Ship Structural Design

Ocean Engineering: A Wiley Series

EDITOR:
MICHAEL E. McCORMICK, Ph.D.
U.S. Naval Academy

ASSOCIATE EDITOR:
RAMESWAR BHATTACHARYYA, D. Ingr.
U.S. Naval Academy

Michael E. McCormick	Ocean Engineering Wave Mechanics
John B. Woodward	Marine Gas Turbines
H. O. Berteaux	Buoy Engineering
Clarence S. Clay and Herman Medwin	Acoustical Oceanography: Principles and Applications
F. W. Wheaton	Aquacultural Engineering
Robert M. Sorensen	Basic Coastal Engineering
Rameswar Bhattacharyya	Dynamics of Marine Vehicles
Stephen C. Dexter	Handbook of Oceanographic Engineering Materials
John B. Woodward	Low Speed Marine Diesel
J. P. Hooft	Advanced Dynamics of Marine Structures
Owen F. Hughes	SHIP STRUCTURAL DESIGN A Rationally-Based, Computer-Aided, Optimization Approach

SHIP STRUCTURAL DESIGN

A Rationally-Based, Computer-Aided, Optimization Approach

OWEN F. HUGHES
Department of Naval Architecture
University of New South Wales
Sydney, Australia

A WILEY-INTERSCIENCE PUBLICATION

JOHN WILEY & SONS, New York · Chichester · Brisbane · Toronto · Singapore

Library of Congress Cataloging in Publication Data:

Hughes, Owen F.
Ship structural design.

(Ocean engineering)
"A Wiley-Interscience publication."
Includes index.
1. Hulls (Naval architecture) 2. Strength of materials. I. Title. II. Series: Ocean engineering (John Wiley & Sons)
VM163.H83 1983 623.8'144 83-1110
ISBN 0-471-03241-7

Printed in the United States of America

10 9 8 7 6 5 4 3 2 1

To my parents,
Earl and Marion

SERIES PREFACE

Ocean engineering is both old and new. It is old in that man has concerned himself with specific problems in the ocean for thousands of years. Ship building, prevention of beach erosion, and construction of offshore structures are just a few of the specialties that have been developed by engineers over the ages. Until recently, however, these efforts tended to be restricted to specific areas. Within the past decade an attempt has been made to coordinate the activities of all technologists in ocean work, calling the entire field "ocean engineering." Here we have its newness.

Ocean Engineering: A Wiley Series has been created to introduce engineers and scientists to the various areas of ocean engineering. Books in this series are so written as to enable engineers and scientists easily to learn the fundamental principles and techniques of a specialty other than their own. The books can also serve as textbooks in advanced undergraduate and introductory graduate courses. The topics to be covered in this series include ocean engineering wave mechanics, marine corrosion, coastal engineering, dynamics of marine vehicles, offshore structures, and geotechnical or seafloor engineering. We think that this series fills a great need in the literature of ocean technology.

MICHAEL E. MCCORMICK, *Editor*
RAMESWAR BHATTACHARYYA, *Associate Editor*

March 1983

PREFACE

For a structure as large and as complex as a ship there are three levels of structural design, the second and most central of which is the subject of this book. Concept design deals with the topology or overall geometry of the structure; preliminary design establishes the scantlings (structural dimensions) of all principal structural members; and detail design is concerned with local aspects such as joints, openings, and reinforcements. Overall structural geometry is generally determined by overall design requirements rather than by structural requirements, while detail design is largely guided and constrained by fabrication methods and requirements. Also, since local structural details are numerous and basically similar among various structures they lend themselves to standardization and to design from handbooks and structural codes. Thus, it is in preliminary design where the structural designer has the largest number of significant decisions and options, and the greatest scope for optimizing the structure so that it best fulfills the objectives and satisfies all of the various constraints and requirements.

Rationally-based design is design from first principles using the tools of modern engineering science: computers and the methods of structural analysis and optimization which computers have made possible. Thus, the rationally-based approach is ideally suited for preliminary structural design, and it is this approach and this level of design that is the subject of this book. As shown by some examples in Section 1.3, this type of design offers substantial benefits to all parties concerned: owner, designer, builder, and operator.

Designing from first principles requires two separate and very extensive analyses: a response analysis to ascertain the true and complete response of the structure to all loads and load combinations, and a limit state analysis to ascertain all of the possible limit or failure values of these responses. Taken together these two analyses are by far the dominant part of rationally-based design, and this is reflected in this text in which 15 of the 17 chapters are devoted to various aspects of analysis. Because of this predominance of analysis, rationally-based design is necessarily computer based and this is the key to many of its benefits: speed, accuracy, thoroughness, economy, easy modification, and so forth. Also, as explained in Section 1.3, the necessary computer programs are already available and the hardware and software costs are quite moderate.

Of the many different topics and aspects in preliminary structural design some are an inherent part of rationally-based design (e.g., the aspects pertaining to response analysis and limit analysis) while others are more distinct and external (e.g., the selection of materials) or are simply constraints in the optimization process (e.g., the avoidance of some natural frequency). One of the advantages of the rationally-based approach is that it unifies and coordinates these many different aspects. Even for the more distinct or external aspects the rationally-based approach provides a framework by which each can be better coordinated with the other aspects.

Because there are so many aspects to preliminary design it is simply not possible to treat all of them in detail in one book. Therefore, the author's plan is that the subject should be covered in two books, of which this is the first. This book deals with the methodology of rationally-based design and with its two major components; response analysis and limit analysis. These aspects are presented in such a way that this first book is complete in itself, and does not rely on or require the second book for understanding.

PREREQUISITES, LEVELS OF STUDY, AND TIME REQUIREMENTS

The material in this text is suitable for either graduate or undergraduate study, or a combination of both. The only prerequisites are knowledge of mechanics of solids, strength of materials, and the basic aspects of matrix algebra and of statistics. If necessary, the latter two could be covered in a few introductory classes or in outside reading. The total time required to cover all of the topics in this book is about eight semester hours. For undergraduates and for persons with only the basic prerequisites a six semester hour course can be achieved by omitting chapters 4, 6, 10, 15, and 17 and skimming chapters 8 and 16. For persons who have already completed some courses in structural analysis (covering, say, columns, plate bending and buckling, frame analysis, and the finite element method) all of the remaining topics could be covered in four semester hours, by omitting chapters 5, 7, 9, 11, and 12 and skimming chapters 6 and 10.

Sydney, Australia
June 1983

OWEN F. HUGHES

ACKNOWLEDGMENTS

I would never have been able to write this book without the kind help and encouragement that I have received from many people—both within and outside of the field of ship structures—through the many years of its preparation. So many persons have helped in various ways that it is simply not possible, in this limited space, to acknowledge and thank each of them individually. To all of them I would like to offer my most sincere and heartfelt thanks. I am deeply grateful for their help, and even more for the kindness and generosity which motivated and accompanied it.

But there are some persons who deserve a very special thanks and acknowledgment. I am deeply grateful to Professor Farrokh Mistree, now at the University of Houston, for his generous and cheerful friendship, for his many valuable suggestions and unstinting help and encouragement, and in particular for the many valuable contributions which he made in the computer implementation of the design method presented herein. Special thanks are also due to Dr. Vedran Žanić of the University of Zagreb, who also contributed in all of these ways. Other academic colleagues to whom I am also grateful for reading drafts of chapters and offering valuable suggestions and corrections are Professors J. R. Paulling (Berkeley), J. B. Caldwell (Newcastle), M. K. Ochi (Florida), and P. Terndrup Pedersen and J. Juncher Jensen (Tech. U. of Denmark).

I am also very grateful to Mr. Stanley Stiansen and the senior research staff of the American Bureau of Shipping for the strong encouragement and support which they gave at a relatively early stage in the development of the design method. This early and sustained support was of immense help.

And finally I wish to offer special thanks to Mrs. Diane Augee for having unfailingly provided such speedy and flawless typing of the manuscript through all of its many drafts and revisions over the past six years. Her outstanding professional skill, and the characteristic thoughtfulness, patience, and cheerful enthusiasm which she has contributed throughout have been of immeasurable help, and constitute a major contribution to the work.

O.F.H

CONTENTS

Ship Structural Design

Chapter

One

RATIONALLY-BASED STRUCTURAL DESIGN

1.1 INTRODUCTION

We are at present in the midst of a gradual but profound change in the philosophy and practice of ship structural design. The growing demand for more efficient marine transportation has already produced significant changes in ship sizes, types, and production methods. For example, since 1960 many new ship types have appeared, such as lng carriers, multihull vessels, barge carriers, ro/ro vessels, surface effect ships, and hydrofoils. This process of change will almost certainly continue because there are so many factors which give rise to it and stimulate it. Some examples are:

- The need for adequate protection against pollution and for adequate safety with respect to dangerous cargoes
- New trade patterns, cargoes or tasks, which introduce special problems or lead to unusual ship geometry or proportions
- The development of designs for standard ships which are to be produced in larger numbers and hence require a greater degree of cost efficiency
- The development of a wide variety of other types of floating structures for the extraction of ocean resources and for other uses of the sea

The complexities of modern ships and the demand for greater reliability, efficiency, and economy require a scientific, powerful, and versatile method for their structural design. In the past, ship structural design has been largely empirical, based on accumulated experience and ship performance, and expressed in the form of structural design codes or "rules" published by the various ship classification societies. These rules provide simplified and easy-to-use formulas for the structural dimensions, or "scantlings," of a ship. This approach saves time in the design office and, since the ship must obtain the approval of a classification society, it also saves time in the approval process.

However, there are several disadvantages to a completely "rulebook" approach to design. First, the modes of structural failure are numerous, complex, and interdependent, and with such simplified formulas the margin against failure remains unknown; thus one cannot distinguish between structural adequacy and overadequacy. Therefore, such formulas cannot give a truly efficient design and in some cases the extra steel may represent a significant cost penalty throughout the life of the ship.

Second, these formulas are only aimed at avoiding structural failure, but there are usually several ways of achieving this and the particular way which is implied in the formula may not be the most suitable in regard to the specific goals of the ship owner or the particular purpose or economic environment of the ship. A true design process must be capable of accepting an objective, of actively moving toward it, and of achieving it to the fullest extent possible.

Third, and most important, these formulas involve a number of simplifying assumptions and can be used only within certain limits. Outside of this range they may be inaccurate, and the history of structural design abounds with examples of structural failures—in ships, bridges, and aircraft—which occurred when a standard, time-honored method or formula was used, unknowingly, beyond its limits of validity.

For these reasons there is a general trend toward "rationally-based" structural design, which may be defined as

> Design which is directly and entirely based on structural theory and computer-based methods of structural analysis and optimization, and which achieves an optimum structure on the basis of a designer-selected measure of merit.

Thus rationally-based design involves a thorough and accurate analysis of all of the factors affecting the safety and performance of the structure throughout its life, and a synthesis of this information, together with the goal or objective which the structure is intended to achieve, to produce that design which best achieves the objective and which provides adequate safety. This process involves far more calculation, but that can be more than offset by using computers. For this reason, rationally-based structural design is necessarily computer-based, semiautomated design.

This type of design was first developed and applied for aircraft and aerospace structures and it continues to have its greatest application in these areas because of the predominant economic significance of structural weight, and hence structural efficiency, coupled with the obvious need for very high structural reliability. In land-based structures the move toward this approach was given strong impetus in the 1970s by a series of structural failures of steel box girder bridges, which showed that for the larger and more slender bridges the existing empirically based design codes were inadequate. In the ocean environment, an elementary form of this approach is already used for the design of offshore structures, partly because there is little or no previous experience on which to rely, and partly because of the high economic stakes involved. In this area, as well as in ship structures, the classification societies have encouraged and contributed greatly to the development of rationally-based methods.

Rationally-based design is definitely not fully automated design, that is, a "black box" process in which the designer's only role is to supply the input data, whereupon the process presents him/her with a finished design. This type of design requires that all design decisions—objectives, criteria, priorities, constraints, and so on—must be made before the design commences, and many of these decisions have to be built into the program, making it difficult for the designer to even be aware of them, much less to have control over them. Rather, of its very nature rationally-based design must be interactive. The designer must always remain "in charge" and be able to make changes—in his/her objectives, criteria, constraints, priorities, and so on—in the light of intermediate results. Therefore, a rationally-based design process should allow the designer to interrupt, go back, make changes, call for more information, skip some steps, and so forth.

From all of the foregoing it is obvious that rationally-based design gives the designer much more scope, capability, and efficiency than ever before. But it is also obvious that it requires more extensive and more sophisticated computational facilities than are required for rule-based design. Hence for conventional ship types the simpler rule-based approach will continue to be used in many instances.* Moreover, as each new application becomes more familiar and more thoroughly understood, it becomes possible to find simplifications and shortcuts, and to gradually develop a new code—a new set of rules—which produces designs that are sufficiently close to those produced by the rationally-based approach. Thus the two approaches are complementary and a good designer will use whichever is more appropriate for a given situation.

Preliminary Design and Detail Design

In a ship structure, as in most structures, the principal dimensions are usually not determined by structural considerations but rather by more general requirements such as beam and draft limitations, required cargo capacity, and so on. For this reason structural design usually begins with the principal dimensions already established, and the designer must determine the complete set of scantlings that provide adequate strength and safety for least cost (or whatever other objective is chosen). Structural design consists of two distinct levels:

1. Preliminary design, which determines the location, spacing, and scantlings of the principal structural members†
2. Detail design, which determines the geometry and scantlings of local structure (brackets, connections, cutouts, reinforcements, etc.)

*In particular for small vessels (say, less than 50 m) there will be many instances in which the benefits of rationally-based design are not worth the extra computation and sophistication.

†For naval vessels this is termed "contract design."

Of these two levels, the rationally-based approach has more relevance and more potential benefits in regard to preliminary design because:

a. This level has the greatest influence on the structural design and hence offers very large potential savings.
b. This level provides the input to detail design and the benefits of the latter are strongly dependent on the quality of this input.

In fact, rationally-based preliminary design offers several kinds of potential benefits. The economic benefits are illustrated by the tanker example quoted in Section 1.3, in which the rationally-based approach gives a 6% savings in ship structural cost compared with current standard designs (which for a large tanker represents a savings of over one million dollars) and an even greater amount of extra revenue from the increased cargo capacity arising from weight savings. Naval vessels can obtain greater mission capability by a reduction of weight and/or vertical center of gravity. Ship designers gain a large increase in design capability and efficiency, and are able to concentrate more on the conceptual and creative (and more far-reaching and rewarding) aspects of design. And finally, there are also substantial benefits to be gained in ship structural safety and reliability.

This is not meant to imply that detail design is less important than preliminary design; it is equally important for obtaining an efficient, safe and reliable ship. Here also there are many benefits to be gained by applying modern methods of engineering science, but the applications are different from preliminary design and the benefits are likewise different. Since the items being designed are much smaller it is possible to do full-scale testing, and since they are more repetitive it is possible to obtain the benefits of mass production, standardization, methods engineering, and so on. In fact, production aspects are of primary importance in detail design.

Also, most of the structural items that come under detail design are similar from ship to ship, and so in-service experience provides a sound basis for their design. In fact, because of the large number of such items it would be inefficient to attempt to design all of them from first principles. Instead it is generally more efficient to use design codes and standard designs which have been proven by experience. In other words, detail design is an area where a rule-based approach is very appropriate, and the rules that are published by the various ship classification societies contain a great deal of useful information on the design of local structure, structural connections, and other structural details.

Aims and Scope of the Book

As indicated in the Preface, this book is the first of a coordinated pair of books on ship structural design, and it presents a methodology for rationally-based design and all of the analysis aspects of such design. Now that we have defined the term "rationally-based" and have noted the distinction between preliminary design and detail design, we can give a more specific statement of the aims and scope of the two books. The aim of the first book is twofold:

> To present the fundamental and essential aspects of rationally-based preliminary ship structural design in a complete and unified treatment that assumes only basic engineering subjects such as mechanics and strength of materials
>
> To present a methodology for rationally-based design that is practical, efficient, and versatile, and that has already been implemented in a computer program, tested, and proven

The aim and approximate scope of the second book is:

> To provide information on other basic aspects of ship structural design such as choice of configuration, design of particular substructures, structural optimization (especially practical design-oriented aspects), material selection, vibration, design for production, welded connections and some aspects of detail design (especially fatigue), and in all of these areas pay particular attention to computer-aided aspects

This first book is entirely self-sufficient and self-contained; that is, it covers all of the basic aspects of rationally-based design that are required by a designer, even aspects as basic as frame analysis and the finite element method. This has been done for two reasons. First, because this book is intended primarily as a textbook, and in the field of ship structures such books are few and far between. For too long, students in this field have had to consult a large number of different books and other sources in order to obtain the theory and methods that they need, and these are often in a form that is not suited for ship applications. Because of the greater complexity and sophistication of rationally-based design, the lack of a unified and comprehensive text would constitute a correspondingly greater difficulty for students, and a serious obstacle to further progress in this field.

The second reason is that rationally-based design, both in general and in the particular method presented herein, is radically different from the traditional rule-based method, and although many of its features are familiar to designers (such as frame analysis, finite element analysis, and elastic buckling) other features are either relatively new (such as nonlinear finite element theory and statistical prediction of wave loads) or totally new (such as the use of a hull module as the basis or subject of the design, some new techniques for structural modeling, and the method for ultimate strength analysis of a hull module).

For this reason the book is also intended for practicing designers who wish to become more familiar with this alternative method of design. In fact, the book's role here is of particular importance because rationally-based design of its very nature requires at least a basic knowledge of its underlying theory and methods. Once this is acquired the method's enormous capability (some of which is demonstrated in Section 1.3 and in the references given there) becomes available to the designer. Moreover, the method's breadth of application and the benefits gained from it increase in proportion to the knowledge of it. It is the author's hope that the presentation of the underlying theory and analysis methods in this text will assist designers to obtain the maximum possible benefits from this new approach.

Also, the author would like to emphasize that the design method presented herein is not the only possible method, at least not in regard to the particular component methods for achieving the basic tasks, such as the structural modeling techniques and the methods of member limit analysis. The methods presented herein have been selected or developed on the basis of their suitability for rationally-based design, but this type of design involves so many different areas that there are bound to be some particular methods and techniques that are as good or better than those given here. Moreover, as further progress is made in such areas as structural theory, numerical methods, and computer hardware and software, still better methods will be developed. But the important point is that now, as the result of many years of effort by many persons and organizations both inside and outside of the field of ship structures, all of the required ingredients for rationally-based design are available.

Applicability to Naval Design

Another important feature of the design method presented herein is that it applies equally well to naval vessels and commercial vessels. Because classification rules are intended for commercial vessels and are not suitable for warships, various navies and naval design agencies have developed their own methods of structural design. Like the classification rules, these methods have evolved over a long period and many of them have been systematized and codified into some form of *design manual*—a sort of naval counterpart to the rules. But because of the need for greater structural efficiency and other special requirements, naval design methods have generally been more thorough and rigorous than rule-based design, and have shown a stronger trend toward a rationally-based approach. Thus in addition to the design manuals many current methods of naval design already include some of the basic features of rationally-based design, although not as a single integrated design method.

Section 1.2 gives a brief overview of the basic features of rationally-based design, including a discussion of the different aims, measures of merit, and design criteria in commercial ships and naval ships. Section 1.3 considers the capabilities, applications, and some sample results. Once these aspects have been treated it becomes apparent that the method presented herein applies equally well to both ship types, and that it matches the needs and challenges of naval design particularly well. Because commercial vessels are much more common, most of the explanations and figures in this text refer to this type. Therefore it seems desirable at this point to briefly summarize the reasons why the rationally-based method presented herein is so well suited for naval design, even though a full appreciation can only be obtained after covering Sections 1.2 and 1.3.

First of all, naval ship structures are subject to many special requirements and constraints. For example, they must be capable of withstanding specified levels of blast, shock, and other special loads. Also, they must be damage tolerant, that is, capable of sustaining some structural damage without loss of main functions. As discussed in Section 1.2 the mission requirements of naval vessels make it extremely important to minimize the weight and vertical center of gravity of the structure to the extent allowed by the various constraints (such as cost, adequate strength and safety, and damage tolerance). Hence there is a paramount need for structural optimization, which is one of the basic features of rationally-based design.

But also, in order to achieve a truly optimal structure, the other aspects of rationally-based design are also required. In particular, rationally-based design explicitly calculates each load effect and each limit value, and imposes the stipulated safety margin for each individual limit state. In contrast, current naval design methods utilize specially defined "design loads" and some typical or representative limit states (which are simplified in order to preserve generality and facilitate computation) together with some general requirements such as maximum permissible stress. Since the actual individual load effects and limit values are not evaluated, the actual safety margins are not known precisely. Hence it is necessary to use larger safety margins in the representative limit states in order to achieve a sufficiently high probability that none of the margins in the actual limit states will be zero or negative. This inevitably results in some excess material. In contrast, since rationally-based design considers each limit state explicitly, it can achieve the same degree of safety without excess material.

Finally, it is worth noting that the ability of rationally based design to deal with both commercial and naval ships can also help to unify the field of ship structural design, which until now has been largely split into these two separate areas.

Applicability to Other Types of Structure

In this text the rationally-based approach is described purely in terms of ships. However, because this approach represents the most fundamental and most general type of engineering design, the material presented herein is also applicable to a wide variety of other steel structures, both fixed and floating.* All of the basic principles and most of the analysis methods for other steel structures are the same as for ships, and the scope of this text could have been extended to include these other structures without requiring any fundamental change of approach. However, this would have required the consideration of additional types of loads and failure modes, plus some new examples to illustrate these other applications, and this would have increased the book's length unduly.

*For example, in Reference 1 the method presented herein was used for the structural optimization of a large steel box girder bridge.

Practicality of the Method

Rationally-based design is necessarily computer-based and this raises a number of practical questions in regard to computer implementation, accuracy, cost-effectiveness, availability, ease of use, documentation, and so on. These are important questions and they are dealt with fully in Section 1.3. But since practicality is so essential in a design method it is appropriate to mention here that this method has been developed over a period of 10 years and that during all this time a computer program based on it has been developed simultaneously. The program is called MAESTRO†, and a cruder and partially completed forerunner was called SHIPOPT. The latter was used for thoroughly testing and demonstrating the basic features, and the results have been documented in the technical literature, showing MAESTRO to be extremely powerful, versatile, cost-effective and easy to use. The published ship design applications include a 14,000 tonne cargo ship, a 96,000 tonne tanker, a 140,000 tonne bulk carrier, and a destroyer. In all cases the program produced an optimum structural design in a matter of minutes and at very low cost. In addition to its use for optimum design, the analysis portion of MAESTRO can be used to evaluate a given design, to assess proposed changes to a design or to an existing ship, or to evaluate the seriousness of damage incurred by a vessel. The program is also a valuable tool for ship structural research and for the teaching of ship structural design. Further details of all these aspects are given in Section 1.3 and in the references cited there.

1.2 BASIC ASPECTS OF STRUCTURAL DESIGN

One of the most familiar and fundamental concepts in engineering is that of a system, which may be anything from a simple device to a vast multilevel complex of subsystems.

A ship is an obvious example of a relatively large and complex system, and in most cases the ship itself is a part of an even larger system. The ship consists of

†*M*ethod for *A*nalysis, *E*valuation and *STR*uctural *O*ptimization.

several subsystems, each essential to the whole system, for example, propulsion subsystem, cargo-handling subsystem. The structure of the ship can be regarded as a subsystem, providing physical means whereby other subsystems are integrated into the whole and given adequate protection and a suitable foundation for their operation.

In general terms the design of an engineering system may be defined as:

> The formulation of an accurate model of the system in order to analyze its response—internal and external—to its environment, and the use of an optimization method to determine the system characteristics that will best achieve a specified objective, while also fulfilling certain prescribed constraints on the system characteristics and the system response.

Translating this to the case of preliminary ship structural design, a rationally-based design procedure can be described as follows:

1. The external loads are predicted as accurately as possible, using statistical methods where appropriate.
2. The load effects and limit values of load effects are calculated accurately throughout the structure for all loadcases.
3. The minimum required margins between the load effects and their limit values are all selected on the basis of a required degree of safety.
4. The resulting strength requirements are expressed in the form of mathematical constraints on the design variables (in most cases nonlinear constraints).
5. The designer is left free to specify the measure of merit of the structure, that is, the criteria that is to be used in achieving the *best* structure, and the influence of each design variable on the measure of merit. Also the designer is able to specify any number of other constraints on the design, of any form whatsoever, in addition to the strength-related constraints.
6. An optimization method automatically and efficiently solves for the values of the design variables which give the maximum value of the measure of merit while also satisfying all of the constraints.

From this description of a rationally-based design procedure it is possible to identify six essential tasks:

1. Calculation of environmental loads.
2. Overall response analysis.
3. Substructure response analysis.
4. Limit state analysis.
5. Formulation of reliability-based structural constraints.
6. Solution of a large nonlinear optimization problem.

Figure 1.1 illustrates the overall design process, consisting of these six tasks. All of them are inherently very large tasks, especially for structures as large as ships, and the principal difficulty or challenge in developing a rationally-based design procedure is to develop methods that can perform these tasks to the required degree of accuracy and thoroughness, within acceptable amounts of total manhours and total computational cost. Each of these tasks is now considered briefly, in order to define it more precisely and to explain broadly what it entails.

Calculation of Environmental Loads

The environmental loads are the loads, both static and dynamic, that come from the ship's environment (mainly due to gravity and fluid pressures) and from its motion. Most of these loads are relatively independent of the structural design, that is, they are not much affected by the structural layout or by the scantlings. Rather they are more a function of hull shape, the type and distribution of cargo, and other nonstructural factors. Therefore, although the calculation of these loads is the first step of structural design, and one of the most crucial aspects of the entire process, it is essentially a separate initial task. Some of the loads can be readily calculated and controlled by the designer (e.g., those arising from cargo distribution). Other loads, particularly wave loads and other hydrodynamic loads (slamming, sloshing, etc.) are sufficiently complex that their calculation is not regarded as part of the designer's task but rather that of hydrodynamicists and other specialists. In contrast, the other two types of calculations—response analysis and limit state analysis—are inherently and totally structural in nature.

Overall Response Analysis

This analysis entails the calculation of the effects of the environmental loads on the overall structure (bending moment, deflection, stress, etc.). For reference the load effects will be represented by the symbol $\mathbf{Q}$ (or Q_i if referring to the ith load effect).

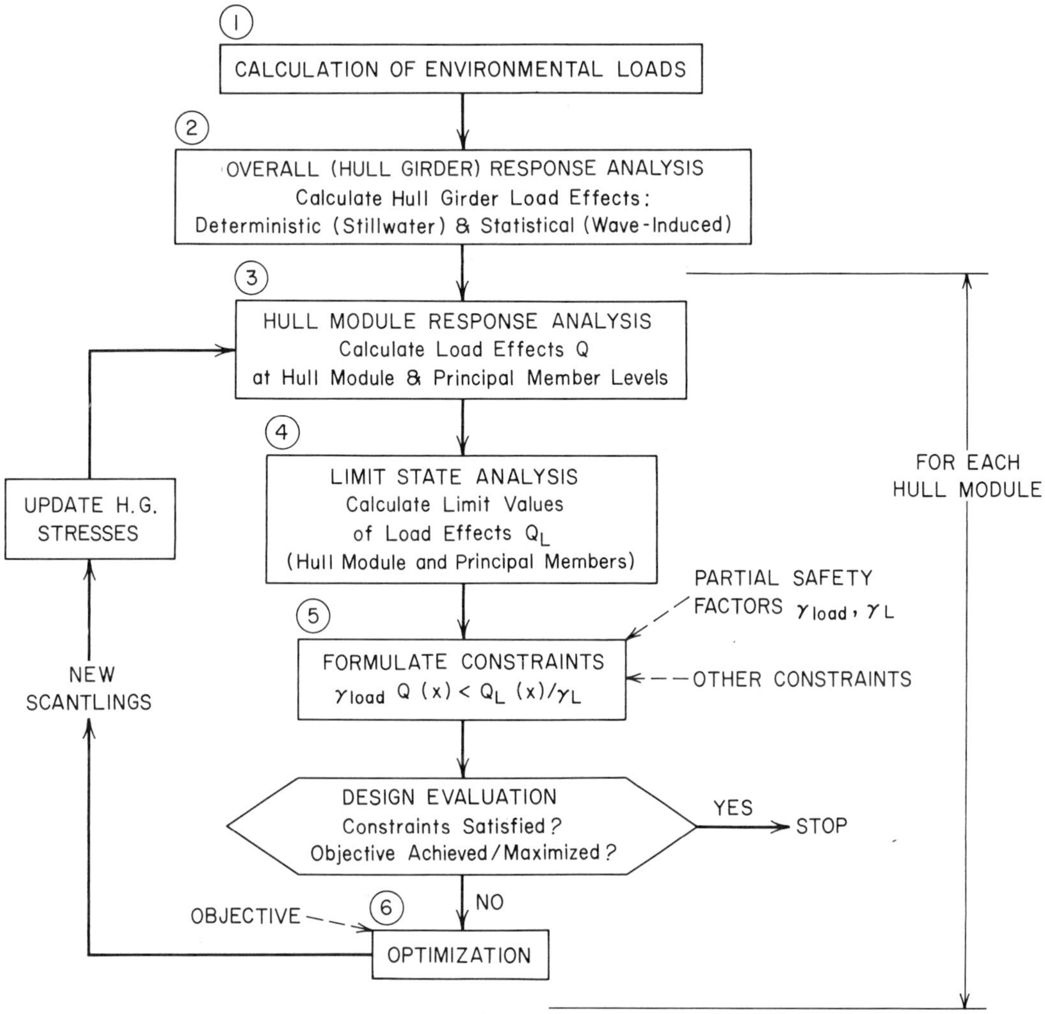

Figure 1.1 Rationally-based structural design.

For all large structures both the loading and the response* are sufficiently complex that the response analysis must be performed in two stages—an analysis of the overall structure followed by separate and more detailed analyses of various substructures. Many of the load effects from the overall analysis constitute the loads and the boundary conditions at the substructure level. For a ship the overall structure is essentially a beam—a floating box girder that is internally stiffened and subdivided—and in which the decks and bottom structure are flanges and the side shell and any longitudinal bulkheads are the webs. The hull girder analysis deals only with those longitudinally integrated forces and moments which are dealt with in beam theory: vertical shear force F_y, longitudinal bending moment in the (ship's) vertical and horizontal planes M_z and M_y, and longitudinal twisting moment M_x. Of these, the most significant is the vertical bending moment M_z, that is, bending about the z-axis in Fig. 1.2. This load effect is caused mainly by the unequal distributions of weight and buoyancy along the length of the ship, accentuated by waves. Horizontal bending (i.e., bending about the y-axis) occurs when the ship is in an inclined condition, due to rolling, and also arises from quartering seas in which wave crests on one side of the ship are in phase with troughs on the other. In most ships the maximum value of M_y is much smaller than the maximum value of M_z (typically 20% or less) but in large tankers and some other cases it can rise to as high as 50%. For simplicity we here consider only

*The "response" of a structure is simply the group of load effects that is caused; the two terms are essentially the same and they are used interchangeably throughout this text.

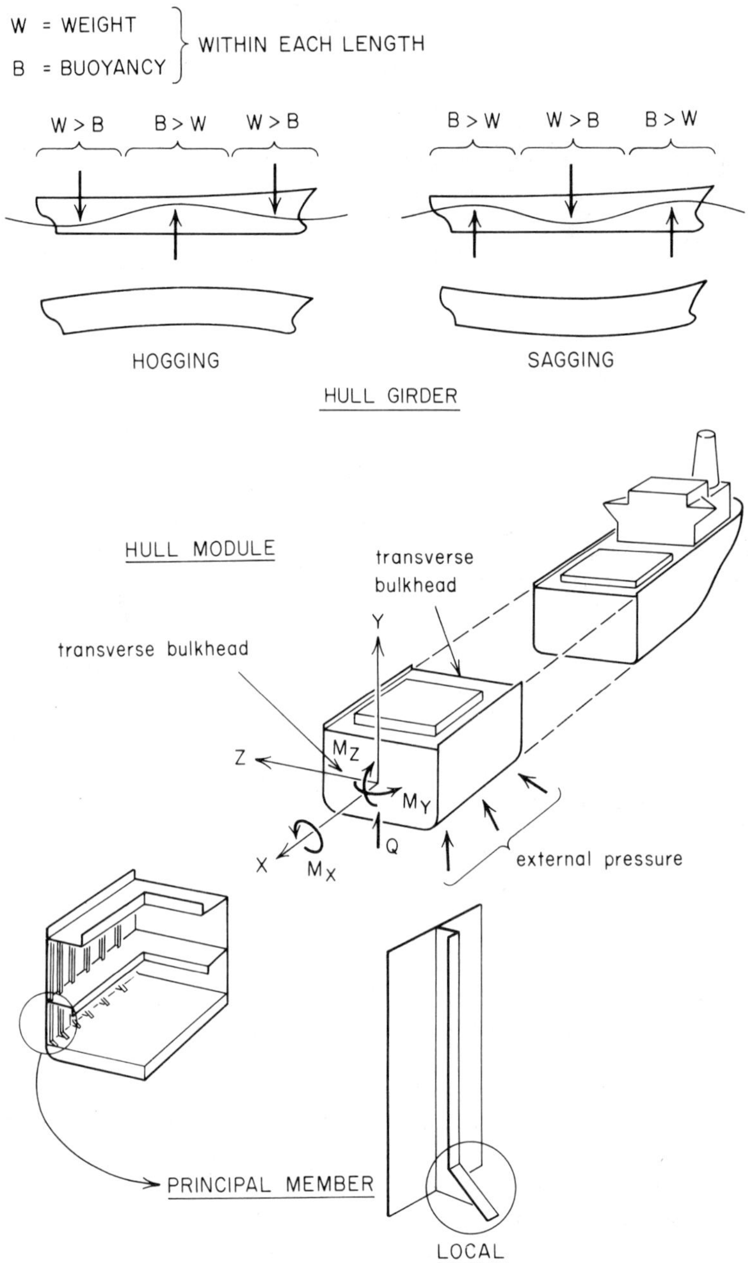

Figure 1.2 Levels of structural analysis.

vertical bending; horizontal bending and its relation to vertical bending is considered in Section 3.6.

The bending moment varies along the length of the ship, being zero at the ends and having a maximum value which usually occurs near the midlength of the ship. The maximum value of hull girder bending moment is the single most important load effect in the analysis and design of ship structures. Hull girder bending is referred to as either "hogging" or "sagging," depending on the sense of curvature which it causes, as shown in Fig. 1.2.

The hull girder analysis assumes that hull girder bending satisfies simple beam theory (Bernoulli-Euler), which implies the following assumptions:

1. Plane cross sections remain plane.
2. The beam is essentially prismatic (no openings or discontinuities).

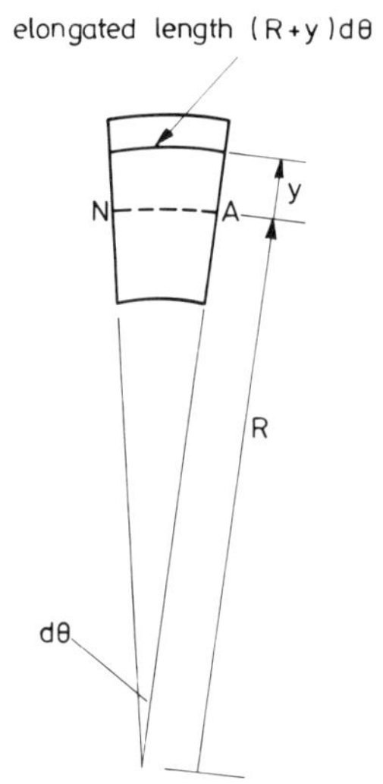

Figure 1.3 Strain distribution in simple beam theory.

3. Other modes of response to the loads (e.g., transverse and longitudinal deflection and distortion, caused by shear and/or torsion) do not affect hull girder bending and may be treated separately.
4. The material is homogeneous and elastic.

The first assumption is illustrated in Fig. 1.3. Under the action of a bending moment a beam undergoes curvature of radius R locally and, if plane cross sections remain plane, the longitudinal strain ε_x varies linearly in the vertical direction and is related to R as follows:

$$\varepsilon_x = \frac{(R + y)\, d\theta - R\, d\theta}{R\, d\theta} = \frac{y}{R}$$

The horizontal surface where y, and hence also the strain, is zero is referred to as either the neutral surface or the neutral axis. The material is assumed to be homogeneous and elastic, and so the longitudinal stress is

$$\sigma_x = E\varepsilon_x = E\frac{y}{R} \tag{1.2.1}$$

If there is no external axial force equilibrium requires

$$\int_A \sigma_x\, dA = 0$$

which reduces to

$$\int_A y\, dA = 0$$

and this indicates that the neutral surface coincides with the horizontal centroidal axis of the cross section.

Equilibrium of moments requires that the external moment M_z is balanced by the moment of the internal stress forces

$$M_z = \int_A y\, \sigma_x\, dA$$

which from (1.2.1) reduces to

$$M_z = \frac{EI}{R} \tag{1.2.2}$$

where I is the moment of inertia of the cross section, defined by

$$I = \int_A y^2\, dA$$

Equation 1.2.1 relates the curvature to the external bending moment and if this is used to eliminate R from (1.2.2) the result is the familiar equation for bending stress at a height y from the neutral axis

$$\sigma_x = \frac{M_z y}{I} \tag{1.2.3}$$

Section Modulus

Equation 1.2.3 indicates that σ_x is greatest when y is greatest, that is, at the extreme upper or lower edge of the section. When y corresponds to one of these extreme values the quantity I/y is called the *section modulus* and is usually denoted by Z. Since the neutral axis is not generally at half-depth there will be two extreme values of y: y_D for the deck and y_K for the keel, and there will thus be two values of Z: $Z_D = I/y_D$ and $Z_K = I/y_K$. The bottom structure is usually sturdier than the deck and so the neutral axis is usually below the half-depth. A height of 0.4 D above the keel is typical, but the location varies widely between different ship types. Thus the largest hull girder stress usually occurs in the deck rather than the bottom. More precisely, it occurs in the uppermost member which is longitudinally effective, that is, which is of sufficient length and has a sufficiently rigid attachment to the rest of the hull girder to act as part of the hull girder. In most cases this is a deck and that deck constitutes the uppermost flange of the hull girder. If the side shell extends up to this deck then it is referred to as the *strength deck*.

Since the hull girder bending moment is essentially independent of the scantlings, the section modulus gives a direct measure of the change in hull girder bending stress arising from a change in scantlings. Therefore in each cycle of the rationally-based design procedure of Fig. 1.1, the new values of hull girder stress are obtained directly by calculating the new location of the neutral axis and the new value of I; there is no need to repeat the hull girder analysis.

Section modulus is also useful whenever it is desired to assess or control the maximum hull girder stress (wave-induced, stillwater, or total) by itself, separately from the stresses arising from hull module and principal member response. For example, because the wave-induced hull girder stress is cyclic it is necessary to restrict its amplitude in order to guard against fatigue failure.

DEPARTURE FROM SIMPLE BEAM THEORY

Equation 1.2.3 states that the stress is constant across horizontal decks and varies linearly in the sides. There are several factors which can cause the actual stress distribution to differ from this idealized distribution. Firstly, because of transverse shear there is some longitudinal distortion of the cross section of the hull girder; this is referred to as "warping" of the cross section. This means that the first assumption is not fulfilled, at least not perfectly. Likewise, the second and third assumptions are not fulfilled because the hull girder is not prismatic, having large openings and other discontinuities and discrete elements (transverse bulkheads). Also, it is a complex assembly of members, transverse as well as longitudinal, and there are several other modes of response, in addition to warping, which affect the hull girder bending response. For ships in which there are no major changes in cross section other than in-line hatches (for which the intermediate portions of deck may be ignored) the longitudinal stress due to hull girder bending generally follows the idealized distribution quite closely (ignoring stress concentrations and other local effects) as shown in Fig. 1.4. For such ships the effects of shear and of other responses, of transverse structure, and even of openings and discontinuities, can be calculated separately (or at least estimated) in order to assess their importance and to apply corrections where necessary. In some cases superposition can be used. This structurally prismatic type of hull girder is considered in Chapter 3. For ships with significant changes in cross section most of the load effects can only be obtained at the hull module level. The boundary conditions for the hull module analysis can be obtained from a nonprismatic hull girder analysis using a "hull segment" element. The latter is presented in Section 6.3.

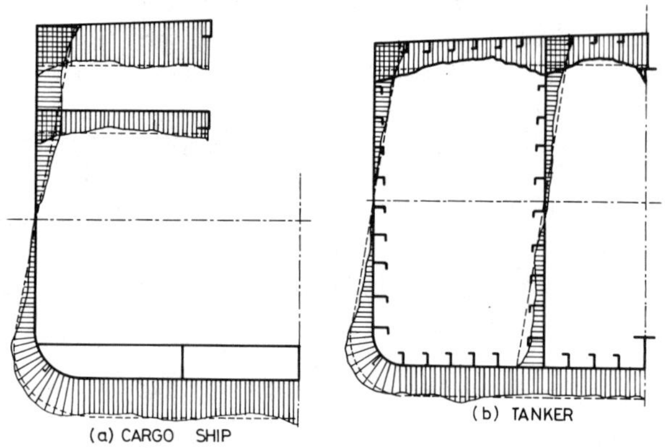

Figure 1.4 Typical hull girder bending stress distribution for structurally prismatic ships.

Shear force can be quite significant with some cargo types and distributions (especially bulk cargoes) and both shear force and twisting moment can be significant if the hull girder has low torsional rigidity. This occurs in vessels with large and numerous deck openings, such as container or barge-carrying vessels. Shear force and its effects are treated in the latter part of Chapter 3. Torsion is dealt with in Chapter 6.

Hull Module

SIGNIFICANCE AND ROLE OF A HULL MODULE

Ideally a structure should be designed as a single unit; that is, all of the steps of rationally-based design should be applied to the structure as a whole, in order to be sure to account for all of the structure's interactions and other characteristics. This is essential if the design is to be truly optimal. However, present day computational means are not yet adequate for the complete procedure to be applied to a structure as large and as complex as a ship. Therefore it is necessary to subdivide the overall structure and to achieve its design by means of a coordinated series of rationally-based designs of these substructures. If these separate designs are to still produce the true optimal overall design it is necessary to continue to account for all of the structure's interactions and other characteristics. Obviously, this includes all types of interactions between the substructures: in response, in limit states, and in overall optimality. Therefore the division into substructures must not prevent, omit, or inaccurately portray any of the structure's responses, limit states, or optimality tradeoffs.

On practical grounds this can only be achieved if the substructures can be defined such that their interactions are few, localized, and easily calculated and accounted for. Hence there are two main requirements which a substructure must meet:

1. It must be sufficiently small, cohesive, and regular as to make its rationally-based design both feasible and practical.
2. It must be sufficiently large and sufficiently autonomous in its response, its limit states, and its effect on the overall optimality, such that no significant characteristic of the overall structure is lost, and such that all significant interactions which do exist can be fully accounted for.

For a ship structure the second requirement can only be satisfied by making each substructure a complete segment of the hull girder. In this text such a substructure is referred to as a *hull module* (see Fig. 1.2). A smaller substructure could never adequately represent all of the characteristics of the overall structure; nor would it have sufficient autonomy in the three required areas: response, limit states, and optimality tradeoffs.

Taking each hull module as being a complete segment largely eliminates interactions in the last two areas. If all of the hull modules are of reasonable length then any form of overall (hull girder) failure will occur wholly within a hull module. And because of the need for longitudinal continuity there can be no appreciable tradeoffs in structural cost (or other measure of optimality) between one hull module and another, especially if they are all reasonably long.

Structural interactions between hull modules cannot of course be eliminated, but they can be minimized by always locating the hull module boundaries at transverse bulkheads. But it is also necessary to be able to analyze each hull module in isolation, and therefore each of them must be defined such that a complete set of boundary conditions can be established for it from the hull girder analysis. It is this need for boundary conditions, together with the need to correctly portray the structural response, which determines the minimum length of a hull module. These two requirements must always be fulfilled by each and every hull module. The actual length that is required depends on the overall layout of the vessel, the arrangement and geometry of its holds or major compartments, and the location and size of all large openings.

In certain rare cases these requirements may preclude subdivision; that is, the ship must be designed as a whole. This would be the case, for example, with a completely open-deck ship because for such a vessel the torsional response is important and the hull girder analysis is not sufficiently detailed to determine accurate boundary conditions anywhere within the ship length. In this case the hull module, being the same size as the hull girder, would have to be modeled more crudely than usual. For a similar reason there are some container ships for which the hull module must extend over all of the cargo holds, from the forepeak bulkhead to the engine room bulkhead. For other container ships it is possible (by means of a special type of hull girder analysis presented in Chapter 6) to obtain suitable boundary conditions at intermediate bulkheads, thereby allowing the use of shorter hull modules. For bulk carriers it will be shown that the hull module must extend over at least three cargo holds. For some other ship types it can be a single hold. A further and more detailed discussion of the minimum length of a hull module is presented in Section 2.3.

Apart from minimum length there are no hard and fast rules for choosing the number, location, and length of the hull modules. The choice depends on many factors and it varies according to the situation and the circumstances surrounding each design. This is just one of many aspects of structural design which must be left to the judgement of the designer. Some examples for particular ship types are given in References 16 and 17. If the vessel consists of a number of basically similar cargo holds it is usually desirable to use identical scantlings for all of them for greater simplicity and economy. In this case the designer defines a hull module which satisfies the minimum length requirement (and which may therefore need to include more than one hold) and he proceeds to design that hull module using a number of different *loadcases* (combinations of loads which occur together). Each loadcase consists of the loads that act on one of the hull modules which make up the hull girder.

In other cases a ship may have very little repetition in its layout so that there is no automatic advantage in using a minimum length hull module. Here the choice of subdivision into hull modules depends mainly on how much the layout and geometry varies among the different portions of the ship, and how much commonality or variation the designer wishes to impose on the design. Perhaps surprisingly, the amount of computation is not a principal factor in this decision, because in the design method presented herein the total amount of computation is only moderately increased if the ship is divided into several short hull modules instead of a few long ones, assuming that in all cases a full design is performed for each hull module.

In fact, although there is some savings if there are fewer hull modules, and although the theoretical ideal is to design the complete structure all at once, there are practical limitations and disadvantages in trying to represent the ship with very few hull modules. From sheer practicality the computer program which implements the design method must necessarily incorporate and rely on some degree of regularity in the geometry and structure of the hull module; otherwise either the programming task becomes too large or the program itself grows too large to be efficient, or even manageable. Because of this regularity in the hull module the inclusion of more and more different portions of the ship in one hull module requires more and more modeling simplifications. These constitute artificial constraints on the design, thus limiting the extent to which it can approach the true optimum. Therefore in general a separate hull module should be defined for each portion of the ship that has substantially different geometry, subject to the minimum length requirement.

However it is not always necessary to perform a complete design for every portion. In some ships there may be a few portions that are of greater economic or operational importance and it may be sufficient to design these first and then obtain the scantlings of the other portions by a judicious choice of intermediate values, or even the same values in order to obtain greater simplicity and regularity of structure and the cost savings which this often brings.

In the design process of Fig. 1.1 the first two steps refer to the ship as a whole and therefore they only need to be performed once, whereas the design cycle consisting of steps 2 to 6 must be performed in its entirety (including however many cycles are needed for convergence) *for each hull module*.

By way of summing up these basic considerations regarding the rationally-based design of ship structures, we may list the three aspects which are crucial for its achievement:

- A suitable choice of the hull modules
- The use of a true rationally-based procedure for the hull module design
- A careful and thorough synthesis of the hull module designs with each other and with the other aspects of structural design

Hull Module Response Analysis

For purposes of structural definition, analysis, and optimization it is necessary to subdivide the hull module into further substructures. In the design method presented herein there are two further levels of substructure: principal member and local structure. These are illustrated in Fig. 1.2. Although we are further subdividing the hull module, it is still analyzed and designed as one unit, and so the manner in which this further subdivision is done does not greatly affect the optimality of the resulting hull module. However, it has a very strong influence on the speed and efficiency of both the analysis and the optimization of the hull module.

The hull girder analysis provides the boundary conditions—both supports and loads—for each of the hull modules. From this point on, each of the hull modules is designed separately, though of course there must be coordination between them and a final synthesis, which often includes standardization. The design process begins with the response analysis of the hull module. We note here that the load effects that are calculated in the hull module analysis become the loads and the boundary conditions for the member analysis, and the load effects for the latter become the loads and boundary conditions for the local structural analysis. Thus the same physical quantity or phenomenon can be referred to as either a "load" or a "load effect," depending on whether one wishes to indicate a cause or an effect.

The importance of the hull module analysis cannot be overemphasized. The purpose of structural response analysis is to calculate all of the significant load effects in all of the principal members, and the main purpose of the hull module analysis is to provide the boundary conditions which are required for the analysis of the principal members. Because of the complex three-dimensional geometry and loading in a hull module, and the consequent high degree of statical indeterminacy, it is not possible to determine the boundary conditions of the principal members except by performing a hull module analysis. Putting it another way, the principal structural members cannot be analyzed or designed in isolation, but rather only in conjunction with a hull module analysis. The point is emphasized here because most earlier methods of ship structural design did not include a hull module analysis, mainly because the computational means were not available.

Another result of the statical indeterminacy of the hull module is that the distribution of loads among the principal members and the pattern of internal forces in

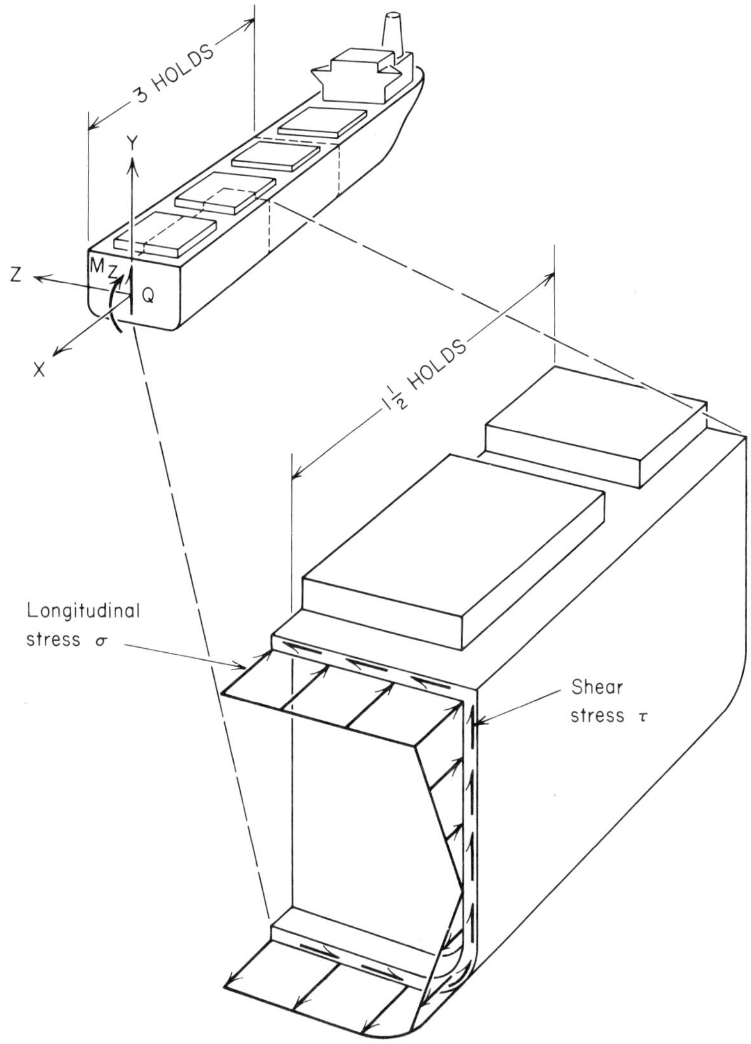

Figure 1.5 Application of hull girder load effects.

each of them is dependent on the structural dimensions of the principal members. As the design proceeds and these dimensions change so also do the internal forces change, and for this reason the design of the hull module and its constituent principal structural members is necessarily an iterative process, as shown in Fig. 1.1.

The loads on the hull module which come from the hull girder analysis are the hull girder stresses in the ship's cross section that are "exposed" by "cutting" the hull module away from the rest of the ship. Unless a special hull girder analysis has been used, this cut must be made at a transverse bulkhead (either just forward or just aft of the bulkhead, so that only the longitudinal material is cut) because the ordinary hull girder analysis assumes that there is no transverse deformation of the cross section. There are two types of applied stresses—longitudinal stress σ which acts normal to the cross section, and shear stress τ which acts in the plane of the cross section. These are illustrated in Fig. 1.5. When applied to the hull module as external forces they account for the effect of the rest of the ship structure and they provide the reaction forces which place the hull module in equilibrium. (It will not be perfect equilibrium and so sufficient restraints must be applied to the module to prevent rigid body motion, but the reaction forces at these support points will be small.) Wherever there is symmetry in the hull module only half of the structure in that direction needs to be modeled (even though the loads may not by symmetric). Hence for a doubly symmetric hull module only a one-quarter model is required, as shown in Fig. 1.5.

This procedure of isolating the hull module and applying the hull girder stresses as loads assumes that these stresses are independent of and unaffected by the structural response which occurs at the hull module level. This assumption is valid because the longitudinal structure of the hull girder is essentially orthogonal to the transverse structure and so the athwartships loads, which are ignored in the hull girder analysis, do not have any appreciable effect in the longitudinal direction (and at the ends of the vessel, where the taper is larger, the stresses themselves are much smaller). Of course, one way in which the hull girder stresses could be affected by the hull module response is that if the combined loads acting on the hull module were to cause buckling or yielding of any longitudinal member, then that member would no longer be longitudinally effective, thereby reducing the hull girder moment of inertia below the value that was used in the hull girder analysis. However, the check for this ineffectiveness occurs in the formulation and evaluation of the constraints (step 5 in Fig. 1.1) in which every member is checked regarding yielding, buckling, and all other possible forms of failure or unserviceability. If a member is found to be inadequate, it is redesigned in step 6. But before the members can be checked the hull module analysis must be performed because there are some longitudinal stresses at this level which must be added to the hull girder stress. Also, the check against buckling requires first that the buckling loads be calculated for all of the relevant members. Thus the structural modifications which are required in order to preserve the validity of the hull girder analysis can only be made after the hull module analysis has been performed. But the design process is necessarily iterative anyway, due to the statical indeterminacy of the structure and the nonlinearities in most of the ultimate strength constraints. As the design proceeds, each of the members will be made to satisfy all requirements and in this way the validity of the hull girder analysis will be guaranteed. Therefore, it is both permissible and logical to analyze the hull module separately and subsequently to the hull girder.

Local Structure

At the most detailed level are those elements of structure which are local in extent: components of principal members, brackets, connections, reinforcements, foundations, fittings, and so on. Basically, a structural element can be classified as local if it does not have any appreciable effect on the load distribution within the hull module; in other words, it is local if it does not affect the magnitude and overall distribution of internal forces in nearby principal members, but rather has only a local effect on the internal force distribution in these members.

Because of the irregular geometry of local structure its analysis usually represents a significant computational task. Obviously the analysis and design of local structure can only be done after the structural dimensions of the principal members have been determined. But as noted previously and as illustrated in Fig. 1.1, the design of the principal members is an iterative process, and it would be inefficient to include the design of local structure as part of this process. Rather, the design of local structure—detail design—is a separate process which comes after the preliminary design process shown in Figure 1.1. Reference 2 contains a great deal of information on detail design, as do the rules and other publications of the classification societies. Moreover, most of the items that come under the heading of local structure are not unique to ships, and there are many design manuals and handbooks for land-based steel structures which contain useful information.

Limit State Analysis

A *limit state* is any condition in which a structure or a structural member has become unfit for one of its intended roles, due to one or more loads and/or load effects.* There are two broad categories of limit states: the *ultimate* or *collapse* limit states, in which the structure or member has failed in its primary, load-carrying role; and the *serviceability* limit states, which involve the deterioration or loss of other, less vital functions. The *limit values* are the values of the loads or load effects which produce or correspond to a limit state. A limit value is denoted by the symbol Q_L. The symbol $\mathbf{Q}_L$ represents the values of a group of loads and/or load effects which produce a limit state. A limit state analysis consists of the calculation of the limit values, perhaps in various combinations and sequences, which correspond to a specified limit state, either in a member or in the overall structure. An ultimate limit state is often referred to as the *ultimate strength* of the

*For the overall structure it is loads, whereas for a member it is usually load effects. For simplicity we often use the term "load", even when the term "load effect" might be more accurate. The symbol Q denotes whatever agent is causing a limit state; hence Q can represent either a load or a load effect. Where a distinction is important the symbol F will be used for a load and the symbol Q for a load effect.

structure or member, and the two terms are used interchangeably throughout this text.

Serviceability limit states arise from the fact that some members are designed on the basis of a form of failure other than structural failure. For example, as shown in Chapter 9, laterally loaded plating is usually designed on the basis of a maximum allowable "permanent set" (plastic dishing). The limit value is the load which causes this limit state, whereas the ultimate load is that value at which the plate can no longer support the load.

There are three basic types of structural failure, plastic deformation, instability, and fracture, and within these there are several different modes of failure, some of which are more serious than others; these are explained in Section 2.4. Moreover, these various failure modes can combine and also interact, depending on the member properties and on the loading. And finally, there are always several different loading arrangements and load combinations which must be considered (hogging and sagging, deep draft and light draft, various distributions of cargo, etc.). Hence for each structural member there are usually several limit states, not all of which have the same degree of seriousness. In general, rationally-based design requires that each and every relevant limit state be examined, and those which interact should be examined together. There should be no *a priori* assumptions as to whether some limit state will or will not govern the design.

Thus, to take a very simple example, the laterally loaded plate referred to above should be examined for both types of limit state. The limit load for ultimate failure is much larger than the limit load for allowable permanent set. But because of the far greater degree of seriousness there must be a much greater margin between the ultimate load and the expected service load, and so either requirement may govern the required thickness of plating. In order to determine which is the governing requirement it is necessary to perform both limit state analyses.

The level of seriousness of a limit state corresponds approximately to the level or extent of structure which has failed: overall, substructure, or local. For the preliminary design of a ship structure the appropriate levels are hull module and principal member. The hull girder level is here redundant because the hull module is a complete segment of the hull girder and failure of the hull module is failure of the overall ship. Hence in this text the terms "overall failure" and "structure failure" refer to failure at the hull module level, unless noted otherwise. Failure of local structure is not sufficiently serious to be included with the other levels. Also, as noted earlier, this level of structure is usually dealt with in detail design rather than preliminary design.

Thus there are two levels of structure that can reach a limit state: the structure as a whole (the hull module) and the principal members. Ideally the limit analysis of the overall structure should include the limit analysis of the individual principal members. However the limit analysis of a hull module is a very large computational task and the total amount of computation can be reduced by performing the two separately: a hull module limit analysis which uses a simplified structural model of the hull module; and a separate limit analysis of each different principal member, for each different load combination which that member faces. The member limit analyses provide the values of a member ultimate strength which are used in the hull module limit analysis, and the latter specifies the load combinations which are to be used in each member limit analysis. The determination of these load combinations is crucial for rationally-based ship structural design. At the member level it is not possible to adequately account for the interaction between members. Hence it is not possible to know the true loads that are acting on each member as the structure approaches collapse. Moreover, most large structures have a high degree of statical indeterminacy, and also alternative paths through which loads can be transmitted. It is unusual—and undesirable—for large structures to have a member that is so vital that collapse of the member would mean collapse of the structure. In most cases structure collapse requires a large number of individual failures in various members. Some of those failures occur within the same members and cause them to collapse; others are more widely distributed and therefore do not cause member collapse. It is even possible for a structure to collapse by a mechanism involving several members, none of which has undergone complete collapse. Hence it is absolutely necessary to examine the strength of the structure as a whole, in order to identify any and all mechanisms which may cause collapse.

An example of member collapse is the collapse of a stiffened panel in the deck of a ship. In this case the load is the hull girder bending stress σ_x. The collapse could be caused by any of the three basic types of failure (for simplicity we here ignore combinations and interactions). Hence there are (at least) three separate limit values of σ_x, and the panel collapses when σ_x reaches the lowest of these three values. The magnitude of each of these limit values is determined by the design of the panel—its geometry, scantlings, material, and so on. Expressing this in more general terms, we say that each limit value Q_L is a function of the design variables **X**, and when we wish to indicate this dependency we shall write $Q_L(\mathbf{X})$. Thus, the limit values are under the control of the designer, and the safety of the structure is achieved mainly by choosing **X** such that each of the limit values $Q_L(\mathbf{X})$ exceeds the corresponding load Q by a satisfactory margin.

Safety, Uncertainty, and Structural Constraints

(A) STRENGTH CONSTRAINTS

Almost every design involves constraints, that is, conditions or requirements which must be satisfied. In structural design the most important constraints are the strength constraints—those which are aimed at providing adequate safety and serviceability. Structural safety is inherently probabilistic, it is the probability that a structure will not fail. The risk of failure arises from the various uncertainties which are involved: uncertainties in loads, load effects, and limit values of load effects, the latter being due to variations in material quality, workmanship, fabrication, and so on. There are two broad types of uncertainty, statistical and nonstatistical. The first type arises from genuine statistical randomness and can therefore be treated using statistical theory. The second type arises from subjective elements and from events which are not truly random and are difficult to predict. Wave loads and material properties are examples of statistical uncertainties; these can be dealt with adequately by statistical methods. Examples of nonstatistical uncertainty are those arising from the operation of the ship, such as operating errors (improper loading, mishandling, etc.) or a change of service.

Obviously, a rationally-based design procedure must be able to deal with both types of uncertainty in such a way that the required degree of safety (which is ultimately decided by society as a whole, through the medium of regulatory authorities and insurance rates) is achieved in a clear and explicit manner. The first type—statistical uncertainty—is dealt with by statistical methods. For all loads, load effects, and limit values which are probabilistic, statistical theory is used to estimate suitable values to be used for design. If the quantity involves a large number of peak values, such as wave-induced bending moment, then the calculation is based on *extreme values* of that quantity, and the particular extreme value which is selected for design is referred to as the *characteristic value*. In this text an extreme value is denoted by the symbol ^ placed above the quantity (e.g., $\hat{M}_w$ for wave-induced bending moment) and the characteristic value is denoted by a subscript c (e.g., $\hat{M}_{w,c}$).

Besides dealing with statistical uncertainty, a rationally-based design procedure must also provide some means whereby the designer (or the regulatory authority) can explicitly allow for the other uncertainties. This is done by specifying a minimum value of the margin between $\hat{Q}_c$ and Q_L. In practice this margin is usually specified in terms of a *safety factor* γ_0 which is the minimum factor by which Q_L must exceed $\hat{Q}_c$. In terms of γ_0 the constraint is of the form

$$\gamma_0 \hat{Q}_c(\mathbf{X}) \leq Q_L(\mathbf{X}) \tag{1.2.4}$$

But in addition to accounting for uncertainty, it is also necessary to utilize some further safety factors in order to allow for the degree of seriousness of each type of failure, both in regard to safety (loss of life) and serviceability (loss of revenue or loss of mission capability). Likewise it is also necessary to apply some factors to account for particular circumstances such as the type of ship (passenger, cargo, naval, hazardous cargo, etc.) and the cost and the operational importance of the ship. These various factors are known as *partial safety factors*. The required degree of safety is provided by the total factor of safety, which is the product of the partial safety factors. Thus in (1.2.4) γ_0 denotes the total factor of safety.

Strength constraints are very often nonlinear, for a variety of reasons. First, two of the three basic types of failure are generally nonlinear: instability in typical ship structural members is usually inelastic, and plastic deformation is inherently nonlinear. Therefore most of the limit value expressions $Q_L(\mathbf{X})$ are nonlinear, and hence most of the structural constraints, even those involving only one load, are nonlinear. Modes of failure that involve more than one load and/or more than one structural member are even more nonlinear. Also in a statically indeterminate structure the load effect in a member $Q(\mathbf{X})$, can be a nonlinear function of the design variables **X**.

Failure Involving Multiple Loads. For simplicity we have thus far considered limit states which involve only one load. For a limit state which involves two or more loads one of the loads is selected as the principal independent variable and an expression for its limit value is obtained as a function, not only of the design variables, but also of the other loads. For example, in the failure of deck plating due to plate buckling, the primary load is the longitudinal compressive stress σ_x. The limit value is the value of σ_x that causes buckling; in buckling terminology this is referred to as the "critical" stress $(\sigma_x)_{cr}$. If there is also a transverse stress σ_y acting on the plate this constitutes a second load and it influences the value of $(\sigma_x)_{cr}$. From plate buckling theory (Chapter 12) one can obtain an expression for this influence; in general form it is

$$(\sigma_x)_{cr} = f(\mathbf{X}, \sigma_y)$$

Hence the constraint equation takes the form

$$\gamma_0 \sigma_x \leq f(\mathbf{X}, \sigma_y)$$

and in the design of the plating the design variables must be such as to satisfy this inequality.

(B) Other (Nonstrength-Related) Constraints

As shown in Fig. 1.1, besides the strength constraints there are other constraints on the structural design arising from (1) operational requirements (e.g., minimum size of hatches, limitations on distortion and on vibration, etc.) and (2) fabrication considerations (e.g., maximum plate thickness for cold rolling, minimum spacing between stiffeners for welding, etc.).

These other constraints are relatively straightforward because they are expressed directly in terms of the structural design variables, being either minimum or maximum values of design variables or of ratios of design variables. An example of the latter is the fabrication constraint that the height of the transverse frame h_f must be sufficiently larger than the height h_s of a stiffener which passes through it, so that the cutout does not interfere with the flange of the frame. Thus, for example, if it were desired to restrict the stiffener height to no more than 80% of the frame height, then the constraint would be $h_s \leq 0.8\, h_f$. Constraints of this type are important, but there is no need to give them any further treatment in this book because they are straightforward and are already contained in structural design manuals and structural codes, such as the rules of the classification societies. Also, since they have a very simple mathematical form (linear inequality) it is a simple matter to incorporate them in any mathematical algorithm or computer program for rationally-based structural design.

Definition of the Objective in Structural Optimization

Rationally-based design, of its very nature, must have a goal or objective, and there must be some measure for assessing the merit of a design vis-a-vis that objective. Hence in a rationally-based design process the designer must be able to nominate and quantify the objective of the design, and the design process must be capable of actively and automatically achieving this objective to the fullest extent possible, subject to the constraints. This in turn means that the design process must include an optimization method which is capable of solving an optimization problem involving a large number of constraints of various types (linear and nonlinear, equality and inequality) and in which the measure of merit is totally flexible. That is, the optimization method should not restrict the measure of merit to linear expressions or to special cases (such as "least weight") since these may not suit the designer's needs.

Mathematical optimization of any kind requires that the measure of merit be defined as a mathematical quantity which is to be maximized (or minimized) and which is expressed as a function of the design variables. The measure of merit is therefore also referred to as the "objective function." In the overall design of a ship the structural design interacts with the other aspects of design, and something which is beneficial from a structural point of view may be detrimental in some other regard. Therefore the structural design objective should reflect the overall goal and the objective function should account for the results of interactions between the structural design and the other aspects of ship design. The goal depends first of all on the basic purpose of the ship, and in this regard the two principal categories are commercial vessels and naval vessels. In this section we consider the measure of merit for a ship structure, first for commercial vessels (while keeping in mind that many of the factors relating to this are also relevant to naval vessels) and then for naval vessels.

(A) Commercial Vessels

For commercial vessels the objective is profitability, either of the ship itself or of some larger system. The principal factors which determine a ship's profitability are shown (in greatly simplified form) in Fig. 1.6, taken from Ref. 3. The quantities that are strongly influenced by the structural design are outlined, and it is clear that the structural design can affect profitability

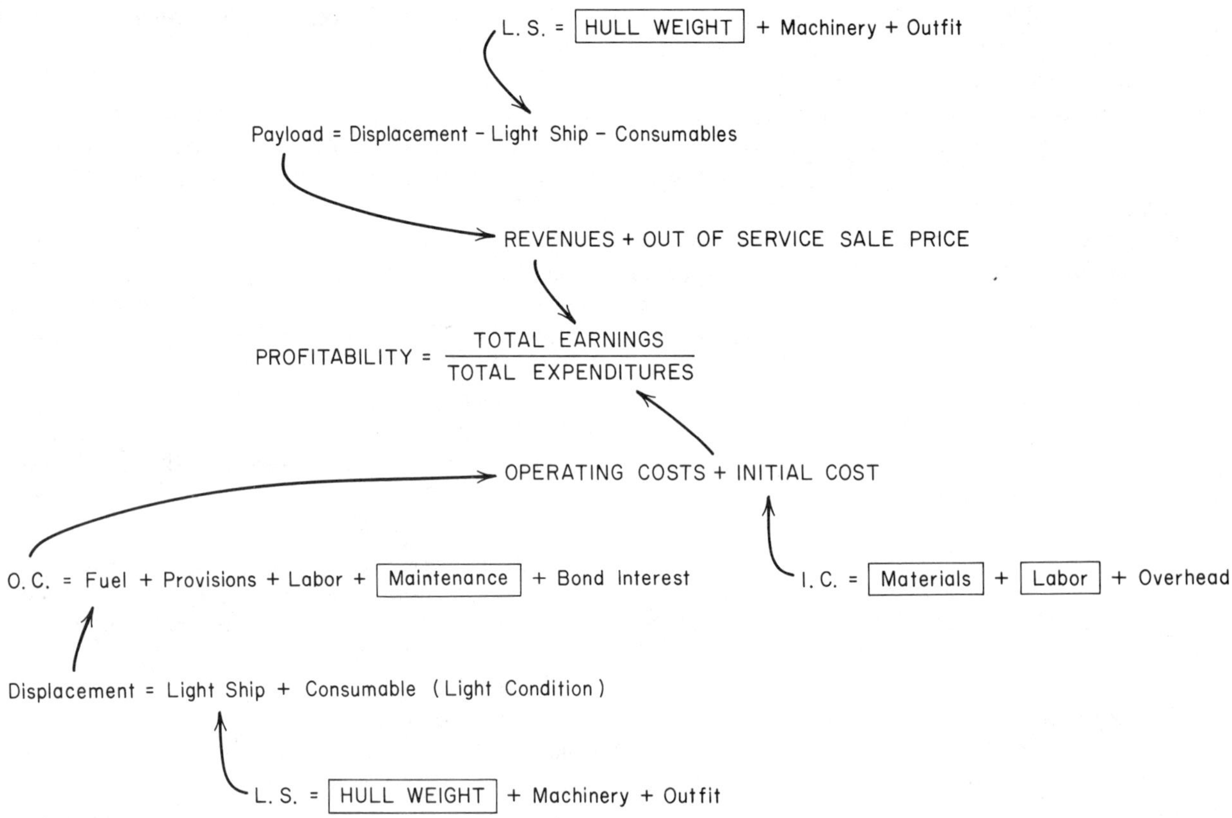

Figure 1.6 Principal factors in ship profitability.

at various levels and in various categories: payload, initial cost, operating cost, and so on.

The choice of the objective function also depends on which person or agency has the authority to decide; that is, it depends on whose behalf the designer is acting. In most cases it is the ship owner, but it may be the ship operator or the shipyard or the controller of some larger system in which the ship is to operate. For example, a shipyard which is responsible for the design as well as the construction would probably give greater importance to initial cost than would a ship owner, whereas the latter would have a greater interest in operational aspects and life-cycle economics.

Also, the factors and influences shown in Fig. 1.6 have different degrees of importance, and not all of them need to be included in the objective function. In many cases the only strong influence which the scantlings have on profitability is their effect on initial cost, and in such cases "least initial cost" is a sufficiently accurate objective.

Alternatively, with small weight-critical vessels such as hydrofoils and surface-effect ships, profitability or performance is determined almost entirely by hull weight because decreased structural weight allows a direct and corresponding increase in the payload. Therefore in these cases the weight of the structure is included in the objective function, as a different type of cost. Moreover, even very large vessels can be weight-sensitive, such that a savings in hull weight gives an increase in payload, as well as reductions in cost. The ways in which this occurs are indicated in Fig. 1.6, from Evans [3].

The question then arises as to what is the proper combination of the two goals of weight reduction and initial cost reduction. This question can only be answered by a careful study of the economic life cycle of each ship, to determine the trade-off between initial cost and increased revenue from weight savings. As an approximate means of allowing for this combination, Caldwell [4] has proposed a useful nondimensional objective function which combines weight and cost in the form

$$U = \nu\frac{W}{W_0} + (1 - \nu)\frac{C}{C_0} \qquad (1.2.5)$$

where W_0 and C_0 are the weight and initial cost of some basis or standard design, W and C are the weight and initial cost of a proposed design, and ν is a number which varies between zero (where least initial cost is the objective) and unity (where weight-saving is the objective). It can be shown that if the weight saved in structure can be taken up by cargo then the best value of the weighting factor ν in (1.2.5) is

$$\nu = \frac{1}{1 + R_c/R_w} \qquad (1.2.6)$$

in which

$$R_w = \frac{\text{weight of structure}}{\text{weight of cargo}}$$

in the basis design and

$$R_c = \frac{\text{annual costs arising from initial cost}}{\text{total annual cost}}$$

in the basis design.

For a large tanker R_w and R_c are typically 0.2 and 0.4, respectively, giving $\nu = \frac{1}{3}$, whereas for a small general cargo vessel the values are typically 0.25 and 0.25, giving $\nu = \frac{1}{2}$. Equations (1.2.5) and (1.2.6) thus indicate that increasing ship size and specialization will increase the importance of initial cost in the objective function. Obviously, more direct factors, such as increased labor costs relative to material costs, will have an even greater effect.

Savings in Fabrication Cost. For this reason, one of the most effective ways of reducing the initial cost of a large structure is to reduce the work content, and this can be achieved by simplifying the design and by increased standardization. There are opportunities for both of these measures at all levels of ship structural design, as shown by the following examples:

1. Overall hull structure.
   a. Simplifying the geometry of the hull. The work content in producing curved panels is many times that for flat panels. This has led to greater use of straight line sheer and camber (or their elimination altogether), greater use of knuckles, chines and flat surfaces, or plating with single, rather than double, curvature.
2. Hull module.
   a. Arranging the major structural elements such as decks, bulkheads, and frames so as to maximize the use of standard panels.
   b. Simplifying the structural arrangements in complicated modules, such as those at the fore and aft ends of the ship. Some examples are fore ends incorporating cylindrical bows, or frames mounted normal to the plating and greater use of flat areas and knuckles near the stern.
3. Principal members (girders, frames, stiffened panels).
   a. Selecting unidirectional stiffening, even when on the basis of the strength : weight ratio cross-stiffening is more "efficient."
   b. Use of corregated plating to avoid stiffening.
   c. Use of straight flanges on the web frames of curved shell plating, even though such frames may protrude further into the hold.
4. Local structure.
   a. Standardization of dimensions of plates and stiffeners to facilitate automated production and to achieve savings due to large batch ordering.
   b. Design of structural connections so as to reduce the work content and perhaps also to facilitate inspection and maintenance.

Decisions about the overall hull structure—the first level—are more a part of the overall design of the ship rather than the structural design as such. And the fourth level—local or detail design—is basically similar for all welded steel structures, and hence there are a number of references and design manuals already available for this level of design. Therefore this book focuses mainly on levels 2 and 3; it is at these levels that rationally-based ship structural design *per se* can provide the greatest benefits.

Savings in Operating Costs. In spite of its importance, initial cost is seldom the only significant cost that is affected by the structural design. Operating cost can also be strongly influenced by the structural design in such areas as:

1. Durability—by which is meant the ability of the structure to maintain its original level of structural efficiency. The design of a structure has an important influence on its resistance to damage, which in turn helps to avoid both lost revenue and the costs of repair.
2. Maintenance—the minimization of corrosion, which is of particular importance for marine structures, requires the provision of good access for inspection, painting, and so on and this must be considered at the design stage.

3. Habitability—noise and vibration levels are directly influenced by structural design.

(B) Naval Vessels

For naval ships the objective is to obtain the maximum possible mission capability over the life of the ship, subject to budget limitations. Thus cost is here a constraint instead of an objective. The two aspects of a structural design which have the greatest influence on mission capability are the weight of the structure and its vertical center of gravity (VCG):

1. Weight savings permits either a higher speed or more mission-related equipment (weapons, sensors, etc.) or increased range and endurance, or some combination of these.
2. A low VCG of the hull structure is of great benefit since most of the important weapons and sensor systems involve large topside weight. In fact, the provision of adequate stability is often the limiting factor on the number or size of such systems, particularly as a vessel gets older and it becomes necessary to fit more modern systems.

These objectives tend to produce a structure that is more intricate and involves less material. Hence for naval vessels the structural cost (i.e., the cost that is attributable to structure and is a function of the structural design variables) is mainly fabrication cost; the material cost is much smaller and, for a given material, it has little influence in determining the final optimum design.* Thus naval design involves a trade-off between weight, VCG, and fabrication cost. The designer seeks to determine the number, arrangement, and size of structural members which will give the lowest possible weight and VCG, subject to cost limitations and to a variety of other constraints requiring satisfactory strength, reliability, endurance, and functioning of the vessel. The constraint on cost is somewhat different from the other constraints. Rather than being an absolute limit it is a somewhat elastic barrier in which the rigidity of the resistance to further increase in cost depends on the cost : benefit ratio, that is, how much benefit the increase in cost will yield. Nevertheless, besides the cost/benefit type of constraint, there can also be an explicit upper limit on total cost.

*The possibility of using a very expensive material such as titanium would be investigated by making a separate optimum design using that material and then judging whether the weight/VCG savings is worth the extra cost.

The situation is illustrated in Fig. 1.7 in which the coordinates are the three main aspects: weight, VCG, and fabrication cost. The curved surface represents the combined and net effect of all of the constraints. It is actually an irregular surface—it is the outer envelope of all of the many individual constraints, each of which is itself a surface consisting of the lowest possible combinations of weight, VCG, and cost which just satisfy the requirements of that particular constraint. In different regions of the design space different constraints are the outermost or "active" constraints. Any point in the design space that does not penetrate below this surface is a feasible design, and any point actually on the surface is an optimum design. The surface indicates what sort of trade-offs are available to the designer.

For example, by raising the height of the strength deck or increasing its thickness the designer may achieve a lower overall weight at no extra fabrication cost, but incur a higher VCG. Alternatively, if he is willing to pay more for a more sophisticated and intricate structural arrangement (more stiffeners and thinner plating) he can achieve a lower weight, or a lower VCG, or both. The choice of which tradeoff to make and the selection of the final optimum combination must be left to the designer because this depends entirely on the particular mission requirements, cost limitations, and other design requirements and priorities.

The objective function U is a combination of weight W and VCG in the form

$$U = \nu \frac{\text{VCG}}{\text{VCG}_0} + (1 - \nu)\frac{W}{W_0} \tag{1.2.7}$$

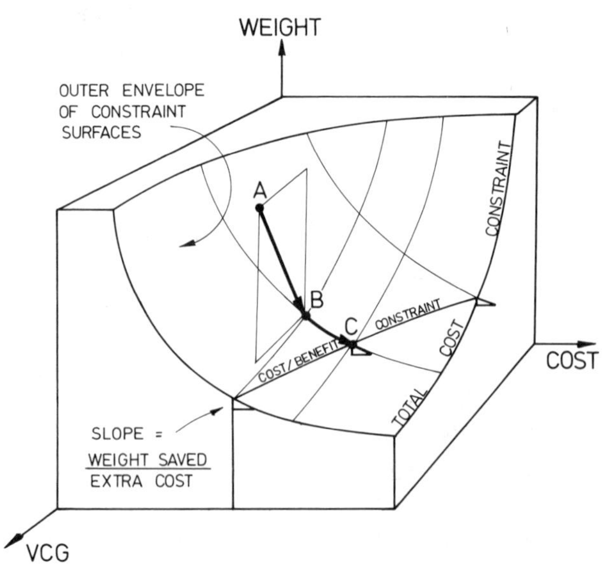

Figure 1.7 Design space.

where W_0 and VCG_o are the weight and VCG of a previous design of approximately the same size, in which the VCG is at a satisfactory height. The value of the parameter ν is then determined by making a few trial optimizations, starting with $\nu = 0.5$ and adjusting ν until the resulting optimal design has a satisfactory VCG. If the previous design was smaller or if it is desired to achieve a lower VCG ν should be increased.

For purposes of explanation the benefits of using a rationally-based optimization approach in the design of naval ship structures may be grouped into three categories. First there is a savings in weight and VCG simply because of the greater accuracy of the rationally-based method. In current practice design both the load effects and the limit values of the load effects are only known approximately, and therefore so also are the margins between them. To help prevent the actual margin from becoming less than the required margin (which allows for genuine and unavoidable uncertainties such as wave loads) the design margin is made larger than is actually required. But in rationally-based design the actual margin is known and so it can be set to the required value rather than an inflated value. The result is illustrated in Figure 1.8 which is an enlargement of Fig. 1.7. Point A represents a current practice design. The calculation and use of actual margins gives an immediate savings in structural weight, and the improved (in fact, optimal) distribution of material gives a reduction in VCG (point B in the figure). Both savings can be achieved with no additional fabrication cost. As indicated earlier, the proportion between the two can be altered to any desired figure by modifying the value of the parameter ν in the objective function.

But these savings are just the start; the designer can also investigate the question of cost. He can ease some or all of the cost constraints, which would then allow more intricate and efficient (but costlier) structural arrangements and proportions (point C in Fig. 1.8). The most common example is an increase in the number of stiffeners and a decrease in plate thickness. And finally, since rationally-based implies computer-based it gives the designer a far greater ability to examine other design alternatives and perform parametric studies, thereby achieving still further savings in weight and VCG. For example he can investigate the use of an improved steel or other material, with a higher ultimate strength or an improved fatigue resistance. These changes would have the effect of moving the corresponding constraint surfaces in Fig. 1.7 closer to the origin, and if one of them had been an active (outermost) constraint for the current optimum design, then the optimal design surface in that region of the design space also moves closer to the origin, and an immediate savings in weight and/or VCG is gained.

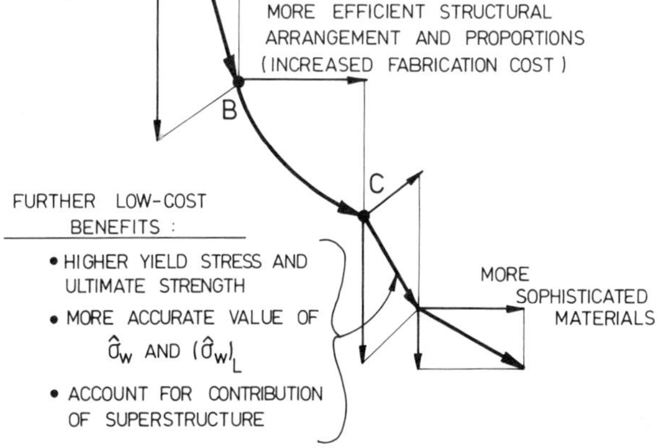

Figure 1.8 Benefits of rationally-based design.

Overall Procedure for Ship Structural Optimization

The final basic aspect of the rationally-based design procedure shown in Fig. 1.1 is the structural optimization process. This aspect consists of a mathematical method which utilizes the information provided by the other aspects and generates the design (i.e., the set of scantlings for all of the principal members) which maximizes the objective and which satisfies all of the constraints. The aim of the present section is twofold: to give an overview of structural optimization and also, now that we have discussed all of the aspects of the overall design procedure of Fig. 1.1, to explain briefly how that procedure works. For this it is not necessary to have a comprehensive or detailed knowledge of mathematical optimization theory. It is sufficient to know the basic features. Therefore, since the primary aim of this text is to present and explain the methodology—both theory and practice—of rationally-based structural design, there is no attempt to cover mathematical optimization theory, or even to give a complete coverage of structural optimization, but rather only those aspects which are needed by a designer. The coverage is in three parts:

1. In this section, a brief summary of the basic features of structural optimization, presented as part of a simple example of the overall design process of Fig. 1.1.
2. In the next section, a brief summary of the broad classes of optimization methods, some comments on these in relation to the requirements of preliminary ship structural design, and references where more detailed information on these methods may be found.
3. In Section 2.7 a summary of a "dual level" optimization strategy which permits the efficient optimization of large structures in which some constraints apply to the structure as a whole.

The second book, which is planned as a companion to this one, will provide further information on practical and design-oriented aspects of optimization, such as computer implementation, postoptimal design studies, and other aids for decision-making in structural design.

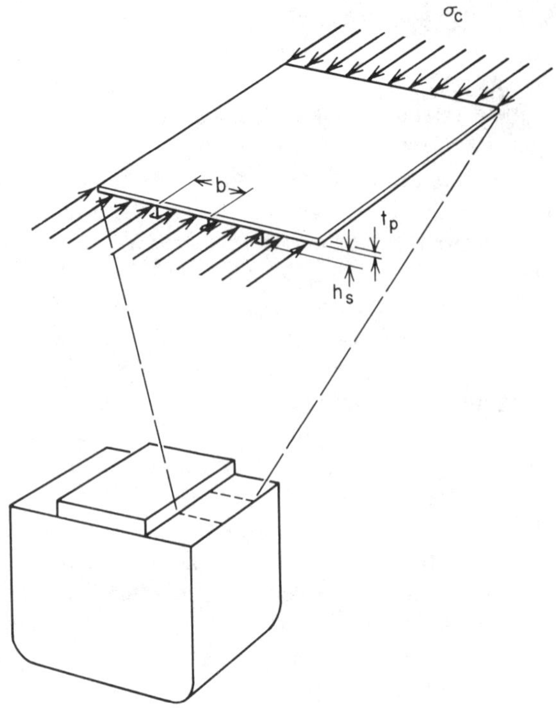

Figure 1.9 Stiffened panel example.

Sample Application of the Procedure for Rationally-Based Preliminary Structural Design

We now present a simple example of the rationally-based design procedure of Fig. 1.1, illustrating the various steps of the procedure and showing how the structural optimization step brings together and utilizes the results of the other steps. The procedure is intended for the structural design of an entire ship, but in order to have a simple example we will here apply it to just one small part of the structure—a stiffened panel in the strength deck of a vessel, as shown in Fig. 1.9. We will also make many simplifying assumptions, none of which are made in the actual procedure. For example, let us say that the panel has only two design variables: the plate thickness t_p and the stiffener height h_s. In reality the panel would have several more design variables: the number of stiffeners (or their spacing) and the web thickness and flange area of the stiffeners. But for this example let us say that the number and type of stiffeners have already been selected, perhaps because of a need to match some existing structure.

Calculation of Environmental Loads

The first step in structural design is to investigate the maximum or extreme value of each load so as to determine a suitable extreme value to be used for design, that is, the *characteristic load* Q_c. For simplicity let us say that the only significant load is hull girder bending moment M and that the only significant load at the panel level is the hull girder bending stress σ. In reality, of course, there would also be other loads acting on the panel (lateral pressure, shear stress, etc.) and the magnitude of these loads could only be obtained by a hull girder analysis, followed by a hull module analysis.

As will be discussed in Chapter 4 the wave-induced portion of the hull girder bending moment M_w is probabilistic and therefore statistical methods must be used to establish a *characteristic extreme value* for it, denoted as $\hat{M}_{w,c}$. For standard types of ships, a value for $\hat{M}_{w,c}$ is available from the classification societies, having been determined by research and by at-sea measurements for that type of vessel (Section 3.5).

Hull Girder Analysis

The next step is the hull girder analysis which consists of the calculation of:

1. The section properties of the hull girder (location of the neutral axis, value of the moment of inertia, etc.).
2. The stillwater bending moment M_s.

3. The combined total hull girder bending stress $\sigma_c = (\hat{M}_{w,c} + M_s)/Z$.

Since there are two values of wave bending moment (hogging and sagging) and several values of still water bending moment (corresponding to different cargo and/or ballast configurations) it is necessary to perform the hull girder analysis for several loadcases.

We note that in order to perform the hull girder analysis it is necessary to have some initial or starting value for the scantlings of all of the longitudinally oriented structure in the hull girder. For the design variables of the panel let us denote the initial values as t_{p1} and h_{s1}, or in vector form, $\mathbf{x}_1$. These initial values are arbitrary; they are required simply because any computer calculation or analysis (such as steps 2, 3, and 4) requires specific numerical values for all quantities. These values do not require any calculation by the designer: he can either select standard values, or arbitrary values, or values from some other design. This will be discussed further when considering the structural optimization step.

Hull Module Analysis

The next step is to perform the hull module analysis, for the current scantlings, $\mathbf{x}_1$. As shown in Fig. 1.1 this step and all of the subsequent steps constitute a cycle (the hull module design cycle) which must be performed repeatedly, because at the end of each cycle the values of the design variables (which comprise all of the scantlings of the hull module) are altered by the optimization process. This alteration makes it necessary to repeat the hull module analysis and the subsequent steps, using the new values of $\mathbf{x}$. The hull girder analysis does not need to be included in this cycle because the bending moment and shear force do not change. All that is required is a recalculation of the section properties of the hull girder, whereupon all of the new hull girder bending stresses can be calculated from the basic beam equation $\sigma_c = M/Z$.

For reference we will denote the current values of $\mathbf{x}$ as $\mathbf{x}_i$ (or t_{pi} and h_{si}); that is, $\mathbf{x}_i$ represents the scantlings which are used during the ith hull module design cycle.

The purpose of the hull module analysis is to obtain all of the values of the individual load effects in each of the principal members, for each loadcase. In this example we are examining only one principal member—a deck panel—and we are saying, purely for simplicity, that the only load effect is the hull girder stress. Thus we are, in effect, ignoring the hull module analysis.

Limit State Analysis

To illustrate this step let us assume that there are five limit states for the panel: three types of compressive collapse, with collapse being initiated by (1) plate buckling, (2) torsional buckling of the stiffeners, or (3) flexural buckling of plating and stiffeners acting together; and also (4) large plastic deformation under tensile load and (5) fracture due to fatigue. Hence the five limit values are the three buckling stresses denoted by $\sigma_{bj}(\mathbf{x})$ with $j = 1, 2,$ and 3, the yield stress σ_Y, and the fatigue-derived limit on wave-induced stress, $(\hat{\sigma}_{w,c})_L$ corresponding to the expected number of hogging and sagging cycles in the life of the ship (which is about 10^8 for a 25-year life).

The purpose of the limit state analysis is to calculate the limit values for the current values of $\mathbf{x}$. In this example two of the limit values are material properties and do not depend on $\mathbf{x}_i$, and so the limit analysis here consists of the calculation of the three buckling stresses, $\sigma_{bj}(\mathbf{x}_i)$. The theory for this is presented in Chapters 13 and 14.

Formulation of Constraints

The next step is to formulate the constraints for the optimization problem. The general form of a constraint was given in (1.2.4). Applying that here gives the following equations for the five panel constraints

$$\gamma_0\sigma_c \leqslant \sigma_{b1}(\mathbf{x}) \qquad \gamma_0\sigma_c \leqslant \sigma_Y$$
$$\gamma_0\sigma_c \leqslant \sigma_{b2}(\mathbf{x}) \qquad \gamma_0\hat{\sigma}_{w,c} \leqslant (\hat{\sigma}_{w,c})_L$$
$$\gamma_0\sigma_c \leqslant \sigma_{b3}(\mathbf{x})$$

As mentioned before, an actual design involves several loadcases, which here would mean several different values of σ_c. Hence for each constraint it is necessary to use whichever value of σ_c is the worst for that type of limit state. Moreover, in reality, each loadcase usually involves several loads acting on the member, in various different combinations, and so the search for the worst combination must be very systematic and thorough.

As explained in Section 1.5 the total factor of safety, γ_0, is made up of several partial safety factors which are chosen in accordance with: (1) the degree of (nonstatistical) uncertainty which exists in regard to both the load and the limit value, and (2) the degree of seriousness of the limit state. For this example the degree of seriousness is about the same for all constraints since they all refer to collapse rather than unserviceability. Although there would be differing degrees of uncertainty in σ_{bj}, σ_Y, and $(\hat{\sigma}_{w,c})_L$ let us say, again purely for simplicity, that γ_0 is the same for all five constraints.

Since the three limit values for buckling are functions of the design variables t_p and h_s, the three buckling constraints are equations in these two variables and can be plotted as curves on a diagram of t_p versus h_s, as shown in Fig. 1.10. In this type of diagram any specific combination of t_p and h_s—that is, any specific panel design—is a particular point on the diagram. For example, point A represents the initial or starting design, corresponding to t_{p1} and h_{s1}. The plane of the diagram represents all possible designs and is referred to as the *design space* (or hyperspace; the concept is extended to any number of design variables). The constraint equations are inequalities and therefore each curve is the boundary between all designs that satisfy that constraint and all that do not. In the figure the permissible side of each constraint is indicated by shading the unpermissable side.

The two constraints corresponding to tensile yield of the panel and fatigue fracture cannot be drawn in the figure because the limit values are material properties and therefore the limit states are essentially independent of the design variables. The only way in which the design variables **x** are involved is through σ_c, because σ_c is inversely proportional to the hull girder section modulus Z. But the plate thickness of a single panel has only a small influence on Z, and h_s has even less influence. That is, even a large change in t_p would cause only a small increase in the section modulus, and hence only a small decrease in σ_c. If either of these constraints was not satisfied, such that it was necessary to reduce σ_c, then this would require increasing the plate thickness of *all* of the deck panels and probably also all of the bottom panels. In other words, these two constraints relate to the overall structure, via section modulus, and not just to this panel. Constraints of this type are discussed in Section 2.5. Also, the determination of the *optimum* combination of thickness changes requires that all of the panels be redesigned together, in a coordinated manner.

We have already seen that a principal member cannot be analyzed in isolation, but only in conjunction with the other principal members, through the medium of the hull module analysis. We now see that in optimization the situation is similar—a principal member cannot be designed in isolation, but only in conjunction with the other principal members. For the present example let us assume that the current values of deck and bottom thicknesses are already sufficient to give a reasonable value of σ_c, such that these two constraints are not violated. In that case they will not become

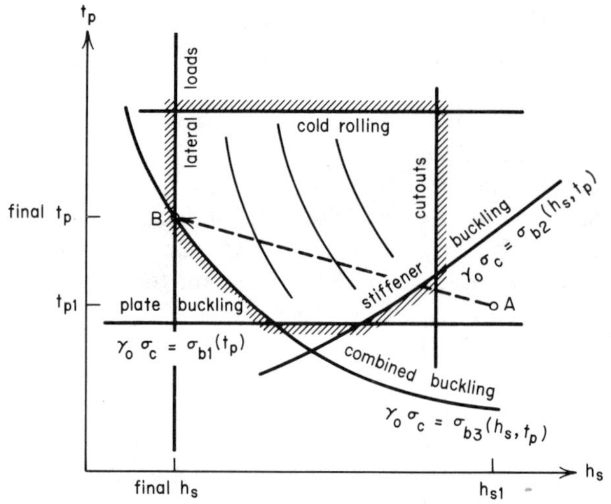

Figure 1.10 Design space for optimum design of a stiffened panel.

violated due to a reduction in thickness of just this one panel, and so we will not consider them further in this example.

Besides the strength constraints there are other constraints on the structural design arising from operational requirements and fabrication considerations. For the deck panel, for example, there might be a maximum plate thickness for cold rolling, a minimum stiffener height (for a given stiffener spacing b) to support lateral loads, and a maximum stiffener height (for a given height of the transverse deck beams h_f) so that the cutout for the stiffener leaves sufficient web area in the beam.

We note that in a more realistic example the stiffener spacing and the frame height would also be design variables, and hence the latter two constraints would only restrict the ratios h_s/b and h_s/h_f; not the absolute value of h_s.

These additional constraints are also drawn on Fig. 1.10. Taken as a group the constraints define a region of the design space in which none of the constraints is violated. This region is known as the *feasible* design space, and corresponds to the shaded outline in the figure.

Objective Function

In this example let us assume that least initial cost has been chosen as the objective for this design. The cost of a panel would depend mainly on the amount of steel used and the cost of fabrication. In the present example t_p and h_s are the only variables and the amount of steel is linearly proportional to each of them, but in general the amount of steel is the product of two design variables; for example, in a stiffener web it is $h_s t_w$, where t_w is the web thickness. The fabrication cost is related to the design variables in a completely different way from the material cost, and therefore the sum of the two costs will be even more nonlinear. Thus the cost function $C(\mathbf{X})$ is, like many of the constraints, a nonlinear function of the design variables. In Fig. 1.10 a typical cost function for the panel is indicated by means of contour lines of constant cost.

Structural Optimization

The figure also shows two particular combinations of t_p and h_s; that is, two specific panel designs. Design A represents the initial design, that is, the starting point for the optimization process. It may correspond to an actual panel in some existing ship, or it may be purely arbitrary. For any good optimization process the latter is sufficient. Thus the starting design need not be feasible (and often will not be, even if it does correspond to an actual design, since the loads and/or limit values for the current design are different). Design A is not only infeasible (it violates the maximum h_s and the stiffener torsional buckling constraint) but also expensive. Design B is the optimum design; it is the point within the feasible region which has the least cost. The task of the mathematical optimization process is to find this optimum point, from any starting point, and to do so with as little computation as possible.

Postoptimality Information

In Fig. 1.10 the optimum design is governed by two constraints: combined buckling of stiffener and plating and the minimum height of stiffeners to support lateral loads. Another feature of a good optimization method is that, if requested, it can inform the designer as to all the circumstances relating to the optimum design, such as what are the governing constraints. Other useful information which should be available includes: how steep or flat is the optimum; that is, how much extra cost would be incurred in moving away from the optimum design point in various directions; what would be the effect on the cost of relaxing or tightening any of the active constraints; and if relaxation were possible, which direction would give the greatest further decrease in cost.

Optimization Methods for Large Structures

Mathematical optimization is purely a mathematical process and therefore it can and should be a fully automated process. That is, in the overall design process of Fig. 1.1 the optimization step is simply a "black box," which performs a specific task: it accepts as input an objective function and a set of constraints, and it returns as output the specific optimal values of the design variables, that is, the values that maximize the objective while satisfying all of the constraints. Since there is only one optimum point in the design space the optimization task is straightforward and unambiguous; there is no need or reason for any intervention by the designer. The only requirement is that the method must find the optimum rapidly and efficiently. The particular manner in which it does this is not important, and any optimization method that meets the requirement can be used in the overall design process of Fig. 1.1, without any need for the designer to have a detailed knowledge of it or of mathematical optimization theory. Therefore in this section we merely describe the broad classes of optimization methods, identify those methods which have been proven to be capable of meeting the requirements of ship structural optimization, and provide references from which detailed information about these methods may be obtained.

Types of Nonlinear Optimization Methods

During the past two or three decades many optimization methods have been developed which have at least some nonlinear capability. The majority may be grouped into three categories.

- Fully nonlinear methods, such as mathematical programming methods and various algorithms for unconstrained minimization.
- Special purpose methods, such as fully stressed design, the optimality criteria methods, and geometric programming.
- Methods based on sequential application of linear programming.

In all there are many different optimization methods and nearly all of them can be used for structural optimization in some application or other. However, most of them tend to be suitable only for a particular type of problem. Reference 5 gives a basic explanation of all of the principal methods and demonstrates their applications. A brief review of various methods with particular reference to their application in ship structural optimization is given in [6].

In general, it may be said that the majority of the methods are suitable either for nonlinear problems which are small (i.e., which have only a few design variables) or for large problems which are only slightly nonlinear. As far as the author is aware, only the third type of method—sequential linear programming—has been shown to be capable of solving the large nonlinear problem which is involved in ship structural optimization with sufficient speed, computational efficiency, and generality. The fully nonlinear methods perform satisfactorily for small problems (say five or six design variables) but the amount of computation increases sharply with problem size, and for a structure as large and complex as a hull module these methods are simply not feasible. The special purpose methods are rapid and efficient but they are too restricted; they cannot handle constraints that are highly nonlinear and/or involve many design variables, and the first two (fully stressed design and optimality criteria) cannot handle an arbitrary (user-specified) nonlinear objective.

Methods Based on Sequential Linear Programming

In this type of method all of the nonlinear functions [the objective function and the limit values $Q_L = f(\mathbf{x})$] are replaced by linear approximations and the linearized problem is solved very rapidly by the well-known method of linear programming, using the Simplex algorithm. This process is repeated sequentially, with the linear approximations being recalculated at each new design point. The original version of sequential linear programming was developed around 1960 by Kellog [7] and by Griffith and Stewart [8]. In this first version the linearized form of each function f was simply the linear terms of its Taylor Series expansion, which involves the various first derivatives of f with respect to each design variable: $\partial f / \partial x_i$. It was found that unless all of the functions were only slightly nonlinear the linearized problem was too different from the actual nonlinear problem and the process would not converge. Hence for many years the method was limited to problems which were only moderately nonlinear. But it was subsequently shown that this limitation can be overcome by using some second-derivative terms in formulating the linearizations. Various second-order methods have been developed. In 1976 Hughes and Mistree presented the SLIP2 method [9], which uses the second derivatives of the nonlinear functions (but only terms of the form $\partial^2 f / \partial x_i^2$; mixed derivatives $\partial^2 f / \partial x_i \partial x_j$ are not required) to obtain a more accurate linearization of both the objective function and the constraints. This method was developed specifically to meet the requirements of rationally-based ship structural design, and it is this method which is used in the MAESTRO computer program. As shown in the examples and references given in Section 1.3, SLIP2 is able to solve problems involving a large number of constraints of various types (linear and nonlinear, equality and inequality) and in which the objective may be any user-specified nonlinear function of the design variables. Most importantly the method is very rapid and cost-effective. The SLIP2 method is not limited to structures; it is a general purpose method which can be used for commercial, industrial, or other optimization applications. The complete mathematical algorithm and logic structure for SLIP2 are given in Ref. 10.

Another second-order version of Sequential Linear Programming was developed by Murtagh and Saunders [11]. In Ref. 12 this method is demonstrated for several large-scale optimization problems in commerce and industry. Although these examples do not include structural optimization it is clear that the method is also suitable for this application.

Pedersen [13] has presented a systematic method for using "move limits" in Sequential Linear Programming which overcome most of the convergence problems referred to earlier.

Coverage and Plan of this Text

Having now had an introductory look at all of the basic aspects of rationally-based design, we can now give a more precise summary of the scope of the book and the organization of topics. The rest of Chapter 1 is devoted to some further basic aspects of structural safety, probabilistic design, and the use of partial safety factors. The specification of safety factors is primarily the responsibility of the classification societies, and therefore this text does not seek to determine or recommend any specific combination of safety factors or any specific values for them. Instead, these are described in general terms, and the combinations and values given are purely for illustration.

Chapter 2 summarizes the four major analysis tasks in rationally-based ship structural design: calculation of loads, of the structure's response to the loads, of the various limit values of each response, and optimization. Thus chapters 1 and 2 give the overall methodology of rationally-based structural design.

Chapters 3 to 8 deal with the structural response of the ship. Chapter 3 presents the hull girder response analysis. This is the oldest and best established aspect of preliminary structural design, and was developed in conjunction with rule-based design. Therefore this text does not cover all of the traditional topics, but only those which continue to be relevant for rationally-based design. Chapter 4 summarizes the theory and techniques for obtaining a more precise estimate of wave bending moment, when account is taken of the probabilistic, dynamic, and nonlinear aspects of wave loads.

Chapters 5 through 8 present the various aspects of hull module analysis, which was earlier shown to be one of the most important and most distinctive features of rationally-based structural design. In the hull module analysis the method of structural modeling is very similar to frame analysis. Moreover much of the development of the finite element method came about as a generalization and extension of frame analysis. For this reason frame analysis provides a good introduction and foundation for the finite element method, particularly for the design-oriented finite element method which is used for the hull module analysis. Therefore the explanation of the hull module analysis is presented in a sequence of four chapters. Chapter 5 presents a summary of frame analysis, whch serves mainly to introduce all of the basic concepts. Chapter 6 presents a special beam element which is the basis for a more accurate and more versatile type of hull girder analysis, permitting the analysis of a non-prismatic hull girder and the calculation of torsional response. Chapter 7 presents the basic features of the finite element method and introduces some basic two-dimensional elements which are of use in hull module analysis. Finally, Chapter 8 presents various aspects of the modeling strategy for the hull module analysis and some further finite elements that are useful for this analysis.

Chapters 9 to 14 represent the largest part of the book and deal with the limit analysis of the principal members: plates, columns, beam-columns, and stiffened panels. In each case the elastic aspects are covered first and then the inelastic. The former consists mainly of material culled from many different sources and augmented or generalized in whatever way was necessary to make it suitable for computer-based analysis. For example, for all of the various types of elastic buckling it is necessary to have either explicit expressions or numerical procedures for calculating the values of the buckling coefficients. In many cases these coefficients were only available in charts or tables, and so suitable expressions were developed. Some methods are new, such as that for clamped beam-columns in Section 11.3, and the generalized orthotropic plate theory which allows for shear distortion, presented in Section 10.2.

For inelastic limit analysis there were only a few cases in which a suitable method was available, that is, a method which had the required combination of computational economy, accuracy, and generality of application. In some cases a suitable method was obtained by combining and adapting existing methods (e.g., the ultimate compressive strength of plates in Section 12.6. In other cases a new method was developed, for example, the estimation of the permanent set of plating (Section 9.4) and the ultimate compressive strength of stiffened panels (Chapter 14).

In regard to the limit analysis of ship panels and girders, mention should be made of the *Ships' Load and Strength Manual,* published by the Norwegian classification society Det norske Veritas [14]. This manual contains a comprehensive set of approximate explicit formulas for the strength of girders and stiffened panels, that is, formulas for the limit values of the forces and stresses in these members corresponding to various types of failure. In general these formulas have been obtained by making suitable engineering approximations to the underlying theory. Formulas of this type are an essential part of any practical method of rationally-based structural design, and the manual provides them in a systematic and concise manner. It is not aimed at giving theory or derivations and it does not consider the hull module, either in regard to response analysis or limit analysis, since neither of these analyses can be achieved by or put in terms of explicit formulas. Therefore the author considers the manual to be a very useful adjunct and complement to this text, both in purpose and in scope.

Chapters 15 to 17 deal with the limit state analysis of the hull module. Because of the structural complexity of a hull module and the complex interaction which often occurs between instability and plastic deformation, this analysis requires an *incremental load-deflection* approach, which traces the history of the collapse. Elastic instability is examined in Chapter 15, which introduces nonlinear (incremental) frame and finite element analysis and shows that for a complex three-dimensional structure, in which the members carry both lateral and axial loads, an incremental approach is essential; the classical bifurcation buckling analysis is not sufficient. Chapter 16 deals with failure by large plastic deformation. It presents a new incremental method of plastic frame analysis and shows that here also the classical method of plastic frame analysis is not adequate for the limit analysis of a hull module, principally because of the interaction between instability and plastic deformation. Finally, in Chapter 17 these two incremental methods for the limit analysis of large structures—one for failure by instability and the other for failure by large plastic deformation—are combined to give a comprehensive method for the limit analysis of a hull module.

1.3 PRACTICALITY AND PERFORMANCE OF THE METHOD

As explained earlier, the aims of this text are to present a method for rationally-based design and to explain all of the basic theory and analysis methods which are required for that method. But the first requirement of a design method is that it be practical, and a method which lacks this characteristic is of no real value, no matter how "rational" it may be. Thus when a method is proposed which has a much larger theoretical content and is entirely computer-based, the questions which immediately arise are: what stage has it reached regarding implementation, availability, and actual use; what are its benefits and its performance characteristics; what does it require; in short: how practical is it? Therefore before proceeding any further with the theory and methodology it is necessary to demonstrate that the method of rationally-based design which is presented herein is truly as powerful, practical, and beneficial as implied in the previous section, and to provide factual information that will answer all of the questions just raised.

This can readily be done because the method has been developed over a long period of time (since 1972) and throughout all of this period the practical aspects have been given just as much attention as the theoretical aspects. As each portion of the method was developed it was computer-implemented and tested before being accepted, and each portion received many further tests as other portions were added or modified. If at any stage some portion was found to be impractical it was promptly discarded and work was begun on a replacement. In the first six years or so there were many discards.

Two milestones in the overall development were the completion of a first version of the method in 1975, for which the computer program was called AUSTROSHIP, and a second version in 1978, for which the program was called SHIPOPT. This was followed by a series of validation tests; the finite element portion was validated against DAISY, a large general purpose finite element program owned by the American Bureau of Shipping, and the structural optimization portion was validated by a series of formal, full-scale design studies involving four ship types: a 14,000 dwt general purpose cargo vessel [15], a 96,000 dwt segregated ballast tanker and a 140,000 dwt bulk carrier [16], and a destroyer [17]. Finally, in 1981 and 1982 the complete method was implemented in the computer program known as MAESTRO.

The practicality and performance of the method, and also of MAESTRO, may be judged from the results of the validation tests and design studies. All of them showed a similar performance, and since Ref. 16 is the most comprehensive study most of the results quoted here are from that reference.*

Since this section deals mainly with the features and performance of a particular computer program—MAESTRO—it is perhaps appropriate to emphasize that the subject matter of the book is not a computer program but rather the underlying theory of, and a general methodology for rationally-based structural design, which is necessarily computer based. The theory and methodology presented in this book can serve as the foundation for various computer programs, some more general and some more specific. It is not limited to one particular program.

Use of MAESTRO for Structural Evaluation

Being a program for rationally-based design, MAESTRO is organized along the lines of the design process of Fig. 1.1. Thus, corresponding to steps 3, 4, and 5 it contains:

1. A special high-speed design-oriented finite element method which calculates the load effects **Q** (deflections and stresses) in all of the principal members for all loadcases.

*Only published studies are quoted so that, if desired, full details may be readily obtained. These studies were all performed using the predecessor program, SHIPOPT; the author has repeated all of them using MAESTRO and found that the computing costs remain the same, in spite of the many new features and capabilities of the latter. For example, MAESTRO accommodates non-prismatic ship sections such as bows and sterns and changes in frame spacing, and it includes overall limit states as well as those of principal members. These and other features will be demonstrated in further design studies.

2. A set of coordinated subroutines which perform limit state analysis, examining all relevant types of failure and calculating $\mathbf{Q}_L$, the limit values of the load effects, for each different principal member and for all loadcases.
3. Other subroutines which formulate the constraints against each type of limit state for each different principal member. This involves searching the values of $\mathbf{Q}$ and $\mathbf{Q}_L$ to find and use the currently worst combinations of these two quantities. The program then makes a note of the corresponding lowest value of the margin of safety for each limit state, and the location and loadcase where each lowest value occurs. If requested it prints all of this information in structured easy-to-read tables (in several optional levels of detail).

These three features make the program a powerful tool for a variety of structural analysis and evaluation purposes, in addition to optimum design. When used in this mode the program executes only one cycle and stops just short of the optimization step. Because the finite element portion is extremely rapid and because a given set of input data is easily modified, the designer can quickly determine the effect of a design change which he has in mind. Moreover, the structural evaluation portions of the program have many valuable applications outside of, or immediately following, the design of a ship. Some examples are:

1. To check the structural adequacy of a proposed design.
2. To investigate proposed structural modifications.
3. To assess the seriousness of structural damage and the degree of urgency of repair.
4. To assess the structure after an actual or a projected corrosion wastage.

A common example of the first application is the checking of a proposed design prior to or as part of the classification approval process. For large and/or nonstandard ships classification societies usually require that a finite element analysis be made of the hull structure in order to check whether the general stress levels are satisfactory and whether there are any particular locations of overstressing. The analysis is usually performed in two stages: a three-dimensional "coarse mesh" analysis of a hull module (much the same as MAESTRO) followed by a series of two-dimensional "fine mesh" analyses of selected areas. MAESTRO uses a level of modeling similar to the "coarse mesh" analysis used with other finite element programs, together with an internal logic developed specifically for high computational speed. As shown in the bulk carrier example, MAESTRO is approximately 12 to 15 times faster than conventional finite element programs. Also, MAESTRO uses more sophisticated finite elements and therefore, in spite of the coarse mesh, it gives all of the important stress values in all of the principal structural members of the ship. Moreover, besides calculating the stresses, MAESTRO also performs the complete limit state analysis just described and produces a series of tables giving a complete assessment of the vessel's structural adequacy, in various optional levels of detail: first, overall adequacy—are there any problems and if so which regions of the ship. Then for each region it gives the location and loadcase of the lowest value of safety margin for each type of limit state. Finally, if desired, it gives a complete report on each principal member in each region: its lowest value of safety margin for each relevant limit state, together with the corresponding loadcase, and the various stresses in each principal member for all of the loadcases.

These features, together with a simple, user-oriented data input, make MAESTRO ideal for obtaining a rapid and yet thorough evaluation of the adequacy of a proposed design. This is useful for two purposes:

1. To assess a design in which there are some nonstandard aspects, but not sufficiently unusual to warrant a detailed (fine mesh) finite element analysis.
2. For designs that are nonstandard, to determine whether there are any regions which require a two-dimensional fine mesh analysis and if so, where and for which loadcase. These detailed analyses can then be performed immediately because MAESTRO provides all the required boundary conditions.

Selection of Design Objective

MAESTRO allows the designer to achieve an optimum ship structure and leaves him free to specify how the optimality is to be measured. For commercial vessels the usual objective is maximum profitability over the ship's lifetime. The factors which most affect profitability are initial cost and operating revenue. Initial cost is a combination of material cost and fabrication cost, and to a first approximation these two may be expressed in terms of the volume (or weight) of material and the combined lengths of all of the girders, frames, and stiffeners. These are the parameters which were used in Ref. 16, but the program can accept any cost algorithm defined by the user. The specific algorithm contained four factors:*

1. Cost per unit volume for stiffened panels.
2. Cost per unit length of stiffening for stiffened panels.
3. Cost per unit volume for web frames and girders.
4. Cost per unit length for web frames and girders.

The other principal aspect of profitability—operating revenue—is determined mainly by cargo capacity. In bulk carriers a savings in hull steel weight gives a corresponding increase in cargo deadweight, and hence revenue. The additional revenue due to weight savings (or, on a cost basis, the lost revenue due to weight increase) can be allowed for by increasing the volumetric cost factors.

The four cost factors are part of the data input and if desired can be different for different regions of the hull module. The lineal (cost per unit length) factors would reflect such items as welding costs and would influence the optimum number of stiffeners in each strake of plating. These factors might vary according to which shipyard is building the ship. For example, one shipyard might have more automatic welding machines so that its cost per unit length of stiffener weld might be cheaper. In this case, the optimum design would probably have more stiffeners and less steel than at another shipyard with higher welding costs. The type of ship could also influence the cost function. In a double bottom bulk carrier, one would want to penalize increased double bottom height since this reduces cargo capacity. In this way, the program would automatically look first at other structural changes before increasing the double bottom height.

Example 1—Segregated Ballast Tanker

The first example from Ref. 16 is a segregated ballast tanker. Since one of the main purposes of this study was to assess the economic benefits of rationally-based design, all of the design specifications (principal dimensions, geometry, loads, etc.) were the same as for an actual, rule-based, manually produced design. As explained in Ref. 16 special steps were taken to avoid any bias in favor of MAESTRO and to remove the rather uncertain question of corrosion allowance from the comparison. In rationally-based design there is a clear distinction between steel that is required for adequate strength and steel that is provided in order to allow for corrosion. MAESTRO provides only the former; the latter must be added on after the optimization.

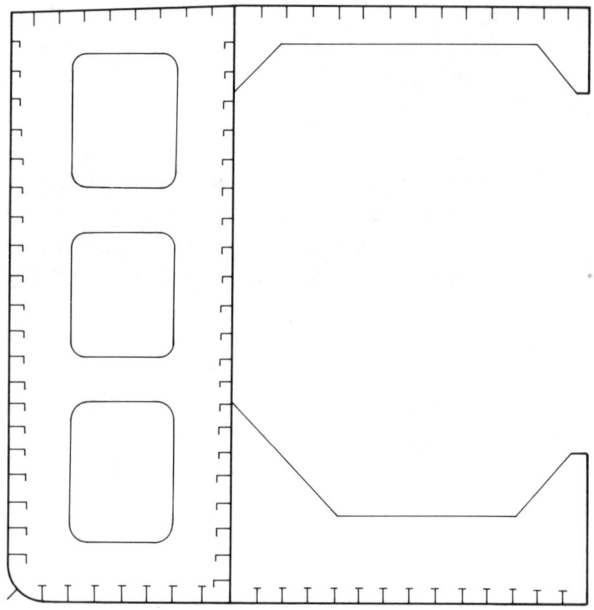

Figure 1.11 Basis ship: 96,000 dwt tanker.

The transverse section of the basis ship is shown in Fig. 1.11. For the three cargo tank lengths which comprised the MAESTRO structural model the cost of the basis design was 9708 cost units, and the structural weight (which is automatically calculated by MAESTRO) was 8050 tons. As mentioned earlier, the initial scantlings for MAESTRO are completely arbitrary and so in this case, in order to provide a direct and graphic comparison between the rule-based design and the MAESTRO design, the scantlings of the former were used as the initial scantlings for MAESTRO. The performance of MAESTRO is shown in Fig. 1.12.† The solution for the optimum design required 12 design cycles, which took about 12 minutes of computer time on an IBM 370/158. The total computing cost was about $130. The resulting optimum design had a total life-cycle savings (in which increased revenue from weight savings is converted to and subtracted from initial cost) of 8477 cost units, which is a 13% improvement on the basis or current practice design. The savings in initial cost was 6%, which for a tanker of this size represents a savings of the order of one million dollars.

*As shown there, the costs are not, and do not need to be, actual dollar costs, but rather cost indicators, or indices, which portray the correct relative proportions of the various costs.

†Including solutions with two other quite different sets of initial scantlings.

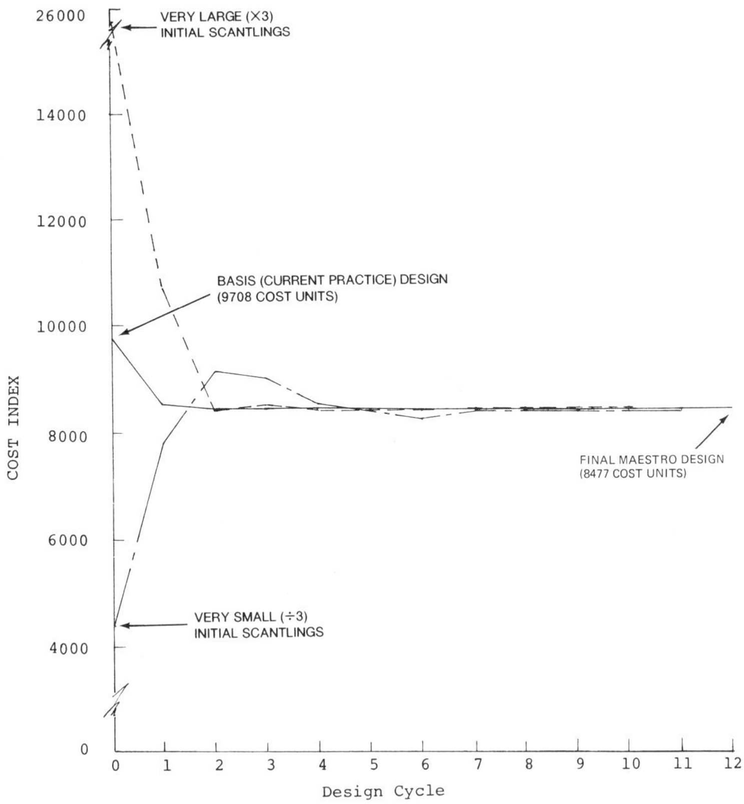

Figure 1.12

Effect of Using Standard Sections

The foregoing savings will be decreased slightly by the need to use standard plate thicknesses and standard rolled sections for the stiffeners. MAESTRO deliberately treats these design variables, and also the stiffener spacing, as continuous variables in order to avoid the enormous computation and complexity of discrete variable optimization. This also has the advantage of allowing the designer himself to decide these questions, and to do so *after* he has seen what the idealized optimum is. This is very helpful information and with it he can make a better decision as to how much standardization he wants. The standard sizes do not all have to be the next larger size; it is usually sufficient to choose the nearest size, unless this would mean decreasing all of the scantlings in a particular stiffened panel. After inserting standard sizes wherever he wishes, the designer then runs MAESTRO for one more cycle with all of these design variables fixed (a standard feature on MAESTRO). The program then automatically adjusts all other scantlings so as to obtain the new (and final) optimum design which has all of the standard sizes and still satisfies all of the constraints. By observing the new optimum cost the designer can immediately see how far from the idealized optimum his selection of standard sizes has taken him. In most cases the departure is small. For example in the tanker the insertion of standard sections reduced the life-cycle savings from 13% to 11% compared to the rule-based design, which still represents a substantial savings. Moreover this guarantees that the 6% savings in initial cost is a real savings, being based on realistic limitations to standard sections.

Ability to Repeat a Preliminary Design

Since the preliminary structural design, by definition, does not examine local effects, it often happens that in the detail design it is necessary to increase some of the scantlings because of local loads, cutouts, and so on. This will happen more frequently when the preliminary design is an optimum design, since there is no excess steel in such a design. In most cases the cost increase is small. However if there are many such loads or other influences, such that the subsequent cost increase is found to be large, then these are not really local effects; they are general effects and the preliminary design should be redone with these effects included. With manual design this is a large and time-consuming task; in many cases the more expensive locally adjusted design would have to be accepted. With MAESTRO it is a relatively easy matter to add the loads or make an approximate modeling of the local geometry, and rerun the program. In that case the extra design requirement will be satisfied in the optimal way, and so the cost increase (over the previous preliminary design) will be much less than the cost increase associated with making local adjustments. It will also be less than that obtained by redoing a manual preliminary design.

A similar situation arises when an important design requirement is changed after a preliminary design has been completed, and to fulfil the new requirement properly would mean repeating the design. A designer using standard manual methods must choose between making another costly and time-consuming design or only partially fulfilling the requirement by making local changes.

Examination of Alternative Structural Configurations

Another principal benefit of the program is that it allows the designer to compare the optimum designs for alternative structural configurations. In order to demonstrate this a study was made of an alternative tanker design having three longitudinal girders in the center tank; that is, three parallel "ring frames" around the inside of the tank in a vertical longitudinal plane, instead of only one ring frame in the ship's centerplane. The data for this alternative configuration took about one man day to prepare. The results showed that the optimum design based on this configuration was slightly more costly than that of the single-girder model, and hence that there was no point in adding the extra girders. To have performed such a study without a program like MAESTRO would have constituted a major research project, occupying several manweeks and requiring a large amount of computer time.

Example 2—Bulk Carrier

The other ship type which is investigated in Ref. 16 is a 140,000 dwt bulk carrier. This also was an actual current practice design, but in this case the design chosen was not the final design but an earlier version in which a coarse mesh finite element stress check (by DAISY, a general-purpose finite element program) had found stresses exceeding yield in the bottom and inner bottom plating. The yielding was caused by a combination of stresses from the local cargo bending moment and the overall hull girder bending moment (see Figs. 1.13 and 2.3). This early version of the design was chosen to illustrate how MAESTRO can be used to check a tentative design and also, if desired, to produce optimal corrections. For this model it was necessary to analyze five hold lengths because of the ballast tank arrangement. Since the model was structurally symmetric longitudinally, it was only necessary to model $2\frac{1}{2}$ hold lengths of the structure.

Structural Evaluation

At first MAESTRO was used in the "evaluation mode" to compare its results with those obtained earlier in the DAISY three-dimensional coarse mesh analysis. To analyze seven loadcases DAISY had required more than 50 minutes of computer time. To analyze five loadcases (and also to perform a complete limit state analysis, and to evaluate the adequacy of the entire structure) MAESTRO took about three minutes, using the same computer and modeling the same portion of the ship ($2\frac{1}{2}$ cargo holds). All of the deflections and all of the stresses given by the MAESTRO analysis were in good agreement with those given by DAISY. Figure 1.13 is an example of a deflection plot produced by MAESTRO. In these plots the deflection is greatly exaggerated for clarity.

Optimization

After the structural evaluation of the existing design MAESTRO was used to produce an optimum (least cost) design while also correcting the deficiencies. The cost factors were basically the same as those used in Ref. 15.

With the height of the double bottom fixed at its original value of 2.3 meters, the optimum design produced by MAESTRO had thicker plating but smaller longitudinals. The resulting cost was 7.5% and the weight 6.4% less than the starting design. Next, the double bottom height was allowed to vary. This produced a design that had a lower double bottom height (1.9 m) and thicker bottom plating (24 mm). Compared to the original design, the cost was reduced 8.8%. These results indicate that the design having a smaller double bottom height is of slightly lower cost. In addition, this design is able to carry more cargo. Because the hull module in this example included more cargo holds than the segregated ballast tank model ($2\frac{1}{2}$ cargo holds versus $1\frac{1}{2}$ cargo holds), the

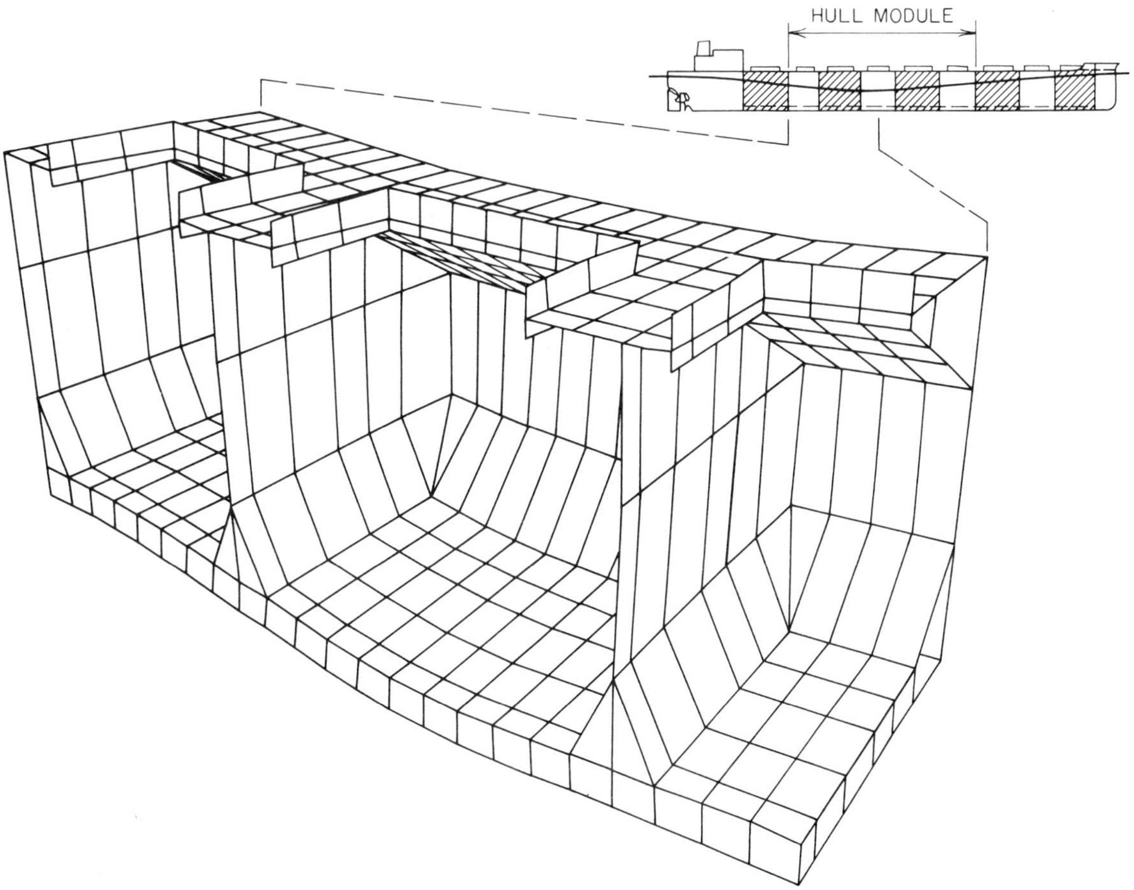

Figure 1.13 MAESTRO deflection plot—full load sagging condition; heavy ore cargo.

computer cost per design cycle was higher. An eight-design-cycle run in this case used 12 minutes of computer time and cost about $150.

The effect of varying the double bottom floor spacing in the bulk carrier was also examined. The original design had a floor spacing of 2.4 m (23 bays in the MAESTRO model). From this original model two others were developed: one with a spacing of 2.208 m (25 bays) and one with a spacing of 2.76 m (20 bays). The overall lengths of all three designs were identical. The length of the cargo holds however did vary due to the differing floor spacing, with the variation not more than 14%. One man-day was required to generate these two additional models.

Running MAESTRO for these three floor spacings without restricting the double bottom height produced the results shown in Fig. 1.14. Because there were no restrictions on double bottom height, each floor spacing resulted in a different double bottom height. From the cost index curve, it appears that the original spacing of 2.4 m was a reasonable choice. To determine the minimum value of the curve, it would be necessary to make additional runs with larger frame spacings. Notice that the weight is lowest at the original frame spacing.

To eliminate the variation in double bottom height as a factor in determining the optimum frame spacing,

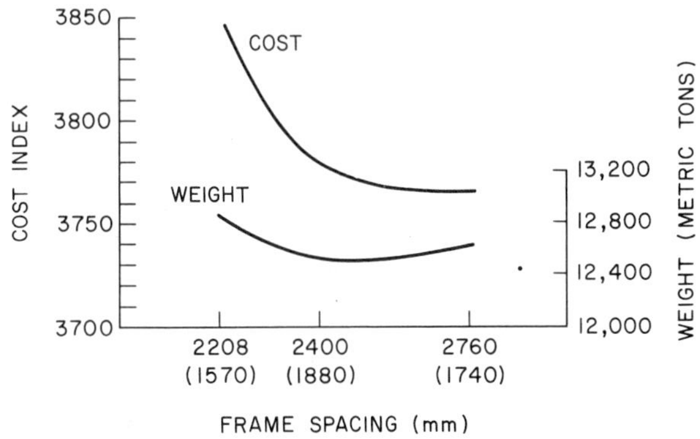

Figure 1.14 Cost and weight versus frame spacing—variable double bottom height. Note: Numbers in parentheses are the double bottom height (mm).

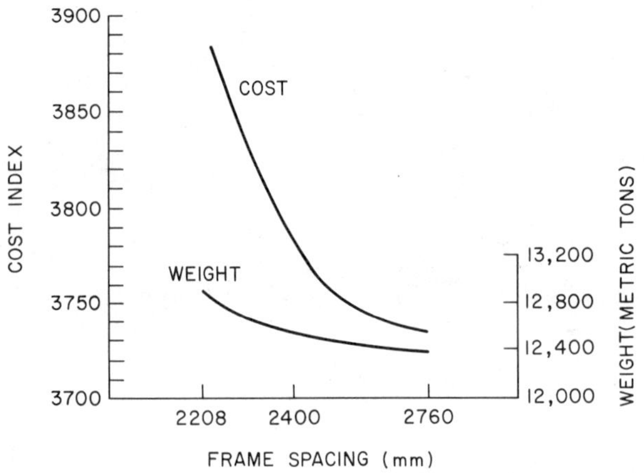

Figure 1.15 Cost and weight versus frame spacing—1880 double bottom height.

the three models were run with the double bottom height fixed at 1.9 m. The weight and cost index curves for these runs are shown in Fig. 1.15. In this case it appears that of the three spacings, a frame spacing of 2.76 m is still the optimum.

Use of MAESTRO in Teaching and Research

Since rationally-based design deals with the actual characteristics of a structure, including the complex and interrelated response of all of its members and the calculation of the various limit values of these responses, and since the program is a mathematical model of these characteristics, it is very helpful in teaching both the theory and the practice of ship structural design. First, because of its structural analysis and structural evaluation features, the program assists in gaining a deeper understanding of the complex and interrelated characteristics of a ship structure, and in learning what types of structural arrangement are more efficient. Second, because it also has an optimization capability the program is in effect a "ship design simulator," in which a designer can try out various ideas and possibilities, and can increase his knowledge in many aspects of structural design, such as the relative cost efficiency of various structural arrangements and the optimum proportions of structural members for various structural arrangements.

The program also has many useful applications in ship structural research. For example it may be used to examine the effect of corrosion on the strength of the hull girder and to develop improved corrosion criteria (initial allowance, permissible wastage—both thickness and extent, etc.). The program may also be used to perform a series of "hindcasting" analyses of various types of ships in order to determine the approximate magnitude of the safety factors that are inherent in current design practice, and to see how the factors compare for various locations in the ship, types of principal structural members, types and modes of failure, and types of ship.

Other Practical Aspects

MAESTRO includes several features that make it versatile and easy to use. For example, there is a pause and restart feature which allows the user to execute several design cycles, stop the program, change any part of his input (e.g., give new or modified optimization constraints), and then execute a few more design cycles with both this new input and the design scantlings from any of the previous design cycles. Thus he is able to monitor and control the design process as closely as desired, and can at any time change any of the design requirements.

The use of the program does not require any specialist knowledge about computing or about optimization theory. There are extensive error-checking and graphics facilities, including three-dimensional perspective plots of the structural model and/or the structure deflections, with or without hidden line removal (Fig. 1.13 is an example.). The program is fully documented, with a User's Manual, a Data Input Manual and an Example Manual. In addition, this book itself serves as a very complete type of Theoretical Manual for the program. The program is written in FORTRAN and contains about 30,000 statements, divided into approximately 150 subroutines. It has been operated successfully on many different computers, including some of the relatively small "super mini" computers such as the DEC VAX 11-750 and the PRIME 750. Distribution and technical support is handled by Giannotti and Associates of Annapolis, MD.

1.4 STRUCTURAL SAFETY

Uncertainty, Risk, and Safety

In the design of ocean structures there are many uncertainties which must be dealt with. First there is the uncertainty of the loads, especially those arising from waves. The ocean environment is severe, complex, and continuously varying. Ocean waves are essentially probabilistic and can be adequately defined only by means of statistics. Second, there are uncertainties regarding material properties such as yield stress, fatigue strength, notch toughness, and corrosion rate. For example, in ordinary steel which has not had special quality control the yield stress can vary by as much as 10%, and it is also dependent on the rate of loading and on the effects of welding. Third, there is inevitably some degree of uncertainty in the analysis of a structure as complex as a ship. Both the response analysis and the limit state analysis necessarily involve assumptions, approximations, and idealizations in formulating mathematical models of the physical environment and of the structure's response to that environment. Fourth, there can be variations and hence uncertainties in the quality of construction, and this factor has a particularly strong influence on the strength of a structure. Finally, there are uncertainties of operation, such as operating errors (improper loading, mishandling, etc.) or a change of service.

Wherever there are uncertainties there is risk of failure. For a structure, the risk of failure is the probability of a load reaching or exceeding its limit value. That is, for each limit state

$$\text{risk} = P_f = \text{prob}(Q \geq Q_L) \qquad (1.4.1)$$

in which, for simplicity, we are here considering a limit state which involves only one load. The safety of a structure is the converse—the probability that it will not fail. Hence

$$\text{safety} = \text{prob}(Q < Q_L) = 1 - P_f \quad (1.4.2)$$

Since there are always some uncertainties, and hence some risk of failure, it is impossible to make a structure absolutely safe. Instead it can only be made "sufficiently safe," which means that the probability of failure can be brought down to a level that is considered by society to be acceptable for that type of structure. Therefore, if a structural design process is to be rationally based, the whole question of safety must be dealt with on a probabilistic basis, and the process must provide the means whereby the designer can ensure that the degree of safety meets or exceeds the required level. The calculation of the probability of a particular type of failure involves the probability density functions of the relevant load and of the limit value of that load. If these probability density functions are denoted by $p_Q(\cdot)$ and $p_{Q_L}(\cdot)$ respectively, then the

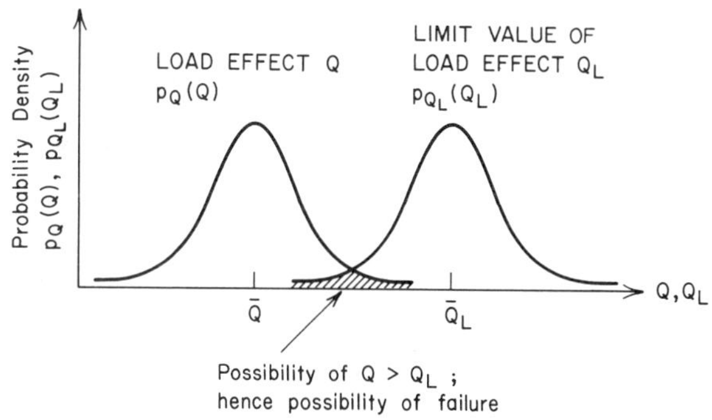

Figure 1.16 Probability distributions of load effect and of limit value of load effect.

probability of this particular type of failure occurring is

$$P_f = \int_0^\infty \left[\int_0^\eta p_{Q_L}(\xi)\, d\xi \right] p_Q(\eta)\, d\eta \quad (1.4.3)^*$$

This is illustrated in Fig. 1.16 which shows that even though $\overline{Q}_L$, the mean of the limit value, is well above the mean load there is still some overlap of the curves and hence some possibility of failure. (Note: The probability of failure is *not* equal to the area of the overlap, but this area nevertheless provides a useful visual and qualitative indication of P_f.) The figure also shows that the important regions of the distribution curves are the tails, because this is where the overlap occurs. Unfortunately it is this portion of a distribution curve which is most difficult to obtain with any precision, mainly because one is dealing with rare events. However it will be shown that there are ways of obtaining satisfactory estimates and upper limits of P_f, even though the tail portions of the distribution curves are not known precisely.

Levels of Safety

The required level of safety varies according to the type of failure and the seriousness of its consequences. Because these levels are ultimately determined by society there are no precise values or exact rules for determining them, but they can be estimated by surveys and by examining the statistics regarding failures, particularly those types in which the failure rate is considered by the public to be generally satisfactory, in the sense that the costs and resource usages that

*This equation assumes that Q and Q_L are independent random variables.

TABLE 1.1 COMPARATIVE ANNUAL PROBABILITY OF DEATH PER 10,000 PERSONS

Activity	Annual Exposure (hr)	Annual Risk per 10^4
Offshore operations (including diving and vessels)		65
Mountaineering (international)	100	27
Distant water trawling (1958–1972)	2900	17
Offshore operations (other than construction, diving, or vessels)		11
Air travel (crew)	1000	12
Coal mining	1600	3.3
Car travel	400	2.2
Construction site	2200	1.7
Air travel (passenger)	100	1.2
Home accidents (all persons)	5500	1.1
Home accidents (able bodied)	5500	0.4
Manufacturing	2000	0.4
Structural failure (land-based structures)	5500	0.001
All causes (England and Wales, 1960–1962)		
Male, age 30	8700	13
Female, age 30	8700	11
Male, age 50	8700	73
Female, age 50	8700	44
Male, age 53	8700	100

would be required to further reduce the failure rate are considered to be unwarranted when balanced against other needs. For example, in regard to occupational risk Ref. 18 reviewed a range of activities and obtained the results shown in Table 1.1. From the results of a study of merchant ship losses the authors of Ref. 19 estimated a value of between 0.003 and 0.006 as the lifetime probability of overall structural failure that has been tacitly accepted over the past 20 years for large ocean-going ships. From studies of this type it would appear that the value of P_{fT}, the total annual failure probability per structure (ship, aircraft, drill rig, etc.) ranges from 10^{-3} or less for failures that have moderately serious consequences (substantial economic loss but no fatalities) to 10^{-5} or less for catastrophic failures, such as the crash of a passenger aircraft.

A different approach to the question of required level of safety is the use of economic criteria [20]. This is particularly appropriate for cases in which loss of life is not involved. For a large number of similar structures the total annual cost C_T of each of them can be formulated as:

$$C_T = C_i + P_f[C_f]_{EQ} \qquad (1.4.4)$$

where C_i = initial cost, converted to an annual depreciation cost
P_f = annual probability of failure
$[C_f]_{EQ}$ = equivalent failure cost in present worth.

The equivalent failure cost involves a discounting of future damage costs to present worth by appropriate interest rates. Failure costs should include all costs involved such as salvage operation, pollution abatement, cleanup, and lost production. It could also include loss of reputation and public confidence. In Fig. 1.17 a typical relationship between annual cost and annual probability of failure is demonstrated. The figure shows that the most important parameter is the ratio ρ between failure cost and initial cost.

Many structures are comprised of, or can be divided into, a set of members each of which is sufficiently important that failure of any one member is regarded as failure of the structure. If the failure modes of the members are independent the total failure probability is

$$P_{fT} = \sum P_{fj} \qquad (1.4.5)$$

where P_{fj} is the failure probability of the jth member. Since the total probability of failure consists of several components there is an opportunity to optimize the way in which this total required level is obtained. Under some simplified assumptions the optimal design

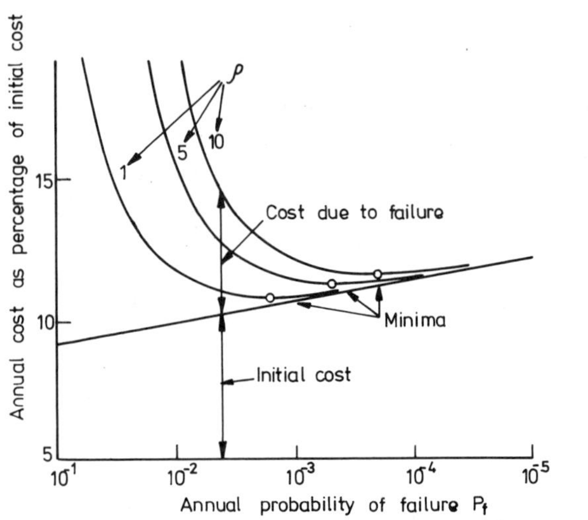

Figure 1.17 Typical relationship between cost and total probability of failure.

will come close to satisfying the following equation [21]:

$$\frac{C_{ij}}{C_i} = \frac{P_{fj}}{P_{fT}} \tag{1.4.6}$$

where C_{ij} = cost of member j. This approach can be used to decide on the relative safety margins of the members.

1.5 PROBABILISTIC DESIGN METHODS

The question of a design procedure based on a probabilistic model for loads and strength has received a great deal of attention in the field of structural engineering. Pugsley [22] and Freudenthal [23] were the pioneers for aircraft and civil structures in the 1940s. They demonstrated how a relationship can be derived between safety factors and probability of failure provided that the statistical distributions are known. In subsequent years these methods were further developed and were increasingly incorporated in structural design codes, both for steel and for concrete. In the latter case this approach has been particularly successful because it accounts for the large variability in the strength of this material. A probabilistic design code is currently being developed for the design of offshore structures, stimulated by the higher risks and the higher economic stakes involved in that field.

In contrast, in the field of ship structures the probabilistic approach is still at the early stages, in spite of the obvious probabilistic nature of wave loads. But in fairness it should be pointed out that the load and response analysis is much more complicated for ships than it is for fixed structures or for aircraft, because it must deal with the exceedingly complex interaction between wave excitation and ship motions merely to compute the loads.

Exact and Approximate Probabilistic Methods

The task of achieving a specified level of safety can be pursued at various levels of mathematical rigor. We shall first show that a fully rigorous method requires the gathering of a great deal of information and is simply not justified in the majority of cases. Then we shall present two approximate methods, with an emphasis on the second one—the Partial Safety Factors method.

(A) Fully Probabilistic Design

The most rigorous and most general type of probabilistic design is that which utilizes the complete probability distribution functions of all relevant quantities (loads, load effects, and limit values) to calculate P_f from (1.4.3) for each load and for each type of failure. These values are then combined into an overall probability of failure which is then adjusted, by making modifications to the design, until it falls within the stipulated acceptable overall risk. This approach requires the determination of all of the probability distributions, either by measurement (of the complete phenomena or of their separate constituent aspects, and either full-scale or model) or by theoretical considerations, all of which is a very large task. Since the most highly probabilistic loads are those arising from waves, and since hull girder bending moment is the most important load effect, research efforts during the past 10 years have been concentrated on obtaining the probability distribution for the extreme value of wave-induced hull girder bending moment. Sufficient data regarding waves have now been collected and processed statistically to produce some approximate probability distributions for this load effect; these are discussed in Section 4.4. The probability distributions of other loads and load effects are less known and much work remains to be done. Likewise, the distributions of limit values are not easy to obtain since they arise from so many separate variations (material properties, accuracy of analysis, and quality of construction), each of which requires the collection of a great deal of statistical information. In areas where this information is not yet available it is necessary to use less rigorous techniques.

Moreover, the availability of information is not the only factor that should be considered; another important question is whether the application really requires the complexity of the fully probabilistic method, because a design method should always be as simple as the circumstances permit. A complex method always introduces more likelihood of errors in its use. Also, greater complexity usually increases both the cost and the time required for the design. Therefore it is important to consider whether the added accuracy of a rigorous but complex method is really justified, in regard to both safety and economy, for the particular application. For aerospace structures it often

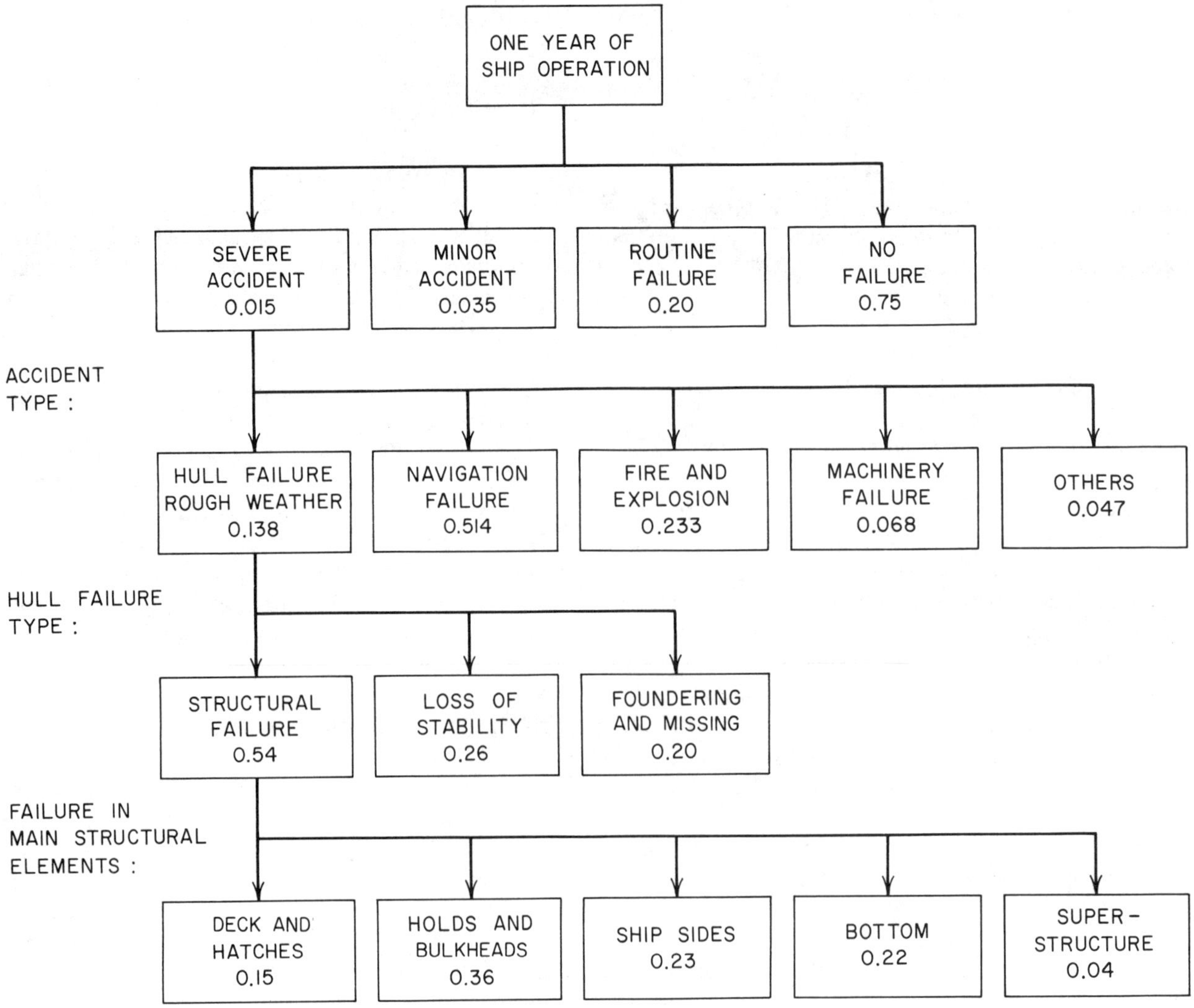

Figure 1.18 Empirical distribution of ship casualties.

is justified, but for ship structures this is less likely.

In this regard it is also relevant to examine what proportion of ship accidents are due to structural failure. Fig. 1.18 from Ref. 24 presents the results of a survey which showed that in a given sample of ship casualties only about 7% (0.138 × 0.54 × 100%) of severe accidents were caused by structural failure. In view of the many other causes of severe accidents and the relative infrequency of structural failure, it is clear that even a large increase in the rigor and accuracy of structural design would not improve the overall risk of casualty very much. Resources used for this purpose could be used more effectively to improve the other risks involved. Hence there is need for moderation in regard to the statistical complexity of the structural design method.

(B) Approximate Probabilistic Methods

The complexity of the fully probabilistic approach has led to the development of simplified methods which retain the basic statistical foundation but which do not require the complete probability distribution curves, but rather only the mean and the variance.

Two alternative approximate methods are available and they are basically similar, having the following two fundamental features:

1. All failure modes which are independent are treated separately. This greatly simplifies the process but it requires that a value of acceptable risk must be defined separately for each type of failure (although in

practice the same value can be used for all types which have the same degree of seriousness) and it precludes the possibility of combining the separate risks. Therefore it requires approximations, which must necessarily be on the conservative side, in order to deal with combinations of loads of differing probability and combinations of interactive modes of failure.

2. All probability distributions are characterized by their first two moments, that is, by the mean and the variance (see Section 4.1. for a brief summary of basic statistical definitions and theorems). For this reason these methods are sometimes referred to as second moment methods. If these two parameters have been established for the load effects and for the limit values, it is then possible for the relevant safety authority to specify the required level of safety in terms of a set of deterministic (i.e., nonstatistical) safety factors from which the designer can immediately calculate the required strength (limit values) which the structure must have.

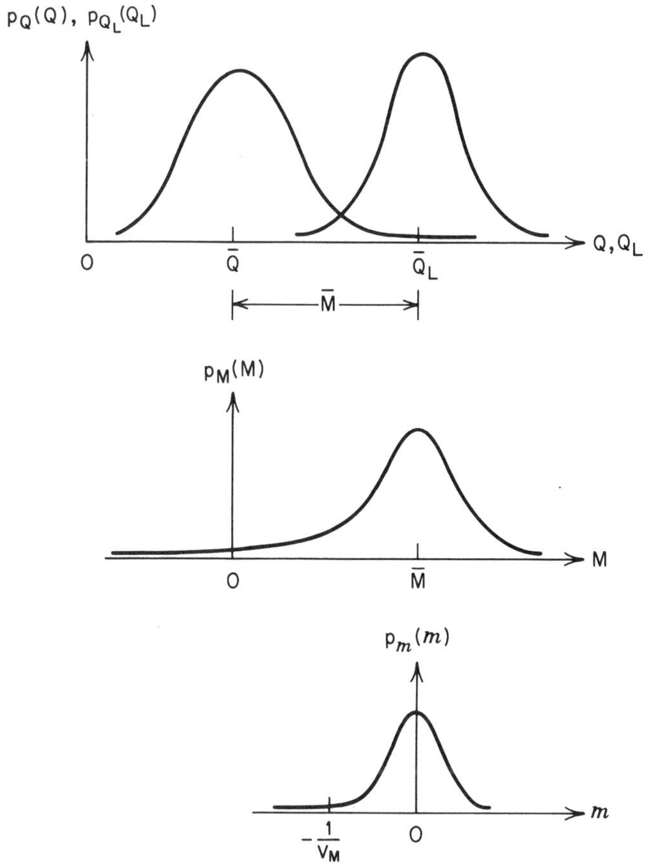

Figure 1.19 Probability distributions of the safety margin.

Safety Index Method

This method is the earlier of the two [25], but it has not been as widely adopted as the second, the Method of Partial Safety Factors. Nevertheless it will be briefly described here because it introduces basic concepts common to both methods and because the Safety Index itself is a very useful tool in establishing suitable values for the partial safety factors.

The degree of safety is directly related to the margin between the actual value of the load effect and the limit value

$$M = Q_L - Q$$

and failure occurs when the margin becomes negative. Since these are both random variables, M will be likewise, having a probability density function $p_M(M)$, as shown in Fig. 1.19. Therefore the degree of safety depends not only on the separation of the two curves as measured by the distance between their mean values

$$\overline{M} = \overline{Q}_L - \overline{Q}$$

but also it depends inversely on the "spread" of the two curves, as measured for example, by their coefficients of variation. Therefore it will also bear some inverse relationship to V_M, the coefficient of variation of M. If V_M is large the degree of safety will be correspondingly less, and vice-versa. The probability of failure is

$$P_f = \text{prob}[M < 0]$$

Subtracting $\overline{M}$ from both sides of the inequality and normalizing by means of the standard deviation s_M gives

$$P_f = \text{prob}\left[\frac{M - \overline{M}}{s_M} < -\frac{\overline{M}}{s_M}\right] \qquad (1.5.1)$$

By definition the coefficient of variation is $V_M = s_M/\overline{M}$ and therefore

$$P_f = \text{prob}\left[\frac{M - \overline{M}}{s_M} < -\frac{1}{V_M}\right] \qquad (1.5.2)$$

The left-hand term within the brackets is the normalized margin, for which the distribution has zero mean and unit variance. Let us denote this normalized margin as $\mathcal{M}$ and let $P_{\mathcal{M}}(\cdot)$ be its probability distribution

$$\mathcal{M} = \frac{M - \overline{M}}{s_M}$$

$$P_{\mathcal{M}}(\cdot) = \text{cumulative probability distribution of } \mathcal{M}$$
$$= \text{prob}[\mathcal{M} < \cdot]$$

Equation 1.5.2 becomes

$$P_f = \text{prob}\left[\mathcal{M} < -\frac{1}{V_M}\right] \qquad (1.5.3)$$

The occurrence of $1/V_M$ on the right-hand side within the brackets confirms that the degree of safety depends on the inverse of the coefficient of variation. We therefore give this quantity the name Safety Index and denote it by the symbol $\mathcal{S}$

$$\mathcal{S} = \frac{1}{V_M}$$

In terms of the Safety Index Equation 1.5.3 is

$$P_f = \text{prob}[\mathcal{M} < -\mathcal{S}]$$
$$= P_{\mathcal{M}}(-\mathcal{S}) \qquad (1.5.4)$$

This last expression shows that there is a direct correspondence between the Safety Index $\mathcal{S}$ and the probability of failure. The larger the Safety Index, the smaller the probability of failure, that is, the safer the structure. If the complete distribution $P_{\mathcal{M}}(\cdot)$ is known (i.e., if the distributions of Q and Q_L are known) then the exact value of P_f corresponding to a given value of $\mathcal{S}$ can be determined.* But even if the exact $P_{\mathcal{M}}(\cdot)$ is not known, designing on the basis of a specified value of $\mathcal{S}$ produces a consistent degree of safety from one design to another, for each type of structure. For ship structures a suitable value of $\mathcal{S}$ can be determined for each type of failure by analyzing the statistics regarding ships which have proven to be reasonably efficient and which also have a satisfactory safety record. This has been done, for example, for extreme hull girder bending moment in Ref. 26 and 27.

*For example, for the normal distribution the Safety Index is the same as the standard deviation.

One of the principal advantages of the Safety Index and also of the Method of Partial Safety Factors is that the provision of adequate safety, which is a probabilistic quantity, is converted and expressed deterministically in terms of a specific "design" value of load and a specific limit value which the structure must possess. The Safety Index method makes use of mean values and in particular it involves their ratio which is referred to as the central safety factor, γ_C

$$\gamma_C = \frac{\overline{Q}_L}{\overline{Q}}$$

It can be seen that γ_C corresponds to the familiar single safety factor of deterministic design.

The relationship between $\mathcal{S}$ and γ_C can be derived as follows, starting with the definition of $\mathcal{S}$:

$$\mathcal{S} = \frac{1}{V_M} = \frac{\overline{M}}{s_M} = \frac{\overline{Q}_L - \overline{Q}}{\sqrt{s_L^2 + s_Q^2}} = \frac{\gamma_C - 1}{\sqrt{\gamma_C^2 V_L^2 + V_Q^2}}$$

Upon rearranging for γ_C

$$\gamma_C = \frac{1 + \mathcal{S}\sqrt{V_L^2 + V_Q^2 - \mathcal{S}^2 V_L^2 V_Q^2}}{1 - \mathcal{S}^2 V_L^2} \qquad (1.5.5)$$

In the Safety Index method of design the appropriate safety authority specifies the values of $\mathcal{S}$, V_Q, and V_L, depending on the type of structure and the degree of seriousness of the limit state. For each limit state the designer calculates the central safety factor from (1.5.5). He then calculates the mean value (best estimate) of the relevant load effect $\overline{Q}$ by performing a response analysis. Alternatively, for those structures for which the loads and load effects are well established the safety authority may provide a formula for a less exact but more universal "design value" of $\overline{Q}$, together with a larger value of $\mathcal{S}$ which must be used with it. Knowing $\overline{Q}$, the designer then applies the factor γ_C to obtain $\gamma_C\overline{Q}$, and he must then design the structure such that $\overline{Q}_L$ equals or exceeds $\gamma_C\overline{Q}$. This requirement constitutes one of the *strength constraints* which the design must satisfy. Stated mathematically the constraint is

$$\gamma_C\overline{Q} < \overline{Q}_L \qquad (1.5.6)$$

This procedure is carried out for each limit state, thus producing the complete set of strength constraints which govern the design.

Partial Safety Factor Method

This method has already been adopted in several areas of structural design, from simple building codes for civil structures to designs for aircraft and aerospace structures. It has two advantages over the Safety Index method. First, it makes a more explicit distinction between *statistical uncertainty*—that which arises purely from genuine statistical randomness and which can therefore be properly and adequately assessed using statistical theory, and *approximational uncertainty*—that which arises from the assumptions, approximations, and judgements that are necessarily involved in any structural design task.* Second, in the partial safety factor method each principal circumstance affecting the seriousness of the failure, and each principal source of approximational uncertainty, is accounted for explicitly by means of a separate safety factor, and this clarifies matters and permits greater precision and consistency.

The following paragraphs give a fuller explanation of statistical and approximational uncertainty and show that the first can be accounted for by using characteristic values (instead of mean values) whereas the second can be accounted for by using safety factors, with a separate factor for each source of uncertainty.

Statistical Uncertainty

The uncertainty that arises because of the randomness of the variable (Q or Q_L) can and should be assessed by means of basic statistical theory. To do this it is necessary to establish what type of distribution (normal, Poisson, etc.) is involved. In some cases this is known from theoretical considerations. In other cases it is possible to determine by observation which basic type most nearly resembles the actual distribution. Once the type is known the uncertainty can be calculated by means of the basic laws and relationships of statistics. A very useful way of dealing with statistical uncertainty is in terms of a *characteristic value,* which is the value corresponding to a specified percentage of the area under the probability density curve, that is, to a specified probability of exceedance. For example, Fig. 1.20*a* illustrates a characteristic value of load Q_c corresponding to a 5% probability of exceedance. Figure 1.20*b* illustrates a *characteristic limit value* of load, in which case the 5% probability refers to *non*-exceedance.

In contrast to the Safety Index method, which uses the mean values $\bar{Q}$ and $\bar{Q}_L$, the partial safety factor method uses characteristic values Q_c and $Q_{L,c}$ thereby automatically accounting for the statistical probability of failure. Thus, if the only uncertainty was purely statistical (i.e., if Q and Q_L exactly followed their assumed distribution function) there would be no need for any safety factors, and the strength constraint would be simply

$$Q_c \leq Q_{L,c} \tag{1.5.7}$$

in which the characteristic values would be selected so as to provide whatever degree of safety was required.

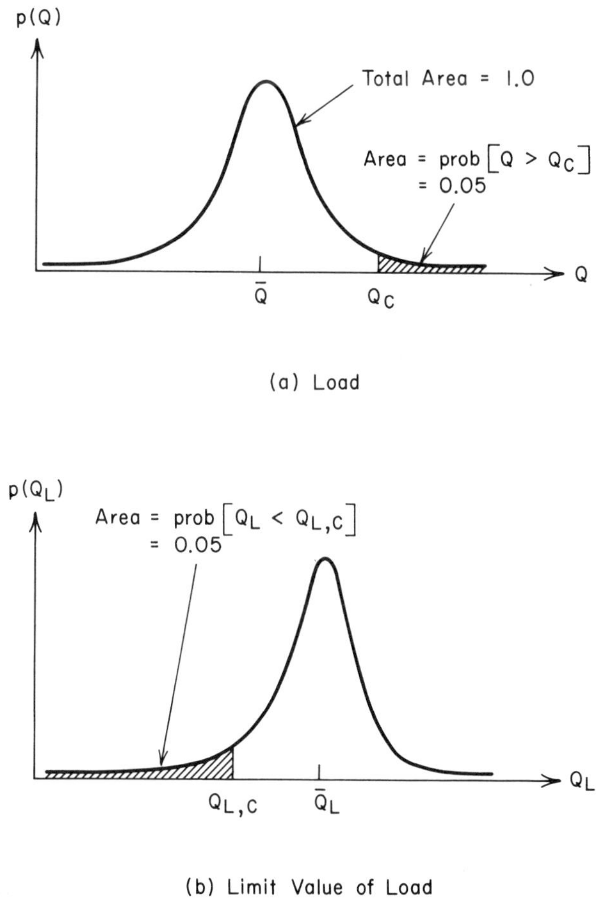

Figure 1.20 Illustration of characteristic values.

*Some authors use the terms *objective* and *subjective* uncertainty, but the latter term has the connotation of a purely personal view, or even of a biased view, whereas in true engineering design the assumptions, approximations, and judgements are made in an objective manner, involving investigation, consultation, and consensus. Estimative or judgemental uncertainty would also be appropriate terms.

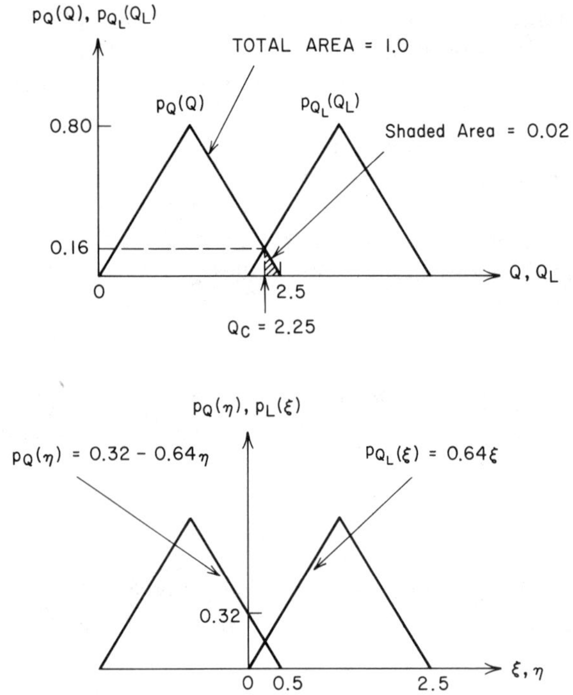

Figure 1.21 Idealized probability distributions.

To illustrate this let us take the idealized triangular probability distribution of Fig. 1.21 for both Q and Q_L. Let us define the characteristic values Q_c and $Q_{L,c}$ as those values which correspond to a 2% probability of exceedance and nonexceedance, respectively. We superimpose the two distributions such that their characteristic values coincide, thereby just fulfilling the constraint which is expressed by (1.5.7). For this case (1.4.3) becomes

$$P_f = \int_0^{0.5} \left[\int_0^{\eta} p_{Q_L}(\xi)\, d\xi \right] p_Q(\eta)\, d\eta \quad (1.5.8)$$

in which the limits of integration and the origin of the dummy variables ξ and η reflect the fact that the integration need only be performed over the length of the overlap. We now take further advantage of this in writing simple expressions for p_Q and p_L which are valid within this range:

$$p_{Q_L}(\xi) = 0.64$$
$$p_Q(\eta) = 0.32 - 0.64\,\eta$$

Substituting into (1.5.8) and integrating gives

$$P_f = \int_0^{0.5} 0.64 \frac{\eta^2}{2} (0.32 - 0.64\,\eta)\, d\eta$$

and integrating again gives

$$P_f = \frac{(0.64)^2}{384} = .00107$$

or a failure probability of about 0.1%. Hence if the required safety corresponded to a failure probability of 0.1%, then the two-percentile characteristic values would be the appropriate values, and the strength constraint of (1.5.7) would give the required safety without needing any factor of safety.

Approximational Uncertainty

In reality, of course, besides the purely statistical uncertainty, there is always some additional uncertainty which is either not statistical in nature (e.g., uncertainty arising from value judgements, from approximations, or from legal, political or other nontechnical influences on design) or which, although statistical, cannot be included in that category because sufficient information is not available. This uncertainty is here called *approximational* because most of it arises from the approximations that are inevitable in structural design. Indeed, even the use of statistical theory to describe ocean waves involves some assumptions and approximations. But in dealing with any statistical aspect in design the goal should be first to obtain sufficient information so as to be able to use statistical theory and to account for most of the uncertainty by this means, even though some approximations are still required, and then to seek to improve the information so as to further reduce the approximational uncertainty. Thus as more information becomes available the amount of approximational uncertainty is reduced. However it will never be entirely eliminated; some approximations will always be necessary, and in addition there will always be some sources of uncertainty for which statistics are not entirely adequate; of that, at least, there is no uncertainty.

Because of approximational uncertainty the characteristic values that account for statistical uncertainty are not sufficient in themselves. It is necessary to further increase the separation of the Q and Q_L curves, by some amount that can only be estimated and that requires judgement, in order to retain the required degree of safety. There are two different ways of doing this, and in describing these we shall use as an example the case of an approximational uncertainty in the load Q. The two approaches are:

1. Artificially increase the variance V_Q. If the margin between Q and Q_L is being specified in terms of mean values and a central safety factor γ_C, as in the Safety Index method, this produces a larger value for γ_C from (1.5.5), which causes the limit value distribution curve to be displaced to the right thereby increasing the margin. If the margin is being specified in terms of characteristic values this produces a larger characteristic load Q_c, and from (1.5.7) this again causes the limit value curve to be displaced to the right.

It is emphasized that because the uncertainty that is being accounted for either is not statistical or has unknown statistical properties, the amount by which the variance should be increased can only be estimated; it is a matter of judgement and is therefore somewhat subjective and arbitrary. This is emphasized because variance is a statistical quantity and so the stratagem of increasing the variance, as is done in the Safety Index method, might make the method appear to be entirely statistical and objective, with no subjective or arbitrary element. Hence the increase in the variance is described as "artificial" at the beginning of this description.

2. Apply a factor of safety γ_0, the value of which is likewise a matter of judgement in exactly the same degree as the increase of variance. However, in this approach statistical aspects remain unchanged, and the approximational uncertainty is accounted for in a more explicit manner. For example, if statistical uncertainty is being accounted for by means of characteristic values, as in the example involving the triangular distributions, then the strength constraint of (1.5.7) would become

$$\gamma_0 Q_c \leq Q_{L,c} \tag{1.5.9}$$

The second approach is the basis for the Partial Safety Factor method, which is presented next. We note here that since the statistical uncertainty is already accounted for by the characteristic values, the magnitude of γ_0 will be much smaller than the value of γ_c in the Safety Index method.

Partial Safety Factors

Regardless of which of the two semiprobabilistic methods is used, and regardless of which technique is used to account for approximational uncertainty, it is absolutely essential to be able to specify different levels of safety for different types of failures, in accordance with the degree of seriousness of the failures. That is, in addition to and quite apart from the need to account for uncertainty, it is also necessary to adjust the separation between the curves of Q and Q_L to account for the degree of seriousness of the particular type of failure which is being considered. Thus there is a need for some simple and explicit method for adjusting the separation between Q and Q_L; the safety factor γ_0 which relates the two characteristic values Q_c and $Q_{L,c}$ provides just such a method. Therefore instead of regarding this factor as a single quantity we will regard it as the product of several *partial safety factors*. These factors can then be used for two main purposes:

1. To account for the degree of seriousness of the particular limit state in regard to safety and serviceability (for a commercial ship the latter refers mainly to the economic consequences of the failure) taking into account any special circumstances (purpose of the ship, type of cargo, interaction of this limit state with others, etc.). Since safety and serviceability are not the same, it is best to use two independent partial safety factors for this task.
2. To account for the approximational uncertainties such as:
   a. Deviation of the probability distribution of the loads from the assumed distribution, due to unforeseen actions or conditions, and consequent deviation of the load effects.
   b. Assumptions and approximations in the response analysis and in the limit state analysis.
   c. Deviation of the limit value from its assumed distribution due to unpredictable factors (e.g., poor workmanship).
   d. Other matters requiring estimation and judgement.

The number of partial safety factors varies from as few as three to eight or more, depending on the type of structure and on what level of detail is preferred for their specification. Factors are sometimes further subdivided if this gives greater precision or consistency. In this text we will use four factors:

γ_{S1} and γ_{S2}, which account for the seriousness, in regard to safety and serviceability, of the type of failure under consideration.

γ_Q, which accounts for the approximational uncertainties in the loads and load effects, including the discrepancy between the structure's actual load effect and the value predicted by the (necessarily) idealized response analysis.

γ_L, which accounts for the approximational uncertainties in the estimated limit value.

The first three of these factors are applied to the load (or load effect) in the same way as γ_0 is applied in (1.5.9). But since the fourth factor refers specifically to the limit value it is customary to transfer it to the denominator of the right-hand side and to apply it to $Q_{L,c}$ as a dividing factor. For this reason γ_{S1}, γ_{S2}, and γ_Q are often referred to as *load factors* while γ_L is called a *limit value reduction factor*, or, in some cases a *usage factor*.

Thus in the Partial Safety Factor method the strength constraints are of the form

$$\gamma_{S1}\gamma_{S2}\gamma_Q Q_c \leq Q_{L,c}/\gamma_L \qquad (1.5.10)$$

The situation is illustrated in Fig. 1.22.

The Partial Safety Factor method has been adopted for civil engineering design codes in nearly all of Europe and in Canada, Australia, and other countries. Recently, it has been adopted by the American Institute of Steel Construction as an alternative to its existing code. These codes are actually a synthesis of the Safety Index method and the Partial Safety Factor method because the former was used to establish suitable values for the partial safety factors. The Safety Index is ideal for measuring and comparing the relative safety of different structures and structural members. Structures or members designed to the same Safety Index will have essentially the same degree of safety. From a survey of past designs it is possible to calculate values of $\mathcal{S}$ for structures that were designed using earlier codes and that have proven satisfactory. For example the AISC found that $\mathcal{S} = 3$ gave a reasonable correlation with its previous code. Once a satisfactory value of $\mathcal{S}$ is established, it is then possible to calculate partial safety factors for the various loads, load effects, and limit values, such that they all have a consistent degree of safety, even though they may have quite different degrees of uncertainty. This process is basically the responsibility of the safety authorities rather than the designers, and so it is beyond the scope of this text. We here merely mention a few basic and simplified aspects.

In general the distribution that is found to give the best fit to measured values of loads, load effects, and limit values is the lognormal distribution, defined in Section 4.1. From the definition of the Safety Index

$$\mathcal{S} = \frac{1}{V_M} = \frac{\overline{M}}{S_M} = \frac{\overline{Q}_L - \overline{Q}}{\sqrt{S_L^2 + S_Q^2}}$$

it may be shown that for the lognormal distribution

$$\mathcal{S} = \frac{\dfrac{\overline{Q}_L}{\overline{Q}}\sqrt{\dfrac{1 + V_Q^2}{1 + V_L^2}}}{\sqrt{ln[(1 + V_Q^2)(1 + V_L^2)]}} \qquad (1.5.11)$$

where V_Q and V_L are the coefficients of variation of the load effect and of the limit value. If V_Q and V_L are 0.3 or less (1.5.11) can be simplified quite accurately to:

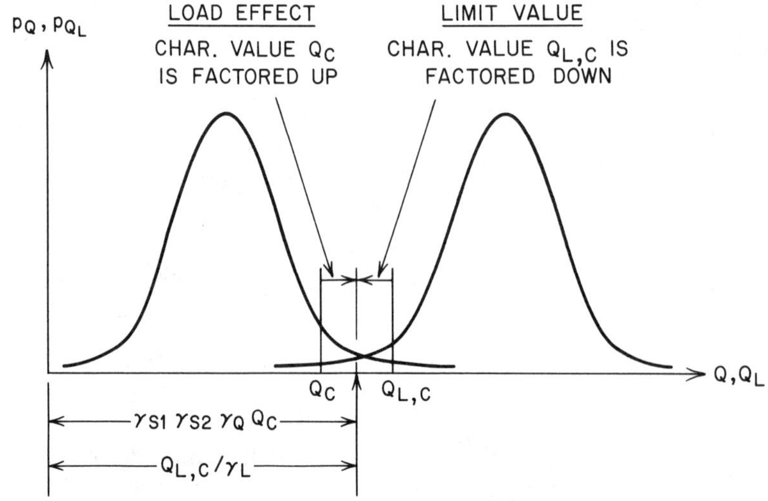

Figure 1.22 Use of partial safety factors.

$$\mathcal{S} = \frac{ln\left(\dfrac{\overline{Q}_L}{\overline{Q}}\right)}{\sqrt{V_Q^2 + V_L^2}} \qquad (1.5.12)$$

This expression can be further simplified by introducing a "splitting" constant, α, and making the approximation

$$\sqrt{V_Q^2 + V_L^2} \cong \alpha(V_Q + V_L)$$

If, for example, V_Q and V_L have about the same magnitude, then a choice of $\alpha = 0.7$ gives a good approximation for a fairly wide range of magnitudes. With this approximation (1.5.12) can be rearranged into the form of a strength constraint

$$\gamma_Q \overline{Q} \leq \overline{Q}_L / \gamma_L \qquad (1.5.13)$$

in which

$$\gamma_Q = \exp(\alpha \mathcal{S} V_Q)$$
$$\gamma_L = \exp(\alpha \mathcal{S} V_L)$$

This equation states that if the mean value (best estimate) of the load effect multiplied by γ_Q is less than the mean (best estimate) of the limit value divided by γ_L, then a Safety Index greater than $\mathcal{S}$ will be achieved. Note that in this formulation the mean values $\overline{Q}$ and $\overline{Q}_L$ are used instead of the characteristic values, and so the factors γ_Q and γ_L account for both statistical and approximational uncertainty. This is done if neither Q nor Q_L are amenable to a statistical representation and nearly all of the uncertainty is approximational.

The establishment of a suitable set of partial safety factors for ship structures is a large task, but it is absolutely essential for progress in this field, and it will be of benefit to all in the shipping world: owners, designers, builders, operators, insurers, and others. The new civil engineering codes and the limited set of partial safety factors in Ref. 14 provide a good start, but there is need for further effort by persons and agencies directly concerned with ship structures. The American Petroleum Institute and other agencies are currently engaged in a similar task for offshore structures.

Distinction Between Stillwater Loads and Wave Loads

The hull girder loading on a ship may be divided into two broad categories—stillwater loads and wave loads. The first relate mainly to cargo loading and other controllable factors and they are, for the most part, specific, deliberate and directly calculable; they are basically deterministic and for the statistical fluctuations which do occur it is a relatively simple matter to calculate characteristic values. Therefore the calculation of the corresponding characteristic load effects is relatively straightforward, requiring only a single response analysis. Wave loads are the opposite: they are essentially probabilistic and the calculation of hull girder bending moment in the at-sea condition is a major task, primarily because there is a quite complicated interaction between the load and the response: the seaway loading causes ship motion and the motion influences the seaway loading. Because of this cross-coupling in the response the characteristic value of load effect (such as hull girder bending moment which is the most important load effect) cannot be obtained by defining a single characteristic value of load and then performing a single, static response analysis. On the contrary, the response analysis for wave-induced loads is quite complex, involving statistical theory, hydrodynamics, and systems analysis, and it can only be done by means of rather sophisticated computer programs that have been developed for the purpose. A brief summary of this analysis is presented in Section 2.2, and it is treated more fully in Chapter 4.

For an individual designer such an analysis represents a complex and time-consuming task, and yet it is the only way of obtaining accurate values for the wave-induced load effects for a specific ship design. However, for ships of standard geometry and proportions there is a much easier method, which is made possible by the following observations:

1. The load effects of the stillwater loads can be calculated separately, and the stillwater loads have only a secondary and indirect influence on the wave loads.
2. The stillwater load effects depend on the particular geometry and cargo distribution, but the wave-induced characteristic load effects, being long-term statistical quantities, are more general and universal, and are applicable to a whole class of ships having the same geometry and proportions.

Over the past several years, ship classification societies and other researchers have utilized computer programs for wave-induced response analysis to calculate the characteristic value of hull girder bending moment for ships of standard geometry and proportions. Then by subtracting the stillwater portion of the bending moment they obtained characteristic values for the wave-induced bending moment. Other information has been obtained from long-term measurements of bending strain in ships. As a result it has been possible to develop expressions for these characteristic values in terms of the principal dimensions of the ship. These expressions are given in the rules of the various classification societies. Some examples are given in Section 3.5. Thus for ships of standard geometry and proportions the designer can concentrate on the still-water response analysis and load effects, which he must do because they are particular to his design.

For ships of nonstandard geometry or proportions the characteristic value of total bending moment should be calculated for the most important cargo configurations using a wave response analysis program. These programs have now been developed to the point where the user need only supply basic information such as the ship's lines (or offsets) and the cargo distribution. They are available at the computer bureaus of various classification societies, and with the development of minicomputers it is now possible to have such programs in individual ship design offices.

1.6 LOAD FACTORS AND DEGREES OF SERIOUSNESS OF FAILURE

As mentioned earlier, three of the four partial safety factors are applied to the load and hence these three are commonly referred to as load factors.* It will be convenient to also define a "total" load factor, γ_{load}, which is simply the product of these three partial load factors.

$$\gamma_{\text{load}} = \gamma_{S1}\,\gamma_{S2}\,\gamma_Q \tag{1.6.1}$$

The first two refer to the degree of seriousness of the failure and in this section we present some qualitative definitions of degrees of seriousness and some sample values of load factors. The fourth factor, γ_L, is applied as a divisor to the limit value of the load effect, and its value is determined mainly by material properties and fabrication considerations.

Degrees of Seriousness of Failure

In order to assess the degree of seriousness of a structural failure we must examine the consequences: what are the losses and how severe are they? We have seen that the two principal attributes by which the fitness of a ship is measured are *safety* and *serviceability*. Accordingly, we may distinguish two different types of losses.

1. Loss of life and other serious and irreparable non-economic losses.
2. Loss of main function which, for a commercial vessel, means economic loss due to lost revenue, cost of repair or replacement, lawsuits, and so on.

The foregoing categories also apply to non-commercial vessels in which the main function is the performance of some mission or service that has no direct relationship with economic factors. For such vessels the performance can be quantified by means of a performance index; in fact, as pointed out in Section 1.2, a design cannot be said to be rationally-based unless the objective is specified and its dependency on the design variables is quantified. The same performance index that serves as the objective function can also be used to assess the degree of seriousness of a failure that adversely affects the performance.

Although safety and serviceability have much in common they are distinct; there are some failures that can cause fatalities without causing loss of main function, and vice-versa. Also they have different relative importance in different situations. For example, in naval vessels the main function is the performance of a mission and therefore serviceability—the accomplishment of the mission—has greater importance relative to safety than it has for commercial vessels.

Since there are two separate attributes for measuring seriousness—safety and serviceability—we use two separate factors, γ_{S1} and γ_{S2}. Even with this distinction between safety and serviceability, the degree of seriousness is still difficult to assess and to quantify. There are any number of degrees of seriousness; it is a continuous rather than a discrete quantity. Nevertheless, for our purpose we must nominate and define a few specific degrees of seriousness. As an example, we will herein distinguish four degrees of seriousness which we will call *extreme, severe, moderate,* and *slight*. These must be defined in terms of their likely consequences in regard to safety and serviceability. For the attribute of safety the degree of seriousness of a failure corresponds to its consequences in regard to loss of life. Table 1.2 describes in general terms the sorts of consequences that are associated with the four degrees of seriousness. Similarly, for the attribute of serviceability the seriousness is measured by loss of main function, and Table 1.3 describes the sorts of

*Strictly speaking, they should be called "partial load factors," but the word "partial" is often omitted.

TABLE 1.2 DEGREES OF SERIOUSNESS OF STRUCTURAL FAILURE IN REGARD TO SAFETY

Level of Structure	Degree of Seriousness of Failure	Consequences in Regard to Loss of Life	Sample Range of $\gamma_{S1}\gamma_Q$	γ_{S1} for $\gamma_Q = 1.1$
Hull Module Collapse	Extreme	Some fatalities likely; may include all personnel if there is another failure or harsh conditions or mismanagement.	1.30–1.54	1.18–1.40
Principal Member Collapse	Severe	Small but definite risk that the failure may cause a few fatalities at occurrence; risk of subsequent fatalities very small unless there is another failure or harsh conditions or mismanagement.	1.20–1.40	1.09–1.27
	Moderate	No appreciable risk of fatalities but the structure is weakened and a slight risk would arise if there is another failure or harsh conditions or mismanagement.	1.10–1.30	1.0–1.18
	Slight	No risk of fatalities, but the resulting local damage constitutes a slight risk of injury (e.g., warped deck plating).	1.0–1.20	1.0–1.09

TABLE 1.3 DEGREES OF SERIOUSNESS OF STRUCTURAL FAILURE IN REGARD TO SERVICEABILITY

Level of Structure	Degree of Seriousness of Failure	Consequences in Regard to Loss of Main Function[a]	Sample Range of γ_{S2}
Hull Module Collapse	Extreme	Complete loss (ship out of service) for a long period. May be permanent loss (i.e., total loss of ship) if there is another failure or harsh conditions or mismanagement.	1.2–1.4
Principal Member Collapse	Severe	Complete loss for short period or partial loss (ship operational but severely handicapped). Repair costly and urgent.	1.1–1.2
	Moderate	Ship operational but inefficient. Loss of some secondary functions. Repair as soon as practicable.	1.05–1.1
	Slight	Main function unimpaired; some inconvenience or inefficiency at the secondary level. Repair as soon as convenient.	1.0–1.05

[a]Revenue earning or mission performance.

consequences that would correspond to these four degrees. It is emphasized that the values of partial load factors given in the tables are merely sample values, given for illustration only. The way in which they were obtained is described in the next section, which also points out the need to take into account particular factors such as the type of ship (passenger, cargo, naval, hazardous cargo, etc.) and the cost and the operational importance of the ship.

Since the primary aim of structural constraints is to provide adequate safety and serviceability, the most important limit state is that of ultimate failure or collapse of the structure (hull module). The other limit states are merely stages toward collapse. The provision of an adequate degree of safety against structure collapse automatically provides a proportional degree of safety against less serious forms of failure, and this is usually sufficient. But the converse is not true; the provision of adequate safety against lesser forms of failure does not necessarily provide sufficient safety at the overall level—which is where it is most required.

In the past, because a true ultimate strength analysis of the overall structure was not possible, the only alternative was to define certain ways in which overall failure could occur, in terms of certain combinations of member failures. This has obvious limitations, the most serious being that structure collapse can in fact be caused by the collapse of one member or of many, depending entirely on the geometry and proportions of the structure and on the loading. The correlation of structural collapse with member collapse requires long-term experience and carefully documented information concerning actual failures of that particular type of structure. If the structure is of a different type, or even of different proportions, it may very well be susceptible to a certain combination of member failures which has not occurred before. This of course is precisely what has happened in many structural failures.

Obviously it is better to determine what the actual forms of failure are for the structure in question, under the various combinations of loads which are expected. Since the ultimate strength testing of a complete hull module is usually out of the question, this can only be done by means of a model—either a physical model or a mathematical model. With modern computers the latter is not only possible but it is also just as reliable and is far easier and more efficient than the former. As is shown briefly in Section 2.5, and in more detail in Chapters 15 to 17, the mathematical calculation of the ultimate strength of a hull module requires an *incremental* or *load-deflection* approach, that is, the formulation of a mathematical model and the determination of the structure's actual load-deflection relationship by an incremental analysis of the individual failures which lead to collapse of the structure. There is no other way of determining the true forms of collapse and the true (i.e., lowest) collapse load for a large structure which is subjected to a variety of load combinations. A load-deflection approach is normally a rather large computational task, but Chapter 17 presents a modeling strategy and a method of analysis that makes it quite economical with present-day computers.

If a true hull module ultimate strength analysis is performed then there is no need to attempt to anticipate the form of the collapse. Hence there is no need to define intermediate stages of failure in terms of numbers and types of member failures, or to classify these according to their degree of seriousness and assign partial safety factors for them. Therefore the ultimate strength analysis of the overall structure logically involves only one level of seriousness—the extreme level. The other levels shown in Tables 1.2 and 1.3 are mainly intended for the ultimate strength analysis of principal members, which we will next consider. But we note in passing that if there is some particular intermediate form of failure for which it is desired to have a specified degree of safety, then these other levels can be used for this purpose.

The other type of limit state analysis is that of the principal members. In most cases these are sufficiently simple and their characteristics are sufficiently well known that it is possible to define all of the possible limit states for each type of member (the member types being distinguished on the basis of the topology, geometry, and material of the member) in terms of the member's structural dimensions and material properties, and also in terms of *the loads and boundary conditions acting on it*. The last point is emphasized because if the member loads and boundary conditions are not known accurately, then a multimember load-deflection approach must be used. But if the limit states can be thus defined, then it is possible to calculate the limit load directly, although some approximations and idealizations may be required.

Since there are several limit states and since some of them may be known to have more serious consequences than others (again because of safety or serviceability or both) it is desirable to have two or three levels of seriousness of member failure, and that is the purpose of the other three categories—severe, moderate, and slight—of Tables 1.2 and 1.3. Each member failure is assigned to one of these categories depending on which of the consequences described in the tables best matches the consequence which that limit state would have on the safety and serviceability of the ship.

Sample Values of Load Factors

At the hull module level the most significant load effect is hull girder bending moment. Because of the complexity of this load effect there is as yet no standard or universally accepted value for the percentile level to be used in calculating the characteristic value of wave bending moment, or for the load factors γ_Q and γ_{S1}. The specification of these values is the prerogative of the classification societies and other ship structure authorities, and these bodies are currently engaged in the determination of suitable values. Therefore in this section we present only some sample values of load factors which are intended merely to illustrate the discussion and to give some indication of the approximate magnitude and range of these factors.

As we have seen, the degree of seriousness of a structural failure, from the point of view of safety, is accounted for by γ_{S1}. But this degree of seriousness depends on many factors and circumstances relating to the nature of the ship and its service. For example, a given type of structural failure is more serious in a vessel carrying passengers, not only because of the added number of people, but also because, unlike a crew, they are uninvolved and untrained. Therefore, although we can define certain degrees of seriousness in general terms as in Table 1.2, it would not be sufficient to assign one specific value of γ_{S1} for each of these degrees. Instead it is necessary to give a set of values that covers all of the important circumstances relating to safety. In order to give at least a qualitative indication of this, Table 1.2 gives a sample range of values of $\gamma_{S1}\gamma_Q$ for each of the defined levels of seriousness. The reason for giving the product instead of the separate load factors is explained later.

The factor γ_{S2} accounts for the degree of seriousness in regard to serviceability, and in this regard the seriousness of a given failure depends on the importance of the ship's main function, or on the economic scale of the ship and of the system in which it is operating. For example, a failure that causes three weeks of ship immobility for repairs is more serious for a fast or expensive or specialized ship than for a slow-speed, low-cost, easily-substituted ship. Thus the choice of γ_{S2} is normally made not by regulatory authorities but by whomever is responsible for the overall system in which the ship is to operate, on the basis of economic and operational criteria. Hence the value of γ_{S2} can vary over an appreciable range, and to illustrate this, Table 1.3 gives a range of values. But here again, as in Table 1.2, the values are for illustration only, and the specified upper and lower values are not intended as limits in any way.

The sample values of $\gamma_{S1}\gamma_Q$ for the "extreme" level of seriousness in Table 1.2 are based on a study by Faulkner and Sadden [27] which analyzed several actual and "rule-based" naval vessels. Working backward from available data (load estimates, strain records, etc.) the authors estimated that the percentile level for the characteristic hull girder bending moment which is implicit in current design practice is of the order of 2%, that is, a wave bending moment which is sufficiently large that there is only a 2% chance of it being exceeded in the ship's lifetime.* However, this figure of 2% is only approximate because there are still some aspects of ship response analysis theory in which greater accuracy is required, such as some of the nonlinear effects of the ship motion response analysis, the spectral distribution function for ocean waves, and the statistical prediction of long-term values.

These are all examples of approximational uncertainty regarding the load and hence they can be accounted for by γ_Q. If the values of the other two load factors were known it would be possible to deduce the value of γ_Q which is implicit in present-day design practice. But, since there are as yet no officially established or standard values for either of these factors the authors of Ref. 27 could only obtain values of the total load factor, γ_{load}, defined in (1.6.1).†

The results indicated that for hull module collapse (specifically, collapse of the strength deck) which corresponds to the "extreme" level of seriousness, the value of γ_{load} varied from 1.56 to 1.83 for the rule-designed naval vessels, for which the values of γ_{S1} would resemble those which are implicit in merchant ship design. Since all of the ships were of the same type, γ_{S2} is constant and so the product $\gamma_{S1}\gamma_Q$ is proportional to the load factor. Although the value of γ_{S2} is unknown it is probably rather low since these were naval vessels. In order to give some indication of the value of the product $\gamma_{S1}\gamma_Q$, let us choose a value of 1.2 for γ_{S2} (this corresponds to the lower end of the sample range of γ_{S2} in Table 1.3). Dividing this into the values of γ_{load} gives a range for $\gamma_{S1}\gamma_Q$ from 1.30 to 1.54, and this is the range shown in the top row of the table. The other three ranges were chosen such that all four are similar in size and have a small overlap.

From the above discussion it is clear that the total load factor is influenced by all three partial factors, each of which has a different purpose and is determined according to different criteria. Since the value of γ_{S2} does not relate to safety and is chosen by persons other than the safety authorities, the values of γ_{S1} and γ_Q must be sufficiently large such that the overall level of safety is satisfactory even when γ_{S2} is 1.0, as it might be for a low-cost, easily replaceable vessel. Obviously, a value of γ_{S2} less than 1.0 cannot be allowed. In principle, this same restriction applies to the other partial safety factors as well. However, circumstances can arise where there might be grounds for allowing a factor to be less than 1.0, such as when a standard characteristic load is being used which is too severe for the vessel in question (e.g., ocean wave bending moment for a vessel intended only for semi-sheltered waters).

*This is not the probability of failure; it is merely the definition of the characteristic value of the load effect, Q_c in Fig. 1.18*b*. The probability of failure is much less than 2% because it also involves the characteristic value of the *limit* bending moment $Q_{L,c}$ which is itself defined in terms of a small percentage chance of non-exceedance (typically 5%), and the margin between Q_c and $Q_{L,c}$ which is provided by the partial safety factors.

†Actually Ref. 27 presented values of the *overall* safety factor, $\gamma_0 = \gamma_{S1}\gamma_{S2}\gamma_Q\gamma_L$, but from the information given in the paper it is possible to deduce that γ_L is approximately 1.08.

REFERENCES

1. O. F. Hughes, F. Mistree, and J. Davies, "Automated Limit State Design of Steel Structures," *Proceedings of the Sixth Australasian Conference on the Mechanics of Structures and Materials,* Christchurch, August 1977, pp. 98–106.
2. R. Taggart, Ed., *Ship Design and Construction,* Society of Naval Architects and Marine Engineers, New York, 1980.
3. J. H. Evans, Ed., *Ship Structural Design Concepts,* Cornell Maritime Press, Cambridge, MD, 1975.
4. J. B. Caldwell, "Theory and Synthesis of Thin-Shell Ship Structures," International Association for Shell Structures, Symposium, Hawaii, 1971.
5. R. H. Gallagher and O. C. Zienkiewicz, Eds., *Optimum Structural Design,* John Wiley & Sons, New York, 1973.
6. O. F. Hughes, "Structural Optimization," Section 4 of Chapter IV.1, Proc. 7th International Ship Structures Congress, Paris, 1979, pp. 38–50.
7. H. J. Kellog, "The Cutting Plane Method for Solving Complex Programs," *SIAM J.,* **8,** 703–712, 1960.
8. R. E. Griffith and R. A. Stewart, "A Nonlinear Programming Technique for the Optimization of Continuous Processing Systems," *Manage. Sci.,* **7,** 379–392, 1961.
9. O. F. Hughes and F. Mistree, "An Automated Ship Structural Optimization Method," *Computer Applications in the Automation of Ship Design,* Jacobssen et al., Eds., North Holland Publishing Co., 1976, pp. 203–212.
10. F. Mistree, O. F. Hughes, and H. B. Phuoc, "An Optimization Method for the Design of Large, Highly Constrained Complex Systems," *Engineer. Optimiz.,* **5** (3), August 1981.
11. B. A. Murtagh and M. A. Saunders, Systems Optimization Laboratory Report 80-1, Stanford, January 1980.
12. B. A. Murtagh and M. A. Saunders, "A Projected Lagrangian Algorithm and its Implementation for Sparse Nonlinear Constraints," Mathematical Programming Study on Constrained Minimization, A. G. Buckley and J. L. Goffin, Eds., North-Holland, New York, in press.
13. P. Pedersen, "Optimal Joint Positions for Space Trusses," *ASCE,* **99,** ST12, 1973, pp. 2459–2476.
14. P. Lersbryggen, Ed., *Ships' Load and Strength Manual,* Det norske Veritas, Hovik, Norway, 1978.
15. O. F. Hughes, F. Mistree, and V. Žanić, "A Practical Method for the Rational Design of Ship Structures," *J. Ship Research,* **24** (2), June 1980, pp. 101–113.
16. D. Liu, O. F. Hughes, and J. E. Mahowald, "Applications of a Computer-Aided, Optimal Preliminary Ship Structural Design Method," *Trans. SNAME,* 1981.
17. O. F. Hughes, W. A. Wood, and R. Janava, "SHIPOPT—A CAD System for Rationally-Based Ship Structural Design and Optimization," *4th International Conference on Computer Applications in Ship Design,* Annapolis, June 1982.
18. A. R. Flint and M. J. Baker, "Risk Analysis for Offshore Structures—the Aims and Methods," Proc. Conf. on Design and Construction of Offshore Structures, *I.C.E.,* 1977.
19. E. V. Lewis et al., "Load Criteria for Ship's Structural Design," SSC Report SSC-240, 1973.
20. Construction Industry Research and Information Association (CIRIA): "Rationalization of Safety and Serviceability Factors in Structural Codes," CIRIA Report 63, July 1977.
21. F. Moses, "Safety and Reliability of Offshore Structures," in *Safety of Structures Under Dynamic Loading,* I. Holand, D. Kavlie, J. Moe, and R. Sigbjornsson, Eds., Tapir Publ., Trondheim, 1978.
22. A. G. Pugsley, "A Philosophy of Aeroplane Strength Factors," Report and Memo No. 1906, British Aeronautical Research Committee, 1942.
23. A. M. Freudenthal, "The Safety of Structures," *Trans. ASCE,* **102,** 1947.
24. S. Gran, "Reliability of Ship Hull Structures," Det norske Veritas Report No. 78-216, 1978.
25. A. M. Freudenthal, "Safety and Probability of Structural Failure," *Trans. ASCE,* **121,** 1956.
26. A. Mansour, "Approximate Probabilistic Method of Calculating Ship Longitudinal Strength," *J. Ship Research,* **18** (3), Sept. 1974, pp. 203–213.
27. D. Faulkner and J. A. Sadden, "Toward a Unified Approach to Ship Structural Safety," *Trans. RINA,* 1979.

Chapter

Two

LOADS, RESPONSE, LIMIT STATES, AND OPTIMIZATION

This chapter takes a closer look at four of the main aspects of rationally-based design in order to identify the principal components of each of them and to help the reader to gain a clear overall picture by classifying these components according to their types, levels, and the relationships between them. The logical place to start is with the loads and so we begin with a brief summary of the principal types of loads which act on a ship.

2.1 LOADS ON SHIPS

One way of classifying the loads on ships is according to the level of structure at which they act, because some loads influence the structure at just one of the four levels mentioned earlier: hull girder, hull module, principal member, and local (see Fig. 1.2). But other loads have an influence at more than one level and the most fundamental load—external pressure on the hull—has an influence at all four levels. Nevertheless the loads can be classified approximately in this way and it is important to have a clear concept of the levels at which the various loads act, or have their principal influence.

Another way of classifying loads is according to how they vary with time: static, slowly varying, or rapidly varying. And in calculating the load effects, there are three types of structural analysis which more or less correspond to these: static, quasi-static and dynamic. In a dynamic analysis the effects of the time variation of the loading are fully accounted for. Almost any irregular dynamic loading can be represented as a combination of regularly varying loads and if the force–displacement relation is linear or only slightly nonlinear then the problem of calculating the load effects can be solved "in the frequency domain," with frequency as the principal independent variable instead of time, which greatly simplifies the calculations. The frequency-based distribution of a load or a load effect (response) is called a *spectrum,* and so we speak of a *wave spectrum* and a *response spectrum.* If the force–displacement relation is nonlinear then the problem must be solved "in the time domain," with time as the independent variable.

A quasi-static analysis is simply a static analysis in which the motions are estimated and their effect on the structure is accounted for approximately by including some inertia forces. Since there is no essential difference between these two we will in this text usually speak of just "static" and "dynamic," and only use the term "quasi-static" when it is desired to emphasize that some motion effects are being allowed for in the static analysis. In most cases this will be clear from the context.

Slowly varying loads are those for which even the shortest component period is appreciably longer than the fundamental (longest) natural period of vibration of the structure. In most cases slowly varying loads can be dealt with by means of static analysis with only a small loss of accuracy, whereas rapidly varying loads usually require a dynamic analysis for sufficient accuracy.

Whenever possible, in order to minimize computation, the static and dynamic analyses are performed separately, with the latter dealing only with the fluctuation in load, that is, departure from the static load. The total response of the structure is then obtained by superimposing the two results.

In terms of the three load types just defined, the principal loads on ships are:

Static (or essentially static) loads

1. All of the "stillwater" loads: external and internal pressures (buoyancy and bulk cargo); all weights.
2. Drydocking loads.
3. Thermal loads

Slowly varying loads

1. The wave-induced dynamic pressure distribution on the hull due to the combination of wave encounter and the resulting ship motion.
2. Sloshing of liquid cargoes.
3. Shipping of green seas on deck.
4. Wave slap on sides and on foredecks.
5. Inertia loads, especially on masts and other elongated structures, and also on decks and frames at the attachment points for containers and other heavy objects.
6. Launching and berthing loads.
7. Ice-breaking loads (at the hull girder level).

Rapidly varying loads

1. Slamming.
2. Forced (mechanical) vibration; pressure pulses from the propeller.
3. Other dynamic loads, as discussed later.

The static loads are relatively straightforward and do not require any special comments or explanation. The calculation of the stillwater hull girder load effects is dealt with in Chapter 3. The slowly varying and rapidly varying loads are more complex, and the following sections define and describe the various loads in these categories.

The treatment here is mainly qualitative, with the aim of covering the main aspects and of giving an overview of the various types of loads and the ways in which they are accounted for. More specific information about loads is presented in the Ship's Load and Strength Manual—Ref. 14 of Chapter 1. This manual provides formulas for characteristic extreme values for wave-induced loads and expressions for calculating various deterministic loads, such as static and dynamic forces from general and dry bulk cargo, and static and dynamic pressures in liquid cargo tanks, both full and partly full.

Slowly Varying Loads

The most important load in this category is the wave-induced dynamic pressure, and the most important effect is the wave-induced hull girder bending moment M_w. The term "wave-induced" means the difference or departure from the stillwater value of that load (pressure, bending moment, shear force, etc.) such that the total value at any point is the sum of the two. Ideally, since it is influenced by the ship's motions, the pressure on the hull should be calculated as part of the ship motion analysis, and the wave bending moment obtained from it by integration. But as mentioned in Chapter 1, recent progress in ship motion analysis and in the statistical description of ocean waves has permitted the development of approximate but sufficiently accurate expressions for the characteristic value of extreme wave-induced (vertical) bending moment $\hat{M}_{w,c}$. The accuracy and applicability of these expressions is being steadily improved by theoretical analyses and by model and full-scale testing. In the hull girder analysis the corresponding wave-induced hull girder (vertical bending) stress σ_w is obtained from the section modulus formula $\sigma_w = \hat{M}_{w,c}/Z$.

In the (static) hull module analysis these wave-induced hull girder stresses, together with the stillwater hull girder stresses, constitute the loads at the ends of the module. The other hull module loads are an equivalent static pressure distribution on the hull representing the actual dynamic pressure distribution, and the various gravity loads (cargo, steel weight, etc.) with an allowance for inertia effects if these are considered important.

Of the other slowly varying loads which were listed previously, loads (2) through (5) should also, ideally, be derived from a ship motion analysis, but the first three are highly nonlinear and hence the computation is very difficult. Sloshing requires a dynamic analysis for satisfactory accuracy, but from a series of such analyses for various tank sizes, geometries, locations, and so on, some approximate equivalent static loads can be determined.* For green seas on deck the downward pressure loading is approximately equal to the

*See, for example, Ref. 14 of Chapter 1.

hydrostatic pressure for that height of water (i.e., negligible dynamic effect). The horizontal pressure (on deckhouse fronts, etc.) has both static and dynamic components, and it can be represented approximately by an equivalent static pressure; a value of 50 kPa is typical. Wave slap is similar and is of the same order of magnitude. Inertia loads require knowledge of the ship's peak acceleration at the point in question; a precise value requires a ship motion analysis but an approximate value can be obtained from the experimental data and ship motion studies which are available in the technical literature. Some information on launching, berthing, and ice-breaking loads can also be found in the literature.

Rapidly Varying Loads

There are several types of load which have short periods and which usually require dynamic response analysis.

Slamming

There are two types of slamming: bottom slamming and bow flare slamming.*

Bottom slamming occurs when, because of pitching, possibly combined with the occurrence of a wave trough, the ship's bottom emerges from the water and subsequently, because of its relatively flat area and the combined speed of ship and wave surface, undergoes a severe hydrodynamic impact on reentry. The impact is sufficiently rapid and intense to generate a high intensity pressure pulse on the bottom plating, of very short duration (typically 0.1 to 1 sec) which is often accompanied by a loud booming or slamming sound.

Bow flare slamming is the plunging of the upper flared portion of the bow deeper into the water. This is a somewhat more gradual phenomenon (usually without any sound unless the flare is very concave) but it also imparts a relatively sudden and intensive force to the forward part of the ship. Circumstantially the two types of slamming are quite different; bottom slamming requires emergence of the ship's forefoot whereas bow flare slamming occurs later in the ship's pitching cycle, when the bow is already partially immersed. They are independent phenomena and either can occur without the other. But in spite of their external differences, these two phenomena arise from the same fundamental cause—the relatively sudden change in the breadth of the immersed cross section† of the ship as it moves downward. This change of breadth causes a change in the momentum of the surrounding water and it does so by changing both the velocity and the mass (i.e., the amount) of water which is involved. The velocity change—both in direction and in magnitude—is easy to understand because the immersed section of the ship is growing in size and changing in shape. But the motion of a body in a fluid involves simultaneous movement of the fluid and this makes the body behave as if it had additional or added mass; more force is required to accelerate the body. If the submerged body grows rapidly in size, as happens in slamming, the added mass also grows rapidly and resists the body's motion, which causes a rapid increase in local pressure.

Slamming has important effects at two different levels of structure:

Hull Girder Level. Slamming causes a sudden vertical acceleration and deflection of the bow and excites flexural vibration of the hull girder, mainly in the fundamental two-node mode but also in higher modes. This hull girder flexural vibration is referred to as "whipping." As shown in Fig. 2.1, the vibratory hull girder bending stress ("whipping stress") is of much higher frequency than the wave-induced stress, and is effectively superimposed on it. The period of the fundamental mode of whipping is usually in the range from 0.5 to 2 sec. The hull girder aspects of slamming are discussed further in Section 4.5 which deals with the nonlinear ship motion and response analysis.

Principal Member Level. The shell plating and its supporting structure (stiffeners, frames, webs, etc.) are subjected to high impact pressure forces, which accelerate and deflect all of this structure and set up vibrations, particularly in the plating. Damage may occur in the form of permanent deformation of the plating and other structure. The analysis of this level of slamming response requires detailed information on the pressure distribution which varies with time and space. The calculation of the permanent deformation is treated in Section 9.5.

Bottom slamming is a particularly complex phenomenon. The magnitude and duration of the impact

*Some authors prefer to use the term slamming only for bottom slamming and to refer to the second type as bow flare impact.

†Strictly speaking it is the waterplane *area* of the immersed *volume* which is changing; that is, the phenomenon is actually three-dimensional. But since a ship is a generally prismatic body the flow is idealized as being two-dimensional.

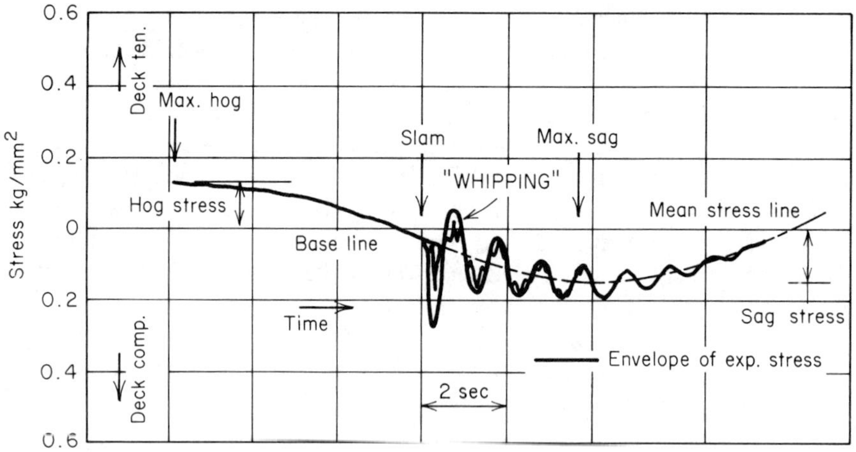

Figure 2.1 Whipping stress due to slamming.

pressure depend strongly on, and are sensitive to, the angle and the relative shape of the ship's bottom and the water surface, and also on their relative speed of approach. This type of slamming has been the subject of an enormous amount of study and research, and the work is continuing. Ochi [1] has presented criteria for predicting its occurrence and severity, Mansour and d'Oliveira [2] have presented a method for calculating the whipping moment, and Ochi and Motter [3], Kawakami et al. [4] and Evans [5] have presented methods for predicting the impact pressure and the whipping stresses due to bottom slamming.

Forced Vibration

At the various substructure levels vibration may be set up by the main or auxiliary engines, by fluctuating hull pressure due to propellers (especially if cavitating) or other sources of excitation. Although these are not severe loads they can influence the structural design in two ways:

By requiring the redesign of a structural member in order to avoid resonance.

By requiring thicker plating to avoid fatigue damage (by reducing the stress level).

Other Dynamic Loads

Ice-impact (at the local or other substructure level).

Underwater explosion (e.g., minehunter).

Certain idealized collision or grounding loads, which may be specified by safety authorities for some ship types, which the ship structure must be capable of absorbing without undergoing certain specified type(s) of failure, such as the failure of cargo tank bulkheads adjacent to those ruptured by the collision.

Springing

For most ships the period of wave encounter is longer than the longest natural period of hull girder vibration. But for ships which are relatively flexible (e.g., some of the very long Great Lakes vessels) the period of two-node vibration is sufficiently long (of the order of 1 sec) that it can be excited by the shorter period components of the encountered waves. This phenomenon is known as "springing." Since it depends on the period of encounter, springing is more likely at higher ship speeds. Springing is undesirable for two reasons. First, it produces, at the fore and aft ends of the ship, a noticeable rise and fall of the deck, with a period of the order of 1 sec, which is distracting and even uncomfortable. It seldom lasts for more than a few cycles because it depends on so many coincident factors, but it is nonetheless an undesirable phenomenon, particularly in ships with accommodation aft. Second, since springing occurs at a higher frequency than ordinary wave-induced bending it increases the number of stress cycles in the ship's lifetime. Thus if it were to occur frequently it would be a contributing factor to hull girder fatigue.

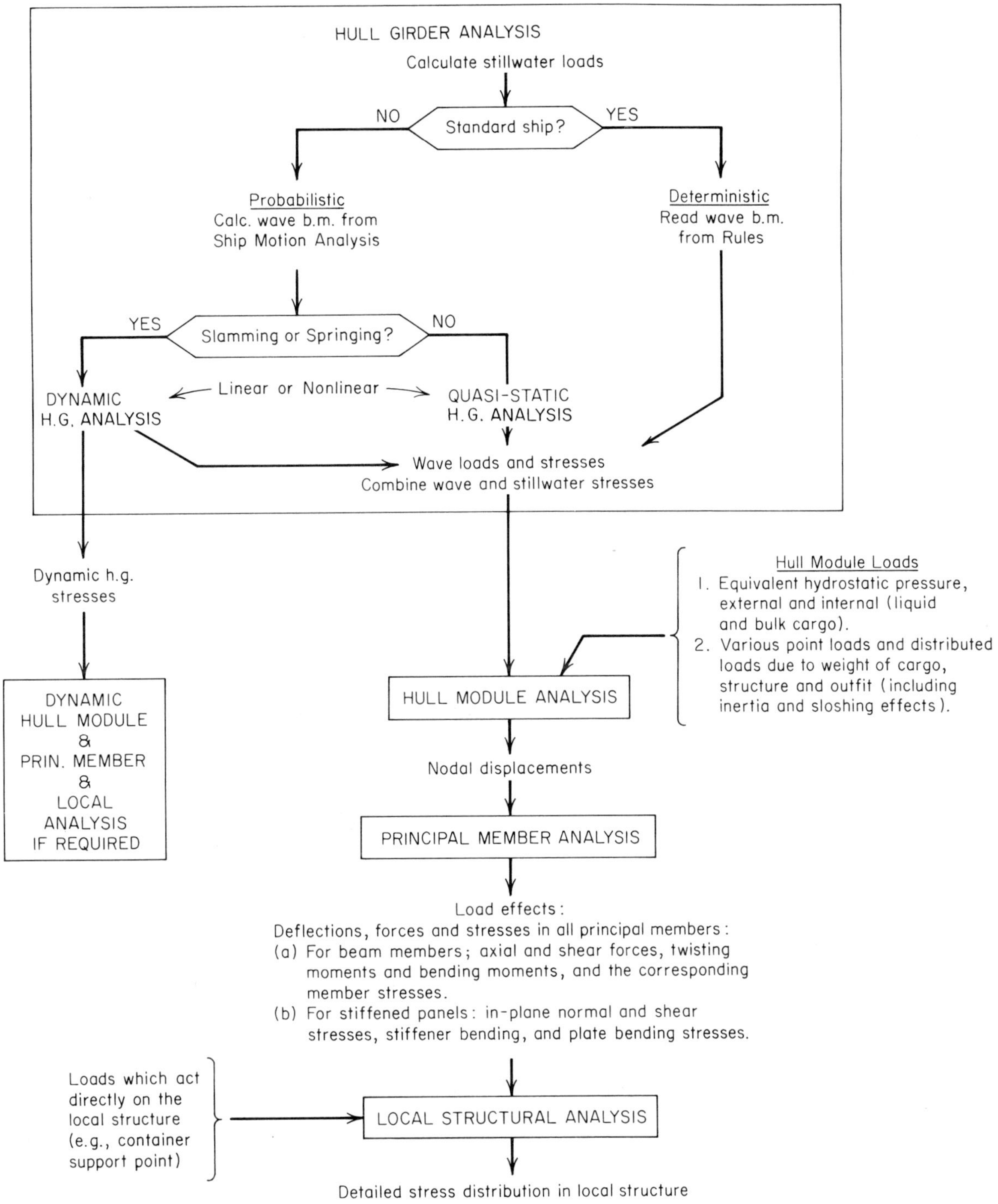

Figure 2.2 Levels of response analysis.

2.2 TYPES OF STRUCTURAL RESPONSE ANALYSIS

Static Only or Static and Dynamic

Figure 2.2 illustrates the principal types of loads and load effects which are involved in the four levels of structural response analysis, from hull girder to local. The figure shows that at each level the analysis may need to include a dynamic structural analysis. The need depends entirely on whether that level of structure is subjected to any significant rapidly varying loads, that is, loads for which the shortest component period is the same order of magnitude or shorter than the longest natural period of that level of structure. Since the latter differs markedly for different levels, with longer periods for larger levels of structure, a load that is slowly varying at a lower level may constitute a rapidly varying load at a higher level (providing that it qualifies as a significant load at that higher level). Springing is an example of this; wave loads have too long a period to cause any excitation at the hull module level but they can cause vibration of the hull girder if the ship is relatively flexible. An impulsive or high-frequency load, if it is large enough, will cause a vibratory response at several levels. Slamming, in particular, can cause a response at all four levels. But most rapidly varying loads have their principal influence at the local and/or principal member level and are not large enough to induce vibration of an entire hull module or hull girder. Of course, vibration can be transmitted from one member to the next and thus extend over large regions of the ship's structure, but that is still member vibration.

In summary we may say that at the hull girder and hull module levels a wave-excited dynamic analysis is not required for most ship types. However a calculation of hull girder natural frequency is nearly always performed. At the principal member and local levels a vibration analysis may be required if there is some significant and unavoidable source of excitation (propellers, machinery, etc.). In most cases the only requirement is to calculate the natural frequencies and to design the structure so as to avoid resonance.

On the other hand, there are some instances when a dynamic analysis is required at the hull girder and/or hull module levels. For relatively flexible ships a dynamic hull girder analysis should be performed to check for springing. Also a dynamic analysis may be required at one or both levels for "high performance" vessels, that is, vessels that have greater structural efficiency and hence are more flexible, particularly at the hull module level, and which at the same time must face significant dynamic loads at these levels because of high-speed operation, unusual hull geometry, or other reasons. Some container ships and some naval vessels are in this category.

Probabilistic or Deterministic

In addition to the choice between static and dynamic, there are also two different types of response analysis depending on whether an explicit statistical approach is used to define the loads and to calculate the load effects:

*Probabilistic**—the characteristic values of load effect are calculated explicitly for the particular structure and load.

Deterministic—the characteristic values are obtained from approximate expressions which have been derived previously by means of a systematic series of probabilistic analyses.

Probabilistic analysis should be used for ships for which the hull girder loads are not already well established, such that there are no prederived characteristic values of wave-induced bending moment. If these are well established and characteristic values are available, then deterministic analysis is sufficient. Most types of cargo ships are in this latter category.

If a probabilistic response analysis is required it is usually required only at the hull girder level because the most uncertain load is usually the wave load. At more detailed levels there are seldom any loads which are so truly random as to permit probabilistic analysis. Hence the analysis at lower levels is usually deterministic, using the characteristic values of load effects obtained from the hull girder analysis and deterministic estimates for the maximum values of all loads that occur at these lower levels.

For limit state analysis the situation is reversed. Here it is the limit value which is uncertain, and the principal source of the (statistical) uncertainty arises at the local level (especially material properties) and the principal member level (especially connections and

*Strictly speaking this should be called semiprobabilistic, but we have seen in Chapter 1 that fully probabilistic analysis is presently not feasible and is probably not justified. Hence in this text the distinction is not required, and from this point on we shall use the terminology defined above.

fittings). Hence the limit analysis is probabilistic at these levels and produces a characteristic value, $Q_{L,c}$, for the relevant limit value Q_L. At higher levels the limit analysis is deterministic, combining these characteristic values together in order to calculate the limit values for the modes of failure which occur at those higher levels.

Linear or Nonlinear

Finally, wave response analysis may also be classified according to whether it is linear or nonlinear. The former is much easier but for severe sea states there are several sources of nonlinearity: the waves themselves, the governing hydrodynamic equations, and the ship geometry. A linear analysis is based on several simplifying assumptions, of which the principal ones are:

1. The irregular wave surface of the ocean can be represented as the linear sum of a large number of individual regular waves of different heights and frequencies.
2. The hydrodynamic forces on a ship hull can be obtained by considering each transverse section of the ship separately and combining the results linearly.
3. The wave force acting on each section is linearly proportional to the "emergence" at that section, that is, the difference between the local wave height and the ship's still waterplane.

The accuracy of these assumptions is discussed in Section 4.5. Suffice to say here that the accuracy of the first two is generally satisfactory and that the third assumption is valid for ships which are approximately wall-sided in the waterplane region. If this is not so, or if there is any other source of nonlinearity, a nonlinear method of response analysis must be used.

From what has been said previously it can be seen that wave response analysis can be quite complex, involving probabilistic, dynamic and nonlinear aspects. These are discussed in Chapter 4, after first covering the deterministic aspects of hull girder response analysis in Chapter 3. But for the purposes of this section, in which we are seeking to present an overall view, it may be helpful to list the principal steps which are involved for the simplest type of wave response analysis, which is based on *linear* ship response theory (see Fig. 4.27).

1. A ship motion response analysis is performed repeatedly for a complete range of deterministic wave loads, each consisting of regular waves of a single specific frequency ω and unit height. This gives the transfer function $\Phi(\omega)$ for the relevant response (e.g., wave-induced hull girder bending moment M_w).
2. A typical ocean storm condition is represented by a group of selected ocean wave spectra (i.e., frequency-based distributions of wave height). For each wave spectrum $S(\omega)$ the corresponding response spectrum [e.g., $M_w(\omega)$] is the product of $|\Phi(\omega)|^2$ and the wave spectrum.
3. The results for the various wave spectra are then combined, according to the proportion in which each spectrum is present in a particular sea state.
4. The analysis is repeated for various sea states (and also various ship headings and speeds). The resulting short-term response spectra are then combined statistically so as to obtain long-term characteristic values of the load effects, such as the value of bending moment for which there is a sufficiently high probability of nonexceedance in the ship's lifetime.

Reference 6 gives a thorough exposition of the linear probabilistic theory of ship dynamics.

2.3 HULL MODULE AND PRINCIPAL MEMBER ANALYSIS

We now focus attention on the second and third levels of elastic response analysis of a ship structure, the hull module analysis and the analysis of the principal members which make up the hull module. These two levels are treated together because they are closely related—so much so that they can be viewed as two stages of a single analysis. The goal of this analysis is to calculate all of the significant load effects in all of the principal members, and the purpose of the hull module analysis is to provide the boundary conditions that are required for the analysis of the principal members. Because of the complex three-dimensional geometry and loading in a hull module, and the consequent high degree of statical indeterminacy, it is not possible to determine the boundary conditions of the principal members except by performing a hull module analysis.

Minimum Length of a Hull Module

We have seen that the two requirements which determine the minimum length of a hull module are the need for known boundary conditions and the require-

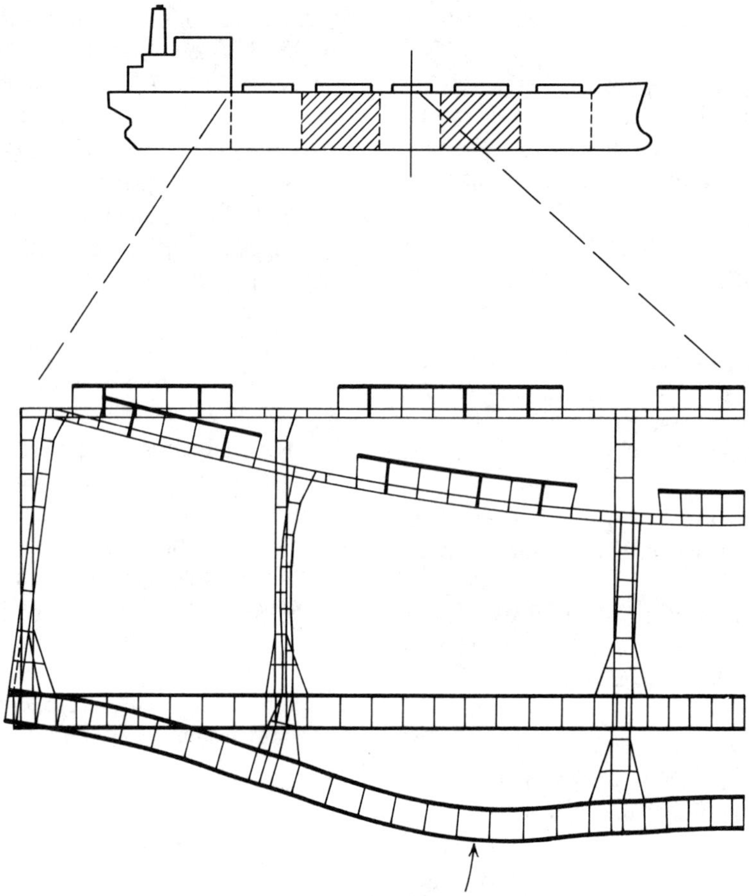

Figure 2.3 Hull module loading and response with a dense cargo.

ment that the hull module must include and acurately portray any form of structural loading or response which is not portrayed at the hull girder level. This often requires that the hull module include several cargo holds (or watertight compartments of a naval vessel) depending on the ship's overall geometry and the nature and arrangement of the cargo or other loading. For example, in a bulk carrier loaded with a very dense cargo such as iron ore, only about half of the internal volume can be filled because of the weight of the cargo. Also, in a bulk carrier each hold should be either full or empty, to prevent cargo shifting. In order to avoid excessive hull girder bending moment the empty holds must be interspersed between the full holds, as shown in Fig. 2.3. This arrangement causes large shear force between the holds which cannot be adequately examined in the hull girder analysis because of the complex interaction between the side shell, double bottom, and transverse bulkhead. Also, as shown in Fig. 2.3, there is a very large amount of inter-bulkhead bending of the double bottom which cannot be examined in isolation (i.e., by considering only one cargo hold) because the boundary support of the double bottom within each hold is provided largely by the double bottom structure in adjacent holds. Unless these adjacent holds are included in the hull module there is no way of knowing and accounting for the correct amount of support. Therefore in such vessels the hull module must extend over at least three cargo holds.

Similarly in tankers and other liquid bulk carriers the cargo tanks are either full or empty (with cargo or ballast) in order to avoid free surface and to minimize tank cleaning. Here again, in order to avoid excessive hull girder bending, the full and empty tanks are interspersed, both longitudinally and transversely, in a checkerboard pattern, as in Fig. 2.4. This arrangement produces a complex pattern of shear force between all of the tanks, in both the longitudinal and transverse bulkheads, and it also produces alternating hogging

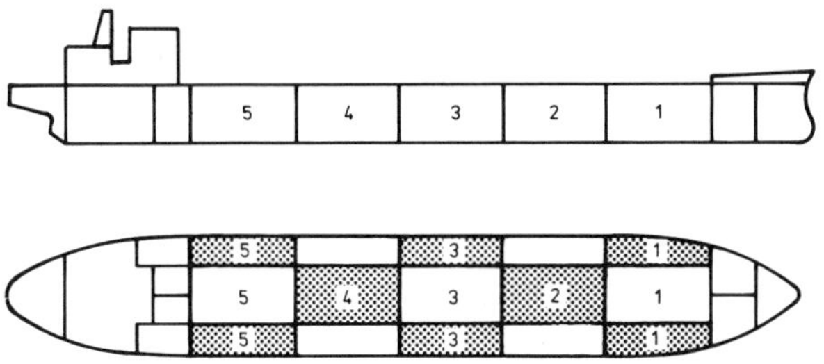

Figure 2.4 Ballasted or partly loaded liquid bulk carrier.

and sagging in each combination of three in-line tanks, both longitudinally and transversely. This inter-tank loading and response is not represented or analyzed adequately in the hull girder analysis. In fact the hull structure of a tanker is completely outside of the scope of simple beam theory because it has two extra "webs"—the longitudinal bulkheads—whose vertical deflection along the ship length is different from that of the side shell. These longitudinal bulkheads have approximately the same in-plane rigidity as the transverse bulkheads and so their vertical deflection relative to the side shell depends almost entirely on the vertical shear distortion of the transverse bulkhead. Therefore the hull module must extend over several cargo holds in order to include the interaction between the longitudinal and transverse bulkheads. Hence for all bulk carriers, dry or liquid, the hull module must extend longitudinally for at least three tanks.

It was mentioned earlier that the hull module boundary conditions can be obtained from the hull girder analysis, providing that the boundaries of the hull module are located at transverse bulkheads. These bulkheads can usually be regarded as a rigid body and the transverse displacements in this plane are all part of a single rigid body displacement. This is the most fundamental assumption of hull girder analysis and hence it must be fulfilled if the hull module boundary conditions are to be obtained from a hull girder analysis. When the transverse bulkheads are not full depth or are almost entirely absent (e.g., ro-ro ships and train and car ferries) the hull can undergo significant transverse distortion, known as "racking" (see Fig. 2.5). For such vessels the hull module should extend over a large portion of the ship length and at least one loadcase should include a combination of loads which represents the "worst case" for racking.

Finally, even if there are relatively rigid transverse bulkheads there may be significant rotation of them if the vessel has low torsional rigidity. This can occur with container ships and other vessels with large deck openings. In this case the correct boundary conditions

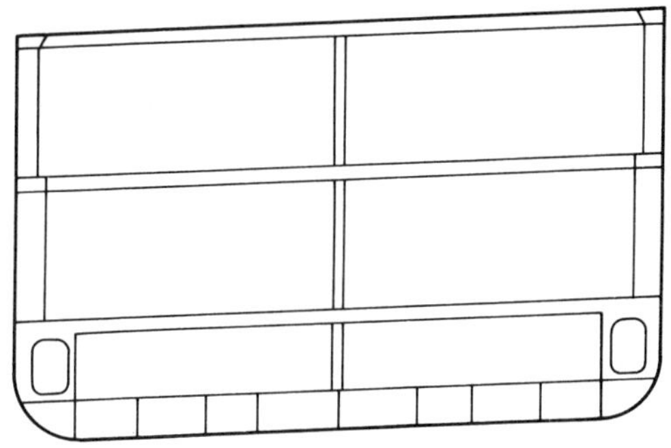

Figure 2.5 Transverse shear or "racking" deformation.

for the hull module analysis can be obtained by a more accurate hull girder analysis, using the hull girder element presented in Chapter 6.

Of course if there is symmetry of structure in any direction, then the hull module (i.e., the mathematical model which is created) need be only half of the length in that direction. This is always true, even when the load is not symmetric in that direction. Thus, since there is nearly always port–starboard symmetry, the hull module is nearly always a half section of the ship, even when the loading is not transversely symmetric, as in torsional loads. The modeling of unsymmetric loads is explained in Chapter 8.

Design-Oriented Finite Element Analysis

The basic method for the hull module analysis is the finite element method. During the past 15 years this method has become one of the most powerful and versatile methods in structural analysis, and most readers of this book will already be familiar with it. However, there are two points which are relevant in this regard. First, up to the present time in most structural fields the finite element method has been used mainly for structural analysis, rather than as an integrated part of structural design. In particular, it is used principally for the analysis of portions of a structure that have complex geometry and for detailed three-dimensional stress analysis. But this type of usage is somewhat more specialized than design, and is of more direct interest to structural analysts and researchers than to structural designers. Because of this, many structural designers, particularly those who are engaged in or are responsible for the overall structural design, have had little need or reason to gain more than a casual acquaintance with the finite element method. Since this book is intended mainly for structural designers, and since one of its principal aims is to show how modern computer-based structural analysis methods can be integrated into structural design, it includes, in Chapter 7, a brief introduction to the finite element method. The presentation is simpler and less extensive than in most books dealing with this method because it concentrates on fundamentals and on those aspects which are of importance in rationally-based structural design.

This leads to the second point, which is that the use of finite element analysis as an integrated part of structural design requires a different type of analysis—a *design-oriented* finite element analysis, with a different modeling strategy and choice of elements from those which are used for detailed stress analysis. Thus even for structural specialists, if they wish to become involved in and contribute to rationally-based structural design, there is a need to learn and to make use of these design-oriented modeling and analysis techniques.

In order to understand the basic relationship between the hull module analysis and the principal member analysis, it is necessary to have a clear idea of the two principal stages in any finite element analysis: structure analysis and member analysis.

1. ***Structure analysis.*** *Use basic elements to set up a mathematical model of the structure and solve this for the structure displacements.*
Using basic structural elements for which the *stiffness matrix* (i.e., the coefficients of the force–displacement relationships) are known, the analyst chooses a combination of element types and sizes which will give an adequate representation of the structure. He then uses a standard mathematical technique (which is straightforward, but is much too tedious to do by hand and is therefore done by computer) to combine these element stiffness matrices mathematically, according to the way in which the elements are arranged in the structure, thus obtaining the force–displacement equations for the structure. These equations are then solved (again using the computer) to obtain the structure displacements.

2. ***Member analysis.*** *For each element, transform the structure displacements at the element's boundary points (nodes) into element coordinates and solve for the element's internal forces.*
The structure displacements at the boundaries of each element provide the necessary boundary conditions for the separate structural analysis of each element, whereby all of its internal forces and stresses can be calculated.

In design-oriented finite element analysis it is desirable to have a close identification between the elements and the members which are being designed. That is, as far as possible the design variables and the analysis variables should be the same, in order to reduce the amount of computation and to simplify the interface between the response analysis, limit analysis, structural evaluation and synthesis steps in the overall design process (steps 2 to 6 in the flow chart of Fig. 1.1). Therefore the hull module analysis makes use of elements that directly correspond to the principal members, and as a result the hull module analysis and the principal member analysis correspond precisely to the two stages just outlined. Of course the principal members of a hull module are relatively large portions of structure: beams, columns, stiffened panels, and so on. In ordinary finite element analysis they would usually be modeled by a number of smaller, general-purpose elements, in order to obtain a complete and detailed picture of the stress distribution throughout the member. But the determination of the structural dimensions of the principal member does not require a complete or a detailed structural analysis. Rather it only requires a knowledge of the internal forces and stresses which are of significance for each particular type of member. The detailed stress distribution is only required for the design of local structure ("detail design") and this is a separate and subsequent phase in the overall design process. Of course there are some local structural requirements which have an influence on the sizes of the principal members, but this can be accounted for by means of constraints, as shown in Section 1.2, in the discussion of step 5. And even though the local analysis may reveal that some of the principal members must be redesigned, nevertheless, because the design of the principal members ("preliminary design") is an iterative process it is still more efficient to perform the local analysis afterwards, as part of the detail design, rather than as part of the preliminary design.

2.4 BASIC TYPES OF STRUCTURAL FAILURE

In contrast to response analysis, which deals with the linear elastic response of a structure to prescribed loads, limit state analysis seeks to determine those levels and combinations of loads which cause structural failure, both of individual members and of the overall structure. Structural failure is nearly always nonlinear—either a geometric nonlinearity (buckling, or any other large deflection) or a material nonlinearity (yielding and plastic deformation). Also, it is possible for both types of nonlinearity to occur together, and so, in general, limit state analysis is much more complex and difficult than response analysis. In fact, it is probably the most complicated aspect of rationally-based structural design, and for this reason the entire second half of this text—chapters 9 through 17—is devoted to limit state analysis. Because of its importance we here give a brief and qualitative review of the basic types of member failure, and the next two sections deal with the basic aspects of limit state analysis at the overall structure level.

For steel members the three basic types of failure and their subdivisions are as follows:

1. Large local plasticity
2. Instability
   - Bifurcation
   - Nonbifurcation
3. Fracture
   - Direct (tensile rupture)
   - Fatigue
   - Brittle

In practice an individual failure in a structural member often involves a combination of these basic types, particularly the first and second types. For example, for sturdy members instability is preceded and accompanied by plasticity. Also the occurrence of local plasticity can seriously diminish the stability of a member and even convert it into a mechanism. Nevertheless each of these three is a distinct type of failure and since we here wish to review only their basic aspects we shall consider each of them separately. Their interactions are considered in Chapters 9 through 17.

To discuss the basic types of failure and to appreciate the difference between them, it is necessary to examine the relationship between load and deflection for each of them. Figure 2.6 presents a sample of load–deflection curves for individual members, illustrating the variety of shapes the curves have, depending on the type of member and the type of loading and support which it receives. The curves are for typical ship structural members, having the following features: (1) they contain local imperfections, including eccentricities and locked-in stresses due to forming and welding; (2) they are made of good quality steel such that premature fracture does not occur and ultimate fracture is preceded by large plastic deformation; and (3) they are of relatively sturdy proportions such that buckling is not purely elastic but rather involves some yielding. Although the curves differ they have some basic aspects in common. In most cases the curve consists of an elastic portion, a region of transition from mainly elastic to mainly plastic behavior, and a plastic region in which the slope becomes small such that the deflection increases greatly for only a small increase in load. The slope of the curve is the instantaneous *stiffness* of the member, indicating its ability to carry further load. It is also a measure of the *stability* of the member. An unstable member is one which can undergo a large increase in deflection with little or no increase in load.

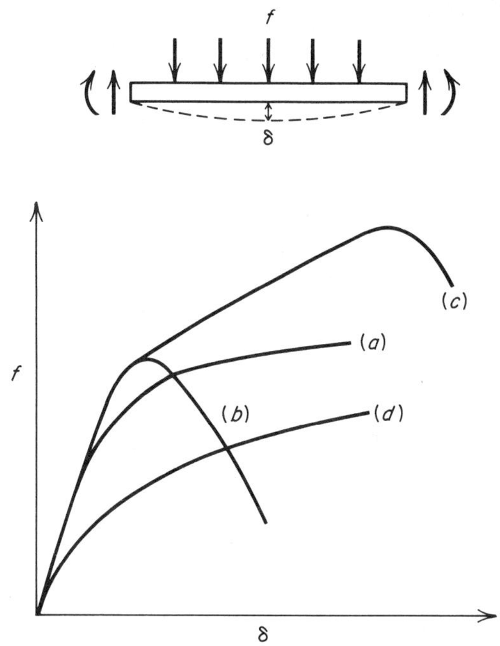

Figure 2.6 Load—deflection curves: (*a*) failure by plastic deformation, (*b*) bifurcation buckling of beams and columns, (*c*) bifurcation buckling of plates, and (*d*) nonbifurcation buckling.

Local Plastic Deformation

We first consider a member that is not susceptible to instability, either because all axial compression loads are small or the member is of very sturdy proportions or it is completely braced and supported. For this type of member the load–deflection curve resembles curve (a) in Fig. 2.6. In the elasto-plastic region of the curve, local regions of plastic deformation occur progressively at the most highly stressed points, and this gradually decreases the member stiffness. In the plastic region of the curve, when this local plastic deformation has grown larger or has occurred at several different points, the stiffness has become quite small and the deflection increases rapidly, eventually becoming so large that the member is considered to have failed; it is no longer fulfilling its main purpose. There is no obvious or precise point on the curve at which failure occurs. The failure load (or collapse load or ultimate strength) is usually taken as the load at the beginning of the plastic region, where the deflection first begins to grow rapidly. Although the member's actual loss of function may not occur until the deflection reaches a much larger value, the reserve of strength is too small to warrant consideration. This type of structural failure is essentially due to material failure, due to high stress levels. Hence this type of failure requires a *stress analysis*.

The foregoing discussion assumes implicitly that the deflection and deformation do not significantly alter either the geometry of the member or the equilibrium equations. Hence it is sufficient to perform all calculations using the initial geometry and the initial equilibrium equations. This is a *first-order stress analysis,* and it is by far the most common type of analysis. In some cases, however, the effect of the deflection and deformation on the geometry may be quite important. The effect can cause either a strengthening or a weakening of the member. For example, as shown in Fig. 2.7, a strengthening effect occurs in a laterally loaded beam whose ends are completely restrained against axial deflection ("pull-in") because the relatively small change in geometry due to the lateral deflection causes part of the load to be carried by the action of membrane tension instead of by bending. In such a case, the stresses in the member have to be calculated with due regard to the effect of this deflection on the equilibrium of the member. This is a *second-order stress analysis*.

More often the change in geometry due to deflection produces weakening effects. For example, in the beam-column of Fig. 2.8 the lateral deflection represents some additional eccentricity and this change of geometry, in conjunction with the axial load f_x, causes additional bending moment in the member. This particular effect is a fundamental characteristic of all geometric *instability* problems; that is, in such problems the lateral deflection always affects the bending moment in the member and hence it alters the equilibrium conditions. Therefore geometric instability is essentially a second-order phenomenon and *stability analysis,* which is distinct from stress analysis, is necessarily a second-order analysis.

Bifurcation Buckling

Instability, or buckling, can occur in any member or part of a member which carries an axial or in-plane compressive load. There are two types of buckling; bifurcation and nonbifurcation. The most common example of bifurcation buckling is the buckling of a simple column. The general shape of the load–deflection curve for such a member is shown in Fig. 2.9. For an elastic column there is some axial load at which an alternative equilibrium position exists, corresponding to a bent shape, and this load is the *buckling* or *bifurcation* load of the member. In a typical col-

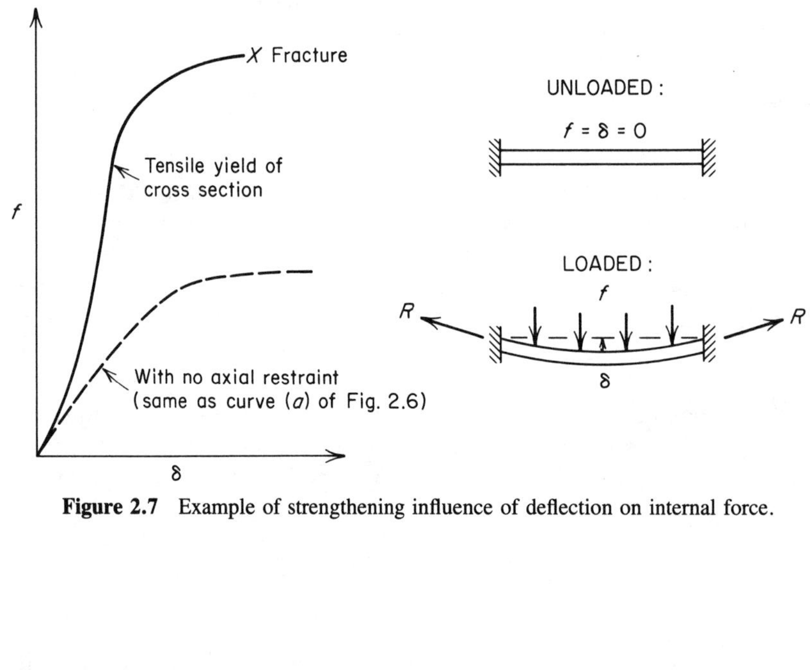

Figure 2.7 Example of strengthening influence of deflection on internal force.

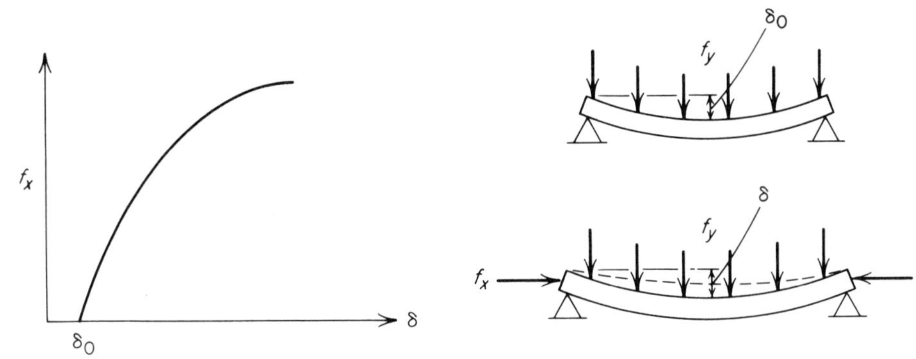

Figure 2.8 Example of weakening influence of deflection on internal force.

umn, which contains some initial eccentricity, the axial load induces bending of the column and this in turn causes some further lateral deflection. For low levels of load this effect is negligible, and the initial portion of the curve is approximately linear. But as the axial load increases the lateral deflection becomes significant and induces more and more bending, which in turn further increases the lateral deflection. The result is a relatively rapid loss of stiffness. Also, the large bending stress, in combination with the axial stress, may cause yielding of the compression flange, and this further decreases the member stiffness. This loss of stiffness causes the curve to reach a peak and then to fall. The falling portion of the curve is only obtained if the load on the member is quickly decreased, so that it follows the curve. If the load is not decreased, then since the member has zero stiffness, the deflection increases suddenly to some very large value, that is, the member collapses.

In addition to ordinary flexural buckling a beam or column of open section, such as an I-beam, can undergo *flexural-torsional buckling* if it is subjected to bending in the plane of its web and if it is not constrained to remain in this plane. The bending can be due to either a distributed lateral load, bending moments applied to the ends, or an eccentric axial compressive load. Initially, at low levels of bending load, the member deflects by bending in the plane of the web. However, at a certain value of load the compression flange may become unstable and buckle laterally, while the tension flange remains stable. Hence the cross section undergoes a combination of twist and lateral deflection, as shown in Fig. 2.10. The curve of load versus vertical deflection is illustrated by curve (a) in Fig. 2.11. Alternatively, if sufficient lateral supports are provided for the compression flange such that twist of the cross section cannot occur, then the load–deflection curve continues upward until failure occurs either by plastic deformation—curve (b)—or possibly, if there is an axial compressive load and if the member is slender, by flexural buckling in the plane of the web—curve (c).

In many cases the prebuckling deflection is relatively small and this permits an idealized approach, known as the *eigenvalue approach,* which greatly simplifies the calculation of the buckling load. In this approach, an *ideal* or *perfect* member and loading conditions are considered. The member is assumed to have no geometric or material imperfections, and to be

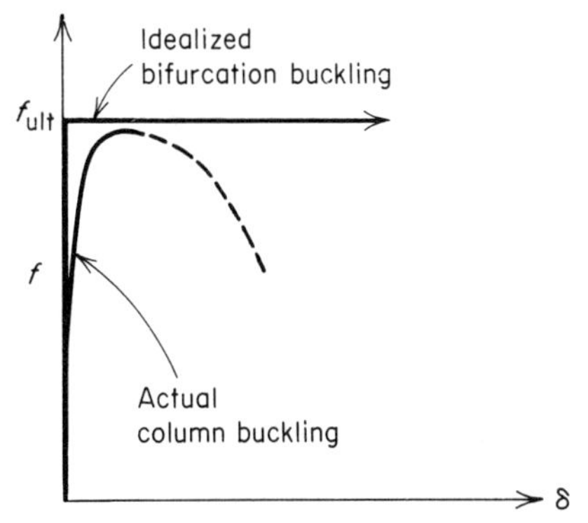

Figure 2.9 Column buckling.

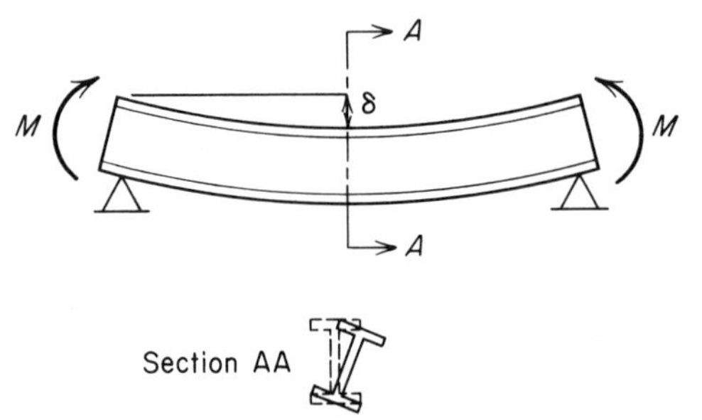

Figure 2.10 Flexural-torsional buckling.

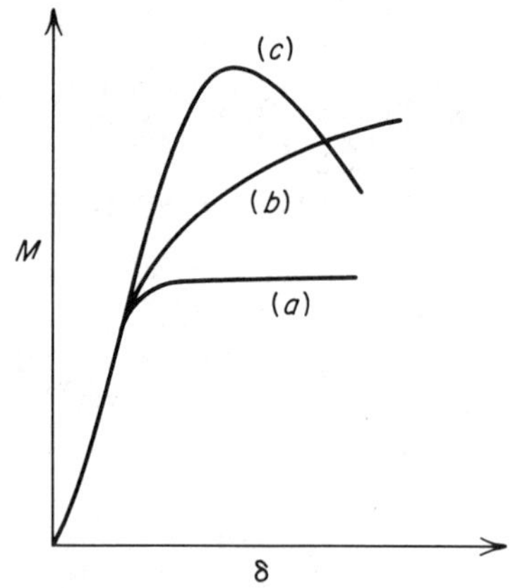

Figure 2.11 Load-deflection curves for a beam-column under various load and support conditions.

loaded in such an ideal manner that the only deflections that occur prior to buckling are those in the direction of the applied loads. That is, in the case of a concentrically loaded column the load does not produce transverse deflection until a load of a unique magnitude, the critical load, is reached. At that point the member has two possible equilibrium states—undeflected and deflected—and it adopts the deflected shape because the strain energy of the undeflected shape is just exceeding the strain energy of the deflected shape. The load–deflection behaviour is that shown by the heavy line in Fig. 2.9. The essential feature of this idealized form of buckling is that the member can undergo indefinite lateral deflection at no increase in axial load. This feature makes it possible to calculate the buckling load or, more precisely, the *bifurcation load,* by means of an eigenvalue analysis.

However, the eigenvalue approach is only appropriate when the buckling is of bifurcation type and when the member fulfills the idealizing assumptions reasonably closely. This is particularly important because in nonideal conditions (eccentric load, eccentric member, local yielding) the buckling load is always *less* than that calculated by an eigenvalue analysis. Hence if the assumptions are not met to a reasonable degree the result is a potentially serious overestimate of the buckling load, or of the stiffness of the member under large load.

In plates and stiffened panels there is some postbuckling reserve of strength; that is, the ultimate or collapse load is greater than the buckling load, as illustrated by curve (c) in Fig. 2.6. The buckling analysis and ultimate strength analysis of these members is taken up in various chapters later in the book.

Nonbifurcation Buckling

In nonbifurcation buckling the load–deflection curve is similar to curve (d) of Fig. 2.6. The lateral deflection commences as soon as the axial load is applied, and it increases progressively, at an increasing rate, causing the member to progressively lose its stiffness from the very beginning of the loading, until finally the stiffness becomes zero. Hence this type of buckling is a gradual phenomenon. It occurs whenever one of the deflections, which is caused by and which increases with the applied load, is such as to have a weakening effect on the member, right from the beginning of the loading. This can occur in members (beam-columns, plates, and stiffened panels) that are subjected to relatively large lateral loads which increase as the axial load increases. A notable example is a beam-column which is subjected to bending in two planes (biaxial bending), as shown in Fig. 16.28.

As can be seen from Fig. 2.6 the load–deflection curve for nonbifurcation buckling resembles that for failure by local plastic deformation. There is no obvious buckling point, or any precise peak value of load. As with failure by plastic deformation, the gradual loss of stiffness may produce such large deflections that the member is deemed to have failed on that account. Hence the failure load is defined as the value which corresponds to some limiting value, either of deflection or of stiffness.

Static Fracture: Ductile and Brittle

The provision of adequate safety against failure by static fracture is achieved by keeping the stress level throughout each member sufficiently below σ_{UTS}, the ultimate tensile strength of the material. Thus the constraint against failure by fracture is

$$\gamma\sigma \leq \sigma_{UTS} \qquad (2.4.1)$$

The partial safety factor is chosen according to the degree of uncertainty of σ (if not already factored for this) and the importance of the member and the seriousness of the consequence of the fracture. The general distribution of stress in each member is known from the hull module and principal member analyses. Thus the static fracture constraint is much simpler than the constraints against plastic deformation and instability because the limit value, instead of depending strongly and in a complicated manner on the member's dimensions and proportions, is virtually independent of them. Once the material has been selected the limit value is known.

The term "brittle fracture" refers to the fact that below a certain temperature the ultimate tensile strength of steel diminishes sharply. The value of this transition temperature depends almost entirely on the chemical composition of the steel and the metallurgical processes by which it is made. For ship structure a good quality steel is absolutely necessary, and in most cases sufficient, to avoid brittle fracture.

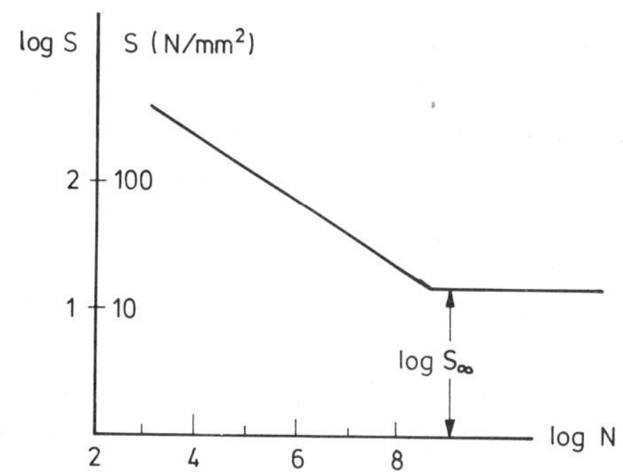

Figure 2.13 Typical S–N diagram.

Fatigue Fracture

In steel and other metals a fluctuating stress can initiate microscopic cracks which gradually increase in size until, after a large number of cycles, the cracks have become so large that fracture occurs. Fatigue fracture is entirely distinct from static tensile fracture. In fatigue the most important parameter is the *stress range*, S, which is the total (peak to trough) variation in the cyclic stress, as shown in Fig. 2.12. Since fatigue damage (crack growth) is cumulative the occurrence of fracture depends on the magnitude and duration (number of cycles) of the individual cyclic loads which act on the structure throughout its life. For an individual cyclic load of constant amplitude S the number of cycles N which is required for fatigue fracture has been established experimentally for each type of steel and other materials. This information is usually presented in "S-N diagrams" of the type shown in Fig. 2.13, in which the horizontal distance to the sloping line is the *fatigue life* (number of cycles to failure) at that level of S for a certain type of structural specimen (geometry, direction of weld, and quality of weld). Each line is obtained by testing to failure a series of identical structural specimens at different levels of S and then performing linear regression analysis on a log-log plot of the data, with say a 95% confidence limit. Thus each line (or "S-N curve," as it is commonly called) represents an exponential relationship of the form

$$S_N = \left(\frac{C}{N}\right)^{1/m} \qquad (2.4.2)$$

which in logarithmic form becomes

$$\log N = \log C - m \log S_N$$

where N = number of cycles to failure for a constant amplitude stress range S_N

m = the negative slope of the log-log plot of the S-N Curve

S_N = the constant amplitude stress range for failure at N cycles

$\log C$ = the life intercept of the S-N curve

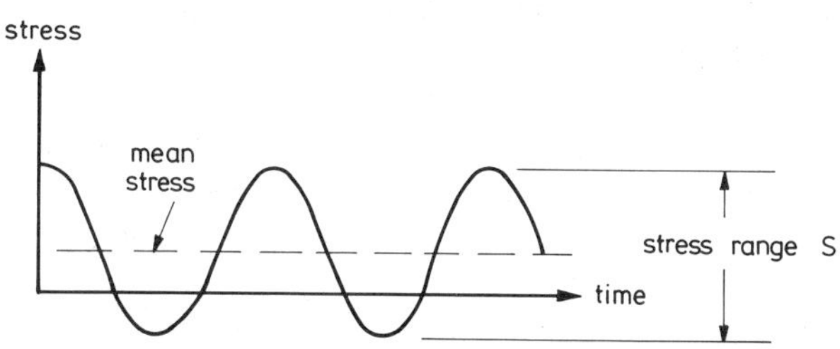

Figure 2.12 Stress range.

For most materials there is a threshold level of stress range S_∞ below which fatigue damage does not occur, regardless of the number of cycles. This is commonly referred to as the *fatigue limit*.*

In a ship structure there are three main sources of cyclic stresses: wave induced loads, especially bending of the hull girder; the alternation between loaded and ballasted conditions which occurs in tankers and some other vessels; and mechanical sources such as the engine and the propellers. The number of wave bending cycles in a ship's life (say, 20 years) is of the order of 10^8.

In general, fatigue failure is prevented by controlling the cyclic stress amplitude, and in most cases the most efficient way to control stress is to either increase the local scantlings and/or modify the local geometry so as to reduce stress concentrations, eccentricities,

*A better term would be "fatigue threshold," since fatigue only commences at this level of S.

and discontinuities. Hence in the overall process of ship structural design the prevention of fatigue falls mainly within the scope of detail design. But for all cyclic stresses that are not locally controllable the matter of fatigue must be dealt with at the preliminary design stage. The principal example is wave-induced hull girder bending stress σ_w which is proportional to section modulus and is relatively unaffected by local changes in scantlings. Therefore, one of the most important tasks in preliminary design is to ensure that σ_w is sufficiently small so that it cannot in itself cause fatigue in any portion of the hull girder, taking into account all unavoidable and non-locally-treatable influences. Obviously one way of preventing fatigue fracture is to totally prevent the occurrence and accumulation of any fatigue damage by keeping the cyclic stresses below the fatigue limit. This approach is adopted when the number of stress cycles is extremely large, as in mechanically produced stresses. Such stresses are usually very local and therefore relatively easy to control. However, this approach would be very inefficient (if not impossible) for the more extensive stresses which arise from full/ballasted cargo variations and hull girder bending. Instead it is necessary to design ships in such a way that over their lifetime they can sustain some cumulative fatigue damage without any appreciable risk of fracture. To do this it is necessary to know how the individual amounts of fatigue damage caused by various sequences of cyclic stresses of different magnitude, duration, and mean value interact and accumulate to finally cause fracture. This is an extremely complex question and many aspects are still only partly resolved.

There are two methods for dealing with fatigue, one based on fracture mechanics and the other based on fatigue tests together with the hypothesis of linear damage accumulation. The fracture mechanics method is more detailed, examining crack growth and calculating the number of load cycles that are needed for small initial defects (which are always present in welds and in the structure adjacent to welds) to grow into cracks large enough to cause fracture. The growth rate is proportional to the stress range, expressed in terms of a *stress intensity factor* K which accounts for stress magnitude, weld and joint geometry, and the current crack size. The equation for crack growth rate is of the form

$$\frac{da}{dN} = C(\Delta K)^m \tag{2.4.3}$$

in which a is the crack depth, C and m are crack propagation parameters which come from fracture mechanics, and ΔK is the range of K corresponding to the stress range.

The other method is based on fatigue test data (S-N curves) together with the hypothesis, commonly known as Miner's Rule, that fatigue damage accumulates linearly.* According to this hypothesis the total fatigue life under a variety of stress ranges is the weighted sum of the individual lives at constant S, as given by the S-N curves, with each being weighted according to the fractional exposure to that level of stress range. To apply this hypothesis the long-term distribution of stress range is replaced by a stress histogram, consisting of a convenient number of constant amplitude stress range blocks ΔS_i and a number of stress cycles n_i. The constraint against fatigue fracture is then expressed in terms of a nondimensional damage ratio η:

$$\sum_{i=1}^{B} \frac{n_i}{N_i} \le \eta_L \tag{2.4.4}$$

where B = number of stress blocks
ni = number of stress cycles in stress block i
N_i = number of cycles to failure at constant stress range S_i
η_L = limit damage ratio

The limit damage ratio η_L depends on the maintainability, that is, the possibility for inspection and repair, as well as the importance of the particular structural detail. For important details which are exposed to seawater a typical value of η_L is 0.3 if there is good access and maintainability, and 0.1 if not.

As stated earlier, fatigue relates mainly to structural details, and fatigue analysis and prevention is primarily a part of detail design, which is outside the scope of this book. The principal exception is hull girder fatigue due to wave-induced bending, and this is dealt with in Section 3.6.

2.5 LIMIT STATES OF THE OVERALL STRUCTURE

Besides the failure of individual members, there are limit states that involve several members and that therefore cannot be dealt with at the member level.

*Originally proposed by Palmgren [7] and later by Miner [8]; it is usually known under the latter's name because he developed it on a logical basis from a consideration of the work done during each loading cycle.

TABLE 2.1 OVERALL LIMIT STATES

Type of Overall Limit State	Description of Limit State Analysis	Overall Design Variables
Collapse (loss of stiffness of hull girder)		
1. Combination of member failures	Hull module ultimate strength analysis	Member limit values involved in the collapse
2. Hull girder fatigue	Calculate minimum required values of Z_D and Z_B	Strake areas
Noncollapse (e.g., excessive flexibility of hull girder)		
1. Springing 2. Hull/machinery incompatibility 3. Excessive angular deflection	Calculate minimum required value of $I{:}I_R$	Strake areas

The most obvious example is collapse of the overall structure, but there are also other types of overall limit states, as shown in Table 2.1.

Types of Collapse

Collapse of a structure is essentially the loss of overall stiffness and load-carrying ability. In shell structures, in which the load is carried almost entirely by membrane compression (aircraft fuselage, submarine hull, etc.), it is possible for collapse to occur by bifurcation buckling of the overall structure. For these structures the limit load can be calculated (or at least estimated closely) by a direct eigenvalue method. But in a ship structure many of the members carry large bending moments and because of their relatively sturdy proportions they undergo extensive yielding both before and during any buckling. A typical load-deflection curve for the structure is illustrated in Fig. 2.14(*a*), in which F and Δ are some characteristic values of load and deflection for the structure as a whole. For example, Figures 2.14(*b*), (*c*) and (*d*) illustrate how overall collapse might occur in a tanker during loading if the transverse framing were not properly designed. (For clarity the deflections are grossly exaggerated.) Let us say that this tank is the only one being loaded, in which case the draft is approximately constant and F is simply the total weight of cargo. As F is increased various principal members undergo failure by yielding or buckling or some combination of the two. In other modes of collapse some members might fail by fracture (again, assuming an inadequate design). These successive failures cause a decrease in overall stiffness—the slope of the F-Δ curve. However, the slope does not become zero immediately because other members can carry a further increase in load, including the load shed by the failed member and the further load it would have taken. As more failures occur (some interactively) the structure suffers a progressive loss of stiffness, a large increase in deflection, and eventually collapse. In this type of collapse, involving combined types of failures and nonlinear interaction among various members, a rigorous and accurate value of the limit load can only be obtained by calcu-

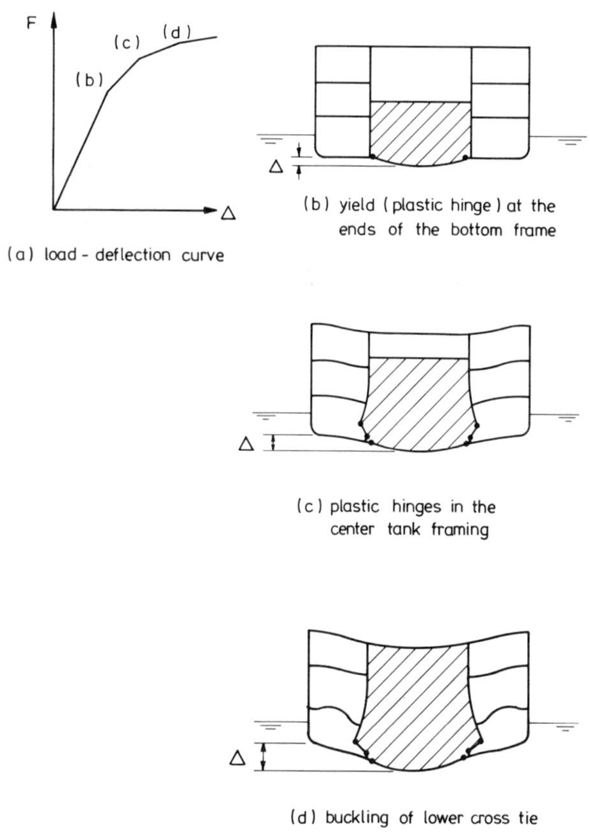

Figure 2.14 Load-deflection relationship during overall collapse.

lating the complete load–deflection relationship, using an incremental or stepwise approach. This consists in applying a sequence of load increments from one failure to the next, and in updating the mathematical analysis model so that it always gives a correct portrayal of the stiffness and other properties of the structure. In order to detect the individual failures it is necessary to continuously monitor all of the member internal forces and deformations, and to compare them with limit values which have been calculated for the various modes of failure, for each type of member. In order to detect any multimember form of instability, up to and including final collapse of the structure, it is necessary to monitor the stiffness of the overall structure.

Another possible type of overall collapse is fatigue fracture of the hull girder due to the cyclic wave-induced hull girder bending stress, σ_w. In order to prevent this the amplitude of σ_w must be kept below a certain limit value. The maximum values of σ_w occur in the deck and bottom and are inversely proportional to the section moduli $Z_D = I/y_D$ and $Z_B = I/y_B$. Hence the fatigue limit state involves the structure as a whole and cannot be dealt with exclusively at the member level. The manner in which the minimum required values of Z are obtained is discussed in Section 3.6.

NonCollapse Overall Limit States

Collapse is not the only limit state which pertains to the overall structure. For example it may be necessary to place a limit on overall (hull girder) flexibility, in order to avoid "springing" (see Section 2.1) or some undesirable hull/machinery interaction. In naval vessels such a limit may be necessitated by the weapons tracking and guidance systems, which have certain upper limits on the relative angular deflection between two points along the hull. The limiting of the hull girder flexibility is achieved by imposing a minimum required value of the hull girder moment of inertia I_R.

2.6 METHODS FOR HULL MODULE ULTIMATE STRENGTH ANALYSIS

Available Methods: Accuracy versus Computational Cost

As may be seen from the foregoing description, the collapse of a ship structure is an extremely complex process and in the past there has been no method available that could provide a complete and accurate limit analysis of such a structure. Therefore it has been necessary to use simplified and approximate approaches, most of which may be grouped into two categories.

Ultimate Strength Analysis of Selected Substructures

By simplifying and idealizing the geometry of a hull module it is possible to identify particular portions of it for which some approximate ultimate strength analysis methods can be developed and applied. For example, the decks and sides of a ship might be regarded as grillages, or the double bottom might be idealized as an orthotropic plate. However, this approach suffers the same problems as isolated member analysis: the boundary conditions are not known accurately; that is, the interactions with adjacent portions of structure are not known, even if those portions are of the same type. Also the geometry of ship structures seldom admits of such simplifications; for example, decks and sides often have large openings, and a double bottom may be connected to the decks above by pillars.

Separate Analysis of Buckling and Plastic Hinge Failure

Another technique which has commonly been used to simplify the analysis is to divide it into two separate parts, each of which examines only one type of failure.

Buckling Analysis. This analysis assumes that collapse is caused purely by bifurcation buckling and the loads are assumed to have a fixed proportion relative to one another. These assumptions convert the equilibrium equations into an eigenvalue problem, for which an immediate and direct solution can be obtained.

Plastic Frame Analysis. In this analysis the collapse mechanism is assumed to be caused entirely by the occurrence of localized regions of large plastic deformation ("plastic hinges") and a systematic procedure is used to find the particular pattern of hinges which corresponds to the collapse load, that is, the lowest intensity of loading (of which the component loads again have fixed proportions) which would produce a mechanism and hence collapse of the structure.

Because they are based on an *a priori* selection of the type of failure which causes the structure's collapse, both of the foregoing analyses are able to deal directly and exclusively with the final or collapse state of the structure. There is no need to examine earlier stages in the structure's progression toward collapse. Hence the amount of computation involved is relatively small. However, this division of ultimate strength analysis into two separate analyses, each involving only one type of failure, is not sufficiently accurate for rationally-based design. For one thing, it ignores the possibility of a structural collapse involving both types of failures, at different locations, and it also ignores the possibility of the two types occurring together and interacting.

In recent years the finite element method has been generalized and used for nonlinear problems involving both types of nonlinearity—geometric and material. This has been achieved by introducing incrementation and iteration, that is, by using a load–deflection approach involving successive increments of load (which must be determined iteratively) and by performing several linear finite element analyses at each load level. This approach can account for the simultaneous and interactive occurrence of both yielding and buckling. Moreover, it can be applied not only to an individual member but, in principle at least, to a complete structure involving multiple and interactive member failures by combined yielding and buckling. Unfortunately, however, this process of incrementation and iteration involves a large amount of computation, and it increases to enormous proportions as the size and complexity of the structure increases. For present day computers the computational expense of nonlinear finite element analysis is too large for the method to be used in a fully rigorous form and for every application. However, there are two main applications—one pertaining to individual structural members and one to the overall structure—for which the superiority of the method justifies its use, even though for the time being it is necessary to make some simplifications in order to reduce the amount of computation. These two applications are discussed briefly in the next two sections.

Ultimate Strength Analysis of Individual Members

At this level the nonlinearities of buckling and plastic deformation can be dealt with by approximate analytical, closed-form methods. Since these methods involve much less computation and yet have sufficient accuracy for structural design, they are preferable to nonlinear finite element analysis for design applications, in which the analysis must be performed repeatedly. However, these methods are only approximate. In most cases they are semiempirical, consisting of a formula that has been derived from the underlying theory (which is always somewhat idealized) together with some numerical terms or coefficients that have been obtained experimentally or numerically. Because of the greater accuracy and generality of nonlinear finite element analysis, its principal and most valuable application at the member level is to check the accuracy and to guide the further development of these approximate analytical methods.

In order to achieve their simplicity these various methods each pertain to a specific type of structural member and type (or combination) of structural failure. Therefore Chapters 9 to 14 are arranged according to type of member: plates, columns, beam columns, and stiffened panels. The aim of these chapters is simply to present the basic theory and some of the resulting methods and formulas (and the underlying assumptions and limitations) for estimating the ultimate strength of these members.

Ultimate Strength Analysis of a Hull Module

As mentioned before, although nonlinear finite element analysis is the only method that can give a truly rigorous and accurate estimate of the ultimate strength of a large structure, its application on this scale would require a prohibitive amount of computation, at least for present day computers. For a completely general three-dimensional structure there are not only a large number of different possible modes of member failure, each of which must be repeatedly checked during the course of a detailed and time-consuming load–deflection analysis (which must be performed for each loadcase), but also the ultimate strength depends on the *sequence* of the individual loads of each loadcase, and so the complete analysis should be performed for each *combination* of individual loads. Moreover, structural design is itself an iterative process, requiring repeated limit state analyses. Hence the total computational task is truly enormous, and only in special circumstances would such an analysis be performed, even for one loadcase. Of course with continued improvements in computational capability this situation will eventually be altered, but for the time being it is necessary to utilize special techniques for reducing the amount of computation. This can be done in two ways:

1. By developing a method that is specifically for a

hull girder, rather than attempting to use a general method.

2. By imposing some special constraints on the design which, although they may prevent the final design from being the true and exact optimum design, nevertheless have the great advantage of reducing the number of possible modes of overall collapse (i.e., combinations of member failures) and of totally eliminating those modes that are computationally most expensive (and/or which are undesirable on other grounds).

As noted in Section 1.2, hull module collapse is synonymous with hull girder collapse, that is, the inability of the hull girder to withstand one or more of the overall loads: vertical and horizontal bending, transverse shear, and torsion. In other words, hull module collapse is the loss of hull girder stiffness in either bending, shear, or torsion. Now a hull module is essentially prismatic and the transverse structure is basically orthogonal to the longitudinal structure. Also the very large in-plane stiffness of the plating means that there is never any "sway" (large deflections of joints) in the longitudinal direction. For this reason the modes of overall collapse (i.e., loss of stiffness) can only occur in the longitudinal direction and/or in the transverse plane of the ship. These two categories of hull module collapse can be summarized as follows:

1. Longitudinal: the collapse of sufficient longitudinally oriented principal members of the hull girder to cause a large decrease in its bending, shear, or torsional stiffness.

2. Transverse: the occurrence of sufficient failures in various principal members to create a collapse mechanism in the transverse framing (as in Fig. 2.14), thereby removing the support to some of the longitudinal principal members which form part of the hull girder.

These two categories are not independent because a collapse in the longitudinal direction may involve one or more transverse frames. Hence there are many possible combinations of longitudinal and transverse member failures (each of which can involve simultaneous buckling and plastic deformation) and if all of these were to be included in the hull module limit analysis the amount of computation would still be prohibitive, at least for present day computers. Therefore it is necessary to impose some special constraints which make the analysis computationally feasible. The simplest way of doing this is to impose the constraint that all of the transverse frame members must be strong enough so that all longitudinal members would undergo interframe collapse before undergoing any type of collapse involving more than one frame space. This greatly simplifies the analysis in two ways:

1. The hull module only needs to be analyzed one segment at a time, each segment being comprised of all of the principal members within one frame space, and since many segments are identical only the more heavily loaded of these need to be analyzed.

2. The longitudinal and transverse modes of collapse become independent and can be analyzed separately, which further reduces the total computation.

For longitudinal overall collapse the predominant load is hull girder bending moment and the principal response is hull girder flexure, which can be expressed in terms of the local curvature of the hull girder, ϕ. All other longitudinal responses can be dealt with as corrections or modifications to this principal response. As illustrated in Fig. 17.2 the ultimate longitudinal strength is the value of bending moment M_{ult} at which the flexural stiffness of the hull girder—the slope of the M–ϕ curve—becomes zero. This occurs when sufficient members (panels, for the most part) within a particular segment of the hull girder (the critical segment) have themselves undergone collapse, either tensile or compressive. During the overall collapse the various panels are all at different stages of collapse, and within each panel the principal load effect is the elongation or shortening which the hull girder bending moment imposes as a result of curvature ϕ.

One of the principal advantages of requiring that no other type of collapse can precede interframe collapse is that with the latter the hull girder continues to function as a beam during the entire collapse process, and therefore the fundamental assumption of beam theory—that plane cross sections remain plane—continues to apply with sufficient accuracy. Hence the longitudinal strain due to hull girder flexure is still given by $\varepsilon = y\phi$, as for elastic hull girder bending. Obviously since each panel is undergoing collapse, involving yielding or buckling or both, the corresponding stress distribution across the section of the critical segment will be highly nonlinear, and even within each panel the relationship between average stress and average strain will vary as the strain increases. But the distribution of strain continues to be approximately linear up to hull girder collapse. Therefore the member limit values which play the biggest role in longitudinal collapse are the values of *ultimate strain* ε_{ult} of stiffened panels. For each panel ε_{ult} is the amount of imposed compressive strain which causes collapse of that panel. In order to calculate $M_{\rm ult}$ it is also necessary to know the *ultimate strength* of each panel—the value of longitudinal compressive load which it carries at collapse. This is usually expressed as $\sigma_{a,\rm ult}$, the average value of longitudinal stress across the panel section at the time of panel collapse. The calculation of $\varepsilon_{\rm ult}$ and $\sigma_{a,\rm ult}$, is dealt with in Chapter 14.

For transverse overall collapse the principal load effect is the bending moment M in the beams that make up the transverse framing—specifically the values of M at the ends and at the midlength of each beam. This load effect has three different limit values: the critical value M_{cr} for failure due to flexural-torsional buckling (see Fig. 2.10) and M_p and M_{fY}, the values which would cause either a *plastic hinge* in the beam (see Figs. 16.3, 4, and 5) or a local *plastic mechanism* in the beam flange, due to flange yield (see Fig. 14.4). Overall collapse occurs when individual member failures have made the transverse framing incapable of providing adequate support to the longitudinal principal members. This usually occurs because of the formation of a mechanism in the transverse framing.

The concepts and methods which form the basis of hull module ultimate strength analysis are presented in Chapters 15 through 17. Chapter 15 presents the basic aspects of nonlinear finite element analysis and its application to structure instability. Chapter 16 deals with plastic frame analysis and shows the need for an incremental approach. Chapter 17 shows how these two fundamental aspects can be integrated to achieve a comprehensive method for hull module limit analysis which involves an acceptable amount of computation.

2.7 OPTIMIZATION OF LARGE STRUCTURES

In nonlinear optimization the amount of computation increases exponentially with the number of design variables which must be optimized simultaneously, and a direct solution to a typical ship structure optimization problem would involve a prohibitive computational cost. If the overall problem could be subdivided into a number of subproblems the computation would be substantially reduced. In fact if each structural member could be optimized separately, the total amount of computation would be relatively small. But in a large structure such as a ship, member-by-member optimization is neither desirable nor possible. It is undesirable because the objective function is a nonlinear function of the design variables, and member-by-member optimization would not allow any tradeoffs between members, which is where a large part of the benefits of optimization come from. And in any case member-by-member optimization is not possible because, as shown in Section 2.5, some limit states relate to the overall structure and cannot be dealt with at the member level.

In a structure as large as a hull module there are typically 100 to 200 design variables, and a nonlinear optimization problem of this size involves far too much computation to be solved as a single problem, even with a rapid and efficient optimization algorithm such as one of the sequential linear programming methods discussed in Section 1.2. Therefore one of the prerequisites for rationally-based ship structural design is a method or strategy for subdividing the overall problem while still retaining true overall optimization and the capability of dealing with overall constraints. In this section we present a brief outline of a *dual level* strategy which meets these requirements.

Subdivision of the Overall Optimization Problem

The subdivision of the overall optimization problem is based on the concept of a submodule, which is a region of structure in which a sufficient number of the scantlings are linked, either by fixed structural geometry or by explicit constraints linking two or more scantlings, such that the region forms a logical entity from an optimization point of view.* The characteristics which most clearly make a region suitable as a submodule are geometric uniformity and identical, repetitive structural members. Such characteristics are frequently imposed on portions of structure in order to simplify the design and to thereby gain increased economy and efficiency in nearly all aspects of the ship's existence: design, fabrication, outfitting, operation, and maintenance. Modularity of cargo is another common reason for imposing structural uniformity. Regardless of the reason, any such uniformity provides an opportunity for reducing the amount of computation, and the choice of submodules should reflect and take full advantage of uniformity. In most cases the choice is obvious and straightforward, because it is a natural extension of the designer's decisions regarding geometric uniformity and member repetition. In making such decisions he is in effect imposing various constraints on the design—constraints which require that some scantlings or geometric features in one member or region should bear a fixed relationship (in most

*The division of the structure into submodules is for optimization purposes only, and is entirely separate and distinct from the subdivision into elements and superelements in the hull module analysis.

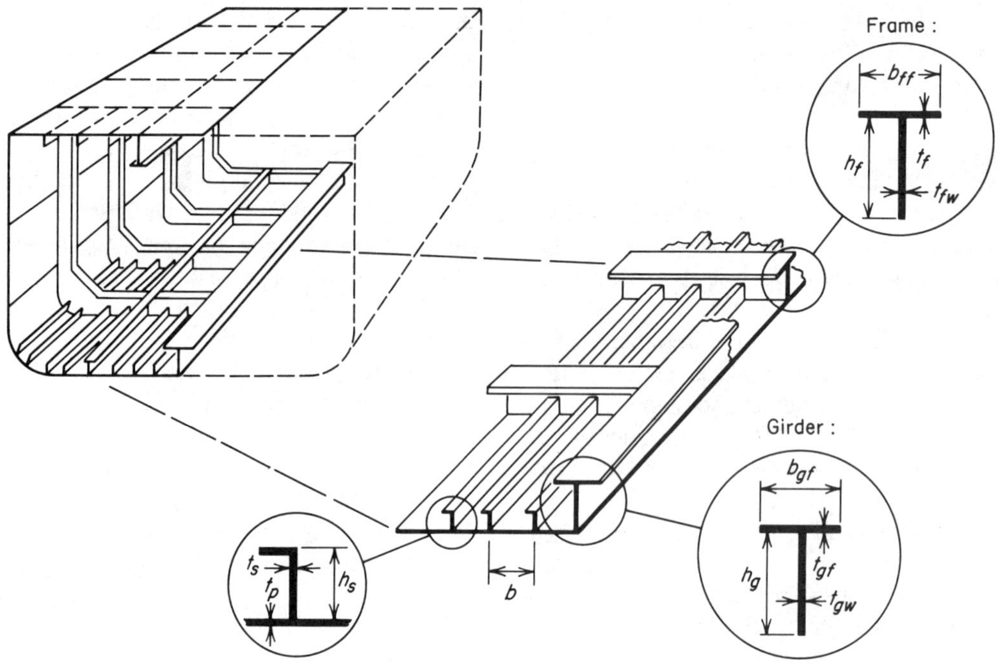

Figure 2.15 Strake submodule.

cases, be identical) to those of another member or region.* For these constraints to be properly accommodated, the optimization problem should extend over and include all of the linked members or regions. Hence the physical size of the submodule depends mainly on the extent of the linking.

Fig. 2.15 shows the most common type of submodule, which for simplicity shall be called a "strake." It consists of a longitudinal row of panels, transverse frame segments, and, if applicable, longitudinal girder segments.† In most ships these members are uniform and repetitive in the longitudinal direction over quite large distances—nearly always over the full length of a hold, and often over several hold lengths (or compartments for a naval vessel). Local changes in geometry or scantlings (cutouts, reinforcements, etc.) are disregarded because they are dealt with in detail design. Whatever the length of the uniformity, that is the length of the submodule. Because the panels, frame segments, and girder segments in the submodule are identical the total number of design variables is quite small; it is the same number as for just one member of each type. In Fig. 2.15 the strake submodule has 12 design variables: four for the stiffened panel, four for the frame segments, and four for the girder segment (in this example, since the stiffener is a rolled section, not all of its dimensions are design variables).

Submodules also reduce the number of limit values that have to be calculated because identical members all have the same value. Hence there is just one constraint for each mode of member failure, and to formulate each constraint it is necessary only to scan for and utilize the largest value of each relevant load effect.

A submodule can also extend in the transverse direction. For example, in a double bottom there is an appreciable amount of geometric uniformity and repetition of scantlings in the transverse direction as well as the longitudinal direction. In the transverse floors the height, thickness, and stiffening arrangement are usually constant in the transverse direction. Also the various strakes of plating in the shell and inner bottom are often of uniform thickness (except for the keel strake, which is usually thicker). If a designer wishes to have this degree of uniformity and repetition in the transverse direction then he should regard the entire double bottom as a single submodule.

Alternatively, he may wish to allow the scantlings to vary from member to member in the transverse

*For simplicity the discussion here is in terms of identity constraint, but the principles apply to any simple and direct type of linking, such as a fixed proportion of scantlings or geometry between members.

†In ship construction terminology a "strake" is a lengthwise strip of plating; e.g. keel strake, sheer strake, etc. The distinction between this and a strake submodule will be clear from the context.

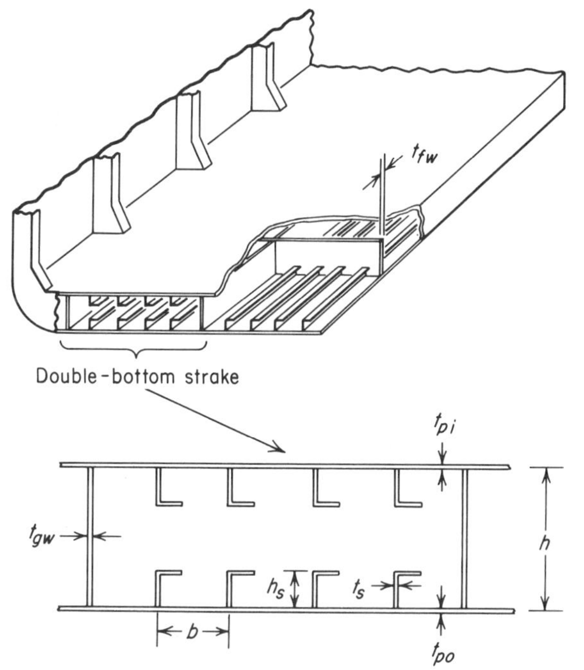

Figure 2.16 Double bottom strake submodule.

direction, with each longitudinal group of panels, floor segments, and girder segments being optimized on the basis of the largest load effects in that row rather than the largest load effects in all members of that type. In this case the submodule would extend only in the longitudinal direction (a "double bottom strake") as shown in Fig. 2.16. Each longitudinal girder would be associated with one or the other submodule, and so there would be eight design variables in each double bottom strake. Hence in a typical double bottom there would be some 16 or 24 design variables. If the designer wanted to further "tailor" the double bottom structure to local requirements he could define the longitudinal rows of panels in the bottom, shell, and side girders as separate strakes. Thus the designer has a great deal of flexibility in his choice of submodules. In principle he could go to the extreme of defining each cell of the double bottom as a separate submodule, but in practice this would be an extremely expensive structure to build.

As mentioned earlier, the amount of computation increases exponentially with the number of design variables. The author has found from experience that the total amount of computation (for the complete hull module) will remain reasonable as long as the number of design variables in each submodule does not exceed 30. Thus the simple and general rule to follow is to make submodules as large as possible, subject to the limit of 30 design variables. If a proposed submodule would have more than 30 design variables it should be divided into two submodules or, alternatively, some further uniformity should be specified. Thus for example a double bottom could not be defined as one submodule unless *most* of the scantlings in the longitudinal direction were constrained to be uniform.

In the last sentence the word "most" is emphasized because not all of the scantlings in a submodule need to be uniform. It frequently happens that in a particular portion of structure it is desired that most members of a certain type should be identical, but there are a few which ought to be allowed to differ, perhaps because they carry special loads. For example, Fig. 2.17 illustrates a hull module that corresponds to one complete cargo hold with, say, eight transverse frames, the second and seventh of which coincide with the forward and aft ends of the hatch. In each of the side and deck strakes the second and seventh frame segments carry larger loads than the other frame segments. If these two segments were included in the strake the final optimum scantlings of the other six frame segments would be dictated by the requirements of these two. Therefore the two hatch end segments should be defined as a separate submodule. The fact that they are not contiguous is no problem because the grouping of members in a submodule is done for optimization purposes only and not for analysis. Hence there is no need for a submodule to be a single or continuous portion of structure.

Figure 2.18 illustrates another example of submodule selection. In a transverse bulkhead the submodules would extend horizontally since the hydro-

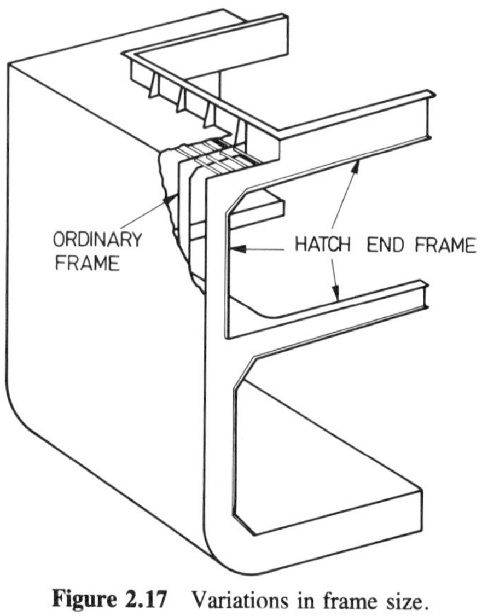

Figure 2.17 Variations in frame size.

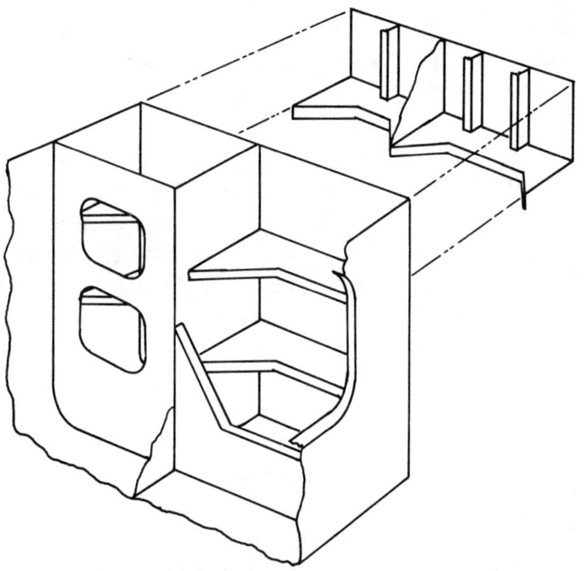

Figure 2.18 Submodule in a transverse bulkhead.

static load is constant and, if desired, the panels and horizontal girder segments (including their brackets) can be constrained to be identical in this direction (if desired, local adjustments could subsequently be made to individual brackets.) Note that this submodule would include the same region of structure on *all* of the bulkheads (or as many as the designer wishes to be identical).

Design Variables in the Overall Optimization Problem

An overall design variable is any mathematical combination or function of the ordinary design variables (the scantlings) which plays the role of an independent variable in an overall (hull girder) limit state. There are two different groups of overall design variables, one for hull girder collapse and one for constraints relating to hull girder section properties Z and I.

Member Limit Values Which Participate in Overall Collapse

The first group of overall design variables consists of those member limit values which are involved in the collapse of the hull girder, either longitudinal or transverse. These will be denoted as $\mathbf{Q}_{L,hg}$.* For longitudinal collapse $\mathbf{Q}_{L,hg}$ consists of the values of ultimate longitudinal strain $\varepsilon_{a,u}$ of those panels which are involved in the collapse. The number of such panels varies, depending on the structural geometry of the ship, but it is usually less than 10. The constraint against overall longitudinal collapse may be expressed symbolically in the form

$$M_{\text{ult}}(\mathbf{Q}_{L,hg}) \geq \gamma_{s,o} M \qquad (2.7.1)$$

in which $\gamma_{s,o}$ is the partial safety factor for overall collapse.

Transverse overall collapse involves a number of different loads, some of which are independent. Hence the load is not a simple scalar quantity as is hull girder bending moment; rather it is a set or group of individual loads which we shall denote as $\mathbf{F}$. The symbol $\mathbf{F}_d$ represents the set of *design loads;* for those loads which are random a *characteristic value* or *design extreme value* is used. As shown in Chapter 17 the basic method for transverse ultimate strength analysis is incremental plastic frame analysis, in which $\mathbf{F}$ constitutes the load vector. In practice, because the components of $\mathbf{F}$ can occur in different proportions and different sequences, there are several load vectors. The magnitude of each vector is characterized by its norm, which we denote as $\overline{F}$

$$\overline{F} = \mathbf{F}^T\mathbf{F}$$

The transverse ultimate strength of the ship $\mathbf{F}_{\text{ult}}$ is the set of values of $\mathbf{F}$ which cause collapse of the transverse framing. Since there are several load vectors $\mathbf{F}_{\text{ult}}$ is always taken as the worst of these, that is, the particular combination and sequence which causes collapse for the smallest value of $\overline{F}$.

As mentioned in Section 2.5 the member limit values which are involved in transverse overall collapse are M_{cr}, M_P, and M_{fY}. Hence for transverse collapse the quantity $\mathbf{Q}_{L,hg}$ just introduced consists of the particular values of M_{cr}, M_P, and M_{fY} that are involved in the collapse (using the worst load vector). Here again the number of such values is usually less than 10. Each member limit value within $\mathbf{Q}_{L,c}$ is for that load effect and that member, considered in isolation. The interaction with other load effects and other members is dealt with as part of the overall ultimate strength analysis. In terms of $\mathbf{Q}_{L,hg}$ the constraint against transverse collapse is

$$\overline{F}_{\text{ult}}(\mathbf{Q}_{L,hg}) \geq \gamma_{s,o} \overline{F}_d \qquad (2.7.2)$$

in which, as with (2.7.1), the parentheses signify the functional dependence of the overall ultimate strength on $\mathbf{Q}_{L,hg}$, the member limit values that are involved in

*In this section boldface indicates a set or group of quantities.

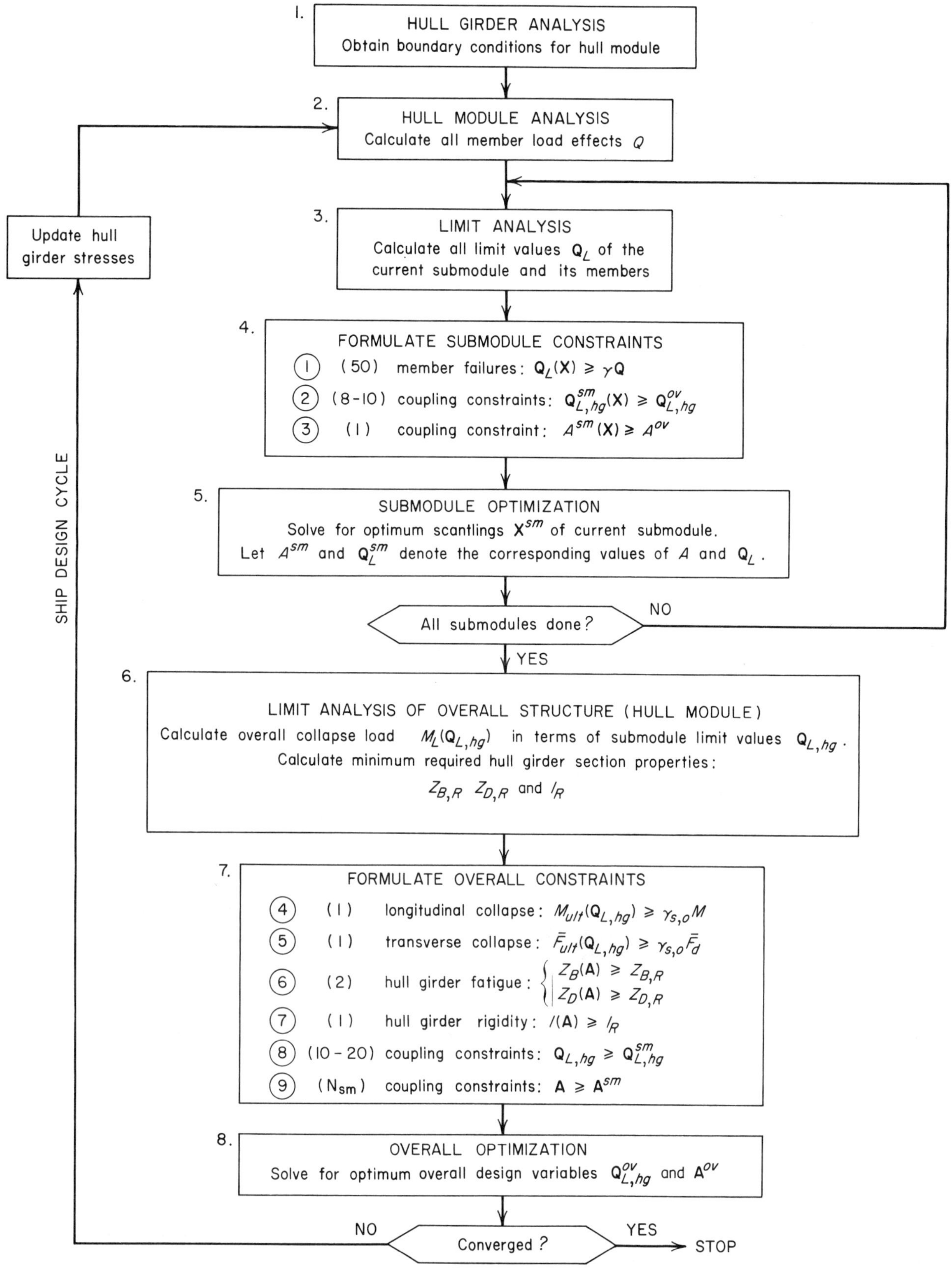

Figure 2.19 Dual level optimization procedure. *Notes:* 1. In the first ship design cycle constraints ② and ③ do not exist. 2. Each constraint in ② only applies if the relevant constraint in ④ or ⑤ was active in the previous ship design cycle and if ⑧ was *not* active. 3. Constraint number ③ only applies if the relevant constraint in ⑥ or ⑦ was active in the previous ship design cycle and if ⑨ was *not* active. 4. Each constraint in ⑧ only applies if the corresponding constraint in group ① was active in the immediately previous optimization of that submodule. 5. N_{sm} is the total number of submodules.

the collapse, and indicate that these latter values are the "design variables" in the overall level of optimization.

SECTION AREAS OF STRAKES

In the overall limit states relating to hull girder fatigue and flexibility the hull girder section modulus Z and moment of inertia I take the place of the load effects. These geometric quantities can be expressed in terms of the cross-sectional areas of the strakes (or other submodules) which comprise the ship's cross section. Therefore these areas are treated as overall design variables, represented by **A**, and the constraints relating to hull girder fatigue and flexibility are then functions of these variables, which are few in number, rather than as a function of the many individual design variables (i.e., scantlings). In symbolic form these constraints are

$$Z_B(\mathbf{A}) \geq Z_{B,R}$$

$$Z_D(\mathbf{A}) \geq Z_{D,R}$$

$$I(\mathbf{A}) \geq I_R$$

in which subscripts B and D stand for bottom and deck, and R signifies the required minimum value of Z and I. The former comes from the consideration of fatigue (Section 3.6). The latter is only required if there is a need to limit the hull girder flexibility, for example, to avoid springing.

Summary of the Dual Level Strategy

The use of submodules and overall design variables makes it possible to divide the optimization problem into several smaller problems, one for each of the submodules and one for the overall structure. In each submodule problem the design variables are the scantlings of the submodule, denoted as **X**. In the overall problem the design variables are $\mathbf{Q}_{L,hg}$ and **A**.

The strategy of dual level optimization is summarized in Fig. 2.19 in the form of a flow chart. Each submodule problem contains all of the constraints relating to that submodule, either as a whole or to its component members. Each constraint is a function of the design variables **X** of that submodule. For reference this group of constraints is labeled ①. In order to give some indication of the problem size the number in parentheses immediately after the label represents a typical number of constraints in that group. For group 1 the figure given (50) is only a rough "order of magnitude" figure for a submodule; the actual number varies widely depending on the type of submodule.

Groups ② and ③ are coupling constraints which ensure that the overall design variables for the submodule, $\mathbf{Q}_{L,hg}^{sm}$ and $\mathbf{A}^{sm}$, are not decreased below the values determined in the overall problem, $\mathbf{Q}_{L,hg}^{ov}$ and $\mathbf{A}^{ov}$. Note that each of these two constraint groups is invoked only if any one of the overall constraints in which they are involved (groups ④ and ⑤ for $\mathbf{Q}_{L,hg}$, and groups ⑥ and ⑦ for **A**) was an active constraint in the previous design cycle. Obviously, constraint groups ② and ③ only exist after the first complete design cycle because the overall optimization is not done until the end of each overall (ship design) cycle.

The solution of each submodule optimization problem gives the optimum scantlings $\mathbf{X}^{sm}$ of that submodule, for the current distribution of load effects in that submodule. Of course, one complete design cycle is not sufficient because the change in the submodule scantlings from initial to optimal affects the distribution of load effects throughout the entire structure (all submodules). The new load effects are obtained when another hull module analysis is performed at the start of the second ship design cycle. These updated load effects are used in solving the various submodule optimization problems, followed by another solution of the overall optimization problem. In the second ship design cycle the changes in **X**, and also in **A** and $\mathbf{Q}_{L,hg}$, will be smaller than in the first cycle, and it has been shown [9] that the overall design process converges quickly, generally in eight or ten design cycles.

2.8 LIMIT STATE ANALYSIS AND STRUCTURAL OPTIMIZATION

This final section contains a few further comments about limit analysis arising from the fact that in rationally-based design the limit states are the basis for the constraints in the mathematical optimization process.

1. In structural optimization the final optimum design is governed by the active constraints. There are many of these (approximately equal to the number of design variables) each corresponding to some limit state. Those constraints in which the limit state is of the collapse type will involve several individual types of failure. It is not known beforehand which particular types of failure will be the "active" ones, or which loadcase is the most severe for each type. Therefore all

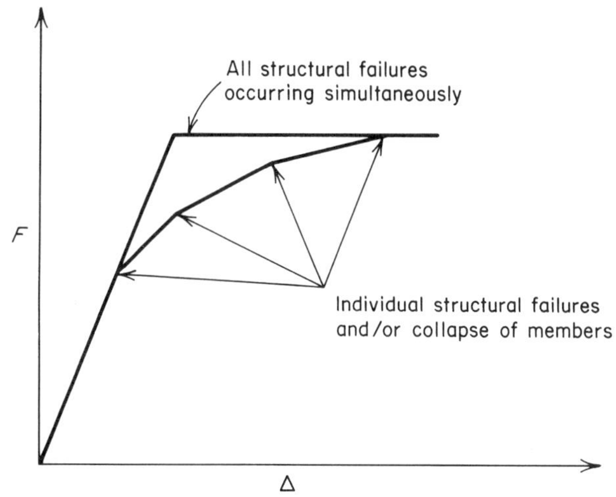

Figure 2.20

of the relevant types of failure must be investigated for each member, for all loadcases.

2. The process of structural optimization tends to produce a design in which a large proportion of the active constraints are of the collapse type. This means that although the collapse load is sufficiently high to provide the required degree of safety, the structure has been proportioned such that at that load it would undergo several simultaneous types of collapse. This makes the collapse more sudden than it would be if the individual structural failures and member collapses occurred sequentially; the contrast is illustrated in Fig. 2.20. Unlike the sequential type of collapse, the sudden collapse provides no natural or inherent warning of the possibility of total collapse. Therefore the degree of seriousness of this type of collapse is greater. However it would be wrong to say that an optimum structure is necessarily less safe than a nonoptimum structure. What is important is the overall degree of safety, that is, the probability of avoiding loss of life, loss of revenue or of mission performance, and so on. Hence what is required is that the probability of avoiding any particular type of collapse must be appropriate for the nature and consequences of that type of collapse. Also the suddenness can be removed by providing a reliable system which monitors the relevant load effects in the key structural members and automatically warns of overloads or other dangerous conditions.

3. Alternatively, with the incremental methods for overall ultimate strength analysis, the designer can control the degree of simultaneity of the structural failures to whatever extent he wishes. He may, for example, specify load margins between failures and he can nominate which types of failure are to occur first, to serve as natural warnings. He can also rule out certain combinations of failures entirely by placing limits on the relative values of design variables and/or load effects, either within members or between members.

4. Finally, although it may be obvious, it bears emphasizing that any effort to design an optimum structure should include all aspects of rationally-based design, and not just optimization. The aspects which are particularly vital are:

a. A careful prediction of the extreme load (obtained and defined statistically wherever possible).

b. An accurate calculation of the ultimate strength of the structure (either allowing explicitly for the interactions of different types of failure or excluding the possibility of interaction by means of special constraints).

c. A proper choice of the target level of safety and the corresponding partial safety factors.

REFERENCES

1. M. K. Ochi, "Prediction of Occurrence and Severity of Ship Slamming at Sea," 5th Symposium on Naval Hydrodynamics, Office of Naval Research, *ACR* 112, 1964.
2. A. Mansour and J. M. d'Oliveira, "Hull Bending Moment Due to Ship Bottom Slamming in Regular Waves," *J. Ship Research,* June 1975.
3. M. K. Ochi and L. E. Motter, "Prediction of Slamming Characteristics and Hull Responses for Ship Design," *Trans. SNAME,* 1973.
4. M. Kawakami, J. Michimoto, and K. Kobashi, "Prediction of Long-Term Whipping Vibration Stress Due to Slamming of Large Full Ship in Rough Seas," *International Shipbuilding Progress,* **24** (272), April 1977, pp. 83–110.
5. J. Harvey Evans, "Preliminary Design Estimation of Hull Girder Response to Slamming," *Trans. SNAME,* 1982.
6. W. G. Price and R. E. D. Bishop, *Probabilistic Theory of Ship Dynamics,* Chapman and Hall, London, 1974.
7. A. Palmgren, "Die Lebenstauer von Kugellagern," *ZVDI,* **68,** 1924, p. 339.
8. M. A. Miner, "Cumulative Damage in Fatigue," *Journal of Applied Mechanics,* **12,** 1945, p. A-159.
9. O. F. Hughes, R. Janava, and W. Wood, "SHIPOPT—A CAD System for Rationally-Based Ship Structural Design and Optimization," *Computer Applications in the Automation of Shipyard Operation and Ship Design,*" IV, North-Holland, 1982.

Chapter

Three

HULL GIRDER RESPONSE ANALYSIS—PRISMATIC BEAM

In this chapter we consider the overall or primary level of ship structural loading and response, in which the ship is idealized as a hollow thin-wall box beam, referred to as the "hull girder." At this level of consideration we can make several simplifying assumptions and approximations, the principal one being that the hull girder acts in accordance with simple beam theory. In Section 3.8 some corrections for these approximations will be given. For the sake of clarity and overall perspective, the principal assumptions are listed hereunder.

1. There is only one independent variable, longitudinal position, and loads and deflections have only a single value at any cross section.
2. The hull girder remains elastic, its deflections are small, and the longitudinal strain due to bending varies linearly over the cross section, about some transverse axis of zero strain (neutral axis).
3. Dynamic effects may be either neglected or accounted for by equivalent static loads. Hence static equilibrium may be invoked.
4. Since the bending strain is linear, the horizontal and vertical bending of the hull girder may be dealt with separately and superimposed. Since they are similar, and since vertical bending (bending in the vertical plane of the hull girder) predominates, we shall deal mainly with it.

3.1 BASIC RELATIONSHIPS: LOAD, SHEAR FORCE, BENDING MOMENT

As shown in Fig. 3.1, overall static equilibrium requires that the total upwards buoyancy force equals the weight of the ship [considered as a force; e.g., megaNewtons (MN), *not* tonnes] and that these two vertical forces coincide; that is, the longitudinal center of buoyancy (l.c.b.) must coincide with the longitudinal center of gravity (l.c.g.). The notation to be used herein is shown in Fig. 3.1. Using this notation the first requirement is

$$\rho g \int_0^L a(x)\,dx = g \int_0^L m(x)\,dx \qquad (3.1.1)$$
$$= g\Delta$$

where $a(x)$ = immersed cross-sectional area
$m(x)$ = mass distribution (mass per unit length)
ρ = mass density of sea water (or fresh water, if appropriate)
g = gravitational acceleration
Δ = displacement.

The factor g is retained on both sides to emphasize that it is forces which are involved.

Similarly, equilibrium of moments requires that

$$\rho g \int_0^L a(x)x\,dx = g \int_0^L m(x)x\,dx \qquad (3.1.2)$$
$$= g\Delta\, l_G$$

where l_G = distance from origin to l.c.g.

Application of Beam Theory

In elastic, small-deflection beam theory the governing equation for the bending moment $M(x)$ is

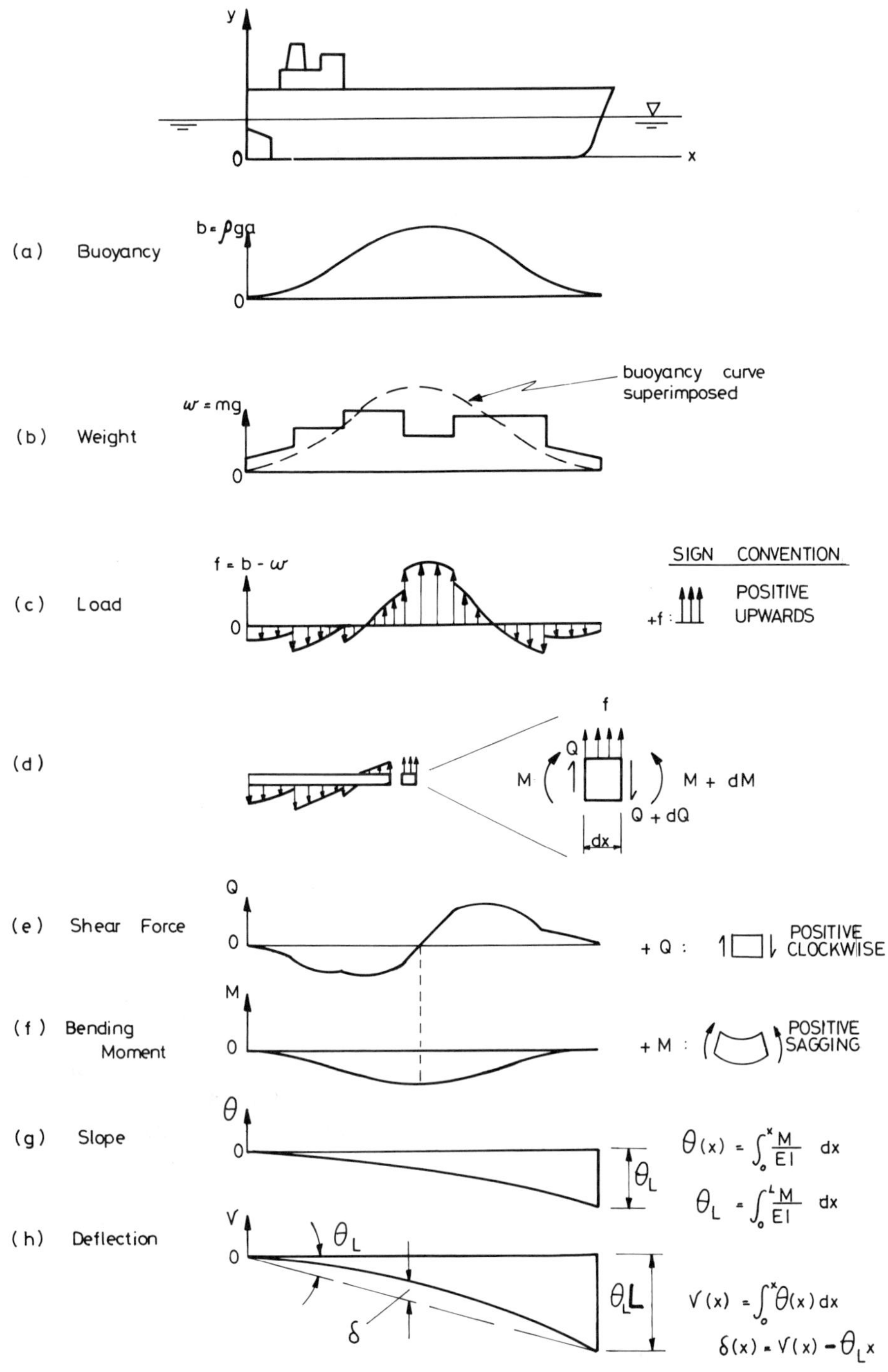

Figure 3.1 Summary of hull girder bending

$$\frac{d^2M}{dx^2} = f(x) \tag{3.1.3}$$

in which the right hand side, $f(x)$, is the loading on the beam, expressed as a distributed vertical force. For a ship this is a *net* distributed force; that is, it is the net resultant of the buoyancy force $b(x)$ and the weight force $w(x)$, as shown in Fig. 3.1*c*. In the sign convention adopted herein forces are positive upward, but an exception is made for the weight force which is conventionally regarded as positive. Hence the net force is $f(x) = b(x) - w(x)$.

The solution for $M(x)$ requires two integrations. The first yields the transverse shear force $Q(x)$, and is obtained by imposing vertical force equilibrium of a differential element considered as a free body, as shown in Fig. 3.1*d*.

$$Q + f dx - Q - dQ = 0$$

or

$$f = \frac{dQ}{dx} \tag{3.1.4}$$

from which

$$Q(x) = \int_0^x f(x)\, dx + C \tag{3.1.5}$$

For ships the constant of integration is always zero because the hull girder is a "free–free" beam, with zero shear force at the ends: $Q(0) = 0$.

As shown in Fig. 3.1*d* equilibrium of moments (say, about the right-hand end of the element, and positive clockwise) yields

$$M + Q\, dx + f\, dx \frac{dx}{2} - M - dM = 0$$

The dx^2 term is of second order and therefore

$$Q = \frac{dM}{dx} \tag{3.1.6}$$

from which

$$M(x) = \int_0^x Q(x)\, dx + C \tag{3.1.7}$$

The sign conventions for shear force and bending moment are shown in Fig. 3.1*e*, *f*. Shear force at any point is positive if the integral, or net accumulation, of load up to that point is positive. If we define a "positive face" as the cross section which one sees when looking in the positive x-direction, then the shear force is positive upward when acting on a positive face, and positive downward when acting on a negative face. As shown in Fig. 3.1*e*, an alternative way of expressing this is to say that positive shear causes clockwise rotation of an element.

In a similar manner, the bending moment at any point is positive if the integral, or net accumulation, of shear force up to that point is positive. It can easily be shown that with this definition, positive bending moment corresponds to beam curvature that is concave upward. This condition or state of bending is referred to as "sagging," and the opposite state, concave downward, is referred to as "hogging."

In discussing a sign convention for shear force and bending moment it is important to realize that we are actually considering each of them as a *state of loading*, rather than a specific force or couple. These two states of loading have their own sign convention, quite apart from the sign convention adopted for specific forces and couples. Thus, for example, the differential element in Figure 3.1*d* is in a state of positive shear, even though one of the two end forces acting on it is downward.

Characteristics of Shear Force and Bending Moment Curves

As we have seen, both the shear force and the bending moment must be zero at the ends. Since the load is the derivative of the shear force Q, a point of zero load corresponds to a local maximum or minimum value of Q, as shown by the dashed lines in Fig. 3.1. In most cases the loading is approximately similar forward and aft of amidships, and under these conditions the shear force is approximately asymmetric, passing through zero somewhere near amidships and having maximum values, positive and negative, near the quarter points.

Similarly, since the shear force and bending moment are related by $Q = dM/dx$, a point of zero Q corresponds to a local maximum or minimum value of bending moment. In general, therefore, the bending moment will be a maximum, positive or negative, near amidships, but if the loading is very unsymmetric the maximum bending moment may be some distance from amidships. Since shear force is zero at both ends, the bending moment curve will have zero slope at both ends, and its value will usually be small forward and aft of the quarter points.

Distinction between Still Water Loading and Wave Loading

In order to calculate the load on the hull girder, it is necessary to first calculate both the distributed buoyancy force and the distributed weight force. In regard to the former it is useful to distinguish between the buoyancy force in still water and the additional, and quite different, buoyancy force that occurs as a result of waves. The still water buoyancy is a completely static quantity, and it depends mainly on the shape of the immersed hull. It therefore most logically forms part of the hydrostatic calculations. The additional buoyancy force due to waves is markedly different from the buoyancy force in still water, being essentially both dynamic and probabilistic. Therefore, in order to simplify the analysis, the buoyancy distribution in waves is calculated separately and is superimposed on the static and deterministic still water buoyancy force.

In order to calculate the still water buoyancy distribution, the location of the still waterline of the vessel must be determined from the two overall equilibrium requirements of (3.1.1) and (3.1.2), and therefore it is also necessary to know the weight (i.e., mass) distribution $m(x)$, or at least the overall weight Δ and the location of the l.c.g. Thus, once the lines of a ship have been specified, the still water buoyancy is fixed and calculable, and hence the still water load, shear force, and bending moment depend entirely on the weight distribution. We therefore begin by examining some steps and techniques for estimating this distribution.

3.2 ESTIMATION OF WEIGHT DISTRIBUTION

The calculation of the longitudinal distribution of weight or mass $m(x)$ is a difficult process, partly because $m(x)$ is made up of discrete items rather than being a continuous and regular curve, and partly because at the design stage many of the individual weights are known only approximately. The calculation cannot be as readily or as thoroughly automated as for the buoyancy distribution, but it can be greatly facilitated and even largely automated by means of a systematic approach and by the use of suitable approximate methods.

The weights in a ship fall into two main categories: those which are relatively unchanging, such as the ship's own structural weight; and those which do change, such as cargo, fuel, stores, and ballast. The first group constitutes the "lightweight" of a ship, that is, the weight when it is without cargo, fuel, and so on (this condition is referred to as the "lightship" condition). The second group is called the "deadweight." The deadweight changes with each different cargo loading, and hence there are usually several loading conditions which need to be investigated. The two most common conditions are "full load" and "ballast."

In general, the information regarding weights is of a discrete nature and must be gathered together and entered into a "Table of Weights" or some other suitable form of information storage. In most cases the following information is specified for each logically distinct item:

1. Total weight.
2. Vertical* and longitudinal center of gravity (l.c.g.).
3. Longitudinal extent.
4. The type of distribution over this extent.

Once this information is available the rest of the calculations can be done by computer: the lightweight curve and, for each loading condition, the deadweight curve, the total weight curve, the displacement, and the l.c.g. Knowing the latter two items makes it possible to calculate the still water buoyancy curve for each loading condition, usually by means of a comprehensive hydrostatics program. Then, for each loading condition the appropriate buoyancy and weight curves are combined to get the load curve, and this is then integrated twice to get the shear force and bending moment curves.

In specifying the extent and distribution of individual weights, it is helpful and even necessary to use some approximations and idealizations. Nearly all items can be represented in terms of one or more of three basic types of distribution: point, uniform distribution, and trapezoidal distribution. Also, for cargo and ballast, an alternative approach is possible. For these items the weight per unit length is related to the cross-sectional area of the relevant cargo or ballast space, and their weight distribution may be taken as the product of the sectional area curve of the relevant space times the mass density of the cargo or ballast.

The idealization of a weight as a trapezoidal, uniform, or even as a point mass does not introduce any appreciable error into the calculation of shear force and bending moment as long as it is done reasonably and carefully. After all, both of these quantities are themselves idealizations: they arise from and are a measure of the overall behavior of the hull girder, not the local response; they have only a single value at any cross section. Moreover, they are obtained by integrating the load, and integration is a smoothing process.

Typical examples of point loads are machinery (one point load at each foundation point), masts, winches,

*For calculation of stability and ship motions.

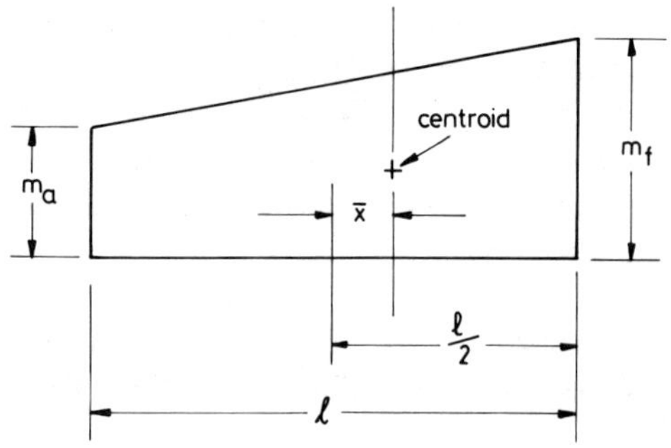

Figure 3.2 Trapezoidal representation of a weight.

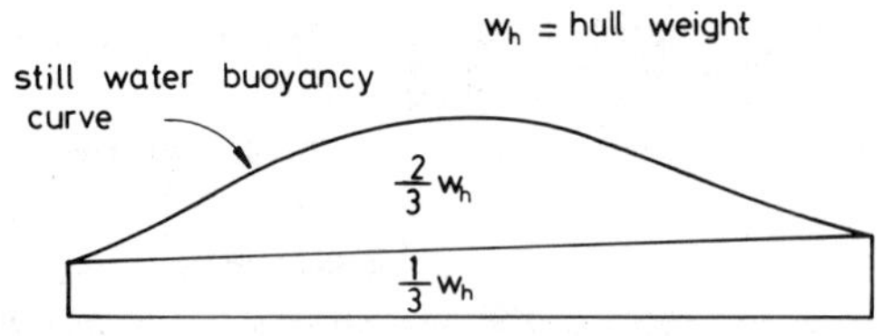

Figure 3.3 Approximation for hull weight distribution.

and transverse bulkheads. Examples of uniform loads are: hull steel within the parallel midbody; and cargo, fuel, ballast, and other homogenous weights within prismatic spaces. Outside of the parallel midbody and particularly toward the ends of the ship a trapezoidal distribution is appropriate, although even here some items can be accurately represented as uniform loads, such as superstructure. For a trapezoid, say of length l, the relevant information may be specified in two different ways: either as total mass, M_0, with a specified position of center of gravity within this length (say a distance $\bar{x}$ from the center; see Fig. 3.2) or in terms of mass per unit length at the forward and after ends: m_f and m_a. The formulas for converting from one form to another are:

$$\left.\begin{aligned} \bar{x} &= \frac{l}{6}\left[\frac{m_f - m_a}{m_f + m_a}\right] \\ M_0 &= \frac{l(m_f + m_a)}{2} \end{aligned}\right\} \qquad (3.2.1)$$

and

$$\left.\begin{aligned} m_a &= \frac{M_0}{l} - \frac{6M_0\bar{x}}{l^2} \\ m_f &= \frac{M_0}{l} + \frac{6M_0\bar{x}}{l^2} \end{aligned}\right\} \qquad (3.2.2)$$

One of the major items of the weight distribution is the hull itself, and this will sometimes be required before detailed structural design of the hull has been completed. A useful first approximation to the hull weight distribution is obtained by assuming that two-thirds of its weight follows the still water buoyancy curve and the remaining one-third is distributed in the form of a trapezoid, with end ordinates such that the center of gravity of the entire hull is in the desired position, as shown in Fig. 3.3. An alternative approach is to use a uniform weight distribution over the parallel midbody portion and two trapezoids for the end portions, with end ordinates again chosen such that the l.c.g. of the hull is in the desired position.

A sample weight curve is given in Fig. 3.4. By commencing with "lightship" components (hull steel, machinery, equipment, and outfit), the designer can determine the distribution of variable weight items (cargo, fuel oil, ballast, etc.) for those loading conditions which will be most serious in both hogging and sagging. With moderate to heavy cargo densities, the average lineal density throughout a ship's machinery space is likely to be less than a cargo space. Hence with a "machinery aft" layout, as in tankers and ore carriers, the sagging moment will probably be greater than the hogging moment.

The most significant loading condition is the "full and down" condition, with the ship loaded with sufficient cargo to float at its "loadline," the water line corresponding to the minimum permissible freeboard. In some cases, beam and draft restrictions may be the overriding constraints. For simplicity, the cargo is usually assumed to be homogeneous and of a density which just produces the "full and down" condition. Lighter-density cargoes reduce displacement (and hence the loading), whereas heavier densities permit flexibility, and hence optimization of load distribution, as the cargo weight remains essentially limited by the statutory draft limitation. Regardless of whether the maximum bending moment in the homogeneous cargo case is hogging or sagging, it can be reduced by

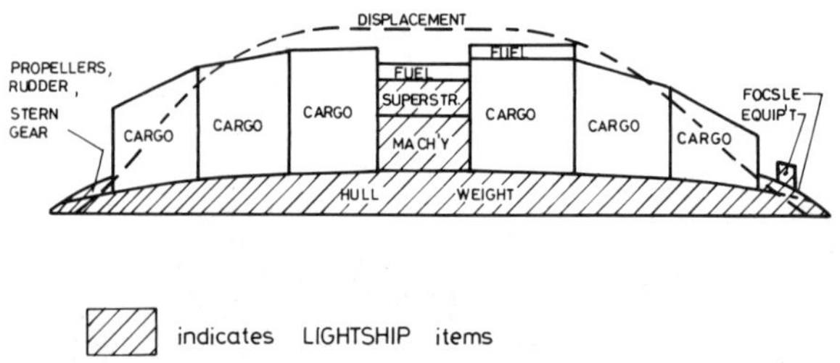

Figure 3.4 Typical weight distribution.

concentrating the more dense cargo either amidships or in the ends respectively.

Because the cargo is the largest item of weight and because there are so many possible variations in its distribution there are often some distributions and combinations that would cause excessive values of bending moment and that therefore must be avoided. This is particularly the case with bulk carriers in which, for various reasons (e.g., tank cleaning, avoidance of free surface, variety of cargo types over a given route) it is more efficient to have the cargo holds or tanks either completely full or completely empty. Given such extreme differences it is important that they be spread out and interspersed, rather than grouped together, because the latter would give excessive shear force and/or bending moment, as shown in Fig. 3.5. Therefore, in addition to restrictions arising from stability requirements, the loading manual of a cargo ship may also contain restrictions on the permissible cargo combinations and distributions, arising from longitudinal strength requirements.

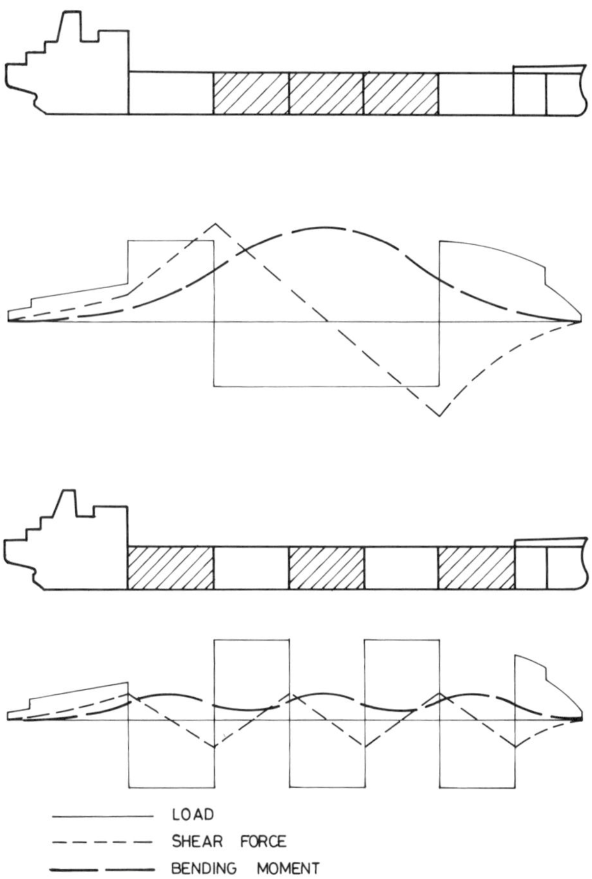

Figure 3.5 Effect of cargo distribution.

3.3 CALCULATION OF STILL WATER BENDING MOMENT

From the foregoing discussion it is clear that the calculation of the still water bending moment is, conceptually at least, a straightforward task; it is simply a double integration of the difference between the buoyancy force and the weight force. There are three main types of methods:

1. Hand methods involving explicit integration of the load curve, using some simple numerical integration formula such as the Trapezoidal Rule or Simpson's One-Third Rule, and performed with the aid of a hand calculator.
2. The simplified hand method of Faresi [1], in which the integrals of buoyancy and of hull weight have already been obtained for a series of hull forms and are presented in the form of charts; values of $Q(x)$ and $M(x)$ at any position x along the ship's length are obtained by a tabular calculation involving readings from the charts combined with the values and first moments of all of the discrete weights, that is, all weights other than the hull weight.
3. A computer program, which is essentially the automation of method 1.

The calculations are both straightforward and tedious, and hence there are many computer programs available for this task. The largest part of the calculation is finding the equilibrium values of draft and trim for each loading condition. Since this is a basic part of the hydrostatic calculations nearly all hydrostatics programs include a routine for calculating the still water load, shear force, and bending moment distributions along the ship length.

Some approximate methods are available for estimating the amidships bending moment $M_{\otimes}$ [2,3] but this is not sufficient for rationally-based design. At the very least, it is necessary to obtain the maximum bending moment M_{max} and in some cases this may not occur at amidships. Some typical cases are:

1. Vessels with unusual internal arrangements (e.g., machinery space three-fourths aft).
2. Tankers and bulk carriers with some empty cargo holds.

For example, in tankers with empty spaces amidships $M_{\otimes}$ may be quite small and there may be two peak values of bending moment, at or near the two quarter points. Also, bulk carriers are commonly loaded in alternate holds and in this case there is a peak value of bending moment in way of each alternate hold. Moreover, with bulk carriers the shear force Q can be quite significant and it will have several peak values. Therefore even in preliminary design it is best to calculate Q and M over the full length of the vessel, or at least between the aftermost and forwardmost peak values of Q.

3.4 CORRECTION FOR CHANGES IN WEIGHT

Because of the wide variety of possible loading conditions, the ship will rarely be in the same condition as was assumed for the still water bending moment calculation. It is therefore important to be able to calculate, as simply as possible, the effect of the addition or removal of weight on the hull girder bending moment. A useful technique for this is to construct an *influence line* diagram. An influence line shows the effect on the *maximum* bending moment of the addition of a unit weight at any position x along the ship length. The height of the line at x represents the effect on M_{max} of the addition of a unit weight at x. Two influence lines are normally drawn, one for the maximum hogging and one for the maximum sagging condition. Influence lines could, of course, be drawn to show the effect of additions on other bending moment values (i.e., other than M_{max}) but these would be of less interest.

Let us take the case of a weight P which is added at a distance x_p forward of amidships, as shown in Fig. 3.6. Other relevant quantities are defined in the figure. As a result of this addition the ship will undergo a parallel sinkage v and a nondimensional trim t (total trim divided by the total length L). If A_W and I_L are the area and the longitudinal moment of inertia of the waterplane about the LCF, then (assuming that the change in the waterplane is small) we have

$$v = \frac{P}{\rho g A_W} \quad \text{and} \quad t = \frac{P(x_P - x_F)}{\rho g I_L}$$

where x is positive forward of amidships. Let R denote the position of maximum bending moment, M_{max}, located at a distance x_R from amidships. The total change in M_{max} can be determined by taking moments about R, either forward or aft. Choosing the forward side, we see that the change in M_{max} is the sum of the moment of buoyancy of the parallel sinkage forward of R plus the moment of buoyancy of the wedge forward of R minus the moment of added weight. (This is consistent

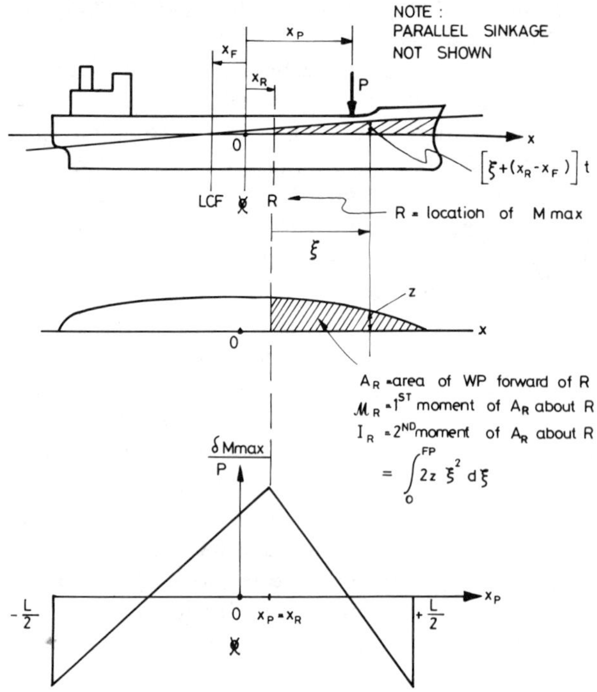

Figure 3.6 Influence line for changes in M_{max} due to added weight.

with the convention that sagging bending moment is positive). These three quantities are:

1. Moment of buoyancy of parallel sinkage forward of R about R

$$= \rho g \, v \, \mathcal{M}_R = \frac{P \mathcal{M}_R}{A_W}$$

where $\mathcal{M}_R$ is the moment about R of the waterplane area forward of R.

2. Moment of buoyancy of wedge forward of R

$$= \rho g \int_0^{FP} 2y\xi[\xi + (x_R - x_F)t \, d\xi$$

$$= \rho g [I_R + \mathcal{M}_R (x_R - x_F)]t$$

$$= \frac{P(x_P - x_F)}{I_L}[I_R + \mathcal{M}_R (x_R - x_F)$$

3. Moment of added weight $= P(x_P - x_R)$.

Note: since we are calculating moments forward of R, this third term is only included if the added weight is forward of R; if not, it is omitted. To assist in remembering this we shall write this third term as $P\langle x_P - x_R \rangle$, in which the pointed brackets indicate that whenever the quantity within the brackets becomes negative its value is taken as zero.

Therefore, the net effect on the bending moment at R (i.e., on $M_{\max}$) due to the addition of a weight P is

$$\delta M_{\max} = P\left\{\frac{\mathcal{M}_R}{A_W} + \frac{(x_P - x_F)}{I_L}[I_R + \mathcal{M}_R(x_R - x_F)] - <x_P - x_R>\right\} \quad (3.4.1)$$

This equation is also valid for negative values of x_P (i.e., for P aft of amidships), provided that, once again, if the expression in pointed brackets is negative it is taken as zero. A discontinuity occurs at R, the position of maximum bending moment. As shown in Fig. 3.6, the influence lines are straight lines which cross the axis at approximately the quarter points of the vessel. Therefore, a weight added within this length causes an increased sagging moment and an added weight outside this length causes an increased hogging moment. To construct an influence line diagram, (3.4.1) should be evaluated for three values of x_P: $-L/2$, x_R, and $+L/2$.

By making suitable approximations it is possible to simplify (3.4.1). It can be shown that if both LCF and R are taken as being at amidships then $x_R = x_F = 0$, $I_R = I_L/2$, and $\mathcal{M}_R = \frac{1}{2}A_W\, x_{(1/2)WP}$, where $x_{(1/2)WP}$ is the distance from amidships to the centroid of the forward half waterplane. With these simplifications (3.4.1) becomes

$$\delta M_{\max} \cong P\left\{\frac{x_{(1/2)WP}}{2} + \frac{x_P}{2} - <x_P>\right\}$$

Thus, if the weight is added forward of amidships the result is

$$\delta M_{\max} \cong \frac{P(x_{(1/2)WP} - x_P)}{2}$$

and this has a direct and relatively simple physical interpretation: it is $-\frac{1}{2}$ times the moment of P about the centroid of the half waterplane area. Moreover, for this simplified case it is possible to define the terms in such a way that a single expression applies, regardless of whether P is forward or aft of amidships. To do this we define l as the distance between the added weight and the *nearer* centroid of the half-waterplane area, as shown in Fig. 3.7, and we adopt the sign convention that if P lies between this centroid and amidships, the moment change is positive (sagging); otherwise negative (hogging). The approximate expression for the moment change is then simply

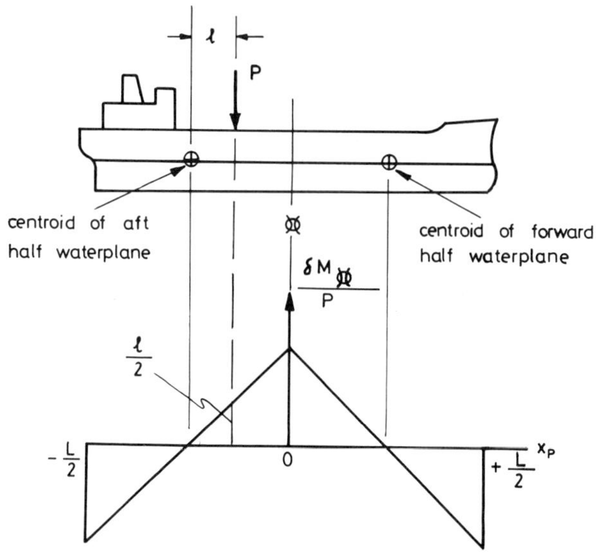

Figure 3.7 Simplified influence line: change in $M_{\otimes}$ due to added weight.

$$\delta M_{\max} \cong \frac{Pl}{2} \quad (3.4.2)$$

Of course, if a weight is *removed* then $-P$ replaces P in all of the above formulas.

In addition to their use in design, influence lines are a helpful tool for the ship operator and are sometimes provided as a part of the loading manual. However, it should be noted that they are intended for small weight changes only; certainly not more than 10% of the displacement. If the change exceeds this amount a new bending moment calculation should be done. Obviously, the approximate formula of (3.4.2) is even less accurate and it should not be used at all if either the LCF or the location of $M_{\max}$ is not close to amidships.

In some cases it may be desired to find the change in bending moment at some other point along the vessel, or perhaps along the entire length, instead of only the change in $M_{\max}$. As before, let $M(x)$ denote the bending moment at some arbitrary position x and let $\delta M(x)$ be the change in $M(x)$ due to the addition of a weight at x_P. The moment change can be calculated by first calculating the change in the shear force as a function of x and then integrating. The change in the shear force is

$$\delta Q(x) = \rho g \int_{-L/2}^{x} 2z(v + tx)\, dx$$

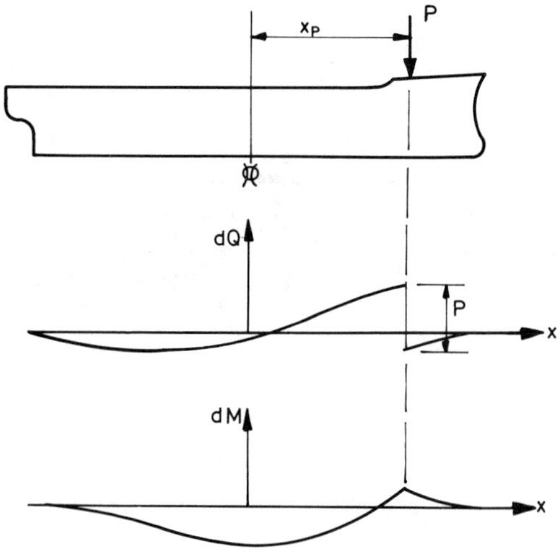

Figure 3.8 Change in shear force and bending moment distribution.

where v and t are the sinkage and trim, as before. Substitution for these gives

$$\delta Q(x) = \frac{2P}{A_W}\int_{-L/2}^{x} z\,dx + \frac{2P(x_P - x_F)}{I_L}\int_{-L/2}^{x} zx\,dx \qquad (x < x_P) \tag{3.4.3}$$

and the change in the bending moment is then

$$\delta M(x) = \int_{-L/2}^{x} [\delta Q(x)]\,dx \qquad (x < x_P) \tag{3.4.4}$$

These integrals must be evaluated by some graphical or numerical technique; the process is illustrated in Fig. 3.8. The expression for $\delta Q(x)$ is valid only between the after end and the location of the weight. Beyond that point, the weight P itself must be added to the shear formula, and this causes a corresponding change in the bending moment expression (see Fig. 3.8).

3.5 APPROXIMATE DESIGN VALUES OF WAVE LOADS

Because of the complexity of ocean waves and of the dynamic interaction between ship and waves, the direct calculation of an appropriate design value of wave loading for a given ship is a very complex task, and one which has occupied the attention of several generations of naval architects. After many years of investigation significant progress has been made, due largely to simultaneous and complementary advances in several fields: ocean wave data collection, statistical theory, system response analysis, free surface hydrodynamics, and above all, computing. Thanks to these advances methods are now available which are more rationally-based, more accurate, and more comprehensive than the methods of only a few years ago. Being computer-based, these new methods also eliminate much of the tedium of the older methods.

But as mentioned in Section 1.5, the design values for wave-induced loads are long-term statistical quantities and are therefore applicable to all ships of a given size, speed, hull geometry, and distribution of mass. Therefore, over the past several years the classification societies have gathered much information concerning wave loads from the computer-based methods, from model studies and from direct full-scale measurements. From this they have developed explicit formulas for the characteristic extreme values for standard ship types, expressed in terms of the principal dimensions of the ship. Because of the complexity of wave loading and because it is probabilistic there is a good deal of scatter in the measured values, and the formulas developed from them are only approximate and somewhat conservative. Nevertheless, the explicit specification of wave loads is of great convenience and usefulness to the designer, and this helps to offset the approximations and possible extra safety margin which it contains.

For special ship types the characteristic extreme values of wave loads should be calculated explicitly, using the computer-based methods. This does not require specialist knowledge or involve an undue amount of calculation because these computer-based methods are currently being incorporated into user-oriented packages, which require for their use only a broad familiarity with the basic concepts of statistics, ship motions, and system response analysis, all of which are presented in Chapter 4.

In the remainder of this section we present a set of characteristic wave loads from Ref. 4 which may be used for the design of standard ships. These characteristic values correspond to a probability of exceedence of the order of 10^{-8}.

Vertical Bending Moment

From 0.1 L aft of amidships to $C_B L/4$ forward of amidships:

$$(M_{w,c})_{\text{sag}} = 125\, L^2 B (C_B + 0.2) h_{w,c} \tag{3.5.1a}$$

$$(M_{w,c})_{\text{hog}} = 165\, L^2 B\, C_B^2\, h_{w,c} \tag{3.5.1b}$$

in units of Nm, where the symbols are, in order of appearance, length between perpendiculars, beam, block coefficient, and the characteristic value of wave height for the vessel, which is defined as follows:

$$\left.\begin{array}{l}\text{For } L < 350 \text{ m } h_{w,c} \text{ is the smallest of:}\\ h_{w,c} = D, \text{ the depth of the hull girder, or}\\ h_{w,c} = 13 - \left(\dfrac{250 - L}{105}\right)^3, \text{ or}\\ h_{w,c} = 13 - \left(\dfrac{L - 250}{105}\right)^3;\\ \text{For } L > 350 \text{ m:}\\ h_{w,c} = \dfrac{227}{\sqrt{L}}\end{array}\right\} \qquad (3.5.2)$$

Forward and aft of the length just specified, the bending moment is assumed to reduce linearly to zero at the fore and aft perpendiculars (FP and AP), as shown in Fig. 3.9. However for ships with small block coefficient, high speed, and large bow flare, the sagging moment in the forward region is first reduced linearly to an intermediate value of

$$\left(\frac{0.1\, C_A}{0.5 - 0.25\, C_B}\right)(M_{w,c})_{\text{sag}}$$

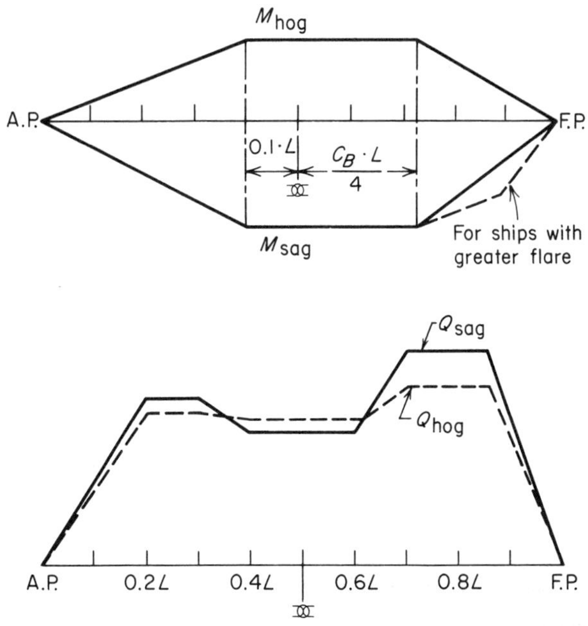

Figure 3.9 Distributions of wave-induced bending moment and shear force.

at 0.1 L from FP, and then linearly to zero. The coefficient C_A is defined as

$$C_A = 0.8\frac{V}{\sqrt{L}} + \frac{15(A_D - A_{WL})}{LB}$$

where V is the contracted trial speed of the ship in knots, A_D is the horizontally projected area of upper deck (including any forecastle deck) forward of 0.2 L from FP, and A_{WL} is the area of waterplane forward of 0.2 L from FP at the moulded summer draft, d.

Horizontal Bending Moment

From 0.1 L aft of amidships to 0.25 C_BL forward of amidships the characteristic value of horizontal wave bending moment on the hull girder, as a combination of the direct horizontal moment and the heeling component of vertical bending moment, is

$$(M_{w,c})_{\text{hor.}} = 1500\sqrt[4]{\frac{L}{200}}\, C_B L^2(d + 0.3B) \qquad (3.5.3)$$

where d is the moulded summer draft. Outside the specified area amidship, a linear reduction to zero at the perpendiculars is assumed.

Torsional Moment

The characteristic value of wave-induced torsional moment at a position x along the hull girder in a quartering sea (waves coming from 20 to 50° from abeam) is:

$$(M_{w,c})_{\text{tor.}} = 130\, LB^2 h_{w,T} C_{WL}^2 \left[\sin^2\left(180°\frac{x}{L}\right) + \alpha \sin\left(360°\frac{x}{L}\right)\right] \qquad (3.5.4)$$

where

$$h_{w,T} = 13 - \left(\frac{250 - 0.7L}{105}\right)^3 \quad \text{(maximum value of 13)}$$

$$\alpha = \frac{3.8\, C_B\, dp}{C_{WL}^2 B^2}$$

x is the length (m) from FP to the section considered for head seas, or the length (m) from AP to the section considered for following seas. C_{WL} is the waterplane coefficient, d is the summer draft, and p is the distance from the shear center to a level $0.6d$ above the baseline.

Equation 3.5.4 is based on the assumption that the shear center position is fairly constant along the loaded length. If not, a more accurate calculation of $(M_{w,c})_{\text{tor.}}$ should be made. For slender ships with deck cargo at great height above the waterline it is advisable to check the torsional moment due to roll. Expressions for characteristic values of roll period, angle, and acceleration are given in Reference 4.

Vertical Shear Force

The wave-induced vertical shear force on the hull girder, in Newtons, is given in the following table, and can be taken as constant within the specified lengths.

Length Forward of AP	$(Q_{w,c})_{\text{sag}}$	$(Q_{w,c})_{\text{hog}}$
(0.2–0.3)L	$\dfrac{(M_{w,c})_{\text{sag}}}{0.32L}$	$\dfrac{(M_{w,c})_{\text{hog}}}{0.34(C_B + 0.2)L}$
(0.4–0.6)L	$\dfrac{(M_{w,c})_{\text{sag}}}{0.39L}$	$\dfrac{(M_{w,c})_{\text{hog}}}{0.37(C_B + 0.2)L}$
(0.7–0.85)L	$\dfrac{(M_{w,c})_{\text{sag}}}{0.25L}$	$\dfrac{(M_{w,c})_{\text{hog}}}{0.29(C_B + 0.2)L}$

Between the specified lengths and to a value of zero at the perpendiculars, a linear variation is assumed, as shown in Fig. 3.9. $(Q_{w,c})_{\text{sag}}$ in the forward half length should be increased for vessels with large bow flare.

3.6 HULL GIRDER BENDING STRESS

Constraint on Hull Girder Stress in Rule-Based Design

For general cargo ships and most other non-bulk carriers, rule-based design does not involve a hull module analysis, and so only the hull girder stresses are explicitly calculated; the additional stresses due to hull module response and principal member response are not calculated. Likewise such design does not include a limit state analysis, either of the hull module or of the principal members, to find the magnitude and combinations of total applied stress which would cause the various types of failure. Therefore, in this type of design it is necessary to place a restriction on the total hull girder stress σ_h (wave-induced plus stillwater) as an approximate means of preventing member failure due to the combination of σ_h and the other stresses in the member.

In contrast, as explained in Section 1.2, rationally-based design specifically examines all types of failure, at both levels (principal member and hull module, and in this context the latter is identical to hull girder) for all loadcases, and generates specific constraints against all of these failures. Since failures of the hull module are more serious they are given a larger factor of safety. If some constraint against hull module failure is not satisfied by the current design then the deficiency is corrected *in whatever way is optimal:* either by making a general increase in scantlings, thereby increasing the section modulus and decreasing σ_h, or by strengthening the principal members that are the source of the deficiency, or by some combination of these two strategies—whatever is optimal, according to the designer's specified measure of merit. In determining the optimal correction it is necessary to take into account all constraints that involve σ_h and all constraints that involve the same design variables as in the violated constraint. Obviously this is more accurate and thorough than a blanket rule specifying a maximum permissible value of σ_h. It involves far more computation, but the overall design methodology presented herein and the associated methods for response analysis, limit state analysis, and optimization are all well within the computational capabilities of current computers. In fact, as shown by the design studies referred to in Section 1.3, the rationally-based method presented herein involves no more time or total cost (computer costs plus man-hour costs) than rule-based design. However, it should also be repeated here that for some design offices and for some types of ships, particularly smaller vessels, rule-based design is more appropriate because it is so much simpler in concept and because it can be done by hand.

Section Modulus Requirement to Prevent Hull Girder Fatigue Failure

The cyclic wave-induced hull girder bending stress, if not kept within certain limits, could lead to fatigue fracture of the hull girder. Hence as shown in Table 2.1 one of the structural constraints at the overall (hull girder) level is the requirement of a sufficiently large section modulus to prevent this. In current rule-based design the minimum value of section modulus is specified as a simple function of ship length, beam, and block coefficient. There is no explicit check of hull girder fatigue because this section modulus requirement is intended to cover a variety of limit states (similar to the restriction on total hull girder stress) and for the more common types and proportions of cargo ships these two requirements generally result in a wave-induced stress that is within satisfactory limits for fatigue. But this is not sufficient for rationally-based design because the process of optimization may produce a design having different proportions from previous ships.* In rationally-based design hull girder fatigue and all other important types of failure are always examined explicitly, and since this type of design is computer-based the computation involved is not a problem.

Limit Value of Wave-Induced Stress

The basic aspects of fatigue analysis were described in Section 2.4, and for this application we use the method based on *S-N* curves and the linear damage hypothesis (Miner's Rule) of (2.4.4). From (2.4.2) the *S-N* relationship is

$$N S_N^m = C \tag{3.6.1}$$

and the damage ratio η which is used in Miner's Rule is

$$\eta = \sum_{i=1}^{B} \frac{n_i}{N_i} \tag{3.6.2}$$

where B = number of stress blocks
n_i = number of stress cycles in stress block i
N_i = number of cycles to failure at constant stress range S_i

The *S-N* curves are for constant amplitude stress cycles whereas in hull girder bending the stress range is a random variable, described by a probability density function $p_S(s)$ (see Fig. 3.10) and quantified by some type of characteristic value S_c corresponding to

*In fact, for long slender ships such as naval vessels the prevention of hull girder fatigue is often one of the active constraints in the structural design.

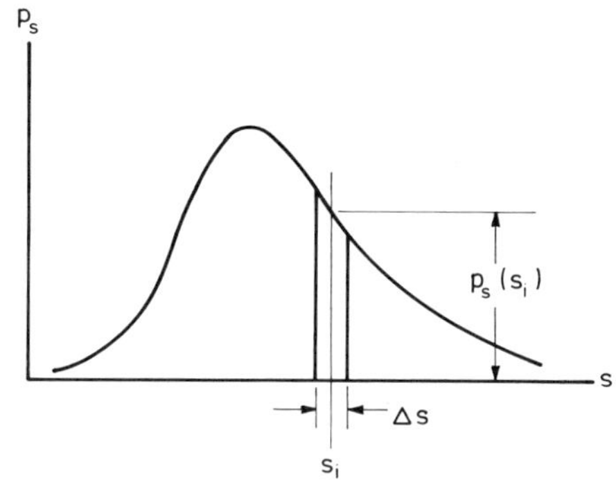

Figure 3.10 Probability density function for wave-induced hull girder bending stress range (long term).

a specified probability of exceedance.† A probability of 10^{-8} is commonly used for offshore structures and can also be used for a hull girder. Because a 20-year life represents about 10^8 cycles this means that S_c is the stress range that is expected to be exceeded once in the life of the structure.

The application of *S-N* test data to a hull girder requires a relationship between the characteristic value of wave-induced random stress and the constant amplitude stress of the S-N curves. This is obtained by applying Miner's Rule for the full lifetime distribution of stress range, as given by the probability density function $p_S(s)$. The distribution is divided into a large number of narrow stress blocks of width Δs, as illustrated in Fig. 3.10. In each block the fractional number of cycles is $p_S(s_i)\Delta s$. If N denotes the total number of cycles in the life of the ship, then

$$\begin{aligned} n_i &= \text{number of cycles in } \Delta s \\ &= N p_S(s_i)\, \Delta s \end{aligned} \tag{3.6.3}$$

Substituting (3.6.1) and (3.6.3) into (3.6.2) gives

$$\eta = \sum \frac{N p_S(s_i)\, \Delta s}{C / S_i^m}$$

and as $\Delta s \to 0$ this becomes

$$\eta = \frac{N}{C} \int_0^\infty s^m p_S(s)\, ds$$

†We here employ the customary probabilistic notation in which the random variable (here the stress range S) is written in lower case when it is an indefinite quantity. Section 4.1 contains a brief review of some of the fundamentals of probability theory.

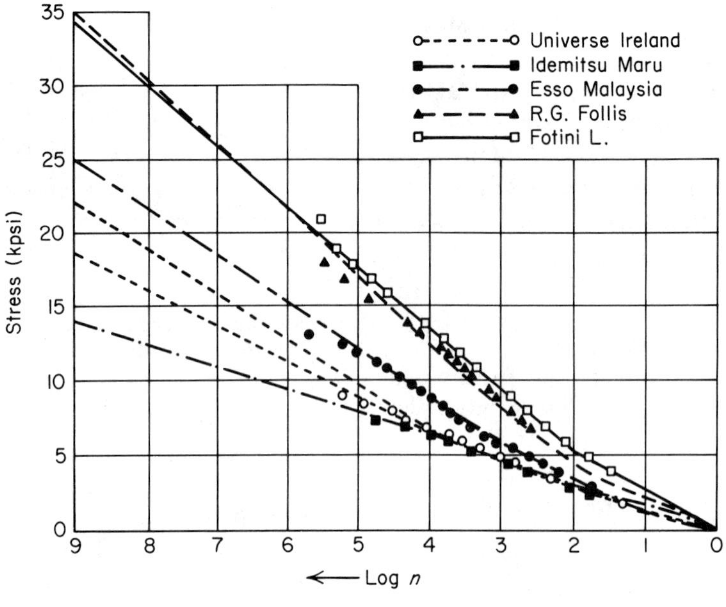

(*a*)

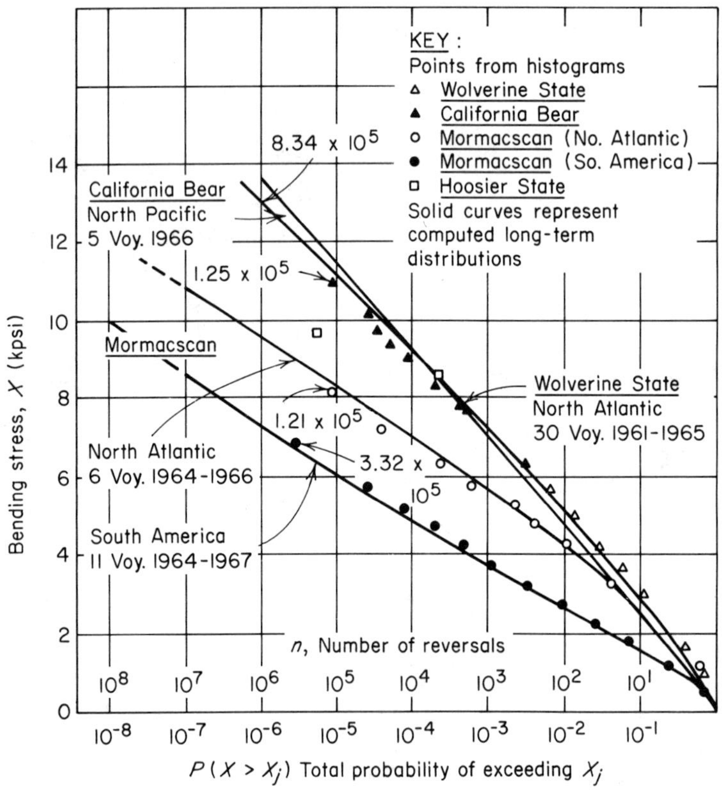

(*b*)

Figure 3.11 Long-term trends in service stresses for (*a*) large bulk carriers [5] and (*b*) dry cargo vessels [6].

The integral expression is the expected value, or mean value, of the random variable S^m (see Section 4.1) which is written as $E(S^m)$. Hence the damage ratio becomes

$$\eta = \frac{N}{C} E(S^m) \tag{3.6.4}$$

where

$$E(S^m) = \int_0^\infty s^m p_S(s)\, ds \tag{3.6.5}$$

Figure 3.11 gives measured values of the cumulative probability distribution of stress range for large tankers and dry cargo vessels [5,6]. From these and other at-sea measurements [7,8] it has been found that $p_S(s)$ can be represented satisfactorily by the two-parameter Weibull probability function. For this function the cumulative probability distribution [which is the integral of $p_S(s)$] is given by

$$P_S(s) = 1 - \exp\left\{-(\ln N)\left(\frac{s}{S_c}\right)^k\right\} \tag{3.6.6}$$

in which $P_S(s)$ = the probability that a stress range of magnitude s will not be exceeded
S_c = the value of S for which the probability of exceedence is 10^{-8}
k = Weibull Shape Parameter

The Weibull distribution is plotted in Fig. 3.12 for various values of k. Table 3.1 from Ref. 9 summarizes the results of the at-sea measurements illustrated in Fig. 3.11, and shows that for large tankers and bulk carriers k ranges from 0.7 to 1.0, while for dry cargo ships and containerships it ranges from 1.0 to 1.3. For design applications k can either be set at the appropriate upper value or it can be estimated more accurately by calculating the ratio S/S_c for some shorter period such as $N = 10^3$, and then using Fig. 3.12. The calculation would be done by means of a ship motion and loads computer program, of which several are now available. In most cases the programs require only the lines of the ship and the approximate weight distribution. The basic principles underlying such programs are presented in Chapter 4, and for most applications a more detailed or specialized knowledge of the theory is not necessary in order to use the programs.

For the Weibull distribution the expected value is

$$E(S^m) = S_c^m (\ln N)^{-m/k}\, \Gamma\left(1 + \frac{m}{k}\right)$$

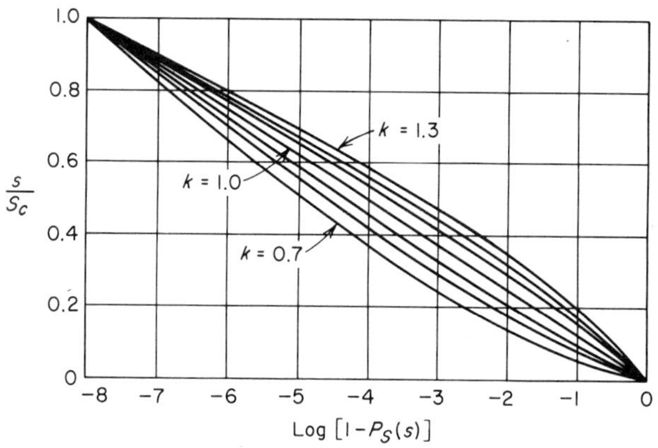

Figure 3.12 Ship loading histories modeled by Weibull distributions.

Substitution into (3.6.4) gives the damage ratio

$$\eta = \frac{N}{C} S_c^m (\ln N)^{-m/k}\, \Gamma\left(1 + \frac{m}{k}\right) \tag{3.6.7}$$

To prevent fatigue fracture this ratio must not exceed 1.0. In practice, because of the various uncertainties in regard to loads, fabrication, and operation, the limit value η_L is made substantially less than 1.0. In offshore structures typical values for principal structural members are in the range from 0.1 to 0.3 [10].

Equation 3.6.7 is suitable for checking an existing structure or the completed design of a structure. In order to bring the fatigue criterion into the design process we need an expression for the limit value of S_c, denoted as $(S_c)_L$, which can then be combined with the characteristic value of wave-induced bending moment to obtain the minimum required value of section modulus. $(S_c)_L$ is the value of S_c obtained from (3.6.7) by substituting η_L for η and taking the $1/m$ power of both sides. We also note from (3.6.1) that the quantity $(N/C)^{1/m}$ is simply S_N, the limit or failure level of constant amplitude stress range corresponding to the ship life. Making these substitutions in (3.6.7) and solving for $(S_c)_L$ gives

$$(S_c)_L = \xi \frac{S_N}{\gamma_F} \tag{3.6.8}$$

in which

$$\xi = (\ln N)^{1/k} \left[\Gamma\left(1 + \frac{m}{k}\right)\right]^{-1/m} \tag{3.6.9}$$

and

$$\gamma_F = (\eta_L)^{-1/m} \tag{3.6.10}$$

TABLE 3.1 SHIP LOADING HISTORIES COMPARED WITH WEIBULL DISTRIBUTIONS

Type of Ship	Name of Ship	Notes	Weibull Distribution Shape Parameter, k	Stress Range at Probability of Exceedance = 10^{-8} S_c (ksi)
Dry Cargo	Wolverine State	1,5	1.2	16.5
	California State	1,5	1.0	18.0
	Mormacscan	1,5,7	1.3	12.0
	Mormacscan	1,5,8	1.0	10.0
Large Tankers	Idemitsu Maru	2,5	1.0	12.3
	R. G. Follis	2,5	0.8	30.0
	Esso Malaysia	2,5	0.8	21.8
	Universe Ireland	2,3,5	0.7	18.7
Bulk Carrier	Fotini L.	2,5	0.9	29.5
SL-7 Containerships	See Note 9	4,6,9	1.2	34.1

Notes:

1. Data from Ref. 6.
2. Data from Ref. 5.
3. Data from Ref. 8.
4. Data from Ref. 7.
5. Load history is for wave-induced loading with dynamic effects filtered.
6. Load history is for wave-induced loading with dynamic effects included.
7. Load history based on North Atlantic voyages.
8. Load history based on South American voyages.
9. Load history based on data collected from eight SL-7 containerships.

The factor ξ is the *random load factor* which relates the limit value S_N from the *S-N* data to the limit value of the characteristic load, $(S_c)_L$. Table 3.2 gives values of ξ for various values of k and m for a life of 10^8 cycles. For any other life N, the values in the table would be multiplied by

$$\frac{(\ln N)^{1/k}}{(18.42)^{1/k}}$$

In (3.6.8) γ_F is the partial safety factor which is implied in the selection of η_L and is given by (3.6.10). For example if $m = 3$, a choice of $\eta_L = 0.3$ is equivalent to using a partial safety factor $\gamma_F = 1.41$.

The final step is to express the characteristic value of applied stress in terms of the characteristic value of wave-induced bending moment and the section modulus, since it is the latter which is required for design. The characteristic bending moment may be either calculated from a ship motion and loads program or, for ships of standard type, simply estimated by an appropriate formula such as (3.5.1). This is related to S_c by the section modulus Z:

$$S_c = \frac{(M_{w,c})_{\text{sag}} + (M_{w,c})_{\text{hog}}}{Z(\mathbf{A})} \tag{3.6.11}$$

in which $Z(\mathbf{A})$ indicates the dependency of Z on the sectional areas $\mathbf{A}$ of the longitudinally continuous members in the hull girder cross section. These sectional areas constitute the overall design variables in the optimization of the structure, as explained in Section 2.7. The constraint which is to be imposed on the overall design variables in order to prevent fatigue

TABLE 3.2 RANDOM LOAD FACTOR ξ BASED ON A LIFE OF 10^8 CYCLES

m	$k = 0.7$	$k = 0.8$	$k = 0.9$	$k = 1.0$	$k = 1.1$	$k = 1.2$	$k = 1.3$
3.0	19.23	14.96	12.12	10.14	8.69	7.60	6.75
3.5	16.35	13.04	10.77	9.14	7.93	7.00	6.27
4.0	14.15	11.53	9.68	8.32	7.30	6.50	5.86
4.5	12.41	10.31	8.79	7.64	6.76	6.07	5.52

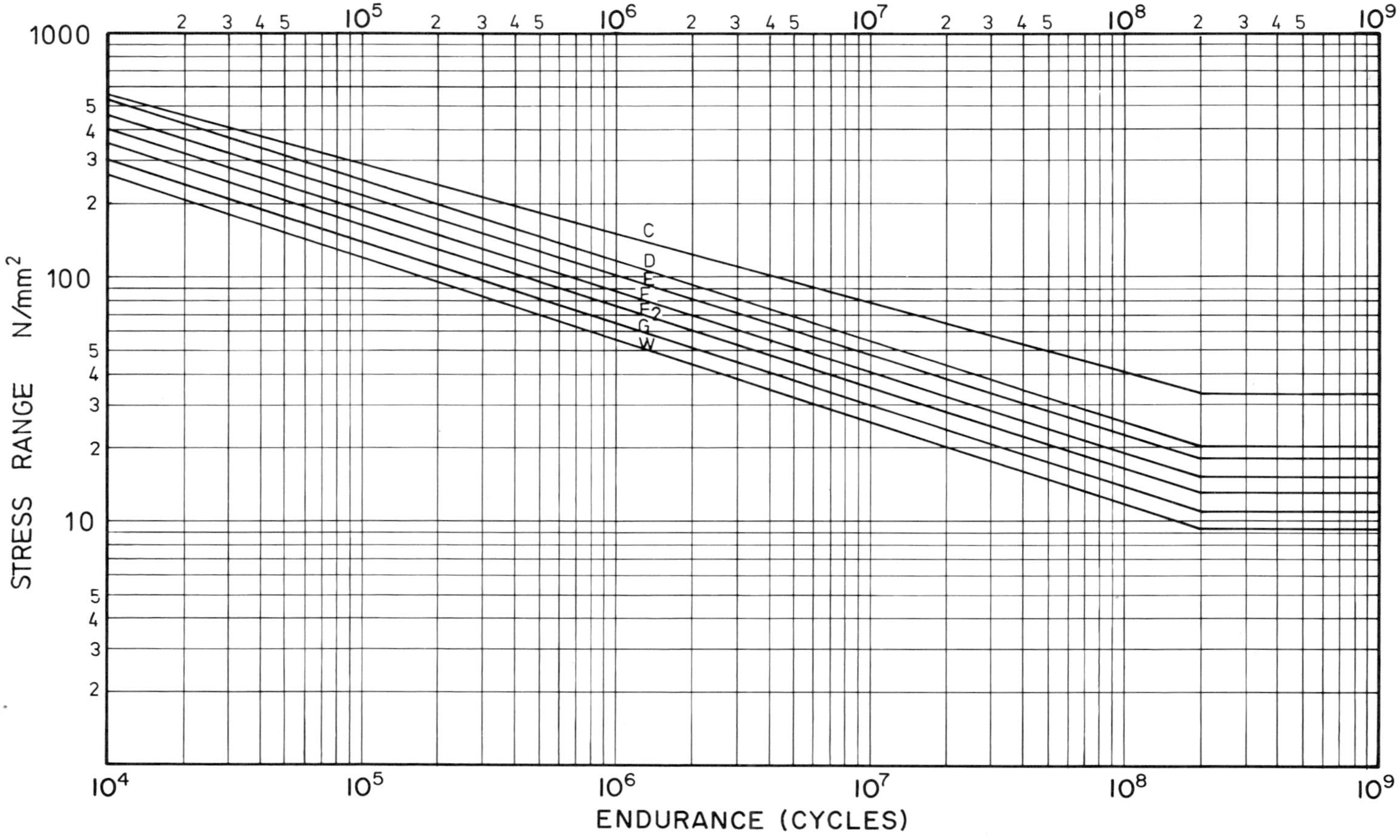

Figure 3.13 S–N curves for hull girder fatigue.

failure of the hull girder is

$$\gamma_F S_c \leq (S_c)_L \qquad (3.6.12)$$

Substituting (3.6.7) and (3.6.10) gives the specific mathematical requirement that the overall design variables **A** must satisfy in order to avoid fatigue failure of the hull girder

$$Z(\mathbf{A}) \geq [(M_{w,c})_{\text{sag}} + (M_{w,c})_{\text{hog}}]\frac{\gamma_F}{\xi S_N} \qquad (3.6.13)$$

There are at least two such constraints, one for the deck and one for the bottom. This constraint should also be imposed on any other part of the hull girder that could possibly be the source of a fatigue failure, either because of a lower class of welding or a larger value of wave-induced stress due to stress raisers (openings, eccentricities, discontinuities, etc.) which are part of the design and cannot be eliminated.

S-N Curves for Hull Girder Fatigue Analysis

Figure 3.13 presents *S-N* curves from Ref. 11 for plate and box girder structure which are used for offshore structures [10], and which can also be used for ship hulls. Table 3.3 describes the conditions (type of joint) for which each curve applies,* and Table 3.4 gives the numerical values of m and C for each curve. The stress range to be associated with Class B to G is the stress range in the parent plate adjacent to the weld. For Class W the stress range is the shear stress range in the weld.

Generally, stress concentration due to the structure geometry, but excluding that for the weld attachment itself, is to be accounted for in the applied stress range.

*Type 6 is for structural details, but is included here for the sake of completeness.

TABLE 3.3 JOINT CLASSIFICATIONS[a]

Type Number, Description, and Notes on Mode of Failure	Class	Explanatory Comments	Examples, Including Failure Modes
TYPE 1 MATERIAL FREE FROM WELDING			
Notes on potential modes of failure			
In plain steel, fatigue cracks initiate at the surface, usually either at surface irregularities or at corners of the cross section. In welded construction, fatigue failure will rarely occur in a region of plain material since the fatigue strength of the welded joints will usually be much lower. In steel with rivet or bolt holes or other stress concentrations arising from the shape of the member, failure will usually initiate at the stress concentration			
1.1 Plain steel			
(a) With all surfaces machined and polished and of uniform or uniformly varying cross-section	A		
(b) In the as-rolled condition, or with cleaned surfaces but with no flame-cut edges or reentrant corners	B	Beware of using Class B for a member which may acquire stress concentrations during its life, for example, as a result of rust pitting. In such an event Class C would be more appropriate	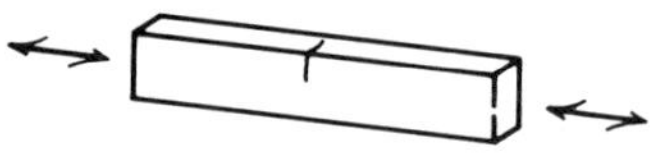
(c) As (b) but with any flame-cut edges subsequently ground or machined to remove all visible sign of the drag lines	B	Any reentrant corners in flame-cut edges should have a radius greater than the plate thickness	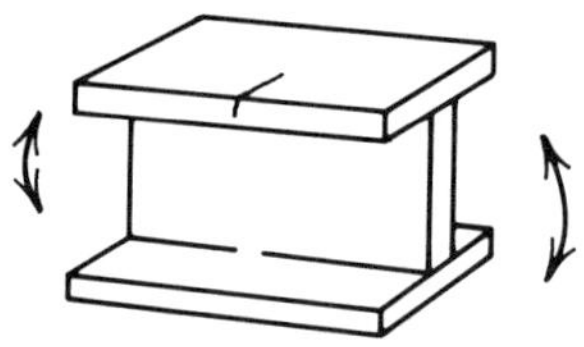
(d) As (b) but with the edges machine flame-cut by a controlled procedure to ensure that the cut surface is free from cracks	C	Note, however, that the presence of a reentrant corner implies the existence of a stress concentration so that the design stress should be taken as the net stress multiplied by the relevant stress concentration factor	

TABLE 3.3 JOINT CLASSIFICATIONS[a] (Continued)

Type Number, Description, and Notes on Mode of Failure	Class	Explanatory Comments	Examples, Including Failure Modes

TYPE 2 CONTINUOUS WELDS ESSENTIALLY PARALLEL TO THE DIRECTION OF APPLIED STRESS

Notes on potential modes of failure

With the excess weld metal dressed flush, fatigue cracks would be expected to initiate at weld defect locations. In the as-welded condition, cracks might initiate at stop-start positions or, if these are not present, at weld surface ripples

General comments

1 Backing strips

If backing strips are used in making these joints: (a) they must be continuous, and (b) if they are attached by welding those welds must also comply with the relevant Class requirements (note particularly that tack welds, unless subsequently ground out or covered by a continuous weld, would reduce the joint to Class F, see joint 6.5)

2 Edge distance

An edge distance criterion exists to limit the possibility of local stress concentrations occurring at unwelded edges as a result, for example, of undercut, weld spatter, or accidental overweave in manual fillet welding (see also notes on joint Type 4). Although an edge distance can be specified only for the "width" direction of an element, it is equally important to ensure that no accidental undercutting occurs on the unwelded corners of, for example, cover plates or box girder flanges. If it does occur it should subsequently be ground smooth

2.1 Full or partial penetration butt welds, or fillet welds

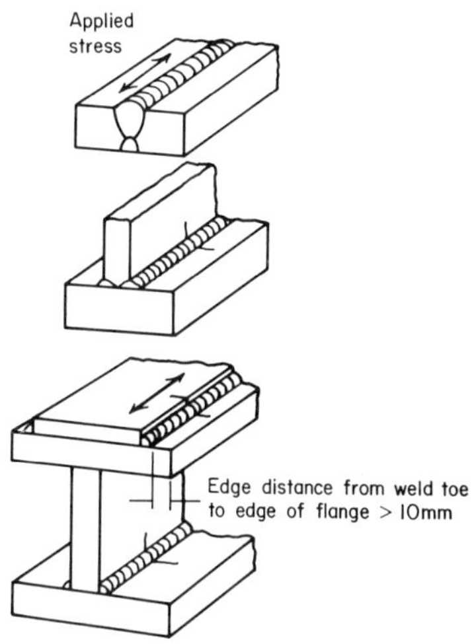

Parent or weld metal in members, without attachments, built up of plates or sections, and joined by continuous welds

Description	Class	Explanatory Comments
(a) Full penetration butt welds with the weld overfill dressed flush with the surface and finish-machined in the direction of stress, and with the weld proved free from significant defects by nondestructive examination	B	The significance of defects should be determined with the aid of specialist advice and/or by the use of fracture mechanics analysis. The NDT technique must be selected with a view to ensuring the detection of such significant defects
(b) Butt or fillet welds with the welds made by an automatic submerged or open arc process and with no stop-start positions within the length	C	If an accidental stop-start occurs in a region where Class C is required remedial action should be taken so that the finished weld has a similar surface and root profile to that intended
(c) As (b) but with the weld containing stop-start positions within the length; this class applies also to manual welding	D	For situation at the *ends* of flange cover plates see joint Type 6.4

TYPE 3 TRANSVERSE BUTT WELDS (i.e., essentially perpendicular to the direction of applied stress)

Notes on potential modes of failure

With the weld ends machined flush with the plate edges, fatigue cracks in the as-welded condition normally initiate at the weld toe, so that the fatigue strength depends largely upon the shape of the weld overfill. If this is dressed flush the stress concentration caused by it is removed and failure is then associated with weld defects. In welds made on a permanent backing strip, fatigue cracks initiate at the weld metal/strip junction, and in partial penetration welds (which should not be used under fatigue conditions), at the weld root

Unless made on a permanent backing (Type 3.2) welds made entirely from one side are not classified for fatigue purposes, since adequate control cannot be exercised over the profile of the root bead which is where fatigue cracks would be likely to initiate

Design stresses

In the design of butt welds of Types 3.1 or 3.2 which are not aligned the stresses must include the effect of any eccentricity. An approximate method of allowing for eccentricity in the thickness direction is to multiply the nominal stress by $(1 + 3e/t)$, where

TABLE 3.3 JOINT CLASSIFICATIONS[a] (Continued)

Type Number, Description, and Notes on Mode of Failure	Class	Explanatory Comments	Examples, Including Failure Modes
e is the distance between centers of thickness of the two abutting members; if one of the members is tapered, the center of the untapered thickness must be used, and			
t is the thickness of the thinner member			
With connections which are supported laterally, for example, flanges of a beam which are supported by the web, eccentricity may be neglected			
3.1 Parent metal adjacent to, or weld metal in, full penetration butt joints welded from both sides between plates of equal width and thickness or where differences in width and thickness are machined to a smooth transition not steeper than 1 in 4		Note that this includes butt welds that do not completely traverse the member, such as circular welds used for inserting infilling plates into temporary holes	
(a) With the weld overfill dressed flush with the surface and with the weld proved free from significant defects by nondestructive examination	C	The significance of defects should be determined with the aid of specialist advice and/or by the use of fracture mechanics analysis. The NDT technique must be selected with a view to ensuring the detection of such significant defects	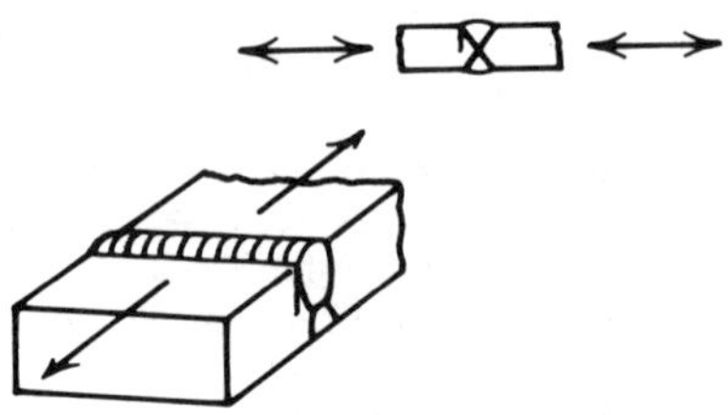
(b) With the welds made in the shop, either manually or by an automatic process other than submerged-arc, provided all runs are made in the flat position	D	In general, welds made positionally, or on site, or by the submerged-arc process tend to have a poor reinforcement shape, from the point of view of fatigue strength. Hence such welds are downgraded from D to E	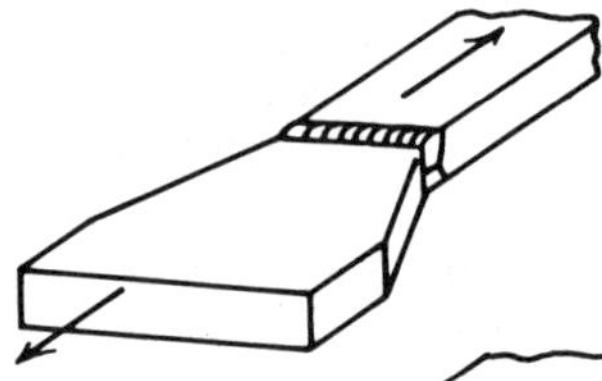
(c) Welds made other than in (a) or (b)	E	In both (b) and (c) the corners of the cross section of the stressed element at the weld toes should be dressed to a smooth profile Note that step changes in thickness are, in general, not permitted under fatigue conditions, but that where the thickness of the thicker member is not greater than 1.15 × the thickness of the thinner member, the change can be accommodated in the weld profile without any machining. Step changes in width lead to large reductions in strength (see joint Type 3.3)	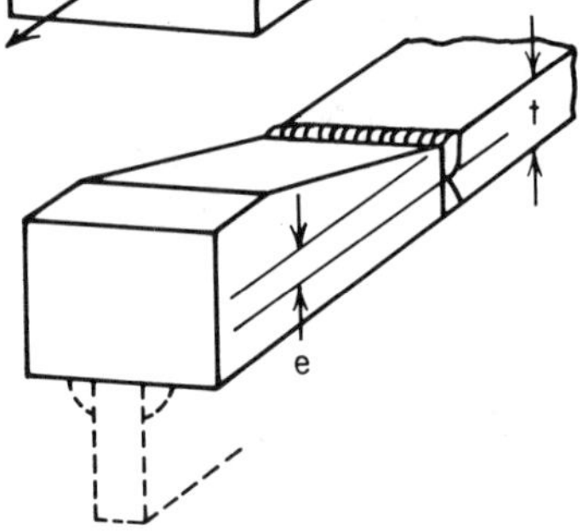
3.2 Parent metal adjacent to, or weld metal in, full penetration butt joints made on a permanent backing strip between plates of equal width and thickness or with differences in width and thickness machined to a smooth transition not steeper than 1 in 4	F	Note that if the backing strip is fillet welded or tack welded to the member the joint could be reduced to Class G (joint Type 4.2)	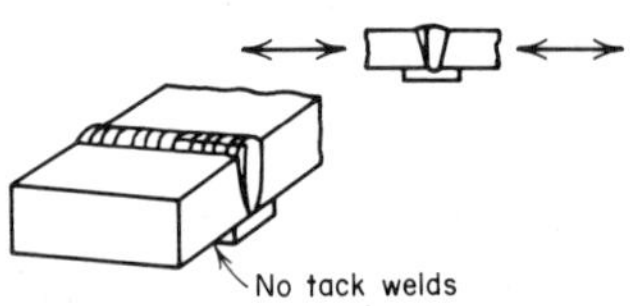
3.3 Parent metal adjacent to, or weld metal in, full penetration butt welded joints made from both sides between plates	F2	Step changes in width can often be avoided by the use of shaped transition plates, arranged so as to enable butt	

TABLE 3.3 JOINT CLASSIFICATIONS[a] (Continued)

Type Number, Description, and Notes on Mode of Failure	Class	Explanatory Comments	Examples, Including Failure Modes
of unequal width, with the weld ends ground to a radius not less than 1.25 times the thickness *t*		welds to be made between plates of equal width Note that for this detail the stress concentration has been taken into account in the joint classification	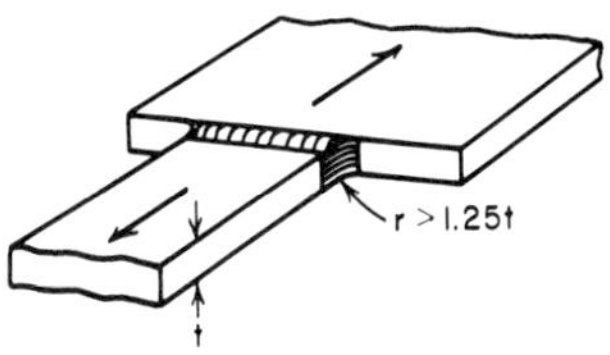

TYPE 4 WELDED ATTACHMENTS ON THE SURFACE OR EDGE OF A STRESSED MEMBER

Notes on potential modes of failure

When the weld is parallel to the direction of the applied stress, fatigue cracks normally initiate at the weld ends, but when it is transverse to the direction of stressing they usually initiate at the weld toe; for attachments involving a single, as opposed to a double, weld cracks may also initiate at the weld root. The cracks then propagate into the stressed member. When the welds are on or adjacent to the edge, of the stressed member the stress concentration is increased and the fatigue strength is reduced; this is the reason for specifying an 'edge distance' in some of these joints (see also note on edge distance in joint Type 2)

Type Number, Description, and Notes on Mode of Failure	Class	Explanatory Comments	Examples, Including Failure Modes
4.1 Parent metal (of the stressed member) adjacent to toes or ends of bevel-butt or fillet welded attachments, regardless of the orientation of the weld to the direction of applied stress, and whether or not the welds are continuous round the attachment		Butt welded joints should be made with an additional reinforcing fillet so as to provide a similar toe profile to that which would exist in a fillet welded joint	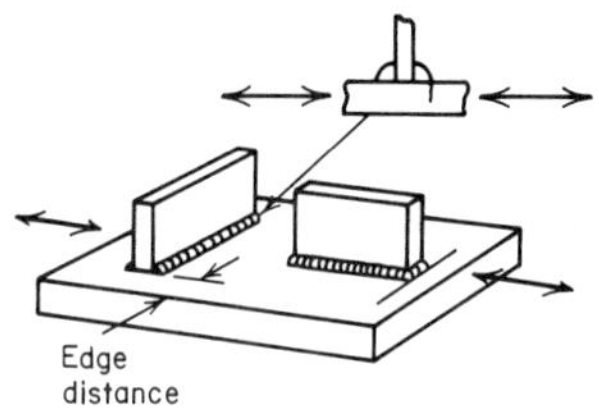
(a) With attachment length (parallel to the direction of the applied stress) $\leq$ 150 mm and with edge distance $\geq$ 10 mm	F	The decrease in fatigue strength with increasing attachment length is because more load is transferred into the longer gusset, giving an increase in stress concentration	
(b) With attachment length (parallel to the direction of the applied stress) > 150 mm and with edge distance $\geq$ 10 mm	F2		
4.2 Parent metal (of the stressed member) at the toes or the ends of butt or fillet welded attachments on or within 10 mm of the edges or corners of a stressed member, and regardless of the shape of the attachment	G	Note that the classification applies to all sizes of attachment. It would therefore include, for example, the junction of two flanges at right angles. In such situations a low fatigue classification can often be avoided by the use of a transition plate (see also joint Type 3.3)	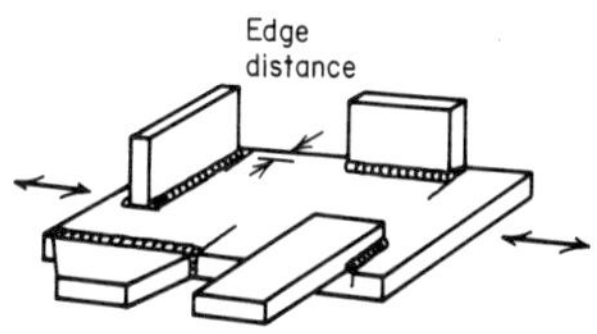
4.3 Parent metal (of the stressed member) at the toe of a butt weld connecting the stressed member to another member slotted through it		Note that this classification does not apply to fillet welded joints (see joint Type 5.1b). However it does apply to loading in either direction (L or T in the sketch)	
(a) With the length of the slotted-through member, parallel to the direction of the applied stress, $\leq$ 150 mm and with edge distance $\geq$ 10 mm	F		
(b) With the length of the slotted-through member, parallel to the direction of the	F2		

TABLE 3.3 JOINT CLASSIFICATIONS[a] (Continued)

Type Number, Description, and Notes on Mode of Failure	Class	Explanatory Comments	Examples, Including Failure Modes
applied stress, > 150 mm and with edge distance ≥ 10 mm			
(c) With edge distance < 10 mm	G		
TYPE 5 LOAD-CARRYING FILLET AND T BUTT WELDS			
Note on potential modes of failure			
Failure in cruciform or T joints with full penetration welds will normally initiate at the weld toe, but in joints made with load-carrying fillet or partial penetration butt welds cracking may initiate either at the weld toe and propagate into the plate or at the weld root and propagate through the weld. In welds parallel to the direction of the applied stress, however, weld failure is uncommon; cracks normally initiate at the weld end and propagate into the plate perpendicular to the direction of applied stress. The stress concentration is increased, and the fatigue strength is therefore reduced, if the weld end is located on or adjacent to the edge of a stressed member rather than on its surface			
5.1 Joint description Parent metal adjacent to cruciform joints or T joints (member marked X in sketches)		Member Y can be regarded as one with a non-load-carrying weld (see joint Type 4.1). Note that in this instance the edge distance limitation applies	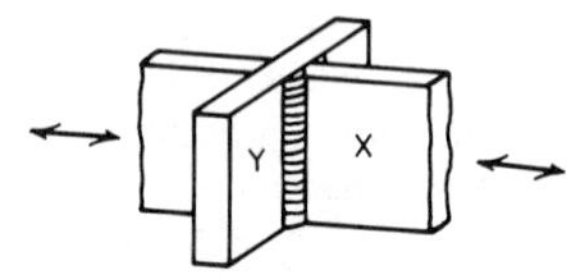
(a) Joint made with full penetration welds and with any undercutting at the corners of the member dressed out by local grinding	F		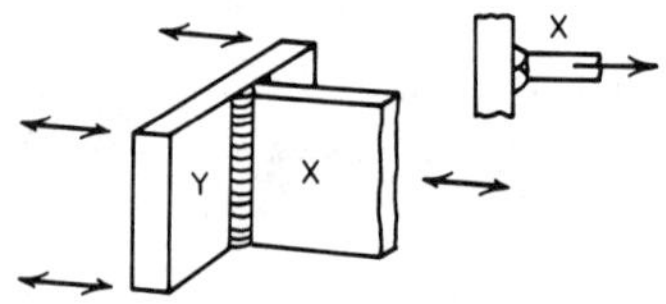
(b) Joint made with partial penetration or fillet welds with any undercutting at the corners of the member dressed out by local grinding	F2	In this type of joint, failure is likely to occur in the weld throat unless the weld is made sufficiently large (see joint Type 5.4)	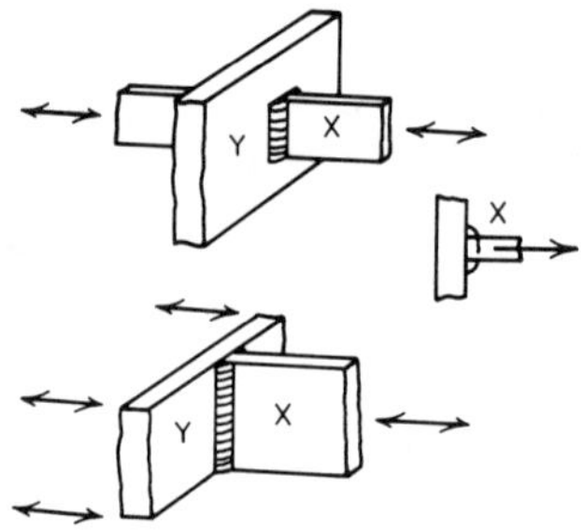
5.2 Parent metal adjacent to the toe of load-carrying fillet welds which are essentially transverse to the direction of applied stress (member X in sketch)		The relevant stress in member X should be calculated on the assumption that its effective width is the same as the width of member Y	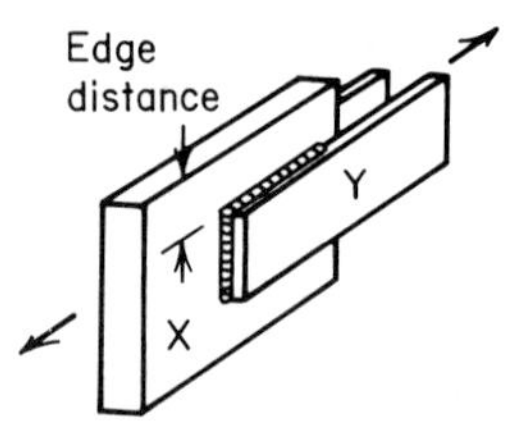
(a) Edge distance ≥ 10 mm	F2	These classifications also apply to joints with longitudinal welds only	
(b) Edge distance < 10 mm	G		
5.3 Parent metal at the ends of load-carrying fillet welds which are essentially parallel to the direction of applied stress, with the weld end on plate edge (member Y in sketch)	G		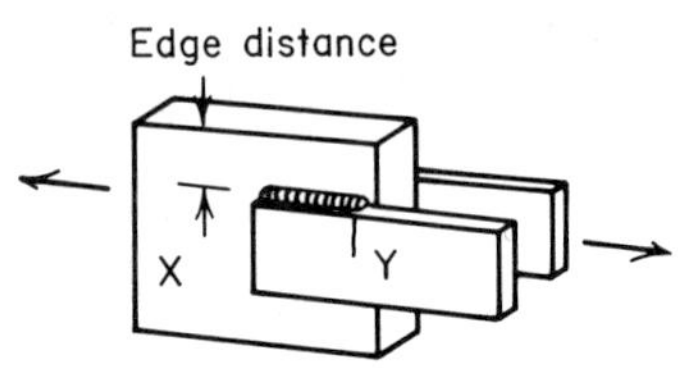
5.4 Weld metal in load-carrying joints made with fillet or partial penetration welds, with the welds either transverse or parallel to the direction of applied stress (based on nominal shear stress on the minimum weld throat area)	W	This includes joints in which a pulsating load may be carried in bearing, such as the connection of bearing stiffeners to flanges. In such examples the welds should be designed on the assumption that none of the load is carried in bearing	

TABLE 3.3 JOINT CLASSIFICATIONS[a] (Continued)

Type Number, Description, and Notes on Mode of Failure	Class	Explanatory Comments	Examples, Including Failure Modes

TYPE 6 DETAILS IN WELDED GIRDERS

Notes on potential modes of failure

Fatigue cracks generally initiate at weld toes and are especially associated with local stress concentrations at weld ends, short lengths of return welds, and changes of direction. Concentrations are enhanced when these features occur at or near an edge of a part (see notes on joint Type 4)

General comment

Most of the joints in this section are also shown, in a more general form, in joint Type 4; they are included here for convenience as being the joints which occur most frequently in welded girders

Type Number, Description, and Notes on Mode of Failure	Class	Explanatory Comments	Examples, Including Failure Modes
6.1 Parent metal at the toe of a weld connecting a stiffener, diaphragm, etc. to a girder flange		Edge distance refers to distance from a free, i.e. unwelded, edge. In this example, therefore, it is not relevant	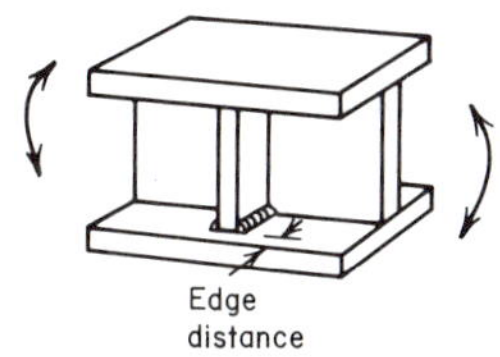
(a) Edge distance ≥ 10 mm (see joint Type 4.2)	F	as far as the (welded) edge of the web plate is concerned. For reason for edge	
(b) Edge distance < 10 mm	G	distance see note on joint Type 2	
6.2 Parent metal at the end of a weld connecting a stiffener, diaphragm, etc. to a girder web in a region of combined bending and shear	E	This classification includes all attachments to girder webs	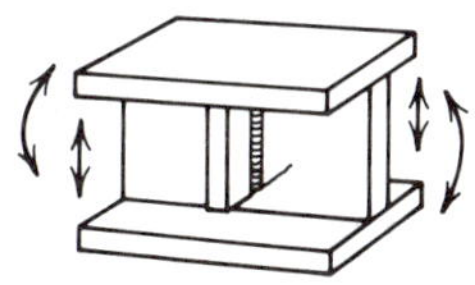
6.3 Parent metal adjacent to welded shear connectors			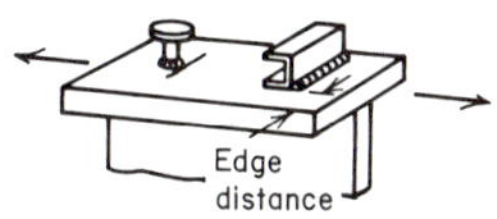
(a) Edge distance ≥ 10 mm	F		
(b) Edge distance < 10 mm (see Type 4.2)	G		
6.4 Parent metal at the end of a partial length welded cover plate, regardless of whether the plate has square or tapered ends and whether or not there are welds across the ends	G	This Class includes cover plates which are wider than the flange. However, such a detail is not recommended because it will almost inevitably result in undercutting of the flange edge where the transverse weld crosses it, as well as involving a longitudinal weld terminating on the flange edge and causing a high stress concentration	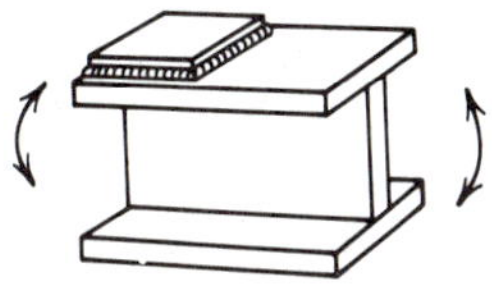
6.5 (a) Parent metal adjacent to the ends of discontinuous welds, e.g. intermittent web/flange welds, tack welds unless subsequently buried in continuous runs	E	This also includes tack welds which are not subsequently buried in a continuous weld. This may be particularly relevant in tack welded backing strips	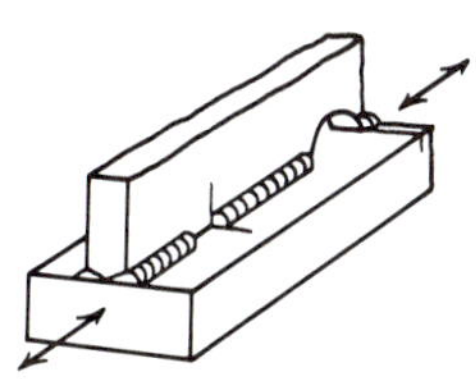
(b) Same as (a) but adjacent to cope holes	F	Note that the existence of the cope hole is allowed for in the joint classification; it should not be regarded as an additional stress concentration	

[a]This table is copied from Welding Institute Research Bulletin, Vol. 17, No. 5, May 1976, with the kind permission of the Welding Institute and of Cambridge University Press.

TABLE 3.4 COEFFICIENTS OF S-N CURVES (UNITS N/mm²)

Class	$\log_{10} C$	m
B	15.01	4.0
C	13.63	3.5
D	12.18	3.0
E	12.02	3.0
F	11.80	3.0
F_2	11.63	3.0
G	11.39	3.0
W	11.20	3.0
X	14.57	4.1

This set of *S-N* curves applies to stiffened plating and other portions of structure which constitute the hull girder and which contain normal amounts of welding. The curves do not apply to complicated joints such as the intersection of tubular cylinders, or to any portion of structure containing a high degree or high density of welding. For local structural details Ref. 9 is recommended.

It should also be noted that these *S-N* curves are for uniaxial stress. If there is a large wave-induced shear stress at the ship section under consideration the principal stress should be used instead of the longitudinal stress. Also these curves presume that efficient protection is provided against corrosion.

Finally, fatigue life is also influenced to some extent by the value of the mean stress and by temperature. The curves presented here are for a moderate level of mean stress (i.e., stillwater bending stress) and a normal seasonal range of ocean temperature. Other conditions would require either a different set of *S-N* curves or some type of correction factor.

Example

We next present a sample calculation of the minimum value of section modulus which is required to prevent hull girder fatigue. For purposes of demonstration the principal particulars of the ship are the same as for the SL-7 containership: length between perpendiculars 268 m, beam 32 m, and block coefficient 0.54. Let us assume that a value of 1.2 has been established for γ_F, and that the lowest class of welding in the primary hull girder structure is class D. From Table 3.1 the value of k for the SL-7 is 1.2. In the design of a new ship the value would either be estimated conservatively as 1.3 or a more precise value would be obtained from a ship motion and loads computer program. Likewise the characteristic value of wave-induced bending moment would be obtained either from design formulas such as (3.5.1) or from a program. For this example we use the former, which gives

$$(M_{w,c})_{\text{sag}} = 125(268)^2(32)(0.74)(13) = 2764 \text{ MNm}$$

$$(M_{w,c})_{\text{hog}} = 165(268)^2(32)(0.54)^2(13) = 1438 \text{ MNm}$$

From Fig. 3.13, for class D at $N = 10^8$, $S_N = 25$ MPa. From Table 3.4 $m = 3$ and hence from Table 3.2 the random load factor is $\xi = 6.75$. Hence from (3.6.13) the minimum required value of section modulus is

$$Z_{\min} = \frac{(4202)1.2}{(6.75)25} = 29.9 \text{ m}^3$$

In the SL-7 ships the section modulus in the midships region is generally in the range from 30 to 40 m³, depending on the particular cross section. Thus there are some cross sections which are at the lower limit of Z based on fatigue,* and any further decrease would be inadvisable. In the rationally-based design process of Fig. 2.19 such a decrease would be prevented by the fatigue constraint (labeled ⑥ in the figure) which corresponds to (3.6.13).

Section Modulus Reduction Away from Amidships

Because the ship will meet a wide variety of stillwater and wave loads during its lifetime, the envelope curve of all of the various bending moment distributions is relatively flat throughout the amidships region. Therefore, the required value of section modulus needs to be maintained at almost its full magnitude for some appreciable length forward and aft of amidships. For simplicity, and in order to avoid having to calculate the envelope curve, it is common procedure to maintain the full value of section modulus throughout the middle 40% or so of the ship length. Outside of this length the scantlings may be reduced but the reduction must be gradual, for example, to one-half of the full value at 0.15 L from the ends.

If the bending moment envelope curve is used, items included in the hull girder section amidships

*For the assumed values of γ_F, weld class and wave-induced bending moment.

may be reduced in size or eliminated as the bending moment decreases, but always in a gradual manner and always such that the maximum permissible stress is not exceeded.

Allowable Area for Section Modulus

In general, the following items are included in the calculation of the section modulus, provided they are longitudinally continuous:

Deck plating (strength deck and other effective decks).
Shell and inner-bottom plating.
Longitudinal bulkheads and girders.
Longitudinal stiffeners.

Some items may be partly effective, depending on their length, manner of attachment, and longitudinal stiffness:

1. Length: To be effective a member must extend over a sufficient length such that some portion of the longitudinal stress field enters the member. An approximate rule-of-thumb is that the longitudinal stress can diffuse at a rate of 1 in 2 in the web and about 1 in 4 in the flanges, that is, angles of divergence of about 30 and 14°. Thus, as shown in Fig. 3.14*a*, a longitudinal bulkhead would become fully effective in its midlength region only if it extended over a length of at least 2 D, where D is the depth of the ship, and at locations closer to its ends it would be only partly effective. Similarly, if at a certain point along the ship length the height of the side shell increases by H, as in a full width superstructure, the additional side plating only becomes fully effective after a length of 2 H. An even longer length is required for the entire superstructure to become fully effective (see Fig. 3.14*c*. This same rationale can also be used to estimate and allow for the ineffective plating immediately forward or aft of an opening. As shown in Figure 3.14*d*, the material which lies within the "shadow area" subtended by the two 14° rays is ineffective. This shadow area rule is useful for dealing with openings that occur close to, but not quite within, the same transverse plane. It also indicates that the material between hatches is seldom very effective; for example, if the spacing of the hatches is equal to their width, only one-fourth of such plating would be effective. Thus if there were other openings, such as sideports, located

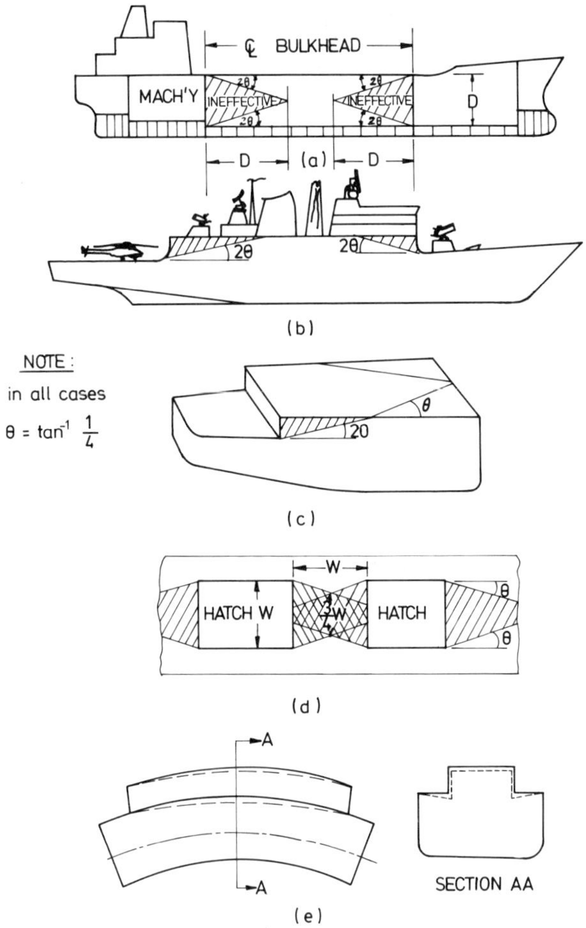

Figure 3.14 Effective longitudinal material.

at other points in the ship's transverse plane between the hatches, it could happen that this plane would be the critical one.

2. Manner of attachment: The attachment must be capable of transmitting shear force without undergoing longitudinal slip. In welded structures this is generally not a problem.

3. Longitudinal stiffness: To be fully effective a member must have the same (or greater) Young's modulus as the rest of the hull girder and must have parity of strain; that is, it must undergo the same elongation or compression as is occurring in the rest of the hull girder at that same elevation. Thus there are two further ways in which a member might be only partly effective: by having a smaller material stiffness (e.g., aluminum) or by deflecting vertically so as to adopt a larger radius of curvature, thereby reducing its longitudinal strain and thus "shirking" part of the stress load it would otherwise carry. This phenomenon can

occur, for example, in superstructures that are not full width, as illustrated in Figure 3.14*e*. The manner of dealing with sections containing more than one material (i.e., different Young's moduli) is taken up in Section 3.6. The question of superstructure is discussed in Section 3.9.

In most cases the critical hull girder cross section will be that section which contains the least amount of effective material—that is, the section containing the largest hatches or other openings—but it also depends on the distance of these from the neutral axis. If there is any doubt, the section modulus should be calculated for all of the potentially critical sections.

In general, the *net* sectional areas of longitudinal members are to be used in the section modulus calculation, except that small isolated openings need not be deducted provided the openings and the shadow area breadths of other openings in any one transverse section do not reduce the section modulus by more than a few percent.

Figure 3.15 is an example of a longitudinal structural section showing (only) the longitudinally effective material. The two quantities to be calculated are the position of the neutral axis of the section and the moment of inertia I about the neutral axis. If done by hand the calculation is best carried out in tabular form, as shown in Table 3.5. The dimensions and the vertical height of the centroid of each item are entered and then the area and the first and second moments of area of each item are calculated about some axis zz. The choice of the axis zz is arbitrary but the keel position is probably the most appropriate because the location of the neutral axis is usually quoted as height above the keel. To obtain the moment of inertia of an item about the zz axis it is necessary to add to ah^2 the moment of

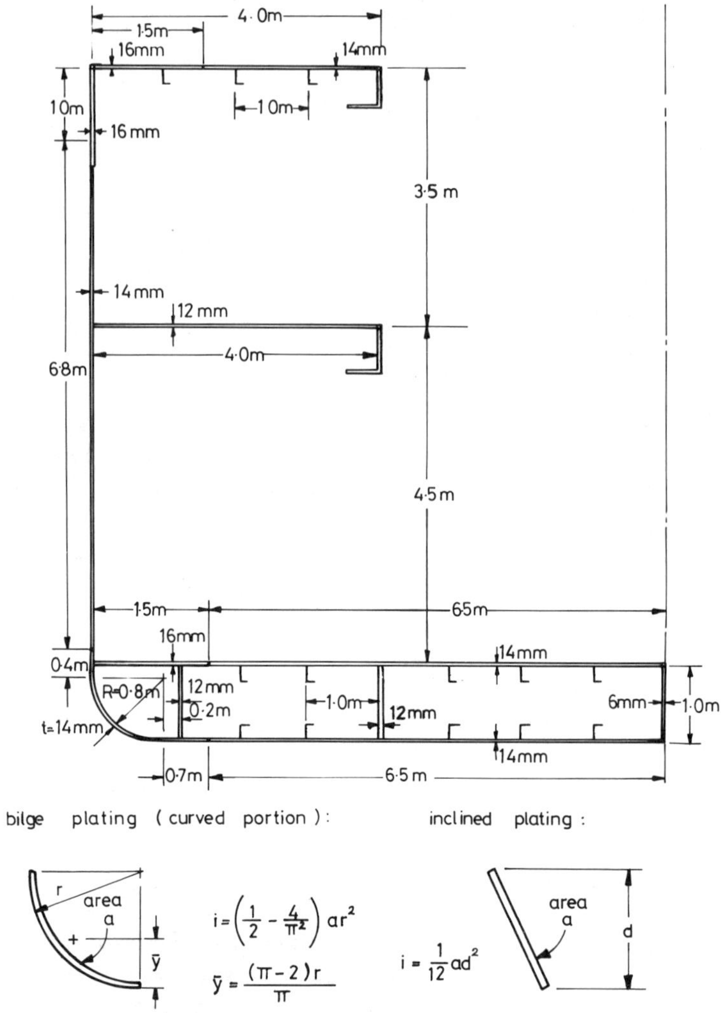

Figure 3.15 Example of longitudinally effective material.

TABLE 3.5 SECTION MODULUS CALCULATION

Item	Scantlings m × mm	Area $a(\mathrm{m}^2)$	Height $h(\mathrm{m})$	Moment ah	2nd Moment ah^2	Local 2nd Moment $i(\mathrm{m}^4)$
Strength Deck Plating	2.5 × 14	0.0350	9.0	0.3150	2.835	
S.D. stringer plate	1.5 × 16	0.0240	9.0	0.2160	1.944	
S.D. longitudinals (mm)	W160 × 14; F40 × 14	0.0084	8.9	0.0748	0.666	
Sheer strake	1.0 × 16	0.0160	8.5	0.1360	1.156	0.001
Side plating	7.2 × 14	0.1008	4.4	0.4435	1.951	0.435
2nd deck plating	4.0 × 12	0.0480	5.5	0.2640	1.452	
Bilge (curved portion)	$R = 0.8$; $t = 14$	0.0176	0.29	0.0051	0.001	0.001
Inner Bottom plating	6.5 × 14	0.0910	1.0	0.0910	0.091	
I.B. margin plate	1.5 × 16	0.0240	1.0	0.0240	0.024	
I.B. longitudinals (mm)	W200 × 10; F66 × 15	0.0150	0.86	0.0129	0.011	
Side girders	1.0 × 12	0.0240	0.5	0.0120	0.006	0.002
Center girder ($\frac{1}{2}$)	1.0 × 6	0.0060	0.5	0.0030	0.001	0.001
Bottom plating	7.2 × 14	0.1008	0	0	0	
Bottom longitudinals (mm)	W200 × 10; F66 × 15	0.0150	0.14	0.0021	0.000	
Upper hatch side girder	W0.5 × 25; F0.4 × 25	0.0225	8.64	0.1944	1.680	
Lower hatch side girder	W0.5 × 25; F0.4 × 25	0.0225	5.14	0.1157	0.595	
Totals (for half section)		0.5706		1.9095	12.413	0.440

Height of neutral axis: $h_{NA} = \dfrac{\Sigma a_i h_i}{\Sigma a_i} = \dfrac{1.9095}{0.5706} = 3.346\ \mathrm{m}$

$$\frac{1}{2}I_{zz} = 12.413 + 0.440 \qquad = 12.853\ \mathrm{m}^4$$

$$\text{Parallel axis term} = -(\Sigma a_i)h_{NA}^2 \qquad = -6.390\ \mathrm{m}^4$$

$$\frac{1}{2}I = 6.463\ \mathrm{m}^4$$

Full values: $A = 1.142\ \mathrm{m}^2$

$I = 12.93\ \mathrm{m}^4$

$$Z_D = \frac{I}{h_D - h_{NA}} = \frac{12.93}{9.000 - 3.346} = 2.287\ \mathrm{m}^3$$

$$Z_K = \frac{I}{h_{NA}} = \frac{12.93}{3.346} = 3.864\ \mathrm{m}^3$$

inertia of the item about its own neutral axis, which is denoted as i. This is entered in the last column. Most of the items can be treated as rectangles so that the moment of inertia is simply $\frac{1}{12}ad^2$ where a is the area and d the depth. Note that, as shown in Fig. 3.15, this expression is also valid for inclined sections of plating.

It will be noted that there are only a few entries in the last column; this is because the moment of inertia of what may be called "horizontal" material about its own neutral axis is sufficiently small to be negligible. If, for example, a panel of plating is of thickness t and width 100 t, the value of i is $10^6 t^4$ if the panel is vertical but is only $10^2 t^4$ if it is horizontal; since the values are to be summed the latter quantity is clearly negligible.

The distance of the neutral axis above the keel is

$$h_{NA} = \frac{\Sigma a_i h_i}{\Sigma a_i}$$

Finally, from the parallel axis theorem, the moment of inertia about the neutral axis is

$$I = I_{zz} - Ah_{NA}^2$$

where I = moment of inertia about the neutral axis
I_{zz} = moment of inertia about some axis zz
A = total area
and h_{NA} = distance from zz to neutral axis.

As shown in the example of Table 3.5, the calculation is usually carried out for one side of the ship only, and therefore the results have to be multiplied by two.

For simple beams, consisting of two flanges and a web, an explicit formula for moment of inertia is available; this is presented in Section 8.3.

Combined Vertical and Horizontal Bending

The calculation just described assumes that the ship is upright and that the bending moment is in the ship's vertical plane. This is referred to as *vertical bending*. If the ship is at an angle of heel due to rolling, it will also be subjected to *horizontal bending*, that is, a bending moment M_y acting in the ship's horizontal plane (see Figure 3.16). For this bending moment the neutral axis is the ship's vertical centerline.

Let us first take the case in which M_y is entirely due to inclination of the vessel, say to an angle θ. In this case M_y and M_z are directly related, being components of the total bending moment M (which acts in the *true* vertical plane):

$$M_y = M \sin\theta$$

$$M_z = M \cos\theta$$

If y and z are the coordinates of any point in the cross section and I_{NA} and I_{CL} are the moments of inertia about the horizontal axis in the upright condition and about the centerline respectively, then the stress at (y,z) is

$$\sigma = \sigma_V + \sigma_H = \frac{M\cos\theta}{I_{NA}}y + \frac{M\sin\theta}{I_{CL}}z \qquad (3.6.14)$$

When $\sigma = 0$ it follows that

$$\frac{y\cos\theta}{I_{NA}} + \frac{z\sin\theta}{I_{CL}} = 0$$

or

$$y = -\frac{I_{NA}}{I_{CL}}\tan\theta\, z$$

This gives the equation of the neutral axis in the inclined condition. It will be seen that this is inclined to the neutral axis in the upright condition at an angle ψ given by

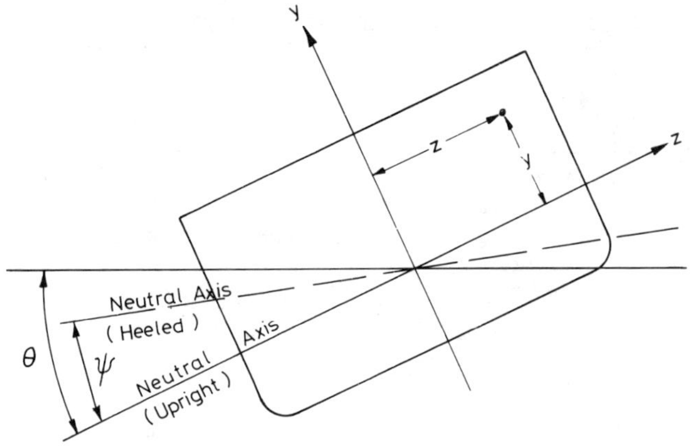

Figure 3.16 Neutral axis with simultaneous horizontal and vertical bending.

$$\tan\psi = \frac{y}{z} = -\frac{I_{NA}}{I_{CL}}\tan\theta$$

If the vessel were such that $I_{NA} = I_{CL}$, then $\tan\psi = -\tan\theta$, and the neutral axis would remain horizontal. In general this is not so, I_{CL} being larger than I_{NA}, and so the neutral axis is inclined to the horizontal. This is a well-known feature of unsymmetric bending. In a ship there is one axis of symmetry, the centerline, and therefore I_{CL} and I_{NA} are the maximum and minimum moments of inertia respectively.

Referring to (3.6.14), the angles of heel at which the greatest and least stresses occur may be obtained by putting $d\sigma/d\theta = 0$.

Hence,

$$\frac{d\sigma}{d\theta} = -\frac{M\sin\theta}{I_{NA}}y + \frac{M\cos\theta}{I_{CL}}z = 0$$

or

$$\tan\theta = \frac{z}{y}\frac{I_{NA}}{I_{CL}} = \frac{Z_{NA}}{Z_{CL}} \qquad (3.6.15)$$

The greatest and least stresses will also be associated with the maximum values of y and z and this means that these stresses will occur at the corners of the section. Figure 3.17 is a qualitative illustration of the stresses at the four corners of a ship section, for which the maximum and minimum stresses occur at a heel angle of about 30°.

In practice the horizontal and vertical bending moments are not so directly coupled and do not necessarily occur simultaneously. Their relationship varies

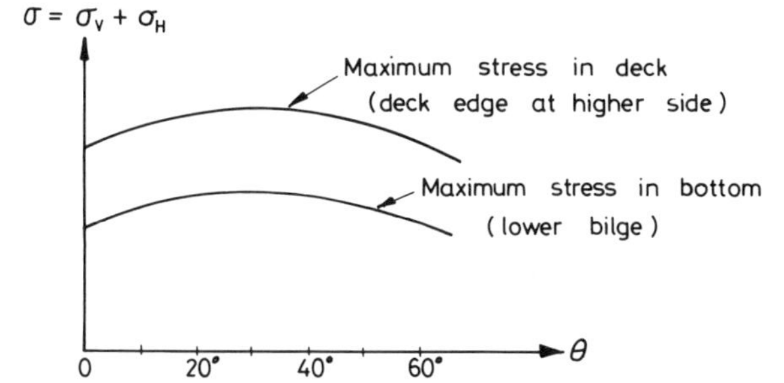

Figure 3.17 Variation of maximum total stress with angle of heel.

with different sea conditions and depends mainly on ship heading. The situation is summarized in Fig. 3.18 which shows the results of theoretical calculations of the characteristic value of wave bending stress in a 300-m tanker [12]. (We recall from Section 1.2 that the characteristic value is the extreme value corresponding to some specified probability of exceedence in the ship's lifetime.) The figure shows the stresses at the deck edge due to horizontal and vertical bending, $\hat{\sigma}_H$ and $\hat{\sigma}_V$, and also the total stress, $\hat{\sigma}$, over a range of ship headings. This type of calculation, and the underlying statistical principles, is discussed in Chapter 4. But to interpret Fig. 3.18 it should be mentioned here that the degree of simultaneity or correlation between two random variables may be measured by a correlation coefficient ε. If the variables are independent then $\varepsilon = 0$; if they are completely linked, such that they always occur simultaneously, then $\varepsilon = 1$. Horizontal and vertical bending stresses are random variables and for this reason their characteristic values cannot be just added together, as is done for deterministic quantities. Instead, the combined characteristic stress depends on the extent to which the variables are correlated. If $\hat{\sigma}_H$ and $\hat{\sigma}_V$ denote the individual values, it can be shown that the magnitude of the combined stress $\hat{\sigma}$ is given by

$$\hat{\sigma} = \sqrt{\hat{\sigma}_H^2 + 2\,\varepsilon\,\hat{\sigma}_H\,\hat{\sigma}_V + \hat{\sigma}_V^2} \qquad (3.6.16)$$

Thus, if the bending moments are independent their stresses will be uncorrelated ($\varepsilon = 0$) and the magnitude of the long-term total stress is

$$\hat{\sigma} = \sqrt{\hat{\sigma}_H^2 + \hat{\sigma}^2{}_V}$$

If on the other hand the bending moments are perfectly correlated, such that they always coincide, then $\varepsilon = 1$ and the total long-term stress is simply the sum of the two separate values:

$$\hat{\sigma} = \hat{\sigma}_H + \hat{\sigma}_V$$

Figure 3.18 presents the values of $\hat{\sigma}_H$ and $\hat{\sigma}_V$, their correlation coefficient ε, and the total stress $\hat{\sigma}$, as given by (3.6.16), plotted for a range of ship headings. For beam seas and quartering seas ε is nearly unity, indicating that $\hat{\sigma}_H$ and $\hat{\sigma}_V$ are closely correlated. This is because for beam seas and quartering seas the horizontal bending is due mainly to heeling of the ship. Likewise the largest value of $\hat{\sigma}_H$ occurs with beam seas because rolling is usually a maximum for this heading. Thus the horizontal and vertical bending moments are, for the most part, components of the same overall bending moment and so their respective stresses are closely correlated. Therefore the total stress calculated from (3.6.16) is nearly the same as the direct sum of $\hat{\sigma}_H$ and $\hat{\sigma}_V$.

For head seas the situation is the reverse; here the horizontal stress has its smallest value and it is nearly independent of the vertical stress. This is because there

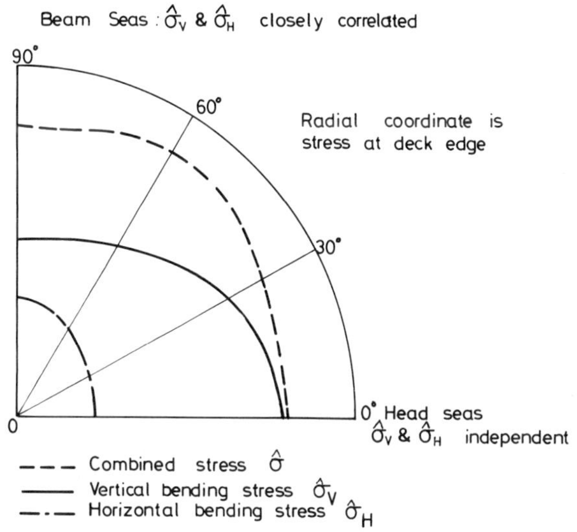

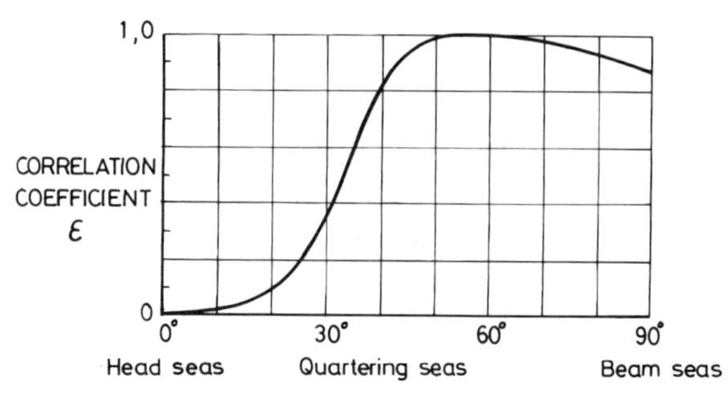

Figure 3.18 Combination of long term vertical and horizontal bending stress.

is less rolling and much of the horizontal bending is caused by irregular seas which give different load distributions along the two sides of the ship. The total stress as given by (3.6.16) is only slightly greater than the vertical stress, and for this heading $\hat{\sigma}_H$ could be ignored.

Figure 3.18 shows that as the heading angle increases, $\hat{\sigma}_V$ diminishes and $\hat{\sigma}_H$ increases. Also, because of the change in their correlation the total stress is approximately constant, with a peak value which is about 12% larger than the maximum (head seas) value of σ_V. This indicates that for design purposes it is sufficient (and conservative) to calculate the two stress values separately, using characteristic values of wave bending moment such as those given by (3.5.1) and (3.5.3), and then to simply add them together. Note that this gives the stress at the deck edge; the stress in the deck at the centerline would be $\hat{\sigma}_V$ because $\hat{\sigma}_H$ is always zero at this location.

Dissimilar Materials: the Composite Beam Technique

In some cases the hull girder may consist of a combination of materials having different values of Young's modulus, E. There are two quite different situations which require completely different approaches, depending on whether the vertical variation of hull girder bending strain in the dissimilar portion has the same linear distribution as in the main portion, $\varepsilon = y/R$. If it does then it is an integral part of the hull girder and the dissimilarity can be dealt with by the "composite beam" technique presented in this section. Unfortunately, the most common instance of material dissimilarity in a hull girder—an aluminum superstructure—does not usually satisfy this requirement. The requirement would only be satisfied in the midlength region of a long, full-width superstructure in which the sides were a direct continuation of the sides of the hull (such as in a car carrier or a large ferry), and providing that there were no expansion joints. This combination is rare because without expansion joints such a superstructure would incur the largest strains in the section, and the scantlings required for adequate strength would partly offset the weight savings of the aluminum. (Nevertheless there is a net savings, and this is important in naval design, but to properly exploit this savings requires a more accurate analysis of both hull and superstructure. This is discussed in Section 3.9.)

As its name implies, the composite beam technique is intended for beams of solid cross section which are fabricated of two or more materials, such as a reinforced concrete beam. Its main application in this text is in Section 14.3, dealing with the ultimate strength of longitudinally stiffened panels. In such a panel each stiffener and its associated width of plating together constitute an unsymmetric beam-column. If this member is loaded up to collapse in such a way that the plate flange is in compression, this flange will have a smaller axial stiffness than the stiffener flange because of plate buckling, and a convenient way of allowing for this is to regard the plate as being of a different material, with a smaller average value of elastic modulus (see Figs. 14.6 and 14.7).

The reason for introducing the composite beam technique at this point is to establish what the stress distribution would be in an aluminium superstructure if it were fully effective, so as to be able to compare other more realistic cases against this idealized case. Of course the technique has direct application to a hull girder if the dissimilar portion is part of the cross section of the main hull; for example, a deck constructed of titanium or other special material such as carbon fiber composites, or a concrete barge with some longitudinally continuous steel portions.

We saw earlier that the fundamental assumption of beam bending theory is that plane cross sections remain plane. We now assume that the dissimilar portion is an integral part of the hull girder, such that the bending strain ε is still linearly proportional to the distance y from the neutral axis; that is

$$\varepsilon = \frac{y}{R} \tag{3.6.17}$$

The corresponding value of the bending stress at any point is $\sigma = E\varepsilon$, and hence it is clear that the value of the stress depends on the local value of E. The next step in the derivation of elementary beam theory is the requirement that there is no net axial force in the beam:

$$\int \sigma\, dA = \int E \frac{y}{R}\, dA = \frac{1}{R} \int Ey\, dA = 0$$

If there are two or more portions of the beam which have different values of E (say E_i, i = 1, 2, . . .) then the integral must be evaluated separately for each portion. An alternative approach, which we shall adopt, is to choose one of the values of E_i (say E_1) as a reference value and relate the other values to it by means of a "transformation factor" T_i which is defined as

$$T_i = \frac{E_i}{E_1} \tag{3.6.18}$$

T_i is a variable which is a function of position within the beam cross section. Within any one material its value is constant, but the value changes abruptly when passing from one material to another. In the reference material its value is unity.

In terms of T_i the foregoing integral becomes

$$\frac{E_1}{R} \int T_i y \, dA = 0$$

or simply

$$\int T_i y \, dA = 0$$

This equation corresponds to that for homogeneous beams which specifies that the neutral axis passes through the centroid of the section. We may preserve this interpretation for the composite beam if we regard the transformation factor as being applied to the *area* dA instead of to E_1. That is, we consider the beam to be of only one material, with modulus E_1, and we account for the nonhomogeneity by transforming the area of the material. The area of each portion is shrunk or expanded according to the local value of T_i, and this results in a distorted or "transformed" section. The process is illustrated in Fig. 3.19 for a beam of solid section and also for a thin-walled beam.

This transformation allows us to calculate the location of the neutral axis of the composite section: it is the horizontal centroidal axis of the transformed section.

It is important to note that the shrinking or expanding must be done in such a way that the vertical position (i.e., the value of y) of each element dA is unchanged. Hence for solid sections it must be done horizontally, as illustrated in Fig. 3.19. For thin-walled sections the transformation can always be applied to the thickness, regardless of whether or not the wall is oriented vertically. This is advantageous, since it reduces the visual or conceptual distortion of the section.

The next step in our review of elementary beam theory is the requirement of equilibrium between the external bending moment acting on the section and the moment of the internal stress forces.

$$M = \int \sigma y \, dA = \frac{1}{R} \int E y^2 \, dA$$

For homogeneous beams, E is brought outside of the integral and the integral is then defined as the "moment of inertia" of the beam—a purely geometric quantity.

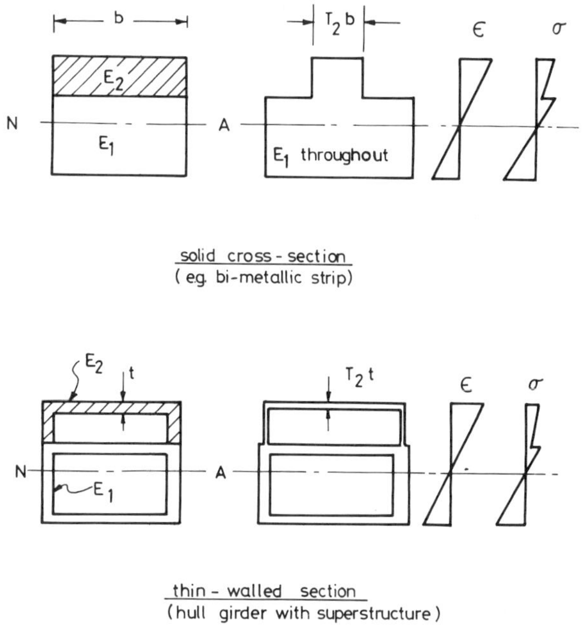

Figure 3.19 Section containing multiple materials.

This leads to the familiar "moment-curvature" relation for homogeneous beams

$$\frac{1}{R} = \frac{M}{EI}$$

For composite beams the modulus is a variable, E_i, and must be left within the integral. However, if we again replace the variable E_i by the product $T_i E_l$ then we may write

$$M = \frac{E_1}{R} \int T_i y^2 \, dA \tag{3.6.19}$$

As noted previously, we are regarding T_i as multiplying dA and giving either a reduced or enlarged area for each nonstandard portion of the cross section. Therefore the integral in (3.6.19) is simply the moment of inertia of the transformed section, which we shall denote as I_{tr}:

$$I_{tr} = \int T_i y^2 \, dA \tag{3.6.20}$$

This definition of I_{tr} means that the "moment-curvature" relation of elementary beam theory continues to hold, as if the beam were a homogeneous beam of modulus E_1, with I_{tr} in place of I:

$$\frac{1}{R} = \frac{M}{E_1 I_{tr}} \tag{3.6.21}$$

We may now use this expression to replace $1/R$ in (3.6.17):

$$\varepsilon = \frac{My}{E_1 I_{tr}} \tag{3.6.22}$$

As before, to find the bending stress σ at any point we multiply ε by E *at that point*: $E_i = T_i E_1$. This gives

$$\sigma = T_i \frac{My}{I_{tr}} \tag{3.6.23}$$

In comparing this to the familiar bending stress equation $\sigma = My/I$, we see that not only must we use I_{tr} in place of I but also, *for the stress formula* (and only for this formula) *we must insert the transformation factor T_i as an extra term* (in addition to its earlier use in creating the transformed section).

The tabular method for calculating neutral axis position and moment of inertia is especially well suited for dealing with composite beams. For each structural member which is made of a material other than the reference material, the member area in the left-hand column of Table 3.5 is simply transformed (multiplied) by the factor T_i for that material; thereafter the calculations proceed as for a homogeneous beam.

The greatest advantage of the transformed section approach is that it provides a way of adapting all of the standard homogeneous beam formulas for use in nonhomogeneous beams: the section is first transformed, neutral axis position and I_{tr} are calculated, and these properties, together with a uniform modulus E_1, are used as the properties of the beam. Thus in all of the formulas for the deflection, rotation, or stiffness of a beam, $E_1 I_{tr}$ is used in place of EI. As we shall see, these same properties can also be used in the formula for shear flow in beams. It is only the formulas for stress (bending stress and shear stress) that require special attention.

Changes to Section Modulus

The provision of the required section modulus is necessarily an iterative process, particularly if it is to be done efficiently, avoiding an overly large value. As the design progresses it will be necessary to add or subtract material in the hull girder cross section. A typical situation is shown in Fig. 3.20 in which an area a is added at a height y above the neutral axis of a vessel having moment of inertia I, total area A, and distances y_D and y_K to the deck and keel. The effect of

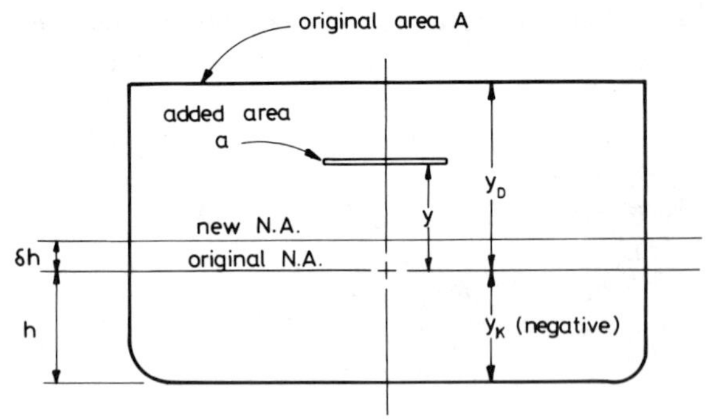

Figure 3.20 Effect of adding area. *Notes:* 1. y values are positive above the original N.A., and negative if below. 2. If material is *removed*, a is negative.

the addition is to raise the neutral axis a distance δh and to increase the moment of inertia to a value $I + \delta I$ (about the new neutral axis). The net effect on the deck and bottom can vary, depending on the location of a. For example, although the addition shown would reduce the deck stress (because it increases I and decreases y_D) it might increase the keel stress because y_K is increased and this might outweigh the increase in I. In addition to this complication, there is also the fact that the calculation of I is a lengthy computation and it would not be desirable to have to repeat it for every change of area, even within a computer-aided design program. Therefore it would be helpful to have some means for quickly estimating the effect of adding material. To this end we first present some expressions which indicate where the material must be added to avoid any increase in stress. Then we will derive a general expression for the stress derivative $\partial\sigma/\partial a$ which is particularly useful for computer-aided design applications.

For a given bending moment, the stress in the keel will not be increased if the section modulus I/y_K is not reduced; that is, if

$$\frac{I + \delta I}{y_K + \delta h} - \frac{I}{y_K} \geq 0$$

or

$$I y_K + y_K\, \delta I - I y_K - I\, \delta h > 0$$

or

$$\frac{\delta I}{I} \geq \frac{\delta h}{y_K} \tag{3.6.24}$$

With the addition of the area a, the rise of the neutral

axis is

$$\delta h = \frac{ay}{A + a} \qquad (3.6.25)$$

and the additional moment of inertia is

$$\delta I = ay^2 + i - (A + a)(\delta h)^2$$

If the material is added below the original neutral axis the value of y is negative. If the material is *removed* then the value of a is negative and also i is negative in the foregoing equation.

In what follows we assume that i, the local moment of inertia of the added area, is negligible. Substitution for δh from (3.6.25) gives

$$\delta I = ay^2 - \frac{a^2y^2}{A + a} = \frac{Aay^2}{A + a} \qquad (3.6.26)$$

There are two cases which require separate consideration: (1) the material is added within the existing cross section, that is, $y < y_D$ (or $|y| < |y_K|$ if the material is added below the neutral axis); and (2) the material is added above the existing section, $y > y_D$, such that the maximum stress now occurs in the added material. This would happen if, for example, a superstructure deck were to become longitudinally effective.

MATERIAL ADDED WITHIN THE SECTION ($y < y_D$)

As just noted, both δh and δI would act to decrease the deck stress. At the keel, the condition expressed in (3.6.24) becomes

$$\frac{Aay^2}{A\rho^2(A + a)} > \frac{ay}{y_K(A + a)}$$

where $\rho = \sqrt{\dfrac{I}{A}}$ = radius of gyration

This reduces to

$$y > \frac{\rho^2}{y_K} \qquad (3.6.27)$$

Thus, to achieve a decrease in stress at the keel as well as in the deck, the material must be added at a height greater than ρ^2/y_K from the neutral axis. For a thin-walled box girder of depth D and beam 2 D, ρ is approximately 0.44 D, and ρ^2/y_K is approximately 0.4 D. For ships, ρ is somewhat smaller because of internal decks, and so on; typical values are from 0.33 D to 0.4 D and values of ρ^2/y_K are generally in the range from 0.3 D to 0.45 D.

From (3.6.25) and (3.6.26) the new section modulus at the deck, Z'_D, is

$$Z'_D = \frac{I + \delta I}{y_D - \delta h} = \frac{I + \dfrac{Aay^2}{A + a}}{y_D - \dfrac{ay}{A + a}} \qquad (3.6.28)$$

and if a is small relative to A this may be approximated by

$$Z'_D \cong Z_D(1 + r) + Ayr \qquad (3.6.29)$$

where $$r = \frac{ay}{(A + a)y_D}$$

The corresponding expressions for the new section modulus at the keel are

$$Z'_K = \frac{I + \delta I}{y_K + \delta h} = \frac{I + \dfrac{Aay^2}{A + a}}{y_K + \dfrac{ay}{A + a}} \qquad (3.6.30)$$

and its simpler approximate form

$$Z'_K \cong Z_K(1 - r) + Ayr \qquad (3.6.31)$$

with $r = ay/[(A + a)y_K]$. It may be seen from these equations that to obtain maximum benefit, the new material should be added on the side where the maximum stress occurs (usually above the neutral axis) and, within the section, as far from the neutral axis as possible (i.e., usually in the strength deck). As mentioned earlier, if material is removed then a is negative in all of the foregoing equations.

MATERIAL ADDED ABOVE THE STRENGTH DECK ($y > y_D$)

If the material is added above the strength deck, the maximum stress now occurs in the added material. The distance of this material from the new position of the neutral axis is

$$y - \delta h = y - \frac{ay}{A + a} = \frac{Ay}{A + a}$$

and the condition that the section modulus to the new material should be not less than Z_D is

$$\frac{I + \delta I}{y - \delta h} - \frac{I}{y_D} \geq 0$$

which becomes, after substitution and rearrangement,

$$a \geq \frac{A\left(\dfrac{y}{y_D} - 1\right)}{\dfrac{y^2}{\rho^2} + 1} \qquad (3.6.32)$$

Therefore, if the material is added above the deck, there is a minimum area at any position y for which there will be a reduction in maximum stress. If the added material has less area than this minimum the stress in the added material will be greater than the original maximum value.

For the keel the new section modulus is as given in (3.6.30) or (3.6.31), and the required condition for the keel stress not to increase, (3.6.27), is usually satisfied because y is large.

Derivative of Hull Girder Stress

For some applications, notably in computer-aided design, a method is needed for calculating the incremental change in hull girder stress due to a change in effective material. If the area a_m of any member m is changed by a small amount δa_m the change in hull girder stress σ_l at any arbitrary location l is given by

$$\delta\sigma_l = \frac{\partial \sigma_l}{\partial a_m}\,\delta a_m \qquad (3.6.33)$$

and so if a general expression were available for the stress derivative, $\partial\sigma_l/\partial a_m$, the change in stress could be readily calculated. We shall therefore derive such an expression and in order for it to be more useful we shall deal with the general case in which the hull girder may contain differing materials, and the member being changed may be of any material. Therefore the expressions will be derived in terms of the transformed section properties A_{tr}, I_{tr}, and h_{NA}, and it is assumed that these have been calculated beforehand using transformation factors. Note that in all cases the area a_m is the *actual* area of the member; its effective area is $T_m a_m$. In the special case in which there is a *change of material* of a member, such that its Young's modulus changes from E_m to E'_m, there is need for some additional terms in (3.6.33). Firstly, the effect of the change in member area, δa_m, will be augmented or diminished by the ratio of the moduli and so this ratio needs to be applied as a multiplying factor. Secondly, changing E_m is equivalent to making another change in area (separate from and simultaneous to δa_m) equal to the original area multiplied by the proportional change in modulus: $(E'_m - E_m)/E_m$. Thus the most general form of (3.6.33) is

$$\delta\sigma_l = \left(\frac{\partial \sigma_l}{\partial a_m}\right)\delta a_m \frac{E'_m}{E_m} + \left(\frac{\partial \sigma_l}{\partial a_m}\right) a_m \frac{E'_m - E_m}{E_m} \qquad (3.6.34)$$

The most common example of a change in area is a change δt in the thickness of a strake of plating. If the width of the strake is b then for this case δa_m is given by $b\ \delta t$. Figure 3.21 illustrates some of the basic parameters relating to the member being changed: area a_m, height of centroid above the keel h_m, distance from neutral axis y_m (positive upward), vertical depth d_m, and transformation factor T_m. A subscript l denotes the location where the stress change is to be calculated. For generality the expression to be derived requires a rigorous sign convention for y, the distance from the neutral axis, and so we adopt the usual choice of positive upward. We thus have $y_l = h_l - h_{NA}$, and $y_m = h_m - h_{NA}$.

In order to obtain an expression for $\partial\sigma_l/\partial a_m$ it will

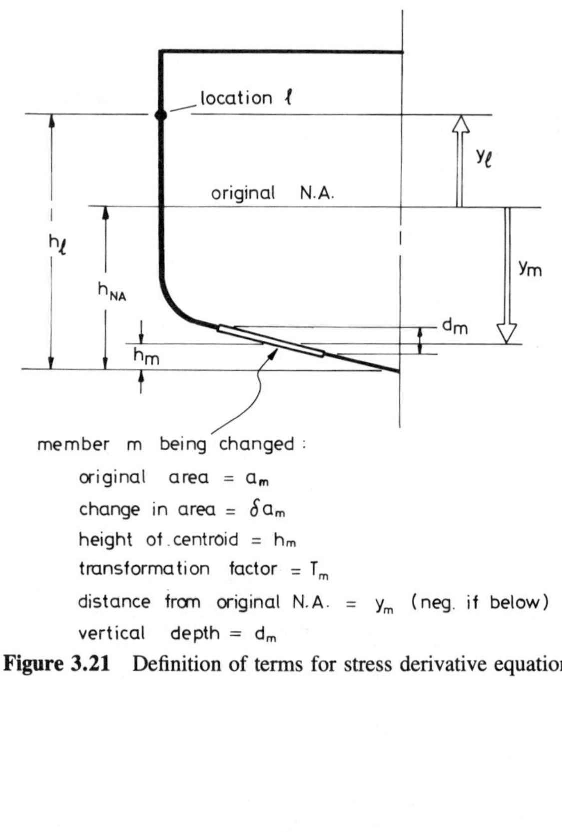

Figure 3.21 Definition of terms for stress derivative equation.

be necessary to have the partial derivatives of h_{NA} and I_{tr} with respect to a_m. These can be obtained from the expressions developed earlier for δh and δI—the changes in h_{NA} and in I due to adding some material. In the present case the added material is δa_m and its effective area is $T_m\ \delta a_m$; hence the latter replaces a in (3.6.25) and (3.6.26). Also, A_{tr} and I_{tr} replace A and I, and for greater generality and accuracy we here retain i_m, the moment of inertia of the added material, which was dropped during the derivation of (3.6.26). This term is usually negligible except when the member m is oriented vertically. In this case, assuming that it is a straight, thin-walled member of vertical depth d_m, i_m is equal to $\delta a_m\ d_m^2/12$ (see Fig. 3.15). With these changes (3.6.25) and (3.6.26) become

$$\delta h_{NA} = \frac{y_m T_m\ \delta a_m}{A_{tr} + T_m\ \delta a_m} \qquad (3.6.35)$$

$$\delta I_{tr} = \frac{A_{tr} y_m^2 T_m\ \delta a_m}{A_{tr} + T_m\ \delta a_m} + T_m \frac{\delta a_m\ d_m^2}{12} \qquad (3.6.36)$$

We may obtain the desired derivatives by dividing these changes by the increment of area δa_m and taking the limit.

$$\frac{\partial h_{NA}}{\partial a_m} = \lim_{\delta a_m \to 0} \left[\frac{\delta h_{NA}}{\delta a_m}\right] = \lim_{\delta a_m \to 0} \left[\frac{T_m y_m}{A_{tr} + T_m\ \delta a_m}\right]$$

$$= \frac{T_m y_m}{A_{tr}} \qquad (3.6.37)$$

$$\frac{\partial I_{tr}}{\partial a_m} = \lim_{\delta a_m \to 0} \left[\frac{\delta I_{tr}}{\delta a_m}\right]$$

$$= \lim_{\delta a_m \to 0} \left[\frac{A_{tr} y_m^2 T_m}{A_{tr} + T_m\ \delta a_m} + \frac{T_m\ d_m^2}{12}\right]$$

$$= T_m\left(y_m^2 + \frac{d_m^2}{12}\right) \qquad (3.6.38)$$

We may now proceed to derive the stress derivative which is required in (3.6.33) and (3.6.34). The hull girder bending stress at y_l is

$$\sigma_l = T_l \frac{M y_l}{I_{tr}} \qquad (3.6.39)$$

and its derivative is

$$\frac{\partial \sigma_l}{\partial a_m} = T_l M \frac{I_{tr}\left(\dfrac{\partial y_l}{\partial a_m}\right) - y_l\left(\dfrac{\partial I_{tr}}{\partial a_m}\right)}{I_{tr}^2}$$

Since $y_l = h_l - h_{NA}$, the first derivative in the numerator is $-\partial h_{NA}/\partial a_m$, which is given by (3.6.37), and the other derivative is given by (3.6.38). Substitution yields

$$\frac{\partial \sigma_l}{\partial a_m} = -T_l \frac{M y_l}{I_{tr}} \left[\frac{T_m y_m}{A_{tr} y_l} + \frac{T_m}{I_{tr}}\left(y_m^2 + \frac{d_m^2}{12}\right)\right]$$

and from (3.6.39) this is

$$\frac{\partial \sigma_l}{\partial a_m} = -\sigma_l T_m \left[\frac{y_m}{A_{tr} y_l} + \frac{y_m^2 + \dfrac{d_m^2}{12}}{I_{tr}}\right] \qquad (3.6.40)$$

In the simplest case when the hull girder is of one material and $d_m \ll y_m$ this equation becomes

$$\frac{\partial \sigma_l}{\partial a_m} = -\sigma_l \left[\frac{y_m}{A y_l} + \frac{y_m^2}{I}\right] \qquad (3.6.41)$$

This expression, in conjunction with (3.6.33), is less cumbersome than using the equations for change in section modulus: (3.6.28) and (3.6.30). However, the latter are more accurate for larger changes in area (a instead of δa; the different notation serves as a reminder of this). Of course, if the change in area is truly large—say, more than 10% of the total area—then the section modulus should be recalculated.

3.7 CALCULATION OF HULL GIRDER SHEAR STRESS

Shear Stress in Open Sections

In the hull girder, as in any beam loaded by transverse vertical forces, there is a vertical shear force Q acting on the cross section. In a thin-walled section such as an I-beam or a box girder it is important to know *how* the total shear force Q is distributed across the section so that the wall thicknesses can be adequately sized. That is, it is necessary to determine the distribution of the shear stress τ around the entire cross section.

Figure 3.22 shows a thin-walled symmetric box girder subjected to a vertical shear force Q. From elementary beam theory it is known that over a differential segment of length dx, Q causes a change of bending moment given by

$$dM = Q\ dx$$

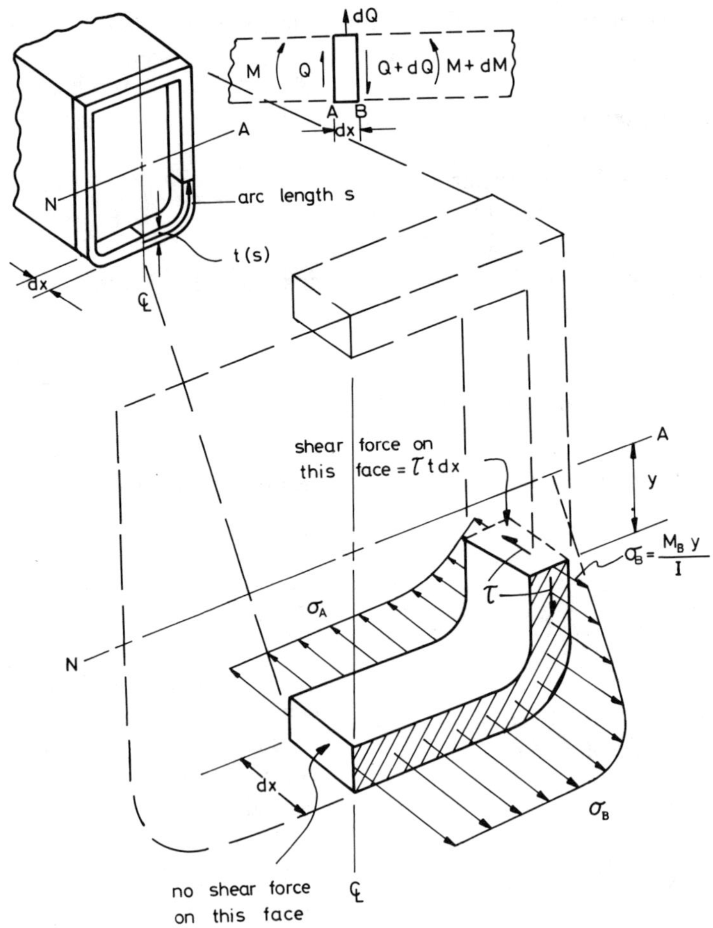

Figure 3.22 Free body diagram for transverse shear.

Due to this change in the bending moment the bending stresses σ_A and σ_B on the two faces of the differential segment are not equal. Therefore, if we isolate a portion of the differential segment by making two cuts, one at the centerline and the other at an arc length s from the centerline, the imbalance in the longitudinal normal stress forces must be counterbalanced by longitudinal shear stress forces across the cut sections. However, because of symmetry, there can be no shear stress in the centerplane cut and hence the balancing force must come entirely from the shear stress τ at the other cut. Longitudinal equilibrium therefore states that

$$\tau t\, dx = \int_0^s \sigma_B t\, ds - \int_0^s \sigma_A t\, ds \qquad (3.7.1)$$

Substituting $\sigma = My/I$ on both faces

$$\tau t\, dx = \frac{M_B - M_A}{I} \int_0^s yt\, ds$$

$$= \frac{dM}{I} \int_0^s yt\, ds$$

and substituting $dM = Q\, dx$ gives

$$\tau t = \frac{Q}{I} \int_0^s yt\, ds \qquad (3.7.2)$$

The integral on the right-hand side is a function of the geometry of the section and of position s around the section. For convenience we assign the symbol m to this quantity:

$$m = \int_0^s yt\, ds \qquad (3.7.3)$$

and we note that m is the first moment about the neutral axis of the cumulative section area starting from the "open" end (shear-stress-free end) of the section.

Substituting for m in (3.7.2) and solving for τ:

$$\tau = \frac{Qm}{tI} \tag{3.7.4}$$

The product τt has special significance in the torsion of thin-walled sections, and has some analogies to the flow of an ideal fluid within a closed pipe. This product is therefore referred to as the "shear flow" and is assigned the symbol q:

$$q = \tau t \tag{3.7.5}$$

The shear flow also a useful quantity in the present case, in which the shear stress is due to a transverse load. From (3.7.4) we observe that the shear flow is given by

$$q = \frac{Qm}{I} \tag{3.7.6}$$

Since Q and I are constants for the entire section, the shear flow is directly proportional to m. In fact, the ratio Q/I may be regarded as simply a scaling factor, and once the distribution of m has been calculated the shear flow distribution is identical to it but with different units. Still another advantage of q is that its value does not vary abruptly with local thickness changes, as does τ.

It may be seen from the derivation of (3.7.4) that the values of Q and I are normally those for the entire hull girder cross section, whereas the calculation of m is performed using a half-section. This is the convention which is usually followed. Obviously, the Q and I values for a half-section could also be used since their ratio is then unchanged. However, *m should always be calculated for a half-section;* if both port and starboard halves had been used in deriving (3.7.4), then there would have been two symmetric and identical cut faces, each carrying an axial shear force of $\tau t\, dx$. In this case the denominator of (3.7.4) would have to be $2tI$. The convention of using a half-section for the calculation of m is also more appropriate because in practice half-sections are used for structural drawings.

The calculation of m is illustrated in Fig. 3.23 for an idealized hull girder. For horizontal portions the moment arm y is constant and m therefore increases linearly with arc length. This occurs in the deck and bottom if there is no camber or deadrise. For instance, in the deck

$$m(s_1) = gt_D s_1$$

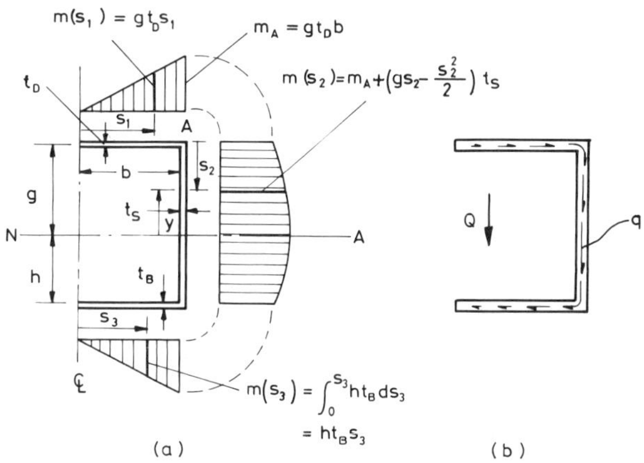

Figure 3.23 Calculation of moment term m by integrating along branches.

and

$$m_A = m(b) = gt_D b$$

In the side shell m is parabolic

$$m(s_2) = m_A + \int_0^{s_2} yt_s\, ds_2 = m_A + \left(gs_2 - \frac{1}{2}s_2^2\right)t_s$$

In ships of normal proportions the parabola is very flat and hence the shear flow q is almost constant vertically.

Because of changes in the orientation and thickness of the plates which make up the hull girder cross section, the integration for m is usually performed in segments. The integration is always commenced at the "open" end of any branch. As shown in Fig. 3.24 this need not be on the centerline; it may be at the edge of a hatch or other opening.

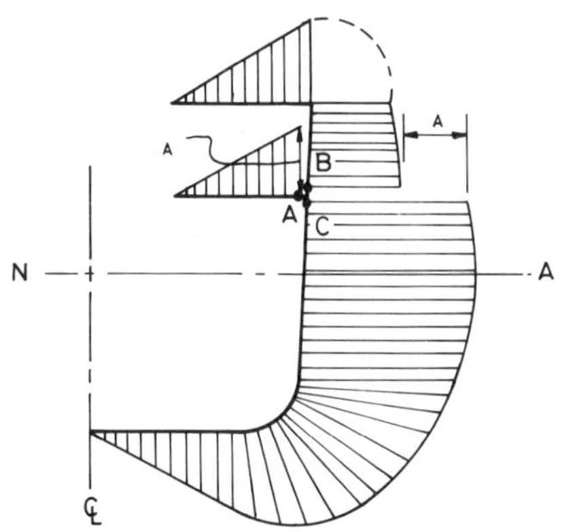

Figure 3.24 Conservation of shear flow at corners and branch points.

Figure 3.24 also shows the effect of multiple branches, for example, additional decks. If an imaginary cut were made at point C the shear force at that point would have to balance the net imbalance in bending stress forces in the second deck and all plating above it. Therefore all of this area must be included in the calculation of m at point C. The new area which is incurred in passing from B to C is the area of the second deck, and therefore the increment in m is equal to the total value of m for the second deck. That is, $m_C = m_A + m_B$ and, since q is directly proportional to m, $q_C = q_A + q_B$. This illustrates one of the reasons for the use of the term "shear flow": at any junction or branchpoint the increment in the shear flow is equal to the flow contributed or taken away by the branch, as shown in Fig. 3.25. It should be noted that since deck and side plating may differ, this rule of continuity of shear flow does not hold for τ. Fig. 3.26 illustrates how τ changes with changes in thickness.

It is well known (e.g., from Mohr's circle) that because of equilibrium the shear stress in a differential element at any point occurs in the form of two equal and opposite stress couples, one positive and the other negative. Since they are equal it makes little difference which is which, and so there is no need for a rigorous sign convention for either m or τ or q. The direction of the shear flow may be determined by inspection because in the webs of the hull girder it has the same direction—upward or downward—as the overall shear force, Q. In "open" sections such as that of Fig. 3.25 the flow is a "straight-through" flow; there is rarely a reversal in the flow direction and so usually it is only the magnitude of m, or of q, which is of interest, and its sign will usually be positive. Also, the integration for m is always commenced at the open end of each branch because that simplifies the computation. Therefore, the moment arm y is always taken as positive at the beginning of each branch, regardless of which side of the neutral axis the branch commences, and it only becomes negative if and when that particular branch crosses the neutral axis. For this reason, it is best to stop at the neutral axis and to finish that branch by starting from the other end. If this is not possible then the integration can proceed across the neutral axis, provided that a negative moment arm is used for all points on the other side.

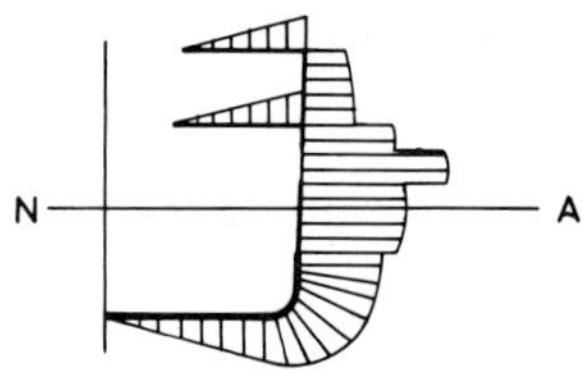

Figure 3.26 Change of τ due to change of thickness.

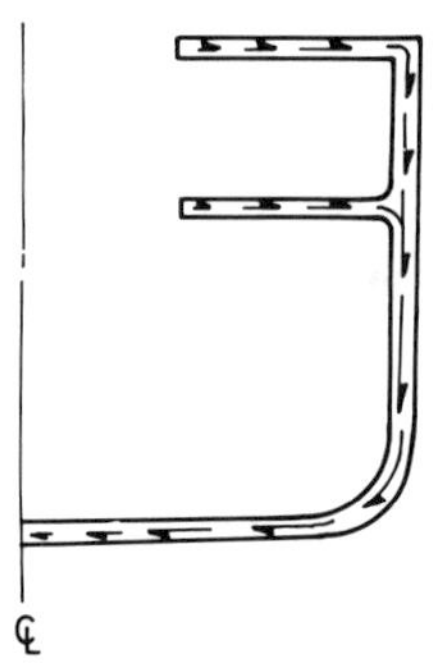

Figure 3.25 Sample diagram indicating direction of shear flow.

Shear Stress in Multicell Sections

In the definition of m in (3.7.3) it was assumed that the line integral was always commenced at a point of zero shear flow. Thus, as noted earlier, if there are any branch points (such as a second deck and side shell connection) then each branch must have zero shear flow at its end. This is equivalent to requiring that there should be no closed loops or "cells" within the overall half-section of the hull girder. Consequently, for the tanker of Fig. 3.27 the value of m can only be calculated along AB and FE; it cannot be calculated anywhere around the perimeter of the wing tank BCDEB. The difficulty arises from the fact that the shear flow divides at point B (and reunites at point E) and the separate components cannot be determined from simple statics. The problem is *statically indeterminate* and, as usual in such problems, additional information must be obtained from a consideration of geometric compatibility. The standard technique consists of two main steps:

1. Remove a sufficient number of restraints so as to render the problem statically determinate, even though this artificial case has displacements at these restraint points which should in fact be zero.
2. For each such restraint point, impose the geometric condition that the displacement should be zero.

In the present case the restraint is not an external support, as it is for beams, but rather an internal self-restraint: one portion of the hull girder imposing a

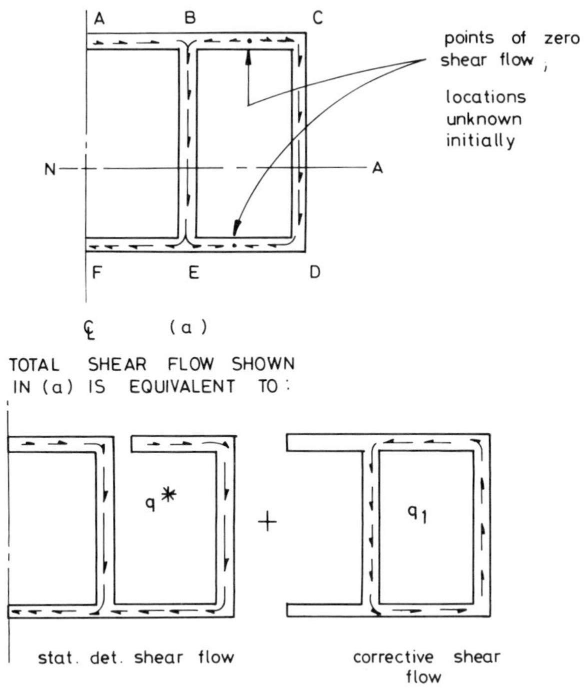

Figure 3.27 Calculation of shear flow in multicell sections.

restraint on an adjacent portion (and the second imposing an equal and opposite restraint on the first).

The foregoing procedure will be illustrated by applying it to the tanker of Fig. 3.27. The shear flow in the hull girder is rendered statically determinate at point B by making a longitudinal cut there, as shown in Figure 3.27*b*. This transforms the hull girder half-section into an open section with three branches: ABE, BCDE, and FE, each of which has zero shear flow at its ends. In the general case, with N closed cells, a total of N cuts would have to be made. The first moment term m can now be calculated from (3.7.3). To signify that this value of m corresponds to an artificially cut section it will be denoted as m^*. Equation 3.7.3 therefore becomes

$$m^*(s) = \int_0^s y(s)t\, ds \qquad (3.7.7)$$

As before, $m^*(s)$ is a function of position within and around the hull girder half-section. The position of any point is specified by the arc length s from the tip of the particular branch within which that point lies.

The corresponding value of shear flow will be denoted as q^* and (3.7.6) therefore becomes

$$q^*(s) = \frac{Q}{I}\, m^*(s) \qquad (3.7.8)$$

It now remains to remove the error which the longitudinal cuts have introduced into the structure. The effect of each cut is to permit longitudinal "slip" to occur between the two cut edges. Hence the geometric condition must be imposed that the slip is actually zero at each cut, and in order to satisfy this condition some additional or correcting shear flow must be added. It is shown in Section 6.1 that equilibrium of an isolated thin-walled section requires constant shear flow within that section. Therefore, the correcting shear flow is introduced in the form of N separate, constant values of shear flow, $q_i (i = 1, \ldots, N)$, one for each of the originally closed cells. The sum of all of these, plus the q^* distribution, must give the correct total shear flow q. That is

$$q = q^* + \sum_{i=1}^{N} q_i \qquad (3.7.9)$$

The situation is illustrated in Fig. 3.27 for the case of a tanker. In this case there is only one closed cell and hence only one correcting shear flow, q_1, as shown in Fig. 3.27*c*.

Now for a prismatic thin-walled section of unit length, the longitudinal slip that would occur due to a cut is equal to the cyclic integral of the longitudinal shear strain from one edge of the cut to the other around any closed path in the section. That is,

$$\text{slip} = \oint \gamma\, ds = \frac{1}{G}\oint \tau\, ds = \frac{1}{G}\oint \frac{q}{t}\, ds \qquad (3.7.10)$$

where γ is the shear strain at any point in the cross section. In the present application the path of integration for each cut is taken to be the perimeter of the closed cell associated with that cut (i.e., the cell which was "opened" by that cut). Therefore the condition of zero slip at each cut is replaced by the condition that for each cell the foregoing cyclic integral is zero. This gives rise to a set of N equations:

$$\oint_{\text{cell } j} \left(\frac{q}{t}\right) ds = 0 \qquad (j = 1, \ldots, N) \qquad (3.7.11)$$

Note that because of the zero on the right-hand side, G has been eliminated.

Substituting for q from (3.7.9) and applying the cyclic integration to each term gives (3.7.11) in expanded form:

$$\left.\begin{aligned}
\oint_{\text{cell 1}} \frac{q_1}{t}\,ds + \oint_{\text{cell 1}} \frac{q_2}{t}\,ds + \cdots + \oint_{\text{cell 1}} \frac{q_N}{t}\,ds &= -\oint_{\text{cell 1}} \frac{q^*}{t}\,ds \\
\oint_{\text{cell 2}} \frac{q_1}{t}\,ds + \oint_{\text{cell 2}} \frac{q_2}{t}\,ds + \cdots + \oint_{\text{cell 2}} \frac{q_N}{t}\,ds &= -\oint_{\text{cell 2}} \frac{q^*}{t}\,ds \\
&\vdots \\
\oint_{\text{cell } N} \frac{q_1}{t}\,ds + \oint_{\text{cell } N} \frac{q_2}{t}\,ds + \cdots + \oint_{\text{cell } N} \frac{q_N}{t}\,ds &= -\oint_{\text{cell } N} \frac{q^*}{t}\,ds
\end{aligned}\right\} \qquad (3.7.12)$$

It should be noted that all of the terms on the left-hand side are of the form

$$\oint_{\text{cell } j} \frac{q_i}{t}\,ds$$

and so for any two cells i and j which are *not adjacent* the corresponding term will be zero, and if they are adjacent the cyclic integral will only yield a nonzero value along their common boundary.

Since the tanker of Fig. 3.27 has only one closed cell there is only one compatibility equation:

$$q_1 \oint_{\text{BCDEB}} \frac{ds}{t} = -\oint_{\text{BCDEB}} \frac{q^*}{t}\,ds$$

and hence the correcting shear flow in the wing tank is

$$q_1 = \frac{-\displaystyle\oint_{\text{BCDEB}} \frac{q^*}{t}\,ds}{\displaystyle\oint_{\text{BCDEB}} \frac{ds}{t}} \qquad (3.7.13)$$

As required by (3.7.9) the total shear flow is obtained by adding the correcting shear flow q_1 to the (artificially) determinate shear flow q^*. This is illustrated in Fig. 3.27. The final expression may be simplified by noting that in (3.7.6) Q and I are constant for the entire section. Therefore these components of q^* may be removed from the integration. The final result for the tanker is then

$$q = \frac{Q}{I}\left[m^* - \frac{\displaystyle\oint \frac{m^*}{t}\,ds}{\displaystyle\oint \frac{ds}{t}}\right] \qquad (3.7.14)$$

SIGN CONVENTION

In applying this theory, some care is needed with regard to sign convention because two different types of line integrals are involved: (1) the integral for the calculation of $m^*(s)$ to obtain $q^*(s)$; and (2) the cyclic integrals of q^*/t and q_i/t around each of the closed cells. These two integrals involve different sign conventions.

1. *Calculation of* m^*—The general direction of the shear flow q^* can be established by inspection once the locations of the cuts are chosen. For this purpose it is usually helpful to make a sketch of the cut section and to indicate the directions of q^* by means of arrows, as in Fig. 3.28*a*. Once again the integration should always be commenced at the open end of each branch, and y is always initially positive and only becomes negative if and when that particular branch crosses the neutral axis.

2. *Cyclic integration of* q^*/t *and* q/t—As noted before, the direction of q^* at any point can be established by inspection once the locations of the cuts are chosen. It is necessary to know the direction of q^* at all points because in all of the cyclic integrals on the right-hand side of (3.7.12) the sign allocated to q^* depends on its local direction. The sign convention used herein is as follows:

> The shear flow q^* at any point is positive if at that point the direction of flow is clockwise with respect to that cell.

Within each equation of (3.7.12) (i.e., within each cell) the cyclic integrations must all be done in the same direction. It may be helpful to always proceed clockwise around each cell because then the definition of positive q^* is easier: q^* is positive if it is flowing

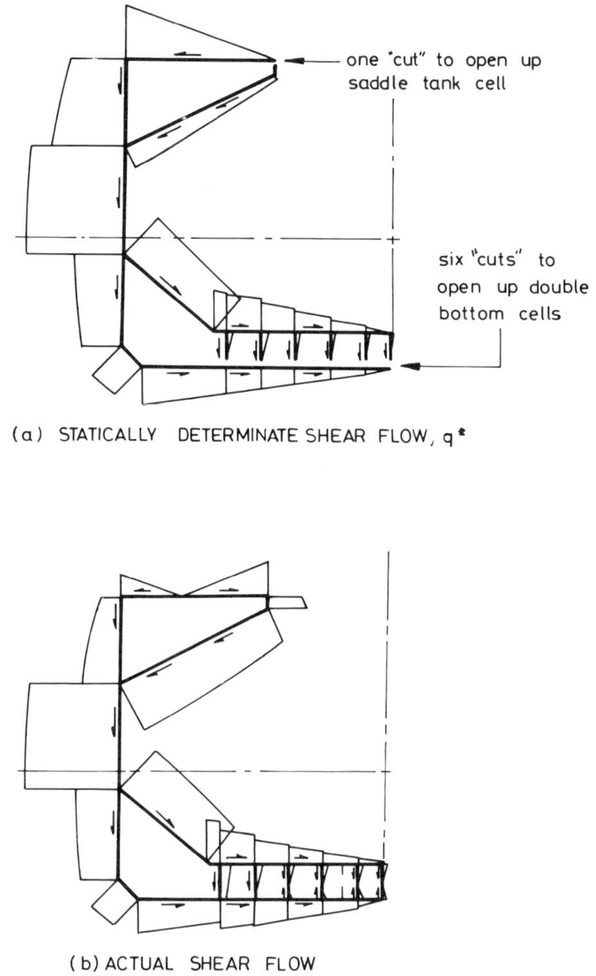

Figure 3.28 Shear flow in a typical multicelled section (bulk carrier).

with the integration and negative if it is flowing *against*.

The same sign convention (positive clockwise) is used for the corrective shear flows, q_i. If any of the q_i values turns out to be negative it simply means that in that particular cell the corrective shear flow is counterclockwise (as happens in the sample problem of Figure 3.27).

The method presented in this section is essentially a hand technique, and it becomes cumbersome for cross sections with many closed cells, such as the bulk carrier in Fig. 3.28. Some computer programs are available which are automated versions of the above method. However, an alternative method is presented in Section 6.4 which is more suitable for computer implementation.

Example of Shear Flow Calculation for a Multicell Section

The shear flow will be calculated for the idealized tanker of Figure 3.29*a*. To better illustrate the method, the thicknesses will be expressed in terms of a reference thickness, t_A, instead of substituting their numerical values.

Step 1. Calculation of m (Fig. 3.30)*

Branch BCDE:

$$\text{BC:} \quad m^* = \int_0^{s_1} 12t_A\, ds_1 = 12t_A s_1, \qquad m_C^* = 120t_A$$

$$\text{CD:} \quad m^* = m_C^* + \int_0^{s_2} (12 - s_2)t_A\, ds_2$$

$$= 120t_A + \left(12s_2 - \frac{s_2^2}{2}\right)t_A, \qquad m_D^* = 160t_A$$

$$\text{DE:} \quad m^* = m_D^* + \oint_0^{s_3} (-8)t_B\, ds_3 = 160t_A - 8t_B s_3,$$

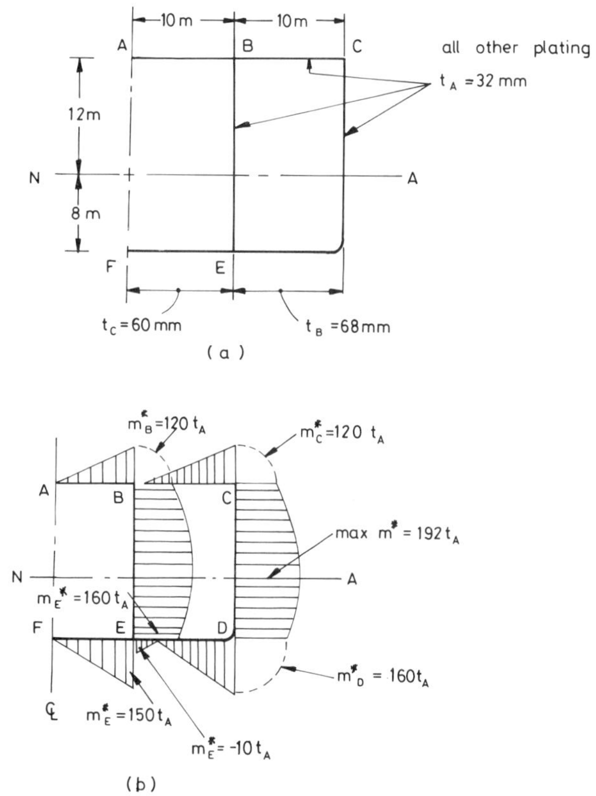

Figure 3.29 Example of shear flow calculation in a multicell section.

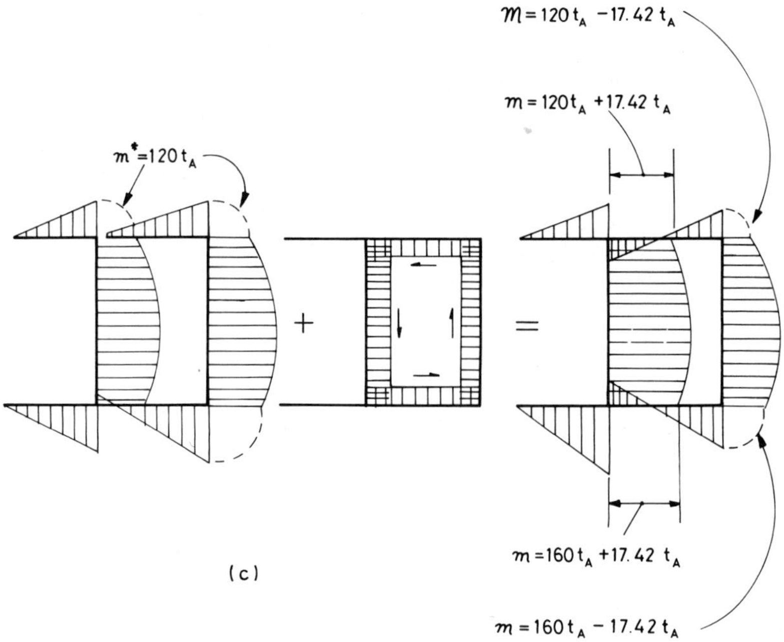

Figure 3.29 (Continued)

$$m_E^* = 160t_A - 80t_B$$
$$= \left(160 - 80\,\frac{68}{32}\right)t_A$$
$$= -10t_A$$

Branch ABE:

AB: $m^* = \int_0^{s_4} 12t_A\,ds_4 = 12t_A s_4, \qquad m_B^* = 120t_A$

BE: $m^* = m_B^* + \int_B^{s_5} (12 - s_5)t_A\,ds_5$

$$= 120t_A + \left(12s_5 - \frac{s_5^2}{2}\right)t_A, \qquad m_E^* = 160t_A$$

Branch FE:

FE: $m^* = \int_0^{s_6} 8t_C\,ds_6 = 8t_C s_6,$

$$m_E^* = 80t_C = 80\,\frac{60}{32}\,t_A = 150t_A$$

Step 2. Cyclic integration $\oint \frac{m^*}{t}\,ds$ *(Fig. 3.31)*

BC: $\int_0^{10} \frac{12t_A s_1}{t_A}\,ds_1 = 600$

CD: $\int_0^{20} \frac{\left(120 + 12s_2 - \frac{s_2^2}{2}\right)t_A}{t_A}\,ds_2$

$$= \left[120s_2 + 6s_2^2 - \frac{s_2^3}{6}\right]_0^{20} = 3467$$

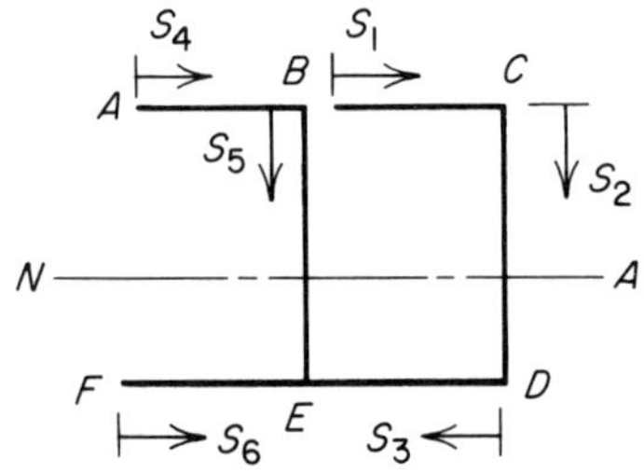

Figure 3.30

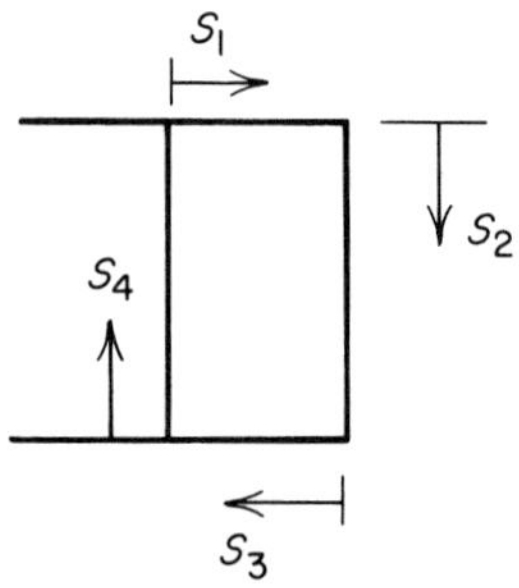

Figure 3.31

DE: $\displaystyle \int_0^{10} \frac{160t_A - 8t_B s_3}{t_B}\, ds_3$

$$= \int_0^{10} \left(160\,\frac{32}{68} - 8s_3\right) ds_3 = 353$$

EB: $\displaystyle \int_0^{20} \frac{-\left(120 + 12s_4 - \frac{s_4^2}{2}\right)t_A}{t_A}\, ds_4$

$$= -\left[120s_4 + 6s_4^2 - \frac{s_4^3}{6}\right]_0^{20} = -3467$$

(negative because the integration is against the flow)

$$\text{Therefore } \oint \frac{m^*}{t}\, ds = 600 + 3467 + 353 - 3467 = 953$$

Step 3. Cyclic integration $\oint \frac{ds}{t}$ *(Fig. 3.31)*

$$\left.\begin{aligned} \text{BCD:}&\quad \int_0^{30} \frac{ds}{t_A} = \frac{30}{t_A} \\ \text{DE:}&\quad \int_0^{10} \frac{ds}{t_B} = \frac{10}{t_B} \\ \text{EB:}&\quad \int_0^{20} \frac{ds}{t_A} = \frac{20}{t_A} \end{aligned}\right\} \oint \frac{ds}{t} = \frac{30 + 10\left(\frac{32}{68}\right) + 20}{t_A} = \frac{54.71}{t_A}$$

Step 4: Calculation of corrective shear flow

$$q_1 = -\frac{Q}{I}\,\frac{\oint \frac{m^*}{t}\, ds}{\oint \frac{ds}{t}} = -\frac{Q}{I}\,\frac{953}{54.71}\, t_A = -\frac{Q}{I}(17.42t_A)$$

Negative value indicates that corrective shear flow is counter clockwise. If $\oint (m^*/t)\, ds$ had been negative, q_1 would be positive.

Step 5. Total shear flow

$$q = q^* + q_1 = \frac{Q}{I}(m^* - 17.42t_A)$$

The resulting distribution of q is plotted in Figure 3.29c.

Shear Flow in Sections Containing Different Elastic Moduli

The foregoing analysis was for a homogeneous cross section and will now be generalized for a section having two or more different values of Young's modulus. The bending stress σ in (3.7.1) will now be given by (3.6.23) which is

$$\sigma = T_i \frac{My}{I_{tr}}$$

and so (3.7.1) becomes

$$\tau t\, dx = \frac{M_B - M_A}{I_{tr}} \int_0^s T_i y t\, ds \tag{3.7.9}$$

Here again if we regard the transformation factor T_i as multiplying the area, or more specifically the thickness t, we obtain an equivalent homogeneous section in which the wall thickness at any point is $T_i t$. That is, the thickness has been scaled up or down in proportion to the local T_i.

As a result of this scaling, the integral in (3.7.9) is the value of m for the transformed section; that is, it is the first moment of the area of the transformed section along an arc length s, taken about the neutral axis of the transformed section. In conformity with our notation for bending, we denote this value of m as m_{tr}.

$$m_{tr} = \int_0^s y T_i t\, ds \tag{3.7.10}$$

and this is the generalized version of (3.7.3). The expression for the shear flow (3.7.6) then becomes

$$q = \frac{Qm_{tr}}{I_{tr}} \tag{3.7.11}$$

To obtain the shear stress at any point we divide the shear flow by the *true* local thickness t:

$$\tau = \frac{Qm_{tr}}{I_{tr} t} \tag{3.7.12}$$

Unlike the bending stress distribution, the shear stress does not undergo an abrupt change at a material interface (unless there is also a branchpoint there, in which case the abrupt change is entirely due to the new branch). The shear stress distribution does, however,

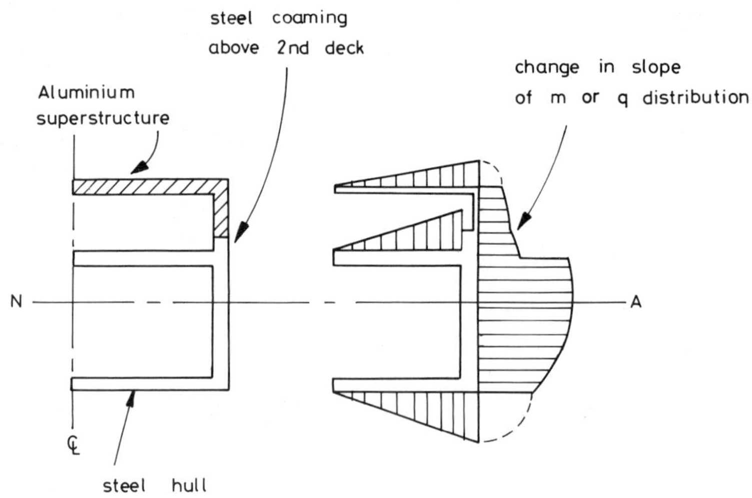

Figure 3.32 Shear flow in a section having two elastic moduli.

undergo an abrupt change of slope at a material interface. A typical distribution of shear stress in the side shell of a composite hull girder is shown in Fig. 3.32.

3.8 SHEAR EFFECTS AND OTHER DEPARTURES FROM SIMPLE BEAM THEORY

Shear Lag

Simple beam theory assumes that plane cross sections remain plane, and that therefore the bending stress is directly proportional to the distance from the neutral axis. Thus in any flange-and-web type of beam the stress should be constant across the flanges. However, in most cases the bending is not caused by the application of a pure couple to the ends of the beam; rather, it is caused by vertical loads and these loads are absorbed by the webs of the beam and not by the flanges. That is, even for a hull girder, in which the vertical loads initially act on the flanges, they are immediately transferred to the webs by transverse beams and frames; the plating of the flanges can only take longitudinal in-plane loads (we are discussing principal loads, not small local loads). Therefore, the vertical loads act on the webs and cause them to deflect to some radius of curvature, thus inducing maximum strain in the flanges. Since they carry maximum strain, and hence maximum stress, the flanges make the largest contribution to the bending stiffness. But, it is important to note that this maximum strain comes initially from the webs and only reaches the flanges by shear. This is illustrated in Fig. 3.33 which shows a portion of a box girder cantilever loaded by a vertical force F. The force is reacted by, or carried by, the webs, which deflect to some radius of curvature such that the upper and lower edges of the web are elongated and shortened. For simplicity, the curvature is not shown; only the change in length. At the upper edge the elongated web pulls the plating with it, through shear forces, and this sets up shear stresses in the flange; these were discussed fully in the previous section. The bending and shear stresses cause stretching and in-plane distortion of the flange. On the left and right sides of the figure an element is shown before and after this stretching and distortion. The shear distortion is such that the inner edge of the element does not have to stretch as much as the outer edge; that is, the longitudinal strain is less at the inner edge and therefore so is the longitudinal stress. This same phenomenon will occur at each element, from the edges to the centerline, although it will diminish steadily and disappear at the centerline because the shear stress diminishes to zero. The overall result is that the flange undergoes in-plane longitudinal distortion and therefore *plane cross sections do not remain plane* when shear stress is present. This distortion is commonly referred to as "warping" of the cross section. The most significant aspect of the shear distortion is that the

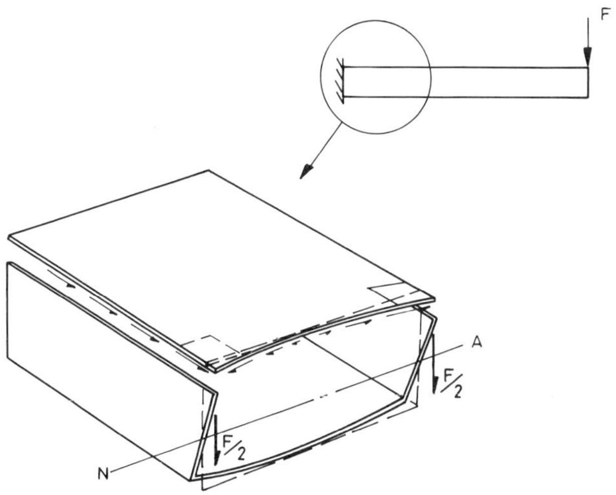

Figure 3.33 Shear lag in box girders.

inner portion of the flange carries less bending stress, and is therefore less effective than the outer portion. That is, due to shear effects, the bending stress remote from a web "lags behind" the stress near the web. The phenomenon is therefore termed the "shear lag" effect. The same effect occurs in the compression flange since here the shear distortion allows fibers that are remote from a web to avoid some of the shortening and hence to carry less compressive stress. Shear lag occurs in any wide-flanged section that carries a lateral load in the manner of a beam. In an open section, such as the standard single web beam of Fig. 3.34*a* it is the outer edges of the flange which are less effective.

The exact distribution of stress in a wide-flanged section can be found using the mathematical theory of elasticity but this analysis, involving the use of stress functions, is too complex for design calculations. The exact analysis shows that the magnitude of the shear lag effect (i.e., the extent to which the distribution of bending stress differs from simple beam theory) is dependent on:

1. The ratio of width to length of the flange.
2. The distribution of lateral loading along the beam.
3. The relative proportions of web and flanges.
4. The type of section (single or multiple web, symmetric or unsymmetric, etc.).
5. The position along the beam. The shear lag effect in general varies from point to point along the length, and is maximum at maximum shear force gradient (concentrated loads).

Shear lag is of importance in beams having very wide flanges and shallow webs, such as aircraft wings. In steel box girders the effect is much smaller, even in box girder bridges [13] which have large concentrated loads due to the point supports. In the hull girder bending of ships, the shear lag effect is usually only a few percent. It is more important in the consideration of the effects of superstructures and of the effective breadth of plating in local strength problems which will be discussed later.

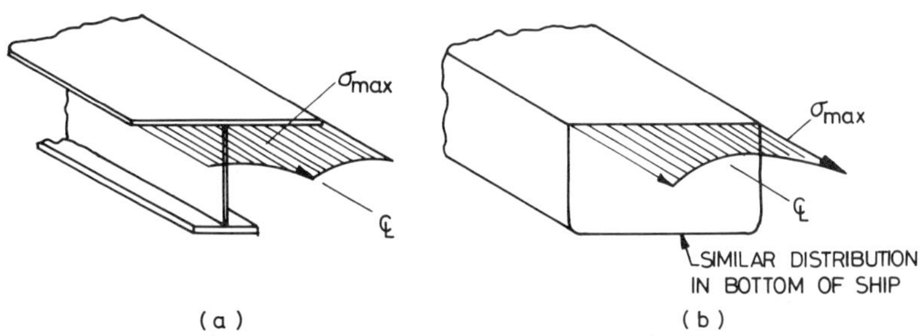

Figure 3.34 Shear lag in beam flanges.

Effective Breadth Due to Shear Lag

Rather than using a mean value of flange stress as a way of allowing for shear lag, it is preferable to retain the value of the maximum stress σ_{max} at the junction of the flange and web, so that it can be allowed for in designing the flange. Therefore the usual way of allowing for shear lag is to use an "effective breadth" of flange, b_e, which is defined as:

> The breadth of plating which, when used in calculating the moment of inertia of the section, will give the correct stress at the junction of web and flange, using simple beam theory.

Thus the effective breadth must be such that the total longitudinal force in the flange is equal in the actual and simplified cases. Equating forces

$$b_e \sigma_{max} = \int_0^b \sigma_x \, dz$$

or

$$b_e = \frac{1}{\sigma_{max}} \int_0^b \sigma_x \, dz \tag{3.8.1}$$

Effective breadth is illustrated in Fig. 3.35. In references 15 and 16, among many others, effective breadths are calculated for a wide variety of structures and loading conditions. These references should be consulted if it is required to investigate shear lag effects in detail.

The main conclusions from such analyses are:

1. The most important parameter which determines effective breadth of plating is the ratio of flange width b to the length L_0 between points of zero bending moment. For simply supported beams L_0 = beam span. A low L_0/b ratio results in a small ratio of b_e/b.
2. Effective breadth varies from point to point along the span of a beam, being smallest at points of concentrated loading on the beam, where there is a discontinuity in the shear force curve. Conversely, there is no shear lag effect in pure bending (no shear).
3. Shear lag occurs in both tension and compression flanges equally, provided that in the latter case buckling does not occur.

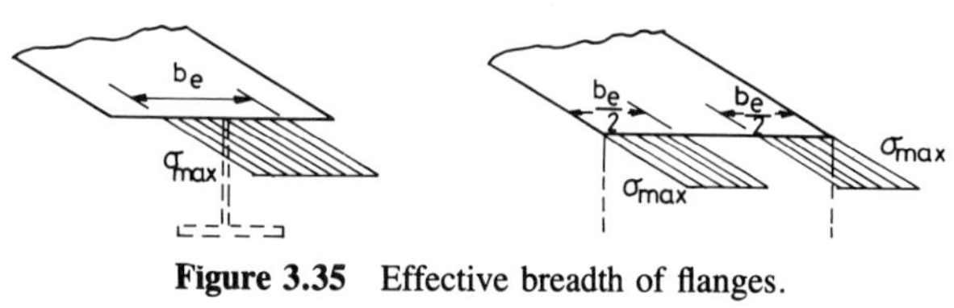

Figure 3.35 Effective breadth of flanges.

For design purposes the effective breadth at the section of maximum bending moment is of most importance. Figure 3.36, derived from Refs. 14 and 15, enables effective breadths to be found at points of maximum bending moment. These breadths should then be used in calculating the effective geometrical properties of the section, and hence the maximum bending stress in the beam.

Modification of Shear Force Curve Due to Double Bottom

By applying simple beam theory to the hull girder we obtained the relationship that the shear force Q is the integral of the net vertical load $f(x)$. However, this assumes that $f(x)$ is applied directly to the vertical shear-carrying webs of the girder, that is, directly to the side shell. This assumption would be accurate if the ship contained a purely transverse framing system such that the vertical loads on the bottom and on each deck were transferred immediately to the side shell. However, if there are longitudinal members in the decks or bottom they will transmit some fraction of the vertical load to the transverse bulkheads, and the bulkheads in turn will transmit this load to the side shell as a vertically distributed shear force (which is equivalent to a point load as far as its effect on Q is concerned). Thus the overall loading $f(x)$ is transmitted to the side shell in two parts: a distributed part due to the direct transfer of load by transverse members, and a concentrated part at each bulkhead. The latter gives rise to a discontinuity in the shear curve at the bulkhead.

This effect occurs, for example, in ships with double bottoms because a double bottom usually contains several full depth longitudinal girders. Also, the longitudinal continuity of these members gives them added effectiveness in transmitting part of the load on the double bottom to the transverse bulkheads. This is illustrated in Fig. 3.37 which shows three adjacent hold spaces, of which the central one is empty and the others full, as is common in ships carrying heavy bulk cargo. Figure 3.37*a* shows the idealized loading which is assumed to implied by simple beam theory, and the corresponding shear force distribution is

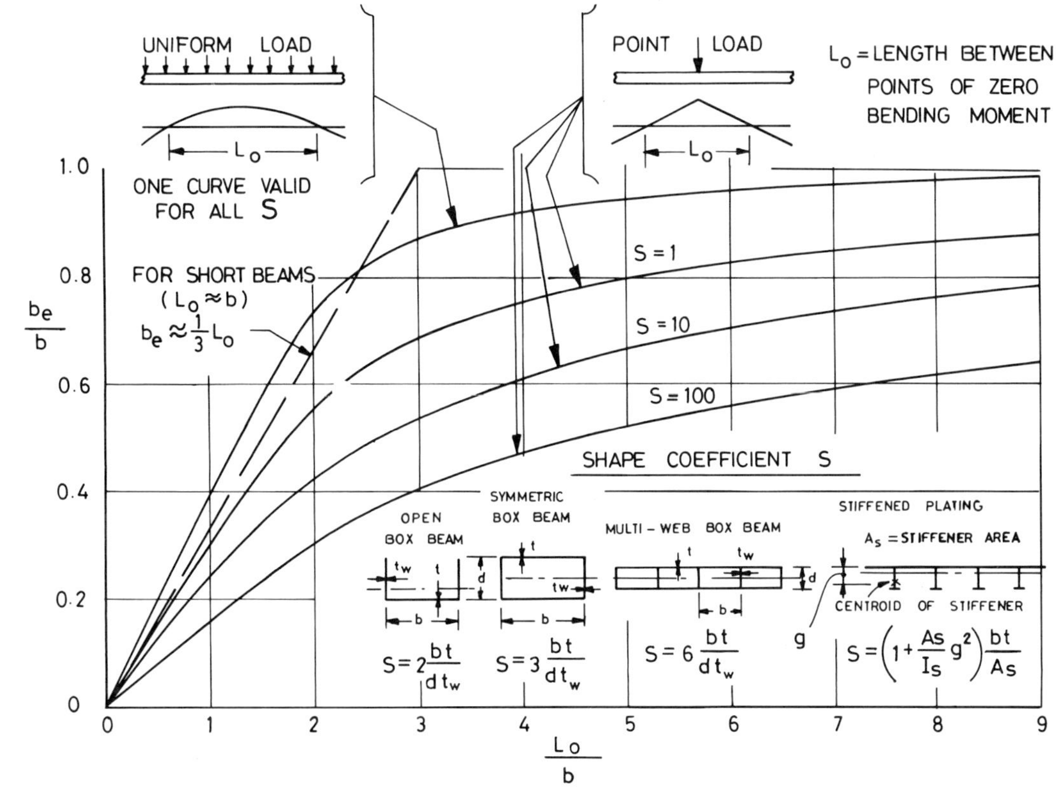

Figure 3.36 Effective breadth at maximum bending moment.

shown below this. Figure 3.37*b* shows the actual force distribution in the hull (still somewhat idealized, of course). The net vertical force (buoyancy minus weight) in each hold space is partly transferred to the bulkheads, and some cancellation occurs there if these transferred forces are of opposite sign, as in this case. The net transferred force then passes to the side shell, where it appears as a vertically distributed shear force F_s. The other portion of the net vertical force acting on the double bottom is transferred directly to the side shell. However, as shown in the figure it does not appear there as a uniformly distributed load because in reality the double bottom is equivalent to a large rectangular orthotropic plate. The behavior of orthotropic plates is dealt with in Chapter 10; it suffices here to simply note that a force acting normal to a rectangular plate is transferred mainly to the midregion of the plate sides and not to the corners. The

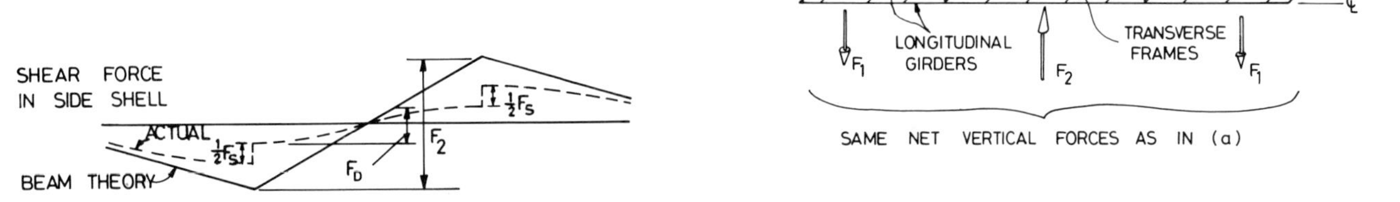

Figure 3.37 Shear force in side shell.

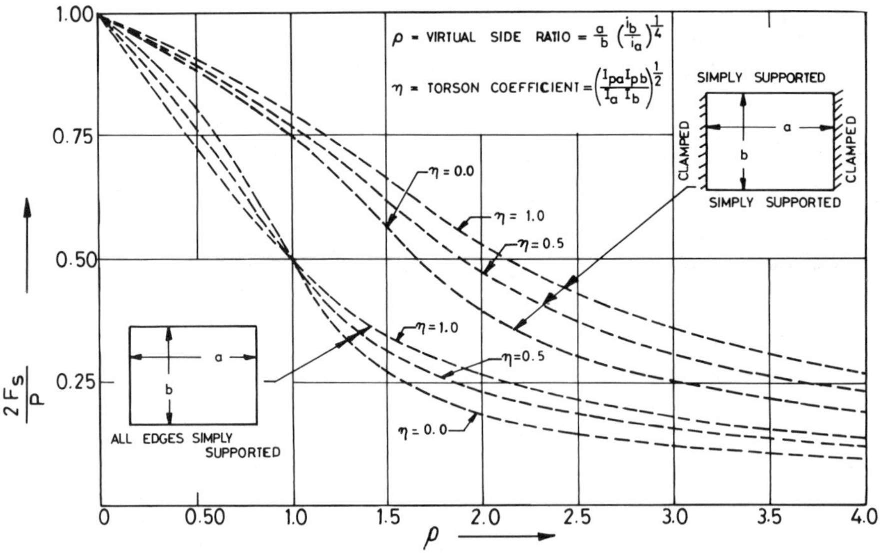

Figure 3.38 Proportion of double bottom load carried by transverse bulkheads. Symbols: F_s = total force carried by *each* bulkhead (across full width of ship); P = total vertical load on the double bottom; $i_a(i_b)$ = unit stiffness in longitudinal (transverse) direction; $I_a(I_b)$ = moment of inertia of a longitudinal girder (transverse frame) comprised of web plus upper and lower flanges (i.e., tank top and bottom plating) of effective width; $I_{pa}(I_{pb})$ = moment of inertia of girder (frame) flanges alone (i.e., no web).

nonuniformity of the force acting on the side shell means that the actual shear force distribution is nonlinear, and the longitudinal "detour" and partial cancellation of some of the vertical force means that the actual shear force in the side shell is less than that predicted by simple beam theory. The decrease is somewhat exaggerated in Fig. 3.37, but it can be sufficiently large to warrant consideration (40% or more, depending on the length and longitudinal rigidity of the double bottom). An approximate value of the reduction can be estimated with the aid of Fig. 3.38, adapted from Ref. 16. For a double bottom in a hold space of length a and breadth b, the figure gives the proportion of the net vertical load on the double bottom which is transmitted to and taken by the transverse bulkheads. The curves were obtained using orthotropic plate theory, and the notation is explained in Chapter 10. Note that the curves give total force (both bulkheads) and so to obtain F_s, the concentrated shear force which acts on the side shell (both sides combined) at *each* bulkhead, the ordinate of the curve should be multiplied by $P/2$. To obtain the corrected shear force distribution $Q(x)$ it is usually sufficiently accurate to subtract a uniform load F_s/a from the original (simple beam theory) load curve $f(x)$, and add the concentrated load F_s at each bulkhead. This is repeated for each hold space and then the corrected load curve is integrated.

This effect also occurs in decks which contain substantial longitudinal girders but it is of lesser magnitude because such decks are singly plated grillages and as such do not have the large torsional rigidity of doubly plated structure. Also, unless there are a large number of such girders orthotropic plate theory is not applicable; Chapter 10 gives some criteria regarding its applicability.

Finally, it should be mentioned that the foregoing discussion does not apply to ships that have internal webs, that is, longitudinal bulkheads. In such vessels it is not only a question of the path by which the net vertical load reaches the webs, but also the proportion of load which the internal bulkheads take relative to the side shell. This depends on many factors, particularly the layout and geometry of the framing system and the relative vertical rigidity of the bulkheads and the shell. Such questions cannot be dealt with at the hull girder level; they require a three-dimensional hull module analysis (Chapters 5–8).

3.9 HULL–SUPERSTRUCTURE INTERACTION

The interaction between hull and superstructure (or deckhouse) is highly three-dimensional and quite complex, and therefore it too cannot be dealt with adequately at the hull girder level, but rather only at the hull module level. Also, since the simple beam theory

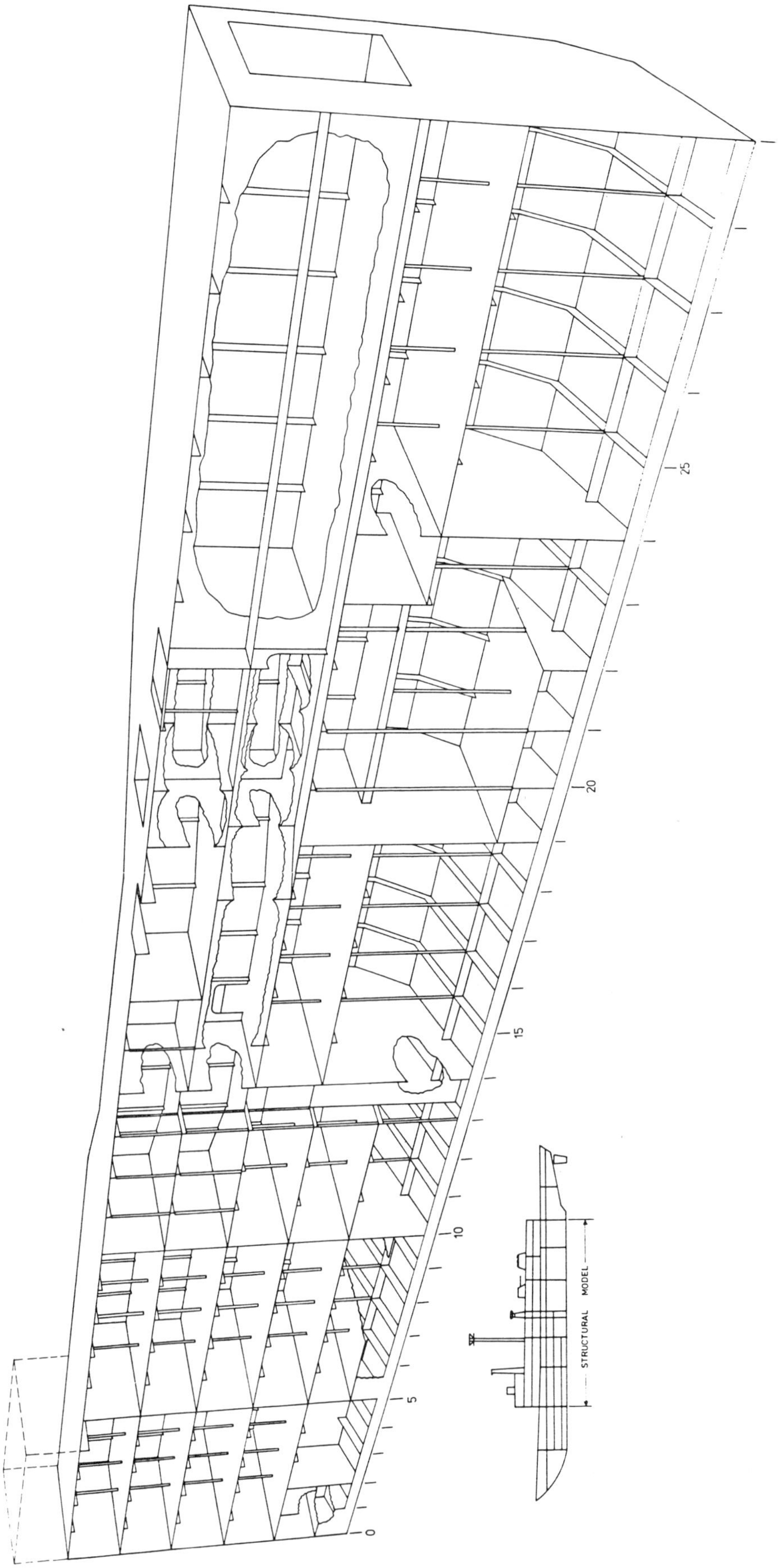

Figure 3.39 Hull/superstructure module for a naval vessel.

of the hull girder analysis cannot provide accurate values of stress or deflection at any transverse section of the superstructure, the hull module should extend over the full length of the superstructure. In most cargo ships the superstructure is simply part of the hull module containing the engine room, and has no influence on the choice of length of the hull module. But in passenger vessels and naval vessels the superstructure can be quite long and hence the need to include all of it will often dictate the required length of the hull module. If desired the initial hull module can be a simplified model which provides the boundary conditions for shorter and more detailed hull modules. But case studies performed by the author for long hull/superstructure modules have shown that with the structural modeling techniques presented in Chapters 7 and 8, the computational cost of the hull/superstructure module analysis remains quite reasonable and does not require a simplified module. For example, for the hull/superstructure module shown in Fig. 3.39 the complete analysis required 10 minutes of CPU time on a CDC Cyber 171, which corresponds to a commerical cost of about $200.

In the design of vessels with relatively long superstructures it is important to appreciate the role of the superstructure in the overall hull girder, particularly the degree to which it participates in the hull girder bending and the relative magnitude of the resulting stresses. Therefore, although the actual analysis of the superstructure is part of the hull module analysis, we here consider briefly and qualitatively the factors that have the greatest influence on the distribution and relative magnitude of hull girder bending stress in the superstructure.

This topic has long been of concern to naval architects but it is only recently, with the advent of the finite element method, that an adequate analysis has been possible. The landmark paper of Paulling and Payer [17] was the first comprehensive analysis, and since then many other contributions have been made, particularly by naval designers [18, 19, 20] because naval vessels have relatively long superstructures while at the same time it is important to keep the weight of the superstructure as small as possible because of the criticality of KG in naval vessels.

House Side Continuous with Ship Side

In essence a superstructure participates in hull girder bending to the extent that its webs (sides) are forced to act as a vertical continuation of the ship's sides, and to undergo in-plane bending to the same radius as the hull. This will be maximal if the house sides are in the same plane as the ship's sides. Figure 3.40*a* illustrates the deflected shape and the distribution of longitudinal strain ε_X for such a case. The bottom fibers of the house side undergo the same extension as the top of the hull. At the midlength the upper deckhouse fibers also contribute and so the axis of zero strain is higher and the curvature $1/R$ is smaller than if the deckhouse was not present. Hence the vertical rate of increase in strain is smaller. But away from the midlength the upper deckhouse fibers undergo progressively less extension and at the ends of the house ε_X will be zero.

Thus even though the bottom of the house side has the same curvature as the hull the vertical distribution of strain is not linear because of the longitudinal shear deformation (warping) of the house side; that is, a plane cross section of house and hull does not remain plane. The figure shows typical strain distributions at the midlength and quarter points, and also the lengthwise variation of ε_X in the upper fibers of the house side. The sharp reentrant corners would be avoided in practice since they cause very high stress concentration.

House Side Offset from Ship Side

Figure 3.40*b* illustrates the effect if the house sides are not in line with the ship's sides. Because of the relative flexibility of the deck beams the house sides are able to adopt a much larger radius of curvature and thereby escape from a good deal of the bending. (No matter how sturdy a deck beam may be, its flexural stiffness is only a small fraction of the in-plane stiffness of plating.) In this situation the strain is small even at the midlength, and the deckhouse is largely independent of the hull in regard to primary bending.

As the ship undergoes hogging and sagging the bottom corners of the deckhouse will alternately pull upwards and push downwards on the deck, and sufficient area of attachment must be provided to keep the cyclic stresses sufficiently small in order to avoid fatigue. Also it is usually advisable to terminate the deckhouse at transverse bulkheads in order to avoid excessive cyclic deflections and cyclic stresses in the deck structure.

Intermediate Transverse Bulkhead (Full Height)

Figure 3.40*c* illustrates the effect of an intermediate transverse bulkhead which continues up into the super-

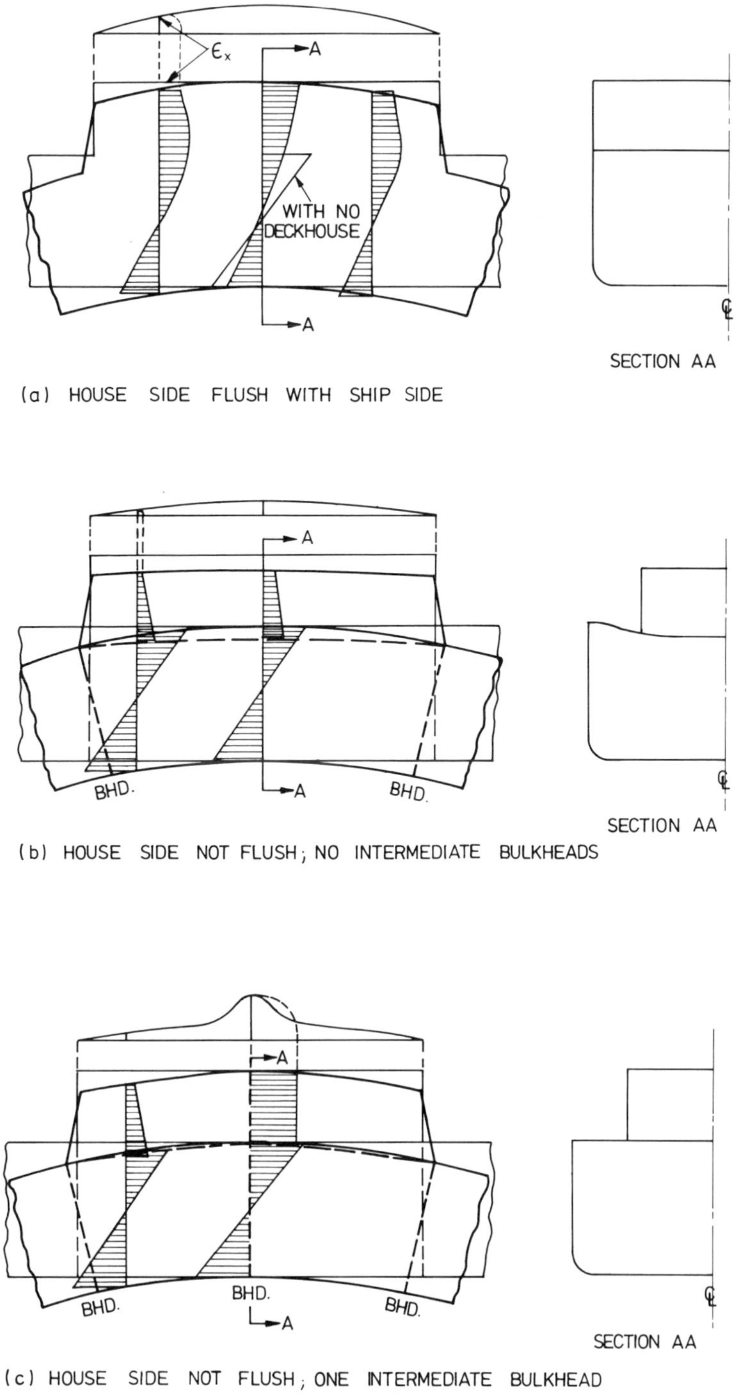

Figure 3.40 Interaction between hull and deckhouse.

structure. This forces the superstructure at that point to undergo the same deflection as the hull and causes it to undergo substantial local bending about its own neutral axis, in addition to the tension imposed by the bending of the hull. Away from the bulkhead the strain quickly diminishes because the regions of superstructure between transverse bulkheads can still maintain a larger radius of curvature than the hull. This configuration causes very large "lift off" forces at the bottom corners of the deckhouse, and special attention must be given to the design of these areas to avoid fatigue.

REFERENCES

1. R. Faresi, "A Simplified Approximate Method to Calculate Shear Force and Bending Moment in Still Water at Any Point of the Ship's Length," American Bureau of Shipping Publication, April 1965.
2. J. M. Murray, "Longitudinal Strength of Tankers," *Trans. NECIES*, **75,** 1958–59, p. 39.
3. W. Muckle, *Strength of Ships' Structures*, Edward Arnold, London, 1967.
4. P. Lersbryggen, Ed., *Ships' Load and Strength Manual*, Det norske Veritas, Hovik, Norway, 1978.
5. R. S. Little, E. V. Lewis, and F. C. Bailey, "A Statistical Study of Wave Induced Bending Moments on Large Ocean-going Tankers and Bulk Carriers," *Trans. SNAME*, 1971.
6. D. Hoffman and E. V. Lewis, "Analysis and Interpretation of Full-Scale Data on Midship Bending Stresses of Dry Cargo Ships," SSC-196, June 1969.
7. R. A. Fain and E. T. Booth, "Results of the First Five 'Data Years' of Extreme Stress Scratch Gauge Data Collected Aboard Sea-Land's SL-7's," SSC-286, March 1979.
8. E. V. Lewis and R. B. Zubaly,"Dynamic Loadings Due to Wave and Ship Motions," *STAR Symposium, SNAME*, 1975.
9. W. Munse, "Fatigue Criteria for Ship Structural Details," *Extreme Loads Response Symposium, SNAME*, pp. 231–247, 1981.
10. *Rules for Design, Construction and Installation of Offshore Structures*, Det norske Veritas, Hovik, Norway, 1977.
11. *Welding Institute Research Bulletin*, **17** (5), May 1976.
12. International Ship Structures Congress, Report of Committee I-3, 6th ISSC, Boston, August 1976.
13. P. J. Dowling, J. E. Harding, and P. A. Frieze, *Steel Plated Structures*, Crosby Lockwood Staples, London, 1977.
14. H. A. Schade, "The Effective Breadth of Stiffened Plating under Bending Loads," *Trans. SNAME*, **59,** 1951.
15. H. A. Schade, "The Effective Breadth Concept in Ship Structure Design," *Trans. SNAME*, **61,** 1953, pp. 410–430.
16. H. A. Schade, "Hull Strength," Chapter 3 of *Ship Design and Construction*, A. M. D'Arcangelo, Ed., *SNAME*, New York, 1969.
17. J. R. Paulling and H. G. Payer, "Hull-Deckhouse Interaction by Finite Element Calculations," *Trans. SNAME*, **76,** 1968, pp. 281–296.
18. G. C. Mitchell, "Analysis of Structural Interaction Between a Ship's Hull and Deckhouse," *Trans. RINA*, **120,** 1978, pp. 121–134.
19. J. D. McVee, "A Finite Element Study of Hull-Deckhouse Interaction," *Computers and Structures*, **12,** 1980, pp. 371–393.
20. J. T. Kamerrer, "A Design Procedure for Determining the Contribution of Deckhouses to the Longitudinal Strength of Ships," Assoc. of Senior Engineers, Bureau of Ships, 3rd Annual Technical Symposium, 1966.

Chapter

Four

WAVE LOADS—STATISTICAL, DYNAMIC, AND NONLINEAR ASPECTS

In this chapter we present a summary of the basic theory and technique for:

1. The mathematical representation of the physical and statistical characteristics of the sea surface elevation, considering it as a random sum of waves of various heights and frequencies, and grouped into "sea states" according to severity.
2. The calculation of the response (both linear and nonlinear) of a given ship to a random sea of specified characteristics, and hence the wave-induced loads on the ship (hull pressure distribution, hull girder bending moment, etc.)
3. The determination of design values of the wave-induced loads, that is, values which have a specified probability of nonexceedence in the ship's lifetime.

This involves various specialized branches of knowledge, including probability theory and extreme value statistical theory, statistical data—both short-term and long-term—regarding ocean storm waves, the nonlinear hydrodynamics of waves and of two-dimensional flow around solid bodies in the presence of a free surface, and the solution of coupled nonlinear fourth-order differential equations. Obviously it is a very complex field and a complete coverage would be far beyond the scope of this book, and also beyond what is normally required for rationally-based structural design of ships. As mentioned earlier, a deterministic approach is often sufficient, in which the wave bending moment is obtained from approximate formulas given by the classification societies. These formulas are the result of a variety of statistical analyses of theoretical and experimental studies and full-scale measurements, and they are adequate for all standard ship types. Nevertheless the estimation of loads is one of the most crucial aspects of structural design and hence it is important for ship structural designers to know at least the basic aspects of the theory and technique for a statistical calculation of wave loads. Moreover, there are some ship types for which it is advisable to perform an explicit statistical estimate of the wave loads. These calculations are performed by means of comprehensive computer programs developed expressly for that purpose, and it is important for the designer to be familiar with at least the basic aspects, so that he can use the programs correctly and to maximum advantage.

Probability theory is becoming an increasingly important tool in many branches of engineering, and many if not most modern engineering curricula include a basic course on probability and statistics. Therefore this chapter contains only a brief summary of the particular topics and formulas that are necessary for the presentation of the overall method. Section 4.2 deals with the basic aspects regarding the frequency analysis of a random process, especially the *spectral density function* of such a process. Section 4.3 presents information regarding the extreme values of a random process, and a formula for the *characteristic extreme value* of a wave-induced load effect, $\hat{Q}_c$, for a given sea state. Section 4.4 deals with the mathematical and statistical representation of ocean waves, showing how a typical sea state may be represented in terms of a family of *spectral density functions,* or more commonly, *wave spectra,* and presents information about the short-term and long-term statistics of these sea states. Section 4.5 deals with the nonlinear theory of ship response to wave excitation. The section includes a brief summary of linear response theory because, although not sufficiently accurate for the severe sea states for which a ship must be designed, the linear theory introduces several basic concepts and techniques, especially that of a transfer function. Because of the complexity of the nonlinear theory, this section deals only with motions in the vertical plane, that is, pitch and heave. It is these components which are of principal interest since they allow the computation of the vertical shear force and bending moment in the hull girder. Two alternative approaches are discussed. If, as is usual, the hull girder is sufficiently stiff that its fundamental period of vibration is much shorter than the mean period of wave excitation, then the waves do not cause any hull girder vibration (springing). Therefore the only significant response to the waves is wave bending moment, and the analysis is quasistatic; that is, the only time-dependent aspect is the ship's rigid body motion. Once the equations of motion have been solved the wave forces on the hull can be evaluated and can then be integrated to obtain shear force and bending moment. The only dynamic response would be whipping, caused by bottom slamming or bow flare impact, and if an analysis of this response is required it can be performed separately, modeling the hull girder as a flexible beam, and superimposed. It need not be performed simultaneously with the ship motion analysis, except for a small ship in which the motion is strongly affected by the impact. However, if the hull girder is relatively flexible, such that springing is possible, then the foregoing two-stage approach must be replaced by a unified method of solution in which a dynamic structural analysis is included as part of the ship motion analysis.

Section 4.6 shows how the methods of the previous sections can be combined to obtain the design value of wave bending moment which was discussed in Chapter 1, that is, the value corresponding to some selected probability of nonexceedence in the ship's lifetime. In brief, for the most probable extreme value of sea severity the wave excitation is calculated for the relevant wave spectra and the nonlinear equations of motion are then solved for the time-dependent response [wave bending moment, $M_w(t)$] and hence the response spectra, as shown in Section 4.5. Finally, the extreme value formulas of Section 4.3 are used to obtain the characteristic wave bending moment, $\hat{M}_{w,c}$. Section 4.6 also presents a method for investigating the full range of sea states over the ship's entire lifetime, which might be done if a particularly thorough investigation of hull girder fatigue were required.

4.1 BASIC ELEMENTS OF PROBABILITY

Probability Density Function and its Moments

In the general theory of statistics a random variable X is an event, or an outcome, among all possible outcomes. If the possible outcomes form a continuous "space," $-\infty < x < \infty$, and each event is some portion of this space, then the probability of an event occurring is simply the probability that X will lie within that portion of x. This probability is specified by a *probability density function* $p_X(x)$ as shown in Fig. 4.1. In terms of this function the probability that X lies between x and $x + dx$ is $p_X(x)dx$. That is

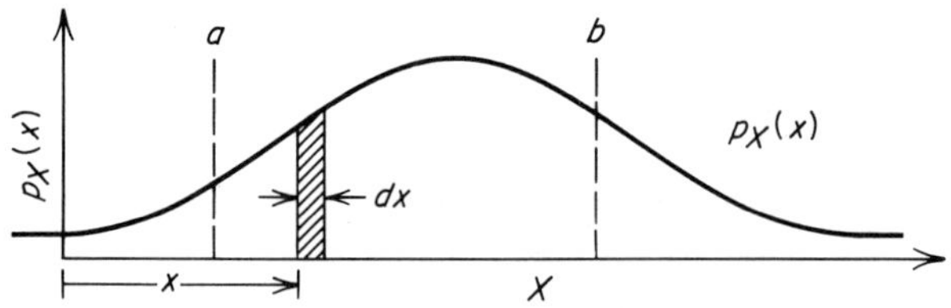

Figure 4.1 Probability density function.

$$\text{Prob}[x \leqslant X \leqslant x + dx] = p_X(x)dx$$

In general, the probability of X lying within any interval is equal to the *area* under the $p_X(x)$ curve within that interval.

$$\text{Prob}[a \leqslant X \leqslant b] = \int_a^b p_x(x)dx \qquad (4.1.1)$$

and hence the total area under the curve is equal to unity:

$$\text{Prob}[-\infty < X < \infty] = \int_{-\infty}^{\infty} p_X(x)dx = 1$$

The *cumulative probability distribution function,* $P_X(x)$, often referred to as simply the "probability distribution," is the indefinite integral of the probability density function

$$P_X(x) = \int_{-\infty}^{x} p_x(x)dx$$

We next define $E[X]$, the *average* or the *expected value,* either of a random variable X or more generally, of any function of X, $f(x)$, as follows:

$$E[X] = \int_{-\infty}^{\infty} x\, p_X(x)dx \qquad (4.1.2)$$

$$E[f(X)] = \int_{-\infty}^{\infty} f(x)p_X(x)dx \qquad (4.1.3)$$

Equation 4.1.3 states that, for any function $f(X)$ of a random variable which has a probability density function $p_X(x)$, the *average of* $f(X)$ is equal to the *moment of* $p_X(x)$, with $f(X)$ taken as the moment arm. For the special case in which $f(X)$ is X itself, the average is simply the *mean,* or *mean value,* of X, that is, the most direct and most familiar type of average. We shall use the symbol μ to indicate this mean value, thus:

$$\mu = E[X] = \int_{-\infty}^{\infty} x\, p_X(x)dx \qquad (4.1.4)$$

We next examine the average or expected value of some simple functions of X, namely, powers of X. In this case the moments are second moments, third moments, and so on, of $p_X(x)$. For example, the average of X^2, or the *mean square* of X, is the expected value of X^2. From (4.1.3) this is

$$E[X^2] = \int_{-\infty}^{\infty} x^2 p_X(x)dx \qquad (4.1.5)$$

which is the second moment of $p_X(x)$, taken about $x = 0$.

In defining the averages for higher powers of X it is convenient to introduce the *deviation* from the mean, $X - \mu$, and to take powers of the deviation instead of X, because this corresponds to taking moments about the mean μ.

With this definition the second moment is a measure of the spread or dispersion of $p_X(x)$ and is known as the *variance*, σ^2.

$$\begin{aligned}\sigma^2 &= E[(X - \mu)^2] \\ &= \int_{-\infty}^{\infty} (x - \mu)^2\, p_X(x)dx \qquad (4.1.6) \\ &= m_2\end{aligned}$$

The symbol m_2 is introduced for this second moment in order to be consistent with the more general relationship to be defined in (4.1.8).

The only difference between the variance and the earlier form of second moment, the mean square, is the use of a different moment arm and hence they are closely related:

$$\begin{aligned}\sigma^2 &= \int_{-\infty}^{\infty} x^2 p_X(x)dx - 2\mu\int_{-\infty}^{\infty} x p_X(x)dx \\ &\qquad + \mu^2\int_{-\infty}^{\infty} p_X(x)dx \\ &= E[X^2] - \mu^2 \qquad (4.1.7)\end{aligned}$$

Since $p_X(x)dx$ is a pure number, the units of variance are those of X^2. This is not convenient for some purposes and therefore the measure of dispersion is usually taken as the positive square root of σ^2. This quantity, σ, is referred to as the *standard deviation*. It is also the root mean square of the deviation.

It may be helpful to note that if $p_X(x)$ were visualized as the mass distribution of a rod, then μ would equal the distance of the center of mass from $x = 0$, and σ^2 would equal the moment of inertia about the center of mass. Also, (4.1.7) is simply a statement of the parallel axis theorem.

We next proceed to define moments of higher order, denoting a moment of order k as m_k:

$$m_k = E[(X - \mu)^k] = \int_{-\infty}^{\infty} (x - \mu)^k p_X(x)dx \quad (4.1.8)$$

Each of these moments is a type of weighted average of $p_X(x)$; that is, each moment is a parameter which helps to characterize $p_X(x)$, and it would be possible to define or describe a probability density function in terms of its various moments. However, a complete and unique definition would in general require moments of all orders. This is impractical and we usually deal only with the first two orders: the mean and the variance. As we shall see, for some particular probability distributions this information is sufficient to completely define the distribution.

Some One-Dimensional Probability Density Functions*

GAUSSIAN (OR NORMAL) DISTRIBUTION

$$p_X(x) = \frac{1}{\sqrt{2\pi}\,\sigma} e^{-(x-\mu)^2/2\sigma^2} \; (-\infty < x < +\infty) \qquad (4.1.9)$$

This is by far the most common distribution; many random processes, be they physical or mathematical, exhibit this relationship. One of the features of this distribution is that it is defined entirely and explicitly in terms of its mean value μ and its variance σ^2. Some typical Gaussian distributions are shown in Fig. 4.2 for $\mu = 0$ and $\sigma = 1$, 2, and 3.

*Where a slash (/) is used in an exponent, all of the terms to the right of the slash are in the denominator, unless indicated otherwise by parentheses.

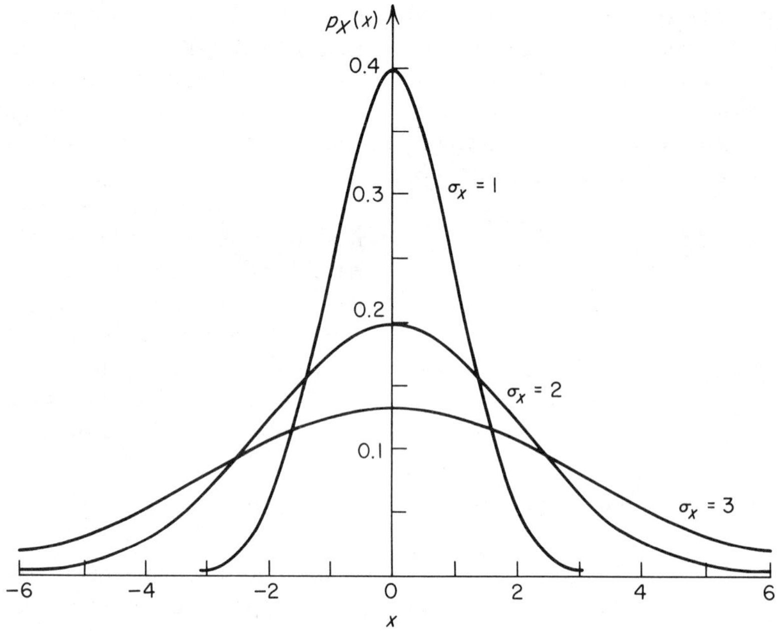

Figure 4.2 Gaussian probability density function, with $\sigma_X = 1, 2$, and 3, and with $\mu_X = 0$.

Log-Normal Distribution

A positive random variable X is said to have a log-normal distribution if its natural logarithm, $Z = \ln X$, has a normal (Gaussian) distribution. That is

$$p_Z(z) = \frac{1}{\sqrt{2\pi}\,\sigma_Z} e^{-(z-\mu_Z)^2/2\sigma_Z^2}$$

where μ_Z and σ_Z^2 are the mean and the variance of $\ln X$. The probability density function of X is then

$$p_X(x) = p_Z(x)\frac{dZ}{dX} = \frac{p_Z(x)}{x}$$

$$= \frac{1}{\sqrt{2\pi}\,\sigma_Z x} e^{-(\ln x-\mu_Z)^2/2\sigma_Z^2} \qquad (x \geqslant 0) \quad (4.1.10)$$

It has been found from ocean wave data that the two parameters which characterize a sea state—significant height and modal period—follow a log-normal distribution. This is discussed in Section 4.4.

Rayleigh Distribution

The surface elevation X at any point in the ocean is a random variable with a Gaussian distribution and a zero mean. In Section 4.2 it will be shown that the *peak values* of X, denoted here as $\tilde{X}$, have a Rayleigh probability density function which is

$$p_{\tilde{X}}(\tilde{x}) = \frac{\tilde{x}}{\sigma_X{}^2} e^{-\tilde{x}^2/2\sigma_X{}^2} \qquad (\tilde{x} > 0) \quad (4.1.11)$$

or in nondimensional form, with $Y = \tilde{x}/\sigma_x$

$$p_Y(y) = p_X(y)\frac{d\tilde{X}}{dY} = \frac{y}{\sigma_X} e^{-y^2/2}\sigma_X$$

$$= y\, e^{-y^2/2}$$

where σ_X^2 is the variance of X. This is illustrated in Fig. 4.3.

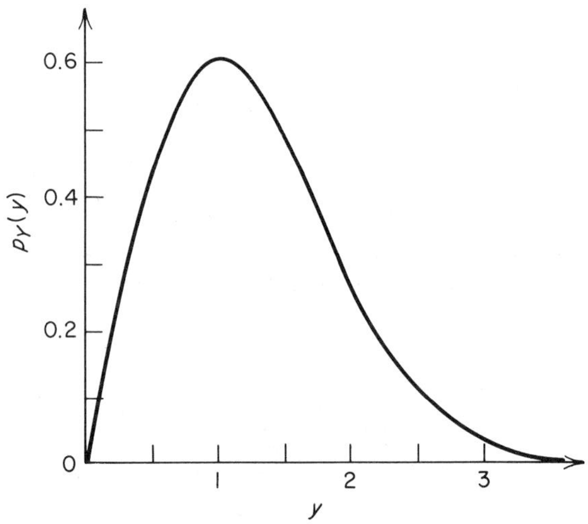

Figure 4.3 The Rayleigh probability density function of the nondimensional random variable $y = \tilde{x}/\sigma_X$.

Note that all three of the foregoing distributions are defined in terms of no more than two parameters: μ and σ^2. In such cases the first two moments, $m_1 = \mu = E[X]$ and $m_2 = \sigma^2 = E[X^2]$, are sufficient to completely define the distribution; if they are known then the full distribution can be created. For other distributions further information would be required.

CENTRAL LIMIT THEOREM

The central limit theorem states that if a random variable Z is the sum of n independent random variables

$$Z = \sum_{i=1}^{n} X_i$$

and if all of the X_i have the same distribution then the probability density function $p_Z(z)$ for Z approaches the Gaussian distribution as n approaches infinity, regardless of the type of distribution of the X_i variables. That is, a random variable which is the sum of similar random effects tends to have a Gaussian distribution.

Probability Density Function for Two Variables

If there are two events or random variables X and Y, the corresponding probability density function expresses the probability of both events occurring together

$$\text{Prob}[(x \leq X \leq x + dx) \cap (y \leq Y \leq y + dy)] = p_{XY}(x, y)dx\,dy$$

A typical probability density function for two variables is shown in Fig. 4.4, illustrating that the joint probability is the volume under that part of the $p_{XY}(x, y)$ surface bounded by the two intervals.

$$\text{Prob}[(a \leq X \leq b) \cap (c \leq Y \leq d)] = \int_a^b \int_c^d p_{XY}(x, y)dy\,dx$$

Also, the total volume under the complete surface is unity.

In the special case when the two random variables are independent, then the joint probability density function is simply the product of the two separate distributions

$$p_{XY}(x, y) = p_X(x)\, p_Y(y) \qquad (4.1.12)$$

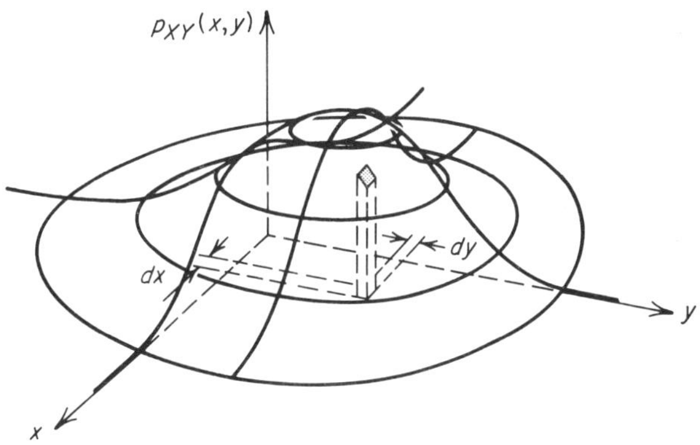

Figure 4.4 Second-order probability density function.

For a joint distribution we can again take moments, as before. The expected value of the product XY corresponds to a mixed second moment

$$E[XY] = \int_{-\infty}^{\infty} \int_{-\infty}^{\infty} xy\, p_{XY}(x, y)dx\,dy \qquad (4.1.13)$$

If we take moments about the respective mean values μ_X and μ_Y (computed independently, from the one-dimensional distributions) we have what is known as the *covariance* C_{XY} of the random variables X and Y

$$\begin{aligned} C_{XY} &= E[(X - \mu_X)(Y - \mu_Y)] \\ &= \int_{-\infty}^{\infty} \int_{-\infty}^{\infty} (x - \mu_X)(y - \mu_Y)p_{XY}(x, y)dx\,dy \\ &= E[XY] - E[X]E[Y] \qquad (4.1.14) \end{aligned}$$

When $X = Y$ the covariance C_{XX} reduces to the variance σ_X^2 of the random variable X.

The *correlation coefficient* ρ_{XY} of the random variable is defined as

$$\rho_{XY} = \frac{C_{XY}}{\sigma_X \sigma_Y} \qquad (4.1.15)$$

It can be shown that the correlation coefficient lies in the range $-1 \leq \rho_{XY} \leq 1$. When $\rho_{XY} = 0$, the random variables X and Y are said to be "uncorrelated." If two random variables X and Y are independent then they are also uncorrelated since

$$\begin{aligned} E[XY] &= \int_{-\infty}^{\infty} \int_{-\infty}^{\infty} xy\, p_{XY}(x, y)dx\,dy \\ &= \int_{-\infty}^{\infty} xp_X(x)dx \int_{-\infty}^{\infty} yp_Y(y)dy \\ &= E[X]E[Y] \end{aligned}$$

and $\quad C_{XY} = E[XY] - E[X]E[Y] = 0 = \rho_{XY}$

Note, however, that uncorrelated random variables are not necessarily independent random variables.

The expected value of a linear combination of random variables $Z = aX + bY$ is

$$\begin{aligned} E[Z] &= E[(aX + bY] \\ &= \int_{-\infty}^{\infty}\int_{-\infty}^{\infty} (ax + by)p_{XY}(x, y)dx\,dy \\ &= a\int_{-\infty}^{\infty} x\int_{-\infty}^{\infty} p_{XY}(x, y)dy\,dx \\ &\quad + b\int_{-\infty}^{\infty} y\int_{-\infty}^{\infty} p_{XY}(x, y)dx\,dy \\ &= a\int_{-\infty}^{\infty} xp_X(x)dx + b\int_{-\infty}^{\infty} yp_Y(y)dy \\ &= a\mu_X + b\mu_Y \end{aligned} \tag{4.1.16}$$

and the variance is

$$\begin{aligned} \sigma_Z^2 &= E[\{(aX + bY) - (a\mu_X + b\mu_Y)\}^2] \\ &= E[\{a(X - \mu_X) + b(Y - \mu_Y)\}^2] \\ &= a^2E[(X - \mu_X)^2] + 2abE[(X - \mu_X)(Y - \mu_Y)] \\ &\quad + b^2E[(Y - \mu_Y)^2] \\ &= a^2\sigma_X^2 + 2abC_{XY} + b^2\sigma_Y^2 \end{aligned} \tag{4.1.17}$$

If X and Y are independent random variables, then the covariance $C_{XY} = 0$ and

$$\sigma_Z^2 = a^2\sigma_X^2 + b^2\sigma_Y^2 \tag{4.1.18}$$

We may extend these results to the set of n independent random variables $X_1, X_2, \ldots, X_n$ with individual expected values μ_i and variances σ_i^2 $(i = 1, 2, \ldots, n)$. It may be shown that the summed random variable

$$z = \sum_{i=1}^{n} X_i$$

has the expected value

$$\mu_Z = \sum_{i=1}^{n} \mu_i \tag{4.1.19}$$

and variance

$$\sigma_Z^2 = \sum_{i=1}^{n} \sigma_i^2 \tag{4.1.20}$$

4.2 RANDOM PROCESSES

A *random process* is a random function of a *time* parameter. Figure 4.5 shows several samples or observations of a random, time-dependent quantity $X(t)$. In each case the time t is the time from the commencement of sampling. Since X is a random variable, each curve $X^{(k)}(t)$ represents just one sample out of an infinite number of possible samples or measurements. The collection of such samples, theoretically infinite in number, is known as an "ensemble."

Of course, there may be more than one independent variable; for example, the process may depend on location as well as time. Whatever the dependency, it is necessary that each sample be taken under identical conditions, at least in regard to the factors which affect $X(t)$, so that $X(t)$ is a truly random process. If variations in some such factor cannot be avoided then that factor must be included as a second independent variable, in addition to t, and the process would then be "two-dimensional." In the case of ocean waves, in which $X(t)$ is the sea surface elevation, it is possible to identify and isolate these factors sufficiently well so that for short term observations (up to a few hours in duration) $X(t)$ is essentially one-dimensional.

Stationary and Ergodic Random Process

If the statistical characteristics of a random process do not change with time t we say that the process is *stationary*. Thus, for example, if the process in Fig. 4.5 were a stationary random process, then the probability density functions at t_1, t_2, and so on would all be the same. Now, the statistical quantities used in the analysis of random processes are averages of $X(t)$ and of various functions of $X(t)$, such as its square, as is done for a simple, non-time-dependent variable X. If the process is stationary then the averages and moments are invariant over time, and they therefore have the same definition and meaning as given earlier for simple, non-time-dependent random variables. That is, the mean, the variance, and the higher moments of the process continue to be given by (4.1.4), (4.1.6), and (4.1.8).

It has been found experimentally that, over the short term (as defined before) the sea surface elevation is a stationary random process. Hence all of the earlier definitions of averages and moments are still applicable, even though the random variable X is now a function of time $X(t)$.

Because a random process is a function of time there are in fact two different ways of calculating aver-

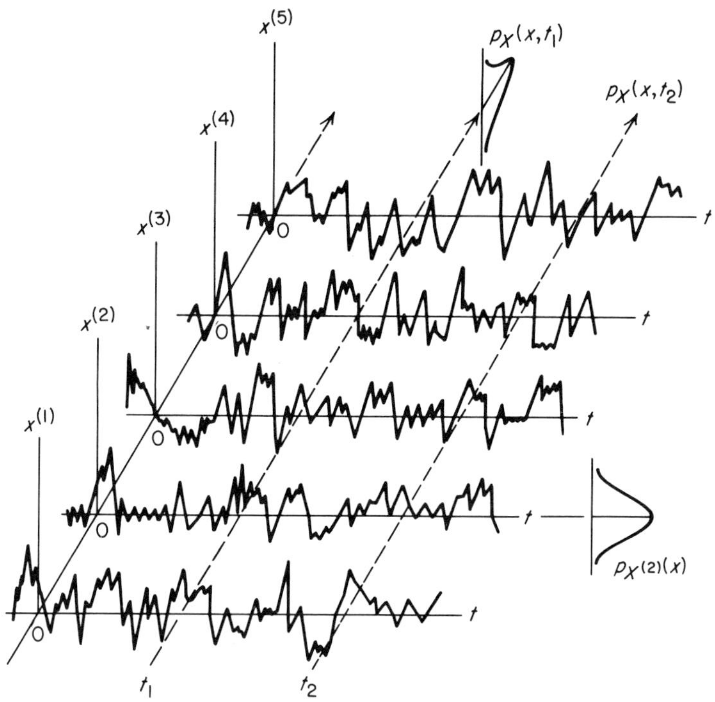

Figure 4.5 Schematic representation of a random process $X(t)$. Each $X^{(j)}(t)$ is a sample function (or, realization) of the ensemble.

ages. The averages may be taken over all of the samples of the ensemble at the same instant of time, say t_1, or they may be taken for a particular sample, say $X^{(1)}(t)$ over all time from $-\infty$ to ∞. These two alternatives are referred to as *ensemble averages* and *temporal* (or *time*) *averages,* respectively. All of the averages defined thus far have been ensemble averages. For instance, the mean μ, the variance σ^2, and the higher moments m_k are in fact the ensemble averages of X, $(X - \mu)$, and $(X - \mu)^k$, respectively. Likewise, the ensemble average of any function $f(X)$ of X is given by (4.1.3).

The temporal averages, on the other hand, are computed for a particular sample $X(t) = X^{(1)}(t)$ [we will omit the (1) superscript from here on] over some length of time—a *long* time for sufficient accuracy. Thus, for example, the *temporal mean* is*

$$\mu = \langle X(t) \rangle = \lim_{T\to\infty} \frac{1}{T} \int_{-T/2}^{T/2} X(t)dt \qquad (4.2.1)$$

Likewise, the *temporal mean square* is

$$\langle X^2(t) \rangle = \lim_{T\to\infty} \frac{1}{T} \int_{-T/2}^{T/2} X^2(t)dt \qquad (4.2.2)$$

*Brackets ⟨ ⟩ indicate a temporal averaging.

It is often useful to represent a time-dependent variable as the sum of a temporal mean value plus a fluctuating component. The fluctuating component is described by the *variance* σ^2 which is from (4.1.6)

$$\sigma^2 = \langle [X(t) - \mu]^2 \rangle$$
$$= \lim_{T\to\infty} \frac{1}{T} \int_{-T/2}^{T/2} [X(t) - \mu]^2 dt \qquad (4.2.3)$$

In practice, of course, some finite length of time must be used in place of the limit $T\to\infty$, and so the above equalities become approximations, or estimates, of the mean, the mean square, and the variance.

In general, these time averages would be different from those obtained by averaging "across the ensemble." However, many random processes, including ocean waves, are such that time averages formed from a single sample over a time interval are, in the limit, equal to the ensemble averages. Such processes are known as *ergodic* processes. In qualitative terms, an ergodic process is one in which a single sample $X(t)$ is sufficiently typical to represent the entire process. Obviously, an ergodic process must be stationary, but a stationary random process is not necessarily ergodic. Ergodicity implies that all of the various expectations are equal to, and may be replaced by, the corresponding temporal averages. This is quite important because, although most of the theoretical relationships are defined or derived in terms of ensemble averages, in practice these are difficult, if not impossible, to obtain, whereas calculating time averages from a single observation is relatively easy. If the process is ergodic these can be used in place of the ensemble averages.

As noted earlier, there are two distinct methods of specifying a random process $X(t)$:

1. Through its probability density function.
2. Through various averages, which correspond to various moments of the probability density function.

The first of these methods is impractical, since it involves an enormous amount of information. The second is practical, providing that the averages can be measured. Hence the importance of ergodicity. With this property, the required averages may be computed from measurements of a single observation. Ergodicity is assumed in virtually all practical applications, particularly in the estimation of parameters of empirical models. Fortunately, it has been established by oceanographers and statisticians that the surface elevation at a given location in the ocean can be considered, at least for engineering purposes, as an ergodic process with a Gaussian distribution. This will be taken up in Section 4.4.

Auto-Correlation Function for an Ergodic Random Process

The averages which are used when analyzing linear systems by statistical methods are those which represent or measure the degree of association between values of the random variable $X(t)$ at times differing by a specified time interval τ. These averages are called *correlation functions*. The most basic of them is the *auto-correlation function*, $R(\tau)$, which is the average, or expected value, of the product of any two values of X: $X_1 = X(t_1)$ and $X_2 = X(t_2) = X(t_1 + \tau)$

$$R(\tau) = E[X(t)X(t+\tau)] = E[X_1X_2] = \int_{-\infty}^{\infty}\int_{\infty}^{\infty} x_1x_2 p_{XX}(x_1, x_2)dx_1dx_2 \qquad (4.2.4)$$

For a stationary process $R(\tau)$ is independent of t since $E[X(t_1)X(t_1+\tau)] = E[X(t_2)X(t_2+\tau)]$. Also, since the product X_1X_2 is commutative we have

$$R(\tau) = E[X_1X_2] = E[X_2X_1] = E[X(t_2)X(t_2-\tau)] = R(-\tau)$$

indicating that $R(\tau)$ is an even function of τ. The value at the origin is simply the mean square

$$R(0) = E[X^2] \qquad (4.2.5)$$

In order to get some idea of the shape of $R(\tau)$, let us consider an ensemble of functions $X(t)$ with zero mean.* If t and τ are such that there is little or no association between $X(t)$ and $X(t+\tau)$, then a particular value of $X(t)$ is just as likely to be associated with a positive value of $X(t+\tau)$ as with negative value. For this reason we can expect the average value of the product $X(t)X(t+\tau)$ to be near zero. On the other hand, the association between the two values may be such that both tend to have the same sign, either positive or negative. In this case the product will have an average which is definitely positive. Again, if $X(t)$ and $X(t+\tau)$ tend to have opposite signs the average of the product will be negative.

To clarify matters let us consider two extreme cases. On the one hand values of a function $X(t)$ of the ensemble at different times may be completely unrelated, however close the times may be. An ensemble of this type, which is called *white noise*, would in general have no continuity; it represents in fact the characteristic of erratic variation in the extreme. In this case $R(\tau) = 0$ for all values of τ except zero, where it equals the mean square.

On the other hand, if each sample $X(t)$ were identical, thus exhibiting perfect correlation, then $R(\tau)$ would be a constant, equal to the mean square.

Usually the situation is intermediate between these extremes. Thus if τ is small, $X(t+\tau)$ can be expected to lie in a range of values that do not differ greatly from $X(t)$. On the other hand, if τ is large there will be very little association between $X(t)$ and $X(t+\tau)$. Consequently, provided that there are no periodic components present in the random process, $R(\tau)$ tends to zero since products will be equally positive and negative. That is, $R(\tau)\rightarrow 0$ as $\tau\rightarrow\infty$. With a nonzero mean, $R(\infty) = \mu^2$. Some typical auto-correlation functions are shown in Fig. 4.6.

Instead of the ensemble average of X_1X_2 we can form the time average; the *temporal auto-correlation* function is

*This restriction is not necessary for the theory developed in this section but since it is characteristic of ocean waves and since it also simplifies the development, it is introduced at this stage.

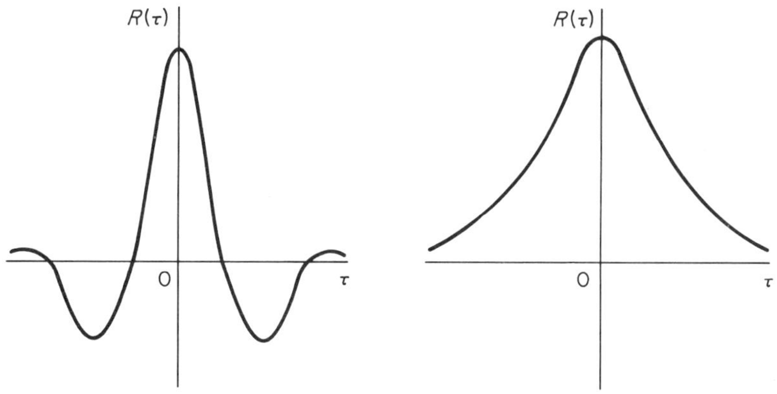

Figure 4.6 Possible forms of the auto-correlation function.

$$R(\tau) = \langle X(t)X(t + \tau)\rangle$$
$$= \lim_{T\to\infty} \frac{1}{T}\int_{-T/2}^{T/2} X(t)X(t + \tau)dt \qquad (4.2.6)$$

For an ergodic process this will be identical to the ensemble average of (4.2.4), and in the case of a physical process the time average is much easier to obtain. Hence (4.2.6) is taken as the definition of $R(\tau)$.

Some further insight into the concept of auto-correlation may be gained by examining a typical method for measuring the auto-correlation of a random signal (for example the heave motion of a buoy for measuring wave heights). Figure 4.7 shows a block diagram for the procedure. The sample function is recorded on a tape, and a tape recorder with *two* heads is used. The spacing between the heads is adjustable. If the speed of the tape is V, the time interval between readings is $\tau = a/V$. The values $f(t)$, $f(t + \tau)$ are then multiplied and averaged over a long time period T. For finite averaging time the measured quantity $y(t)$ fluctuates slowly. In the limit $T\to\infty$ the reading y approaches the temporal auto-correlation function $R(\tau)$ and, if the process is ergodic, this is also equal to the ensemble auto-correlation function. In order to obtain the auto-correlation function as a function of τ it is necessary to perform the average just described for a whole range of values of τ, that is, a whole range of values of a.

Frequency Analysis of an Ergodic Process: Spectral Density Function

In dealing with fluctuating time variations it is usually more convenient to work in terms of frequency. This

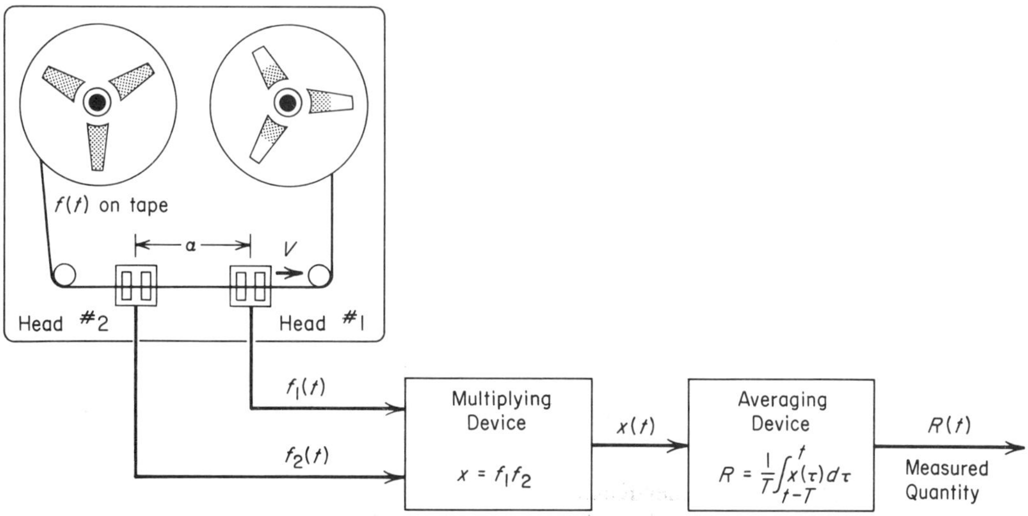

Figure 4.7 Schematic diagram for measurement of auto-correlation function $R(\tau)$ for sample function $f(t)$. The adjustable distance a between tape heads is proportional to lag τ.

is obviously true for *periodic* variations, but it is also true for random variations, particularly if the process is ergodic.

Before considering a random function $X(t)$, it is useful to recall the Fourier expansion of a *deterministic* time dependent function $g(t)$. When $g(t)$ is periodic with period T, or frequency $\omega_0 = 2\pi/T$, we have

$$g(t) = \sum_{n=-\infty}^{\infty} g(n)e^{in\omega_0 t}$$

where $$g(n) = \frac{1}{T}\int_{-T/2}^{T/2} g(t)e^{-in\omega_0 t}\,dt$$

The quantity $g(n)$ is referred to as the "Fourier transform" of the periodic function $g(t)$. The function $g(t)$ and the quantity $g(n)$ are said to constitute a "Fourier transform pair."

When $g(t)$ is a nonperiodic function it may, under fairly general conditions, still be represented in a transformed manner as a "Fourier integral":

$$g(t) = \int_{-\infty}^{\infty} G(\omega)e^{i\omega t}\,d\omega$$

in which $$G(\omega) = \frac{1}{2\pi}\int_{-\infty}^{\infty} g(t)e^{-i\omega t}\,dt$$

is the Fourier transform of the function $g(t)$. Here again the functions $g(t)$ and $G(\omega)$ are a Fourier transform pair, constituting inverses of each other. For deterministic time-dependent functions the transformation to the "frequency domain" usually simplifies and facilitates the analysis.

For a random process $X(t)$ a similar frequency-based or spectral representation can be achieved. Although it is not possible to transform $X(t)$ directly, it can be shown that the auto-correlation function $R(\tau)$ fulfills all of the requirements for defining a Fourier transform. The Fourier transform of $R(\tau)$ and its inverse for a stationary random process $X(t)$ are respectively

$$S(\omega) = \frac{1}{2\pi}\int_{-\infty}^{\infty} R(\tau)\cos\omega\tau\,d\tau \tag{4.2.7}$$

$$R(\tau) = \int_{-\infty}^{\infty} S(\omega)\cos\omega\tau\,d\omega \tag{4.2.8}$$

These two equations are called the "Wiener-Khintchine Relations." The quantity $S(\omega)$ is called the "spectral density function" because its domain is the spectrum $-\infty < \omega < \infty$. Like the auto-correlation function, the spectral density function is a measure, or at least an indicator, of the repetitiveness or harmonic content of the random process $X(t)$, and of its distribution over the frequency range. If $X(t)$ exhibits some approximate repetition having an approximate frequency ω_0, then $S(\omega)$ will have a local peak in the vicinity of ω_0.

The full name of this function is the "*mean square* spectral density function," or the variance spectrum for $E[x(t)] = 0$. It is called this because the area under the $S(\omega)$ curve is the mean square value of $X(t)$. This may be shown from (4.2.8) in which, if we set $\tau = 0$, the right-hand side becomes the area under the $S(\omega)$ curve, and the left-hand side becomes $R(0)$, which is simply the mean square of the process, as was shown earlier in (4.2.5). That is,

$$\begin{aligned}\text{Area under } S(\omega) = \int_{-\infty}^{\infty} S(\omega)d\omega &= R(0)\\ &= E[X^2(t)]\\ &= \langle X^2(t)\rangle\end{aligned} \tag{4.2.9}$$

Hence, in an analogous manner to the probability density function, the area $S(\omega)d\omega$ which lies within the interval or "band" from ω to $\omega + d\omega$ is the contribution which is made to the mean square by components whose frequencies lie within that band.

Other properties of $S(\omega)$ may be deduced from its definition. Since $R(\tau)$ is real and even, it follows that $S(\omega)$ will be a real and even function of ω. Also, since $S(\omega)$ is the spectral density of the mean square, it is nonnegative. An example of a typical spectral density function is shown in Fig. 4.8. It should be noted that in defining $S(\omega)$ we adopted, simply for mathematical convenience, the range or spectrum $-\infty < \omega < \infty$. Since we wish to apply this theory to physical processes, it is necessary to use a modified, one-sided spectrum: $\omega \geqslant 0$. For this purpose we define the one-sided mean square spectral density function

$$S^+(\omega) = \begin{cases} 2\,S(\omega) & \text{for } \omega \geqslant 0\\ 0 & \text{otherwise}\end{cases} \tag{4.2.10}$$

as shown in Fig. 4.8. Since both the auto-correlation function and the spectral density function are real and even, we have

$$R(\tau) = \int_0^{\infty} 2S(\omega)\cos\omega\tau\,d\omega$$

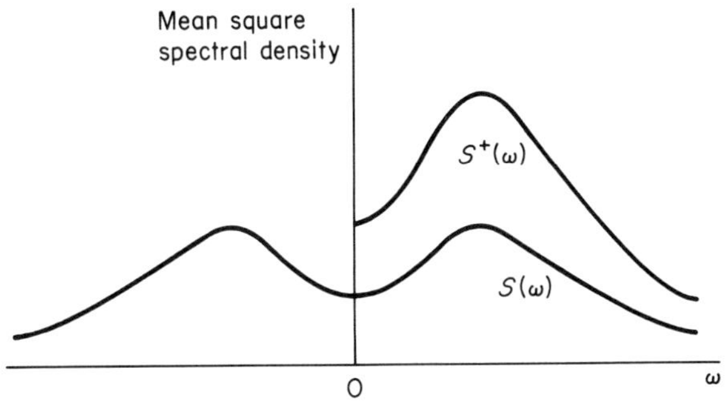

Figure 4.8 One- and two-sided mean square spectral density functions of a random process.

$$= \int_0^\infty S^+(\omega) \cos \omega\tau \, d\omega \qquad (4.2.11)$$

and so for a one-sided spectrum $S^+(\omega)$ bears the same relationship to $R(\tau)$ as did $S(\omega)$ for the two-sided spectrum. Since S and S^+ are so closely related, and since the latter is used in all practical applications, we will not bother using the $+$ sign to distinguish them. We will use the two-sided spectrum only when it is more convenient mathematically, and its use in such instances will be clear from the context.

Since the spectral density function is the mathematical transform of the auto-correlation function of $X(t)$, the units of $S(\omega)$ depend on the units of $X(t)$. If $X(t)$ is wave height in meters then $S(\omega)$ has units of m^2 sec. It may be shown that for any kind of wave—gravitational, electromagnetic, and so on—the area $S(\omega_0)\,\delta\omega$ within a bandwidth $\delta\omega$ is directly proportional to the total energy of all of the components which lie within the band $(\omega_0 - \frac{1}{2}\delta\omega, \omega_0 + \frac{1}{2}\delta\omega)$. Because of this direct relationship between energy and the mean square spectral density function, and because of the attraction of a shorter name, the terms "energy spectrum" or "wave spectrum" are commonly used in place of the full name. Sometimes this is even further shortened to simply "spectrum," although strictly speaking this word means a range of frequencies rather than a function defined within that range.

Moments of the Spectral Density Function

One of the principal advantages of the spectral density function is that all of the basic characteristics of a random process can be expressed in terms of moments of this function:

$$m_n = \int_0^\infty \omega^n S(\omega) d\omega$$

where n can be any integer. We have already seen in (4.2.9) that the mean square value is the "zero-order moment. That is

$$\langle X^2(t) \rangle = m_0 = \int_0^\infty S(\omega) d\omega \qquad (4.2.12)$$

Some other basic characteristics which may be expressed in terms of spectral moments are the following:

Average mean period

$$\overline{T}_p = 2\pi \frac{m_0}{m_1} \qquad (4.2.13)$$

Average zero-crossing period

$$T_{p0} = 2\pi \sqrt{\frac{m_0}{m_1}} \qquad (4.2.14)$$

Average crest-to-crest period:
For a general random process

$$T_{pc} = 4\pi \left(\frac{\sqrt{1-\varepsilon^2}}{1+\sqrt{1-\varepsilon^2}} \right) \sqrt{\frac{m_o}{m_2}} \qquad (4.2.15)$$

and for a narrow-band random process

$$T_{pc} = 2\pi \sqrt{\frac{m_2}{m_4}} \qquad (4.2.16)$$

Broadness (or "bandwidth")

$$\varepsilon = \sqrt{1 - \frac{m_2^2}{m_0 m_4}} \qquad (4.2.17)$$

Properties of a Narrow-Band Random Process

Figure 4.9 shows two extreme types of random process, and their corresponding energy spectra. The two types are:

1. *Narrow-Band Process*—A process which is made up of components whose frequencies lie within a narrow band or range, whose width is small compared with the magnitude of the center frequency of the band, ω_0. This periodicity produces regularly spaced peaks in the auto-correlation function and a single narrow peak at ω_0 in the energy spectrum.

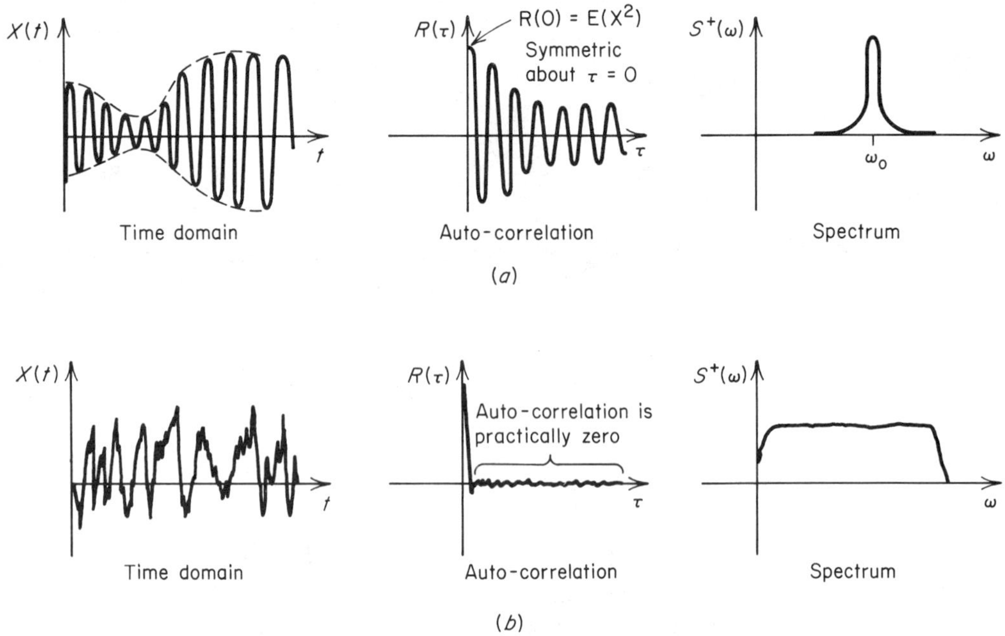

Figure 4.9 Narrow-band (*a*) and wide-band (*b*) processes.

2. *Wide-Band Process*—In this case the process contains components of many different frequencies, so that there is little or no periodicity and the auto-correlation function is almost zero. The energy spectrum is therefore quite wide.

Analysis of ocean wave data has shown that for a fully developed, wind-generated, mid-ocean sea state (i.e., no growth or decay, no coastal effects, and no swell) the wave spectrum is relatively narrow-banded. Of course high-frequency wave components do occur but they correspond to waves that are small—both in height and in length—and which have little effect on the ship. In fact, the ship acts as a filter, such that the spectra of ship motions and of the hull girder load effects are even more narrow-banded than the wave spectrum. Also, like the waves, these various responses have distributions which are Gaussian and stationary (in the short term, i.e., for a given sea state). This is an important point because a process of this type is much easier to analyze than a general, wide-band process. In this section we present some of the principal characteristics of a narrow-band process, $X(t)$; these will have application not only to waves but also to the various wave-induced load effects in ships, especially hull girder bending moment.

For simplicity we shall assume that $X(t)$ has a zero mean. This is, of course, true for the ocean surface elevation and it is also true for wave-induced motions and load effects if they are calculated from the linearized theory presented in Section 4.5. The more accurate nonlinear motions and load effects have nonzero mean values, but this in itself does not present any difficulty. Rather, the difficulty with nonlinear processes is that a frequency-based response analysis is not possible; the analysis must be performed in the time domain. This is discussed in Section 4.5.

Distribution of Peak Values

For design purposes we are interested in the peak values, $\tilde{X}$, rather than the full range of values of $X(t)$. The peak values of a random process are a special subgroup and they therefore have a probability density function of their own, different to that of $X(t)$. If we denote this function as $p_{\tilde{X}}(\tilde{x})$, the probability that a peak value, chosen at random, will exceed the value $\tilde{X}$ is

$$\text{Prob}\{\text{peak exceeds } \tilde{X}\} = \int_{\tilde{X}}^{\infty} p_{\tilde{X}}(x)dx$$

This is illustrated in Fig. 4.10.

As shown originally by Rice [1] and in standard texts such as Ref. 2, $p_{\tilde{X}}(\tilde{x})$ may be derived by examining the probability of "positive crossings" of $X = \tilde{x}$, that is, the average frequency $\omega_{\tilde{x}}$ with which X exceeds a specified magnitude $\tilde{x}$. If $X(t)$ is Gaussian the fre-

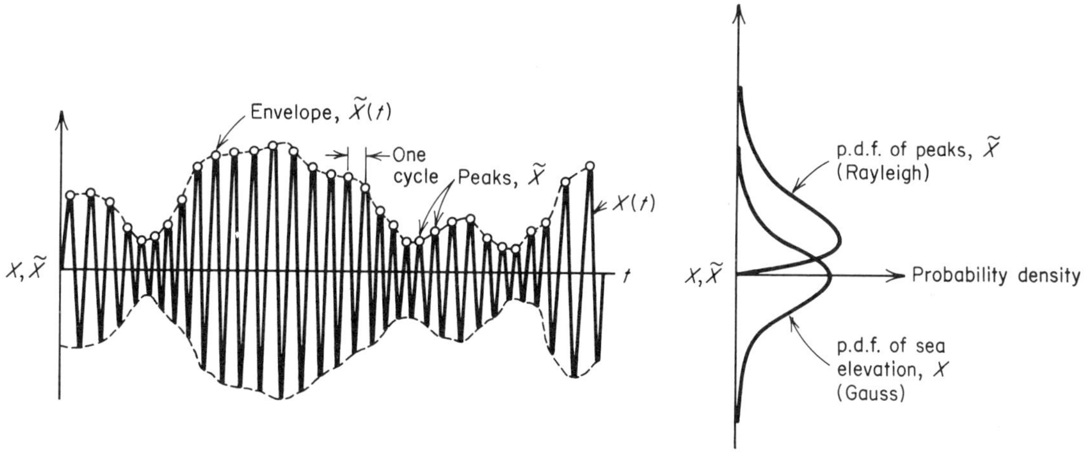

Figure 4.10 Distribution of peak values in a narrow-band process.

quency is

$$\omega_{\tilde{x}} = \frac{1}{2\pi}\frac{\sigma_{\dot{X}}}{\sigma_X} e^{-\tilde{x}^2/2\sigma_X^2} \quad (4.2.18)$$

where σ_X and $\sigma_{\dot{X}}$ are the standard deviations of $X(t)$ and $\dot{X}(t)(=dX/dt)$, respectively. Setting $\tilde{x} = 0$ gives the average frequency for the process, ω_0. In a narrow band process, $X(t)$ crosses the axis before and after each peak, with only rare exceptions. Therefore the expected number of positive peak values is approximately equal to the number of cycles. In time T there will be, on average, $\omega_0 T$ cycles. Of these, the number which have peak values exceeding $\tilde{x}$ will be $\omega_{\tilde{x}}T$. Hence the probability that any peak value, chosen at random, exceeds $\tilde{x}$ is equal to $\omega_{\tilde{x}}/\omega_0$. That is

$$\frac{\omega_{\tilde{x}}}{\omega_0} = \int_{\tilde{x}}^{\infty} p_{\tilde{x}}(x)dx \quad (4.2.19)$$

Substitution of (4.2.18) and differentiation with respect to $\tilde{x}$ gives

$$p_{\tilde{X}}(\tilde{x}) = \frac{\tilde{x}}{\sigma_X^2} e^{-\tilde{x}^2/2\sigma_X^2} \quad (4.2.20)$$

which is the Rayleigh distribution of (4.1.11). We have thus shown that for a random process which is Gaussian, ergodic, and narrow-banded, the peak values follow a Rayleigh distribution. For a process with zero mean the variance σ_X^2 is equal to the mean square, and from (4.2.12) this is equal to m_0, the area under the spectral density function of the process. In terms of m_0, (4.2.20) is

$$p_{\tilde{X}}(\tilde{x}) = \frac{\tilde{x}}{m_0} e^{-\tilde{x}^2/2m_0} \quad (4.2.21)$$

The cumulative probability distribution function is

$$P_{\tilde{X}}(\tilde{x}) = \int_0^{\tilde{x}} p_{\tilde{X}}(x)dx = 1 - e^{-\tilde{x}^2/2m_0} \quad (4.2.22)$$

Thus, the probability of $\tilde{X}$ exceeding a specified value $\tilde{X}_p$ is

$$\text{Prob}\{\tilde{X} > \tilde{X}_p\} = 1 - P_{\tilde{X}}(\tilde{X}_p) = e^{-\tilde{X}_p^2/2m_0}$$

By inverting this we can define any value $\tilde{X}_p$ in terms of its probability of being exceeded. For example, as illustrated in Fig. 4.11*a*, we can define $\tilde{X}_{1/3}$ as the value of $\tilde{X}$ for which there is a probability of exceedance of $\frac{1}{3}$. This implies

$$\text{Prob}\{\tilde{X} \geqslant \tilde{X}_{1/3}\} = e^{-\tilde{X}_{1/3}^2/2m_0} = \frac{1}{3}$$

from which we obtain

$$\tilde{X}_{1/3} = \sqrt{2m_0 \ln 3} = 1.482\sqrt{m_0}$$

The average of all of the values above the one-third value is called the *significant value*, and is denoted $\tilde{X}_S$. It is the horizontal distance to the centroid of the area under $p_{\tilde{x}}(\tilde{x})$ beyond $\tilde{x} = \tilde{X}_{1/3}$, which is given by

$$\tilde{X}_S = 3\int_{\tilde{X}_{1/3}}^{\infty} \tilde{x}\, p_{\tilde{x}}(x)dx$$

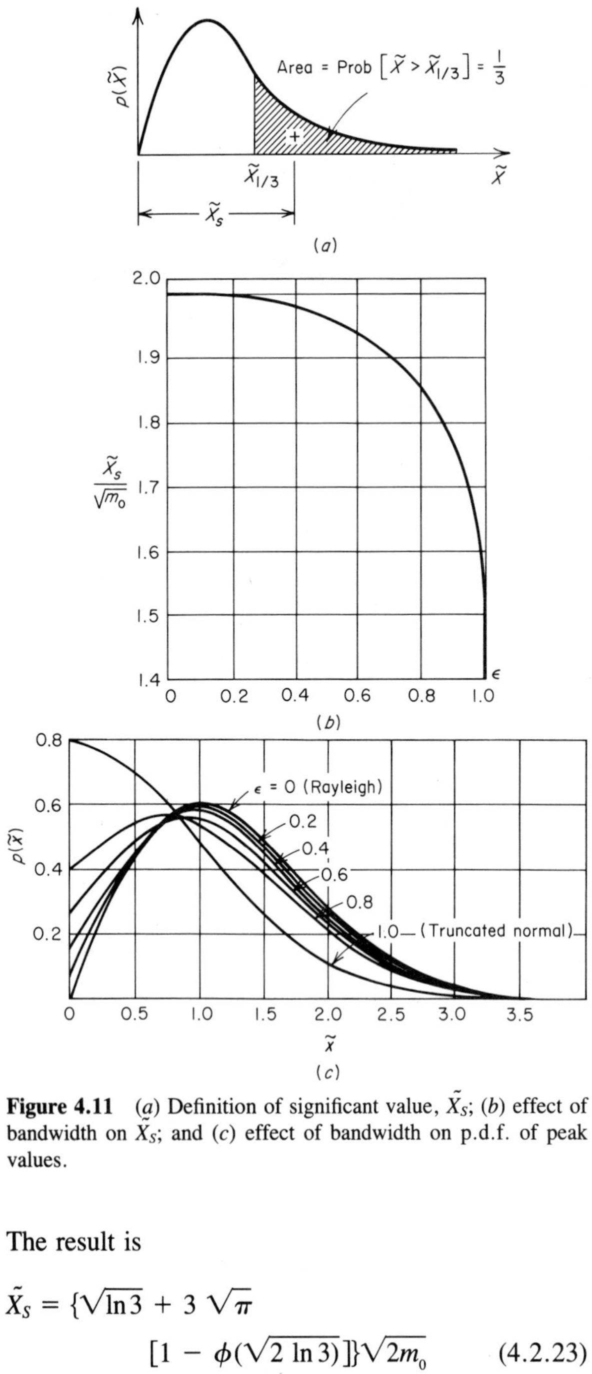

Figure 4.11 (*a*) Definition of significant value, $\tilde{X}_S$; (*b*) effect of bandwidth on $\tilde{X}_S$; and (*c*) effect of bandwidth on p.d.f. of peak values.

The result is

$$\tilde{X}_S = \{\sqrt{\ln 3} + 3\sqrt{\pi}\,[1 - \phi(\sqrt{2\ln 3})]\}\sqrt{2m_0} \tag{4.2.23}$$

which to three significant figures is

$$\tilde{X}_S = 2.00\sqrt{m_0},$$

where

$$\phi(\xi) = \frac{1}{2}\left[1 + \text{erf}\left(\frac{\xi}{\sqrt{2}}\right)\right]$$

and $\text{erf}(\xi/2^{1/2})$ is the error function or the probability integral

$$\text{erf}\left(\frac{\xi}{\sqrt{2}}\right) = \frac{2}{\sqrt{\pi}}\int_0^{\xi/\sqrt{2}} e^{-\xi^2}\,d\xi$$

Similarly, the average of the highest "one-*n*th" values of a random process which has a Rayleigh distribution is

$$\overline{X}_{1/n} = \{\sqrt{\ln(n)} + n\sqrt{\pi}\,[1 - \phi\sqrt{\ln(n)}]\}\sqrt{2m_0}$$

When n becomes large, this reduces to

$$\overline{X}_{1/n} = \sqrt{2m_0 \ln(n)} \tag{4.2.24}$$

The average $1/n$th value provides information about the magnitude of the larger peaks of a random process, and hence it characterizes the "severity" of the process. In particular the *significant value*, $\tilde{X}_S$ is used for measuring and specifying the severity of sea states. In this case the random variable is the trough-to-crest wave height, $\tilde{H}$, which is twice the amplitude of $\tilde{X}$. Therefore the preceding expression must be multiplied by 2, giving

$$\tilde{H}_S = 4.01\sqrt{m_0} \tag{4.2.25}$$

The statistical properties of a general non-narrow-band process have been investigated by Cartwright and Longuet-Higgins [3] and by Huston and Skopinski [4]. Fig. 4.11*b* from Ref. 5 shows the results of computations by Ochi of significant wave heights (in dimensionless form) for various values of the bandwidth parameter ε, defined in (4.2.17). As can be seen in the figure, there is no appreciable difference in the significant wave heights for ε up to 0.5. The difference only becomes significant for ε greater than 0.8. Since the ε-value for ocean waves is generally within the range from 0.4 to 0.8, the significant wave height evaluated on the assumption of a narrow-band process is only slightly overestimated, of the order of 5% or so.

Fig. 4.11*c* shows the probability density function of the peak values of various non-narrow-band random processes, as a function of ε. For a totally wide-band process the function is simply the Gaussian curve.

4.3 EXTREME VALUES: PREDICTION OF THE CHARACTERISTIC EXTREME VALUE

As explained in Chapter 1 the rationally-based design of ships requires the consideration of the largest value (extreme value) of the wave loading, especially the wave-induced bending moment, which is expected to occur within the ship's lifetime, and in particular the prediction of a *characteristic value* which is associated with a certain probability of nonexceedance in that time. In order to achieve this we must first examine the concept of an *extreme value* of a random process. This topic was first treated by Fisher and Tippett and was later systematized by Gumbel [6].

In defining a design value it might seem sufficient to use the $1/n$th value, for which the probability of being exceeded is $1/n$, and to simply choose an appropriate value of n. It is true that the probability of exceeding the value $\tilde{Q}_{1/n}$ is, on the average, once in n observations (peaks).* But there is no assurance that $\tilde{Q}_{1/n}$ will definitely occur once in n observations. For design purposes we need information concerning the *extreme value,* that is, the largest peak value that will occur in the life of the ship, and also the probability (or risk) that this largest value will exceed a specified magnitude. The *characteristic value* is then that magnitude of extreme value which has an appropriate probability of exceedance. Since the statistical properties of negative peak values are essentially the same as those of positive peak values, only the latter are considered in the following discussion (see Fig. 4.12).

There are three different methods for calculating extreme values, depending on whether or not the probability density function of the peak values is known analytically. In discussing extreme value statistics this function is referred to as the *initial* probability density function and its integral is referred to as the *initial* probability distribution. If the former is known analytically then an exact analytical expression can be derived for the probable extreme value by applying *order statistics*. This situation exists, for example, in regard to the short-term peak values of wave height, since the initial probability density function for this process is the Rayleigh distribution.

If the initial probability density function is not known analytically and the only information available is some measured (or observed) data then there are two other possibilities, depending on the nature of the data.

1. If enough data is available to allow an approximate initial probability distribution $P_{\tilde{Q}}(\tilde{Q})$ to be constructed, then it will be shown that the extreme value for a longer period of time can be established by plotting the distribution in the form $1/[1 - P_{\tilde{Q}}(\tilde{Q})]$ and extrapolating the curve. Obviously this method possesses the usual uncertainties associated with any process of extrapolation. It is used, for example, to estimate the most severe sea state likely to be encountered over a very long period of time.

2. If the available data consists of a large number of short-term extreme values (e.g., the largest wave measured or observed each day over a period of many years) then the extreme value likely to be encountered over a longer period can be estimated by means of the asymptotic distributions that were originally developed by Fisher and Tippet and later systematized by Gumbel.

*The symbol Q is chosen because in the majority of applications of the results obtained here, the random process $\tilde{Q}(t)$ is a response or a load effect, which in this text is denoted by the symbol Q.

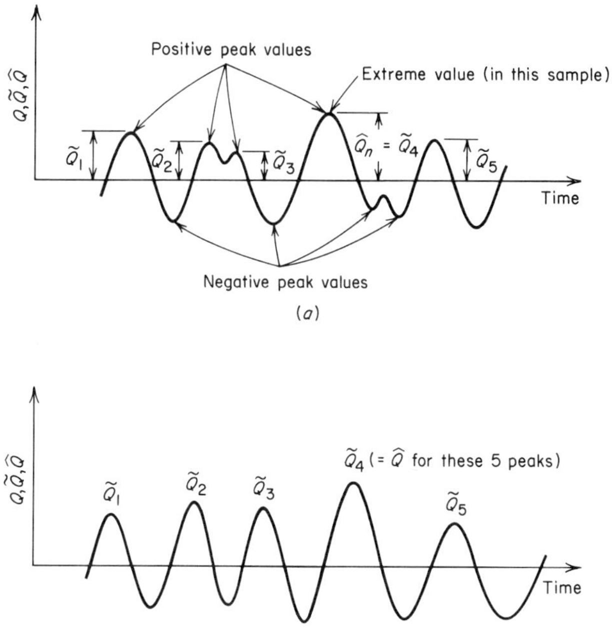

Figure 4.12 Random processes. (*a*) Nonnarrow-band, (*b*) narrow-band.

Exact Evaluation of Extreme Values

Let $p_{\tilde{Q}}(\tilde{Q})$ be the probability density function which governs the positive peak values $\tilde{Q}$ of a random process $Q(t)$, and let $P_{\tilde{Q}}(\tilde{Q})$ be the corresponding cumulative probability function, as shown in Fig. 4.13. The function $1 - P_{\tilde{Q}}(\tilde{Q}_0)$ gives the probability that any one peak value, selected at random, will exceed a specified magnitude, $\tilde{Q}_0$. But for design purposes we wish to know the probability that, out of a total of n peak values that are encountered, at least one peak will exceed a specified magnitude. We will derive this in two stages. First we will obtain the probability density function for the *extreme value*, that is, the largest peak

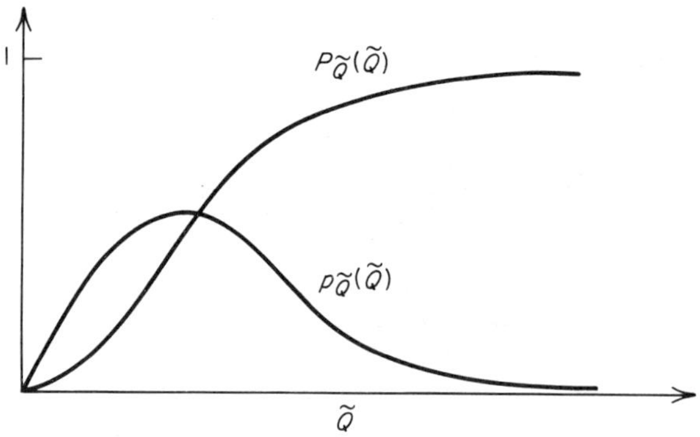

Figure 4.13 Distributions of peak values.

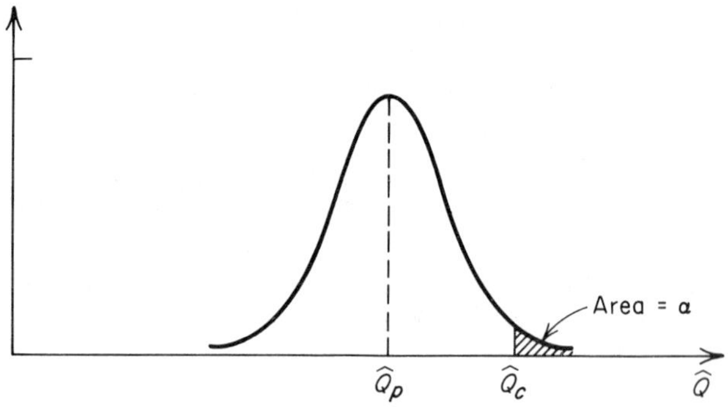

Figure 4.14 Typical p.d.f. of extreme values.

value among a sample of n peak values, and then we will use this to relate the magnitude of the extreme value to its probabiilty of occurrence.

Let $(\tilde{Q}, \tilde{Q}_2, \ldots \tilde{Q}_n)$ be a typical sample of size n, where each $\tilde{Q}_i$ is an observed peak value of $Q(t)$, as in Fig. 4.12. Since none of the members of the sample are specially defined or distinguished in any way, all of them have the same probability density function, $p_{\tilde{Q}}(\tilde{Q})$. Let us now arrange this random sample into an ordered sample $(\hat{Q}_1, \hat{Q}_2, \ldots \hat{Q}_n)$ in ascending order of magnitude such that $\hat{Q}_1 < \hat{Q}_2 < \cdots < \hat{Q}_n$. We choose a new symbol, $\hat{Q}$, in order to emphasize that each of these random variables now has its own separate definition (largest peak, second largest peak, etc.) and therefore each has its own probability density function $p_{\hat{Q}}(\hat{Q}_i)$, different from $p_{\tilde{Q}}(\tilde{Q}_i)$. The variable of interest is the extreme value, that is, the largest peak value in the ordered sample. Since the other values are not of interest we will dispense with the subscript and denote the extreme value as $\hat{Q}$. When we wish to associate $\hat{Q}$ with the sample size we will write $\hat{Q}_n$.

We wish to obtain the probability density function of $\hat{Q}$. Now for any one member to be the *largest* requires that all of the other $n - 1$ peak values must be less than $\hat{Q}$, and the probability of this is $[P_{\tilde{Q}}(\hat{Q})]^{n-1}$ (assuming the peaks are statistically independent). But since there are n members, any one of which could be the largest, we must multiply this probability by n. Hence the probability density function of $\hat{Q}$ is

$$p_{\hat{Q}}(\hat{Q}) = p_{\tilde{Q}}(\hat{Q}) n [P_{\tilde{Q}}(\hat{Q})]^{n-1}$$

A typical extreme value probability density function is illustrated in Fig. 4.14. The value of $\hat{Q}$ for which the probability density function $p_{\hat{Q}}(\hat{Q})$ is maximum is the extreme value that is most likely to occur in n observations, and this is called the "probable extreme value" denoted by $\hat{Q}_p$. The probable extreme value has some usefulness since the extreme value which actually occurs in n peak values of wave loading will be very close to this value. It is obtained as the solution of the following equation.

$$\frac{d}{d\hat{Q}}[p_{\hat{Q}}(\hat{Q})] = 0$$

which can be expanded to

$$p'_{\tilde{Q}}(\hat{Q}) P_{\tilde{Q}}(\hat{Q}) + (n - 1)[p_{\tilde{Q}}(\hat{Q})]^2 = 0 \qquad (4.3.1)$$

For a narrow-band process the initial distributions are the Rayleigh distributions

$$p_{\tilde{Q}}(\tilde{Q}) = \frac{2\tilde{Q}}{R} e^{-\tilde{Q}^2/R} \qquad (4.3.2a)$$

and

$$P_{\tilde{Q}} = 1 - e^{-\tilde{Q}^2/R} \qquad (4.3.2b)$$

$$\text{where } R = \begin{cases} 2\sigma_Q^2 = 2m_0 \text{ for amplitude} \\ 8\sigma_Q^2 = 8m_0 \text{ for height (crest-to-trough)} \end{cases}$$

Substituting these into (4.3.1) gives

$$2\frac{\hat{Q}^2}{R}(ne^{-\hat{Q}^2/R} - 1) - (e^{-\hat{Q}^2/R} - 1) = 0$$

The second term may be neglected for large n, giving

$$\hat{Q}_p = \sqrt{R \ln(n)} \qquad (4.3.3)$$

Ochi [5] has shown that for a random process having a bandwidth parameter $\varepsilon \leq 0.9$ (which easily applies to wave-induced bending moment and most other wave-induced loads) the solution to (4.3.1) is

$$\hat{Q}_p = \sqrt{2m_0 \ln\left\{\frac{2\sqrt{1-\varepsilon^2}}{1+\sqrt{1-\varepsilon^2}}\, n\right\}} \qquad \text{for } \varepsilon \leq 0.9 \tag{4.3.4}$$

However, the probability that the largest peak value may exceed the probable extreme value is quite large, and hence it is not appropriate to use this value for engineering design. For example, it may be shown [5] that for a perfectly narrow-banded process ($\varepsilon = 0$) the probability that the extreme value will exceed $\hat{Q}_p$ is $1 - e^{-1} = 0.632$. For purposes of structural design, we must obtain an extreme value for which the probability of being exceeded is some acceptably small value, α, chosen by the designer—a typical value is $\alpha = 0.01$. As explained in Chapter 1, α is not the probability of failure, since that also involves the probability distribution function of the limit value of the wave-induced load or load effect. Rather, α is a *risk parameter* by means of which the designer has explicit control over the probability, or risk, of the design load being exceeded in a total of n peak values. Hence we wish to obtain the extreme value $\hat{Q}_c$, which we refer to as the *characteristic value*, for which there is a probability α of being exceeded. That is, $\hat{Q}_c$ must be such that

$$\text{Prob}\{\text{extreme value among } n \text{ peak values} > \hat{Q}_C\} = \alpha$$

An equivalent requirement is that there must be a probability of $1 - \alpha$ that all of the n peak values will be less than or equal to $\hat{Q}_c$. For each peak value this probability is $P_{\tilde{Q}}(\hat{Q}_c)$, and the joint probability of this being satisfied for all peak values simultaneously is the product of the n individual probabilities, that is, $[P_{\tilde{Q}}(\hat{Q}_c)]^n$. Thus we have

$$[P_{\tilde{Q}}(\hat{Q}_c)]^n = 1 - \alpha$$

Considering that α is small and n is large, we have

$$\begin{aligned} P_{\tilde{Q}}(\hat{Q}_c) &= (1-\alpha)^{1/n} \\ &\approx 1 - \frac{\alpha}{n} + \mathrm{O}(\alpha^2) \end{aligned} \tag{4.3.5}$$

where $\mathrm{O}(\alpha^2)$ indicates terms of the order of α^2. By developing further the results of Ref. 3 Ochi derived an expression for $P_{\tilde{Q}}(\tilde{Q})$ which, in combination with (4.3.5), gives the following expression for $\hat{Q}_c$:

$$\hat{Q}_c = \sqrt{2m_0 \ln\left(\frac{\sqrt{1-\varepsilon^2}}{1+\sqrt{1+\varepsilon^2}}\,\frac{2n}{\alpha}\right)} \qquad \text{for } \varepsilon \leq 0.9 \tag{4.3.6}$$

This equation expresses the design value as a function of the number of peak values, n, of the random process under consideration, such as the wave-induced bending moment. For practical purposes it is preferable to express the design value in terms of time rather than as a function of the number of peak values. Ochi has further shown that the number of peak values can be expressed as a function of time by

$$n = (60)^2\left(\frac{T}{4\pi}\right)\left(\frac{1+\sqrt{1-\varepsilon^2}}{\sqrt{1-\varepsilon^2}}\right)\sqrt{\frac{m_2}{m_0}} \tag{4.3.7}$$

where T is time in hours. Substituting in (4.3.6) gives

$$\hat{Q}_c = \sqrt{2m_0 \ln\left\{\frac{1800T}{\pi\alpha}\sqrt{\frac{m_2}{m_0}}\right\}} \tag{4.3.8}$$

It may be seen that the characteristic value $\hat{Q}_c$ expressed in terms of time T is no longer a function of the bandwidth parameter ε but is a function only of the area m_0 and the second moment m_2 of the spectral density function of the random process. Therefore (4.3.8) applies for any Gaussian process, regardless of its bandwidth.

From (4.3.4) and (4.3.7), the most probable extreme value, $\hat{Q}_p$, can be also expressed as a function of time by

$$\hat{Q}_p = \sqrt{2m_0 \ln\left\{\frac{1800T}{\pi}\sqrt{\frac{m_2}{m_0}}\right\}} \tag{4.3.9}$$

Estimation of Extreme Values by Extrapolation of Measured Data

The previous section presented an analytical expression for extreme values when the initial probability distribution of peak values is known analytically. In practice, the initial distribution is often not known with such precision. For example, the probability distribution that is applicable for the long-term sea severity (as measured by significant wave height H_s)* is not known analytically. The cumulative probability distribution $P(H_s)$ is derived empirically from analysis of observed (or measured) data accumulated over a certain period of time, and there seems to be no single mathematically defined probability distribution function which applies to this quantity. In fact, there is probably no way of theoretically deriving such a distribution, as contrasted with the Rayleigh probability function applicable for the height of individual waves in a given sea. For design purposes we need to estimate extreme values of H_s, particularly the probable extreme value $H_{s,p}$ corresponding to a certain period, and so we must use an approximate method for this purpose.

For long-term statistics it is possible to estimate the general asymptotic form of the cumulative probability distribution and to use the theory of asymptotic distribution of extreme values, which is applicable to any probability function if certain conditions are met [7]. In this method the initial cumulative probability distribution is assumed to be of the form

$$P_{\tilde{Q}}(\tilde{Q}) = 1 - e^{-f(\tilde{Q})} \tag{4.3.10}$$

The function $f(\tilde{Q})$ specifies the precise manner in which $P_{\tilde{Q}}(\tilde{Q})$ approaches its asymptote. The probability density function is the derivative of $P_{\tilde{Q}}(\tilde{Q})$

$$p_{\tilde{Q}}(\tilde{Q}) = \frac{d}{d\tilde{Q}}[1 - e^{-f(\tilde{Q})}]$$
$$= e^{-f(\tilde{Q})}f'(\tilde{Q})$$

In our intended application $\tilde{Q}$ is significant wave height H_s but for generality we retain the symbol $\tilde{Q}$. Likewise the symbol for the extreme value of $\tilde{Q}$ in a sample of size n is $\hat{Q}$. As shown in the previous section the probable extreme value $\hat{Q}_p$ is the solution of (4.3.1). In the present case this equation becomes

$$\frac{f''(\hat{Q})}{[f'(\hat{Q})]^2}[1 - e^{-f(\hat{Q})}] + ne^{-f(\hat{Q})} - 1 = 0$$

For large n the first term is small in comparison with the other terms, and the equation becomes

$$e^{f(\hat{Q})} \cong n \tag{4.3.11}$$

and $\hat{Q}_p$ is the particular value of $\hat{Q}$ which satisfies this equation. Equation 4.3.10 holds for any value of $\tilde{Q}$, including the probable extreme value, $\hat{Q}_p$. When (4.3.11) is substituted back into (4.3.10) the assumed functional expression for the asymptote drops out, leaving simply

$$\frac{1}{1 - P_{\tilde{Q}}(\hat{Q}_p)} = n \tag{4.3.12}$$

This result shows that the probable extreme value expected to occur in n observations can be evaluated from the initial cumulative distribution function and the number of observations. Although the initial cumulative distribution function $P_{\tilde{Q}}(\tilde{Q})$ is not known analytically, it can be constructed from the observed data. Then the left-hand side of (4.3.12) can be plotted versus $\hat{Q}_p$, and the extreme value for a number of occurrences larger than n can be estimated by extrapolating the curve. The left-hand side of (4.3.12) is known as the *return period*, because it corresponds to the number of occurrences in between extreme values of a certain magnitude. The use of this technique to estimate the probable extreme sea state is shown in Section 4.4.

Estimation of Extreme Value from Observed Short-Term Extreme Values

The observed short-term extreme value considered here is the largest value that is either measured or observed during a certain period of time. As an example, let us assume that the data of the largest wave height measured every day are available. If the number of measured data is large, then it is possible to estimate wave height expected to occur in a much longer period of time, for example 20 or 40 years. The estimation is based on the asymptotic distribution of extreme values developed by Gumbel. This theory shows that if the initial distributions are of exponential type, which means that they satisfy the following relationship

$$\frac{p_{\tilde{Q}}(\tilde{Q})}{1 - P_{\tilde{Q}}(\tilde{Q})} = -\frac{p'_{\tilde{Q}}(\tilde{Q})}{p_{\tilde{Q}}(\tilde{Q})}, \quad \tilde{Q} \to \infty$$

(which is satisfied, for example, by the Rayleigh dis-

*Strictly speaking significant wave height is a peak value quantity and in the notation of the previous section would be written as $\tilde{H}_s$. The symbols ˜ and ˆ for peak value and extreme value are necessary when discussing the completely general case, but in most applications the fact that a quantity is a peak value or an extreme value is obvious from the context. For example, a *significant value* is always a peak value, by definition. Therefore significant wave height will generally be written as H_s; the symbol ˜ is not required. Likewise since the subscripts p and c indicate the probable extreme value and the characteristic extreme value, there is no need to also use the symbol ˆ for these quantities. Therefore the probable extreme significant wave height will be denoted as $H_{s,p}$.

tribution and by (4.3.10)) then a suitable asymptotic approximation to the exact cumulative probability function of the extreme value is given by

$$p_{\hat{Q}}(\hat{Q}) = e^{-e^{-\kappa(\hat{Q}-\hat{Q}_p)}} \qquad (4.3.13)$$

where

$$\kappa = \frac{\pi}{\sqrt{6\,\mathrm{Var}[\hat{Q}]}}$$

$$\hat{Q}_p = E[\hat{Q}] - \frac{\gamma}{\kappa}$$

in which $\mathrm{Var}[\hat{Q}]$ and $E[\hat{Q}]$ are the variance and the expected value of the available data (i.e., the observed short-term extreme values) and γ is Euler's constant, 0.577.

Following Gumbel [6] Loukakis and Grivas [8] have shown that if the initial distribution is the Rayleigh distribution [see (4.3.2)] then the parameter κ in (4.3.13) is given by

$$\kappa = 2\sqrt{\frac{\ln(n)}{R}} \qquad (4.3.14)$$

4.4 STATISTICAL REPRESENTATION OF THE SEA SURFACE

The surface of the ocean—that is, the pattern of surface elevation—is highly irregular and totally random (nonrepeating) even under relatively calm conditions. Fortunately for naval architects, oceanographers have found that the irregularity of the ocean surface can be represented as the superposition of a large number of regular waves having different heights, lengths, directions, and random phase differences. This finding is very important because it allows the ocean surface to be described mathematically and also allows the use of statistical methods to predict the maximum wave loads in a ship's lifetime.

Superposition of Regular Waves

The first major contributions were made by Pierson [9] and Pierson et al. [10] who proposed that the completely irregular and unrepeating pattern of the ocean surface could be represented as the sum of an infinite number of regular sinusoidal waves, of all frequencies, each of which satisfies the governing hydro-

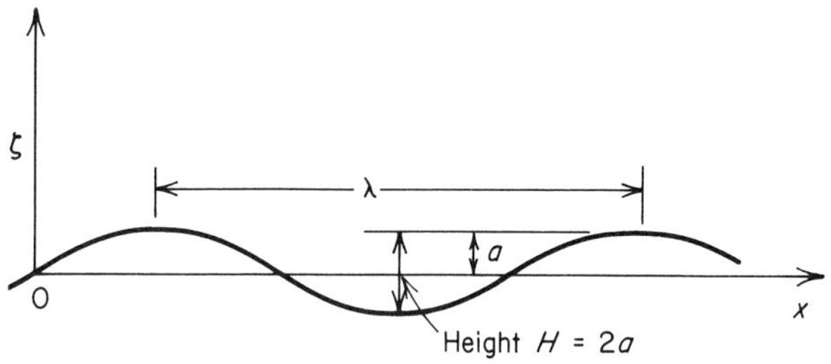

Figure 4.15 Geometry of a regular wave.

dynamic equation for gravitational waves. For simplicity we begin with a "long-crested" sea, that is, a sea in which the waves are all parallel. The surface profile of a typical component wave is

$$\zeta(x, t) = a\cos(-kx - \omega t + \theta) \qquad (4.4.1)$$

These and other quantities and terms are defined as follows (see Fig. 4.15):

a = *wave amplitude*, measured from the mean water surface, which is also the location of the x-axis.

λ = *wave length*, the horizontal distance between successive crests or troughs.

$k = 2\pi/\lambda$ = *wave number*

T = *wave period*, the time taken for two successive crests to pass a fixed point on the x-axis or the time taken by a crest to travel a distance equal to one wavelength.

$\omega = 2\pi/T$ = *wave frequency*

θ = *phase angle*

For deep water the wave number and the wave frequency are related by

$$k = \frac{\omega^2}{g} \qquad (4.4.2)$$

The total energy (kinetic and potential) per unit area of water surface is

$$\Omega = \frac{1}{2}\rho g a^2 \qquad (4.4.3)$$

It was Pierson et al. [10] who first proposed that the surface elevation $h(x, t)$ of an irregular sea could be represented as

$$h(x, t) = \lim_{n\to\infty} \sum_{i=1}^{n} a_i \cos \psi_i \qquad (4.4.4)$$

where

$$\psi_i = -k_i x - \omega_i t + \theta_i$$

in which the phase angle θ_i is a random variable, equiprobably distributed over the range $(0, 2\pi)$.

Each component wave must satisfy the governing differential equation for gravitational waves. This equation is most conveniently expressed in terms of a velocity potential $\phi(x, y, t)$ defined such that its derivatives correspond to velocities:

$$u = -\frac{\partial \phi}{\partial x}, \qquad v = -\frac{\partial \phi}{\partial y}$$

In terms of the potential function the governing equation is Laplace's equation

$$\nabla^2 \phi = 0$$

and the pressure at any point is

$$p(x, y, t) = \rho\left[\frac{\partial \phi}{\partial t} + gy + \frac{1}{2}(\nabla\phi)^2\right]$$

The problem is nonlinear because the free surface boundary condition is $p = 0$ and the pressure $p(x, y, t)$ is a nonlinear function of ϕ. The problem is further complicated by the fact that the boundary—the free surface—is an undulating time-dependent surface instead of a fixed boundary. If the wave height is small in comparison to the wave length then the problem may be linearized, and it may be shown that the solution for the linearized problem is the velocity potential

$$\phi(x, y, t) = \frac{a\omega}{k} e^{ky} \sin \psi \qquad (4.4.5)$$

If we wish to discuss the surface elevation at a particular location we may set $x = 0$

$$h(t) = \lim_{n\to\infty} \sum_{i=1}^{n} \zeta_i = \lim_{n\to\infty} \sum_{i=1}^{n} a_i \cos(-\omega_i t + \theta_i) \qquad (4.4.6)$$

It can easily be verified that each of the component random processes $\zeta_i(t)$ is stationary and hence $h(t)$ is stationary. Also, since these processes are independent with zero mean values, terms of the type $E[\zeta_i(t_1)\zeta_j(t_2)]$ are all zero for $i \neq j$ and are only nonzero when $i = j$, so that the mean square wave height is given by

$$E[h^2(t)] = E\left[\sum_{i=1}^{n} \zeta_i^2(t)\right] = \sum_{i=1}^{n} E[\zeta_i^2(t)] = \frac{1}{2}\sum_{i=1}^{n} a_i^2 \qquad (4.4.7)$$

and the auto-correlation function of the summed process is

$$\begin{aligned} R(\tau) &= E[h(t_1)h(t_2)] \\ &= \sum_{i=1}^{n}\sum_{j=1}^{n} E[\zeta_i(t)\zeta_j(t+\tau)] \\ &= \sum_{i=1}^{n} R_i(\tau) \end{aligned} \qquad (4.4.8)$$

Because the surface elevation $h(t)$ is the sum of a large number of independent variables it will, by the central limit theorem (Section 4.1) have a Gaussian, or normal, probability density function. For the case of a zero mean this is

$$p(h) = \frac{1}{\sigma\sqrt{2\pi}} e^{-(h/\sigma)^2/2} \qquad (4.4.9)$$

and since the mean is zero, the variance equals the mean square:

$$\sigma^2 = E[h^2]$$

We note that, because $h(t)$ is Gaussian, the variance is sufficient to uniquely define the entire process.

As indicated earlier, it is advantageous to describe a random process in terms of its mean square spectral density function $S(\omega)$ which is the Fourier transform of $R(\tau)$, given by (4.2.11). For all practical applications we use the one-sided form of the function. As we have seen, the area under this curve is the mean square, or in this case, the variance:

$$\sigma^2 = \int_0^\infty S(\omega)d\omega \qquad (4.4.10)$$

From (4.4.3) the energy of each component wave is $\Omega_i = \frac{1}{2}\rho g a_i^2$ and hence the total energy is

$$\Omega = \Sigma\Omega_i = \tfrac{1}{2}\rho g \Sigma\, a_i^2$$

Using (4.4.7) and noting that the variance equals the mean square, the foregoing can be rewritten as

$$\Omega = \rho g \sigma^2$$

and hence, as noted earlier, the spectral density function is often referred to as the *energy spectrum* because the area under this curve is directly proportional to the total energy of the waves per unit area of surface:

$$\Omega = \rho g \int_0^\infty S(\omega) d\omega \qquad (4.4.11)$$

Again, for each component wave the relationship is

$$\tfrac{1}{2}\rho g a_i^2 = \rho g S(\omega_i)\delta\omega$$

and so the relationship between the wave spectrum and the amplitude of each component wave is

$$a_i = \sqrt{2S(\omega_i)\delta\omega} \qquad (4.4.12)$$

In terms of $S(\omega)$ the surface elevation is

$$h(t) = \lim_{\substack{n\to\infty \\ \delta\omega\to 0}} \sum_{i=1}^{n} \sqrt{2S(\omega_i)\delta\omega}\,\cos(-\omega_i t + \theta_i) \qquad (4.4.13)$$

This relationship is illustrated in Fig. 4.16 for $n = 14$.

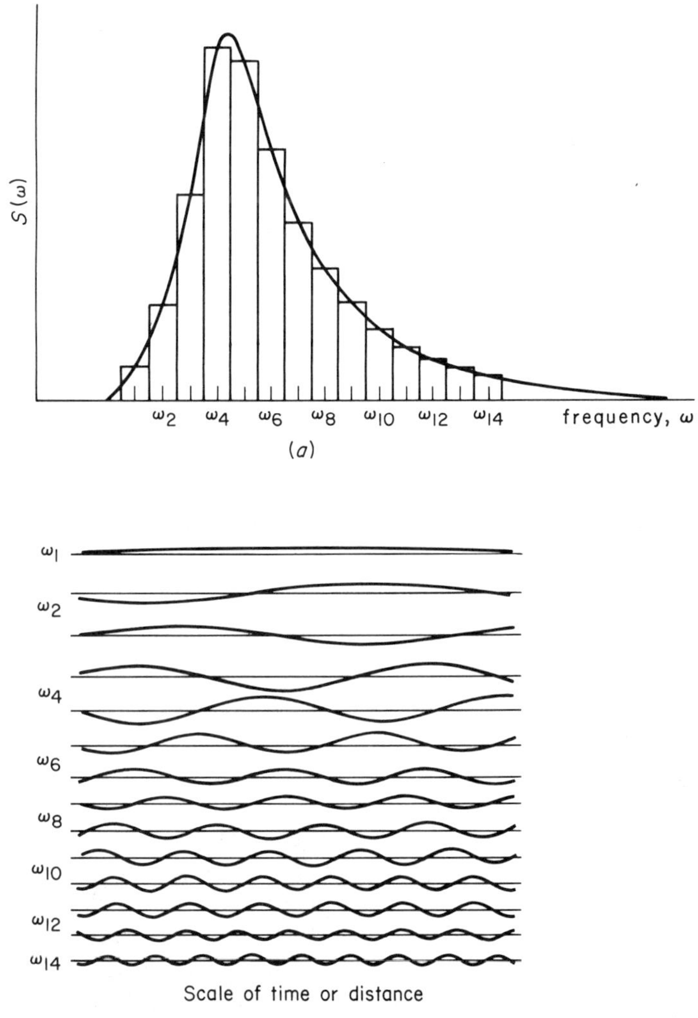

Figure 4.16 Typical energy spectrum showing approximation by a finite sum of components. (*a*) Spectrum. (*b*) Component waves, length of each = $1.56T^2 = 61.6/\omega^2$, amplitude = $[2S(\omega_n)\delta\omega]^{1/2}$.

Mathematical Representation of Ocean Wave Spectra

Over the space of many years oceanographers and other researchers have gathered and tabulated both visual and measured data concerning ocean waves in various parts of the world. One of the most comprehensive sets of data is that presented by Hogben and Lumb [11] in which data for 50 ocean areas around the world (but with primary emphasis on the North Atlantic) are tabulated for different seasons and different wave directions for wave periods varying roughly from 0.6 to 22.5 sec over 10 increments. These data are based on almost two million sets of observations reported from ships at sea over a period of eight years. Further data on the North Atlantic have been presented by Roll [12] and Walden [13]. Data for the North Pacific are provided by Yamanouchi et al. [14]. All of these data are visual observations and hence they are not entirely consistent, being based on judgements made by many different observers. Measured wave data are more precise but this type of data is limited in quantity and geographical location when compared with the vast accumulation of visual observations. Most of the measured data has been obtained from British weather ships in the North Atlantic. Some additional wave data has been generated by a hindcast technique [15]. Reference [16] is a very comprehensive summary of standardized wave and wind conditions.

Two-Parameter Formulas for Wave Spectra

Bretschneider [18] was the first to propose that the wave spectrum for a given sea state could be represented in terms of two parameters which were characteristic of that sea state such as average wave height, $\overline{H}$, and average wave period, $\overline{T}$. Various other formulas in addition to Bretschneider's have been proposed such as those of Pierson-Moskowitz, the International Towing Tank Committee, and the International Ship Structure Congress. In some cases the average frequency $\overline{\omega}(= 2\pi/\overline{T}$, if expressed in radians/sec) is used instead of $\overline{T}$. In terms of $\overline{H}$ and $\overline{\omega}$ the general form of these two-parameter formulas is

$$S(\omega) = A\left(\frac{\overline{\omega}}{\omega}\right)^k \frac{\overline{H}^2}{\omega} e^{-B(\overline{\omega}/\omega)^l} \tag{4.4.14}$$

where the coefficients A and B and the exponents k and l are selected to fit the data and the system of units. Some formulas make use of an alternative pair of parameters: significant wave height, H_s (shown as X_s in Fig. 4.11*a*) and modal wave frequency, ω_m, which is the frequency at which the wave spectrum has its maximum height. It may be shown that for a narrow-band spectrum these two parameters are related to the first two as follows:

$$H_s = \sqrt{\frac{8}{\pi}}\,\overline{H} = 1.60\,\overline{H}$$

$$\omega_m = \frac{(0.8)^{1/4}}{\Gamma\left(\frac{3}{4}\right)}\,\overline{\omega} = 0.77\,\overline{\omega}$$

where $\Gamma(\ )$ denotes the Gamma function. In the work to follow we will use a two-parameter formula, now adopted by the ITTC, which is a modified version of Bretschneider's formula, expressed in terms of H_s and ω_m:

$$S(\omega) = \frac{5}{16}\left(\frac{\omega_m}{\omega}\right)^4 \frac{H_s^2}{\omega} e^{-1.25(\omega_m/\omega)^4} \tag{4.4.15}$$

For coastal waters the JONSWAP spectrum [17] is used.

Relationship between Wave Frequency and Significant Wave Height

Both of these parameters are random processes and to use any of the two-parameter formulas it is necessary to know the probability density function and the interdependence, if any, of the two parameters. H_s is customarily chosen as the primary parameter and it then becomes necessary to account for the dependency of ω_m (or $\overline{\omega}$) on it. Until recently this was done by attempting to fit a deterministic expression for ω_m as a function of H_s, but as pointed out by Ochi [19], this is not appropriate because both quantities are random processes, and ω_m is neither totally dependent nor totally independent of H_s. The dependency can only be accounted for by establishing the statistical relationship between these two parameters. Specifically, what is needed is the probability density function of the wave frequency for a given value of H_s, that is, the *conditional* probability density function of ω. From an analysis of North Atlantic data Ochi has shown that H_s follows the log-normal probability density function given in (4.1.11) (and not the Weibull distribution, as others have proposed). Only the very large values of H_s (the uppermost 1%, which corresponds roughly to values exceeding 10 m) depart from this law, and for such extreme values the probability density function is obtained in a different manner, by means of the asymptotic approximation of the cumulative probability distribution function, as discussed in Section

4.3. Ochi also showed that if the data are expressed in terms of wave period instead of wave frequency, the modal period $T_m (= 2\pi/\omega_m)$ also follows a log-normal probability density function. Therefore since both H_s and T_m are log-normally distributed, the wave period for a given wave height follows the conditional log-normal probability density function which is

$$p_T(T_m \mid H_s) = \frac{1}{\sigma_T^* \sqrt{2\pi}\, T_m} e^{-\frac{1}{2}[(\ln T_m - \mu_T^*)/\sigma_T^*]^2} \quad (4.4.16)$$

where

$$\mu_T^* = \mu_T + \rho \frac{\sigma_T}{\sigma_H} (\ln H_s - \mu_H)$$

$$\sigma_T^* = \sqrt{1 - \rho^2}\, \sigma_T$$

In this expression the two pairs of parameters—μ_H, σ_H and μ_T, σ_T—are the mean and the standard deviation of the probability density function for $\ln(H_s)$ and $\ln(T_m)$ respectively, and ρ is the correlation coefficient between the two random variables, H_s and T_m. Table 4.1 gives the values of these five parameters which Ochi obtained by averaging the values calculated from the data obtained at seven weather stations in the North Atlantic.

It was mentioned previously that most of the available data on wave height and wave period are visually observed values, H_v and T_v. For example the data in Ref. 13 are values observed at nine locations in the North Atlantic over a period of 10 years. The observations were made regularly by trained oceanographers at $3\frac{1}{2}$-hour intervals, as an average. Wave heights were visually estimated, and wave periods were counted by a stopwatch from visually observed wave crests. Therefore, the observed wave height H_v represents neither the significant nor the average wave height. Also, the observed wave period T_v represents neither the zero-crossing nor the average wave period. Hence, in order to use this data it is necessary to convert it to the statistical quantities $\overline{H}$ and T_0, or H_s and T_m. Having shown that both the visually observed data and the measured data follow a log-normal distribution, Ochi obtained the following relationships between them

$$H_s = H_v^{1.08} \quad (4.4.17)$$

$$T_m = 2.99 T_v^{0.73} \quad (4.4.18)$$

Short-Crested Waves

The mathematical spectral formulas and wave spectra previously discussed are spectra of the sea at a fixed point, or one-dimensional spectra. This can be thought of as describing a long-crested irregular sea. A more complete representation of the sea is given by a two-dimensional *directional spectrum*, $S(\omega, \Theta)$, as shown in Fig. 4.17, which indicates the direction Θ as well as the frequencies of the wave components and thus accounts for the typical short-crestedness of ocean storm waves. The most common method for approximating $S(\omega, \Theta)$ is to use the form

$$S(\omega, \Theta) = S(\omega) f(\Theta) \quad (4.4.19)$$

The function $f(\Theta)$ is referred to as the *spreading function*. A function commonly used is

$$f(\Theta) = \begin{cases} \frac{2}{\pi} \cos^2 \Theta & \left(-\frac{\pi}{2} \le \Theta \le \frac{\pi}{2}\right) \\ 0 & \text{otherwise} \end{cases} \quad (4.4.20)$$

Figure 4.17 shows a contour plot of $S(\omega, \Theta)$ after the spreading function has been used.

TABLE 4.1 PARAMETERS ASSOCIATED WITH BIVARIATE LOG-NORMAL PROBABILITY DISTRIBUTION FOR VARIOUS LOCATIONS IN THE NORTH ATLANTIC

Weather Station	A	B	C	D	I	J	K
Significant height							
μ_H	0.946	0.910	1.024	0.968	1.112	1.053	0.748
σ_H	0.619	0.588	0.571	0.578	0.562	0.565	0.680
Modal period							
μ_T	2.505	2.462	2.494	2.483	2.588	2.594	2.600
σ_T	0.218	0.218	0.216	0.209	0.142	0.147	0.174
Correlation coefficient							
ρ	0.498	0.594	0.578	0.586	0.358	0.339	0.331

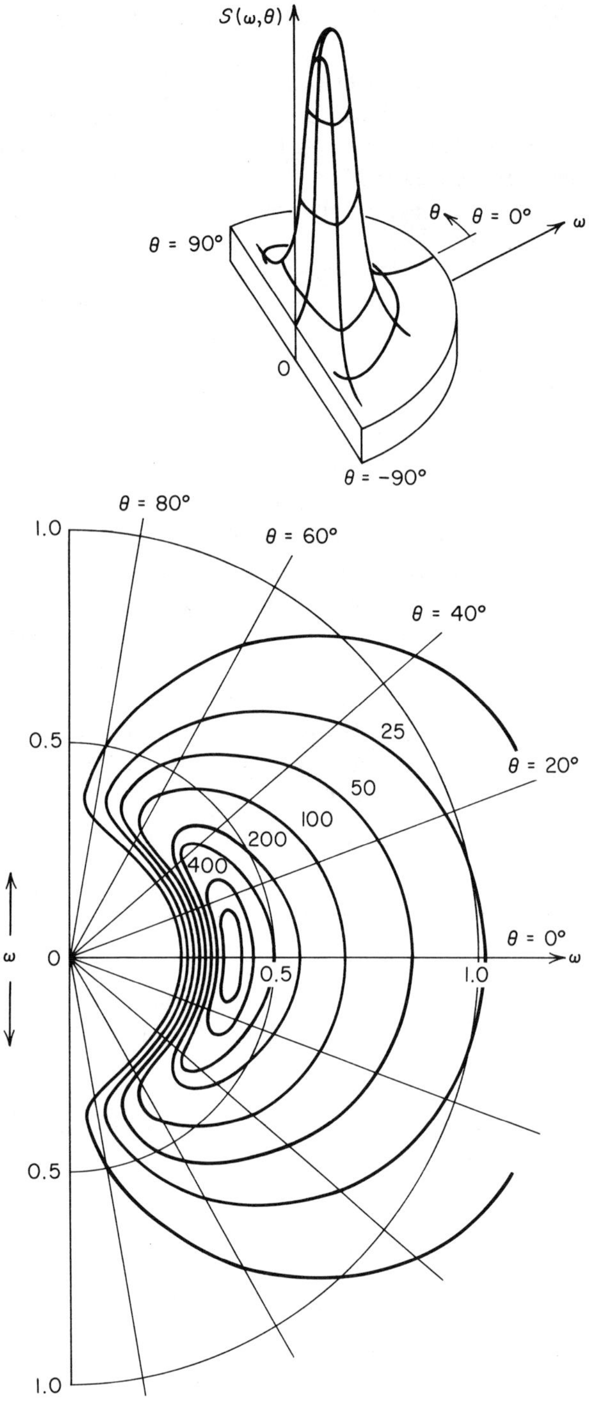

Figure 4.17 Idealized directional sea spectrum.

Families of Wave Spectra

The spectra given by a two-parameter formula such as (4.4.15) are very idealized and simplified. In reality, the shape of wave spectra observed in the ocean varies considerably (even for the same significant wave height) depending on the geographical location, duration and fetch of wind, stage of growth or decay of a storm, and existence of swell. Since a ship encounters an infinite variety of wave conditions, and since the magnitude of the response is significantly influenced by the shape of the wave spectrum, it is necessary to have a method that accounts for this variety of wave spectra. For this purpose researchers have developed families of wave spectra, which consists of groups of some 10 or 12 spectra for each of several levels of sea state severity, that is, for each of several significant wave heights. Examples are the H-family of Lewis [20], the extension of it by Hoffman [21], and the two-parameter and six-parameter families of Ochi [22]. Ochi's approach is chosen for illustration because it is based on the probabilistic relationship between ω_m and H_s expressed in (4.4.16). The six-parameter family is more comprehensive and versatile, but involves more computation. Both families are briefly summarized here.

Family of Two-Parameter Wave Spectra

There are nine members of this family, each of which is generated from (4.4.15) for a different value of ω_m (or equivalently, T_m). Since the probability density function of T_m for any given H_s is known, the nine values of T_m can be chosen so as to give a complete and balanced representation of the variation in wave period which occurs in each level of sea state. This is achieved by choosing the most probable value of T_m and four pairs of values on either side of it, corresponding to confidence coefficients 0.95, 0.85, 0.75, and 0.50. The resulting expressions for ω_m as a function of H_s are given in Table 4.2, which also gives the

TABLE 4.2 MODAL FREQUENCIES FOR THE (MEAN) NORTH ATLANTIC WAVE SPECTRA AS A FUNCTION OF SIGNIFICANT WAVE HEIGHT (ω_m in rps, H_s in meters)

Confidence Coefficients	Value of ω_m	Weighting Factor
Lower ω_m		
0.95	$0.048(8.75 - \ln H_s)$	0.0500
0.85	$0.054(8.44 - \ln H_s)$	0.0500
0.75	$0.061(8.07 - \ln H_s)$	0.0875
0.50	$0.069(7.77 - \ln H_s)$	0.1875
Most probable	$0.079(7.63 - \ln H_s)$	0.25
Upper ω_m		
0.50	$0.099(6.87 - \ln H_s)$	0.1875
0.75	$0.111(6.67 - \ln H_s)$	0.0875
0.85	$0.119(6.65 - \ln H_s)$	0.0500
0.95	$0.134(6.41 - \ln H_s)$	0.0500

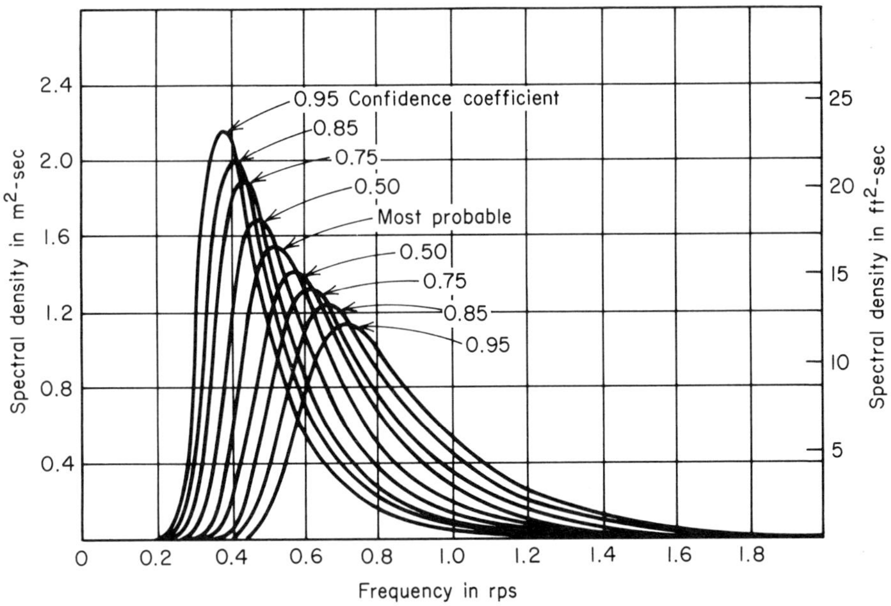

Figure 4.18 Family of two-parameter wave spectra for significant wave height 3.0 m (9.8 ft).

weighting function by which each of the nine wave spectra is multiplied to reflect the differing probabilities of occurrence. Figure 4.18 shows the family of two-parameter wave spectra for a significant wave height of 3 m.

Family of Six-Parameter Wave Spectra

Ochi's six-parameter family accounts for two additional sources of variation in ocean wave spectra. The first is the shape, or degree of sharpness, of the spectrum peak. To account for this feature Ochi added a shape parameter, $\lambda (0 \leq \lambda \leq \infty)$ to the basic two-parameter formula of (4.4.15); the generalized formula is

$$S(\omega) = \frac{1}{4\Gamma(\omega)}\left[\frac{4\lambda + 1}{4}\left(\frac{\omega_m}{\omega}\right)^4\right]^{\lambda}\frac{H_s^2}{\omega}\, e^{-[(4\lambda+1)/4](\omega_m/\omega)^4} \tag{4.4.21}$$

where $\Gamma(\,)$ is the Gamma function. The shape of the wave spectrum becomes sharper with increasing λ. For $\lambda = 1$, (4.4.21) reduces to the two-parameter wave spectrum formula of (4.4.15). The derivation of (4.4.21) is given in Ref. 22.

Another characteristic of actual wave spectra is that there often exists a plateau, or even a second peak, at a higher frequency, as shown in Fig. 4.19. This may arise because of swell coexisting with wind-generated waves, or because of the growth or decay of a storm. Although the wave energy at the higher frequencies is usually much less than that at the lower frequencies, its contribution to ship response may be significant depending on the ship size and speed.

Thus, it is highly desirable to represent the shape of the entire spectrum as closely as possible, and this may be achieved by separating the spectra into two parts, one which includes primarily the lower-frequency components of the energy and the second which covers primarily the higher-frequency components of the energy, as shown in Fig. 4.19. This gives a six-parameter spectral formula:

$$S(\omega) = \frac{1}{4}\sum_{j=1,2}\frac{1}{\Gamma(\lambda_j)}\left[\frac{4\lambda_j + 1}{4}\left(\frac{\omega_{mj}}{\omega}\right)^4\right]^{\lambda_j}\frac{H_{sj}^2}{\omega}\, e^{-[(4\lambda_j+1)/4](\omega_{mj}/\omega)^4} \tag{4.4.22}$$

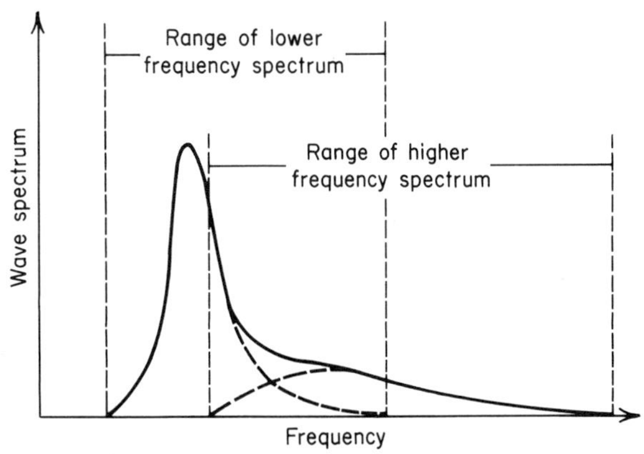

Figure 4.19 Decomposition of wave spectra.

where $j = 1,2$ stands for the lower and higher frequency components, respectively.

DETERMINATION OF THE SIX PARAMETERS

In order to obtain expressions for the six parameters, Ochi first grouped the 800 observed North Atlantic wave spectra from Refs. 23–25 into ten groups, according to severity, and then for each group performed a statistical analysis, of which the principal steps are given next. In this analysis it was possible to deal with only five parameters by working in terms of the ratio of the two significant wave heights: $r_H = H_{s1}/H_{s2}$. The steps are:

1. Probability density functions were established for each parameter. For example, it was found that the parameters λ_1 and λ_2 both follow the gamma probability law for all 10 groups.
2. Three values were determined from the probability density function for each parameter, namely, the modal value and upper and lower values corresponding to a confidence coefficient of 0.95.
3. For each value of a parameter, values of each of the other parameters were determined from the original data by taking their respective averages in the region of ±5%. Thus a total of 15 spectra were established for a given sea severity.
4. Of these 15 spectra, five are associated with the modal value of the five parameters. It was found, however, that these five spectra had nearly the same shape; therefore, the spectrum associated with the modal value of the parameter r_H was chosen as representative, and this spectrum is called the most probable spectrum for a given sea severity. Thus, a total of 11 spectra was derived as a family of wave spectra for a specified sea severity.

In Table 4.3 the values of the six parameters for this family are expressed as functions of significant wave height H_s and from these expressions the complete family of spectra for the desired sea can be generated from (4.4.22).

The weighting factor for each member of the family is

Most probable spectrum	0.50
All other spectra	(each) 0.05

The weight given to the most probable wave spectrum is much higher than that for the other spectra. This is because the most probable spectrum also represents four other spectra associated with the modal value of the parameters, as stated above.

Figure 4.20 shows the family of six-parameter spectra for a significant wave height of 3.0 m, and may be compared with the family of two-parameter spectra shown in Fig. 4.18. It can be seen that the members of the six-parameter family have a wider variety of shapes than the members of the two-parameter family. Some members of the six-parameter family have double peaks, and the majority of the spectra have sharper peaks than those of the two-parameter family.

Duration of Sea States

The extreme value of a response is a function of the number of peak values encountered, and hence it is necessary to know the duration of each sea state. Ideally we wish to know the length of time in which a ship is exposed to a given sea condition. In many instances a ship will encounter a particular sea state some time after that sea state has arisen or will depart from the relevant area of ocean before the sea state has subsided. However it is always in the midst of some sea

TABLE 4.3 VALUES OF SIX PARAMETERS AS A FUNCTION OF SIGNIFICANT WAVE HEIGHT (H_s in meters)

	H_{s1}	H_{s2}	ω_{m1}	ω_{m2}	λ_1	λ_2
Most probable spectrum	$0.84\ H_s$	$0.54\ H_s$	$0.70\ e^{-0.046\ H_s}$	$1.15\ e^{-0.039\ H_s}$	3.00	$1.54\ e^{-0.062\ H_s}$
0.95 confidence spectra	$0.95\ H_s$	$0.31\ H_s$	$0.70\ e^{-0.046\ H_s}$	$1.50\ e^{-0.046\ H_s}$	1.35	$2.48\ e^{-0.102\ H_s}$
	$0.65\ H_s$	$0.76\ H_s$	$0.61\ e^{-0.039\ H_s}$	$0.94\ e^{-0.036\ H_s}$	4.95	$2.48\ e^{-0.102\ H_s}$
	$0.84\ H_s$	$0.54\ H_s$	$0.93\ e^{-0.056\ H_s}$	$1.50\ e^{-0.046\ H_s}$	3.00	$2.77\ e^{-0.112\ H_s}$
	$0.84\ H_s$	$0.54\ H_s$	$0.41\ e^{-0.016\ H_s}$	$0.88\ e^{-0.026\ H_s}$	2.55	$1.82\ e^{-0.089\ H_s}$
	$0.90\ H_s$	$0.44\ H_s$	$0.81\ e^{-0.052\ H_s}$	$1.60\ e^{-0.033\ H_s}$	1.80	$2.95\ e^{-0.105\ H_s}$
	$0.77\ H_s$	$0.64\ H_s$	$0.54\ e^{-0.039\ H_s}$	0.61	4.50	$1.95\ e^{-0.082\ H_s}$
	$0.73\ H_s$	$0.68\ H_s$	$0.70\ e^{-0.046\ H_s}$	$0.99\ e^{-0.039\ H_s}$	6.40	$1.78\ e^{-0.069\ H_s}$
	$0.92\ H_s$	$0.39\ H_s$	$0.70\ e^{-0.046\ H_s}$	$1.37\ e^{-0.039\ H_s}$	0.70	$1.78\ e^{-0.069\ H_s}$
	$0.84\ H_s$	$0.54\ H_s$	$0.74\ e^{-0.052\ H_s}$	$1.30\ e^{-0.039\ H_s}$	2.65	$3.90\ e^{-0.085\ H_s}$
	$0.84\ H_s$	$0.54\ H_s$	$0.62\ e^{-0.039\ H_s}$	$1.03\ e^{-0.030\ H_s}$	2.60	$0.53\ e^{-0.069\ H_s}$

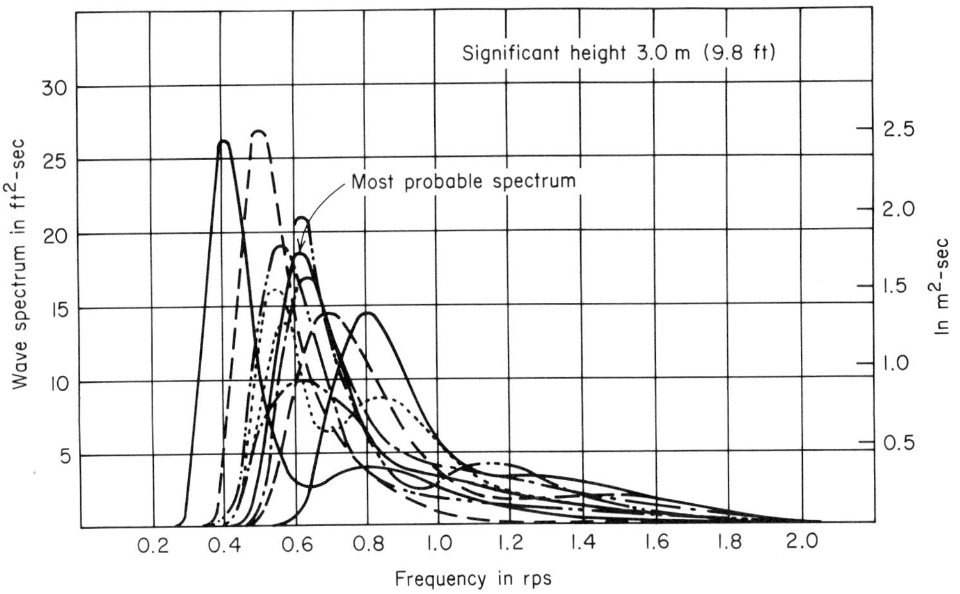

Figure 4.20 Family of six-parameter wave spectra for significant wave height 3.0 m (9.8 ft).

state and the transfers from one to another largely cancel out. Hence we can assume that the exposure time to a given sea state is equal to the duration of that state. This is in fact slightly conservative because with modern forecasting and long-range weather radar a ship can often avoid the worst regions of a storm area. Figure 4.21 taken from Ref. 26 is an envelope curve of the longest recorded durations of every 1.5 m interval of significant wave height, estimated from analysis of the data given in Ref. 23. For example, significant wave heights between 6 and 7.5 m can be expected to persist for a maximum of 40 hours. At the upper and lower extremities of sea severity the maximum duration tends toward a constant value of approximately 3 hours and 45 hours, respectively.

Estimation of Most Probable Extreme Sea State

The second method of Section 4.3 can be used to estimate the probable extreme value of significant wave height, $H_{s,p}$ from available wave data. Ochi [19]

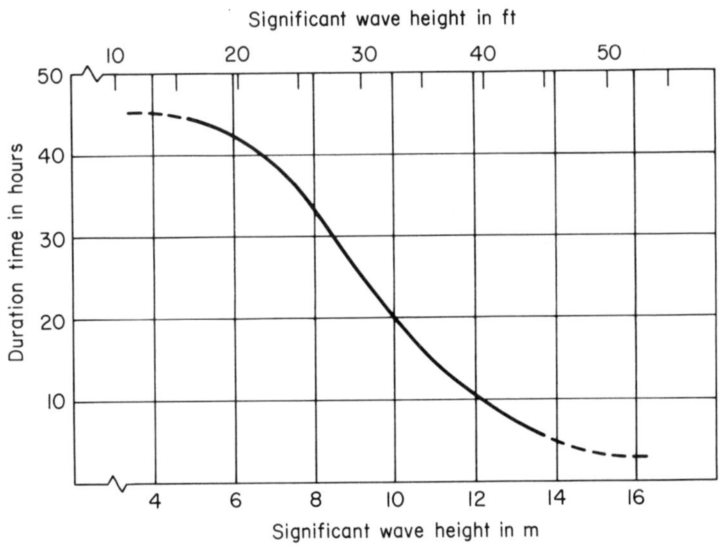

Figure 4.21 Significant wave height and its persistence in the North Atlantic every 1.52-m interval.

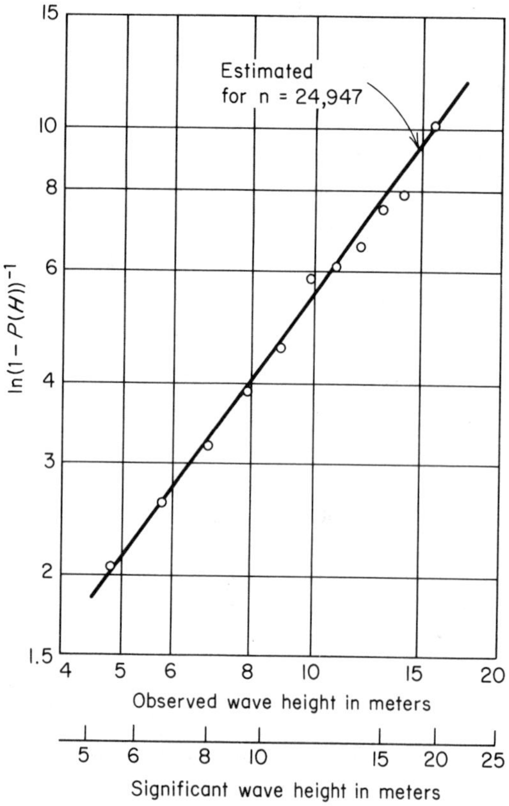

Figure 4.22 Probable extreme significant wave height in 10 years at Station J (data from Ref. 13).

has done this, using the visually observed data from 10 weather stations, collected over a period of 10 years. As an example, Fig. 4.22 shows the results for Station J, which generally encountered the most severe conditions of the 10 stations. The cumulative distribution of H_s, $P(H_s)$, was evaluated from the data, and then the left-hand side of (4.3.12) was plotted in logarithmic form. The resulting points lie approximately along a straight line, showing that the long-term distribution of H_s is approximately exponential; that is, that $f(H_s)$ in (4.3.10) is approximately linear in H_s.

In estimating the probable extreme values from the line drawn in the figure, the number of observations involved in the data has to be considered since the magnitude of the extreme value depends on this number. For Station J the number of observations in 10 years was $n = 24{,}947$. Hence the right-hand side of (4.3.12) is $\ln(24{,}947) = 10.12$, and this point on the line corresponds to a visually observed wave height of 15.4 m. Hence this is the probable extreme value for a 10-year period. Using the conversion given in (4.4.17), the probable extreme significant wave height expected to occur in 10 years is 19.2 m. Ochi also showed that these estimates obtained from visually observed wave data agree well with the estimates calculated from measured values of H_s, providing that the comparison is made for the same sample size, that is, the same value of n.

Figure 4.23 gives the combined results for all 10 stations and covers the full range of H_s, including the smaller values [below $P(H_s) = 0.99$] which follow the log-normal probability density function as discussed earlier.

The data used for Fig. 4.22 cover a 10-year period, whereas we wish to know the probable extreme significant wave height expected in the lifetime of the ship, which is say 20 years. If we assume that the statistical characteristics of the extreme values expected to occur in 20 years are the same as those observed in the data accumulated in 10 years, then the value for 20 years may be obtained by extending the line to a point corresponding to a sample size which is twice as large. The ordinate of Fig. 4.22 is $\ln[\{1 - P(H_{s,p})\}^{-1}]$, and from (4.4.26) this is equal to the natural logarithm of n. Hence the new ordinate is

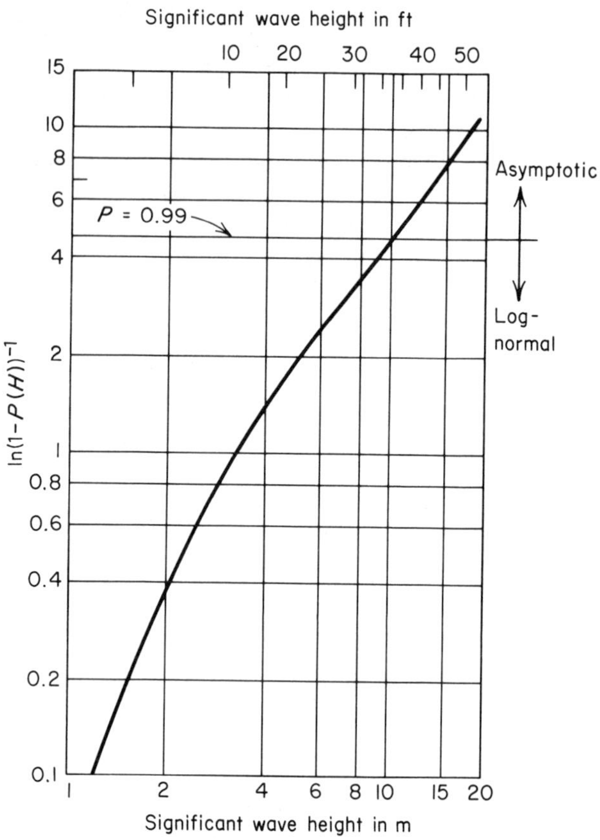

Figure 4.23 Cumulative distribution of significant wave height for the (mean) North Atlantic.

ln(2n) which is 10.82. By extrapolating the straight line up to this point we obtain a value of 20.2 m for $H_{s,p}$. Thus the 20-year value is about 5% greater than that corresponding to 10 years.

Frequency of Occurrence of Various Sea States

As will be discussed in Section 4.6, it is necessary or at least advisable to perform a response analysis for not only the probable extreme significant wave height $H_{s,p}$, but also for one or two smaller values of H_s. For these values it is necessary to decrease the value of the risk parameter in proportion to their more frequent occurrence, in order to maintain a uniform risk level. Hence we need to know the frequency of occurrence of the various sea states. This is even more necessary if a response analysis is being performed for the full range of H_s and over the complete life of the ship, as might be done if a rigorous hull girder fatigue calculation is required.

The frequency of occurrence of various sea severities can be obtained from the available wave data. For small and moderate sea states the governing probability density function is the log-normal function. This applies up to a cumulative probability distribution of 0.99 (i.e., values of H_s that have at least a 1% likelihood of occurring). For severe sea states the calculation must be based on the asymptotic extreme distribution. Ochi performed the calculations for the average North Atlantic data and obtained the results shown in Table 4.4, which gives the frequencies for each one-meter interval of significant wave height.

TABLE 4.4 LONG-TERM FREQUENCY OF OCCURRENCE OF VARIOUS SEA STATES IN THE (MEAN) NORTH ATLANTIC

Significant Wave Height (Meters)	Frequency of Occurrence	Significant Wave Height (Meters)	Frequency of Occurrence
< 1	0.0503	9–10	0.0079
1–2	0.2665	10–11	0.0054
2–3	0.2603	11–12	0.0029
3–4	0.1757	12–13	0.0016
4–5	0.1014	13–14	0.00074
5–6	0.0589	14–15	0.00045
6–7	0.0346	15–16	0.00020
7–8	0.0209	16–17	0.00012
8–9	0.0120	17 <	0.00009

4.5 CALCULATION OF SHIP RESPONSE TO A RANDOM SEA

We now examine the interaction between the ship and the waves so as to be able to calculate the wave-induced structural loading on the ship. In this regard it will be helpful to consider the interactive process between the wave and the ship as a system, as illustrated in Fig. 4.24. The input to the system is the irregular and randomly varying elevation of the sea surface. We have seen that the usual practice is to represent this irregular sea surface as the linear superposition of a large number of regular waves of various amplitude and frequencies, and that the resulting combination is conveniently described in terms of a wave spectrum. The output of the system is whichever form of ship response we are interested in, such as ship motion, hull

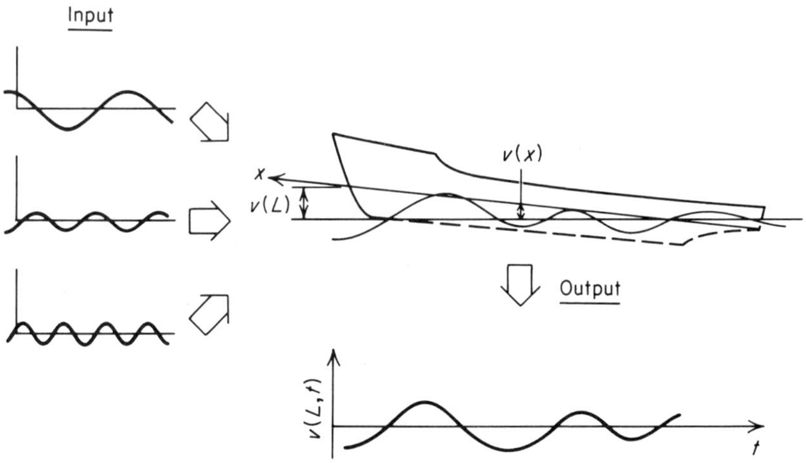

Figure 4.24 Wave–ship system.

pressure distribution, and bending moment. In the preceding four sections we have obtained all of the necessary information about the input which, being a random process, requires the use of statistical methods for its determination. The first use of such methods to predict ship motions in a realistic seaway was in the landmark paper of St. Denis and Pierson [27].

We now turn our attention to the wave–ship system, and we consider the physical aspects of the problem—the mechanics and dynamics of the fluid flow and of the ship motion and deformation—in order to establish the equations governing the interaction between the waves and the ship, and thus to be able to calculate the ship response to a specified wave excitation. In addition to the ship's "body forces" (weight and inertia force) the external forces acting on the ship may for simplicity be grouped into three categories:

1. The dynamic pressure on the ship's hull due to the incident and diffracted waves.
2. Inertia forces arising from the acceleration of the fluid.
3. Damping forces arising from wave radiation from the ship (and also from viscosity but this is relatively small).

All three of these are important, and the wave-induced structural loading cannot be idealized as simply a redistribution of buoyancy as the ship moves through the waves. Rather, because of the strong mutual dependency between the forces and the ship's motion, the wave-induced structural loads on a ship can only be determined as part of the complete ship motion problem. To make the solution possible, several simplifications are introduced in the theory of ship motion. The principal simplification is that the ship is subdivided into a number of prismatic segments, and the various forces are calculated separately for each segment using two-dimensional flow theory, thus neglecting the longitudinal component of relative velocity and any type of interaction between adjacent segments. The shear force and bending moment along the ship are then obtained by integration of the vertical forces on the various segments. Since the ship is being represented as a series of prismatic strips of hull, this approach is usually referred to as *strip theory* or the *strip method*.

Even with these simplifications the problem is still nonlinear because all three of the forces listed are nonlinear functions of the instantaneous draft of each prismatic segment, which in turn depends on both the wave motion and the ship motion. In the face of this complexity most analysts working in this area have chosen to linearize the problem. This involves two further and much more restrictive simplifications. The first is the linearization of the boundary conditions of the differential equation governing the fluid flow, as discussed in Section 4.4. As noted there, this approximation is only valid if the wave height is small relative to the wave length. The second approximation is to linearize the equations of motion of the ship. This means ignoring the dependency of the three forces on the instantaneous draft, which implies a wall-sided ship and small amplitudes of motion. With these simplifications the wave-ship system becomes a linear system, which is far easier to analyze than a nonlinear system. In particular a linear system is both *additive* and *homogeneous*, which means that the output to a sum of inputs is equal to the sum of the outputs produced separately by each input, and that the output produced by a scaled input $cX(t)$ is equal to $cY(t)$, where $Y(t)$ is the output produced by $X(t)$ alone. This means that the total (irregular) response can be obtained by calculating the response to each (regular) component wave separately and then superimposing these separate responses. This greatly simplifies the analysis because the response to an individual regular wave is also regular. Linearity also allows the problem to be solved in terms of frequency rather than of time, and this greatly reduces the amount of computation.

However, in reality, the wave–ship interaction is nonlinear and the foregoing approximations can only be made for small ship motions. For design purposes, in regard to both seakeeping and structural safety, it is the severe sea states which are of interest. Hence the linear theory of ship motions is not adequate for rationally-based ship structural design.

However, the general theory regarding the response of linear systems is of great value and has application in other areas of structural design such as vibration analysis. Also, in the field of ship motions some of the terminology is based on the linear theory, and a good deal of the available data is expressed in terms of quantities which are defined by the linear theory. In particular the notion of a frequency response function, or transfer function, is of fundamental importance. Therefore, before presenting the nonlinear theory we give a brief summary of linear response analysis and the definition and use of a transfer function. A comprehensive coverage is given in References 28 and 29, and in Reference 6 of Chapter 2.

Linear Response Theory: Transfer Function

We have seen that time-varying processes can be represented in terms of a spectral density function. There are many areas of engineering, such as vibration theory and communications, which deal with linear systems in which the input and the output are time-varying functions, say $X(t)$ and $Y(t)$. We will now show that in all such cases the response analysis is greatly simplified if it is performed using the spectral representation of $X(t)$ and $Y(t)$ because, if the system is linear, the two spectra are directly related by a *frequency response function* $\Phi(\omega)$, commonly referred to as a *transfer function*. The transfer function depends on the characteristics of the system, and it can be determined either by mathematical analysis (providing the governing equations for the system are known) or by experiment, or by a combination of the two. To establish the basic properties of this function let us consider the response to a sinusoidal wave input. If the input is a constant amplitude cosine wave of fixed frequency

$$x(t) = x_0 \cos \omega t$$

then it can easily be shown that the output will also be a steady state cosine wave of constant amplitude, the same frequency ω and a phase difference θ. That is

$$y(t) = y_0 \cos(\omega t - \theta) \qquad (4.5.1)$$

Information about the amplitude ratio y_0/x_0 and the phase angle θ defines the transmission characteristics or transfer function of the system at the fixed frequency ω. Instead of thinking of amplitude ratio and phase angle as two separate quantities, it is customary to use a single complex number to represent both quantities. This is called the *(complex) frequency response function* $\Phi(\omega)$ which is defined such that its magnitude is equal to the amplitude ratio and the ratio of its imaginary part to its real part is equal to the tangent of the phase angle. If

$$\Phi(\omega) = A(\omega) - iB(\omega) \qquad (4.5.2)$$

where $A(\omega)$ and $B(\omega)$ are real functions of ω, then

$$|\Phi(\omega)| = \sqrt{A^2 + B^2} = \frac{y_0}{x_o} \qquad (4.5.3)$$

and

$$\frac{\text{Imaginary part}}{\text{Real part}} = \frac{B}{A} = \tan\theta \qquad (4.5.4)$$

Using complex exponential notation, we can now say that if the input is a harmonic wave of amplitude x_0,

$$\begin{aligned} x(t) &= x_0 \cos \omega t = x_0[\text{the real part of } e^{i\omega t}] \\ &= x_0 \operatorname{Re}\{e^{i\omega t}\}, \end{aligned}$$

then the corresponding harmonic output will be

$$y(t) = x_0 \operatorname{Re}\{\Phi(\omega)e^{i\omega t}\} \qquad (4.5.5)$$

We now wish to determine the transfer function for a general input, $X(t)$. Let $\Upsilon(t - \tau)$ be the response $Y(t)$ of the system for the special case when the input is a unit impulse at time $t = \tau$; that is $\Upsilon(t - \tau)$ is the response to $X(t) = \delta(\tau)$, as shown in Fig. 4.25. To obtain the response for a general input function $X(t)$, the latter may be divided into a series of impulses, as shown in Fig. 4.26, and the response from each impulse superimposed

$$Y(t) = \sum_i [X(\tau_i)\Delta\tau]\Upsilon(t - \tau_i)$$

In the limit, as $\Delta\tau \to 0$

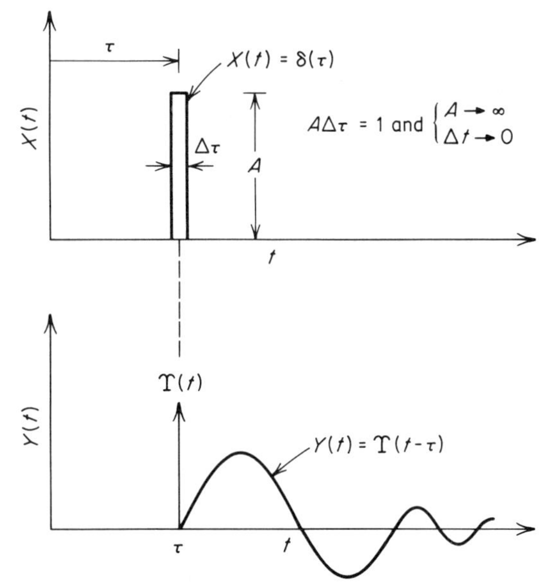

Figure 4.25 Response to a unit impulse.

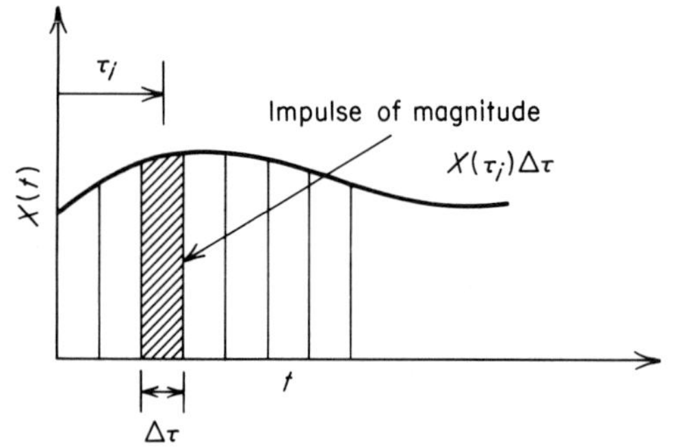

Figure 4.26 Impulse representation of a general input.

$$Y(t) = \int_{-\infty}^{\infty} X(\tau)\Upsilon(t - \tau)d\tau$$

This integral is known as the Duhamel or convolution integral. By an appropriate change of variables it may be written in an alternative form which is more convenient for mathematical purposes

$$Y(t) = \int_{-\infty}^{\infty} X(t - \tau)\Upsilon(\tau)d\tau \qquad (4.5.6)$$

It may be shown that the frequency response function $\Phi(\omega)$ and the impulse response function $\Upsilon(t)$ constitute a "Fourier transform pair" and are related by

$$\Phi(\omega) = \frac{1}{2\pi}\int_{-\infty}^{\infty} \Upsilon(t)e^{-i\omega t}\, dt \qquad (4.5.7)$$

and

$$\Upsilon(t) = \int_{-\infty}^{\infty} \Phi(\omega)e^{i\omega t}\, d\omega \qquad (4.5.8)$$

We have seen that the analysis of a fluctuating process such as ocean surface elevation is simplified by using a spectral representation, that is, by describing the process in terms of its frequencies, and we are therefore led to rewrite the relationship between $X(t)$ and $Y(t)$ of (4.5.6), expressing it in terms of their spectral density functions $S_X(\omega)$ and $S_Y(\omega)$. The details of this step are available in standard texts such as Ref. 2. The result is the strikingly simple relationship

$$S_Y(\omega) = |\Phi(\omega)|^2 S_X(\omega) \qquad (4.5.9)$$

Thus for a linear system the spectral density of the response is simply the spectral density of the input multiplied by a single scalar function of ω: the square of the amplitude of the transfer function. Once $S_Y(\omega)$ is known the averages and expected maxima can be computed from its moments, using the formulas presented in Section 4.2. $|\Phi(\omega)|^2$ is often called the *response amplitude operator* (RAO) of the system. We have seen that $|\Phi(\omega)|$ is the amplitude of the steady state response to a unit sinusoidal input of frequency ω. Thus one way of determining the transfer function for ship motions and loads is to subject a ship model to a series of sinusoidal waves of unit amplitude and various frequencies and to measure the response amplitude for each frequency. However, this is only valid for small waves and small response amplitudes, so that the wave–ship system is linear.

Thus if the wave–ship system is linear the calculation of wave bending moment M_w or other wave-induced response is both simple and rapid. For later reference we here summarize the procedure for the calculation of the response for a given wave spectrum $S_w(\omega)$, and a given ship heading and ship speed (and hence a given wave encounter frequency ω_e):

1. Calculate the transfer function $\Phi(\omega)$ by performing a ship motion analysis for the response to regular waves of unit height over a complete range of wave encounter frequency.
2. Express the wave spectrum in terms of encounter frequency and apply the transfer function to obtain the response spectrum $S_M(\omega_e)$, as in (4.5.9).
3. By calculating the appropriate moments m_0, m_2, and so on, of $S_M(\omega_e)$ determine whatever response values are required, such as

bandwidth: $\varepsilon = \sqrt{1 - \dfrac{m_2^2}{m_0 m_4}}$

significant value: $M_{w,s} = 2\sqrt{m_0}$

probable extreme value:

$$\hat{M}_{w,p} = \sqrt{2 \ln\left(\frac{1800T}{\pi}\sqrt{\frac{m_2}{m_0}}\right)}$$

The use of this procedure to calculate the characteristic extreme value of bending moment is illustrated schematically in Fig. 4.27.

Loukakis and Grivas [8] present evidence to show that, providing the wave–ship system is linear, the peak values of the response can be considered to follow a Rayleigh distribution, even though the response is not a narrow-band process, and that the effect of bandwidth can be accounted for by using a

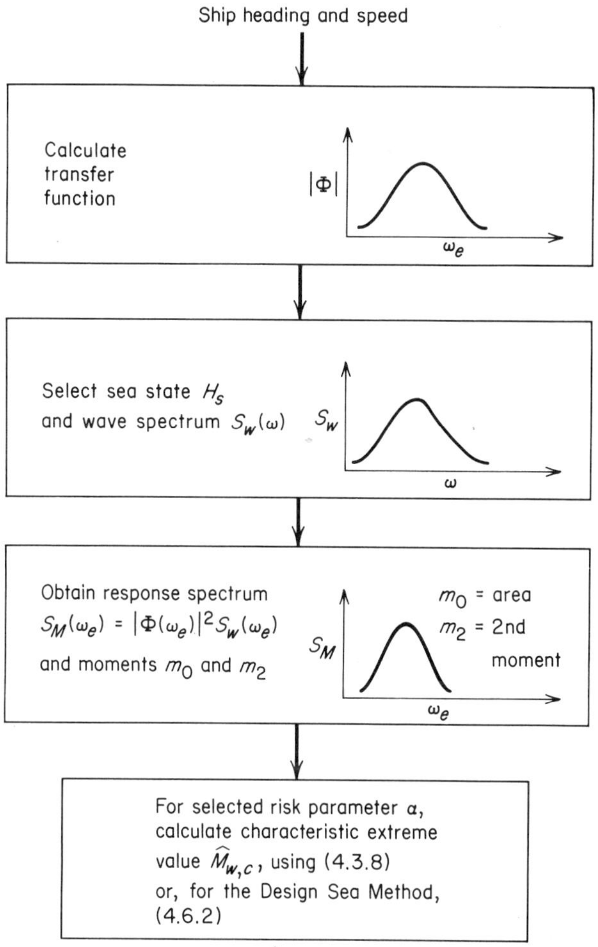

Figure 4.27 Calculation of $\hat{M}_{w,c}$ for a linear wave–ship system.

modified Rayleigh parameter [see (4.3.2)] $R' = 2m_0(1 - \varepsilon^2/2)$, in place of the usual parameter, $R = 2m_0$. In this case the probable extreme value is, from (4.3.3),

$$\hat{M}_{w,p} = \sqrt{R' \ln(n)} = \sqrt{2m_0(1 - \varepsilon^2/2)\ln(n)} \tag{4.5.10}$$

However this expression is only approximate and it differs from (4.3.4) which was presented by Ochi [6] and which was derived in a more rigorous manner.

Calculation of Nonlinear Wave-Induced Forces

We turn now to a more realistic nonlinear model of the wave–ship system, and we begin by examining the forces that act on a vessel due to waves. As explained briefly at the beginning of Section 4.5, nearly all of these forces are nonlinear functions of the instantaneous draft, which in turn depends on both the wave motion and the ship motion. Thus all of the various nonlinearities are due to the fluid dynamics and the shape of the ship, and not to the structural aspects. Therefore the task of calculating the wave-induced forces lies entirely within the field of hydrodynamics rather than structures, and since most of these fluid-related forces are nonlinear it is a quite complex task, requiring an extensive knowledge of hydrodynamics. Obviously this is well outside the normal role of a ship structural designer, even in rationally-based design, and hence it is also beyond the scope of this book. Properly speaking, it is the task of hydrodynamicists to develop methods (necessarily computer-based) for evaluating these forces and for calculating ship motions and the wave-induced hull girder loads. However, the ship structural designer does need to be familiar with the basic aspects so as to be able to make proper use of the information and methods which the hydrodynamicist provides. Hence in this section we give a brief overview of the calculation of the wave-induced forces on a ship.

Figure 4.28*a* shows the coordinate systems and some of the notation. The $\mathcal{X}$, $\mathcal{Y}$, $\mathcal{Z}$ axes are fixed in space, with the $\mathcal{X}$-axis at the still water level. The ship is assumed to move forward at speed U in the $\mathcal{X}$-direction. The x, y, z-axes are attached to the ship, with the origin at the AP in the ship's centerplane and the x-axis along the ship's keel. $H(x, t)$ is the instanta-

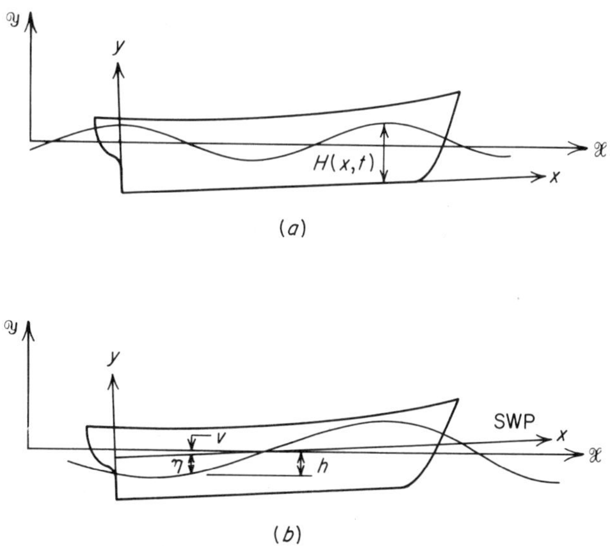

Figure 4.28 Coordinate system for ship response.

neous height of the local free surface above the keel, and this can be chosen as the primary variable. The external force acting on the ship is made up of the three components listed at the beginning of Section 4.5. In terms of $H(x, t)$ and other variables explained below, the total vertical force per unit length at location x is

$$F(x, t) = -\left[\frac{D}{Dt}\left\{m(x, H)\frac{DH}{Dt}\right\} + N(x, H)\frac{DH}{Dt} + \int_0^H B(x, y)\left(\frac{\partial p}{\partial \mathcal{Y}}\right)dy\right] \quad (4.5.11)$$

where m is the added mass per unit length and N is the damping force per unit length. The operator D/Dt is the total derivative with respect to time t; that is

$$\frac{D}{Dt} \equiv \frac{\partial}{\partial t} - U\frac{\partial}{\partial x}$$

The symbol p represents the local Froude-Krylov fluid pressure. The "body forces"—that is, the weight per unit length and the product of ship mass per unit length times the local acceleration—are not included in the foregoing equation; they will be accounted for later.

All three terms in the foregoing equation are nonlinear; the added mass m and the damping coefficient N are nonlinear functions of H, and the pressure distribution within the fluid is a nonlinear function of depth and of fluid velocity.

At the present time a good deal of research is being devoted to obtaining more complete and more accurate solutions to the foregoing equation, and steady progress is being made. This area of ship research includes many specialized aspects and a variety of solution methods, and in a brief overview such as this it is simply not possible to describe all of these various aspects and methods.

At the present time the most comprehensive method to have been published appears to be the second-order method of Jensen and Pedersen [30]. Therefore in this section we present a brief summary of their approach. For this method it is more convenient to locate the x-axis in the plane of the ship's waterline in still water, as shown in Fig. 4.28*b*. This plane, which is fixed relative to the ship, will be referred to as the SWP. For simplicity the waves are taken to be long-crested waves, moving toward the ship (head seas).

In order to allow for component waves whose amplitude is not small relative to their length, Jensen and Pedersen employed the second-order wave theory of Longuet-Higgins [31]. In this approach the wave potential is represented as the sum of a first-order term, ϕ_1 and a second term ϕ_2, and the surface profile $h(x, t)$ is similarly the sum of two components, h_1 and h_2. In each case the first terms correspond to the linearized wave theory presented in Section 4.4, and the second terms constitute (approximate) corrections to the linear terms. For the surface elevation we have

$$h \cong h_1 + h_2 \quad (4.5.12)$$

where

$$h_1(\mathcal{X}, t) = \sum_{i=1}^{n} a_i \cos \psi_i$$

with

$$\psi_i = -k_i\mathcal{X} - \omega_i t + \theta_i$$

which is the same linear expression for h that was presented earlier, in (4.4.4). The second-order term is

$$h_2(\mathcal{X}, t) = \frac{1}{4}\sum_{i=1}^{n}\sum_{j=1}^{n} a_i a_j [(k_i + k_j)\cos(\psi_i + \psi_j) - |k_i - k_j|\cos(\psi_i - \psi_j)] \quad (4.5.13)$$

Similarly the total velocity potential is

$$\phi \cong \phi_1 + \phi_2 \quad (4.5.14)$$

where

$$\phi_1(\mathcal{X}, \mathcal{Y}, t) = \sum_{i=1}^{n} \frac{a_i\omega_i}{k_i} e^{k_i\mathcal{Y}} \sin \psi_i \quad (4.5.15)$$

in which each term within the summation is the linearized potential which was presented earlier, in (4.4.5). The second-order contribution to ϕ is

$$\phi_2(\mathcal{X}, \mathcal{Y}, t) = \frac{1}{2}\sum_{i=1}^{n}\sum_{j=1}^{n} a_i a_j \max(-\omega_i, \omega_j) e^{|k_i - k_j|\mathcal{Y}} \sin(\psi_i - \psi_j) \quad (4.5.16)$$

in which

$$\max(-\omega_i, \omega_j) \equiv \begin{cases} -\omega_i \text{ if } \omega_i > \omega_j \\ \omega_j \text{ if } \omega_i < \omega_j \end{cases}$$

We now rewrite (4.5.10) for the external vertical force per unit length as follows.

$$F(x, t) = -\left[\frac{D}{Dt}\left\{m(x, \eta)\frac{D\eta}{Dt}\right\} + N(x, \eta)\frac{D\eta}{Dt} + \int_{-T}^{-\eta} B(x, y)\left(\frac{\partial p}{\partial \mathcal{Y}}\right)_{y+v} dy\right] \qquad (4.5.17)$$

In this expression $v(x, t)$ is the instantaneous vertical displacement of the hull girder at location x and η is the difference between this local hull girder displacement and the local height of the ocean surface. That is

$$\eta(x, t) = v(x, t) - h(x, t) \qquad (4.5.18)$$

Thus η is the "emergence" at location x; it is the local height of the ship's SWP above the local sea surface. $T(x)$ is the still water draft along the length of the ship and hence the instantaneous local draft $H(x, t)$ and the emergence $\eta(x, t)$ are directly related: $H(x, t) = T(x) - \eta(x, t)$. The emergence η is more convenient to use because it is the perturbation or departure of the instantaneous local draft from its still water equilibrium value $T(x)$. The local breadth of the ship is $B(x, y)$ and hence $B(x, -\eta)$ is the breadth at the instantaneous local fluid surface.

When the η dependence of m and N, and the y dependence of B are neglected, (4.5.17) reduces to the linear strip theories proposed by Lewis [32] and Gerritsma and Beukelman [33].

Jensen and Pedersen evaluated $F(x, t)$ by a perturbational method, taking into account linear and quadratic terms in the relative displacement η and thereby in the local displacement of the hull girder v and the local wave surface elevation h. In order to do this they used the following linear approximations for the added mass m and the damping coefficient N, constructed at $\eta = 0$, and for B, constructed at $y = 0$.

$$m(x, \eta) \cong m(x, 0) + \eta\left(\frac{\partial m}{\partial \eta}\right)_{\eta=0} \equiv m_0(x) + \eta(x, t)m_1(x) \qquad (4.5.19)$$

$$N(x, \eta) \cong N(x, 0) + \eta\left(\frac{\partial N}{\partial \eta}\right)_{\eta=0} \equiv N_0(x) + \eta(x, t)N_1(x) \qquad (4.5.20)$$

$$B(x, y) \cong B(x, 0) + y\left(\frac{\partial B}{\partial y}\right)_{y=0} \equiv B_0(x) + yB_1(x) \qquad (4.5.21)$$

Froude-Krylov Buoyancy Force

The buoyancy force F_B on a prismatic hull segment of unit length at position x can be expressed as

$$F_B(x, t) = -\int_{-T}^{-\eta} B(x, y)\left(\frac{\partial p}{\partial \mathcal{Y}}\right)_{y+v} dy \qquad (4.5.22)$$

where the pressure p is given by Bernoulli's equation

$$p(\mathcal{Y}, x, t) = \rho\left(\frac{\partial \phi}{\partial t} + g\mathcal{Y} + \frac{1}{2}(\nabla\phi)^2\right) \qquad (4.5.23)$$

in which ρ is the fluid density, ϕ the velocity potential of the wave as expressed in (4.5.14), and g the acceleration of gravity. In terms of the two components of ϕ we can approximate p as the sum of three terms

$$p \cong p_0 + p_1 + p_2$$

where

$$p_0 = -\rho g\mathcal{Y} \qquad (4.5.24)$$

$$p_1 = -\rho\frac{\partial \phi_1}{\partial t} \qquad (4.5.25)$$

$$p_2 = -\rho\left[\frac{\partial \phi_2}{\partial t} + \frac{1}{2}(\nabla\phi_1)^2\right] \qquad (4.5.26)$$

in which the contribution of $\nabla\phi_2$ in the third term is neglected.

Substituting for the components of ϕ into the expressions for the first- and second-order contributions to the pressure in (4.5.25) and (4.5.26) gives

$$p_1 = \rho g \sum_{i=1}^{n} a_i e^{k_i \mathcal{Y}} \cos \psi_i \qquad (4.5.27)$$

$$p_2 = -\frac{1}{2}\rho \sum_{i=1}^{n}\sum_{j=1}^{n} a_i a_j [\max(\omega_i, \omega_j)|\omega_i - \omega_j| e^{|k_i - k_j|\mathcal{Y}} + \omega_i\omega_j e^{(k_i+k_j)\mathcal{Y}}] \cos(\omega_i - \omega_j) \qquad (4.5.28)$$

We now rewrite (4.5.17), separating the hydrostatic contribution to the buoyancy force, due to p_0, from the hydrodynamic contributions due to p_1 and p_2, and then further subdividing the hydrostatic contribution into the portion from the keel to the still water plane (SWP) and the portion from the SWP to the local fluid surface height.

$$F_B = \int_{-T}^{0} B(x, y) \frac{\partial p_0}{\partial \mathcal{Y}} dy + \int_{0}^{-\eta} B(x, y) \frac{\partial p_0}{\partial \mathcal{Y}} dy - \int_{-T}^{-\eta} B(x, y) \left(\frac{\partial (p_1 + p_2)}{\partial \mathcal{Y}} \right)_{y+v} dy$$

The first term is the still water buoyancy of the segment, and when all segments have been considered this force will be equal and opposite to the weight of the ship. These two forces eventually cancel and they are responsible only for the stillwater loads, which we prefer to analyze separately from the wave-induced loads. Therefore from this point on we omit both the static buoyancy and the weight (the inertia effects of the ship's mass will be included when writing the equations of motion). Therefore we delete the first term of the foregoing equation. The next step is to substitute for the three components of pressure from (4.5.24), (4.5.25), and (4.5.26). The result is

$$\begin{aligned} F_B = {} & \rho g \int_{0}^{-\eta} B(x, y)\, dy \\ & - \rho g \sum_{i=1}^{n} a_i k_i \cos \psi_i \int_{-T}^{-\eta} B(x, y) e^{k_i(y+v)}\, dy \\ & + \frac{1}{2} \rho g \sum_{i,j=1}^{n} a_i a_j \cos(\psi_i - \psi_j) \int_{-T}^{-\eta} B(x, y) \\ & [\max(\sqrt{k_i}, \sqrt{k_j}) \\ & \times (\sqrt{k_i} - \sqrt{k_j})(k_i - k_j) e^{|k_i - k_j|\mathcal{Y}} \\ & + \sqrt{k_i k_j}\,(k_i + k_j) e^{(k_i + k_j)\mathcal{Y}}]\, dy \end{aligned}$$

We then substitute for $B(x, y)$ from (4.5.21), perform the three integrations with respect to y and then substitute for η from (4.5.18). In performing the latter step we also make a second-order approximation for v, as done earlier for h. That is we say

$$v \cong v_1 + v_2$$

and therefore the expression which is substituted for η is

$$\eta \cong v_1 - h_1 + v_2 - h_2$$

Finally, the entire expression is expanded and only first- and second-order terms are kept, that is terms in v_1, v_1^2, v_2, h_1, h_1^2, and h_2. If the first- and second-order terms are grouped the resulting expression for the distributed buoyancy force is

$$F_B \cong B_{B1} + F_{B2} \tag{4.5.29}$$

where

$$F_{B1}(x, t) = -\rho g B_0 \left[v_1 - \sum_{i=1}^{n} a_i \kappa_i \cos \psi_i \right] \tag{4.5.30}$$

and

$$\begin{aligned} F_{B2}(x, t) = {} & -\rho g B_0 v_2 + \frac{1}{2} \rho g B_1 v_1^2 \\ & - \rho g v_1 \sum_{i=1}^{n} a_i \cos \psi_i (B_1 - B_0 k_i \kappa_i) \\ & + \frac{1}{4} \rho g \sum_{i,j=1}^{n} a_i a_j [\cos(\psi_i + \psi_j) B_1 \\ & + \cos(\psi_i - \psi_j)(B_1 + 2B_0 \\ & \{[\sqrt{k_i k_j} - \max(k_i, k_j)]\kappa_{|i-j|} \\ & - \sqrt{k_i k_j}\ \kappa_{|i+j|}\})] \end{aligned} \tag{4.5.31}$$

in which κ_i and $\kappa_{|i \pm j|}$ are the Smith correction factors, defined as

$$\kappa_i(x) = 1 - \frac{k_i}{B_0(x)} \int_{-T}^{0} B(x, y) e^{k_i y}\, dy \tag{4.5.32a}$$

$$\kappa_{|i \pm j|}(x) = 1 - \frac{|k_i \pm k_j|}{B_0(x)} \int_{-T}^{0} B(x, y) e^{|k_i \pm k_j| y}\, dy \tag{4.5.32b}$$

Hydrodynamic Forces

We next consider the hydrodynamic part of the force components, that is, the first two terms in (4.5.17)

$$F_H(x, t) = -\frac{D}{Dt}\left(m \frac{D\eta}{Dt} \right) - N \frac{D\eta}{Dt} \tag{4.5.33}$$

Substitution of the linearized expressions for m and N, together with a second-order representation of the emergence, $\eta \cong \eta_1 + \eta_2$, gives the following second-order expression for F_H:

$$F_H(x, t) \cong F_{H1} + F_{H2} \tag{4.5.34}$$

where

$$F_{H1} = -m_0 \frac{D^2\eta_1}{Dt^2} - n_0 \frac{D\eta_1}{Dt} \tag{4.5.35}$$

$$F_{H2} = -m_0 \frac{D^2\eta_2}{Dt^2} - n_0 \frac{D\eta_2}{Dt} - \eta_1\left(m_1 \frac{D^2\eta_1}{Dt^2} + n_1 \frac{D\eta_1}{Dt}\right) - m_1\left(\frac{D\eta_1}{Dt}\right)^2 \tag{4.5.36}$$

and where

$$n_0 = n_0(x) = N_0 - U \frac{\partial m_0}{\partial x}$$

$$n_1 = n_1(x) = N_1 - U \frac{\partial m_1}{\partial x}$$

Now introducing into (4.5.35) and (4.5.36) $\eta = v - h^*$, where h^* is the wave height including second-order terms with the amplitudes reduced by the Smith correction factor, we finally get

$$F_{H1} = -m_0 \frac{D^2 v_1}{Dt^2} - n_0 \frac{Dv_1}{Dt} - \sum_{i=1}^{n} a_i \omega_i \kappa_i (m_0 \omega_i \cos\psi_i - n_0 \sin\psi_i) \tag{4.5.37}$$

$$\begin{aligned} F_{H2} = &-m_0 \frac{D^2 v_2}{Dt^2} - n_0 \frac{Dv_2}{Dt} - v_1\left(m_1 \frac{D^2 v_1}{Dt^2} + n_1 \frac{Dv_1}{Dt}\right) - m_1\left(\frac{Dv_1}{Dt}\right)^2 \\ &+ \sum_{i=1}^{n} a_i \kappa_i \left[\cos\psi_i\left(m_1 \frac{D^2 v_1}{Dt^2} + n_1 \frac{Dv_1}{Dt} - \omega_i^2 m_1 v_1\right) + \sin\psi_1\left(n_1 \omega_i v_1 + 2m_1 \omega_i \frac{Dv_1}{Dt}\right)\right] \\ &+ \frac{1}{4} \sum_{i=1}^{n} \sum_{j=1}^{n} a_i \kappa_i a_j \kappa_j \\ &\{\cos(\psi_i + \psi_j)(\omega_i + \omega_j)^2[m_1 - (k_i + k_j)m_o] \\ &+ \cos(\psi_i - \psi_j)(\omega_i - \omega_j)^2(m_1 + |k_i - k_j|m_0) \\ &+ \sin(\psi_i + \psi_j)(\omega_i + \omega_j)[-n_1 + (k_i + k_j)n_0] \\ &+ \sin(\psi_i - \psi_j)(\omega_i - \omega_j)(-n_1 - |k_i - k_j|n_0)\} \end{aligned} \tag{4.5.38}$$

This completes the determination of a set of first- and second-order contributions to the external vertical force per unit length acting on the hull of the ship in an irregular sea.

Calculation of Structural Response

The degree of complexity of the hull girder structural response analysis depends mainly on whether it is necessary to include a dynamic hull girder response analysis, and if so whether it needs to be performed as part of the ship motion analysis. For ships of standard types and proportions a dynamic analysis is generally not necessary, although the hull girder vibration frequencies should be calculated to check against resonance with main engines, and so on. If a dynamic analysis is desired (for example, to investigate the response to slamming) it can be performed separately, after the wave-induced loads have been calculated for the ship. Therefore, in general, the ship motion analysis can be performed on the basis of a rigid ship. However there are two main instances in which the hull girder should be modeled as a flexible beam and the ship's dynamic structural response should be calculated as part of the ship motion analysis:

1. If the hull girder is relatively flexible such that its natural two-node frequency of vibration is low enough for springing to be a possibility (below, say, 4 Hz, although this depends on the ship's operating speed).
2. If the response to slamming is to be examined and the vessel is so small that its motion is significantly affected by slamming.

In the following a brief summary is given of both types of hull girder response analysis—one in which the ship is regarded as rigid and only its rigid body response is calculated (mainly bending moment) and one in which the ship is modeled as a flexible beam and its flexural vibration (whipping) is calculated in addition to its rigid body response.

Flexible Ship

The structural response, whether it be low frequency (wave-induced bending) or high frequency (whipping) has no significant effect on the forces which are acting, or on the ship's overall motion. Therefore all of the expressions for the various wave-induced forces presented in the previous section are applicable for both a flexible ship and a rigid ship. The governing differential equation for the vertical deflection of a flexible beam subjected to a dynamic distributed load $F(x, t)$ is

$$EI\frac{\partial^4 v}{\partial x^4} + m_s\frac{\partial^2 v}{\partial t^2} - m_s r^2\frac{\partial^4 v}{\partial t^2 \partial x^2} = F(x, t) \quad (4.5.39)$$

in which $I(x)$ is the moment of inertia for vertical bending, $m_s(x)$ is the ship mass per unit length, and r is the radius of gyration of the sectional mass m_s in rotation about a horizontal transverse axis through the section's center of mass. Hence $m_s r^2(x)$ is the sectional or local mass moment of inertia. This term is small relative to the others and is usually neglected in beam vibration analysis. Shear deflection may also be accounted for but is omitted here for simplicity.

Jensen and Pedersen [30] showed that for the second-order representation of $F(x, t)$ given in the previous section, a solution can be obtained for the hull girder deflection $v(x, t)$ as the sum of its two components, $v_1(x, t)$ and $v_2(x, t)$, and that each of these is of the form

$$v_l(x, t) = \sum_{r=0}^{n} [u_l(t)]_r f_r(x) \qquad (l = 1, 2)$$

where $f_r(x)$ is the eigenfunction (deflected shape) for the rth mode of beam vibration (including the two rigid body modes: heave and pitch), $[u_1(t)]_r$ is given by a series of the form

$$[u_1(t)]_r + \sum_{i=1}^{n} C_{1ri}\cos\omega_i t + C_{2ri}\sin\omega_i t$$

and $[u_2(t)]_r$ is of the form

$$\begin{aligned}[u_2(t)]_r = \sum_{i,j=1}^{n} [&C_{3ri}\cos(\omega_i + \omega_j)t \\ &+ C_{4ri}\cos(\omega_i - \omega_j)t \\ &+ C_{5ri}\sin(\omega_i + \omega_j)t \\ &+ C_{6ri}\sin(\omega_i - \omega_j)t]\end{aligned}$$

and the terms which comprise $C_{1ri}, \ldots, C_{6ri}$ are determined from the amplitudes and phase angles of the n component waves which comprise the given wave spectrum. Once the hull girder deflection is known the wave-induced forces can be calculated and then integrated to give the hull girder shear force and bending moment. Like the forces, the bending moment is the sum of a first-order term and a second-order term

$$M(x, t) \cong M_1(x, t) + M_2(x, t)$$

and each of these is expressed as a series of order n in which the terms are explicit analytical functions of x and t. The functions of t are similar to those in the series given above for u_1 and u_2, and the functions of x are obtained by a double integration of the forces along the hull girder. Details are given in Ref. 30.

Rigid Ship

If the ship is rigid then the local deflection $v(x)$ is everywhere defined by the vertical position V and rotation γ of the ship's center of gravity. Hence we introduce the coordinates $\bar{x}, \bar{y}, \bar{z}$ which are fixed in the ship, with the origin at the ship's center of gravity. In these coordinates the expressions that replace $v_1(x)$, $v_2(x)$ and their derivatives are

$$\left.\begin{aligned} v_l(x) &\Leftarrow V_l + \bar{x}\gamma_l \\ \dot{v}_l(x) &\Leftarrow \dot{V}_l + \bar{x}\dot{\gamma}_l \\ \ddot{v}_l(x) &\Leftarrow \ddot{V}_l + \bar{x}\ddot{\gamma}_l \end{aligned}\right\} (l = 1, 2) \qquad (4.5.40)$$

where γ is the pitch angle, and $\dot{V}$ and $\ddot{V}$ are time derivatives of V. The equations of motion for the ship become

$$M_s\ddot{V} = \int_L F(\bar{x}, t)\, d\bar{x} \qquad (4.5.41)$$

$$I_z\ddot{\gamma} = \int_L \bar{x}F(\bar{x}, t)\, d\bar{x} \qquad (4.5.42)$$

where M_s is the mass of the ship and I_z is the moment of inertia of the SWP about the $\bar{z}$-axis. $F(\bar{x}, t)$ is obtained by substituting $\bar{x} + x_G$ and $\bar{y} + y_G$ in place of x and y, and by substituting the new expressions for $v_1(x)$, $v_2(x)$, and their derivatives in all of the terms of $F(x, t)$. The only remaining quantities which depend on x (or $\bar{x}$) are m_0, m_1, N_0, N_1, n_0, n_1, B_0, and B_1, all of which are properties of the ship's form and which can be established by strip theory. Therefore the integration of the forces in (4.5.41) and (4.5.42) can be performed. If the first- and second-order terms are then grouped separately the result is two pairs of coupled differential equations, one for $V_1(t)$ and $\gamma_1(t)$, and the other for $V_2(t)$ and $\gamma_2(t)$. Upon solving these the values of $V_1(t)$, $V_2(t)$, and their derivatives are substituted back into $F(\bar{x}, t)$ and the shear force and bending moment can then be obtained by integration. Once again the expressions will involve summations of terms that are a function of the amplitudes and phase angles of the n component waves of the given wave spectrum.

Ship Speed in a Seaway

The response of a ship is dependent on the ship speed, and for the prediction of the design value of response the speed should be the maximum likely value corresponding to the selected sea state, that is, the design speed minus the speed reductions which are entailed by that sea state. There are two different types of speed reduction in a seaway: one due to the added resistance and reduced propulsive efficiency associated with waves and ship motions, and the other due to voluntary slowdown by the ship's master in order to ease severe ship motions. For severe seas the latter is more significant. For a rough estimate of speed reduction for design purposes the following formula derived by Aertssen [34, 35] from full-scale trials may be used.

$$\Delta U = \frac{V}{100}\left(\frac{m}{L} + n\right) \tag{4.5.43}$$

where ΔU = loss of speed in a seaway in knots
U = ship speed in calm water in knots
L = ship length in meters
m, n = constants depending on ship heading and sea severity [34]

Results and Conclusions Regarding Nonlinear Response Analysis

Nonlinearity Arising from Hull Shape

It has been known for some time, from strain measurements made at sea, that in ships with a large bow flare and/or a relatively small block coefficient and/or a relatively high operating speed (e.g., containerships and warships) the sagging bending moment is appreciably larger than the hogging bending moment, primarily because of the significant departure from wall-sidedness, especially at the bow where the hydrodynamic forces are largest. In addition there are several important hydrodynamic differences between the entry and the exit of a body through the free surface of a fluid, and between the crest and trough of a wave when it is nonsinusoidal. The linear theory of ship motion ignores all of these effects and consequently predicts the same value of wave bending moment for both hogging and sagging.

In Ref. 36 Jensen and Pedersen applied their perturbational method to the nonlinear response analysis of a VLCC tanker and a containership, both operating in the fully loaded condition in a moderate sea state (H_s = 6 m). Solutions were obtained for both an elastic hull and a rigid hull. The containership was of typical form, with a large bow flare, relatively small block coefficient and a relatively high operating speed. This study was one of the first truly nonlinear analyses to be performed on such vessels. Figure 4.29*a* shows the theoretical results for the containership. The figure is a logarithmic plot of the probability distribution $P(\hat{M}_w)$ of the nondimensional extreme values of hogging and sagging wave bending moment over a short term (approximately one hour). It can be seen that the sagging bending moment is approximately 20% larger than the value predicted by the linear theory, and the hogging bending moment is approximately 10% smaller. The figure also shows some measured values of short-term extreme values of wave bending moment, and it can be seen that the theory agrees well with these values.

Figure 4.29*b* is a plot of wave bending moment over the length of the ship and shows how the increase in sagging bending moment extends over the forward half of the ship and moves the peak value forward by about 8% of the ship length.

These results illustrate the reason for specifying a larger value of sagging wave bending moment for ships of this type. They also show clearly that for a wave-induced response analysis for a ship of this type a nonlinear method should be used.

In contrast, the results for the VLCC showed that for this type of vessel, at deep draft, the nonlinear effects are very small and the linear theory is adequate.

The results also showed that for both the containership and the tanker the principal effect of hull flexibility is to increase the variance (but not the magnitude) of wave bending moment by about one-third compared to the value for a rigid hull. For the containership 40% of the increase came from the nonlinear terms, whereas for the tanker only 8% of the increase came from these terms. This indicates that in an investigation of springing, in which the item of principal interest is the ship's flexible response to the waves, the nonlinear ship motion theory should be used for flared or high-performance vessels, whereas the linear theory appears to be sufficient for a tanker in the deep draft

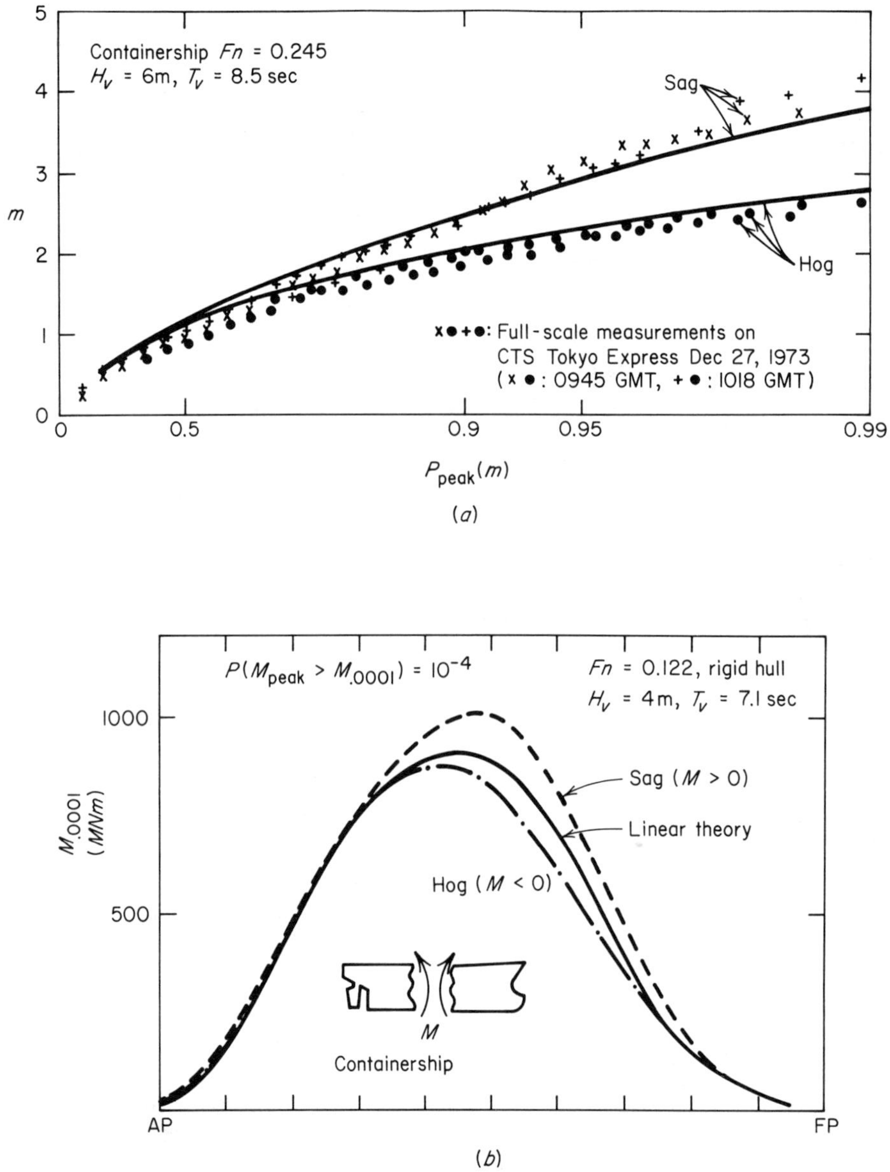

Figure 4.29 Nonlinear response of a container ship. (*a*) Probability distribution function $P_{peak}(m)$ for the peak values of the nondimensional midship bending moment $m = M/\lambda$, where λ is the standard deviation of the first-order contribution (λ_{calc} = 451 MNm). (*b*) Longitudinal distribution of the most probable largest bending moment amplitudes during 10,000 maxima. $Fn = 0.122$, H_v = 4m, and T_v = 7.1 sec. The effects of springing are neglected (rigid hull).

condition. For a tanker in the ballast condition the nonlinear theory is preferable because the shallow draft forward introduces nonlinear effects.

In obtaining these results Jensen and Pedersen used a special mathematical technique that allows $P(\hat{M}_w)$ to be calculated directly, without need of working in the time domain, even though the problem is nonlinear. However, in this technique the maximum value of $P(\hat{M}_w)$ that can be considered becomes smaller as the degree of nonlinearity increases. Therefore if the ship response becomes increasingly nonlinear with increasing sea state (as it does for the containership) the technique is limited to moderate sea states and can only obtain short-term extreme values. To obtain long-term extreme values for severe sea states requires a solution for M_w as a function of time, followed by random sampling. This is discussed later in this section.

Bottom Slamming and Bow Flare Slamming

These two phenomena were described in Section 2.1. As explained there they are fundamentally similar; an impulsive and intensive pressure on the bottom or flared bow caused mainly by a rapid increase in the added mass m as the bottom or flared section crosses the local fluid surface. In the equations of motion the relevant term is contained in the first term of (4.5.33); it is the time derivative of m.

From the description of slamming in Section 2.1 and from the brief summary of the relevant ship motion equations earlier in this section, it is clear that slamming is a complex and highly nonlinear phenomenon. As mentioned in Section 2.1, slamming causes effects at both the hull girder level and the principal member level of structure. We are here concerned only with the first. Fortunately, at this level there are two factors which simplify the treatment of slamming. First, the structure response—whipping—usually occurs at a much higher frequency than do the wave-induced responses such as wave bending moment, which occur at the frequency of wave encounter. Second, for large and medium-sized ships the overall mass and the overall damping are sufficiently large that the heaving and pitching motions of the ship are not appreciably affected by slamming. Therefore the ship's motion and all of its wave-induced responses can be calculated without considering the extra complexities of slamming. If slamming is to be investigated this can be done separately, using the information given by the ship motion analysis regarding relative positions, angles, and velocities of the ship and the water surface. Then the whipping response can be superimposed on the wave-induced response, calculated for a rigid ship.

For a smaller vessel the ship motion is strongly influenced by slamming and consequently if an accurate investigation of slamming is desired (as it would be, for example, in a naval patrol boat) it can only be done as part of a ship motion analysis in which the highly nonlinear slamming terms are included.

Various theoretical and experimental studies such as Ref. 37 have confirmed that the maximum bending moment due to slamming can be calculated with sufficient accuracy by using rigid ship theory to calculate the ship motions and other parameters affecting slamming, then performing a dynamic analysis of the elastic hull to obtain the whipping response to slamming, and then superimposing this on the wave-induced response. This is illustrated in Figs. 4.30 and 4.31 for a tanker in ballast with a shallow draft forward, and in Fig. 4.32 for a fully loaded fast container ship with large bow flare. In all cases the total wave bending moment was calculated for an elastic ship, and the wave-induced bending moment was calculated for a rigid ship. The figures illustrate how the whipping component of the former oscillates around the latter. It may be seen that in the tanker forebody (Fig.

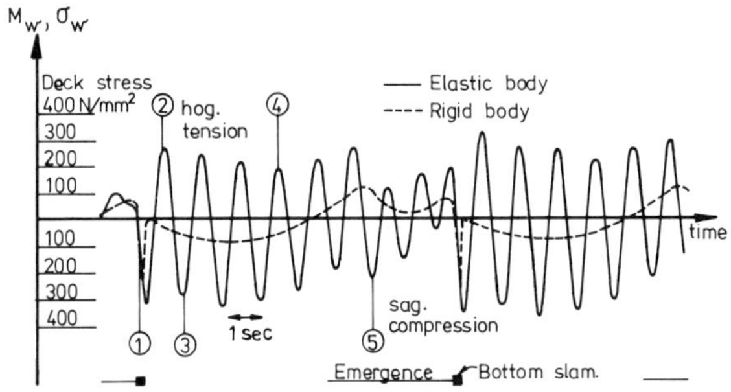

Figure 4.30 Time history of bending moment at midships; tanker in ballast.

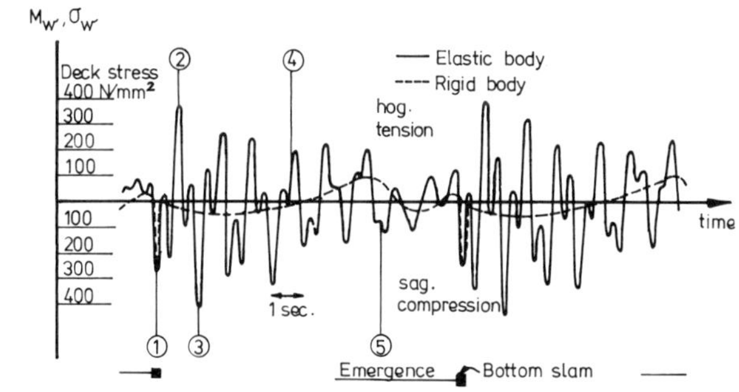

Figure 4.31 Time history of bending moment at station $7^1/_2$; tanker in ballast.

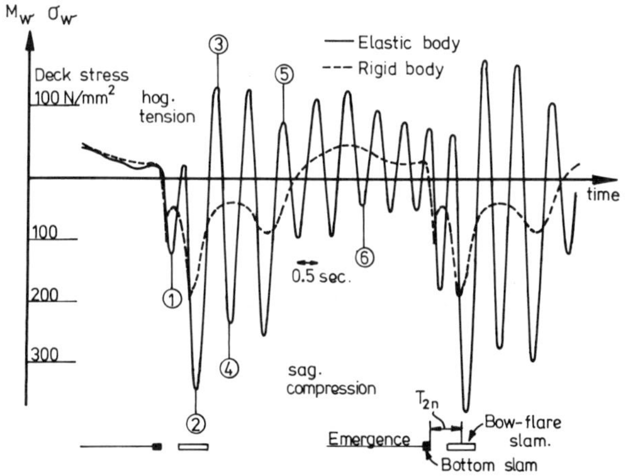

Figure 4.32 Time history of bending moment at midships; containership, Fn = 0.261.

4.31) the three-node whipping mode becomes significant.

Figs. 4.33 and 4.34 show the longitudinal distribution of wave bending moment at various instants of time in the response history. The circled numbers correspond to the times indicated by the vertical lines in the time histories.

A containership often experiences both bottom slamming and bow flare slamming, and in such cases the dynamic response depends crucially on the time interval τ between these two excitations, which in turn depends mainly on ship speed. The case shown in Fig. 4.32 corresponds to a relatively high speed (Froude no. = 0.261) and illustrates the worst case, in which the time interval is approximately equal to the ship's natural period of two-node vibration T_{2n} such that the effects of bow-flare slamming are superimposed on those of bottom slamming. In contrast, the vibrations caused by bottom and bow-flare slamming are canceled out in Fig. 4.35, when $\tau \approx 1.5\,T_{2n}$. In Fig. 4.36, at an even lower ship speed, the impact force decreases significantly, because the relative velocity between the ship and the wave surface becomes low. This illustrates the well-known fact that reduction of the ship's speed is the most effective means of preventing serious structural damage due to slamming.

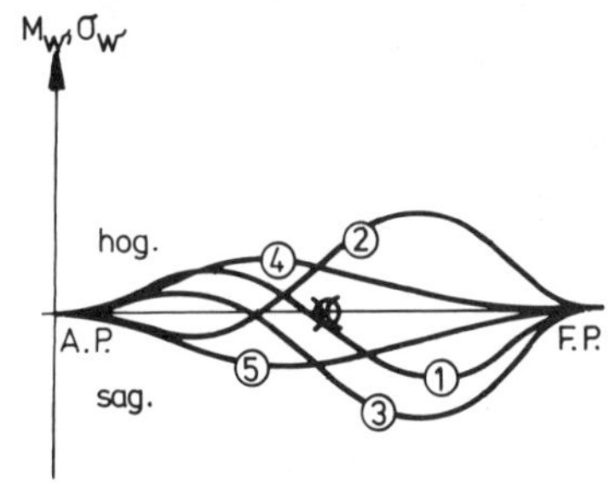

Figure 4.33 Longitudinal distribution of bending moment; tanker in ballast.

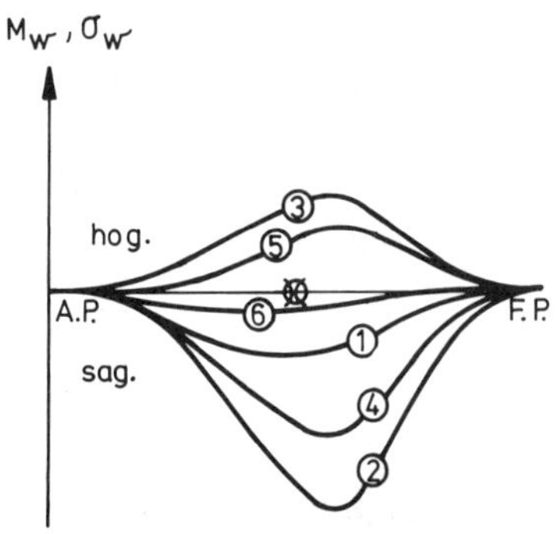

Figure 4.34 Longitudinal distribution of bending moment; containership.

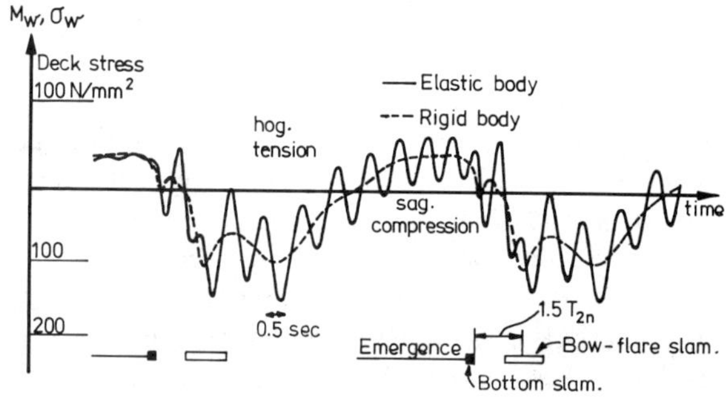

Figure 4.35 Time history of bending moment at midships; containership, Fn = 0.15.

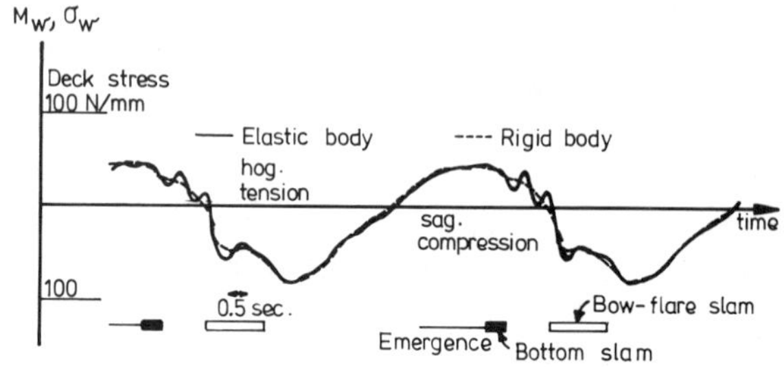

Figure 4.36 Time history of bending moment at midships; containership, Fn = 0.07.

Finally we note that Fig. 4.34 illustrates once again the fact that for a high-speed ship with flared bow the maximum sagging wave bending moment greatly exceeds the maximum hogging wave bending moment.

Calculation of Extreme Value of a Nonlinear Response

For ship design it is the more severe sea states which are of greatest interest and it is precisely these conditions for which the wave–ship system usually becomes nonlinear. The only exceptions are vessels that are of deep draft and are almost perfectly wall-sided.* We have also seen that for severe conditions the wave-induced bending moment has a nonzero mean value and, most difficult of all, is non-Gaussian.

The nonlinearity means that it is not possible to solve for the response in the frequency domain, by means of a transfer function. Instead we must perform a stepwise solution in the time domain for the response

*For this reason floating platforms generally have cylindrical legs, and for these structures the linear theory is often sufficient, even for severe seas.

[$M_w(t)$, say] and then construct a histogram. The latter must consist of *random* samples of peak values of $M_w(t)$ and the number of samples n must be large, which means that the solution for $M_w(t)$ must be performed for a large number of time steps. The equations to be solved are either (4.5.39) for a springing analysis (flexible hull), or (4.5.41) and (4.5.42) for ordinary response analyses (rigid hull). If the method of Jensen and Pedersen is used the former analysis involves the solution of a number of coupled nonlinear second-order differential equations, and the latter analysis involves only two pairs of coupled nonlinear second-order differential equations. This method has the advantage of providing an explicit expression for the time derivative of $M_w(t)$, which assists in calculating the peak values.

Since the response is non-Gaussian the distribution of peak values will not be the Rayleigh distribution. Instead as shown by Ochi [38] it follows the generalized Gamma distribution which is

$$p(x) = \frac{g}{\Gamma(m)} \lambda^{gm} x^{gm-1} e^{-(\lambda x)^g} \qquad (4.5.44)$$

where $\Gamma(\,)$ is the Gamma function. The parameters g, m, and λ can be evaluated by calculating certain moments of the data (i.e., of the histogram). An asymptotic estimate of the characteristic extreme value $\hat{M}_{w,c}$ is obtained as the solution of

$$\lambda(\hat{M}_{w,c})^{g(m-1)} e^{-(\lambda \hat{M}_{w,c})^g} = \frac{\alpha}{n} \Gamma(m) \qquad (4.5.45)$$

where α is the risk parameter (i.e., the selected probability of the extreme value exceeding $\hat{M}_{w,c}$), and n is the total number of samples of M_w. Further details are given in Ref. 38.

The foregoing procedure is summarized schematically in Fig. 4.37.

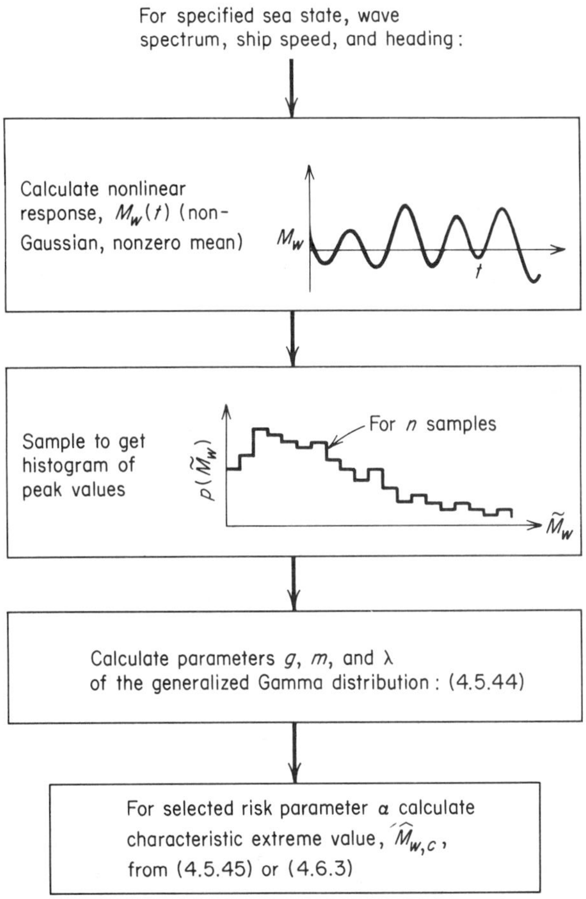

Figure 4.37 Calculation of characteristic extreme value of response for a nonlinear wave–ship system for specified short-term conditions.

4.6 CALCULATION OF DESIGN VALUE OF SHIP RESPONSE

We next consider the method for predicting the design value of the response. For this we need a method that takes into account all of the factors and correctly combines them to obtain a suitable design value of wave bending moment or other wave-induced response. As we have seen there are four factors:

1. The severity of the sea state, as measured by H_s, and the frequency of occurrence and the duration of each level of severity, so as to determine the ship's exposure time to each.
2. For each sea state, a family of wave spectral shapes which are of sufficient variety to represent the range of sea conditions that a ship meets in its lifetime.
3. Ship heading to the waves in a given sea state.
4. Ship speed for a given heading and sea state.

Also, it is necessary to select a value for the risk parameter α which is used in (4.3.8) when calculating the design value of the response from the moments of the response spectrum. The overall aim is to determine the largest response value, resulting from the worst combination of these factors, for which the probability of being exceeded in the life of the ship is α. There are two ways of achieving this, depending on the way in which the first factor, H_s, is accounted for. One approach is to regard H_s as a continuous variable and to perform the response analysis for the entire range of sea states, including light and moderate seas, and to then take a weighted sum of all of these responses, with each weighted according to the frequency of occurrence of that sea state throughout the ship's lifetime. The other method is to use long-term ocean wave statistics to determine the probable extreme value of significant wave height $H_{s,p}$ and then to perform the response analysis for that value and, as a check, for at least one or two other values of H_s. For convenience we will refer to these two methods as the Lifetime Weighted Sea method and the Design Sea method, respectively.

The Lifetime Weighted Sea method considers every cycle of wave encounter throughout the ship's lifetime, regardless of the magnitude of the response. Such a complete and even exhaustive accounting is required only if the effect of the response is cumulative, and if even low levels of response contribute to the effect, that is, fatigue. Hence this method should be used for vessels in which a specially accurate fatigue analysis is desired. For most ships, however, fatigue failure due to wave loads is prevented by simply requiring that the average wave-induced stress not exceed a certain magnitude, depending on the number of cycles, the material properties, and other local factors. If greater accuracy is required it is necessary to estimate the expected number of cycles for various stress levels, but this information can be obtained by the Design Sea method. Hence even for fatigue it is usually not necessary to examine the lifetime ship response. This is even more the case with other types of failure, which are not cumulative and which involve only one (or a few) specific limit value(s) of response. Also, most types of vessels spend only a small fraction of their operating time—perhaps 5 or 10%—in seas that are sufficiently severe to be considered for design. Hence the calculation of response values in mild seas and their inclusion in predicting the extreme value for design would represent unnecessary computation for most ship types. Furthermore, Ochi [19] has shown that both methods give similar results; that is, the Lifetime Weighted Sea method, in spite of its greatly increased computation, gives no more accuracy than the Design Sea method. However the former may be required when there is some special susceptibility to cumulative damage, and so we shall give a brief summary of both methods.

Design Sea Method

This method consists of calculating the characteristic extreme value of response (e.g., $\hat{M}_{w,c}$) for the probable extreme sea state $H_{s,p}$ corresponding to the lifetime of the vessel, and a few other selected values of H_s. For each sea state values of $\hat{M}_{w,c}$ are calculated for all combinations of the other three factors (spectra, heading, speed) and the largest value is entered on a plot of $\hat{M}_{w,c}$ versus H_s (it may be instructive to plot all of the values). This process is performed for each of the selected sea states so as to produce an envelope curve of largest $\hat{M}_{w,c}$ versus H_s, as shown in Fig. 4.38. Such a figure gives a clear picture of the relationship between $\hat{M}_{w,c}$ and H_s, especially for severe seas, and permits the designer to choose the most appropriate design value of $\hat{M}_{w,c}$. In most cases this will be the value corresponding to $H_{s,p}$, but there could be exceptions. For example, there could be some anomaly in the vessel's response, such that $\hat{M}_{w,c}$ is larger for a smaller value of H_s. Also it might be found that in the vicinity of $H_{s,p}$ the rate of increase of $\hat{M}_{w,c}$, that is, $d\hat{M}_{w,c}/dH_s$ is still large, so that a smaller risk factor α might be advisable. For a high-performance vessel a wider range of sea states should be examined because operation at speed in medium seas may be worse than the hove-to condition in extreme seas.

In calculating $\hat{M}_{w,c}$ for sea states of lesser severity than the probable extreme level, it is necessary to allow for the larger number of encounters which the ship will have with those sea states. For the probable extreme sea state the probable number of encounters is, by definition, one. For values of H_s less than $H_{s,p}$ the probable number of encounters is larger, say k, and

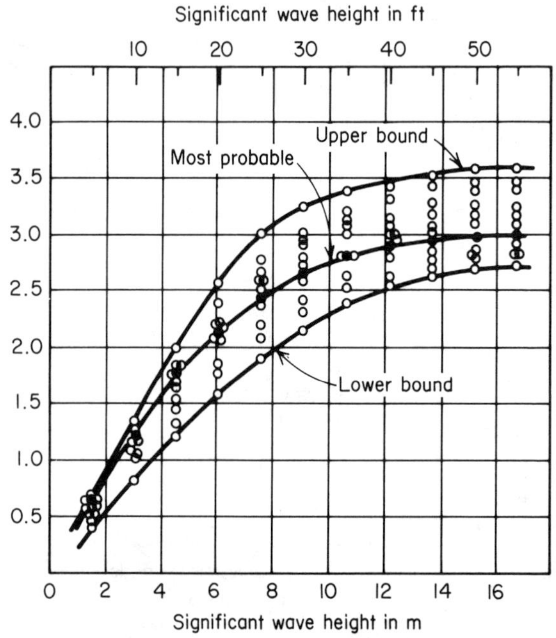

Figure 4.38 Envelope curves of $\hat{M}_{w,c}$ vs. H_s.

for each such sea state it is necessary to divide the risk parameter α by k in order that the true total risk is the same value for all sea states. The value of k for each level of H_s may be estimated from the information regarding the return period of H_s, such as Fig. 4.23. If n_p represents the number of observed wave peaks corresponding to $H_{s,p}$ and n_i represents the corresponding number for a sea state of lesser severity, then the probable number of encounters with the latter is

$$k = \frac{n_p}{n_i} \tag{4.6.1}$$

In the equations for the characteristic extreme value of response—either (4.3.8) for a linear wave–ship system or (4.5.45) for a nonlinear system—α must be replaced by α/k. For reference the modified equations are given here:

linear system:

$$\hat{Q}_c = \sqrt{2m_0 \ln\left(\frac{1800T}{\pi\alpha/k}\sqrt{\frac{m_2}{m_0}}\right)} \tag{4.6.2}$$

nonlinear system:

$\hat{Q}_c$ is obtained from

$$\lambda(\hat{Q}_c)^{g(m-1)}e^{-(\lambda\hat{Q}_c)^g} = \frac{\alpha}{kn}\Gamma(m) \tag{4.6.3}$$

The procedure is illustrated schematically in Fig. 4.39. The principal steps of the procedure are as follows:

1. Specify the next condition:
   a. Select the next sea state H_s; this will be one of the following:
      (1) The probable extreme sea state for the ship's lifetime, $H_{s,p}$, obtained from long-term wave statistics such as Fig. 4.23.
      (2) A few other values of $H_{s,p}$ below and perhaps above H_s. The number and range of values will depend on the type of ship and on the results obtained from the first few values. Estimate the longest duration of this sea state from available information, such as Fig. 4.21. Likewise, estimate the number of encounters with this sea state in the ship's lifetime. For $H_{s,p}$ the number is 1; values for lesser H_s may be obtained from (4.6.1).
   b. Select the next spectrum of the family (e.g., each of the 11 six-parameter spectra of (4.4.22), for which the parameter values are given in Table 4.3.
   c. Select the next heading.
   d. Select the next ship speed, allowing for speed reduction (see equation 4.5.43).
2. For the chosen risk parameter α calculate the characteristic extreme value of response $\hat{M}_{w,c}$ for the specified condition, using either linear theory (Fig. 4.27) or nonlinear theory (Fig. 4.37).
3. Repeat from Step 1, b, c, or d for all headings, speeds, and wave spectra, and note the largest value of $\hat{M}_{w,c}$. It may also be desirable to plot other values so as to obtain cross curves or to observe trends (e.g., the effect of ship speed).
4. Based on the relationship between $\hat{M}_{w,c}$ and H_s, select the final *design value* of $\hat{M}_{w,c}$.

Lifetime Weighted Sea Method

During the 20 years or so of the average working lifetime of a ship a wide range of weather conditions, and hence of sea states, will be encountered. This total time span may be regarded as a large number of short intervals, each of a few hours duration, during which the sea state remains constant. The values of H_s computed for such intervals throughout the lifetime of a ship will be characterized by some kind of distribution, or probability density function, over the lifetime of the ship. Likewise, the total lifetime response history of the ship may be thought of as a series of short-term response episodes, and in the Lifetime Weighted Sea method all of these short-term responses are combined by a procedure that takes into account the relative amount of exposure to the various levels of sea severity.

We first present the general form of the Lifetime Weighted Sea method, and then a simplified version which can be used when the wave–ship system is essentially linear.

General Form

In its general form the Lifetime Weighted Sea method requires the complete probability density function of the response for each specified condition, that is, each combination of speed, heading, wave spectrum, and sea state. Figure 4.40 is a schematic outline of the method. The probability density function of peak values of response for the ship's lifetime $[p_{\tilde{Q}}(\tilde{Q})]_L$ is a weighted sum of the various short-term probability density functions, each of which is for a particular

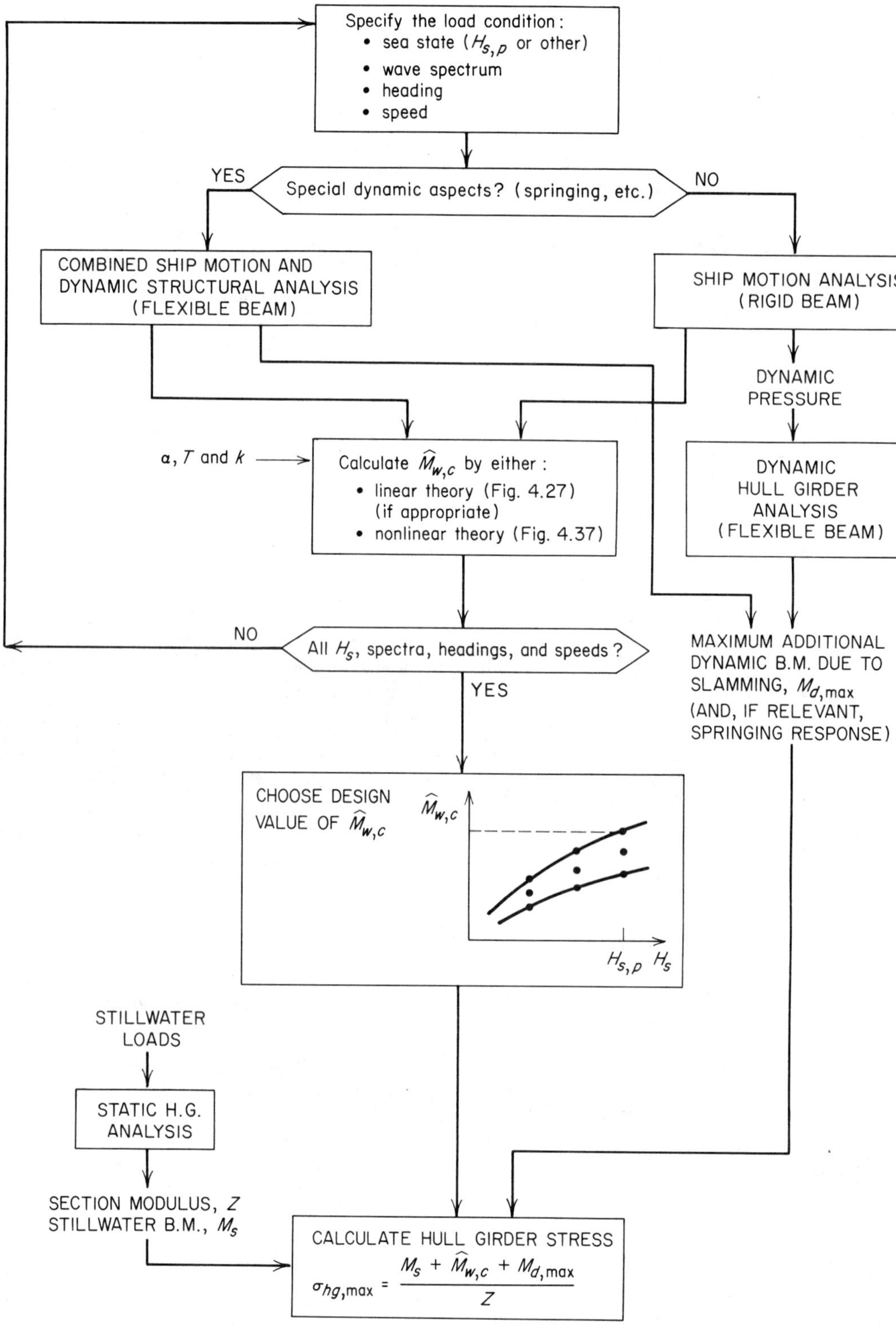

Figure 4.39 Design Sea Method.

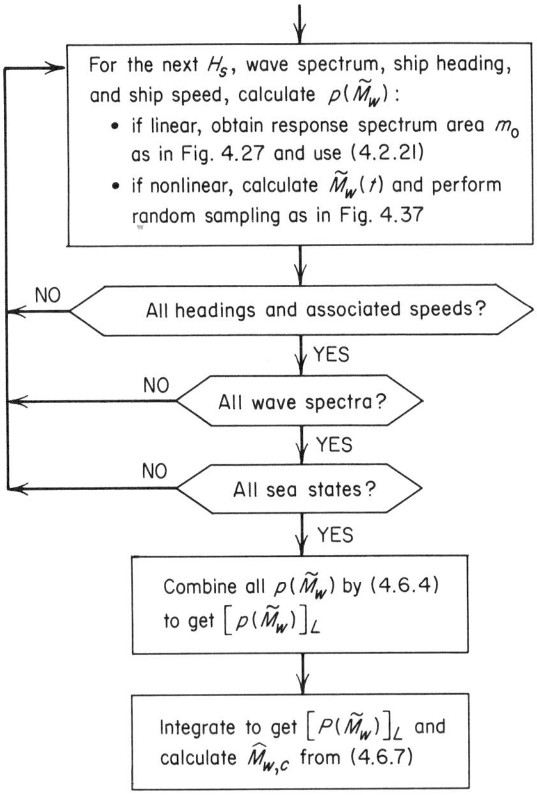

Figure 4.40 Principal steps in the Lifetime Weighted Sea Method.

combination of the four factors and carries four weighting factors to account for the relative frequency of that particular combination. The expression is

$$[p_{\tilde{Q}}(\tilde{Q})]_L = \frac{\sum_i \sum_j \sum_k \sum_l \bar{n} f_i f_j f_k f_l [p_{\tilde{Q}}(\tilde{Q})]_{ijkl}}{\sum_i \sum_j \sum_k \sum_l \bar{n} f_i f_j f_k f_l} \qquad (4.6.4)$$

where $[p_{\tilde{Q}}(\tilde{Q})]_{ijkl}$ = probability density function for the peak values of the short-term response, that is, the response for the given H_s, wave spectrum, heading, and speed

$\bar{n}$ = average number of responses per second for the short-term response

$= \dfrac{1}{2\pi}\sqrt{\dfrac{m_2}{m_0}}$

m_0 = area under the short-term response spectrum

m_2 = second moment of the short-term response spectrum

f_i = weighting factor for sea state, from Table 4.4

f_j = weighting factor for wave spectrum; these are 0.5 and 0.05 for the family of six-parameter spectra

f_k = weighting factor for heading to waves in a given sea state*

f_l = weighting factor for speed in a given sea state and heading

Figure 4.41*a* shows a typical lifetime probability density function for the peak values of wave-induced bending moment.

The total number of responses expected in the lifetime of the ship is

$$n_L = \left(\sum_i \sum_j \sum_k \sum_l \bar{n} f_i f_j f_k f_l\right) \times T \times (60)^2 \qquad (4.6.5)$$

where T is the total exposure time to the sea, in hours.

For fatigue analysis the number of cycles for each range of response magnitude is equal to the product of the average value of $[p_{\tilde{Q}}(\tilde{Q})]_L$ for that range times n_L.

After the lifetime probability density function has been calculated the next step is to use the second method from Section 4.3 to estimate the characteristic extreme value of response over the ship's lifetime, $\hat{Q}_c$. Equation 4.4.26 provides an estimate of the probable extreme value $\hat{Q}_p$ in terms of $[P_{\tilde{Q}}(\tilde{Q})]_L$, the cumulative distribution of peak values for the ship's lifetime:

$$\frac{1}{1 - [P_{\tilde{Q}}(\hat{Q}_p)]_L} = n_L \qquad (4.6.6)$$

and $P_{\tilde{Q}_L}(\bullet)$ can be obtained by integrating (4.6.4) with respect to $\tilde{Q}$.

Figure 4.41*b* shows a typical lifetime probability distribution for wave-induced bending moment $\tilde{M}_w$, and Fig. 4.41*c* shows a plot of the return period $1/[1 - P_{\tilde{M}}(\tilde{M}_w)]$. Equation 4.6.6 states that the probable extreme value $\hat{M}_{w,p}$ can be obtained by entering the diagram at a return period equal to n_L and reading the corresponding value of $\tilde{M}_w$.

For design purposes we wish to introduce a risk parameter α and obtain the characteristic extreme value $\hat{Q}_c$, which has a probability α of being exceeded in the lifetime of a ship (i.e., in n_L response cycles). This is equivalent to the probable extreme value for a

*For most sea states all headings are equi-probable, but for very severe seas a ship's master will alter course in order to reduce the severity of the response.

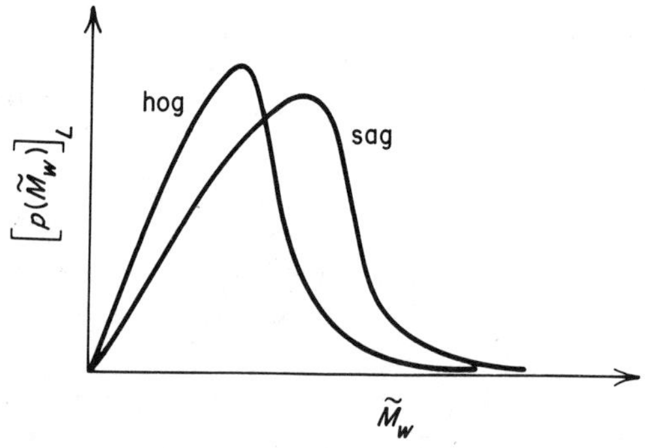

(*a*) Lifetime probability density function

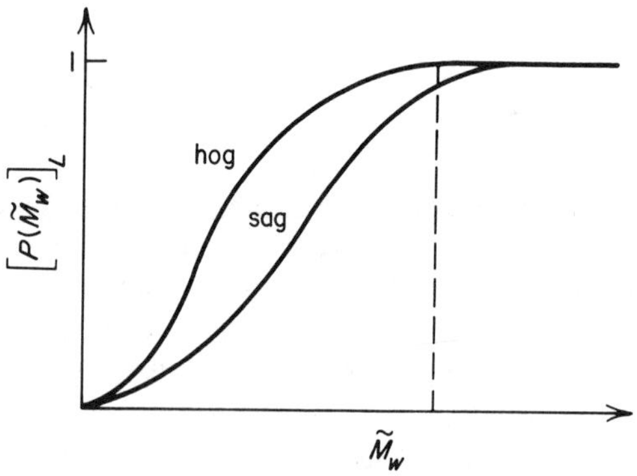

(*b*) Lifetime probability distribution

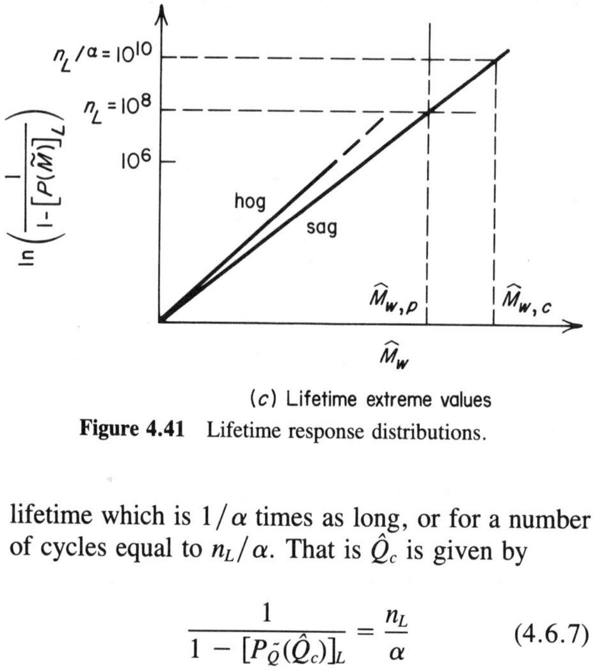

(*c*) Lifetime extreme values

Figure 4.41 Lifetime response distributions.

lifetime which is $1/\alpha$ times as long, or for a number of cycles equal to n_L/α. That is $\hat{Q}_c$ is given by

$$\frac{1}{1 - [P_{\tilde{Q}}(\hat{Q}_c)]_L} = \frac{n_L}{\alpha} \tag{4.6.7}$$

Figures 4.22 and 4.23 showed that the long-term cumulative distribution of the peak values of H_s is approximately exponential. From Fig. 4.41*c* the long-term distribution of the response appears to also be approximately exponential, since the plot of $\ln[1/\{1 - P_{\tilde{M}}(\hat{M}_w)\}]$ against $\hat{M}_w$ is approximately a straight line.

Simplified Method for Linear Systems

We have seen that the general Lifetime Weighted Sea method requires the calculation of the complete probability density function for all magnitudes of response, and this involves an enormous amount of computation. A simplified method has been proposed by Loukakis and Grivas [8] for use when the wave–ship system is sufficiently linear that the peak values of response can be assumed to follow the Rayleigh distribution, with the modified parameter R' which was introduced in Ref. 8. In this case the probable extreme value $(\hat{Q}_p)_{ijkl}$ for each specified short-term condition is given approximately by (4.5.10). Knowing $(\hat{Q}_p)_{ijkl}$, the corresponding short-term extreme value distribution $[P_{\hat{Q}}(\hat{Q})]_{ijkl}$ may be expressed by Gumbel's asymptotic approximation, given in (4.3.13), with R' in place of R in (4.3.14) for κ. As in the general method the lifetime distribution (in this case, of extreme values) consists of a weighted sum of all of the short-term distributions. Therefore the probability of the response $\tilde{Q}$ exceeding a specified characteristic extreme value $\hat{Q}_c$ is

$$P_e = \text{Prob}[\tilde{Q} > \hat{Q}_c]$$
$$= \frac{\sum_i \sum_j \sum_k \sum_l \bar{n} f_i f_j f_k f_l \{1 - [P_{\hat{Q}}(\hat{Q}_c)]_{ijkl}\}}{\sum_i \sum_j \sum_k \sum_l \bar{n} f_i f_j f_k f_l} \tag{4.6.8}$$

in which $\bar{n}$ and the weighting factors are as defined previously. P_e is precisely the same as the risk parameter α introduced earlier. Thus the final step is to specify P_e and solve (4.6.8) for the corresponding value of $\hat{Q}_c$. The solution must be performed inversely, that is, P_e is calculated for two (or more) trial values of $\hat{Q}_c$ and the correct value of $\hat{Q}_c$ is then obtained by interpolation.

Loukakis and Grivas found that a value of $10^{-2.5}$ for P_e (or α) gave a value of $\hat{M}_{w,c}$ approximately equal to the design value of wave bending moment (sagging) specified in the A.B.S. Rules. Using this value they then calculated $\hat{M}_{w,c}$ for various ratios of principal ship dimensions (length/beam, beam/draft, etc.) and found that the variations were in general agreement with those implied by the A.B.S. formula.

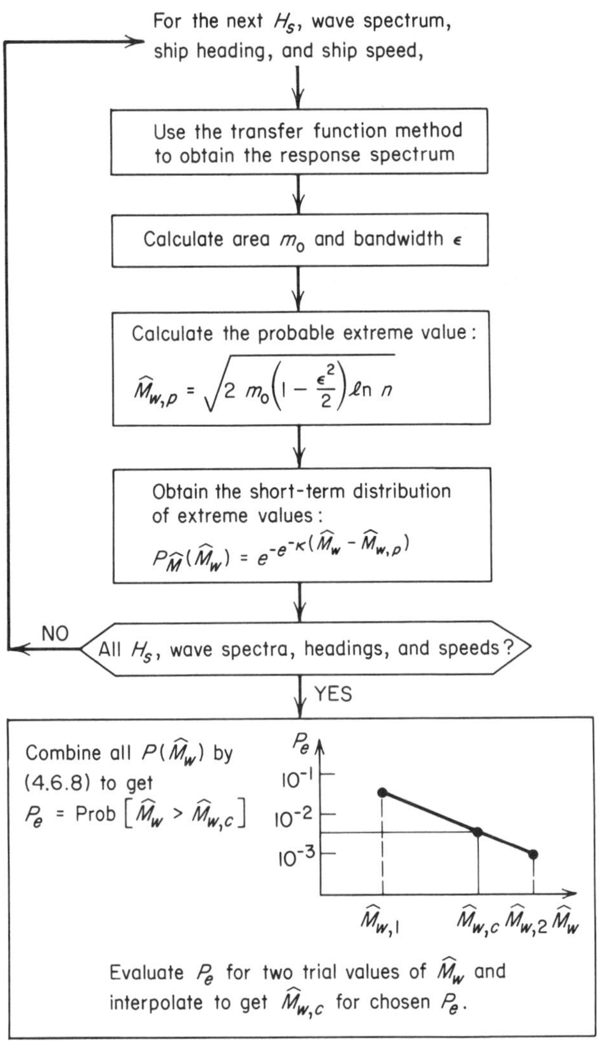

Figure 4.42 Lifetime Weighted Sea Method—simplified version.

The simplified method is illustrated schematically in Fig. 4.42. Although it involves much less computation than the general method it still requires substantially more computation than the Design Sea method because it investigates all sea states. Also it is limited to applications in which the response is linear, which precludes its use for most of the more challenging ship applications (high-performance, unusual geometry, etc.).

Summary

The foregoing sections have shown that the Lifetime Weighted Sea method is computationally expensive and that its use is only justified if fatigue is of special importance. If the ship's response is known to be linear then the simplified method can be used. However, as pointed out by Ochi [16], all lifetime methods have still another disadvantage. A ship spends a very large percentage of its life in small or moderate seas and therefore an inaccuracy in estimating the characteristics of small or moderate sea states, or in calculating the ship response at these levels, could introduce an appreciable error into the estimates of the characteristic value of response. For this reason, as well as for the savings in computation, the Design Sea method is more appropriate for the majority of ship design applications.

Finally, in an effort to give a concise overall summary of this entire chapter, Fig. 4.43 on page 178 brings together and summarizes the various levels, tasks, options, and analysis methods in the overall process of calculating a design value of a wave-induced load effect, such as hull girder bending moment. The figure also indicates in general terms which combinations of methods and options are appropriate, or necessary, for various design applications.

REFERENCES

1. Rice, O., "Mathematical Analysis of Random Noise," *Bell System Tech. Journal* **23,** 1944, and **24,** 1945.
2. Crandall, S. and Mark, W., "Random Vibration in Mechanical Systems," Academic Press, 1963.
3. Cartwright, D. E. and Longuet-Higgins, M. S., "The Statistical Distribution of the Maxima of a Random Function," *Proc. Roy. Soc.,* Series A, **237,** 1956, pp. 212–232.
4. Huston, W. B. and Skopinski, T. H., "Probability and Frequency Characteristics of Some Flight Buffet Loads," *NACA Tech. Note* 3733, 1956.
5. M. K. Ochi, "Stochastic Analysis and Probabilistic Prediction of Random Seas," *Advances in Hydroscience,* **13,** Academic Press, 1982, pp. 217–375.
6. Gumbel, E., "Statistics of Extremes," Columbia University Press, New York, 1966.
7. David, H. A., *Order Statistics,* John Wiley & Sons, New York, 1970.
8. Loukakis, T. A. and Grivas, S. B., "A Method for Establishing Ship Design Wave Bending Moment and Its Comparison with Classification Societies' Rules," *Ocean Eng.* **7,** 1980, pp. 357–371.
9. Pierson, W. J., "A Unified Mathematical Theory for the Analysis (of) Propagation and Refraction of Storm-Generated Ocean Surface Waves, Parts I and II," New York University, College of Engineering, Engineering Research Division, Dept. of Meteorology and Oceanography, 1952, 461 pp.
10. Pierson, W. J., Neumann, G., and James, R. W., "Practical Methods for Observing and Forecasting Ocean Waves by

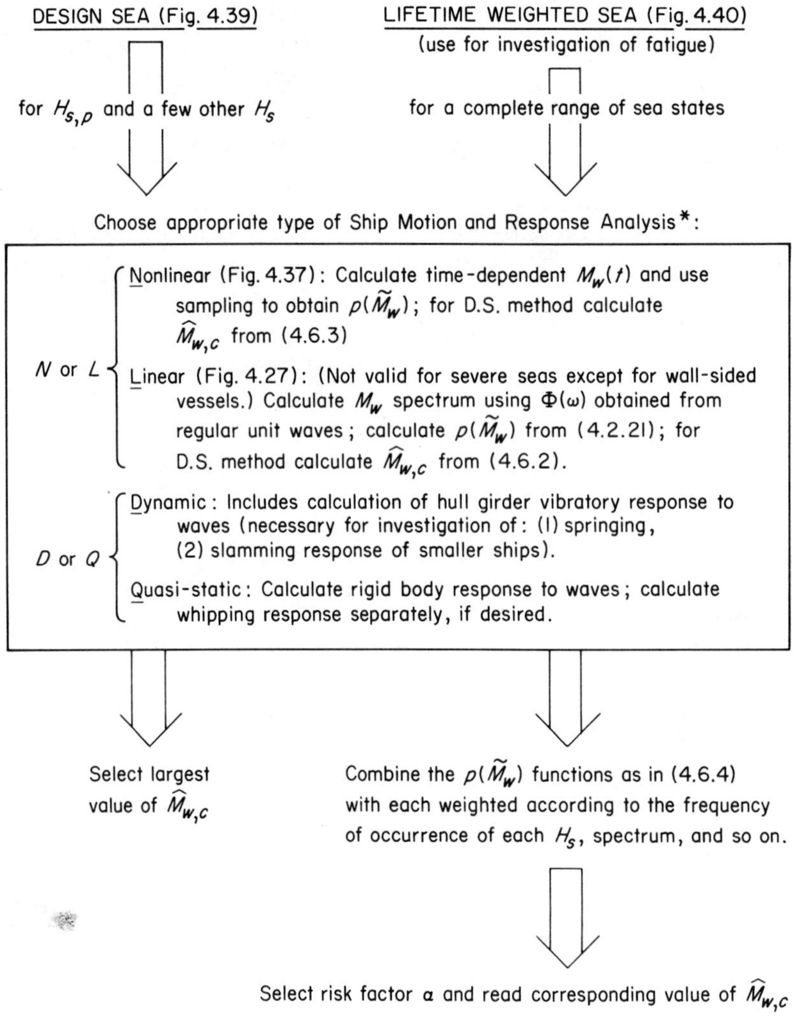

Figure 4.43 Summary of methods for calculating design value of wave bending moment $\hat{M}_{w,c}$. *Combinations: L Q, easiest and most common; L D, generally unsuitable or inappropriate; N Q, current best method for high performance ships; N D, required for springing and small ship whipping, but presently limited to deterministic applications (too wide-banded and too much computation for probabilistic analysis).

Means of Wave Spectra and Statistics," Hydrographic Office Publication No. 603, 1955.

11. Hogben, N. and Lumb, F. E., *Ocean Wave Statistics,* National Physical Laboratory, H. M. Stationery Office, London, 1967.
12. Roll, H. U., "Height, Length and Steepness of Seawaves in the North Atlantic and Dimensions of Seawaves as Functions of Wind Force," *T&R Bull.* No. 1–19, *SNAME,* 1953.
13. Walden, H., "Die Eigenschaften der Meerswellen in Nordatlantischen Ozean," *Deutscher Wetterdienst, Einzelveröffentlichungen,* No. 41, 1964 (in German).
14. Yamanouchi, Y. et al., "Statistical Investigations on Wind and Waves in North Pacific Ocean and Coastal Seas of Japan," *Shipbuilding Res.* **7** (2), November 1965.
15. Chen, Henry, Chen, H. H., and Hoffman, D., "The Implementation of the 20-Year Hindcast Wave Data in the Design and Operation of Marine Structure," Offshore Technology Conference, Houston, 1979.
16. Bales, S. L., Lee, W. T., and Voelker, J. M., "Standardized Wave and Wind Environments for NATO Operational Areas, "DTNSRDC Report SPD-0919-01, July 1981.
17. Ewing, J. A., "Wave Prediction: Progress and Applications," Proc. of 6th International Ship Structures Congress, Boston, 1976.
18. Bretschneider, C. L., "Wave Variability and Wave Spectra for Wind-Generated Gravity Waves," Beach Erosion Board, U.S. Army Corps of Engineers, Technical Memorandum No. 118, 1959.
19. Ochi, M. K., "Wave Statistics for the Design of Ships and Ocean Structures," *Trans. SNAME,* **86,** 1978, pp. 47–76.
20. Lewis, E. V., "Predicting Long-Term Distribution of Wave-Induced Bending Moments on Ship Hulls," SNAME Spring Meeting, Montreal, 1967.
21. Hoffman, D., "Wave Data Application for Ship Response Predictions," Webb Institute of Naval Architecture, October 1975.

22. Ochi, M. K. and Hubble, E. N., "On Six-Parameter Wave Spectra," *Proceedings,* 15th Coastal Engineers Conference, American Society of Civil Engineers, 1976.
23. Moskowitz, L., Pierson, W. J., and Mehr, E., "Wave Spectra Estimated from Wave Record Obtained by *OWS Weather Explorer* and *OWS Weather Reporter,* New York University, College of Engineering, (I) 1962, (II) 1963, (III) 1965.
24. Bretschneider, C. L. et al., "Data for Wave Conditions Observed by the *OWS Weather Reporter* in December 1959," *Deutsche Hydrographische Zeitschrift,* Band 15 Heft 6, 1962.
25. Miles, M., "Wave Spectra Estimates from a Stratified Sample of 323 North Atlantic Wave Records," Report LTR-SH-118A, Division of Mechanical Engineering, National Research Council, Ottawa, Ontario, Canada, 1972.
26. Ochi, M. K. and Motter, L. E., "Prediction of Extreme Ship Responses in Rough Seas of the North Atlantic," *Proceedings,* Symposium on the Dynamics of Marine Vehicles and Structures in Waves, 1974.
27. St. Denis, M. and Pierson, W. J., "On the Motions of Ships in Confused Seas," *Trans. SNAME,* **61,** 1953.
28. Newman, J. N. *Marine Hydrodynamics,* M.I.T. Press, Cambridge, Mass., 1977.
29. Bhattacharyya, R., *Dynamics of Marine Vehicles,* Wiley-Interscience, New York, 1978.
30. Jensen, J. J. and Pedersen, P. T., "Wave-Induced Bending Moments in Ships—A Quadratic Theory," Supplementary Papers of *R.I.N.A.,* **121,** 1979, pp. 151–165.
31. Longuet-Higgins, M. S., "The Effect of Non-Linearities on Statistical Distributions in the Theory of Sea Waves," *J. Fluid Mech.* **17,** 1963, pp. 459–480.
32. Lewis, E. V., "Motions of Ships in Waves," *Principles of Naval Architecture,* J. P. Comstock, Ed., *SNAME,* Chapter IX, 1967.
33. Gerritsma, J. and Beukelman, W., "The Distribution of the Hydrodynamic Forces on a Heaving and Pitching Ship Model in Still Water," Fifth Symposium on Naval Hydrodynamics, 1964.
34. Aertssen, G., "Service Performance and Trials at Sea," 12th International Towing Tank Conference, 1969.
35. Aertssen, G., "The Effect of Weather on Two Classes of Container Ships in the North Atlantic," *Roy. Inst. Naval Arch.,* 1975.
36. Jensen, J. J. and Pedersen, P. T., "Bending Moments and Shear Forces in Ships Sailing in Irregular Waves," *J. Ship Research,* **25** (4), December 1981, pp. 243–251.
37. Yamamoto, Y., Fujino, M., and Fukasawa, T., "Motion and Longitudinal Strength of a Ship in Head Sea and the Effects of Non-Linearities," *Naval Arch. Ocean Eng,* Soc. Nav. Arch. of Japan, **18,** 1980, pp. 91–100.
38. Ochi, M. K., "Generalization of Rayleigh Probability Distribution and Its Application, *J. Ship Res.* **22** (4), December 1978, pp. 259–265.

Chapter

Five

MATRIX STIFFNESS ANALYSIS: FRAMES AND GRILLAGES

In this chapter we consider structures which are made up of one-dimensional beam or bar elements connected together at their ends. The connection may be either pinned or rigid. Such structures are referred to as frames and the connection points are referred to as *nodes*. Frames are probably the most common class of structure. In a ship, in spite of the plating, the three-dimensional assemblage of deck beams, side frames, and longitudinal girders constitutes a framework, especially in regard to the transverse loads because these act normal to the plating and are carried mainly by the framing system.

5.1 BASIC CONCEPTS

Frame Analysis: Nodal Displacements

Frame analysis is a well established technique in the field of structural analysis. In this technique everything is expressed in terms of what happens at the nodes. External forces are applied at the nodes (and only there), and the displacements of the structure are expressed entirely in terms of nodal displacements. The starting point is to determine the response characteristics of each individual member, that is, the relationship between nodal forces and nodal displacements. This relationship is taken to be linear, which means that each nodal displacement u_j is linearly proportional to each nodal force f_i: $f_i = k_{ij} u_j$. The coefficient k_{ij} is referred to as a *stiffness coefficient*. The complete relationship between all nodal forces and all nodal displacements in a member is a system of linear equations, and such systems are best expressed in matrix notation, $\mathbf{f} = \mathbf{ku}$, in which $\mathbf{f}$ and $\mathbf{u}$ are vectors containing the nodal forces and displacements, and $\mathbf{k}$ is a square matrix containing all of the stiffness coefficients; this is the *stiffness matrix* of the member.

Fundamental Laws

In structural analysis there are three fundamental "laws" or relationships which must be satisfied:

1. Equilibrium of forces (within each member and between members).
2. Compatibility of displacements (within each member and between members).
3. Law of material behavior (stress-strain law) of each member.

As noted, the first two of these must be satisfied at two levels: within each member and also for the structure as a whole (i.e., between members). For one-dimensional members such as bars and beams, which we deal with in this chapter, all three laws are exactly satisfied within each member because they are implicit in the member's nodal force–nodal displacement relationship, that is, the member stiffness. Therefore, this chapter deals mainly with the first two of the laws applied to the structure as a whole. In other words, there must be equilibrium of forces between the external loads and the various member forces, and there must be compatibility in the deformations of the members, such that they continue to fit together.

For a statically determinate structure the equilibrium requirement is sufficient because the member forces can be calculated directly, whereupon the member displacements and internal forces can be calculated directly. But a frame structure is statically indeterminate, and so in frame analysis the equilibrium and compatibility requirements are imposed at each node of the structure. These requirements, together with the force-displacement relationships of each member (as embodied in its stiffness matrix **k**) produce a system of equations for the nodal displacements. After solving for these displacements, some previously established relationships between them and the member's internal forces and deformations (that is, stresses and strains) are used to solve for the latter.

In a typical three-dimensional frame structure there may be a large number of nodal displacements, several thousand perhaps, but the solution of a large system of linear equations is a routine task for a computer. Because there are a large number of equations, any discussion of the underlying theory and of the setting up of the equations is greatly facilitated by using matrix notation; hence this field is often referred to as matrix frame analysis or matrix stiffness analysis.

For one-dimensional members, imposing the requirements of equilibrium and compatibility at the nodes is sufficient to ensure that they are satisfied everywhere, both within members and between members, because the members are connected only at the nodes. For members that are two- or three-dimensional in extent it is necessary to define a relationship between nodal displacements and internal deformation in such a way that these two laws are satisfied to sufficient accuracy both within the member and between members. This is the key step in finite element analysis, and it is the only major difference between the basic finite element method and frame analysis. Hence much of the material of this chapter serves as a foundation for the presentation of the finite element method in Chapter 7.

Stiffness Matrix of a Structure

The device most directly associated with stiffness is a simple elastic spring, such as that shown in Fig. 5.1. It can also be regarded as an example of a structural member because it has the same basic structural characteristics as a pin-ended bar: it undergoes axial displacements, transmits axial forces, and exhibits a linear internal (or material) behavior; that is, there is a linear relationship between internal force $\mathcal{F}$ (axial

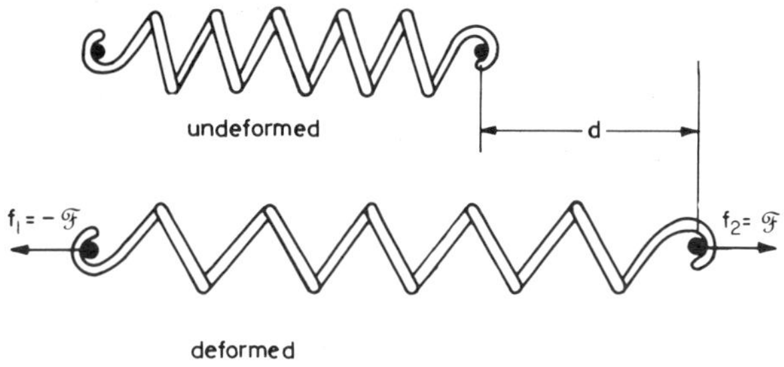

(a)

(*a*) Nodal forces f_1 (negative; $f_1 = -\mathcal{F}$) and f_2 (positive: $f_2 = \mathcal{F}$) (*b*) Internal force $\mathcal{F}$. Magnitude of $\mathcal{F}$ = magnitude of a pair of equal and opposite nodal forces reqired to cause a specified deformation, *d*. Sign of $\mathcal{F}$: positive for elongation.

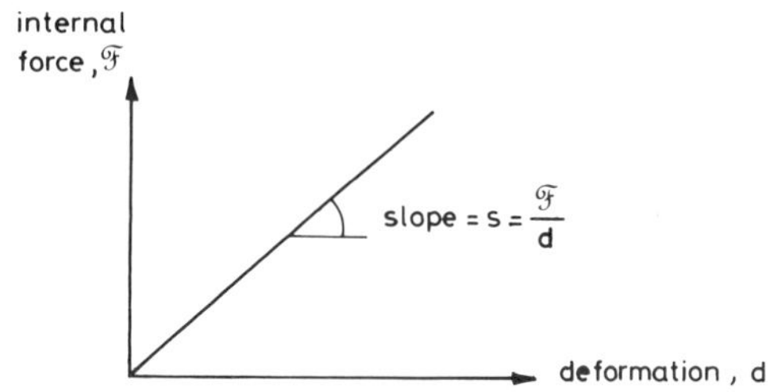

(b)

Figure 5.1 Internal force–deformation relationship for an elastic spring element.

force) and internal deformation d (elongation or shortening) which is of the form

$$\mathcal{F} = sd$$

The quantity s is termed the *stiffness* of the spring and corresponds to the slope of the internal force–deformation diagram (Fig. 5.1).*

Besides being a structural member the same spring could be considered as a structure—a one-member structure which is supported at one node and loaded at the other by an applied load F (see Fig. 5.2). The load

*Note that there is a subtle but important distinction between the internal force in a member and the applied external forces f_1 and f_2. Although their definitions are such that they always have the same magnitude, they are conceptually quite different. The internal force $\mathcal{F}$ is a state or situation inside of the member, and it is defined in terms of the internal deformation. In the present case it is the magnitude which a *pair* of equal and opposite forces must have in order to cause a specified elongation. It is a scalar quantity (no direction), it is diffused throughout the member, and its sign depends on the type of deformation: positive for elongation. In contrast the external forces are directional and can be regarded as vectors (although here they have only one component each), they act at specific locations, and their sign depends on their direction.

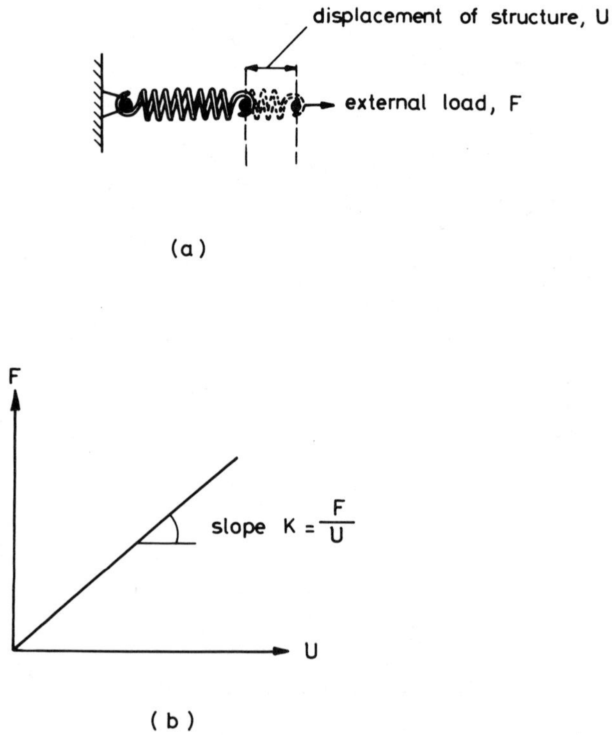

Figure 5.2 Load–displacement relationship for an elastic spring structure.

causes a structure displacement U which is linearly proportional to F,

$$F = KU \tag{5.1.1}$$

and the constant of proportionality K is termed the stiffness of the structure. In this case, since the structure is identical to the member, the two stiffnesses are the same: $K = s$. Knowing the value of this stiffness and of the applied load, (5.1.1) may be inverted to give the displacement

$$U = \frac{1}{K}F \tag{5.1.2}$$

For this structure the response consists of just one displacement. For more realistic structures such as the pin-jointed frame shown in Fig. 5.3, it is necessary to determine the displacements of nodes C, D, E, and F in order to be able to evaluate the member deformations (axial strains) and hence the internal forces (stresses) in the members.

Thus, the most basic task of structural analysis is to determine the relationship between the nodal loads and the nodal displacements of the structure. This relationship will be analogous to (5.1.1) for the simple spring structure, but there will be many interconnecting members and therefore many interrelated nodal displacements. Likewise, there can be many external loads because these may occur at any node. Therefore a general linear relationship between all of the external loads and all of the nodal displacements will consist of a system of simultaneous equations, and such systems are best expressed in matrix notation. The load–displacement relation is therefore of the form

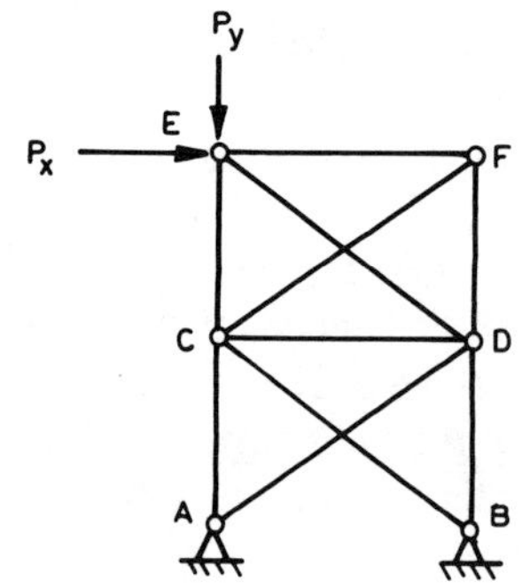

Figure 5.3 Statically indeterminate pin-jointed frame.

$$\mathbf{F} = \mathbf{KU} \tag{5.1.3}$$

in which **F** and **U** are vectors of nodal loads and nodal displacements respectively, and the matrix **K** is termed the "stiffness matrix" of the structure, since it consists of the coefficients relating **F** to **U**. For the very simple structure of Fig. 5.2, which has only one possible displacement, this matrix is of order 1 × 1 and the vectors **F** and **U** contain only one term each.

It can now be seen that the basic task of determining the load–displacement relationship of a structure consists essentially in obtaining the stiffness matrix of the structure because once **K** has been determined, the solution for **U** follows immediately. As will now be shown, the load–displacement relationship for each member or *element* (we will use the latter term from now on) can also be expressed in terms of an element stiffness matrix $\mathbf{k}^e$. Next, it will be shown that the structure stiffness matrix **K** is in fact obtained ("assembled") by a systematic superposition of the element stiffness matrices, taking account of the particular way in which the elements are arranged and connected in the structure.

Stiffness Matrix of a Spring (or Bar) Element

For simplicity we again consider the simple spring of Fig. 5.1 and we now regard it as a basic element, forming part of a larger structure. In this case each of its two nodes would be connected to other elements in

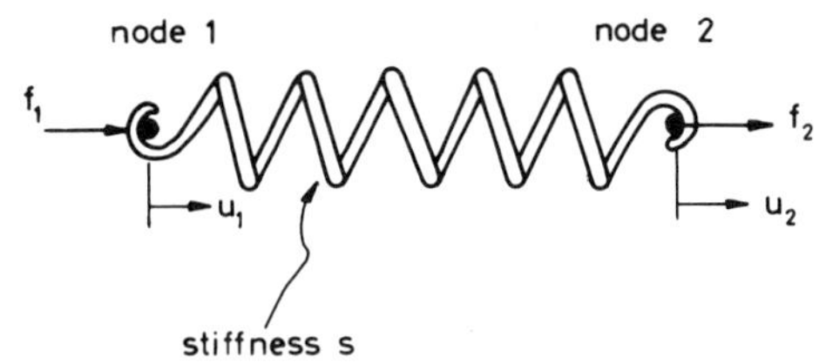

Figure 5.4 Element nodal forces and displacements for the spring element.

the structure and so in general each node would be subjected to a nodal force and would undergo a nodal displacement, as shown in Fig. 5.4. For a pin-jointed bar element the stiffness s is AE/L. Positive forces and displacements are as shown in the figure. The vector of nodal forces is

$$\mathbf{f} = \begin{Bmatrix} f_1 \\ f_2 \end{Bmatrix}$$

and the vector of nodal displacements is

$$\mathbf{u} = \begin{Bmatrix} u_1 \\ u_2 \end{Bmatrix}$$

Therefore the stiffness matrix for the spring element is of order 2 × 2 and the relationship between nodal forces and nodal displacements is of the form

$$\begin{Bmatrix} f_1 \\ f_2 \end{Bmatrix} = \begin{bmatrix} k_{11} & k_{12} \\ k_{21} & k_{22} \end{bmatrix} \begin{Bmatrix} u_1 \\ u_2 \end{Bmatrix} \quad \text{or} \quad \mathbf{f} = \mathbf{k}^e\mathbf{u} \quad (5.1.4)$$

in which the superscript e indicates that this stiffness matrix is for a single element. We now seek to determine the individual terms of $\mathbf{k}^e$. These can be obtained by considering each nodal displacement in isolation (i.e., keeping the other nodal displacement zero) and using the law of material behavior to evaluate the two nodal forces. For example, let us consider nodal displacement u_1 and keep $u_2 = 0$, as shown in Fig. 5.5*a*. From equilibrium the two nodal forces must be equal and opposite, and from the definition of $\mathscr{F}$ their magnitude must be the same as that of $\mathscr{F}$. Since $u_2 = 0$ the deformation is $d = u_2 - u_1 = -u_1$, and hence the material law of the spring requires that $\mathscr{F} = s(-u_1)$. Therefore $f_1 = -\mathscr{F} = su_1$ and $f_2 = -f_1 = -su_1$. If these values are substituted into (5.1.4) the result is

$$\begin{Bmatrix} su_1 \\ -su_1 \end{Bmatrix} = \begin{bmatrix} k_{11} & k_{12} \\ k_{21} & k_{22} \end{bmatrix} \begin{Bmatrix} u_1 \\ u_2 = 0 \end{Bmatrix}$$

from which

$$su_1 = k_{11}u_1 \quad \text{or} \quad k_{11} = s$$

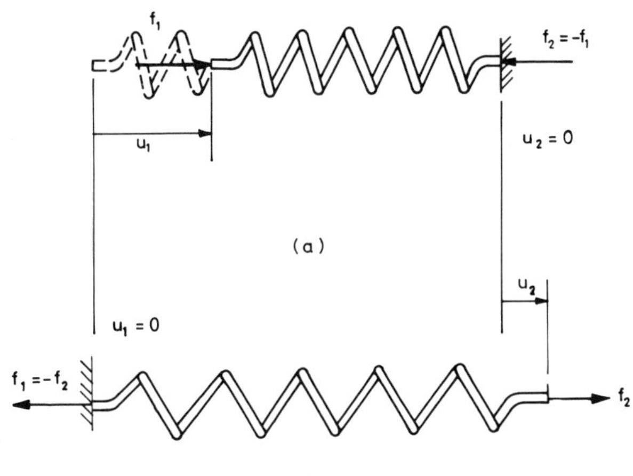

Figure 5.5 Calculation of $\mathbf{k}^e$ by imposing individual nodal displacements.

and

$$-su_1 = k_{21}u_1 \quad \text{or} \quad k_{21} = -s$$

We have thus obtained the terms in the first column of $\mathbf{k}^e$. The terms in the second column are obtained in a similar manner by setting $u_1 = 0$ (Fig. 5.5*b*) and by again observing that from equilibrium and the definition of the spring constant s the nodal forces are $f_1 = -su_2$ and $f_2 = su_2$. Substituting these values into (5.1.4) and again noting that $u_1 = 0$ gives

$$-su_2 = k_{12}u_2 \quad \text{or} \quad k_{12} = -s$$

together with

$$su_2 = k_{22}u_2 \quad \text{or} \quad k_{22} = s$$

Therefore the complete element stiffness matrix for the spring element is

$$\mathbf{k}^e = \begin{bmatrix} s & -s \\ -s & s \end{bmatrix} \quad (5.1.5a)$$

Hence the relationship between element nodal forces and element nodal displacements is

$$\begin{Bmatrix} f_1 \\ f_2 \end{Bmatrix} = \begin{bmatrix} s & -s \\ -s & s \end{bmatrix} \begin{Bmatrix} u_1 \\ u_2 \end{Bmatrix} \quad \text{or} \quad \mathbf{f} = \mathbf{k}^e\mathbf{u} \quad (5.1.5b)$$

It may be seen that the stiffness matrix is symmetric. This is true of all stiffness matrices, whether for a single element or for an entire structure. The symmetry comes from (or is an alternative way of stating) Maxwell's reciprocal theorem. Further examination of

$\mathbf{k}^e$ would also reveal that it is singular; the reason for this will be explained subsequently.

It may also be noted that in obtaining $\mathbf{k}^e$ we made use of the laws of material behavior and of equilibrium within the element. For such a simple element the law of compatibility hardly arises but it *is* satisfied; the spring's internal deformation (stretching or shortening) is linear and continuous.

Assembling the Structure Stiffness Matrix

The next point to consider is how the stiffness matrix for a structure can be obtained from the element stiffness matrices of its constituent elements. To achieve this we make use of the element force–displacement relationship, (5.1.5), of which the element stiffness matrix is the crucial part, together with the two laws that must be met at the structure level—those of equilibrium and of compatibility. Both laws are to be applied at the nodes of the structure (because everything is being described in terms of nodal values) and so we need a structure node numbering system, quite apart from the element node numbering system of Fig. 5.4. We will also need a sign convention for the applied loads $\mathbf{F}$ and for the structure nodal displacements $\mathbf{U}$.

Consider, for example, the two-element structure of Fig. 5.6. The structure nodes are labeled A, B, and C (letters are chosen instead of numbers to emphasize the distinction between the element system and the structure system) and both the loads $\mathbf{F}$ and the displacements $\mathbf{U}$ are positive to the right. Since there are three nodes there may be up to three applied (external) loads, as shown in the figure, but there could just as well be only one or two.*

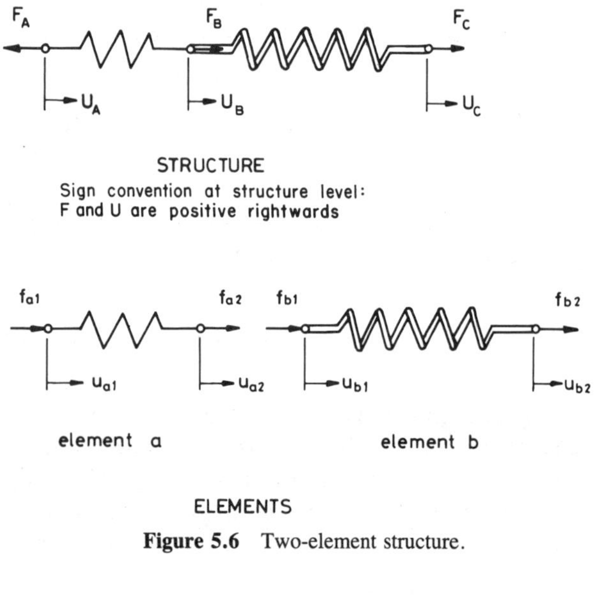

Figure 5.6 Two-element structure.

Also, each load will act in a particular direction. In the example loads F_B and F_C act towards the right and load F_A to the left. Hence the numerical value of the latter would be negative.

The law of equilibrium requires that at each node there must be equilibrium between the external applied load at that node (if any) and the sum of the element forces at that node. Therefore (see Fig. 5.6)

$$\begin{aligned} F_A &= f_{a1} \\ F_B &= f_{a2} + f_{b1} \\ F_c &= f_{b2} \end{aligned}$$

Next, we use the element force–displacement relationship, that is, the element stiffness expressions of (5.1.5), to express the element nodal forces $\mathbf{f}$ in terms of the element nodal displacements $\mathbf{u}$. The foregoing equations then become

$$\begin{aligned} F_A &= s_a u_{a1} - s_a u_{a2} \\ F_B &= -s_a u_{a1} + s_a u_{a2} + s_b u_{b1} - s_b u_{b2} \\ F_C &= -s_b u_{b1} + s_b u_{b2} \end{aligned}$$

The next step is to impose interelement compatibility of the *structure* nodal displacements. This consists essentially in relating the element displacements $\mathbf{u}$ to the structure displacements $\mathbf{U}$. In this example the relationships are

$$\begin{aligned} u_{a1} &= U_A \\ u_{a2} &= u_{b1} = U_B \\ u_{b2} &= U_C \end{aligned}$$

Substituting these gives

$$\begin{aligned} F_A &= s_a U_A - s_a U_B \\ F_B &= -s_a U_A + (s_a + s_b) U_B - s_b U_c \\ F_C &= - s_b U_B + s_b U_C \end{aligned}$$

which are the desired equations relating the applied loads to the structure displacements. Therefore the coefficients of these equations constitute the individual terms of the structure stiffness matrix $\mathbf{K}$, introduced in

*For the structural analysis the applied loads are known quantities, although it will often be the task of the structural analyst to calculate them before commencing the analysis.

(5.1.3). In matrix form the above equations are

$$\begin{Bmatrix} F_A \\ F_B \\ F_C \end{Bmatrix} = \begin{bmatrix} s_a & -s_a & 0 \\ -s_a & s_a + s_b & -s_b \\ 0 & -s_b & s_b \end{bmatrix} \begin{Bmatrix} U_A \\ U_B \\ U_C \end{Bmatrix}$$

or

$$\mathbf{F} = \mathbf{KU} \tag{5.1.6}$$

It may be seen that $\mathbf{K}$ is an assemblage which includes each and every one of the individual terms of $\mathbf{k}^a$ and $\mathbf{k}^b$, assembled together in a systematic manner. Hence the laws of equilibrium and compatibility are equivalent to a rule or procedure by which the terms of the $\mathbf{k}^e$ matrices are assembled to form $\mathbf{K}$. If we can specify this procedure in a suitably general and practical form we can then obtain $\mathbf{K}$ by a direct assembly process, without having to explicitly impose these laws. To achieve this we first show that the assembly process is essentially a superposition of the element stiffness matrices, and then we present a procedure for assembling $\mathbf{K}$ which is computationally practical and efficient.

To demonstrate the superposition of the element stiffness matrices we first write down the full set of equilibrium equations for each element.

Element a | Element b

$$\begin{Bmatrix} f_A \\ f_B \end{Bmatrix} = \begin{bmatrix} s_a & -s_a \\ -s_a & s_a \end{bmatrix} \begin{Bmatrix} U_A \\ U_B \end{Bmatrix} \qquad \begin{Bmatrix} f_B \\ f_C \end{Bmatrix} = \begin{bmatrix} s_b & -s_b \\ -s_b & s_b \end{bmatrix} \begin{Bmatrix} U_B \\ U_C \end{Bmatrix}$$

Although the two $\mathbf{k}^e$ matrices are of the same order they may not be added directly since they relate to different sets of displacements. However, by inserting rows and columns of zeros, both may be expanded in such a way that they each relate to the three displacements U_A, U_B, and U_C.

$$\begin{Bmatrix} f_A \\ f_B \\ f_C \end{Bmatrix} = \begin{bmatrix} s_a & -s_a & 0 \\ -s_a & s_a & 0 \\ 0 & 0 & 0 \end{bmatrix} \begin{Bmatrix} U_A \\ U_B \\ U_C \end{Bmatrix}$$

$$\begin{Bmatrix} f_A \\ f_B \\ f_C \end{Bmatrix} = \begin{bmatrix} 0 & 0 & 0 \\ 0 & s_b & -s_b \\ 0 & -s_b & s_b \end{bmatrix} \begin{Bmatrix} U_A \\ U_B \\ U_C \end{Bmatrix}$$

We now impose equilibrium of forces which says that the sum of the element forces at each node is equal to the external load at that node (or equals zero if there is no external load). Performing the summation gives

$$\begin{Bmatrix} F_A \\ F_B \\ F_C \end{Bmatrix} = \begin{bmatrix} s_a & -s_a & 0 \\ -s_a & s_a + s_b & -s_b \\ 0 & -s_b & s_b \end{bmatrix} \begin{Bmatrix} U_A \\ U_B \\ U_C \end{Bmatrix}$$

From this we can see that the way in which the individual terms of $\mathbf{k}^a$ and $\mathbf{k}^b$ are combined together (assembled) to form $\mathbf{K}$ depends entirely on how the physical elements are arranged (or assembled) in the actual structure. In explaining the assembly procedure in more detail we shall identify each of the individual terms in a stiffness matrix by its row and column, and we note here that in $\mathbf{K}$ these rows and columns correspond to structure nodes, whereas in $\mathbf{k}^a$ and $\mathbf{k}^b$ they correspond to element nodes. Also, the arrangement of the elements in the structure may be described by specifying the structure node where each element node is located. That is, for each node of each element the corresponding structure node is specified, and this summarizes the basic layout of the structure. For this example the structure node corresponding to each element node is shown in Fig. 5.7 immediately below the element stiffness matrices.

The foregoing summation of matrices shows that each term in a particular row and column position of $\mathbf{K}$ (corresponding to structure nodes I and J, say) is the sum of all of the $\mathbf{k}^e$ terms whose row and column numbers (i.e., whose element node numbers) correspond to (are located at) structure nodes I and J. Thus a particular $\mathbf{k}^e$ term, say k_{lm}, is added into the K_{IJ} position of $\mathbf{K}$, where I and J are the structure nodes at which the element nodes l and m are located in the structure.

With this rule it is no longer necessary to expand each of the element matrices up to the full structure size and then perform matrix addition. Instead, we may think of each of the element matrices as being "broken up" and its individual terms being inserted directly into their proper position in $\mathbf{K}$, according to an "assembly plan" which is simply a list of the structure nodes corresponding to the element nodes of that element. With this procedure, the only information which is required is the list of structure node numbers for each element. This list is usually referred to as the *location vector*. There is one of these for each element and it is usually a very short vector since its length equals the number of nodes in the element. The location vectors for the above example are shown in Fig. 5.7.

Because each element stiffness matrix is symmetric the resulting structure stiffness matrix is likewise always symmetric. Therefore only half of it needs to be assembled.

In the simple example presented here the elements are all in-line and there is only one component of displacement at each node. With more practical elements there are several components of displacement at each node and so in later sections of this chapter, and

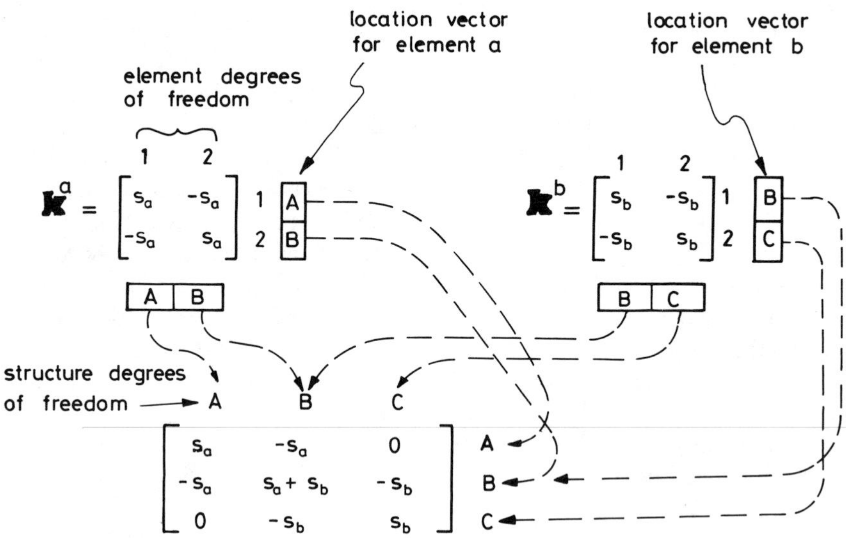

Figure 5.7 Location vectors and assembly of the structure stiffness matrix.

also in Chapter 7, the foregoing definitions will need to be generalized. Nevertheless these simple in-line spring elements illustrate all of the basic principles of matrix stiffness analysis and provide a useful introduction.

Solution Procedure

After the structure stiffness matrix has been fully assembled an apparent paradox arises—in every case the matrix is singular! This would seem to indicate that the equations for the structure's nodal displacements cannot be solved; that the displacements are indeterminate. This apparent paradox has a simple explanation—the structure has not yet been tied down or given any points of support, and in this situation the applied loads would cause the entire structure to move as a rigid body, thus indeed giving indeterminate displacements. Obviously, then, it is necessary to provide sufficient restraints on the structure to prevent this rigid body motion. These restraints are applied by specifying zero values for some of the structure's displacements. In practice structures often have more restraints than the minimum number. If so, they must all be specified because otherwise the structure would not be accurately represented. The full set of restraints is referred to as the support conditions. In the simple two-element structure of Fig. 5.6 there is only one possible type of rigid body motion—horizontal movement—and hence only one restraint is required for this structure. For example, let us say that the structure is supported at node A, so that $U_A = 0$. Then (5.1.6) can be rewritten in partitioned form as follows:

$$\begin{Bmatrix} F_A \\ \hline F_B \\ F_C \end{Bmatrix} = \left[\begin{array}{c|cc} s_a & -s_a & 0 \\ \hline -s_a & s_a + s_b & -s_b \\ 0 & -s_b & s_b \end{array}\right] \begin{Bmatrix} U_A = 0 \\ \hline U_B \\ U_C \end{Bmatrix} \qquad (5.1.7)$$

The system of equations contains two unknown displacements U_B and U_C, and an unknown reaction F_A. F_B and F_C are known applied loads. Using standard matrix manipulation we have

$$\{F_A\} = s_a\{U_A\} + [-s_a \quad 0]\begin{Bmatrix} U_B \\ U_C \end{Bmatrix}$$

$$\begin{Bmatrix} F_B \\ F_C \end{Bmatrix} = \begin{Bmatrix} -s_a \\ 0 \end{Bmatrix}\{U_A\} + \begin{bmatrix} s_a + s_b & -s_b \\ -s_b & s_b \end{bmatrix}\begin{Bmatrix} U_B \\ U_C \end{Bmatrix}$$

and since U_A is zero we have

$$\{F_A\} = [-s_a \quad 0]\begin{Bmatrix} U_B \\ U_C \end{Bmatrix} \qquad (5.1.8)$$

together with

$$\begin{Bmatrix} F_B \\ F_C \end{Bmatrix} = \begin{bmatrix} s_a + s_b & -s_b \\ -s_b & s_b \end{bmatrix}\begin{Bmatrix} U_B \\ U_C \end{Bmatrix} \qquad (5.1.9)$$

Equation 5.1.9 consists of two equations in the two unknowns U_B and U_C and may be solved for these values which, when substituted in (5.1.8) then give the value of the unknown reaction F_A.

It should be noted that (5.1.9) may be obtained directly from (5.1.7) simply by deleting the rows and columns of **K** corresponding to zero displacements.

Once the displacements have been obtained, the internal forces in the elements may be determined from the law of material (or internal) behavior of the element, that is, the relationship between internal force and internal deformation. For a spring this relationship is simply $\mathcal{F} = sd$ where $\mathcal{F}$ is the internal force in the spring and d is the change in length of the spring. Hence for the two elements we have

$$\begin{aligned} \mathcal{F}_a &= s_a(u_{a2} - u_{a1}) \\ \mathcal{F}_b &= s_b(u_{b2} - u_{b1}) \end{aligned} \qquad (5.1.10)$$

Since the elements are all in-line the required element displacements may be obtained by simply substituting the appropriate structure displacements.

$$\begin{aligned} \mathcal{F}_a &= s_a(U_B - U_A) \\ \mathcal{F}_b &= s_b(U_C - U_B) \end{aligned}$$

This completes the solution process. It will be helpful at this point to summarize the principal steps in matrix stiffness analysis because these remain basically the same regardless of the type of structure and the types of element which are used. The seven principal steps are given in Table 5.1. The transformation referred to in Step 2 was not necessary in the foregoing example. It is introduced and explained in the next section. Step 7 will be explained in Section 5.3.

TABLE 5.1 PRINCIPAL STEPS IN MATRIX STIFFNESS ANALYSIS

1 Define the structural model (structure coordinate system, nodes, node numbers, element types, and loads).

2 Determine each element stiffness matrix $\mathbf{k}^e$ and transform it to structure coordinates (whereupon it is denoted $\mathbf{K}^e$).

3 Assemble the structure stiffness matrix **K** from the individual element stiffness matrices $\mathbf{K}^e$ thus obtaining the system of equilibrium equations.

4 Apply the support conditions to the system of equations.

5 Solve for the displacements and then, if desired, the reaction forces at the supports.

6 For each element, obtain the nodal displacements and then use the internal force (or stress) matrix of that element to obtain the internal forces (or stresses).

7 If equivalent nodal loads were used to represent distributed loads, superimpose the fixed-end forces from the solution of the "internal problem."

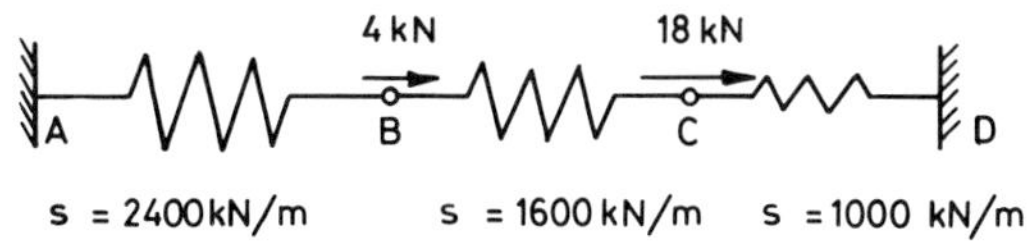

Figure 5.8 Numerical example, three-element structure.

Numerical Example

A simple numerical example is given to illustrate the procedure. The structure shown in Fig. 5.8 consists of three springs and is supported at nodes A and D. If axial loads of 4 kN and 18 kN are applied at nodes B and C respectively as shown, determine the displacements at nodes B and C and the reaction forces at A and D.

The element stiffness matrices and location vectors are:

$$\text{Element a:}\quad \mathbf{K}^a = \begin{array}{c} \begin{array}{cc} \text{A} & \text{B} \end{array} \\ \begin{bmatrix} 2400 & -2400 \\ -2400 & 2400 \end{bmatrix} \end{array}$$

$$\text{Element b:}\quad \mathbf{K}^b = \begin{array}{c} \begin{array}{cc} \text{B} & \text{C} \end{array} \\ \begin{bmatrix} 1600 & -1600 \\ -1600 & 1600 \end{bmatrix} \end{array}$$

$$\text{Element c:}\quad \mathbf{K}^c = \begin{array}{c} \begin{array}{cc} \text{C} & \text{D} \end{array} \\ \begin{bmatrix} 1000 & -1000 \\ -1000 & 1000 \end{bmatrix} \end{array}$$

Assembling these matrices to form the structure stiffness matrix:

$$\mathbf{K} = \begin{array}{cccc} \text{A} & \text{B} & \text{C} & \text{D} \end{array}$$
$$\mathbf{K} = \begin{bmatrix} 2400 & -2400 & & \\ -2400 & \begin{matrix}2400\\+1600\end{matrix} & -1600 & \\ & -1600 & \begin{matrix}1600\\+1000\end{matrix} & -1000 \\ & & -1000 & 1000 \end{bmatrix} \begin{matrix} \text{A} \\ \text{B} \\ \text{C} \\ \text{D} \end{matrix}$$

The boundary conditions are $U_A = U_D = 0$. Hence these two rows and columns of **K** can be deleted as shown by the dashed lines in the above matrix. Also, the corresponding forces and displacements are excluded from the equilibrium equations, which are then

$$\begin{Bmatrix} 4 \\ 18 \end{Bmatrix} = \begin{bmatrix} 4000 & -1600 \\ -1600 & 2600 \end{bmatrix} \begin{Bmatrix} U_B \\ U_C \end{Bmatrix}$$

The solution to these equations is $U_B = 0.005$ m and $U_C = 0.010$ m. Knowing the displacements, we can now solve for each of the unknown reaction forces by using the equilibrium equation corresponding to that force. These are the same equations which were deleted from the original system of equations. They are

$$F_A = 2400U_A - 2400U_B + 0U_C + 0U_D$$

and

$$F_D = 0U_A + 0U_B - 1000U_C + 1000U_D$$

Substituting the displacements gives $F_A = -12$ kN and $F_D = -10$ kN. Knowing all of the forces permits an equilibrium check to be made, so as to verify the solution (not an essential step, but highly recommended):

$$F_A + F_B + F_C + F_D = -12 + 4 + 18 - 10 = 0$$

5.2 PIN-JOINTED FRAMES

Transformation of Coordinates

At the beginning of this chapter it was shown that a pin-ended bar is equivalent to a spring. Its law of internal force-deformation is

$$\mathcal{F} = \frac{AE}{L} d$$

where $\mathcal{F}$ is the tension or compression in the bar and d is the change in length of the bar. The physical stiffness is AE/L and so from (5.1.5b) the relationship between nodal forces acting on the element and nodal displacements of the element is

$$\begin{Bmatrix} f_1 \\ f_2 \end{Bmatrix} = \frac{AE}{L} \begin{bmatrix} 1 & -1 \\ -1 & 1 \end{bmatrix} \begin{Bmatrix} u_1 \\ u_2 \end{Bmatrix} \quad \text{or} \quad \mathbf{f} = \mathbf{k}^e\mathbf{u} \qquad (5.2.1)$$

which establishes $\mathbf{k}^e$, the element stiffness matrix for this element. In this equation the forces and displacements are always axial regardless of how the bar may be orientated in the structure. That is, (5.2.1) is expressed in "local" or "element" coordinates (x,y) in which the x-axis is always aligned with the element as shown in Fig. 5.9. In this text all quantities expressed in element coordinates are denoted by a lower case letter ($\mathbf{f}$, $\mathbf{k}$, $\mathbf{u}$, etc.).

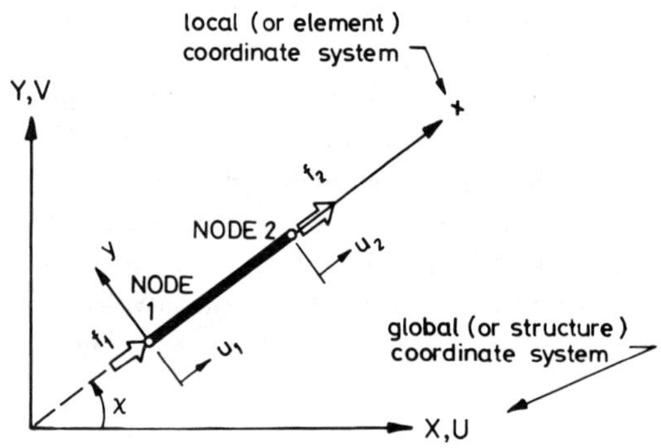

Figure 5.9 Coordinate systems.

In structural frameworks the members occur at various angles to one another and it is necessary to allow for this. In particular, in order to assemble the structure stiffness matrix it is first necessary to express the stiffness matrix of each element of the structure, not in terms of its own element coordinates, but in terms of a "global" or "structure" coordinate system, that is, a coordinate system that is used as the basic reference system for the structure as a whole.

In Fig. 5.9 a pin-ended bar element is inclined at an angle χ to the global system, where χ is positive when measured counterclockwise from the global X-axis to the element x-axis. Axes x and y are the local or element coordinates and X and Y are the structure coordinates. The respective displacements are u and v and U and V, and the respective forces are f_x, f_y, and F_X, and F_Y. Since an axial displacement u of an element generally possesses both a U and V component in the structure coordinate system, it is necessary to expand (5.2.1)

$$\begin{Bmatrix} f_{x1} \\ f_{y1} \\ f_{x2} \\ f_{y2} \end{Bmatrix} = \frac{AE}{L} \begin{bmatrix} 1 & 0 & -1 & 0 \\ 0 & 0 & 0 & 0 \\ -1 & 0 & 1 & 0 \\ 0 & 0 & 0 & 0 \end{bmatrix} \begin{Bmatrix} u_1 \\ v_1 \\ u_2 \\ v_2 \end{Bmatrix} \qquad (5.2.2)$$

or

$$\mathbf{f} = \mathbf{k}^e\boldsymbol{\delta}$$

in which the symbols $\mathbf{f}$ and $\boldsymbol{\delta}$ represent the vectors of nodal forces and nodal displacements. That is

$$\mathbf{f} = \begin{Bmatrix} f_{x1} \\ f_{y1} \\ f_{x2} \\ f_{y2} \end{Bmatrix} \quad \text{and} \quad \boldsymbol{\delta} = \begin{Bmatrix} u_1 \\ v_1 \\ u_2 \\ v_2 \end{Bmatrix} \qquad (5.2.3)$$

Since a pin-ended bar can only carry an axial load, $f_{y1} = f_{y2} = 0$. It may be seen from Fig. 5.10 that the local and global system of forces at node 1 are related by the expressions

$$f_{x1} = F_{X1} \cos\chi + F_{Y1} \sin\chi$$
$$f_{y1} = -F_{X1} \sin\chi + F_{Y1} \cos\chi$$

and similar expressions apply for node 2. Hence if we define

$$\mu = \sin\chi; \qquad \lambda = \cos\chi \qquad (5.2.4)$$

then the full relationship between the two systems of forces is

$$\begin{Bmatrix} f_{x1} \\ f_{y1} \\ f_{x2} \\ f_{y2} \end{Bmatrix} = \begin{bmatrix} \lambda & \mu & 0 & 0 \\ -\mu & \lambda & 0 & 0 \\ 0 & 0 & \lambda & \mu \\ 0 & 0 & -\mu & \lambda \end{bmatrix} \begin{Bmatrix} F_{X1} \\ F_{Y1} \\ F_{X2} \\ F_{Y2} \end{Bmatrix} \qquad (5.2.5)$$

or

$$\mathbf{f} = \mathbf{TF}$$

where $\mathbf{T}$ is called the transformation matrix. It may be shown that this transformation matrix has the very useful property that its inverse is equal to its transpose, that is,

$$\mathbf{T}^{-1} = \mathbf{T}^T$$

For orthogonal systems the transformation between local and global displacements is the same as that between the two sets of forces, namely

$$\boldsymbol{\delta} = \mathbf{T}\boldsymbol{\Delta} \qquad (5.2.6)$$

where

$$\boldsymbol{\Delta} = \begin{Bmatrix} U_1 \\ V_1 \\ U_2 \\ V_2 \end{Bmatrix}$$

A general transformation involves rotations of three axes x, y, z relative to the structure axes X, Y, Z. The x-axis is at angles θ_{xX}, θ_{xY}, and θ_{xZ} measured from the axes X, Y, and Z as shown in Fig. 5.11 and the cosines of these three angles are the direction cosines of χ with respect to X, Y, and Z. They are denoted by λ_{11}, λ_{12}, and λ_{13}. Similarly, the direction cosines of y and z are λ_{21}, λ_{22}, λ_{23} and λ_{31}, λ_{32}, λ_{33} respectively. Thus, the nodal transformation matrix $\boldsymbol{\lambda}$ for transforming forces or deflections at each node from structure coordinates to element coordinates is

$$\boldsymbol{\lambda} = \begin{bmatrix} \cos\theta_{xX} & \cos\theta_{xY} & \cos\theta_{xZ} \\ \cos\theta_{yX} & \cos\theta_{yY} & \cos\theta_{yZ} \\ \cos\theta_{zX} & \cos\theta_{zY} & \cos\theta_{zZ} \end{bmatrix} \qquad (5.2.7)$$

which would be used as follows

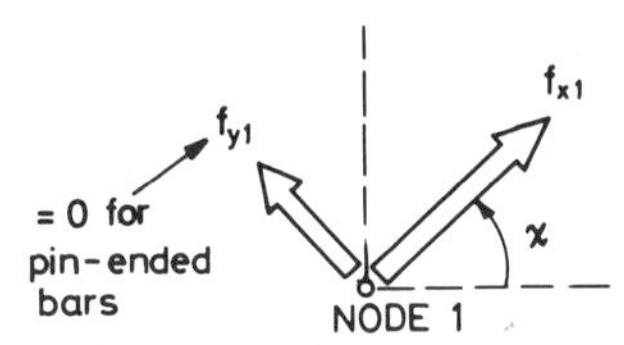

(a) ELEMENT NODAL FORCES AT NODE 1 EXPRESSED IN ELEMENT COORDINATES

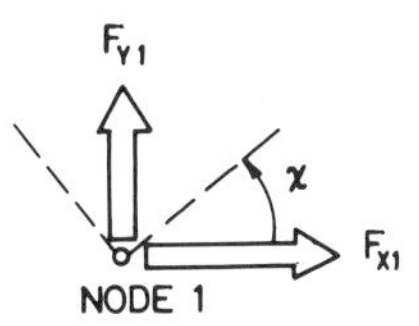

(b) ELEMENT NODAL FORCES AT NODE 1 EXPRESSED IN STRUCTURE COORDINATES

Figure 5.10 Transformation of nodal forces.

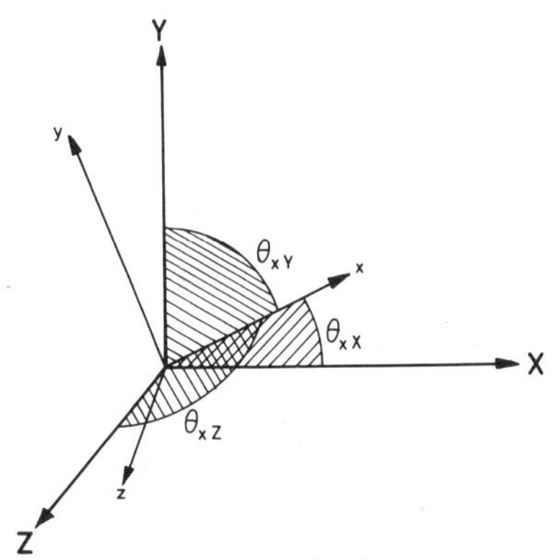

Figure 5.11 General transformation of axes.

$$\mathbf{f}_i = \boldsymbol{\lambda}\mathbf{F}_i \quad \text{and} \quad \boldsymbol{\delta}_i = \boldsymbol{\lambda}\boldsymbol{\Delta}_i$$

where i denotes the node number.

As with $\mathbf{T}$ we have $\boldsymbol{\lambda}^{-1} = \boldsymbol{\lambda}^T$, and therefore the reverse transformations are

$$\mathbf{F}_i = \boldsymbol{\lambda}^T\mathbf{f}_i \quad \text{and} \quad \boldsymbol{\Delta}_i = \boldsymbol{\lambda}^T\,\boldsymbol{\delta}_i$$

The transformation matrix $\boldsymbol{\lambda}$ applies at each node and hence for a pin-jointed bar element the full transformation matrix is

$$\mathbf{T} = \begin{bmatrix} \boldsymbol{\lambda} & \mathbf{O} \\ \mathbf{O} & \boldsymbol{\lambda} \end{bmatrix}$$

If the bar element lies in the x,y plane in element coordinates, and in the X,Y plane in structure coordinates, then $\boldsymbol{\lambda}$ reduces to the simple 2×2 matrix given in (5.2.5)

$$\boldsymbol{\lambda} = \begin{bmatrix} \cos\theta_{xX} & \cos\theta_{xY} \\ \cos\theta_{yX} & \cos\theta_{yY} \end{bmatrix}$$

$$= \begin{bmatrix} \cos\theta_{xX} & \cos\left(\frac{\pi}{2} - \theta_{xX}\right) \\ \cos\left(\frac{\pi}{2} + \theta_{xX}\right) & \cos\theta_{xX} \end{bmatrix}$$

$$= \begin{bmatrix} \cos\chi & \sin\chi \\ -\sin\chi & \cos\chi \end{bmatrix}$$

$$= \begin{bmatrix} \lambda & \mu \\ -\mu & \lambda \end{bmatrix}$$

Using the transformation matrix $\mathbf{T}$ we can derive an expression for the element stiffness matrix in global coordinates, $\mathbf{K}^e$. The basic force–displacement relationship for the element is given by (5.2.2). If we substitute (5.2.5) on the left-hand side and (5.2.6) for $\boldsymbol{\delta}$ on the right-hand side we obtain

$$\mathbf{TF} = \mathbf{k}^e\mathbf{T}\boldsymbol{\Delta}$$

Premultiplying both sides by $\mathbf{T}^{-1}$ and using the fact that $\mathbf{T}^{-1} = \mathbf{T}^T$ gives

$$\mathbf{F} = \mathbf{T}^T\mathbf{k}^e\mathbf{T}\boldsymbol{\Delta} = \mathbf{K}^e\boldsymbol{\Delta}$$

and therefore the element stiffness matrix in local coordinates $\mathbf{k}^e$ is transformed to the element stiffness matrix in global coordinates $\mathbf{K}^e$ by the compound transformation

$$\mathbf{K}^e = \mathbf{T}^T\mathbf{k}^e\mathbf{T} \tag{5.2.8}$$

The result of this double multiplication is

$$\mathbf{K}^e = \frac{AE}{L}\begin{bmatrix} \lambda^2 & \lambda\mu & -\lambda^2 & -\lambda\mu \\ \lambda\mu & \mu^2 & -\lambda\mu & -\mu^2 \\ -\lambda^2 & -\lambda\mu & \lambda^2 & \lambda\mu \\ -\lambda\mu & -\mu^2 & \lambda\mu & \mu^2 \end{bmatrix} \tag{5.2.9}$$

It may be seen that $\mathbf{K}^e$ is a symmetric matrix consisting of four sub-matrices κ^e

$$\mathbf{K}^e = \begin{bmatrix} \kappa^e & -\kappa^e \\ -\kappa^e & \kappa^e \end{bmatrix}$$

where

$$\kappa^e = \frac{AE}{L}\begin{bmatrix} \lambda^2 & \lambda\mu \\ \lambda\mu & \mu^2 \end{bmatrix}$$

Unless the elements are all in-line the transformation of the element stiffness matrix from $\mathbf{k}^e$ to $\mathbf{K}^e$ of (5.2.9) must always be performed before the structure stiffness matrix $\mathbf{K}$ can be assembled. Since the arrangement of elements differs for each structure an element stiffness matrix, if it is to be of general use, must first be expressed in terms of its own local coordinates. Also, since the various elements in a structure generally have different sizes and orientations the transformation to structure coordinates must be performed element by element. Hence in the principal steps of matrix stiffness analysis given in Table 5.1, step 2 consists of the transformation of each element stiffness matrix from $\mathbf{k}^e$ to $\mathbf{K}^e$, that is, from local to global coordinates. For all standard elements a general expression for $\mathbf{k}^e$ will be available and so the first part of step 2—determining $\mathbf{k}^e$—is merely a matter of substituting that element's dimensions and physical properties.

After assembling and solving the complete system of equations (steps 3 to 5 in Table 5.1) the next step in a structural analysis is the calculation of the internal forces in each element. $\mathcal{F} = sd$, where s is the element's material stiffness and d is the deformation. For this it is necessary to convert back to element coordinates in order to determine the element deformation d. For a pin-ended bar the material stiffness is the axial stiffness AE/L. Hence (5.1.10) becomes

$$\mathcal{F}_e = \left(\frac{AE}{L}\right)_e (u_2 - u_1)_e \tag{5.2.10}$$

where u_1 and u_2 are the nodal displacements in element coordinates for element e. Thus before transforming the nodal displacements of the structure from structure coordinates to element coordinates it is first necessary to allocate these displacements to the appropriate ends (nodes) of each element. In a complicated structure this would be done with the aid of the location vectors. After this the transformation from $\mathbf{\Delta}$ to $\boldsymbol{\delta}$ can be performed (separately for each element) and the internal forces can be evaluated. If desired, the transformation can be incorporated into the expression for internal force so that the latter is expressed directly in terms of the structure displacements. In expanded form (5.2.6) is

$$\begin{Bmatrix} u_1 \\ v_1 \\ u_2 \\ v_2 \end{Bmatrix}_e = \begin{bmatrix} \lambda & \mu & 0 & 0 \\ -\mu & \lambda & 0 & 0 \\ 0 & 0 & \lambda & \mu \\ 0 & 0 & -\mu & \lambda \end{bmatrix} \begin{Bmatrix} U_1 \\ V_1 \\ U_2 \\ V_2 \end{Bmatrix}$$

and therefore

$$\begin{aligned} u_2 - u_1 &= (\lambda U_2 + \mu V_2) - (\lambda U_1 + \mu V_1) \\ &= [\lambda \quad \mu] \begin{Bmatrix} U_2 - U_1 \\ V_2 - V_1 \end{Bmatrix} \end{aligned}$$

Substituting in (5.2.10) gives

$$\mathscr{F}_e = \left(\frac{AE}{L}\right)_e [\lambda \quad \mu]_e \begin{Bmatrix} U_2 - U_1 \\ V_2 - V_1 \end{Bmatrix} \qquad (5.2.11)$$

Degrees of Freedom: Demonstration of Complete Method

The complete process will now be demonstrated for the simple three-bar pin-jointed frame illustrated in Fig. 5.12. The frame is supported at A and C and all members have the same cross-sectional area A and Young's modulus E. The frame carries vertical and horizontal loads as shown in the figure. Since there are three nodes and two independent displacements U and V at each node, the vectors of applied loads and displacements each contain six terms as follows:

$$\mathbf{F} = \begin{Bmatrix} F_{XA} \\ F_{YA} \\ F_{XB} \\ F_{YB} \\ F_{XC} \\ F_{YC} \end{Bmatrix} \quad \text{and} \quad \mathbf{\Delta} = \begin{Bmatrix} U_A = 0 \\ V_A = 0 \\ U_B \\ V_B \\ U_C = 0 \\ V_C = 0 \end{Bmatrix}$$

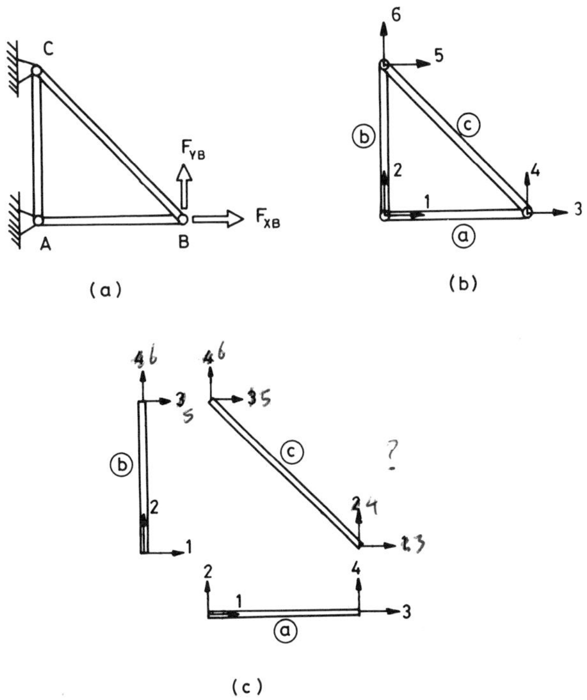

Figure 5.12 Three-member pin-jointed frame. (*a*) Structure; (*b*) structure degrees of freedom; and (*c*) element degrees of freedom.

At this point it is appropriate to introduce the notion of "degrees of freedom." The degrees of freedom of a structure, or of an element, are basically the same as the nodal displacements, but they are slightly more general in concept. Specifically, the degrees of freedom of a structure or element are all of the independent nodal parameters that are used to describe all of the particular ways or modes in which the structure or element can deflect. There are two reasons for introducing this new term:

1. To allow for additional and more general types of "displacement." The most common example is a degree of freedom which is a derivative of an ordinary displacement. For example, in the case of a beam it could be the slope dv/dx or the curvature d^2v/dx^2 at the end (or node) of the beam; these additional "displacements" are derivatives of the beam's ordinary lateral displacement v. The use of this form of displacement will be discussed in the next section.

2. To provide a suitable terminology for the continuous sequential numbering of all of the displacements in a structure (or element), as an alternative to the "two-level" system used thus far which specifies first the node number and then the particular nodal force or displacement at that node. The sequential numbering system based on degrees of freedom is more in harmony with standard matrix notation and is

also more suitable for the practical computer implementation of frame analysis and finite element analysis. Nevertheless, the two-level node numbering is often more suitable for explanatory purposes and we shall continue to make frequent use of it.

The degree of freedom approach is illustrated in Fig. 5.12*b* which shows that the structure degrees of freedom are numbered sequentially within each node, moving from one node to another. Within each node the numbering should follow some systematic and unchanging sequence (e.g., the X-direction first, then the Y-direction) whereas the nodes may be taken in any sequence. However, we shall see in the next section that the amount of computation can be reduced by a careful choice of node sequence.

Using the numbering system based on structure degrees of freedom, the vectors of applied loads and structure displacements become

$$\mathbf{F} = \begin{Bmatrix} F_1 \\ F_2 \\ F_3 \\ F_4 \\ F_5 \\ F_6 \end{Bmatrix} \quad \text{and} \quad \boldsymbol{\Delta} = \begin{Bmatrix} \Delta_1 \\ \Delta_2 \\ \Delta_3 \\ \Delta_4 \\ \Delta_5 \\ \Delta_6 \end{Bmatrix}$$

The expression for $\mathbf{K}^e$ for a typical element is given by (5.2.9) and to evaluate this the direction cosines λ and μ for each element must be determined. Recalling that χ is measured counter-clockwise from the structure X-axis to the element x-axis, these values are

Member	χ	λ	μ	λ^2	μ^2	$\lambda\mu$
a	0	1	0	1	0	0
b	90	0	1	0	1	0
c	135°	−.707	.707	$\frac{1}{2}$	$\frac{1}{2}$	$-\frac{1}{2}$

Since the terms in $\mathbf{K}^e$ involve $\lambda^2\,\mu^2$ and $\mu\lambda$, a change in angle from χ to $(\chi + \pi)$, which causes both μ and λ to change sign, does not affect the signs of the terms in the stiffness matrix. For example, with respect to member c, $\chi = 135°$ if measured counter-clockwise at node B, or 315° (or −45°) if measured counter-clockwise at node C. Either way, in substituting into (5.2.9) the resulting $\mathbf{K}^e$ is the same.

Once the structure degrees of freedom have been defined the location vectors can be determined by inspection. Sometimes a diagram such as Fig. 5.12*c* is of assistance. For each element, one proceeds sequentially through the element degrees of freedom, expressed in structure coordinates, and records the corresponding structure degree of freedom. In a hand solution it is helpful to write the location vector above and to the right of the transformed stiffness matrix for that element. For the three elements of the sample problem the $\mathbf{K}^e$ matrices and the location vectors are as follows:

Element a:

$$\begin{matrix} & & 1 & 2 & 3 & 4 & \\ \mathbf{K}^a = \dfrac{AE}{L} & \Bigg[& \begin{matrix}1\\0\\-1\\0\end{matrix} & \begin{matrix}0\\0\\0\\0\end{matrix} & \begin{matrix}-1\\0\\1\\0\end{matrix} & \begin{matrix}0\\0\\0\\0\end{matrix} & \Bigg] \begin{matrix}1\\2\\3\\4\end{matrix} \end{matrix}$$

Element b:

$$\begin{matrix} & & 1 & 2 & 5 & 6 & \\ \mathbf{K}^b = \dfrac{AE}{L} & \Bigg[& \begin{matrix}0\\0\\0\\0\end{matrix} & \begin{matrix}0\\1\\0\\-1\end{matrix} & \begin{matrix}0\\0\\0\\0\end{matrix} & \begin{matrix}0\\-1\\0\\1\end{matrix} & \Bigg] \begin{matrix}1\\2\\5\\6\end{matrix} \end{matrix}$$

Element c:

$$\begin{matrix} & & 3 & 4 & 5 & 6 & \\ \mathbf{K}^c = \dfrac{AE}{2\sqrt{2}L} & \Bigg[& \begin{matrix}1\\-1\\-1\\1\end{matrix} & \begin{matrix}-1\\1\\1\\-1\end{matrix} & \begin{matrix}-1\\1\\1\\-1\end{matrix} & \begin{matrix}1\\-1\\-1\\1\end{matrix} & \Bigg] \begin{matrix}3\\4\\5\\6\end{matrix} \end{matrix}$$

These matrices are each "broken up" and their terms are assembled together to form the stiffness matrix for the structure. However, before we can do so it is necessary to redefine the assembly process in more general terms. In the initial presentation, in Section 5.1, the example contained only one displacement at each node and so it was possible to use node numbers to identify the rows and columns of the stiffness matrices—both element and structure. It has now been shown that practical problems have multiple displacements (degrees of freedom) at each node, and therefore it is the degree of freedom number which defines the rows and columns of the stiffness matrices—both element and structure. Hence the rule given earlier for performing the assembly process is easily generalized by substituting the term "degree of freedom" in place of "node number." Also, in the earlier example the elements were all in-line and so there was no need to transform $\mathbf{k}^e$ *to* $\mathbf{K}^e$. Hence to obtain the general form of the rule it is also necessary to replace $\mathbf{k}^e$ by $\mathbf{K}^e$. The result is the following general rule for assembling the structure stiffness matrix $\mathbf{K}$:

> For each transformed element stiffness matrix in turn, say $\mathbf{K}^e$, each term of $\mathbf{K}^e$, say K^e_{lm}, is added into the K_{IJ} position of $\mathbf{K}$, where I and J are the structure degrees of freedom corresponding to the element degrees of freedom l and m. The correspondence between I and l and between J and m is indicated by the (previously constructed) location vector for that element.

Applying this rule for the present example gives:

$$\mathbf{K} = \begin{bmatrix} a_{11}+b_{11} & a_{12}+b_{12} & a_{13} & a_{14} & b_{13} & b_{14} \\ a_{21}+b_{12} & a_{22}+b_{22} & a_{23} & a_{24} & b_{23} & b_{24} \\ a_{31} & a_{32} & a_{33}+c_{11} & a_{34}+c_{12} & c_{13} & c_{14} \\ a_{41} & a_{42} & a_{43}+c_{21} & a_{44}+c_{22} & c_{23} & c_{24} \\ b_{31} & b_{32} & c_{31} & c_{32} & b_{33}+c_{33} & b_{34}+c_{34} \\ b_{41} & b_{42} & c_{41} & c_{42} & b_{43}+c_{43} & b_{44}+c_{44} \end{bmatrix}$$

where a_{ij} represents the term in the ith row and the jth column of $\mathbf{K}^a$, and similarly for b_{ij} and c_{ij}. For each term in an element stiffness matrix the numbers in the location vector above and to the right of that term indicate the column and row in $\mathbf{K}$ where it is to be inserted. For example, the term b_{34} of $\mathbf{K}^b$ is inserted into the $\mathbf{K}_{56}$ location of $\mathbf{K}$. Later, when $\mathbf{K}^c$ is being processed, the term c_{34} is also inserted into the $\mathbf{K}_{56}$ location, being added to b_{34}. In terms of the actual numerical values the fully assembled structure stiffness matrix is

$$\mathbf{K} = \frac{AE}{L}\begin{bmatrix} 1 & 0 & -1 & 0 & 0 & 0 \\ 0 & 1 & 0 & 0 & 0 & -1 \\ -1 & 0 & 1.35 & -0.35 & -0.35 & 0.35 \\ 0 & 0 & -0.35 & 0.35 & 0.35 & -0.35 \\ 0 & 0 & -0.35 & 0.35 & 0.35 & -0.35 \\ 0 & -1 & 0.35 & -0.35 & -0.35 & 1.35 \end{bmatrix}$$

where for brevity 0.35 is written in place of $1/[2(2)^{1/2}] = 0.35355$.

The next step is to impose the boundary conditions and to reduce the system of equations accordingly. For this structure both displacements are zero at node A and at node C. These correspond to degrees of freedom 1, 2, 5, and 6, and therefore these four rows and columns of $\mathbf{K}$, and the corresponding terms in $\mathbf{F}$ and $\mathbf{\Delta}$, should be deleted. The reduced system is

$$\begin{Bmatrix} F_3 \\ F_4 \end{Bmatrix} = \frac{AE}{L}\begin{bmatrix} 1.35 & -0.35 \\ -0.35 & 0.35 \end{bmatrix}\begin{Bmatrix} \Delta_3 \\ \Delta_4 \end{Bmatrix}$$

For typical structures the numerical values of F_3 and F_4 would now be inserted and the structure displacements would be obtained by a numerical elimination (or "matrix triangularization") scheme, taking full advantage of the symmetry of $\mathbf{K}$. For such a small system as this one the solution may be obtained in terms of general loads F_3 and F_4 by using matrix inversion. The result is

$$\begin{Bmatrix} \Delta_3 \\ \Delta_4 \end{Bmatrix} = \frac{L}{AE}\begin{bmatrix} 1 & 1 \\ 1 & 3.828 \end{bmatrix}\begin{Bmatrix} F_3 \\ F_4 \end{Bmatrix}$$

If desired, the reaction forces at nodes A and C can now be obtained by substituting Δ_3 and Δ_4 into the equations corresponding to the deleted rows of $\mathbf{K}$.

$$\begin{Bmatrix} F_1 \\ F_2 \\ F_5 \\ F_6 \end{Bmatrix} = \begin{bmatrix} -1 & 0 \\ 0 & 0 \\ -0.35 & 0.35 \\ 0.35 & -0.35 \end{bmatrix}\begin{bmatrix} 1 & 1 \\ 1 & 3.828 \end{bmatrix}\begin{Bmatrix} F_3 \\ F_4 \end{Bmatrix}$$

$$= \begin{bmatrix} -1 & -1 \\ 0 & 0 \\ 0 & 1 \\ 0 & -1 \end{bmatrix}\begin{Bmatrix} F_3 \\ F_4 \end{Bmatrix} = \begin{Bmatrix} -F_3 - F_4 \\ 0 \\ F_4 \\ -F_4 \end{Bmatrix}$$

The final step is to calculate the internal forces in the elements. In order to do this the structure displacements $\mathbf{\Delta}$ must first be distributed amongst the appropriate element nodes, using the location vectors. For the sample problem the correspondence between the displacements is summarized in Table 5.2.

TABLE 5.2 DISTRIBUTION OF STRUCTURE DISPLACEMENTS Δ TO ELEMENT NODES

	Element Nodal Displacements (in structure coordinates)			
Element	U_1	V_1	U_2	V_2
a	Δ_1	Δ_2	Δ_3	Δ_4
b	Δ_1	Δ_2	Δ_5	Δ_6
c	Δ_3	Δ_4	Δ_5	Δ_6

With this information (5.2.11) may now be evaluated for each element in turn. For example, for element a:

$$\mathcal{F}_a = \left(\frac{AE}{L}\right)_a [\lambda \quad \mu]_a \begin{Bmatrix} U_2 - U_1 \\ V_2 - V_1 \end{Bmatrix}_a$$

$$= \frac{AE}{L}[1 \quad 0]\begin{Bmatrix} \Delta_3 - \Delta_1 \\ \Delta_4 - \Delta_2 \end{Bmatrix}$$

$$= \frac{AE}{L}[1 \quad 0]\frac{L}{AE}\begin{Bmatrix} F_3 + F_4 - 0 \\ F_3 + 3.828F_4 - 0 \end{Bmatrix}$$

$$= F_3 + F_4$$

Similarly, for element b

$$\mathcal{F}_b = \frac{AE}{L}[0 \quad 1]\begin{Bmatrix} \Delta_5 - \Delta_1 \\ \Delta_6 - \Delta_2 \end{Bmatrix}$$

$$= \frac{AE}{L}[0 \quad 1]\frac{L}{AE}\begin{Bmatrix} 0 - 0 \\ 0 - 0 \end{Bmatrix}$$

$$= 0$$

and for element c

$$\mathcal{F}_c = \frac{AE}{\sqrt{2}L}[-.707 \quad .707]\begin{Bmatrix} \Delta_5 & \Delta_3 \\ \Delta_6 & \Delta_4 \end{Bmatrix}$$

$$= \frac{AE}{2L}[-1 \quad 1]\frac{L}{AE}\begin{Bmatrix} 0 - F_3 - F_4 \\ 0 - F_3 - 3.828F_4 \end{Bmatrix}$$

$$= -1.414F_4$$

Numbering Sequence of Nodes

If two displacements (or degrees of freedom) in the structure, Δ_l and Δ_m, are linked together by a bar element this will give rise to terms in the structure stiffness matrix in locations K_{ll}, K_{mm}, K_{lm} and K_{ml}, as shown in Fig. 5.13. If these two degrees of freedom have widely differing numbers then the equilibrium equations corresponding to F_l will have a wide "bandwidth" w_b, as shown in the figure. In the numerical solution of a system of equations the amount of computation depends on the product $N\overline{w}_b^2$, where N is the total number of degrees of freedom and $\overline{w}_b$ is the average bandwidth. For a given structure N is a constant and so for the present discussion we may ignore the effect of N. Therefore in the degree of freedom numbering the order in which the nodes are considered should be such that the average difference in the degree of freedom numbers is as small as possible. The rule for achieving this is simple: the numbering should always proceed across the shortest width of the structure, thus only gradually moving along the length of the structure. This is illustrated in the 13-bar pin-jointed structure of Fig. 5.14. To simplify the discussion we can use node numbers instead of degrees of freedom, because this will not alter the conclusion in any way. It simply means that since there are two degrees of freedom at each node, each square in the stiffness matrix diagram represents four terms. In the first case (Fig.5.14a) this rule is not followed and $\overline{w}_b$ is 3.9, whereas if the rule is followed (Fig. 5.14b) the value is 2.7. Thus the first node sequence requires $(3.9/2.7)^2 = 2.09$ times as much computation as the second sequence.

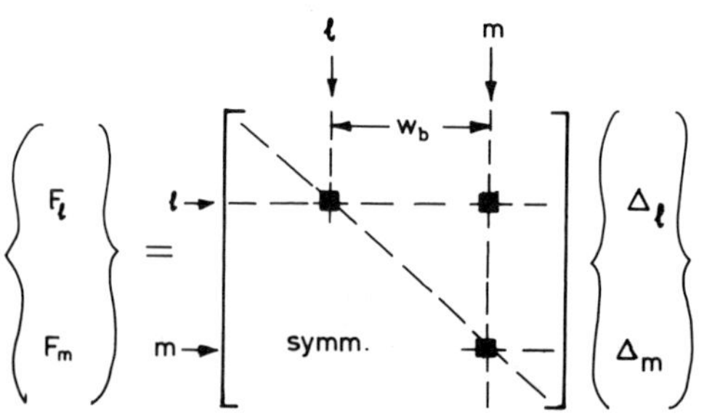

Figure 5.13 Bandwidth of the structural stiffness matrix.

5.3 BEAM ELEMENT: RIGID-JOINTED FRAME ANALYSIS

Flexure-Only Beam Element

For several obvious reasons the two- and three-dimensional frameworks found in ships are rigid-jointed rather than pin-jointed: the need for watertightness, the use of welding, and the large lateral loads which must be carried. The basic element for a rigid-jointed frame is the beam element shown in Fig. 5.15. The beam is loaded by forces and moments at each node and is assumed to be of uniform flexural rigidity EI.

By definition, the nodes of a beam element are the points where the external forces and moments act, and these are assumed to be in equilibrium with the internal forces which act along the beam's neutral axis. Therefore, unless special provision is made for nodal eccentricity, the nodes of a beam element are located at the centroid of the beam's cross section.

The element stiffness matrix $\mathbf{k}^e$ may again be obtained by following the same procedure as for the bar element, that is, by imposing each of the nodal displacements in turn, determining the nodal forces set up, and then superimposing the results obtained from each individual case. Alternatively, since the element is a prismatic beam, we may make use of beam theory to obtain the required relationships between the nodal forces and displacements. Specifically, these relationships may be obtained from the well-known slope-deflection equations of beam theory. For the element shown in Fig. 5.15 the slope-deflection equations are

$$M_1 = \frac{6EI}{L^2}v_1 + \frac{4EI}{L}\theta_1 - \frac{6EI}{L^2}v_2 + \frac{2EI}{L}\theta_2$$

$$M_2 = \frac{6EI}{L^2}v_1 + \frac{2EI}{L}\theta_1 - \frac{6EI}{L^2}v_2 + \frac{4EI}{L}\theta_2$$

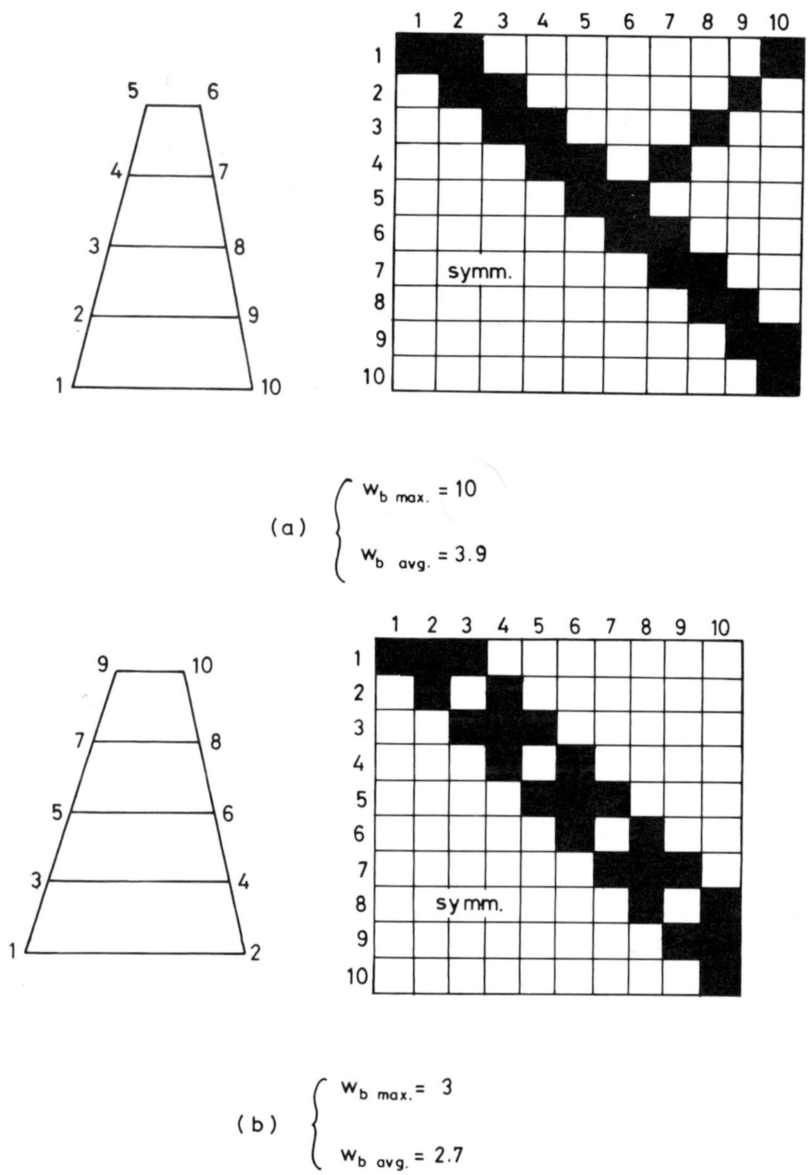

Figure 5.14 Effect of structure node numbering on bandwidth.

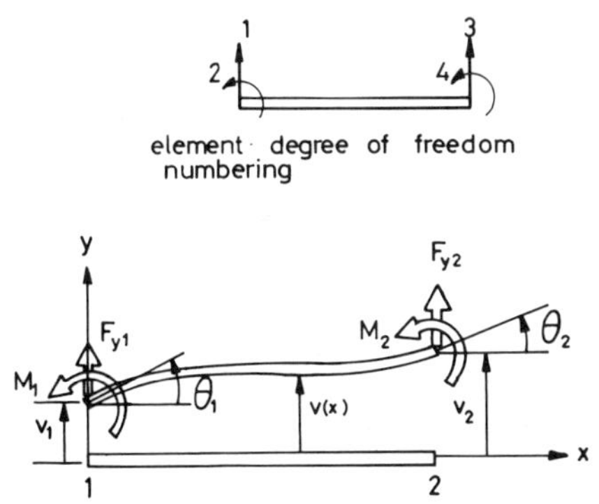

Figure 5.15 Flexure-only beam element.

From vertical force equilibrium

$$F_{y1} = -F_{y2} = \frac{M_1 + M_2}{L}$$

$$= \frac{12EI}{L^3} v_1 + \frac{6EI}{L^2} \theta_1 - \frac{12EI}{L^3} v_2 + \frac{6EI}{L^2} \theta_2$$

In matrix notation these equations are

$$\begin{Bmatrix} F_{y1} \\ M_1 \\ F_{y2} \\ M_2 \end{Bmatrix} = \frac{EI}{L^3} \begin{bmatrix} 12 & 6L & -12 & 6L \\ 6L & 4L^2 & -6L & 2L^2 \\ -12 & -6L & 12 & -6L \\ 6L & 2L^2 & -6L & 4L^2 \end{bmatrix} \begin{Bmatrix} v_1 \\ \theta_1 \\ v_2 \\ \theta_2 \end{Bmatrix} \tag{5.3.1}$$

or

$$\mathbf{f} = \mathbf{k}^e \boldsymbol{\delta}$$

which defines the nodal force vector $\mathbf{f}$, the nodal displacement vector $\boldsymbol{\delta}$, and the stiffness matrix $\mathbf{k}^e$ for a beam element in which only the flexural response is considered. It is important to note that in this element axial forces and axial deformation are deliberately ignored. Hence this element can only be used for applications in which the loads are carried purely by beam bending, such as in laterally loaded flat grillages, or in which the axial displacements and flexural displacements are completely independent, such as in orthogonal frameworks, providing also that the structure does not undergo "sway," that is, translation of joints so as to cause axial rigid body motion of any element. Hence this element cannot be used for trusses or for other frame structures in which a given load is carried partly by axial stiffness and partly by bending stiffness, such as a pitched grillage (e.g., a ship's bottom that has some deadrise). In fact, about the only types of structure for which this element can be used are multispan beams and flat grillages. The latter are particularly common in ships, and this application is discussed in the next section. The element is introduced here at this point because it allows a simpler explanation of some of the principles and techniques of frame analysis.

The nodal displacements given above include the rotation θ at the end of the beam. It should be noted that in small-deflection beam theory the quantity that is actually involved is the *slope* of the beam, that is, the derivative $v' = dv/dx$; for small deflections this is equivalent to the angle of rotation and the latter term is usually chosen because a rotation is easier to visualize as a form of displacement. However, for basic derivations the derivative term is preferable because it shows that within the element the deflected state of the beam is described completely by the general displacement function $v(x)$; the slope at any point—including the ends—is simply the derivative of $v(x)$ at that point. Thus the four nodal "displacements" are in reality the nodal values of $v(x)$ and of $v'(x)$, and this illustrates why the name "degree of freedom" is often used instead of "displacement."

For an arbitrarily orientated beam element it is again necessary to expand $\mathbf{k}^e$ to allow for the element displacements v (which are always normal to the beam) to be transformed into U and V displacements in the global coordinate system. The nodal bending moments are unaffected. Consequently $\mathbf{T}$ takes the form

$$\mathbf{T} = \begin{bmatrix} \lambda & \mu & 0 & 0 & 0 & 0 \\ -\mu & \lambda & 0 & 0 & 0 & 0 \\ 0 & 0 & 1 & 0 & 0 & 0 \\ 0 & 0 & 0 & \lambda & \mu & 0 \\ 0 & 0 & 0 & -\mu & \lambda & 0 \\ 0 & 0 & 0 & 0 & 0 & 1 \end{bmatrix} \tag{5.3.2}$$

As shown in (5.2.8) the transformation of the element stiffness matrix from local to global coordinates is $\mathbf{T}^T\mathbf{k}^e\mathbf{T}$. For the beam element the result is

$$\mathbf{K}^e = \frac{EI}{L^3} \begin{bmatrix} \boldsymbol{\kappa}_1 & \boldsymbol{\kappa}_2^T \\ \boldsymbol{\kappa}_2 & \boldsymbol{\kappa}_3 \end{bmatrix} \tag{5.3.3}$$

where

$$\boldsymbol{\kappa}_1 = \begin{bmatrix} 12\mu^2 & & \text{symm.} \\ -12\lambda\mu & 12\lambda^2 & \\ -6\mu L & 6\lambda L & 4L^2 \end{bmatrix}$$

$$\boldsymbol{\kappa}_3 = \begin{bmatrix} 12\mu^2 & & \text{symm.} \\ -12\lambda\mu & 12\lambda^2 & \\ 6\mu L & -6\lambda L & 4L^2 \end{bmatrix}$$

and

$$\boldsymbol{\kappa}_2 = \begin{bmatrix} -12\mu^2 & 12\lambda\mu & 6\mu L \\ 12\lambda\mu & -12\lambda^2 & -6\lambda L \\ -6\mu L & 6\lambda L & 2L^2 \end{bmatrix}$$

From this point on, the use of this element for the structural analysis of a rigid-jointed frame would proceed along the remaining steps of Table 5.1: the overall stiffness matrix $\mathbf{K}$ is assembled, the support conditions are applied, and the resulting set of equations solved. However, before proceeding further in regard to the application of the beam element, we shall use it as an example to demonstrate the general method for defining an element and deriving its stiffness matrix.

General Method for Deriving An Element Stiffness Matrix

In presenting the general method we choose the flexure-only beam element as the example because this element is quite simple and yet is sufficiently general to illustrate all of the basic steps in the definition of an element and the derivation of its stiffness matrix. In Chapter 7 it will be shown that in finite element theory this same method is used to derive the element stiffness matrix for a wide variety of two- and three-dimensional finite elements. Hence this method constitutes one of the basic foundations of matrix structural analysis. In terms of the seven steps of matrix structural analysis given in Table 5.1, this method provides the material which is essential for step 2, that is, the complete definition of the element properties, and especially the element stiffness matrix $\mathbf{k}^e$. The method consists of five steps, and in order to distinguish them from the seven basic steps of matrix structural analysis we shall identify them by means of Roman numerals.

STEP I: SELECT SUITABLE DISPLACEMENT FUNCTION

The first step is to specify a displacement function which uniquely defines the state of displacement at all points within the element in terms of the nodal degrees of freedom. For a beam the displacement within the element is $v(x)$. This displacement can conveniently be represented by a polynomial expression and since the aim is to express $v(x)$ in terms of the nodal degrees of freedom $\boldsymbol{\delta}$, the assumed polynomial must contain one unknown coefficient for each degree of freedom possessed by the element. As defined previously the beam element possesses four degrees of freedom

$$\boldsymbol{\delta} = \begin{Bmatrix} v_1 \\ v_1' \\ v_2 \\ v_2' \end{Bmatrix} \tag{5.3.4}$$

and therefore we choose a cubic polynomial to represent $v(x)$, since this has four coefficients.

$$v(x) = C_1 + C_2x + C_3x^2 + C_4x^3 \tag{5.3.5}$$

In matrix form this is

$$v(x) = \mathbf{H}(x)\mathbf{C} \tag{5.3.6}$$

where $\mathbf{H}(x) = [1 \quad x \quad x^2 \quad x^3]$

and $\mathbf{C}^T = [C_1 \quad C_2 \quad C_3 \quad C_4]$

Equation (5.3.4) defines the vector of nodal displacements $\boldsymbol{\delta}$. We now construct the *generalized displacement vector* $\boldsymbol{\delta}(x)$ which expresses the values of the displacements $v(x)$ and $v'(x)$ at any point within the element, in terms of the four (as yet undetermined) coefficients. Differentiating (5.3.5) gives

$$v'(x) = C_2 + 2C_3x + 3C_4x^2$$

and by writing this in matrix form and combining it with (5.3.6) we obtain

$$\boldsymbol{\delta}(x) = \begin{Bmatrix} v(x) \\ v'(x) \end{Bmatrix} = \begin{bmatrix} 1 & x & x^2 & x^3 \\ 0 & 1 & 2x & 3x^2 \end{bmatrix} \begin{Bmatrix} C_1 \\ C_2 \\ C_3 \\ C_4 \end{Bmatrix} \tag{5.3.7}$$

STEP II: RELATE GENERAL DISPLACEMENTS WITHIN THE ELEMENT TO ITS NODAL DISPLACEMENTS

The coefficients of the displacement function $\mathbf{C}$ are now expressed in terms of the nodal displacements $\boldsymbol{\delta}$ and hence by substituting into (5.3.7) the displacements at any point within the element are obtained in terms of the nodal displacements $\boldsymbol{\delta}$.

Since $\boldsymbol{\delta}(x)$ represents the displacement at any point x the nodal displacements can be obtained from it by simply substituting the appropriate nodal coordinates into (5.3.7).

At node 1, $x = 0$ and therefore

$$v_1 = C_1 \quad \text{and} \quad v_1' = C_2$$

At node 2, $x = L$ and therefore

$$v_2 = C_1 + C_2L + C_3L^2 + C_4L^3$$
$$v_2' = C_2 + 2C_3L + 3C_4L^2$$

Writing these in matrix form

$$\begin{Bmatrix} v_1 \\ v_1' \\ v_2 \\ v_2' \end{Bmatrix} = \begin{bmatrix} 1 & 0 & 0 & 0 \\ 0 & 1 & 0 & 0 \\ 1 & L & L^2 & L^3 \\ 0 & 1 & 2L & 3L^2 \end{bmatrix} \begin{Bmatrix} C_1 \\ C_2 \\ C_3 \\ C_4 \end{Bmatrix}$$

or $\boldsymbol{\delta} = \mathbf{AC}$

Since the matrix $\mathbf{A}$ is known, this equation may be solved for the vector of unknown coefficients $\mathbf{C}$

$$\mathbf{C} = \mathbf{A}^{-1}\,\boldsymbol{\delta} \tag{5.3.8a}$$

and it may be shown that

$$\mathbf{A}^{-1} = \begin{bmatrix} 1 & 0 & 0 & 0 \\ 0 & 1 & 0 & 0 \\ \dfrac{-3}{L^2} & \dfrac{-2}{L} & \dfrac{3}{L^2} & \dfrac{-1}{L} \\ \dfrac{2}{L^3} & \dfrac{1}{L^2} & \dfrac{-2}{L^3} & \dfrac{1}{L^2} \end{bmatrix} \quad (5.3.8b)$$

For future use we note here that (5.3.6) and (5.3.8) can be combined to obtain an expression for the displacement within the element, $v(x)$, in terms of the nodal displacements $\boldsymbol{\delta}$. Combining these two equations gives

$$v(x) = \mathbf{H}(x)\mathbf{A}^{-1}\,\boldsymbol{\delta}$$

and the result of the matrix multiplication is

$$v(x) = \mathbf{N}(x)\,\boldsymbol{\delta} \quad (5.3.9a)$$

in which $\mathbf{N}(x)$ is known as the *shape function* and is given by

$$\mathbf{N}(x) = [(1 - 3\xi^2 + 2\xi^3) \quad x(1 - 2\xi + \xi^2) \quad (3\xi^2 - 2\xi^3) \quad x(-\xi + \xi^2)] \quad (5.3.9b)$$

where $\xi = x/L$.

STEP III: EXPRESS THE INTERNAL DEFORMATION IN TERMS OF THE NODAL DISPLACEMENTS

The definition of "deformation" of an element depends on which particular field of solid mechanics is involved. In plane elasticity the deformation is strain $\varepsilon(x, y)$, whereas for beam bending it is the curvature of the beam. In each case the deformation is some form of derivative of the displacement; in plane elasticity it is a first derivative ($\varepsilon_x = \partial u/\partial x$, etc.) whereas the curvature of a beam is (for small deflections) the second derivative of the beam displacement, $v''(x) = d^2v/dx^2$. Therefore, from (5.3.5)

$$v''(x) = 2C_3 + 6C_4x$$

or in matrix form

$$v''(x) = [0 \quad 0 \quad 2 \quad 6x]\begin{Bmatrix} C_1 \\ C_2 \\ C_3 \\ C_4 \end{Bmatrix}$$

From (5.3.8a)

$$v''(x) = [0 \quad 0 \quad 2 \quad 6x]\mathbf{A}^{-1}\,\boldsymbol{\delta}$$

and substitution for $\mathbf{A}^{-1}$ followed by matrix multiplication gives

$$v''(x) = \mathbf{B}\,\boldsymbol{\delta}$$

where (5.3.10)

$$\mathbf{B} = \left[\frac{-6}{L^2} + \frac{12x}{L^3} \quad \frac{-4}{L} + \frac{6x}{L^2} \quad \frac{6}{L^2} - \frac{12x}{L^3} \quad \frac{-2}{L} + \frac{6x}{L^2}\right]$$

STEP IV: EXPRESS THE INTERNAL FORCE IN TERMS OF THE NODAL DISPLACEMENTS, USING THE ELEMENT'S LAW OF ELASTIC BEHAVIOR

As with internal deformation, the definition of internal force depends on the class of problem being considered. For problems in two- and three- dimensional elasticity it consists of the components of stress, whereas for beam bending it is the internal bending moment in the beam, $\mathcal{M}_z(x)$*. In this step the goal is to express the internal force in terms of the nodal

*The use of script symbols for internal forces is especially helpful in the case of bending moment, $\mathcal{M}$, because it is necessary to distinguish it from an applied (nodal) moment M; these two quantities are not the same and they have a different sign convention. The term "bending moment" is used to describe the internal state of the beam, that is, its response to the external load. The deformation response is a curvature and so the sign of the internal force response (i.e., bending moment $\mathcal{M}$) is determined by the curvature of the beam: $\mathcal{M}_z(x)$ is positive if $v''(x)$ is positive. A nodal moment M is an applied load, a certain type of force (a couple) that acts at a particular location. It may be either an external load acting on the structure or an element nodal moment acting on a particular element. Either way, its sign is determined by the coordinate system for forces and deflections (using the right hand rule), either at the structure level or at the element level. Thus, M_z is a moment acting about the z-axis and it is positive if its rotation would give an advance in the $+z$-direction. Of course, *at a node* (say element node 1) the internal bending moment $\mathcal{M}_z(0)$, or $\mathcal{M}_{z1}$, would be equal *in magnitude* to the element nodal moment acting at that node, M_{z1}, but the sign could be different, and this could be a source of confusion and error if these two are not clearly distinguished. In small problems which are solved by hand the possibility of confusion and error is perhaps not so serious but with large problems and for computer implementation it is important to distinguish between these two quantities. In addition to using different symbols, in this text the phrase "bending moment" shall be used only for $\mathcal{M}$, and never for M. Other terms, such as (element) nodal moment, generalized nodal force (or load), applied moment, and so on, which do *not* contain the phrase "bending moment," will be used for M.

displacements. Since a relationship between the internal deformation and the nodal displacements $\boldsymbol{\delta}$ is already known, the internal force can be related to the nodal displacements providing that the element's internal force–deformation relationship is known. For elasticity problems this relationship is Hooke's Law. For beam bending it is the moment–curvature relationship

$$\mathcal{M}_z(x) = EIv''(x) \tag{5.3.11}$$

Therefore the expression for the internal force in terms of nodal displacements is

$$\mathcal{M}_z(x) = EI\mathbf{B}\ \boldsymbol{\delta}$$

From the expression for $\mathbf{B}$ in (5.3.10) it can be seen that in the beam element the bending moment varies linearly along the beam.

For any element the expression for internal force is used not only in deriving the element stiffness matrix but also in any structural analysis which makes use of that element, because step 6 of matrix structural analysis is the calculation of the internal forces in the various elements by transforming the nodal displacements back into element coordinates and substituting them into the expression for internal force. From (5.3.10) the expression is

$$\mathcal{M}_z(x) = \mathfrak{b}[-6L + 12x \quad -4L^2 + 6Lx \quad 6L - 12x \quad -2L^2 + 6Lx]\begin{Bmatrix} v_1 \\ v_1' \\ v_2 \\ v_2' \end{Bmatrix} \tag{5.3.12a}$$

in which $\mathfrak{b} = \dfrac{EI}{L^3}$.

In many cases it is desired, or is sufficient, to obtain the internal forces only at the nodal points. In the beam element, for example, the linear distribution of $\mathcal{M}_z(x)$ means that it is sufficient to calculate $\mathcal{M}_z$ only at the nodes. An expression for these nodal values is easily obtained by simply substituting $x = 0$ and $x = L$ into the foregoing expression. The result is

$$\begin{Bmatrix} \mathcal{M}_z(0) \\ \mathcal{M}_z(L) \end{Bmatrix} = \mathbf{S}\ \boldsymbol{\delta} \tag{5.3.12b}$$

where

$$\mathbf{S} = \frac{EI}{L^3}\begin{bmatrix} -6L & -4L^2 & 6L & -2L^2 \\ 6L & 2L^2 & -6L & 4L^2 \end{bmatrix}$$

STEP V: OBTAIN THE ELEMENT STIFFNESS MATRIX $\mathbf{k}^e$ BY RELATING NODAL FORCES TO NODAL DISPLACEMENTS

In this step the internal forces are replaced by statically equivalent nodal forces $\mathbf{f}$ and hence the nodal forces are related to the nodal displacements $\boldsymbol{\delta}$, thereby defining the required element stiffness matrix $\mathbf{k}^e$.

The principle of virtual work is used to determine the set of nodal forces that is statically equivalent to the internal forces. The condition of equivalence may be expressed as follows: during any virtual displacement imposed on the element, the total external work done by the nodal forces must equal the total internal work done by the internal forces. While the virtual displacement is being imposed all actual forces, both nodal and internal, are constant. The arbitrary virtual nodal displacements are represented by the symbol $\boldsymbol{\delta}^*$; that is

$$\boldsymbol{\delta}^* = \begin{Bmatrix} v_1^* \\ v_1'^* \\ v_2^* \\ v_2'^* \end{Bmatrix}$$

We begin with the external virtual work. If the above virtual nodal displacements are imposed on an element in which the actual nodal forces are

$$\mathbf{f} = \begin{Bmatrix} f_1 \\ M_1 \\ f_2 \\ M_2 \end{Bmatrix}$$

then the virtual work done externally is

$$\begin{aligned} W_{\text{ext}} &= v_1^* f_1 + v_1'^* M_1 + v_2^* f_2 + v_2'^* M_2 \\ &= \boldsymbol{\delta}^{*T}\mathbf{f} \end{aligned}$$

Next we evaluate the internal virtual work. If an arbitrary virtual curvature $v''(x)^*$ is imposed on a beam of length L in which the actual internal bending moment is $\mathcal{M}_z(x)$, the internal virtual work is

$$W_{\text{int}} = \int_0^L (v''(x)^*)^T \mathcal{M}_z(x)\, dx$$

If the virtual curvature is imposed indirectly, in terms of virtual nodal displacements $\boldsymbol{\delta}^*$, the corresponding value of $v''*$ is

$$(v''(x)*)^T = (\mathbf{B}\ \boldsymbol{\delta}*)^T = \boldsymbol{\delta}*^T\mathbf{B}^T$$

In matrix algebra, expanding the transpose of a product causes the terms to be reversed. The actual internal bending moment can be expressed in terms of the actual curvature $v''(x)$ by means of (5.3.11) and so the internal work resulting from the virtual nodal displacement is

$$W_{\text{int}} = \int_0^L \boldsymbol{\delta}*^T\mathbf{B}^T EIv''(x)\ dx$$

Finally, from (5.3.10) the actual curvature $v''(x)$ can be expressed in terms of the actual nodal displacement $\boldsymbol{\delta}$. The result is

$$W_{\text{int}} = \int_0^L \boldsymbol{\delta}*^T\mathbf{B}^T EI\mathbf{B}\ \boldsymbol{\delta}\ dx$$

Equating the internal and external work and setting the arbitrary virtual displacement $\boldsymbol{\delta}*$ to unity (we can do this because the value is arbitrary) gives

$$\mathbf{f} = \left[\int_0^L \mathbf{B}^T EI\mathbf{B}\ dx\right]\boldsymbol{\delta}$$

Therefore since this equation relates nodal force $\mathbf{f}$ to nodal displacement $\boldsymbol{\delta}$ the quantity in brackets must be the element stiffness matrix $\mathbf{k}^e$. Substituting for $\mathbf{B}$ and performing the integration produces the same stiffness matrix which was derived earlier using the slope-deflection equations and was presented in (5.3.1).

Ordinary Beam Element

For general frame structures in which the members are not orientated orthogonally the axial stiffness and the flexural stiffness are not independent and the former must be included in the element stiffness matrix. Hence the usual beam element has six degrees of freedom as shown in Fig. 5.16. These are usually taken in the following order.

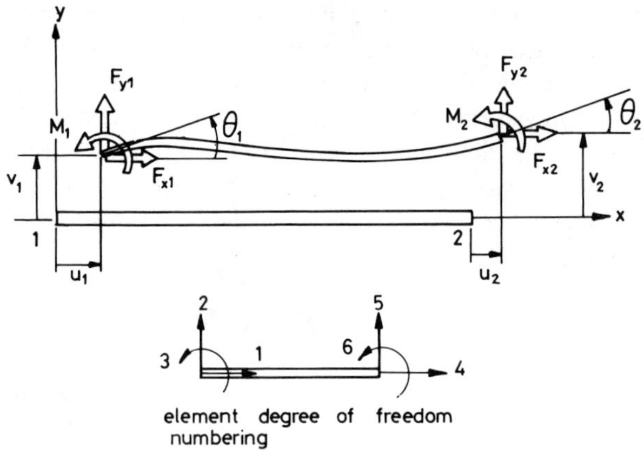

Figure 5.16 Ordinary beam element.

$$\boldsymbol{\delta} = \begin{Bmatrix} u_1 \\ v_1 \\ \theta_1 \\ u_2 \\ v_2 \\ \theta_2 \end{Bmatrix} \tag{5.3.13}$$

Accordingly, in the 6 × 6 element stiffness matrix the axial stiffness terms occur in the first and fourth rows and columns, and they are the same as for the pin-ended bar element. For convenience we define the following axial and flexural parameters

$$\mathfrak{a} = \frac{AE}{L} \quad \text{and} \quad \mathfrak{b} = \frac{EI}{L^3} \tag{5.3.14}$$

In terms of $\mathfrak{a}$ and $\mathfrak{b}$ the stiffness matrix of the ordinary beam element is

$$\mathbf{k}^e = \begin{bmatrix} \mathfrak{a} & 0 & 0 & -\mathfrak{a} & 0 & 0 \\ 0 & 12\mathfrak{b} & 6\mathfrak{b}L & 0 & -12\mathfrak{b} & 6\mathfrak{b}L \\ 0 & 6\mathfrak{b}L & 4\mathfrak{b}L^2 & 0 & -6\mathfrak{b}L & 2\mathfrak{b}L^2 \\ -\mathfrak{a} & 0 & 0 & \mathfrak{a} & 0 & 0 \\ 0 & -12\mathfrak{b} & -6\mathfrak{b}L & 0 & 12\mathfrak{b} & -6\mathfrak{b}L \\ 0 & 6\mathfrak{b}L & 2\mathfrak{b}L^2 & 0 & -6\mathfrak{b}L & 4\mathfrak{b}L^2 \end{bmatrix} \tag{5.3.15}$$

which can be recognized as the "flexure-only" stiffness matrix plus the axial stiffness terms.

As before, the element stiffness matrix in structure coordinates is $\mathbf{T}^T\mathbf{k}^e\mathbf{T}$ with $\mathbf{T}$ defined in (5.3.2). The result is

$$\mathbf{K}^e = \begin{bmatrix} \mathfrak{a}\lambda^2 + 12\mathfrak{b}\mu^2 & & & & \text{symmetric} & \\ (\mathfrak{a}-12\mathfrak{b})\lambda\mu & \mathfrak{a}\mu^2 + 12\mathfrak{b}\lambda^2 & & & & \\ -6\mathfrak{b}L\mu & 6\mathfrak{b}L\lambda & 4\mathfrak{b}L^2 & & & \\ -\mathfrak{a}\lambda^2 - 12\mathfrak{b}\mu^2 & -(\mathfrak{a}-12\mathfrak{b})\lambda\mu & 6\mathfrak{b}L\mu & \mathfrak{a}\lambda^2 + 12\mathfrak{b}\mu^2 & & \\ -(\mathfrak{a}-12\mathfrak{b})\lambda\mu & -\mathfrak{a}\mu^2 - 12\mathfrak{b}\lambda^2 & -6\mathfrak{b}L\lambda & (\mathfrak{a}-12\mathfrak{b})\lambda\mu & \mathfrak{a}\mu^2 + 12\mathfrak{b}\lambda^2 & \\ -6\mathfrak{b}L\mu & 6\mathfrak{b}L\lambda & 2\mathfrak{b}L^2 & 6\mathfrak{b}L\mu & -6\mathfrak{b}L\lambda & 4\mathfrak{b}L^2 \end{bmatrix} \tag{5.3.16}$$

The expression for the internal forces is simply a combination of (5.2.10) for the axial force and (5.3.12) for the bending moment. In the former we shall divide by the beam cross-sectional area A to obtain axial stress σ_x. If desired, an expression for the shear force can also be included since the shear force is simply the derivative of the internal bending moment: $\mathcal{Q}(x) = (d/dx)[\mathcal{M}_z(x)]$. From (5.3.10) and (5.3.11)

$$\mathcal{Q}(x) = \frac{EI}{L^3}[12 \quad 6L \quad -12 \quad 6L]\,\boldsymbol{\delta} \qquad (5.3.17)$$

Thus the shear force is constant along the beam element, as is the axial force. Therefore it is sufficient to have an expression for the forces only at the nodes. This is

$$\begin{Bmatrix} \sigma_{x1} \\ \mathcal{Q}_1 \\ \mathcal{M}_{z1} \\ \sigma_{x2} \\ \mathcal{Q}_2 \\ \mathcal{M}_{z2} \end{Bmatrix} = \begin{bmatrix} \dfrac{-E}{L} & 0 & 0 & \dfrac{E}{L} & 0 & 0 \\ 0 & 12\mathfrak{b} & 6\mathfrak{b}L & 0 & -12\mathfrak{b} & 6\mathfrak{b}L \\ 0 & -6\mathfrak{b}L & -4\mathfrak{b}L^2 & 0 & 6\mathfrak{b}L & -2\mathfrak{b}L^2 \\ \dfrac{-E}{L} & 0 & 0 & \dfrac{E}{L} & 0 & 0 \\ 0 & 12\mathfrak{b} & 6\mathfrak{b}L & 0 & -12\mathfrak{b} & 6\mathfrak{b}L \\ 0 & 6\mathfrak{b}L & 2\mathfrak{b}L^2 & 0 & -6\mathfrak{b}L & 4\mathfrak{b}L^2 \end{bmatrix} \begin{Bmatrix} u_1 \\ v_1 \\ \theta_1 \\ u_2 \\ v_2 \\ \theta_2 \end{Bmatrix} \qquad (5.3.18)$$

Note that although these are nodal values the script form is used for $\mathcal{M}$ and $\mathcal{Q}$ in order to indicate that they are internal forces.

Restraints and Specified Displacements

Earlier, when discussing boundary conditions and restraints, it was shown that the imposition of zero displacement at a node can be achieved by simply deleting that row and column from the structure equilibrium equations. This is quite suitable if the equations are being solved by hand (any reduction in the number of equations is then most valuable) but it is not so suitable for a computer solution. Deletion of rows and columns from $\mathbf{K}$ would involve repacking it within the computer, and this is time-consuming. A more efficient technique is to replace the diagonal terms of $\mathbf{K}$ corresponding to the zero displacement by unity and replace the rest of the terms in the corresponding row and column by zeros.

A second method is available which requires only one multiplication for each degree of freedom that is to be eliminated. In this method each diagonal term of $\mathbf{K}$ that corresponds to a zero displacement is multiplied by a large number such as 10^{25}. For example, suppose the displacement which is degree of freedom number I is to be zero. After the multiplication the equilibrium equation corresponding to that degree of freedom is

$$F_I = K_{I1}\Delta_1 + K_{I2}\Delta_2 + \cdots + 10^{25}K_{II}\Delta_I + \cdots + K_{IN}\Delta_N$$

and thus

$$\Delta_I = \frac{F_I - (K_{I1}\Delta_1 + K_{I2}\Delta_2 + \cdots + K_{IN}\Delta_N)}{10^{25}K_{II}}$$

Since the diagonal term is now so much greater than the off-diagonal terms, the value of the displacement Δ_I is very close to zero.

If a displacement Δ_I is required to have some specified nonzero value d then this effect too may be allowed for using either of these methods, together with an adjustment of the load vector. If the first method is used, then after placing the 1 and the zeros in $\mathbf{K}$ the F_I term in the force vector is replaced by the value of the prescribed displacement d. Alternatively, using the second method, the F_I term is replaced by the diagonal term multiplied by the prescribed displacement. Thus writing out the equation for row I

$$10^{25}K_{II}d = K_{I1}\Delta_1 + K_{I2}\Delta_2 + \cdots + 10^{25}K_{II}\Delta_I + \cdots + K_{IN}\Delta_N$$

Consequently

$$\Delta_I = \frac{10^{25}K_{II}d - (K_{I1}\Delta_1 + K_{I2}\Delta_2 + \cdots + K_{IN}\Delta_N}{10^{25}K_{II}}$$

Since the diagonal term $10^{25}K_{II}$ is so much larger than the off-diagonal terms, then to a good approximation $\Delta_I = d$.

Distributed Loads: Equivalent Nodal Loads

In frame analysis, and also in finite element analysis, all quantities are specified or represented in terms of their values at the nodes. Until now we have assumed that the loads occur only at nodes whereas in most frame structures, and certainly in ships, the loads are distributed loads. The technique for dealing with such loads is to divide the analysis into two parts, a local analysis which deals only with the internodal loading and response of the loaded beam, and an overall analysis which deals with the response of the overall structure. A separate solution is obtained for each, and then the solutions are superimposed. For this purpose the distributed load is divided into two parts: an internal load which is applied within the beam and which accounts for the internodal bending; and an external load which is applied at the boundaries of the loaded beam (i.e., at the nodes) and which acts on the entire structure. Since it is a nodal load the usual frame analysis method may be used for the solution.* Obviously each load must give an accurate solution within its own field, and the two loads taken together must match the original load. In the case of beam bending this is achieved by defining the internal load as the original distributed load plus equilibrium end forces and moments that correspond to complete fixity at the boundary, that is, zero deflection and zero rotation at the ends (or nodes) of the loaded beam. In this way the internal problem satisfies equilibrium and may be dealt with in isolation from the overall solution. Thus the internal problem is simply the given distributed load acting on a clamped beam, and for standard types of distributed loading the solution is already available. In particular, the bending moments at the ends of the beam are simply the "fixed-end moments" for that loading. Figure 5.17 gives the fixed-end reaction forces and moments for a variety of distributed loads commonly occuring in ship structures.

The external load must satisfy three requirements:

1. It must consist of nodal forces and moments.
2. In regard to its effect on the overall solution it must be statically equivalent to the actual distributed load.
3. The combination of the internal and external loads must be the actual load.

From (3) the external load must be equal and opposite to the clamping forces and moments which were added to the distributed load in the internal problem. Since these clamping forces and moments equilibrated the distributed load they were statically equivalent to it but opposite in sign. Hence the forces and moments in the external problem, being of opposite sign to these clamping forces and moments, will be exactly statically equivalent to the distributed load. Thus in any frame analysis the statically equivalent nodal loads which are used in place of a distributed load are exactly the same, but opposite in sign, as the fixed-end reaction forces and moments corresponding to that load.

Although the equivalent nodal loads are an adequate approximation of the distributed loads in regard to obtaining the structural (nodal) displacements, they are not sufficient in themselves to obtain the nodal forces and moments. Since these equivalent loads have been added artificially at the nodes, the nodal values of the element internal forces (shear force and bending moment) which are calculated in step 6 of Table 5.1 will be too large by this amount, and the excess must be subtracted. Therefore if an equivalent load has been used for any element, the fixed-end forces and moments must be added to the nodal forces and moments obtained from step 6, regardless of whether the full internal distribution of shear force and bending moment is required or not. This correction constitutes step 7 in the overall procedure of matrix stiffness analysis.

Example

The portal frame of Fig. 5.18 will be used as an example to demonstrate several aspects of matrix stiffness analysis which have been discussed previously:

1. The modeling of symmetry in a structure.
2. The use of equivalent nodal loads to represent a distributed load.
3. The relative magnitude of axial and flexural stiffness and the order of accuracy attainable when neglecting the axial degree of freedom in orthogonal frames.

*This technique is used in several branches of continuum mechanics. For example, in fluid mechanics a flow is often represented by the combination of a "near field" flow which accounts for the local aspects (e.g., flow around a body) and a "far field" flow which accounts for the effects in the rest of the field.

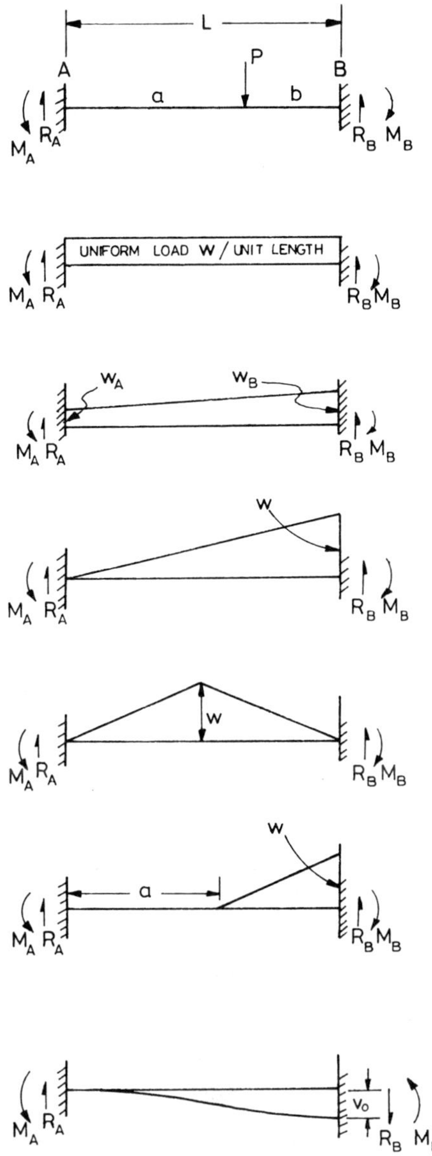

$$R_A = \frac{Pb^2(3a+b)}{L^3} \qquad R_B = \frac{Pa^2(a+3b)}{L^3}$$

$$M_A = \frac{Pab^2}{L^2} \qquad M_B = -\frac{Pa^2b}{L^2}$$

$$R_A = \frac{wL}{2} \qquad R_B = \frac{wL}{2}$$

$$M_A = \frac{wL^2}{12} \qquad M_B = -\frac{wL^2}{12}$$

$$R_A = \frac{L}{20}(7w_A + 3w_B) \qquad R_B = \frac{L}{20}(3w_A + 7w_B)$$

$$M_A = \frac{L^2}{60}(3w_A + 2w_B) \qquad M_B = \frac{-L^2}{60}(2w_A + 3w_B)$$

$$R_A = \frac{3}{20}\,wL \qquad R_B = \frac{7}{20}\,wL$$

$$M_A = \frac{wL^2}{30} \qquad M_B = \frac{-wL^2}{20}$$

$$R_A = \frac{wL}{4} \qquad R_B = \frac{wL}{4}$$

$$M_A = \frac{5wL^2}{96} \qquad M_B = \frac{-5wL^2}{96}$$

$$R_A = \frac{w(L-a)^3(3L+2a)}{20L^3}$$

$$R_B = \frac{w(L-a)}{20L^3}\left[10L^3 - (L-a)^2(3L+2a)\right]$$

$$M_A = \frac{w}{60L^2}(L-a)^3(2L+3a)$$

$$M_B = \frac{-w}{60L^2}(L-a)^2(3L^2+4aL+3a^2)$$

$$R_A = -R_B = \frac{12\,EI\,v_o}{L^3}$$

$$M_A = M_B = \frac{6\,EI\,v_o}{L^2}$$

Figure 5.17 Fixed-end reaction forces and moments.

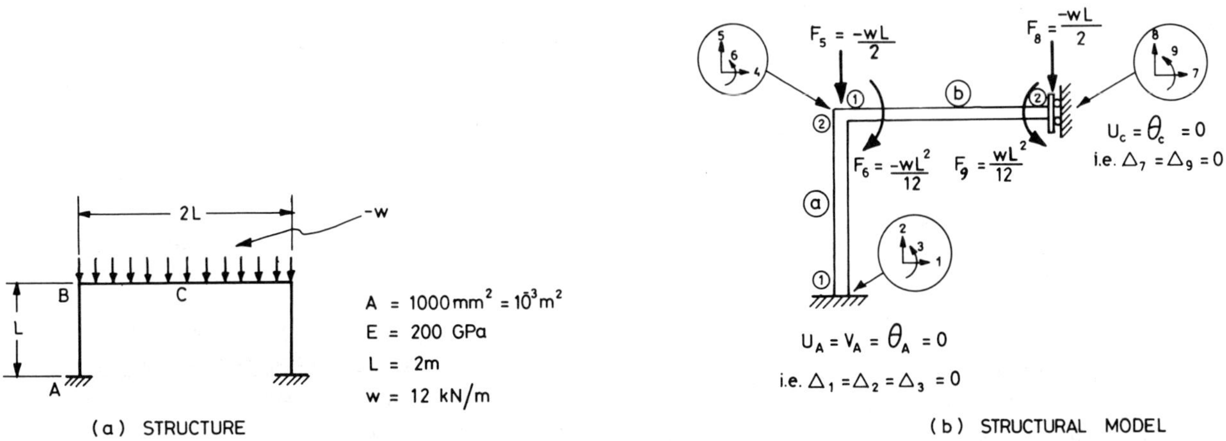

Figure 5.18 Sample problem with distributed load.

The solution will be presented in terms of the seven steps of Table 5.1.

Step 1. Define the Structural Model

The structural modeling is illustrated in Fig. 5.18*b*. Because of symmetry only half of the structure is modeled. To permit this a node is introduced at point C, even though there is no joint there. The geometric conditions arising from symmetry are: zero horizontal displacement and zero slope at the point of symmetry. The clamped support at A means that the three displacements at that point are also zero. Although the frame is orthogonal the solution will be performed using the ordinary beam element, so that the axial degree of freedom is included in the modeling. The result will then be compared with a solution obtained using the flexure-only beam element. The structure degrees of freedom are numbered from 1 to 9 as shown. The uniform distributed load of intensity w is represented by means of the statically equivalent loads $wL/2$ and $wL^2/12$. It may be seen that F_9 will not actually be applied to the structure because degree of freedom 9 is eliminated due to symmetry. The axial and flexural stiffness parameters are $\mathfrak{a} = AE/L$ and $\mathfrak{b} = EI/L^3$. Their ratio is

$$\frac{\mathfrak{a}}{\mathfrak{b}} = \frac{EA}{L}\frac{L^3}{EA\rho^2} = \left(\frac{L}{\rho}\right)^2$$

where ρ = radius of gyration of the beam cross section. The ratio L/ρ is the *slenderness ratio* of the beam and for typical flange-and-web beams its value is between 10 and 40. For illustrative purposes we take a deliberately low value of 10, which then means that $\mathfrak{a}/\mathfrak{b} = 100$. The numerical values of all dimensions and physical properties are given in the figure.

Step 2. Calculate Each Element Stiffness Matrix in Structure Coordinates

The general expression for the beam element stiffness matrix at an arbitrary orientation χ is given in (5.3.16). The values of λ and μ for the two elements are as follows.

Member	χ	λ	μ
ⓐ	90°	0	1
ⓑ	0°	1	0

The element stiffness matrix for element ⓐ is given. It is suggested that the reader should evaluate at least a few of the terms himself and then compare them with the following.

$$\begin{array}{cccccc} \boxed{1} & \boxed{2} & \boxed{3} & \boxed{4} & \boxed{5} & \boxed{6} \end{array}$$

$$\mathbf{K}^{ⓐ} = \begin{bmatrix} 12\mathfrak{b} & 0 & -6\mathfrak{b}L & -12\mathfrak{b} & 0 & -6\mathfrak{b}L \\ 0 & \mathfrak{a} & 0 & 0 & -\mathfrak{a} & 0 \\ -6\mathfrak{b}L & 0 & 4\mathfrak{b}L^2 & 6\mathfrak{b}L & 0 & 2\mathfrak{b}L^2 \\ -12\mathfrak{b} & 0 & 6\mathfrak{b}L & 12\mathfrak{b} & 0 & 6\mathfrak{b}L \\ 0 & -\mathfrak{a} & 0 & 0 & \mathfrak{a} & 0 \\ -6\mathfrak{b}L & 0 & 2\mathfrak{b}L^2 & 6\mathfrak{b}L & 0 & 4\mathfrak{b}L^2 \end{bmatrix}$$

$$\begin{array}{cccccc} \boxed{\times} & \boxed{\times} & \boxed{\times} & \boxed{4} & \boxed{5} & \boxed{6} \end{array}$$

$$= \begin{bmatrix} & & & & & \\ & & & & & \\ & & & & & \\ & & & 12 & 0 & 12 \\ & & & 0 & 100 & 0 \\ & & & 12 & 0 & 16 \end{bmatrix} \begin{array}{c} \boxed{\times} \\ \boxed{\times} \\ \boxed{\times} \\ \boxed{4} \\ \boxed{5} \\ \boxed{6} \end{array}$$

The location vector is written above the matrix. In a hand solution such as this it is often worthwhile to anticipate which degrees of freedom are to be eliminated and to not bother calculating these terms in $\mathbf{K}^e$. For greater clarity the foregoing matrix has been written out in full initially but in substituting the numerical values this short-cut has been used. The eliminated degrees of freedom are indicated by ×. The stiffness matrix for element ⓑ is

$$\begin{array}{cccccc} \boxed{4} & \boxed{5} & \boxed{6} & \boxed{7} & \boxed{8} & \boxed{9} \end{array}$$

$$\mathbf{K}^{ⓑ} = \begin{bmatrix} \mathfrak{a} & 0 & 0 & -\mathfrak{a} & 0 & 0 \\ 0 & 12\mathfrak{b} & 6\mathfrak{b}L & 0 & -12\mathfrak{b} & 6\mathfrak{b}L \\ 0 & 6\mathfrak{b}L & 4\mathfrak{b}L^2 & 0 & -6\mathfrak{b}L & 2\mathfrak{b}L^2 \\ -\mathfrak{a} & 0 & 0 & \mathfrak{a} & 0 & 0 \\ 0 & -12\mathfrak{b} & -6\mathfrak{b}L & 0 & 12\mathfrak{b} & -6\mathfrak{b}L \\ 0 & 6\mathfrak{b}L & 2\mathfrak{b}L^2 & 0 & -6\mathfrak{b}L & 4\mathfrak{b}L^2 \end{bmatrix}$$

$$\begin{array}{cccccc} \boxed{4} & \boxed{5} & \boxed{6} & \boxed{\times} & \boxed{8} & \boxed{\times} \end{array}$$

$$= \begin{bmatrix} 100 & 0 & 0 & -100 & 0 & 0 \\ 0 & 12 & 12 & 0 & -12 & 12 \\ 0 & 12 & 16 & 0 & -12 & 8 \\ -100 & 0 & 0 & 100 & 0 & 0 \\ 0 & -12 & -12 & 0 & 12 & -12 \\ 0 & 12 & 8 & 0 & -12 & 16 \end{bmatrix} \begin{array}{c} \boxed{4} \\ \boxed{5} \\ \boxed{6} \\ \boxed{\times} \\ \boxed{8} \\ \boxed{\times} \end{array}$$

Steps 3 and 4. Assemble the Structure Stiffness Matrix and Apply the Boundary Conditions

In this problem these two steps are combined because the elimination of the degrees of freedom is here being achieved by simply not assembling these terms in the structure stiffness matrix. If a given structure is to be analyzed repeatedly with the same boundary conditions (as, for example, in structural optimization) then this technique is more efficient.

The result of the assembly process is given next. Here again it is suggested that the reader should perform the assembly process himself and then compare his result with that given.

$$\mathbf{K} = \begin{array}{c|ccccccccc|}
 & 1 & 2 & 3 & 4 & 5 & 6 & 7 & 8 & 9 \\
\hline
1 & & & & & & & & & \\
2 & & & & & & & & & \\
3 & & & & & & & & & \\
4 & & & & 112 & 0 & 12 & & 0 & \\
5 & & & & 0 & 112 & 12 & & -12 & \\
6 & & & & 12 & 12 & 32 & & -12 & \\
7 & & & & & & & & & \\
8 & & & & 0 & -12 & -12 & & 12 & \\
9 & & & & & & & & & \\
\end{array}$$

Thus the equilibrium equations are a 4×4 system of equations. The load vector must be expressed in units which are consistent with **K**. In this problem the units of $\mathfrak{a}$ and $\mathfrak{b}$, and hence of **K**, are MN/m. Therefore the magnitudes of the equivalent nodal loads are

$$\frac{wL}{2} = \frac{12 \times 10^{-3}(2)}{2} = 12 \times 10^{-3}\ \text{MN}$$

and

$$\frac{wL^2}{12} = \frac{12 \times 10^{-3}(4)}{12} = 4 \times 10^{-3}\ \text{MNm}$$

Hence the system of equations is

$$\begin{Bmatrix} 0 \\ -12 \\ -4 \\ -12 \end{Bmatrix} \times 10^{-3} = \begin{bmatrix} 112 & 0 & 12 & 0 \\ 0 & 112 & 12 & -12 \\ 12 & 12 & 32 & -12 \\ 0 & -12 & -12 & 12 \end{bmatrix} \begin{Bmatrix} \Delta_4 \\ \Delta_5 \\ \Delta_6 \\ \Delta_8 \end{Bmatrix}$$

Step 5. Solve for Displacements and Reactions
The solution to this system is

$$\begin{aligned}
\Delta_4 &= 0.0917 \times 10^{-3}\ \text{m} \\
\Delta_5 &= -0.2400 \times 10^{-3}\ \text{m} \\
\Delta_6 &= -0.8550 \times 10^{-3}\ \text{rad} \\
\Delta_8 &= -2.0950 \times 10^{-3}\ \text{m}
\end{aligned}$$

Reaction Forces. If desired, the support reactions could now be calculated from the appropriate (excluded) equilibrium equations. For example F_3, the reaction moment at A, is given by the third excluded equation

$$\begin{aligned}
F_3 &= \mathfrak{b}[-6L \quad 0 \quad 4L^2 \quad 6L \quad 0 \quad 2L^2 \quad 0 \quad 0 \quad 0]\,\boldsymbol{\Delta} \\
&= -5.74 \times 10^{-3}\ \text{MNm}
\end{aligned}$$

Steps 6 and 7. Calculate the Element Internal Forces

A. *Element* ⓐ

1. *Element Nodal Displacements*

For a simple structure such as this the element nodal displacements may be obtained from the structure displacements simply by inspection, but for a more complex structure, or in a computer program, it would be necessary to use the location vector for the allocation of the $\boldsymbol{\Delta}$ values and then to use the transformation matrix $\mathbf{T}$ to obtain the element nodal displacements in element coordinates. For example, for element ⓐ

$$\boldsymbol{\delta}_{ⓐ} = \mathbf{T}_{ⓐ}\boldsymbol{\Delta}_{ⓐ} = \begin{bmatrix} \lambda & \mu & 0 & 0 & 0 & 0 \\ -\mu & \lambda & 0 & 0 & 0 & 0 \\ 0 & 0 & 1 & 0 & 0 & 0 \\ 0 & 0 & 0 & \lambda & \mu & 0 \\ 0 & 0 & 0 & -\mu & \lambda & 0 \\ 0 & 0 & 0 & 0 & 0 & 1 \end{bmatrix} \begin{Bmatrix} \Delta_1 \\ \Delta_2 \\ \Delta_3 \\ \Delta_4 \\ \Delta_5 \\ \Delta_6 \end{Bmatrix}$$

and since $\lambda = 0$ and $\mu = 1$ the result is

$$\boldsymbol{\delta}_{ⓐ} = \begin{Bmatrix} u_1 \\ v_1 \\ \theta_1 \\ u_2 \\ v_2 \\ \theta_2 \end{Bmatrix}_{ⓐ} = \begin{Bmatrix} \Delta_2 \\ -\Delta_1 \\ \Delta_3 \\ \Delta_5 \\ -\Delta_4 \\ \Delta_6 \end{Bmatrix} = \begin{Bmatrix} 0 \\ 0 \\ 0 \\ -0.24 \\ -0.0916 \\ -0.8550 \end{Bmatrix} \times 10^{-3}$$

2. *Element Internal Forces*

For the beam element the internal forces are axial stress, shear force, and bending moment. Since the first two are constant and the third is linear within the element it is sufficient to calculate these forces only at the nodes. The expression for this, in terms of nodal displacements, is given by (5.3.18). The results are

Axial stress

$$\begin{aligned}
\sigma_{x1} = \sigma_{x2} &= \frac{E}{L}(-u_1 + u_2) \\
&= \frac{200 \times 10^3}{2}(0 - .24 \times 10^{-3}) = -24\ \text{MPa}
\end{aligned}$$

Shear force

$$Q_1 = Q_2 = \mathfrak{b}[0 \quad 12 \quad 6L \quad 0 \quad -12 \quad 6L]\,\boldsymbol{\delta}_{\textcircled{a}}$$
$$= -9.16 \times 10^{-3}\ \text{MN} = -9.16\ \text{kN}$$

Bending moment

$$\begin{Bmatrix} \mathcal{M}_z(0) \\ \mathcal{M}_z(L) \end{Bmatrix} = \mathfrak{b}\begin{bmatrix} 0 & -6L & -4L^2 & 0 & 6L & -2L^2 \\ 0 & 6L & 2L^2 & 0 & -6L & 4L^2 \end{bmatrix}\boldsymbol{\delta}_{\textcircled{a}}$$

which gives

$$\mathcal{M}_z(0) = 5.74\ \text{kNm}$$
$$\mathcal{M}_z(L) = -12.58\ \text{kNm}$$

The distribution of $\mathcal{M}_z(x)$ is plotted in Fig. 5.19*a* to the left of the element.

3. *Element Nodal Forces*

Usually there is no need to calculate the element nodal forces **f** but we will do so in this example to illustrate the distinction between nodal forces F_{y1}, F_{y2} and shear force Q, and between nodal moments M_{z1}, M_{z2} and bending moment $\mathcal{M}_z(x)$. It will also be demonstrated that each element is in equilibrium. The quantity that relates element nodal forces and element nodal displacements is the element stiffness matrix, since $\mathbf{f} = \mathbf{k}^e\,\boldsymbol{\delta}$. Hence from (5.3.15)

$$F_{x1} = \mathfrak{a}(u_1 - u_2) = 100(0 + 0.24 \times 10^{-3})$$
$$= 24\ \text{kN}$$
$$F_{y1} = \mathfrak{b}(12v_1 + 6L\theta_1 - 12v_2 + 6L\theta_2)$$
$$= -9.16\ \text{kN}$$
$$M_{z1} = \mathfrak{b}(6Lv_1 + 4L^2\theta_1 - 6Lv_2 + 2L^2\theta_2)$$
$$= -5.74\ \text{kNm}$$
$$F_{x2} = \mathfrak{a}(-u_1 + u_2) = -24\ \text{kN}$$
$$F_{y2} = \mathfrak{b}(-12v_1 - 6L\theta_1 + 12v_2 - 6L\theta_2)$$
$$= +9.16\ \text{kN}$$
$$M_{z2} = \mathfrak{b}(6Lv_1 + 2L^2\theta_1 - 6Lv_2 + 4L^2\theta_2)$$
$$= -12.58\ \text{kNm}$$

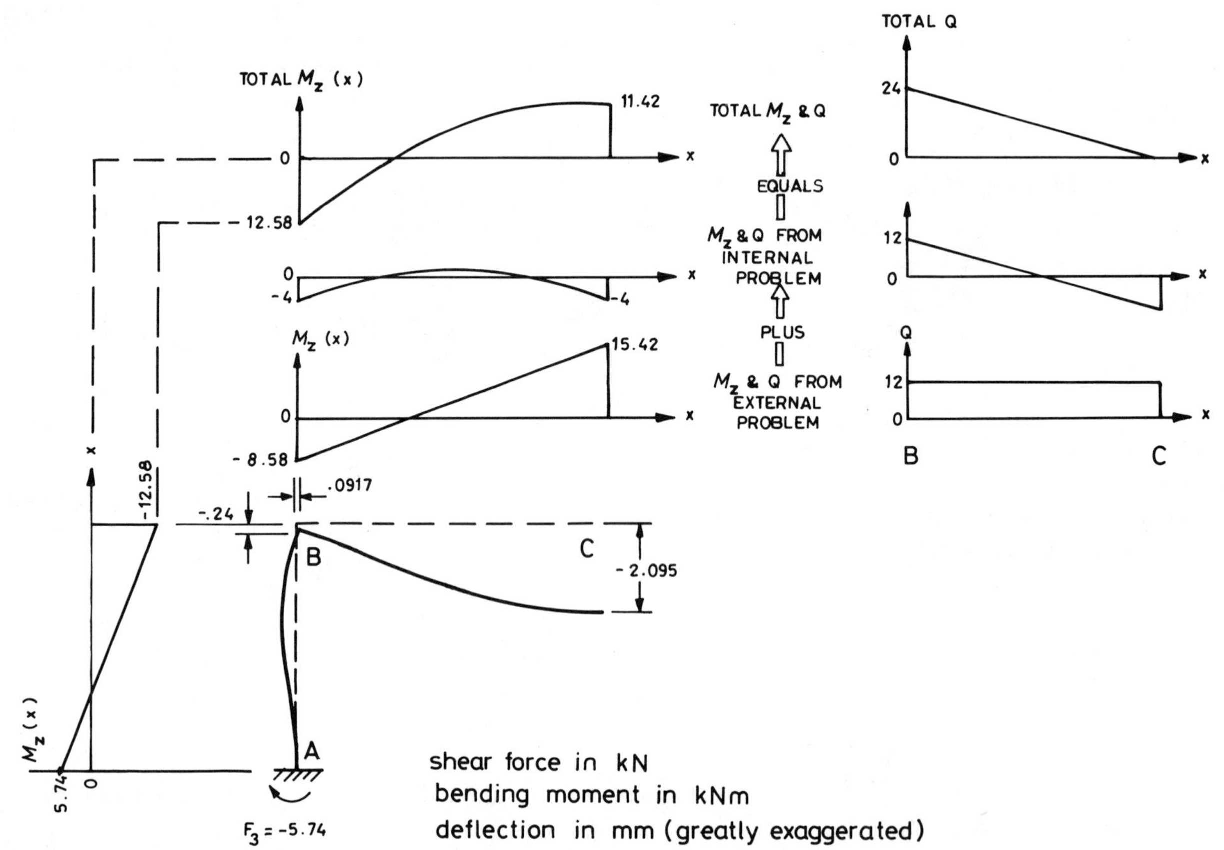

(a)

Figure 5.19*a* Deflection, shear force, and bending moment.

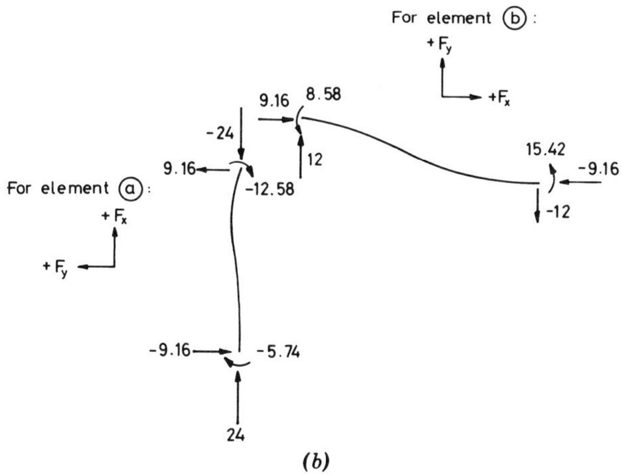

(b)

Figure 5.19*b* Element nodal forces.

These forces and moments are illustrated in Fig. 5.19*b*. Equilibrium of forces is obvious, and equilibrium of moments may be verified by taking moments about point A:

$$\sum M_A = -5.74 - 12.58 + 9.16L = 0$$

B. *Element* ⓑ

1. *Element Nodal Displacements*

In a similar fashion the nodal displacements of element ⓑ are

$$\boldsymbol{\delta}_{ⓑ} = \begin{Bmatrix} u_1 \\ v_1 \\ \theta_1 \\ u_2 \\ v_2 \\ \theta_2 \end{Bmatrix}_{ⓑ} = \begin{Bmatrix} \Delta_4 \\ \Delta_5 \\ \Delta_6 \\ \Delta_7 \\ \Delta_8 \\ \Delta_9 \end{Bmatrix} = \begin{Bmatrix} 0.0916 \\ -0.2400 \\ -0.8550 \\ 0 \\ -2.0950 \\ 0 \end{Bmatrix} \times 10^{-3}$$

2. *Element Internal Forces*

From (5.3.18)

Axial Stress

$$\sigma_{x1} = \sigma_{x2} = \frac{E}{L}(-u_1 + u_2)$$

$$= \frac{200 \times 10^3}{2}(-0.0916 \times 10^{-3} + 0)$$

$$= -9.16 \text{ MPa}$$

Shear force

$$Q_1 = Q_2 = \mathfrak{b}[0 \quad 12 \quad 6L \quad 0 \quad -12 \quad 6L]\, \boldsymbol{\delta}_{ⓑ}$$

$$= 12 \times 10^{-3} \text{ MN} = 12 \text{ kN}$$

Bending moment

$$\begin{Bmatrix} \mathcal{M}_z(0) \\ \mathcal{M}_z(L) \end{Bmatrix} = \mathfrak{b}\begin{bmatrix} 0 & -6L & -4L^2 & 0 & 6L & -2L^2 \\ 0 & 6L & 2L^2 & 0 & -6L & 4L^2 \end{bmatrix} \boldsymbol{\delta}_{ⓑ}$$

$$= \begin{Bmatrix} -8.58 \\ 15.42 \end{Bmatrix} \text{kNm}$$

The distributions of Q and of $\mathcal{M}_z(x)$ are plotted in Fig. 5.19*a*, immediately above the element. Since this element has a distributed load, these two internal forces must be corrected by superimposing the corresponding forces from the solution of the internal problem. In the present example the internal problem is simply a uniformly loaded, clamped beam. The corresponding shear force and bending moment distributions are shown just above those obtained from the frame analysis. The final and complete distributions, which are the sum of these two, are drawn at the top of the figure. Note that the total value of $\mathcal{M}_z$ at point B, -12.58, agrees with the value which was calculated in element ⓐ. This is another useful check on the solution.

3. *Element Nodal Forces*

Here again, although not necessary for the solution, we shall calculate the element nodal forces in order to point out some of the relationships and differences between these and other quantities. As before, $\mathbf{f}_{ⓑ} = \mathbf{k}^e\, \boldsymbol{\delta}_{ⓑ}$ with $\mathbf{k}^e$ given by (5.3.15). Substitution of $\boldsymbol{\delta}_{ⓑ}$ gives

$$F_{x1} = -F_{x2} = 9.16$$

$$F_{y1} = -F_{y2} = 12$$

$$M_{z1} = 8.58; \qquad M_{z2} = 15.42$$

These forces and moments are illustrated in Fig. 5.19*b*. Once again, equilibrium of forces is immediately evident and equilibrium of moments can be verified by taking moments about point B:

$$\sum M_B = 8.58 + 15.42 - 12L = 0.$$

This summation of moments illustrates the fact that frame analysis takes no account of any eccentricity of axial loads arising from the beam's deflection. This procedure is valid as long as the deflections are small, which they are in this example because the largest deflection is 2.095 mm, which is only 0.1% of the beam length L. The small-deflection assumption is usually valid for the elastic analysis of ships and other plated steel frame structures, because the overall stiffness of such structures is usually sufficiently large such that any load which does not cause yielding produces only small deflections. However, the stiffness of a structure depends on the material stiffness, on the member proportions and scantlings (thickness, etc.) and on the arrangement of the members. Hence when using frame analysis (or any other elastic, small-deflection technique) a check should always be made on the size of the deflections relative to the dimensions of the structure.

Another observation which may be made in the foregoing example is that at point B the net vertical force of -12 kN is equal to the applied load at that point, which is the equivalent load, $F_4 = -wL/2$. Similarly, the net moment of -4 kNm is equal to the applied moment, $F_6 = -wL^2/12$. However, the net moment at point C (15.42 kNm) is not equal to the equivalent moment, $F_9 = wL^2/12$, because the latter was never applied to the structure; it was excluded by the zero-slope boundary condition at point C.

SOLUTION USING FLEXURE-ONLY BEAM ELEMENT

In the foregoing example the ratio of $\mathfrak{a}$ and $\mathfrak{b}$ is 100. In more slender frames and particularly in plated frameworks such as occur in ships the ratio is much larger still, often being 500 or more. If such frames are completely orthogonal, such that axial loads and displacements are independent (or nearly independent) of lateral loads and displacements, and if axial forces and displacements are not of primary interest, then the axial degree of freedom can be ignored and the flexure-only beam element may be used. The use of this element is equivalent to regarding the ratio $\mathfrak{a}/\mathfrak{b}$ as being infinity, such that there is no shortening or elongation of any beam. If several of these elements are coaxial and if any one of the nodes is externally restrained against axial displacement, this restraint will extend to all nodes on that axis. Moreover, if another beam which is similarly restrained is attached to one of these nodes and is orthogonal to it, then the orthogonal deflection of that node is also restrained. Thus in an orthogonal framework there may be many nodes in which one or more translational degrees of freedom are restrained; the number of such restraints depends on the arrangement of the elements and on the extent of external restraint which is imposed on the structure. All of these restrained degrees of freedom can be eliminated, and this can give a significant reduction in the size of the overall problem. In an *ad hoc* solution formulated for a specific structure (as opposed to using a general-purpose frame analysis program) these restrained degrees of freedom can be omitted from the structural model right from the beginning, and this is particularly helpful in such solutions, especially if a quick and simple estimate is being sought.

For example, in the above portal frame the vertical restraint at A, in conjunction with nonshortening of element ⓐ, causes point B to also have vertical restraint. Likewise, the horizontal restraint at C is transmitted to point B by element ⓑ. Hence at point B there is only one degree of freedom: rotation. The structural model is therefore reduced to the extremely simple, two-degree-of-freedom problem shown in Fig. 5.20. The solution is

$$\Delta_1 = -0.8 \times 10^{-3} \text{ rad}$$

$$\Delta_2 = -1.8 \times 10^{-3} \text{ m}$$

These agree reasonably well with the displacements from the first solution ($-.855 \times 10^{-3}$ and -2.095×10^{-3}). The difference in the values of Δ_2 is almost entirely accounted for by the axial shortening of element ⓐ (-0.24×10^{-3}). In frame structures the response that is usually of greatest interest is bending moment. For the flexure-only element this is given by (5.3.12) and for the given structure the results are:

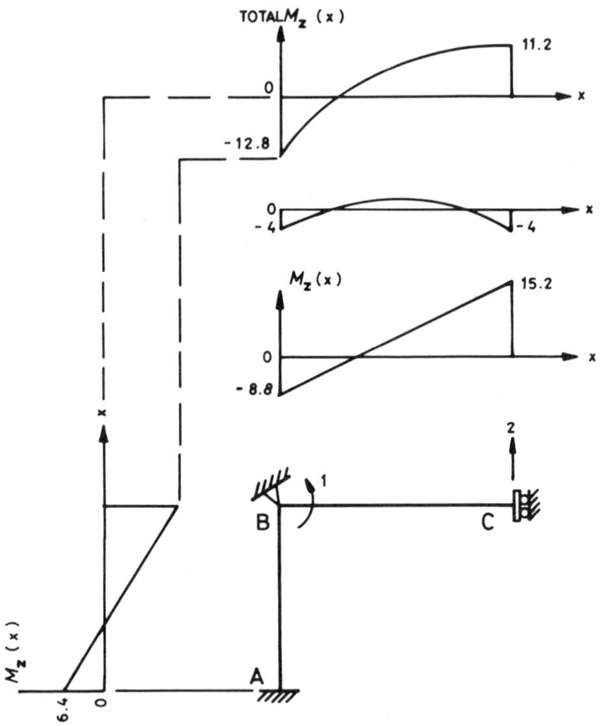

Figure 5.20 Solution using flexure-only beam element.

Element ⓐ

$$\begin{Bmatrix} \mathcal{M}_z(0) \\ \mathcal{M}_z(L) \end{Bmatrix} = \mathbf{S}\boldsymbol{\delta}_{ⓐ}$$

$$= \mathfrak{b}\begin{bmatrix} -6L & -4L^2 & 6L & -2L^2 \\ 6L & 2L^2 & -6L & 4L^2 \end{bmatrix} \begin{Bmatrix} 0 \\ 0 \\ 0 \\ -0.8 \end{Bmatrix}$$

$$= \begin{Bmatrix} 6.4 \\ -12.8 \end{Bmatrix}$$

Element ⓑ

$$\begin{Bmatrix} \mathcal{M}_z(0) \\ \mathcal{M}_z(L) \end{Bmatrix} = \mathbf{S}\boldsymbol{\delta}_{ⓑ} = \mathbf{S}\begin{Bmatrix} 0 \\ -0.8 \\ -1.8 \\ 0 \end{Bmatrix}$$

$$= \begin{Bmatrix} -8.80 \\ 15.20 \end{Bmatrix}$$

This result for element ⓑ is not complete because this element carries a distributed load and the above bending moment corresponds only to the external problem. The internal problem is the same as before—a uniformly loaded clamped beam. The final solution is the sum of the two and this is shown in Fig. 5.20. The values of bending moment agree very well with those of the earlier, more exact solution, particularly in those locations where the bending is large, as can be seen from the following comparison:

Location	Ordinary Beam Element	Flexure-Only Beam Element	% Difference
A	5.74	6.4	11.5
B	12.58	12.8	1.8
C	11.42	11.2	1.9

In this example a relatively low value of 100 was chosen for the ratio $\mathfrak{a}/\mathfrak{b}$. With larger values the accuracy of the flexure-only beam element would be still greater.

5.4 APPLICATION TO GRILLAGES AND LATERALLY LOADED PANELS

A grillage is a plane structure of intersecting beams, with or without plating, which carries a lateral load through the action of beam bending. Most grillages are orthogonal; that is, they are made up of two sets of parallel beams which intersect at right angles. Throughout this text the term "grillage" implies a flat grillage. It also implies that in-plane loads are sufficiently small such that all in-plane effects (both load and response) are completely independent of the lateral loads and the grillage bending response. If this condition is not satisfied then the structure will be referred to as a cross-stiffened panel. The response under in-plane loads is dealt with in Chapters 13 and 14. Some examples of grillages are platform decks and decks near the ship's neutral axis, transverse bulkheads, ramps, and hatch covers.

A grillage may be either regular or irregular. A regular grillage is rectangular in shape, the beams in each direction are of uniform size and spacing and each beam is prismatic and continuous. For regular grillages several special analysis methods are available which take advantage of the regularity; see, for example, Refs. 1 and 2.

If a grillage cannot be idealized as two orthogonal sets of identical full-length prismatic beams intersecting each other, then it is irregular and it must be modeled as a laterally loaded plane frame made up of many individual flexure-only beam elements which are connected at nodes. Each of these elements may be different and they need not be orthogonal. Thus, because of the great versatility of frame analysis even quite complex grillages are no more than straightforward applications of this method. For this reason, if an efficient frame analysis computer program is readily available it is often easier and more efficient (because of familiarity with the program) to use it for all grillages—regular as well as irregular.

Allocation of Distributed Load

In stiffened panels the plating transmits the lateral load to the beams and in cross-stiffened panels it is necessary to estimate the proportion of the load which is taken by each set of beams. When one set of beams (say the stiffeners) is more closely spaced than the other (the girders) it is reasonable to assume that all of the load is transmitted to the stiffeners. The girders then support the stiffeners, and so the loads on the girders are the reaction forces at each stiffener. For a uniform pressure p and a stiffener spacing b, the distributed load q on each stiffener is $q = pb$.

When the spacings are more nearly equal it is necessary to apportion the load between the two sets of beams. An empirical but sufficiently accurate allocation is given by Faulkner's equation [3]:

$$\left.\begin{aligned} &\text{load on each } \textit{short} \text{ adjacent beam} = pb^2 \\ &\text{load on each } \textit{long} \text{ adjacent beam} = pb^2\left(\alpha - \frac{1}{2}\right) \end{aligned}\right\} \quad (5.4.1)$$

in which $\alpha = a/b$

a = longer dimension of the plating between the beams

b = shorter dimension of the plating between the beams

Strictly speaking, the distribution of these loads along each of the beams would give a triangular and a trapezoidal load curve, respectively, but these are usually simplified to uniform distributions q_a and q_b:

$$\left.\begin{aligned} q_a &= pb\left(1 - \frac{1}{2\alpha}\right) \\ q_b &= \frac{pb}{2} \end{aligned}\right\} \quad (5.4.2)$$

Structural Modeling of a Grillage

The analysis of grillages is simplified by the fact that the load, and hence the displacements and rotations, all occur in a plane normal to the structure. By definition all in-plane effects are absent or are small enough to be calculated separately and superimposed. Therefore the flexure-only beam element is ideal for grillage analysis. If the grillage is orthogonal, and most are, then there are only three degrees of freedom at each node: vertical deflection and two rotations. Also, if the orientation of the element in the structure is properly chosen then the transformation from element to structure coordinates is not required; that is, each element stiffness matrix $\mathbf{k}^e$ can be inserted directly into $\mathbf{K}$, as indicated by the location vector. This is made possible by choosing the orientation of each element (i.e., the direction of the element x-axis) such that the four degrees of freedom of the element act in the same directions as the corresponding degrees of freedom of the structure.

The deck grillage of Fig. 5.21*a* will be used as an example to illustrate the principal steps in this type of application of frame analysis. We will also take advantage of this example to illustrate several useful modeling techniques and approximations:

1. As noted in Section 5.3, the nodes of beam elements are located at the neutral axis of the beam. In the structural model of Fig. 5.21*c* the nonalignment of the element is ignored and they are joined directly together, as if the neutral axes of all elements were in the same plane. This vertical repositioning of members is valid only if there are no (appreciable) axial loads, because otherwise the eccentricity of the connections would give rise to further moments which have not been allowed for. Therefore let us assume that the deck in Fig. 5.21 is a "platform deck" (not longitudinally continuous).

2. Torsional stiffness of the beams is ignored because this type of stiffness is relatively small for thin-wall beams of open cross section. This simplifies the analysis because it reduces the number of degrees of freedom in the structure. Also, the slight error which is incurred lies on the conservative side.

3. Support of the deck beams at the side shell is treated as simple support because the side frames provide relatively little rotational restraint. Again, the

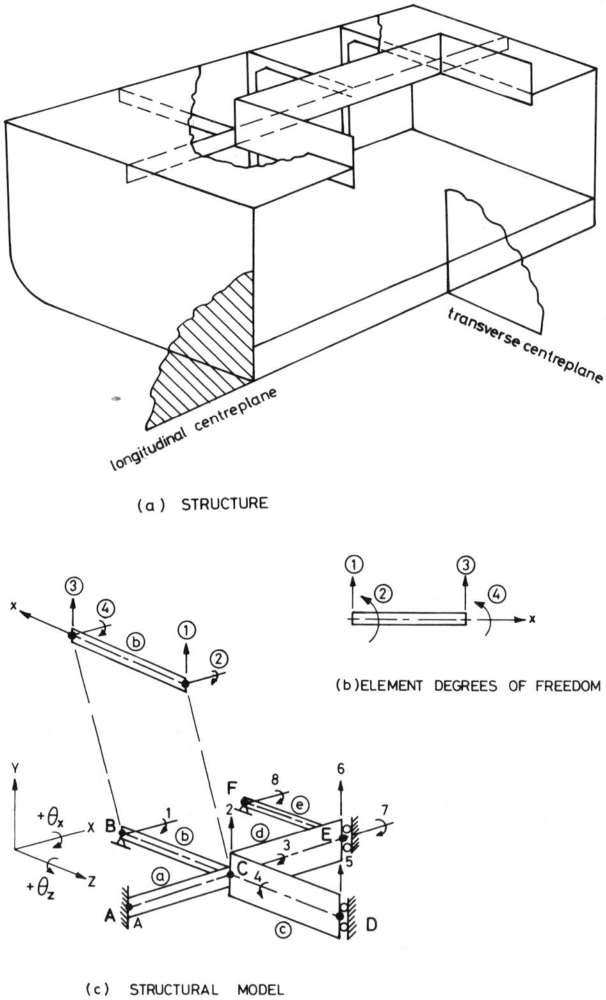

Figure 5.21 Grillage example.

error in this approximation lies on the conservative side. If desired the side frames could be included in the grillage because their vertical orientation does not affect their flexural properties.

4. Support of the longitudinal hatch side girder at the transverse bulkhead is treated as clamped because the hatch side girder is longitudinally continuous through the bulkhead. This is done only for simplicity; if desired the grillage could extend over several cargo holds.

A typical flexure-only beam element is shown in Fig. 5.21*b*. In this example the element degrees of freedom are circled in order to distinguish them from the structure degrees of freedom. The structural model for the grillage is shown in Fig. 5.21*c*, showing the orientation of the various elements in the structure. The orientation of each element is chosen such that the element rotational degrees of freedom have the same positive rotational direction as the rotational degrees of freedom of the structure. Thus, for example, element ⓑ must be inserted into the structure such that its x-axis points in the $-Z$-direction in order that its rotational degrees of freedom (numbers ② and ④) may act in the same direction as the corresponding structure degrees of freedom (numbers 3 and 1) for which the positive direction is θ_X, as shown in the figure.

The elimination of structural degrees of freedom due to boundary conditions (structural supports and symmetry) is done as part of the setting up of the structural model. The simple support at the side shell eliminates vertical deflection and twist at B and F; the clamped support at the bulkhead eliminates all deflections and rotations at A; transverse symmetry requires zero slope at D and hence the bending rotation is eliminated there, and longitudinal symmetry eliminates bending rotation at E. In addition, since the torsional stiffness is not being modeled, it is permissible to eliminate the torsional degree of freedom at points D and E. This may at first seem to be an artificial strengthening of the structure but this is not so because the absence of any torsional stiffness in the beam elements means that there is no way for this restraint to be transmitted into the structure; it has no effect whatever. In fact, the absence of torsional stiffness in the element makes it mandatory to eliminate the torsional degree of freedom at points D and E, and also at points A, B, and F, because none of the beams can offer any resistance to a torsional displacement and so the structure stiffness matrix would be singular, having zeros on the main diagonal at these degree of freedom positions.

After these eliminations have been made there are a total of eight structural degrees of freedom and these are numbered as shown in Fig. 5.21*c*. The numbering is chosen with a view to keeping the bandwidth of the equilibrium equations as narrow as possible, but the chosen numbering sequence is not unique; there are several others which would have been equally satisfactory.

The need to place the rotational degrees of freedom of the elements in alignment with those of the structure means that for elements ⓐ and ⓓ the x-axis points in the $+X$-direction, and for elements ⓑ, ⓒ and ⓔ the x-axis points in the $-Z$-direction. Therefore the location vectors for the five elements are

	①	②	③	④
ⓐ	0	0	2	4
ⓑ	2	3	0	1
ⓒ	5	0	2	3
ⓓ	2	4	6	0
ⓔ	6	7	0	8

These vectors indicate precisely where each term of each element stiffness matrix $\mathbf{k}^e$ is to be inserted into $\mathbf{K}$. For brevity we adopt the symbol a_{ij} to represent the term $k_{ij}^{ⓐ}$ in the element stiffness matrix for element ⓐ, and a similar notation for elements ⓑ, ⓒ, and ⓓ. The resulting structure stiffness matrix is

$$
\mathbf{K} = \begin{array}{c} \\ 1 \\ 2 \\ 3 \\ 4 \\ 5 \\ 6 \\ 7 \\ 8 \end{array}
\begin{array}{cccccccc}
1 & 2 & 3 & 4 & 5 & 6 & 7 & 8 \\
\hline
b_{44} & b_{41} & b_{42} & & & & & \\
b_{14} & \begin{array}{c} a_{33}+b_{11} \\ +c_{33}+d_{11} \end{array} & c_{34}+b_{12} & a_{34}+d_{12} & c_{31} & d_{13} & & \\
b_{24} & c_{43}+b_{21} & c_{44}+b_{22} & & c_{41} & & & \\
 & a_{43}+d_{21} & & a_{44}+d_{22} & & d_{23} & & \\
 & c_{13} & c_{14} & & c_{11} & & & \\
 & d_{31} & & d_{32} & & d_{33}+e_{11} & e_{12} & e_{14} \\
 & & & & & e_{21} & e_{22} & e_{24} \\
 & & & & & e_{41} & e_{42} & e_{44}
\end{array}
$$

The next step would be to insert the loads and then solve the equilibrium equations. Thus by means of frame analysis the analysis of a relatively complex grillage structure has been reduced to a symmetric, banded 8×8 system of equations, which is small enough to be solved on a hand-held programmable calculator.

5.5 GENERAL BEAM ELEMENT

Thus far, for simplicity, we have considered beam bending in only one plane and we have ignored torsion and the effect of shear deflection. In this section these additional features of beam bending are discussed and are combined to form a stiffness matrix for a general beam element which is suitable for most three-dimensional frames, especially the plated steel frameworks found in ships.

The beam is assumed to be a straight beam of uniform cross section capable of resisting axial forces, bending moments about the two principal axes in the plane of its cross section, and twisting moments about its centroidal axis. The nodal forces acting on the beam are: axial forces f_1 and f_7; shearing forces f_2, f_3, f_8, and f_9; moments f_5, f_6, f_{11}, and f_{12}; and twisting moments (torques) f_4 and f_{10}. The location and positive direction of these forces are shown in Fig. 5.22. The corresponding displacements $\delta_1, \ldots, \delta_{12}$ will be taken, as before, to be positive in the positive directions of the forces.

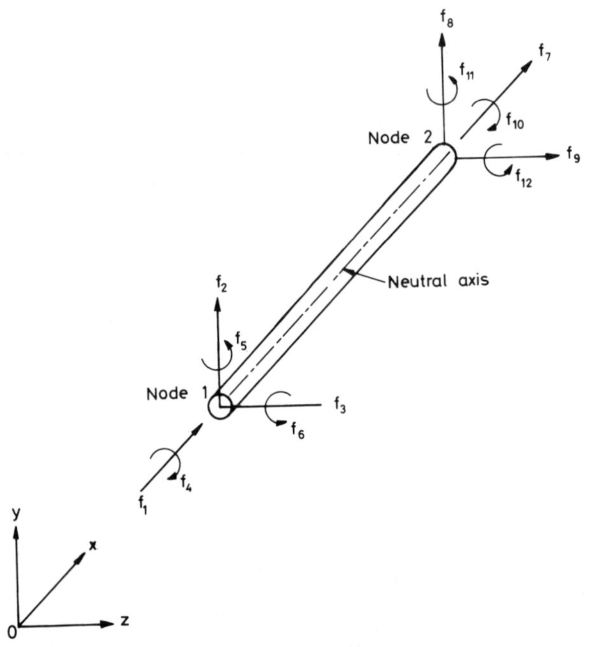

Figure 5.22 General beam element.

As is well known, the axial forces f_1 and f_7 depend only on their corresponding displacements; the same is true of the torques f_4 and f_{10}. However, for arbitrary orientation of the y- and z-axes the bending moments and shearing forces in the xy-plane would depend not only on their corresponding displacements but also on the displacements corresponding to the forces in the xz-plane. Only if the x-axis coincides with the centroidal axis of the beam and the xy- and xz-planes coincide with the principal axes of the cross section can the bending and shear in the two planes be considered independently of each other. Hence we adopt this orientation for the element axes.

Another advantage of using principal axes is that this allows the element stiffness matrix (which is of order 12×12) to be expressed in a more compact and banded form, in terms of four small submatrices each of which refers exclusively to a particular mode of displacement: axial displacement, u; y displacement, v; z displacement, w; and torsional displacement, θ. This bandedness bears no relationship to the speed of solution because the element stiffness matrix is merely a record of the element's properties, and the order of terms within it is arbitrary. However, the compactness makes this form more convenient for derivations and for observing relationships between the various modes of displacement. The compact form is obtained by using a different numbering system for the element degrees of freedom. Instead of numbering all of the degrees of freedom at one node before proceeding to the other node they are divided into the above four groups (u, v, w, and θ displacements) and each group is numbered fully before proceeding to the next group. These alternative degrees of freedom are related to the usual degrees of freedom $\boldsymbol{\delta}$ of Fig. 5.22 as follows:

$$\left.\begin{array}{l} \mathbf{u} = \begin{Bmatrix} u_1 \\ u_2 \end{Bmatrix} = \begin{Bmatrix} \delta_1 \\ \delta_7 \end{Bmatrix} \\[2ex] \mathbf{v} = \begin{Bmatrix} v_1 \\ v_1' \\ v_2 \\ v_2' \end{Bmatrix} = \begin{Bmatrix} \delta_2 \\ \delta_6 \\ \delta_8 \\ \delta_{12} \end{Bmatrix} \qquad \mathbf{w} = \begin{Bmatrix} w_1 \\ w_1' \\ w_2 \\ w_2' \end{Bmatrix} = \begin{Bmatrix} \delta_3 \\ \delta_5 \\ \delta_9 \\ \delta_{11} \end{Bmatrix} \\[2ex] \boldsymbol{\theta} = \begin{Bmatrix} \theta_1 \\ \theta_2 \end{Bmatrix} = \begin{Bmatrix} \delta_4 \\ \delta_{10} \end{Bmatrix} \end{array}\right\} \quad (5.5.1)$$

Shear deflection occurs in the v and w directions, and since these two are quite similar we shall concentrate on the former and derive the generalized element stiffness (sub)matrix $\mathbf{k}_v$ for the v and v' degrees of freedom (i.e., for deflection in the y-direction and bending about the z-direction) taking into account the effect of shear deflection in the y-direction.

Effect of Shear Deflection

If a beam has a large value of bending stiffness parameter b ($b = EI/L^3$) because of large flanges or because it is short or because it has a deep web (say one-fifth of the length or more) then the bending deflection is small and the vertical shear deflection of the web (which is ignored in elementary beam theory) can be of the same order as the bending deflection, such that it cannot be ignored. We therefore seek to derive an expression for the beam's stiffness in regard to vertical (i.e., y-directed) displacement which will account for this effect. To do so we will use the same technique as for the bar element in Section 5.1; that is, impose each of the nodal displacements in turn while all of the others are held at zero and determine all of the nodal forces which are set up. Each of the nodal forces is then divided by the single nodal displacement and the quotients thus obtained are the terms in one complete column of the element stiffness matrix—the column corresponding to that displacement.

VERTICAL DISPLACEMENT AND ROTATION (v AND v')

The vertical deflection v of a beam subjected to end (nodal) forces and moments, as shown in Fig. 5.23*a*, is given by

$$v = v_b + v_s$$

where v_b is the vertical deflection due to bending strains and v_s is the additional deflection due to shearing strains, such that

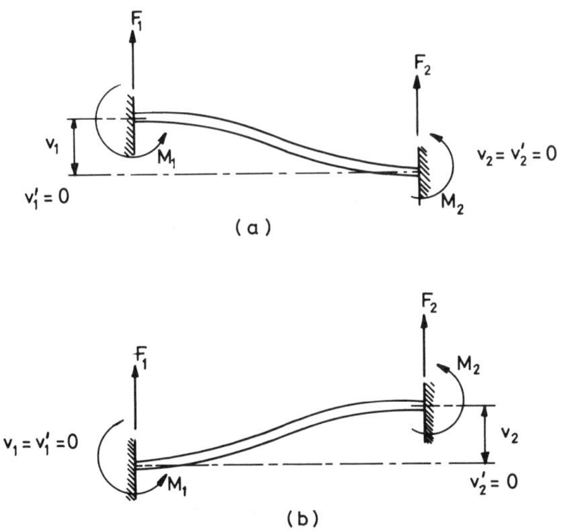

Figure 5.23 Nodal displacements imposed individually.

$$\frac{dv_s}{dx} = \frac{-F_1}{GA_s} \tag{5.5.2}$$

with A_s representing the beam cross-sectional area effective in shear. The bending deflection for the beam shown in Fig. 5.23*a* is governed by the differential equation

$$EI_z\frac{d^2v_b}{dx^2} = F_1x - M_1 \tag{5.5.3}$$

From integration of (5.5.2) and (5.5.3) it follows that

$$EI_zv = \frac{F_1x^3}{6} - \frac{M_1x^2}{2} + \left(C_1 - \frac{F_1EI_z}{GA_s}\right)x + C_2 \tag{5.5.4}$$

where C_1 and C_2 are the constants of integration. The component of slope due to shear, dv_s/dx, is not a rotation in the ordinary sense and is totally unrelated to the nodal moments. Therefore the second or rotational degree of freedom is the slope due to bending only: $v' = dv_b/dx$. Hence the boundary condition for a clamped end is $dv_b/dx = 0$, and the total slope at this point is nonzero and is equal to dv_s/dx.

Therefore the boundary conditions in Fig. 5.23*a* are:

$$\frac{dv}{dx} = \frac{dv_s}{dx} = \frac{-F_1}{GA_s} \qquad \text{at } x = 0,\ x = L$$

and

$$v = 0 \qquad \text{at } x = L$$

Equation 5.5.4 becomes

$$EI_zv = \frac{F_1x^3}{6} - \frac{M_1x^2}{2} - \frac{F_1\Phi xL^2}{12} + (1 + \Phi)\frac{L^3F_1}{12} \tag{5.5.5}$$

where

$$\Phi = \frac{12EI_z}{GA_sL^2}$$

and the forces are related by

$$M_1 = \frac{F_1L}{2}$$

The remaining forces acting on the beam can be determined from the equations of equilibrium:

$$\left.\begin{aligned} F_2 &= -F_1 \\ \text{and} \qquad M_2 &= -M_1 + F_1 L \end{aligned}\right\} \tag{5.5.6}$$

Now at $x = 0$, $v = v_1$ and hence from (5.5.5)

$$v_1 = (1 + \Phi)\frac{F_1 L^3}{12EI_z}$$

Using the foregoing equations the following results are obtained

$$k_{11} = \frac{F_1}{v_1} = \frac{12EI_z}{(1 + \Phi)L^3}$$

$$k_{21} = \frac{M_1}{v_1} = \frac{F_1 L}{2v_1} = \frac{6EI_z}{(1 + \Phi)L^2}$$

$$k_{31} = \frac{E_2}{v_1} = \frac{-12EI_z}{(1 + \Phi)L^3}$$

$$k_{41} = \frac{M_2}{v_1} = \frac{-M_1 + F_1 L}{v_1} = \frac{6EI_z}{(1 + \Phi)L^2}$$

Similarly, if the left-hand end of the beam is clamped, as shown in Fig. 5.23*b*, then by use of the differential equations for the beam deflections or the condition of symmetry it can be demonstrated that

$$k_{33} = k_{11} = \frac{12EI_z}{(1 + \Phi)L^3}$$

$$k_{43} = -k_{21} = \frac{-6EI_z}{(1 + \Phi)L^2}$$

In order to determine the stiffness coefficients associated with the nodal rotations v_1' and v_2' (which are the derivatives of v_b, evaluated at the nodes) the beam is

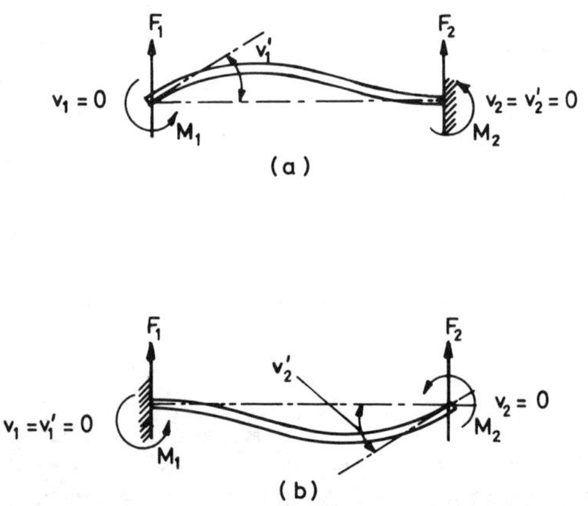

Figure 5.24 Nodal rotations imposed individually.

given a rotation at each end, as shown in Fig. 5.24, and the associated nodal forces and moments are calculated. The force-deflection relationship is again given by (5.5.4), but the constants C_1 and C_2 in these equations must now be evaluated from a different set of boundary conditions. The boundary conditions of Fig. 5.24*a* are

$$v = 0 \qquad \text{at } x = 0,\ x = L$$

$$\text{and} \qquad \frac{dv}{dx} = \frac{dv_s}{dx} = -\frac{F_1}{GA_s} \qquad \text{at } x = L$$

Equation 5.5.4 becomes

$$EI_z v = \frac{F_1}{6}(x^3 - L^2 x) + \frac{M_1}{2}(Lx - x^2)$$

and the forces are related by

$$F_1 = \frac{6M_1}{(4 + \Phi)L} \tag{5.5.7}$$

As before, the remaining forces acting on the beam can be determined from the equations of equilibrium, that is, (5.5.6).

Now at $x = 0$

$$\frac{dv_b}{dx} = \frac{dv}{dx} - \frac{dv_s}{dx} = v_1'$$

so that

$$v_1' = \frac{M_1(1 + \Phi)L}{EI_z(4 + \Phi)} \tag{5.5.8}$$

Hence, from (5.5.6), (5.5.7), and (5.5.8)

$$k_{22} = \frac{M_1}{v_1'} = \frac{(4 + \Phi)EI_z}{(1 + \Phi)L}$$

$$k_{32} = \frac{F_2}{v_1'} = \frac{-F_1}{v_1'} = -\frac{6EI_z}{(1 + \Phi)L^2}$$

$$k_{42} = \frac{M_2}{v_1'} = \frac{-M_1 + F_1 L}{v_1'} = \frac{(2 - \Phi)EI_z}{(1 + \Phi)L}$$

The deflected shape in Fig. 5.24*b* is the same as in Fig. 5.24*a*; the figure is simply rotated 180°. Hence it follows that

$$k_{44} = k_{22} = \frac{(4 + \Phi)EI_z}{(1 + \Phi)L}$$

We now summarize these results for the stiffness submatrix $\mathbf{k}_v$ corresponding to nodal forces and moments in the xy-plane. For this purpose we define a generalized bending stiffness parameter $\bar{\mathfrak{b}}_z$ which includes the effect of shear

$$\left.\begin{aligned} \bar{\mathfrak{b}}_z &= \frac{EI_z}{(1 + \Phi_y)L^3} \\ \text{where} \qquad \Phi_y &= \frac{12EI_z}{GA_{sy}L^2} \end{aligned}\right\} \tag{5.5.9}$$

The subscript z indicates that the bending is about the z-axis and the subscript y indicates that the deflection, including the shear deflection, is in the y-direction. A_{sy} is the proportion of the cross-sectional area over which the y-directed shear force is assumed to act. In terms of these quantities the stiffness submatrix is

$$\mathbf{k}_v = \bar{\mathfrak{b}}_z \begin{bmatrix} 12 & & & \text{symm.} \\ 6L & (4+\Phi_y)L^2 & & \\ -12 & -6L & 12 & \\ 6L & (2-\Phi_y)L^2 & -6L & (4+\Phi_y)L^2 \end{bmatrix} \tag{5.5.10}$$

Horizontal Displacement and Rotation (w and w')

The stiffness coefficients associated with the displacements in the xz-plane, w and w', can be derived directly from foregoing results. To distinguish the forces and moments in the two planes the more complete notation F_{y1}, F_{y2}, M_{z1}, and M_{z2} will be used for the nodal forces just derived, that is, forces in the y-direction and moments about the z-direction. Likewise the symbols F_{z1}, F_{z2}, M_{y1}, and M_{y2} will be used for the nodal forces in the xz-plane. It should be observed, however, that with the sign convention of Fig. 5.22 the directions of the positive nodal moments in the yx- and zx-planes are different. This is illustrated in Fig. 5.25 which shows that the positive direction of the nodal moments M_{y1} and M_{y2} is opposite to that of M_{z1} and M_{z2}. From this it can be shown that the following terms in $\mathbf{k}_w$ are of opposite sign to the corresponding terms in $\mathbf{k}_v$: k_{21}, k_{41}, k_{32}, and k_{43}. All other terms have the same sign as in $\mathbf{k}_v$ and therefore

$$\mathbf{k}_w = \bar{\mathfrak{b}}_y \begin{bmatrix} 12 & & & \text{symm.} \\ -6L & (4+\Phi_z)L^2 & & \\ -12 & 6L & 12 & \\ -6L & (2-\Phi_z)L^2 & 6L & (4+\Phi_z)L^2 \end{bmatrix} \tag{5.5.11}$$

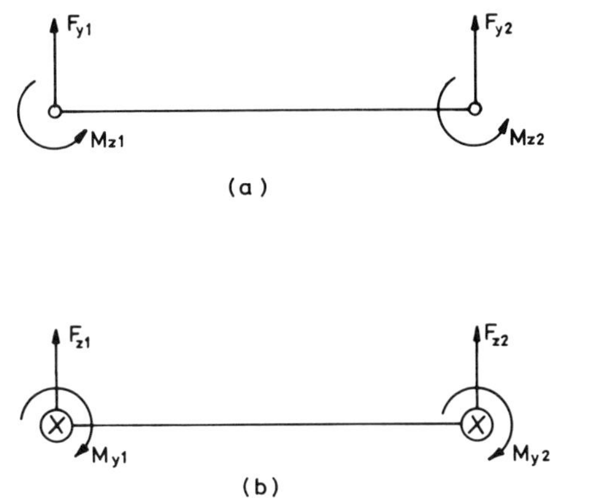

Figure 5.25 Sign convention for nodal forces and moments.

$$\left.\begin{aligned} \text{where} \qquad \bar{\mathfrak{b}}_y &= \frac{EI_y}{(1 + \Phi_z)L^3} \\ \text{and} \qquad \Phi_z &= \frac{12EI_y}{GA_{sy}L^2} \end{aligned}\right. \tag{5.5.12}$$

Torsional and Axial Stiffness

The next chapter deals with the torsional loading and response of the hull girder. It includes a summary of the principal aspects of the torsion of prismatic thin-wall beams, and it presents a special beam element that is suitable for structures in which warping of the cross section of the beam is restrained. However, in a free-standing framework made up of slender beams there is little restraint against warping of the beam cross section. In this case the torsional rigidity of the beam is simply GJ, where the torsional stiffness parameter J depends on the shape of the cross section (see Section 6.1 for methods of calculating J). Hence, for the present purpose of deriving a beam element for frame analysis it is appropriate to adopt this simpler expression. Since warping restraint increases the torsional rigidity of a beam any error incurred is on the conservative side. Also, if the frame structure is such that warping restraint could be significant then the hull girder element of Chapter 6 can be used for the analysis; it is not restricted to hull girder applications.

With no warping restraint the differential equation for the twist θ of a beam is

$$M_x = -GJ\frac{d\theta}{dx}$$

Integration gives

$$M_x x = -GJ\theta + C_1$$

By imposing a twisting moment M_{x1} at node 1, giving a rotation there of θ_1, together with the boundary condition $\theta = 0$ at $x = L$, the constant of integration C_1 is found to be

$$C_1 = M_{x1} L$$

Since $\theta = \theta_1$ at $x = 0$, it follows that

$$M_{x1} = \frac{GJ}{L} \theta_1$$

The equilibrium condition for the twisting moment is

$$M_{x2} = -M_{x1}$$

Hence $$k_{11} = \frac{M_{x1}}{\theta_1} = \frac{GJ}{L}$$

and $$k_{21} = \frac{M_{x2}}{\theta_1} = \frac{-GJ}{L}$$

From symmetry of the beam element, $k_{22} = k_{11}$. Therefore the stiffness submatrix for the two torsional degrees of freedom θ_1 and θ_2 is

$$\mathbf{k}_\theta = \frac{GJ}{L}\begin{bmatrix} 1 & -1 \\ -1 & 1 \end{bmatrix} \qquad (5.5.13)$$

Finally, from (5.2.1) the axial stiffness submatrix is

$$\mathbf{k}_u = \alpha\begin{bmatrix} 1 & -1 \\ -1 & 1 \end{bmatrix} \qquad (5.5.14)$$

where $$\alpha = \frac{AE}{L}$$

Stiffness Matrix for the General Beam Element

Having obtained the four submatrices we can now combine them to form the stiffness matrix for the general beam element as follows

$$\mathbf{k}^e = \begin{bmatrix} \mathbf{k}_u & & & \text{symm.} \\ 0 & \mathbf{k}_v & & \\ 0 & 0 & \mathbf{k}_w & \\ 0 & 0 & 0 & \mathbf{k}_\theta \end{bmatrix} \qquad (5.5.15)$$

in which the four submatrices are given by (5.5.14), (5.5.10), (5.5.11), and (5.5.13).

If it is desired to express $\mathbf{k}^e$ using the degrees of freedom of Fig. 5.22, the various terms of the four submatrices would be inserted into $\mathbf{k}^e$ in the positions indicated by (5.5.1). The resulting stiffness matrix is given in (5.5.16), on page 217.

Coordinate Transformation for a Generally Orientated Beam Element

It is necessary to provide for the case in which the beam has a completely general orientation in three dimensions relative to the structure axes X,Y,Z. For the general beam element the displacement vector at each node contains two sets of components—translation and rotation—and the nodal force vector at each node contains both forces and moments. For this general case the transformation between element axes and structure axes is more conveniently expressed if the degrees of freedom are grouped together at each of the two nodes, as in Fig. 5.22. The transformation matrix is

$$\mathbf{T} = \begin{bmatrix} \boldsymbol{\lambda} & & & \\ & \boldsymbol{\lambda} & & \\ & & \boldsymbol{\lambda} & \\ & & & \boldsymbol{\lambda} \end{bmatrix}$$

where $\boldsymbol{\lambda}$ is the 3×3 nodal transformation matrix given in (5.2.7).*

*If the principal moments of inertia for the beam elements are different it is also necessary to rotate about the element x-axis.

REFERENCES

1. J. Clarkson, *The Elastic Analysis of Flat Grillages*, Cambridge University Press, 1965.
2. Pin-Yu Chang, A simple method for elastic analysis of grillages, *J. Ship Res.*, **12** (2), June 1968, pp. 153–159.
3. D. Faulkner, Chapter 22 of *Ship Structural Design Concepts*, J. Harvey Evans, Ed., Cornell Maritime Press, 1975.

$$
\mathbf{k}^e = \begin{bmatrix}
\mathfrak{a} & & & & & & & & & & & \text{symmetric} \\
0 & 12\bar{\mathfrak{b}}_z & & & & & & & & & & \\
0 & 0 & 12\bar{\mathfrak{b}}_y & & & & & & & & & \\
0 & 0 & 0 & \mathfrak{t} & & & & & & & & \\
0 & 0 & -6\bar{\mathfrak{b}}_y L & 0 & (4+\Phi_z)\bar{\mathfrak{b}}_y L^2 & & & & & & & \\
0 & 6\bar{\mathfrak{b}}_z L & 0 & 0 & 0 & (4+\Phi_y)\bar{\mathfrak{b}}_z L^2 & & & & & & \\
-\mathfrak{a} & 0 & 0 & 0 & 0 & 0 & \mathfrak{a} & & & & & \\
0 & -12\bar{\mathfrak{b}}_z & 0 & 0 & 0 & -6\bar{\mathfrak{b}}_z L & 0 & 12\bar{\mathfrak{b}}_z & & & & \\
0 & 0 & -12\bar{\mathfrak{b}}_y & 0 & 6\bar{\mathfrak{b}}_y L & 0 & 0 & 0 & 12\bar{\mathfrak{b}}_y & & & \\
0 & 0 & 0 & -\mathfrak{t} & 0 & 0 & 0 & 0 & 0 & \mathfrak{t} & & \\
0 & 0 & -6\bar{\mathfrak{b}}_y L & 0 & (2-\Phi_z)\bar{\mathfrak{b}}_y L^2 & 0 & 0 & 0 & 6\bar{\mathfrak{b}}_y L & 0 & (4+\Phi_z)\bar{\mathfrak{b}}_y L^2 & \\
0 & 6\bar{\mathfrak{b}}_z L & 0 & 0 & 0 & (2-\Phi_y)\bar{\mathfrak{b}}_z L^2 & 0 & -6\bar{\mathfrak{b}}_z L & 0 & 0 & 0 & (4+\Phi_y)\bar{\mathfrak{b}}_z L^2
\end{bmatrix} \tag{5.5.16}
$$

where $\mathfrak{a} = \dfrac{EA}{L}$; $\quad \bar{\mathfrak{b}}_z = \dfrac{EI_z}{(1+\Phi_y)L^3}$; $\quad \bar{\mathfrak{b}}_y = \dfrac{EI_y}{(1+\Phi_z)L^3}$; $\quad \mathfrak{t} = \dfrac{GJ}{L}$

Chapter

Six

HULL GIRDER RESPONSE ANALYSIS—NONPRISMATIC BEAM

From earlier discussions in Chapter 3 and elsewhere it is clear that for some ship types the ordinary hull girder analysis presented in Chapter 3 is not sufficiently accurate. This analysis treats the hull girder as a single prismatic beam. This is a drastic simplification and if there are large hatches over an extended length or if there is any major discontinuity in the ship's cross section then this assumption is grossly in error. Also, the ordinary hull girder analysis does not investigate torsion and for ships with large deck openings, such as containerships, the provision of adequate torsional strength is important. As we shall see, these ships are necessarily nonprismatic because in order to possess adequate torsional strength they contain closed cellular structure and transverse box girders in between the hatches. Thus there are some short segments of the hull that have a radically different cross section from the other segments.

Ships of this type require a more accurate form of hull girder analysis, capable of dealing with the torsional response of a nonprismatic structure. The hull module analysis is sufficiently accurate and detailed for this, but this analysis requires that the hull module extend over whatever length is needed to establish known boundary conditions. The simplest and most direct way of fulfilling both of these requirements is to replace the hull girder analysis with a full-length hull module analysis, with whatever modeling simplifications may be required. Even with these simplifications this analysis will give response values of sufficient accuracy to establish the boundary conditions for smaller hull modules, which can have the usual modeling accuracy. These are then analyzed by an ordinary hull module analysis, separate from the one which was used to establish the boundary conditions.

If only a portion of the ship length has a reduced torsional rigidity, there is a special form of hull girder analysis which is capable of dealing with the torsional response of a nonprismatic structure, and which can provide suitable boundary conditions for the hull modules, including a hull module that corresponds to the torsionally flexible portion of the ship. In essence this improved hull girder analysis consists of dividing the hull girder into prismatic segments, each corresponding to a portion of the hull girder for which the cross section is constant, however short that may be. In the structural model each segment is represented by a special beam element which includes extra degrees of freedom to allow it to represent the torsional response. The stiffness matrix and other properties of this element are obtained from the theory of torsion of thin-walled beams.

Torsion of noncircular sections causes longitudinal deformation of the cross section, called *warping*. In solid sections this has a negligible effect and the assumption may be made that a plane cross section remains plane during torsion. But in thin-walled beams the warping deformation can induce considerable longitudinal normal stresses, called warping stresses. Although these are self-balancing, they cannot be considered as local stresses. The theory of torsion of thin-walled beams was developed by numerous authors, mainly in Germany. Recent comprehensive texts include Refs. 1–3. This theory is based on three assumptions:

1. Since the beam is everywhere thin-walled, the warping and the shear flow are assumed to be essentially constant through the thickness.
2. During torsion the beam cross section undergoes rotation and longitudinal warping, but its *shape* is assumed to remain unchanged; that is, there is no distortion in the tranverse or cross-sectional plane.
3. The longitudinal warping causes additional (or "secondary") stresses of both types, normal stress and shear stress, but the theory ignores the effect of the latter; that is, it ignores the additional shear deformation caused by the secondary shear stress.

This theory has been utilized for the analysis of containerships by several authors, notably deWilde [4] and Hapel [5]. The basic features of the hull segment method were developed by Haslum and Tonnessen [6] and by Kawai [7]. The former used the solution to the governing differential equation for torsion in order to describe the torsional response within each element, and used a transfer matrix approach to connect the elements together. Kawai used the ordinary finite element approach based on a stiffness matrix, both at the element level and at the structure level.

Although the hull segment method is very useful, it is still a very simplified representation of the hull girder. The basic source of inaccuracy is the lack of compatibility in the warping displacements between two adjacent elements which have different cross sections. This incompatibility is inherent in the method and cannot be removed; it can only be minimized and/or compensated for. Haslum and Tonnessen made an approximate allowance for it by adjusting the magnitude of the warping such that equilibrium between adjacent elements was satisfied. Gunnlaugsson and Pedersen [8] presented a more accurate method in which the warping displacement is coupled to the bending so as to give a least-squares minimization of the discontinuity. Westin [9] adopted the method of Haslum and Tonnessen and added an iteration feature so as to calculate and account for the additional shear deformation caused by the secondary (warping-induced) shear stress, thus eliminating assumption 3 of the basic theory and avoiding this source of inaccuracy. He applied this improved hull segment method to a series of beams of mixed open and closed cross section which, although idealized, were representative of the hull structure of containerships. From these solutions and from finite element solutions by Ulfvarson [10], Westin showed that in such beams the cross section does *not* retain its shape, even when some transverse bulkheads are fitted, and that *this in-plane distortion of the cross section is the major source of inaccuracy in the hull segment method.* Assumption 3 of the theory—neglecting the extra shear deformation caused by warping—also introduces some inaccuracy, but not nearly as much.

The thin-walled torsional theory was derived for prismatic beams, and in such beams the assumption that the cross section does not change its shape is fulfilled satisfactorily. Thus it is not the theory which is at fault but rather its application to beams which are markedly nonprismatic. If each prismatic beam segment in the hull girder acted in isolation its longitudinal warping and other responses would be as predicted by the theory. But when two segments which have totally different warping response in isolation are connected rigidly together, further deformation must occur in order to achieve geometric compatibility. Some of this deformation is longitudinal and some is transverse; the precise distribution depends entirely on the geometry of the two segments. In any case this additional deformation is quite outside the scope of the theory and cannot be predicted by it.

Improved versions of the hull segment method have been developed by Pedersen [11] and by Kawai [12]. Pedersen's approach retains the classical differential equation formulation and introduces specific discontinuity boundary conditions to account for abrupt changes of section. A brief description is given in Section 6.3, after covering the basic theory of torsion. Kawai's method includes the effect of shear deformation as well as warping. Both methods give quite good overall accuracy (certainly as much as could be reasonably expected from beam theory) and are ideally suited for obtaining the boundary conditions for a hull module torsional analysis. However they cannot provide accurate values of stress in the vicinity of the changes of section, which is where the largest values occur. Therefore they are not equivalent to a hull module analysis, but they can be sufficient if the ship is of a standard well-proven geometry and there is no need to calculate actual stress values.

The stress inaccuracy extends some distance in each direction from the junction, and so in a short segment of hull girder which differs markedly from the segments fore and aft of it, the inaccuracy is present throughout the segment. Therefore in using the hull segment method to obtain the boundary conditions for the hull module analysis, it is not sufficient to locate the boundaries of the hull module at transverse bulkheads; they must also be located well away from any junction of two widely different segments. This means they can only be located at transverse bulkheads which have similar prismatic segments on both sides, and both segments must extend for an appreciable distance from the bulkhead (at least half of the ship's beam, for each segment). For most container ships this requirement is only fulfilled near the ends of the vessel and therefore the best procedure is to use a simplified full-length hull module analysis instead of the hull segment method. However, if only a portion of the ship has low torsional rigidity, then if desired a separate hull module can be used for the design of that portion and the boundary conditions for it may be obtained from the hull segment method. The only requirement is that the hull module should be bounded by transverse bulkheads which are well beyond (outside of) the ends of that portion.

The aim of this chapter is threefold: (1) to present the main aspects of the theory of torsion of prismatic thin-walled sections, with an emphasis on warping and its effects; (2) to give a broad outline of the hull segment method; and (3) to show how the concept of warping leads to a computer-oriented method for calculating hull girder shear stress in multicell sections. Section 6.1 presents the basic theory of torsion of thin-walled sections. Mathematical derivations and details are not given since these are available in the references cited. Section 6.2 presents a stiffness matrix developed by Kawai [7] for representing the torsional response. This matrix is based on the same displacement function as the flexure-only beam element of Chapter 5, which simplifies the presentation and reinforces the concepts presented in that chapter. Section 6.3 gives a brief summary of how a sequence of hull segment elements may be assembled together to represent a hull girder. This is the essential feature of the hull segment method, and the details of two alternative versions are given in References [11] and [12]. Section 6.4 shows how the concept of warping and a finite element representation of its leads to a new method for calculating the shear stress due to transverse shear loads (rather than torsion) in multicell sections. This method is more suitable for computer applications than the method presented in Chaper 3.

6.1 REVIEW OF TORSION OF PRISMATIC THIN-WALLED BEAMS

For thin-walled beams the torsional characteristics differ markedly, depending on whether the section is open or closed. Open sections have much less torsional stiffness than closed sections; that is, they undergo more rotation for a given twisting moment. Also, they exhibit much more warping, that is, nonuniform axial deformation u such that an initially plane cross section is no longer plane (see Fig. 6.1). In fact, the low rotational stiffness of open sections is directly linked to their ability to warp, and if warping of any section is prevented or partly restrained then that section becomes effectively stiffer. Therefore, in our discussion of torsion we will deal separately with open sections and with closed sections, and within each of these we will first consider the case of free warping and then the effect of warping restraint.

As noted previously, the theory deals only with prismatic members and it assumes that during the twisting the cross section retains its shape, that is, that it rotates in rigid body fashion about a point, which is referred to as the center of twist, and does not deform in its own plane. Although such deformation does occur due to transverse shear, its magnitude relative to the size of the cross section is very small (in prismatic members). Under these conditions the displacements of any point (y,z) in the cross section are (see Fig. 6.1)

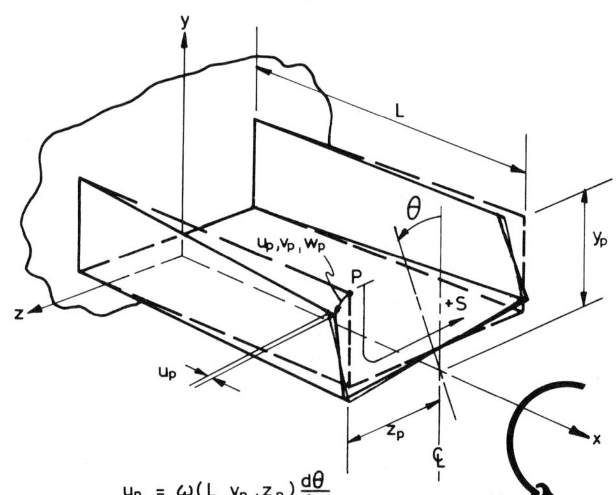

$$u_p = \omega(L, y_p, z_p)\frac{d\theta}{dx}$$
$$v_p = -\theta z_p$$
$$w_p = \theta y_p$$

Figure 6.1 Torsional deflection of a prismatic thin-wall beam of open section.

$$u = \omega_n(y,z)\theta' = \omega_n(s)\theta' \qquad (6.1.1)$$
$$v = -\theta z$$
$$w = \theta y$$

where $\theta' = d\theta/dx$ = rate of twist.

That is, the distribution of the warping around the cross section is given by a (normalized) *warping function* $\omega_n(y,z)$ which does not change in the axial direction x. Also, as will be shown, the magnitude of the warping is directly proportional to the rate of twist. In thin-wall sections the warping is constant through the thickness and hence it can be expressed as a function of arc length s instead of y and z. If the cross section has an axis of symmetry then the center of twist* lies on this axis and the warping is antisymmetric about this axis.

If there is no warping restraint, then torsion of any section—open or closed—is governed by the first-order differential equation

$$M_x = GJ\frac{d\theta}{dx}$$

where M_x = the twisting moment
G = the shear modulus
J = St. Venant's torsional constant

Figure 6.1 shows the sign convention which is used herein. The twisting moment M_x, the twist angle θ, and the arc length s are all positive in the clockwise direction when looking along the $+x$-axis. As noted earlier, open and closed sections differ markedly in regard to both their stress distribution and the degree of warping. We shall now describe each of these briefly, before going into detail.

Figure 6.2 shows the shear stress distribution in a member having an open section. The shear stress varies linearly through the thickness and is zero at the mid-thickness. Such a member, regardless of its shape, has essentially the same stress distribution as a flat slab of width b and thickness t, where b is the total arc length of the open cross section of the member. Thus in reality the member has a very slender "solid" cross section, and the shear stress can "circulate" (thus balancing the applied twisting moment) only within

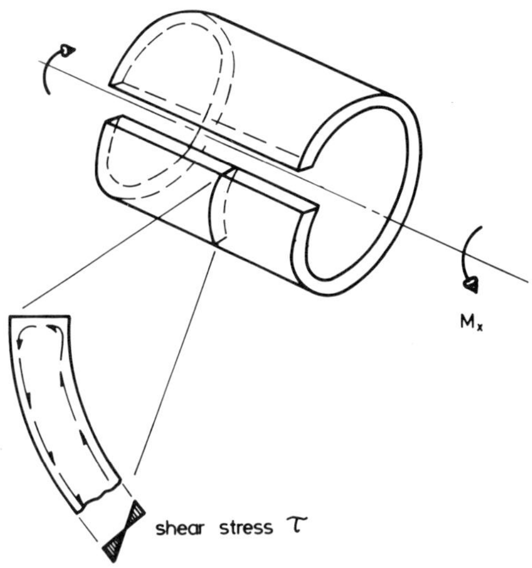

Figure 6.2 Shear stress in open sections.

this confined area. Also, the shear stress is maximum along the outside edges and much lower everywhere inside, and this is a quite inefficient distribution. The main problem is that because of the slenderness of the member the moment of these internal stresses is small, and so equilibrium between this internal moment and the applied twisting moment necessarily involves a large twisting angle and large peak values of shear stress.

In a closed cross section (see Fig. 6.3) the "circulation" of shear stress occurs around the closed path formed by the section, and this has two advantages: the stress is constant through the thickness, thus

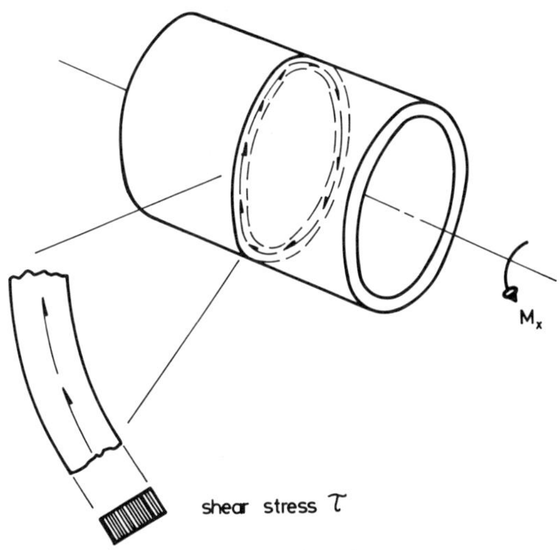

Figure 6.3 Shear stress in closed sections.

*The center of twist of any section—open or closed—is identical to the shear center of that section.

giving a better utilization of material, and it has a much larger moment arm about the center of twist. The result is that a closed section involves much less rotation and shear stress for a given twisting moment. Also, as will be shown subsequently, in a closed section the *shear flow* $q(q = \tau t)$ is constant around the section.

We turn now to the question of warping of the two types of sections. In an open section, warping is not caused by shear. In fact, as we have seen, the mean value of shear stress (averaged through the thickness) is everywhere zero. As a result the warping (as long as it is not restrained) is caused entirely by rigid body rotation of the member wall within its own plane. This rotation is the direct kinematic complement to the axial rotation of the member. Figure 6.4 shows that for a differential member of length dx the tangential displacement of any point around the section is $hd\theta$, where h is the distance from the center of twist to the tangent to the point in question. This distance shall be considered positive when, in conjunction with the positive direction of the arc length coordinate s, it would correspond to a positive twist. The tangential displacement causes a "spiral" angle $\psi = hd\theta/dx = h\theta'$. Since the net shear stress is zero through the wall thickness there is no in-plane shear deformation of the wall. Instead, the wall rotates in rigid body fashion and so the cross section rotates out of its plane by this same angle, ψ, and this causes a local warping displacement $du = -\psi\, ds = -h\theta'\, ds$. Therefore, the warping displacement of any point in the cross section, as a function of arc length s around the section, is proportional to $\omega(s)$ which is defined as

$$\omega(s) = \int_0^s h(s)\, ds \tag{6.1.2}$$

This function is related to the normalized warping function by

$$\omega_n(s) = -\omega(s) + \omega_0 \tag{6.1.3}$$

The quantity ω_0 corresponds to a constant of integration. As will be shown subsequently, it is determined by axial equilibrium requirements.

In a closed section warping is necessarily much smaller because regardless of how it may vary as a function of arc length around the (closed) cross section, it must return to the same value after a complete cycle. Hence for a circular section there is no warping whatever. The underlying reason for the reduction in warping is the presence of shear stress. As we have seen, in a closed section the shear stress τ is nonzero and is essentially constant through the thickness. The shear strain $\gamma = \tau/G$ causes shear deformation of the member walls which, as shown in Fig. 6.5, corresponds to a local rotation of the cross section which is opposite to the rigid body rotation ψ. Hence the net warping of a differential element of arc length ds is

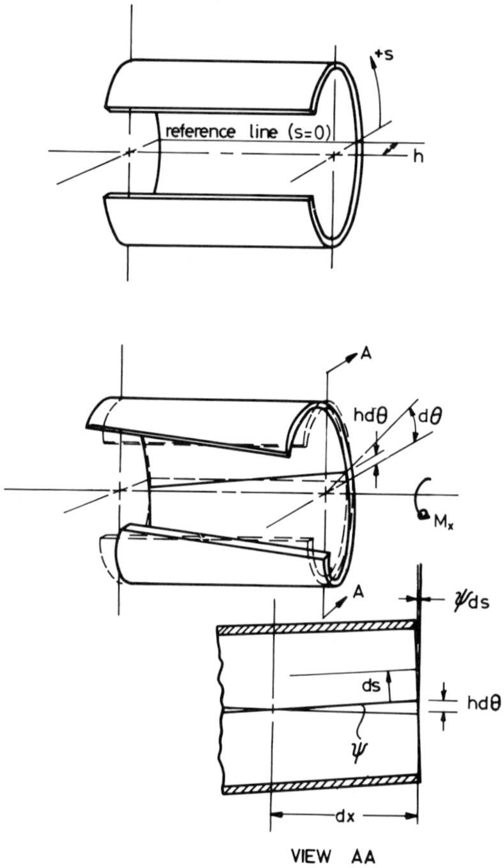

Figure 6.4 Warping in an open section.

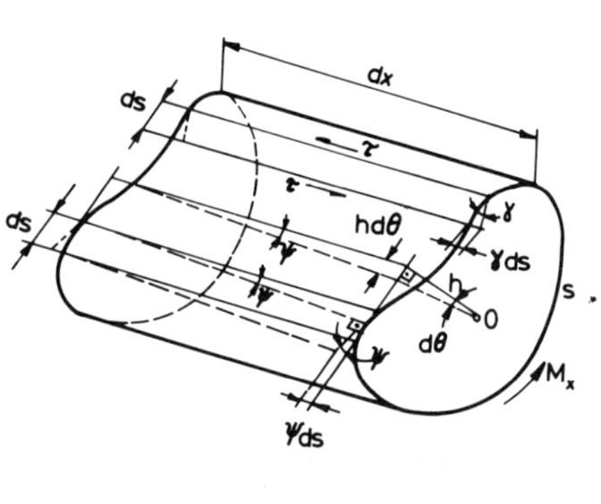

Figure 6.5 Warping in a closed section.

$$du = (\gamma - \psi)\, ds = \left(\frac{\tau}{G} - h\theta'\right) ds \qquad (6.1.4)$$

There is always some cancellation between these two components and hence there is less warping with closed sections.

If warping is restrained then there is some additional resistance to the twisting moment, and it will be shown that for all sections—open or closed—this gives rise to an additional term in the differential equation such that it becomes

$$M_x = GJ\frac{d\theta}{dx} - EI_\omega \frac{d^3\theta}{dx^3}$$

in which I_ω is the warping torsional stiffness of the section, to be defined subsequently.

Open Sections: Free Warping

Shear Stress

As noted earlier, for free warping the relationship between twisting moment and rate of twist is

$$M_x = GJ\theta'$$

and it may be shown (see, for example, Ref. 2) that for open sections St. Venant's torsional constant is given by

$$J = \frac{1}{3}\int_0^b t^3\, ds \qquad (6.1.5a)$$

where b is the total arc length of the cross section. For a section composed of straight portions a more convenient form of equation is

$$J = \frac{1}{3}\sum_{i=1}^{n} t_i^3 b_i \qquad (6.1.5b)$$

where b_i and t_i are the breadth and thickness of the individual portions of the section.

The shear stress at any point is maximum at the outer surfaces and zero at the mid-thickness. The maximum value is given by

$$\tau_{\max} = \frac{M_x t}{J}$$

Warping

As explained earlier, warping in open sections is due entirely to the transverse rotation of the cross section which necessarily accompanies the axial rotation so as to give rigid body rotation because the net shear stress is zero everywhere around the section. The distribution of the warping around the cross section is given by the normalized warping function defined in (6.1.2) and (6.1.3). The value of ω_0 and the location of the shear center are determined from the conditions that the net warping must be zero (since otherwise there would be extension of the member) and that the net first moments of the warping, both horizontal and vertical, about the center of twist must be zero (because to be otherwise would require a horizontal or vertical bending moment acting on the member; see Ref. 2 for details). Stated mathematically these three conditions are

$$\int_0^b \omega_n(s)t\, ds = 0 \qquad (6.1.6a)$$

$$\int_0^b y\omega_n(s)t\, ds = 0 \qquad (6.1.6b)$$

$$\int_0^b z\omega_n(s)t\, ds = 0 \qquad (6.1.6c)$$

In the absence of symmetry the procedure to evaluate ω_0 and to determine the location of the center of twist is to define a set of reference axes ξ, η with the origin at the centroid, as shown in Fig. 6.6*a*, and then calculate the following quantities

$$\left.\begin{aligned} \omega_c(s) &= \int_0^s h_c\, ds \\ I_\xi &= \int_0^b \eta^2 t\, ds \\ I_\eta &= \int_0^b \xi^2 t\, ds \end{aligned}\right\} \qquad (6.1.7)$$

Then the value of ω_0 is given by

$$\omega_0 = \frac{1}{A_s}\int_0^b \omega(s)t\, ds \qquad (6.1.8)$$

in which A_s is the total area of the cross section:

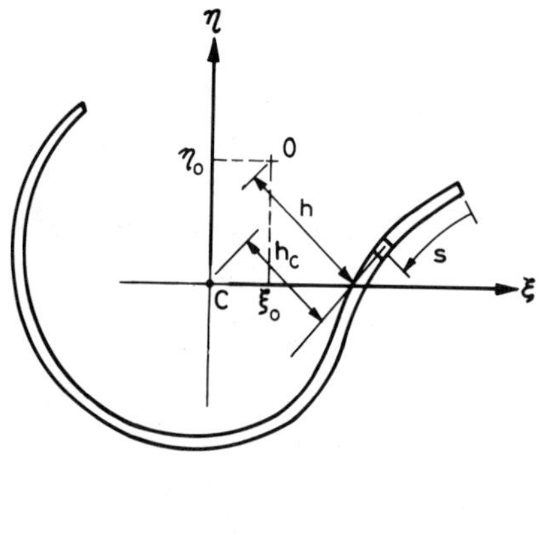

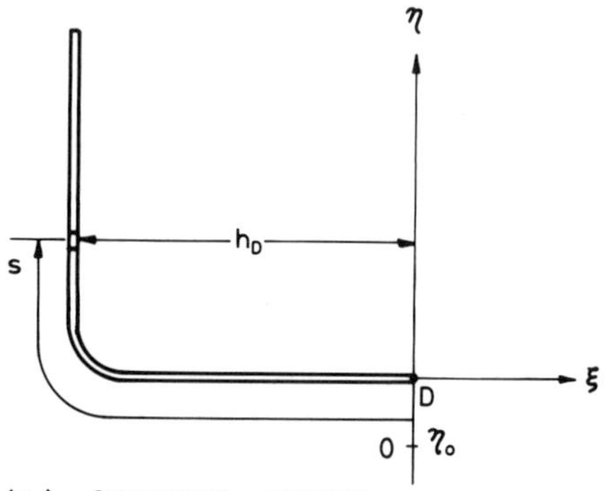

Figure 6.6 Calculation of center of twist.

$$A_s = \int_0^b t\, ds$$

Thus, ω_0 is the mean value of the general warping function $\omega(s)$.

The location of the center of twist is

$$\left.\begin{aligned} \xi_0 &= \frac{1}{I_\xi}\int_0^b \eta\omega_c(s)t\, ds \\ \eta_0 &= \frac{-1}{I_\eta}\int_0^b \xi\omega_c(s)t\, ds \end{aligned}\right\} \qquad (6.1.9)$$

If the section has an axis of symmetry then the center of twist lies on that axis and the foregoing calculations may be simplified. Since ships are usually symmetric about their vertical centerplane, we shall describe the calculations for the case when the y-axis is an axis of symmetry. First, when there is symmetry it is not necessary to place the origin of the ξ, η axes at the centroid. Instead it may be placed at any convenient point on the axis of symmetry, say D. If the ship has a flat bottom it is advantageous to place the origin there because then h_D, the distance to the tangent line of any point, is zero for all points in the bottom. Also, s should be measured from the centerline because this greatly simplifies matters. Note that s is still positive clockwise when facing in the $+x$-direction. With these conventions the resulting equations are

$$\omega_D(s) = \int_0^s h_D\, ds \qquad (6.1.10)$$

$$\eta_0 = \frac{-1}{I_\eta}\int_0^b \xi\omega_D(s)t\, ds \qquad (6.1.11)$$

$$\omega_n(s) = -\omega_D(s) - \eta_0\xi \qquad (6.1.12)$$

Open Sections: Warping Restrained

If a beam is composed of a series of prismatic but differing segments, each segment will have a different warping response and hence will interfere to some extent with the warping of adjacent segments. When warping in a prismatic member is restrained, axial stresses σ_x are set up in the member and, as we shall see, these stresses are not constant in the axial direction. Moreover, since they are not constant they are accompanied by a secondary distribution of shear stress τ_2 over and above the primary distribution due to the (free warping) torsion. Since the warping is constant through the thickness, so also are σ_x and τ_2, and therefore it will be more convenient to deal with shear flow, $q_2 = \tau_2 t$. Note that in this regard the secondary shear stress differs from the primary shear stress τ_1 which (for open sections) varies linearly through the thickness. The total shear stress in the member is the sum of τ_1 and τ_2.

We may obtain expressions for q_2 and σ_x by considering axial equilibrium of a differential element, as shown in Fig. 6.7.

$$\left(\sigma_x t + \frac{\partial\sigma_x t}{\partial x}dx\right) ds - \sigma_x t\, ds + \left(q_2 + \frac{\partial q_2}{\partial s}\right) dx - q_2\, dx = 0$$

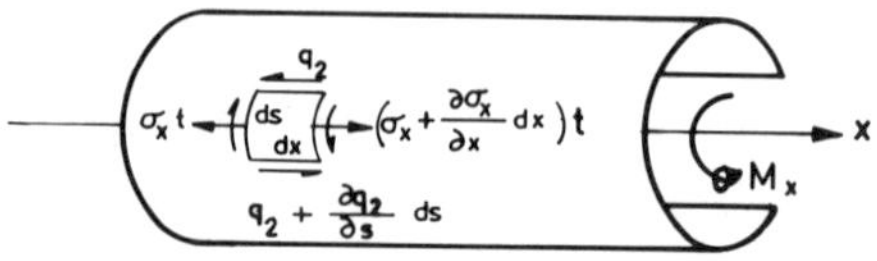

Figure 6.7 Stresses due to warping restraint.

or

$$\frac{\partial q_2}{\partial s} + t\frac{\partial \sigma_x}{\partial x} = 0$$

Therefore

$$q_2(s, x) = -\int_0^s \frac{\partial \sigma_x}{\partial x} t\, ds + (q_2)_0 \qquad (6.1.13)$$

The constant of integration $(q_2)_0$ will be zero providing that $q_2 = 0$ at $s = 0$; this will be satisfied as long as the integration is commenced at one of the free edges of the member. Moreover, it is automatically satisfied if the integration is performed over the entire section, as it will be in some of the derivations that follow.

The value of σ_x is obtained as follows:

$$\sigma_x = E\varepsilon_x = E\frac{\partial u}{\partial x} = E\frac{\partial}{\partial x}[\omega_n(s)\theta']$$

or

$$\sigma_x = E\omega_n \frac{d^2\theta}{dx^2} \qquad (6.1.14)$$

Substituting this into (6.1.13) gives

$$q_2(s, x) = -\int_0^s E\frac{d^3\theta}{dx^3}\omega_n t\, ds$$

$$= -E\frac{d^3\theta}{dx^3}\int_0^s \omega_n t\, ds \qquad (6.1.15)$$

with s measured from one of the free edges of the member. The secondary shear flow q_2 sets up a secondary twisting moment M_{x2} which may be evaluated as follows:

$$M_{x2} = \int_0^b q_2 h\, ds$$

and from (6.1.15) this is

$$M_{x2} = -EI_\omega \frac{d^3\theta}{dx^3} \qquad (6.1.16)$$

where $$I_\omega = \int_0^b \left(\int_0^s \omega_n t\, ds\right) h\, ds$$

The latter may be simplified by rearranging as follows:

$$I_\omega = \int_0^b \omega_n t\, ds \int_0^b h\, ds - \int_0^b \int_0^s (h\, ds)\omega_n t\, ds$$

and observing that the first integral is zero from (6.1.6a). Also, from (6.1.2) and (6.1.3) we see that the inner integral of the second term is

$$\int_0^s h\, ds = \omega_0 - \omega_n$$

Hence the expression becomes

$$I_\omega = \int_0^b (\omega_n - \omega_0)\omega_n t\, ds$$

$$= \int_0^b \omega_n^2 t\, ds - \omega_0 \int_0^b \omega_n t\, ds$$

and again (6.1.6a) means that the second integral is zero. Therefore we finally obtain

$$I_\omega = \int_0^b [\omega_n(s)]^2 t\, ds \qquad (6.1.17)$$

This quantity is referred to as the *warping torsional stiffness*. The product EI_ω which occurs in (6.1.16) is the *warping torsional rigidity* and, as the equation indicates, the additional resistance to twisting which arises when warping is restrained is directly proportional to EI_ω and to the third derivative of the twisting angle. The general equation governing the torsion of open sections is

$$M_x = GJ\frac{d\theta}{dx} - EI_\omega \frac{d^3\theta}{dx^3} \qquad (6.1.18)$$

As we shall now see, this equation is also valid for closed sections, but the expression for J is different.

Closed Sections: Free Warping

Shear Stress

In a closed section the shear stress τ is constant through the thickness. We seek now to determine how τ varies around the section. Figure 6.8b shows an element of varying thickness and from the requirement that the net axial force is zero we have

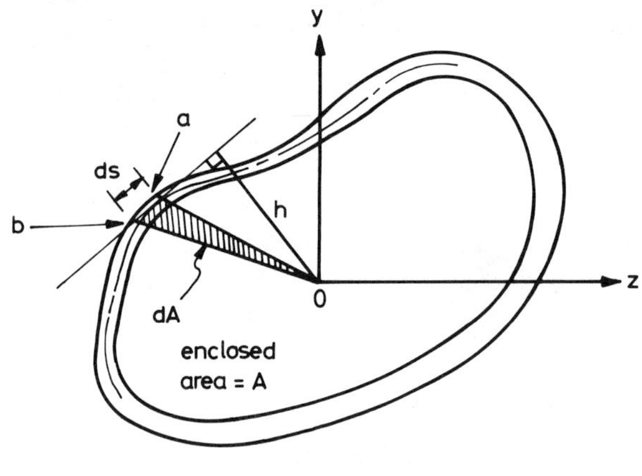

(a) GEOMETRY

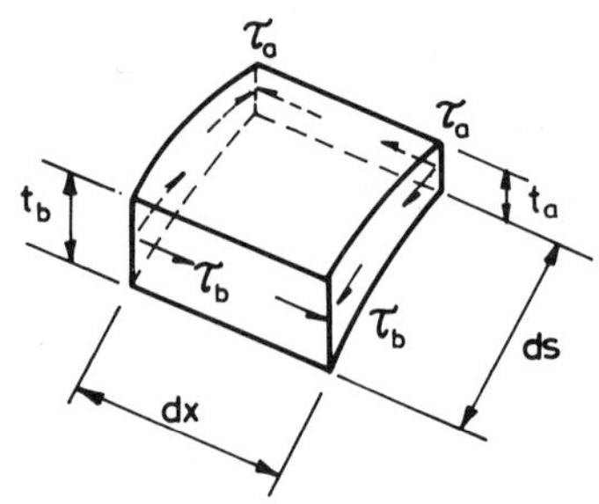

(b) EQUILIBRIUM OF STRESSES

Figure 6.8 Geometry and equilibrium of stresses in a closed section.

$$(\tau_b t_b - \tau_a t_a)\, dx = (q_b - q_a)\, dx = 0$$

and therefore the shear flow q is constant around the section.

Let us now consider overall equilibrium between the applied twisting moment M_x and the moment of the internal shear stress. Figure 6.8*a* shows that the force due to an element ds is $q\, ds$ and the moment it exerts about the center of twist is

$$dM_x = hq\, ds$$

From the geometric properties of a triangle the sector area subtended by ds (shown shaded in Fig. 6.8*a*) is

$$dA = \frac{1}{2} h\, ds$$

and so dM_x can be expressed in terms of this sector area

$$dM_x = 2q\, dA$$

The total moment is obtained by integrating around the full section which gives

$$M_x = 2qA \qquad (6.1.19)$$

where A is the total area enclosed by the perimeter of the closed section. The shear flow around the section is then

$$q = \frac{M_x}{2A} \qquad (6.1.20)$$

For a typical member segment of length dx, we now invoke the principle of strain energy, which states that the work done by M_x equals the internal strain energy:

$$\frac{1}{2} M_x\, d\theta = dx \oint \frac{1}{2} \frac{\tau^2}{G}\, t\, ds$$

and therefore, from (6.1.20)

$$M_x = GJ \frac{d\theta}{dx} = GJ\theta' \qquad (6.1.21)$$

where

$$J = \frac{4A^2}{\oint \frac{ds}{t}} \qquad (6.1.22)$$

This value of J for closed sections differs markedly from the value for open sections given by (6.1.5). For most section shapes it greatly exceeds the latter. For example, in a square box section of side $600t$ (typical of ships) the ratio of the two values is 270,000! Hence in a closed section which has attached open section members, such as longitudinal stiffeners on a hull, the latter's St. Venant torsional stiffness J may be ignored.

Warping

As discussed earlier, and as shown in Fig. 6.5, the warping of a closed section involves the combination and partial cancellation of the shear deformation and the "rotational" or rigid body warping. The net warping distribution is

$$u(s) = \int_0^s \left(\frac{\tau}{G} - h\theta'\right) ds + \omega_0 \theta' \qquad (6.1.23a)$$

By means of (6.1.19), (6.1.20), and (6.1.21) this may be converted to

$$u(s) = [\omega_n(s)]\theta' = \left(\frac{J}{2A}\int_0^s \frac{ds}{t} - \int_0^s h\, ds + \omega_0\right)\theta' \qquad (6.1.23b)$$

Closed Sections: Warping Restrained

Here again, as for open sections, any restraint on the warping will introduce a nonuniform axial stress distribution $\sigma_x(s, x)$ and an accompanying secondary shear flow $q_2(s, x)$. The derivation of q_2 exactly parallels the derivation given for open sections earlier in this Section, with the warping function now given by (6.1.23). The only difference is that the constant of integration $(q_2)_0$ in (6.1.13) can no longer be automatically set to zero because there is now no free edge at which q_2 is known to be zero, and which would provide a convenient starting point for the integration. Thus the equation corresponding to (6.1.15) is

$$q_2(s, x) = -E\frac{d^3\theta}{dx^3}\int_0^s \omega_n t\, ds + (q_2)_0 \qquad (6.1.24)$$

In order to evaluate $(q_2)_0$ we make use of the fact that since the section is closed, $q_2(s)$ must be such that the net total warping displacement which it causes over one full cycle of the section is zero. Therefore we make a cyclic evaluation of the warping caused by q_2, and to do so we divide q_2 into two parts: one part, say q_2^*, which corresponds to the first term of (6.1.24) and for which the value at the starting point of the integration is taken to be zero; and a second part which is the true, but unknown, starting value $(q_2)_0$. The latter is constant and hence the foregoing condition is

$$\frac{1}{G}\oint \frac{q_2^*}{t}\, ds + \frac{(q_2)_0}{G}\oint \frac{ds}{t} = 0$$

from which

$$(q_2)_0 = -\frac{\oint \frac{q_2^*}{t}\, ds}{\oint \frac{ds}{t}} \qquad (6.1.25)$$

After solving for $(q_2)_0$, the additional resisting moment which results from q_2 is found in the same way as for an open section

$$M_{x2} = \oint q_2 h\, ds = -EI_\omega \frac{d^3\theta}{dx^3}$$

where $$I_\omega = \int_0^b \omega_n^2 t\, ds$$

and ω_n is given by (6.1.23*b*). The foregoing equations are the same as (6.1.16) and (6.1.17) and so these equations apply to both open and closed sections, with the warping function being given by (6.1.3) or (6.1.23*b*) respectively. The integration for I_ω may be done piecewise, in any order and in any direction, such that each portion of the section is traversed once only. Therefore (6.1.18) is the governing differential equation for both types of sections, with J being calculated from either (6.1.5) or (6.1.22) as appropriate.

Multiple Cell and Mixed Sections

Multiple Cell: Free Warping

In a section containing n cells, each cell comprises a closed section and so the overall shear flow consists of the superposition of n separate circulating shear flows, one in each cell, each of which is constant around the perimeter of its cell. Along the interface between any two cells the shear flow is the algebraic sum of the two circulating shear flows. From (6.1.19) the torque transmitted by each cell is

$$M_{xi} = 2A_i q_i$$

and the total torque transmitted by the section is the sum of these values

$$M_x = \sum_{i=1}^{n} 2A_i q_i \qquad (6.1.26)$$

Since there are n unknown values of q, the problem is statically indeterminate and we must invoke geometric compatibility. In a closed section the geometric condition is that there must be zero net warping in one complete circuit of each cell. From (6.1.23a) this condition is

$$\frac{1}{G}\oint \tau\, ds - \theta' \oint h\, ds = 0$$

in which ω_0 does not appear because the integral is exactly one cycle. Rearrangement gives

$$\frac{1}{\theta'}\oint \frac{q}{t}\, ds = 2AG$$

As the integration proceeds around the ith cell the shear flow along each wall of the cell will be the value associated with that cell, q_i, plus the flow from the adjacent cell, if any. For example, in Fig. 6.9 there are two walls which adjoin other cells, designated $i - 1$ and $i + 1$. For convenience we regard all of the circulating shear flows as circulating in the positive (i.e., clockwise) direction. If the solution gives a negative value for any particular shear flow this indicates that the shear flow around that cell is anticlockwise. With this convention the contribution of adjacent cells will always be opposite to q_i and must be subtracted. Thus for the ith cell we have the equation

$$\frac{q_i}{\theta'}\oint \frac{ds}{t} - \frac{q_{i-1}}{\theta'}\int_{i-1,i} \frac{ds}{t} - \frac{q_{i+1}}{\theta'}\int_{i+1,i} \frac{ds}{t} = 2A_i G$$

Since there is no transverse deformation all cells undergo the same rate of twist, θ'. Therefore although θ' is an unknown in the foregoing equation, it will have the same value in all such equations and so it can be regarded as a normalizing factor which is applied to all of the unknown values of q. Therefore we define a normalized shear flow

$$\overline{q}_i = \frac{q_i}{\theta'} \tag{6.1.27}$$

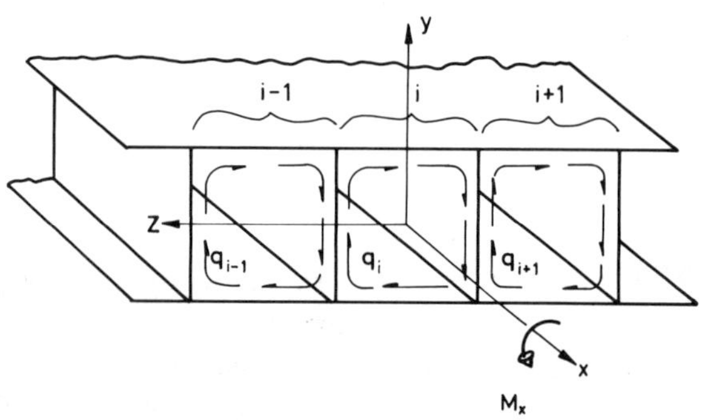

Figure 6.9 Shear flow in a multi-cell section.

Also, we note that the integrals in the equation may all be evaluated from the given geometry and so they constitute known coefficients, which will be represented by the symbol C. The equation then becomes

$$C_i\overline{q}_i - C_{i-1}\overline{q}_{i-1} - C_{i+1}\overline{q}_{i+1} = 2A_iG$$

This equation is written for all cells, resulting in a system of n equations

$$[C]\{\overline{q}\} = 2G\{A\} \tag{6.1.28}$$

which can be solved for $\{\overline{q}\}$. The total torque is, from (6.1.26),

$$M_x = 2\{A\}^T\{q\} = 2\theta'\{A\}^T\{\overline{q}\}$$

and therefore

$$\theta' = \frac{M_x}{2\{A\}^T\{\overline{q}\}}$$

From the latter equation we see that the denominator is the torsional rigidity of the overall section, corresponding to GJ for a single closed section. That is, for a multicell section

$$GJ = 2\{A\}^T\{\overline{q}\} \tag{6.1.29}$$

After solving for θ', the values of q may be calculated from the normalized values.

Multiple Cell: Warping Restraint

The effect of warping restraint is accounted for by the warping torsional stiffness, $I_\omega = \int_0^b \omega_n^2 t\, ds$, and for multiple cells the integration must be done in such a way that no wall is traversed more than once. Some typical examples are illustrated in Fig. 6.10. The figure also shows that if the cells are slender relative to the overall dimensions, such as a double wall side shell, a double bottom, or an isolated "torsion box," then for the calculation of I_ω it is sufficiently accurate to combine the two walls and to lump other areas such as cross webs. Stiffeners and other small members may be ignored.

Mixed Sections

If the overall cross section contains both closed portions and open portions, the total St. Venant stiffness is the sum of J for these two portions, and the total warping stiffness is the sum of I_ω for the two portions.

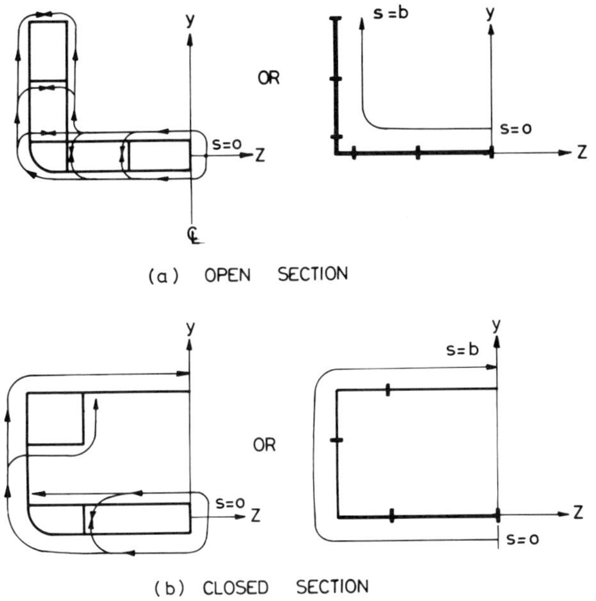

Figure 6.10 Integration paths for calculating I_ω.

As mentioned earlier, the value of J for open sections is negligible compared to that for closed sections, unless the open section is very long or very thick-walled. In fact, in a mixed section the only significant contribution of open section portions is their contribution to the warping torsional stiffness I_ω. This can be estimated with sufficient accuracy by assuming that the warping is constant within that portion and is equal to the value in the closed section at the point of attachment to the latter. With this assumption the area of the open portion may be lumped at the point of attachment, as shown in Fig. 6.10*b*.

Figure 6.11 illustrates some modeling assumptions and idealizations which may be used, depending on whether the overall section is open or closed. Because of symmetry the calculations are usually performed for a half section, as shown in the figure.

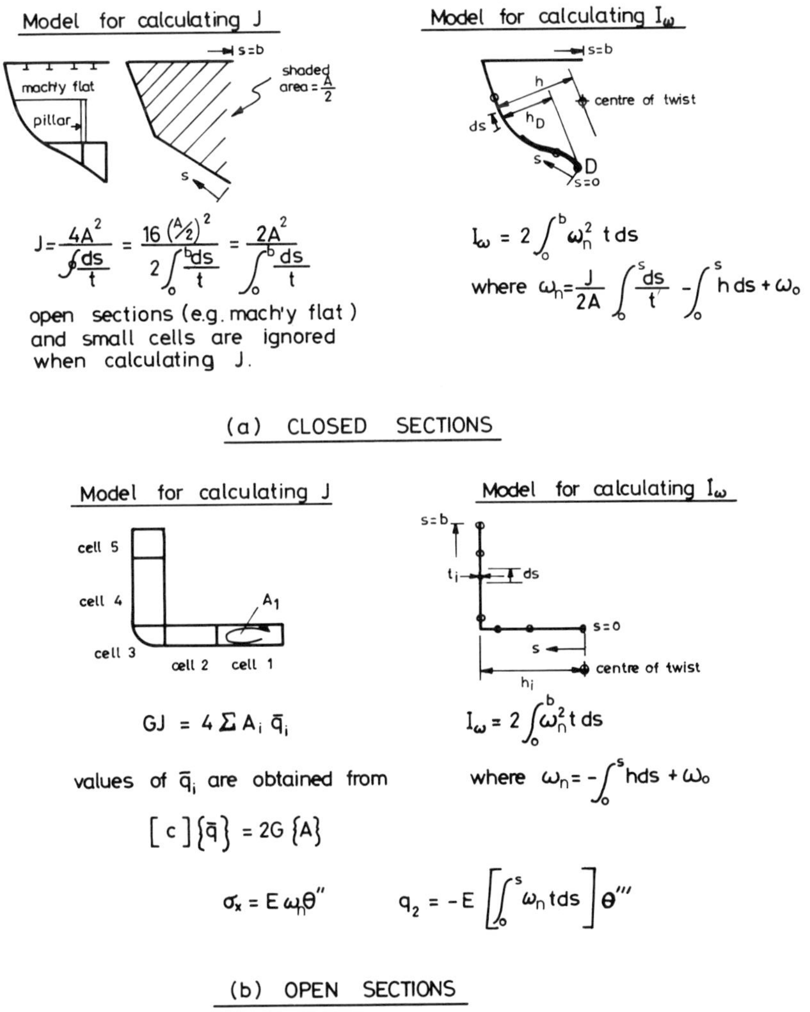

Figure 6.11 Structural models for calculation of torsional stiffnesses.

Numerical Calculation of the Warping Function

For hand calculation $\omega_n(s)$ is calculated from the equations just derived but for automated or computer-based applications a numerical approach is more suitable. As shown by Herrmann [13] the distribution of $\omega_n(s)$ can be represented in a discretized manner by defining "nodes" at which ω_n has a specific value and by assuming a linear variation between nodes. Since the value of ω_0 is not known beforehand, the problem is formulated in terms of the general warping function ω, rather than ω_n. The method produces a system of linear equations for the nodal values of ω. After solving for these the value of ω_0 is calculated by numerical integration (within the same computer program) and added to each value, thus giving ω_n. The warping torsional stiffness I_ω can then be computed, again using numerical integration. In this approach the overall cross section is modeled as a series of straight-line "elements," each of constant thickness and of unit depth in the axial (x) direction. Along the element the warping function varies linearly as shown in Fig. 6.12. The elements can be any length and so it is possible to model quite complicated cross sections with good accuracy. In terms of the nodal values ω_1 and ω_2, the warping within an element is

$$\omega(s) = \omega_1 + (\omega_2 - \omega_1)\frac{s}{b}$$

in which the $+s$-direction for the element is the same as for the overall section. From (6.1.4) the shear strain in the element is

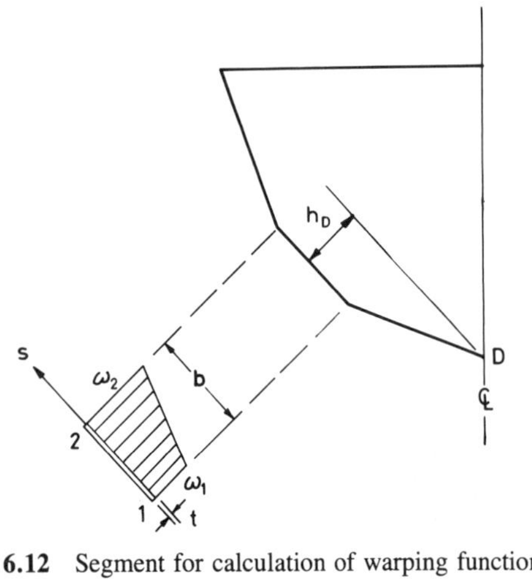

Figure 6.12 Segment for calculation of warping function distribution.

$$\gamma = \frac{\partial u}{\partial s} + h\theta'$$

and from the definition of the warping function we have $u = (-\omega + \omega_0)\theta'$ and therefore

$$\gamma = \left(-\frac{\partial \omega}{\partial s} + h\right)\theta'$$

In terms of strain the potential energy of element e is

$$U_e = \frac{Gt_eL}{2}\int_0^{b_e}[\gamma(s)]^2\,ds$$

where L is the length of the element in the x-direction that is, the length of the overall prismatic member. From the above equation for γ the expression for U_e becomes

$$U_e = \frac{Gt_eL}{2}\int_0^{b_e}\left[-\left(\frac{\partial\omega}{\partial s} + h_e\right)\theta'\right]^2 ds$$
$$= \frac{Gt_eL}{2}(\theta')^2\int_0^{b_e}\left[\frac{\omega_1 - \omega_2}{b_e} + h_e\right]^2 ds$$

The total potential energy of the structure is the sum of that for each of the elements, taken over all of the elements, say N_e in number.

$$U_{\text{total}} = \frac{GL}{2}(\theta')^2\sum_{e=1}^{N_e} t_e\int_0^{b_e}\left[\frac{\omega_1 - \omega_2}{b_e} + h_e\right]^2 ds$$

The overall distribution of ω, as defined by the set of nodal values $(\omega_1, \ldots, \omega_i, \ldots, \omega_N)$ must be such as to minimize the total potential energy.* That is

$$\frac{\partial U_{\text{total}}}{\partial \omega_i} = 0 \qquad i = 1, \ldots, N$$

which yields a system of N equations, one for each node. The differentiation can be done inside the integral and the common terms $(GL/2)(\theta')^2$ can be dropped. The result is

$$\sum_{e=1}^{N_e} t_e\int_0^{b_e} 2\left[\frac{\omega_1 - \omega_2}{b_e} + h_e\right]\left(\frac{\pm 1}{b_e}\right) ds = 0$$

*The work done by the load is independent of ω, and therefore does not need to be included in the total energy.

in which the $\pm$ arises from the differentiation. The plus sign applies if ω_i corresponds to ω_1 of element e, and the minus sign applies if it corresponds to ω_2. Integrating along the element length gives

$$\sum_{e=1}^{N_e} \pm t_e\left(\frac{\omega_1 - \omega_2}{b_e} + h_e\right) = 0 \qquad (6.1.30)$$

There is one such equation for each nodal value ω_i. In each equation the summation need not be performed over all N_e elements but only on the M elements which are attached to node i. Let us use the subscript m to denote each of these attached elements, such that $m = 1, 2, \ldots, M$. Also, for each element, let us denote the node remote from i by the subscript r. Then as the summation process at node i proceeds, taking each of the M elements in turn, the values of ω_1 and ω_2 of each element will either be ω_i and $\omega_{r,m}$ or the reverse, depending on the orientation of element m. If i corresponds to node 1, the term in brackets in (6.1.30) would be $(\omega_i - \omega_{r,m} + h_m b_m)$*. If i corresponds to node 2 the term would be

$$-(\omega_{r,m} - \omega_i + h_m b_m)$$
$$= (\omega_i - \omega_{r,m} - h_m b_m)$$

Hence the system of equation may be written as

$$\sum_{m=1}^{M} t_m(\omega_i - \omega_{r,m} \pm h_m b_m) = 0 \qquad i = 1, \ldots, N \qquad (6.1.31)$$

in which the minus sign applies whenever i corresponds to node 2 of the mth element.

In most cases the location of the center of twist is not known beforehand and so in the foregoing solution the values of h are measured relative to an assumed location D (which we are here presuming to be on the vertical centerline) and the result is ω_D, as in (6.1.10). In this case the normalized warping function ω_n can be calculated directly from (6.1.11) and (6.1.12) using numerical integration; there is no need to calculate ω_0. The warping torsional stiffness I_ω can then be calculated from (6.1.17), again using numerical integration.

*After first multiplying (6.1.30) through by b_e.

6.2 TORSIONAL STIFFNESS MATRIX FOR A THIN-WALLED PRISMATIC BEAM ELEMENT

In the past the torsional analysis of ships has been based on representing the open length of the hull girder (i.e., the portion containing the hatches) as a single prismatic member, thus temporarily neglecting the cross decks and other nonprismatic features, so as to apply the differential equation of (6.1.18) to the ship as a whole. The procedure is to estimate an average or effective value of J and of I_ω, substitute these into the differential equation, impose suitable boundary conditions to account for the other portions of the hull girder, apply the twisting moment M_x, and solve for $\theta(x)$. Some further approximate analysis is then required to allow for the cross decks, and so on. This approach is analogous to the traditional single beam hull girder analysis for calculating the hull girder bending response. However it is not as satisfactory because ships that have large open-section portions are inherently less prismatic and the structure which must be provided to give them sufficient torsional strength makes them still less prismatic. Therefore a more accurate method of analysis is required and this is achieved by representing the ship as a series of "hull segment" elements, each prismatic within its own length and with appropriate features (degrees of freedom, stiffness matrix, etc.) so as to accurately represent the structural characteristics of that particular segment of the hull girder. Most of these characteristics can be adequately represented by the usual beam element properties, but the torsional characteristics require the development of a special stiffness matrix which can account for both warping and warping restraint. The element must satisfy the governing differential equation for torsion, and its stiffness matrix must reflect both St. Venant and warping torsional stiffnesses. Kawai [7] showed that a suitable element may be obtained by using a third-order polynomial to represent the twist $\theta(x)$

$$\theta(x) = \mathbf{H}(x)\mathbf{C}$$

where

$$\mathbf{H}(x) = [1 \quad x \quad x^2 \quad x^3]$$

and

$$\mathbf{C}^T = [C_1 \quad C_2 \quad C_3 \quad C_4]$$

and by using two degrees of freedom θ and θ', at each end of the element. That is, the nodal displacement vector $\boldsymbol{\theta}$ is

$$\boldsymbol{\theta} = \begin{Bmatrix} \theta_1 \\ \theta_1' \\ \theta_2 \\ \theta_2' \end{Bmatrix} \qquad (6.2.1)$$

The length of the element is L and the origin of element coordinates is at the left end. Thus, there is a close parallel between this element and the "flexure-only" beam element of Chapter 5, with θ and θ' in place of v and v'. The nodal load corresponding to θ is, of course, the twisting moment M_x. The nodal load corresponding to θ' will be the load which would cause this type of deflection, that is, a change in the *rate* of twist at a point without affecting the twist angle at that point. As shown in Fig 6.13*a* this load is a pair of transverse twisting moments, referred to as a "bi-moment," M_ω. It is also termed a "warping moment" since it warps the cross section out of its plane. Fig 6.13*b* shows a simplified slab-shaped element subjected to two point loads and Fig. 6.13*c* shows the equivalent nodal loads: twisting moments M_{x1} and M_{x2}, and bi-moments, $M_{\omega 1}$ and $M_{\omega 2}$. The vector of element nodal forces for torsion will be denoted as $\mathbf{f}_\theta$.

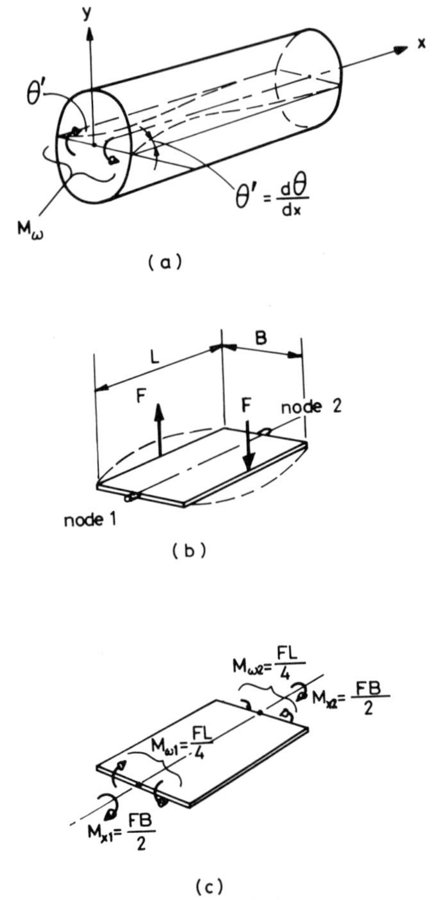

Figure 6.13 Nodal forces for torsion.

$$\mathbf{f}_\theta = \begin{Bmatrix} M_{x1} \\ M_{\omega 1} \\ M_{x2} \\ M_{\omega 2} \end{Bmatrix} \qquad (6.2.2)$$

The next step in the derivation is to express the general twist angle at any location within the element in terms of the nodal values (step II of the procedure of Section 5.3). The result is

$$\theta(x) = \mathbf{H}(x)\mathbf{A}^{-1}\boldsymbol{\theta} \qquad (6.2.3)$$

$$\text{where } \mathbf{A}^{-1} = \begin{bmatrix} 1 & 0 & 0 & 0 \\ 0 & 1 & 0 & 0 \\ -\dfrac{3}{L^2} & -\dfrac{2}{L} & \dfrac{3}{L^2} & -\dfrac{1}{L} \\ \dfrac{2}{L^3} & \dfrac{1}{L^2} & -\dfrac{2}{L^3} & \dfrac{1}{L^2} \end{bmatrix}$$

The torsional differential equation involves the first and third derivatives of θ and from (6.2.3) these are

$$\left.\begin{aligned} &\theta'(x) = \mathbf{B}_1\boldsymbol{\theta} \\ &\text{where } \mathbf{B}_1 = \left[\frac{-6}{L}(\xi - \xi^2) \quad 1 - 4\xi + 3\xi^2 \right. \\ &\qquad\qquad \left. \frac{6}{L}(\xi - \xi^2) \quad -2\xi + 3\xi^2\right] \end{aligned}\right\} \qquad (6.2.4)$$

with $\xi = x/L$, and

$$\left.\begin{aligned} &\theta'''(x) = \mathbf{B}_2\boldsymbol{\theta} \\ &\text{where} \quad \mathbf{B}_2 = \frac{6}{L^2}\left[\frac{2}{L} \quad 1 \quad \frac{-2}{L} \quad 1\right] \end{aligned}\right\} \qquad (6.2.5)$$

Matrices $\mathbf{B}_1$ and $\mathbf{B}_2$ relate the internal deformation to the nodal displacements (step III). The element internal force–deformation relationship (step IV) is the torsional differential equation. For convenience we divide this equation into two parts corresponding to St. Venant stiffness and warping stiffness

$$M_x(x) = M_{x1}(x) + M_{x2}(x)$$

where

$$M_{x1}(x) = GJ\theta'(x) \tag{6.2.6}$$

and

$$M_{x2}(x) = -EI_\omega\theta'''(x)$$

From (6.2.4) and (6.2.5) we can express these internal twisting moments in terms of the nodal displacements

$$M_{x1}(x) = GJ\mathbf{B}_1\boldsymbol{\theta}$$
$$M_{x2}(x) = -EI_\omega\mathbf{B}_2\boldsymbol{\theta}$$

From the definitions of $\mathbf{B}_1$ and $\mathbf{B}_2$ we observe that $M_{x1}(x)$ varies parabolically along the element whereas $M_{x2}(x)$ is constant along the element.

The next step (step V) is to determine the element torsional stiffness matrix $\mathbf{K}_\theta$, relating the nodal forces $\mathbf{f}_\theta$ to the nodal displacements $\boldsymbol{\theta}$.

$$\mathbf{f}_\theta = \mathbf{K}_\theta\boldsymbol{\theta} \tag{6.2.7}$$

As shown in Section 5.3, $\mathbf{K}_\theta$ is obtained by expressing the internal work in terms of the nodal displacements $\boldsymbol{\theta}$ and equating it to the external work done by the nodal forces $\mathbf{f}_\theta$ during a virtual displacement $\boldsymbol{\theta}^*$ of the element. In the resulting equation $\mathbf{K}_\theta$ is the coefficient matrix multiplying $\boldsymbol{\theta}$.

If an arbitrary twist rate θ' is imposed on an element in which the internal twisting moments are $M_{x1}(x)$ and $M_{x2}(x)$ the internal work done is

$$W_{\text{int}} = \int_0^L [\theta' M_{x1}(x) + \theta' M_{x2}(x)]\, dx$$

$$= \int_0^L [\theta' GJ\theta' + \theta'(-EI_\omega)\theta''']\, dx$$

If virtual nodal displacements $\boldsymbol{\theta}^*$ are imposed the corresponding value of θ' is $\theta' = (\mathbf{B}_1\boldsymbol{\theta}^*)^T = \boldsymbol{\theta}^{*T}\mathbf{B}_1^T$ (the transposed form is required for the next step). The actual internal twisting moments can be expressed in terms of the actual displacements $\theta'(x)$ and $\theta'''(x)$ by means of (6.2.6) and so the internal work resulting from these virtual nodal displacements is

$$W_{\text{int}} = \int_0^L [\boldsymbol{\theta}^{*T}\mathbf{B}_1^T GJ\theta' + \boldsymbol{\theta}^{*T}\mathbf{B}_1^T(-EI_\omega)\theta''']\, dx$$

From (6.2.4) and (6.2.5) the general displacements θ' and θ''' can be expressed in terms of the nodal displacements. The result is

$$W_{\text{int}} = \int_0^L [\boldsymbol{\theta}^{*T}\mathbf{B}_1^T GJ\mathbf{B}_1\boldsymbol{\theta} + \boldsymbol{\theta}^{*T}\mathbf{B}_1^T(-EI_\omega)\mathbf{B}_2\boldsymbol{\theta}]\, dx$$

The external work done by the nodal forces is

$$W_{\text{ext}} = \boldsymbol{\theta}^{*T}\mathbf{f}_\theta$$

Equating these and setting the virtual displacement $\boldsymbol{\theta}^*$ to unity gives

$$\mathbf{f}_\theta = \left[\int_0^L \mathbf{B}_1^T GJ\mathbf{B}_1\, dx\right]\boldsymbol{\theta} + \left[\int_0^L \mathbf{B}_1^T(-EI)\mathbf{B}_2\, dx\right]\boldsymbol{\theta}$$

and hence the terms in brackets are the two parts of the torsional stiffness matrix, corresponding to St. Venant torsional stiffness and warping torsional stiffness. Denoting these as $\mathbf{k}_J$ and $\mathbf{k}_\omega$ respectively we have

$$\mathbf{k}_\theta = \mathbf{k}_J + \mathbf{k}_\omega \tag{6.2.8}$$

Substituting for $\mathbf{B}_1$ and $\mathbf{B}_2$ and performing the integrations gives

$$\mathbf{k}_J = \frac{GJ}{L}\begin{bmatrix} \frac{6}{5} & & \text{Symm.} & \\ \frac{L}{10} & \frac{2L^2}{15} & & \\ -\frac{6}{5} & \frac{L}{10} & \frac{6}{5} & \\ \frac{L}{10} & \frac{-L^2}{30} & \frac{-L}{10} & \frac{2L^2}{15} \end{bmatrix} \tag{6.2.9}$$

and

$$\mathbf{k}_\omega = \frac{EI_\omega}{L^3}\begin{bmatrix} 12 & & \text{Symm.} & \\ 6L & 4L^2 & & \\ -12 & -6L & 12 & \\ 6L & 2L^2 & -6L & 4L^2 \end{bmatrix} \tag{6.2.10}$$

Once the nodal displacements $\boldsymbol{\theta}$ have been determined, the shear flow arising from warping restraint, q_2, can be calculated from (6.1.24)

$$q_2(s, x) = -E \int_0^s \omega t \, ds \, \mathbf{B}_2\boldsymbol{\theta} + (q_2)_0$$

$$= -E \int_0^s \omega t \left(\frac{12M_{x1}}{L^3} + \frac{6M_{\omega 1}}{L^2} - \frac{12M_{x2}}{L^3} + \frac{6M_{\omega 2}}{L^2} \right) ds + (q_2)_0$$

For open sections $(q_2)_0$ is zero. Similarly, from (6.1.14) the axial stress arising from the warping restraint is

$$\sigma_x = E\omega\theta''(x)$$

and by differentiating (6.2.4) we obtain an expression for $\theta''(x)$ in terms of the nodal displacements

$$\theta''(x) = \mathbf{B}_3\boldsymbol{\theta}$$

$$\text{where } \mathbf{B}_3 = \left[\frac{-6}{L}(1 - 2\xi) \quad -4 + 6\xi \quad \frac{6}{L}(1 - 2\xi) \quad -2 + 6\xi \right]$$

Hence in the torsion element derived herein the stress distribution in the lengthwise direction is constant for q_2 and linear for σ_x. This is entirely adequate for representing a short prismatic segment of a hull—typically the length of a cargo hatch. Moreover, if the segment is long it can be represented by two or more elements.

6.3 STIFFNESS MATRIX FOR HULL SEGMENT ELEMENT

Having obtained a suitable stiffness matrix for torsion of a thin-wall beam, we now proceed to combine this with the general beam stiffness matrix which accounts for axial and lateral deflection and lateral bending. As the main reference coordinates for the element we choose the centroidal axes (x,y,z) because all of the bending properties of a beam (moment of inertia, bending moment axis, etc.) are defined in terms of these axes. However, the element nodes are not placed at the centroid because in a beam with twisting the rotation θ of the cross section is defined with respect to the axis of twist, and so it is necessary to place the

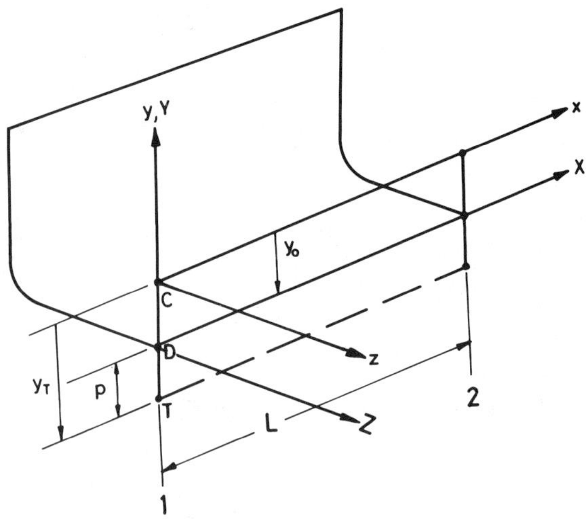

Figure 6.14 Coordinate system for hull girder element.

beam nodes on this axis.* Since the cross section does not deform, the lateral deflection due to bending has the same value over the entire cross section and so placing the node at the center of twist does not affect the v and w bending displacements.

As shown in Fig. 6.14 the center of twist is assumed to be at a distance y_T from the centroid (negative if below). Note that we are taking advantage of port/starboard symmetry because it greatly simplifies the analysis. If required, the equations for the general unsymmetric case may be found in Ref. 7.

To avoid any ambiguity the element nodal displacements in the vertical and the horizontal directions are denoted $\mathbf{v}_T$ and $\mathbf{w}_T$. This is important because in the axial direction the nodal displacement is *not* the displacement of the center of twist. Instead, in order to satisfy axial equilibrium, the axial nodal displacement must be defined as the *average* axial displacement of the cross section, and this is not the same as the axial displacement of the center of twist. The axial nodal displacements are denoted as $\mathbf{u}$.

With this choice of element coordinates and nodal displacements, there is no coupling between the four response modes: axial, vertical bending, horizontal bending, and twist. Therefore the generalized beam stiffness matrix in element coordinates is obtained by

*Note that we are not saying that the element always rotates about its center of twist; the element as a whole, including the axis of twist, can deflect sideways thus altering the effective center of rotation.

simply grouping together the four stiffness matrices as follows

$$\begin{Bmatrix} \mathbf{f}_u \\ \mathbf{f}_v \\ \mathbf{f}_w \\ \mathbf{f}_\theta \end{Bmatrix} = \begin{bmatrix} \mathbf{k}_u & 0 & 0 & 0 \\ 0 & \mathbf{k}_v & 0 & 0 \\ 0 & 0 & \mathbf{k}_w & 0 \\ 0 & 0 & 0 & \mathbf{k}_\theta \end{bmatrix} \begin{Bmatrix} \mathbf{u} \\ \mathbf{v}_T \\ \mathbf{w}_T \\ \boldsymbol{\theta} \end{Bmatrix} \quad \text{or} \quad \mathbf{f} = \mathbf{k}_{hg}\mathbf{d} \tag{6.3.1}$$

in which the individual terms are defined by

$$\begin{Bmatrix} F_{x1} \\ F_{x2} \end{Bmatrix} = \frac{EA_s}{L} \begin{bmatrix} 1 & -1 \\ -1 & 1 \end{bmatrix} \begin{Bmatrix} u_1 \\ u_2 \end{Bmatrix} \quad \text{or} \quad \mathbf{f}_u = \mathbf{k}_u\mathbf{u} \tag{6.3.2}$$

$$\begin{Bmatrix} F_{y1} \\ M_{z1} \\ F_{y2} \\ M_{z2} \end{Bmatrix} = \frac{EI_z}{L^3} \begin{bmatrix} 12 & & & \text{Symm.} \\ 6L & 4L^2 & & \\ -12 & -6L & 12 & \\ 6L & 2L^2 & -6L & 4L^2 \end{bmatrix} \begin{Bmatrix} v_{T1} \\ v'_{T1} \\ v_{T2} \\ v'_{T2} \end{Bmatrix}$$

$$\text{or} \quad \mathbf{f}_v = \mathbf{k}_v\mathbf{v}_T \tag{6.3.3}$$

$$\begin{Bmatrix} F_{z1} \\ M_{y1} \\ F_{z2} \\ M_{y2} \end{Bmatrix} = \frac{EI_y}{L^3} \begin{bmatrix} 12 & & & \text{Symm.} \\ 6L & 4L^2 & & \\ -12 & -6L & 12 & \\ 6L & 2L^2 & -6L & 4L^2 \end{bmatrix} \begin{Bmatrix} w_{T1} \\ w'_{T1} \\ w_{T2} \\ w'_{T2} \end{Bmatrix}$$

$$\text{or} \quad \mathbf{f}_w = \mathbf{k}_w\mathbf{w}_T \tag{6.3.4}$$

$$\begin{Bmatrix} M_{x1} \\ M_{\omega 1} \\ M_{x2} \\ M_{\omega 2} \end{Bmatrix} = \begin{bmatrix} \mathbf{k}_J + \mathbf{k}_\omega \end{bmatrix} \begin{Bmatrix} \theta_1 \\ \theta'_1 \\ \theta_2 \\ \theta'_2 \end{Bmatrix} \quad \text{or} \quad \mathbf{f}_\theta = \mathbf{k}_\theta\boldsymbol{\theta} \tag{6.3.5}$$

and $\mathbf{k}_J$ and $\mathbf{k}_\omega$ are defined in (6.2.9) and (6.2.10).

Since the hull girder is not prismatic the axis of twist of each hull girder element will be at a different location and hence we must define a global coordinate system (X,Y,Z) with global nodes located at some appropriate position y_0 so that adjacent elements can be connected. As shown in Fig. 6.14 the bottom centerline position is a suitable location. Then, in order to assemble the global stiffness matrix we must transform the element stiffness matrix to the global coordinates. The local and global nodes are separated by a vertical distance $p = y_0 - y_T$ and it may easily be shown that the displacements along the axes of the global system are related to the displacements in element coordinates as follows

$$\begin{aligned} U(x) &= u(x) - y_0 v'_T(x) \\ V(x) &= v_T(x) \\ W(x) &= w_T(x) + p\theta(x) \\ V'(x) &= v'_T(x) \\ W'(x) &= w'(x) + p\theta'(x) \end{aligned}$$

The rotational degrees of freedom $\theta(x)$ and $\theta'(x)$ are unaffected by the transformation and so a new symbol for them is not required.

For the transformation of $\mathbf{k}_{hg}$ in element coordinates to $\mathbf{K}_{HG}$ in global coordinates we need to invert the foregoing equations, which gives

$$\begin{aligned} u(x) &= U(x) + y_0 U'(x) \\ v_T(x) &= V(x) \\ w_T(x) &= W(x) - p\theta(x) \\ v'(x) &= V'(x) \\ w'(x) &= W'(x) - p\theta'(x) \end{aligned}$$

Therefore the transformation matrix $\mathbf{T}$ is

$$\begin{Bmatrix} \mathbf{u} \\ \mathbf{v}_T \\ \mathbf{w}_T \\ \boldsymbol{\theta} \end{Bmatrix} = \begin{bmatrix} \mathbf{I} & \mathbf{y}_0 & 0 & 0 \\ 0 & \mathbf{I} & 0 & 0 \\ 0 & 0 & \mathbf{I} & -p\mathbf{I} \\ 0 & 0 & 0 & \mathbf{I} \end{bmatrix} \begin{Bmatrix} \mathbf{U} \\ \mathbf{V} \\ \mathbf{W} \\ \boldsymbol{\theta} \end{Bmatrix} \quad \text{or} \quad \boldsymbol{\delta} = \mathbf{T}\boldsymbol{\Delta} \tag{6.3.6}$$

where

$$\mathbf{y}_0 = \begin{bmatrix} 0 & 0 & y_0 & 0 \\ 0 & 0 & 0 & y_0 \end{bmatrix}$$

Using this transformation matrix the stiffness matrix of the hull girder element expressed in global coordinates is $\mathbf{K}_{HG} = \mathbf{T}^T\mathbf{k}_{hg}\mathbf{T}$. The resulting $\mathbf{K}_{HG}$ and the overall force–deflection relationship are

$$\begin{Bmatrix} \mathbf{F}_u \\ \mathbf{F}_v \\ \mathbf{F}_w \\ \mathbf{F}_\theta \end{Bmatrix} = \begin{bmatrix} \mathbf{K}_u & \mathbf{y}_0\mathbf{K}_u & 0 & 0 \\ \mathbf{y}_0\mathbf{K}_u & \mathbf{y}_0^T\mathbf{K}_u\mathbf{y}_0 + \mathbf{K}_v & 0 & 0 \\ 0 & 0 & \mathbf{K}_w & -p\mathbf{K}_w \\ 0 & 0 & -p\mathbf{K}_w & p^2\mathbf{K}_w + \mathbf{K}_\theta \end{bmatrix} \times \begin{Bmatrix} \mathbf{U} \\ \mathbf{V} \\ \mathbf{W} \\ \boldsymbol{\theta} \end{Bmatrix} \quad \text{or} \quad \mathbf{F} = \mathbf{K}_{HG}\boldsymbol{\Delta} \tag{6.3.7}$$

This relationship reveals that vertical bending of a ship is coupled only with axial displacement and that torsion is coupled only with horizontal bending. Hence if the hull girder element is used in a vertical bending analysis of a nonprismatic hull, each element requires only the usual six degrees of freedom of a standard beam element.

Each prismatic segment of the hull girder with a different cross section has a different warping function. Hence at a node joining two different segments the common value of θ' produces two different and incompatible distributions of warping, implying a discontinuity of longitudinal displacement. In the real structure there is of course no such discontinuity; instead there is a very localized transition from one warping distribution to the other. Hence it is θ' that could better be regarded as discontinuous at the node, rather than the warping. To allow for this, Pedersen [11] introduced a coupling coefficient s_1 relating the two values of θ' at each node; for example at the ith node

$$\theta'_{i+} = s_{1i}\theta'_{i-}$$

where the subscripts + and − refer to the two segments that are connected at the node—one forward of the node and the other aft.

Since horizontal bending is strongly coupled to torsion Pedersen also introduced a discontinuity in the W' rotation, expressed in terms of another coupling coefficient s_{2i}:

$$W'_{i+} = W'_{i-} + s_{2i}\theta'_{i-}$$

By imposing a suitable restriction on the discontinuity in the warping displacement he obtained explicit formulas for s_1 and s_2 in terms of the sectorial coordinate* properties of the two cross sections. It is then possible to assemble the global stiffness matrix and solve for the nodal displacements. However, Pedersen has also shown that a semianalytical solution can be obtained, using the more general Vlasov-Timoshenko beam theory [1] to account for those effects which vary only gradually such as flexure and shear deflection. This approach is well suited for dynamic analysis as well as static. It produces a set of four coupled differential equations of boundary value type. The abrupt changes in cross-sectional properties which occur at the junction of open and closed hull segments and at transverse portions of deck between hatches are introduced as discontinuity conditions on the solution, in addition to the usual boundary conditions. For static loads the system of differential equations is solved numerically by a series of integrations. Natural frequencies may be calculated by an iterative solution of the set of differential equations. Forced vibrations are solved by a modal approach. Details of these solution methods are given in Ref 11.

*Sectorial coordinates are an alternative means of deriving and expressing torsional properties; see for example Refs. 1 and 2.

6.4 PRACTICAL CALCULATION OF HULL GIRDER SHEAR EFFECTS

In Section 3.7 the shear stress due to transverse shear loads was calculated by considering equilibrium. For a multicell section it was necessary to invoke geometric compatibility which requires that the net warping of each closed cell must be zero. This approach requires careful consideration of the connectivity of the section and of the pattern of the shear flow—its direction in the various cell walls and the way in which it divides at branch points. Such a method is not well suited for computer implementation and so in this section we present an alternative method which is formulated not in terms of shear flow but rather in terms of the warping due to transverse shear. This approach is simpler because warping is everywhere single-valued, even at branch points.

We have now seen that when transverse shear is present plane cross sections do not remain plane, as assumed in simple beam theory, but undergo deformation out of this plane. This type of deformation is referred to as warping, and occurs both in torsion and in transverse shear. These two types of warping are distinct, having separate causes. If there is both twisting and transverse shear both types of warping will exist, and if the warping is restrained the two types will interact. For simplicity this section deals only with warping due to transverse shear.

In a prismatic beam which is not subject to twist the shear strain in the cross section is simply the derivative of the warping:

$$\gamma_{xy} = \frac{\partial u}{\partial y}; \qquad \gamma_{xz} = \frac{\partial u}{\partial z}$$

In a thin-walled beam it is more convenient to use the tangential or arc length coordinates within the cross section, for which the relationship is;

$$\gamma = \frac{\partial u}{\partial s} \tag{6.4.1}$$

and there is no need for subscripts. This relationship may also be obtained by setting the twist angle ψ to zero in (6.1.4). If end effects are ignored the warping in a prismatic beam maintains the same cross-sectional pattern along the length of the beam, and the magnitude at any section is proportional to the transverse shear force $Q(x)$ at that section. It is therefore convenient to introduce a warping function $\omega_Q(s)$ which describes the transverse distribution of warping, and to express the warping as a product

$$u(s, x) = \omega_Q(s)Q(x) \tag{6.4.2}$$

Since the shear strain γ (and hence also the shear stress, $\tau = G\gamma$) is simply the derivative of the warping, the distribution of transverse shear stress can be determined by first solving for the warping distribution and then differentiating. This approach was developed by Mason and Herrmann [14] and is similar to the latter's method for calculating the torsional warping function which was presented in Section 6.1. It was first applied to ship hulls by Kawai [7]. The method is essentially an application of the classical Rayleigh-Ritz technique. First the distribution of ω_Q is represented in a discretized manner by defining "nodes" at which ω_Q has a specific value and by assuming a linear variation between nodes. Then the total potential energy of the system Π is expressed in terms of the nodal values $\boldsymbol{\omega}_Q$, and the derivative of Π with respect to each value of $\boldsymbol{\omega}_Q$ is set equal to zero, thus obtaining a system of equations for the $\boldsymbol{\omega}_Q$ values. After solving for these the complete distribution of shear stress can be readily calculated.

For use in conjunction with the hull segment method the derivation is presented here for a typical hull segment of length L. As shown in Fig. 6.15, a hull segment may consist of any number of flat rectangular elements of constant thickness t_e, breadth b_e, and length L. In Refs. 7 and 13 the transverse shear force Q was taken as constant along the length of the element, but for greater generality we here allow for a linear variation, as occurs if the overall load on the hull girder is a uniform distributed load. If the magnitude of this load is f_y (see Fig. 6.15) the shear force at any section is:

$$Q(x) = Q_A\left(1 + \phi\frac{x}{L}\right) \tag{6.4.3}$$

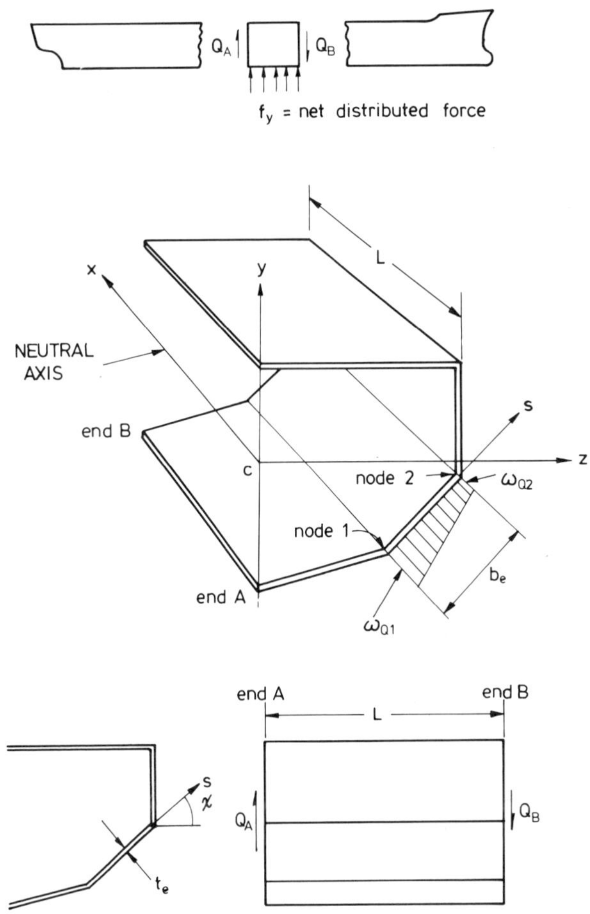

Figure 6.15 Warping due to transverse shear.

where

$$\phi = \frac{f_y L}{Q_A}$$

and Q_A is the shear force at end A of the segment. The warping is then:

$$u(s, x) = \omega_Q(s)Q_A\left(1 + \phi\frac{x}{L}\right) \tag{6.4.4}$$

Since the longitudinal variation of ω_Q is independent of the transverse variation, the problem can be formulated entirely in terms of the warping at end A, and so the problem is essentially a two-dimensional shear flow problem, in which the shear flow $q(s)$ (at end A) is to be determined.

The assumption of a linear variation of $\omega_Q(s)$ within each element implies constant shear flow in the element, whereas it is known from Section 3.7 that the shear flow varies linearly in horizontal portions of the ship's cross section and parabolically in all other portions. It will be shown that the constant value of shear flow which is obtained for each element is the mean value for that element, and that the other part of the total shear flow (a linear or parabolic distribution, having a zero mean value) can be obtained separately for each element and superimposed on the mean value.* Thus the assumption of a linear distribution of warping within the element does not constitute an approximation, and so a single element can be used for any straight, constant thickness portion of the cross section, even if its breadth b_e is quite large.

In terms of nodal values ω_{Q1} and ω_{Q2} the warping within an element is:

$$\omega_Q(s) = \omega_{Q1} + (\omega_{Q2} - \omega_{Q1})\frac{s}{b_e} \qquad (6.4.5)$$

in which the $+s$-direction is from element node 1 to element node 2. The orientation of the element is defined by the angle χ which the $+s$-axis makes with the horizontal axis of the hull girder, as shown in Fig. 6.15.

In terms of strain the potential energy of an element is:

$$U_e = \frac{Gt_e}{2}\int_0^L\int_0^{b_e}[\gamma(s,x)]^2\,ds\,dx$$

Substituting from (6.4.1) and (6.4.4) and integrating along the length gives:

$$U_e = \frac{1}{2}C_Q Q_A^2 G t_e L \int_0^{b_e}\left(\frac{\partial \omega_Q}{\partial s}\right)^2 ds \qquad (6.4.6)$$

where

$$C_Q = 1 + \phi + \frac{1}{3}\phi^2 \qquad (6.4.7)$$

Substituting the linear expression for $\omega_Q(s)$ and summing over all the elements (say N_e in number):

$$U = \frac{C_Q Q_A^2 GL}{2}\sum_{e=1}^{N_e}\int_0^{b_e}\left(\frac{\omega_{Q2}^e - \omega_{Q1}^e}{b_e}\right)^2 t_e\,ds \qquad (6.4.8)$$

The other component of the total potential energy is the negative of the work done by the loads as a result of the warping. For the differential element in Fig. 6.16 the work done is the product of the net longitudinal force in the element

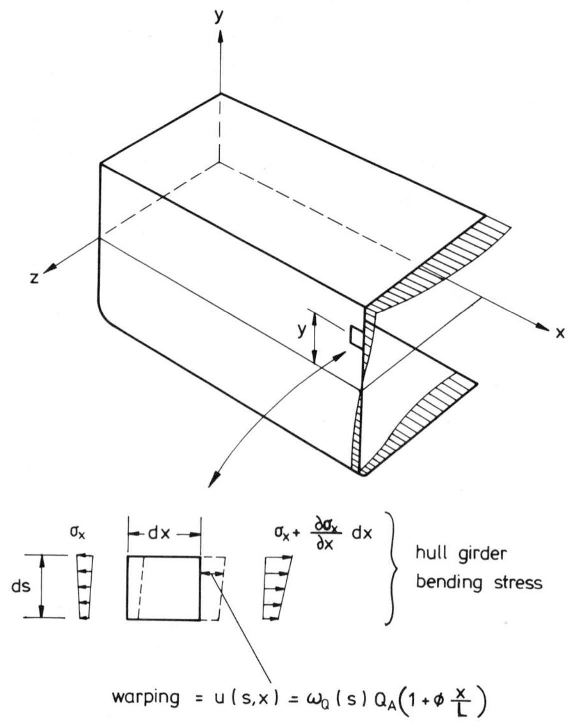

Figure 6.16 Distribution of warping.

$$dF = \frac{\partial \sigma_x}{\partial x}\,dx\,t_e\,ds$$

times the distance through which the element moves, which is the local warping displacement $u(s, x)$. That is, the work done throughout the element is:

$$W = \int_{\text{vol}} u(s,x)\,dF \qquad (6.4.9)$$

(There is no $\frac{1}{2}$ factor because the stress is caused by the external loads and not by the warping.) The longitudinal stress can come either from hull girder bending or from a restraint on the *torsional* warping if the segment is undergoing twist. For bending stress dF may be related to the change in bending moment and hence to the transverse shear force Q, as was done in deriving (3.7.1) and (3.7.2) (and as illustrated in Fig. 3.22). The steps are:

$$dF = \frac{\partial}{\partial x}\left(\frac{My}{I}\right)dx\,t_e\,ds$$

$$= \frac{\partial M}{\partial x}\frac{y}{I}\,dx\,t_e\,ds$$

$$= \frac{Q(x)}{I}\,y\,dx\,t_e\,ds$$

*This approach was suggested to the author by Dr. Vedran Žanić of the University of Zagreb.

After substituting this expression and (6.4.4) and (6.4.5) into (6.4.9), integrating along the length of the element, and finally summing over all of the elements, the result is:

$$W_Q = \frac{C_Q Q_A^2 L}{I} \sum_{e=1}^{N_e} t_e \int_0^{b_e} \left[\omega_{Q1}^e + \frac{(\omega_{Q2}^e - \omega_{Q1}^e)}{b_e} \right] y(s)\, ds \tag{6.4.10}$$

where C_Q is defined in (6.4.7). The value of $y(s)$ can be expressed in terms of the value at element node 1 and the orientation angle χ (see Fig. 6.15)

$$\left. \begin{aligned} y(s) &= y_1 + s\mu \\ \text{where} \qquad \mu &= \sin\chi \end{aligned} \right\} \tag{6.4.11}$$

We have seen in Section 6.1 that if a section that is undergoing twist is even partially restrained from warping, this causes a longitudinal stress σ_x which varies in the axial direction. From (6.1.4)

$$\sigma_x = E\omega_n \frac{d^2\theta}{dx^2}$$

and so the net axial force in an element of width ds and length dx is:

$$\begin{aligned} dF &= \frac{\partial \sigma_x}{\partial x} dx\, t_e\, ds \\ &= E\omega_n \frac{d^3\theta}{dx^3} dx\, t_e\, ds \end{aligned}$$

Again substituting into (6.4.9) and integrating lengthwise, the following expression is obtained for the work done by the *torsional* warping stress:

$$W_\theta = EQ_A t_e \int_0^L \frac{d^3\theta}{dx^3} \int_0^{b_e} (\omega_n \omega_Q\, ds) \left(1 + \phi \frac{x}{L}\right) dx \tag{6.4.12}$$

Thus if there is torsion and if the torsional warping is restrained so as to give a nonzero $d^3\theta/dx^3$, then the two types of warping interact. For most ships transverse shear Q is a much more severe load than torsion, and the shear flow q is due almost entirely to Q. Therefore the remainder of this section deals with transverse shear only.

The total potential energy of the system is $\Pi = U - W$, and the system of equations for the nodal values of ω is obtained by requiring that Π must be a minimum with respect to each ω_i, as was done in Section 6.1 for torsional warping. That is

$$\frac{\partial \Pi}{\partial \omega_i} = 0 \qquad i = 1, \ldots, N_n$$

where N_n is the number of nodes. The steps are straightforward and the resulting system of equations is

$$G \sum_{m=1}^{M} t_m \left(\frac{\omega_{Q,i} - \omega_{Q,r}^m}{b_m} \right) = \frac{1}{2I} \sum_{m=1}^{M} t_m b_m \left(y_{1m} + \frac{n_m}{3} \mu_m b_m \right) \qquad i = 1, \ldots, N_n \tag{6.4.13}$$

In these equations the subscript e has been replaced by m to indicate that since each equation refers to one node (the ith node) the summation over the elements need only include the M elements that touch that node. For each of these M elements the subscript r denotes the node which is *remote* from node i, and the symbol n_m is the *element* node number (either 1 or 2) which corresponds to node i. The symbols y_1 and μ were defined in (6.4.11).

It can be shown that axial equilibrium requires that $\omega_Q = 0$ at all points which lie on the neutral axis. This requirement can be imposed by placing nodes at all such points and setting these values of ω_Q equal to zero.* This effectively divides the system of equations into two independent subsystems.

In order to demonstrate the simplicity of this method and to show how the final distribution of shear flow is obtained, it will be applied to the same idealized tanker as in Section 3.7. For brevity the calculation will be performed only for the upper portion. As shown in Fig. 6.17 there are only three nodal values of ω_Q to be determined, and only four elements are required. For elements a and b the parameter values are:

$$t = 0.032, \quad b = 10, \quad y_1 = 12, \quad \chi = 0^\circ \quad \mu = 0$$

while for elements c and d they are:

$$t = 0.032, \quad b = 12, \quad y_1 = 0, \quad \chi = 90^\circ, \quad \mu = 1$$

*If this is not done the system of equations is singular.

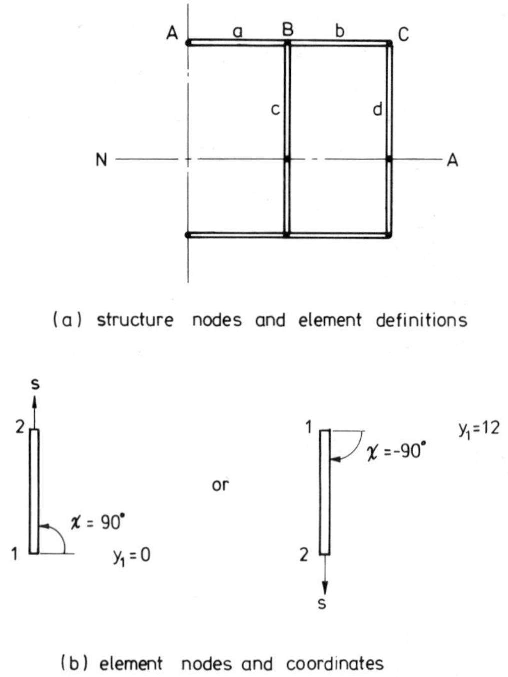

Figure 6.17 Sample calculation of warping due to transverse shear.

These values correspond to a choice of the lower end as element node 1. It would be equally valid to choose the upper end, as shown in Fig. 6.17*b*.

At node A only element a is involved and so (6.4.13) becomes:

$$t_a \frac{\omega_{Q,A} - \omega_{Q,B}}{b_a} = \frac{1}{2GI}\, t_a b_a \left(y_{1a} + \frac{1}{3}\, \mu_a b_a \right)$$

Substituting the foregoing values gives:

$$0.0032\omega_{Q,A} - 0.0032\omega_{Q,B} = \frac{0.32}{2GI}\,(12)$$

At node B elements a, b, and c are involved and the equation (with zero values already inserted) is

$$\left(\frac{t_a}{b_a} + \frac{t_b}{b_b} + \frac{t_c}{b_c} \right) \omega_{Q,B} - \frac{t_a}{b_a}\, \omega_{Q,A} - \frac{t_b}{b_b}\, \omega_{Q,C} - \frac{t_c}{b_c}\, 0$$

$$= \frac{1}{2GI} \Big[t_a b_a (y_{1a} + 0)$$

$$+ t_b b_b (y_{1b} + 0) + t_c b_c \left(0 + \frac{2}{3}\, b_c \right) \Big]$$

which becomes:

$$0.009067\omega_{Q,B} - 0.0032\omega_{Q,A} - 0.0032\omega_{Q,C}$$

$$= \frac{10.752}{2GI}$$

At node C elements b and d are involved and the equation is:

$$\left(\frac{t_b}{b_b} + \frac{t_c}{b_c} \right) \omega_{Q,C} - \frac{t_b}{b_b}\, \omega_{Q,B} - \frac{t_d}{b_d}\, 0$$

$$= \frac{1}{2GI} \left[t_b b_b (y_b + 0) + t_d b_d \left(0 + \frac{2}{3}\, b_d \right) \right]$$

which becomes:

$$0.005867\omega_{Q,C} - 0.0032\omega_{Q,B} = \frac{6.912}{2GI}$$

The solution to these three equations is:

$$GI\omega_{Q,A} = 2827.8$$

$$GI\omega_{Q,B} = 2227.8$$

$$GI\omega_{Q,C} = 1804.2$$

The mean value of shear stress in each element is given by:

$$\bar{\tau}_e = G \frac{\partial u}{\partial s} = GQ \frac{\partial \omega_Q^e}{\partial s} = GQ \frac{\omega_2^e - \omega_1^e}{b_e} \qquad (6.4.14)$$

Since Q and I are constant for a given section it is convenient to deal with scaled values of τ which correspond to a unit value of Q/I, and then apply this factor at the end. Also, the solution in Section 3.7 is given in terms of m/t, which from (3.7.4) is seen to be the same scaled value of shear stress.

Thus for element a the mean value of the (scaled) shear stress is:

$$\bar{\tau}_a = \frac{\omega_{Q,A} - \omega_{Q,B}}{b_a} = \frac{2827.8 - 2227.8}{10} = 60$$

and reference to Fig. 3.29*c* shows that this is indeed the correct mean value of shear stress in this portion of the ship section. The other values obtained from (6.4.14) are:

$$\bar{\tau}_b = 42.35, \qquad \bar{\tau}_c = 185.65, \qquad \bar{\tau}_d = 150.35$$

Once the mean value of τ is known in any element the local variation can be found from (3.7.2), which gives the local or incremental distribution of τ in any element, over and above the value at the end where the integration is started (node 1). We require a local distribution which has a zero *mean* value, instead of a zero value at node 1, and this can be obtained by subtracting the mean value. Thus for each element the local distribution $\tau_e^l(s)$ which is to be superimposed on the mean value $\overline{\tau}_e$ obtained from a warping solution is given by:

$$\tau_e^l(s) = \frac{Q}{t_e I}\left[m_e(s) - \overline{m}_e\right]$$

where $m(s)$ is the first moment of area, defined in (3.7.3). Expressing $y(s)$ in terms of y_1 and μ as in (6.4.11), this equation becomes, after integration:

$$\tau_e^l(s) = \frac{Q}{I}\left[y_{1e}\left(s - \frac{b_e}{2}\right) + \frac{\mu}{2}\left(s^2 - \frac{b_e^2}{3}\right)\right] \quad (6.4.15)$$

Since this expression comes from the shear flow equations of Section 3.7, it follows the same rules in regard to sign convention. That is, it assumes that the s-axis is in the direction of *increasing* shear flow (or shear stress). If this is not so, then the sign of $\tau_e^l(s)$ must be reversed.

In the tanker example the resulting local distributions for elements a and c are:

$$\tau_a^l(s) = \frac{Q}{I}\left[12(s - 5)\right]$$

$$\tau_c^l(s) = -\frac{Q}{I}\left[0 + \tfrac{1}{2}\left(s^2 - \frac{144}{3}\right)\right] = \frac{Q}{I}\left[24 - \tfrac{1}{2}s^2\right]$$

Note that for $\tau_c^l(s)$ the sign has been reversed because element node 1 is at the neutral axis ($y_{1c} = 0$) and the s-axis points upward ($\mu = 1$), which is the direction of *decreasing* τ. The maximum value of τ occurs at the neutral axis and for element c this value is

$$\begin{aligned}\tau(y = 0) &= \overline{\tau}_c + \tau_c^l(0)\\ &= (185.6 + 24)Q/I\\ &= 209.6Q/I\end{aligned}$$

which again agrees with the earlier solution.

REFERENCES

1. V. Z. Vlasov, *Thin-Walled Elastic Beams,* National Science Foundation, U.S. Dept. of Commerce, 1961; also Oldbourne Press, London, 1961.
2. C. F. Kollbrunner and K. Basler, *Torsion in Structures,* Springer-Verlag, Berlin, 1969.
3. C. F. Kolbrunner and N. Hajdin, *Dünnvandige Stäbe,* Band 1 and 2, Springer-Verlag, 1972 and 1975.
4. G. deWilde, "Structural Problems in Ships With Large Hatch Openings," Int'l. Shipbuilding Progress, Jan.-Feb. 1967, pp. 7–33.
5. K.-H. Hapel, *Torsion von Containerschiffen,* Technische Universität Berlin—Institut für Schiffstechnik, Berlin, 1975.
6. K. Haslum and A. Tonnessen, "An Analysis of Torsion in Ship Hulls," European Shipbuilding, No. 5/6, pp. 67–90, 1972.
7. T. Kawai, "The Application of Finite Element Methods to Ship Structures," *Computers and Structures,* **3,** 1973, pp. 1175–1194.
8. G. A. Gunnlauggson and P. T. Pedersen, *A Finite Element Formulation for Beams with Thin-Walled Cross Sections,* Computers and Structures **15,** 6, 1982, pp. 691–699.
9. H. Westin, "Torsion of Non-Prismatic Beam Girders With Special Applications to Open Ships," TRITA-SKP-1055, Div. of Naval Arch., Royal Inst. of Technology, Stockholm, 1980.
10. A. Ulfvarson, "Studies on the Structural Behavior of Open Ships Under Torsion," Report E-46, Div. of Naval Arch., Royal Inst. of Technology, Stockholm, 1975.
11. P. Terndrup Pedersen, "A Beam Model for the Torsional Bending Response of Ship Hulls," *Trans. RINA,* 1982.
12. T. Kawai, "Studies on the Ultimate Strength Analysis of a Ship Structure by Means of a New Discrete Thin-Walled Beam Element," *J. Soc. Nav. Arch. of Japan,* **150,** pp. 405–412.
13. L. R. Herrmann, "Elastic Torsional Analysis of Irregular Shapes," J. of Engg. Mechanics Division, A.S.C.E., no. EM6, December 1965, pp. 11–19.
14. W. E. Mason, Jr. and L. E. Herrmann, "Elastic Shear Analysis of General Prismatic Beams," J. of Engg. Mechanics Division, A.S.C.E., no. EM4, August 1968, pp. 965–983.

Chapter

Seven

BASIC ASPECTS OF THE FINITE ELEMENT METHOD

7.1 AIM, APPROACH, AND SCOPE

The aim of this chapter is to explain as simply and as directly as possible the basic ideas and features of the finite element method, with specific regard to its applications in ship structural analysis. In its totality the finite element method is very broad and powerful; it includes numerous special features and techniques and it has a great variety of applications—both structural and nonstructural. Structural applications include not only the small-deflection analysis of two- and three-dimensional elastic continua, but also structural stability (buckling) and structural dynamics (e.g., vibration analysis). Also, with the aid of special iterative techniques, the method can be used in nonlinear solid mechanics, dynamics, and stability problems. Nonstructural applications are also numerous because, although the method was originally derived for structures, it has subsequently been given a broader and more general foundation by means of variational theory. As a result it can be used for a wide variety of problems for which the solution corresponds to the minimization of a functional. In structural applications the functional is strain energy, but in the broader fields of continuum mechanics (fluid and solid), the functional can be related to pressure, temperature, fluid velocity, and so on. Likewise it can be the potential in the various aspects of potential field theory (electrostatics, magnetism, gravity, etc.).

However, such generality would not be helpful or appropriate in a book dealing exclusively with ship structures. Several books such as Refs. 1, 2, 3, and 4 are available which present the finite element method in varying degrees of generality, breadth of application, and level of detail, and these are recommended for readers wishing to explore the method more thoroughly. Reference 4 is a good introductory text, having a clear and simple presentation and dealing only with elastic structural analysis of a two-dimensional continuum. Of the four references, 1 and 3 are more comprehensive in that they include nonlinear and nonstructural applications.

The scope and level of treatment of this chapter are intended to explain merely the basic aspects of the method as it applies to ship structures. This is a middle course between two extremes: that of attempting to give a general and mathematically rigorous treatment and to cover a broad spectrum of features and applications; and that of omitting the method entirely on the grounds that the material can be found in other texts. Some specific advantages of this course are:

1. It makes the method easier to learn and does not require a knowledge of variational calculus.
2. It avoids requiring the reader to have a previous knowledge of the method.
3. It leads to a more unified and cohesive coverage of ship structural analysis and design, especially in regard to overall philosophy, underlying assumptions, and methodology.
4. It avoids the need for an additional and separate text, with the consequent problems of differing notation, approach, and emphasis.

Furthermore, although the finite element method has now been given a broader and more fundamental theoretical basis, it was originally developed mainly from matrix structural analysis. Hence it is possible to present the method as a generalization and extension of matrix frame analysis to continuum structures such as plates and shells. Although not all of the historical development actually came from frame analysis, most of the concepts and terminology are similar. Hence, with the aim of providing greater unity and simplicity of presentation, the method is here presented as an extension of matrix frame analysis. Consequently the principal basis for the present chapter is the material in Chapter 5.

The key feature of the finite element method and the point of departure from frame analysis is the use of two- and three-dimensional elements for the discrete representation of a continuum. The first paper to clearly and fully implement this idea was presented in 1956 by Turner et al. [5]. It should however be mentioned that some of the basic features of the method had appeared before this in papers by Courant [6], McHenry [7], Hrenikoff [8], and others. There followed many other contributions by many authors, including a series of papers on matrix structural analysis by Argyris and his collaborators, beginning in 1954 and culminating in a monograph by Argyris and Kelsey [9] in 1960. The first elements were for plane stress applications. After this, finite elements were developed for three-dimensional solids, plates in bending, thin shells, thick shells, and other structural forms. Once these had been established for linear elastic analysis, work was directed toward more complex areas such as dynamic response, buckling, and material and geometric nonlinearities. This required the introduction of iterative finite element analysis.

In addition to these developments in the structural field, in the early 1960s the method began to be recognized as a form of the Ritz method, and in 1964 Zienkiewicz and Cheung [10] showed that it is applicable to all field problems that can be cast in variational form.

For structural applications a number of general-purpose finite element computer programs are now available and this, together with the inherent versatility of the finite element method, has meant that the number and variety of practical applications of the method have grown enormously in recent years.

7.2 FUNDAMENTALS OF THE METHOD

As mentioned, the basic concept of the finite element method is the same as in matrix frame analysis, namely, that the structure can be represented as an assemblage of individual structural elements interconnected at a discrete number of nodes. This manner of representation is a natural one for frame structures since they consist of discrete beam members connected together at a number of joints. But in a continuum structure such as a panel of plating, a corresponding natural subdivision does not exist and therefore it is necessary to divide the continuum artificially into a number of elements, connected at their nodes. The artificial elements, or "finite elements," are usually either rectangular or triangular in shape, and the nodes are usually but not necessarily at the corners. For example, Fig. 7.1 shows the web of a deep beam subdivided into triangular elements.

To use matrix methods, the essential requirement is that the structural continuum must be represented in terms of a finite number of discrete variables. These variables are the nodal displacements and, in some cases, their derivatives. If the latter are included we speak of "degrees of freedom" instead of nodal displacements, but this does not arise for the simple elements which are presented in this chapter. In terms of nodal displacements the essential requirement is that the internal displacements of the elements must be related to the nodal displacements, and all of the interactions between elements must be expressed in the terms of the nodal displacements. In this way the only unknowns in the problem are the nodal displacements and so the problem becomes discrete instead of continuous. Although there may be a large number of nodal displacements they are nevertheless a finite number of discrete variables which are interrelated by linear equations and therefore they can be conveniently handled by matrix methods.

From the discussion at the beginning of Chapter 5, it is clear that to achieve an exact solution the finite element representation would have to satisfy the requirements of equilibrium and geometric compatibility, and would have to satisfy them everywhere, both within each of the elements and also between elements. This amounts to four separate requirements. By way of illustration let us consider the requirement of interelement compatibility. In a continuous struc-

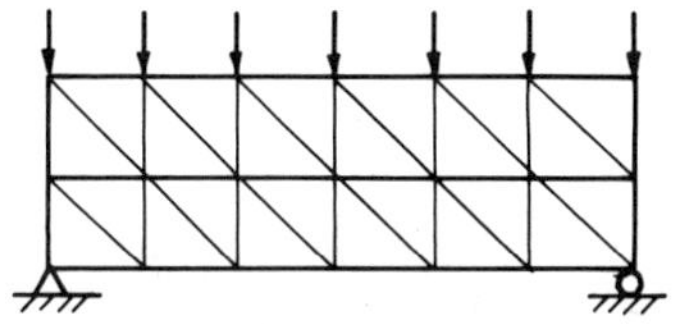

Figure 7.1 Representation of a beam web using triangle elements.

ture such as the plate web of Fig. 7.1, there is complete continuity of displacement along the common boundaries of the elements. Hence in the finite element model it would not be sufficient to satisfy continuity of displacement only at the nodes, and to place no conditions on the displacements along the common boundaries. This would make the model much more flexible than the structure since it would allow gaps and overlaps to occur, as shown in Fig. 7.2. One way of reducing the error is to use smaller and more numerous elements because then there are more nodes and hence more points at which compatibility is satisfied.

However, of its very nature, a discrete model can never give an exact representation of a continuum, regardless of the number of discrete variables which it employs. There is always some error, although it may be made negligibly small and localized. Hence in the finite element model it is not possible to fulfill all four of the foregoing requirements exactly, even though the elements may be made small and numerous.

However, besides using smaller and more numerous elements, it is also possible to reduce the error by carefully choosing and defining suitable element properties. This specification of element properties is the most fundamental feature of the finite element method and is the means whereby it can satisfy the four requirements sufficiently well without having to use extremely small elements. The element behavior is described by a specially chosen set of functions which represent either the stresses or the displacements in that element. In other words, each element is constrained to deform in a specific pattern. The result is that even though equilibrium and compatibility are explicitly enforced only at the nodes, the specified internal behavior pattern of the elements ensures that intra- and interelement equilibrium and compatibility are all satisfied to a sufficient degree. For example, if the displacement within each element is specified as being linear, this linearity would also apply along the element edges. Therefore, since two adjacent elements share a common displacement at each of their two common nodes, they must also have equal displacements all along their common boundary. Hence the specification of linear displacement would cause the requirement of interelement compatibility to be satisfied.

Thus there are two principal features of the finite element method which require discussion: the subdivision of the structure into elements and, more importantly, the choice of the internal behavior function. We first consider some aspects and criteria regarding subdivision. Then in order to explain and illustrate the choice of the internal behavior function we consider some specific elements which are simple, commonly used, and suitable for ship structure applications.

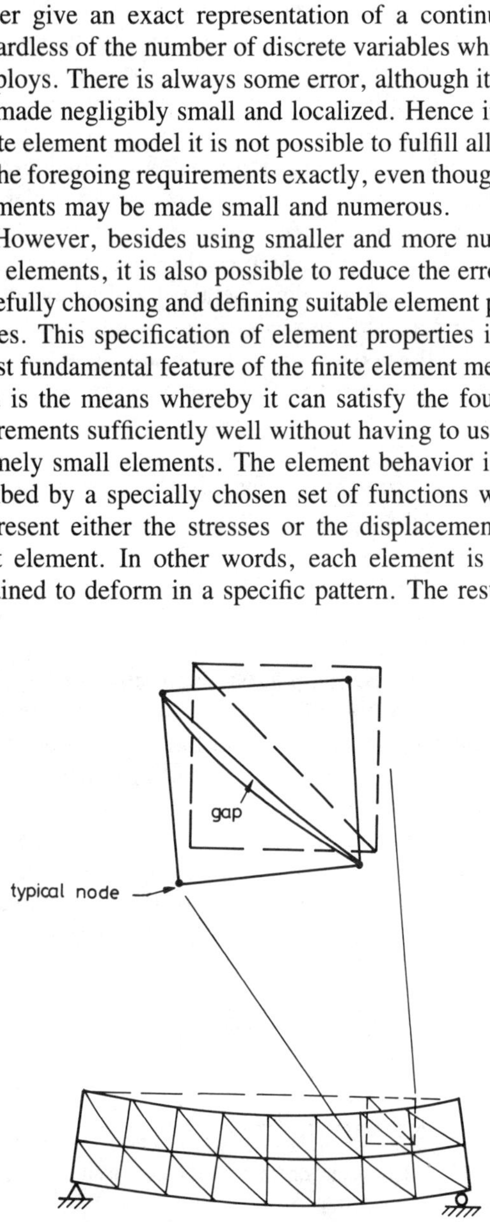

Figure 7.2 Deformation of elements.

Subdivision of Structure

As mentioned previously, the accuracy of the finite element method increases with the number of elements used. However, increasing the number of elements also increases the computer time required for a solution and hence it increases the cost.

In many cases a "graded mesh," that is, a gradation of element sizes, is adopted to provide a more detailed modeling of regions of the structure where there are stress concentrations such as in the vicinity of openings and cut-outs or near the point of application of concentrated loads, as shown in Fig. 7.3. Mesh grading is a very efficient technique and can give a significant savings in solution time without any loss of accuracy. However, because of the enormous variety of structures and of loads it is not possible to give a general rule as to the number or size of elements or the

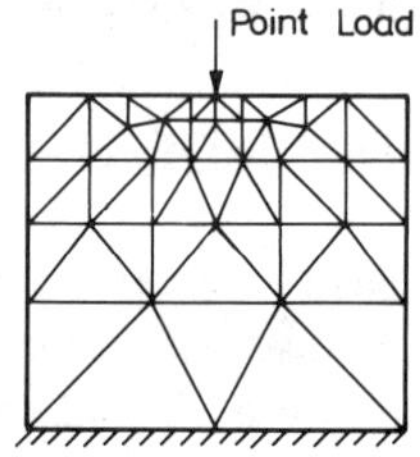

Figure 7.3 Mesh grading.

type of mesh grading which is required to provide sufficient accuracy. In each case the subdivision should be based on experience with similar structures. If this is not possible, then a series of appropriate test problems should be solved using different mesh sizes in order to observe the rate of convergence and thereby determine a suitable mesh size for the particular problem. Some examples are given later in this chapter.

The external loads acting on the structure must be represented by an equivalent system of point loads applied at the element nodes. In the case of concentrated loads the obvious course is to choose the mesh such that there is a node at the point of application of the load. For distributed loading, statically equivalent nodal point loads must be used. Most plane stress elements do not have a rotational degree of freedom and hence there is no possibility of using equivalent nodal moments. This is of little consequence because with the close proximity of nodes the equivalent nodal forces are in themselves a good approximation to the distributed loading. For example, Fig. 7.4 shows a deep web of the type considered earlier, with a length of 6 m and a uniform distributed load of 10 kN/m. The choice of equivalent nodal loads is obvious and the loads can be obtained by inspection, as shown in the figure.

Elements for Ship Structural Analysis and Design

In this chapter we consider only two-dimensional plane stress elements because it will be shown that these are sufficient both for hull module analysis and for the analysis of individual panels of plating, whether stiffened or unstiffened. Three-dimensional elements are only required for very local and detailed stress analysis, and such analysis does not fall within the scope of this book because it is neither a principal nor a unique part of ship structural analysis and design. Nor are plate bending elements included because those aspects of plate bending that are relevant to ship structures can be dealt with by plate theory (elastic and inelastic, as covered in Chapter 11) together with stress concentration theory, if required. Shell elements and other more complicated elements are also not included because they are rarely needed for hull module analysis. For panels and shells that have only slight curvature, ordinary plane stress elements are sufficient. Special plane stress elements are available for axisymmetric and spherical shells, and there are general shell elements which can deal with the combination of plane stress and plate bending. However such elements would only be required for a hull module or a panel with highly curved plating or with irregular geometry, loading, or boundary conditions, such that ordinary shell theory is inapplicable. These situations are less common and in any case the element stiffness matrices and other properties of general shell elements, and of other specialized elements, are available in various books which are devoted exclusively to the finite element method. Hence there is little purpose in including all of that information here.

For all of these reasons this chapter deals only with plane stress elements. Three such elements are presented, a triangular element and two rectangular elements. One of the rectangular elements is particularly well suited for use in a hull module analysis, and it also serves as the basis for a stiffened panel element which is presented in Section 8.4. All three elements, but especially the triangular element, are also suitable for most local or detailed plane stress analyses.

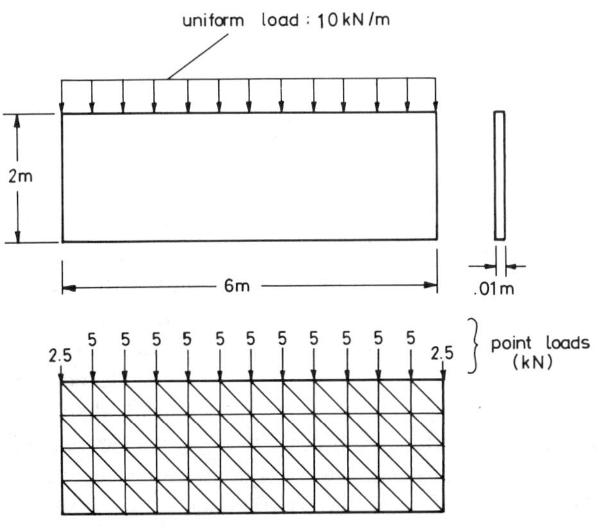

Figure 7.4 Equivalent nodal point loads.

7.3 CONSTANT STRESS TRIANGLE (CST) ELEMENT

As the first example of a finite element we take the case of a flat triangular-shaped element of constant thickness t and isotropic material properties. This element is of primary importance because of its versatility of geometry; almost any two-dimensional shape can be represented by an assemblage of triangles. Moreover, three-dimensional curved surfaces can also be modeled with this element, providing that the curvature is everywhere gradual and that the height of the surface (or the ratio of true surface area to projected area) is small.

In Section 5.1 the five basic steps in the derivation of the stiffness matrix of an element were presented for the beam element. We now show how the same five steps can be used to derive the stiffness matrix for finite elements, beginning with the triangle element.

The coordinate system and the node numbering are shown in Fig. 7.5. The positions of the nodes are (x_1,y_1), (x_2,y_2) and (x_3,y_3). Since we are dealing with displacements in a plane the element has two degrees of freedom at each node, as shown in Fig. 7.6, giving a total of six degrees of freedom.

$$\boldsymbol{\delta} = \begin{Bmatrix} \boldsymbol{\delta}_1 \\ \boldsymbol{\delta}_2 \\ \boldsymbol{\delta}_3 \end{Bmatrix} = \begin{Bmatrix} u_1 \\ v_1 \\ u_2 \\ v_2 \\ u_3 \\ v_3 \end{Bmatrix} \tag{7.3.1}$$

The corresponding forces are

$$\mathbf{f} = \begin{Bmatrix} \mathbf{f}_1 \\ \mathbf{f}_2 \\ \mathbf{f}_3 \end{Bmatrix} = \begin{Bmatrix} f_{x1} \\ f_{y1} \\ f_{x2} \\ f_{y2} \\ f_{x3} \\ f_{y3} \end{Bmatrix} \tag{7.3.2}$$

Both $\boldsymbol{\delta}$ and $\mathbf{f}$ contain six terms, and therefore the element stiffness matrix $\mathbf{k}^e$ is a 6×6 matrix since by definition it is the coefficient matrix in the element nodal force–displacement relationship

$$\mathbf{f} = \mathbf{k}^e\,\boldsymbol{\delta}$$

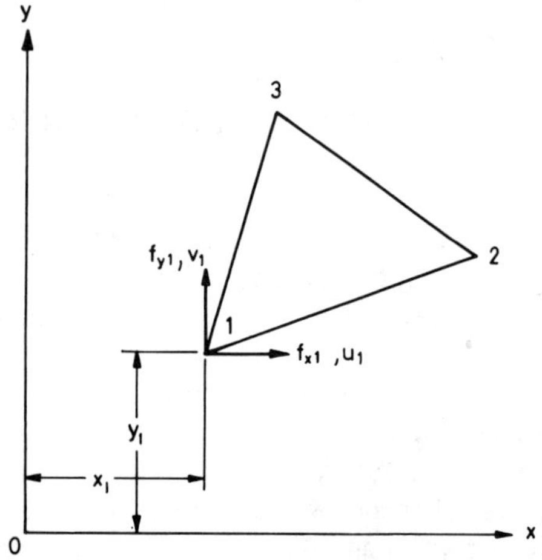

Figure 7.5 Coordinates and node numbering for triangle element.

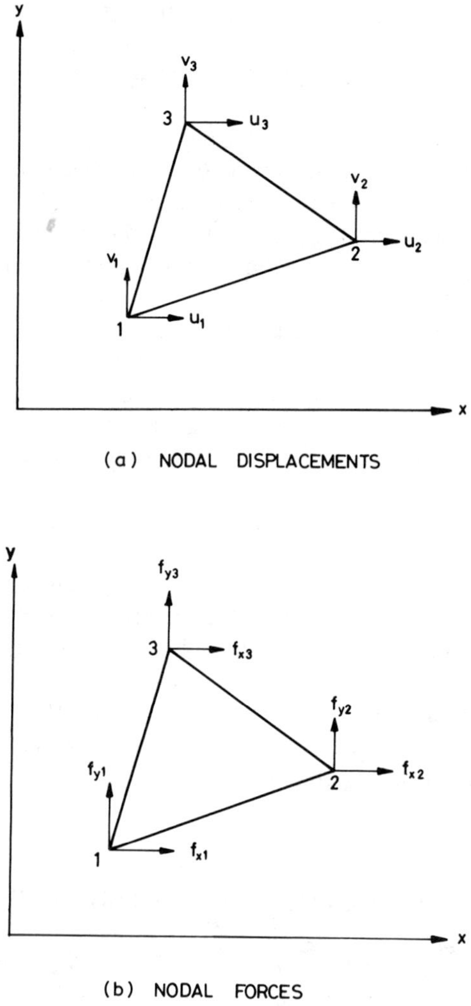

Figure 7.6 Nodal displacements and forces.

STEP I. SELECT SUITABLE DISPLACEMENT FUNCTION

As mentioned at the beginning of this chapter, the key aspect of the finite element method is the specification of the displacement pattern of an element in terms of some mathematical function. A polynomial is a function that is both simple and versatile, since it can be multidimensional (for plane stress only two dimensions, x and y, are required) and it can contain any number of independent parameters (the coefficients) depending on the order of the polynomial. Hence we choose a polynomial to describe the internal displacements in the triangle. Since there are six degrees of freedom we must choose a polynomial having six independent parameters or coefficients. Also, since no bias should be shown in either the x- or the y-direction, we should use three coefficients for the u displacement and three for the v displacement. Hence the logical choice is to represent each of these displacement components by a linear polynomial in x and y:

$$\begin{aligned} u &= C_1 + C_2 x + C_3 y \\ v &= C_4 + C_5 x + C_6 y \end{aligned} \tag{7.3.3}$$

With this choice the displacement varies linearly in x and y. Hence it varies linearly along each boundary of each element, and since the displacement of adjoining elements must agree at their common nodes, the displacements must also agree along the entire boundary. Hence as mentioned earlier the choice of a linear displacement function, in conjunction with corner nodes, ensures continuity of displacement along all interelement boundaries. As shown subsequently, it does *not* provide perfect equilibrium along the boundaries, but rather only at the nodes. Thus there will inevitably be some error but it can be reduced to an acceptable level by making the elements sufficiently small.

Equation 7.3.3 can be written in matrix form as

$$\boldsymbol{\delta}(x,y) = \begin{Bmatrix} u \\ v \end{Bmatrix} = \begin{bmatrix} 1 & x & y & 0 & 0 & 0 \\ 0 & 0 & 0 & 1 & x & y \end{bmatrix} \begin{Bmatrix} C_1 \\ C_2 \\ C_3 \\ C_4 \\ C_5 \\ C_6 \end{Bmatrix} \tag{7.3.4}$$

or

$$\boldsymbol{\delta}(x,y) = \mathbf{H}(x,y)\mathbf{C}$$

STEP II. RELATE THE GENERAL DISPLACEMENT WITHIN THE ELEMENT TO THE NODAL DISPLACEMENTS

This step is achieved by substituting the values of the nodal coordinates into (7.3.4), thereby obtaining expressions that can be solved for the unknown coefficients. For example at node 1

$$\boldsymbol{\delta}_1 = \boldsymbol{\delta}(x_1,y_1) = \mathbf{H}(x_1,y_1)\mathbf{C} = \begin{bmatrix} 1 & x_1 & y_1 & 0 & 0 & 0 \\ 0 & 0 & 0 & 1 & x_1 & y_1 \end{bmatrix} \mathbf{C}$$

and for nodes 2 and 3

$$\boldsymbol{\delta}_2 = \begin{bmatrix} 1 & x_2 & y_2 & 0 & 0 & 0 \\ 0 & 0 & 0 & 1 & x_2 & y_2 \end{bmatrix} \mathbf{C}$$

$$\boldsymbol{\delta}_3 = \begin{bmatrix} 1 & x_3 & y_3 & 0 & 0 & 0 \\ 0 & 0 & 0 & 1 & x_3 & y_3 \end{bmatrix} \mathbf{C}$$

Substituting these three matrices into (7.3.1) yields

$$\boldsymbol{\delta} = \begin{Bmatrix} \boldsymbol{\delta}_1 \\ \boldsymbol{\delta}_2 \\ \boldsymbol{\delta}_3 \end{Bmatrix} = \begin{bmatrix} 1 & x_1 & y_1 & 0 & 0 & 0 \\ 0 & 0 & 0 & 1 & x_1 & y_1 \\ 1 & x_2 & y_2 & 0 & 0 & 0 \\ 0 & 0 & 0 & 1 & x_2 & y_2 \\ 1 & x_3 & y_3 & 0 & 0 & 0 \\ 0 & 0 & 0 & 1 & x_3 & y_3 \end{bmatrix} \begin{Bmatrix} C_1 \\ C_2 \\ C_3 \\ C_4 \\ C_5 \\ C_6 \end{Bmatrix} \tag{7.3.5}$$

The matrix of nodal coordinates is known as the *connectivity matrix*. For future reference we denote it as **A** and this equation then becomes

$$\boldsymbol{\delta} = \mathbf{AC}$$

The unknown polynomial coefficients **C** are now determined from (7.3.5) by inverting the connectivity matrix.

$$\mathbf{C} = \mathbf{A}^{-1}\,\boldsymbol{\delta} \tag{7.3.6}$$

For the small matrix in this example, the inversion of **A** can be done algebraically, but for larger matrices the inversion must be performed numerically by the computer. Algebraic inversion is tedious, but it only needs to be done once, when the element is first derived, and then never has to be repeated. This is of advantage in a design context in which the proportions of the structure—and hence of the elements—are changing, thus requiring repeated calculation of the element stiffness matrix. The inverse matrix $\mathbf{A}^{-1}$ for the triangle element is given in (7.3.7).

$$\mathbf{A}^{-1} = \frac{1}{2A_{123}} \begin{bmatrix} x_2y_3 - x_3y_2 & 0 & -x_1y_3 + x_3y_1 & 0 & x_1y_2 - x_2y_1 & 0 \\ y_2 - y_3 & 0 & y_3 - y_1 & 0 & y_1 - y_2 & 0 \\ x_3 - x_2 & 0 & x_1 - x_3 & 0 & x_2 - x_1 & 0 \\ 0 & x_2y_3 - x_3y_2 & 0 & -x_1y_3 + x_3y_1 & 0 & x_1y_2 - x_2y_1 \\ 0 & y_2 - y_3 & 0 & y_3 - y_1 & 0 & y_1 - y_2 \\ 0 & x_3 - x_2 & 0 & x_1 - x_3 & 0 & x_2 - x_1 \end{bmatrix} \tag{7.3.7}$$

$$\text{where } 2A_{123} = \det \begin{vmatrix} 1 & x_1 & y_1 \\ 1 & x_2 & y_2 \\ 1 & x_3 & y_3 \end{vmatrix}$$

$$= (x_2y_3 - x_3y_2) - (x_1y_3 - x_3y_1) + (x_1y_2 - x_2y_1)$$

$$= 2 \times \text{area of triangle } 123$$

Equation 7.3.6 expresses the six unknown coefficients $\mathbf{C}$ in terms of the nodal displacements $\boldsymbol{\delta}$. From (7.3.4) the general displacement $\boldsymbol{\delta}(x,y)$ at any point (x,y) within the element can now be expressed in terms of the nodal displacements $\boldsymbol{\delta}$ by substituting for $\mathbf{C}$. The result is

$$\boldsymbol{\delta}(x,y) = \mathbf{H}(x,y)\mathbf{A}^{-1}\,\boldsymbol{\delta} \tag{7.3.8}$$

STEP III. EXPRESS THE INTERNAL DEFORMATION (STRAIN) IN TERMS OF THE NODAL DISPLACEMENTS

In general the internal deformation of a member is defined as the derivative of its displacement. The most common example is strain, which is the first derivative (or local rate) of displacement. For rigid body deflections the displacement (translation or rotation) is constant and hence the strain is zero. In Chapter 5 the internal deformation for a one-dimensional bar element was the elongation (or shortening), but this could have been converted to axial strain by dividing by the length. For the beam element the internal deformation was curvature rather than strain because in that case the internal "force" (bending moment) is proportional to the second derivative of lateral displacement. Also, in both cases the member was one-dimensional and therefore the internal deformation was a single quantity, that is, a scalar.

To deal with elastic continua we must revert to the more general measure of internal deformation, that is, strain. For a two-dimensional continuum there are three components of strain and so we can define a *strain vector*

$$\boldsymbol{\varepsilon}(x,y) = \begin{Bmatrix} \varepsilon_x \\ \varepsilon_y \\ \gamma \end{Bmatrix} \tag{7.3.9}$$

in which ε_x and ε_y are the direct strains and γ is the shear strain. From the theory of elasticity the relationship between the strain $\boldsymbol{\varepsilon}$ and the displacements u and v is

$$\begin{aligned} \varepsilon_x &= \frac{\partial u}{\partial x} \\ \varepsilon_y &= \frac{\partial v}{\partial y} \\ \gamma &= \frac{\partial u}{\partial y} + \frac{\partial v}{\partial x} \end{aligned} \tag{7.3.10}$$

Substituting for u and v from eq. (7.3.3)

$$\begin{aligned} \varepsilon_x &= \frac{\partial}{\partial x}(C_1 + C_2x + C_3y) \\ &= C_2 \\ \varepsilon_y &= \frac{\partial}{\partial y}(C_4 + C_5x + C_6y) \\ &= C_6 \\ \gamma &= \frac{\partial}{\partial y}(C_1 + C_2x + C_3y) + \frac{\partial}{\partial x}(C_4 + C_5x + C_6y) \\ &= C_3 + C_5 \end{aligned}$$

Thus

$$\boldsymbol{\varepsilon}(x,y) = \begin{Bmatrix} \varepsilon_x \\ \varepsilon_y \\ \gamma \end{Bmatrix} = \begin{Bmatrix} C_2 \\ C_6 \\ C_3 + C_5 \end{Bmatrix}$$

We note that because of the linearity of the chosen displacement function—(7.3.3) or (7.3.4)—all three components of strain are constant within the element. This will be discussed further at the end of the derivation. Although the strain happens to be constant we shall continue to denote the strain matrix as $\boldsymbol{\varepsilon}(x,y)$ in order to indicate that it expresses the strain throughout the element, rather than only at the nodes.

In terms of the full coefficient matrix the above equation is

$$\boldsymbol{\varepsilon}(x,y) = \begin{Bmatrix} \varepsilon_x \\ \varepsilon_y \\ \gamma \end{Bmatrix} = \begin{bmatrix} 0 & 1 & 0 & 0 & 0 & 0 \\ 0 & 0 & 0 & 0 & 0 & 1 \\ 0 & 0 & 1 & 0 & 1 & 0 \end{bmatrix} \begin{Bmatrix} C_1 \\ C_2 \\ C_3 \\ C_4 \\ C_5 \\ C_6 \end{Bmatrix} \qquad (7.3.11)$$

or

$$\boldsymbol{\varepsilon}(x,y) = \mathbf{GC}$$

By substituting for **C** from (7.3.6) we obtain the desired expression for strain in terms of nodal displacements

$$\boldsymbol{\varepsilon}(x,y) = \mathbf{GA}^{-1}\,\boldsymbol{\delta}$$

and if we define a *strain coefficient matrix*

$$\mathbf{B} = \mathbf{GA}^{-1} \qquad (7.3.12)$$

the foregoing expression becomes

$$\boldsymbol{\varepsilon}(x,y) = \mathbf{B}\,\boldsymbol{\delta} \qquad (7.3.13)$$

Because $\mathbf{A}^{-1}$ has been obtained explicitly, the strain coefficient matrix **B** can be obtained by performing the matrix multiplication of (7.3.12). The result is

$$\mathbf{B} = \frac{1}{2A_{123}} \times \begin{bmatrix} y_2 - y_3 & 0 & y_3 - y_1 & 0 & y_1 - y_2 & 0 \\ 0 & x_3 - x_2 & 0 & x_1 - x_3 & 0 & x_2 - x_1 \\ x_3 - x_2 & y_2 - y_3 & x_1 - x_3 & y_3 - y_1 & x_2 - x_1 & y_1 - y_2 \end{bmatrix} \qquad (7.3.14)$$

where, as defined earlier, in (7.3.7), A_{123} is the area of the element.

STEP IV. EXPRESS THE INTERNAL FORCE (STRESS) IN TERMS OF THE NODAL DISPLACEMENTS, USING THE ELEMENT'S LAW OF ELASTIC BEHAVIOR

In an elastic continuum the internal force is stress, and for two-dimensional members there are three components of stress: σ_x, σ_y, and τ. Accordingly the stress vector is

$$\boldsymbol{\sigma} = \begin{Bmatrix} \sigma_x \\ \sigma_y \\ \tau \end{Bmatrix} \qquad (7.3.15)$$

For plane stress the relationship between stress and strain is

$$\varepsilon_x = \frac{\sigma_x}{E} - \frac{\nu\sigma_y}{E}$$

$$\varepsilon_y = \frac{-\nu\sigma_x}{E} + \frac{\sigma_y}{E}$$

$$\gamma = \frac{\tau}{G} = \frac{2(1+\nu)}{E}\,\tau$$

where E is Young's modulus, G is the shear modulus and ν is Poisson's ratio. In matrix notation the relationship is

$$\boldsymbol{\varepsilon}(x,y) = \begin{Bmatrix} \varepsilon_x \\ \varepsilon_y \\ \gamma \end{Bmatrix} = \frac{1}{E}\begin{bmatrix} 1 & -\nu & 0 \\ -\nu & 1 & 0 \\ 0 & 0 & 2(1+\nu) \end{bmatrix} \begin{Bmatrix} \sigma_x \\ \sigma_y \\ \tau \end{Bmatrix} \qquad (7.3.16)$$

For our purpose we require the inverse of this relationship which is

$$\boldsymbol{\sigma}(x,y) = \begin{Bmatrix} \sigma_x \\ \sigma_y \\ \tau \end{Bmatrix} = \frac{E}{1-\nu^2}\begin{bmatrix} 1 & \nu & 0 \\ \nu & 1 & 0 \\ 0 & 0 & \dfrac{1-\nu}{2} \end{bmatrix} \begin{Bmatrix} \varepsilon_x \\ \varepsilon_y \\ \gamma \end{Bmatrix} \qquad (7.3.17)$$

or

$$\boldsymbol{\sigma}(x,y) = \mathbf{D}\boldsymbol{\varepsilon}(x,y)$$

Finally, upon substituting for $\boldsymbol{\varepsilon}(x,y)$ from (7.3.13), the desired expression for the element stresses in terms of the nodal displacements is obtained.

$$\boldsymbol{\sigma}(x,y) = \mathbf{DB}\,\boldsymbol{\delta} \qquad (7.3.18)$$

The principal steps in matrix structural analysis (i.e., both frame analysis and the finite element method) were outlined in Chapter 5 and are summarized in Table 5.1. As shown there, the element stiffness matrix is used in the setting up (assembly) of the system of equations for the structure displacements. After solving this system the next step (step 6) is to calculate the stresses within each element, and to do this we need an expression for the stresses in terms of the element nodal displacements. This is given by (7.3.18) and since **B** and **D** are both known explicitly for the triangle element we can perform the matrix multiplication and thus obtain an explicit *stress matrix* $\mathbf{S} = \mathbf{DB}$ which provides the stresses directly from the element nodal displacements. The stress matrix for the triangle element is given in (7.3.19).

$$\boldsymbol{\sigma}(x,y) = \mathbf{S}\boldsymbol{\delta}$$

where

$$\mathbf{S} = \frac{E}{2A_{123}(1-\nu^2)} \begin{bmatrix} y_{23} & -\nu x_{23} & -y_{13} & \nu x_{13} & y_{12} & -\nu x_{12} \\ \nu y_{23} & -x_{23} & -\nu y_{13} & x_{13} & y_{12} & -x_{12} \\ \frac{-(1-\nu)x_{23}}{2} & \frac{(1-\nu)y_{23}}{2} & \frac{(1-\nu)x_{13}}{2} & \frac{-(1-\nu)y_{13}}{2} & \frac{-(1-\nu)x_{12}}{2} & \frac{(1-\nu)y_{12}}{2} \end{bmatrix} \tag{7.3.19}$$

and $x_{ij} = x_i - x_j$

and $y_{ij} = y_i - y_j$

STEP V. OBTAIN THE ELEMENT STIFFNESS MATRIX BY RELATING NODAL FORCES TO NODAL DISPLACEMENTS

In this step the element is given virtual nodal displacements $\boldsymbol{\delta}^*$ and the external work, involving actual nodal forces $\mathbf{f}$, is equated to the internal work, involving the virtual strain $\boldsymbol{\varepsilon}^*$ (expressed in terms of $\boldsymbol{\delta}^*$) and the actual stress $\boldsymbol{\sigma}$ (expressed in terms of the actual displacements $\boldsymbol{\delta}$, corresponding to $\mathbf{f}$). The virtual nodal displacements are then set to unity and the resulting matrix relating $\mathbf{f}$ to $\boldsymbol{\delta}$ is the required element stiffness matrix.

The external virtual work is

$$W_{\text{ext}} = \boldsymbol{\delta}^{*T}\,\mathbf{f}$$

and the internal virtual work is

$$W_{\text{int}} = \int_{\text{vol}} [\boldsymbol{\varepsilon}^*(x,y)]^T \boldsymbol{\sigma}(x,y)\; d\text{vol}$$

where vol is the volume of the element.

Substituting for $\boldsymbol{\varepsilon}^*(x,y)$ from (7.3.13) and for $\boldsymbol{\sigma}(x,y)$ from (7.3.17) gives

$$W_{\text{int}} = \int_{\text{vol}} [\mathbf{B}\boldsymbol{\delta}^*]^T \mathbf{D}\boldsymbol{\varepsilon}(x,y)\; d\text{vol}$$

$$= \int_{\text{vol}} \boldsymbol{\delta}^{*T}\mathbf{B}^T\mathbf{D}\mathbf{B}\boldsymbol{\delta}\; d\text{vol}$$

Setting $\boldsymbol{\delta}^*$ equal to unity and equating the internal and external work gives

$$\mathbf{f} = \left[\int_{\text{vol}} \mathbf{B}^T\mathbf{D}\mathbf{B}\; d\text{vol}\right]\boldsymbol{\delta}$$

and therefore the element stiffness matrix is the quantity in the square brackets. The matrices $\mathbf{B}$ and $\mathbf{D}$ contain only constant terms and hence they can be taken outside the integral, leaving only $\int d\text{vol}$. For an element of constant thickness this equals the area of the triangle multiplied by its thickness t. Therefore

$$\mathbf{k}^e = \mathbf{B}^T\mathbf{D}\mathbf{B}A_{123}t$$

where, as before

$$2A_{123} = \det \begin{vmatrix} 1 & x_1 & y_1 \\ 1 & x_2 & y_2 \\ 1 & x_3 & y_3 \end{vmatrix}$$

For the triangle element $\mathbf{B}$ has been obtained explicitly and so an explicit expression for $\mathbf{k}^e$ can be obtained by matrix multiplication. For convenience of presentation the result is given in two components

$$\mathbf{k}^e = \mathbf{k}_\varepsilon + \mathbf{k}_\gamma \tag{7.3.20a}$$

which correspond to the direct strain terms and the shear strain term of the $\mathbf{D}$ matrix of (7.3.17). The two components are

$$\mathbf{k}_\varepsilon = \frac{Et}{4A_{123}(1-\nu^2)} \times$$

$$\begin{bmatrix} y_{32}^2 & & & & & \\ -\nu y_{32}x_{32} & x_{32}^2 & & & \text{Symmetric} & \\ -y_{32}y_{31} & \nu x_{32}y_{31} & y_{31}^2 & & & \\ \nu y_{32}x_{31} & -x_{32}x_{31} & -\nu y_{31}x_{31} & x_{31}^2 & & \\ y_{32}y_{21} & -\nu x_{32}y_{21} & -y_{31}y_{21} & \nu x_{31}y_{21} & y_{21}^2 & \\ -\nu y_{32}x_{21} & x_{32}x_{21} & \nu y_{31}x_{21} & -x_{31}x_{21} & -\nu y_{21}x_{21} & x_{21}^2 \end{bmatrix}$$

and (7.3.20b)

$$\mathbf{k}_\gamma = \frac{Et}{8A_{123}(1+\nu)} \times$$

$$
\begin{bmatrix}
x_{32}^2 & & & & & \\
-x_{32}y_{32} & y_{32}^2 & & & \text{Symmetric} & \\
-x_{32}x_{31} & y_{32}x_{31} & x_{31}^2 & & & \\
x_{32}y_{31} & -y_{32}y_{31} & -x_{31}y_{31} & y_{31}^2 & & \\
x_{32}x_{21} & -y_{32}x_{21} & -x_{31}x_{21} & y_{31}x_{21} & x_{21}^2 & \\
-x_{32}y_{21} & y_{32}y_{21} & x_{31}y_{21} & -y_{31}y_{21} & -x_{21}y_{21} & y_{21}^2
\end{bmatrix}
$$

where, as before, $x_{ij} = x_i - x_j$ and $y_{ij} = y_i - y_j$.

For the analysis of an individual panel or a group of co-planar panels, the x,y axes are the structure axes and so there is no need for a transformation between element and structure coordinate systems. However, in a hull module analysis there may be some panels that are not parallel to one of the planes of structure axes, thus requiring a transformation of coordinates. If the hull module is longitudinally prismatic the x,y axes of the triangle element will usually be parallel to at least one of the structure axes and so the transformation from structure coordinates to element coordinates is similar to that of (5.2.5). In the general case, when the plating being modeled is not parallel to any of the structure axes, the general transformation of (5.2.7) must be used. Since the element is two-dimensional and has no z-directed degree of freedom, the nodal transformation matrix is

$$\boldsymbol{\lambda} = \begin{bmatrix} \cos\theta xX & \cos\theta xY \\ \cos\theta yX & \cos\theta yY \end{bmatrix} \tag{7.3.21}$$

and the complete transformation matrix is

$$\mathbf{T} = \begin{bmatrix} \boldsymbol{\lambda} & \mathbf{0} & \mathbf{0} \\ \mathbf{0} & \boldsymbol{\lambda} & \mathbf{0} \\ \mathbf{0} & \mathbf{0} & \boldsymbol{\lambda} \end{bmatrix}$$

which applies for both displacements and forces:

$$\boldsymbol{\delta} = \mathbf{T}\boldsymbol{\Delta} \quad \text{and} \quad \mathbf{f} = \mathbf{T}\mathbf{F}$$

while the stiffness matrix transformation is

$$\mathbf{K}^e = \mathbf{T}^T \mathbf{k}^e \mathbf{T}$$

We now examine the question as to whether the requirements of equilibrium and compatibility are satisfied within the element and between elements. First, since the assumed displacement field is a continuous function, compatibility is clearly satisfied within the element. Second, in regard to internal equilibrium, the equilibrium equations for a two-dimensional stress field are

$$\begin{aligned} \frac{\partial \sigma_x}{\partial x} + \frac{\partial \tau}{\partial y} &= 0 \\ \frac{\partial \sigma_y}{\partial y} + \frac{\partial \tau}{\partial x} &= 0 \end{aligned} \tag{7.3.22}$$

The element strains are constant because they are obtained by differentiation of the linear displacement field. The stresses, being related to the strains by elastic constants, are also constant, and therefore these equations are satisfied everywhere within the element.

We have already seen that the assumption of a linear displacement field ensures compatibility between elements. Thus there only remains the question of equilibrium of stresses across the boundary of adjacent elements. Figure 7.7 illustrates such a boundary between elements A and B. The stress matrix for the triangle element was presented in (7.3.19). From this equation it may be seen that each stress component is a function of the displacements at all three nodes of an individual element. Consequently, although both $\boldsymbol{\sigma}_A$ and $\boldsymbol{\sigma}_B$ are functions of u_1, u_2, v_1, and v_2, $\boldsymbol{\sigma}_A$ is a function of u_3 and v_3, whereas $\boldsymbol{\sigma}_B$ is a function of u_4 and v_4, and therefore the boundary stresses will not in general be equal.

Thus, for the triangular plane stress element both the equilibrium and compatibility requirements are satisfied within the element, and the requirement of interelement compatibility is also satisfied along the element boundaries. However, interelement equilibrium is not satisfied exactly along the boundaries. Of course, equilibrium between elements is satisfied in a discrete and less detailed manner at the nodes, by means of the structure (or global) equilibrium equations which enforce equilibrium of element nodal forces throughout the structure. Therefore, as the elements are made smaller and more numerous the solution error due to this localized lack of equilibrium will become progressively smaller and more negligible. This is shown in the following example.

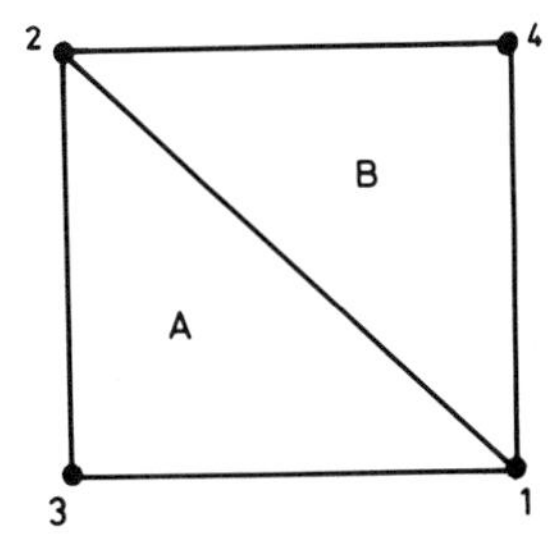

Figure 7.7 Adjacent triangular elements.

Sample Application of the Triangle Element

The deep web beam of Fig. 7.4 will be used as an example to illustrate the use of the triangle element. For this simple example there is an exact solution available from theory of elasticity, and this will allow us to assess the accuracy of this element and the reduction of error with decreasing mesh size. Figure 7.8 shows two different mesh patterns which could be used. Numerical studies [11] have shown that the first pattern gives better accuracy. It also has the advantage that automatic generation of the mesh can be accomplished more easily. Another important point regarding mesh creation is that the accuracy of the element is significantly reduced if any of the vertex angles are small. Hence in choosing the mesh pattern and layout (whether regular or irregular) one should always avoid elongated triangles.

Figure 7.9 shows three finite element models for which solutions have been obtained in Ref. 4. The element size decreases in the proportion 1.0 : 0.8 : 0.4, and the corresponding number of elements increases from 96 to 150 to 600. The simply supported end condition is modeled by specifying zero vertical displacement of all the nodes at the ends of the beam. Figure 7.10, adapted from Ref. 4, shows the vertical deflection of the beam for the three cases, together with the solution from simple beam theory and the exact solution from the theory of elasticity. It may be seen that in each case the finite element values are less than the exact values, but as the mesh size is reduced the results approach the exact solution. The fact that the deflections are always smaller than the exact values is because in matrix stiffness analysis the use of an assumed displacement function to satisfy compatibility always causes the finite element model to be stiffer than the actual structure. Since the error is thus one-sided the possibility exists of reducing the error by relaxing the compatibility requirement slightly. This technique is illustrated in Section 7.5.

Figure 7.11 shows the vertical distribution of longitudinal stress at the midlength of the beam for the same five cases. Since the stresses obtained by the finite element method are constant within each triangle element, each stress value is plotted at the height corresponding to the centroid of the element. Other techniques are possible, such as calculating nodal stress values by averaging the stresses in elements meeting at the node.

Finally, in concluding this presentation of the triangle (or CST) element, it bears mentioning once again that a group of triangles of various shapes and sizes

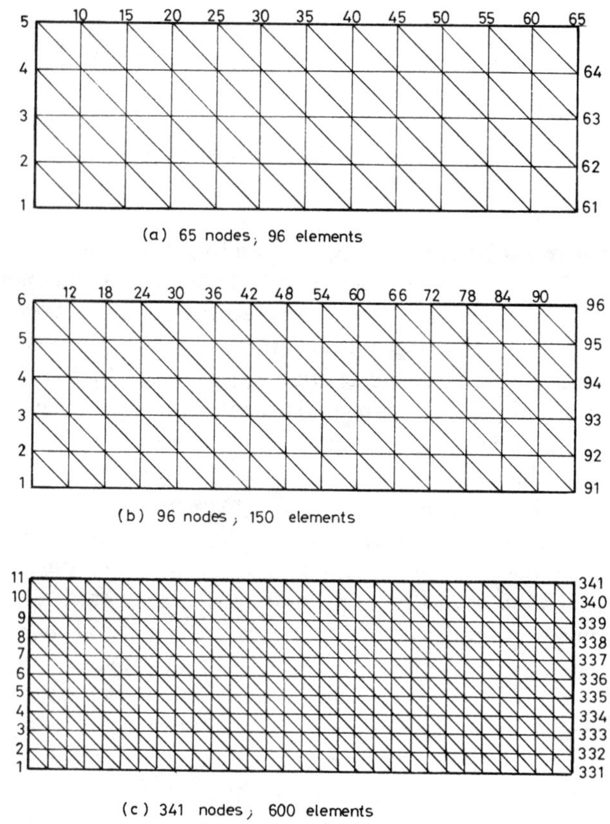

Figure 7.9 Typical finite element models (from Ref. 4).

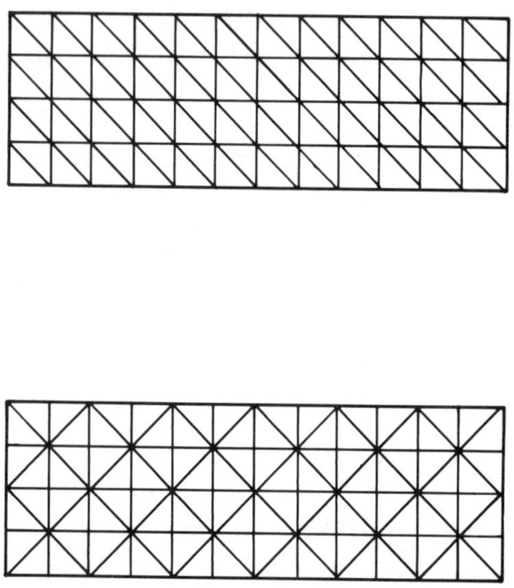
Figure 7.8 Alternative mesh patterns.

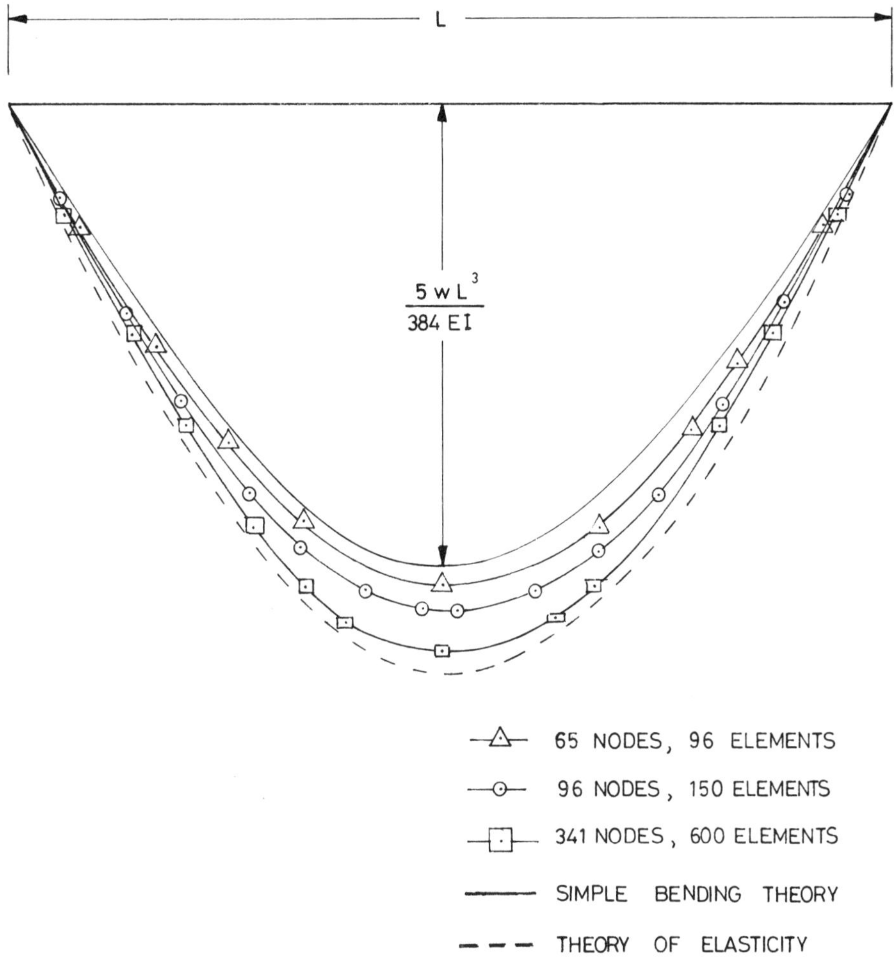

Figure 7.10 Deflection of simply supported deep beam with uniform loading (from Ref. [4]).

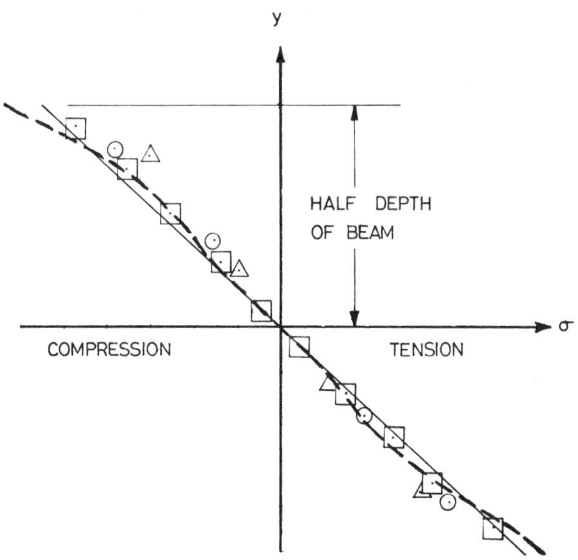

Figure 7.11 Vertical distribution of longitudinal stress, σ (from Ref. 4).

can represent almost any irregularly shaped region. This very useful feature, together with the simplicity of the element, has made it one of the most widely used of elements and this in turn has produced a well-documented relationship between error and mesh size; see for example Ref. 11.

7.4 LINEAR STRAIN RECTANGLE (LSR) ELEMENT

In this section we again use the five basic steps to derive the stiffness matrix of a rectangular plane stress element. The rectangle has sides a and b and thickness t as shown in Fig. 7.12a. The node numbering system is also shown in the figure. Since there are two degrees of freedom at each node the element has eight degrees of freedom. Fig. 7.12b shows the eight nodal displace-

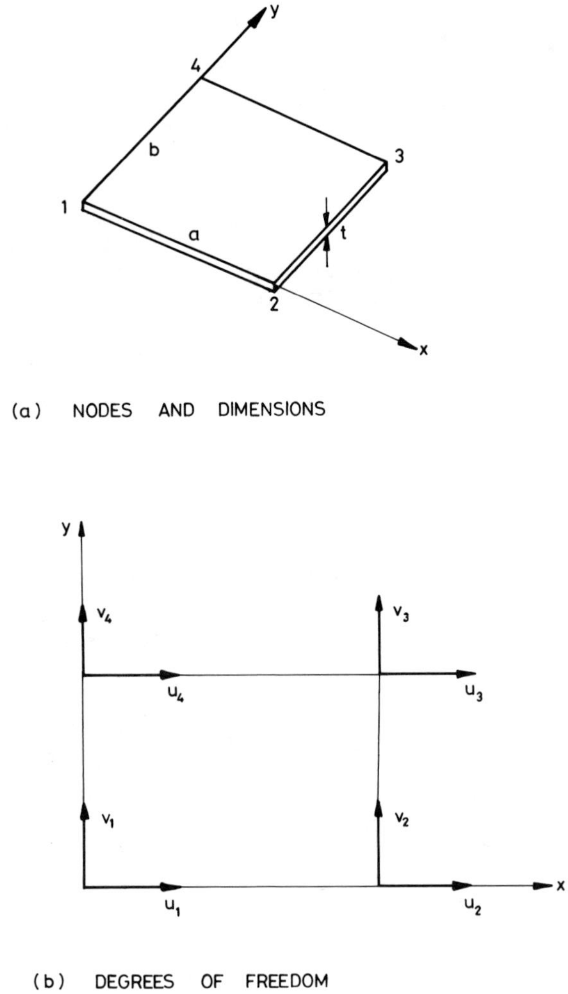

Figure 7.12 Linear strain rectangle.

ments which in matrix form are

$$\boldsymbol{\delta} = \begin{Bmatrix} u_1 \\ v_1 \\ u_2 \\ v_2 \\ u_3 \\ v_3 \\ u_4 \\ v_4 \end{Bmatrix}$$

STEP I. SELECT SUITABLE DISPLACEMENT FUNCTION

Since the element has eight degrees of freedom, eight unknown coefficients must be involved in the polynomial representing the displacement pattern. To avoid bias in the x- or y-direction we use four coefficients for the u displacement and four for the v displacement, and we take an equal number of x and y terms in each. The result is

$$\begin{aligned} u(x,y) &= C_1 + C_2x + C_3y + C_4xy \\ v(x,y) &= C_5 + C_6x + C_7y + C_8xy \end{aligned} \qquad (7.4.1)$$

It may be seen that when x is a constant both u and v vary linearly with y, and similarly when y is a constant both displacements vary linearly with x. The displacements thus vary linearly along each side of the element. Since the displacements of two adjacent elements must be equal at their common nodes, the displacements will also agree along the entire common boundary, that is, interelement compatibility is satisfied.

Writing (7.4.1) in matrix form

$$\begin{Bmatrix} u \\ v \end{Bmatrix} = \begin{bmatrix} 1 & x & y & xy & 0 & 0 & 0 & 0 \\ 0 & 0 & 0 & 0 & 1 & x & y & xy \end{bmatrix} \begin{Bmatrix} C_1 \\ C_2 \\ C_3 \\ C_4 \\ C_5 \\ C_6 \\ C_7 \\ C_8 \end{Bmatrix} \qquad (7.4.2)$$

or

$$\boldsymbol{\delta}(x,y) = \mathbf{H}(x,y)\mathbf{C}$$

STEP II. RELATE THE GENERAL DISPLACEMENT WITHIN THE ELEMENT TO THE NODAL DISPLACEMENTS

As before, this step is achieved by substituting the values of the nodal coordinates into (7.4.2) four times, once for each node, and then solving for **C**. The substitution gives

$$\boldsymbol{\delta} = \begin{bmatrix} 1 & 0 & 0 & 0 & 0 & 0 & 0 & 0 \\ 0 & 0 & 0 & 0 & 1 & 0 & 0 & 0 \\ 1 & a & 0 & 0 & 0 & 0 & 0 & 0 \\ 0 & 0 & 0 & 0 & 1 & a & 0 & 0 \\ 1 & a & b & ab & 0 & 0 & 0 & 0 \\ 0 & 0 & 0 & 0 & 1 & a & b & ab \\ 1 & 0 & b & 0 & 0 & 0 & 0 & 0 \\ 0 & 0 & 0 & 0 & 1 & 0 & b & 0 \end{bmatrix} \mathbf{C} \qquad (7.4.3)$$

or

$$\boldsymbol{\delta} = \mathbf{AC}$$

It may be shown that the solution for **C** is

$$\mathbf{C} = \frac{1}{ab}\begin{bmatrix} ab & 0 & 0 & 0 & 0 & 0 & 0 & 0 \\ -b & 0 & b & 0 & 0 & 0 & 0 & 0 \\ -a & 0 & 0 & 0 & 0 & 0 & a & 0 \\ 1 & 0 & -1 & 0 & 1 & 0 & -1 & 0 \\ 0 & ab & 0 & 0 & 0 & 0 & 0 & 0 \\ 0 & -b & 0 & b & 0 & 0 & 0 & 0 \\ 0 & -a & 0 & 0 & 0 & 0 & 0 & a \\ 0 & 1 & 0 & -1 & 0 & 1 & 0 & -1 \end{bmatrix} \boldsymbol{\delta} \tag{7.4.4}$$

or

$$\mathbf{C} = \mathbf{A}^{-1}\,\boldsymbol{\delta}$$

From (7.4.2) the expression for the general displacement in terms of the nodal displacements is

$$\boldsymbol{\delta}(x,y) = \mathbf{HA}^{-1}\,\boldsymbol{\delta} \tag{7.4.5a}$$

Although it is not required for the derivation of the element stiffness matrix, we next evaluate the product $\mathbf{HA}^{-1}$ to obtain an explicit expression for the element internal displacement $\boldsymbol{\delta}(x,y)$ in terms of the nodal displacements, in order to examine the displacement field of the element more closely. Substituting for **H** and $\mathbf{A}^{-1}$ and multiplying them gives

$$\begin{aligned} u &= (1-\xi)(1-\eta)u_1 + \xi(1-\eta)u_2 + \xi\eta u_3 \\ &\qquad + (1-\xi)\eta u_4 \\ v &= (1-\xi)(1-\eta)v_1 + \xi(1-\eta)v_2 + \xi\eta v_3 \\ &\qquad + (1-\xi)\eta v_4 \end{aligned} \tag{7.4.5b}$$

where $\xi = x/a$ and $\eta = y/b$.

Figure 7.13 shows the displacement field which results when a unit horizontal displacement is imposed at node 3 ($u_3 = 1$; $v = 0$) and all other nodes are pinned: $u_i = v_i = 0$, $i = 1, 2, 4$. A similar pattern would occur if unit displacements were imposed at other nodes. The displacement pattern is extremely simple, which is the reason why interelement compatibility is satisfied, but this simplicity brings with it a large degree of artificiality in the way the element deforms. For example, the pattern implies very large local shear deformation in the vicinity of nodes 2 and 3. The actual deformation of an element of plating subjected to the prescribed nodal displacements of Fig. 7.13 would depend on the total boundary conditions on all of the edges, but the deformation would certainly include some in-plane bending and curved edges, as shown later in Fig. 7.15. Since the LSR element lacks this degree of versatility it will inevitably be too stiff and this brings with it some error, in spite of the fact that interelement compatibility is satisfied exactly. As we shall see in the next section, elements can be derived which have this versatility and have greater accuracy, even when compatibility is not satisfied exactly.

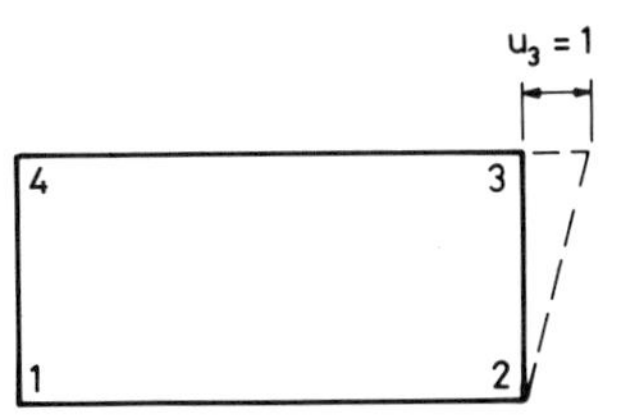

Figure 7.13 Typical unit nodal displacement.

Step III. Express the Internal Deformation (Strain) in Terms of the Nodal Displacements

The definition of strain in a two-dimensional continuum was given in (7.3.10). Substituting for u and v from (7.4.2) and performing the required differentiation gives

$$\boldsymbol{\varepsilon}(x,y) = \begin{bmatrix} 0 & 1 & 0 & y & 0 & 0 & 0 & 0 \\ 0 & 0 & 0 & 0 & 0 & 0 & 1 & x \\ 0 & 0 & 1 & x & 0 & 1 & 0 & y \end{bmatrix} \mathbf{C} \tag{7.4.6}$$

or

$$\boldsymbol{\varepsilon}(x,y) = \mathbf{GC}$$

As before, we substitute for **C** from (7.4.4) and introduce the strain matrix $\mathbf{B} = \mathbf{GA}^{-1}$, in terms of which (7.4.6) becomes

$$\boldsymbol{\varepsilon}(x,y) = \mathbf{B}\,\boldsymbol{\delta} \tag{7.4.7}$$

On performing the required matrix multiplication the strain matrix is found to be

$$\mathbf{B} = \frac{1}{ab} \times \begin{bmatrix} -b+y & 0 & b-y & 0 & y & 0 & -y & 0 \\ 0 & -a+x & 0 & -x & 0 & x & 0 & a-x \\ -a+x & -b+y & -x & b-y & x & y & a-x & -y \end{bmatrix} \tag{7.4.8}$$

From this expression it may be seen that for any given set of nodal displacements $\boldsymbol{\delta}$ the ε_x strain is constant in the x-direction and varies linearly with y. Similarly, the ε_y strain is constant in the y-direction, and varies linearly with s. Finally, the shear strain γ varies linearly with both x and y.

STEP IV. EXPRESS THE INTERNAL FORCE (STRESS) IN TERMS OF THE NODAL DISPLACEMENTS, USING THE ELEMENT'S LAW OF ELASTIC BEHAVIOR

Stress is obtained from strain by $\boldsymbol{\sigma} = \mathbf{D}\boldsymbol{\varepsilon}$, with the $\mathbf{D}$ matrix for plane stress given by (7.3.17). Therefore this step consists in simply substituting for $\boldsymbol{\varepsilon}(x,y)$ from (7.4.7) to obtain

$$\boldsymbol{\sigma}(x,y) = \mathbf{DB}\boldsymbol{\delta}$$

For the derivation of the element stiffness matrix this expression is all that is required. If desired, the explicit stress matrix $\mathbf{S}(x,y) = \mathbf{DB}$ could be calculated at this point, so that it would be available when performing a structural analysis with the element. The expression for $\mathbf{S}(x,y)$ would show that all stress components in the rectangle vary linearly in the x- and y-directions. Substitution of these stresses into the equilibrium relations of (7.3.22) would reveal that, for an arbitrary set of nodal displacements, equilibrium is not satisfied exactly. It would, however, be satisfied for uniform extension ($u_1 = u_4$, $u_2 = u_3$, $v_1 = v_2$, $v_3 = v_4$), which corresponds to uniform direct strain: ε_x = constant and ε_y = constant. The departure from internal equilibrium is therefore proportional to the degree of shear that is present, and so the element can be expected to have less accuracy when large shear stresses are present.

Since the stress distribution within the element depends on all four nodal displacements, two adjacent elements will not have identical stress distributions along their common boundary, and so once again interelement equilibrium is not satisfied to an exact and detailed degree, but only in an overall node-by-node manner.

Since the stresses vary linearly within the element, the stress matrix $\mathbf{S}(x,y)$ does not give specific numerical values of stress as it did for the triangle element. Hence in order to calculate actual stress values it is necessary to choose some specific location. The usual practice is to choose the nodes, much the same as in the calculation of nodal values of bending moment in a beam element. The nodal stress matrix $\mathbf{S}$ is obtained by making four successive substitutions of the four nodal coordinates into the $\mathbf{B}$ matrix, giving an 8×12 matrix.

$$\mathbf{S} = \begin{bmatrix} \mathbf{DB}(0,0) \\ \mathbf{DB}(a,0) \\ \mathbf{DB}(a,b) \\ \mathbf{DB}(0,b) \end{bmatrix}$$

The detailed results are not given here because in the next section we present another rectangle element which has better accuracy than the simple LSR element presented here.

Since the stress distribution in each element is constrained to be of the form specified by $\mathbf{S}(x,y)$, it gives only an approximate representation of the stress distribution in the actual structure. Hence at a common node shared by two or more elements the respective nodal stress values will not be exactly equal. However, the discrepancy is usually small, especially if a small mesh size is used, and a good approximation can be obtained by averaging all of the stress values at a particular node.

STEP V. OBTAIN THE ELEMENT STIFFNESS MATRIX BY RELATING NODAL FORCES TO NODAL DISPLACEMENTS

This step is identical to that of the previous section and hence the element stiffness matrix is again given by

$$\mathbf{k}^e = \left[\int_{\text{vol}} \mathbf{B}^T\mathbf{DB}\, d\text{vol}\right]$$

or, for an element of constant thickness

$$\mathbf{k}^e = t\left[\int_0^b \int_0^a \mathbf{B}^T\mathbf{DB}\, dx\, dy\right]$$

with $\mathbf{B}$ defined by (7.4.8).

Unlike the triangle element, in which $\mathbf{B}$ contained only constant terms, $\mathbf{B}$ here contains both x and y terms and cannot be taken outside the integral. Hence the product $\mathbf{B}^T\mathbf{DB}$ must be evaluated first, and then the terms of the resulting matrix must be integrated over the area of the element. As with the triangle element, it is convenient to present the result as the sum of two parts:

$$\mathbf{k}^e = \mathbf{k}_\varepsilon + \mathbf{k}_\gamma \qquad (7.4.9a)$$

where

$$\mathbf{k}_\varepsilon = \frac{Et}{12(1-\nu^2)} \times \begin{bmatrix} 4/\alpha & & & & & & & \\ 3\nu & 4\alpha & & & & & & \\ -4/\alpha & -3\nu & 4/\alpha & & & & & \\ 3\nu & 2\alpha & -3\nu & 4\alpha & & & & \\ -2/\alpha & -3\nu & 2/\alpha & -3\nu & 4/\alpha & & & \\ -3\nu & -2\alpha & 3\nu & -4\alpha & 3\nu & 4\alpha & & \\ 2/\alpha & 3\nu & -2/\alpha & 3\nu & -4/\alpha & -3\nu & 4/\alpha & \\ -3\nu & -4\alpha & 3\nu & -2\alpha & 3\nu & 2\alpha & -3\nu & 4\alpha \end{bmatrix} \tag{7.4.9b}$$

in which $\alpha = a/b$ and

$$\mathbf{k}_\gamma = \frac{Et}{24(1+\nu)} \times \begin{bmatrix} 4\alpha & & & & & & & \\ 3 & 4/\alpha & & & & & & \\ 2\alpha & 3 & 4\alpha & & & & & \\ -3 & -4/\alpha & -3 & 4/\alpha & & & & \\ -2\alpha & -3 & -4\alpha & 3 & 4\alpha & & & \\ -3 & -2/\alpha & -3 & 2/\alpha & 3 & 4/\alpha & & \\ -4\alpha & -3 & -2\alpha & 3 & 2\alpha & 3 & 4\alpha & \\ 3 & 2/\alpha & 3 & -2/\alpha & -3 & -4/\alpha & -3 & 4/\alpha \end{bmatrix} \tag{7.4.9c}$$

If it is required to transform coordinates the transformation matrix is

$$\mathbf{T} = \begin{bmatrix} \boldsymbol{\lambda} & & & \mathbf{0} \\ & \boldsymbol{\lambda} & & \\ & & \boldsymbol{\lambda} & \\ \mathbf{0} & & & \boldsymbol{\lambda} \end{bmatrix}$$

with $\boldsymbol{\lambda}$ given by (7.3.21).

7.5 CONSTANT SHEAR STRESS RECTANGLE (CSSR) ELEMENT

We now present another rectangle element which is generally superior to the LSR element. It is actually one of the oldest elements, having been introduced in the landmark paper by Turner et al. [5]. Instead of using an assumed displacement function to define the internal behavior of the element, we do this by specifying a stress distribution within the element. We choose a simple distribution in which the direct stresses vary linearly and the shear stress is constant (the latter characteristic furnishing the name of the element)

$$\begin{aligned} \sigma_x &= C_1 + C_2 y \\ \sigma_y &= C_3 + C_4 x \\ \tau &= C_5 \end{aligned} \tag{7.5.1}$$

where $C_1, \ldots, C_5$ are constant coefficients, as yet unknown. Figure 7.14 shows the local coordinate system and other details.

This simple stress distribution, unlike that of the LSR element, satisfies the stress equilibrium relationships of (7.3.22) within the rectangle. On the other hand, as we shall see, the resulting displacement distribution does not satisfy interelement compatibility of boundary displacements.

From the stress–strain relationship for plane stress given in (7.3.16) we have

$$\varepsilon_x = \frac{\partial u}{\partial x} = \frac{1}{E}(C_1 + C_2 y - \nu C_3 - \nu C_4 x) \tag{7.5.2}$$

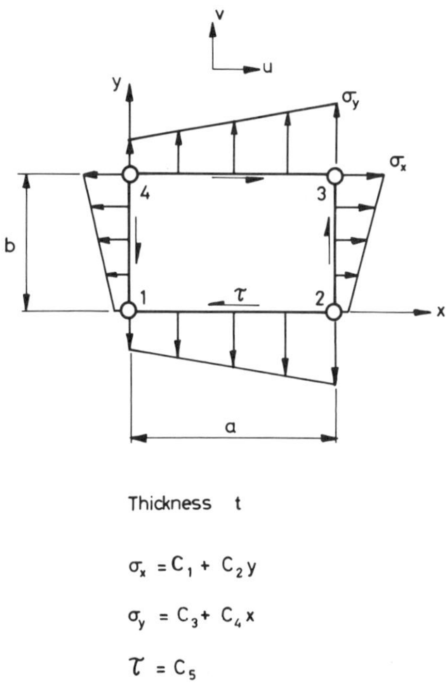

Figure 7.14 Constant shear stress rectangle (CSSR) element.

which, when integrated, becomes

$$u = \frac{1}{E}\left(C_1x + C_2xy - \nu C_3x - \frac{\nu C_4x^2}{2}\right) + \frac{f(y)}{E} \tag{7.5.3}$$

where $f(y)$ is an arbitrary function of y. Similarly, starting with the strain ε_y we can show that

$$v = \frac{1}{E}\left(C_3y + C_4xy - \nu C_1y - \frac{\nu C_2y^2}{2}\right) + \frac{g(x)}{E} \tag{7.5.4}$$

where $g(x)$ is a function of x only. Also, from the equation for the shear strain we have

$$\gamma = \frac{\partial u}{\partial y} + \frac{\partial v}{\partial x} = \frac{\tau}{G} = 2(1 + \nu)\frac{C_5}{E} \tag{7.5.5}$$

Substituting (7.5.3) and (7.5.4) into (7.5.5) and rearranging, we have

$$\frac{\partial f}{\partial y} + C_4y = 2(1 + \nu)C_5 - \left[\frac{\partial g}{\partial x} + C_2x\right] = C_6 \tag{7.5.6}$$

where C_6 represents a constant, which is the only possible condition that will satisfy (7.5.6). Solving for $f(y)$ and $g(x)$, we obtain

$$f(y) = -\frac{C_4y^2}{2} + [(1 + \nu)C_5 + C_6]y + C_7 \tag{7.5.7}$$

$$g(x) = -\frac{C_2x^2}{2} + [(1 + \nu)C_5 - C_6]x + C_8 \tag{7.5.8}$$

The constants of integration C_7 and C_8 represent rigid-body translations, while the previously introduced constants C_5 and C_6 define rigid-body rotation.

If we substitute (7.5.7) and (7.5.8) into (7.5.3) and (7.5.4) we obtain the general displacement in terms of the unknown coefficients

$$\boldsymbol{\delta}(x,y) = \begin{Bmatrix} u \\ v \end{Bmatrix} = \mathbf{H}(x,y)\mathbf{C} \tag{7.5.9}$$

where

$$\mathbf{H} = \frac{1}{E}\begin{bmatrix} x & xy & -\nu x & -\frac{1}{2}(\nu x^2 + y^2) & (1+\nu)y & y & 1 & 0 \\ -\nu y & -\frac{1}{2}(\nu y^2 + x^2) & y & xy & (1+\nu)x & -x & 0 & 1 \end{bmatrix}$$

Equation 7.5.9 is the displacement function which results from the chosen stress distribution and which fulfills the requirement of geometric compatibility within the element. Hence the foregoing analysis is an alternative method of accomplishing Step I in the standard procedure for deriving the stiffness matrix of an element. From here on the remaining four steps are the same as for the previous two elements.

STEP II. RELATE THE GENERAL DISPLACEMENT WITHIN THE ELEMENT TO THE NODAL DISPLACEMENTS

The unknown coefficients $\mathbf{C}$ can now be expressed in terms of the element nodal displacements $\boldsymbol{\delta}$ by the usual technique of substituting each pair of nodal coordinates into $\mathbf{H}(x,y)$ in (7.5.9) to give

$$\boldsymbol{\delta} = \mathbf{AC}$$

where

$$\mathbf{A} = \begin{bmatrix} \mathbf{H}(0,0) \\ \mathbf{H}(a,0) \\ \mathbf{H}(a,b) \\ \mathbf{H}(0,b) \end{bmatrix}$$

and then solving for $\mathbf{C}$ by inverting $\mathbf{A}$

$$\mathbf{C} = \mathbf{A}^{-1}\boldsymbol{\delta} \tag{7.5.10}$$

In the present case the inverse matrix is

$$\mathbf{A}^{-1} = \frac{E}{a}\begin{bmatrix} -\mathcal{C} & -\mathcal{A}\alpha & \mathcal{C} & -\mathcal{A}\alpha & \nu\mathcal{A} & \mathcal{A}\alpha & -\nu\mathcal{A} & \mathcal{A}\alpha \\ \frac{1}{b} & 0 & \frac{-1}{b} & 0 & \frac{1}{b} & 0 & \frac{-1}{b} & 0 \\ -\mathcal{A} & -\mathcal{C}\alpha & \mathcal{A} & -\nu\mathcal{A}\alpha & \mathcal{A} & \nu\mathcal{A}\alpha & -\mathcal{A} & \mathcal{C}\alpha \\ 0 & \frac{1}{b} & 0 & \frac{-1}{b} & 0 & \frac{1}{b} & 0 & \frac{-1}{b} \\ -\mathcal{B}\alpha & -\mathcal{B} & -\mathcal{B}\alpha & \mathcal{B} & \mathcal{B}\alpha & \mathcal{B} & \mathcal{B}\alpha & -\mathcal{B} \\ \frac{-3}{4}\alpha & \frac{3}{4} & \frac{1}{4}\alpha & \frac{-3}{4} & \frac{-1}{4}\alpha & \frac{1}{4} & \frac{3}{4}\alpha & \frac{-1}{4} \\ a & 0 & 0 & 0 & 0 & 0 & 0 & 0 \\ 0 & a & 0 & 0 & 0 & 0 & 0 & 0 \end{bmatrix}$$

where

$$\mathcal{A} = \frac{\nu}{2(1 - \nu^2)} \qquad \mathcal{B} = \frac{1}{4(1 + \nu)}$$

$$\mathcal{C} = \frac{2 - \nu^2}{2(1 - \nu^2)} \qquad \alpha = \frac{a}{b}$$

At this point we digress momentarily from the derivation of $\mathbf{k}^e$ in order to examine the displacement field of this element more closely. For this purpose we calculate the explicit expression for the general displacement $\boldsymbol{\delta}(x,y)$ in terms of the nodal displacements. From (7.5.9) and (7.5.10) the matrix expression is

$$\boldsymbol{\delta}(x,y) = \begin{Bmatrix} u \\ v \end{Bmatrix} = \mathbf{H}(x,y)\mathbf{A}^{-1}\,\boldsymbol{\delta}$$

and after substituting for $\mathbf{A}^{-1}$ and performing the reqired matrix multiplication the final result is

$$\begin{aligned} u &= (1 - \xi)(1 - \eta)u_1 + \xi(1 - \eta)u_2 + \xi\eta u_3 \\ &\quad + (1 - \xi)\eta u_4 \\ &\quad + \frac{1}{2}\left[\nu\alpha(\xi - \xi^2) + \frac{1}{\alpha}(\eta - \eta^2)\right] \times \\ &\quad (v_1 - v_2 + v_3 - v_4) \\ v &= (1 - \xi)(1 - \eta)v_1 + \xi(1 - \eta)v_2 + \xi\eta v_3 \\ &\quad + (1 - \xi)\eta v_4 \\ &\quad + \frac{1}{2}\left[\alpha(\xi - \xi^2) + \frac{\nu}{\alpha}(\eta - \eta^2)\right] \times \\ &\quad (u_1 - u_2 + u_3 - u_4) \end{aligned} \tag{7.5.11}$$

where $\xi = x/a$, $\eta = y/b$, and $\alpha = a/b$.

Fig. 7.15 illustrates the deformation of the element for a unit value of u_3 and with all other nodal displacements held at zero, as was done for the LSR element in Fig. 7.13. Along the top edge ($0 \le \xi \le 1$; $\eta = 1$) the displacements are

$$u = \xi; \qquad v = \frac{\alpha}{2}(\xi - \xi^2)$$

and along the bottom edge ($0 \le \xi \le 1$; $\eta = 0$) they are

$$u = 0; \qquad v = \frac{\alpha}{2}(\xi - \xi^2)$$

In addition to the elongation of the top edge there is some sympathetic curvature of the top and bottom edges, which makes the deformed shape more realistic because it avoids the large local shear deformation which is implied in the LSR element. In fact, by

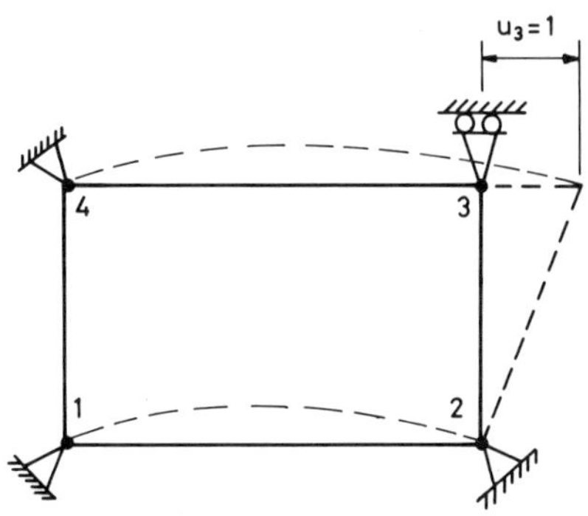

Figure 7.15 Deformation of CSSR element due to displacement u_3.

definition, the shear stress in the CSSR element is uniform throughout the element.

The v displacement of the upper and lower edges is a quadratic expression, which requires three coefficients to define uniquely, whereas there are only two nodal values of v along each edge. Hence interelement compatibility is not entirely satisfied by this element.

It would be possible to have quadratically varying edge displacements and to also satisfy interelement compatibility by adding midside nodes along each edge because with three nodes the quadratic would be uniquely defined. Another possibility is that elements may have more than two degrees of freedom per node (e.g., they could involve the derivatives of u and v) thereby involving more coefficients and hence higher order displacement functions. However, midside nodes cause complications in mesh generation, and both of these approaches increase the total number of degrees of freedom and hence the solution time and cost. In many cases these higher order elements give simply using more elements. For example, a rectangle with midside nodes has a total of 16 degrees of freedom, whereas using four quarter size ordinary rectangle elements would involve 18 degrees of freedom, which is only a marginal increase and which gives a more detailed solution.

The remaining three steps in the derivation of $\mathbf{k}^e$ for the CSSR element are the same as for the LSR element and therefore only the results are presented. The stiffness matrix and the equations for nodal values of stress are given in (7.5.12) and (7.5.13).

$$\mathbf{k}^e = \frac{Et}{4(1-\nu^2)} \begin{bmatrix} \mathcal{A}\alpha+\mathcal{D}/\alpha & & & & & & & \\ \mathcal{B} & \mathcal{A}/\alpha+\mathcal{D}\alpha & & & & & & \\ \mathcal{A}\alpha-\mathcal{D}/\alpha & \mathcal{C} & \mathcal{A}\alpha+\mathcal{D}/\alpha & & & & & \\ -\mathcal{C} & -\mathcal{A}/\alpha+\mathcal{E}\alpha & -\mathcal{B} & \mathcal{A}/\alpha+\mathcal{D}\alpha & & & & \\ -\mathcal{A}\alpha-\mathcal{E}/\alpha & -\mathcal{B} & -\mathcal{A}\alpha+\mathcal{E}/\alpha & \mathcal{C} & \mathcal{A}\alpha+\mathcal{D}/\alpha & & & \\ -\mathcal{B} & -\mathcal{A}/\alpha-\mathcal{E}\alpha & -\mathcal{C} & \mathcal{A}/\alpha-\mathcal{D}\alpha & \mathcal{B} & \mathcal{A}/\alpha+\mathcal{D}\alpha & & \\ -\mathcal{A}\alpha+\mathcal{E}/\alpha & -\mathcal{C} & -\mathcal{A}\alpha-\mathcal{E}/\alpha & \mathcal{B} & \mathcal{A}\alpha-\mathcal{D}/\alpha & \mathcal{C} & \mathcal{A}\alpha+\mathcal{D}/\alpha & \\ \mathcal{C} & \mathcal{A}/\alpha-\mathcal{D}\alpha & \mathcal{B} & -\mathcal{A}/\alpha-\mathcal{E}\alpha & -\mathcal{C} & -\mathcal{A}/\alpha+\mathcal{E}\alpha & -\mathcal{B} & \mathcal{A}/\alpha+\mathcal{D}\alpha \end{bmatrix} \tag{7.5.12}$$

where

$$\mathcal{A} = \frac{1-\nu}{2}; \quad \mathcal{B} = \frac{1+\nu}{2}; \quad \mathcal{C} = \frac{1-3\nu}{2}$$

$$\mathcal{D} = \frac{4-\nu^2}{3}; \quad \mathcal{E} = \frac{2+\nu^2}{3}; \quad \alpha = \frac{a}{b}$$

$$\begin{aligned}
(\sigma_x)_1 = (\sigma_x)_2 &= \frac{E}{1-\nu^2}\left[\frac{-u_1+u_2}{a} + \frac{\nu^2(u_1-u_2+u_3-u_4)}{2a} + \frac{\nu(-v_1-v_2+v_3+v_4)}{2b}\right] \\
(\sigma_x)_3 = (\sigma_x)_4 &= \frac{E}{1-\nu^2}\left[\frac{u_3-u_4}{a} + \frac{\nu^2(-u_1+u_2-u_3+u_4)}{2a} + \frac{\nu(-v_1-v_2+v_3+v_4)}{2b}\right] \\
(\sigma_y)_1 = (\sigma_y)_4 &= \frac{E}{1-\nu^2}\left[\frac{-v_1+v_4}{b} + \frac{\nu^2(v_1-v_2+v_3-v_4)}{2b} + \frac{\nu(-u_1+u_2+u_3-u_4)}{2a}\right] \\
(\sigma_y)_3 = (\sigma_y)_2 &= \frac{E}{1-\nu^2}\left[\frac{-v_2+v_3}{b} + \frac{\nu^2(-v_1+v_2-v_3+v_4)}{2b} + \frac{\nu(-u_1+u_2+u_3-u_4)}{2a}\right] \\
\tau &= \frac{E}{4(1+\nu)}\left[\frac{-u_1-u_2+u_3+u_4}{b} + \frac{-v_1+v_2+v_3-v_4}{a}\right]
\end{aligned} \tag{7.5.13}$$

Accuracy and Application of the CSSR Element

The assumed stress distribution function of the CSSR element and its ability to represent in-plane bending make it especially suitable for the sides and longitudinal bulkheads of a ship because these are the webs of the hull girder and they are subjected to in-plane bending and nearly uniform shear stress. The element can also be used to represent panels of deck plating because, although the shear stress in a deck varies a lower ratio of accuracy to cost than is achieved by linearly in the transverse direction, the use of a local average value within each panel provides an adequate representation. In fact, the CSSR element has been found to be sufficiently accurate and versatile to be able to represent relatively large panels of plating. As shown in Fig. 8.1 it can be used to model the rectangular panels bounded by the major transverse members (frames, floors, or deck beams) and major longitudinal boundaries (bilge, upper deck edge, intersections of decks with longitudinal girders, bulkheads or side shell, etc.).

The suitability of the CSSR element for representing large panels and its superiority over the LSR element are shown by the test results reported in Refs. 12 and 13 which are presented in Figs. 7.16 and 7.17. The test structure is a deep web cantilever beam subjected to a vertical end load. Thus there is a large amount of shear as well as bending, and the proportions are such that the web of the cantilever is analogous to the side shell of a ship. Figure 7.16 shows that even when only four elements are used ($m = 4$; $n = 1$) such that each element extends over the full depth, the error for the CSSR element is only a few percent, compared to about 30% for the LSR element. A more stringent and more relevant criterion is stress rather than deflection, because displacement-based finite elements usually give more accurate deflections than stresses, whereas it is stresses which are of primary interest in design. Values of maximum stress for this test problem have been obtained by Davies [13] and are presented in Fig. 7.17. The figure shows that the CSSR element is again more accurate for coarse mesh modeling. Also, it shows a consistent convergence to the exact value as mesh size decreases, whereas the curve of error for the LSR element overshoots the zero axis.

Of course, the large panels of plating of the type we are here considering would nearly always be stiffened panels, and hence it is necessary to allow for the effect of the stiffeners on both the deflection and the stress. This is treated in Section 8.4, where it is shown that for stiffened panels the CSSR element can again be used to represent the plating and that, moreover, one of the most efficient techniques for modeling the stiffeners (for purposes of a hull module analysis) is a technique which is also based on the assumed stress field of the CSSR element.

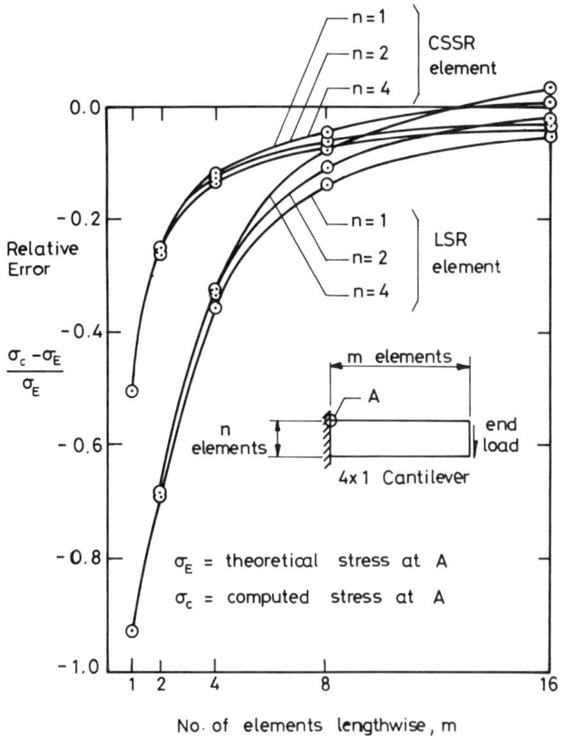

Figure 7.17 Accuracy of stress using the LSR and CSSR elements.

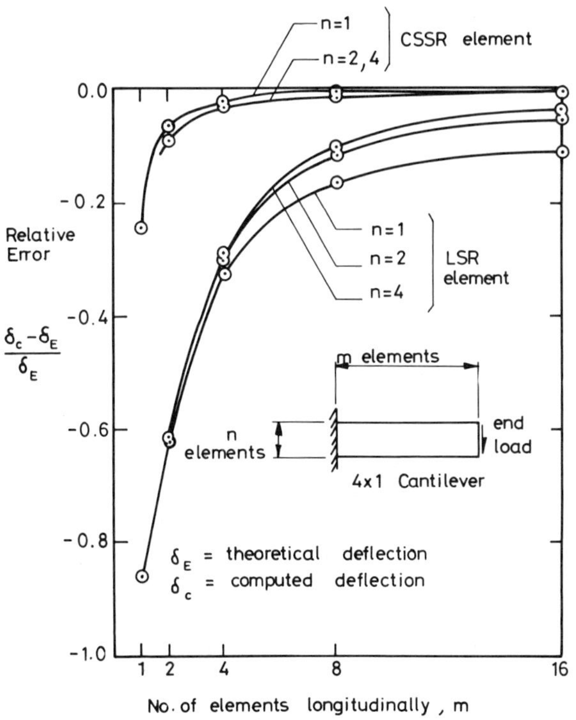

Figure 7.16 Accuracy of deflections using LSR and CSSR elements.

7.6 SHAPE FUNCTIONS AND EQUIVALENT NODAL FORCES

Step II in the derivation of an element stiffness matrix leads to an expression for the displacements $u(x, y)$ and $v(x, y)$ anywhere within the element in terms of the nodal displacements. For example, for the LSR element the expression is given in (7.4.5b) which we now rewrite in matrix notation

$$\boldsymbol{\delta}(x, y) = \mathbf{N}^e(\xi, \eta)\boldsymbol{\delta}$$

or in expanded form

$$\begin{Bmatrix} u(x, y) \\ v(x, y) \end{Bmatrix} = \begin{bmatrix} \mathbf{N}(\xi, \eta) & \\ & \mathbf{N}(\xi, \eta) \end{bmatrix} \begin{Bmatrix} u \\ v \end{Bmatrix} \tag{7.6.1}$$

where

$$\mathbf{N}(\xi, \eta) = [(1 - \xi)(1 - \eta) \;\; \xi(1 - \eta) \;\; \xi\eta \;\; (1 - \xi)\eta]$$

and in which $\xi = x/a$ and $\eta = y/b$.

The vector $\mathbf{N}(\xi, \eta)$ is referred to as the *shape function* because it completely specifies the pattern or shape of the displacement everywhere within the element, given the nodal displacements. We will now show how the shape function can be used to obtain a general expression for the equivalent nodal forces for the two types of distributed external loads: "body forces" **b** (forces on a body that are proportional to the body's volume, such as weight and inertia) and "surface forces" **p** (forces which act on the surface, either normally—pressure—or tangentially—shear or "traction"). To do so we will again utilize the Principle of Virtual Work, but we now apply it *to the overall structure*, rather than to just one element. The expression is

$$\boldsymbol{\Delta}^{*T}\mathbf{P} + \int \boldsymbol{\Delta}^*(\mathbf{X})^T\mathbf{b}(\mathbf{X})dV + \int \boldsymbol{\Delta}^*(\mathbf{S})^T\mathbf{p}(\mathbf{S})dA = \int \boldsymbol{\epsilon}^*(\mathbf{X})^T\boldsymbol{\sigma}(\mathbf{X})dV \tag{7.6.2}$$

The left-hand side is the virtual work done by point loads **P**, body forces **b**, and surface forces **p**. V and A refer to the volume and external surface area of the structure, and **X** and **S** are coordinates indicating location within this volume and on this surface, respectively. The right-hand side is the internal virtual work.

We now make the two substitutions that lie at the heart of the finite element method:

1. We divide the structure volume V and the structure boundary S into elements and use integration by parts; that is, the integrals over the structure are each replaced by a sum of integrals over the elements.
2. We represent the displacement within each element by means of shape functions $\mathbf{N}(\mathbf{X})$ which specify the displacement in terms of the nodal values, $\boldsymbol{\Delta}$:

$$\boldsymbol{\Delta}(\mathbf{X}) = \mathbf{N}(\mathbf{X})\boldsymbol{\Delta} \tag{7.6.3}$$

With these substitutions (7.6.2) becomes*

$$\boldsymbol{\Delta}^{*T}\mathbf{P} + \sum_e \left[\int_{V^e} (\mathbf{N}\boldsymbol{\Delta}^*)^T\mathbf{b}(\mathbf{X})dV^e + \int_{A^e} (\mathbf{N}\boldsymbol{\Delta}^*)^T\mathbf{p}(\mathbf{S})dA^e \right] = \sum_e \int_{V^e} (\mathbf{B}\boldsymbol{\Delta}^*)^T\mathbf{D}\mathbf{B}\,\boldsymbol{\Delta}dV^e$$

or

$$\boldsymbol{\Delta}^{*T}\left[\mathbf{P} + \sum_e \int_{V^e} \mathbf{N}^T\mathbf{b}(\mathbf{X})dV^e + \sum_e \int_{A^e} \mathbf{N}^T\mathbf{p}(\mathbf{S})dA^e\right] = \boldsymbol{\Delta}^{*T}\left[\sum_e \int_{V^e} \mathbf{B}^T\mathbf{D}\mathbf{B}dV^e\right]\boldsymbol{\Delta} \tag{7.6.4}$$

in which the virtual displacements $\boldsymbol{\Delta}^*$ cancel. The right-hand side is seen to be $\mathbf{K}\boldsymbol{\Delta}$, and hence the left-hand side is the complete load vector, in which the second and third terms are the equivalent discrete nodal forces, in structure coordinates, that are used to represent the distributed body forces and surface forces that act on the structure. These equivalent nodal forces are calculated element by element and summed over all of the structure degrees of freedom, in much the same manner as the structure stiffness matrix is assembled element by element.

7.7 QUADRILATERAL AND ISOPARAMETRIC ELEMENTS

In this text the process of defining and deriving each new finite element has usually commenced with the selection of a displacement function $\mathbf{H}(x)$ or $\mathbf{H}(x, y)$, whereupon the shape function $\mathbf{N}$ is obtained from $\mathbf{N} = \mathbf{H}\mathbf{A}^{-1}$. But an alternative (and more general) approach is to begin by specifying the shape function. Once this is done all of the element's properties (stiffness matrix, stress matrix, equivalent nodal forces,

*In reality the shape function is defined in element coordinates and (7.6.3) implies a prior transformation to structure coordinates.

etc.) can be established uniquely. As shown in Refs. 1–4 and other standard texts, many elements are now available for various specialized applications, and many of these have been developed through the use of shape functions. This is a very large topic and all of it is beyond the scope of this text. But there is one such element that is of particular relevance to a hull module: a general quadrilateral that is sufficiently accurate for modeling large nonrectangular panels of plating; for example, panels of shell plating near the bow or stern. Therefore in this section we merely identify the element and, in doing this, make brief mention of the use of shape functions and also of numerical integration in the definition of such elements. The derivation of the element and a demonstration of its accuracy are given in Refs. 14 and 15.

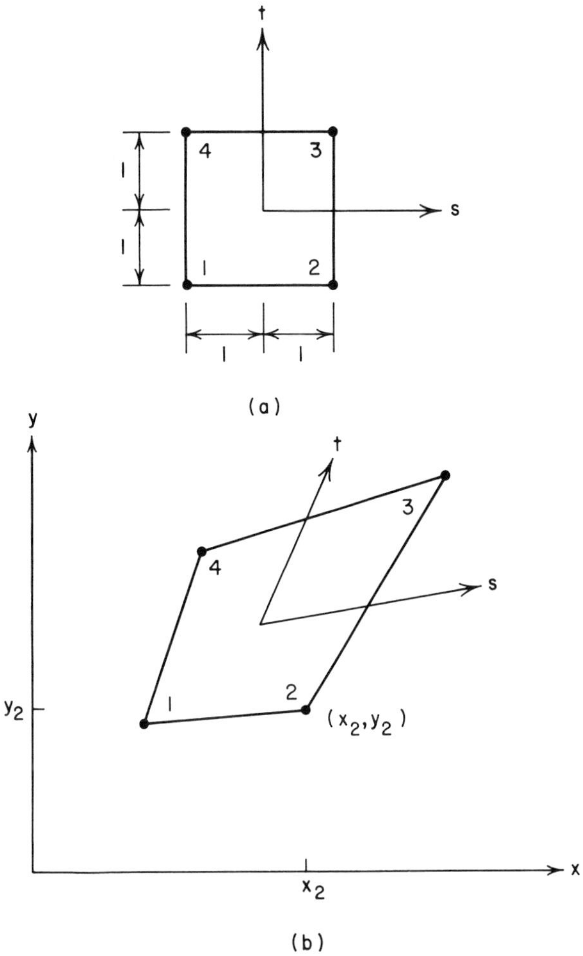

Figure 7.18 (a) Natural coordinates; (b) Isoparametric quadrilateral element.

Natural Coordinates

For a four-sided element, shape functions are conveniently defined in terms of "natural" coordinates s, t which have a unit value at the nodes, as shown in Fig. 7.18a. A set of four shape functions is defined as follows:

$$N_i(s, t) = \tfrac{1}{4}(1 + s_i s)(1 + t_i t) \qquad (i = 1, \cdots, 4) \tag{7.7.1}$$

in which s_i and t_i are the values of s and t at node i (either $+1$ or -1). Thus each $N_i(s, t)$ is a smoothly varying interpolating function which has a unit value at node i and is zero at the other three nodes. The internal displacement can therefore be expressed as the sum of these four functions, with each being multiplied by its associated nodal displacement:

$$\boldsymbol{\delta}(s, t) = \sum_{i=1}^{4} \boldsymbol{\delta}_i N_i(s, t) \tag{7.7.2}$$

Non-Conforming Displacement Modes

The element displacement defined by (7.7.1) and (7.7.2) is in fact the same shape function as that of the LSR element given in (7.6.1); the only difference is the use of natural coordinates. Section 7.5 showed that the displacement function of the CSSR element is preferable to that of the LSR element because it contains quadratic terms that allow the element to accommodate in-plane bending. Shape functions provide a means for adding such higher order terms explicitly and selectively, in order to allow an element to accommodate a particular mode of deformation. For example, this technique may be used to explicitly add some quadratic terms to the LSR displacement, and the resulting element is found to be identical to the CSSR element. This method of derivation was used by Wilson et al. [14], but it was not recognized until some time later that the element was actually the same as the CSSR element. When the quadratic terms are added to the basic shape function of (7.7.2) the resulting displacement function is

$$\boldsymbol{\delta}(s, t) = \sum_{i=1}^{4} \boldsymbol{\delta}_i N_i(s, t) + \sum_{i=5}^{6} \boldsymbol{\delta}_i \overline{N}_i(s, t) \tag{7.7.3}$$

where the additional shape functions are

$$\overline{N}_5(s, t) = 1 - s^2$$

and

$$\overline{N}_6(s, t) = 1 - t^2$$

As noted in the discussion following the derivation of the CSSR element, the presence of the quadratic terms means that interelement compatibility is not perfectly satisfied. Such elements are referred to as "nonconforming," and, if this feature is used properly, it gives greater accuracy because a conforming element is always "too stiff."

Isoparametric Elements: Quadrilateral

Besides their use in describing displacements within elements, shape functions can also be used to define the basic geometry of the elements themselves. This technique allows the development of a wide variety of two- and three-dimensional elements known collectively as *isoparametric* elements. We here consider only one of these—the straight-sided quadrilateral shown in Figure 7.18b. For conceptual purposes this quadrilateral may be regarded as the result of a smoothly varying, in-plane "distortion" of the basic square. The four nodes are moved from their unit coordinates in the square to the points defined by the quadrilateral's nodal coordinates $\mathbf{x}_i = (x_i, y_i)$. The latter may be regarded as scaling factors which are applied to the original unit coordinates. To obtain the complete quadrilateral these scale factors are applied to the unit shape functions, as follows:

$$\mathbf{x} = \sum_{i=1}^{4} \mathbf{x}_i N_i(s, t)$$

References 14 and 15 use this shape function approach to derive a quadrilateral element, called the QM6 element, that is similar to the CSSR element and that reduces to it when the quadrilateral is a rectangle. The element is also summarized in Section 11.3 of Ref. 1. Like the CSSR element, it is well-suited for representing panels of plating in a hull module, and can be used in place of the CSSR element in nonprismatic modules such as bow or stern or other distinctly tapered portions of a ship.

Numerical Integration

Because of the greater complexity of the QM6 element in both external shape and internal displacement, the integration involved in obtaining the element stiffness matrix must be performed numerically rather than analytically. Therefore, neither the stiffness matrix nor the stress matrix can be obtained in explicit algebraic form, but this is not a problem; the matrices are generated by the computer as required. Although this involves more computation the increase is quite acceptable when balanced against the great versatility of the element. Also, the use of numerical integration brings with it the possibility of "optimal sampling," a technique which gives improved accuracy in the final stress values. This is briefly summarized in Section 11.5 of Ref. 1.

7.8 DYNAMIC ANALYSIS BY THE FINITE ELEMENT METHOD

The finite element method can also be used for dynamic structural analysis, such as calculating a structure's response to a given dynamic load or determining its natural frequencies of vibration. For a dynamic system such as the spring and point mass of Fig. 7.19 Newton's law

$$m\ddot{u} = P - ku$$

can be written

$$ku = P - m\ddot{u}$$

The term $-m\ddot{u}$ can be considered an "inertial force," acting oppositely to the direction of acceleration; this is d'Alembert's Principle. This is an example of a body force, as discussed in Section 7.6, and so in a finite element analysis it can be accounted for in the same manner. Thus if an element is undergoing accelerations $\ddot{u}(x, y)$ and $\ddot{v}(x, y)$, the distributed body force $\mathbf{b}$ has components $b_x = -\rho\ddot{u}(x, y)$ and $b_y = -\rho\ddot{v}(x, y)$, where ρ is the mass density of the element. The equivalent nodal forces are given by the second term of (7.6.4).

$$\mathbf{F}_{eq} = \sum_e \int_{V^e} \mathbf{N}^T \mathbf{b}(\mathbf{X}) dV^e \qquad (7.8.1)$$

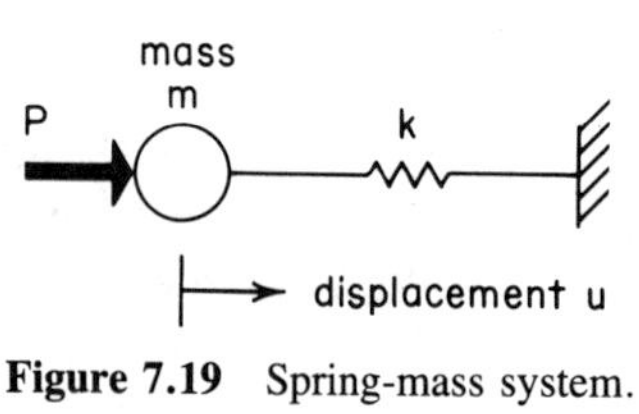

Figure 7.19 Spring-mass system.

For a linear elastic body the natural frequency of vibration may be calculated from Newton's Law with the applied loads set to zero. For the undamped point mass of Fig. 7.19 the equation is

$$m\ddot{u} + ku = 0 \tag{7.8.2}$$

The solution to this differential equation is the simple harmonic vibration

$$u = a \sin \omega t$$

In all such cases the acceleration is a scalar multiple of u

$$\ddot{u} = -\omega^2 a \sin \omega t$$

and ω is the natural frequency, given by

$$\omega = \left(\frac{k}{m}\right)^{1/4}$$

In an extended body such as say, a two-dimensional finite element, the accelerations $\ddot{u}$ and $\ddot{v}$ continue to have the same general form as the displacements u and v. Therefore the same shape function that describes the element's internal displacement in terms of nodal displacements also gives its internal acceleration in terms of the nodal accelerations, $\ddot{u}$ and $\ddot{v}$; that is

$$\ddot{u}(x, y) = N(x, y)\ddot{u}$$

$$\ddot{v}(x, y) = N(x, y)\ddot{v}$$

Transforming to structure coordinates and substituting into (7.8.1) gives

$$\mathbf{F}_{eq} = -\sum_e \left[\int_{V^e} \mathbf{N}^T \boldsymbol{\rho} \mathbf{N} dV^e\right] \ddot{\boldsymbol{\Delta}} \tag{7.8.3}$$

in which for two-dimensional motion

$$\boldsymbol{\rho} = \begin{bmatrix} \rho & 0 \\ 0 & \rho \end{bmatrix}$$

In (7.8.3) the expression in square brackets is a square matrix which accounts for the inertial effects (i.e., the mass) of each element. It is therefore known as the *element mass matrix*, $\mathbf{M}^e$, and the assemblage of these is the *structure mass matrix*: $\mathbf{M} = \Sigma_e \mathbf{M}^e$. Thus, for a complete structure the equation corresponding to (7.8.2) is

$$\mathbf{M}\ddot{\mathbf{U}} + \mathbf{K}\mathbf{U} = 0$$

The general solution is again of the form

$$\mathbf{U} = \overline{\mathbf{U}} \sin \omega t$$

and after taking second derivative and substituting for $\ddot{\mathbf{U}}$ the equation becomes the *eigenvalue* or *characteristic value* problem.

$$[-\omega^2\mathbf{M} + \mathbf{K}]\overline{\mathbf{U}} = 0$$

for which a nonzero solution requires that the determinant of the coefficients be zero.

$$\det\,[-\omega^2\mathbf{M} + \mathbf{K}] = 0$$

The values of ω that are obtained from this are the natural frequencies of the structure.

This extremely brief discussion of dynamic structural analysis is intended simply to show that the finite element method is a powerful tool for such analysis, and that it greatly unifies static analysis and dynamic analysis, allowing dynamic aspects (such as the natural frequencies of a hull module) to be calculated as part of the response analysis. This in turn allows dynamic constraints (such as the avoidance of certain natural frequencies) to become an integral part of the design process, just as much as the static design constraints. The only significant additional steps are the assembly of the structure mass matrix and the calculation of the eigenvalues, for which various computer routines are available. As noted in Section 1.1, it is intended to cover dynamic aspects of design in a companion text to this.

REFERENCES

1. O. C. Zienkiewicz, *The Finite Element Method*, Third Edition, McGraw-Hill, 1977.
2. H. C. Martin and G. F. Carey, *Introduction to Finite Element Analysis*, McGraw-Hill, 1973.
3. R. H. Gallagher, *Finite Element Analysis*, Prentice-Hall, 1975.
4. K. C. Rockey, H. R. Evans, D. W. Griffiths, and D. A. Nethercot, *The Finite Element Method*, Crosby Lockwood Staples, 1975.
5. M. Turner, R. Clough, H. Martin, and L. Topp, "Stiffness and Deflection Analysis of Complex Structures," *J. Aero. Sci.*, **23** (9), Sept. 1956, pp. 805–823.
6. R. Courant, "Variational Methods for the Solution of Problems of Equilibrium and Vibration," *Bull. Am. Math. Soc.*, **49,** 1943, pp. 1–43.

7. D. McHenry, "A Lattice Analogy for the Solution of Plane Stress Problems," *J. Inst. Civil Eng.*, **21,** 1943, pp. 59–82.
8. A. Hrenikoff, "Solution of Problems in Elasticity by the Framework Method," *J. Appl. Mech.*, **8,** 1941, pp. 169–175.
9. J. Argyris and S. Kelsey, *Energy Theorems and Structural Analysis,* Butterworth Scientific Publications, London, 1960.
10. O. C. Zienkiewicz and Y. K. Cheung, "Finite Elements in the Solution of Field Problems," *The Engineer,* **220,** 24 Sept. 1964, pp. 507–510.
11. J. E. Walz, R. E. Fulton, and N. J. Cyrus, "Accuracy and Convergence of Finite Element Approximations," Proceedings of 2nd Conference on Matrix Methods in Structural Mechanics (AFFDL-TR-68-150), Wright-Patterson AFB, Ohio, 1968.
12. R. H. Gallagher et al., "A Correlation Study of Methods of Matrix Structural Analysis," AGARDograf 69, Pergamon (distrib. by Macmillan, New York) 1964.
13. J. D. Davies, "A Finite Element to Model the In-Plane Response of Ribbed Rectangular Panels," M. Eng. Sc. thesis, University of New South Wales, Sydney, 1976.
14. E. L. Wilson, R. L. Taylor, W. P. Doherty and T. Ghabussi, "Incompatible Displacement Models," in *Numerical and Computer Methods in Structural Mechanics,* ed. S. T. Fenves *et al.*, Academic Press, 1973, pp. 43–57.
15. R. L. Taylor, P. J. Beresford and E. L. Wilson, "A Non–conforming Element for Stress Analysis," *Int. J. Num. Meth. Eng.*, **10,** 1976, pp. 1211–1220.

Chapter

Eight

HULL MODULE MODELING AND ELASTIC ANALYSIS

This chapter presents the main principles and some practical aspects of the structural modeling of a hull module, with special emphasis on the particular elements and modeling techniques which the author has found to be suitable for achieving rapid, efficient design-oriented finite element analysis.

8.1 CONSIDERATIONS REGARDING THE STRUCTURAL MODEL

Modeling of a Stiffened Panel

A hull module is essentially a three-dimensional framework of beams and stiffened panels, and the hull module analysis consists of the calculation of all of the principal responses of these members. The phrase "principal response" is meant to exclude those responses that are purely local and that can be analyzed independently of the hull module analysis. In a stiffened panel there are two very common responses of this type: bending of the stiffeners due to a lateral load (pressure, wheel loads, etc.) and bending of the plating between stiffeners, also due to a lateral load. In a hull module most of the lateral loads act initially on the plating. Then, through the action of plate bending, the plating transmits the lateral load to the nearby stiffeners. The stiffeners are essentially small beams and they in turn transmit the load to the nearby major beams, that is, the transverse frames and longitudinal girders. The important point is that both of these load transmissions, one by plate bending and the other by stiffener bending, are local phenomena. That is, they do not affect the next higher level of loading response and their boundary conditions can be estimated with sufficient accuracy to allow these phenomena to be analyzed separately. Therefore it is not necessary to model either the plate bending stiffness or the bending stiffness of the stiffeners. This means that a stiffened panel can be represented by an element that has only membrane stiffness. On the other hand, it *is* necessary to model the in-plane or axial stiffness of the stiffeners and so the element which represents the stiffened panels must have this capability. A suitable element for representing stiffened panels is presented in Section 8.4.

Analysis of Laterally Loaded Panels

The foregoing discussion only applies to plate stiffeners; all of the larger beams in a hull module should be modeled explicitly, including the beams that support the stiffeners. Hence in the separate analysis of the bending response of the stiffened panels the stiffening is uniaxial, and in such panels the stiffeners are essentially independent and can be analyzed as separate beams. Where they are identical and equally spaced only one analysis is required. The flexural properties may be calculated by means of (8.3.6). Under a distributed pressure load the ends can be taken as clamped; under more local types of load which may act on only one panel the end rotational restraint is relatively small because the supporting beams are nearly always of open section and such members have relatively low torsional rigidity. If the stiffening extends over several supporting beams in both directions the maximum bending moment values for a point load P are $Pa/12$ at the ends of the panel and $Pa/6$ at the midlength.

If the panel is stiffened in both directions it may be

analyzed by one of the many grillage analysis methods which are available; see Refs. 1 and 2 of Chapter 5.

Mesh Orthogonality and Mesh Size

In regard to its overall tranverse dimensions, a hull module is usually prismatic or, if it is not so, it may usually be idealized as such; in other words the taper of the hull girder, if any, can usually be neglected. In this case the longitudinal mesh of nodal points is always orthogonal to the transverse plane. This greatly facilitates the generation of the three-dimensional nodal mesh and other aspects of the structural model. Also, it allows rectangular elements to be used for all of the panel elements that are oriented in the longitudinal direction. In regard to individual members, the longitudinal beams are usually prismatic and the stiffened panels in a longitudinal row usually have identical geometry because of the need for longitudinal continuity of structure and the desirability of simple, regular structural geometry.

To minimize the computation required, it is desirable to make the elements in the hull module as large as possible. Since the stiffened panels are represented by rectangular membrane elements, the structural model only requires nodes at the corners of these elements and also at the junctions of beam elements if such junctions occur at other locations. Figure 8.1 shows some typical elements in a hull module. It can be seen that the stiffened panel elements are indeed quite large. For example, in the deck they extend from the side shell to the hatch side girder and from one transverse deck beam to the next. In the side shell they extend a full deck height and from one major frame to the next. If there are small closely spaced subframes they can be treated as stiffeners (i.e., their bending can be examined separately) and so only their axial stiffness needs to be modeled and this can be incorporated in the stiffened panel element.

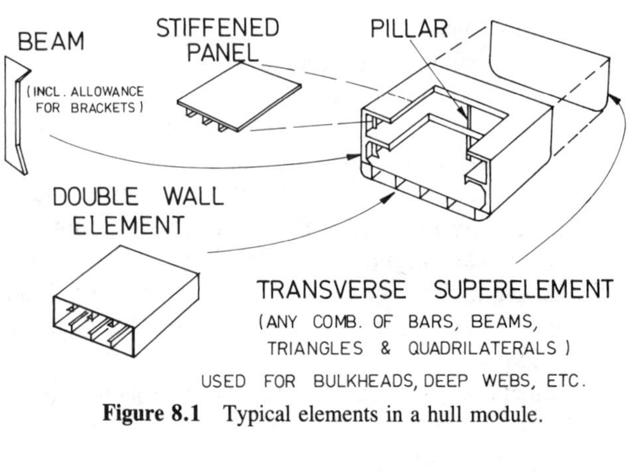

Figure 8.1 Typical elements in a hull module.

Section 8.3 explains how the beam element allows for the contribution which the plating makes to the beam's flexural rigidity. Section 8.6 presents techniques for representing the extra flexural rigidity at the end of a beam, due either to a deep web in the adjacent orthogonal beam or to a bracket, and for representing the shear distortion of joints, which is quite significant for deep beams.

As shown in the figure the bulkhead is usually modeled as a *superelement,* which is a combination of bar, beam, triangle, and quadrilateral elements that have been joined together to form a single element, with its own stiffness matrix. Nodes that are wholly within the superelement and are not required for attaching it to the rest of the structure may be deleted; if the superelement occurs repeatedly this deletion provides a savings in computation. Superelements may also be used to represent large brackets and any other portion of structure which does not correspond to a standard element. The superelement technique is explained in Section 8.7.

The double bottom, or any other double-walled region, may be modeled by a combination of beam elements and stiffened panel elements using a special technique that accounts for the extra torsional rigidity of closed cellular structures. With this technique only one node is required at each intersection of floors and longitudinal girders. The technique is explained in Section 8.8.

Figure 8.2 from Ref. 1 shows a typical hull module for a naval vessel, consisting of three compartments—the engine room and the adjacent compartments forward and aft of it. The first bulkhead is not drawn in order to show the internal structure.

Figure 8.3 shows the division of the hull module into "strakes" (i.e., strake submodules; see Fig. 2.15) denoted by circled numbers. Because the engine room has less internal structure, the shell and deck scantlings there will need to be heavier than in the other two compartments. Therefore in the structural optimization process of Fig. 2.19 each strake would be further subdivided into three separate portions, corresponding to the three compartments, and optimized separately.

In this example the double bottom is modeled using ordinary strakes instead of double wall strakes (Fig. 2.16). This allows more variation in geometry and yields more detailed information about stresses, but requires more degrees of freedom.

In this hull module there are approximately 1500 elements altogether, of four types: bar, beam, CST, and SP (the stiffened panel version of the CSSR element of Section 7.5; this is presented in Section 8.4). There are five different superelements. Figure 8.4 shows the bulkhead superelement. Because the bulk-

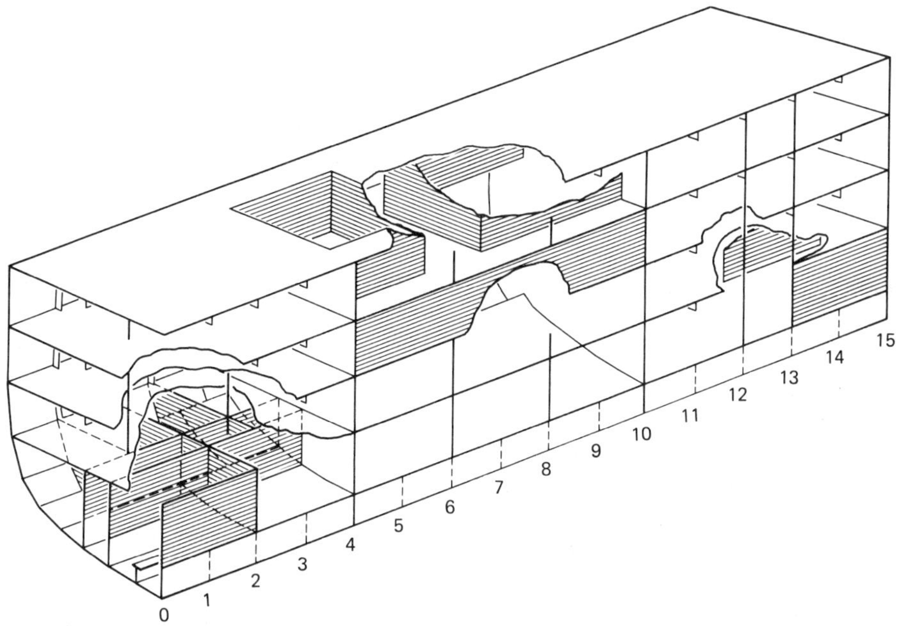

Figure 8.2 Hull module of a naval vessel.

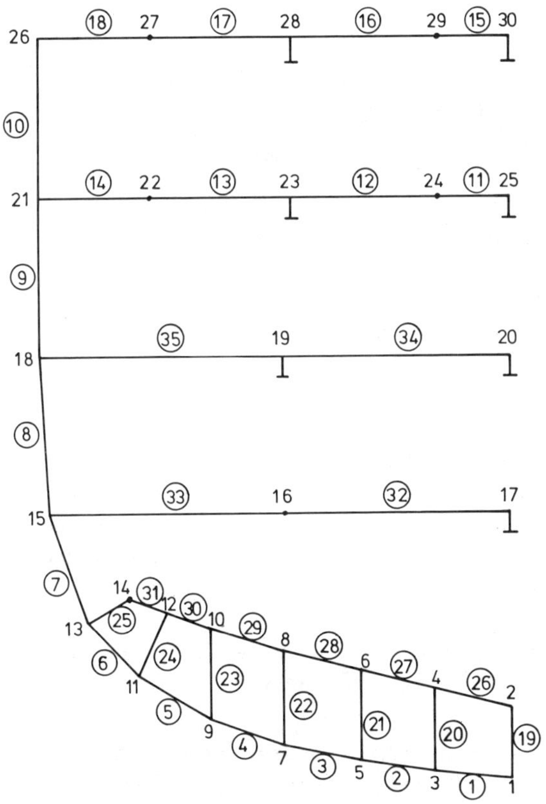

Figure 8.3 Allocation of strakes.

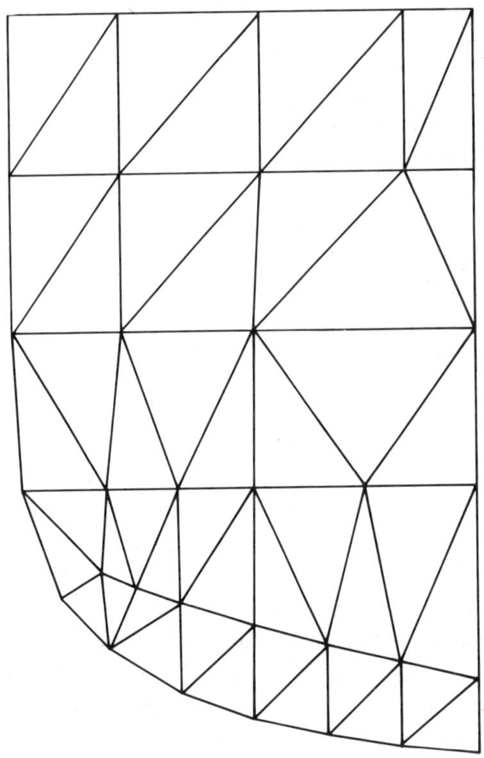

Figure 8.4 Typical bulkhead superelement.

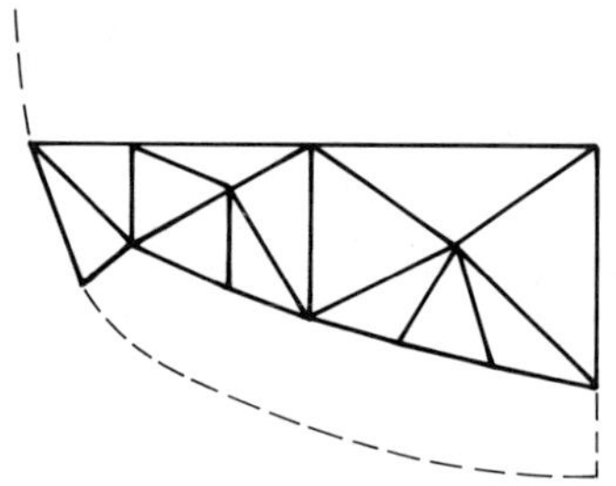

Figure 8.5 Tank wall at frame 2.

heads are identical, only one superelement needs to be assembled and its stiffness matrix is then inserted into the global stiffness matrix four times. A second superelement is used to model the transverse floors, which are 12 in number. The other three superelements are used to model the transverse walls of tanks, as shown in Fig. 8.5.

The complete model has a total of 1242 nodes and 4212 degrees of freedom.

8.2 SYMMETRY OF STRUCTURE AND OF LOADING

In nearly every case the hull module has at least transverse (i.e., port–starboard) symmetry and in this section the discussion focuses on how to take full advantage of this symmetry, even when the load does not possess symmetry. There is no need to discuss the opposite case because if the structure does not have at least transverse symmetry then there would be no possibility of a savings in computation; the full structure would have to be modeled, even if the loading were symmetric.

The important responses at the overall structure (hull girder) level are vertical bending and torsion; horizontal bending is similar to vertical bending and is therefore rarely examined explicitly, except as part of a torsional analysis. In addition to transverse symmetry the structure of a hull module often possesses longitudinal symmetry as well, that is, symmetry about the transverse midlength plane. Even when the structure is not longitudinally symmetric it may be close enough to be idealized as such. On the other hand, there are many instances in which the hull module is definitely not longitudinally symmetric and so both the symmetric and the nonsymmetric cases will be discussed. In addition to symmetry of the structure, there is also the question of symmetry of the loading. Again the symmetry is of two types: transverse and longitudinal, but the latter only gives an advantage if the structure is likewise longitudinally symmetric. Table 8.1 summarizes the various possible combinations. To make the table serve as a convenient guide and to facilitate reference between it and the text, each of the following sections which discusses a specific combination is headed by the row and column identifiers which correspond to that combination in the table.

In general the hull module analysis must be performed for several combinations of cargo loading conditions and other loads. In the following discussion each of these combinations will be referred to as a "loadcase." In principle, all loadcases should be investigated because it will not be known in advance which loadcase will prove to be critical for particular types of failure in particular locations of the hull module. In practice it is often possible to synthesize two or more loadcases into an artificial composite loadcase, combining the worst features of the separate loadcases, if these different features occur in different parts of the hull module and are known to be nearly independent. This synthesis reduces the number of loadcases and hence the amount of computation.

I. Longitudinal and Transverse Symmetry of Load

IA. Longitudinal and Transverse Symmetry of Structure

In a finite element analysis the solution of the equilibrium equations is by far the largest task within the overall analysis. The amount of storage required is proportional to N and the amount of computation required is proportional to Nw_b^2, where N is the total number of degrees of freedom and w_b is the semi-bandwidth (see Section 5.2). It is therefore very important to always take advantage of symmetry to the fullest extent possible. If both the load and the structure are doubly symmetric then only one-fourth of the hull module needs to be modeled. The value of N will be $\frac{1}{4}$ and the value of w_b will be approximately $\frac{1}{2}$ of the value of the full structure.* Thus only $\frac{1}{4}$ of the storage and approximately $\frac{1}{16}$ of the computation would be required compared to the full structure. The boundary conditions for restraints which are applicable to the quarter model for this case of double symmetry are shown in Fig. 8.6. There are two main groups, corresponding to the two planes of symmetry:

$U = \theta_Y = \theta_Z = 0$ at all nodes in the midlength transverse plane;

*Assuming the nodes are numbered first in the transverse direction, starting in the plane of the end bulkhead, repeating at each frame.

TABLE 8.1 STRUCTURAL MODELING FOR HULL MODULE ANALYSIS

Symmetry of Loadcase	Symmetry of Structure: A. Longitudinal and Transverse Symmetry USE HALF-LENGTH MODEL	B. Transverse Symmetry Only USE FULL-LENGTH MODEL
I. Longitudinal and transverse symmetry	One load vector per loadcase; b.c.[a] as in Fig. 8.6	One load vector per loadcase; b.c. as in Fig. 8.7
II. Transverse symmetry only	Divide loadcase into two load vectors (Fig. 8.8) and calculate two responses: R_s for a doubly symmetric vector; b.c. in Fig. 8.6; R_a for a longitudinally antisymmetric vector (S.S.)[b]	Same as above
III. No symmetry	Divide loadcase into four load vectors (Fig. 8.9) and calculate four responses: R_{ss} for a doubly symmetric vector; b.c. in Fig. 8.6; R_{as} for a longitudinally antisymmetric, transversely symmetric vector (S.S.); R_{sa} for a longitudinally symmetric, transversely antisymmetric vector (S.S.); R_{aa} for a doubly antisymmetric vector, b.c. in Fig. 8.10 (S.S.)	Divide loadcase into two load vectors (Fig. 8.11) and calculate two responses: R_s for a transversely symmetric vector; b.c. in Fig. 8.7; R_a for a transversely antisymmetric vector; b.c. in Fig. 8.12 (S.S.)

[a]b.c. Boundary conditions.
[b]S.S. Separate solution. These boundary conditions are different from those of the first load vector, and so the solution for R must be performed separately, after first modifying $[K]$ so that it corresponds to the new boundary conditions.

$W = \theta_X = \theta_Y = 0$ at all nodes in the longitudinal centerplane.

In addition, it is necessary to prevent rigid body motion of the overall structure. The foregoing restraints eliminate all such motion except vertical translation. Hence we must somewhere apply the further restraint that $V = 0$; for example at the top of the side shell.

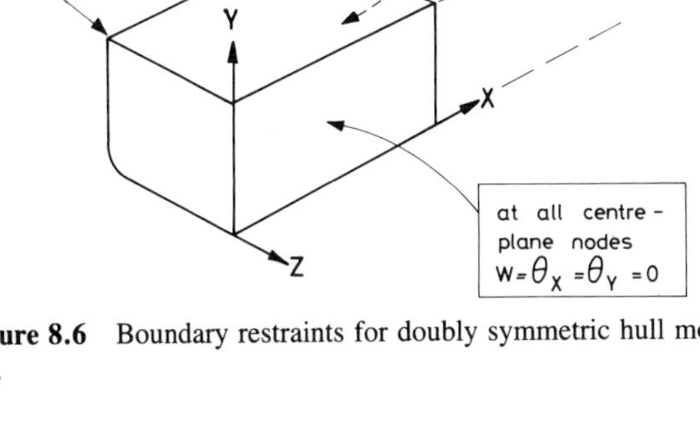

Figure 8.6 Boundary restraints for doubly symmetric hull module.

IB. Structure with Transverse Symmetry Only

If there is no longitudinal symmetry of structure then the full length of the hull module must be included in the structural model, and no advantage can be taken of the longitudinal symmetry of the load. The model and the boundary restraints are shown in Fig. 8.7. The restraints in the longitudinal centerplane are the same as before. Additional restraints must be added to prevent rigid body translation in the X- and Y-directions and rigid body rotation in a plane parallel to the XY plane. The restraints chosen are located in the side shell:

$U = 0$ at the node nearest the neutral axis,

$V = 0$ at the two ends

It is desirable that the two points where $V = 0$ should be widely spaced to avoid having large equilibrating forces which might cause spurious distortions

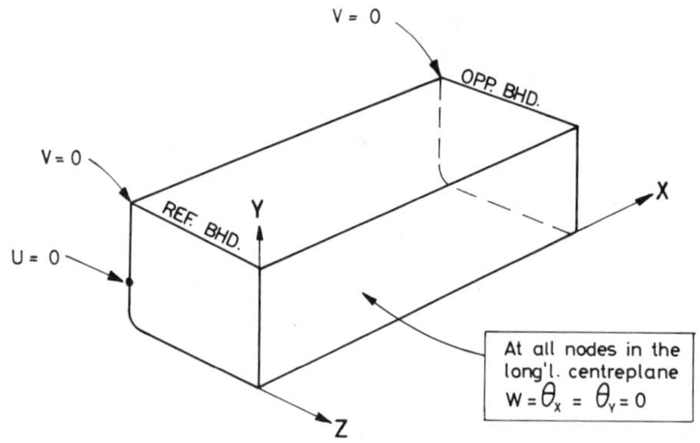

Figure 8.7 Boundary conditions for transversely symmetric loading.

in the model. The total set of restraints is shown in Fig. 8.7, but these are by no means the only ones possible; there are other equally valid combinations. In a full-length model it is helpful to have some system for distinguishing between the two end bulkheads; we shall adopt the nomenclature shown in Fig. 8.7 in which the bulkhead which includes the origin is called the "reference" bulkhead and the other is called the "opposite" bulkhead.

II. Load with Transverse Symmetry Only
IIA. Doubly Symmetric Structure

In general, if a structure has a certain type of symmetry but the load does not, then to take advantage of the structural symmetry it is necessary to divide the load into two parts: a symmetric part and an antisymmetric part. Figure 8.8 illustrates the process for the present case, in which the structure is longitudinally symmetric but the load is not. Because of the symmetry of the structure it will be convenient to introduce a second coordinate system (x, y, z) which has its origin at the midlength of the hull module, as shown in Fig. 8.8. This system is nearly identical to (X, Y, Z); the only change is that $x = X - L/2$, where L is the length of the module.

A typical loadcase would consist of two different values of shear force and bending moment at the two ends of the hull module, together with an unsymmetric, longitudinally distributed vertical load $f(x)$. All of these loads would have been obtained from the hull girder analysis. In order to use the load division technique the distributed load can only be made up of point loads, uniform distributed loads, or loads which vary linearly along the length of the module. This limitation is not all serious because $f(x)$ is the net difference between the longitudinal distributions of bouyancy and weight and it can nearly always be represented satisfactorily by some combination of these three load types. The boundary loads of hull girder shear force and bending moment at the bulkheads are in reality distributed forces which act in the girthwise and longitudinal directions. These are applied to the model in the form of equivalent nodal forces at each of the bulkhead nodes. The calculation of the equivalent nodal forces can be incorporated into the computer program for the finite element analysis, using the principles given in Section 5.3. For simplicity, these separate forces are represented in Fig. 8.8 by their total values F_{Y1} and M_{Z1} at the reference bulkhead, and F_{Y2} and M_{Z2} at the opposite bulkhead. It will also be necessary to have the values of shear force and bending moment at the midlength of the module. When considered as external forces applied to the half model, these two quantities are denoted as F_{Y0} and M_{Z0} respectively (see Fig. 8.8). In the actual finite element solution, the overall forces and moments F_Y and M_Z are load vectors, made up of the individual nodal forces. Likewise, the distributed load $f(x)$ is in fact a collection of nodal forces, and these three groups of loads—F_Y, M_Z, and $f(x)$—together make up the overall load vector $\overline{P}$.* In the work which is to follow the notation $\overline{P}(x)$ means all of the loads (nodal forces) at those nodes which lie within a transverse plane at a distance x from the origin. Thus, for example, $\overline{P}(-L/2)$ would refer to all of the nodal forces at the nodes in the reference bulkhead. These would consist of all of the X-directed forces which make up M_{Z1}, plus all of the Y- and Z-directed forces which make up F_{Y1} plus all other nodal forces which act at the reference bulkhead nodes in the Y- or Z-direction, arising from the distributed load $f(x)$. (Nodal forces in the Z-direction arise from shear and from hydrostatic pressure on the ship sides; these forces are assumed to be symmetric transversely). All computations which are described later in terms of these overall forces and moments would in practice be performed at each node, using the nodal forces. In Fig. 8.8 the sign convention for the M_Z moments is positive anticlockwise because the Z-axis is outward from the page.

Having defined the foregoing quantities, we may now proceed to show that whenever a longitudinally unsymmetric load acts on a longitudinally symmetric hull module, the overall load vector $\overline{P}$ should be replaced by two other load vectors:

*In this section a bar is placed above P to indicate that it is a vector, containing either the complete set, or a subset, of the external nodal loads.

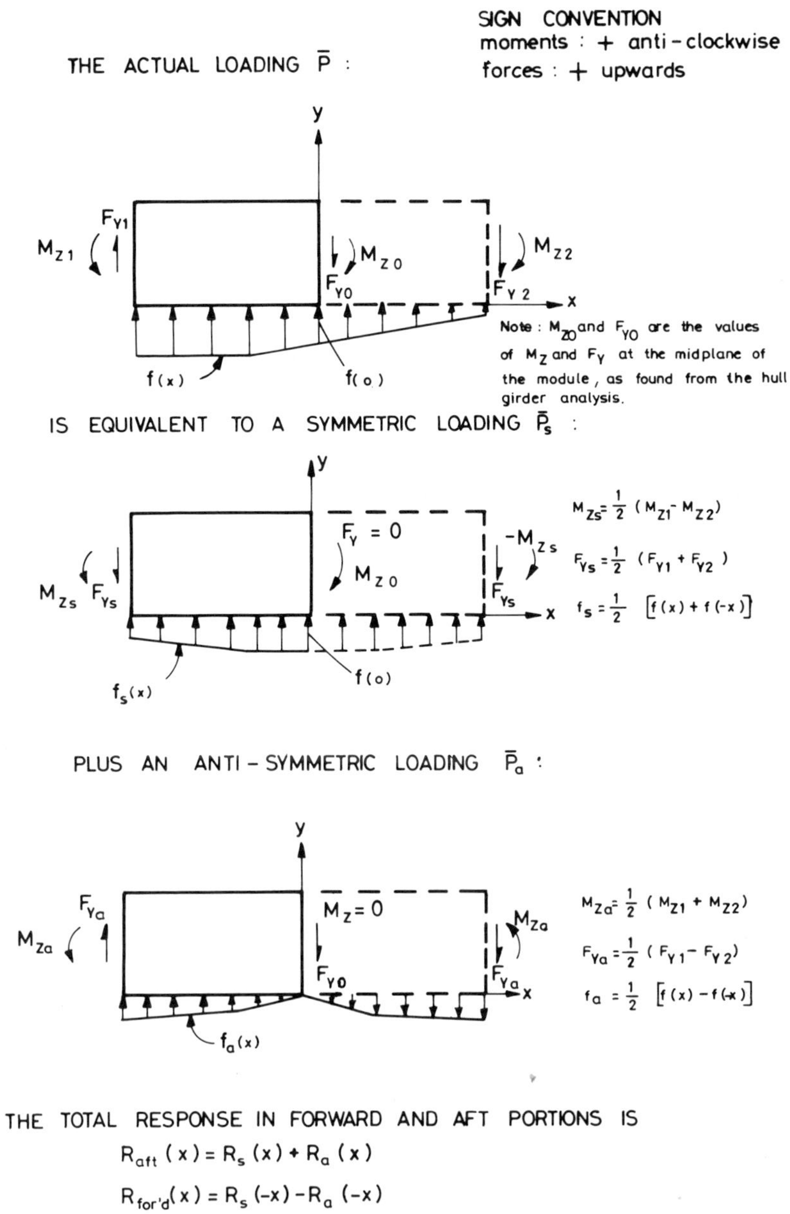

Figure 8.8 Division of a longitudinally unsymmetric load into symmetric and antisymmetric components.

1. A symmetric load vector $\bar{P}_s$ which is obtained by averaging the loads on the forward and aft portions of the module

$$\bar{P}_s = \tfrac{1}{2}[\bar{P}(x) + \bar{P}(-x)] \qquad (8.2.1)$$

2. An antisymmetric load vector $\bar{P}_a$ which is the difference between $\bar{P}$, the original set of loads, and $\bar{P}_s$. Thus the combination, or superposition, of $\bar{P}_s$ and $\bar{P}_a$ is equivalent to the original loadcase $\bar{P}$. The relationships are

$$\begin{aligned}\bar{P}_a &= \bar{P} - \bar{P}_s \\ &= \bar{P}(x) - \tfrac{1}{2}[\bar{P}(x) + \bar{P}(-x)] \qquad (8.2.2) \\ &= \tfrac{1}{2}[\bar{P}(x) - \bar{P}(-x)]\end{aligned}$$

The third line of (8.2.2) shows that $\bar{P}_a$ is exactly antisymmetric; that is, if x is replaced by $-x$ the resulting load has the same magnitude but opposite sign. $\overline{P}_a$ is illustrated in Fig. 8.8.

The foregoing definitions of $\overline{P}_s$ and $\overline{P}_a$ are for quantities that are continuous functions of x such as the distributed loading $f(x)$ or the hull girder bending moment if considered as a function of x, say $BM(x)$. For the specific end forces and moments $F_{Y1,2}$ and $M_{Z1,2}$ we require special definitions. The full set of symmetric forces is therefore defined as follows:

$$(F_{Ys})_{1,2} = \tfrac{1}{2}(F_{Y1} + F_{Y2}) \tag{8.2.3a}$$

$$(M_{Zs})_1 = \tfrac{1}{2}(M_{Z1} - M_{Z2}) \tag{8.2.3b}$$

$$(M_{Zs})_2 = -\tfrac{1}{2}(M_{Z1} - M_{Z2}) \tag{8.2.3c}$$

$$f_s(x) = \tfrac{1}{2}[f(x) + f(-x)] \tag{8.2.3d}$$

The separate definitions of the two M_{Zs} values are required because we are averaging the two end values of hull girder bending moment $BM(x)$, and this quantity has a different sign convention (e.g., positive sagging) from a "moment of forces" such as M_Z (positive anticlockwise). The set of nodal forces corresponding to a given bending moment will have opposite directions, depending on whether it acts on the reference bulkhead or the opposite bulkhead.

For the case we are considering, the structure is doubly symmetric and the load $\overline{P}$ is transversely symmetric. Thus with the longitudinally symmetric load vector $\overline{P}_s$ we have double symmetry and so the model need only be one-quarter of the actual hull module. The boundary restraints will be those derived earlier for the case of double symmetry; they are shown in Fig. 8.6. The $U = 0$ restraint at the midlength plane provides whatever F_X reaction forces are required there. The total moment of these will be the reaction moment which will (or ought to) equal the original bending moment at this point, M_{Z0}; this fact could be used as part of the equilibrium check after the solution. Because the loading is symmetric the shear force will be zero at the midlength plane so there is no need to apply any nodal forces to represent it at this boundary. However, the symmetric distributed load $f_s(x)$ will have a value at this plane; from (8.2.3d) it may be seen that its value is equal to the actual distributed load at this position; that is $f_s(0) = f(0)$. This is illustrated in Fig. 8.8. Thus, the only nodal forces which need to be applied in the midlength plane are the statically equivalent forces corresponding to $f(0)$.

Turning now to the antisymmetric load $\overline{P}_a$, we expand its definition given previously by writing the following explicit expressions for the various parts of this load vector:

$$(F_{Ya})_1 = \tfrac{1}{2}(F_{Y1} - F_{Y2}) \tag{8.2.4a}$$

$$(F_{Ya})_2 = -\tfrac{1}{2}(F_{Y1} - F_{Y2}) \tag{8.2.4b}$$

$$(M_{Za})_{1,2} = \tfrac{1}{2}(M_{Z1} + M_{Z2}) \tag{8.2.4c}$$

$$f_a(x) = \tfrac{1}{2}[f(x) - f(-x)] \tag{8.2.4d}$$

A separate expression is required for $(F_{Ya})_2$ in order to obtain the required antisymmetry because there is no continuous function of x which would allow (8.2.2) to be used. Also, the positive sign in (8.2.4c) is required because of the sign convention for M_Z, as discussed previously.

From (8.2.4d) it may be seen that at the midlength transverse plane the distributed load in the antisymmetric case is zero. Similarly, the general expression for the antisymmetric bending moment $BM_a(x)$ would be

$$BM_a(x) = \tfrac{1}{2}[BM(x) - BM(-x)]$$

and so at $x = 0$ the two terms are identical and the resultant value is zero. Therefore, in the antisymmetric case there is no hull girder bending at the midlength plane and so there is no need to apply any nodal forces to represent such a loading. On the other hand, since all of the vertical forces in the two halves of the structure are equal and opposite, the shear force will not be zero at the midlength plane. It may be shown that its value is equal to the actual value of shear force in this plane, F_{Y0}, that is, the same value as in the original loading $\overline{P}$.

Calculation of Response

After splitting an unsymmetric load into two load vectors and obtaining the deflections for each of them, these two sets of deflections must then be combined so as to give the actual deflections of the structure under the original load. Since the original load is unsymmetric longitudinally the deflections will be different in the forward and aft portions of the hull module. However, since the structure itself is symmetric the results which were obtained for one-half of the module (the aft half, if the X-axis points in the forward direction) apply also to the other half. For purposes of discussion let us say it is the aft half which is modeled and analyzed (this is the case illustrated in Fig. 8.8, assuming that the x-axis points in the forward direction). Within the aft half itself the total value of any deflection component (say V) is simply equal to the sum of the two separately calculated values: V_s and V_a. However, in the forward half the total deflection at any point (x, Y, Z) would be the sum of $V_s(-x, Y, Z)$ (i.e., the symmetric result applies equally to both halves) and $-V_a(-x, Y, Z)$ (i.e., the antisymmetric deflection in the forward half will be simply the negative of that which was obtained for the aft half, because V_a is exactly antisymmetric in x). This holds true for all deflection components and consequently for all member responses (bending moment, axial force, shear force) as long as they are linear functions of the nodal deflections (which they are in an elastic, small deflection analysis). To indicate that the foregoing combination technique for obtaining deflections also permits the evaluation of all other responses, we shall write the combination formula in terms of a general response R. If R_s and R_a denote response values corresponding to the symmetric and antisymmetric solutions the total response is given by

$$\begin{aligned} R_{\text{aft}}(x) &= R_s(x) + R_a(x) \\ R_{\text{for'd}}(x) &= R_s(-x) - R_a(-x) \end{aligned} \tag{8.2.5}$$

IIB. Structure and Load with Transverse Symmetry Only

As in case IB, if there is no longitudinal symmetry of structure then the full length of the hull module must be modeled. In fact, the modeling and boundary restraints of the present case are exactly the same as for case IB because although that case had a doubly symmetric load no advantage could be taken of its longitudinal symmetry.

III. Load With No Symmetry

We now consider the most general case, in which the load has neither longitudinal nor transverse symmetry. Transverse unsymmetry can occur due to unsymmetric cargo stowage or irregular seas, but even in the worst combination of circumstances the degree of unsymmetry is small relative to the magnitude of the load, except for unusual ship types (e.g., heavy lift ships). However, the effect of this unsymmetry of load is to cause a twisting moment and if the ship has low torsional rigidity (e.g., containerships) then the response to such loads must be investigated. We note in passing that a twisting moment could not occur if the load were symmetric longitudinally. Hence the case of a load that is longitudinally symmetric and transversely unsymmetric would rarely be of practical interest and in any case could be treated as an example of the present case by simply disregarding the longitudinal symmetry.

IIIA. Doubly Symmetric Structure

In this case it is necessary to split the given load into four load vectors, as shown in Fig. 8.9. For simplicity, only the bottom pressure loading is shown. This would be the net value of external hydrostatic pressure minus internal cargo loading. The numbers are arbitrary, and serve only to indicate the relative sizes and shapes of the four load components. As illustrated in the figure, the original load surface need not be a single plane; it only needs to be piecewise linear. The four load vectors are:

1. A doubly symmetric load vector $\overline{P}_{ss}$ which is obtained by averaging the loads in the four quarters of the doubly symmetric hull module

$$\overline{P}_{ss}(x,Z) = \tfrac{1}{4}[\overline{P}(x,Z) + \overline{P}(-x,Z) + \overline{P}(x,-Z) + \overline{P}(-x,-Z)] \tag{8.2.6}$$

2. A longitudinally antisymmetric, transversely symmetric load vector

$$\overline{P}_{as}(x,Z) = \tfrac{1}{4}[\overline{P}(x,Z) - \overline{P}(-x,Z) + \overline{P}(x,-Z) - \overline{P}(-x,-Z)] \tag{8.2.7}$$

3. A longitudinally symmetric, transversely antisymmetric load vector

$$\overline{P}_{sa}(x,Z) = \tfrac{1}{4}[\overline{P}(x,Z) - \overline{P}(x,-Z) + \overline{P}(-x,Z) - \overline{P}(-x,-Z)] \tag{8.2.8}$$

4. A doubly antisymmetric load vector

$$\overline{P}_{aa}(x,Z) = \tfrac{1}{4}[\overline{P}(x,Z) - \overline{P}(-x,Z) - \overline{P}(x,-Z) + \overline{P}(-x,-Z) \tag{8.2.9}$$

Load vectors $\overline{P}_{ss}$ and $\overline{P}_{as}$ are the same as those for Case IIA and hence the boundary restraints are the same; the first is shown in Fig. 8.6.

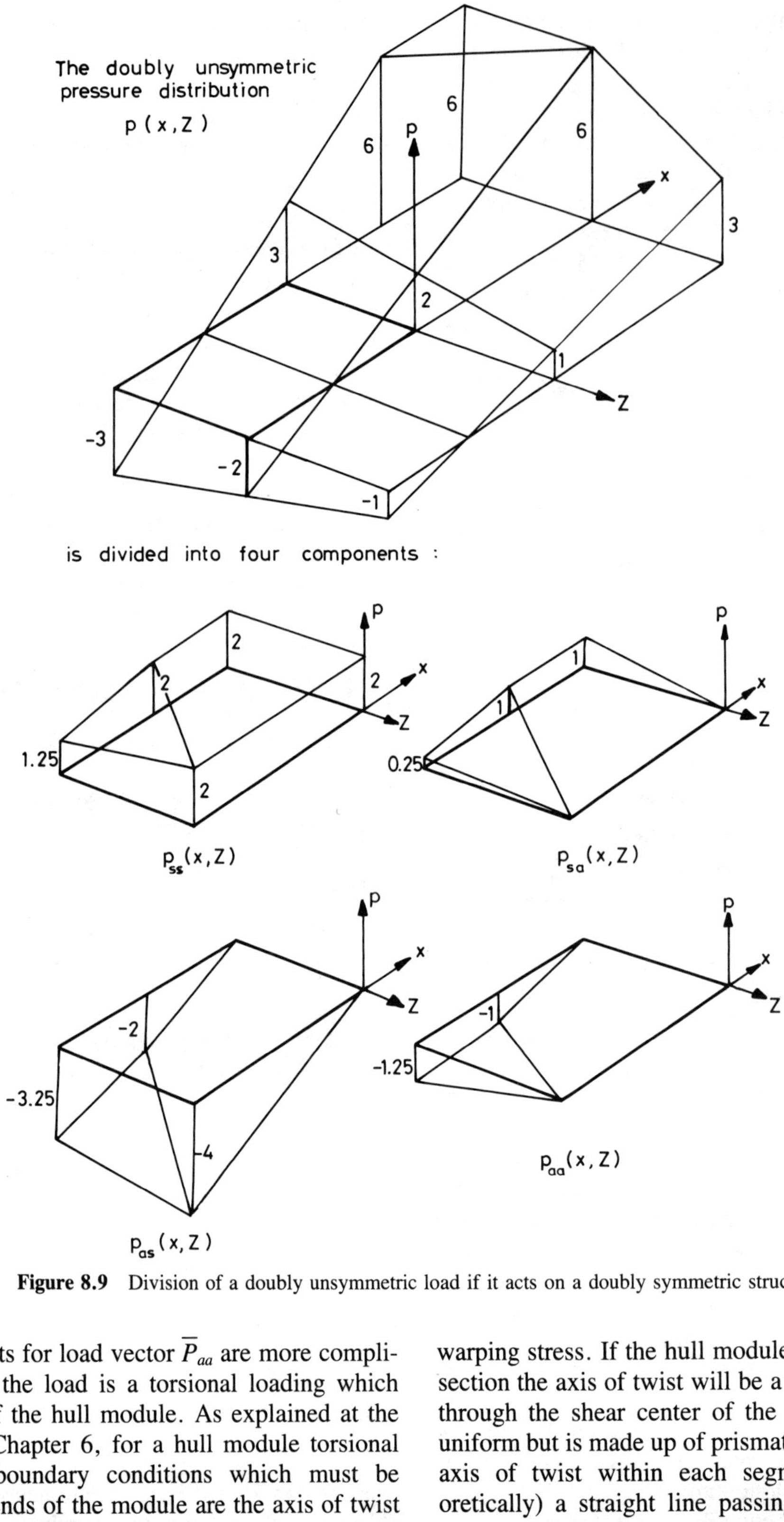

Figure 8.9 Division of a doubly unsymmetric load if it acts on a doubly symmetric structure.

The restraints for load vector $\overline{P}_{aa}$ are more complicated because the load is a torsional loading which causes twist of the hull module. As explained at the beginning of Chapter 6, for a hull module torsional analysis the boundary conditions which must be known at the ends of the module are the axis of twist at that section and the distribution of the longitudinal warping stress. If the hull module is of uniform cross section the axis of twist will be a straight line passing through the shear center of the section. If it is not uniform but is made up of prismatic segments then the axis of twist within each segment is again (theoretically) a straight line passing through the shear center of that segment. But that segment itself will have undergone both translation and rotation, and the location of the shear center differs for different cross sections—markedly so between open and closed sections. Therefore the axis of twist is far from straight; it changes with each segment and is further altered by the (cumulative) translation and rotation of each segment. Thus it is constantly in a state of transition, and it only coincides with the local shear center near the midlength of very long prismatic segments (which themselves have undergone translation and rotation). Therefore, to know the location of the axis of twist at the ends of a hull module it is necessary for the ends to be located well inside the length of a prismatic segment, at a distance equal to at least half the beam of the vessel. The longitudinal stress due to warping is calculated from a special hull girder analysis which is based on the hull segment element, as explained in Chapter 6. Another requirement is that there must not be any appreciable distortion of the shape of the cross section at the ends of the hull module, because this would invalidate the hull girder analysis. Therefore the ends must coincide with a transverse bulkhead. If these two requirements cannot be met then the hull girder analysis must be performed by means of a simplified, full-length hull module analysis. This is the only way to determine the axis of twist and the distribution of warping stress in a cross section that does not fulfill the above requirements. Having determined these values it is then possible to proceed with the torsional analysis of a hull module, as described in this section.

It often happens that a cargo hold of a containership is doubly symmetric but a problem arises in that the transverse plane of symmetry occurs at the midlength of the hold, where there is no transverse bulkhead. Hence there will be some distortion in the shape of the cross section and there is no *a priori* way of knowing how significant its effect is. To avoid introducing what might be a significant error there are only two possible procedures: either determine the actual warping stresses in the midlength plane by means of a simplified hull module analysis over a longer portion of the hull girder for which the boundary conditions are known, or ignore the symmetry and model the full length of the hold. Of course, if the next hold is identical to the current hold then this full-length model can take advantage of symmetry at one end. This situation is illustrated in Fig. 8.10. In the terminology of this section we would say that this is a one-quarter model and that the complete hull module is two cargo holds in length. The boundary conditions are shown in the

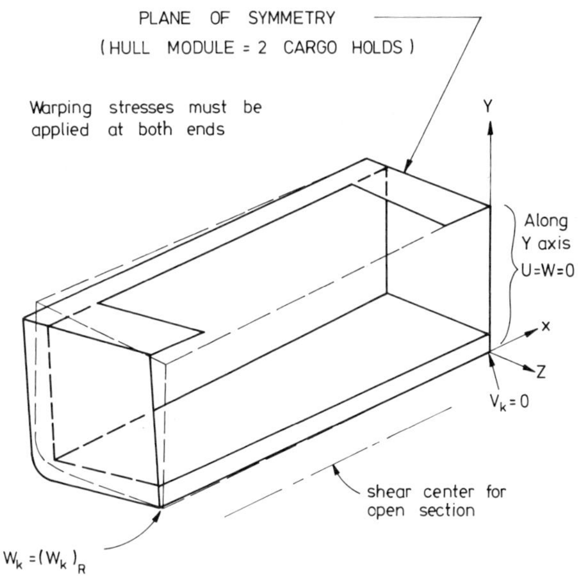

Figure 8.10 Boundary conditions for doubly antisymmetric load $\bar{P}_{aa}$ if structure is doubly symmetric.

figure. The $V = 0$ restraint may be applied at any node; it need not be at the keel.

Calculation of Response

After splitting the original unsymmetric load into the four load vectors, the next step is to obtain the four corresponding solution vectors (displacements). Each solution involves different restraints. Therefore the set of equilibrium equations is slightly different (which is to say the global stiffness matrix is slightly different). But the physical structure is still the same, and therefore the finite element program should be written in such a way that the global stiffness matrix is only modified and not completely reassembled.

The actual displacements, and hence all other responses which are linearly related to displacements, are obtained by combining these four idealized solutions. For the quarter of the hull module that was actually modeled the final solution is simply the sum of the four idealized solutions, and for each of the other three quadrants the final solution is a different combination of these four. If, for example, it was the port aft quarter which was modeled (as in all of the figures if both x and X are taken as pointing forward) the final responses in the four quarters of the hull module are:

$$R_{\text{aft,port}}(x,Z) = R_{ss}(x,Z) + R_{as}(x,Z) + R_{sa}(x,Z) + R_{aa}(x,Z) \quad (8.2.10)$$

$$R_{\text{aft,stbd}}(x,Z) = R_{ss}(x,-Z) + R_{as}(x,-Z) - R_{sa}(x,-Z) - R_{aa}(x,-Z) \quad (8.2.11)$$

$$R_{\text{for'd,port}}(x,Z) = R_{ss}(-x,Z) - R_{as}(-x,Z) + R_{sa}(-x,Z) - R_{aa}(-x,Z) \quad (8.2.12)$$

$$R_{\text{for'd,stbd}}(x,Z) = R_{ss}(-x,-Z) - R_{as}(-x,-Z) - R_{sa}(-x,-Z) + R_{aa}(-x,-Z) \quad (8.2.13)$$

IIIB. Structure with Transverse Symmetry Only

If the hull module is not symmetric about its midlength then the full length of the module must be used for the structural model, and the longitudinal symmetry of the load gives no advantage. However, because of the module's transverse symmetry it is again possible to use a half-width model, provided that the load $\overline{P}$ is split into two load vectors which are symmetric and antisymmetric transversely, as shown in Fig. 8.11.

$$\begin{aligned} \overline{P}_s(Z) &= \tfrac{1}{2}[\overline{P}(Z) + \overline{P}(-Z)] \\ \overline{P}_a(Z) &= \tfrac{1}{2}[\overline{P}(Z) - \overline{P}(-Z)] \end{aligned} \quad (8.2.14)$$

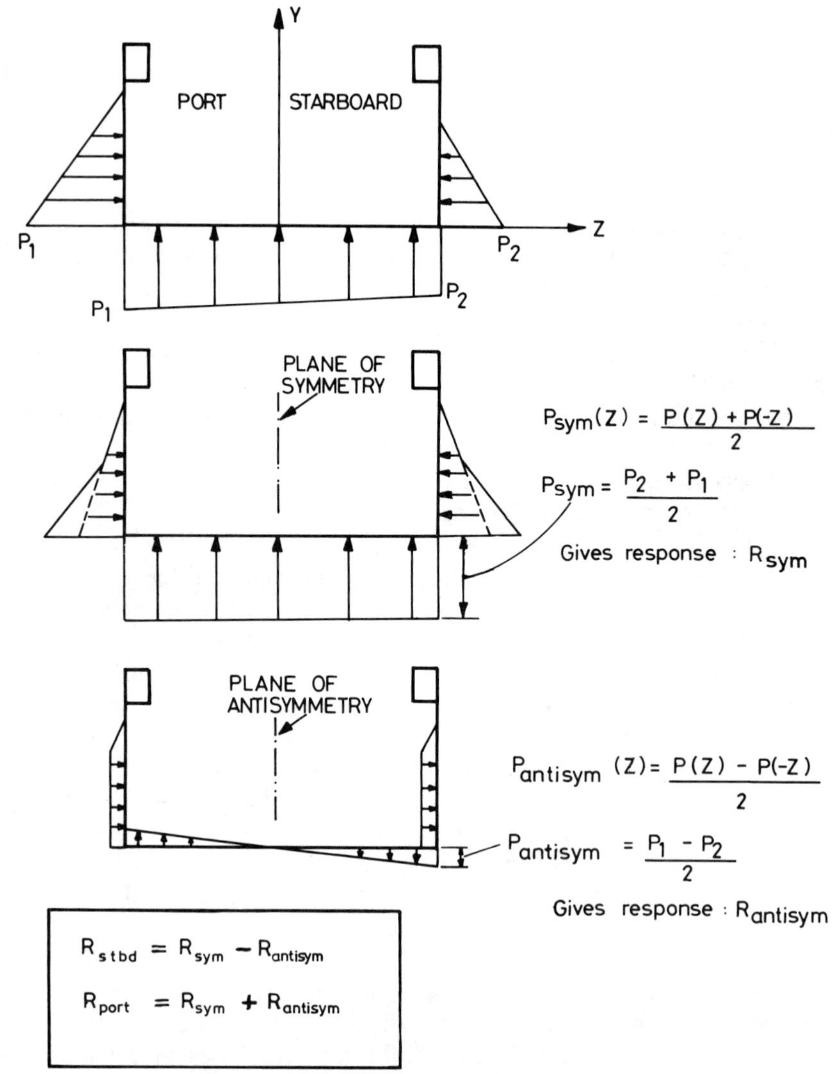

Figure 8.11 Use of symmetric and antisymmetric loads.

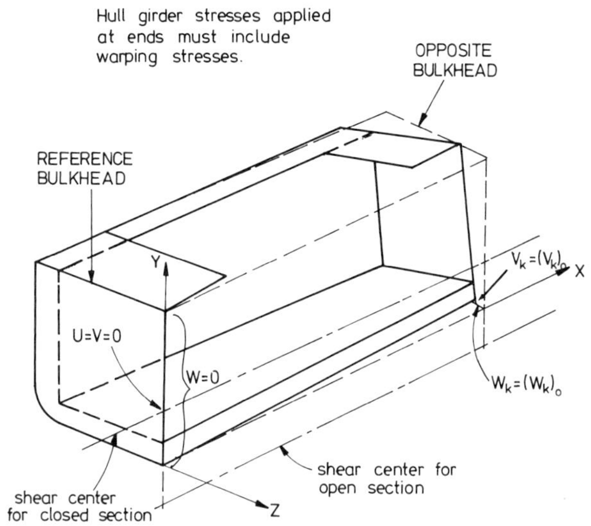

Figure 8.12 Boundary conditions for transversely antisymmetric loading $\overline{P}_a$.

The boundary restraints for a transversely symmetric load and full-length model were derived earlier, in case IB, and are shown in Fig. 8.7. The boundary restraints for a transversely antisymmetric load were also derived earlier, in case IIIA (Fig. 8.10) for a half-length model. For a full-length model they are as shown in Fig. 8.12. The rotational restraint is here applied at the reference bulkhead and comes entirely from the special hull girder analysis based on the hull segment method. The reference rotation may be zero, as shown in the figure, or it may be some finite value, say $(\theta_X)_R$. In the first case the boundary conditions are $W = 0$ along the vertical centerline and $U = V = 0$ at the local axis of twist, which will be the shear center of the hull segment at this end of the hull module, providing that the segment is of adequate length. If the reference bulkhead has a nonzero rotation this total *accumulated* rotation will in general not be about the local shear center, and so the latter will have translated horizontally by some amount. The boundary condition $U = 0$ is still applied at the shear center, but the conditions on V and W no longer have any special relationship to the local shear center. The restraint on vertical movement can be specified at any convenient point on the centerline such as the keel: $V_k = (V_k)_R$ where $(V_k)_R$ comes from the hull girder analysis. The restraint on rotation is expressed by specifying the known values of W at all nodes on the centerline, as obtained from the hull girder analysis.

At the opposite bulkhead restraints are only required in regard to vertical and horizontal translation. Again this simply means specifying V and W at some convenient centerline node such as the keel, as shown in the figure, for which the values $(V_k)_0$ and $(W_k)_0$ come from the hull girder analysis.

Calculation of Response

Here again, because we have split the load into symmetric and antisymmetric load vectors, the total response at any point will be some combination of the two separate responses, R_s and R_a, taking due account of the exactly antisymmetric nature of the latter. If in Fig. 8.12 we again regard the X-axis as pointing in the forward direction then the structural model is the port half of the hull module. Therefore the total response in either of the two halves of the hull module is given by

$$R_{\text{port}}(Z) = R_s(Z) + R_a(Z) \tag{8.2.16}$$

$$R_{\text{stbd}}(Z) = R_s(-Z) - R_a(-Z) \tag{8.2.17}$$

Table 8.2 summarizes the calculation of response for all of the cases considered in this section.

TABLE 8.2 CALCULATION OF TOTAL HULL MODULE RESPONSE FROM RESPONSES TO SYMMETRIC AND ANTISYMMETRIC LOAD VECTORS

	Symmetry of Structure	
Symmetry of Loadcase	A. Longitudinal and Transverse Symmetry (HALF-LENGTH MODEL)	B. Transverse Symmetry Only (FULL-LENGTH MODEL)
II. Transverse symmetry only	$R_{\text{aft}}(x) = R_s(x) + R_a(x)$ $R_{\text{for'd}}(x) = R_s(-x) + R_a(-x)$	No combination required
III. No symmetry	$R_{\text{aft,port}}(x,Z) = R_{ss}(x,Z) + R_{as}(x,Z) + R_{sa}(x,Z) + R_{aa}(x,Z)$ $R_{\text{aft,stbd}}(x,Z) = R_{ss}(x,-Z) + R_{as}(x,-Z) + R_{sa}(x,-Z) + R_{aa}(x,-Z)$ $R_{\text{for'd,port}}(x,Z) = R_{ss}(-x,Z) - R_{as}(-x,Z) + R_{sa}(-x,Z) - R_{aa}(-x,Z)$ $R_{\text{for'd,stbd}}(x,Z) = R_{ss}(-x,-Z) - R_{as}(-x,-Z) - R_{sa}(-x,-Z) + R_{aa}(-x,-Z)$	$R_{\text{port}}(Z) = R_s(Z) + R_a(Z)$ $R_{\text{stbd}}(Z) = R_s(-Z) - R_a(-Z)$

8.3 MODELING OF BEAMS ATTACHED TO PLATING

In modeling a hull module almost the entire structure can be represented in terms of two basic types of elements: a beam element and a rectangular stiffened panel element. For the most part the main purpose of the beams is to carry the transverse loads which act on the plate surface, thereby maintaining this thin-walled surface or "envelope" of plating in its proper position and shape, so that it can in turn carry the in-plane loads which exist in the structure. Hence nearly all of the beams in a ship structure are attached to plating and the important question arises as to what is the best way of modeling this type of structural member. Afterward in Section 8.4, we take up the question of how best to model stiffened panels.

Interaction Between Beams and Plating

The important requirement in modeling beams and panels is to provide an adequate representation of their interaction. Two modes of loading and of response are involved:

1. Lateral loads and the corresponding beam bending response.
2. Axial loads in the beam and in-plane loads in the plating which cause corresponding axial and in-plane strains.

As noted at the beginning of Section 8.1, panel bending due to lateral load is sufficiently independent from beam bending and from in-plane response that it can be dealt with separately. Methods for plate and panel bending analysis are given in Chapters 9 and 10.

The interaction of beams and plating is an interaction between the foregoing two modes of loading and response. The interaction takes place in both directions and we now consider each of these separately.

EFFECT OF BEAM BENDING ON IN-PLANE RESPONSE OF PLATING

In a beam that is attached to plating, the plating forms one flange of the beam. Therefore, in addition to carrying the direct or primary in-plane stress $(\sigma_a)_1$ arising from hull girder loads, the plating also incurs some in-plane stress due to bending, $(\sigma_{x,b})_p$, as shown in Fig. 8.13. The plate flange is usually much wider than the beam's own flange and hence the neutral axis is usually close to the plating. As a result $(\sigma_{x,b})_p$ is usually less than $(\sigma_a)_1$ and is sometimes referred to as the secondary stress. The beam bending stress in the beam flange, $(\sigma_{x,b})_f$, is larger and may be the dominant stress in the beam flange. Due to shear lag the plate flange will have an effective breadth b_e which is less than the full spacing b between the beams, as explained in Section 3.8. As shown in Fig. 3.36 the value of b_e usually lies in the range $0.5 < b_e/b < 0.8$. Because of the larger width of the plate flange the beam is unsymmetric; a formula for calculating the neutral axis position and moment of inertia of unsymmetric beams is given in (8.3.6).

EFFECT OF AXIAL OR IN-PLANE LOADS ON BEAM BENDING

The axial or in-plane load is usually an applied stress, $(\sigma_a)_1$, such as the stress arising from hull girder response, in which case it acts equally on the beam and on the plating (assuming that the beam is rigidly at-

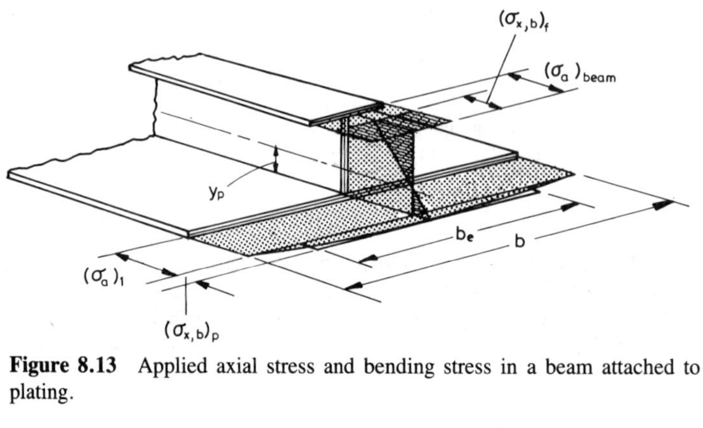

Figure 8.13 Applied axial stress and bending stress in a beam attached to plating.

tached to the plating and is sufficiently long). In this case the applied stress in the beam, $(\sigma_a)_{\text{beam}}$, is the same as that in the plate, $(\sigma_a)_1$. If the stress in the plating is nonuniform then $(\sigma_a)_1$ is simply an average stress and it would differ slightly from $(\sigma_a)_{\text{beam}}$. In this case the overall load does not pass exactly through the combined centroid of the two members and this induces some bending of the beam, but the effect is quite small and can either be ignored or be dealt with subsequently as part of the third level of analysis (individual members). In some special cases the induced bending can be significant (e.g., a short beam with an axial load applied only to the ends of the beam and not to the plating). In such cases the eccentric moment can be calculated explicitly (by taking the moment of the load about the combined centroid) and applied to the ends (nodes) of the beam as an additional load. Therefore, in general, the effect of axial loading on beam bending need not be considered or specially allowed for in the hull module analysis, apart from the inclusion of local bending moments where required.

It is of course true that with compressive axial loads there is the possibility of buckling, either of the beam or of the plating, and this can cause quite significant eccentric bending. As the beam approaches its buckling load, its lateral deflection is magnified and this increases the bending. This phenomenon is dealt with in Section 11.1. Alternatively, if the plate were to buckle before the beam the effective centroid (i.e., the action line of the internal axial force) would move away from the plate and the resulting eccentricity would cause beam bending. This is discussed fully in Chapter 14.

Nevertheless, in spite of the importance of buckling, it is not directly relevant to the hull module analysis. In a rational design process as outlined in Chapter 1, the investigation of member buckling is part of the evaluation phase of the design process, and this must come after the analysis phase so that the loads on the members are known. The important point is that buckling of beams and plates is indeed investigated in each cycle of the overall design process and is immediately eliminated in the redesign phase by redesign of the members, before the next cycle is commenced. Therefore the investigation of member buckling need not and does not form part of the hull module analysis.

From the foregoing discussion it may be concluded that the effect of axial load on beam bending is not important for the hull module analysis and is more properly dealt with as a part of the evaluation of individual members, which is the next step after the hull module analysis. Hence in regard to axial stiffness the beam and the plating are independent and for each of them the axial stiffness may be accounted for in the usual way, as part of the element stiffness matrix $\mathbf{k}^e$. For the beam the relevant tems in $\mathbf{k}^e$ will be $\pm\mathfrak{a}$, where $\mathfrak{a} = AE/L$ and A is the area of the beam only (i.e., the plate flange is not included). The panels of plating, if unstiffened, could be represented by the CSSR (Constant Shear Stress Rectangle) element presented in Section 7.5, with $\mathbf{k}^e$ given by (7.5.12). If stiffened they could be represented by means of the rectangular stiffened panel element which is presented in the next section.

With regard to the effect of beam bending on in-plane response, the situation is quite different because it was shown previously that this interaction is significant and must be accounted for as part of the hull module analysis. That is, it must somehow be incorporated into the beam bending characteristics and this requires a modified type of beam element having suitable bending stiffness characteristics.

We now present two such elements and draw a comparison between them.

Elements for Modeling Beams Attached to Plating

In allowing for the effect of beam bending on in-plane response, two different approaches are possible, resulting in two different beam elements:

Hybrid Beam Element

The element is termed "hybrid" because the axial stiffness and the bending stiffness are based on different cross sections. The axial stiffness parameter is $\mathfrak{a} = AE/L$ in which the area $\mathfrak{a}$ is the area of only the beam itself (flange and web). The bending stiffness parameter is $\mathfrak{b}_e = EI_e/L^3$ in which the neutral axis position and the moment of inertia I_e are calculated for the combined section formed by the beam plus a plate flange of effective breadth b_e. The element nodes are located at the neutral axis of the combined section, as shown in Fig. 8.14. Thus, in the hybrid element the plate's contribution to the beam bending stiffness is accounted for in the calculation of I_e, and the expression for the beam stiffness matrix remains unchanged, that is, (5.5.15) or some parts of it. The only interaction between plating and beams that occurs explicitly in the structural model is the combining of the beam's axial stiffness terms with some of the plate's in-plane stiffness terms which occurs when the struc-

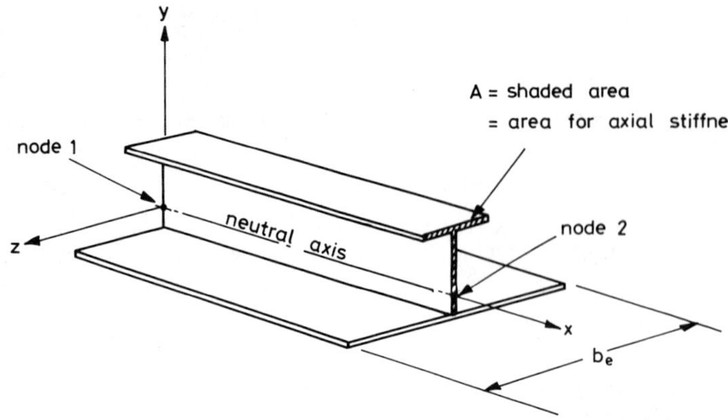

Figure 8.14 Hybrid beam element.

ture stiffness matrix is assembled. With the hybrid element the structural modeling is such that the structure nodes are located at the neutral axis of the combined section. Thus the plate elements, which have only in-plane stiffness, are located (mathematically, in the model) such that their stiffness acts in the plane of the combined neutral axis, as illustrated schematically in Fig. 8.15. This is a correct representation because in addition to the plate element contributing its in-plane stiffness, the beam element contributes its axial stiffness to make up the total in-plane stiffness, and the effective "plane of action" of this total stiffness is the plane passing through the combined centroid. Locating the plate element at the combined centroid does not affect the bending stiffness because this stiffness is accounted for entirely by the beam element.

Eccentric Beam Element

This element consists of only the beam itself (no plate flange). That is, both the area A (and the parameter $\mathfrak{a} = AE/L$) and the moment of inertia I (and the parameter $\mathfrak{b} = EI/L^3$) refer to the beam alone, and I is calculated about the beam's own centroid, the centroidal axes being denoted as $\bar{x}$, $\bar{y}$. However, the element nodes are located not along the beam's centroidal axes but rather at the toe of the web as shown in Fig. 8.16. That is, the origin of the element axes (x,y) is located at a distance g below the origin of the centroidal axes, such that $y = \bar{y} + g$. Therefore in this type of modeling the plate elements are located at their true position, in the plane of the plating. However, the axial force in the beam continues to be proportional to the change in length of the neutral axis, and this is displaced from the element x-axis. Thus, for the element of Fig. 8.17 (in which, for simplicity, nodal displacements v_1 and v_2 are kept at zero) the net change in length of the neutral axis is

$$d = (u_2 - g\theta_2) - (u_1 - g\theta_1)$$

and therefore the internal axial force $\mathcal{F}$ is

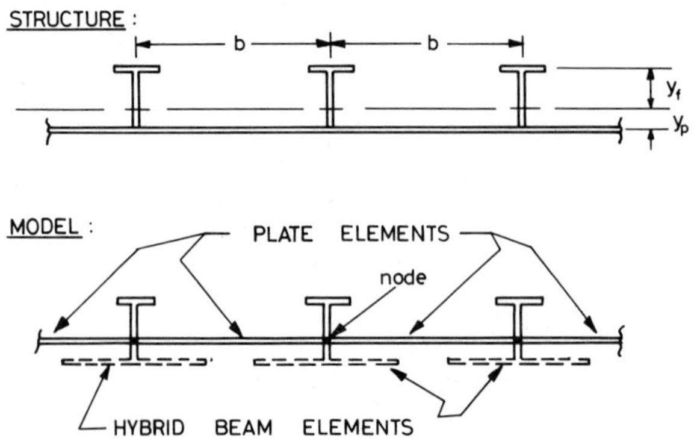

Figure 8.15 Use of the hybrid beam element.

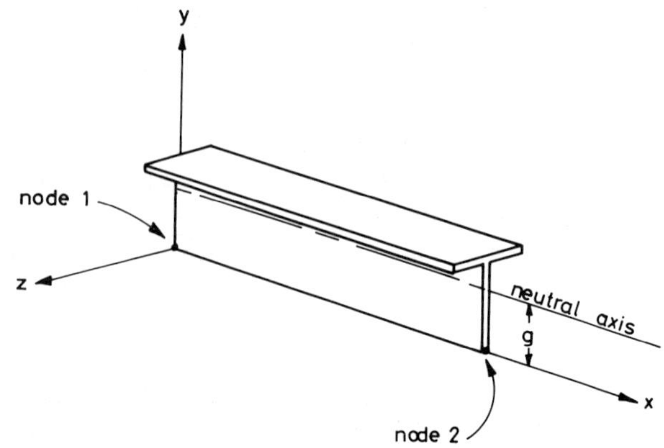

Figure 8.16 Eccentric beam element.

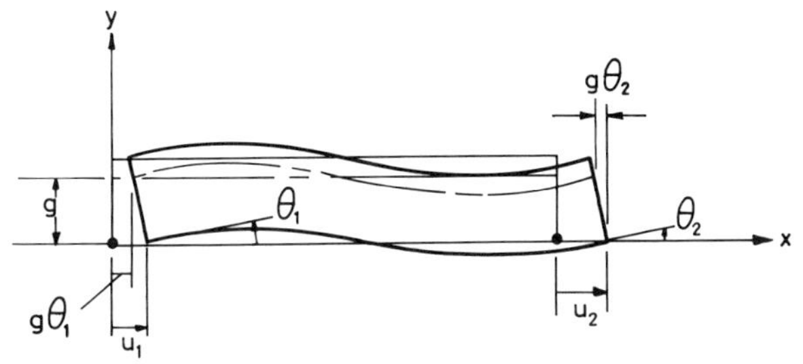

Figure 8.17 Sample nodal displacements of the eccentric beam element.

$$\mathcal{F} = \mathfrak{a}d$$
$$= \mathfrak{a}(-u_1 + g\theta_1 + u_2 - g\theta_2)$$

where $\mathfrak{a} = AE/L$.

From equilibrium the nodal forces are $f_{x1} = -\mathcal{F}$ and $f_{x2} = \mathcal{F}$. Hence the equilibrium equation corresponding to f_{x1} is

$$f_{x1} = \mathfrak{a}(u_1 - g\theta_1 - u_2 + g\theta_2) \qquad (8.3.1)$$

Thus, the placement of the nodes in the plane of the plating, away from the neutral axis, means that in this beam element an axial nodal force causes nodal rotation and vice-versa.

In a similar manner, it could be shown that for the nodal displacements of Fig. 8.17 the nodal moment M_{z1} is

$$M_{z1} = -g\mathfrak{a}u_1 + (4\mathfrak{b}L^2 + g^2\mathfrak{a})\theta_1 + g\mathfrak{a}u_2 + (2\mathfrak{b}L^2 - g^2\mathfrak{a})\theta_2 \qquad (8.3.2)$$

in which, as before, $\mathfrak{b} = EI/L^3$.

The change to the equilibrium equations corresponds to a transformation of the stiffness matrix of the ordinary beam element, $\mathbf{k}^e$, as given in (5.3.15). The latter is for centroidal coordinates, in which the x-axis is the neutral axis. The result of moving the origin downwards by a distance g is that in $\mathbf{k}^e$ there are now some new terms which were formerly zero and which represent two new interactions. First, as we have seen, there is now an interaction between axial force and nodal rotation. This was illustrated by (8.3.1) which shows that the new terms in the first row of $\mathbf{k}^e$ are $k_{13} = -k_{16} = -g\mathfrak{a}$. Second, there is now an interaction between nodal moment and axial displacement as illustrated by (8.3.2), which corresponds to the third row of $\mathbf{k}^e$ and contains the new terms $k_{31} = -k_{34} = -g\mathfrak{a}$. Besides these new interactions the relationship between nodal moment and nodal rotation is changed by the addition of two new terms in the third and sixth rows of $\mathbf{k}^e$. For example, from (8.3.2) it can be seen that k_{33} contains the new term $+g^2\mathfrak{a}$. The complete element stiffness matrix for the eccentric beam element is

$$\mathbf{k}^e_{ecc} = \begin{bmatrix} \mathfrak{a} & & & & & \\ 0 & 12\mathfrak{b} & & & & \\ -g\mathfrak{a} & 6\mathfrak{b}L & 4\mathfrak{b}L^2 + g^2\mathfrak{a} & & & \\ -\mathfrak{a} & 0 & g\mathfrak{a} & \mathfrak{a} & & \\ 0 & -12\mathfrak{b} & -6\mathfrak{b}L & 0 & 12\mathfrak{b} & \\ g\mathfrak{a} & 6\mathfrak{b}L & 2\mathfrak{b}L^2 - g^2\mathfrak{a} & -g\mathfrak{a} & -6\mathfrak{b}L & 4\mathfrak{b}L^2 + g^2\mathfrak{a} \end{bmatrix} \qquad (8.3.3)$$

The expressions for the internal forces (axial stress and bending moment) are also affected by the change of axes. The new expressions are

$$\sigma_x = \frac{E}{L}[-1 \quad 0 \quad g \quad 1 \quad 0 \quad -g]\boldsymbol{\delta} \qquad (8.3.4)$$

$$\left.\begin{aligned} \mathcal{M}_z(0) &= [g\mathfrak{a} \quad -6\mathfrak{b}L \quad -(4\mathfrak{b}L^2 + g^2\mathfrak{a}) \quad -g\mathfrak{a} \quad 6\mathfrak{b}L \quad -(2\mathfrak{b}L^2 - g^2\mathfrak{a})]\boldsymbol{\delta} \\ \mathcal{M}_z(L) &= [g\mathfrak{a} \quad 6\mathfrak{b}L \quad (2\mathfrak{b}L^2 - g^2\mathfrak{a}) \quad -g\mathfrak{a} \quad -6\mathfrak{b}L \quad (4\mathfrak{b}L^2 + g^2\mathfrak{a})]\boldsymbol{\delta} \end{aligned}\right\} \qquad (8.3.5)$$

Comparison of Elements

It will now be shown that, of the two methods for modeling beam and plate combinations, the hybrid beam element is more accurate than the eccentric beam element. Figure 8.18*a* shows a structure consisting of beams attached to plating. For simplicity, let us assume that each beam is simply supported and carries a central point load acting in the positive (upward) direction. Fig. 8.18*b* shows the structural model that would result from using the eccentric beam element. Each beam is modeled using two eccentric beam elements in order to provide a node for the load.

Before examining this case let us first consider the situation if there were no plating at all, as shown in Fig. 8.19*a*. The deformed state of the beam element would then be as shown. Since there are no horizontal loads the length of the neutral axis remains the same. That is, the nodal displacements of the beam would be such that $u_1 = g\theta_1$. From (8.3.4) the resulting axial stress is zero, as indeed it must be. Also, as can easily be verified, v_2 and θ_1 would bear the relationship $v_2 = \frac{2}{3}L\theta_1$. Hence from (8.3.5) the bending moment at the left end would be, after substituting for u_1 and v_2,

$$\begin{aligned} M_z(0) &= g\mathfrak{a}(g\theta_1) - (4\mathfrak{b}L^2 + g^2\mathfrak{a})\theta_1 \\ &\quad + 6\mathfrak{b}L\left(\frac{2}{3}L\theta_1\right) \\ &= 0 \end{aligned}$$

as of course it must be for a simply supported beam.

We now add the plating as shown in Fig. 8.19*b*. Since plate bending is a more local and independent phenomenon, the plating would be modeled by a membrane element, which contributes only in-plane stiffness. Because of its large extent the plate has a relatively large stiffness. Hence it would greatly decrease u_1 and it would also decrease θ_1. However, there is no reason at all for supposing that u_1 and θ_1 would continue to be related by $u_1 = g\theta_1$. Their relative values would depend entirely on the relative axial stiffnesses of the beam and the plating. Therefore (8.3.4) would give a nonzero axial stress in the beam (tensile, in the present case). Likewise, (8.3.5) would give a nonzero bending moment at the left end of the beam element, as shown in the figure.

These spurious values of axial stress and bending moment arise because adding the axial stiffness of the

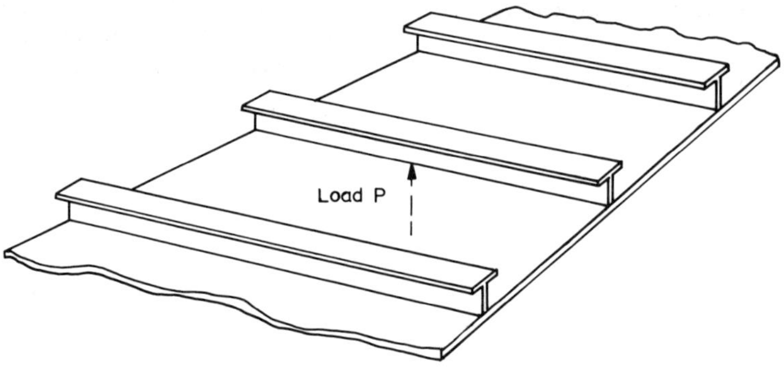

(a) BEAMS ATTACHED TO PLATING

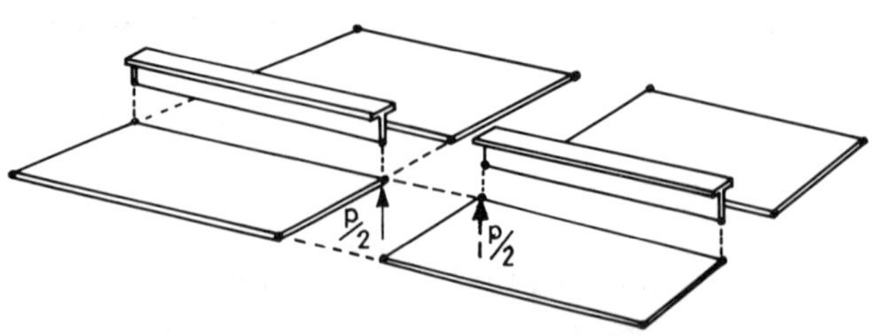

(b) STRUCTURAL MODEL USING ECCENTRIC BEAM ELEMENT

Figure 8.18 Use of eccentric beam element.

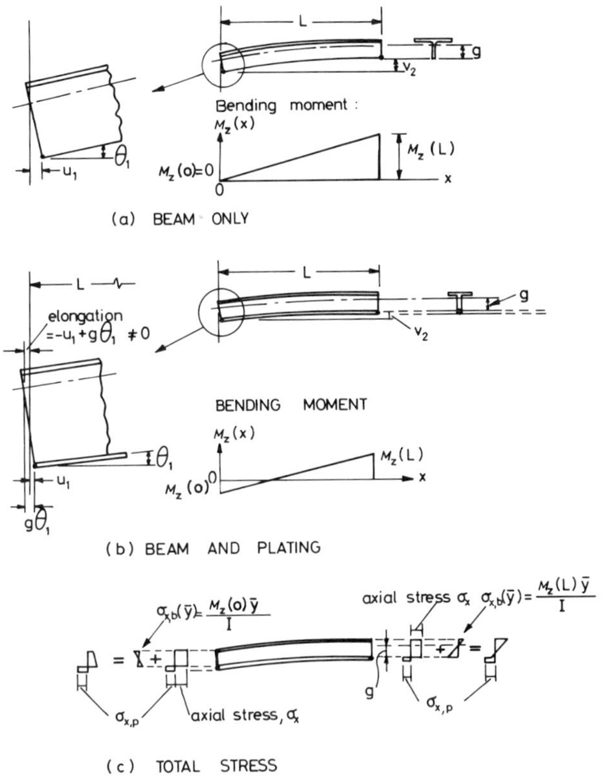

Figure 8.19 Stress distribution using the eccentric beam element.

plate is a relatively crude modeling of the plate's action as the compression flange of the combined section, and the vertical distribution of stress in the cross section does not match the actual distribution.

To see this, let us examine the stress distribution in some detail, as shown in Fig. 8.19*c*. The total stress in the beam is the sum of the axial stress σ_x and the bending stress, $\sigma_{x,b} = \mathcal{M}_z(x)\bar{y}/I$. The stress in the plate, $\sigma_{x,p}$, would be obtained from the plate internal stress equation; in the present case it would be a compressive stress. The total stress distribution in the combined cross section is shown in Fig. 8.19*c*. At the left end of the beam the distribution is obviously wrong, since the stress at a simply supported end should be zero. If the stresses were integrated the net axial force would be small and hence axial force equilibrium is approximately satisfied, but the stresses themselves are completely spurious. At the right end of the beam the stress distribution is such that the point of zero stress would occur approximately at the neutral axis of the combined section and the net axial force would again be approximately zero, but the shape of the stress distribution is only a crude approximation to the actual distribution.

By way of comparison the hybrid element modeling of this structure is shown in Fig. 8.20. Since this

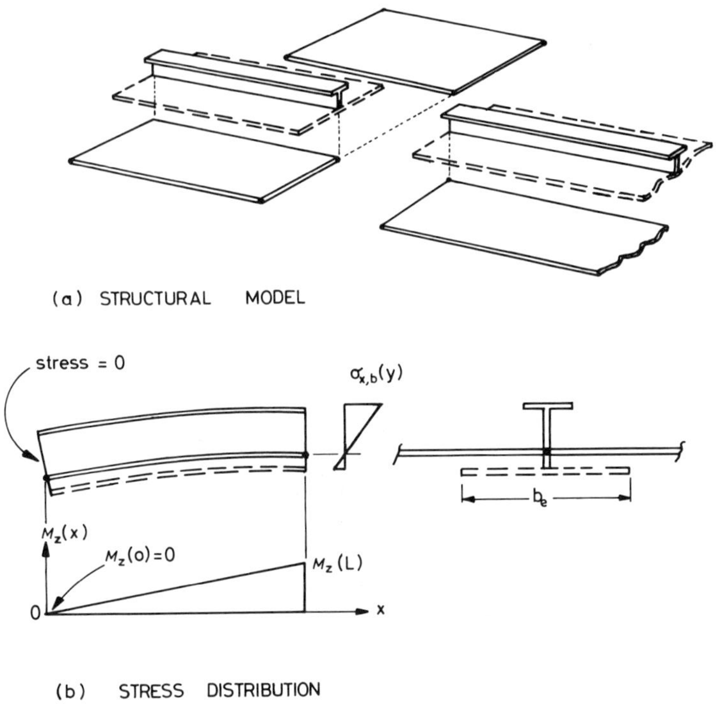

Figure 8.20 Use of hybrid beam element.

element uses the ordinary stiffness matrix (with an effective moment of inertia I_e in place of I) it gives zero axial stress in the beam and zero bending moment at the left end, as required. Also, the stress distribution in the cross section is of the correct linear form. Figure 8.21 illustrates the more general case, in which there are in-plane loads as well as lateral loads. After solving for the nodal displacements, the first line of (5.3.18) would be used to calculate the axial stress in the beam, $\sigma_x(\sigma_x = \sigma_{x1} = \sigma_{x2})$. This calculated axial stress would be virtually identical to the applied axial stress, $(\sigma_a)_{\text{beam}}$, of Fig. 8.13, but since the beam and plating are part of a larger structure the known loads are at points further away and so this applied stress is not known explicitly. Similarly, the plate element's internal force (stress) equation [e.g., (7.5.13) if the CSSR element has been employed] would give the in-plane stress in the plating, $\sigma_{x,p}$. Here again, this resultant stress would be virtually identical to the applied stress in the plating, $(\sigma_a)_1$ of Fig. 8.13. If the latter is the hull girder stress then its value would be known, but $\sigma_{x,p}$ will not equal $(\sigma_a)_1$ exactly because of averaging effects and other modeling approximations.

The bending moment $\mathcal{M}_z(x)$ at any point along the beam is given by (5.3.12a) and the bending stress in the section is then

$$\sigma_{x,b}(y) = \frac{\mathcal{M}_z(x)y}{I_e}$$

In the plane of the plate the bending stress is

$$(\sigma_{x,b})_p = \frac{\mathcal{M}_z(x)y_p}{I_e}$$

and this stress acts uniformly along the effective plate flange of width b_e, as shown in the figure.

Although the stress distribution of the hybrid beam element is of the proper shape, its accuracy depends on the choice of b_e. The eccentric beam element does not involve a choice of b_e, and this might appear to be an advantage. This is not so because the purpose in using b_e is to allow for shear lag, but in the eccentric beam element method of modeling, shear lag is represented only if the plate element is a high-order (many-degree-of-freedom) element, and this increases computation time. If an ordinary element such as the CSSR element is used, the plating is being modeled as fully effective (unless more than one plate element is used between the beams, but this again increases the computation).

Thus for the hull module analysis the hybrid element is preferable, and its accuracy is quite sufficient for this level of analysis as long as care is taken to choose an effective width ratio b_e/b which is appropriate for the given loading and boundary conditions. In an individual member analysis or a local stress analysis more accuracy might be required. This could be obtained by modeling both the beam and the plating in more detail, using a mesh of membrane elements. We shall not deal with this here because it is a simple and well-known example of plane stress analysis performed by the finite element method. This sort of analysis is common to many types of structures and is discussed in a number of basic finite element texts.

Alternative Method of Stress Calculation

For the hybrid beam element there is an alternative method of calculating the stresses which in some circumstances may be preferable to the method given earlier in the section. Instead of calculating the two components of plate flange stress separately and then combining them, it is possible to first transform the

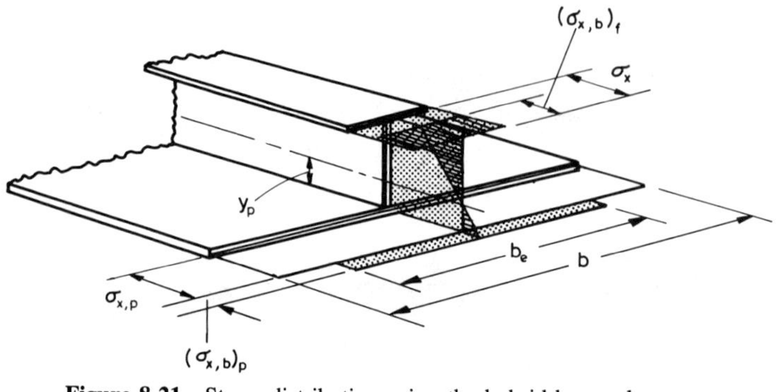

Figure 8.21 Stress distribution using the hybrid beam element.

beam nodal displacements $\boldsymbol{\delta}$ to plate element nodal displacements and to then use the plate element stress equation [e.g., (7.5.13) for the CSSR element] to obtain the (total) stress in the plate. In this case there is no need for a final summation.

The two approaches give similar results, the main difference being that the second approach involves a combining of the nodal displacements of adjacent pairs of beam elements. This requires more "bookkeeping" because the nodal displacements of each beam are used twice and it is important to ensure that the displacements are combined correctly and that there is no unnecessary duplication in the calculations. It may be said that this second approach is plate-oriented whereas the approach described previously is beam-oriented. Also, since the second approach always involves two beams, it has the effect of averaging or smoothing the beam bending stress. Thus, if one beam has a larger bending moment than adjacent beams (due, for example, to a point load) the second approach would smooth out the peak value of bending stress and diffuse some of it to adjacent beams.

Geometric Properties of Beams Attached to Plating

In using the hybrid beam element, and in other calculations concerning beams attached to plating, it is necessary to calculate the neutral axis location and the moment of inertia of the combined section, including a plate flange of effective width b_e.

These properties may be found in one of three ways:

1. If standard sizes of beams and plating are being used, the properties may sometimes be available in a design office manual (if one has been compiled) or, more rarely, in structural steel handbooks.
2. By direct calculation, dividing the cross section into its component areas. If the beam is standard its properties may be found in structural steel handbooks.
3. By means of the following formulas, which are sufficiently accurate for design use provided that the section is a thin-wall section. We define the following quantities (see Fig. 8.22)

A_p = effective plate area = $b_e t_p$

A_w = web area

A_f = flange area

and A_T = total area = $A_p + A_f + A_w$

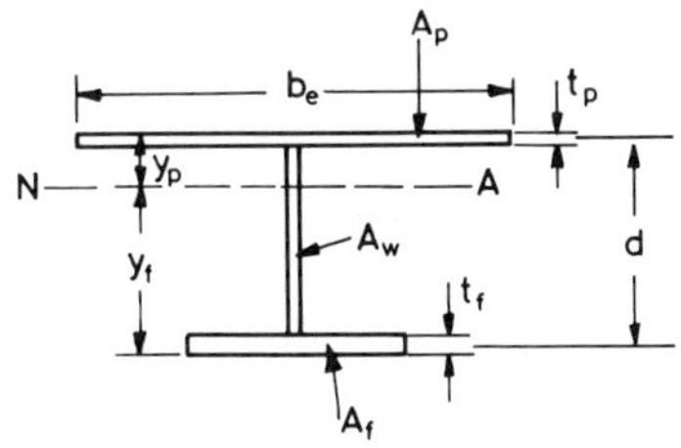

Figure 8.22 Geometric properties of beams attached to plating.

Also t_p = plate thickness

t_f = flange thickness

d = distance from center of plate to center of flange

Then it can be shown that:

$$\left.\begin{aligned} y_f &= \frac{1}{2} t_f + dC_2 \\ y_p &= \frac{1}{2} t_p + d(1 - C_2) \\ I_e &= A_T d^2 C_1 \end{aligned}\right\} \tag{8.3.6}$$

in which

$$C_1 = \frac{A_w\left(\dfrac{A_T}{3} - \dfrac{A_w}{4}\right) + A_f A_p}{A_T^2}$$

$$C_2 = \frac{\dfrac{A_w}{2} + A_p}{A_T}$$

8.4 MODELING OF STIFFENED PANELS

In this section we present an element that models the in-plane stiffness of a flat, rectangular panel of stiffened plating having any number of parallel stiffeners in either direction, or in both directions. The element is primarily intended for representing the rectangular expanse of stiffened plating between the frames and girders of a ship structure, as illustrated in Fig. 8.1. Such panels can be quite large, and so some care has to be taken to ensure that the element has sufficient accuracy to represent such a large portion of structure. For example, the hull girder stress in the side shell varies linearly in the vertical direction, and so the panel element must provide for at least a linear variation in normal stress. It was shown in Chapter 7 that the CSSR element has good accuracy and is a suitable choice for modeling large rectangular panels of plating, and that the QM6 element is suitable for non-rectangular panels. However, large panels nearly always have stiffeners attached to the plating because of lateral loads and/or because of in-plane compressive loads which could cause buckling. Thus the question arises as to what kind of element should be used to represent a stiffened panel.

As a starting point we again take note of the fact that plate bending and stiffener bending need not form part of the hull module analysis and so it is only the in-plane stiffness of the panels which needs to be modeled. Another point is that at the hull module level it would certainly not be appropriate to represent each stiffener individually because this would involve intermediate nodes along the panel edges and consequently excessive amounts of storage and computation. Hence the requirement is for a method of representing the in-plane stiffness of a stiffened panel using only corner nodes. There are three different ways of achieving this:

1. "Lumping" the stiffeners, that is, representing the total effect of the stiffeners by means of two pin-jointed bar elements, one on each side of the panel.
2. Using the orthotropic plate approach, in which the stiffeners are blended with the plating and the stiffened panel is represented as a plate having different stress-strain properties in the two directions. This approach is also discussed in Chapter 10 in connection with panel bending.
3. Using a separate "stiffener element," that is, representing the stiffeners by means of a special membrane element which extends over the same area as the plating and has the same corner nodes and degrees of freedom. In this approach the stiffeners are assumed to follow the same displacement pattern (in the x-direction) as the plating. This allows the strain energy of the stiffeners to be computed and a stiffness matrix to be derived for the stiffeners in the same manner as for the plating. The stiffness matrix of the panel is then the sum of the two separate stiffness matrices.

Lumping is the simplest approach but it is less accurate than the other two and will not be considered in this section. The orthotropic plate approach is the method which has been used most frequently and so we shall give a brief summary of it. However we shall adopt the third approach, which uses a separate stiffener element having the same displacement function as the plating, because we can then choose the CSSR displacement function for the stiffeners as well as for the plating. In this way the complete stiffened panel retains the good accuracy of that element. Another advantage of this approach is that it gives separate stiffness matrices for the plating and for the stiffener, and both are of the form $\mathbf{k} = d\mathbf{k}_0$ in which the design variables of greatest interest appear in the external scalar factor d. As will be shown, this is of significant advantage in rationally-based computer-aided design.

Orthotropic Modeling of a Stiffened Panel

In the orthotropic approach the stiffened panel is represented as an unstiffened but orthotropic plate, that is, a plate that has different stress-strain relationships in the two orthogonal directions. Figure 8.23a shows a rectangular plate with elastic modulus E and Poisson's ratio ν, and with stiffeners parallel to the x-direction.

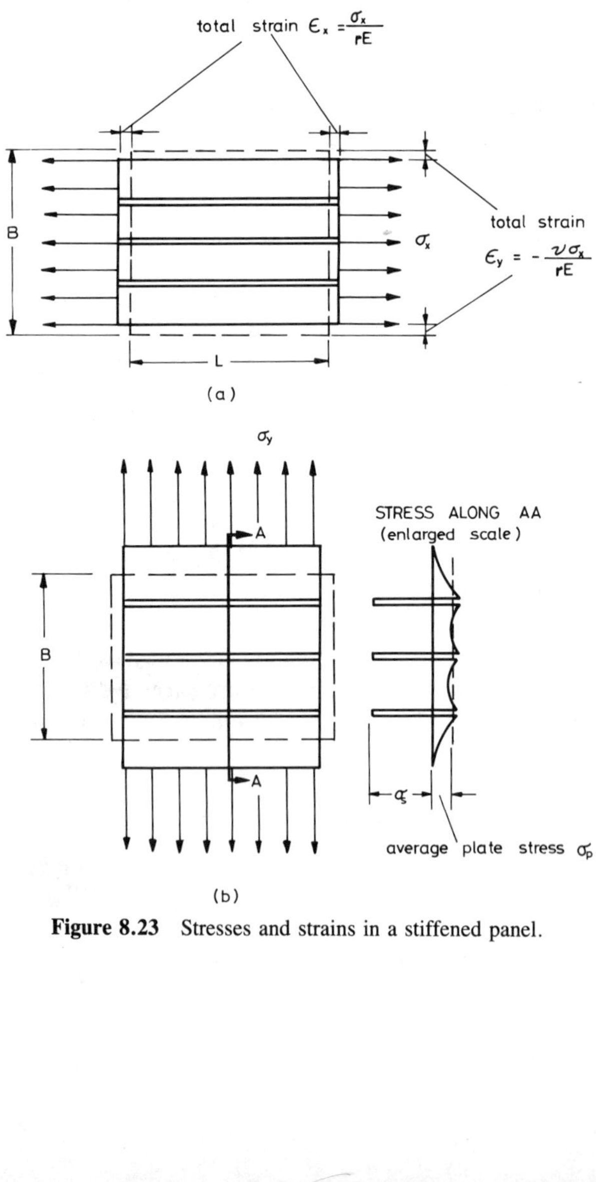

Figure 8.23 Stresses and strains in a stiffened panel.

A uniform stress σ_x causes a strain $\varepsilon_x = \sigma_x/rE$, where r is the ratio of total cross-sectional area to plate area Bt. The related Poisson strain is $\varepsilon_y = -\nu\sigma_x/rE$.

A uniform tensile stress σ_y causes the plate to contract in the x-direction due to the Poisson effect. As the stiffeners can sustain only x-directed stresses, they cannot contract by a σ_y-induced Poisson effect. Thus compatibility of the plate and stiffener deflections gives rise to internal stresses—a compressive stress σ_s in the stiffener and an average tensile stress σ_p in the plate (assuming that the stiffeners are sufficiently close to permit the use of an average plate stress). For equilibrium

$$\sigma_p Bt = \sigma_s(r - 1)Bt$$

The plate length L changes by δ_p and the stiffener length changes by δ_s where

$$\delta_p = \frac{L}{E}(\sigma_p - \nu\sigma_y); \qquad \delta_s = \frac{L}{E}\sigma_s$$

For compatibility $\delta_p = \delta_s$ which implies that

$$\frac{\sigma_p}{\sigma_y} = \frac{\nu(r - 1)}{r}; \qquad \frac{\sigma_s}{\sigma_y} = \frac{-\nu}{r}$$

$$\frac{\varepsilon_x}{\left(\dfrac{-\nu\sigma_y}{E}\right)} = \frac{1}{r}; \qquad \frac{\varepsilon_y}{\left(\dfrac{\sigma_y}{E}\right)} = 1 - \frac{\nu^2(r - 1)}{r}$$

In both the stiffened and unstiffened cases a uniform shear stress τ causes a shear strain $\gamma = \tau/G$. From these results the stress-strain relation for the stiffened plate can be written as

$$\begin{Bmatrix} \varepsilon_x \\ \varepsilon_y \\ \gamma \end{Bmatrix} = \frac{1}{rE}\begin{bmatrix} 1 & -\nu & 0 \\ -\nu & r - \nu^2(r-1) & 0 \\ 0 & 0 & 2r(1+\nu) \end{bmatrix}\begin{Bmatrix} \sigma_x \\ \sigma_y \\ \tau \end{Bmatrix} \tag{8.4.1}$$

or $\boldsymbol{\varepsilon} = \mathbf{D}^{-1}\boldsymbol{\sigma}$

Although this reflects the required orthotropic properties its derivation contains two simplifying assumptions: that the stiffeners are uniformly distributed over the plate, and that the σ_x stress in the stiffened panel is some average of the plate stress σ_p and the stiffener stress σ_s.

Having established the **D** matrix, a stiffness matrix for the orthotropic panel could now be derived by the standard procedure beginning with any suitable displacement function, for example, the linear strain function which was used for the LSR element—(7.4.1)—or the nonlinear displacement function of (7.5.3) and (7.5.4) corresponding to the CSSR element. In each case it would be found that the orthotropic factor r occurs in many different terms of the resulting stiffness matrix and cannot be factored out of the matrix. Therefore if the orthotropic plate approach is used for computer-aided design or for structural optimization, each alteration to the basic design variables (e.g., plate thickness, or any of the stiffener properties) would require a complete recalculation of the stiffness matrix. For this reason the orthotropic plate approach will not be pursued further in this section. It is, however, discussed further in Chapter 10 in regard to bending of double wall panels, and again in Chapter 13 in regard to buckling of cross-stiffened panels.

There is one final point worth mentioning which concerns the stress calculation when using the orthotropic plate approach. It can be shown that the use of orthotropic plate theory (which implies material orthotropy) to represent the effect of stiffeners (structural orthotropy) requires that a final correction be made to the σ_x stress as calculated from the theory. This may be illustrated by considering the panel of Fig. 8.23*b*. A uniform stress σ_x corresponds to a force $F_x = Bt\sigma_x$. The addition of stiffeners changes the area to rBt so that for constant F_x the stress drops to σ_x/r. Since a finite element analysis deals with force–displacement relations rather than stress-strain relations, the stress σ_x found by the orthotropic model must be corrected by a multiplying factor $1/r$.

Special Element for Representing Stiffeners

We now proceed to derive a special rectangular membrane element for representing the stiffeners. This approach was proposed almost simultaneously by Tingleff [2] and by Colville and Abbasi [3] for representing steel reinforcing in concrete. However, the former dealt only with the Constant Strain Triangle element which is not relevant here, and the latter used the LSR displacement function, which has been shown in Figs. 7.16 and 7.17 to be inferior to the CSSR element. Reference 4 appears to be the first to adopt the CSSR displacement for the stiffeners in a rectangular panel. This choice gives an element stiffness matrix which has several advantages over that of Ref.

3: (1) it is simpler; (2) it provides the same improved accuracy for the stiffened panel as the CSSR element provides for plating; and (3) it yields a stiffness matrix in which the design variables appear as explicit scalar factors. After deriving this separate "stiffener element" we shall present a brief summary of some test results which demonstrate its superiority.

The basic technique presented in Refs. 2 and 3 allows each of the stiffeners to be of any length and to have any orientation within the panel. In the case of ship panels the stiffeners are nearly always orthogonal and we will deal mainly with that case, but the basic derivation will be done for the general case in order to show how this approach could be used for modeling diagonal or even nonparallel stiffening. Figure 8.24 shows a completely general stiffener, say the ith stiffener, which begins at a point (x_i, y_i) within the rectangle, makes an angle θ to the x-axis and has length l.

Since we wish the displacement to be compatible with that of the CSSR element we begin with (7.5.9) which gives the displacements at any location, and we adopt it for the stiffeners. In this equation $\mathbf{C}$ is the vector of generalized displacement coefficients, eight in number. Following the usual procedure outlined in Section 7.3, we write the nodal displacement $\boldsymbol{\delta}$ in terms of $\mathbf{C}$ by means of a connectivity matrix $\mathbf{A}$ (Step II):

$$\boldsymbol{\delta} = \mathbf{AC}$$

where

$$\mathbf{A} = \begin{bmatrix} \mathbf{H}(L,B) \\ \\ \mathbf{H}(0,B) \\ \\ \mathbf{H}(0,0) \\ \\ \mathbf{H}(L,0) \\ \\ \end{bmatrix} = \frac{1}{E}\begin{bmatrix} L & LB & -\nu L & \frac{-(\nu L^2+B^2)}{2} & (1+\nu)B & B & 1 & 0 \\ -\nu B & \frac{-(\nu B^2+L^2)}{2} & B & LB & (1+\nu)L & -L & 0 & 1 \\ 0 & 0 & 0 & \frac{-B^2}{2} & (1+\nu)B & B & 1 & 0 \\ -\nu B & \frac{-\nu B^2}{2} & B & 0 & 0 & 0 & 0 & 1 \\ 0 & 0 & 0 & 0 & 0 & 0 & 1 & 0 \\ 0 & 0 & 0 & 0 & 0 & 0 & 0 & 1 \\ L & 0 & -\nu L & \frac{-\nu L^2}{2} & 0 & 0 & 1 & 0 \\ 0 & \frac{-L^2}{2} & 0 & 0 & (1+\nu)L & -L & 0 & 1 \end{bmatrix} \tag{8.4.2}$$

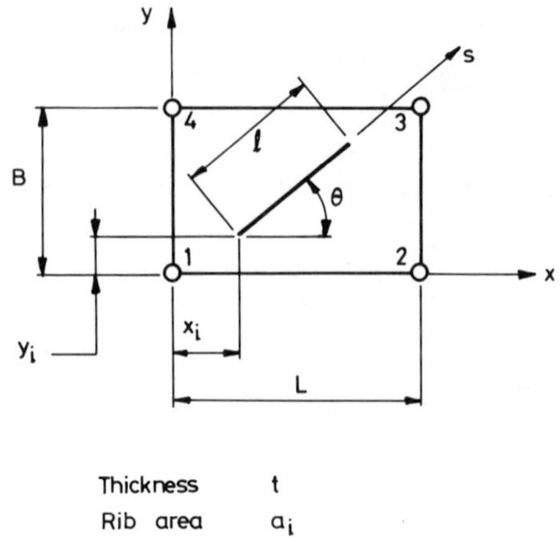

Figure 8.24 Stiffened panel element with stiffeners at arbitrary orientation.

By inverting $\mathbf{A}$ the coefficients $\mathbf{C}$ can be expressed in terms of the nodal displacements

$$\mathbf{C} = \mathbf{A}^{-1}\boldsymbol{\delta} \tag{8.4.3}$$

We next obtain the strain function, in terms of the nodal displacements

$$\varepsilon(x,y) = \frac{\partial}{\partial x}u(x,y)$$

$$= \frac{1}{E}[1 \quad y \quad -\nu \quad -\nu x \quad 0 \quad 0 \quad 0 \quad 0]\mathbf{C}$$

which we shall denote as

$$\varepsilon(x,y) = \mathbf{GC} \tag{8.4.4}$$

From (8.4.3) this is

$$\varepsilon(x,y) = \mathbf{GA}^{-1}\boldsymbol{\delta}$$

which serves to define the strain coefficient vector **B**:

$$\mathbf{B} = \mathbf{GA}^{-1} \tag{8.4.5}$$

such that

$$\varepsilon(x,y) = \mathbf{B}\boldsymbol{\delta} \tag{8.4.6}$$

The stress-strain relationship is

$$\begin{aligned}\boldsymbol{\sigma} &= \mathbf{D}\boldsymbol{\varepsilon}\\ &= E\boldsymbol{\varepsilon}\end{aligned}$$

in which the matrix **D** is here simply the scalar E because the strain is uniaxial. The component of strain in the θ direction comes from the theory of small strains and is

$$\boldsymbol{\varepsilon}_\theta = \boldsymbol{\lambda}\boldsymbol{\varepsilon}$$

where $\boldsymbol{\lambda} = [\cos^2\theta \quad \sin^2\theta \quad \sin\theta\cos\theta]$

The corresponding stress component is

$$\boldsymbol{\sigma}_\theta = E\boldsymbol{\varepsilon}_\theta$$

We next carry out Step V of the basic finite element derivation process, in which the element is given virtual nodal displacements $\boldsymbol{\delta}^*$ and the external work, involving actual nodal forces **f**, is equated to the internal work, involving the virtual strain $\boldsymbol{\varepsilon}^*$ (expressed in terms of $\boldsymbol{\delta}^*$) and the actual stress $\boldsymbol{\sigma}$ (expressed in terms of the actual displacements $\boldsymbol{\delta}$, corresponding to **f**). The virtual nodal displacements are then set to unity and the resulting matrix relating **f** to $\boldsymbol{\delta}$ is the required element stiffness matrix.

The external virtual work is

$$W_{\text{ext}} = \boldsymbol{\delta}^{*T}\mathbf{f}_s + \boldsymbol{\delta}^{*T}\mathbf{f}_p$$

and the internal virtual work is

$$W_{\text{int}} = \int_{R_s} [\boldsymbol{\varepsilon}^*(x,y)]^T\boldsymbol{\sigma}(x,y)\, dR_s + \int_{R_p} [\boldsymbol{\varepsilon}^*(x,y)]^T\boldsymbol{\sigma}(x,y)\, dR_p$$

where R_s and R_p are the stiffener and the plating regions. Thus both W_{ext} and W_{int} consist of two separate components, one for the plating and one for the stiffeners. The plating component leads to the stiffness matrix of the CSSR element, and so from this point on we concentrate our attention exclusively on the stiffener component.

$$\begin{aligned}W_{\text{int},s} &= \int_{R_s} \boldsymbol{\varepsilon}_\theta^{*T}\boldsymbol{\sigma}_\theta\, dR_s\\ &= E\int_{R_s} \boldsymbol{\varepsilon}^{*T}\boldsymbol{\lambda}^T\boldsymbol{\lambda}\boldsymbol{\varepsilon}\, dR_s\end{aligned}$$

Substituting from (8.4.6)

$$W_{\text{int},s} = \mathrm{E}\int_{R_s} \boldsymbol{\delta}^{*T}\mathbf{B}^T\boldsymbol{\lambda}^T\boldsymbol{\lambda}\mathbf{B}\boldsymbol{\delta}\, dR_s$$

Setting $\boldsymbol{\delta}^*$ equal to unity and equating the internal and external work gives

$$\mathbf{k}_s = E\int_{R_s} \mathbf{B}^T\boldsymbol{\lambda}^T\boldsymbol{\lambda}\mathbf{B}\, dR_s$$

This may be evaluated by integration in the x,y coordinate system but it is simpler to integrate along the stiffener axis s. The coordinate transformation needed is

$$\begin{aligned}x &= x_i + s\cos\theta\\ y &= y_i + s\sin\theta\end{aligned}$$

From this point on the derivation is straightforward, but for the completely general stiffener which we have thus far considered it involves tedious multiplications and integrations, and it results in long expressions. Hence we shall now concentrate on the case of predominant interest, which is that of n parallel stiffeners which extend the full length of the panel. For maximum generality we shall continue to allow the stiffeners to be at irregular spacings. Thus for all stiffeners $\theta = 0$, $x_i = 0$, and $l = L$, and the ith stiffener is of cross-sectional area a_i, located at a height y_i from the bottom of the panel. With these simplifications the element stiffness matrix becomes

$$\mathbf{k}_s = E \int_{R_s} \mathbf{B}^T\mathbf{B}\, dR_s$$

Substituting from (8.4.5) and noting that $\mathbf{A}^{-1}$ is not a function of x or y we can write

$$\mathbf{k}_s = E(\mathbf{A}^{-1})^T \mathbf{k}_s' \mathbf{A}^{-1} \tag{8.4.7}$$

where $\mathbf{k}_s'$ is referred to as the "kernal stiffness matrix" and is given by

$$\mathbf{k}_s' = \int_{R_s} \mathbf{G}^T\mathbf{G}\, dR_s$$

In most elements the region of integration is continuous but in the present case there are n discrete prismatic stiffeners of area a_i and constant height y_i extending from $x = 0$ to $x = L$. Recognizing that $\mathbf{k}_s'$ is essentially related to the summation of strain energies, the integration in the vertical direction can be replaced by a summation. Upon substituting for $\mathbf{G}$ from (8.4.4) and integrating along the length of each stiffener the resulting kernel stiffness matrix is

$$\mathbf{k}_s' = \frac{L}{E}\left[\begin{array}{cccc|c}
\sum_n a_i & \sum_n a_i y_i & -\nu \sum_n a_i & -\nu \frac{L^2}{2} \sum_n a_i & \\
 & \sum_n a_i y_i^2 & -\nu \sum_n a_i y_i & -\nu \frac{L^2}{2} \sum_n a_i y_i & \\
 & & \nu^2 \sum_n a_i & \nu^2 \frac{L^2}{2} \sum_n a_i & 0 \\
\text{symm.} & & & \nu^2 \frac{L^3}{3} \sum_n a_i & \\
\hline
 & & 0 & & 0
\end{array}\right]$$

in which blanks are zero and

$$\mathcal{L} = \sum_n a_i y_i^2$$

$$\mathcal{M} = -\sum_n a_i y_i (B - y_i)$$

$$\mathcal{N} = \sum_n a_i (B - y_i)^2$$

$$\mathcal{P} = \frac{\nu^2 L^2}{12} \sum_n a_i$$

That the foregoing stiffness matrix accurately reflects the behavior of various stiffener distributions can be shown with a single stiffener of area a. For $y = 0$ the stiffness matrix is

$$\mathbf{k}_s = \frac{Ea}{12L}\begin{bmatrix}
0 & & & & & & & \text{symm.} \\
 & p & & & & & & \\
0 & & 0 & & & & & \\
 & -p & & p & & & & \\
0 & & 0 & & 12 & & & \\
 & p & & -p & & p & & \\
0 & & 0 & & -12 & & 12 & \\
 & -p & & p & & -p & & p
\end{bmatrix}$$

where $p = (\nu L/B)^2$

This is the same as a bar element of area a joining nodes 1 and 2, plus some small stiffness in the y-direction which is the Poisson effect arising from the lengthwise restraint which the stiffener exerts on the bottom edge of the plating.

The final step is to calculate the element stiffness matrix from (8.4.7). The result is

$$\mathbf{k}_s = \frac{E}{LB^2}\begin{bmatrix}
\mathcal{L} & & & & & & & \text{symmetric} \\
 & \mathcal{P} & & & & & & \\
-\mathcal{L} & & \mathcal{L} & & & & & \\
 & -\mathcal{P} & & \mathcal{P} & & & & \\
\mathcal{M} & & -\mathcal{M} & & \mathcal{N} & & & \\
 & \mathcal{P} & & -\mathcal{P} & & \mathcal{P} & & \\
-\mathcal{M} & & \mathcal{M} & & -\mathcal{N} & & \mathcal{N} & \\
 & -\mathcal{P} & & \mathcal{P} & & -\mathcal{P} & & \mathcal{P}
\end{bmatrix} \tag{8.4.8}$$

Moving the stiffener to $y = B/4$ gives the stiffness matrix

$$
\mathbf{k}_s = \frac{Ea}{16L}\begin{bmatrix}
1 & & & & & & & \text{symm.} \\
 & p' & & & & & & \\
-1 & & 1 & & & & & \\
 & -p' & & p' & & & & \\
-3 & & 3 & & 9 & & & \\
 & p' & & -p' & & p' & & \\
3 & & -3 & & -9 & & 9 & \\
 & -p' & & p' & & -p' & & p'
\end{bmatrix}
$$

where $p' = \frac{4}{3}p = \frac{4}{3}(\nu L/B)^2$.

This shows that the element is more compliant between nodes 3 and 4 than between nodes 1 and 2. Because the stiffener is closer to the latter nodes this result is reasonable.

For $y = B/2$ the stiffener is at the midheight and $\mathbf{k}_s$ is

$$
\mathbf{k}_s = \frac{Ea}{12L}\begin{bmatrix}
3 & & & & & & & \text{symm.} \\
 & p & & & & & & \\
-3 & & 3 & & & & & \\
 & -p & & p & & & & \\
-3 & & 3 & & 3 & & & \\
 & p & & -p & & p & & \\
3 & & -3 & & -3 & & 3 & \\
 & -p & & p & & -p & & p
\end{bmatrix}
$$

Since the displacement of the stiffeners is the same as the u displacement of the attached plating, stress in the stiffeners is given by the equation for σ_x in the CSSR element:

$$
\left.\begin{aligned}
&\sigma_x = C_1 + C_2 x \\
&\text{where} \\
&\frac{C_1}{E} = \frac{u_4 - u_3}{L(1-\nu^2)} + \frac{\nu^2(u_1 - u_2 + u_3 - u_4)}{2L(1-\nu^2)} \\
&\qquad + \frac{\nu(v_1 + v_2 - v_3 + v_4)}{2B(1-\nu^2)} \\
&\frac{C_2}{E} = \frac{u_1 - u_2 + u_3 - u_4}{LB}
\end{aligned}\right\} \qquad (8.4.9)
$$

Uniformly Stiffened Panels

The foregoing discussion assumes that the size, number, and location of the stiffeners are known, and that the type of analysis being performed is such that this degree of detail is appropriate and is desired. This could be the case, for example, in the analysis of a local region of structure of an existing ship or of a nearly finalized design. However, such details do not affect the gross response of a structure as large as a hull module. Moreover, such detail is generally not known during the greater part of a rationally-based structural design process, and averaged stiffener properties would be used instead. And finally, for cost-efficient design and fabrication it is desirable to have identical, evenly spaced stiffeners within a given panel (and, for that matter, from panel to panel, as far as possible). In this case the stiffness matrix can be further simplified, being expressed in terms of the total stiffener cross-sectional area A_s. The result is

$$
\mathbf{k}_s = \frac{EA_s}{12L}\begin{bmatrix}
4 & & & & & & & \\
 & p & & & & & & \\
-4 & & 4 & & & & & \\
 & -p & & p & & & & \\
-2 & & 2 & & 4 & & & \\
 & p & & -p & & p & & \\
2 & & -2 & & -4 & & 4 & \\
 & -p & & p & & -p & & p
\end{bmatrix} \qquad (8.4.10)
$$

where, as before, $p = (\nu L/B)^2$.

For brevity and ease of reference, we shall use the initials USR (Uniform Stiffener Rectangle) to refer to this element, and DRS (Discrete Stiffener Rectangle) to refer to the element of (8.4.8).

The stiffness matrix for the complete stiffened panel element is simply the sum of $\mathbf{k}_s$ for the stiffeners and $\mathbf{k}_p$ for the plating, the latter being the CSSR element stiffness matrix given in (7.5.12). For reference purposes we denote the resulting stiffness matrix for the stiffened panel as $\mathbf{k}_{sp}$

$$
\mathbf{k}_{sp} = \mathbf{k}_s + \mathbf{k}_p \qquad (8.4.11)
$$

For stiffeners and for plating the structural properties which most affect the strength and which are most easily and frequently adjusted during the design of the panel are, respectively: the total amount of stiffening as represented by the total stiffener area A_s, and the plate thickness t. In each of the foregoing matrices these design variables appear only once, as an external scalar factor. This is of significant benefit in computer-aided design and/or structural optimization applications because these variables are repeatedly being changed as the design proceeds. Therefore it is preferable to treat the stiffeners and the plating as two separate elements (the CSSR element and, usually, the USR element) each with its own element stiffness matrix ($\mathbf{k}_s$ and $\mathbf{k}_p$) and to combine them only when assembling the global stiffness matrix, as is done with all other elements. Nevertheless, for ease of reference and simplicity of notation we will often refer to the combination as just one element, the "stiffened panel element" (or SP element, for brevity) and we will use the symbol $\mathbf{k}_{sp}$ to refer to the summed stiffness matrix for this single composite element.

For hull module analysis the stiffener element is the USR element, which uses only the total stiffener area A_s, and in this case the stiffness properties of the composite SP element do not differ appreciably from that of the orthotropic plate element. The real advantage of the SP element is the occurrence of A_s and t as explicit scalar factors. In contrast, we have seen that the DSR element accounts for all of the stiffeners individually, each with its proper area and location. Therefore in applications where the stiffeners are nonuniform, either in size or in spacing, the DRS element should be used for the stiffeners, thus giving the discrete stiffened panel (DSP) element. For all such panels the DSP element is distinctly superior to the orthotropic plate element. In fact, the next section shows that even for uniformly stiffened panels the DSP element is superior. It also demonstrates that the ordinary SP element, using the USR element for the stiffeners, is sufficiently accurate for modeling large stiffened panels, and is therefore suitable for a hull module analysis. Cross-stiffened panels may be modeled by combining two USR elements with a CSSR element. For non-rectangular panels the isoparametric quadrilateral "QM6" element of Section 7.7 may be used in place of the CSSR element. As mentioned there, this element is a generalization of the CSSR element and reduces to it when the quadrilateral is a rectangle.

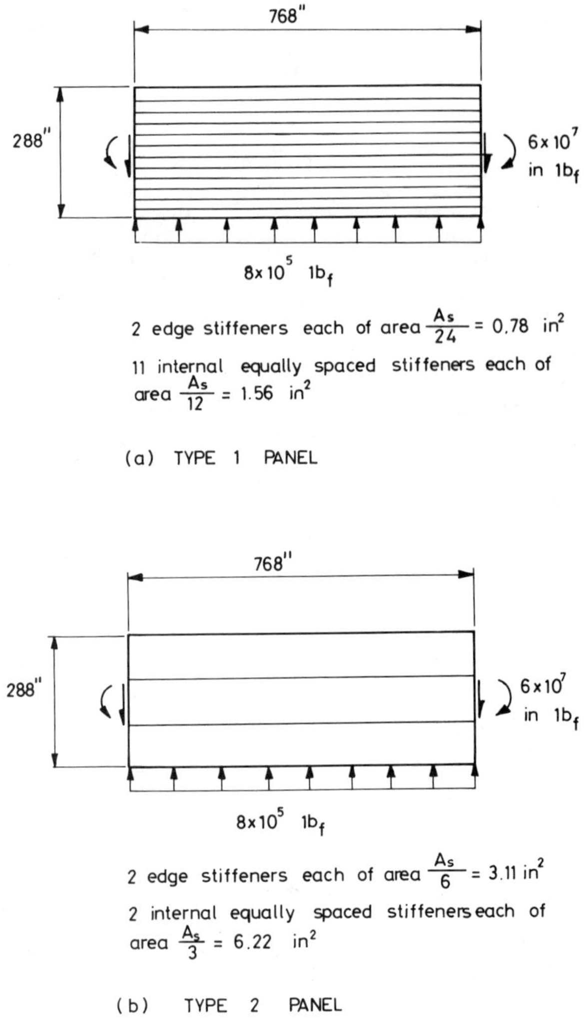

Figure 8.25

Accuracy of Stiffened Panel Elements

Comparison of Various Elements

The performance of a stiffened panel element is not easy to gauge because few stiffened panel cases have analytic solutions and those that do are trivial. The test problem shown in Fig. 8.25 resembles the side plating of a ship in the hogging condition. Four different cases are examined. For a constant total stiffener area of 18.72 in^2, two stiffener distributions were modeled, as shown in the figure, and for each of these two different plate thicknesses were used (0.0833 in and 0.375 in) thus giving two different ratios of total area to plate area ($r = 1.78$ and $r = 1.173$). While this problem is simple enough to allow analytic insights it is too difficult for a complete analytic solution. Therefore a fine-mesh finite element solution using individual bar elements for each stiffener is used as the reference solution. Details of the various models are shown in Fig. 8.26. The reference model and the "lumped stiffener" model use the LSR element to represent the rectangles of plating between the stiffeners.

The errors in maximum deflection and stress are given in Table 8.3 for the various stiffened panel elements. The results indicate that the Lumped element, the SP element, and the Orthotropic element all have approximately the same overall degree of accuracy—one being slightly better than the others depending on the particular stiffener distribution and on whether it is deflection error or stress error. Stress values are generally of greater interest, and these three elements give

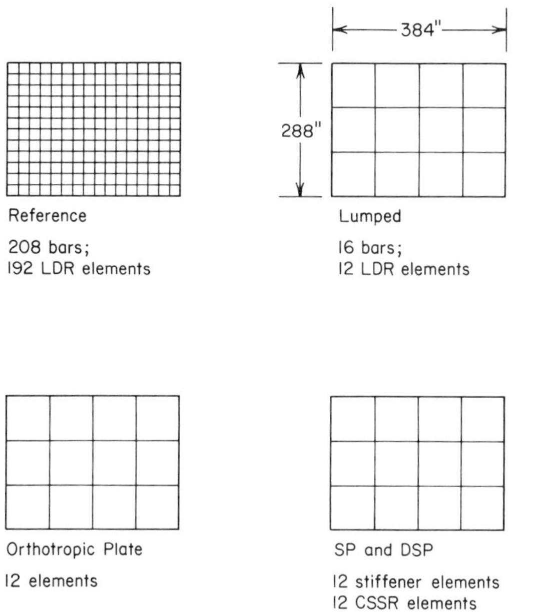

Figure 8.26 Stiffened panel finite element models. Note: Due to symmetry, the structural models represent only one-half of each panel.

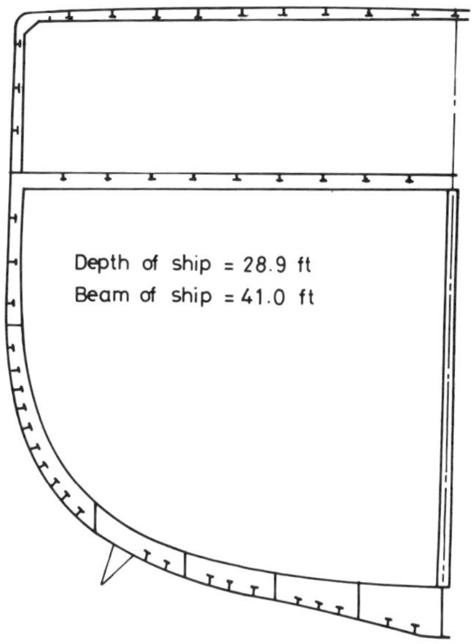

Figure 8.27 Frigate section in way of boiler room.

stress values which are within ±15% of the fine mesh solution, as measured by the rms value of the discrepancies. In contrast, the DSP element is accurate to within ±3%, and it also gives more consistent values as shown by the standard deviations.

Modeling of Large Panels Using the SP Element

We next examine the performance of the SP element when it is used for its intended role—the modeling of large stiffened panels in a hull module analysis. For this purpose the element was used for the analysis of the boiler room compartment of a Whitby class frigate, as shown in Fig. 8.27. The hull module was subjected to hogging bending moment, hydrostatic loading, and deck loads. The details of the problem are not given here but the form of the structure may be recognized from the figures. A fine mesh solution which modeled each stiffener individually was used as the reference solution.

Figure 8.28 shows the longitudinal stresses near the center of the boiler room. The good agreement between the two solutions parallels that seen in the previous test. This applies even to the large panel elements in the deck. Figure 8.29 shows the transverse stresses near the center of the boiler room. The SP elements perform as well as can be expected of elements that cannot follow linearly varying transverse stresses. Figure 8.30 shows the shear stresses near the boiler room bulkhead. Here again there are some stress gradients which the SP element cannot follow. However, the significant stress levels are modeled sufficiently well for a hull module analysis. As expected, the element cannot model stress concentrations caused by local discontinuities (such as those in the decks near the centerline) but this is not the task of a hull module analysis.

We conclude that the in-plane stiffness of rectangular stiffened panels is adequately modeled by the two types of stiffened panel element presented herein. Unlike the usual orthotropic plate and lumped stiffener models, the DSP element gives a realistic account of nonuniform stiffening. Also, the simpler SP element is sufficiently accurate to represent large stiffened panels, as is required in a hull module analysis. Moreover, the principal design variables occur as scalar factors

TABLE 8.3 ERRORS IN MAXIMUM DEFLECTION AND STRESS FOR VARIOUS STIFFENED PANEL ELEMENTS

	% Error					
	Deflection			Stress		
Element Type	rms	Mean	s.d.*	rms	Mean	s.d.*
Lumped	5.5	−4.7	3.4	4.5	−2.5	4.4
Orthotropic	5.3	3.2	5.0	7.5	5.2	6.2
SP	3.3	0.2	3.8	6.2	3.7	5.8
DSP	2.6	−2.5	0.5	1.4	0.2	1.7

*Standard deviation.

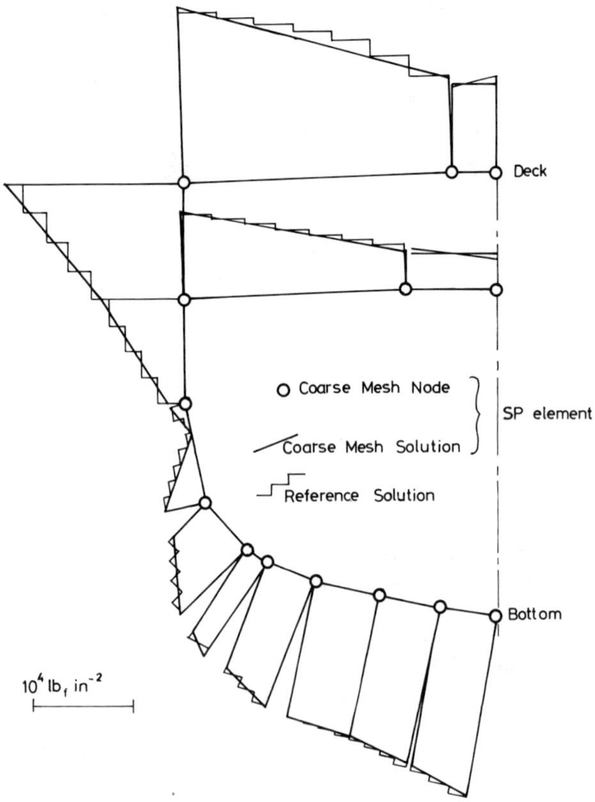

Figure 8.28 Longitudinal stresses near center of boiler room.

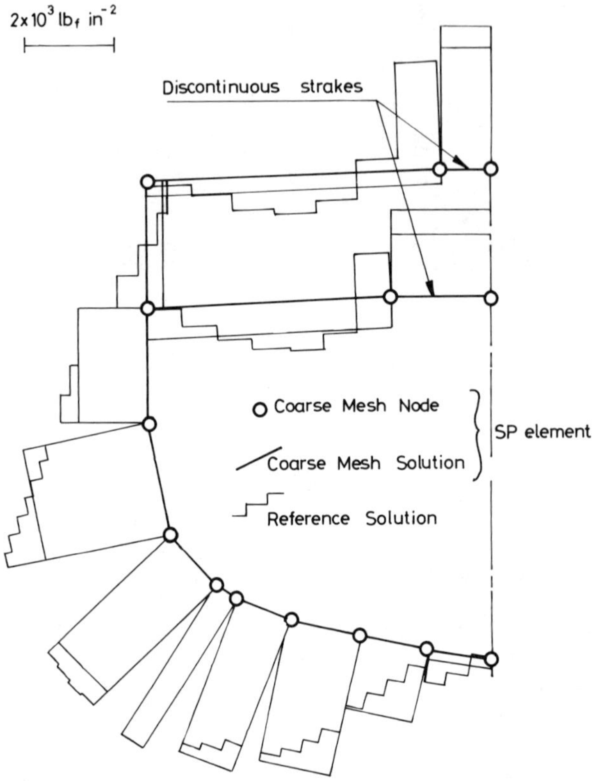

Figure 8.30 Shear stresses near bulkhead.

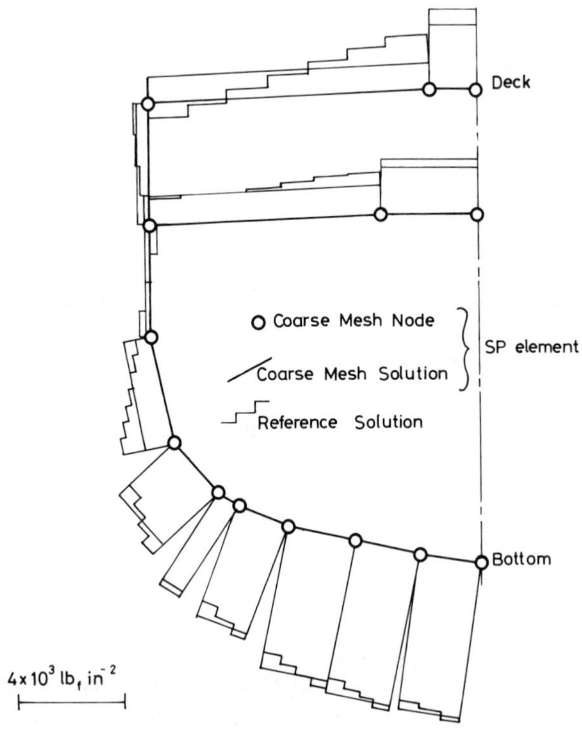

Figure 8.29 Transverse stresses near center of boiler room.

outside the stiffness matrix. Both of these features are of significant benefit in the computer-aided design of ship structures.

8.5 STRUCTURAL MODELING OF A HULL MODULE

There are many basic aspects of the geometry and structural characteristics of a hull module which are common to a wide variety of ship types and which permit the adoption of some simplifying assumptions and the development of modeling and analysis techniques which provide the following benefits:

1. They simplify the data input and allow some degree of automated data generation.
2. They reduce the amount of computation either in the finite element analysis of the hull module or in the subsequent structural evaluation of the principal members, or in both of these areas.
3. They reduce the length and simplify the logic structure of the overall computer program in one or

more of the following phases: hull module analysis, principal member analysis, and the formulation of the (nonlinear) structural constraints (see Fig. 1.1).

This section and the next three sections present the main principles and techniques for structural modeling in finite element analysis, with particular reference to the structural modeling of a hull module. No attempt has been made to cover *all* of the modeling techniques which may be relevant for this application. Rather the intention is merely to present those aspects which are most basic and relevant, together with some particular modeling techniques which the author considers to be of value and which have been used in the MAESTRO computer program for rationally-based preliminary ship structural design, described in Section 1.3. Some of these features are quite common, being used in a variety of finite element programs for various types of structures; others, such as the representation of deep web joints in Section 8.6, and the simplified double bottom modeling techniques in Section 8.8, are of more particular relevance to ship structures.

A hull module is a combination of beam and membrane elements. Beam elements are used to represent the various beams and girders in the bottom, sides, and decks. As explained in Section 8.3, better results are achieved by using the hybrid beam element, in which the position of the neutral axis allows for a plate flange of effective width b_e. In the plane of each transverse frame these beam members interact in the manner of a frame structure, but there are often some portions of structure in the transverse plane which are not beams and which can only be modeled by membrane elements. In particular beam elements should not be used if the beam web is very deep relative to the length. For an ordinary beam element the web height should not exceed about one-fourth of the length. However, if the element contains the shear deflection correction terms of Section 5.5 and the allowances for adjacent beam webs, brackets, and the shear distortion of joints which are presented in this chapter then the web height can be larger. In the author's experience, an approximate upper limit is a web height equal to the flexible length of the beam. Beyond this only membrane elements should be used. Thus for example, in the segregated ballast tanker of Fig. 8.31 the web depth of the upper beam in the wing tank is about at the limit of beam element representation, and the bottom portion of the wing tank structure can only be modeled by membrane elements.

Since a hull module involves both beam and membrane elements many difficult questions arise such as the selection of elements, the choice of nodal positions, and the method of connecting these two very different types of element. The remaining sections of this chapter provide information on these topics and present some specific modeling techniques which allow good accuracy to be achieved with the elements that have been described, despite their relatively large size.

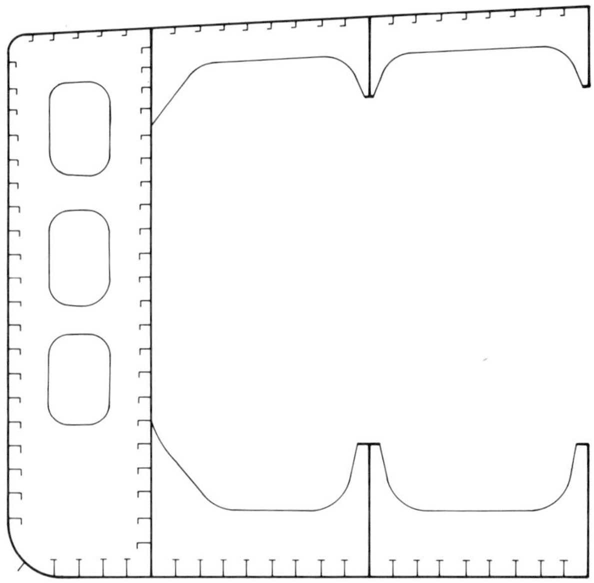

Figure 8.31 Frame structure requiring combined beam and membrane element representation.

Mesh Systems

A hull module can nearly always be idealized as either prismatic or linearly tapering in the longitudinal direction. This makes it relatively simple to generate three-dimensional gridworks or mesh systems, which are useful in specifying various aspects of a structural model such as the location of nodes, the position, size, and orientation of members, and the position of neutral axes. For these applications we shall define two different three-dimensional mesh systems: a reference mesh and a neutral axis mesh.

Reference Mesh

This three-dimensional mesh defines the basic topology (overall dimensions and connectivity) of the hull module. It lies in the planes of the plating (and of any other members represented by membrane elements) both transversely and longitudinally. It therefore re-

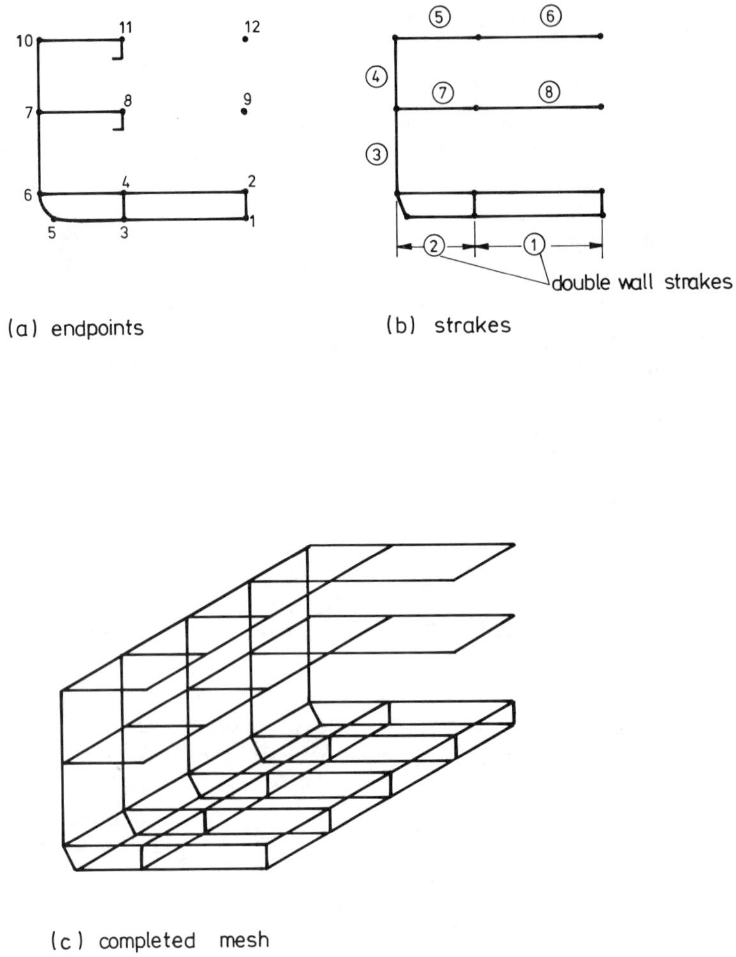

Figure 8.32 Reference mesh.

lates mainly to the membrane elements in the hull module. It is generated in four steps, as shown in Fig. 8.32.

1. In the ship's cross section, nominate endpoints in the plane of the longitudinal plating corresponding to important points in the cross-sectional profile, such as shell and deck intersections. In Fig. 8.32*a* the end points are numbered for reference purposes.
2. Define the location of each "strake" (see Fig. 2.15) by nominating the two endpoint numbers which it connects. The cross-sectional profile formed by the strakes is the approximation to the ship's actual cross section. In Fig. 8.32*b* the strakes are identified by circled numbers. The first two strakes are double wall strakes (see Fig. 2.16).
3. In the plane of each transverse frame insert the set of mesh lines corresponding to the strake cross-sectional profile.
4. Generate longitudinal mesh lines at each end point extending the full length of the hull module and connecting the cross-sectional profiles. The resulting mesh is shown in Fig. 8.32*c*. Portions of strakes 6 and 8 have been deleted to allow for hatches.

The reference mesh has three applications which arise from its very nature:

1. Because it corresponds to the flanges and webs of the hull girder, it is used (at least implicitly) for the hull girder calculations.
2. Because it lies in the plane of the plating it gives the dimensions of the shell and of all internal tanks. It is therefore used for calculating the various hydrostatic loads which act on the hull module.
3. Because it gives the dimensions of each of the membrane elements it is used to calculate all of the structural properties of these elements.

NEUTRAL AXIS MESH

As illustrated in Fig. 8.33 this three-dimensional mesh consists of the neutral axis lines of all of the transverse beam elements (which include a plate flange wherever appropriate) and longitudinal lines passing through and connecting each point where two neutral axis lines intersect. If there are any superelements (Section 8.7) the mesh will include the junction nodes, which are the points at which the beam neutral axes intersect the boundary of the superelement.

Of the two types of mesh the reference mesh is the more basic and more useful, having several different applications in the rationally-based design process of Fig. 1.1. For example the overall dimensions of the hull module are required for the updating, as the design proceeds, of the hull girder sectional area and moment of inertia, and of all loads on the hull module which relate to the dimensions of the membrane elements (e.g., the weight of the plating, and loads arising from lateral pressure acting on plating). Also, as mentioned previously, the reference mesh provides the membrane element dimensions which are required for the calculation of all of their structural properties, including their stiffness matrices. In contrast the neutral axis mesh is generally used only for calculating the structural properties of the beam elements.

Another point of comparison between the two mesh systems is that the reference mesh is easier to construct, since it is based on the actual dimensions. It is also easier to draw and the resulting drawing resembles the actual structure more closely because it is the membrane elements which give the basic shape and subdivision of the structure. The neutral axis mesh is primarily a representation of the beam elements, which of their very nature are simple straight line segments. This mesh could therefore be regarded as the *skeleton* of the hull module. Its construction requires the calculation of the neutral axis locations in all of the transverse beams.

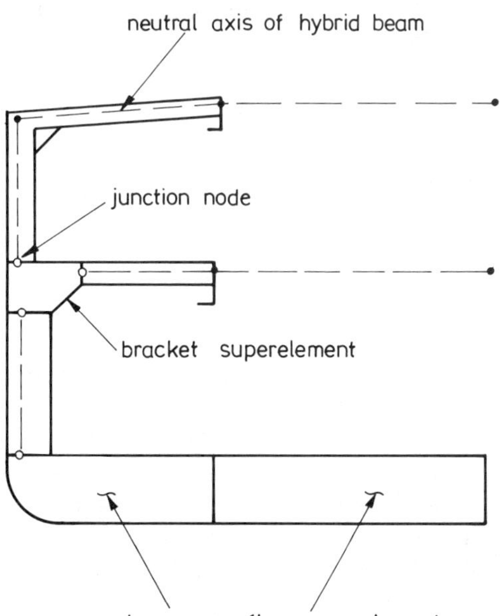

Figure 8.33 Neutral axis mesh.

Location of Structural Nodes

We come now to the question of where the structure nodes should be placed. There are three possibilities:

1. At the reference mesh points.
2. At the neutral axis mesh points.
3. Use the reference mesh for the membrane elements and the neutral axis mesh for the beam elements, and link the two together by means of short, very rigid bar elements.

The third approach is the most accurate, but doubling the number of nodes doubles both N and w_b (see Section 5.2) and causes an eight-fold increase in the computation. We therefore examine the other two approaches to see if either of them might be sufficiently accurate. The possibility of using just one set of nodes for all elements, both flexural and membrane, arises from the fact that in a plated frame the neutral axis of the beam members is very close to the plating. The question is whether there is some method whereby, in one approach or the other, the error can be kept to an acceptable level (which also requires some means for assessing the error).

The key to achieving this lies in a subtle but important difference between the mathematical model of a structure and a physical or graphical model (i.e., a drawing). The latter two models are physical analogues of the structure and therefore all geometric aspects are necessarily and inherently consistent within the model. For example, in a physical model the length of a beam is necessarily equal to the distance between its ends. But the ends are, by definition, the nodes. Therefore there is an automatic and inherent agreement between the member geometry and the geometry of the nodal mesh. To give another example, in a physical model the angle of orientation of a beam is necessarily the same as the angle between the mesh points which constitute the ends of the beam.

In contrast, in a mathematical model there is no *automatic* or *necessary* agreement between the member geometry and the nodal locations. The lengths and orientation angles of members can be obtained by calculation from the declared nodal positions, in which case there is agreement, but they can also be *assigned* a value and this value can be different from the calculated value. This greater flexibility of definition can be used to great advantage in mathematical structural modeling. In particular it allows the element properties (dimensions, orientation, etc.) and other information to be obtained from whichever mesh is appropriate, regardless of which mesh is used to locate the nodes.

Placing the Nodes at the Reference Mesh Points

Adopting the reference mesh points as the structure nodes is equivalent to moving the beams outward, and in a drawing this would give the impression that the beam spans are being increased. But as long as all calculations (of stiffness matrices, etc.) are performed using the true (neutral axis) lengths—obtained from the neutral axis mesh—then there is no increase in span. In fact as far as flexure is concerned there is no change whatever. The only error in this method of modeling is that the axial stiffness of the beam is placed in the plane of the plating instead of along the beam's own centroidal axis. If the beams are in the deck or bottom this outward shift produces a small artificial increase in the moment of inertia of the overall hull module in whichever direction the beams lie—transverse or longitudinal or both. In the transverse direction the overestimate can be ignored because transverse bending of the hull module is not a primary load effect.* In the longitudinal direction the effect is an artificial increase in the hull girder moment of inertia, and it could be significant if there are large longitudinal beams. However this presents no problem because the error that would be introduced can easily be estimated beforehand and if it is too large there is an alternative modeling technique which avoids it. From (3.6.20) it may be shown that for each longitudinal beam the artificial increase in the hull girder moment of inertia is $2ayh_c$, where a is the area of the beam, y is the distance from the hull girder neutral axis to the centroid of the beam (always positive, regardless of whether the beam is above or below the neutral axis), and h_c is the distance from the centroid of the longitudinal beam to the plating midthickness. Hence for a total of N longitudinal beams this form of modeling produces a relative overestimate in the hull girder moment of inertia (excess value divided by the actual value) given by:

$$\frac{\delta I}{I} = \frac{2}{I} \sum_{i=1}^{N} a_i y_i h_{ci} \tag{8.5.1}$$

If the longitudinal beams are too large, such that this error is unacceptable, then the remedy is to define each longitudinal beam to be an additional strake of plating and to model it using a series of panel elements.

Thus the use of the reference mesh points as the structure nodes does not cause any appreciable error providing that: (1) the neutral axis mesh is used for establishing the beam lengths; and (2) a membrane type of modeling is used for very deep beams, and is also used for longitudinal beams in the deck or bottom if warranted by (8.5.1).

Placing the Nodes at the Neutral Axis Mesh Points

Another possible technique for modeling a hull module which is also economical in regard to the total number of degrees of freedom is to choose the neutral axis mesh points as the structure nodes. For a *skeletal* structure, that is, a structure that consists almost entirely of beam members, this is clearly the most appropriate choice. But a hull module is a plated frame and in fact most of the cross-sectional area in the transverse and longitudinal directions consists of plating. As we have seen, the plating and the beams have interdependent and complementary roles. The role of the beams is two-fold: to maintain the overall shape of the structure, especially in regard to the cross section, and to provide normal support to the stiffened panels and the pressure loads which they carry. The first role requires some form of transverse framing system extending around the girth of the ship (a "ring frame") and also extending across most if not all of the internal decks and longitudinal bulkheads. In the special case when the length/breadth ratio of a cargo hold is relatively small, as for example in the center tanks of segregated ballast tankers, then a ring frame might also be placed in the longitudinal vertical plane. However, even here the transverse framing still performs most of the first role. The second role—supporting the

*If the vessel is very wide or if for any other reason transverse bending is significant then the procedure given for the longitudinal direction should also be used for the transverse direction.

stiffened panels—is also performed mainly by the transverse beams. Longitudinal beams are only fitted for this purpose if the longitudinal span is smaller or if there is some particular reason for them, such as hatch side girders.

Hence in nearly all ships the principal beams are those in the transverse plane, and a typical hull module contains many more transverse beams than longitudinal beams. For this reason the neutral axis mesh is usually determined by the neutral axes of the transverse beams, which in turn means that placing the nodes at the neutral axis mesh points will naturally give good results as far as the transverse response is concerned. But in the longitudinal direction the longitudinal beams in the "flanges" and "webs" of the hull girder (deck, bottom, sides, etc.) may constitute a smaller proportion of the cross-sectional area than do the beams in the transverse direction. If so then the longitudinal centroidal plane of each hull girder flange and web will be closer to the plane of the plating; that is, it will be *outside* the neutral axis mesh as determined by the transverse beams. Hence the placement of the nodes at these mesh points causes the cross-sectional area of these portions of the hull girder cross section to be moved inward. Therefore the hull girder moment of inertia—both vertically and horizontally—is less than the true value, which is exactly the opposite problem to that which arises when using the reference mesh. The error is greatest when there are *no* longitudinal beams, and in this case the nodes should definitely be placed in the reference mesh.

If the longitudinal beams are deeper than the transverse beams and if they contribute a significant fraction of the total cross-sectional area, then the neutral axis mesh could be defined in terms of the neutral axes of these longitudinal beams rather than the transverse beams. This would provide good accuracy for the hull girder moment of inertia, with only a slight error arising from the fact that the plating is usually more fully effective in hull girder bending than it is when acting as a plate flange in the bending of individual longitudinal beams. Defining the nodes in this way would cause the hull module bending stiffness in the transverse direction to be smaller than its true value, but this form of hull module response is seldom significant. Also the framework formed by the transverse beams would be reduced slightly in size, but this too would not be significant (and of course the stiffness matrices, etc., of the transverse beams should be calculated in terms of their true effective lengths, measured along their own neutral axes).

Choice of Method

From the foregoing discussion it is clear that each of the two alternative methods for placing the nodes—in the plane of the plating or in the plane of the neutral axes—has advantages and disadvantages. The former is more appropriate for membrane structures; the latter is more appropriate for skeletal structures, and a hull module is a combination of these. But for most applications the former method is better for two reasons: first, it is somewhat easier because the reference mesh is more basic and is already used for several other purposes, and second, it is somewhat more accurate, as long as any very deep longitudinal beams are modeled as separate strakes, using panel elements. Hence the recommended method is to use the reference mesh for specifying the nodal locations and the neutral axis mesh to calculate the lengths and angles of orientation of all beam elements. This provides almost the same accuracy as if a double node system had been used, but without the severe computational penalty which that would involve.

Graphical Representation of the Structural Model

Unfortunately, departing from the limitations of a physical or a graphical model does create a problem regarding the graphical representation of the model. The mathematical model contains an inconsistency (from the physical point of view) which cannot be fully resolved in a physical or graphical model. If the beam elements are drawn on the basis of the nodal locations they appear—incorrectly—out of their proper position, and if they are drawn in their proper position they appear to be unconnected to the membrane elements. The first alternative is visually more misleading, since it displaces an entire system of members away from their true positions. It also produces a more cluttered drawing because the beam elements and the longitudinal membrane elements all coincide. In the second method the inconsistency occurs only at nodes and it is somewhat easier to allow for when looking at a drawing. This can be facilitated by using some graphical means to indicate that the beam node and the membrane node are in fact the same node. An example is shown in Fig. 8.34, which depicts the structural model of the tanker of Fig. 8.31. The deep longitudinal girders are modeled using stiffened panel elements. The other symbols are explained in subsequent sections.

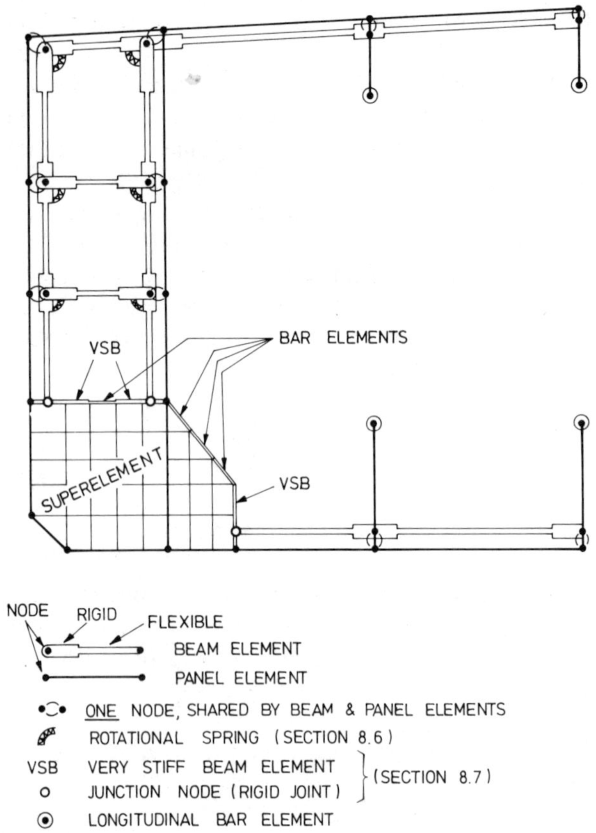

Figure 8.34 Structural model of a tanker.

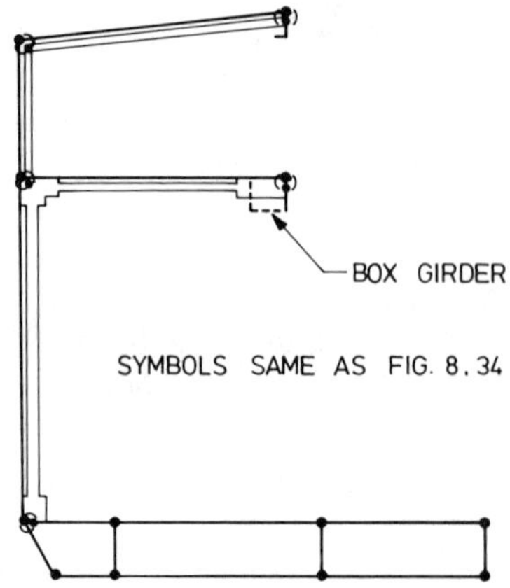

Figure 8.35 Structural model of a general cargo ship.

Figure 8.35 illustrates a structural model of a ship in which the longitudinal beams (the hatch side girders) are not very deep and can therefore be modeled using beam elements rather than membrane elements. For the elements representing the box girder the torsional stiffness is computed from (6.1.22) and inserted into (5.5.13). A superelement is used to represent each transverse floor in the double bottom.

8.6 REPRESENTATION OF JOINTS AND BRACKETS

Use of Rigid Lengths for Flexural Rigidity of Joints

We have noted before that most of the beams in a hull module are attached to plating and consequently their neutral axes lie relatively close to the plating. Therefore regardless of which mesh is used for the structural nodes the beam lengths will often include, at their ends, the web heights of adjacent orthogonal beams. As far as bending is concerned these end portions may be regarded as totally rigid; that is, they remain straight and simply rotate as a rigid body about the support points (the nodes). If the beams are slender, such that these rigid ends are short relative to the total length (less than, say, 10%), the extra rigidity has negligible effect and the end portions can be taken as part of the beam. Note that these end regions must be included in some fashion—either as part of the beam or as separate rigid lengths—because the support points do not change, and the beam length is always the full length between nodes. It would be incorrect to just subtract the end portions and take the remaining length as the length of the beam; this would make the beam too stiff.

If the beams are short and deep such as in the transverse framing of a tanker (see Fig. 8.36) then the effect of the rigid ends becomes important and must be accounted for. This can be done without requiring any additional degrees of freedom and with only a small amount of additional computation. Figure 8.37 shows a typical rigid length modeling. Each beam consists of two rigid end portions of length r_1 and r_2 and an ordi-

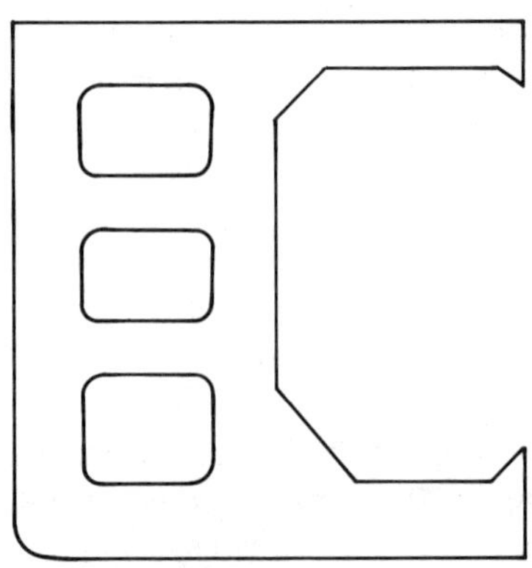

Figure 8.36 Transverse beams in a tanker.

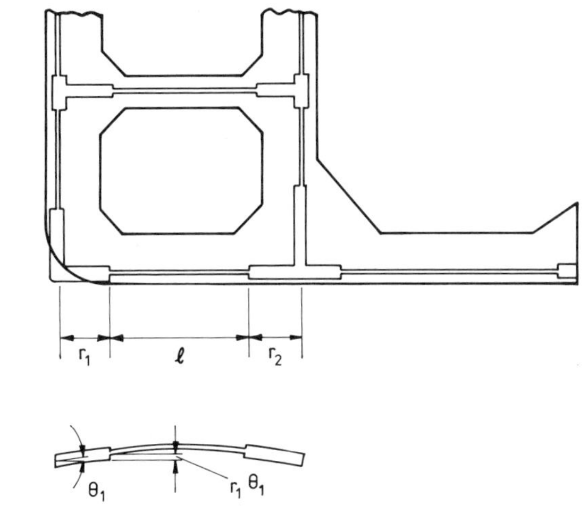

Figure 8.37 Rigid ends in a beam element.

nary, flexible center portion. In the present application the rigid lengths occur only within the beam webs, but it will be shown in the next section that if the joints contain small or medium size brackets it is possible to allow for their presence by extending the rigid length part way along the bracket length. Hence for generality we here use the symbol r for rigid length.

The insertion of the rigid lengths means that in addition to the external nodes there are two internal nodes, and for reference purposes the slope and deflection at the latter are indicated by the symbol ˇ. Since the end portions remain perfectly straight there is a unique mathematical relationship between the slope and deflection at the internal nodes and the slope at the external nodes:

$$\begin{aligned} \check{v}_1 &= v_1 + r_1\theta_1 \\ \check{v}_2 &= v_2 - r_2\theta_2 \end{aligned} \qquad (8.6.1)$$

In matrix notation the transformation from external nodal displacements to internal node displacements is

$$\begin{Bmatrix} \check{v}_1 \\ \check{\theta}_1 \\ \check{v}_2 \\ \check{\theta}_2 \end{Bmatrix} = \begin{bmatrix} 1 & r_1 & & \\ 0 & 1 & & \\ & & 1 & -r_2 \\ & & 0 & 1 \end{bmatrix} \begin{Bmatrix} v_1 \\ \theta_1 \\ v_2 \\ \theta_2 \end{Bmatrix} \qquad (8.6.2)$$

or

$$\check{\boldsymbol{\delta}} = \mathbf{T}\boldsymbol{\delta}$$

Thus, since the internal displacements can be expressed entirely in terms of the external displacements, the complete response of the entire beam can be characterized by the values of slope and deflection at the external nodes, and hence there are *no additional degrees of freedom*. The beam element which is associated with the internal nodes is of length l, given by*

$$l = L - r_1 - r_2 \qquad (8.6.3)$$

This element is an ordinary flexure-only beam element, for which the stiffness matrix is given by (5.3.1) with l in place of L. Since it corresponds to the internal nodes we here denote it as $\check{\mathbf{k}}$.

$$\check{\mathbf{k}} = \frac{EI}{l^3}\begin{bmatrix} 12 & 6l & -12 & 6l \\ 6l & 4l^2 & -6l & 2l^2 \\ -12 & -6l & 12 & -6l \\ 6l & 2l^2 & -6l & 4l^2 \end{bmatrix} \qquad (8.6.4)$$

In external coordinates the corresponding stiffness matrix is obtained by the usual rule:

$$\mathbf{k} = \mathbf{T}^T\check{\mathbf{k}}\mathbf{T}$$

Upon performing these multiplications it is found that the resulting matrix can be expressed as the original matrix, $\check{\mathbf{k}}$, plus a matrix of additive terms which account for the rigid lengths. That is

$$\mathbf{k} = \check{\mathbf{k}} + \mathbf{k}_r \qquad (8.6.5)$$

and the additional rigid length terms are

$$\mathbf{k}_r = \frac{12EI}{l^3}\begin{bmatrix} 0 & & & \text{symm.} \\ r_1 & r_1(l+r_1) & & \\ 0 & -r_1 & 0 & \\ r_2 & \rho l & -r_2 & r_2(l+r_2) \end{bmatrix} \qquad (8.6.6)$$

where $$\rho = \frac{r_1 r_2}{l} + \frac{r_1 + r_2}{2}$$

For general use we require this matrix expressed in structure coordinates, with the beam at an arbitrary orientation angle χ. This is obtained by expanding (8.6.6) to a 6×6 matrix (such that rows and columns 1 and 4 contain zeros) and then applying the general transformation formula $\mathbf{K} = \mathbf{T}^T\mathbf{k}\mathbf{T}$, where $\mathbf{T}$ is defined in (5.3.2). The result is

*Note that the introduction of rigid lengths and the transformation from that nodal system to the full length beam, applies only to the flexural degrees of freedom, v and θ, in the plane of the beam web. In regard to sideways bending the beam is considered to be entirely prismatic and of length L.

$$\mathbf{K}_r = 12\frac{EI}{l^3} \times \begin{bmatrix} 0 & & & & & \text{symm.} \\ 0 & 0 & & & & \\ -\mu r_1 & \lambda r_1 & r_1(l+r_1) & & & \\ 0 & 0 & \mu r_1 & 0 & & \\ 0 & 0 & -\lambda r_1 & 0 & 0 & \\ -\mu r_2 & \lambda r_2 & \rho l & \mu r_2 & -\lambda r_2 & r_2(l+r_2) \end{bmatrix} \tag{8.6.7}$$

For a beam containing a rigid length this matrix is added to that of (5.3.16); in the latter equation l must be substituted in place of L.

After the structure equilibrium equations have been assembled and solved, and the resulting structure nodal displacements $\mathbf{\Delta}$ have been transformed to element nodal displacements $\boldsymbol{\delta}$, the bending moment within the flexible portion of the beam is obtained by using (8.6.2) to obtain the internal nodal displacements $\breve{\boldsymbol{\delta}}$ and inserting these into the element bending moment equation (5.3.12a) with l in place of L.

Extension of Rigid Length for Small Brackets

Brackets can play a very important role in a frame structure—much more important than their size might indicate. There are several reasons for using brackets:

1. They provide more cross-sectional area and moment of inertia at the ends of beams, thereby reducing the transverse shear stress and the normal bending stress, respectively. Also they act as gussets or fillets in alleviating local stress concentration.
2. They increase the degree of fixity at the ends of individual beams, thereby stiffening the frame and also possibly decreasing the midspan bending moment.
3. They facilitate construction by allowing more tolerance in the positioning of the members joined.

The first reason—stress reduction—is usually the most significant because in a rigid-jointed frame which carries distributed loads it is precisely at the ends of each beam where both the bending moment and the shear force are maximum. It should be noted that a bracket does not decrease the end bending moment; in fact it usually increases it because of the increased end fixity. But the bending stress is reduced because the local moment of inertia, and hence also the local section modulus, is increased. For a simple flat plate bracket the section modulus is proportional to the square of the bracket height. Figure 8.38*a* illustrates the benefits of a bracket by superimposing the curves of bending moment (factored and absolute value) and of local section modulus for a uniformly loaded beam which, for simplicity, is taken as clamped. The section modulus curve is multiplied by the yield stress to convert it to a limit value: the bending moment that would cause yield, M_Y. Figure 8.38*b* is the factored load ratio, $\gamma M(x)/[M_Y(x)]$, and it shows that a proper sizing of the bracket provides sufficient section modulus in the end regions to keep the stress there at an acceptable level so that the beam itself can be sized to withstand the smaller bending moment at the center of the beam. Figure 8.39 illustrates the similar benefits in regard to transverse shear stress. Either the bending stress or the shear stress could be the determining factor in choosing the bracket size, depending on the type of load and the end fixity.

Since a bracket increases the local moment of inertia it increases the flexural rigidity EI, and this is related to the curvature ϕ and the bending moment M by $M = EI\phi$. Thus for a constant bending moment the increase in the local value of I produces a corresponding decrease in the local curvature. In other words the beam is flexurally more rigid in way of the bracket. The ordinary prismatic beam element cannot account

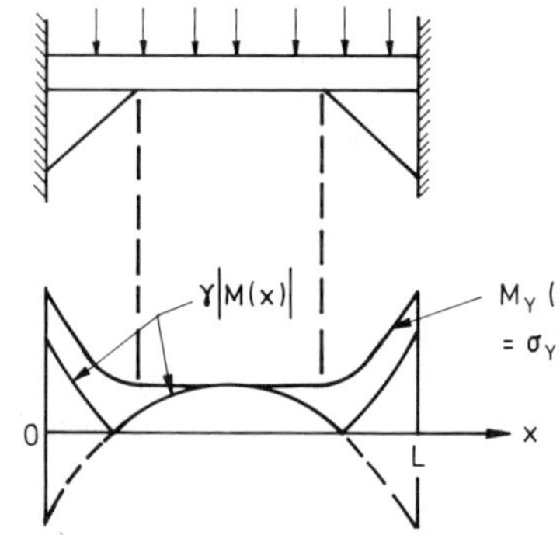

(a) BENDING MOMENT : FACTORED AND LIMIT VALUES

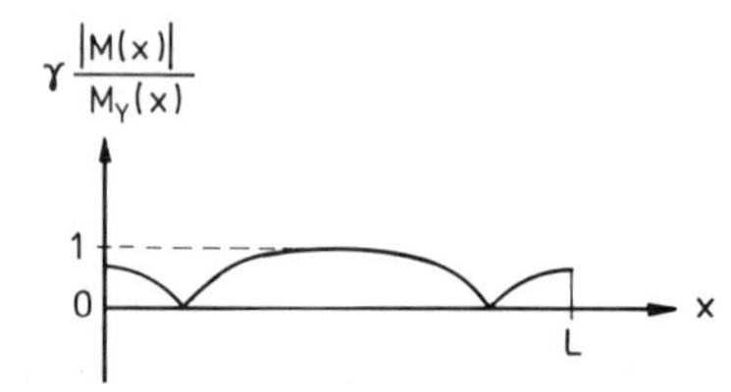

(b) FACTORED LOAD RATIO

Figure 8.38 Use of brackets to reduce end bending stress.

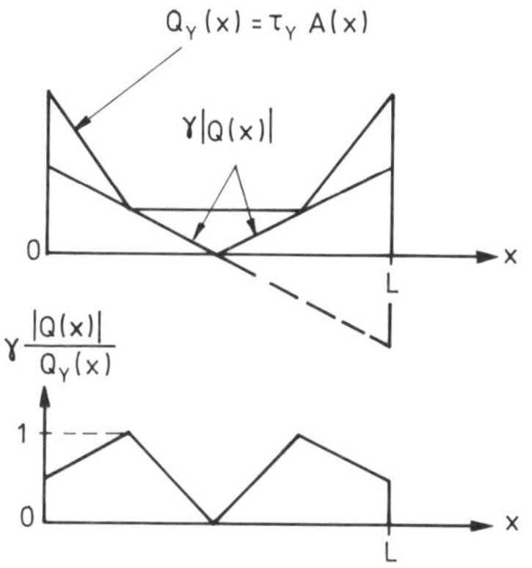

Figure 8.39 Use of brackets to reduce end shear stress.

for this effect, since it has uniform flexural rigidity. It would be possible to account for it by using three beam elements for each beam, with the two end beam elements being made deeper in order to represent the bracketed portions of the beam. However this approach is seldom used since it would triple the number of elements and would increase the total computation by an even greater factor. A much simpler approach, which is widely used because it involves little extra computation, is to idealize the brackets as short rigid end portions or, if rigid end portions are already being used, to simply extend them part way along the bracket length. The extra length should not be the full length since the bracket does have some bending flexibility. In fact the bracket's degree of flexibility can be quite significant. Although the ordinary beam element can represent only a linear variation of bending moment, the beams in frame structures which are subjected to distributed loads have a quadratically varying bending moment, with large peak values at the ends, just where the brackets occur. Thus even though the beam has a larger local value of moment of inertia within the bracket length, it can still undergo significant bending within this length. This is accounted for by regarding only a portion of the bracket length as rigid, say d_r, and the other portion, d_f, as part of the uniform beam, as shown in Fig. 8.40.

There are several approximate formulas for estimating d_r, depending on the shape of the bracket. For simple triangular brackets the formula most commonly used is (see Fig. 8.40)

$$d_r = a_1\left(1 - \frac{h_1}{h_1 + a_2}\right) \tag{8.6.8a}$$

and the flexible portion is then

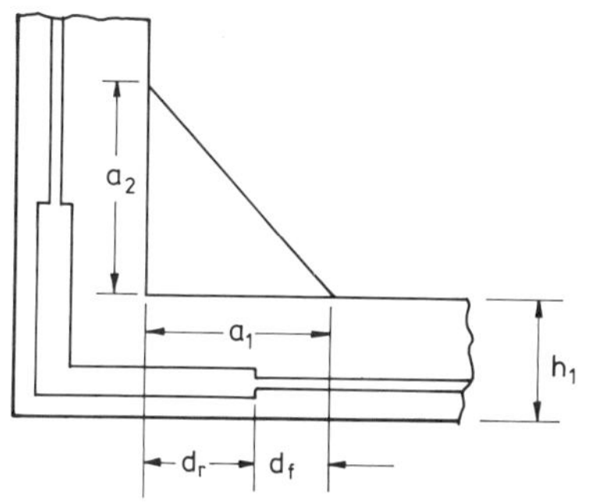

Figure 8.40 Rigid length representation of a bracket.

$$d_f = a_1 - d_r = \frac{h_1 a_1}{h_1 + a_2} \tag{8.6.8b}$$

If the brackets are large the rigid length representation becomes invalid because it concentrates all of the beam's flexure into the central portion and if this portion is short the flexure is seriously exaggerated. Also, the rigid length representation does not provide any information regarding stresses within the bracketed portion of the beam or within the bracket itself. For large brackets or for the calculation of stresses within brackets it is necessary to model the brackets explicitly, using several small finite elements of appropriate types: bars, triangular and rectangular membranes, and so on. The most efficient way of achieving this is to use the superelement technique which is presented in Section 8.7.

But before considering brackets in any detail we shall first consider the phenomenon of shear distortion in the joints of frame structures (bracketed or otherwise) because this has a strong influence on the bending moment in the vicinity of the joints—much more than the flexural rigidity of joints. In fact, for short deep beams it influences the bending moment distribution throughout the frame. Although its importance has been well demonstrated, as for example in Ref. 5, it is still ignored in many frame analysis programs, possibly because of the notion that it would greatly increase the computation. But in the next section we present a simple method for including this phenomenon which requires only one additional degree of freedom at each joint for a plane frame and three at each joint for a three-dimensional frame.

Representation of Shear Distortion of Joints

As shown in Fig. 8.41, when a bending moment is transmitted between two orthogonal beams the rectan-

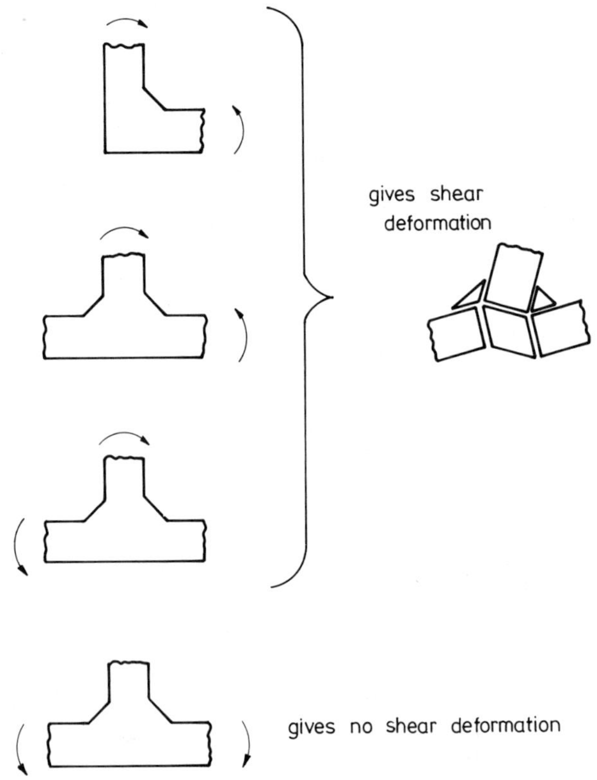

Figure 8.41 Shear deformation of joints due to bending.

gular web region of the joint is subjected to forces which tend to distort it in shear. This shear distortion causes a discrepancy between the angles of rotation of the two beam ends, thus failing to fulfill the assumption of a unique value of nodal rotation, which is a basic assumption of ordinary frame analysis. To a first approximation the magnitude of the discrepancy in the angles is equal to the overall or mean value of shear strain in the web region, γ_w. This will also be true of bracketed joints, although the bracket will of course decrease the magnitude of the shear distortion. If the beams are relatively short the joint flexibility arising from shear distortion is a significant part of the overall flexibility of the frame, and must be accounted for in order to obtain accurate values of bending moments, especially in the vicinity of the joints. Joint flexibility due to shear is a separate and distinct phenomenon from bending flexibility (or its inverse, bending rigidity). Therefore the allowance for joint shear distortion requires a specific modeling technique, distinct from (and harmonious with) the use of end portions to model the flexural rigidity of the web region of joints. In particular, it is not possible to compensate for joint shear distortion by using shorter rigid lengths.

This section presents a simple method for representing joint distortion, with or without brackets. To simplify matters we present the method for the most common type of joint, involving two orthogonal or near orthogonal beams. The extension to a T-shaped joint is straightforward. We will first consider an unbracketed joint and then generalize the result to include the effect of brackets.

The method consists of the insertion of a very stiff rotational spring connecting the two beams which provides a linear elastic moment resisting any increase or decrease in the angle of the joint. This spring is not an "element" in the usual sense because it has no physical size, but it does connect two independent degrees of freedom (θ_i and θ_j in Fig. 8.42) and it supplies a rotational stiffness between them. It therefore has a 2×2 stiffness matrix having the same form as that for a linear spring in (5.1.5a):

$$\mathbf{k}_J = k_J \begin{bmatrix} 1 & -1 \\ -1 & 1 \end{bmatrix} \tag{8.6.9}$$

where k_J is the spring stiffness of the joint, defined as the net total moment acting on the joint divided by the

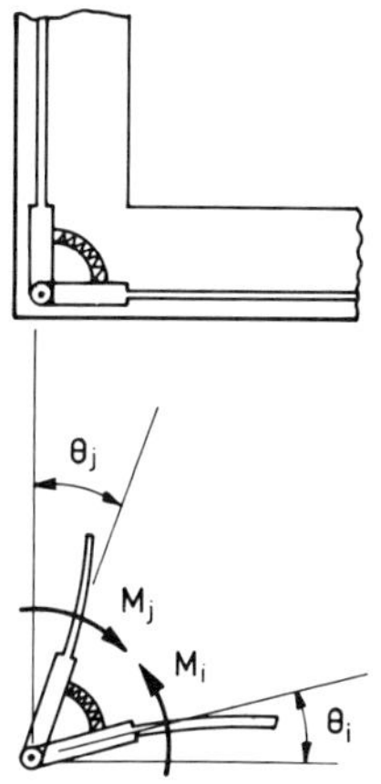

Figure 8.42 Representation of joint shear deformation by an elastic rotational spring.

resulting angle of opening or closing of the joint.* Because web regions at the ends of a beam are being regarded as flexurally rigid (no bending) the joint stiffness is due entirely to the elastic shear strain which occurs in the rectangular region formed by the intersection of the beam webs. The exact distribution of shear strain would depend on the local geometry. We can however obtain an approximate qualitative estimate of the joint stiffness on the basis of the mean value of shear stress within the joint region.

Milchert [5] has shown that for a joint subjected to applied moments M_i and M_j as shown in Fig. 8.43, for which the net moment tending to open or close the joint is

$$\overline{M} = \frac{M_i - M_j}{2} \qquad (8.6.10)$$

the average shear strain in the web region is given approximately by

$$\gamma_w = \frac{\overline{M}}{2Gt_w}\left[\frac{1}{h_i^2} + \frac{1}{h_j^2}\right] \qquad (8.6.11)$$

Therefore the joint stiffness is

$$k_J = \frac{\overline{M}}{\gamma_w} = \frac{2Gt_w h_i h_j}{\alpha_w + \dfrac{1}{\alpha_w}} \qquad (8.6.12)$$

*In Fig. 8.42 and elsewhere the joint stiffness is represented as a curved spring. When there are also rigid ends at the joint it would be possible to group all three into a type of "joint element." This is useful for visualizing the structure, and in drawings of the finite element model this interpretation seems more natural. But in the mathematics the rigid ends are part of the beams, and the only node is the one at the corner.

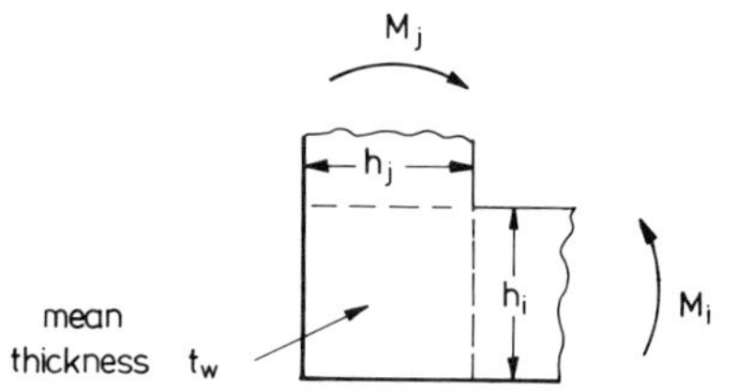

Figure 8.43 Inner web region of a joint.

where α_w is the aspect ratio of the inner web region: $\alpha_w = h_i/h_j$.

This expression for k_J is only a qualitative estimate but we can quantify it by adding an empirically derived coefficient κ:

$$k_J = \kappa \frac{2Gt_w h_i h_j}{\alpha_w + \dfrac{1}{\alpha_w}} \qquad (8.6.13)$$

To evaluate κ a detailed finite element analysis was performed for a typical unbracketed rectangular joint. Milchert's work included a finite element analysis of three *bracketed* joints (Fig. 8.44) and so the beam sizes and the applied moment were made the same as in his work. For an applied moment of 10 MNm the angle of rotation of the web region was $\gamma_w = 140 \times 10^{-5}$ radians. Therefore the stiffness of the joint itself, with no brackets, is

$$k_J = \frac{\overline{M}}{\gamma_w} = \frac{10}{140 \times 10^{-5}} = 7143 \text{ MNm}$$

The beam webs were 2.75 m high and 15 mm thick. Therefore $h_i = h_j = 2.75$ and $t_w = 0.015$. Substituting these, together with $G = 81$ GN/m^2, into (8.6.13) gives a value of 0.78 for κ. Therefore an approximate expression for the rotational stiffness of a joint is

$$k_J = \frac{1.56Gt_w h_i h_j}{\alpha_w + \dfrac{1}{\alpha_w}} \qquad (8.6.14)$$

Effect of Brackets

We now seek to extend the notion of the rotational stiffness of a joint to include the effect of a bracket. The overall flexibility of a bracketed joint may be characterized by the total angular deformation of the bracketed region for a given applied load, that is, the total change of angle γ_T from the node to the toe of the

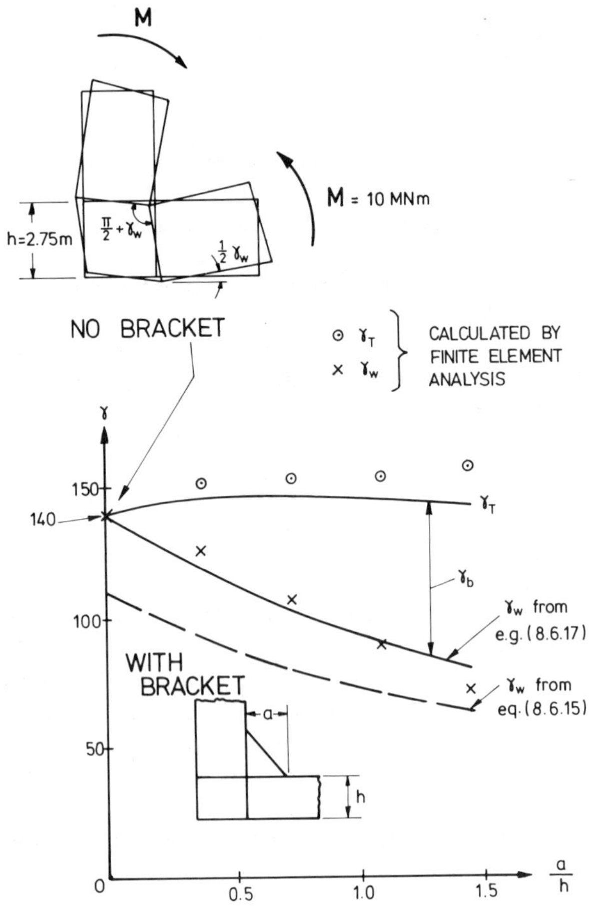

Figure 8.44 Joint shear deformation versus bracket size.

bracket, as illustrated in Fig. 8.45. Figure 8.44 shows values of γ_w and γ_T calculated by finite element analysis for brackets of various sizes, all of which have the same thickness and are subjected to the same load. The results show that bracket size has very little effect on the overall flexibility of the joint. This result may at first seem surprising but is easily explained. When a bracket is added to a joint the most direct result is that the forces which tend to open or close the joint—that is, the forces that cause the shear strain—are absorbed by the bracket in the form of direct tension or compression, instead of shear strain in the inner web region. Hence, as shown in Fig. 8.44 γ_w diminishes with increased bracket size, but the bracketed length of the beam undergoes sufficient bending to offset this and the net result is that γ_T, the total change of angle (measured to the end of the bracket) remains approximately constant.

In this section we wish to continue using the simplest method for representing a bracket, which is to extend the rigid length partly along the bracket, by an amount d_r, and to regard the remaining length as iden-

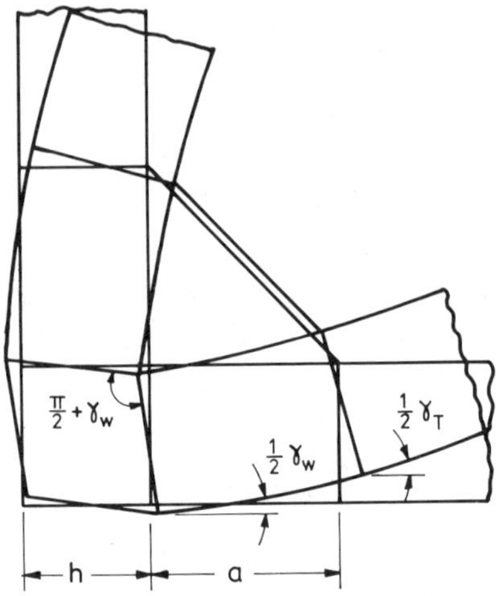

Figure 8.45 Total angular deformation of inner web and bracket.

tical to the unbracketed beam. This choice of modeling allows us to retain the same basic approach, that is, to represent the elastic rotation of the joint entirely in terms of γ_w, the shear strain of the inner web region. With this approach the effect of the bracket is summarized and expressed in terms of how much it diminishes γ_w. For a bracketed joint, Milchert presented the following formula for the average shear strain in the inner web region, in terms of the bracket dimensions a_i and a_j

$$\gamma_w = \frac{\overline{M}}{Gt_w}\left[\frac{1}{h_i(2h_i + a_i)} + \frac{1}{h_j(2h_j + a_j)}\right] \qquad (8.6.15)$$

This expression for γ_w as a function of bracket size is plotted in Fig. 8.44 and is seen to lie below the finite element values. For an unbracketed joint ($a_i = a_j = 0$) Eq. 8.6.15 reduces to (8.6.11). We have seen that this expression for γ_w is only approximate and that the expression for k_J based on it requires a correction factor $\kappa = 0.78$. The corresponding corrected expression for γ_w without brackets is

$$\gamma_w = \frac{1}{0.78}\frac{\overline{M}}{2Gt_w}\left(\frac{1}{h_i^2} + \frac{1}{h_j^2}\right) \qquad (8.6.16)$$

Since (8.6.15) is a generalization of (8.6.11) and reduces to it as the bracket size diminishes, it is clear that this general equation will require a similar correction. We therefore rewrite it as

$$\gamma_w = \frac{1}{0.78}\frac{\overline{M}}{Gt_w}\left[\frac{1}{h_i(2h_i + a_i)} + \frac{1}{h_j(2h_j + a_j)}\right] \qquad (8.6.17)$$

This expression is also plotted in Fig. 8.44 and is seen to agree well with the values of γ_w obtained by a finite element analysis.

The hinge stiffness parameter k_J is, by definition

$$k_J = \frac{\overline{M}}{\gamma_w}$$

and the substitution of (8.6.17) for γ_w gives a general expression for k_J which accounts for the presence of a bracket.

$$k_J = 0.78 \frac{Gt_w}{\dfrac{1}{h_i(2h_i + a_i)} + \dfrac{1}{h_j(2h_j + a_j)}} \tag{8.6.18}$$

The other effect of a bracket is to increase the flexural rigidity at the end of a beam, which we are allowing for by extending the rigid length by a distance d_r within the bracket length. Equation 8.6.7 was proposed as the definition of d_r, and now that we are able to account for the shear deformation of a bracketed joint we can examine the accuracy of this equation. At any section $x (0 \leq x \leq d_f)$ within the bracketed length (see Fig. 8.46*a*) the total angular deformation of the joint consists of two components: the shear distortion of the inner web, γ_w, and an additional angle $\gamma_b(x)$, caused by bending of the bracketed length. We are idealizing γ_b as occurring only along the flexible portion of the bracketed length, $d_f = a - d_r$, and we are regarding this portion as having a constant moment of inertia I, equal to that of the beam. Therefore the value of γ_T, the total angle of rotation at the toe of the bracket is (see Fig. 8.46*c*)

$$\begin{aligned} \gamma_T &= \gamma_w + \gamma_b(x = d_f) \\ &= \gamma_w + \int_0^{d_f} \frac{M(x)}{EI}\, dx \end{aligned}$$

This integral has been evaluated, using (8.6.8b) for d_f, for four different bracketed joints (Milchert's three plus one with $a = 4$ m). These values of γ_b were then added to the values of γ_w given by (8.6.17). The resulting curve of γ_T is shown in Fig. 8.44 and is seen to agree reasonably well with the finite element result. Figure 8.46*b* and *c* further illustrate the method of modeling proposed herein and show the correspondence in the joint distortion angles γ_w and γ_T between the structure and the model.

In summary, these results show that for small and medium size brackets (8.6.17) gives correct values of the angular deformation of the inner web portion of the joint, from the node to the heel of the bracket, and that the use of rigid lengths, with d_f defined by (8.6.8b), gives correct values of angular deformation at the toe of the bracket. We therefore conclude that the combined effects of the flexural rigidity of joints and the shear deformation of joints, either bracketed or unbracketed, are adequately represented by using rigid length modeling and by introducing an extra rotational degree of freedom at each joint, with the corresponding stiffness terms defined by (8.6.9) and (8.6.18).

For large brackets or for the calculation of the stresses within brackets it is necessary to model the brackets explicitly, using the technique presented in Section 8.7.

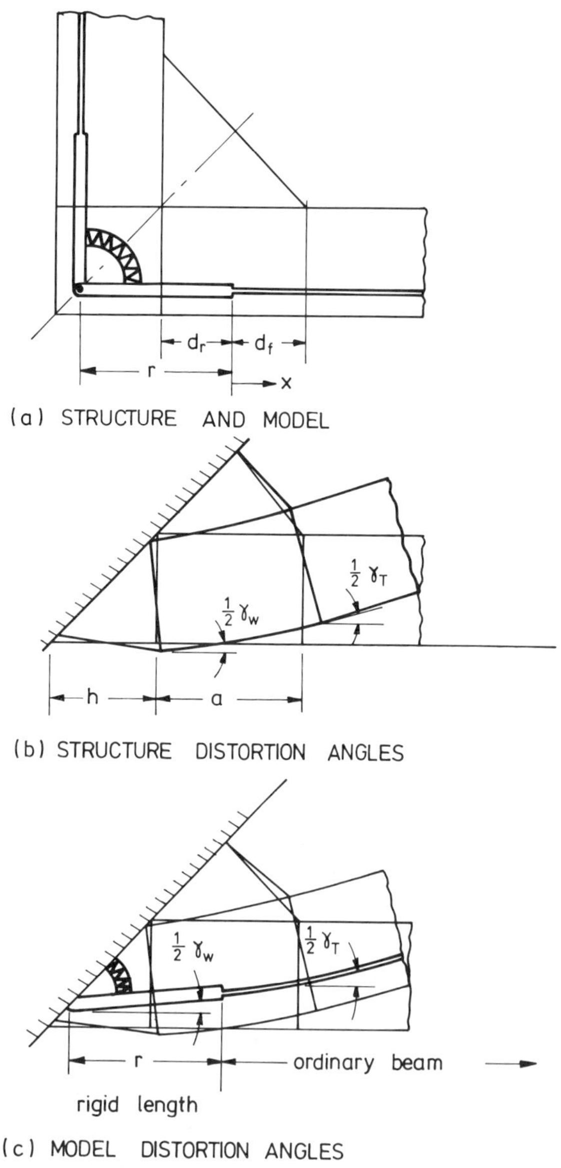

Figure 8.46 Joint distortion angles.

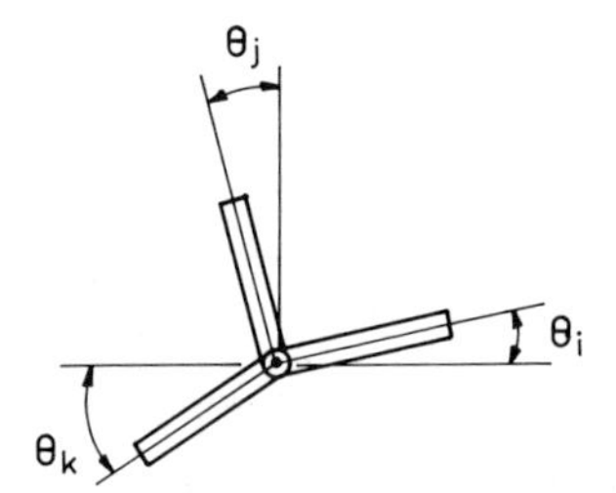

Figure 8.47 Joint rotations including bending.

APPLICATION TO T-SHAPED JOINTS

In a T-shaped joint there might at first seem to be three independent rotations θ_i, θ_j, and θ_k as shown in Fig. 8.47, and three corresponding hinge moments M_i, M_j, and M_k. But, as shown in Fig. 8.41 the hinge between the two in-line beams does not correspond to *shear* stiffness, but rather only bending stiffness. We are here only interested in modeling shear stiffness because the bending stiffness of a joint (with or without brackets) is already accounted for by the use of rigid and flexible lengths. Therefore we can set $\theta_k = \theta_i$; this is indicated in Fig. 8.48 by drawing the in-line beams as continuous and placing the hinge just outside of them.

In regard to shear deformation a T-shaped joint is basically the same as an L-shaped joint because there is still only one inner web region (see Fig. 8.41). Therefore, as shown in Fig. 8.48, there is still only one rotational spring, of stiffness k_J, and the stiffness matrix is again given by (8.6.9). It makes no difference which side of the joint the spring is drawn, the effect is the same in either case.

The presence of a second bracket will influence k_J by altering γ_w, but the effect is probably not large. At the time of this writing, an expression for γ_w for a bracketed T-shaped joint does not seem to have been developed. In this situation we can only obtain a rough estimate of the value of γ_w and hence of k_J, by generalizing (8.6.17) in a manner which seems appropriate.

First, the fact that one of the beams is continuous means that the width h_j of the inner web region is bounded on both sides and the diffusion of shear stress into the new adjacent web material (see Fig. 8.49) is roughly equivalent to making h_j larger. We therefore define an effective value h_j'

$$h_j' = h_j + 0.25h_{i2}$$

where h_{i2} is the height of the new beam. The factor of 0.25 is chosen arbitrarily as a rough estimate of the relative amount of shear stress diffusion.

Second, it seems reasonable to assume that the effect of the second bracket will be similar to that of the first, and therefore (8.6.17) will be generalized by inserting the lengths of the second bracket, a_{i2} and a_{j2}, into the equation in the same manner as the lengths of the first bracket. The generalized equation is

$$\gamma_w = \frac{1}{0.78} \frac{\overline{M}}{Gt_w} \times \left[\frac{1}{h_{i1}(2h_{i1} + a_{i1} + a_{i2})} + \frac{1}{h_j'(2h_j' + a_{j1} + a_{j2})} \right] \tag{8.6.20}$$

and the spring stiffness is then

$$k_J = \frac{\overline{M}}{\gamma_w} \tag{8.6.21}$$

To see what sort of results are produced by these generalizations we shall apply (8.6.20) to the T-shaped joint of Fig. 8.50 in which for simplicity the

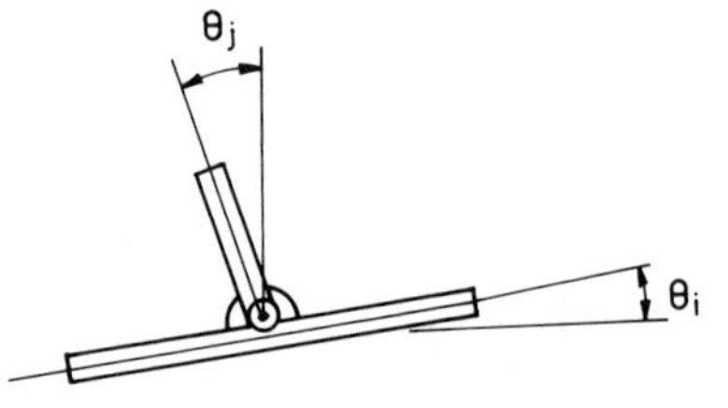

Figure 8.48 Joint rotations excluding bending.

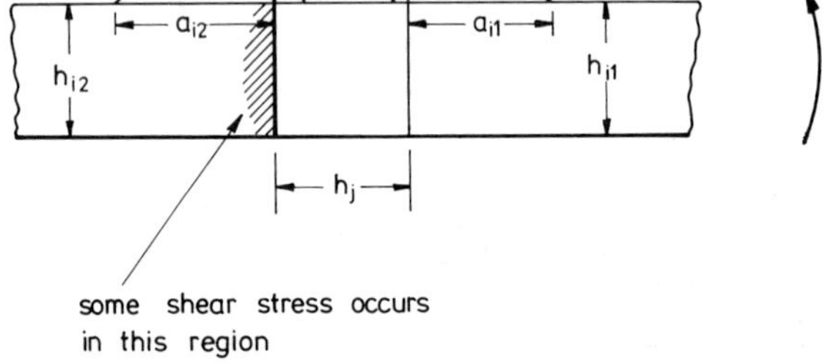

Figure 8.49 Shear deformation in a T-shaped joint.

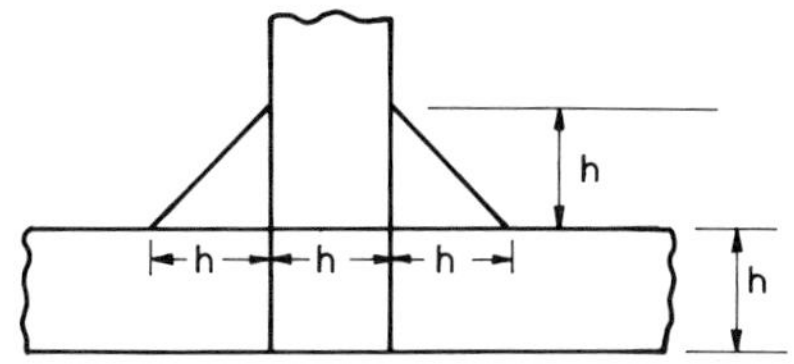

Figure 8.50 Sample bracketed joint.

web heights and the bracket lengths are all made equal. First, in order to obtain reference values, we shall calculate the spring stiffness for an L-shaped joint of the same proportions, with and without a bracket. For a joint with no bracket the shear strain would be, from (8.6.16)

$$(\gamma_w)_{L,0} = \frac{1}{0.78} \frac{\overline{M}}{Gt_w h^2} \tag{8.6.22}$$

and so the spring stiffness is

$$(k_J)_{L,0} = \frac{\overline{M}}{(\gamma_w)_{L,0}} = 0.78Gt_w h^2$$

With one bracket the inner web shear strain is, from (8.6.17)

$$\gamma_w = \frac{1}{0.78} \frac{\overline{M}}{Gt_w} \left[\frac{2}{h(2h + h)} \right] = \frac{2}{3} (\gamma_w)_{L,0}$$

and so k_J is

$$(k_J)_{L,1} = 1.5(k_J)_{L,0}$$

For the T-shaped joint with no brackets, (8.6.20) gives

$$\gamma_w = \frac{1}{0.78} \frac{\overline{M}}{Gt_w} \left[\frac{1}{h(2h)} + \frac{1}{1.25h(2.5h + h)} \right]$$
$$= 0.82(\gamma_w)_{L,0}$$

and therefore

$$(k_J)_{T,0} = 1.22(k_J)_{L,0}$$

For one bracket the value of γ_w is

$$\gamma_w = \frac{1}{0.78} \frac{\overline{M}}{Gt_w} \left[\frac{1}{h(2h + h)} + \frac{1}{1.25h(2.5h + h)} \right]$$
$$= 0.56(\gamma_w)_{L,0}$$

and therefore

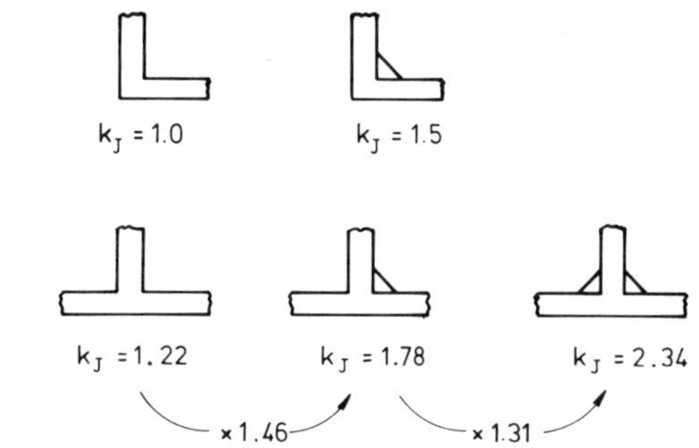

Figure 8.51 Relative spring stiffnesses.

$$(k_J)_{T,1} = 1.78(k_J)_{L,0}$$

And finally for two brackets we have

$$\gamma_w = \frac{1}{0.78} \frac{\overline{M}}{Gt_w} \left[\frac{1}{h(2h + 2h)} + \frac{1}{1.25h(2.5h + 2h)} \right]$$
$$= 0.43(\gamma_w)_{L,0}$$

and

$$(k_J)_{T,2} = 2.34(k_J)_{L,0}$$

These results are summarized in Fig. 8.51. They appear reasonable but it is emphasized that (8.6.20) is only a rough estimate and when a more accurate expression is developed for the shear strain in the inner web region of a T-shaped joint, as a function of beam and bracket dimensions, it should be used in place of (8.6.20).

There are two types of T-shaped joint, depending on whether the middle beam is attached to plating. If it is, then the neutral axis lies close to one side of the web and hence the node is located near the edge of the web. This type of joint is shown in Fig. 8.52. If the middle beam is not attached to plating, as for example one of the horizontal cross ties in the tanker frame of Fig. 8.36, then the neutral axis of the middle beam lies near the beam centerline, as shown in Fig. 8.53. How-

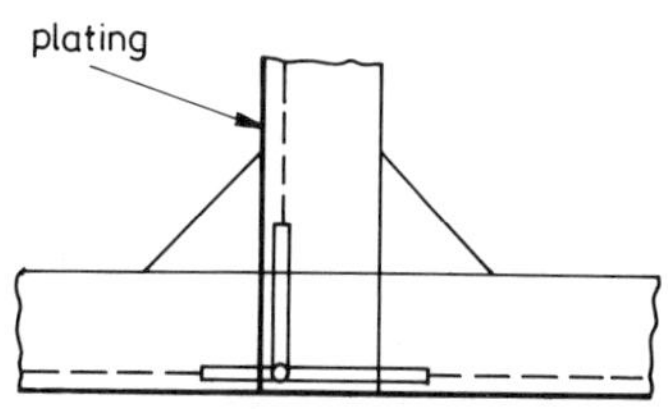

Figure 8.52 Unsymmetric T-shaped joint.

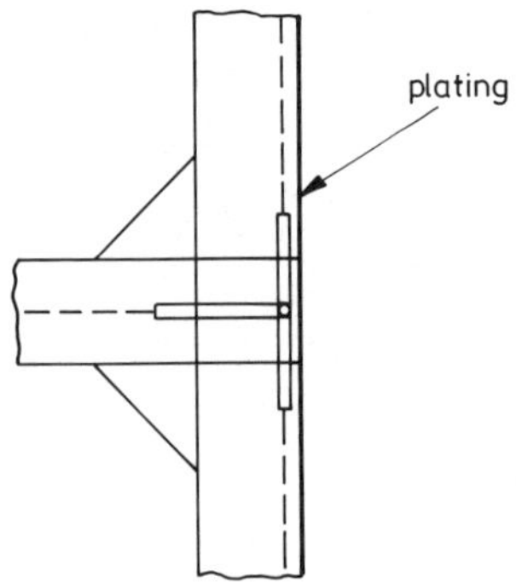

Figure 8.53 Symmetric T-shaped joint.

ever, the inner web region still acts essentially as one unit and so (8.6.20) can also be used for this type of joint.

Sample Application in Hull Module Analysis

In the foregoing sections we have presented improved modeling techniques which account for deep beams, brackets, and the shear deformation of joints—in effect an improved method of frame analysis—because all of these factors are present in a hull module. This section presents a sample application of these techniques and demonstrates their accuracy by comparing the results with those obtained by a more detailed finite element analysis. It also shows the absolute necessity to use these techniques in a hull module analysis by comparing the results with those obtained by ordinary frame analysis, which assumes entirely flexible beams and entirely rigid joints. And finally, since the beams in a hull module usually carry distributed loads, this section also shows how the technique for dealing with such loads, which was presented in Section 5.3 for ordinary beams, is applied to beams which contain a rigid length.

For purposes of comparison we take as our example the same portal frame which was used as an example in Section 5.3, except that now the joints are bracketed. In order to make the example as clear and as simple as possible we will ignore the axial stiffness and use the flexure-only beam element. It was shown in Section 5.3 that for such a simple type of frame the latter element is satisfactory, at least for purposes of demonstration.* In an actual hull module analysis the axial stiffness would, of course, be included.

*This will be even more true in the present example since the ratio of axial to flexural stiffness $\mathfrak{a}/\mathfrak{b}$ is here even larger, being about 470.

Since we now wish to analyze the frame more closely we must specify more information regarding its cross section. The dimensions chosen are shown in Figs. 8.54 and 8.55. In order to make the frame resemble those in ships it is assumed that the structure is plated. Because of the plating it is not possible to achieve a value of I such that $\mathfrak{b} = EI/L^3 = 1$, as in the previous solution (except by adopting unrealistic proportions). Instead, to allow a comparison with the results of Section 5.3 the dimensions have been chosen to give a value of 2 (MN/m) for $\mathfrak{b}$ and the load has been doubled. Therefore the solution to this new problem which would be obtained by ordinary frame analysis would have the same deflection (maximum value 1.8 mm) but the bending moment would be double the old value (e.g., 22.4 kNm at midspan).

Since we are ignoring axial stiffness the constraint against vertical deflection at node A also applies to node B, and the constraint against horizontal deflection at node C likewise applies to node B. Hence the structure contains only three degrees of freedom, Θ_{B1}, Θ_{B2}, and V_C, as shown in Fig. 8.54*b*.

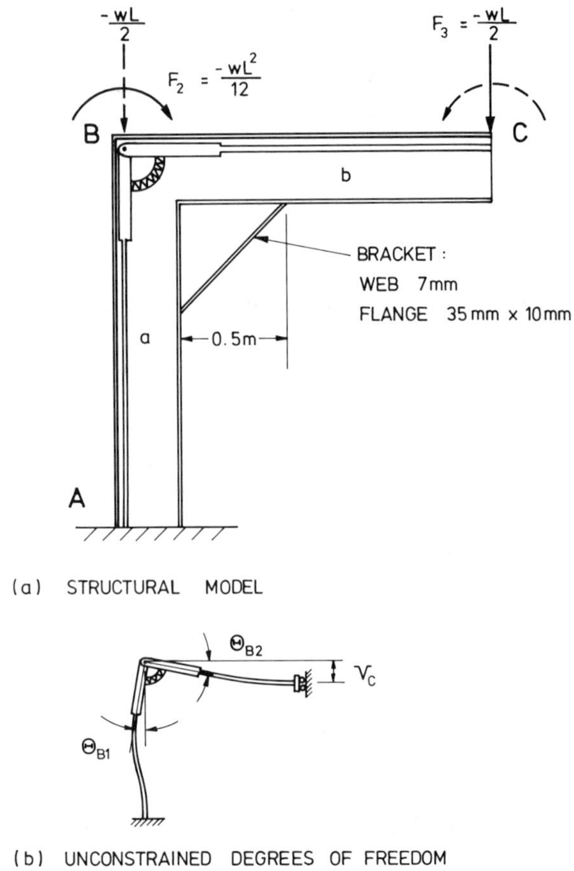

Figure 8.54 Portal frame example.

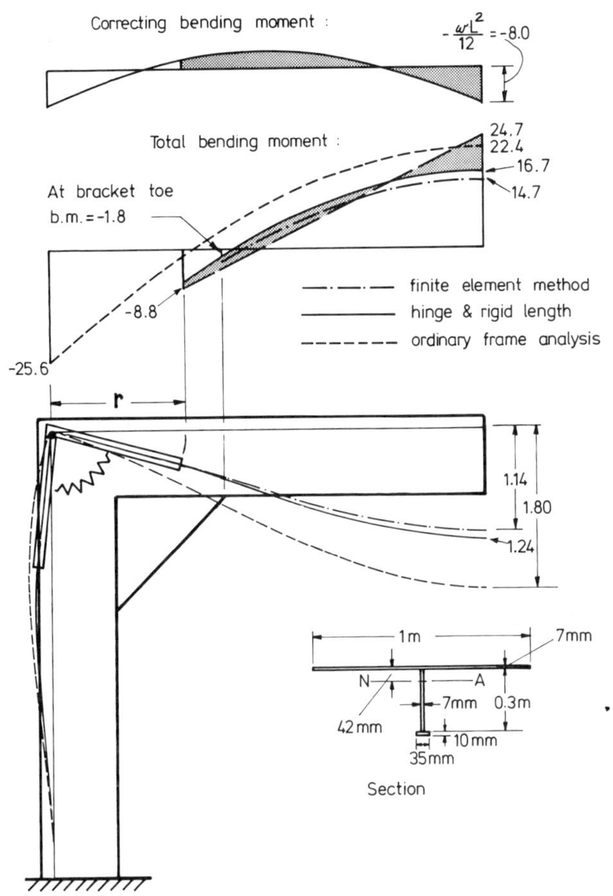

Figure 8.55 Comparison of analysis methods for bracketed frames.

The solution using the joint rotational spring and rigid lengths is presented in five steps.

1. *Calculate the joint stiffness k_J.*

From (8.6.18) the joint stiffness is

$$k_J = 0.78\,\frac{Gt_w}{\left[\dfrac{2}{h(2h+a)}\right]} = \frac{0.78(200/2.6)(.007)}{\left[\dfrac{2}{0.3(0.6+0.5)}\right]}$$

$$= 68.9 \text{ MNm}$$

2. *Calculate the extra rigid length to account for the brackets.*

From (8.6.8a) the extra rigid length is

$$d_r = a\left(1 - \frac{h}{h+a}\right) = 0.5\left(1 - \frac{0.3}{0.8}\right) = 0.31 \text{ m}$$

and therefore the total rigid length is

$$r = h + d_r = 0.30 + 0.31 = 0.61 \text{ m}$$

and the flexible length of the beam is therefore

$$l = L - r = 2 - 0.61 = 1.39 \text{ m}$$

3. *Calculate the element stiffness matrix in structure coordinates.*

From (8.6.5) the element stiffness matrix in element coordinates is

$$\mathbf{k} = \check{\mathbf{k}} + \mathbf{k}_r$$

where $\check{\mathbf{k}}$ is the ordinary flexure-only stiffness matrix for the flexible length l, given in (8.6.4), and $\mathbf{k}_r$ is the matrix which corrects for the rigid length, given in (8.6.6). The formal procedure would be to combine these two matrices and then transform to structure coordinates by the general rule: $\mathbf{K} = \mathbf{T}^T\mathbf{k}\mathbf{T}$. However, it is advantageous to treat the two matrices separately since we have general expressions for each of them: (5.3.16) for $\check{\mathbf{k}}$ (with l in place of L) and (8.6.7) for $\mathbf{k}_r$. The former equation refers to the ordinary beam element, and since we are here applying it for a flexure-only beam element we omit the terms containing α.

For member a the transformation angle is 90° and so the transformation factors are $\lambda = 0$ and $\mu = 1$. Substituting in (5.3.16) and (8.6.7) (with $r_1 = 0$ and $r_2 = r$) and combining the results gives

$$\mathbf{K}^a = 2\left(\frac{L}{l}\right)^3 \times \begin{bmatrix} 12 & & & \text{symm.} \\ -6 & 4l^2 & & \\ -12 & 6l & 12 & \\ -6(L+r) & 2l(L+2r) & 6(L+r) & 4l^2 + 12rL \end{bmatrix}$$

where the external multiplier is

$$\frac{EI}{l^3} = \left(\frac{EI}{L^3}\right)\left(\frac{L}{l}\right)^3$$

and in this example the first factor is 2 (MN/m).

Because of its orientation, member b does not require transformation, and the total element stiffness matrix in structure coordinates is obtained by combining (8.6.4) and (8.6.6), with $r_1 = r$ and $r_2 = 0$. This gives

$$\mathbf{K}^b = 2\left(\frac{L}{l}\right)^3 \begin{bmatrix} 12 & & & \text{symm.} \\ 6(L+r) & 4l^2 + 12rL & & \\ -12 & -6(L+r) & 12 & \\ 6l & 2l(L+2r) & -6l & 4l^2 \end{bmatrix}$$

Since the joint stiffness matrix only contains rotational degrees of freedom it never requires transformation. Hence the stiffness matrix in structure coordinates is

structure d.o.f. →	Θ_{B1}	Θ_{B2}
element d.o.f. →	1	2

$$\mathbf{K}^j = 68.9 \begin{bmatrix} 1 & -1 \\ -1 & 1 \end{bmatrix}$$

In this expression the location vector has been placed above the matrix to indicate where the terms are to be inserted in the structure stiffness matrix.

We now proceed to substitute the numerical values for the terms in $\mathbf{K}^a$ and $\mathbf{K}^b$ and to assemble the structure stiffness matrix. In a small problem such as this the solution can be simplified by introducing the boundary conditions at this stage, and ignoring the element degrees of freedom which correspond to the constrained structure degrees of freedom. We indicate this symbolically by writing the location vector above the stiffness matrices and entering the symbol × for the constrained structure degrees of freedom. None of the terms in these rows and columns requires evaluation; the only terms which need to be evaluated are outlined with a dashed line. Following this convention the situation for element a is as follows

structure d.o.f.	×	×	×	Θ_{B1}
element d.o.f.	1	2	3	4

$$\mathbf{K}^a = 2\left(\frac{L}{l}\right)^3 \begin{bmatrix} & & & \\ & & & \\ & & & \boxed{} \end{bmatrix}$$

and we see that only the final diagonal term needs to be evaluated.

$$K_{44}^a = 2\left(\frac{L}{l}\right)^3 \left[4l^2 + 12rL\right]$$

$$= 2\left(\frac{2}{1.39}\right)^3 [4(1.39)^2 + 12(0.61)2]$$

$$= 133.2$$

For element b the situation is summarized as follows

×	Θ_{B2}	V_C	×
1	2	3	4

$$\mathbf{K}^b = 2\left(\frac{L}{l}\right)^3 \begin{bmatrix} & & & \\ & \boxed{} & & \\ & & & \end{bmatrix}$$

and the required terms are

$$K_{22}^b = 2\left(\frac{L}{l}\right)^3 [4l^2 + 12rL] = 133.2$$

$$K_{23}^b = K_{32}^b = 2\left(\frac{L}{l}\right)^3 [-6(L + r)] = -93.3$$

$$K_{33}^b = 2\left(\frac{L}{l}\right)^3 [12] = 71.5$$

4. *Assemble the structure stiffness matrix and solve for the displacements*

From the results of the previous step the structure stiffness matrix can be written immediately, as follows

$$\mathbf{K} = \begin{bmatrix} 133.2 + 68.9 & -68.9 & 0 \\ -68.9 & 68.9 + 133.2 & -93.3 \\ 0 & -93.3 & 71.5 \end{bmatrix} \quad (\text{columns: } \Theta_{B1}, \Theta_{B2}, V_C)$$

As shown in Fig. 8.54 the nodal loads are

$$F_1 = 0$$

$$F_2 = -\frac{wL^2}{12} = -8 \times 10^{-3} \text{ MNm}$$

$$F_3 = -\frac{wL}{2} = -24 \times 10^{-3} \text{ MN}$$

We note that the vertical force at node B does not appear because the assumption of infinite axial rigidity implies that this load is transmitted directly to the rigid support at the bottom of the structure. Likewise, the moment at node C does not appear because it is immediately absorbed by the rigid rotational restraint at this point. The solution to the structure equilibrium equations is

$$\Theta_{B1} = -0.236 \times 10^{-3} \text{ rad}$$

$$\Theta_{B2} = -0.691 \times 10^{-3} \text{ rad}$$

$$V_C = -1.24 \text{ mm}$$

We note that Θ_{B2} is almost three times as large as Θ_{B1}, whereas in ordinary frame analysis, with rigid joints, they are assumed to be the same angle. This example shows that shear distortion in joints is a significant phenomenon, even when the beam depth is only 10% or so of the length (in this example the proportions are 7.5% and 15%) and even though the joint contains a relatively large bracket.

The deflection is plotted in Fig. 8.55, together with the result from a detailed finite element analysis, as explained subsequently. For comparison the deflection obtained in Section 5.3, using ordinary frame analysis (rigid joint, totally flexible beam) is also plotted. The figure shows clearly that the combination of deep webs and brackets greatly stiffens the frame, and that both of these effects must be included, in addition to shear distortion, to achieve a realistic solution.

5. *Calculate the bending distribution in the beams and the shear stress in the joint*

Since it is only the flexible portions of the beams which undergo bending we must first calculate the element nodal displacements at the internal nodes, using (8.6.1). For example, for the horizontal beam (member b) the vertical displacement at the end of the rigid length is

$$\breve{v}_1 = v_1 + r\theta_1$$

The member nodal displacements are obtained directly from the structure displacements: $v_1 = 0$, $v_2 = V$, $\theta_1 = \Theta_2$, and $\theta_2 = 0$. Therefore the displacements at the ends of the flexible beam are

$$\breve{v}_1 = 0 + (0.61)(-0.691 \times 10^{-3}) = -0.421 \text{ mm}$$

and

$$\breve{v}_2 = v_2 = -1.24 \text{ mm}$$

These nodal displacements produce a linear distribution of bending moment within the beam, given by (5.3.12), with l in place of L. Since the bending moment distribution is linear it is sufficient to calculate the nodal values, using (5.3.12b):

$$\begin{Bmatrix} \breve{\mathcal{M}}_1 \\ \breve{\mathcal{M}}_2 \end{Bmatrix} = \frac{EI}{l^3}\begin{bmatrix} -6l & -4l^2 & 6l & -2l^2 \\ 6l & 2l^2 & -6l & 4l^2 \end{bmatrix}\begin{Bmatrix} \breve{v}_1 \\ \breve{\theta}_1 \\ \breve{v}_2 \\ \breve{\theta}_2 \end{Bmatrix}$$

The resulting values are

$$\breve{\mathcal{M}}_1 = -8.8 \times 10^{-3} \text{ MNm}$$

$$\breve{\mathcal{M}}_2 = 24.7 \times 10^{-3} \text{ MNm}$$

and the corresponding linear distribution of bending moment is shown as a dashed line in Fig. 8.55. This bending moment corresponds to the artificial equivalent nodal forces which were applied at the structure nodes B and C to represent the distributed load. We must now remove the artificiality and account for the internal distribution of the load by superimposing the bending moment for that same load when the ends of the beam are clamped. Note that the beam in question is the full length beam, including the rigid portion. When such a beam is clamped and subjected to a uniform load the bending moment distribution is as shown in Fig. 8.55, with an end value equal to $-wL^2/12 = -8 \times 10^{-3}$ MNm. The total bending moment in the beam is obtained by superimposing the correcting bending moment distribution on the uncorrected linear distribution. The values of bending moment of greatest interest are the maximum values within the prismatic (unbracketed) length of the beam, since these are required in order to design the beam. One of these is the value of bending moment at the toe of the bracket. Since the shape of the correcting bending moment is known (e.g., with a uniform load it is a parabola) this value can easily be calculated. In this example the resulting total bending moment is -1.8 kNm. Since this is less than the center span value of 16.7 kNm the design of the prismatic portion of the beam would be governed by the latter.

In order to demonstrate the accuracy of this improved method of frame analysis the figure also shows the results from a more detailed solution in which the entire bracketed portion is modelled by membrane and bar elements, as shown in Fig. 8.56. The agreement is good and indicates that the methods presented in this section are sufficiently accurate for hull module analysis, even when the hull module contains deep beams and relatively large brackets.

For comparison the bending moment distribution corresponding to ordinary frame analysis (rigid joint, totally flexible beam) is also plotted. The result shows that this method is not sufficiently accurate for a hull module analysis.

8.7 DEFINITION AND USE OF SUPERELEMENTS

We come now to a finite element modeling technique which is of primary importance in the analysis of a large complex three-dimensional structure such as a hull module, and which becomes even more important when the analysis forms part of a rationally-based

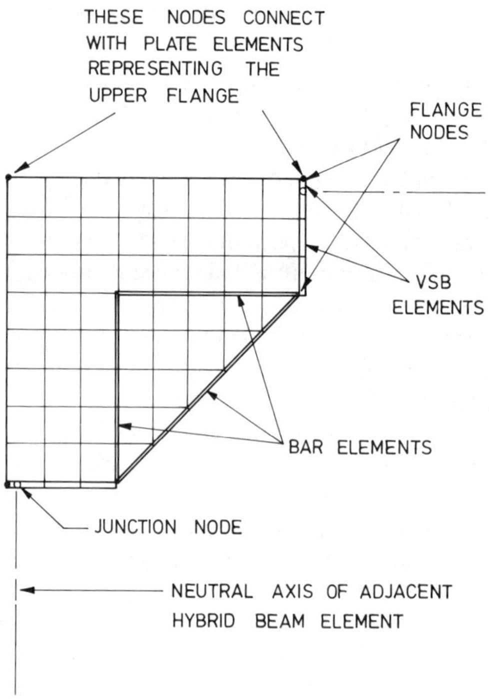

Figure 8.56 Finite element model of bracket.

structural design process. The superelement technique is a method for treating a portion of the structure as if it were a single member having a small number of nodes, even though it actually contains many nodes and is made up of many individual (and different) elements. We shall first briefly summarize the applications of superelements in a hull module analysis, and then explain the way in which a superelement is created.

Applications of Superelements in Hull Module Analysis

In a hull module analysis there are three main uses of superelements:

To save computation when a portion of structure occurs repeatedly.

To preserve functional or behavioral identity of a portion of structure.

To facilitate the generation of the hull module nodal system.

Computational Savings from Repeated Portions of Structure

In large structures it frequently happens that a certain portion of structure containing several nodes occurs repeatedly, in various locations. This repetition provides an important opportunity for more efficient modeling and analysis. We begin by noting that any portion of structure, even though it may be made up of many elements and contain many nodes, can be regarded as a single element—a superelement. The stiffness matrix for this superelement would be obtained by assembling the element stiffness matrices of the individual elements, in the same way as is done for the complete structure. We shall refer to this stiffness matrix as the original or unreduced superelement stiffness matrix. By definition the superelement contains many nodes, but we can distinguish two types: those which serve as connection points between the superelement and the rest of the structure, and those which are wholly internal to the superelement. The former are referred to as boundary nodes and the latter as internal nodes. The internal nodes do not have any role outside of the superelement; they simply account for the internal connectivity and interactions among the various nodes. More specifically, the terms in the original superelement stiffness matrix corresponding to the internal degrees of freedom specify the effect which a force or a displacement at an internal node has on the immediately surrounding nodes. In practice the only interactions that are really required are the interactions between the boundary nodes; if these were known then the internal nodes would become superfluous and so would their associated degrees of freedom. Eliminating these internal degrees of freedom would reduce the size of the superelement stiffness matrix and hence also of the structure stiffness matrix, thereby reducing the computation required to solve the structure equilibrium equations. If the superelement occurs repeatedly the reduction becomes substantial and more than offsets whatever extra computation is required at the start to eliminate the internal degrees of freedom of the superelement—a computation which is performed once only. Thus it would be advantageous if all of the local or "short distance" interactions involving the internal degrees of freedom were combined (or *condensed*) into a set of direct "long distance" interactions involving only the boundary nodes. The procedure for doing this is known as *static condensation* and is described later.

Preservation of Functional or Behavioral Identity

The superelement technique permits any degree of modularization in finite element analysis and helps to achieve the modeling strategy and other characteristics of *design-oriented* finite element analysis which were discussed in Section 2.2. In particular it was noted there that in a hull module analysis the elements should, as far as possible, correspond to the principal structural members. But a hull module often contains some principal structural members that can only be modeled as an assemblage of several small elements, perhaps of differing types. The superelement technique allows these elements to be combined into a single element. In a computer program for rationally-based design this type of modeling is beneficial even if the member occurs only once, because it simplifies the program logic in several aspects. It is especially suitable if the calculation of the member buckling load requires iterative finite element analysis (Section 15.2) which in turn requires the member stiffness matrix. This would be the case, for example, in a large panel with an irregular shape or stiffener arrangement.

Some portions of structure are ideally suited for superelement representation because they not only have a distinct role and type of behavior, but also they occur repeatedly. The most common examples are deep transverse web frames and brackets, especially transverse brackets. These two applications are discussed subsequently in further detail.

Facilitating the Generation of the Hull Module Nodal System

The superelement technique is also useful for representing a portion of structure that has a different geometry or orientation from the rest of the structure (or at least from the surrounding region) or that interrupts an otherwise regular region of structure. A transverse bulkhead is an example of such a structure. In order to be modeled with sufficient accuracy a transverse bulkhead must contain many more nodes than are present in other transverse planes of the hull module. The automated generation of the hull module nodal mesh is greatly simplified if the pattern of nodes in the transverse plane at each frame is the same. The bulkhead contains many extra nodes within its interior, but these are not involved in attaching the bulkheads to the decks, sides, or bottom. Therefore they can be removed by static condensation, leaving only the boundary nodes (and any internal nodes that are points of attachment to longitudinal structure). If this is done then the remaining nodes in the transverse bulkhead coincide longitudinally with the nodes in the transverse plane at each frame, and the automated generation of the three-dimensional nodal system is thereby greatly simplified.

Static Condensation

The process of static condensation is based on the nature of the forward reduction portion of the elimination method for solving a system of linear equations. The reduction process consists in taking each equation in turn (say the ith equation) and progressively eliminating the corresponding unknown variable (the ith variable) from the succeeding equations, while at the same time modifying the succeeding equations to account fully for whatever interactions the eliminated variable had with the other variables in those equations. This is achieved by first normalizing the ith equation (dividing through by its ith coefficient) and then combining the normalized ith equation with each of the succeeding equations in turn, in such a way as to make the coefficient of the ith term zero in all of the succeeding equations. The result is that the ith variable disappears from the succeeding equations, and it has a unit coefficient in the ith equation.

In solving a system of equations this forward reduction process is performed for all equations, resulting in a "triangularized" coefficient matrix containing zeros in all locations below the diagonal. It is then possible to solve for each variable in turn, starting with the last (back substitution).

The forward reduction process can be used to create a superelement. The first step is to assemble the stiffness matrices of the constituent elements into a single stiffness matrix, in the same way that they would be assembled into the structure stiffness matrix if a superelement was not desired. The matrix thus formed is the coefficient matrix of a system of equations describing all of the various interactions between the various degrees of freedom in the superelement—internal and boundary. To create the superelement proper it is necessary to eliminate some of the variables in this system of equations—namely, the internal degrees of freedom, $\boldsymbol{\delta}_i$—and to make the required modification to the system of equations so that their effect is accounted for, even though they are not present. This can be achieved by performing elimination on only part of the system—the part corresponding to $\boldsymbol{\delta}_i$—and then stopping at that point. In order to be eliminated the $\boldsymbol{\delta}_i$ terms must occur at the beginning of each equation, and so it is first necessary to rearrange each equation such that the $\boldsymbol{\delta}_i$ terms occur first, before the terms corresponding to the boundary degrees of freedom. At the same time it is necessary to rearrange the equations themselves in order to maintain the symmetry of the coefficient matrix.

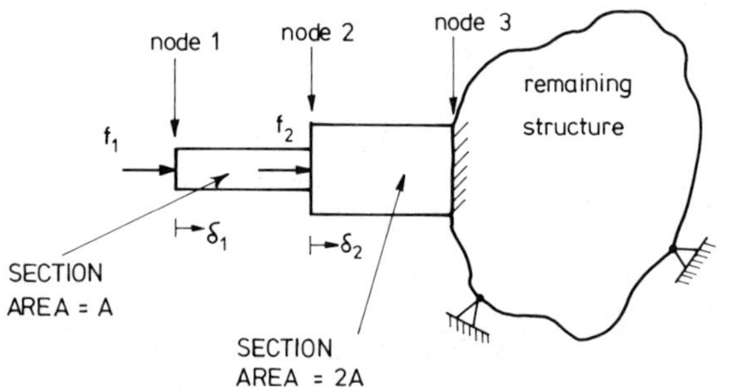

Figure 8.57 Sample structure to illustrate static condensation.

Example

In order to illustrate the process of static condensation as simply as possible, let us consider the two simple bar elements shown in Fig. 8.57. In a more typical case these would form only a part of a larger structure, as shown at the right of the figure. But for simplicity in this example we will subsequently place a boundary support at the right end and treat the two elements as being the complete structure. Two external forces act on the structure as shown. The middle node is not directly connected to the rest of the structure; as far as these two elements are concerned it is an *internal node*. Therefore if desired it can be removed by static condensation, thus converting the two separate elements into a superelement, with nodes only at its boundaries where it connects with the rest of the structure.*

The force–displacement equations for this two-element "member" may be obtained by the usual assembly process. The result is

$$\begin{Bmatrix} f_1 \\ f_2 \\ 0 \end{Bmatrix} = \frac{EA}{L}\begin{bmatrix} 1 & -1 & 0 \\ -1 & 3 & -2 \\ 0 & -2 & 2 \end{bmatrix}\begin{Bmatrix} \delta_1 \\ \delta_2 \\ \delta_3 \end{Bmatrix}$$

Before proceeding with the example we first make note of some information which will be used later to verify the static condensation process. We note that if we temporarily place a rigid support at node 3, so as to convert the two bars into a separate structure, the solution for the nodal displacements would be

$$\delta_1 = \frac{3f_1 + f_2}{2}\frac{L}{EA}$$
$$\delta_2 = \frac{f_1 + f_2}{2}\frac{L}{EA} \tag{8.7.1}$$

We now proceed to demonstrate how the two bar elements can be converted into a superelement by eliminating the middle node. In order to perform static condensation on the middle degree of freedom, we must first rearrange each equation so as to make that degree of freedom be the first term in each equation, and we rearrange the equations so that the equation for the middle node (i.e., the second equation) becomes the first equation. In other words the original 3×3 stiffness matrix of the two combined elements must be rearranged such that the second row and column become the first row and column. In the present example we achieve this by interchanging the first row and column with the second row and column. The resulting equations are

$$\begin{Bmatrix} f_2 \\ f_1 \\ 0 \end{Bmatrix} = \frac{EA}{L}\begin{bmatrix} 3 & -1 & -2 \\ -1 & 1 & 0 \\ -2 & 0 & 2 \end{bmatrix}\begin{Bmatrix} \delta_2 \\ \delta_1 \\ \delta_3 \end{Bmatrix}$$

We now perform one cycle of the Gaussian forward reduction process. This converts the system of equations to

$$\begin{Bmatrix} \dfrac{f_2}{3} \\ f_1 + \dfrac{f_2}{3} \\ \dfrac{2f_2}{3} \end{Bmatrix} = \frac{EA}{L}\begin{bmatrix} 1 & -\frac{1}{3} & -\frac{2}{3} \\ 0 & \frac{2}{3} & -\frac{2}{3} \\ 0 & -\frac{2}{3} & \frac{2}{3} \end{bmatrix}\begin{Bmatrix} \delta_2 \\ \delta_1 \\ \delta_3 \end{Bmatrix}$$

We have now normalized the first equation and we have eliminated the first variable, δ_2, from the succeeding equations. The succeeding equations constitute a reduced or condensed system which is

$$\begin{Bmatrix} f_1 + \dfrac{f_2}{3} \\ \dfrac{2f_2}{3} \end{Bmatrix} = \frac{2}{3}\frac{EA}{L}\begin{bmatrix} 1 & -1 \\ -1 & 1 \end{bmatrix}\begin{Bmatrix} \delta_1 \\ \delta_3 \end{Bmatrix} \tag{8.7.2}$$

and the portion of the original system which has been eliminated is

*Although there is no adjacent structure at the left end of the superelement, the left node cannot be removed because a one-node bar element is meaningless.

$$\frac{f_2}{3} = \frac{EA}{L}\begin{bmatrix} 1 & -\frac{1}{3} & -\frac{2}{3} \end{bmatrix}\begin{Bmatrix} \delta_2 \\ \delta_1 \\ \delta_3 \end{Bmatrix} \qquad (8.7.3)$$

This portion (which here is only one equation) is saved because it will be required later for calculating the internal forces within the superelement.

We see from (8.7.2) that the two end nodes are now directly connected, and that the two bars have in effect been combined into one bar—a superelement—for which the stiffness matrix is

$$\mathbf{k}_{se} = \frac{2}{3}\frac{EA}{L}\begin{bmatrix} 1 & -1 \\ -1 & 1 \end{bmatrix}$$

We also note that the condensation has produced a modified load vector for the superelement—the left-hand side of (8.7.2).

If the two bars were part of a larger structure the superelement stiffness matrix $\mathbf{k}_{se}$ would be assembled into the structure stiffness matrix, as would the stiffness matrix of all the other elements and superelements, and the superelement load vector would be added to the structure load vector. The use of superelements makes the structure stiffness matrix smaller than it would otherwise be, and so the solution of the structure system of equations involves less storage. It also involves less calculation but an overall savings in calculation is only achieved if the superelements occur repeatedly. If each superelement occurred only once there would be no net savings in calculation because the effort required to eliminate the internal degrees of freedom, and thereby create the superelement, is exactly the same as the effort which is later saved in the solution of the system equations. When a superelement occurs repeatedly there is also a savings in the assembly process, because the element stiffness matrix needs to be calculated only once.

In most cases it is desired to calculate the stresses within the superelement and for this purpose the portion of the original system which was removed by condensation should be saved so that the displacements at the internal nodes of the superelement can be calculated.

We can verify the correctness of (8.7.2) without having to consider the rest of the structure shown in Fig. 8.56. For simplicity we can now imagine that the superelement is itself the structure and give it sufficient boundary support to be a structure, namely $\delta_3 = 0$, as shown in the figure. After allowing for this boundary support, the "structure" equations are

$$\begin{Bmatrix} f_1 + \dfrac{f_2}{3} \\ R_3 \end{Bmatrix} = \frac{2}{3}\frac{EA}{L}\begin{bmatrix} 1 & -1 \\ -1 & 1 \end{bmatrix}\begin{Bmatrix} \delta_1 \\ 0 \end{Bmatrix}$$

in which the force at node 3 is now a reaction force R_3. Solving for δ_1 we obtain

$$\delta_1 = \frac{3f_1 + f_2}{2}\frac{L}{EA}$$

which agrees with the answer given in (8.7.1) obtained without using a superelement. The nodal displacement at the internal node of the superelement may be obtained from the eliminated portion of the system of equations, (8.7.3). Substituting the foregoing value of δ_1 (and setting $\delta_3 = 0$ because of the boundary condition) gives

$$\delta_2 = \frac{f_1 + f_2}{2}\frac{L}{EA}$$

which also agrees with the result obtained earlier. Once the internal displacements are known the stresses can be calculated in each of the elements in the superelement.

Use of Superelements for Representing Brackets

A hull module often contains several identical brackets. If the brackets are large, or if it is desired to calculate the stress within them, they must be modeled explicitly rather than as rigid lengths. Because of their geometry the structural modeling requires many individual elements, as shown in Figs. 8.58 and 8.59. The bracket web is a flat membrane and its inner profile is either at an angle or curved; therefore it requires several small membrane elements, some of which may be rectangular and others triangular; the web stiffeners and the bracket flange (or face plate) are usually represented by bar elements; if the flange is curved it requires several such elements. Thus if there are several identical brackets a superelement representation gives significant computational savings.

A superelement modeling becomes even more appropriate if the bracket design is to be included as part of the preliminary design of the ship because the bracket must then fulfill the role of a principal structural member; that is, it must:

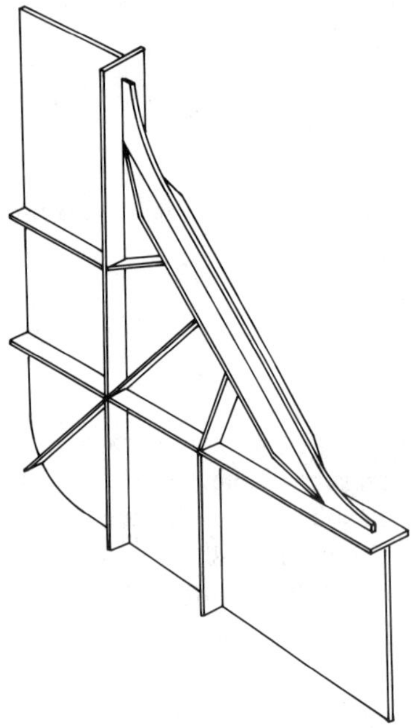

Figure 8.58 Typical large bracket.

1. Incorporate and reflect the effect of the principal structural dimensions of the bracket (a_i, a_j, t).
2. Provide sufficient information regarding the bracket response (deflection, stresses, etc.) to allow the structural assessment of the current bracket design and the formulation of the structural constraint equations which are required for the optimization portion of the design process.

The superelement in Fig. 8.59 is essentially a pure membrane element, with only the translational degrees of freedom, u and v, at each node, whereas the beam elements that are attached to it have three degrees of freedom: u, v, and θ, and it is necessary to have some method for linking these two different types of element. Figure 8.59 illustrates one method which uses very stiff beam (VSB) elements. These are simply ordinary beam elements which are given a very large value of moment of inertia, such as $I_x = 10^{25}$. At each junction between a membrane element and an adjacent beam element a row of VSB elements is inserted connecting the junction node with two nodes in the mem-

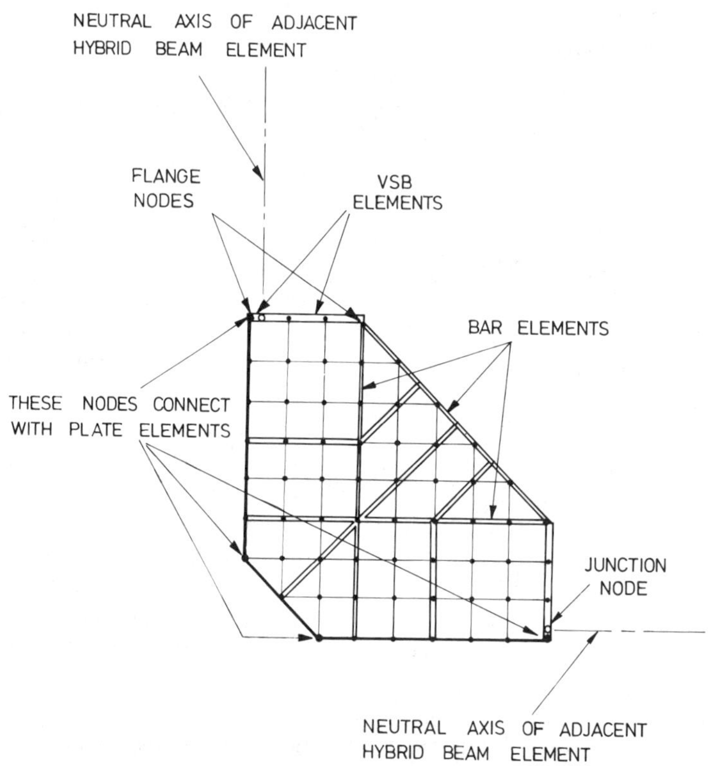

Figure 8.59 Bracket superelement before static condensation.

brane element at the top and bottom of the beam web. These two nodes, which we shall refer to as *flange nodes*, are given three degrees of freedom in order to connect with the VSB elements. The junction node should be located at the (hybrid) beam neutral axis, that is, the centroid of the cross section at the junction. The large flexural rigidity of the VSB element automatically converts the rotation at the beam node into translation of the flange nodes.

It is essential that the beam node should not be one of the mesh nodes within the superelement because then the axial force from the beam would be concentrated on that node, causing severe and erroneous local deformation. Making the beam node distinct from the internal nodes means that the axial force is shared by the two flange nodes, (and perhaps other internal nodes) and since the bracket will also have flanges the two forces will be smoothly and realistically absorbed and diffused into the bracket.

The axial stiffness of the VSB elements should not be a very large value because that would constrain the two flange nodes to remain a fixed distance apart, causing an incorrect strain distribution in the adjacent membrane element. The axial stiffness should be the same for all VSB elements, in spite of their unequal length, in order for the total shear force at the beam node to be distributed evenly among the membrane nodes. The axial stiffness could be expressed as EA_{vsb}/h_i, where h_i is the web height of beam i and A_{vsb} is some representative cross-sectional area of the VSB element. A correct order of magnitude for this (which is all that is required) would be obtained by choosing $A_{vsb} = th_i$, where t is the web thickness. Then the axial stiffness of each VSB element is simply Et. If the bracket contains a web stiffener at the junction, the stiffener's own axial stiffness may be assigned to the VSB elements.

Figure 8.60 shows the same superelement after the internal nodes have been eliminated. If a flange node is not connected to any structure outside of the superelement it may be treated as an internal node and eliminated from the superelement. But if the bracket is attached to plating (as is often the case in a hull module) then the flange node on that side must be retained. In this case there are two closely spaced but separate nodes at the junction: the junction node which connects the superelement to the adjacent beam element, and the flange node which connects it to the adjacent panel elements. Both nodes are required in order to obtain the correct interaction among the elements. This situation is illustrated in Fig. 8.61.

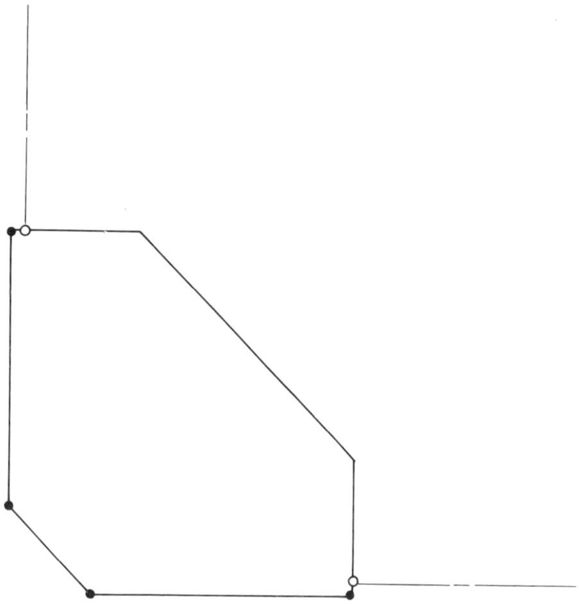

Figure 8.60 Bracket superelement after static condensation.

8.8 APPROXIMATE MODELING OF DOUBLE WALL STRUCTURES

By means of the hybrid beam element and the stiffened panel element it is possible to model various portions of a ship's structure having a wide variety of arrangements and proportions. However, a doubly plated grillage or other closed cellular structural arrangement has special characteristics and requires special mod-

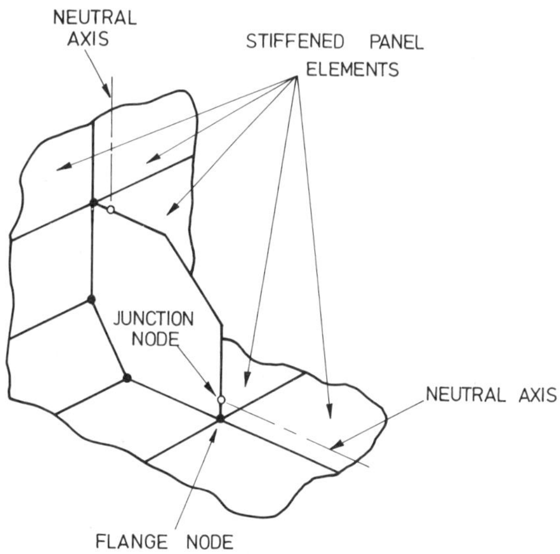

Figure 8.61 Connection to adjacent panel elements.

eling techniques in order to achieve sufficient accuracy. Such structures are common in ships, not only in double bottoms but also in the cellular side walls of container ships and other cellular regions which are intended to provide adequate torsional stiffness, or which may be present for other reasons; cofferdams, collision protection, special liquid storage, extra ballast space, and so on. If the region in question is irregular in shape, such as ballast tanks in a highly curved portion of the hull, then the structure must be modeled entirely with stiffened panel or other membrane elements. But in many cases, particularly in the prismatic portion of a ship's length, a double-walled structure is essentially a double-plated orthogonal gridwork of beams, and the beams are usually at regular spacings. Under these circumstances it is possible to use a simplified method of modeling in which the flexural aspects of the double wall structure are preserved, rather than being converted into membrane strain by a more detailed subdivision of the structure. The principal advantage of this method of modeling is that it requires only a single "layer" of nodes and therefore the same nodal mesh which is used for beams and stiffened panels in singly plated regions of structure can also be used for double wall regions. This mesh commonality greatly simplifies the hull module analysis.

However, although a double wall structure usually contains many beams in two orthogonal directions, and although its resistance to a lateral load comes largely through its flexural rigidity in these two directions, it is nevertheless fundamentally different from an open or singly plated grillage because it contains (in fact it consists of) closed cells which, as we have seen in Section 6.1, have a much larger torsional rigidity than open sections. As a result the lateral load is carried jointly by beam bending and by closed-section torsion, whereas in an open or singly plated grillage the lateral load is carried purely through the action of beam bending.

In fact, a double wall structure involves four different types of response: in-plane (biaxial), bending (about two orthogonal axes), transverse shear (of the beam webs), and closed-cell torsion (which is in-plane shear in the beam webs and in the plating). In the double wall modeling to be presented each of these modes of response is modeled explicitly, using whichever method of modeling is the most efficient for that mode. In particular the torsional rigidity is accounted for simply and explicitly, by means of a specially derived torsional stiffness parameter.

Use of Hybrid Beam and Stiffened Panel Elements

The modeling technique for a double wall structure is illustrated schematically in Fig. 8.62. It may be regarded as a combination of "double-plate-flange" hybrid beam elements and rectangular stiffened panel elements, plus some extra stiffness which is not illustrated but which accounts for the torsional rigidity. With this additional stiffness the modeling technique presented herein has the same degree of accuracy as a more detailed modeling, having nodes at the upper and lower corners of all of the webs and with the two panel elements located at their true positions. In explaining the technique we shall first discuss the attributes and stiffness contributions of the above two elements, then show that some further stiffness is required to account for the closed-cell torsional rigidity, and then proceed to derive a double wall stiffness parameter $\mathfrak{d}$ and a double wall stiffness submatrix $\mathbf{k}_{dw}$ which supplies the required stiffness.

Double-Plate-Flange Hybrid Beam Element

This element is similar to the hybrid beam element of Section 8.3 but it has two plate flanges instead of one. The term "hybrid" again refers to the fact that the element has two different cross sections: one for the axial stiffness and another for the beam bending stiffness. The axial stiffness is that of the web only. Hence for beam i of Fig. 8.63 the axial stiffness parameter is

$$\alpha_i = \frac{Eht_{wi}}{L_i} \tag{8.8.1}$$

where h, t_{wi}, and L_i are shown in the figure.

For the bending stiffness the cross section is a doubly flanged beam in which the flanges are of effective width b_e and equivalent (or "smeared") thickness $\bar{t}$. The latter accounts for the cross-sectional area of the stiffeners and is used for those beams which are parallel to the stiffeners. For example, for beam i in Fig. 8.63 the equivalent thicknesses of the two plate flanges are

$$\left.\begin{aligned} \bar{t}_u &= \frac{L_j t_u + A_s^u}{L_j} \\ \bar{t}_l &= \frac{L_j t_l + A_s^l}{L_j} \end{aligned}\right\} \tag{8.8.2}$$

where L_j, t_u, A_s^u, and so on, are shown in the figure. For

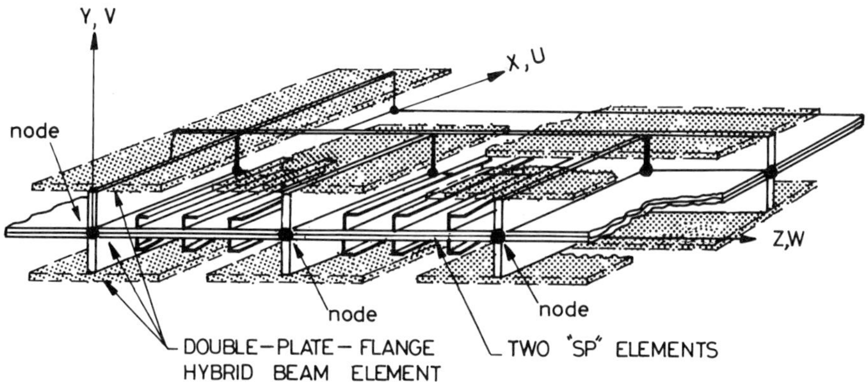

Figure 8.62 Schematic illustration of modeling technique for double wall structures.

beams in the other direction the plate flanges have the same thickness as the plating. Thus for example plate flanges in beam j are t_u and t_l. This illustrates the fact that in a given beam the two flanges need not be the same; each of them has properties according to the panel to which it corresponds. The effective moment of inertia I_e of each beam can be calculated from (8.3.6).

Experiments such as that described in Ref. 6 have shown that for double wall structures the effective breadth is generally quite large ($b_e/b > 0.9$) providing that the web spacings are not excessive ($0.5 < L_i/L_j < 2.0$, say). The neutral axis location may be calculated from (8.3.6). The nodes are located at the mean height of the two neutral axes (which will invariably be very close to each other).

Stiffened Panel Element

As explained earlier, plate bending and stiffener bending need not form part of the hull module analysis and so it is only the in-plane stiffness of the panels which needs to be modeled, using the stiffened panel element presented in Section 8.4. The total in-plane stiffness is the combined in-plane stiffness of the panels plus the axial stiffness of the web. The effective plane of action of the combination is the centroidal plane of the web and panels, which is exactly where the nodes are lo-

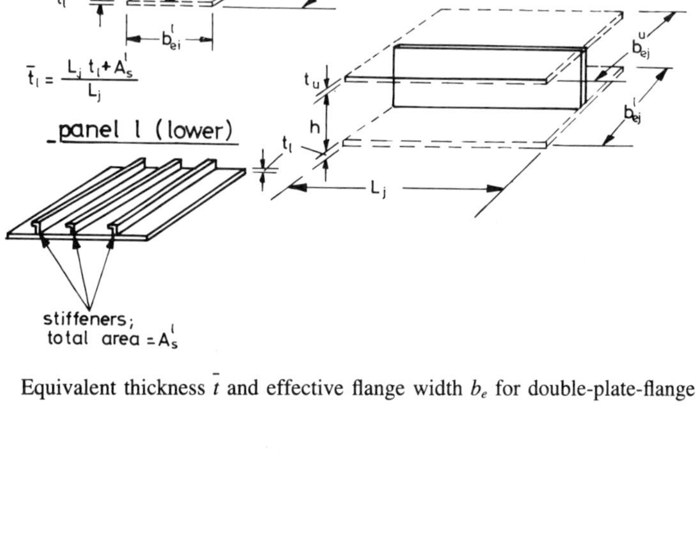

Figure 8.63 Equivalent thickness $\bar{t}$ and effective flange width b_e for double-plate-flange hybrid beam element.

cated, and hence the individual stiffness matrices of the webs, panels, and beams may be combined directly. The combination may either be done in element coordinates, after which the fully assembled stiffness matrix is transformed to global coordinates, or each element stiffness matrix may first be transformed and then added to the global stiffness matrix. In practice the latter procedure is usually chosen because it is more efficient in regard to storage. However, the method of allowing for the closed-cell torsional rigidity is more easily derived and explained in element coordinates and so we here adopt the former approach.

Allowance for Closed-Cell Torsional Rigidity

Before deriving an expression for the torsional stiffness terms of a double wall structure we first demonstrate the need for making a specific allowance for the torsional rigidity of such a structure, over and above the axial and beam bending stiffnesses in the two directions. We do this by demonstrating that a serious deficiency in stiffness still exists when a double wall structure has been modeled as just described, even though all axial and bending stiffnesses have been accounted for in both directions.

In a typical double wall structure each of the box-shaped cells is subjected to some amount of torsion about each of the two in-plane axes, the magnitude of which depends on the location of the cell and on the nature of the external loading. Even if the external loading is a purely lateral load and does not include any direct twisting moment, most cells will be subject to a nonuniform transverse shear load, and this would exert some torsion.* Figure 8.64 shows such a load consisting of two pairs of equal and opposite lateral forces acting at the corners of the box. The forces are such as to impose a unit value of vertical deflection at each corner. From equilibrium it may be shown that all four forces must have the same magnitude regardless of whether the structure is square or not. Figure 8.65 is a side view of the structural model as it responds to this pattern of forces. The figure shows one of the beam elements of the model; it would have a total axial stiffness concentrated along the neutral axis and its bending stiffness would include the effect of the two plate flanges. If the torsional stiffness GJ of the two orthogonal beams at the ends of the given beam is ignored (and it may be, since it is much smaller than

*This is discussed in more detail in Section 9.1, taking the case of a differential element in a laterally loaded plate.

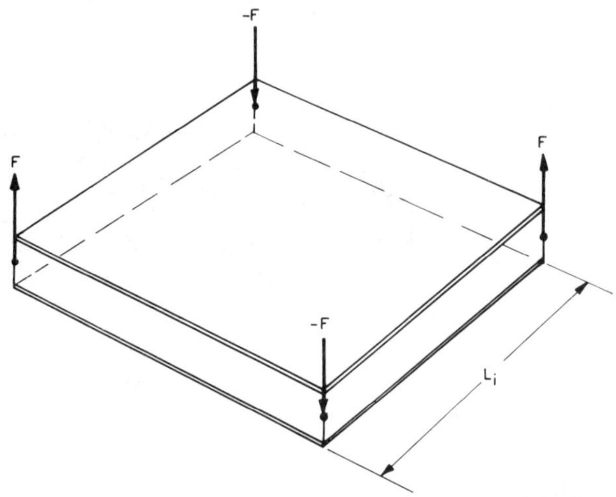

Figure 8.64 Twisting load on a double wall box.

any of the axial or bending stiffness terms) then these adjacent orthogonal beams cannot impose bending moments at the ends of the given beam. Hence the latter remains perfectly straight and undergoes a rigid body displacement such that there is no bending within the beam and the angle of rotation of the ends is $\theta = 2/L_i$. Now, it can easily be verified that if these nodal displacements ($u_1 = u_2 = 0$; $v_2 = -v_1 = 1$; $\theta_2 = \theta_1 = 2/L$) are applied to an ordinary beam element the resultant nodal forces, as calculated from the element stiffness matrix of (5.3.15), all turn out to be zero. Likewise, if shear deflection were allowed for, then the terms of the general beam element stiffness matrix, (5.5.10), would likewise give a zero result. That is, as far as axial and beam bending stiffness is concerned, the structure of Fig. 8.64 has no means of resisting this particular combination of external forces applied at the corners. The only resistance is the relatively small torsional stiffness GJ, and this is why an open gridwork of beams is particularly susceptible to an overall twisting type of load as represented by these four corner forces.

Since a closed box in fact possesses a very large torsional rigidity, there must obviously be some additional stiffness, over and above the axial and bending stiffnesses. As explained in Section 6.1 the large tor-

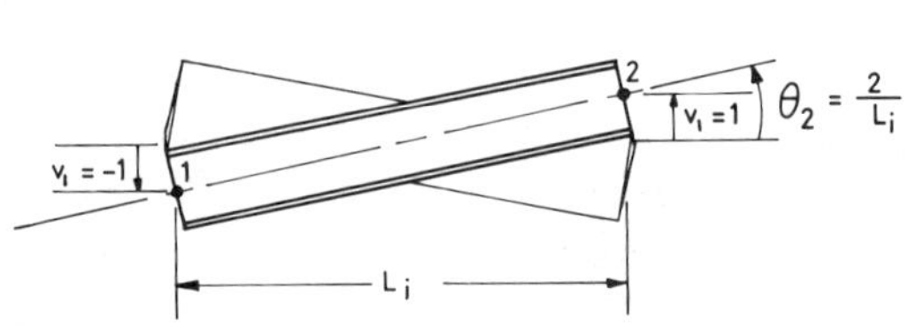

Figure 8.65 Response (unit displacements) of a double wall box.

sional rigidity of a closed section is due to the shear stresses being distributed around a closed path, which means they all act together in the same rotational direction to equilibrate the external torque without necessitating a large amount of twist. If a double wall structure is modeled entirely by membrane elements this continuity of shear flow would be provided automatically, and there would be no need for any special allowance. Indeed, this is the usual way of modeling cellular structures, but we are here endeavoring to achieve a modeling that is both more efficient and more compatible with the modeling of other portions of a hull module.

Let us again consider the case of the overall twisting load consisting of the four corner forces and examine the resulting deformation in more detail. Figure 8.66*a* is an overall view and Fig. 8.66*b* is a plan view of the displacement of the upper edges of the sides of the box. If the panels had no shear rigidity these edges would distort from the original rectangular outline in a pure shear strain type of deformation. The bottom edges would deform in a similar but opposite manner. Hence the panels of plating are subjected to a pure shear type of load, and because of their large in-plane shear rigidity they greatly restrict this deformation. Neither of the two elements discussed provides this in-plane shear rigidity and this is the reason for the observed deficiency in stiffness. The upper panel element of Fig. 8.66*b* shows the degree of freedom numbering and the nodal displacements $\boldsymbol{\delta}^u$ which are caused by the four corner loads. Since the panel has two axes of symmetry the eight displacements fall into two groups

$$\left.\begin{aligned} \delta_1^u &= \delta_3^u = -\delta_5^u = -\delta_7^u \\ \delta_2^u &= -\delta_4^u = -\delta_6^u = \delta_8^u \end{aligned}\right\} \quad (8.8.3)$$

These displacements are related by the panel element nodal forces $\mathbf{f}^u$ which are given by

$$\mathbf{f}^u = -\mathbf{k}_{sp}^u \boldsymbol{\delta}^u \quad (8.8.4)$$

The minus sign is required because we are dealing with the forces exerted *by* the panel rather than the forces *on* the panel.

Figure 8.66*c* is a free-body diagram of one of the webs (web *i*) of the double wall structure. For generality of application we allow all four of the webs to be of different thickness. The free-body diagram does not include the plate flanges or any of the axial or bending characteristics because it was shown above that for this

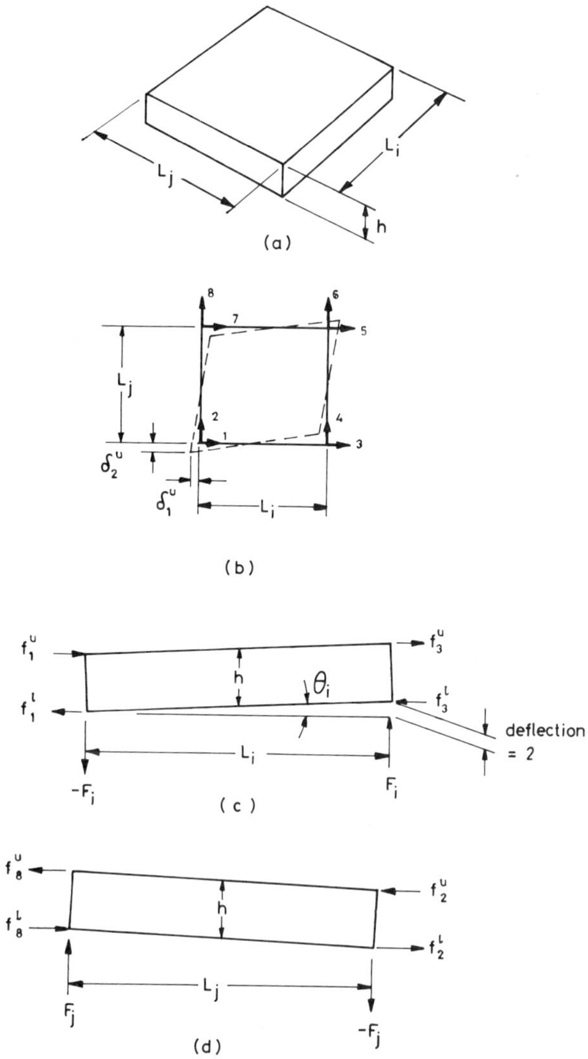

Figure 8.66 Panel and web interactions.

applied load neither the axial nor the beam bending stiffness has any effect. The forces acting at the corners of the web are those exerted by the upper and lower panels and are distinguished from each other by the superscript *u* and *l*. As mentioned earlier, the two panels may be different. However, since each panel has two axes of symmetry and since each web is assumed to be prismatic the forces for each panel fall into two groups, as did the displacements. That is

$$f_1^u = f_3^u = -f_5^u = -f_7^u$$
$$f_2^u = -f_4^u = -f_6^u = f_8^u$$

and similarly for the lower panel. This could be verified by substituting for $\mathbf{k}_{sp}^u$ in (8.8.4) and multiplying by the $\boldsymbol{\delta}$ values of (8.8.3). Therefore within each panel it is possible to represent all of the forces in terms of two representative forces, say f_1^u and f_2^u for the upper panel and f_1^l and f_2^l for the lower panel.

Each of the vertical forces F_i is that portion of F which is absorbed by web i and which is ultimately balanced by the panel forces at the four corners of web i. The other portion of each of the external forces is absorbed by the adjoining web, and is therefore denoted F_j as shown in Fig. 8.66*d*. (Later it will be shown that $F_i = F_j$, but it would complicate matters to attempt to show this now.) Equilibrium of moments in web i yields

$$\begin{aligned} F_i L_i &= \frac{h}{2}(f_1^u - f_1^l) + \frac{h}{2}(f_3^u - f_3^l) \\ &= h(f_1^u - f_1^l) \end{aligned} \tag{8.8.5}$$

Similarly, in web j equilibrium of moments yields

$$\begin{aligned} F_j L_j &= \frac{h}{2}(f_2^u - f_2^l) + \frac{h}{2}(f_8^u - f_8^l) \\ &= h(f_2^u - f_2^l) \end{aligned} \tag{8.8.6}$$

Therefore, since the total vertical force at any corner is $F = F_i + F_j$, we have

$$F = \frac{h}{L_i}(f_1^u - f_1^l) + \frac{h}{L_j}(f_2^u - f_2^l) \tag{8.8.7}$$

We now employ (8.8.4) to express each of the four representative panel forces in terms of the panel corner displacements. From (8.8.3) the latter can also be represented by four representative values: δ_1^u, δ_2^u, δ_1^l and δ_2^l. The results are

$$\begin{aligned} f_1^u &= (k_{11}^u + k_{13}^u - k_{15}^u - k_{17}^u)\delta_1^u \\ &\quad + (k_{12}^u - k_{14}^u - k_{16}^u + k_{18}^u)\delta_2^u \\ f_1^l &= (k_{11}^l + k_{13}^l - k_{15}^l - k_{17}^l)\delta_1^l \\ &\quad + (k_{12}^l - k_{14}^l - k_{16}^l + k_{18}^l)\delta_2^l \\ f_2^u &= (k_{21}^u + k_{23}^u - k_{25}^u - k_{27}^u)\delta_1^u \\ &\quad + (k_{22}^u - k_{24}^u - k_{26}^u + k_{28}^u)\delta_2^u \\ f_2^l &= (k_{21}^l + k_{23}^l - k_{25}^l - k_{27}^l)\delta_1^l \\ &\quad + (k_{22}^l - k_{24}^l - k_{26}^l + k_{28}^l)\delta_2^l \end{aligned}$$

The values of the individual stiffness terms of $\mathbf{k}_{sp}^u$ and $\mathbf{k}_{sp}^l$ are given in two parts in (8.4.10) and (7.5.12), corresponding to the stiffeners and plating, respectively. When these are substituted into the previous four expressions a great deal of simplification occurs. First, all of the stiffener terms cancel. For example, in the expression for f_1^u the two groups of four stiffener terms of $\mathbf{k}_{sp}^u$ are:

$$\frac{EA_s}{12L_i}[(4 - 4 + 2 - 2)\delta_1^u + (0 - 0 - 0 + 0)\delta_2^u]$$

The second simplification is that the plating stiffness terms in the upper and lower panels are exactly the same except for the plate thickness, and there is also a good deal of cancellation among the four stiffness terms in each of the parentheses. For example, for f_1^u and f_1^l the substitution from (7.5.12) yields

$$\begin{aligned} f_1^u &= \frac{Et_u}{4(1-\nu^2)}\{[2\mathscr{A}\alpha + 2\mathscr{A}\alpha]\delta_1^u \\ &\quad + [\mathscr{B} + \mathscr{C} + \mathscr{B} + \mathscr{C}]\delta_2^u\} \\ &= \frac{Et_u}{4(1-\nu^2)}\{[2\alpha(1-\nu)]\delta_1^u + [2(1-\nu)]\delta_2^u\} \end{aligned}$$

and

$$f_1^l = \frac{Et_l}{4(1-\nu^2)}\{[2\alpha(1-\nu)]\delta_1^l + [2(1-\nu)]\delta_2^l\}$$

where $\alpha = L_i/L_j$. Substituting these expressions for the first two forces which appear in (8.8.7) gives

$$f_1^u - f_1^l = \frac{Et_u t_l}{(1+\nu)(t_u + t_l)}[\alpha(\delta_1^u - \delta_1^l) + (\delta_2^u - \delta_2^l)]$$

Similarly, for the second two forces the corresponding expression is

$$f_2^u - f_2^l = \frac{Et_u t_l}{(1+\nu)(t_u + t_l)}\left[(\delta_1^u - \delta_1^l) + \frac{1}{\alpha}(\delta_2^u - \delta_2^l)\right]$$

Substituting these expressions into (8.8.7) yields

$$\left.\begin{aligned} F &= \frac{E\hat{t}h}{2(1+\nu)}\left[\frac{\alpha(\delta_1^u - \delta_1^l)}{L_i} + \frac{(\delta_2^u - \delta_2^l)}{L_i} + \frac{(\delta_1^u - \delta_1^l)}{L_j} + \frac{(\delta_2^u - \delta_2^l)}{\alpha L_j}\right] \\ \text{where} \quad \hat{t} &= \frac{2t_u t_l}{t_u + t_l} \end{aligned}\right\} \tag{8.8.8}$$

The next step is to relate the panel displacements to the unit deflections of the double wall box. Referring to the deflected web of Fig. 8.66*c* it will be recalled that in this deflection there is no axial or bending deformation. Therefore, if shear deformation of the web is neglected, the ends of the web are inclined to the vertical by the angle θ_i. Hence the total horizontal displacement of the left-hand nodes, $(\delta_1^u - \delta_1^l)$, is simply $h\theta_i$. For long webs ($L \gg h$) this would be a satisfactory approximation, but if the web is short and deep then some allowance should be made for the decrease in end rotation caused by shear deformation. In terms of shear strain γ the total horizontal displacement is

$$(\delta_1^u - \delta_1^l) = h(\theta_i - \gamma_i) \qquad (8.8.9)$$

The shear force in the web is constant and is equal to F_i. Therefore the shear strain is also constant and is given by

$$\gamma_i = \frac{F_i}{Ght_{wi}} \qquad (8.8.10)$$

At this point we must choose some system for dealing with a series of adjacent double wall boxes, each of which shares a common web in one or even two directions, in order to ensure that the thickness of the web t_{wi} and any other quantities are neither omitted nor double counted. Since we wish to allow each web to be of a different thickness and since the basic element node numbering is that of the panels, we choose to regard each of the webs of Fig. 8.66 as being half of the actual thickness. The other half thickness will then be accounted for when the adjacent box is processed. This system is also convenient for symmetric double wall structures, provided that the axis of symmetry coincides with a web. It would, however, be necessary to use the full thickness for the end webs of a double wall structure. Provision for this could be incorporated in the computer procedure which assembles the global stiffness matrix.

The allowance for shear deflection, plus allowing the four webs to be different thicknesses, means that there is no longer strict equality within the two groups of displacements of (8.8.3). However, the differences between the shear deflections (which is the only source of error) would be of second order. Hence we can ignore these differences and use an average value of shear deflection. Moreover, in most double wall structures the webs of each set of parallel beams are all of the same thickness. In this case, as long as we distinguish between the two sets (as denoted by i and j) and perform the averaging separately within each set then there is no error at all, and even if there is some difference in thickness the error involved is very small. Hence in (8.8.10) t_{wi} is to be interpreted as the average web thickness of the two "i" beams. Since we have chosen to use a half thickness, we must insert a $\frac{1}{2}$ in the denominator of (8.8.10). Hence the values of shear strain in the two sets of webs are

$$\gamma_i = \frac{2F_i}{Ght_{wi}}$$

$$\gamma_j = \frac{2F_j}{Ght_{wj}}$$

The angle of inclination of each pair of webs is $\theta_i = 2/L_i$ and $\theta_j = 2/L_j$. With the aid of (8.8.9) these expressions for θ and γ can be substituted into the four net horizontal displacements in (8.8.8). The result is

$$F = \frac{Eh^2\hat{t}}{2(1+\nu)}\left[\frac{2\alpha(1-\kappa_iF_i)}{L_i^2} + \frac{2(1-\kappa_jF_j)}{L_iL_j} + \frac{2(1-\kappa_iF_i)}{L_iL_j} + \frac{2(1-\kappa_jF_j)}{\alpha L_j^2}\right]$$

$$= \frac{2Eh^2\hat{t}}{(1+\nu)L_iL_j}(2 - \kappa_iF_i - \kappa_jF_j)$$

where $\kappa_i = \dfrac{L_i}{Ght_{wi}}$ and $\kappa_j = \dfrac{L_j}{Ght_{wj}}$

In this equation we observe that interchanging i and j makes no difference whatever and since F is the sum of F_i and F_j the symmetry indicates that $F_i = F_j = F/2$. Solving for F we obtain

$$F = \frac{4Eh^2\hat{t}}{(1+\nu)L_iL_j}R_s \qquad (8.8.11)$$

in which R_s is a reduction factor due to shear and is given by

$$R_s = \cfrac{1}{1 + \cfrac{2h\hat{t}\Upsilon}{\bar{t}_{wi}\bar{t}_{wj}} + \cfrac{\alpha^2 + 1/\alpha^2}{8(1+\nu)}}$$

where $\hat{t} = \dfrac{2t_u t_l}{t_u + t_l}$

$$\Upsilon = \frac{\bar{t}_{wi}}{L_i} + \frac{\bar{t}_{wj}}{L_j}$$

and $\alpha = \dfrac{L_i}{L_j}$

and in which $\bar{t}_{wi}$ and $\bar{t}_{wj}$ are reduced web thicknesses to allow for cutouts. A method for estimating this is given later, in (8.8.14).

Now that we have evaluated F we can calculate the matrix of additional stiffness terms which accounts for the closed-cell torsional rigidity and which must be added to the global stiffness matrix for each double wall box that occurs in the overall structure. Figure 8.67 shows the 20 degrees of freedom which are involved in the structural model of a typical box. For convenience, in order to relate more directly with the foregoing discussions, the eight in-plane displacements are numbered first. Next are the eight nodal rotations and finally the four vertical displacements of the nodes. With this choice of degree of freedom numbering, the terms which account for torsional rigidity all occur in the lower right-hand corner of the composite stiffness matrix, as shown in Fig. 8.68. For reference we denote this 4 × 4 submatrix as $\mathbf{k}_{dw}$.

The force F at each corner of the box is the result of a unit vertical displacement at each of the four nodes. It has been shown that for this imposed displacement the axial and bending stiffness terms produce no nodal force whatever. Therefore the desired torsional rigidity stiffness terms are uniquely related to the four vertical nodal forces. The force–displacement relationship $\mathbf{f} = \mathbf{k}_{dw}\boldsymbol{\delta}$ for this case is

$$\begin{Bmatrix} -F \\ F \\ -F \\ F \end{Bmatrix} = \begin{bmatrix} & & & \\ & \mathbf{k}_{dw} & & \\ & & & \end{bmatrix} \begin{Bmatrix} -1 \\ 1 \\ -1 \\ 1 \end{Bmatrix}$$

and from these four equations all of the terms of $\mathbf{k}_{dw}$ can be obtained.

After substituting for F from (8.8.11) the result is

$$\mathbf{k}_{dw} = \mathfrak{d} \begin{bmatrix} 1 & & & \text{symm.} \\ -1 & 1 & & \\ 1 & -1 & 1 & \\ -1 & 1 & -1 & 1 \end{bmatrix} \qquad (8.8.12)$$

in which $\mathfrak{d}$, the double wall stiffness parameter, is given by

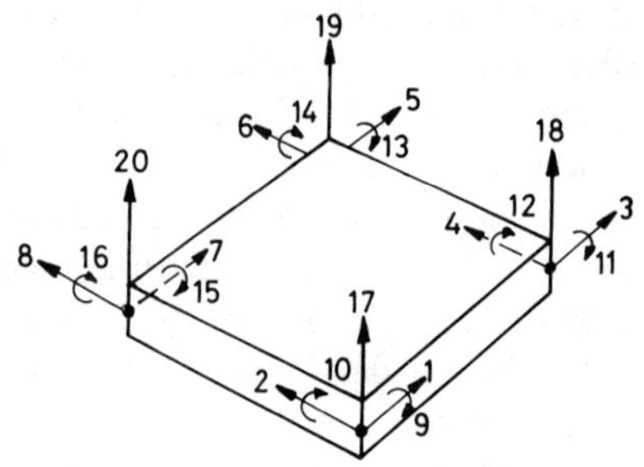

Figure 8.67 Degrees of freedom in a double wall box.

$$\mathfrak{d} = \frac{Eh^2\hat{t}}{(1 + \nu)L_iL_j} R_s \qquad (8.8.13)$$

The double wall stiffness parameter $\mathfrak{d}$ is analogous to the other stiffness parameters introduced earlier: $\mathfrak{a}$ for axial stiffness and $\mathfrak{b}$ for beam bending stiffness.

For webs with regularly spaced openings the reduced web thicknesses $\bar{t}_{wi}$ and $\bar{t}_{wj}$ can be estimated by a formula proposed by Kuhn [7] which has been shown [6] to be valid for a ship's double bottom structure:

$$\frac{\bar{t}_w}{t_w} = \left(1 - \frac{d_o}{e_o}\right)\left[1 - \left(\frac{d_o}{h}\right)^3\right] \qquad (8.8.14)$$

where d_o = diameter of openings
and e_o = spacing between openings.

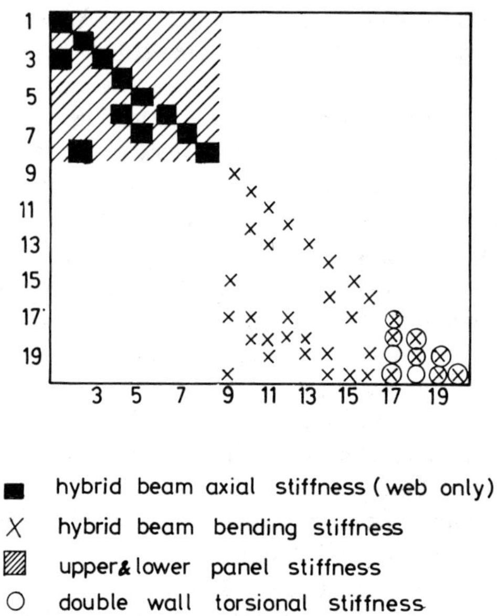

Figure 8.68 Stiffness components in a double wall box.

Calculation of Stress

After solving for the global nodal displacements $\mathbf{\Delta}_{hm}$ of the hull module, the next step is to calculate the stresses in the double wall structure. Therefore the relevant values of the global displacements are transformed to $\mathbf{\Delta}$, the nodal displacements expressed in double wall (X,Y,Z) coordinates. For each double wall box there would be four relevant vectors of nodal displacements, one for each of the corner nodes of the box. We have seen that a double wall structure involves two different types of element: the double-plate-flange hybrid beam element and the stiffened panel element. Each stiffened panel has three different roles which have been modeled separately: (1) it contributes direct, in-plane, biaxial stiffness, (2) it acts as a flange of each of the adjacent beam elements in each direction, and (3) it provides extra torsional rigidity. There are, accordingly, two different methods for calculating the stresses in a double wall structure: either retain the hybrid beam modeling, with its separate treatment of the three roles of the stiffened panel, and superimpose the three corresponding components of stress; or, transform the beam nodal displacements to stiffened panel nodal displacements and use the stiffened panel stress equation. This second method corresponds to the second method for hybrid beam elements which was given in Section 8.3. In that case the two methods were comparable, with the first method being somewhat simpler. However, for double wall structures the second method is decidedly more efficient, as shown by the following descriptions of the two.

Retain the Hybrid Beam Modeling and Calculate Three Separate Stress Components

In this approach the first two components of stress are calculated as in the ordinary hybrid beam modeling of Section 8.3. Then the extra shear stresses due to the torsional rigidity are calculated and the three stress components are summed. Accordingly there are four steps in this method:

1. For each double-plate-flange beam element, use its nodal location vector to select from $\mathbf{\Delta}$ the beam element nodal displacements $\boldsymbol{\delta}$ (in beam element coordinates). Then calculate the beam internal forces at the nodes from (5.3.18), of which there are three parts:
   a. Axial stress in the web.
   b. Shear force in the web.
   c. Bending moment in the flanges.

   The bending moment $\mathcal{M}_z$ is then converted to normal stress in the flanges: $(\sigma_{x,b})_u = (\mathcal{M}_z y_{p,u})/I_e$; $(\sigma_{x,b})_l = (\mathcal{M}_z y_{p,l})/I_e$.
2. Calculate the direct in-plane stresses in the neutral plane of the double wall structure. These stresses are calculated from the CSSR stress equation, (7.5.13). The in-plane nodal displacements u and v that are required in this equation are the U and W components of $\mathbf{\Delta}$, that is, the components in the neutral plane of the double wall structure (see Fig. 8.62).
3. Calculate the extra shear stress by transforming the beam nodal displacements $\mathbf{\Delta}$ to stiffened panel nodal displacements $\boldsymbol{\delta}^u$ and $\boldsymbol{\delta}^l$, and then use (7.5.13) to calculate the shear stress in the panel.
4. Determine the total stress in the stiffened panels by adding the bending stress in the flanges from step 1 to the other two stresses. For each double wall box there are eight groups of stresses to be summed in the upper panel and eight in the lower panel. However, half of these are in common with adjacent boxes in each direction and so to avoid duplication of effort the summation must be done in a systematic fashion.

Transform Directly to Panel Element Coordinates and Calculate Stresses from the SP Stress Equation

In this approach there are only two steps:

1. Calculate the axial stress and the shear force in each of the webs by the same technique as in the first method.
2. Transform the double wall nodal displacements $\mathbf{\Delta}$ to the stiffened panel nodal displacements $\boldsymbol{\delta}^u$ and $\boldsymbol{\delta}^l$, and use the CSSR stress equation, (7.5.13), to obtain the (total) stresses in each panel. This will automatically include whatever shear stress is present due to the extra torsional rigidity. Thus with this approach there is no need for a final summation of stresses.

It is obvious from the foregoing description that the second method is more efficient. The two methods give similar results, the main difference being that in the second method there is some smoothing of the beam bending stress because it is linked with that of the adjacent beams by the panel stress equation. But in fact this is more accurate because in double wall structures there is significantly more stress diffusion than in singly plated panels, mainly because of the greater torsional rigidity. For example, if there is a point load acting on one of the beams of a double wall structure, the stress level in the upper and lower panels near that beam is only moderately larger than the panel stress at the adjacent beams. This smaller and more gradual variation of stress is reproduced more accurately by the second of the above methods. This, coupled with the reduction in computation, makes the second method clearly superior.

To conclude this section, we give some further details of step 2 of the method, the calculation of the panel stresses. These are calculated for each pair of panel elements, upper and lower. It is assumed that as part of the global assembly process, nodal location vectors were defined for each pair of stiffened panel elements, relating the panel element nodes to the global structure nodes. For example, Fig. 8.69 shows a portion of a double wall structure which includes a pair (say, the ith pair) of stiffened panel elements. For these elements the nodal location vector would be

1	2	3	4
Q	R	B	A

The procedure for calculating panel stresses is:

1. Use the nodal location vector of panel i to select the nodal displacements at the global nodes which correspond to the element nodes of panel i and transform them to $\mathbf{\Delta}_i$ (in double wall coordinates X, Y, Z), grouping them into four displacement vectors corresponding to the four panel degrees of freedom, thus

$$\mathbf{\Delta}_i = \begin{Bmatrix} \overline{\mathbf{U}}_i \\ \overline{\mathbf{W}}_i \\ \overline{\boldsymbol{\theta}}_{Xi} \\ \overline{\boldsymbol{\theta}}_{Zi} \end{Bmatrix}$$

where

$$\overline{\mathbf{U}}_i = \begin{Bmatrix} U_Q \\ U_R \\ U_B \\ U_A \end{Bmatrix}; \qquad \overline{\mathbf{W}}_i = \begin{Bmatrix} W_Q \\ W_R \\ W_B \\ W_A \end{Bmatrix}; \text{ etc.}$$

2. Calculate the panel nodal displacements $\overline{\boldsymbol{\delta}}^u$ and $\overline{\boldsymbol{\delta}}^l$ from:

$$\overline{\boldsymbol{\delta}}^u = \begin{Bmatrix} \overline{\mathbf{u}} \\ \overline{\mathbf{v}} \end{Bmatrix}^u = \begin{bmatrix} \mathbf{I} & 0 & 0 & -Y_{p,u}\mathbf{I} \\ 0 & -\mathbf{I} & -\mathbf{I} & 0 \end{bmatrix} \mathbf{\Delta}_i$$

$$\overline{\boldsymbol{\delta}}^l = \begin{Bmatrix} \overline{\mathbf{u}} \\ \overline{\mathbf{v}} \end{Bmatrix}^l = \begin{bmatrix} \mathbf{I} & 0 & 0 & -Y_{p,l}\mathbf{I} \\ 0 & -\mathbf{I} & -\mathbf{I} & 0 \end{bmatrix} \mathbf{\Delta}_i$$

where

$$\overline{\mathbf{u}} = \begin{Bmatrix} u_1 \\ u_2 \\ u_3 \\ u_4 \end{Bmatrix} \quad \text{and} \quad \overline{\mathbf{v}} = \begin{Bmatrix} v_1 \\ v_2 \\ v_3 \\ v_4 \end{Bmatrix}$$

and $\mathbf{I}$ is the identity matrix (4×4). In the double wall coordinate system of Fig. 8.69, $Y_{p,l}$ would be a negative quantity.

3. Substitute the components of $\overline{\boldsymbol{\delta}}^u$, and then of $\overline{\boldsymbol{\delta}}^l$, into (7.5.13) to obtain the stresses in the upper and lower panels, respectively.

REFERENCES

1. D. Edinberg and W. A. Wood, "A SHIPOPT Analysis of the CG-49 Cruiser," Giannotti and Assoc. Report, Annapolis, MD, Oct. 1982.
2. Tingleff, "A Method for Simulating Concentrated Forces and

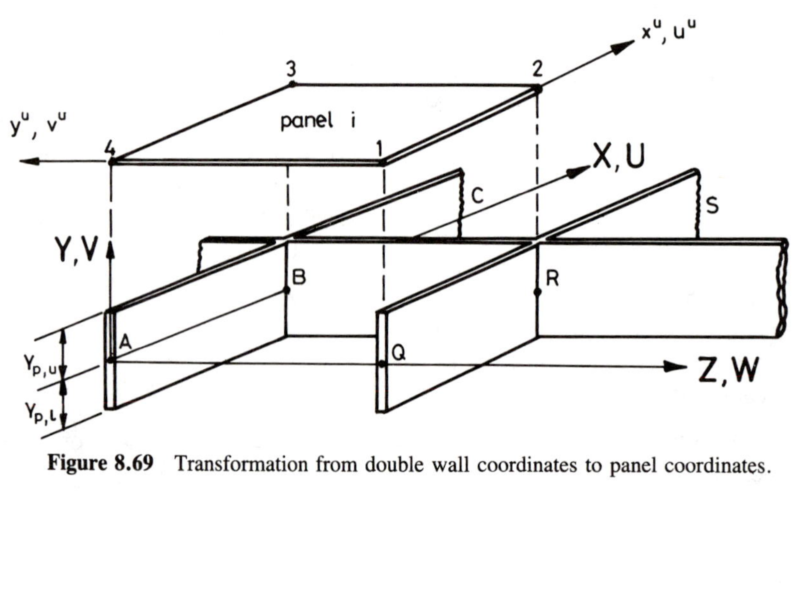

Figure 8.69 Transformation from double wall coordinates to panel coordinates.

Local Reinforcements in Stress Computation," *Mathematics of Finite Elements and Applicatons,* J. Whiteman, Ed., Academic Press, 1973, pp. 463–470.

3. J. Colville and J. Abbassi, "Plane Stress Reinforced Concrete Finite Elements," *J. Struc. Div., ASCE,* **100** (ST5), May 1974, pp. 1067–1083.
4. J. D. Davies, O. F. Hughes, and F. Mistree, "Finite Elements in the Design of Optimum Ship Structures," Report NavArch/76/2, Univ. of New South Wales, May, 1976.
5. T. Milchert, "Further Studies of Web Frames and their Beam Connections," Report No. E-42, Royal Institute of Technology, Div. of Naval Architecture, Stockholm, Dec. 1972.
6. D. G. Williams and J. C. Chapman, "Tests on a One-Eighth Scale Model of the Double Bottom Structure," *Trans. RINA,* 1974, pp. 329–346.
7. P. Kuhn, "The Strength and Stiffness of Shear Webs with Lightening Holes Having 45° Flanges," NACA Report No. ARR/323, 1942, Washington D.C.

Chapter

Nine

PLATE BENDING

9.1 SMALL DEFLECTION THEORY

"Long" Plates (Cylindrical Bending)

Unlike a beam, in which bending occurs only along the length, the bending in a plate usually occurs in two orthogonal directions. An equation relating the deflections to the loading can be developed for the plate, as for the beam. To show the similarity (and the differences) let us begin with the case of a plate that is bent about one axis only (cylindrical bending) as occurs for long plates ($a \gg b$). An elemental strip of such plating of width da is shown in Fig. 9.1. If this strip were an isolated beam its transverse section would deform as shown by the dashed lines, due to Poisson's ratio effects. This is termed "anticlastic" curvature, and it may be shown that this radius of curvature is $1/\nu$ times the primary or bending radius of curvature. However, in plating this transverse deformation does not occur because such a deformation would require that the plate take on a saddle shape, which would mean considerable stretching of the neutral surface and would require enormous strain energy. The prevention of this transverse strain ($\varepsilon_y = 0$) gives rise to a transverse stress $\sigma_y = \nu\sigma_x$, as may be seen from the strain equations:

$$\varepsilon_x = \frac{\sigma_x}{E} - \nu\frac{\sigma_y}{E}$$

$$\varepsilon_y = \frac{\sigma_y}{E} - \nu\frac{\sigma_x}{E} = 0$$

From the second equation we have $\sigma_y = \nu\sigma_x$ and

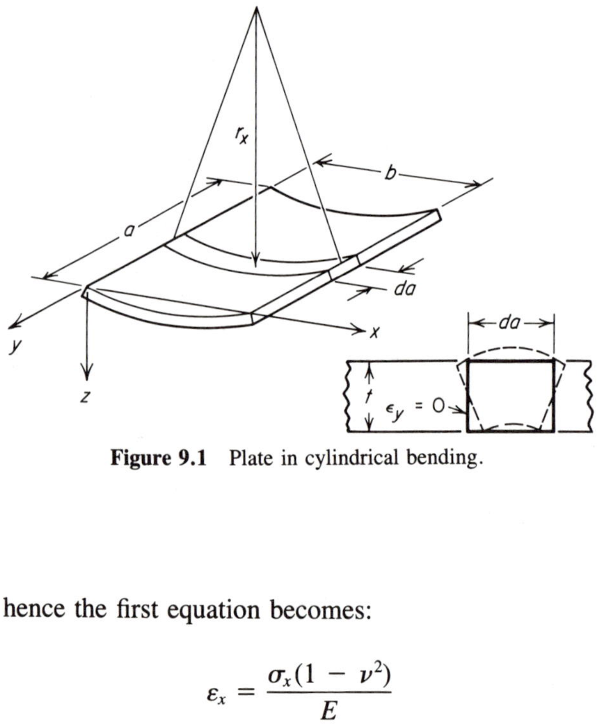

Figure 9.1 Plate in cylindrical bending.

hence the first equation becomes:

$$\varepsilon_x = \frac{\sigma_x(1 - \nu^2)}{E}$$

and

$$\sigma_x = \frac{E}{1 - \nu^2}\,\varepsilon_x \qquad (9.1.1)$$

The latter equation, when compared to the case of a beam ($\sigma_x = E\varepsilon_x$) suggests the definition of a quantity $E' = E/(1 - \nu^2)$, which could be regarded as an "effective" modulus of elasticity. Obviously, this effective modulus is always greater than E, and it may be concluded that a plate is always "stiffer" than a row of beams. The effective modulus is a useful parameter because for a long prismatically loaded plate the effect of this extra stiffness may be fully accounted for by using E' in place of E in all of the various beam deflection formulas (the width of the beam being taken as unity). Thus, for example, the expression for maximum deflection of the long, simply supported plate of Fig. 9.1 due to a lateral pressure p is

$$w_{max} = \frac{5pb^4}{384E'I} = \frac{5pb^4(1-\nu^2)}{32Et^3} \tag{9.1.2}$$

and for the clamped case

$$w_{max} = \frac{pb^4}{384E'I} = \frac{pb^4(1-\nu^2)}{32Et^3} \tag{9.1.3}$$

As with beams, the moment–curvature relation may be obtained by imposing equilibrium of moments over the cross section of the strip of plating (again taking the width as unity).
External bending moment = moment of stress forces.

$$M = \int_{-t/2}^{t/2} \sigma_x z \, dz = \int_{-t/2}^{t/2} \frac{E}{1-\nu^2} \varepsilon_x z \, dz$$

The bending strain is $\varepsilon_x = z/r_x$ and hence

$$M = \int_{-t/2}^{t/2} \frac{E}{1-\nu^2}\left(\frac{z^2}{r_x}\right) dz = \frac{Et^3}{12(1-\nu^2)}\left(\frac{1}{r_x}\right)$$
$$= \frac{D}{r_x} \tag{9.1.4}$$

where

$$D = \frac{Et^3}{12(1-\nu^2)} \tag{9.1.5}$$

Thus, for a plate, the constant of proportionality between moment and curvature is D; this is referred to as the *flexural rigidity* of the plate, and is analogous to the quantity EI in beam theory. This may also be seen by substituting E' for E and noting that for a strip of plating of unit width, $I = t^3/12$.

The radius of curvature of the plate can be expressed in terms of the deflection w of the plate thus:

$$-\frac{\partial^2 w}{\partial x^2} = \frac{1}{r_x} \tag{9.1.6}$$

so that from (9.1.4) the relation between the bending moment and the deflection is

$$-D\frac{\partial^2 w}{\partial x^2} = M$$

and this corresponds exactly to the expression relating bending moment and deflection for a beam, with D substituted for EI.

In a beam, once the maximum bending moment has been determined, the corresponding maximum stress is given by $\sigma = Mc/I$, and this will also be true for a unit strip in a long plate. The section modulus I/c for a unit strip of plating is

$$\frac{\frac{t^3}{12}}{\frac{t}{2}} = \frac{t^2}{6}$$

For a uniform pressure p the maximum bending moment in the unit strip of plating is proportional to pb^2, and hence the maximum stress is proportional to $pb^2/(t^2/6)$. This is usually expressed in the form

$$\sigma_{max} = kp\left(\frac{b}{t}\right)^2 \tag{9.1.7}$$

and the value of the coefficient k depends on the boundary conditions. For simply supported edges $k = \frac{3}{4}$ and for clamped edges $k = \frac{1}{2}$. As we shall see, this same form of equation is used for all plates, whether they are long or not, and the coefficient k also accounts for the effect of aspect ratio a/b.

Derivation of the Plate Bending Equation

In this section we derive the basic equation governing the behavior of panels of plating under lateral load. This equation was first derived by Lagrange in 1811; the outline which follows is adapted from Timoshenko [1] and Jaeger [2]. This theory is only applicable if:

1. Plane cross sections remain plane.
2. The deflections of the plate are small (w_{max} not exceeding $\frac{3}{4}t$).
3. The maximum stress nowhere exceeds the plate yield stress (i.e., the material remains elastic).

In general, a panel of plating will have curvature in two directions at right angles. Let the radii of curvature

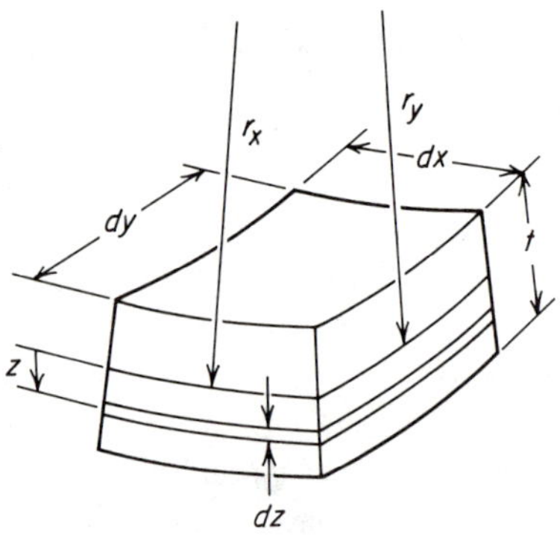

Figure 9.2 Differential element of plating.

in these two directions be r_x and r_y respectively. It follows that if a small element of length dx and breadth dy is considered, as shown in Fig. 9.2, there will be distributed moments m_x and m_y (moments per unit length) along the edges.

The values of these moments can be obtained as follows:

$$\varepsilon_x = \frac{\sigma_x}{E} - \frac{\nu\sigma_y}{E}$$

$$\varepsilon_y = \frac{\sigma_y}{E} - \frac{\nu\sigma_x}{E}$$

Multiplying the second of these equations by ν and adding gives

$$\frac{\sigma_x}{E}(1 - \nu^2) = \varepsilon_x + \nu\varepsilon_y$$

Assumptions 1 and 2 stated previously give rise to the strain–curvature relations

$$\varepsilon_x = -z\left(\frac{\partial^2 w}{\partial x^2}\right) \quad \text{and} \quad \varepsilon_y = -z\left(\frac{\partial^2 w}{\partial y^2}\right)$$

Hence,

$$\sigma_x = \frac{E}{1 - \nu^2}(-z)\left(\frac{\partial^2 w}{\partial x^2} + \nu\frac{\partial^2 w}{\partial y^2}\right)$$

Equilibrium of internal and external moments gives

$$m_x = \int_{-t/2}^{t/2} \sigma_x z\, dz$$

$$= -\int_{-t/2}^{t/2} \frac{E}{1 - \nu^2}\left(\frac{\partial^2 w}{\partial x^2} + \nu\frac{\partial^2 w}{\partial y^2}\right)z^2\, dz$$

$$= -D\left(\frac{\partial^2 w}{\partial x^2} + \nu\frac{\partial^2 w}{\partial y^2}\right) \qquad (9.1.8a)$$

Similarly,

$$m_y = -D\left(\frac{\partial^2 w}{\partial y^2} + \nu\frac{\partial^2 w}{\partial x^2}\right) \qquad (9.1.8b)$$

Suppose now that the intensity of lateral load per unit area on the plate is p; then the load on the element will be $p\ dx\ dy$. This load is carried by the distributed shear forces q acting on the four edges of the element, as shown in Fig. 9.3. In the general case of bending, twisting moments will also be generated on all the four faces, these being denoted by m_{xy} and m_{yx} (again, per unit length). The complete system of forces and moments on the element is shown in Fig. 9.3. Note that all forces and moments are *per unit length* along the edges of the element.

From equilibrium of vertical forces, it may be seen that

$$\left(q_x + \frac{\partial q_x}{\partial x}dx\right)dy - q_x\, dy + \left(q_y + \frac{\partial q_y}{\partial y}dy\right)dx - q_y\, dx + p\, dx\, dy = 0$$

or

$$\frac{\partial q_x}{\partial x} + \frac{\partial q_y}{\partial y} + p = 0 \qquad (9.1.9)$$

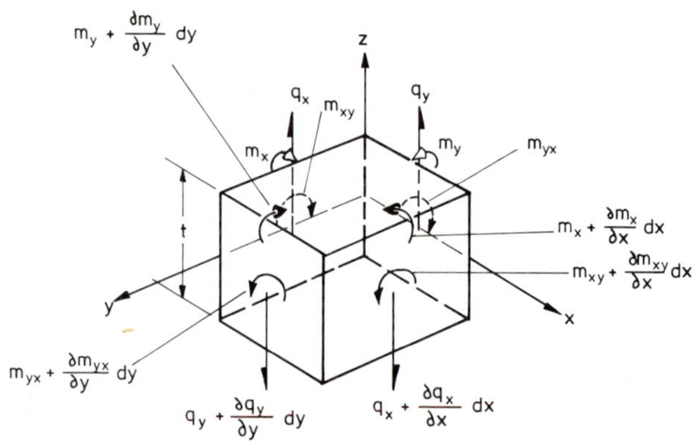

Figure 9.3 Forces and moments in a plate element.

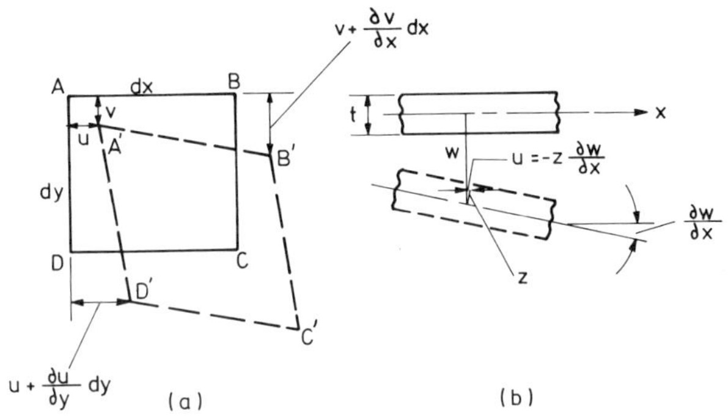

Figure 9.4 Shear strain and plate deflection.

Taking moments parallel to the x-axis,

$$\left(m_{xy} + \frac{\partial m_{xy}}{\partial x}\,dx\right)dy - m_{xy}\,dy + m_y\,dx$$
$$- \left(m_y + \frac{\partial m_y}{\partial y}\,dy\right)dx + q_y\,dx\,dy = 0$$

or

$$\frac{\partial m_{xy}}{\partial x} - \frac{\partial m_y}{\partial y} + q_y = 0 \qquad (9.1.10)$$

Similarly, taking moments parallel to the y-axis leads to

$$\frac{\partial m_{yx}}{\partial y} + \frac{\partial m_x}{\partial x} - q_x = 0 \qquad (9.1.11)$$

Because of the principle of complementary shear stress, it follows that $m_{yx} = -m_{xy}$, so that (9.1.11) becomes

$$-\frac{\partial m_{xy}}{\partial y} + \frac{\partial m_x}{\partial x} - q_x = 0 \qquad (9.1.12)$$

Substituting for q_x and q_y in (9.1.9) from (9.1.10) and (9.1.12)

$$\frac{\partial}{\partial x}\left(-\frac{\partial m_{xy}}{\partial y} + \frac{\partial m_x}{\partial x}\right) + \frac{\partial}{\partial y}\left(\frac{\partial m_y}{\partial y} - \frac{\partial m_{xy}}{\partial x}\right) + p = 0$$
$$\frac{\partial^2 m_x}{\partial x^2} - 2\,\frac{\partial^2 m_{xy}}{\partial x \partial y} + \frac{\partial^2 m_y}{\partial y^2} + p = 0 \qquad (9.1.13)$$

It is now necessary to determine the twisting moment m_{xy} in terms of the deflection. In Fig. 9.4a it will be seen that, if a point A in the plate a distance z from the neutral surface is displaced a distance v in the y-direction, the displacement of a nearby point at $x + dx$ will be $v + (\partial v/\partial x)dx$, so that the change in slope of the line AB will be

$$\frac{v + \dfrac{\partial v}{\partial x}\,dx - v}{dx} = \frac{\partial v}{\partial x}$$

Similarly, the change in slope of the line AD is $\partial u/\partial y$.

The rectangular element ABCD is then changed in shape to a parallelogram A′B′C′D′ and the shear strain is

$$\gamma = \frac{\partial v}{\partial x} + \frac{\partial u}{\partial y}$$

The corresponding shear stress is

$$\tau = G\left(\frac{\partial v}{\partial x} + \frac{\partial u}{\partial y}\right) \qquad (9.1.14)$$

Now from Fig. 9.4b it will be seen that

$$u = -z\,\frac{\partial w}{\partial x}$$

and similarly

$$v = -z\,\frac{\partial w}{\partial y}$$

Making these substitutions in (9.1.14) gives

$$\tau = -2Gz\,\frac{\partial^2 w}{\partial x \partial y}$$

The twisting moment per unit length can now be obtained from equilibrium:

$$m_{xy} = -\int_{-t/2}^{t/2} -2G\,\frac{\partial^2 w}{\partial x \partial y}\,z^2\,dz = \frac{Gt^3}{6}\,\frac{\partial^2 w}{\partial x \partial y}$$

and substituting

$$G = \frac{E}{2(1 + \nu)}$$

gives

$$m_{xy} = \frac{Et^3}{12(1+\nu)}\frac{\partial^2 w}{\partial x \partial y} = \frac{Et^3(1-\nu)}{12(1+\nu)(1-\nu)}\frac{\partial^2 w}{\partial x \partial y}$$

$$= D(1-\nu)\frac{\partial^2 w}{\partial x \partial y} \qquad (9.1.15)$$

It is now possible to substitute for m_x, m_y, and m_{xy} in (9.1.13) using (9.1.8) and (9.1.15)

$$\frac{\partial^2}{\partial x^2}\left[-D\left(\frac{\partial^2 w}{\partial x^2} + \nu\frac{\partial^2 w}{\partial y^2}\right)\right] - 2\frac{\partial^2}{\partial x \partial y}\left[D(1-\nu)\frac{\partial^2 w}{\partial x \partial y}\right] + \frac{\partial^2}{\partial y^2}\left[-D\left(\frac{\partial^2 w}{\partial y^2} + \nu\frac{\partial^2 w}{\partial x^2}\right)\right] = -p$$

or finally

$$\frac{\partial^4 w}{\partial x^4} + 2\frac{\partial^4 w}{\partial x^2 \partial y^2} + \frac{\partial^4 w}{\partial y^4} = \frac{p}{D} \qquad (9.1.16)$$

This is the equation of equilibrium for the plate. It assumes that the deflection is small and that there is no stretching of the middle plane, that is, that membrane effects are absent. Strictly speaking, this implies that the edges of the plate are free to slide in the plane of the plate. However, if the plate edges are restrained from moving, (9.1.16) will still be applicable provided the plate deflection is sufficiently small for the resulting membrane tensions to be neglected. This is usually the case when the maximum deflection does not exceed $\frac{3}{4}t$. Equation (9.1.16) conforms to a type of differential equation known as the biharmonic equation and is often abbreviated as $\nabla^4 w = p/D$.

Boundary Conditions

The solution of (9.1.16) must be such as to satisfy the boundary conditions and therefore the basic task is to find an expression for $w(x,y)$ which will satisfy these conditions. The types of restraint around the boundary of a plate can be idealized as follows:

1. Simply supported—edges free to rotate and to move in the plane of the plate.
2. Pinned—edges free to rotate but not free to move in the plane of the plate.
3. Clamped but free to slide—edges not free to rotate but free to move in the plane of the plate.
4. Rigidly clamped—edges not free to rotate or move in the plane of the plate.

The prevention of in-plane movement (no "pulling in" of the edges) of conditions 2 and 4 can only occur if the structure supporting the plate is very rigid in this direction. In most plated frame structures, and particularly in vehicles and free-standing structures, the individual panels of plating receive relatively little restraint against edge pull-in because such restraint would have to come ultimately from the frames at the edges of the overall stiffened panel, and beam bending stiffness is generally insufficient to provide such rigid in-plane support. This has been verified for ship panels by Clarkson's extensive experimental work [3]. Therefore conditions 1 and 3 are generally applicable. The former would be appropriate when a load acts on a single panel of plating because neither the stiffeners nor the surrounding panels would provide much rotational restraint. Condition 3 would be appropriate for a distributed pressure loading that extends over several panels.

Solution of Special Cases

Simply Supported Plates

The earliest solution to (9.1.16) is that of Navier, who solved the simply-supported case by using a Fourier Series to represent the load $p(x,y)$. The general expression for the load is

$$p(x,y) = \sum_{m=1}^{\infty}\sum_{n=1}^{\infty} A_{mn} \sin\frac{m\pi y}{a} \sin\frac{n\pi x}{b} \qquad (9.1.17)$$

and the coefficient A_{mn} can be obtained by Fourier analysis for any particular load condition. For example, for the case of a uniform pressure p_0 it may be shown that the coefficient A_{mn} is given by

$$A_{mn} = \frac{16p_0}{\pi^2 mn} \qquad (9.1.18)$$

It may be readily shown that for the biharmonic equation a sinusoidal load distribution produces a sinusoidal deflection. That is, the general solution to the foregoing case, satisfying both (9.1.16) and the simply supported boundary conditions, is

$$w = \sum_{m=1}^{m=\infty} \sum_{n=1}^{n=\infty} B_{mn} \sin \frac{m\pi y}{a} \sin \frac{n\pi x}{b} \quad (9.1.19)$$

To find the value of the coefficient in (9.1.19) this, together with (9.1.17) and (9.1.18), is substituted in (9.1.16). This gives, on dropping the common $(\sin m\pi y/a)(\sin n\pi x/b)$ term,

$$B_{mn}\left(\frac{\pi^4 m^4}{a^4} + \frac{2\pi^4 m^2 n^2}{a^2 b^2} + \frac{\pi^4 n^4}{b^4}\right) = \frac{16 p_0}{\pi^2 Dmn}$$

and it will thus be seen that

$$B_{mn} = \frac{16 p_0}{\pi^6 Dmn\left(\dfrac{m^2}{a^2} + \dfrac{n^2}{b^2}\right)^2} \quad (9.1.20)$$

The deflection is then given by

$$w = \sum_{m=1}^{m=\infty} \sum_{n=1}^{n=\infty} \frac{16 p_0}{\pi^6 Dmn\left(\dfrac{m^2}{a^2} + \dfrac{n^2}{b^2}\right)^2} \sin \frac{m\pi y}{a} \sin \frac{n\pi x}{b} \quad (9.1.21)$$

in which m and n, due to the symmetry of the problem, need only take odd values. Values of w_{max} are given in Fig. 9.5 for various values of aspect ratio.

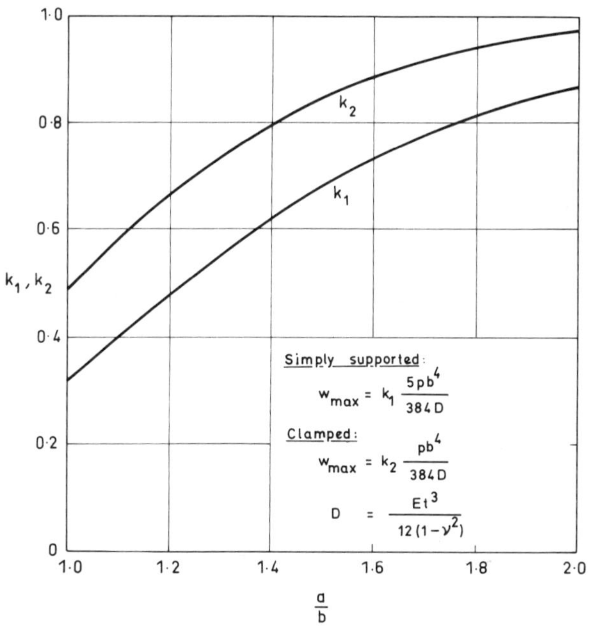

Figure 9.5 Maximum deflection of rectangular plates under uniform pressure (elastic small-deflection theory).

To determine the bending stress in the plate it is first necessary to calculate the bending moments m_x and m_y. From (9.1.21)

$$\frac{\partial^2 w}{\partial x^2} = -\sum_{m=1}^{m=\infty} \sum_{n=1}^{n=\infty} \frac{16 p_0}{\pi^6 Dmn\left(\dfrac{m^2}{a^2} + \dfrac{n^2}{b^2}\right)^2} \frac{\pi^2 n^2}{b^2} \sin \frac{m\pi y}{a} \sin \frac{n\pi x}{b}$$

$$= -\sum_{m=1}^{m=\infty} \sum_{n=1}^{n=\infty} \frac{16 p_0 n}{\pi^4 Dmb^2\left(\dfrac{m^2}{a^2} + \dfrac{n^2}{b^2}\right)^2} \sin \frac{m\pi y}{a} \sin \frac{n\pi x}{b}$$

$$\frac{\partial^2 w}{\partial y^2} = -\sum_{m=1}^{m=\infty} \sum_{n=1}^{n=\infty} \frac{16 p_0 m}{\pi^4 Dna^2\left(\dfrac{m^2}{a^2} + \dfrac{n^2}{b^2}\right)^2} \sin \frac{m\pi y}{a} \sin \frac{n\pi x}{b}$$

Then, for example,

$$m_x = -D\left(\frac{\partial^2 w}{\partial x^2} + \frac{\nu \partial^2 w}{\partial y^2}\right)$$

$$= \sum_{m=1}^{m=\infty} \sum_{n=1}^{n=\infty} \frac{16 p_0}{\pi^4 mn\left(\dfrac{m^2}{a^2} + \dfrac{n^2}{b^2}\right)^2} \left(\frac{n^2}{b^2} + \nu \frac{m^2}{a^2}\right) \sin \frac{m\pi y}{a} \sin \frac{n\pi x}{b}$$

The curvature, and hence the bending moment, will always be greater across the shorter span. By convention, the symbol b is used for the shorter dimension (making the aspect ratio a/b always greater than or equal to unity). Thus, in the present case m_x is the larger of the two moments and has its maximum value in the center of the plate. Values of maximum stress for this case and for the case of a clamped plate are given in Fig. 9.6.

Clamped Plates

The solution for a clamped plate is more complicated. The usual methods are the energy (or Ritz) method and the method of Levy. The energy method, while giving

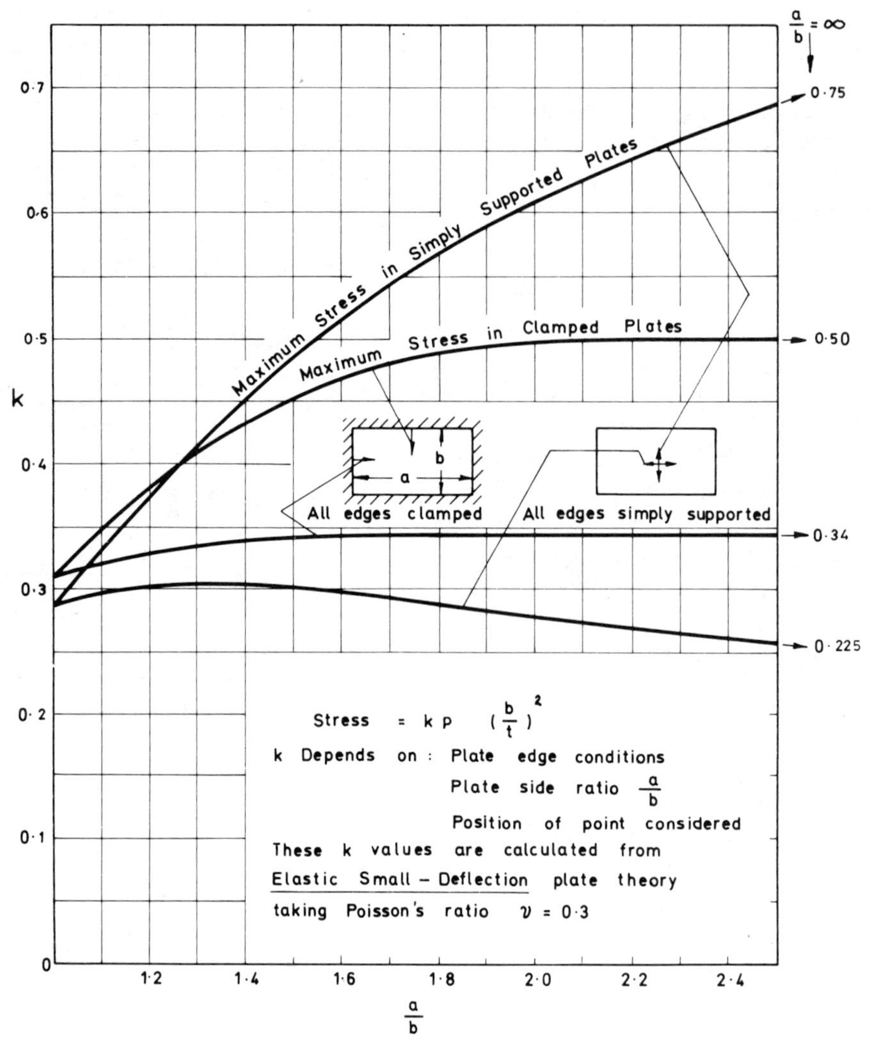

Figure 9.6 Stresses in rectangular plates under uniform lateral pressure.

only approximate results, has the advantage that the assumed deflected surface need not satisfy the governing differential equation while it must satisfy the boundary conditions. The Levy type solution, as described by Timoshenko [1] is achieved by the superposition of three loading systems applied to a simply supported plate: (1) uniformly distributed hydrostatic pressure, (2) moments distributed along the short edges, and (3) moments distributed along the long edges. If w_1, w_2, and w_3 denote the deflections along the appropriate edges for these three cases, the boundary conditions are:

$$\text{short edges:} \left(\frac{\partial w_1}{\partial y}\right)_{y=\pm a/2} + \left(\frac{\partial w_2}{\partial y} + \frac{\partial w_3}{\partial y}\right)_{y=\pm a/2} = 0$$

$$\text{long edges:} \left(\frac{\partial w_1}{\partial x}\right)_{x=\pm b/2} + \left(\frac{\partial w_2}{\partial x} + \frac{\partial w_3}{\partial x}\right)_{x=\pm b/2} = 0$$

where $x = 0$, $y = 0$ is at the center of the plate.

The various deflection components are each represented by trigonometric series, and the foregoing boundary conditions result in two systems of equations each having m unknowns, where m is the number of terms kept in the trigonometric series. The mathematics is quite involved but the answers are well established. The maximum deflection for the case of a uniform load is given in Fig. 9.5. The maximum bending moment occurs at the edge of the plate, and the absolute maximum occurs at the midpoint of the long side. Thus, in terms of the foregoing coordinate system the greatest bending moment is m_x at $x = \pm b/2$; $y = 0$.

The corresponding values of maximum stress at these points are given in Fig. 9.6.

It may be seen from Fig. 9.6 that the effect of aspect ratio is smaller for clamped plates, and beyond about $a/b = 2$ such a plate behaves essentially as a clamped strip and the influence of aspect ratio is negligible.

9.2 COMBINED BENDING AND MEMBRANE STRESSES—ELASTIC RANGE

Large-Deflection Plate Theory

In many cases the use of small-deflection theory for the design of laterally loaded plating leads to thicknesses which are excessive. Small-deflection theory fails to allow for membrane stresses that arise when the deflection becomes large and/or when the edges are prevented from pulling in. With large deflections the plate is no longer a developable surface and the deflection requires in-plane stretching and compression of the plate. If the edges are prevented from pulling in, membrane action becomes significant when the lateral deflection (whether initial deflection or due to load) exceeds approximately half of the plate thickness. If the edges are free to pull in (which is the appropriate assumption for ship plating) membrane action becomes significant somewhat later, typically at a deflection $w > t$. As the deflection increases, an increasing proportion of the load is carried by this membrane action. This situation, in which the lateral load is supported by both bending and membrane action, requires a more comprehensive plate theory, usually referred to as "large-deflection" plate theory. The differential equations for large-deflection plate theory were formulated by von Karman. One of these, the equation relating load to deflection, is a generalization of (9.1.16) and may be derived by considering vertical equilibrium of an element $dx\, dy$, as was done in deriving (9.1.9). Let us denote the membrane tensions and membrane shearing force per unit width as N_x, N_y, and N_{xy}. Taking the x-direction, it can be seen from Fig. 9.7 that at side (1) the vertical component of the ten-

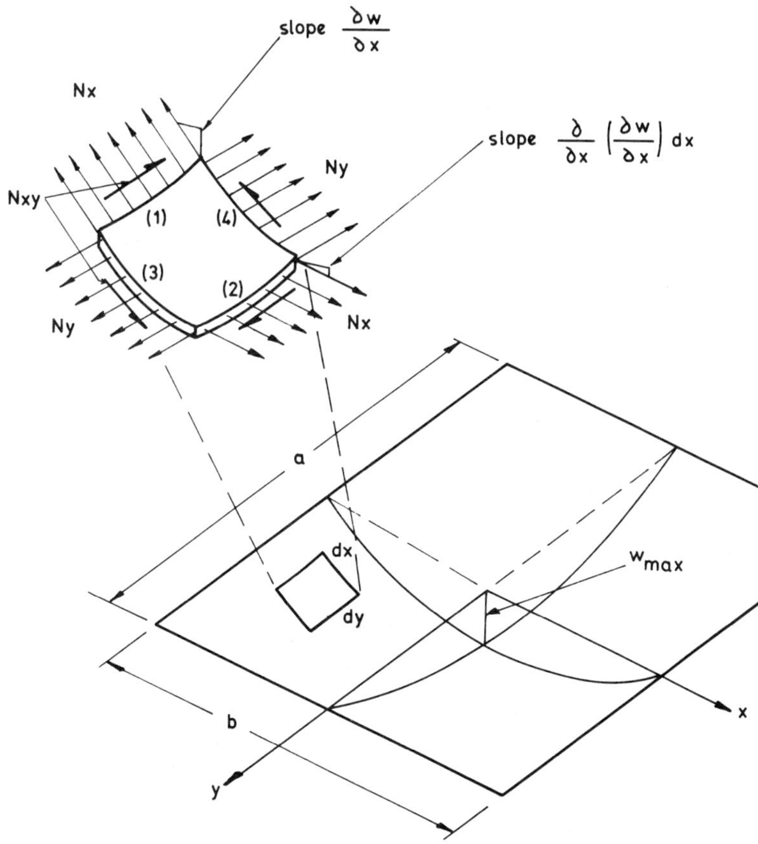

Figure 9.7 Plating with membrane stress.

sion force is $-N_x\, dy\, (\partial w/\partial x)$. The vertical force at side (2) would be the same magnitude (but opposite sign) if the slope $\partial w/\partial x$ were the same, but in general it is not. Therefore we write this component as

$$+ N_x\, dy \left[\frac{\partial w}{\partial x} + \frac{\partial}{\partial x}\left(\frac{\partial w}{\partial x}\right) dx \right]$$

Then the net vertical force is $N_x(\partial^2 w/\partial x^2)\, dx\, dy$. Similarly, in the y-direction the net vertical force is $N_y(\partial^2 w/\partial y^2)\, dx\, dy$.

The vertical component of the shear force along sides (3) and (4) is

$$N_{xy}\, dx \left[\frac{\partial w}{\partial x} + \frac{\partial}{\partial y}\left(\frac{\partial w}{\partial x}\right) dy \right] - N_{xy}\, dx \left(\frac{\partial w}{\partial x}\right)$$

and a similar expression is obtained for the other two sides. Collecting results, we find that the new terms to be added to (9.1.9) are

$$\left(N_x \frac{\partial^2 w}{\partial x^2} + 2N_{xy} \frac{\partial^2 w}{\partial x \partial y} + N_y \frac{\partial^2 w}{\partial y^2} \right) dx\, dt$$

Thus (9.1.16) becomes

$$\nabla^4 w = \frac{1}{D}\left(p + N_x \frac{\partial^2 w}{\partial x^2} + 2N_{xy} \frac{\partial^2 w}{\partial x \partial y} + N_y \frac{\partial^2 w}{\partial y^2} \right) \tag{9.2.1}$$

In general, N_x, N_y, and N_{xy} are functions of x and y and hence (9.2.1) does not provide sufficient information for solution except in special cases where these membrane forces can be separately calculated. The other equation formulated by von Karman is given in Section 101 of Ref. 1. With these two equations, a complete solution can be achieved. However, except for a few simple cases, precise mathematical solutions to these equations are very difficult. Reference 4 presents solutions to a variety of cases.

In ship structures the principal application of (9.2.1) is for plating that is subjected to an in-plane compressive load N_x, which brings with it the possibility of buckling. In this case since N_x is an applied load it is a known quantity. The same may be true of N_y and N_{xy}, or they may continue to be membrane forces arising from large deflections and boundary restraints. In many practical situations N_y and N_{xy} may be taken as zero and then the solution of (9.2.1) presents little problem. The whole question of plate buckling is dealt with in Chapter 12.

Membrane Tension (Edges Restrained Against Pull-in)

The relative magnitude of membrane effects depends on two things: the degree of lateral deflection or "curvature" of the plate surface, whether present initially or due to the action of the lateral load, and the degree to which the edges are restrained from pulling in. Both factors cause the development of membrane tension in the plate, and if the edges are restrained then at large deflections ($w > 1.5t$, say) a high proportion of the lateral load is supported by this means rather than by plate bending. On the other hand, so long as a plate is perfectly flat there are no membrane effects whatever because it is only the normal (or lateral) component of the membrane tension which can carry a portion of the normal (or lateral) pressure load. Likewise, if the plate edges are free to pull in, then no significant membrane tension can arise, and the only membrane effect is the in-plane strain (tensile in the central portion and circumferential compression at the plate edges) which arises because of geometric compatibility. This type of membrane effect only becomes significant for medium to large deflections ($w > t$, say). Note that it cannot arise at all for a very long plate because a cylindrical deflected shape is a developable surface, having no geometric incompatibilities.

As explained in Section 9.1, in ship plating there is relatively little restraint against edge pull-in as long as the deflections are not large. For large deflections such restraint does become significant but in many cases the load required to cause such large deflections would have already caused failure in the stiffeners or beams which support the plating. Nevertheless there are some situations in which these large deflections can be permitted, and in these cases the use of membrane tension can give substantial weight savings. Therefore it is important to have at least some idea of the relative magnitude of membrane tension which can arise if the edges are restrained from pulling in. This can be obtained by examining the case of a unit-width strip of laterally loaded plating with edges prevented from approaching. If $w(x)$ is the deflection of the strip, the difference δ between the arc length of the deflected strip and the original straight length is

$$\delta = \int_0^b \left(\sqrt{1 + \left(\frac{dw}{dx}\right)^2} - 1 \right) dx$$

$$\delta \cong \int_0^b \frac{1}{2}\left(\frac{dw}{dx}\right)^2 dx \tag{9.2.2}$$

In this case the membrane force is unidirectional ($N_y = N_{xy} = 0$). We shall also assume that it is constant over the length. This is a reasonable approximation, and becomes more accurate as deflection increases and the plate becomes more of a "pure membrane." If we call this constant membrane tension T, the extension due to tension is Tb/AE where A, the cross-sectional area per unit width, is equal to t. It follows from (9.2.2) that

$$\frac{Tb}{tE} = \int_0^b \frac{1}{2}\left(\frac{dw}{dx}\right)^2 dx$$

or

$$T = \frac{Et}{2b}\int_0^b \left(\frac{dw}{dx}\right)^2 dx \qquad (9.2.3)$$

Let us assume the strip to be pinned and, in order to investigate the effect of initial deflection, let us assume that there is an initial deflection w_0. Let us further assume that both w_0 and the deflection w_1 due to the load are sinusoidal, so that the total deflection in the loaded condition is

$$w(x) = (w_0 + w_1)\sin\frac{\pi x}{b}$$

From (9.2.2) the total change in length is

$$\delta = \int_0^b \frac{1}{2}(w_0 + w_1)^2 \frac{\pi^2}{b^2}\cos^2\frac{\pi x}{b}\, dx$$

$$= \frac{\pi^2}{4b}(w_0 + w_1)^2$$

Similarly, the change in length due to the initial deflection is

$$\frac{\pi^2 w_0^2}{4b}$$

and the change in length due to the loading is the difference between these two:

$$\delta_1 = \frac{\pi^2(2w_0w_1 + w_1^2)}{4b}$$

Using this value, Muckle [5] obtained a solution for w_1 using an energy method as follows:

Strain energy due to *bending*

$$= \int_0^b \frac{E'I}{2}\left(\frac{d^2w}{dx^2}\right)^2 dx$$

$$= \int_0^b \frac{E'I}{2}(w_1)^2\frac{\pi^4}{b^4}\sin^2\frac{\pi x}{b}\, dx$$

$$= \frac{\pi^4 E'I(w_1)^2}{4b^3}$$

$$\text{Strain energy due to } \textit{tension} = \frac{1}{2}T\delta_p = \frac{1}{2}\frac{AE}{b}\delta_1^2$$

In general, for a deflection $w(x)$ due to a load $p(x)$ per unit length, the work done is

$$W = \int_0^b \frac{1}{2}pw\, dx$$

In the present case, p is a constant and the deflection due to the load is $w_1 \sin(\pi x/b)$. Therefore,

$$W = \frac{pw_1b}{\pi}$$

The work done by the load is equal to the total strain energy, so that

$$\frac{pw_1b}{\pi} = \frac{AE\delta_1^2}{2b} + \frac{\pi^4E'I(w_1)^2}{4b^3}$$

$$= \frac{\pi^4}{32}\frac{AE}{b^3}(2w_0w_1 + w_1^2)^2 + \frac{\pi^4E'I(w_1)^2}{4b^3}$$

Hence

$$\frac{\pi^4AEw_1^3}{32b^3} + \frac{\pi^4AEw_0w_1^2}{8b^3} + \left(\frac{\pi^4AEw_0^2}{8b^3} + \frac{\pi^4E'I}{4b^3}\right)w_1 - \frac{pb}{\pi} = 0 \qquad (9.2.4)$$

For the strip $A = t$, $E' = E/(1-\nu^2)$ and $I = t^3/12$ per unit width.

On substituting these in (9.2.4), the expression becomes

$$w_1^3 + 4w_0w_1^2 + \left(4w_0^2 + \frac{2}{3}\frac{t^2}{1-\nu^2}\right)w_1 - \frac{32pb^4}{\pi^5Et} = 0 \qquad (9.2.5)$$

Effect of Initial Deformation

For plates which have no initial deflection $w_0 = 0$ and (9.2.5) becomes

$$w_1^3 + \frac{2}{3}\frac{t^2}{1-\nu^2}w_1 = \frac{32pb^4}{\pi^5 Et} \qquad (9.2.6)$$

For the initial stages of loading the deflection w_1 will be small relative to the thickness, and hence the first term may be neglected. This gives

$$\frac{w_1}{t} = \frac{48(1-\nu^2)p}{\pi^5 E}\left(\frac{b}{t}\right)^4 \qquad (9.2.7)$$

$$= 0.143\frac{p}{E}\left(\frac{b}{t}\right)^4$$

and this agrees almost exactly with (9.1.2) which was the result obtained from small-deflection theory, ignoring membrane action:

$$\frac{w_1}{t} = \frac{5(1-\nu^2)p}{32E}\left(\frac{b}{t}\right)^4$$

$$= 0.142\frac{p}{E}\left(\frac{b}{t}\right)^4$$

This illustrates the point made earlier that membrane action requires some deflection, either initial or due to load, and if there is no initial deflection then membrane action does not become significant until the deflection due to load approaches the plate thickness. This is further illustrated in Fig 9.8 which shows the stresses in a strip of initially flat ($w_0 = 0$) plating 0.75 m wide and 15 mm thick. It will be seen that as the load increases, the bending stress begins to depart from the value obtained when neglecting membrane effects. There is less deflection, and hence less bending (the plate is straighter, due to the tension T) and therefore the mean magnitude of the bending stress is always lower. However, on the tension side of the plate the membrane stress adds to the bending stress and thus the maximum tensile stress in the plate is slightly greater than the value obtained when membrane effects are neglected. Nevertheless the overall effect is small, being about 5% for a stress level of 100 MPa.

In contrast to this the figure also shows the stresses for the case of an initial deflection $w_0 = 7.5$ mm, that is, half the thickness of the plating. The mean magnitude of the bending stress is greatly reduced and it is obvious that even a small initial deformation allows membrane effects to occur as soon as the load is applied.

In stiffened plating there will always be some degree of inwards dishing of the plate panels bounded by the stiffeners and frames because of the shrinkage of the fillet welds along these boundaries, which pulls the plating inwards. If the plating is relatively thin this causes prominent parallel ridges on the outisde of the plating which are reminiscent of a gaunt rib cage, and the effect is commonly referred to as the "hungry horse" look. A similar type of initial deformation may be caused by previous loading which produced "permanent set," that is, a lateral deflection which involved plastic bending strain along the boundaries.

Although such initial deformation may be undesirable from consideration of appearance, fluid resistance and so on, it can have a beneficial influence on the elastic strength of plates, providing that the edges are restrained from pulling in. Defining "elastic strength" as the applied uniform loading which will just cause inelastic behavior, Table 9.1 taken from Clarkson [6] demonstrates the beneficial effect of initial deformation. The figures quoted are values of the constant C in the equation Elastic Strength $= C \times$ Nondimensional pressure coefficient.

TABLE 9.1 BENEFICIAL EFFECTS OF INITIAL DEFORMATION

Type of Initial Deformation	Elastic Strength	Source of Increase in Elastic Strength
Flat plate	1.59	—
Initial deflection (stress-free) equal to plating thickness	2.58	Membrane action
Initially flat plate dished to a permanent set equal to plating thickness	4.50	Membrane action plus residual stresses

As shown in Table 9.1, if a plate has already acquired some degree of permanent set (and hence "locked-in" or residual stresses) the pressure to cause *further* yielding may be significantly greater than for the *initial* yielding of a plate with a stress-free deflection. However, since the magnitude of such residual stresses will be unknown in practice, it is safer to ignore their effect and work with data for a stress-free plate.

It should be emphasized that initial lateral deformation is beneficial only when the plate edges are at

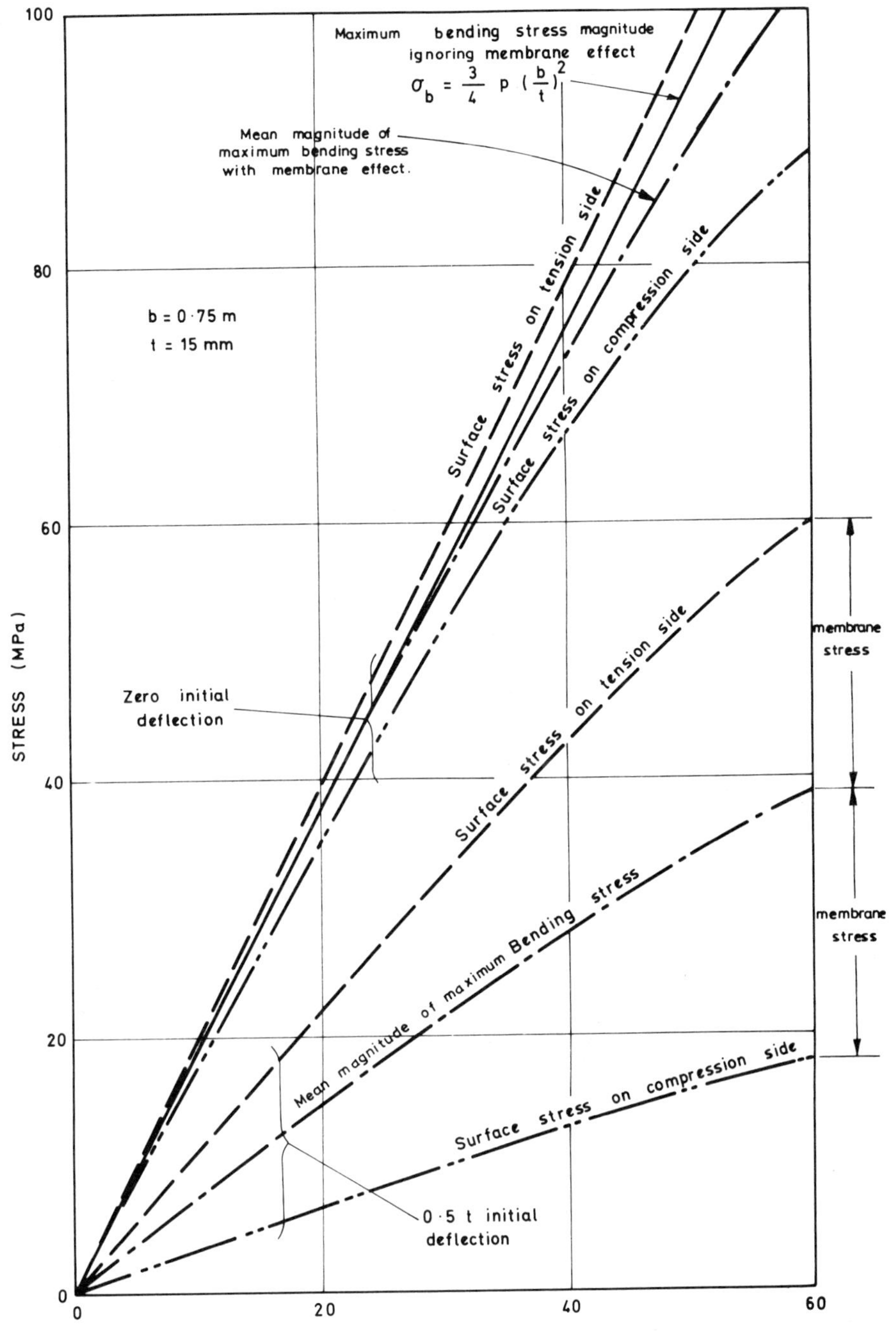

Figure 9.8 Membrane effect for long simply-supported plates.

least partly restrained from pulling in, thus allowing the development of in-plane tension. Also, when there is in-plane compression the lateral deformation can be quite detrimental if the lateral pressure causes a dished profile in line with the in-plane load because the latter acts to increase the deformation, which increases the bending stress and also induces buckling failure. This is examined in Section 12.5.

On the other hand, for transverse bulkheads, platform decks, and other such plating a larger amount of lateral deformation is permissible and such plating is usually designed on the basis of not exceeding a specified value of permanent set. The choice of a maximum permissible value of permanent set depends upon the application. For example, in the case of a collision bulkhead a very large value would be chosen. The design of plating on this basis is discussed in Section 9.4.

9.3 PLATES LOADED BEYOND THE ELASTIC LIMIT

The theory presented so far has been elastic theory which applies only if the material does not yield. However, the pressure applied to a plate which results in the onset of yield does not represent the limit of pressure that the plate can support. The plate may withstand a pressure several times greater than this before it fails in any significant way, or before the deformation becomes unacceptably large. In fact, for continuous plating supported by stiffeners, true "ultimate failure" of the plating almost never occurs because the stiffeners usually have a much lower load capacity than the plating. When the plate deflection becomes very large there does arise some restraint against pull-in, and hence the plate gradually becomes a fully plastic membrane, for which the rupture load is enormous. This extreme level of load and deformation is relevant in some special design applications such as icebreakers and protection against blast or collision, but in general the design of plating in which lateral loads predominate is governed by unserviceability rather than by ultimate failure. In most cases the governing type of unserviceability is a maximum allowable permanent set w_p rather than a maximum stress level. An exception would be plates subject to cyclic loading where fatigue considerations may impose a limit on the level of the working stress. Also, permanent set does have some effect on ultimate strength in the case of wide or approximately square plates that are subjected to in-plane compression and lateral loading, as would occur in a transversely framed strength deck or bottom. This is dealt with in Chapter 12.

Of course, elastic plate bending theory is much simpler and easier to use than inelastic theory and so it has been common practice in plate design to use elastic plate theory in conjunction with an artificial failure stress, for example, twice-yield stress, even though the true failure criterion is deflection, not stress, and even though the accompanying stress is well beyond the yield stress. For a given application, experience can provide the proportion between yield stress and some acceptable level of permanent set. However, in order to achieve a rationally-based design method which can be used for a wide variety of applications it is essential to determine the relationship between load and permanent set and to express this relationship in a form that is suitable for design, combining simplicity and accuracy. For hand design it is desirable to have either charts or simple formulas; for computer-aided design a mathematical formulation is required, either in the form of explicit expressions or a numerical algorithm.

Some examples of serviceability requirements which may dictate the permissible value of w_p are the operation of fork lift trucks or other vehicles, the overall flexural stiffness of the deck or panel, the rigidity of support to attached fittings, the robustness of the plating against damage (dents, etc.), and the avoidance, perhaps on aesthetic grounds, of the "hungry horse" look. As this list indicates it is often a non-structural consideration which determines the maximum permissible w_p. Because of the variety of serviceability requirements no single value or expression for the maximum permissible w_p will be universally suitable. Hence the approach which is taken herein is to use a combination of basic elasto-plastic theory and experimental results to derive a simple, semiempirical mathematical relationship between lateral load and w_p. This approach is a compromise between the complexity of a full elasto-plastic analysis and the approximate "rule of thumb" formulas based on elastic theory. The latter are usually intended for specific applications, whereas the relationship presented herein can be used for a variety of applications. Also it is sufficiently simple to be implemented on a small programmable calculator. Some design charts are also provided herein for even simpler applications.

Theories relating to the behavior of the plate following the onset of yield are called elasto-plastic theories. If the maximum permissible w_p is large, then an accurate analysis requires a combined large deflection and elasto-plastic theory. This is quite complex and so, before proceeding further, it will be helpful to consider which types and levels of theory are generally required for laterally loaded ship plating, as a function of the plate characteristics, the loading and support conditions, and the purpose of the analysis, that is, whether it is for calculating service stresses, or maximum permanent set (and of what magnitude), or possibly ultimate failure, as in the special cases mentioned previously.

To begin, we first consider what are the most basic characteristics of plating and how best to express these. From plate deflection equations such as (9.2.7) it is clear that the deflection depends strongly on the ratio b/t, which is a measure of the slenderness of the plating and hence of its bending flexibility. Ship plating subjected to large lateral loads (e.g., deck and shell plating) is generally within the range $30 < b/t < 80$, whereas superstructure paneling and the stiffened plating in smaller lightweight vessels is usually more slender, having $b/t > 80$. Equation 9.2.7 also shows that the elastic flexibility depends inversely on E. Another fundamental property that characterizes a plate is its yield stress since this measures its elastic range. Hence in the treatment which is to follow we will see that the most useful nondimensional parameter for plating is the plate *slenderness parameter* β which is defined as

$$\beta = \frac{b}{t}\sqrt{\frac{\sigma_Y}{E}} \tag{9.3.1}$$

and which combines all four of the foregoing plate characteristics. As we shall see, the $\frac{1}{2}$ power which is applied to σ_Y/E arises from elasto-plastic theory.

For our purposes in this chapter, plates may be divided into two broad categories; slender plates ($\beta > 2.4$ approximately), and sturdy plates ($\beta < 2.4$). For *sturdy* plates, with a small b/t ratio, the relative deflection w/t will generally be small since it is proportional to $(b/t)^4$. This will apply to both initial deflection and deflection under load. Because of this, and because of the occurrence of edge pull-in, membrane effect in sturdy plates is generally quite small, even after yielding has occurred. Hence small deflection elasto-plastic theory is sufficient for unserviceability analysis. For *slender* plates the deflections will be larger for a given load and hence membrane effects may become significant; in this case, large-deflection theory would be required for unserviceability analysis.

Ship plating which is subjected mainly to lateral loads (e.g., platform decks, tween decks, and shell plating forward) is generally within the range $30 < b/t < 80$, and for mild steel $(\sigma_Y/E)^{1/2}$ is approximately 0.03, giving β values between 0.9 and 2.4. Therefore large-deflection theory is generally not required for the unserviceability analysis of such plating, but elasto-plastic theory usually is required if the analysis is to be rationally-based and versatile in its application.

Application of Elasto-plastic Theory to Laterally Loaded Plates

This section presents a simplified explanation of the formation and growth of permanent set, taken from Ref. 7. The purpose is to establish the relative importance of the various parameters, and to derive the general form of the relationship between load and permanent set. This general form then provides the basis for the design formulas which are presented in the next section. For the load, we take the case of a uniform pressure and assume that this pressure also acts on adjacent panels of plating. In this case the appropriate boundary condition is "clamped; free to pull in." Therefore, yielding will first occur at the boundaries of the plate, at the upper and lower surfaces, and this marks the beginnings of inelasticity and permanent set in the plate. As the load is further increased, the yielding gradually penetrates through the thickness and the permanent set increases. Eventually a *plastic hinge** forms and spreads along each of the plate boundaries and from then on the permanent set increases rapidly.

In order to determine the basic features of the growth of permanent set we begin with the case of an infinitely long plate, thus removing the effect of aspect ratio. Such a plate would undergo cylindrical bending and a typical transverse strip may be treated as a beam of unit width, as was done in the introduction to elastic plate theory. Hence from (9.1.7) the surface stress at the edges is

$$\sigma_x = \frac{1}{2}p\left(\frac{b}{t}\right)^2 \tag{9.3.2}$$

and the stress in the orthogonal direction is $\sigma_y = \nu\sigma_x$. The third principal stress is zero and therefore, using the Hencky-von Mises yield criterion, the value of σ_x at which yield occurs is

$$(\sigma_x)_Y = \frac{\sigma_Y}{\sqrt{1 - \nu + \nu^2}}$$

This result shows that, due to the σ_y stress, yielding does not occur until after σ_x has exceeded σ_Y by about 13%. The section modulus for a strip of plating of unit

*The notion of a plastic hinge and related quantities (such as plastic section modulus) are used mainly in plastic frame analysis, and so the definition and explanation of these terms are given in Sections 16.1 and 16.2. If necessary, these sections may be consulted at this point; they require nothing more than basic mechanics.

width is $Z = t^2/6$. Therefore, if M_Y denotes the bending moment corresponding to initial yield, the value of M_Y must be such that $M_Y/Z = \sigma_Y$. That is

$$M_Y = \frac{\sigma_Y}{\sqrt{1 - \nu + \nu^2}} \frac{t^2}{6} \tag{9.3.3}$$

From (9.3.2) the pressure which causes initial yielding is

$$p_Y = \frac{2\sigma_Y}{\sqrt{1 - \nu + \nu^2}} \left(\frac{t}{b}\right)^2 \tag{9.3.4}$$

By defining a nondimensional load parameter

$$Q = \frac{pE}{\sigma_Y^2} \tag{9.3.5}$$

and by making use of the plate slenderness ratio β, (9.3.4) may be put into the form

$$Q_Y = \frac{2}{\sqrt{1 - \nu + \nu^2}} \left(\frac{1}{\beta}\right)^2 \tag{9.3.6}$$

Likewise, the formation of edge plastic hinges in the strip of plating will follow the same pattern as that shown in Section 16.2 for a beam, assuming elastic-perfectly-plastic material behavior. The plastic section modulus for a strip of plating of unit width is $Z_P = t^2/4$ and hence the bending moment required to form a plastic hinge is

$$M_P = \frac{\sigma_Y}{\sqrt{1 - \nu + \nu^2}} \frac{t^2}{4} = 1.5\, M_Y \tag{9.3.7}$$

In Section 16.2 it is shown that for a beam which is clamped but free to pull in and which is subjected to a uniform load (in this case the pressure p) the load which causes plastic hinges at the ends is

$$p_{EH} = \frac{12 M_P}{b^2}$$

Combining the foregoing relations gives

$$p_{EH} = \frac{3\sigma_Y}{\sqrt{1 - \nu + \nu^2}} \left(\frac{t}{b}\right)^2$$

or, in terms of the nondimensional parameters,

$$Q_{EH} = \frac{3}{\sqrt{1 - \nu + \nu^2}} \left(\frac{1}{\beta}\right)^2 \tag{9.3.8}$$

in which the subscript EH indicates "edge hinge."

Figure 9.9 shows the theoretical load–deflection curve and the corresponding load–permanent set curve for an infinitely long plate. Complete collapse requires the formation of a third plastic hinge, but we are more interested in events prior to this and we want information about the growth of permanent deflection. The simple "hinge mechanism" analysis gives no information about deflection, and so we must examine the formation of the plastic hinges in more detail. As shown in Fig. 9.9 the commencement and early growth of permanent set is entirely due to the edge hinges and this continues to be the case until the edge hinges have nearly been completed. We therefore examine the growth of plasticity at the edge and the relation between the end moment M_e and permanent set. Figure 9.10a is the moment–curvature diagram and we focus attention on the region from $M_e = M_Y$ to $M_e = M_P$. If the bending moment is increased by a differential amount dM_e the plastic region grows and penetrates into the thickness t. The yielded portion cannot take any further bending moment and hence the additional moment dM_e is resisted only by the inner elastic layer of thickness t_r, as shown in Fig. 9.11a. The increase of curvature $d\phi$ is therefore dM/EI_r, where $I_r = t_r^3/12$. That is, EI_r is the local slope of the M–ϕ curve, as shown in Fig. 9.10a. If the moment were removed the curvature would decrease, approximately by an amount M/EI, leaving some permanent local curvature ϕ_p. The permanent curvature is drawn separately in Fig. 9.10b. In the later stages of the hinge formation the decrease in ϕ (the "spring-back" after load removal) would be somewhat less than M/EI but the effect is slight and ignoring it is conservative.

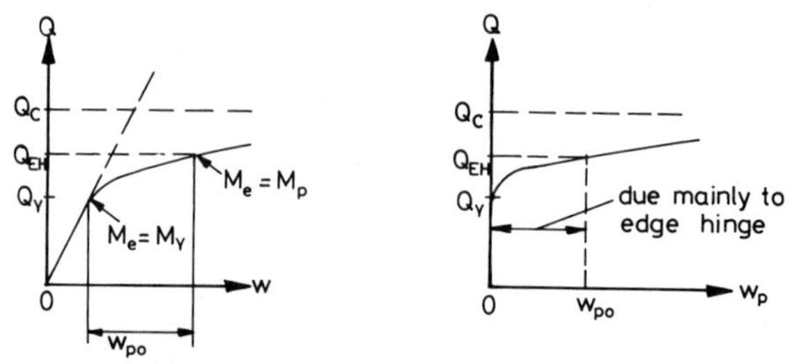

Figure 9.9 Load-deflection and load-permanent set curves for an infinitely long plate.

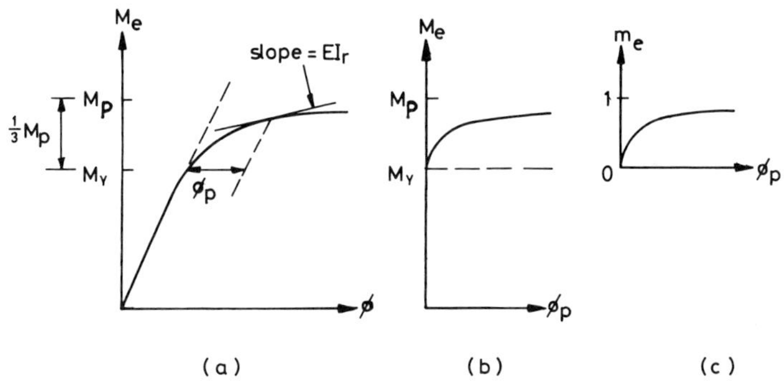

Figure 9.10 Moment-curvature diagrams for an infinitely long plate.

Hence we may write

$$\phi_P(M_e) = \int_{M_Y}^{M_e} \left(\frac{1}{EI_r} - \frac{1}{EI}\right) dM_e \qquad (9.3.9)$$

We define a nondimensional post-yield moment m_e as shown in Fig. 9.10c

$$m_e = \frac{M_e - M_Y}{M_P - M_Y} = \frac{M_e - M_Y}{\dfrac{M_P}{3}} \qquad (9.3.10)$$

In terms of m_e (9.3.9) becomes

$$\phi_P(m_e) = \frac{4M_P}{Et^3} \int_0^{m_e} \left(\frac{t}{t_r^3} - 1\right) dm_e \qquad (9.3.11)$$

after substituting for I and I_r. Figure 9.11a shows the stress distribution after yielding has commenced. It may be shown that equilibrium requires

$$M_e = \left(\frac{t_r^2}{6} + \frac{t^2 - t_r^2}{4}\right)\sigma_Y$$

in which the first term corresponds to the elastic portion. The bending moment at first yield is $M_Y = (\sigma_x)_Y t^2/6$. Substituting M_e and M_Y into the definition of m_e gives

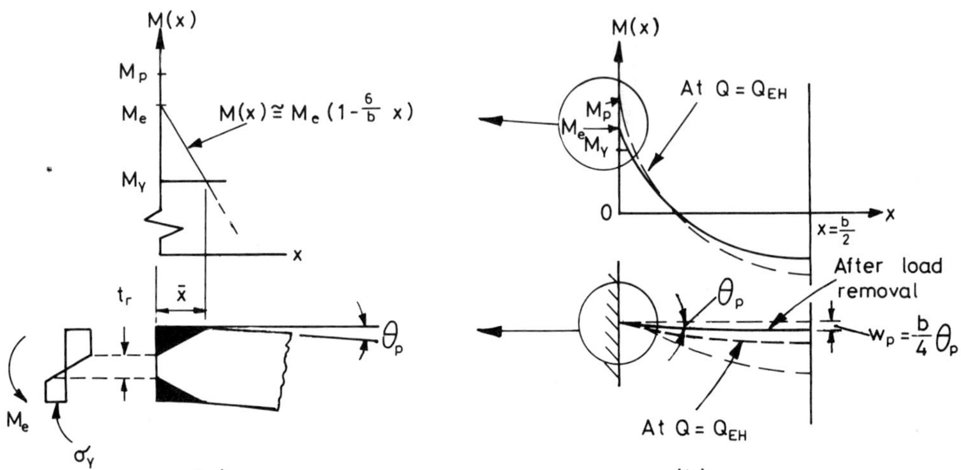

Figure 9.11 Conditions at the edge of a clamped plate.

$$m_e = 1 - \frac{t_r^2}{t^2}$$

Hence

$$\frac{t^3}{t_r^3} = \frac{1}{(1 - m_e)^{3/2}}$$

and (9.3.11) integrates to

$$\phi_p(m_e) = \frac{4M_P}{Et^3}\left[\frac{2}{\sqrt{1 - m_e}} - 2 - m_e\right] \qquad (9.3.12)$$

As shown in Fig. 9.9 the permanent set increases sharply during the formation of the edge hinge. As the load approaches Q_{EH} the plate becomes effectively simply supported with respect to the additional load. Equations 9.1.2 and 9.1.3 derived earlier for a plate strip show that the *rate* of increase in the total deflection would eventually become five times the pre-hinge value. Because of the plastic condition of the plate edges most of this deflection would be locked in and would therefore constitute permanent set. Hence the plate quickly reaches typical permissable values of permanent set and there is no need to consider the formation of a third hinge in the center.

Throughout the range of loading up to $Q = Q_{EH}$, the deflected shape of the plate remains a smooth curve since there is as yet no hinge at the center. If the load is removed after the formation of the edge hinges the plate relaxes to an approximately parabolic shape, with some "locked-in" bending moment and with permanent edge rotation θ_p, as shown in Fig. 9.11*b*. The permanent set w_p is therefore approximately $\theta_p b/4$, because the occurrence of "spring-back" in the edge hinges has already been allowed for. In practice, the stored elastic energy in the plate would cause some slight additional spring-back, but the effect is small and is ignored because our goal here is simply to obtain an understanding of the basic nature of the process and the functional relationships among the variables.

The (permanent) angle of rotation of the plate edge θ_p is obtained by integrating the (permanent) curvature of the plate ϕ_p. This curvature occurs within the length of the plastic zone, say $\bar{x}$ (see Fig. 9.11*a*) and this length is dependent on the magnitude of the load, as characterized by the end moment m_e. Also, the bending moment within the plate is a function of position: $m = m(x)$. Since the length $\bar{x}$ of the plastic zone is small relative to the plate width b we take a linear approximation for $m(x)$:

$$m(x) = m_e - 6\frac{x}{b} \qquad (9.3.13)$$

The length of the plastic zone is obtained by setting $M(x) = M_Y$, or $m(x) = 0$, which gives

$$\bar{x} = \frac{b}{6} m_e \qquad (9.3.14)$$

Adopting the above estimate for w_p we have

$$w_p \approx \frac{b}{4}\theta_p = \frac{b}{4}\int_0^{\bar{x}} \theta_p[m(x)]\,dx \qquad (9.3.15)$$

in which the brackets indicate functional dependence. Equation 9.3.12 may now be generalized to express the curvature at an arbitrary point wthin the plastic zone where the bending moment is $m(x)$:

$$\phi_p(m) = \frac{4M_P}{Et^3}\left[\frac{2}{\sqrt{1 - m}} - 2 - m\right] \qquad (9.3.16)$$

To facilitate the integration of (9.3.15) we regard $m(x)$ as the primary variable, giving it the symbol X:

$$X = m(x) = m_e - \frac{6}{b}x$$

from which

$$dx = -\frac{b}{6}\,dX$$

At $x = 0$, $X = m_e$, and at $x = \bar{x}$, $X = 0$. With these substitutions, together with (9.3.16), (9.3.15) becomes (after reversing the order of the integration)

$$w_p = \frac{4M_p}{Et^3}\frac{b}{24}\int_0^{m_e}\left(\frac{2}{\sqrt{1 - X}} - 2 - X\right)dX$$

$$= \frac{M_P b^2}{6Et^3} F(m_e)$$

where

$$F(m_e) = 4(1 - \sqrt{1 - m_e}) - 2m_e - \frac{m_e^2}{2}$$

Substituting for M_P from (9.3.7) and introducing the plate slenderness β from (9.3.1) gives

$$\frac{w_p}{t} = \frac{\beta^2}{24\sqrt{1 - \nu + \nu^2}} F(m_e) \qquad (9.3.17)$$

At the completion of the edge hinge $m_e = 1$ and the corresponding value of F is 1.5. Hence the value of w_p at this stage of the loading, which was denoted as w_{p0} (see Fig. 9.9) is

$$\frac{w_{p0}}{t} = \frac{1.5\beta^2}{24\sqrt{1 - \nu + \nu^2}} = 0.07\beta^2 \qquad (9.3.18)$$

For later use we define R_w as the ratio of the permanent set w_p and the edge hinge value w_{p0}. From (9.3.17) and (9.3.18) this is

$$R_w = \frac{w_p}{w_{p0}} = \frac{2}{3} F(m_e)$$
$$= \frac{2}{3}\left[4(1 - \sqrt{1 - m_e}) - 2m_e - \frac{m_e^2}{2}\right]$$

The right-hand side of this equation is plotted in Fig. 9.12 and it can be seen that the following expression is a close approximation to it:

$$R_w(m_e) = \frac{w_p}{w_{p0}} \cong 1 - (1 - m_e^3)^{1/3} \qquad (9.3.19)$$

This approximate expression for R_w is chosen because it can be solved explicitly for the bending moment m_e. For later reference the solution is written in terms of a nondimensional function $T(R_w)$ as follows:

$$m_e = T(R_w)$$

$$\text{in which } \begin{cases} T(R_w) = [1 - (1 - R_w)^3]^{1/3} \text{ for } R_w \leq 1 \\ T(R_w) = 1 \text{ for } R_w > 1 \end{cases} \qquad (9.3.20)$$

$$\text{with } R_w = w_p/w_{p0}$$

and with w_{p0} as defined in (9.3.18).

The separate definition for $R_w > 1$ is required because the foregoing analysis does not apply to this range. From the definition of m_e in (9.3.10) it can be

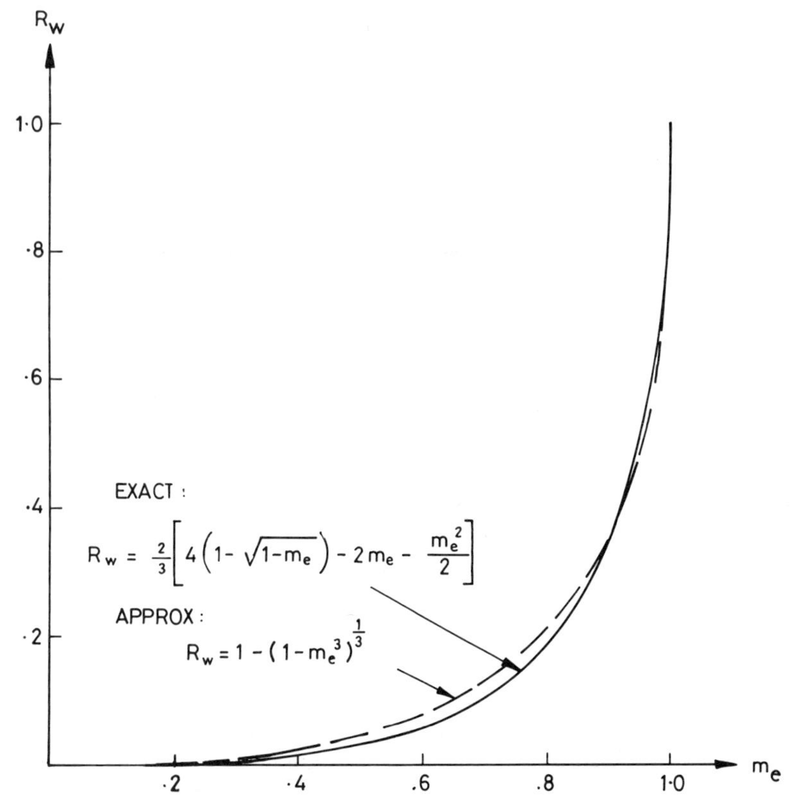

Figure 9.12 Exact and approximate expressions for the permanent set ratio $R_w = w_p/w_{po}$.

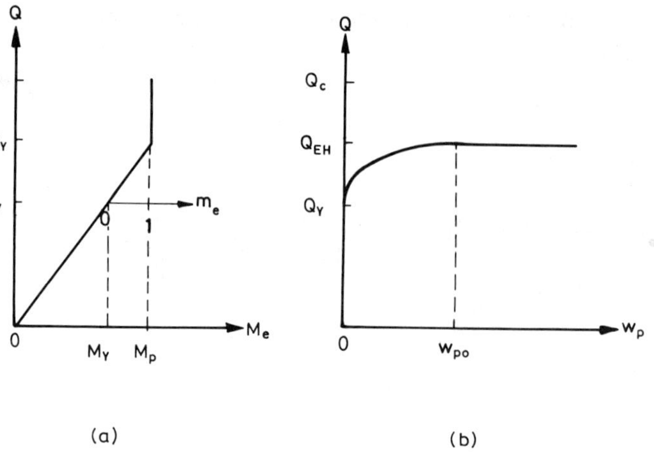

Figure 9.13 Load, edge moment, and permanent set for plates of infinite aspect ratio.

seen that m_e is linearly proportional to the increment of bending moment from M_Y to M_P, and Fig. 9.13*a* shows that within this range M_e is proportional to the increment of load from Q_Y to Q_{EH}. Therefore

$$Q = Q_Y + (Q_{EH} - Q_Y)T(R_w) \qquad (9.3.21)$$

and since $Q_{EH} = 1.5Q_Y$ this becomes

$$Q = Q_Y + \frac{Q_Y}{2}T(R_w)$$

with $T(R_w)$ given by (9.3.20). This expression is plotted in Fig. 9.13*b* which shows that as the edge hinges begin to form, the load–permanent-set curve turns through 90°, forming a "knee," and the permanent set increases rapidly to the value w_{p0} given by (9.3.18), at which point the curve is horizontal. Beyond this point the foregoing analysis is no longer valid because the central hinge begins to form and this absorbs more load. Nevertheless the foregoing analysis shows that for an infinitely long plate (and hence with no aspect ratio effect) the permanent set begins to increase rapidly before the completion of the edge hinges, and the theoretical asymptote of the load–permanent-set curve is a line passing through the edge hinge load Q_{EH}, rather than the collapse load Q_c. The knee of the curve, that is, the region $w_p < w_{p0}$ (or $R_w < 1$), will be referred to as the "transition zone" and it is defined by the transition function $T(R_w)$ of (9.3.20).

For plates of finite aspect ratio, yielding and hinge formation occur at four edges instead of two, and therefore the load required to cause edge hinges will be larger than for a long plate of the same area. Also, the yielding and hinge formation is a more gradual process, beginning at the midpoint of the long sides and gradually extending along the sides, and this makes the transition zone larger than for a long plate. In addition, some membrane effect will be present due to the requirement of geometric compatibility. Nevertheless, the basic mechanism of permanent set is the same as for a long plate and experiments [8] have verified that the formation of edge hinges marks the upper bound of the loading range within which permanent set begins to increase rapidly.

It has also been observed experimentally that once Q_{EH} is exceeded the growth of w_p is very nearly linearly proportional to further increases in load. The main reason for this is that once plastic hinges have formed at the boundaries there is no further change in the basic nature of the boundary conditions with further increase of load. The linearity in the growth of permanent set is illustrated in Fig. 9.14. The figure also shows how the foregoing analysis is applied to plates of finite aspect ratio. The commencement of permanent set occurs at the initial yield load Q_Y, which can be calculated from elastic theory. The subsequent load–permanent-set relationship contains two parts: a curved "transition" portion and a subsequent straight portion. These are defined in terms of two parameters, ΔQ_0 and ΔQ_1, each of which corresponds to a particular increment of load, as shown in Fig. 9.14. The increment ΔQ_0 defines the location of the intercept of the straight portion of the curve (which corresponds to the formation of edge hinges) and Q_1 is the further increment of load at the end of the transition zone and this serves to define the slope of the straight portion.

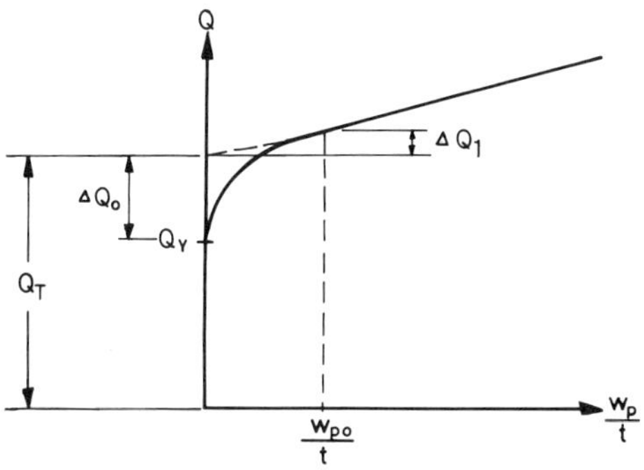

Figure 9.14 Load versus permanent set.

In mathematical terms

$$Q = Q_Y + T(R_w)\left[\Delta Q_0\left(\frac{a}{b}, \beta\right) + \Delta Q_1\left(\frac{a}{b}, \beta\right)R_w\right] \quad (9.3.22)$$

in which the parentheses (but not the square brackets) indicate functional dependence. By definition $R_w = w_p/w_{p0}$, and w_{p0} is given by (9.3.18). As shown in the next section, experiments indicate that the Q versus R_w relationship remains essentially linear up to relatively large values of permanent set. Ultimately, of course, as the load increases, plasticity spreads throughout the central region of the plate and some membrane straining begins to occur because of the large deflections. This latter effect would cause the slope of the curve to increase. At this stage small-deflection theory is no longer valid. However, for typical ship plates there is no need to carry the analysis any further because the magnitude of the permanent set has already exceeded typical serviceability limits.

9.4 DESIGN OF PLATING BASED ON ALLOWABLE PERMANENT SET

Because permanent set involves complicated elasto-plastic behavior, there is no analytical, closed-form method available for the direct calculation of the load required to cause a specified level of permanent set. Some approximate analytical solutions have been obtained for special cases, as in the previous section, but an accurate and general solution requires the use of numerical techniques such as incremental finite element analysis or the solution of the nonlinear plate equations by finite differences. These methods involve too much computation for ordinary design applications. For this purpose the designer requires a rapid and simple method—ideally a single formula—for estimating the load which would produce a given level of permanent set, and since the choice of that level is somewhat arbitrary and does not involve ultimate failure, the calculation does not require the accuracy of these sophisticated numerical techniques.

Plating Subjected to Uniform Pressure

In view of the undesirable complexities which exist on the analytical side and the availability of suitable experimental data obtained by Clarkson [8], an empirical expression was derived in Ref. 7, based on the general form of (9.3.22). The expression is

$$Q = Q_Y + T(R_w)[\Delta Q_0 + \Delta Q_1 R_w] \quad (9.4.1)$$

in which

Q = the load parameter for the pressure load, p

$= \dfrac{pE}{\sigma_Y^2}$

Note: If, as is usual, the maximum value of pressure is subject to some uncertainty, then the pressure load p should be a *factored* load.

$$Q_Y = \frac{2}{\sqrt{1 - \nu + \nu^2}\,\beta^2}\left[1 + 0.6\left(\frac{b}{a}\right)^4\right]$$

$$\Delta Q_0 = \frac{1 + 0.5\beta\dfrac{b}{a}\left[1 + \dfrac{b}{a}\left(3.3 - \dfrac{1}{\beta}\right)\right]}{\sqrt{1 - \nu + \nu^2}\,\beta^2} \quad (9.4.2)$$

$$\Delta Q_1 = 0.95\left(\frac{b/a}{\sqrt{\beta}}\right)^{1.5} \quad (9.4.3)$$

The first term of ΔQ_0, that is, $1/[(1 - \nu + \nu^2)^{1/2}\beta^2]$, is the increase of load above Q_Y which would cause edge hinges in an infinitely long plate, as given by (9.3.6). The second term accounts for the aspect ratio, a/b and for the membrane effect due to compatibility which occurs in plates of finite aspect ratio and which increases with plate slenderness. Equation 9.4.3 gives the further increment in the load parameter at the end of the transition zone, as shown in Fig. 9.14. Figure 9.15 shows that the foregoing empirical expression represents the experimental results of Clarkson with satisfactory accuracy. It therefore provides a simple and rapid means for estimating the load corresponding to a specified level of permanent set.

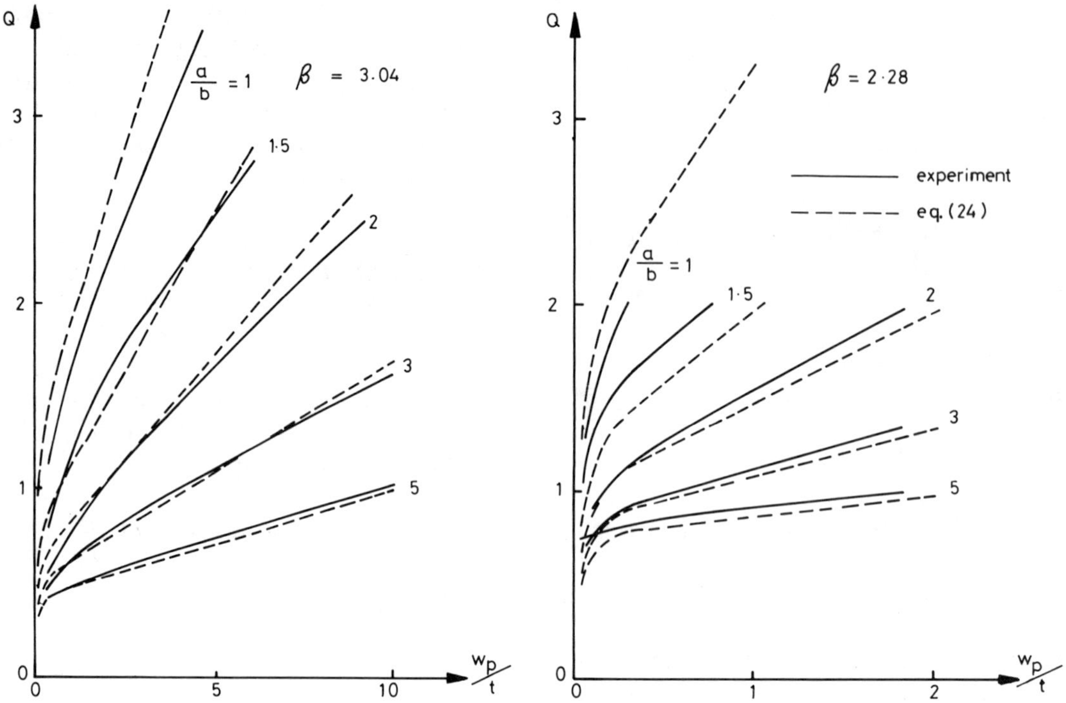

Figure 9.15 Load versus permanent set; laterally loaded plating.

Clarkson's data is for plating which is initially flat, and therefore in (9.4.1) the value of w_p (within R_w) is that due to the load only. In practice, plates are seldom initially flat; they usually have some initial deformation, the most common being that due to welding. There are two types of initial deformation: that caused by previous loads which involves plastic deformation and locked-in stresses, and a "stress-free" initial deformation. Both cases are illustrated in Fig. 9.16. Let us first consider a stress-free initial deformation w_{pi} and define the total permanent set as $w_{pt} = w_{pi} + w_p$. The permanent set ratio is then

$$R_w = \frac{w_{pt} - w_{pi}}{w_{p0}} \tag{9.4.4}$$

If the plate edges are free to pull in and if w_{pi} is not large (say, $w_{pi} < t$) then the presence of w_{pi} does not substantially alter the load–deflection curve; it merely shifts it to the right by an amount w_{pi}, and the curve of Q versus w_p/t is likewise shifted rightward by an amount w_{pi}/t. Thus the load–permanent-set relation of (9.4.1) can be used for plates with stress-free initial permanent set w_{pi} by first subtracting w_{pi} from the total allowable permanent set $(w_{pt})_{max}$.

If the initial permanent set w_{pi} is not stress-free but is due to a previous load (curve OAC in Fig. 9.16), the load–deflection curve would commence at $w = w_{pi}$ and would rise linearly until it almost regained its previous point; only then would the permanent set begin to increase further (curve CAB). Thus if $(w_{pt})_{max}$ denotes the maximum permissible value of total permanent set and if Q_1 denotes the corresponding capa-

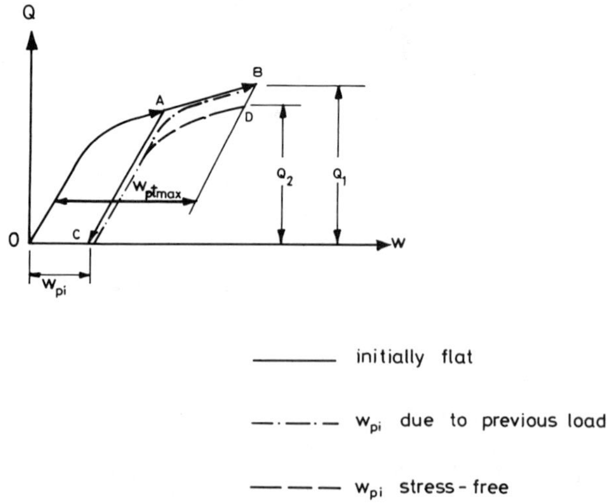

Figure 9.16 Load–deflection curves for various initial conditions.

bility of an initially flat plate (curve OAB) then as shown in the figure an initial permanent set due to a previous load does not reduce the capability of the plate; it is still Q_1. In contrast to this, when the initial permanet set w_{pi} is stress-free the curve of Q versus w is curve CD. As long as w_{pi} is not extreme, this curve has the same shape as for an initially flat plate (curve OAB) but is shifted to the right by an amount w_{pi}. That is, the initial permanent set is not "absorbed" within the subsequent values, but rather is additive. Hence for the same value of maximum permissible permanent set $(w_{pt})_{max}$ the plate's capability will be Q_2, slightly less than Q_1. Eventually, at very large loads, the curve does reach the "initially flat" curve but this is usually well beyond the design range.

In the method presented herein, the initial permanent set due to welding is treated as if it were stress-free. In reality, of course, a welded plate is not stress-free because welding causes large locked-in stresses. Therefore the curve of Q versus w for a welded plate lies somewhere between the "stress-free" curve (CD) and the "initially loaded" curve (CAB). The exact position would be difficult to ascertain and such exactness is not warranted because the prediction of weld-induced deformation, which is considered subsequently, is only approximate. Hence in the method presented here w_{pi} is simply subtracted from the total allowable permanent set, as in (9.4.4). Within the normal design range the error involved in this approximation is small and it lies on the conservative side.

The application of the design formula is as follows. For given trial values of plate thickness and stiffener spacing, the values of ΔQ_0 and ΔQ_1 are calculated and are substituted into (9.4.1), together with the net value of allowable permanent set ratio, calculated from (9.4.4). These trial values are then altered in such a way that the right-hand side, which represents the capability, is equal to or greater than the load parameter for the factored load. Since it involves only a single formula the entire calculation, including the iteration, can easily be programmed on a hand-held programmable calculator. In a rationally-based, optimum design context (9.4.1) constitutes the constraint against unserviceability due to excessive permanent set.*

In order to facilitate the use of the formula in non-computer-aided situations some design curves are given in Fig. 9.17 for $w_p/\beta t = 0.2, 0.4, 0.6, 0.8$, and 1.0. To use the curves, the designer computes Q for

*In this case the equals sign in (9.4.1) would become $\leq$.

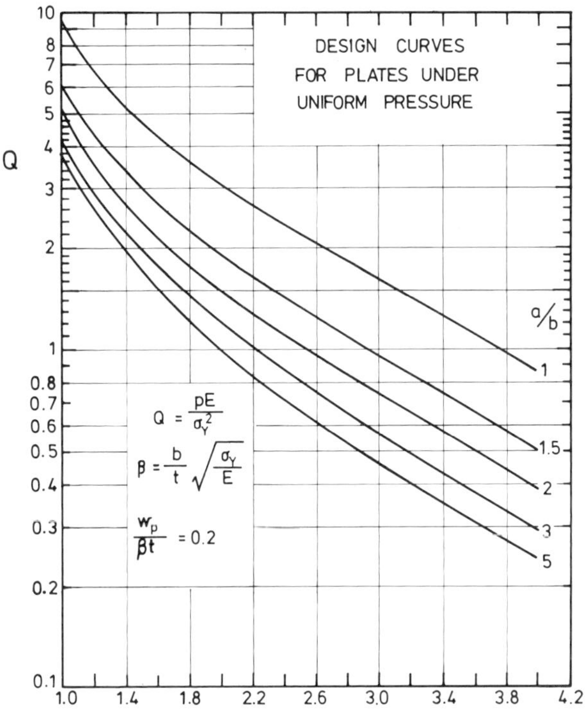

Figure 9.17a

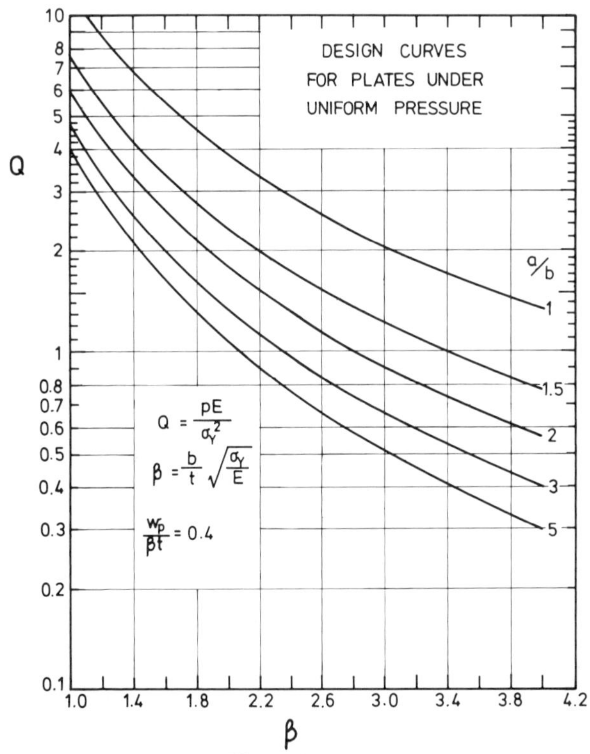

Figure 9.17b

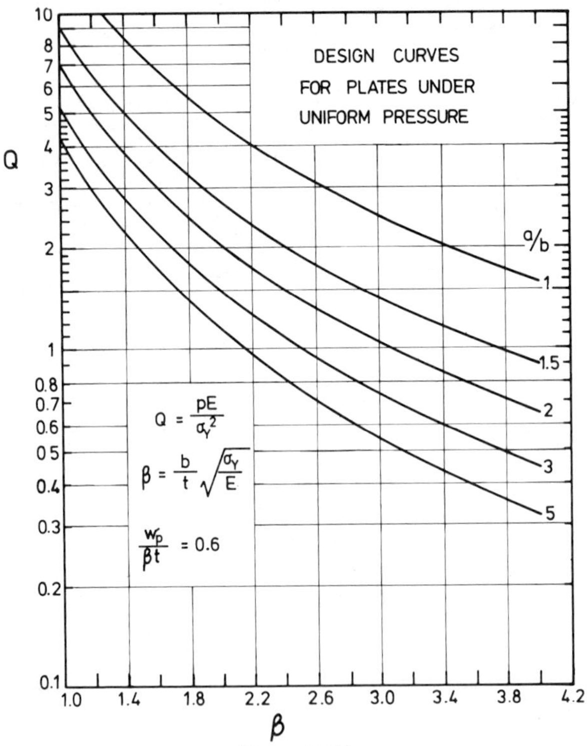

Figure 9.17c

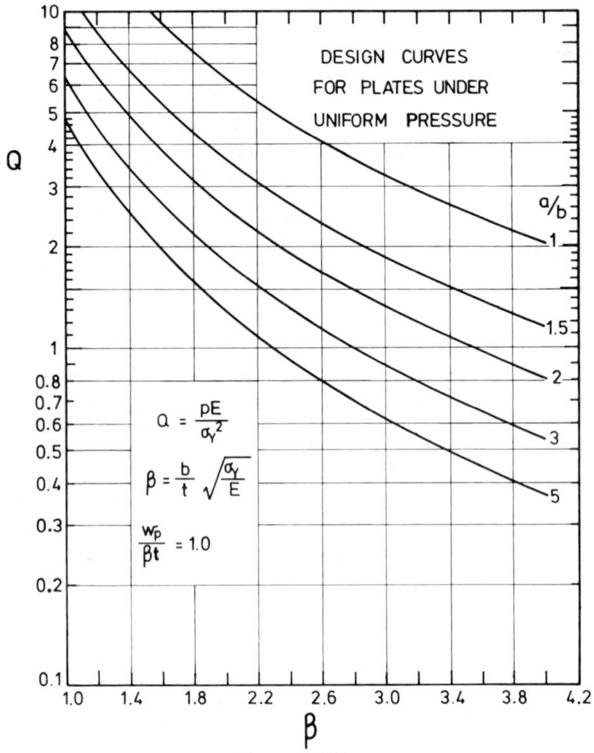

Figure 9.17e

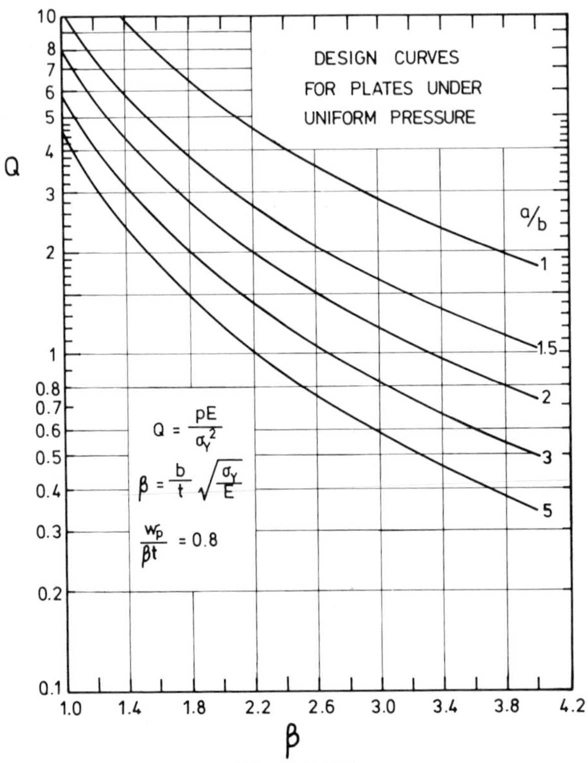

Figure 9.17d

the given design load (factored, if appropriate) and then for a selected stiffener spacing b (and hence aspect ratio a/b) he reads the value of β, from which the thickness may be obtained. If this t is not satisfactory for some reason, the value of b, or perhaps of σ_Y, may be varied as required.

Initial Permanent Set Due to Welding

Antoniou [9] has presented the results of a regression analysis of over 2000 values of w_{pi}/t measured on newly built ships over a number of years, showing that the significant parameters are β and the ratio t_w/t, where t_w is the thickness of the stiffener web, and that the following expression gives a satisfactory fit to the data

$$\frac{w_{pi}}{t} = 0.073\beta^{1.65}\left(\frac{t_w}{t}\right)^{0.42} \qquad (\beta \leq 2.5) \qquad (9.4.5)$$

Typical ship plates have $0.6 < t_w/t < 0.9$ and therefore the mean value is approximately

$$\frac{w_{pi}}{t} \cong 0.065\beta^{1.65}$$

Also, for ship plates the slenderness ratio is usually in the range $1.5 < \beta < 2.5$ and hence the initial welding distortion is usually in the range $0.13 < w_{pi}/t < 0.3$. Similarly from (9.3.18) the amount of permanent set in the transition zone is usually in the range $0.16 < w_{po}/t < 0.44$. Thus the initial welding deformation w_{pi} is about equal to the transition amount. This means that if the total allowable permanent set is more than twice the as-welded value (which is often the case) then the allowable permanent set due to the load is beyond the transition zone and (9.4.1) is simplified because the transition function T is unity.

Antoniou [9] also proposed the following formulas for the maximum permissible value of initial weld-induced permanent set

$$\begin{aligned} \text{for } t \geq 14 \text{ mm:} \quad \left(\frac{w_{pi}}{t}\right)_{\max} &= 0.014\frac{b}{t} - 0.32 \\ \text{for } t < 14 \text{ mm:} \quad \left(\frac{w_{pi}}{t}\right)_{\max} &= 0.018\frac{b}{t} - 0.55 \end{aligned} \tag{9.4.6}$$

and showed that this requirement is quite reasonable, being more than two standard deviations above the mean value given by (9.4.5), and is consistent with other formulations and with general shipbuilding practice.

The question remains as to what total level of permanent set is to be considered acceptable. This depends on the particular circumstances and therefore must be left to the designer. For naval vessels, where weight saving is important, a large value would be appropriate, such as $w_{pt} = b/50$. For cargo vessels a lower value would generally be preferred, such as $b/100$. This question will be considered further at the end of the next section, after dealing with concentrated loads, because in many cases the design load is of this type rather than a uniform pressure. Also many areas of ship plating are subjected to both types of loads and both must be investigated because it will not be known *a priori* which type would cause the larger value of permanent set.

The influence of permanent set on the in-plane compressive strength of plating is discussed in Section 12.7. In brief, for longitudinally stiffened panels the permanent set caused by welding and by lateral pressure does not diminish the ultimate compressive strength; it even enhances it slightly. For transversely stiffened panels even a moderate amount of permanent set causes higher stresses (see Section 12.5) and a lower ultimate compressive strength, and a large permanent set causes a drastic reduction in strength. Since some permanent set is nearly always present, simply due to welding, transverse stiffening should not be used where good in-plane compressive strength is required, and if it is used in any panel that panel should not be counted in the hull girder calculations.

Plating Subjected to Concentrated Loads

Ship plating is also subjected to concentrated loads of various types, such as wheel loads, fenders, pallets, and falling objects. In Ref. 10 it is shown that for design purposes there are two main types of concentrated loads, depending on the number of different locations where they can occur in the panel: single location and multiple location. Here the word "multiple" is not meant to imply simultaneous, but simply any load that will occur several times over the life of the ship and probably in different locations. The distinction between single and multiple is important because, as will be shown subsequently, these two types require different design methods.

Single location loads are either deliberate, such as the weight of a fixed piece of equipment, or accidental such as the dropping of a heavy object. Deliberate single location loads do not usually influence plate design because they can either be placed over a stiffener, or additional stiffeners or other supporting structure can be provided. This is nearly always better than having thicker plating. Although the position of accidental loads may be random, for design purposes the worst position—midway between the stiffeners—must be assumed. Note that here the term accidental implies rare: if it is expected to occur several times in the life of a ship and at various locations in a panel (for example, heavy landing of a helicopter) it is classified as multiple location. Obviously, wheel loads and all other moveable loads are classified as multiple location.

Concentrated loads may be either static, quasi-static, or dynamic. Dynamic loads are those which have such a short duration that the inertia and dynamic response of the plating becomes significant. About the only example of a dynamic concentrated load is projectile impact, and in this case it is necessary to investigate the possibilties of puncture and of plate rupture, in addition to large permanent set. These two limit states are not considered herein. Quasi-static loads are those in which the motion of the load can be accounted for by means of static inertia forces. Ship speeds are such that collision loads fall in this category, but here also it is necessary to investigate puncture and rupture as well as permanent set. Some common examples of quasi-static loads are the heavy landing of an aircraft or helicopter, and wheel loads of vehicles (or any other static weight-related load) when the ship is in heavy seas.

When the load magnitude has a high degree of variability, such as heavy landing of a helicopter, the design load is usually a *design extreme value,* which either corresponds to some probability of occurrence or is associated with some identifiable result (e.g., collapse of the landing gear). Alternatively, in commercial vessels there is sometimes the possibility of controlling and limiting the maximum load by means of operating regulations, such as maximum permissible axle load, wheel load, and tire pressure. If this approach is adopted the maximum permissible load is derived from the maximum allowable permanent set, and this load (again factored to allow some margin for accidental overload and for change of service of the ship) becomes the design load.

Parameters for Describing Concentrated Loads

For simplicity the concentrated load is taken to be rectangular in shape, of dimension $e \times f$, with e parallel to b, as shown in Fig. 9.18. The size or extent of the load is measured by the geometric average of its two dimensions

$$e_m = \sqrt{ef} \tag{9.4.7}$$

This is done because over a period of time a multiple location load will occur in all orientations to the plate and therefore the final cumulative value of permanent set will be independent of the load aspect ratio e/f. Hence both dimensions are given equal weight. The geometric mean is preferable to the arithmetic mean because it relates directly to the area under the load: e_m^2. For wheel loads, e_m is related to the load P and the tire pressure p_t by

$$e_m = \sqrt{P/p_t} \tag{9.4.8}$$

It is also necessary to have a parameter which measures the degree of concentration of the load. For this purpose we define

$$\lambda = e_m/b \tag{9.4.9}$$

And finally the load parameter which is most commonly used for concentrated loads is

$$Q_P = \frac{PE}{b^2\sigma_Y^2} \tag{9.4.10}$$

Design for Multiple Location Loads

Since a wheel load is the most obvious example of a multiple location load, the discussion herein will be in terms of this type of load, but the results apply to all types of multiple location loads.

When a concentrated load which is large enough to produce permanent set moves across a panel from one side to the other, it causes plastic deformation along its entire path and the resulting permanent set is larger than if the same load were applied only at the center. Since wheel loads can occur anywhere in the panel, eventually every part of the panel will have undergone at least some plastic deformation, though it may only be near the plate surfaces. Also, the direction of travel will vary, at least from time to time. There will also be variations in tire diameter and width, and in the wheel arrangement, with twin wheels being especially common. With wide tires or closely spaced twin wheels the load will occasionally be applied along the edge of a panel, straddling a stiffener, as shown in Fig. 9.19. This will cause some plastic rotation of the panel edges—more than would have been caused by loads crossing the panel—and so the permanent set will again be increased. Although not all wheel loads will equal the full design load, over a period of time this load will occur a number of times, in various locations, direction, and wheel arrangements. As a result the distribution and arrangement of plasticity in the panel gradually reaches a final stationary pattern

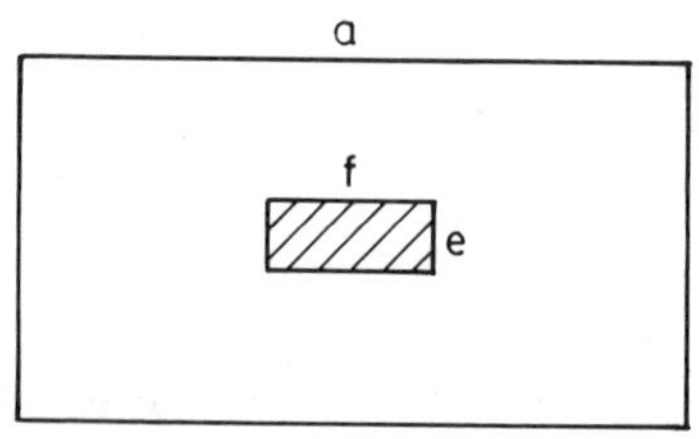

Figure 9.18 Geometry of single location concentrated load.

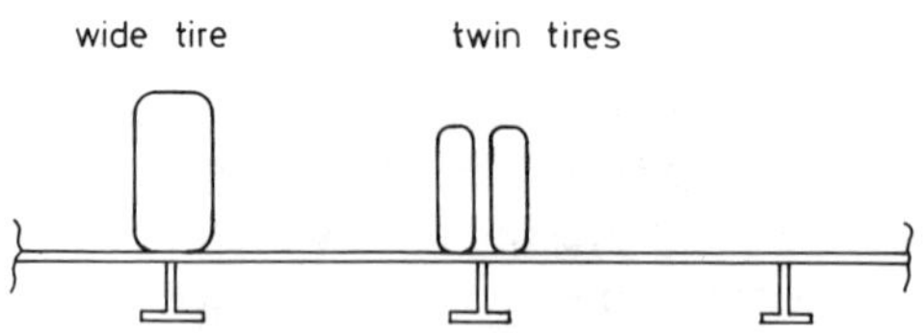

Figure 9.19 Loads causing edge rotation.

and the maximum permanent set reaches a final value corresponding to the design load. It is only this final cumulative value which is of interest to the designer. Moreover, this value is a function only of the panel parameters α and β and the load concentration parameter λ; it is independent of the load shape and load aspect ratio. In Ref. 10 it is shown that because of the cumulative nature of the plastic deformation the final distribution of plasticity is basically similar to that caused by a uniform pressure load. This is true not only for wheel loads but also for all multiple location loads because even if they do not "move" they occur in different locations over the course of time. Because the distribution of plasticity is similar to that for a uniform load then the relationship between load parameter and permanent set is also similar. Hence for a given design load parameter Q_P the eventual stationary value of permanent set is essentially the same as that caused by an equivalent uniform pressure p_e, with load parameter Q_e given by

$$Q_e = \frac{p_e E}{\sigma_Y^2} \tag{9.4.11}$$

The value of Q_e may be expressed in terms of the ratio $r = Q_e/Q_P$. That is

$$Q_e = rQ_P \tag{9.4.12}$$

The relationship between the load P and the resulting permanent set is then given by (9.4.1) with Q_e in place of Q.

There are two ways in which the overall plastic deformation caused by concentrated loads can differ from that caused by uniform loads: in geometry (the pattern of the plasticity) and in magnitude (the value of permanent set). Reference 10 showed that the geometry is basically similar and that the magnitude of the deformation bears a regular and well-behaved relationship with that caused by a uniform load, and that this relationship depends almost entirely on the load concentration ratio λ. It is only slightly dependent on α and β because their effect is mainly accounted for by the formula for uniform load.

We can establish the general features of the function $r(\lambda)$ by imagining an experiment in which a pneumatic tire carrying a constant load P is applied over the entire surface of a panel in every direction so as to produce the final cumulative value of permanent set. In order to establish the behavior of r as λ becomes large, let us imagine a very large balloon type of tire as in Fig. 9.20, such that λ is relatively large. As the

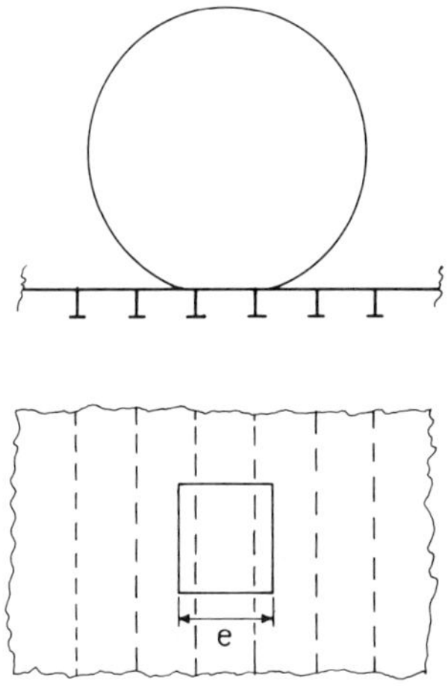

Figure 9.20

tire pressure is decreased the load width increases and if this were continued indefinitely the tire would become equivalent to a uniform pressure load and the equivalent pressure would be equal to the tire pressure. Hence

$$r = \frac{Q_e}{Q_P} = \frac{p_t b^2}{P}$$

and from (9.4.8) this becomes

$$r = \frac{b^2}{e_m^2} = \frac{1}{\lambda^2} \tag{9.4.13}$$

as illustrated in Fig. 9.21.

To establish the behavior of r as λ approaches zero let us imagine a series of experiments with relatively small tires, as in Fig. 9.22, and with progressively larger tire pressures. The width of the tire print e_m, given by (9.4.8) will be smaller each time, and therefore λ will be approaching zero. Since P is constant, Q_P is also constant and so the function $r(\lambda)$ will behave

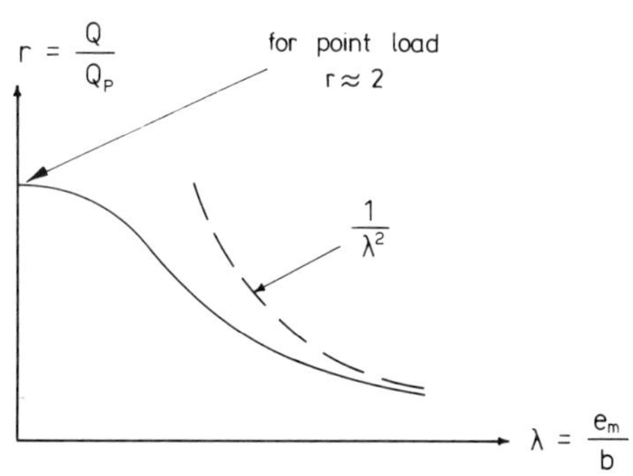

Figure 9.21

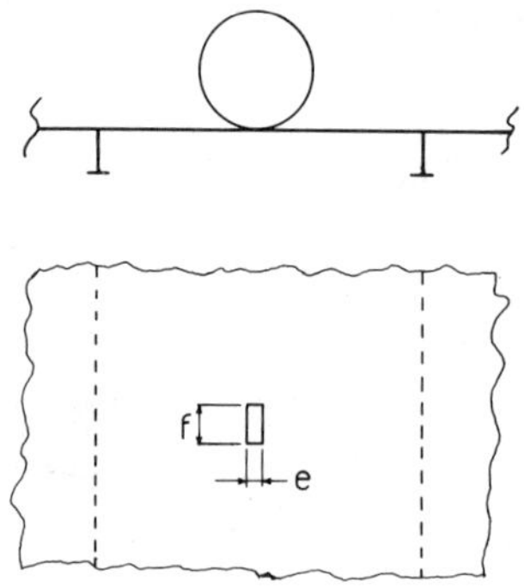

Figure 9.22

in the same manner as Q_e. If the pressure is quadrupled, e_m is halved and so is λ. Since the width of the load is already quite small, halving it does not alter the basic geometry, and since the load P is being kept constant the final amount of permanent set will only be slightly larger than before. Hence the equivalent pressure and its load parameter Q_e will only increase slightly. In the limit, as the width becomes very small, there is no change in geometry and hence Q_e becomes constant. Therefore as λ approaches zero, the value of r approaches a constant value, as shown in Fig. 9.21. In Ref. 10 it is shown that this value is approximately 2 and this information, together with experimental data obtained by Sandvik [11], is used to obtain an approximate expression for $r(\lambda)$

$$r = \frac{0.88}{0.44 + \lambda^2} \tag{9.4.14}$$

which is illustrated in Fig. 9.23.

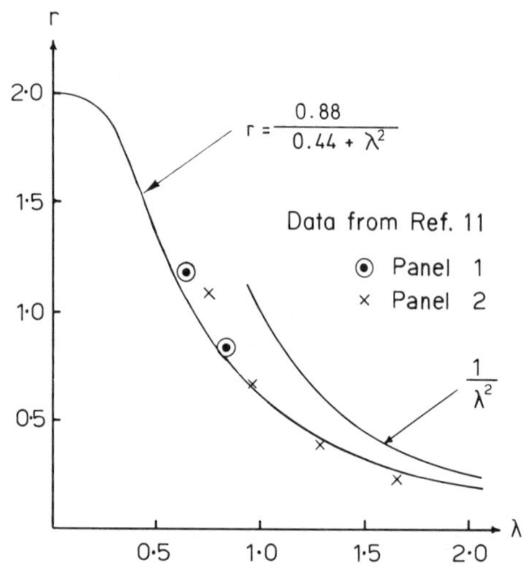

Figure 9.23 Theoretical and experimental values of r.

SELECTION OF LOAD WIDTH e FOR TWIN WHEELS

With twin wheels the question arises as to what value of load width e should be used in order to best represent the load. This depends on the ratio of the tire width d and the wheel spacing s. It also depends on the relative values of the width of the total load envelope parallel to the axle and b, the shorter dimension of the panel; that is, it depends on the ratio (see Fig. 9.24)

$$\chi = \frac{s + d}{b} \tag{9.4.15}$$

The shorter panel dimension is used because the plate bending moment is larger in this direction.

To begin let us consider a situation in which χ is small (say less than 0.4) such that both wheels would be either within the same panel or straddling a stiffener but not very far from the stiffener. In Ref. 12 measurements of the contact pressure across the width of a tire showed that the pressure is largest near the tire walls. Therefore if the wheels are closely spaced (small s/d) the effective width e can be taken as the total width: $e = d + s$. However, as the spacing increases this method of representing the load gradually becomes unrealistic and inaccurate because it artificially increases e_m and hence λ, which in turn causes r, and hence Q_e, to become too low. Thus the error is on the unsafe side. A corollary of this is that making a small or moderate increase in the spacing of twin wheels does not give any substantial decrease in permanent set. Therefore when dealing with moderately spaced twin wheels the best course is to ignore the spacing and use $e = 2d$.

Of course, as the spacing is further increased there

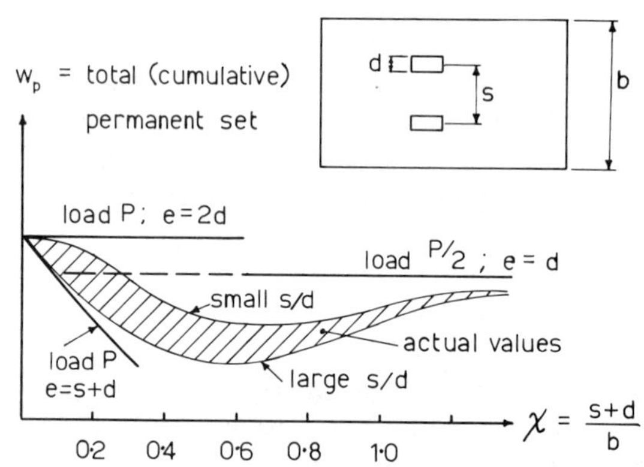

Figure 9.24 Effective width of twin wheels.

is some benefit, as shown (qualitatively only) in Fig. 9.24. However in most cases there is so much variety and uncertainty regarding the spacing and other aspects of the wheel loads which a ship is likely to encounter in its lifetime that it is probably best to adopt the conservative course: $e = 2d$.

For cases in which the wheels are widely spaced relative to b—that is, values of χ aproaching or exceeding 1—the worst loading condition will be when the wheels are acting on different panels, and the loading should be represented as one wheel of width $e = d$ carrying a load $P/2$. This is illustrated in Fig. 9.24 showing that with values of χ above 0.8 or so the maximum value of permanent set corresponds to the single wheel $P/2$ condition. The value of permanent set for this condition will not be as large as when the wheels are close together. However it will not be dramatically less than this because although Q_P has been halved so also has λ, and this causes r, and hence Q_e, to increase. The net effect depends on the width of the tire and other factors. The value of χ at which the single wheel condition can be assumed in place of the $e = 2d$ condition depends on many factors and it would be very difficult to formulate an expression that would cover all cases. After examining a number of design situations with helicopter undercarriages, the authors of Ref. 12 recommended a value of 0.6. Hence the best procedure would seem to be as follows: for $\chi < 0.6$ calculate w_p for both cases (load P with $e = 2d$, and load $P/2$ with $e = d$) and take the larger. For $\chi > 0.6$, take the latter case. This procedure is illustrated in Fig. 9.24.

Permissible Levels of Permanent Set

As mentioned earlier, it is not possible to give general rules for deciding on what amount of permanent set is to be specified as the maximum allowable amount. This depends entirely on the particular circumstances, such as the type of ship, the location and purpose of the plating, and the type of loads. Even when these main factors are specified there may be other factors which influence the decision. Consider, for example, the plating in the vehicle deck of a ro-ro ship. Some types of cargo handling vehicles are operated at relatively high speeds and the permissible value of permanent set may be dictated by the need to avoid a driving hazard. A larger value may be permitted in a nontraffic or parking area. Another factor may be the desire to avoid having pools of water on the deck.

Design for Single Location Loads

As noted earlier, concentrated loads which are expected to occur only rarely should be taken as occurring in the center of the panel. With such loads the pattern of plasticity in the plate is even more complicated than for uniform loads and it would be extremely difficult to derive an expression for permanent set purely from theoretical considerations. Hence for this case also we present an approximate semiempirical design formula obtained in Ref. 10 by combining some basic theory with some experimental results from Ref. 12. The experiments showed that for this type of load the panel aspect ratio α has little effect on the permanent set and that the only effect of load aspect ratio e/f is that loads which are approximately square cause slightly more permanent set than elongated loads. The formula accounts for the latter by means of a simple correction factor Φ.

The range of values of λ covered in the experiment is $0.24 < \lambda < 0.79$. Therefore the design formula may not be accurate for values of λ less than 0.24, and the formula *should not be used* for values of λ greater than 0.8. The formula is

$$Q_P = \Phi\left\{\frac{C_1/\beta^2}{\sqrt{2} - \lambda} + \left[C_2 + \frac{C_3}{\beta^k} + C_4\left(\frac{\lambda}{1-\lambda}\right)^l\right]\left(\frac{w_p}{\beta t}\right)^m\right\}(\lambda < 0.8) \qquad (9.4.16)$$

in which

$$C_1 = 5.0 \qquad k = 1.6$$
$$C_2 = 0.34 \qquad l = 0.8$$
$$C_3 = 3.56 \qquad m = 1.1$$
$$C_4 = 0.23$$

and the correction factor for load aspect ratio is

$$\Phi = 1 - 0.8\left(\frac{ef}{e^2 + f^2}\right)^2 \qquad (9.4.17)$$

A set of design curves based on (9.4.16) is given in Fig. 9.25.

In many cases a given piece of plating is exposed to two or even all three types of load–uniform pressure and single and multiple location concentrated loads. In

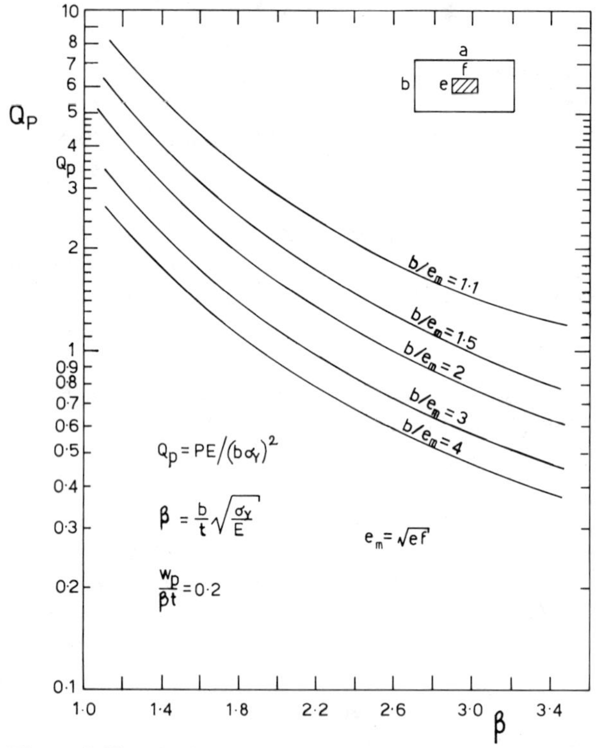

Figure 9.25a Design curves for single location concentrated loads.

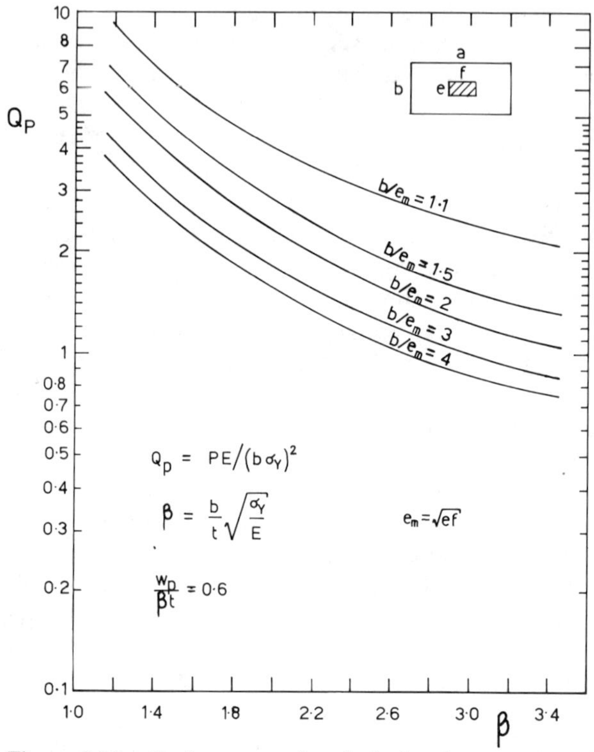

Figure 9.25c Design curves for single location concentrated loads.

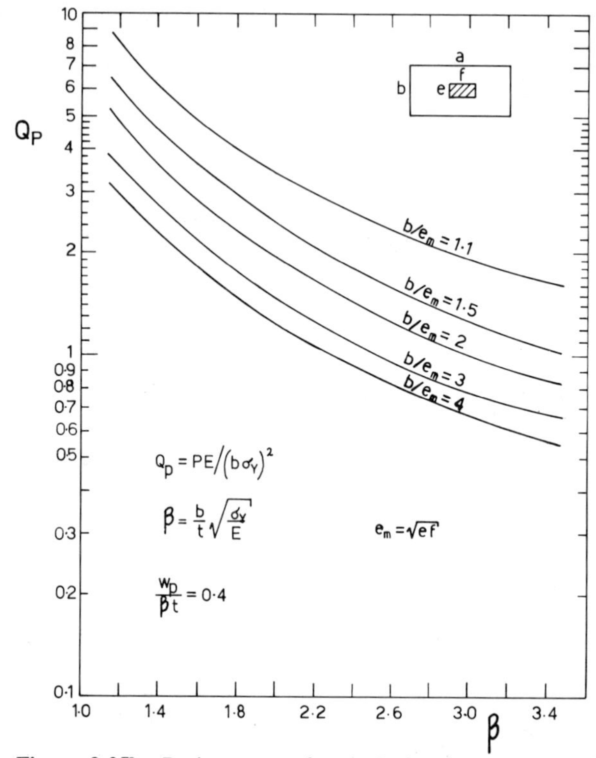

Figure 9.25b Design curves for single location concentrated loads.

the rationally-based, computer-aided structural optimization process of Fig. 1.1 this simply means that in regard to permanent set there are two or three constraints instead of one; there is no need to know (or to try to guess) beforehand which of them will govern the plate thickness. Indeed, it may turn out that none of them do, but rather one of the constraints arising from in-plane strength requirements.

9.5 LARGE DEFLECTION ANALYSIS

As mentioned earlier, there are some design applications which require the calculation of the large deflection response and the ultimate strength of laterally loaded plating. *Elastic* large-deflection theory has only limited application for ship steels because loads severe enough to cause large deflections also cause yielding. Therefore, elastic large-deflection theory is not treated here. Reference 4 presents a summary of the theory, and a series of nondimensional design curves for various boundary conditions.

Elasto-plastic large-deflection theory is quite complicated because there are three separate sources of nonlinearity:

1. Yielding.
2. Large deflections (compatibility; the membrane effect).
3. Restraint from edge pull-in, which appears and becomes significant as deflections become truly large.

Because of the complexity, there is no direct analytical method which is able to yield accurate results for realistic cases. The calculation generally requires computer-based numerical techniques that are too specialized to be dealt with here. Several of the commercially available finite element programs now have elasto-plastic capability and some also have large-deflection capability, and these will undoubtedly increase in number and become more available and economical.

Rigid-Plastic Formulas for Static Pressure Loads

There is, however, one approach which yields simple and explicit load–deflection formulas, although these are only approximate. This is the rigid-plastic hinge-line method originally developed in civil engineering for the design of concrete slabs. The basic approach was presented by Wood [13] and was extended to large deflections by Sawczuk [14]. The principal features of the method, as applied to rectangular plates, are as follows:

1. The edges are assumed to be completely restrained from pulling in, so that at large deflections membrane stresses dominate.
2. Elastic deflections are ignored, and the material is assumed to be rigid-perfectly plastic, that is, zero strain until $\sigma = \sigma_Y$, then unlimited strain with $\sigma = \sigma_Y$.
3. The plate is divided into four rigid regions separated by straight line hinges so as to form a kinematically admissable collapse mechanism, as shown in Fig. 9.26.
4. Yielding is assumed to be governed by the maximum normal stress yield criteria.

With these assumptions the theory results in a pair of alternative equations, depending on whether the permanent set is less than or greater than the plate thickness.

$$\frac{p}{p_c} = \frac{Q}{Q_c} = 1 + \frac{1}{3}W^2\left[\frac{\zeta_0 + (3 + 2\zeta_0)^2}{3 - \zeta_0}\right] \qquad W \le 1 \tag{9.5.1}$$

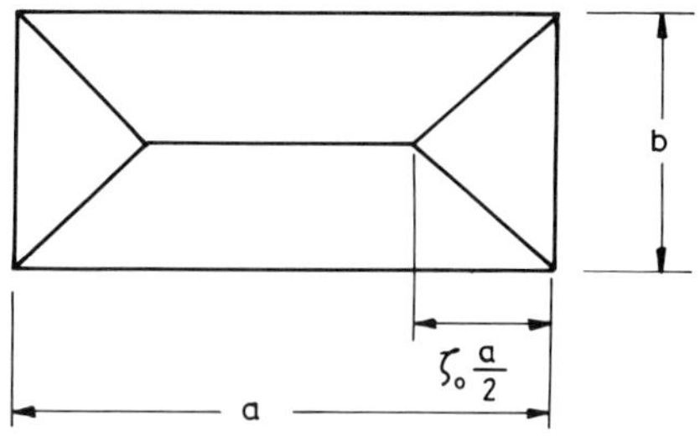

Figure 9.26

$$\frac{p}{p_c} = \frac{Q}{Q_c} = 2W\left[1 - \frac{\zeta_0(2 - \zeta_0)}{3 - \zeta_0}\left(1 - \frac{1}{3W^2}\right)\right] \qquad W > 1 \tag{9.5.2}$$

$$\text{where } p_c = \frac{48M_P}{b^2(\sqrt{3 + \psi^2} - \psi)^2}, \qquad Q_c = \frac{p_c E}{\sigma_Y^2} \tag{9.5.3}$$

$$\zeta_0 = \psi(\sqrt{3 + \psi^2} - \psi) \tag{9.5.4}$$

$$\psi = \frac{b}{a}$$

$$\text{and} \quad W = \frac{w_p}{t}$$

As shown in Fig. 9.26, ζ_0 specifies the location of the interior hinge lines. Equations 9.5.1 and 9.5.2 can be used for either the pressure p or the load parameter $Q = pE/\sigma_Y^2$. Since rigid-plastic theory ignores elastic deflection, Young's modulus E plays no part and so pressure is more appropriate and more convenient. The load parameter is included in order to facilitate comparison with the results of earlier sections. Rigid-plastic theory also ignores the semielastic transition phase which occurs at the beginning of permanent set. Instead, the theory postulates a threshold pressure p_c (or a threshold load parameter Q_c) at which permanent set suddenly begins. The threshold pressure is assumed to be the pressure that would cause sufficient hinge lines so as to form a kinetically admissable collapse mechanism, assuming small deflections, as in Fig. 9.26. From the discussion in Section 9.3, it is clear that this assumption is too optimistic; that is, the threshold pressure of (9.5.3) is too large, as illustrated in Fig. 9.27. The order of the discrepancy can be ascertained by considering the case of a long plate; if ψ is set to zero in (9.5.3), p_c becomes $16M_P/b^2$, which is the pressure to cause three hinges, whereas in Section 9.3 it was shown that a better choice for a thresh-

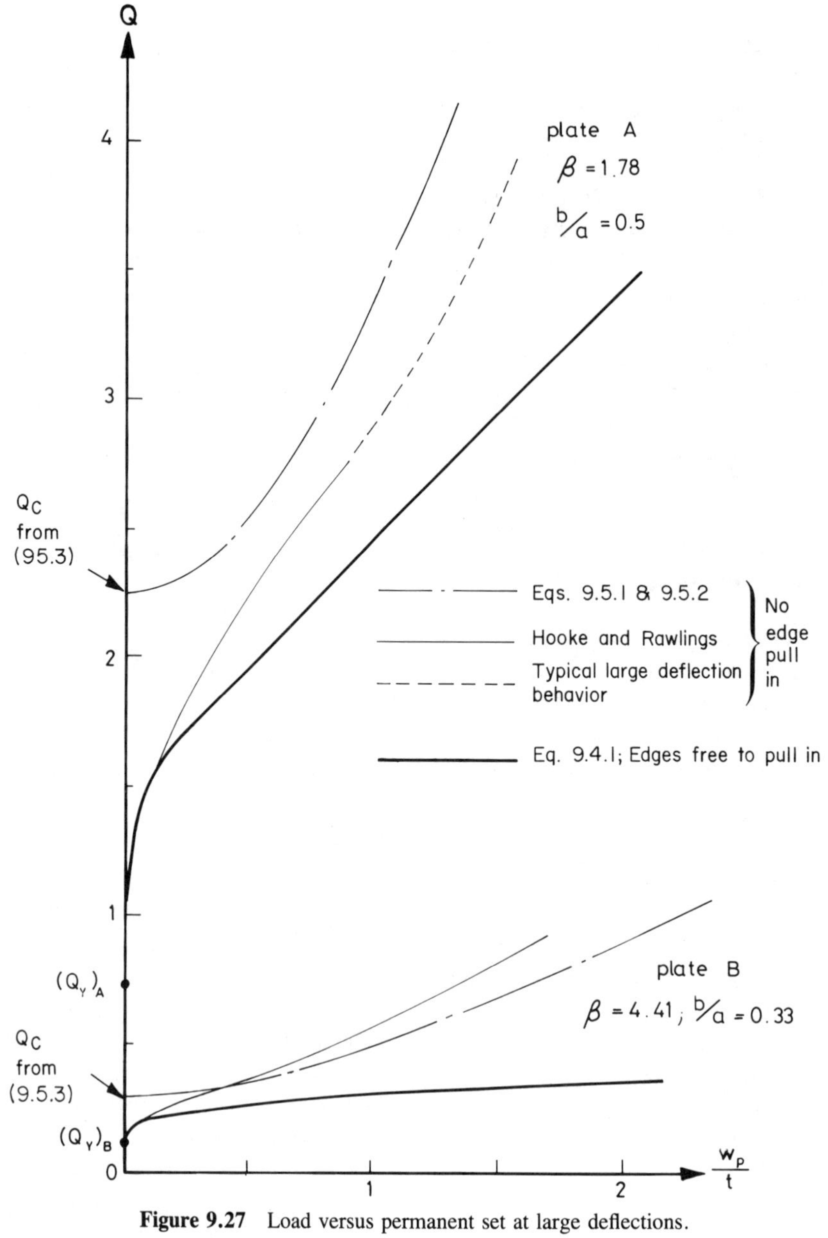

Figure 9.27 Load versus permanent set at large deflections.

old load is the pressure to cause edge hinges only: $12M_P/b^2$. Likewise, the rigid-plastic threshold load parameter, Q_c, is $\frac{4}{3}$ of the true value, Q_{EH} (see Fig. 9.9).

The principal limitation of rigid-plastic theory is the assumption that membrane effects dominate. There are two ways in which this can occur; either because the plate edges are physically restrained from pulling in or because the deflections have become so large that in-plane restraint arises from the nonuniform distribution of in-plane strain due to the nonuniform lateral deflection, such that portions of plating with less deflection provide some restraint to portions with greater deflection. The amount of deflection, and hence of in-plane restraint, depends entirely on the plate slenderness β. For slender plates ($\beta > 2.4$) the lateral deflection grows quickly with load, and consequently some in-plane self restraint arises in the plating. Hence it can be expected that rigid-plastic theory will give better results for slender plating.

Hooke and Rawlings [15] have presented experimental results for laterally loaded plating in which the

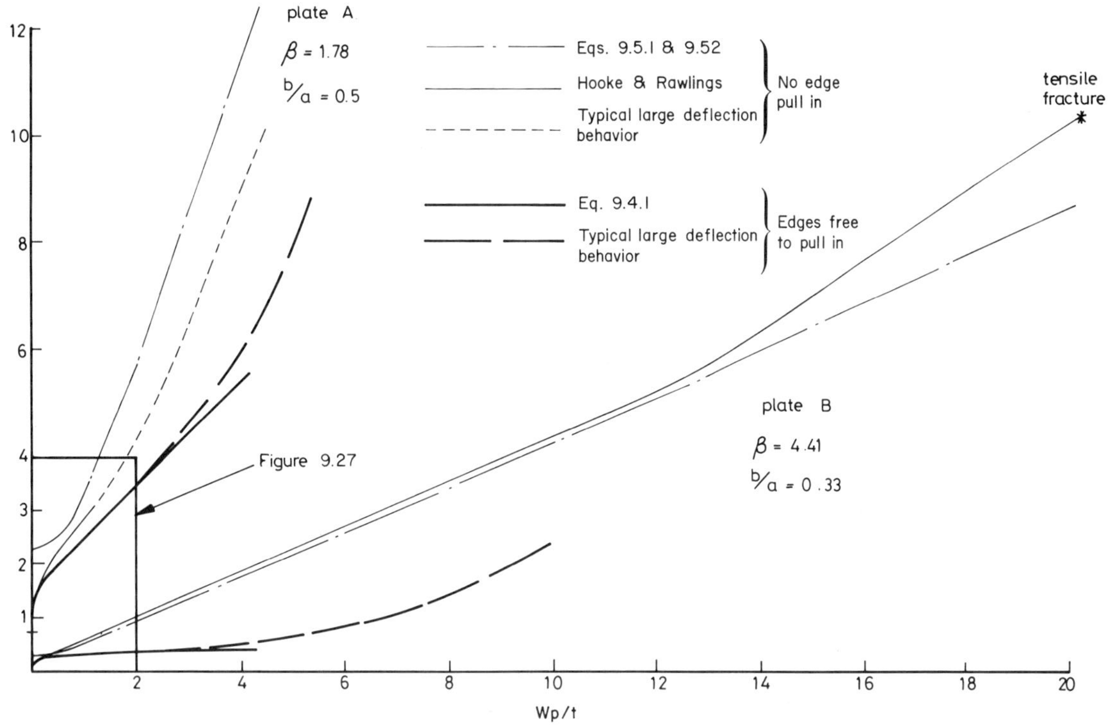

Figure 9.28 Load versus permanent set at large deflections.

plate edges were rigidly bolted to a nondeflecting frame. From these results Jones and Walters [16] have shown that if there is such complete or near-complete restraint, and if the plating is truly slender ($\beta > 2.4$) then the rigid-plastic theory gives good results; for example, plate B in Fig. 9.27. However, the figure also shows that for the relatively sturdy plates which are typical of ship structures the agreement is not as good, especially for $w_p < t$. This was not brought out in Ref. 16 in which a comparison was given only for the three most slender plates tested by Hooke and Rawlings ($\beta = 2.76$, 4.41, and 5.36) and not for the fourth (plate A, $\beta = 1.78$). It is only this fourth plate which has a slenderness that is typical of ship structures.

Moreover, it is important to note that these experimental results are for plates which were physically prevented from pulling in, and this is not a common condition.* In ships, as in any free-standing plated structure, such restraint does not occur (for static loading) until the deflections and the permanent set have become very large. Hence the rigid-plastic theory is not suitable for serviceability-based design (i.e., design based on maximum permanent w_p). In particular, (9.5.1), which is intended for small values of w_p ($w_p < t$) is too optimistic for ship plates of typical slenderness.

In contrast, if there is in-plane restraint and it is maintained throughout the load range, then the rigid-plastic theory gives excellent results for large deflections and large values of w_p. In the experiments of Hooke and Rawlings, one plate (plate B of Fig. 9.27) was loaded to tensile facture. Figure 9.28 shows that, apart from some strain hardening near the ultimate load, the rigid-plastic prediction agrees very well with the experimental result.

We can now summarize, with the aid of Fig. 9.28, the respective areas of application, for static loading, of the semiempirical plate formulas of Section 9.4 and the rigid-plastic formulas. In the figure, the heavy dashed lines indicate qualitatively the behavior of typical ship plating. For values of permanent set within the usual design range (say, $w_p/\beta t < 3$) the edges are essentially unrestrained from pulling in. Hence the semiempirical formulas are sufficiently accurate, whereas the rigid-plastic formulas are too optimistic and are not appropriate for this type of design. More-

*In fact, for static loading, complete restraint against edge pull-in is very difficult to achieve at any stage of the loading, even when a deliberate attempt is made to provide it, as in the previously mentioned experiments. In spite of the careful measures which they took to achieve complete restraint, Hooke and Rawlings found that for the sturdier plates ($\beta < 2$) significant edge pull-in occurred and this explains the decreased slope of their experimental curve compared to the rigid-plastic solution for the case of $\beta = 1.78$ in Fig. 9.27.

over, even for somewhat larger values of allowable permanent set the empirical formulas have the advantage that their error lies on the conservative side rather than the optimistic side. On the other hand, as the edge slope, which is measured by w_p/b or $w_p/\beta t$, becomes large (say $w_p/\beta t > 3$) membrane restraint grows steadily and takes over more and more of the load, and the load–permanent-set curves begin to approach the "edge restrained" curves (see Fig. 9.28).

Eventually, if the material is sufficiently ductile, the plate becomes a fully plastic membrane and if there are no other sources of failure (stress concentration, welding flaws, etc.) it will finally fail when the membrane stress reaches the ultimate tensile stress. Therefore, rigid-plastic theory *is* applicable in the ultimate failure range. Specifically, (9.5.2) can be used for estimating the loads which correspond to very large deflections. Unfortunately this equation does not give the value of the ultimate failure load, but only the load–deflection relationship which occurs as the plate becomes a fully plastic membrane. This is because in a fully plastic membrane, as the load is increased the deflection of the membrane increases so as to carry this increased load, while the membrane stress itself remains constant and equal to the yield stress. However, it is possible to obtain an estimate for the ultimate load by first estimating the membrane strain ε_M from the maximum deflection ($\approx w_p$) as in Section 9.2, and then equating ε_M to the ultimate tensile strain.

Rapidly Varying Loads: Slamming, Collision

Ochi [17] has shown that the maximum impact pressure p_s that acts on the bottom of a ship during a slam can be expressed in the form

$$p_s = kV_s^2$$

where k is a constant which must be determined experimentally for a given body section shape, and V_s is the velocity at the moment of impact. In order for slamming to occur, it is necessary for $V_s \geq V_T$, where V_T is the minimum or threshold relative velocity between a wave and a ship's bow necessary for a slam.

If τ, the duration of the pressure pulse, is small relative to T_p, the natural period of vibration of the plate, then the dynamic aspects (kinetic energy of the load and inertia of the plate) must be taken into account. Elastic structural dynamics theory shows that the natural period of a clamped elastic square plate is

$$T_p = \frac{\pi b^2}{9}\sqrt{\frac{3\rho(1-\nu^2)}{Et^2}}$$

The theory also shows that the response is essentially static whenever τ is appreciably larger than T_p. For example, Biggs [18] has shown that for a one-degree-of-freedom elastic-spring–mass system subjected to an impulsive load, the maximum displacement calculated for the same load applied statically agrees with the corresponding maximum dynamic displacement to within approximately 17% when $\tau > 1.75\ T_p$. Analysis of full-scale measurements [19, 20] and reduced-scale experiments [21, 22] indicates that slamming loads are of sufficient duration to be treated as static. The same also applies to ship collisions, but not, of course, to blast loads or any other truly high-speed loads. Moreover, even if the duration τ is somewhat less than T_p, the quasi-static assumption is conservative because a pressure which is applied for only a brief interval will cause less permanent set than if applied over a longer period. At present, there does not seem to be any simple method for estimating or predicting slam duration, and therefore the safest course is to treat slamming as a static load. Jones [23] has shown that (9.5.1) and (9.5.2) compare well with the experimental results of Refs. 19 to 22. In fact, both equations give better results for moderately dynamic (quasi-static) loads than for loads which are truly static. The contrast is particularly marked for (9.5.1) which neglects the transition phase of static permanent set and which adopts p_c as the threshold pressure. The main reason for the improved accuracy is the fact that with a dynamic load there is more complete restraint against edge pull-in, as is assumed by these equations. The plate can deflect laterally much faster than it can deflect within its own plane and the result is that the ratio of in-plane stiffness to lateral stiffness is much greater in the dynamic case than in the static case. In addition, there are two particular reasons for the marked improvement in the accuracy of (9.5.1). First, as shown by Symonds [24], the difference between elastic-plastic and rigid-plastic response is small if τ/T_p is small and if the total dynamic energy is much larger than the maximum amount of elastic strain energy that the system can absorb. Second, the strengthening effect due to the inertia of the plating is greatest in the initial instants of the loading and this effectively raises the threshold load, so that p_c becomes a more accurate estimate of it. Thus, for example, Jones [23] has shown that for the plates tested in Ref. 19, for which there was no permanent set at a peak slamming pressure of 295 psi, (9.5.3) predicts a threshold pressure of 346 psi, whereas in (9.4.1) the threshold load corresponds to $Q_Y + \Delta Q_0$ with T set equal to unity, and this gives a pressure of 245 psi. This is clearly too low since in reality the plating had not yet undergone any permanent set even at a pressure of 295 psi. Therefore, the rigid-plastic approach is appropriate for quasi-static loads such as slamming, and (9.5.1) and (9.5.2) may be used for the design of plating such that some acceptable amount of permanent set w_p would result from a slamming pressure p_s which itself is chosen on the basis of having an acceptably small risk of occurrence. The latter may be estimated by the methods of Ochi and Motter [25] and Stavovy and Chuang [26]. To facilitate the calculations (9.5.1) and (9.5.2) are presented here in inverted form:

$$\frac{w_p}{t} = \left[\frac{3(3-\zeta_0)}{\zeta_0 + (3-2\zeta_0)^2}\left(\frac{p_s}{p_c} - 1\right)\right]^{1/2} \quad \text{when } \frac{w_p}{t} \le 1 \tag{9.5.5}$$

$$\frac{w_p}{t} = \frac{\dfrac{p_s}{2p_c}\left\{\left(\dfrac{p_s}{2p_c}\right)^2 - \dfrac{4\zeta_0(2-\zeta_0)}{3(3-\zeta_0)}\left[1 - \dfrac{\zeta_0(2-\zeta_0)}{3-\zeta_0}\right]\right\}^{1/2}}{2\left[1 - \dfrac{\zeta_0(2-\zeta_0)}{3-\zeta_0}\right]} \tag{9.5.6}$$

when $\dfrac{w_p}{t} > 1$

in which the threshold pressure p_c is given by (9.5.3). If $p_s < p_c$ there is no permanent set and hence, if there is no particular need to keep shell plating thickness to a minimum, a less sophisticated approach may be used in which the design requirement is simply that $p_c \ge p_s$. Equations 9.5.3 and 9.3.7 then give the following explicit equation for the plate thickness, assuming $\nu = 0.3$:

$$t = 0.272b\left(\sqrt{3 + \left(\frac{b}{a}\right)^2} - \frac{b}{a}\right)\sqrt{\frac{p_s}{\sigma_Y}} \tag{9.5.7}$$

Dynamic Loading and Response

The most useful application of the rigid-plastic approach is in estimating the response of plates to dynamic loads. A principal contributor in this area is Jones, who generalized and extended the hinge line approach to deal with dynamic plastic response [27]. This approach takes into consideration the kinetic energy of the load and the inertia of the plate. When inertia terms are retained in the basic equations it is found that a structure can support a load which is larger than the static limit load, provided that it is removed after a small interval of time. In this circumstance the motion of a structure eventually ceases and reaches a permanent deformed state after expending all the external kinetic energy. It was shown in Ref. 27 that the maximum permanent set w_p of a rigid-plastic fully clamped rectangular plate, which is subjected to a uniformly distributed pressure of constant magnitude p_0 [$p_0 \ge p_c$, where p_c is defined by (9.5.3)] for a duration τ, is

$$\frac{w_p}{t} = (3-\zeta_0)\left[\frac{\{1 + 2\eta(\eta-1)[1-\cos(\gamma\tau)]\}^{1/2} - 1}{2\{1 + (2-\zeta_0)(1-\zeta_0)\}}\right] \tag{9.5.8}$$

where $\gamma^2 = \dfrac{96M_P}{\mu t b^2(3-2\zeta_0)}\left(1 - \zeta_0 + \dfrac{1}{2-\zeta_0}\right)$

$\eta = \dfrac{p}{p_c}$

and $\mu = \rho t$ = mass per unit area (ρ = density).

If $p_0 = p_c$, (9.5.8) gives $w_p/t = 0$ because the theory assumes that either the pressure is sufficiently large or the duration is sufficiently small so that inertia plays a part; if the pressure is merely p_c then the theory implies a very small duration and hence the equation states that no permanent set would occur.

If a rectangular-shaped external pressure–time history has a large magnitude ($p_0 \gg p_c$) and a very short duration τ, then the dynamic loading can be idealized as impulsive. The impulse is the product $p_0\tau$ and is equal to the momentum μV_0, where V_0 is the initial impulsive velocity of the plate. Equation 9.5.8 becomes

$$\frac{w_p}{t} = \frac{(3-\zeta_0)[(1+\Gamma)^{1/2} - 1]}{2[1 + (\zeta_0 - 1)(\zeta_0 - 2)]} \tag{9.5.9}$$

where

$$\Gamma = \frac{\lambda}{6}\left(\frac{b}{a}\right)^2(3-2\zeta_0)\left(1 - \zeta_0 + \frac{1}{2-\zeta_0}\right)$$

and

$$\lambda = \frac{\mu V_0^2 a^2}{4M_P t} = \frac{\rho V_0^2 a^2}{\sigma_Y t^2}$$

which is a nondimensional form of the initial kinetic energy.

Dynamic Concentrated Loads

Since the rigid-plastic approach has been found to be well suited for dynamic uniform pressure loads, it is probably also suitable for dynamic concentrated loads. However it must be borne in mind that this approach examines only permanent set and does not investigate or even allow for the possibility of puncture or tensile fracture, both of which are strong possibilities with loads of this type.

Multiple Location Loads. Here the procedure is the same as for nondynamic multiple location loads except that (9.5.8) is used in place of (9.4.1).

Single Location (Central) Loads. The following relationship between load and permanent set was derived by Kling [28] using rigid-plastic theory and assuming the plate edges to be simply supported.

$$Q_P = Q_{P,T}\left[1 + \left(\frac{w_P}{t}\right)^2 \frac{1 + 2\xi}{3}\right] \qquad \text{for } W < 1$$

$$Q_P = \frac{Q_{P,T}}{2}\left[\frac{3(W-1)^3 + 1}{W}(1 - \xi) + 1 + \frac{W^2}{3}(2 + 7\xi)\right] \qquad \text{for } W > 1 \qquad (9.5.10)$$

where $Q_{P,T}$ is given by (9.4.16) and

$$\xi = \frac{1}{\alpha}\left[\frac{e(b-e) + f(a-f)}{b(b-e) + a(a-f)}\right]$$

and $\quad W = w_p/t$

The load dimensions e and f are shown in Fig. 9.18.

REFERENCES

1. S. Timoshenko and S. Woinowsky-Krieger, *Theory of Plates and Shells,* 2nd ed., McGraw-Hill, 1959.
2. L. G. Jaeger, *Elementary Theory of Elastic Plates,* Pergamon, 1964.
3. J. Clarkson, "Tests of Flat Plated Grillages under Uniform Pressure," *Trans. RINA,* **105,** 1963, pp. 467–484.
4. B. Aalami and D. G. Williams, *Elastic Design of Thin Plates, Vol. 1, Plates Under Transverse Load,* Crosby Lockwood, 1975.
5. W. Muckle, *Strength of Ships' Structures,* Edward Arnold, London, 1967.
6. J. Clarkson, "Strength of Approximately Flat Long Rectangular Plates," *Trans. North East Coast Inst. Eng. Shipbuild.,* **74,** 1958, pp. 21–40.
7. O. F. Hughes, "Design of Laterally Loaded Plating—Uniform Pressure Loads," *J. Ship. Res.,* **25** (2), June 1981, pp. 77–89.
8. J. Clarkson, "Uniform Pressure Tests of Plates with Edges Free to Slide Inwards," *Trans. RINA,* **104,** 1962, pp. 67–76.
9. A. C. Antoniou, "On the Maximum Deflection of Plating in Newly Built Ships," *J. Ship Res.* **24** (1), March 1980, pp. 31–39.
10. O. F. Hughes, "Design of Plating Under Concentrated Lateral Loads," *J. Ship. Res.* **27** (4), 1983.
11. P. C. Sandvik, "Deck Plates Subject to Large Wheel Loads," Report SK/M 28, Univ. of Trondheim, January, 1974.
12. R. L. Jackson and P. A. Frieze, "Design of Deck Structures Under Wheel Loads," *Trans. RINA,* **122,** 1980.
13. R. H. Wood, *Plastic and Elastic Design of Slabs and Plates,* Ronald Press, 1961.
14. A. Sawczuk, "On Initiation of the Membrane Action in Rigid-Plastic Plates," *J. Mécanique,* **3** (1), 1964, pp. 15–23.
15. R. Hooke and B. Rawlings, "An Experimental Investigation of the Behavior of Clamped Rectangular Mild Steel Plates Subjected to Uniform Transverse Pressure", *Proc., Inst. Civil Eng.,* **42,** 1969, pp. 75–103.
16. N. Jones and R. M. Walters, "Large Deflections of Rectangular Plates," *J. Ship Res.* **15** (2) June 1971, pp. 164–171.
17. M. K. Ochi, "Prediction of Occurrence and Severity of Ship Slamming at Sea," 5th Symp. on Naval Hydrodynamics, Office of Naval Research, ACR 112, 1964.
18. J. M. Biggs, *Introduction to Structural Dynamics,* McGraw-Hill, 1964.
19. J. E. Greenspon, "Sea Tests of the U.S.C.G.C. Unimak, Part 3," DTMB Report 978, March 1956.
20. J. W. Wheaton, C. H. Kano, P. T. Diamant, and F. C. Bailey, "Analysis of Slamming Data from the S.S. *Wolverine State,*" Ship Structure Committee Report SSC-210, 1970.
21. R. L. Clevenger and L. C. Melberg, "Slamming of a Ship Structural Model," M.I.T. Engineer's Thesis, May 1963.
22. J. J. Goodwin and J. W. Kime, DTMB Technical Note SML-760-61, July 1963.
23. N. Jones, "Slamming Damage," *J. Ship Res.,* **17** (2), June 1973, pp. 80–86.

24. P. S. Symonds, "Survey of Methods of Analysis for Plastic Deformation of Structures Under Dynamic Loads," Brown University, Report NSRDC/1-67, June 1967.

25. M. K. Ochi and L. E. Motter, "Prediction of Slamming Characteristics and Hull Responses for Ship Design," *Trans. SNAME,* **81,** 1973, pp. 144–176.

26. A. Stavovy and S. L. Chuang, "Analytical Determination of Slamming Pressures for High Speed Vehicles in Waves," *J. Ship Res.,* **20,** (4), December, 1976, pp. 190–198.

27. N. Jones, "A Theoretical Study of the Dynamic Plastic Behavior of Beams and Plates with Finite Deflections," *Int'l J. Solids Struct.,* **7,** 1971, pp. 1007–1029.

28. M. Kling, "Large Deflections of Rectangular Plates Subject to Concentrated Loads," Master's Thesis, Dept. of Ocean Engineering, M.I.T., Cambridge, Mass., June 1980.

Chapter Ten

ORTHOTROPIC PLATE BENDING AND APPLICATION TO DOUBLE WALL PANELS

In many materials, whether natural or man-made, the elastic properties are not *isotropic*; that is, they are not the same in all directions. Examples are wood, glass- and carbon fiber-reinforced materials, laminated materials, and sandwich panels. Even steel plate, which is commonly regarded as isotropic, can exhibit different elastic behavior in different directions, mainly due to the rolling process. In general, such materials are referred to as *anisotropic*.

In the majority of structural materials the anisotropy is orthogonal; that is, the material properties differ only in three (or two) orthogonal directions and are uniform within each direction. In other words, the material has three planes of symmetry with respect to its elastic properties; such materials are called *orthotropic*.

Section 5.4 demonstrated the use of frame analysis for grillages and cross-stiffened panels, in which the beams are modeled individually and the conditions of equilibrium and compatibility are imposed at each intersection. If a cross-stiffened panel is a double wall panel, or if the beams are uniform, numerous, and closely spaced in each direction, then the two sets of beams have a high degree of interaction and an alternative approach becomes possible. In the "orthotropic plate" approach the beams are assumed to interact to such a degree that they can be "smeared" or "blended" into the plating, thus allowing the entire panel to be treated as a single plate. Since the two sets of beams will generally not be identical, the equivalent plate will have two different values of effective bending rigidity in the two orthogonal directions. Hence in regard to bending rigidity the equivalent plate is *orthotropic*. The bending of an orthotropic plate is described by a fourth-order differential equation which is a generalization of the equation for isotropic plates given in (9.1.16). In this method of modeling a stiffened panel, both equilibrium and compatibility are automatically satisfied since they are embodied in the governing orthotropic plate equation.

10.1 ACCURACY AND APPLICABILITY OF THE ORTHOTROPIC PLATE APPROACH

In the past, before computer-based grillage analysis methods were available, the orthotropic plate method was the principal method for the analysis of grillages and stiffened panels. Because of its relative simplicity it was used for a wide variety of panel types. The pioneering work in applying this method to ship panels was done by Schade [1, 2]. As would be expected, the accuracy obtained with this approach depends entirely on the degree to which the stiffened panel resembles a uniform orthotropic plate. The accuracy is greatest for double wall cross-stiffened panels such as a double bottom. A smaller but still satisfactory degree of accuracy is achieved with singly plated cross-stiffened panels if there is a large number of uniform beams in each direction. However, the accuracy diminishes as the number of beams in either direction diminishes. For example, Ref. 3 shows that for a 9×3 grillage the error in maximum bending stress is 15%. In fact, of its very nature a singly plated grillage under a lateral load is represented more accurately by the discrete beam approach, by means of either a grillage method such as the Distributed Reaction method or, for greatest accuracy, by frame analysis. Therefore, now that computer programs which implement these methods are readily available, there is less reason to use the orthotropic plate method for the analysis of singly plated panels. However, it will be shown that for doubly plated panels the much simpler orthotropic plate method is quite accurate and hence for this application it is a viable and often useful alternative to computer-based discrete beam methods.

The orthotropic plate approach can also be used for cross-stiffened panels under in-plane compression; that is, buckling analysis, and for combined lateral and in-plane loads. The panel response is governed for small deflections by an orthotropic version of (9.2.1), with N_x being the applied load. The use of orthotropic plate theory for in-plane and for combined lateral and in-plane loads is discussed more fully in Chapter 13, which deals with stiffened panels subjected to these loads.

The present chapter deals only with laterally loaded orthotropic plates, and it examines only one application: double wall panels. The reasons for this specialization are given in this first section. But first of all, to assist in obtaining a better overall perspective, we list some of the main points regarding the analysis of stiffened plating under various loads, even though some of these points are not discussed or explained in this chapter, but rather in Chapters 13 and 14. There are six main points:

1. Besides its superiority for grillages (i.e., laterally loaded cross-stiffened panels) the discrete beam approach is also more accurate than the orthotropic plate approach for all *singly plated* stiffened panels, that is, for all types of loads and for unidirectional and cross-stiffening. For *doubly plated* panels the two methods have approximately the same accuracy.
2. For *doubly plated* panels the elastic buckling stress is very large—far greater than the yield stress—and hence the usual (elastic) orthotropic plate theory cannot be used to calculate the buckling load of a double wall structure.
3. The analysis of plating under combined lateral and in-plane loads can be dealt with at three different levels of accuracy:
   a. *Linear small-deflection theory,* based on (9.1.16), which ignores any interaction between the lateral and in-plane loads and simply superimposes the responses (stress and deflection) from these two loads. In the present chapter we give the orthotropic equivalent of (9.1.16).
   b. *Corrected small-deflection theory,* which is of two types. In *first-order theory* the response to the lateral load is calculated separately, the same as in (a), but in combining the responses to the two loads allowance is made for the nonlinear effect of the in-plane load (by means of a magnification factor which, for isotropic plates, is calculated from (9.2.1), with $p = N_{xy} = N_y = 0$ and with N_x being the applied load). In *second-order theory* the initial lateral deflection (due to eccentricity or to lateral load) is included in the differential equation, thus allowing interaction with N_x, but no other interaction between lateral and in-plane effects is considered.
   c. *Large-deflection theory,* which for unstiffened plates is summarized by a coupled pair of differential equations.
4. Orthotropic plate theory has the same three levels, and for the analysis of stiffened panels various authors have presented solutions, usually in the form of curves, based on linear small-deflection theory [5, 6], corrected small-deflection theory [7, 8, 9], and large-deflection theory [10, 11, 12]. In the main these solutions were developed for *singly plated* panels, which require the more accurate levels of theory when there are combined lateral and in-plane loads, and when either one or both are of large magnitude.
5. Because of the large bending stiffness of a double wall panel, the deflection under a lateral load is very small relative to the "thickness" of the panel: $w \ll h$. Therefore method (a) is sufficiently accurate for calculating the response of a double wall structure to a lateral load. This has been confirmed experimentally [4].
6. These same tests have shown that for double wall panels method (a) is also sufficiently accurate for analysis under combined loads. That is, the total stress acting on individual panels of the plating on either face of a double wall structure is equal to the sum of the applied in-plane stress σ_a (e.g., the hull girder stress) plus the stress due to bending of the double wall structure by the lateral load (only) as calculated by *linear, small-deflection, orthotropic plate* theory; moreover, this linear superposition remains valid until the summed stress reaches the value which causes collapse of the individual plate panels (and which thereby causes collapse of the double wall panel).

From these points and from the preceding discussion, three conclusions may be drawn:

1. For singly plated cross-stiffened panels, discrete beam methods in the form of simple computer programs are more versatile and more accurate than the orthotropic plate approach.
2. For doubly plated panels in ship structures, orthotropic plate theory has approximately the same (and sometimes better) accuracy as the discrete beam method and hence for this application it constitutes a valid hand-solution alternative to the latter, which is necessarily computer-based.
3. For double wall panels, *small-deflection* orthotropic plate theory is sufficient, both for bending analysis and for combined loads. Hence the results of a bending analysis can be used directly as part of an ultimate strength analysis.

Thus, double wall panels are not only amenable to analysis by small-deflection orthotropic plate theory, but in fact they constitute the most appropriate application for that theory. As noted previously, several solutions have been presented for laterally loaded stiffened panels based on this theory, but these were intended mainly for singly plated panels. For doubly plated panels there are two significant differences which must be taken into account:

1. In most singly plated panels the deflection due to shear is insignificant compared to the bending deflection and can be ignored. But in a doubly plated panel the bending stiffness is much larger relative to the shear stiffness; hence the shear deflection becomes significant and cannot be ignored. For example, in the double bottom structure which is analyzed in Section 10.3 the shear deflection is 18% of the total deflection. However, shear deflection continues to have only a slight effect on the bending response and so it can be omitted from the bending analysis and calculated separately.
2. It will be shown in the next section that for double wall panels, distortion of the webs causes a direct and significant reduction in the torsional rigidity of the panel, whereas for a singly plated panel this effect is small. For this reason the various expressions for torsional rigidity which have been presented heretofore either do not allow for this effect [5,7,8,9,11] or make an allowance which is only applicable to singly plated panels [6,10,12].

Therefore the principal aim of this chapter is to present an improved small-deflection orthotropic plate method which calculates shear deflection (separately) and which takes account of web distortion. The method possesses both simplicity and accuracy, and it permits a hand analysis of a doubly plated cross-stiffened panel subject to lateral loads. Since small-deflection theory is also valid for double wall panels under combined lateral and in-plane loads, this analysis method can also be used as part of the ultimate strength analysis of double wall panels.

10.2 ANALYSIS OF ORTHOTROPIC PLATE BENDING WITH SHEAR DEFLECTION

The stress-strain equations for an orthotropic material are of the form:

$$\begin{aligned}
\varepsilon_x &= S_{11}\sigma_x + S_{12}\sigma_y + S_{13}\sigma_z \\
\varepsilon_y &= S_{21}\sigma_x + S_{22}\sigma_y + S_{23}\sigma_z \\
\varepsilon_z &= S_{31}\sigma_x + S_{32}\sigma_y + S_{33}\sigma_z \\
\gamma_{xy} &= S_{44}\tau_{xy} \\
\gamma_{yz} &= S_{55}\tau_{yz} \\
\gamma_{zx} &= S_{66}\tau_{zx}
\end{aligned}$$

For a plate in a state of plane stress these equations reduce to:

$$\begin{aligned}
\varepsilon_x &= S_{11}\sigma_x + S_{12}\sigma_y \\
\varepsilon_y &= S_{21}\sigma_x + S_{22}\sigma_y \\
\varepsilon_z &= S_{31}\sigma_x + S_{32}\sigma_y \\
\gamma_{xy} &= S_{44}\tau_{xy}
\end{aligned}$$

and for plates, we need consider only σ_x, σ_y, and τ (the xy subscript is no longer needed). The first two of the equations can be rewritten as

$$\sigma_x = \frac{1}{S_{11}S_{22} - S_{12}S_{21}}(S_{22}\varepsilon_x - S_{12}\varepsilon_y)$$

$$\sigma_y = \frac{1}{S_{11}S_{22} - S_{12}S_{21}}(-S_{21}\varepsilon_x + S_{11}\varepsilon_y)$$

and the fourth equation is

$$\tau = \frac{1}{S_{44}}\gamma$$

From symmetry of strain energy it may be shown that

$$S_{12} = S_{21}$$

Therefore these three equations are of the form

$$\sigma_x = C_1\varepsilon_x + C_2\varepsilon_y$$
$$\sigma_y = C_3\varepsilon_y + C_2\varepsilon_x$$
$$\tau = C_4\gamma$$

Thus we see that the stress-strain equations involve four coefficients which represent material properties. The last of these is obviously the shear modulus G, the same as for isotropic materials. There are two different forms for expressing the other three coefficients, each being a way of attaching some physical meaning to them. In the first form all three are presented as effective values of Young's Modulus

$$\sigma_x = E'_x\varepsilon_x + E^*\varepsilon_y$$
$$\sigma_y = E'_y\varepsilon_y + E^*\varepsilon_x \qquad (10.2.1)$$
$$\tau = G\gamma$$

In the second form two of the coefficients are presented as effective values of Poisson's ratio

$$\sigma_x = \frac{1}{1 - \nu_x\nu_y}[E_x\varepsilon_x + \nu_y E_x\varepsilon_y]$$
$$\sigma_y = \frac{1}{1 - \nu_x\nu_y}[E_y\varepsilon_y + \nu_x E_y\varepsilon_x] \qquad (10.2.2)$$
$$\tau = G\gamma$$

together with the symmetry requirement that $\nu_y E_x = \nu_x E_y$. Thus there are again four independent coefficients. In previous applications of orthotropic plate theory to stiffened panels the equivalent orthotropic plate has been obtained by choosing one of these forms of the material stress-strain relationship and then using various analogies or behavior models to relate the material coefficients to the overall structural characteristics (axial, bending, and shear rigidities). However since a double wall panel in a ship structure contains only one type of material it seems preferable to begin with the isotropic stress-strain relationships and to allow for the geometric orthotropy at the structural level rather than at the local stress-strain level. Hence we adopt the usual isotropic relations

$$\sigma_x = \frac{E}{1 - \nu^2}(\varepsilon_x + \nu\varepsilon_y)$$
$$\sigma_y = \frac{E}{1 - \nu^2}(\varepsilon_y + \nu\varepsilon_x) \qquad (10.2.3)$$
$$\tau = G\gamma$$

The derivation of the governing differential equation for a laterally loaded orthotropic plate follows the same steps as for an isotropic plate in Section 9.1. The nomenclature and sign convention are the same, and the various forces and moments are shown in Fig. 9.3.

The first step is to consider equilibrium of the element as a whole. The distributed forces and moments acting on the element are the same as for the isotropic case and the widths s_x and s_y drop out in the same way as did dx and dy. Therefore the resulting equations are the same as (9.1.9), (9.1.10), and (9.1.12), namely:

equilibrium of vertical forces

$$\frac{\partial q_x}{\partial x} + \frac{\partial q_y}{\partial_y} + p = 0 \qquad (10.2.4)$$

equilibrium of moments parallel to the x-axis

$$\frac{\partial m_{xy}}{\partial_x} - \frac{\partial m_y}{\partial_y} + q_y = 0 \qquad (10.2.5)$$

and equilibrium of moments parallel to the y-axis

$$\frac{\partial m_{xy}}{\partial_y} - \frac{\partial m_x}{\partial_x} + q_x = 0 \qquad (10.2.6)$$

Direct and Indirect Effects of Shear

The next step is to obtain the basic compatibility relationship between local deformation (strain) and overall plate displacement (deflection, and/or its derivatives). For beams and for plates this is the strain–curvature relationship; for example, in the x-direction $\varepsilon_x = -z(\partial^2 w/\partial x^2)$. However, in a beam or plate subjected to transverse shear force there is additional deflection due to shear strain γ, and this has the effect of increasing the curvature for a given load. Figure 10.1 shows an element of plating of unit width and thickness t subjected to a pressure p. For simplicity the element chosen is at a horizontal edge (clamped but with no restriction against warping of the cross-section). If shear deflection is ignored the plate takes on a curvature $(\partial^2 w/\partial x^2)_b$ in the x direction. If

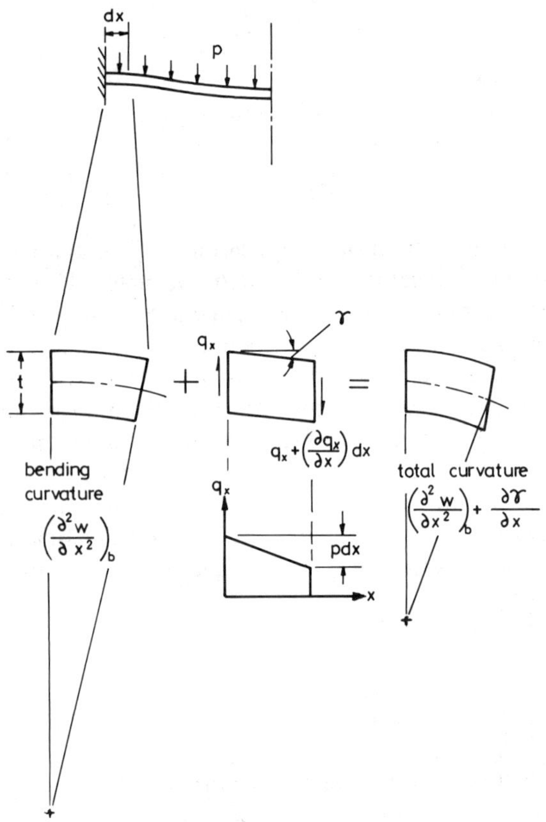

Figure 10.1 Effect of shear on curvature.

the plate is of a sandwich or cellular (flange-web) construction, then the vertical shear force is taken mainly by the core or by the webs, and hence the shear strain γ is essentially constant through the core thickness or the web height:

$$\gamma = \frac{q_x}{Ga_x} \tag{10.2.7}$$

where a_x is the area (per unit width) that is effective in carrying the shear force (per unit width) q_x. The strain γ represents an increase of slope, over and above the slope $\partial w / \partial x$ due to bending. Therefore the total curvature (rate of change of slope) is

$$\frac{\partial}{\partial x}\left(\frac{\partial w}{\partial x} + \gamma\right)$$

Substituting from (10.2.7) the total curvature becomes

$$\frac{\partial^2 w}{\partial x^2} = \left(\frac{\partial^2 w}{\partial x^2}\right)_b + \frac{1}{Ga_x}\frac{\partial q_x}{\partial_x} \tag{10.2.8}$$

The shear distributions $q_x(x,y)$ and $q_y(x,y)$ are not known, and for an exact treatment of the role of shear in plate bending these functions have to be determined as part of the solution. As will be shown, this leads to a sixth-order differential equation for w. However, to deal adequately with the role of shear in laterally loaded double wall panels, it is essential to realize that shear has two different and distinct effects, both of which result in extra deflection:

1. It directly causes an additional vertical deflection, over and above the deflection caused by bending and twisting. As noted previously, this "direct" shear deflection is the cumulative effect of vertical shear strain and it results in extra "sag" or curvature of the panel.
2. It causes distortion of the webs, which in turn reduces the torsional rigidity of the panel, and this causes a larger deflection than would otherwise have occurred.

The first of these effects—the "direct" shear deflection—is well known because it occurs in all structures that carry a transverse load. This shear deflection is distinct from bending and twisting deflections, and within the limits of small-deflection theory it can be calculated separately and superimposed. Also, for double wall panels subjected to a constant pressure loading, this direct shear has only a small effect on bending moment. This is shown in Ref. 4 in which the bending moment computed by an approximate solution which ignores the effect of shear is shown to agree closely with the value given by the sixth-order method. In fact, the approximate method gives a slightly *larger* maximum bending moment and so the error is on the conservative side. The situation is analogous to that in a hull girder, in which the transverse shear causes only a slight departure from the bending stress predicted by simple beam theory (Section 3.8).

The second of the preceding two effects of shear, which results in an "indirect" shear deflection, is a special feature of double wall panels of flange-and-web construction (cellular, honeycomb, etc.) and for all such structures this second effect is decidedly more significant than the first. The torsional rigidity D_{xy} of an orthotropic plate is the ratio of an applied twisting moment to the resulting twist: $D_{xy} = m_{xy}/(\partial^2 w / \partial x \partial y)$. In the case of material orthotropy this rigidity depends mainly on the shear modulus G. But in a double wall panel the orthotropy is due to the internal structural arrangement and therefore D_{xy} depends also on the dimensions of the flanges and webs. In particular, as will be shown, D_{xy} is strongly influenced by the amount of shear distortion of the webs, and this in turn depends on the shear rigidity of the webs Ga_x (and Ga_y in the y-direction) which was introduced in (10.2.7). Thus for a double wall panel, D_{xy} is strongly influenced by a_x and a_y. A reduction in these areas causes a reduction in D_{xy} and this in turn causes a larger deflection for any given load. The precise pattern of the extra deflection depends on the particular load, but the basic relationship between the shear area and D_{xy} is independent of the load. This relationship can be expressed as a reduction factor R_s which is a function of a_x and a_y and which multiplies D_{xy}. In previous applications of orthotropic plate theory to stiffened panels this relationship was not explicitly allowed for in the expressions derived for D_{xy}. For singly plated panels this is acceptable because such panels do not have the large shear stress which comes from closed cell torsion; hence, the amount of shear distortion is quite small, and so the dependency of D_{xy} on a_x and a_y is negligible. But for doubly plated panels this dependency is of prime importance because the torsional rigidity D_{xy} is large *and* the level of shear distortion is large; therefore the influence of the latter on the former is crucial. As will be shown subsequently, for a typical double bottom the reduction factor R_s is about $\frac{1}{3}$; that is, the value of D_{xy} after allowing for web distortion is about one-third of the value when this distortion is ignored. If D_{xy} is determined from tests this effect is automatically included, but there do not seem to be any available test results of a direct measurement of D_{xy} for double wall panels.

The foregoing discussion may be summarized in two main points regarding the bending analysis of double wall panels:

1. The direct shear deflection does not need to be included in the analysis; it can be calculated separately and superimposed, with only a slight loss of accuracy.
2. The decrease in the torsional rigidity D_{xy} due to shear distortion of the webs must be included in the analysis, and it can be expressed as a reduction factor which is a function of the structural dimensions of the panel and which multiplies D_{xy}.

These two considerations point the way toward a simplified and yet accurate method of analysis, which constitutes an alternative to the sixth-order method. The simplified method consists of two separate analyses: an ordinary (fourth-order) orthotropic plate bending analysis which ignores the direct shear deflection but which accounts for the reduction in D_{xy} due to web distortion, and a separate analysis which calculates the direct shear deflection. Apart from having a different value of D_{xy}, the first analysis is the same as previous orthotropic plate analyses and so solutions are already available. The second analysis need only be approximate and it will be shown that for constant pressure loading it consists of a second-order differential equation (Poisson's equation), for which solutions are again readily available.

In the following sections the sixth-order method is presented first, then the simplified method, and then the latter is demonstrated by applying it to the double bottom test structure of Ref. 4. The results are then compared with the test results and with the solutions obtained by other methods, including the sixth-order method.

Exact (Sixth-Order) Method

From (10.2.8) the bending curvature in the x-direction is

$$\left(\frac{\partial^2 w}{\partial x^2}\right)_b = \frac{\partial^2 w}{\partial x^2} - \frac{1}{Ga_x}\frac{\partial q_x}{\partial_x}$$

and from the strain-curvature relationship the strain in the x-direction is

$$\begin{aligned}\varepsilon_x &= -z\left(\frac{\partial^2 w}{\partial x^2}\right)_b \\ &= -z\left(\frac{\partial^2 w}{\partial x^2} - \frac{1}{Ga_x}\frac{\partial q_x}{\partial x}\right)\end{aligned}$$

A similar result is obtained for the strain in the y-direction. Therefore, from (10.2.3)

$$\begin{aligned}\sigma_x &= \frac{-Ez}{1-\nu^2}\left(\frac{\partial^2 w}{\partial x^2} + \nu\frac{\partial^2 w}{\partial y^2} - \frac{1}{Ga_x}\frac{\partial q_x}{\partial x} - \frac{\nu}{Ga_y}\frac{\partial q_y}{\partial y}\right) \\ \sigma_y &= \frac{-Ez}{1-\nu^2}\left(\frac{\partial^2 w}{\partial y^2} + \nu\frac{\partial^2 w}{\partial x^2} - \frac{1}{Ga_y}\frac{\partial q_y}{\partial y} - \frac{\nu}{Ga_x}\frac{\partial q_x}{\partial x}\right)\end{aligned} \tag{10.2.9}$$

The next step is to impose equilibrium between the external bending moments and the moments of the internal stresses on the x and y faces of the equivalent orthotropic plate element. It is at this point that we allow for the nonhomogeneous distribution of material in a double wall structure, which is the cause of the orthotropy. Figure 10.2 shows the dimensions and structural parameters and Fig. 10.3 shows a typical element of the structure, which is the same basic box element which was used in Section 8.8 in deriving a

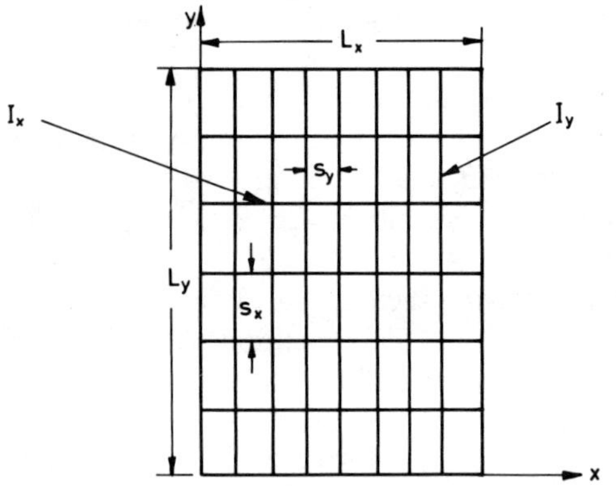

Figure 10.2 Notation for orthotropic plate model of a cross-stiffened panel.

finite element representation of the double bottom. To allow for partial loss of effectiveness of the flanges, due to shear lag and out-of-plane distortion, the flanges of the two half-width beams in the x-direction have a combined effective width b_{ex}, which may be less than s_x. Similarly, the two beams along the y edges have a combined effective flange width b_{ey}, which may be less than s_y. On the x and y faces of the element the areas over which the vertical shear forces act are the combined effective cross-sectional area of the webs of the edge beams, $h\bar{t}_{wx}$ and $h\bar{t}_{wy}$ respectively, where $\bar{t}_{wx}$ and $\bar{t}_{wy}$ are reduced web thicknesses to allow for cutouts. A method for estimating this was given previously, in (8.8.14).

In the current notation the widths of the element are s_x and s_y, whereas in the derivation of the isotropic plate bending equation the element had widths dy and dx. In that case, since the element was homogeneous, the widths dropped out of the equilibrium relationship. For example, in (9.1.8a) the external moment on the left-hand side is actually $m_x\, dy$ and on the right-hand side the cross-sectional area of the element is actually $dy\, dx$, but since dy occurs on both sides it was not written. In the present case the corresponding external moment is $m_x\, s_x$ and the cross-sectional area of the basic element is A_e, the area of the effective flanges and of the two half-thickness webs. Therefore, the equilibrium equation is

$$m_x s_x = \int_{A_e} \sigma_x z\, dA_e$$

Substituting for σ_x from (10.2.9) gives

$$m_x s_x = \frac{-E}{1-\nu^2}\int_{A_e}\left(\frac{\partial^2 w}{\partial x^2} + \nu\frac{\partial^2 w}{\partial y^2} - \frac{1}{Ga_x}\frac{\partial q_x}{\partial x} - \frac{\nu}{Ga_y}\frac{\partial q_y}{\partial y}\right) z^2\, dA_e$$

which can be written as

$$m_x = \frac{-EI_{xe}}{(1-\nu^2)s_x}\left(\frac{\partial^2 w}{\partial x^2} + \nu\frac{\partial^2 w}{\partial y^2} - \frac{1}{Ga_x}\frac{\partial q_x}{\partial x} - \frac{\nu}{Ga_y}\frac{\partial q_y}{\partial y}\right)$$

where

$$I_{xe} = \int_{A_e} z^2\, dA_e$$

is the combined moment of inertia of the two half-thickness webs and the two plate flanges (together with their stiffeners) of total width b_e.

The external factor $1/(1-\nu^2)$ in the foregoing equation comes from the isotropic stress-strain equations and it applies only to the portions of the cross section in which transverse strain is prevented [see, for

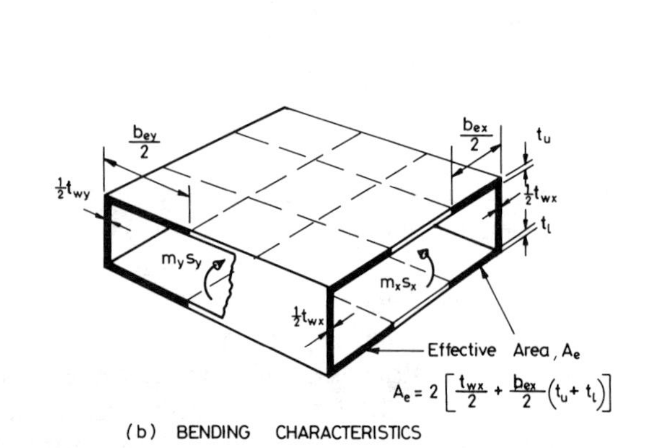

Figure 10.3 Shear and bending characteristics of a double wall panel.

example, the derivation of (9.1.1)]. Thus it applies to the plate flanges, but not to the stiffener or the web portions of the cross section. Therefore in computing I_{xe} the contribution of the web and stiffeners should be multiplied by $(1 - \nu^2)$, to counteract the effect of the external factor. By analogy with isotropic plate theory it is evident that in the orthotropic plate representation the plate flexural rigidity in the x-direction is

$$D_x = \frac{EI_{xe}}{(1 - \nu^2)s_x}$$

A similar result would be obtained by considering equilibrium between m_y and σ_y. The combined results are

$$m_x = -D_x\left(\frac{\partial^2 w}{\partial x^2} + \nu\frac{\partial^2 w}{\partial y^2} - \frac{1}{Ga_x}\frac{\partial q_x}{\partial x} - \frac{\nu}{Ga_y}\frac{\partial q_y}{\partial y}\right)$$

$$m_y = -D_y\left(\frac{\partial^2 w}{\partial y} + \nu\frac{\partial^2 w}{\partial x^2} - \frac{1}{Ga_y}\frac{\partial q_y}{\partial y} - \frac{\nu}{Ga_x}\frac{\partial q_x}{\partial x}\right) \tag{10.2.10}$$

where

$$\left.\begin{aligned} D_x &= \frac{Ei_x}{1 - \nu^2} \quad \text{and} \quad D_y = \frac{Ei_y}{1 - \nu^2} \\ i_x &= \frac{I_{xe}}{s_x} \quad \text{and} \quad i_y = \frac{I_{ye}}{s_y} \end{aligned}\right\} \tag{10.2.11}$$

The next step is to express the twisting moment m_{xy} in terms of w. As shown by Libove and Batdorf [13], the general expression is

$$m_{xy} = \tfrac{1}{2}D_{xy}\left(\frac{\partial^2 w}{\partial x \partial y} - \frac{1}{Ga_y}\frac{\partial q_y}{\partial x} + \frac{\partial^2 w}{\partial x \partial y} - \frac{1}{Ga_x}\frac{\partial q_x}{\partial_y}\right) \tag{10.2.12}$$

in which D_{xy} is the torsional stiffness coefficient. This can be defined in terms of the basic element of Fig. 10.3. If that element were subjected to a distributed twisting moment m_{xy} (acting on all four faces of the element) and if there were no other forces or moments acting, then the distortion would be a pure twist $\partial^2 w/\partial x \partial y$, and D_{xy} is defined as the ratio of twisting moment to twist

$$D_{xy} = \frac{m_{xy}}{\dfrac{\partial^2 w}{\partial x\, \partial y}} \tag{10.2.13}$$

D_{xy} can be determined theoretically or by experiment. Within the field of ship structures the principal theoretical contribution was made by Schade [1,2,5]. However, Schade's expressions do not include the effect of shear distortion of the webs and hence the presentation and discussion of them will be deferred until the next section. Libove and Batdorf [13] give a procedure for determining D_{xy} experimentally (which would automatically include the effect of web distortion) but for analysis and design a mathematical expression is required. Hence we now proceed to derive such an expression for a double wall panel, based on the information that was obtained in Section 8.8 regarding the twisting of a double wall box element. In that earlier study a twist was imposed by means of four unit deflections. Since this deflection pattern does not involve bending, the four webs remain straight and therefore the derivative $\partial^2 w / \partial x\, \partial y$ can be evaluated by first-order finite difference expressions

$$\begin{aligned} \frac{\partial^2 w}{\partial x\, \partial y} &= \frac{1}{s_y}\left[\left(\frac{\partial w}{\partial x}\right)_{34} - \left(\frac{\partial w}{\partial x}\right)_{12}\right] \\ &= \frac{1}{s_y}\left[\frac{w_3 - w_4}{s_x} - \frac{w_2 - w_1}{s_x}\right] \\ &= \frac{w_1 - w_2 + w_3 - w_4}{s_x\, s_y} \\ &= \frac{-4}{s_x\, s_y} \end{aligned} \tag{10.2.14}$$

In the derivation of (8.8.11) in Section 8.8, the twisting moments corresponding to this imposed twist are $-F_iL_i$ and $-F_jL_j$, in the notation of Figs. 8.64, 8.65, and 8.66. The minus signs occur because the imposed twist is negative. Hence the twisting moment per unit length, m_{xy}, is equal to $-F_i$ (or $-F_j$). The derivation shows that the value of this twisting moment is the same in each direction and is given by

$$m_{xy} = -\frac{F}{2} \tag{10.2.15}$$

where F is given by (8.8.11), which allows for the effect of shear distortion of the webs. Substitution of (10.2.14) and (10.2.15) into (10.2.13) gives

$$D_{xy} = \frac{m_{xy}}{\dfrac{\partial^2 w}{\partial x\, \partial y}} = \frac{\dfrac{-F}{2}}{\dfrac{-4}{(s_x\, s_y)}} = \frac{F\, s_x\, s_y}{8}$$

and substitution from (8.8.11) gives (after changing L_i and L_j to s_x and s_y)

$$D_{xy} = \frac{Eh^2 \hat{t} R_s}{2(1+\nu)} \tag{10.2.16}$$

where

$$R_s = \frac{1}{1 + \dfrac{2h\hat{t}\Upsilon}{\bar{t}_{wx}\bar{t}_{wy}} + \dfrac{\alpha^2 + \dfrac{1}{\alpha^2}}{8(1+\nu)}}$$

$$\hat{t} = \frac{2\, t_u\, t_\ell}{t_u + t_\ell}$$

$$\Upsilon = \frac{\bar{t}_{wx}}{s_x} + \frac{\bar{t}_{wy}}{s_y}$$

$$\alpha = \frac{s_x}{s_y}$$

Definitions of t_u, t_ℓ, $\bar{t}_{wx}$, and $\bar{t}_{wy}$ are given in Fig. 8.63 and in (8.8.14).

Now that we have obtained expressions for m_x, m_y, and m_{xy}, these can be substituted into the three equilibrium equations, (10.2.4), (10.2.5), and (10.2.6). Since all three of these moments involve functions of q_x and q_y, we have three coupled differential equations which must be solved simultaneously for w, q_x, and q_y. As shown by Libove and Batdorf [13] they can be combined into a single differential equation of sixth order

$$-\frac{1}{2}\frac{D_{xy}D_x}{Ga_y}\frac{\partial^6 w}{\partial x^6} - \left[\frac{1}{2}\frac{D_{xy}D_x}{Ga_x} + \frac{(1-\nu^2)D_xD_y - \frac{\nu}{2}D_{xy}(D_x+D_y)}{Ga_y}\right]\frac{\partial^6 w}{\partial x^4 \partial y^2}$$

$$- \left[\frac{1}{2}\frac{D_{xy}D_y}{Ga_y} + \frac{(1-\nu^2)D_xD_y - \frac{\nu}{2}D_{xy}(D_x+D_y)}{Ga_x}\right]\frac{\partial^6 w}{\partial x^2 \partial y^4}$$

$$-\frac{1}{2}\frac{D_{xy}D_y}{Ga_x}\frac{\partial^6 w}{\partial y^6} + D_x\frac{\partial^4 w}{\partial x^4} + 2H\frac{\partial^4 w}{\partial x^2\partial y^2} + D_y\frac{\partial^4 w}{\partial y^4} = p \tag{10.2.17}$$

where $H = D_{xy} + \frac{\nu}{2}(D_x + D_y)$ and D_x, D_y, and D_{xy} are given by (10.2.11) and (10.2.16).

The solution of this equation requires the specification of three boundary conditions on each edge, instead of the usual two, and in most cases the third condition can be obtained by considering the shear force. For example, for a simply supported edge (parallel to x) the conditions are

$$w = 0, \quad m_x = 0, \quad \text{and} \quad q_y = 0$$

while for a clamped edge they are

$$w = 0, \quad \frac{\partial w}{\partial x} - \frac{q_x}{Ga_x} = 0, \quad \text{and} \quad q_y = 0$$

and for a free edge they are

$$m_x = 0, \quad m_{xy} = 0, \quad \text{and} \quad q_x = 0.$$

For an elastically restrained edge (say parallel to x) the boundary conditions are expressed in terms of the stiffness k_w, k_{xx}, and k_{xy} of the restraints against vertical deflection, edge rotation, and edge twisting, respectively:

$$q_x = k_w w,$$

$$m_x = k_{xx}\left[\frac{\partial w}{\partial x} - \frac{q_x}{Ga_x}\right],$$

$$\text{and } m_{xy} = k_{xy}\left[\frac{\partial w}{\partial y} - \frac{q_y}{Ga_y}\right]$$

Unfortunately, there does not appear to be any comprehensive set of solutions to (10.2.17) available in the literature. Hence we now present an alternative and simplified method in which the direct shear deflection is calculated separately, thus retaining the usual fourth-order differential equation of orthotropic plate theory, for which solutions are readily available.

Simplified Method

We have seen that the two principal effects of shear are that it causes a direct shear deflection and that it reduces the torsional rigidity D_{xy}. In the simplified method which we now present, the first effect is removed and is dealt with separately, while the second is again accounted for by the reduction factor R_s which is included in the definition of D_{xy} in (10.2.16).

Without shear deflection, the bending moments m_x and m_y of (10.2.10) and the twisting moment m_{xy} become

$$\left.\begin{aligned} m_x &= -D_x\left[\frac{\partial^2 w}{\partial x^2} + \nu\frac{\partial^2 w}{\partial y^2}\right] \\ m_y &= -D_y\left[\frac{\partial^2 w}{\partial y^2} + \nu\frac{\partial^2 w}{\partial x^2}\right] \\ m_{xy} &= D_{xy}\frac{\partial^2 w}{\partial x\,\partial y} \end{aligned}\right\} \quad (10.2.18)$$

These three moments can be substituted into the three equilibrium relationships—(10.2.4), (10.2.5), and (10.2.6). Also, since none of the moment expressions contain q_x or q_y, (10.2.5) and (10.2.6) can be regarded as explicit expressions for q_x and q_y and can be used to eliminate these variables from (10.2.4), thus giving a single fourth-order differential equation for w. The result is

$$\left.\begin{aligned} &D_x\frac{\partial^4 w}{\partial x^4} + 2H\frac{\partial^4 w}{\partial x^2\,\partial y^2} + D_y\frac{\partial^4 w}{\partial y^4} = p \\ &\text{where again} \quad H = D_{xy} + \frac{\nu}{2}(D_x + D_y) \end{aligned}\right\} \quad (10.2.19)$$

and D_x, D_y, and D_{xy} are given by (10.2.11) and (10.2.16).

This equation has been solved by various authors using a variety of methods (Fourier series, Ritz, Galerkin, and finite difference methods) and solutions have been obtained for several different boundary conditions. However, none of these previous solutions allow for the effect of shear distortion on the torsional rigidity D_{xy} (or at least not an explicit allowance) and so this parameter should be evaluated by (10.2.16) rather than by any of the expressions given in these previous solutions.

The earliest solution of (10.2.19) seems to have been that of Huber [14], who showed that the solution involves two nondimensional parameters that are combinations of the plate aspect ratio and the rigidity parameters D_x, D_y, and H.

$$\left.\begin{aligned} \rho &= \frac{L_x}{L_y}\left[\frac{D_y}{D_x}\right]^{1/4} = \text{virtual aspect ratio} \\ \eta &= \frac{H}{2\sqrt{D_x D_y}} = \text{torsional stiffness parameter} \end{aligned}\right\} \quad (10.2.20)$$

To facilitate the evaluation of these parameters we express them in terms of the double wall structural variables using (10.2.11) and (10.2.16). The result is

$$\left.\begin{aligned} \rho &= \frac{L_x}{L_y}\left[\frac{i_y}{i_x}\right]^{1/4} \\ \eta &= \frac{1}{2\sqrt{i_x i_y}}\left[h^2\hat{t}(1-\nu)R_s + \nu(i_x + i_y)\right] \end{aligned}\right\} \quad (10.2.21)$$

It is of interest to compare this expression for η with that proposed by Schade [1, 2, 5] more than 40 years ago and which has been commonly used for ship panels [7, 11]. Schade's expression is meant to apply both to double wall panels (and particularly for ships' double bottoms) and also to single wall panels. In deriving it Schade did not give an expression for the torsional rigidity D_{xy} in terms of the structural variables. Instead, he developed the following approximate expression for the coefficient H in the governing differential equations, (10.2.19)

$$H = \frac{E}{1-\nu^2}\left[t_u\, r_{ux}\, r_{uy} + t_\ell\, r_{\ell x}\, r_{\ell y}\right]$$

where the r values are the distances from the neutral axes (in the x- and y-directions) to the upper and lower flanges of the stiffened panel. In singly plated panels there is a plate flange and a stiffener flange; the terms "upper" and "lower" are used in the foregoing definition because we are dealing here only with double wall panels. After substituting the above expression for H into (10.2.20) to obtain η, Schade then made some final simplifications to obtain

$$\eta = \sqrt{\frac{i_{px}\, i_{py}}{i_x\, i_y}} \quad (10.2.22)$$

where i_{px} and i_{py} are the moments of inertia (per unit width) of the plating only, about the neutral axes in the x- and y-directions. This will be referred to as Schade's simplified expression for η. Williams and Chapman [4] reviewed Schade's work and found that, prior to the final simplifications, his derivation implies the following expression for D_{xy}

$$D_{xy} = \frac{E(i_{px} + i_{py})}{2(1+\nu)}$$

from which

$$H = D_{xy} + \frac{\nu}{2}(D_x + D_y)$$

$$= \frac{E}{2(1 - \nu^2)}\left[(i_{px} + i_{py})(1 - \nu) + \nu(i_x + i_y)\right]$$

Therefore Schade's "more rigorous" torsional stiffness parameter is

$$\eta = \frac{(i_{px} + i_{py})\,(1 - \nu) + (i_x + i_y)}{2\sqrt{i_x\, i_y}} \qquad (10.2.23)$$

Both of these expressions for η differ significantly from that of (10.2.21) in that, although they account for the role of the plating in providing torsional rigidity, there is no allowance for web distortion. In fact, the web thickness does not even appear in the expression whereas in reality, if the web thicknesses were made smaller and smaller the panel would gradually lose all of its torsional rigidity. Thus these expressions for η do not give a correct portrayal of the effect which a change in web thickness would have on the torsional rigidity. In fact, the effect predicted by these expressions is opposite to the actual effect. For example, taking Schade's simplified expression, if the web thicknesses were decreased, i_x and i_y would decrease slightly but i_{px} and i_{py} would be unchanged. Therefore η would become slightly larger, as if the torsional rigidity had increased, which is obviously incorrect.

Marguerre and Woernle [6] have presented a comprehensive range of solutions to (10.2.19) for various boundary conditions, of which the following are reproduced here:

1. All edges simply supported.
2. One pair of edges clamped, other pair of edges simply supported.
3. All edges clamped.
4. One pair of edges elastically rotationally restrained; other pair simply supported.
5. One pair of edges on elastically deflecting cross beams but free to rotate; other pair simply supported.

The solutions for these five cases are presented in Figs. 10.4 to 10.22, in terms of the nondimensional parameters which were derived by Huber and some nondimensional coefficients for maximum deflection and maximum bending moment. The deflection coefficient A is defined by

$$w_{\max} = \frac{5}{384}\,\frac{pL_x^2L_y^2}{\sqrt{D_xD_y}}\,A \qquad (10.2.24)$$

and the bending moment coefficients B_1 and B_2 are defined in the figures. In all of these figures the panel edges marked "a" are of length L_x and those marked "b" are of length L_y.

These curves give only the primary bending moments m_{x0} and m_{y0} defined by

$$m_{x0} = -D_x\frac{\partial^2 w}{\partial x^2}$$

$$m_{y0} = -D_y\frac{\partial^2 w}{\partial y^2} \qquad (10.2.25)$$

The total values of bending moment are obtained from the primary values by adding the other terms from (10.2.10). The result is

$$m_x = m_{x0} + \nu\frac{i_x}{i_y}m_{y0}$$

$$m_y = m_{y0} + \nu\frac{i_y}{i_x}m_{x0} \qquad (10.2.26)$$

For ease of plotting, the curves are presented in terms of the inverse of the virtual aspect ratio

$$\zeta = \frac{1}{\rho} = \frac{L_y}{L_x}\left(\frac{D_x}{D_y}\right)^{1/4} \qquad (10.2.27)$$

Curves are given for $0 < \zeta < 1$, and so the directions of the x- and y-axes must be chosen such that $\zeta < 1$. Thus, the subscript x is assigned to the larger value of "virtual length," which is the true length divided by the fourth root of the rigidity in that direction. Hence a trial evaluation of the virtual aspect ratio ρ must be made before the subscripts x and y can be definitively assigned.

It remains now to give some means for estimating the direct shear deflection $w_s(x, y)$. An exact solution is not required because, for double wall panels having prismatic cross sections and for constant pressure loading, the shear distribution and the resulting direct shear deflection $w_s(x, y)$ are smoothly varying functions. Equation 10.2.8 gives the second derivative of w_s in terms of these derivatives

$$\frac{\partial^2 w_s}{\partial x^2} = \frac{1}{Ga_x}\frac{\partial q_x}{\partial x} \qquad (10.2.28)$$

$$\frac{\partial^2 w_s}{\partial y^2} = \frac{1}{Ga_y}\frac{\partial q_y}{\partial y} \qquad (10.2.29)$$

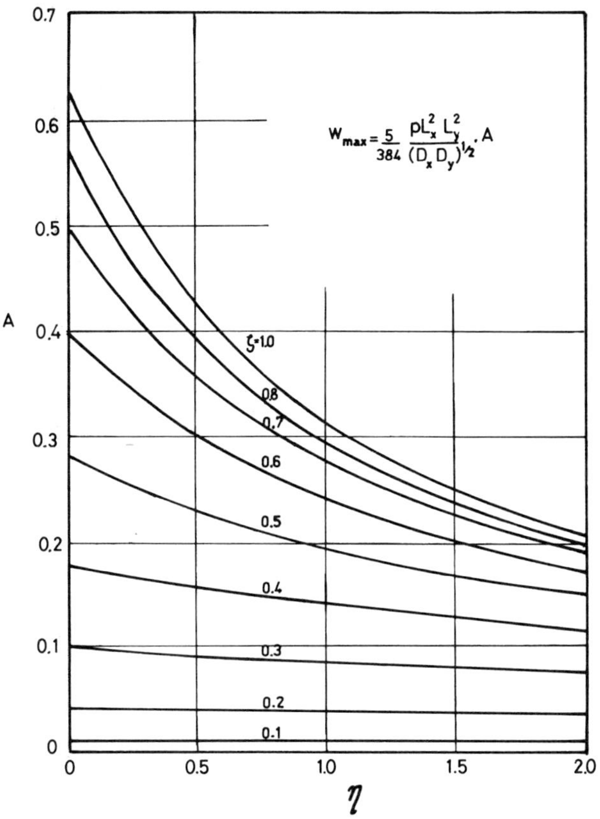

Central Deflection Coefficient—Torsional Rigidity

Simply Supported Plate

Figure 10.4

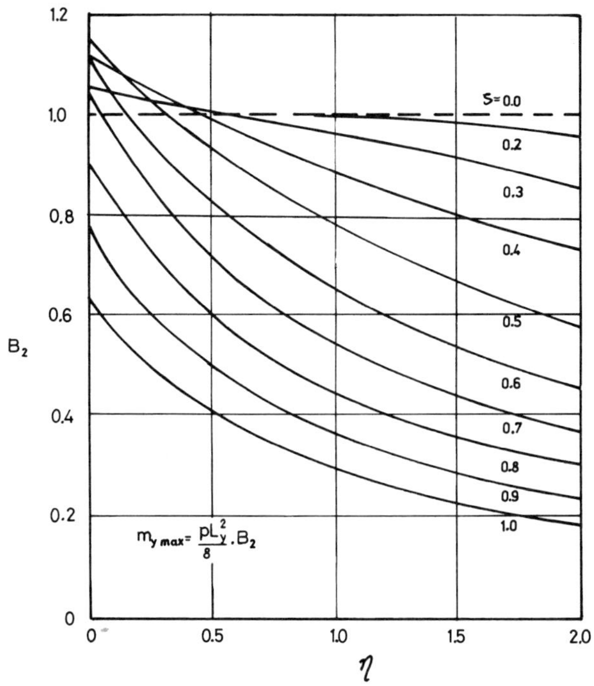

Bending Moment Coefficient—Torsional Rigidity

Simply Supported Plate

Figure 10.6

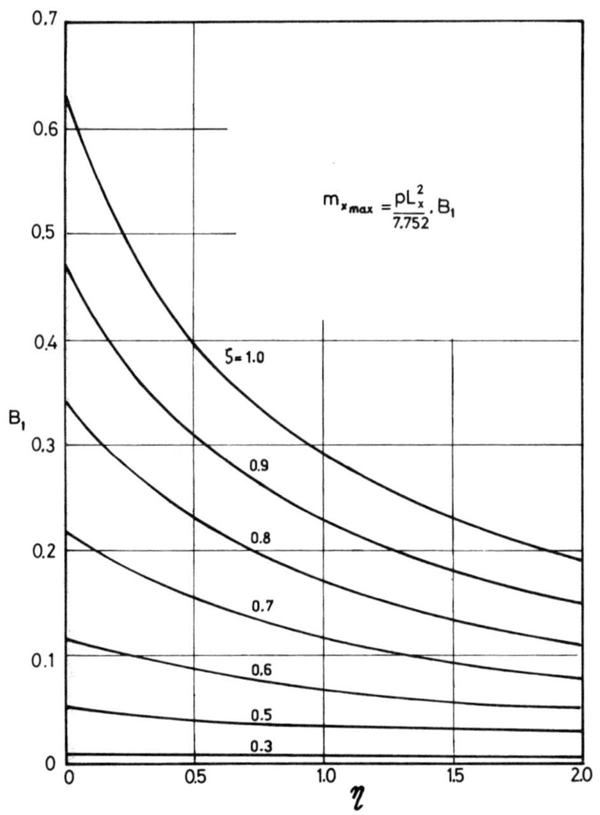

Bending Moment Coefficient—Torsional Rigidity

Simply Supported Plate

Figure 10.5

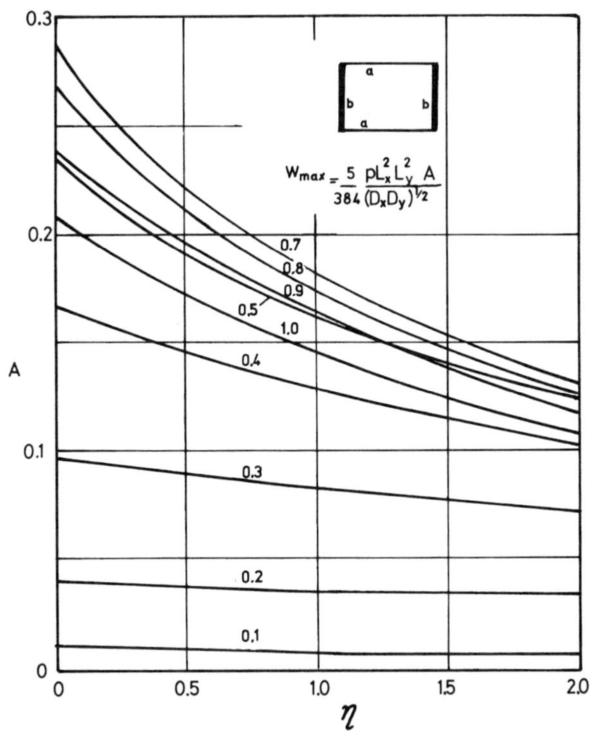

Central Deflection Coefficient—Torsional Rigidity

One Pair of Edges Clamped

Figure 10.7

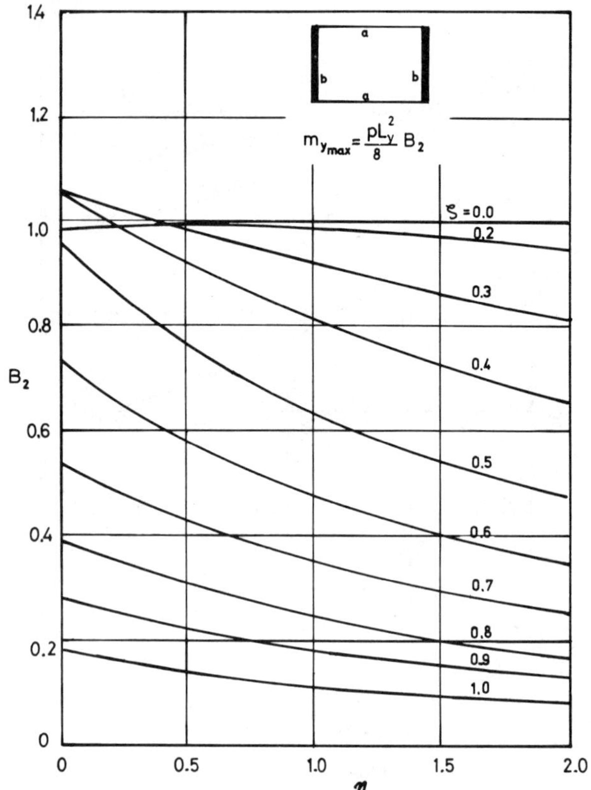

Bending Moment Coefficient — Torsional Rigidity

One Pair of Edges Clamped

Figure 10.8

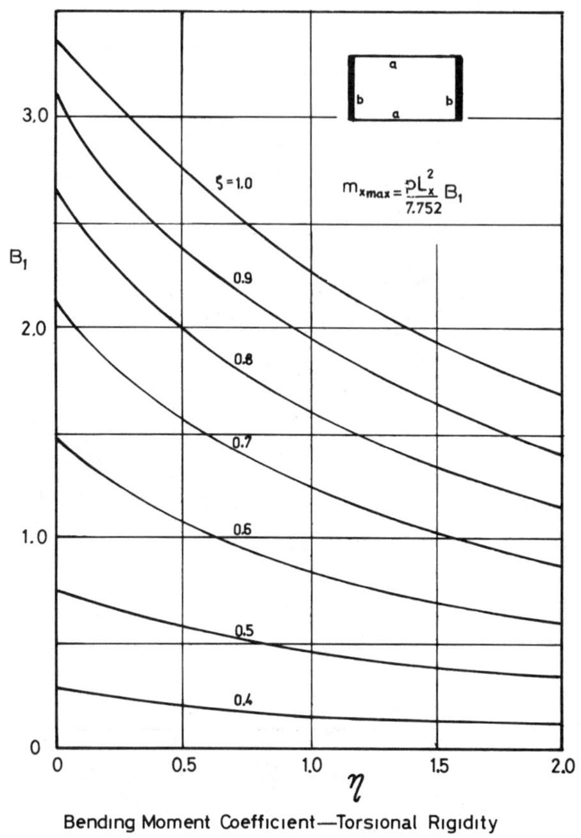

Bending Moment Coefficient—Torsional Rigidity

One Pair of Edges Clamped

Figure 10.9

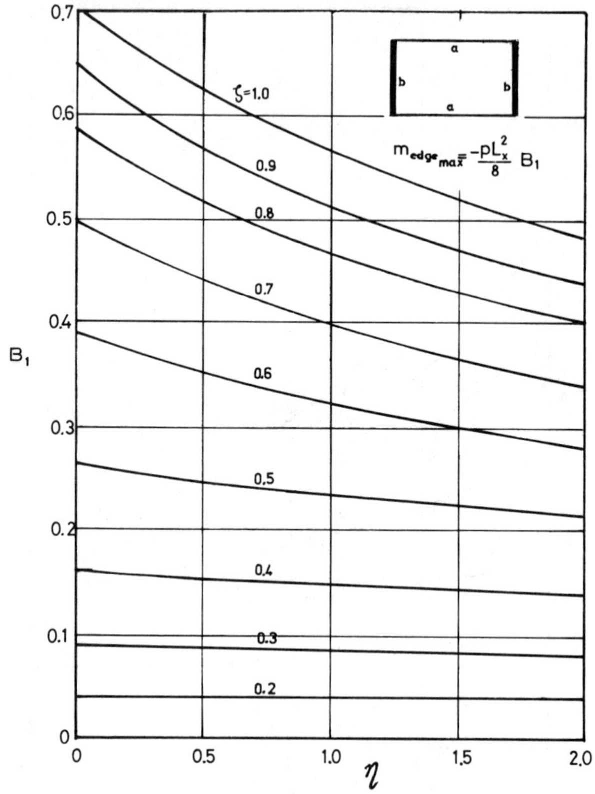

Bending Moment Coefficient — Torsional Rigidity

One Pair of Edges Clamped

Figure 10.10

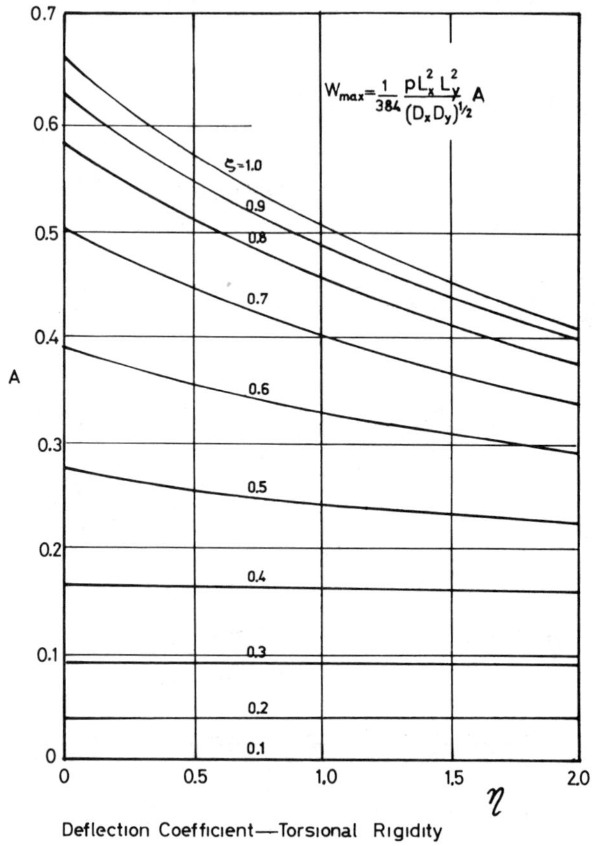

Deflection Coefficient—Torsional Rigidity

Clamped Edges

Figure 10.11

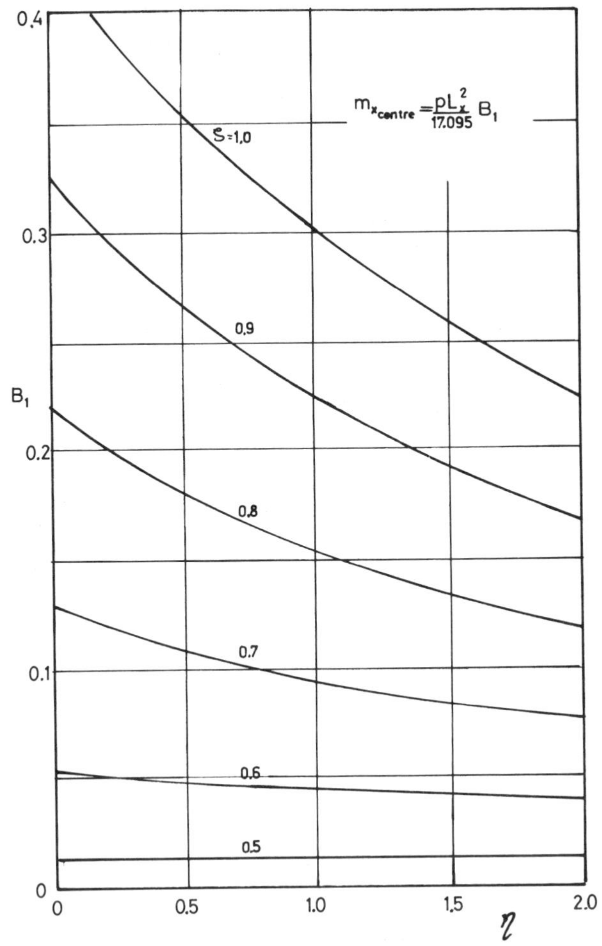

Bending Moment Coefficient—Torsional Rigidity

Clamped Plate

Figure 10.12

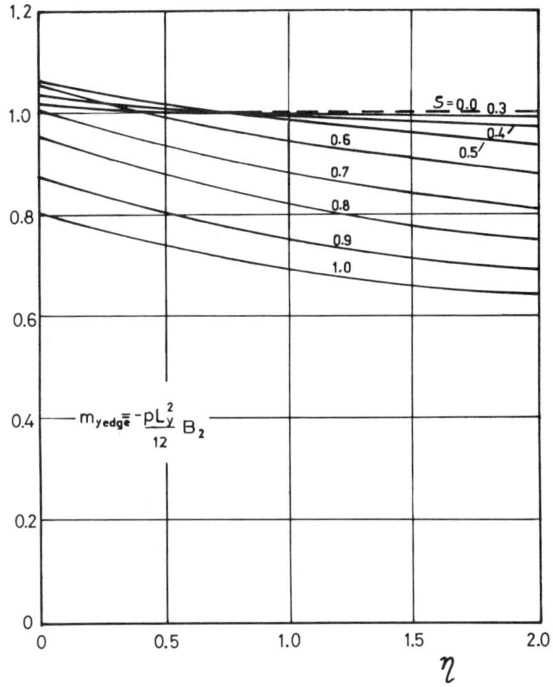

Bending Moment Coefficient—Torsional Rigidity

Clamped Plate

Figure 10.14

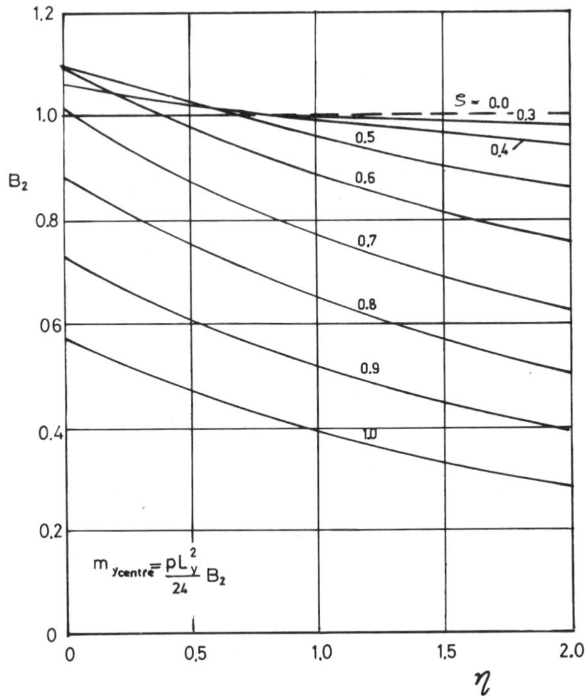

Bending Moment Coefficient—Torsional Rigidity

Clamped Plate

Figure 10.13

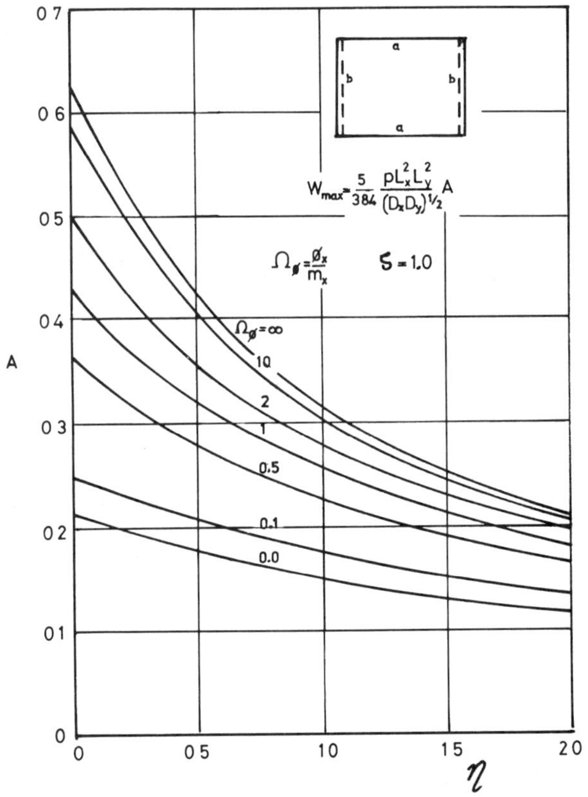

Central Deflection Coefficient—Torsional Rigidity

One Pair of Edges Elastically Restrained

Figure 10.15

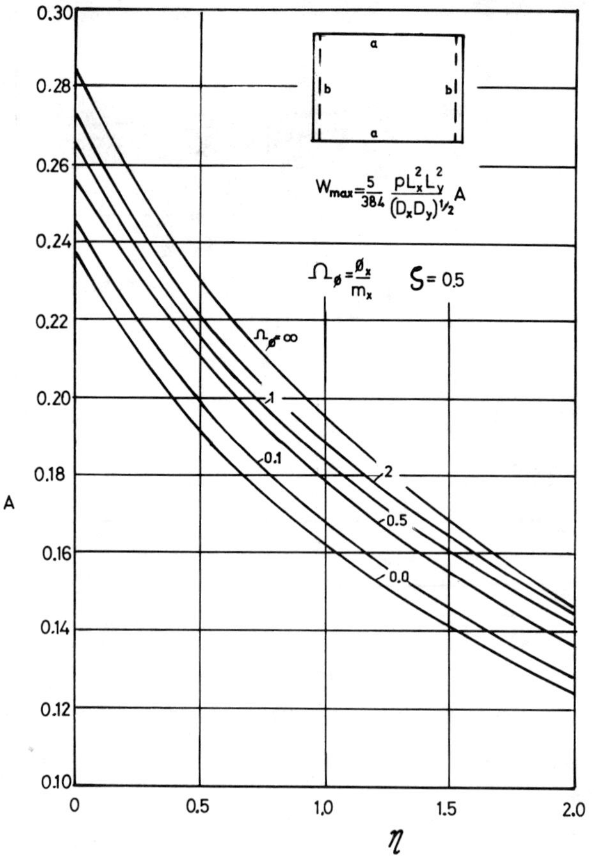

Central Deflection Coefficient—Torsional Rigidity

One Pair of Edges Elastically Restrained

Figure 10.16

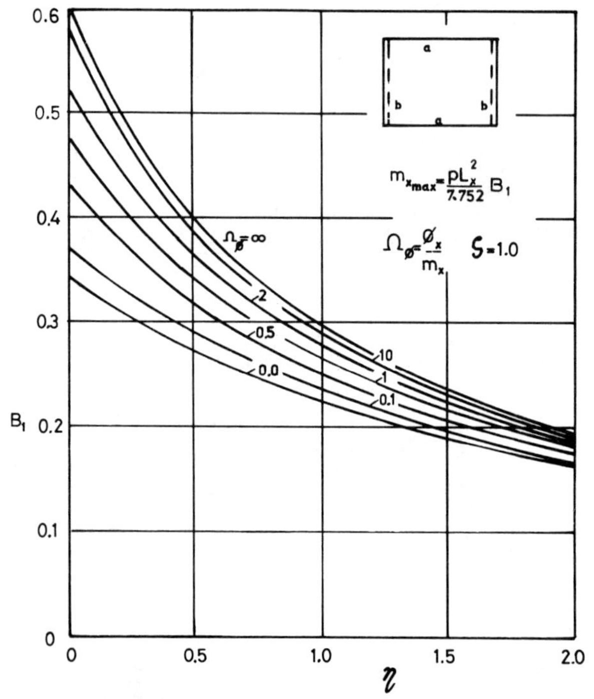

Bending Moment Coefficient—Torsional Rigidity

One Pair of Edges Elastically Restrained

Figure 10.17

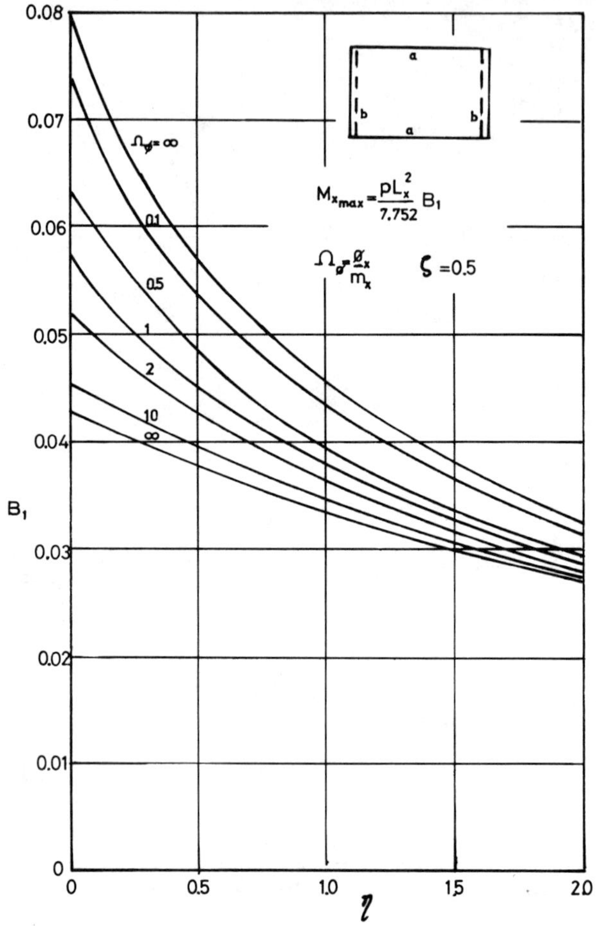

Bending Moment Coefficient—Torsional Rigidity

One Pair of Edges Elastically Restrained

Figure 10.18

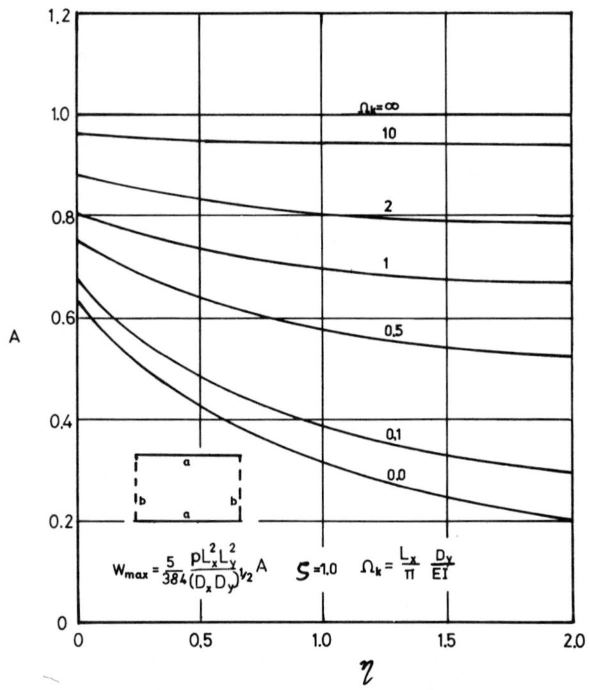

Central Deflection Coefficient—Torsional Rigidity

One Pair of Edges on Elastically Deflecting Beams

Figure 10.19

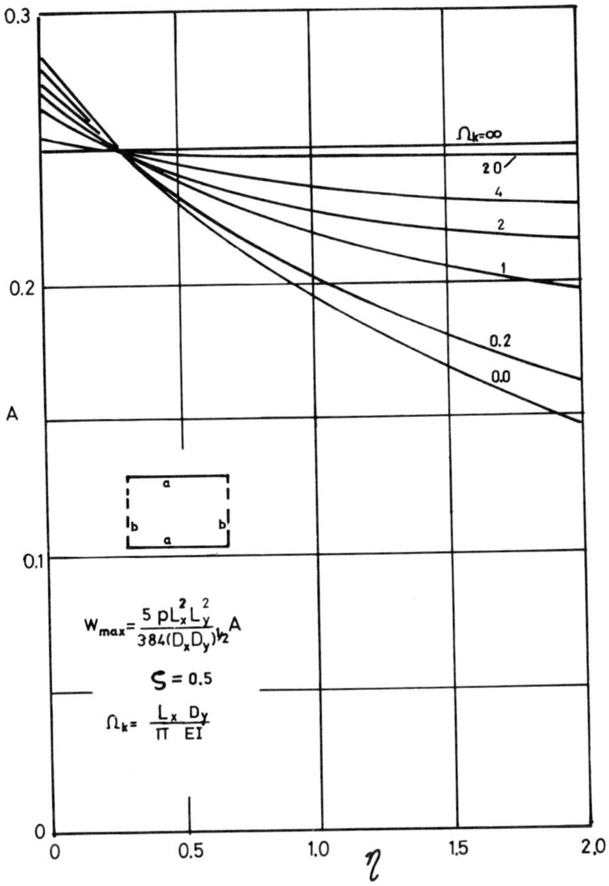

Central Deflection Coefficient —Torsional Rigidity

One Pair of Edges on Elastically Deflecting Crossbeams

Figure 10.20

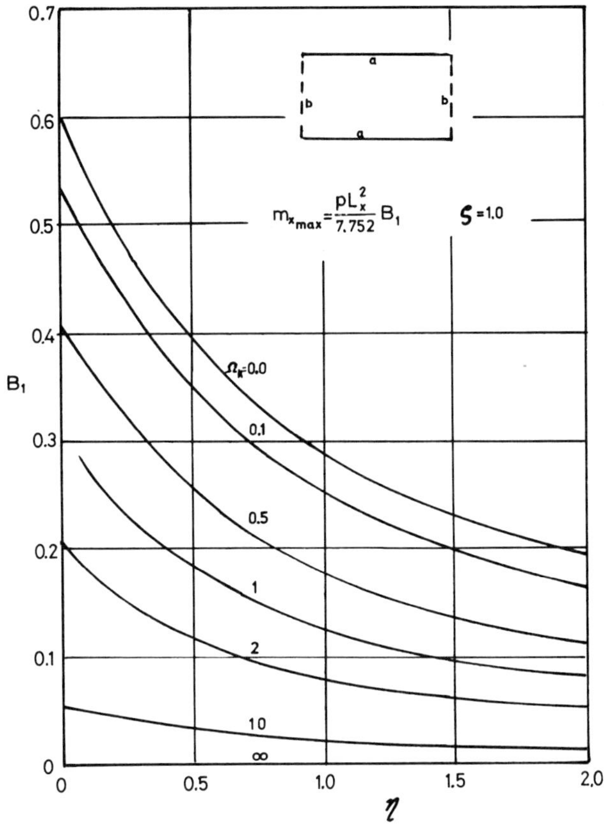

Bending Moment Coefficient— Torsional Rigidity

One Pair of Edges on Elastically Deflecting Crossbeams

Figure 10.21

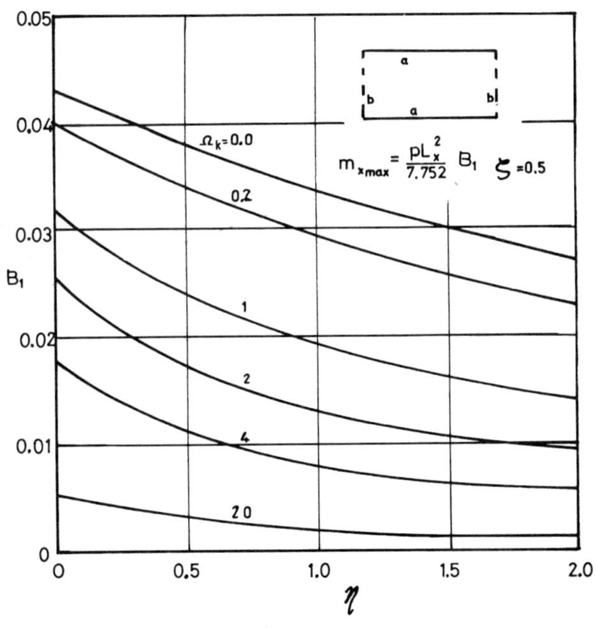

Bending Moment Coefficient—Torsional Rigidity

One Pair of Edges on Elastically Deflecting Crossbeams

Figure 10.22

Also, the shear derivatives must satisfy the requirement of vertical equilibrium as expressed in (10.2.4),

$$\frac{\partial q_x}{\partial x} + \frac{\partial q_y}{\partial y} + p = 0 \qquad (10.2.30)$$

The relative magnitude of the two derivatives will depend on:

1. The relative shear rigidity of the webs in the two directions.
2. The location (x, y) within the panel.

In regard to the first, for our application the material of the panel is isotropic and the difference in shear rigidity is due to a difference in effective shear-carrying area in the two directions. Hence the two derivatives will be approximately proportional to a_x and a_y respectively. As a result of this, the right-hand side terms in the foregoing equations for curvature will tend to be approximately equal, even though the shear areas a_x and a_y may be unequal. For example, if a_x is larger this would reduce the right-hand side of the first equation, but the increased value of $\partial q_x/\partial x$ would largely counteract this reduction. Also, the sum of the two derivatives must always equal $-p$ and therefore, if we sum the two equations, any difference that may exist between the right-hand side values would be further smoothed out. Hence it seems reasonable that in performing this summation we can replace a_x and a_y by their average value $a_s = (a_x + a_y)/2$. The result is

$$\frac{\partial^2 w_s}{\partial x^2} + \frac{\partial^2 w_s}{\partial y^2} = \frac{1}{Ga_s}\left(\frac{\partial q_x}{\partial x} + \frac{\partial q_y}{\partial y}\right)$$

or

$$\nabla^2 w_s = \frac{-p}{Ga_s} \qquad (10.2.31)$$

This is simply Poisson's equation, as occurs in the torsion of solid sections, membrane deflection, two-dimensional irrotational flow, and so on, and solutions are readily available in the literature. In most cases only the maximum deflection is required, and an expression for this can be obtained by expressing the constant pressure loading as a Fourier series

$$p(x, y) = \frac{16\,p_0}{\pi^2}\left(\sum_{m\text{ odd}} \frac{1}{m}\sin\frac{m\pi x}{L_x}\right)\left(\sum_{n\text{ odd}} \frac{1}{n}\sin\frac{n\pi y}{L_y}\right)$$

and seeking a solution of the form

$$w(x, y) = \sum_{m\text{ odd}} \sum_{n\text{ odd}} a_{mn} \sin\frac{m\pi x}{L_x} \sin\frac{n\pi y}{L_y} \qquad (10.2.32)$$

which automatically satisfies the boundary conditions. Substituting $w(x, y)$ into (10.2.31) and solving for a_{mn} gives

$$a_{mn} = \frac{16\,p_0}{Ga_s\pi^4 mn}\left(\frac{1}{\dfrac{m^2}{L_x^2} + \dfrac{n^2}{L_y^2}}\right)$$

and substituting this into (10.2.32) and setting $x = L_x/2$ and $y = L_y/2$ gives

$$(w_{\max})_s = \frac{16\,p_0 L_x^2}{Ga_s\pi^4} \sum_{m\text{ odd}} \sum_{n\text{ odd}} \frac{(-1)^{(m-1)/2}(-1)^{(n-1)/2}}{mn(m^2 + \Pi^2 n^2)} \qquad (10.2.33)$$

where $\Pi = L_x/L_y$.

10.3 COMPARISON OF METHODS WITH TEST RESULTS

To demonstrate the simplified method and also to verify its accuracy, we now use it to analyze the one-eighth scale model of a double bottom structure which was tested by Williams and Chapman [4]. The test structure is shown in Fig. 10.23 and the structural dimensions and parameters are as follows (units are pounds and inches):

$L_x = 130.5$ $\qquad$ $L_y = 88$
$s_x = 22$ $\qquad$ $s_y = 14.5$
$t_u = 0.064$ $\qquad$ $t_\ell = 0.080$
$(b_e/b)_x = 0.85$ $\qquad$ $(b_e/b)_y = 0.795$*
$I_{xe} = 26.06$ in.^{4}** $\qquad$ $I_{ye} = 14.65$ in.^{4}**
$E = 30.2 \times 10^6$ $\qquad$ $h_w = 5.5$
$h = 5.57$ (to plate midthicknesses)
$t_{wx} = 0.064$ $\qquad$ $t_{wy} = 0.064$
$\bar{t}_{wx} = 0.052$ $\qquad$ $\bar{t}_{wy} = 0.045$
Heights from neutral axes to inner bottom plating upper surface:
$r_{ux} = 3.07$ $\qquad$ $r_{uy} = 2.89$

*This includes a factor of 0.85 to allow for out-of-plane distortion which, as pointed out by the authors, should not have been applied for the inner bottom because it is in tension. The correct procedure has been used herein, and hence the value of I_{ye} and i_y differ slightly from Ref. 4.

**In computing I_{xe} and I_{ye}, the contributions of the stiffeners (for I_{xe}) and of the webs have been multiplied by $(1 - \nu^2)$.

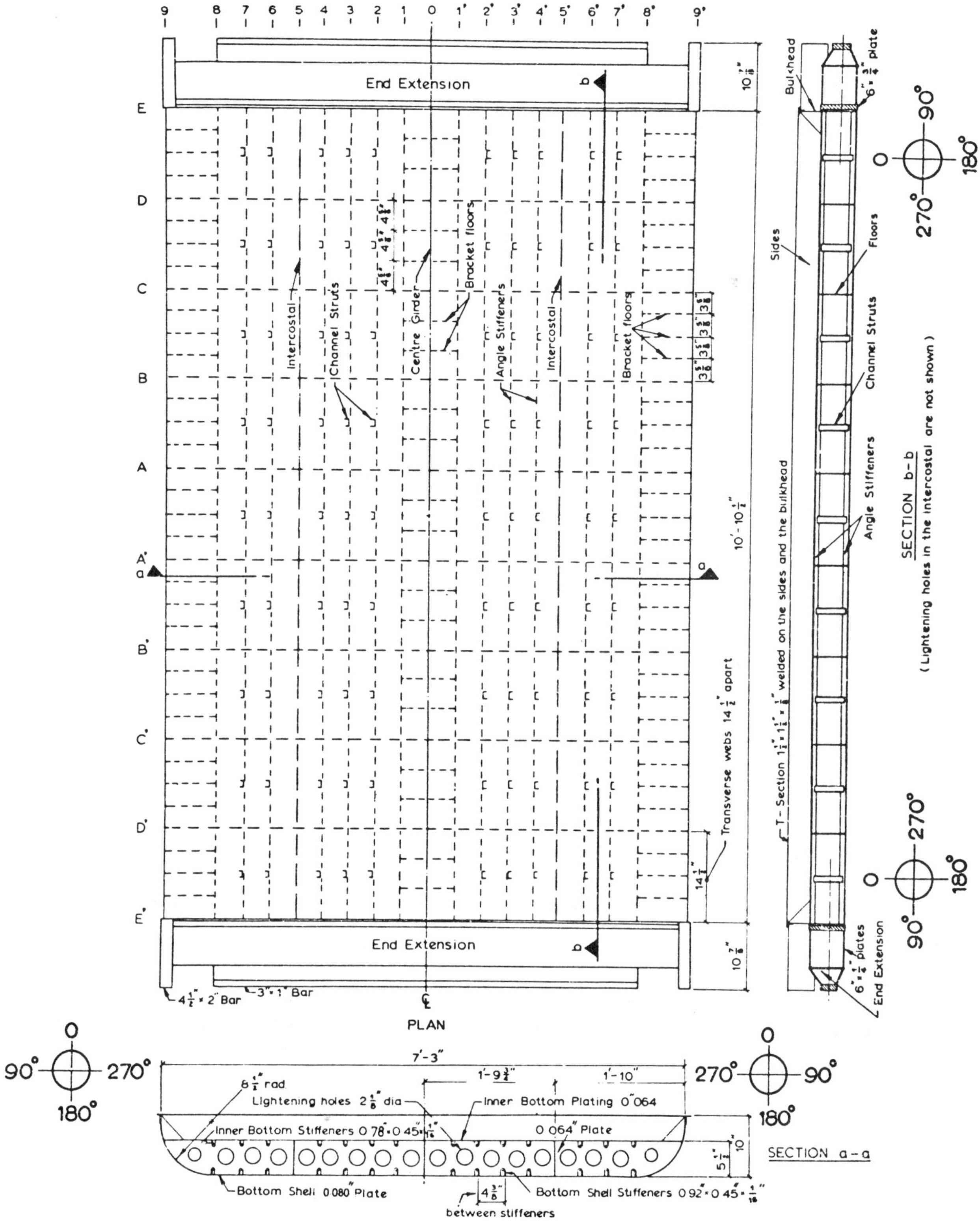

Figure 10.23 Principal dimensions of double-bottom model.

Based on these value the orthotropic parameters are:

$$a_x = 0.0130 \qquad a_y = 0.0171$$

$$i_x = \frac{26.06}{22} = 1.18 \text{ in.}^3 \qquad i_y = \frac{14.65}{14.5} = 1.01 \text{ in.}^3$$

$$D_x = 39.2 \times 10^6 \text{ lb in.}^\dagger \qquad D_y = 33.5 \times 10^6 \text{ lb in.}^\dagger$$

$$\hat{t} = \frac{2t_u t_\ell}{t_u + t_\ell} = 0.0711 \qquad Y = \frac{\bar{t}_{wx}}{s_x} + \frac{\bar{t}_{wy}}{s_y} = 0.00547$$

$$R_s = \frac{1}{1 + \dfrac{2(5.57)(0.0711)(0.00547)}{(0.052)(0.045)} + \dfrac{\alpha^2 + \dfrac{1}{\alpha^2}}{8(1 + \nu)}} = 0.319$$

$$\rho = \frac{130.5}{88}\left(\frac{1.01}{1.18}\right)^{1/4} = 1.426 \qquad \zeta = \frac{1}{\rho} = 0.70$$

$$\eta = \frac{(5.57)^2(0.0711)(0.7)(0.319) + 0.3(1.18 + 1.01)}{2\sqrt{(1.18)(1.10)}} = 0.527$$

From Figs. 10.4, 10.5, and 10.6

$$A = 0.355; \qquad B_1 = 0.155; \qquad B_2 = 0.71$$

Therefore the maximum deflection due to bending is

$$(w_{\max})_b = \frac{5}{384} \frac{pL_x^2 L_y^2}{\sqrt{D_x D_y}} A$$

$$= \frac{5}{384} \frac{(1.0)(130.5)^2 (88)^2}{\sqrt{39.2 \times 33.5 \times 10^6}} (0.355)$$

$$= 0.0169 \text{ in.}$$

In the test the pressure bag acted only on the flat portion of the bottom, which represents only the middle 80% of the width between supports (L_y). In a simply supported beam a load of this type causes a maximum deflection and a maximum bending moment which are 0.952 and 0.960 of the values for a full length load, respectively. Hence the corrected maximum deflection is

$$(w_{\max})_b = 0.0169 \times 0.952 = 0.0160 \text{ in.}$$

Unlike the bending deflection, the shear deflection will be reduced to an 80% value. Therefore from (10.2.33) the maximum deflection due to shear is

$$(w_{\max})_s = 0.8 \frac{16(1.0)(130.5)^2}{(11.62 \times 10^6)(0.015)\pi^4}(0.28)$$

$$= 0.0035 \text{ in.}$$

Therefore the total deflection is

$$(w_{\max})_T = 0.0195 \text{ in.}$$

For comparison, the experimental value and the values obtained by other analysis methods are shown in Table 10.1. It may be seen that the foregoing value agrees well with the experimental value (0.019 in.), as also does the value given by the sixth-order method. The calculated shear deflection of 0.0035 in. represents approximately 18% of the total measured deflection. The table also shows the results of a finite element solution which was obtained using the double bottom modeling technique presented in Section 8.8. This solution includes shear deflection and the effect of shear distortion of the webs (including the effect of web cutouts).

For this double bottom structure the web distortion reduction factor R_s is 0.319, which means that the torsional rigidity D_{xy} is approximately one-third of what it would be if the webs underwent no shear distortion. Accordingly, (10.2.21) gives a value of 0.527 for η. The resulting calculated bending deflection of

† In the notation of Ref. 4, the symbols D_x and D_y include another factor, which is approximately $(1 - \nu^2)$. Hence the numerical values for these symbols are larger than those quoted in Ref. 4.

TABLE 10.1 COMPARISON OF ANALYSIS METHODS FOR DOUBLE WALL PANELS

Method	η and Source	$(w_{max})_b$ [% error]	$(w_{max})_T$	$\varepsilon_x \times 10^6$ [% error]	$\varepsilon_y \times 10^6$ [% error]
Experiment [4]	—	0.0155[a]	0.019	28	43
Finite element solution	—	—	0.0195	30.3	65.0
Average of experimental and finite element values				29.2	54.0
Sixth-order, including shear deflection[b]	—	—	0.020	22 [−25%]	61 [+13%]
Fourth-order, with shear deflection calculated separately					
With web distortion factor R_s	0.527 (10.2.21)	0.0160 [+ 3%]	0.0195[c]	26.7 [− 9%]	56.9 [+ 5%]
Without allowance for web distortion	0.98 Schade (10.2.23)	0.0130 [−16%]	0.0165[c]	19 [−35%]	53 [− 2%]
	0.89 Schade (10.2.22)	0.0135 [−13%]	0.0170[c]	21.5 [−26%]	45.7 [−15%]

[a]Obtained by subtracting $(w_{max})_s$ from the measured total deflection.
[b]As reported in Ref. 15.
[c]Obtained be adding the shear deflection $(w_{max})_s = 0.0035$, calculated from (10.2.33).

0.0160 in. is 3% larger than the experimental value of 0.155 in. (obtained by subtracting the shear deflection of 0.0035 in. from the measured total deflection).

Williams and Chapman calculated the bending deflection by the fourth order method using Schade's more rigorous expression—(10.2.23)—for η. This equation, which does not contain any allowance for web distortion, gives a much larger value of η (0.98), resulting in a bending deflection of 0.0130 in., which is 16% less than the experimental value.

For double bottom structures the expression most commonly used for η is Schade's simplified expression, and so we include this in the comparison. For the test structure $i_{px} = 0.962$ and $i_{py} = 0.978$. Therefore

$$\eta = \sqrt{\frac{(0.962)(0.978)}{(1.18)(1.01)}} = 0.89$$

For $\eta = 0.89$ and $\zeta = 0.70$, Fig. 10.4 gives a value of 0.29 for the deflection coefficient A, and the resulting value of bending deflection (including the correction factor of 0.952) is 0.0135 in., which is approximately 13% less than the experimental value.

Thus for a double bottom of normal proportions and scantlings, Schade's approach, which ignores web distortion, gives an error in predicted deflection of from 13 to 16%, whereas the method presented in Section 10.2, which accounts for web distortion by means of (10.2.21), has an error of approximately 3%.

Moreover, as noted earlier, if web distortion is not allowed for, then the effect or influence of the web scantlings is wrongly represented: the predicted effect of changing the web scantlings is opposite to the actual effect. In rationally-based design, and particularly in structural optimization, it is essential that the expressions for predicting structural response give at least a qualitatively correct portrayal of the effect of changing a design variable. Therefore it is important to include the effect of web distortion, and for this reason (10.2.21) is preferable to any expression that does not allow for this effect.

Continuing the solution of the sample problem, and the comparison of the various methods with experimental and finite element results, we next examine and compare the values of the primary bending moments. The method presented in Section 10.2 gives

$$m_{x0} = \frac{pL_x^2}{7.752}B_1 = \frac{(1.0)(130.5)^2}{7.752}(0.155)$$
$$= 340.5 \text{ lb in./in.}$$

$$m_{y0} = (0.96)\frac{pL_y^2}{8}B_2 = (0.96)\frac{(1.0)(88)^2}{8}(0.71)$$
$$= 659.8 \text{ lb in./in.}$$

in which a correction factor of 0.96 is applied to m_{y0} to allow for the decreased width of load. The total maximum bending moments are, from (10.2.26),

$$m_x = 340.5 + 0.3\frac{1.18}{1.01}659.8 = 571.8 \text{ lb in./in.}$$

$$m_y = 659.8 + 0.3\frac{1.01}{1.18}340.5 = 747.2 \text{ lb in./in.}$$

The first line of Table 10.1 gives the experimentally measured strains, ε_x and ε_y, in the inner bottom plating, at the middle of the double bottom structure. The inner bottom is chosen as the location for the comparison because, unlike the bottom plating, the inner bottom was not subject to the direct application of normal pressure.* The measured strains for the inner bottom were obtained from [15]. In order to compare these measured strains with the values predicted by the method of Section 10.2, it is necessary to convert the foregoing bending moments to strains. This may be done using (10.2.25)

$$\varepsilon_x = -r_{ux}\left(\frac{\partial^2 w}{\partial x^2}\right)_b = -r_{ux}\frac{-m_{x0}}{D_x} = 3.07\frac{340.5}{39.2 \times 10^6}$$
$$= 26.7 \times 10^{-6}$$

$$\varepsilon_y = (2.89)\frac{659.8}{33.5 \times 10^6} = 56.9 \times 10^{-6}$$

The values of bending moment from the other analysis methods have likewise been converted to strains and are listed in the table. It may be seen that the values from the method of Section 10.2 agree well with the experimental values and also with the values given by the finite element solution. In view of the possible sources of experimental error which have been mentioned, it is felt that the finite element results are just as reliable as the experimental results. Therefore, in order to smooth out errors which may exist in each of these, their values have been averaged and these average values, which are given in line 3 of the table, are used as the benchmark values. This procedure is further justified by the fact that for deflection, which is not subject to the sources of error mentioned previously, the agreement between the two is excellent.

The values obtained using Schade's approach are also shown in the table. For $\eta = 0.98$ the values are obtained from Ref. 4. For $\eta = 0.89$, Figs. 10.5 and 10.6 give $B_1 = 0.125$ and $B_2 = 0.570$. Therefore

$$m_{x0} = \frac{pL_x^2}{7.752}(0.125) = 274.6 \text{ lb in./in.}$$

$$m_{y0} = (0.96)\frac{pL_y^2}{8}(0.570) = 529.7 \text{ lb in./in.}$$

from which

$$\varepsilon_x = r_{ux}\frac{m_{x0}}{D_x} = 21.5 \times 10^{-6}$$

$$\varepsilon_y = r_{uy}\frac{m_{y0}}{D_y} = 45.7 \times 10^{-6}$$

These values differ from the benchmark values by 26% and15%, compared to differences of 9% and 5% for the method of Section 10.2. Surprisingly, the latter results are also better than those obtained from the sixth-order method, as quoted in Ref. 15, but this may be due to assumptions regarding boundary conditions.

*In the outer bottom the measured value of ε_x was inconsistent with the values from all of the various methods, including the finite element solution, while at the same time there was good correlation among all of these calculated values. Hence the discrepancy is probably due to some local effect. For example, the direct pressure may have caused additional membrane strain due to edge restraint against pull-in, or the effect of residual welding strain may not have been fully accounted for.

REFERENCES

1. H. A. Schade, "Bending Theory of Ship Bottom Structure," *Trans. SNAME,* **46,** 1938.
2. H. A. Schade, "The Orthogonally Stiffened Plate Under Uniform Lateral Load," *J. Appl. Mech.,* **7** (4), Dec. 1940.
3. J. Clarkson, *The Elastic Analysis of Flat Grillages,* Cambridge University Press, 1965.
4. D. G. Williams and J. C. Chapman, "Tests on a One Eighth Scale Model of the Double-Bottom Structure," *Trans. RINA,* 1974, pp. 329–345.
5. H. A. Schade, "Design Curves for Cross-Stiffened Plating Under Uniform Bending Load," *Trans. SNAME,* **49,** 1941, p. 154.

6. K. Marguerre and H. Woerlne, *Elastic Plates,* Blaisdell Publishing Co., 1969.
7. A. Mansour, "Ship Bottom Structure under Uniform Lateral and Inplane Loads," *Schiff und Hafen,* **5,** (1967) 19 Jahrgang, pp. 323–339.
8. B. H. Falconer and J. C. Chapman, "Compressive Buckling of Stiffened Plates," *The Engineer,* **195,** 1953, pp. 789–791 and 822–825.
9. J. C. Chapman and J. E. Slatford, "Design of Stiffened Plating in Compression," *The Engineer,* **207,** 1959, pp. 292–294.
10. H. G. Schultz, "Zur Tragfahigkeit druckbeanspruchter orthotroper Platten," *Stahlbau,* **33** (4), April 1964, pp. 123–126.
11. A. Mansour, "Charts for Buckling and Post-Buckling Analyses of Stiffened Plates under Combined Loading," *SNAME T & R Bull.* **2-22,** 1976.
12. B. Aalami and J. C. Chapman, "Large Deflection Behaviour of Rectangular Orthotropic Plates under Transverse and In-Plane Loads," *Proc. Inst. Civ. Eng.,* **42,** March 1969, p. 347.
13. C. Libove and S. B. Batdorf, A General Small Deflection Theory for Flat Sandwich Plates, NACA Report No. 899, 1948.
14. M. T. Huber, *Probleme der Statik Technisch Wichtiger Orthotroper Platten,* Gebethner and Wolff, Warsaw, 1929.
15. D. G. Williams, Analysis of a Doubly Plated Grillage under In-Plane and Normal Loading, Ph.D. Thesis, University of London (Imperial College), 1969.

Chapter

Eleven

BUCKLING AND ULTIMATE STRENGTH OF COLUMNS

11.1 REVIEW OF BASIC THEORY

Ideal Columns

The "ultimate load" P_{ult} is the maximum load that a column can carry, and depends on initial eccentricity of the column, eccentricity of the load, transverse loads, end conditions, local or lateral buckling, inelastic action, and residual stresses. The Euler buckling load P_E, on the other hand, is an idealized quantity which does not take any of the foregoing factors into account (except for end rotational restraint, which can be accounted for by using an effective length L_e):

$$P_E = \frac{\pi^2 EI}{L_e^2} \tag{11.1.1}$$

The Euler buckling load is the load for which an ideal column will first have an equilibrium deflected shape. Mathematically, it is the eigenvalue in the solution to Euler's differential equation

$$\frac{d^2w}{dx^2} + \frac{Pw}{EI} = 0$$

Due to the factors just mentioned, the ultimate load of a practical column will be less than the Euler buckling load. In fact, strictly speaking, buckling—the sudden transition to a deflected shape—only occurs in the case of "ideal" columns (columns with no residual stress or eccentricity). Moreover, even in ideal columns the buckling load will be less than the Euler load if the compressive stress in the column exceeds the proportional limit stress, because the diminished slope of the stress-strain curve represents an effective weakening of the column. Shanley [1] has shown that for an ideal column (no eccentricity or residual stress) buckling will occur at the *tangent modulus* load, given by

$$P_t = \frac{\pi^2 E_t I}{L_e^2}$$

in which E_t is the slope of the stress-strain curve corresponding to the level of the compressive stress P_t/A in the column. Since E_t depends on P_t the calculation of P_t is generally an iterative process. Figure 11.1 shows the load–deflection behavior for practical columns, containing residual stress and eccentricity. For such columns it is more convenient to deal in terms of stress rather than load because this allows the effects of yielding and residual stress to be included. The Euler buckling stress is

$$\sigma_E = \frac{P_E}{A} = \frac{\pi^2 E}{\left(\dfrac{L_e}{\rho}\right)^2} \tag{11.1.2}$$

where ρ is the radius of gyration ($I = \rho^2 A$) and L_e/ρ is the *slenderness ratio*. Similarly, the *ultimate strength* of a column is defined as the average applied stress at collapse:

$$\sigma_{ult} = \frac{P_{ult}}{A}$$

For ideal columns the ultimate strength would correspond to the tangent modulus load, that is,

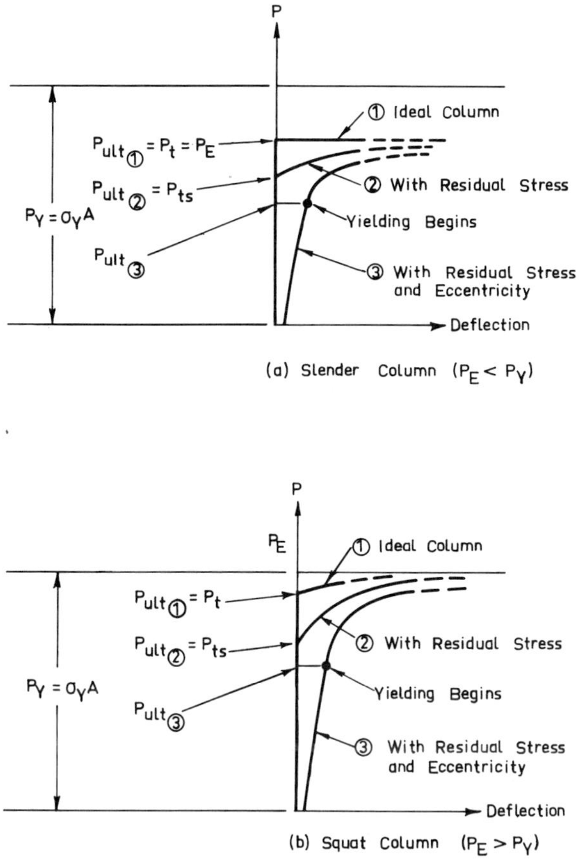

Figure 11.1 Typical load deflection diagrams.

$$(\sigma_{\text{ult}})_{\text{ideal}} = \frac{\pi^2 E_t}{\left(\dfrac{L_e}{\rho}\right)^2} \qquad (11.1.3)$$

The ideal ultimate strength curve, i.e. the relationship between $(\sigma_{\text{ult}})_{\text{ideal}}$ and L_e/ρ, is shown in Fig. 11.2, together with a typical stress-strain curve and the corresponding tangent modulus curve. The figure also shows typical experimental values for essentially straight rolled sections and welded sections. The marked discrepancy is almost entirely due to the compressive residual stress σ_r which exists in rolled, welded, and even cold-worked structural shapes, and this will now be discussed.

Residual Stress

Rolled Sections

In rolled sections the uneven cooling between flange root and flange tip produces tensile residual stresses in the former and compressive residual stresses in the latter, as shown in Fig. 11.3. The effect depends mainly on geometry, and is most pronounced in wide flange sections, in which the residual compressive stress in the flange tips can be as high as 80 MPa for mild steel. The effect is diminished for steels having a higher yield strength; for instance, for $\sigma_Y = 350$ MPa typical values of σ_r range from 20 to 50 MPa. Residual stress due to rolling can usually be eliminated by using quenched and tempered steels or by annealing, but this is costly and often impractical.

The parts of the cross section that have a compressive residual stress will commence yielding when the average applied stress has reached a value of $\sigma_Y - \sigma_r$, and this greatly reduces σ_{ult}. The effect is particularly detrimental because it occurs in the flange tips. The material in the column is no longer homogeneous, and hence the simple tangent modulus approach of (11.1.3) is no longer valid. However, a comparatively simple solution can be achieved for the buckling strength in the primary direction (i.e., bending about the xx-axis in Fig. 11.3) by assuming an elastic-perfectly plastic stress-strain relationship. In this case the progressive loss of bending stiffness, due to progressive yielding of the flanges, is linearly proportional to the extent of the yielded zone. This linearity permits the use of an average value of tangent modulus (averaged over the entire cross section). This

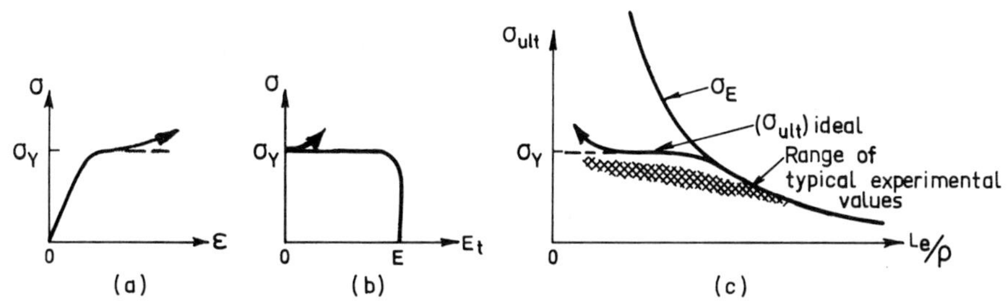

Figure 11.2 Tangent modulus values of ultimate strength for ideal columns.

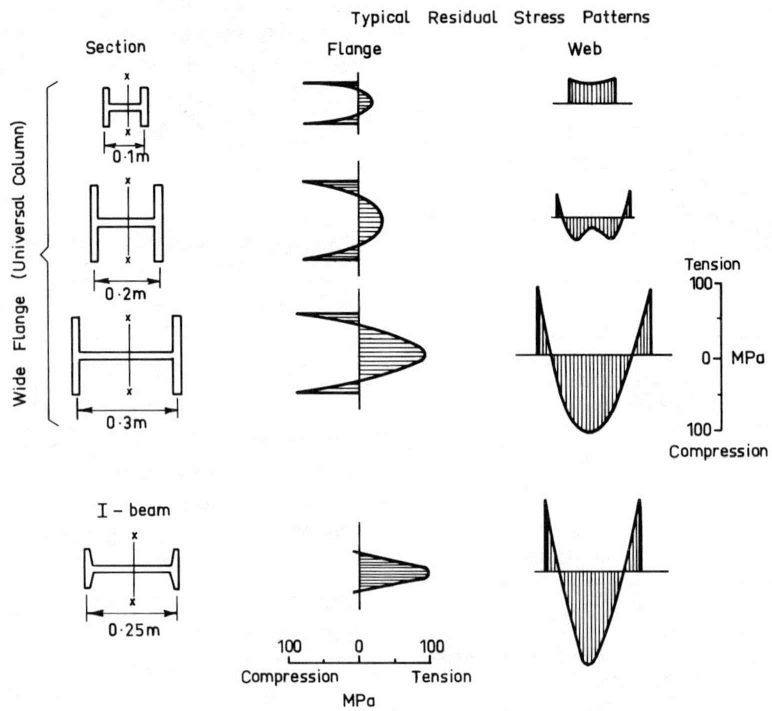

Figure 11.3 Typical residual stresses in rolled sections.

average value is referred to as the *structural tangent modulus, E_{ts}*, and is obtained from stub column tests. A typical stub column stress-strain curve and corresponding structural tangent modulus diagram are shown in Fig. 11.4. For materials that have a definite yield plateau, one of the most useful representations of the structural tangent modulus is the Ostenfeld-Bleich [2] parabola:

$$\frac{E_{ts}}{E} = \frac{\sigma_{av}(\sigma_Y - \sigma_{av})}{\sigma_{spl}(\sigma_Y - \sigma_{spl})} \quad \text{for } (\sigma_{spl} < \sigma_{av} < \sigma_Y) \tag{11.1.4}$$

where σ_{spl} is the *structural proportional limit*. Note that (11.1.4) is only valid in the range $\sigma_{spl} < \sigma_{av} < \sigma_Y$.

Substituting E_{ts} from (11.1.4) in place of E_t in (11.1.3) gives

$$\frac{\sigma_{ult}}{\sigma_Y} = 1 - \frac{\sigma_{spl}}{\sigma_Y}\left[1 - \frac{\sigma_{spl}}{\sigma_Y}\right]\lambda^2 \tag{11.1.5}$$

where

$$\lambda = \frac{L_e}{\rho\pi}\sqrt{\frac{\sigma_Y}{E}} \tag{11.1.6}$$

For rolled, "universal column," (wide flange) sections, a typical value of σ_{spl} is $\frac{1}{2}\sigma_Y$. In this case (11.1.5)

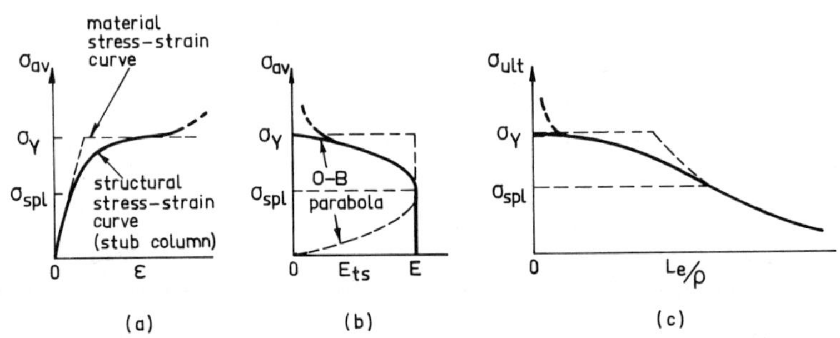

Figure 11.4 Column strength curve based on structural tangent modulus.

becomes

$$\frac{\sigma_{\text{ult}}}{\sigma_Y} = 1 - \frac{\lambda^2}{4} \qquad (11.1.7)$$

This equation is known as the Johnson parabola and is illustrated in Fig. 11.5. Beyond $\lambda = (2)^{1/2}$ the Euler curve applies ($\sigma_{\text{ult}}/\sigma_Y = 1/\lambda^2$). The Johnson parabola has been recommended by the Column Research Council [3] and by the A.I.S.C. [4] for calculating the ultimate strength of a "basic column" (essentially straight; pinned end) for the case of rolled steel sections. This "basic column" value is then used as the basis for design curves for practical, rolled steel columns.

Welded Sections

The distribution of residual stress due to welding is generally quite different from that due to hot rolling, even though both are caused by time-dependent temperature gradients. During welding the metal is in a perfectly plastic state. At any time during and after the instant of welding, some plastic fibers in the cross section will be cooling and other elastic fibers will be heating. The action of the loads on the elastic fibers due to changes in temperature is not counteracted by the plastic fibers until the plastic fibers have cooled sufficiently to become elastic. The interaction between the different fibers in the cross section results in a locked-in tensile stress in and near the weld which is approximately equal to the yield stress of the material. A typical residual stress distribution is shown in Fig. 11.6 together with an idealized distribution. The extent of the tension yield zone is generally from three to six thicknesses out from the weld on each side, and depends mainly on the total heat input. Other principal factors are the cross-sectional area of the weld deposit, the type of welding, and the welding sequence. This locked-in tension gives rise to a compressive residual stress σ_r in the remaining areas of the section. If ηt denotes the width of the tension yield zone and b the total flange width, then from equilibrium

$$\frac{\sigma_r}{\sigma_Y} = \frac{2\eta}{\dfrac{b}{t} - 2\eta} \qquad (11.1.8)$$

Obviously, for narrow thick sections residual stress will be high, and this will seriously diminish the

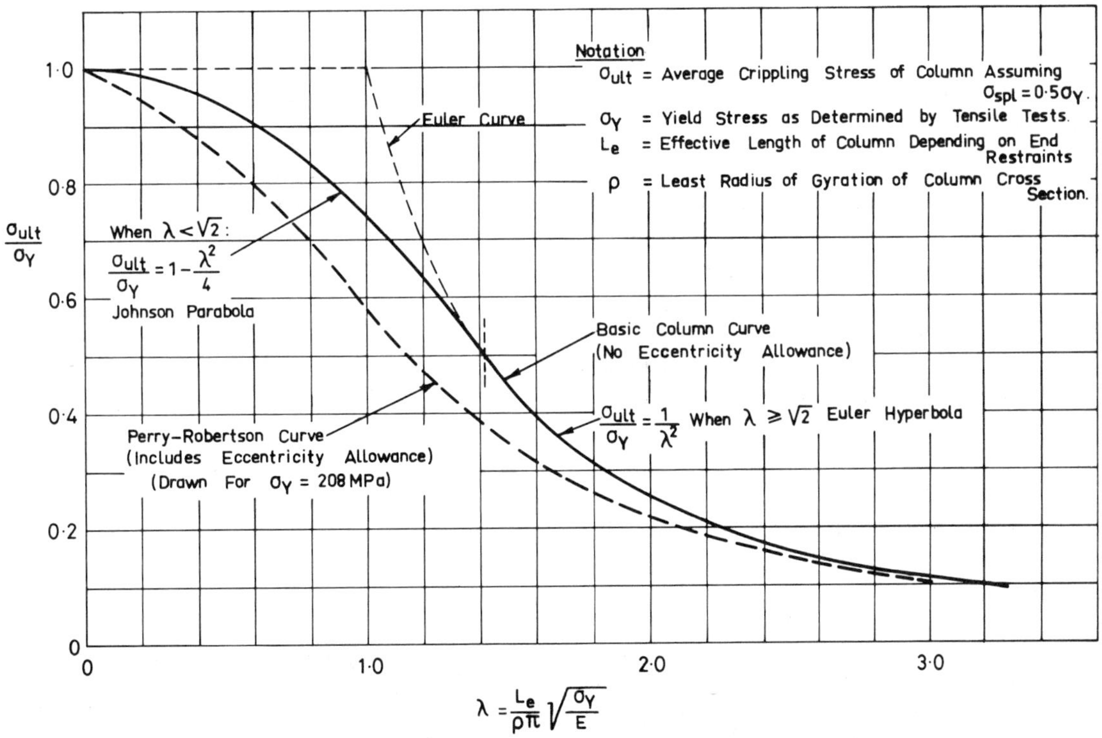

Figure 11.5 Basic column curve (Johnson parabola) and Perry-Robertson curve.

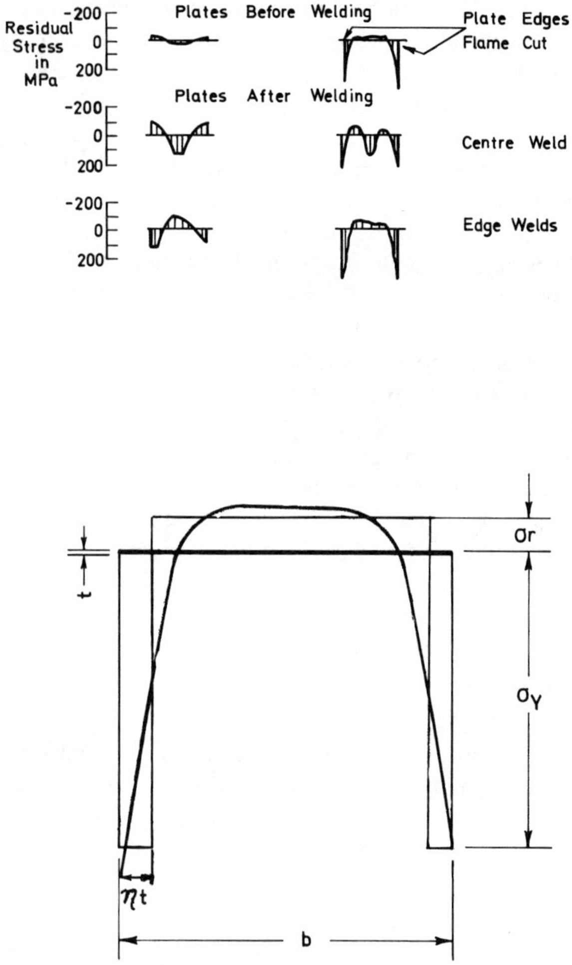

Figure 11.6 Typical residual stresses in welded sections.

strength of the section. It should be noted, however, that the effect will be much worse for open sections than for box sections, since in the latter the tensile stress in the corners delays the final collapse of this region, which is at the extremity of the section. Another point worth noting is that flame-cutting, which is usual ship fabrication practice, creates conditions similar to welding and thus causes high tensile stresses at the edges of flanges, an effect which has been shown to be moderately beneficial [5]. Finally, it should be noted that the effect of residual stress is somewhat diminished for higher yield steels due to a narrower tension zone.

Another important feature of welded sections is that because of the welding distortions the eccentricity is generally larger than for rolled shapes, and this can also seriously diminish the ultimate strength of such sections. This point will be taken up in the next section.

For typical welded universal column (wide flange) sections, the average residual stress is of the order of $\frac{1}{4}\sigma_Y$. Hence $\sigma_{\rm spl} = \sigma_Y - \sigma_r = \frac{3}{4}\sigma_Y$. For stockier sections (such as an I section) and when the bending is in the "strong direction," σ_r is typically $\frac{1}{2}\sigma_Y$, giving $\sigma_{\rm spl} = \frac{1}{2}\sigma_Y$. This again gives the Johnson parabola of (11.1.7), but it should be emphasized that this equation does not allow for eccentricity. This effect will now be discussed.

Eccentricity: Magnification Factor

In practice, columns are seldom perfectly straight, nor is the loading perfectly axial. Figure 11.7 shows a pinned column with an initial deflection $\delta(x)$ and an additional deflection $w(x)$ due to the axial load. For small deflections the governing equation is

$$\frac{d^2w}{dx^2} + \frac{P}{EI}(\delta + w) = 0 \qquad (11.1.9)$$

in which the combination of the total deflection and the axial force gives rise to a bending moment $P(\delta + w)$, which begins to act as soon as the load P is applied. This causes the deflection to increase further, and the deflection continues to grow and be magnified as long as P increases. Thus there is no static equilibrium configuration and no sudden buckling. The solution of (11.1.9) will depend on the nature of δ. Let δ be represented by a Fourier series

$$\delta = \sum_{n=1}^{\infty} \delta_n \sin\frac{n\pi x}{L}$$

and assume that the additional deflection due to the bending is also a Fourier series.

$$w = \sum_{n=1}^{\infty} w_n \sin\frac{n\pi x}{L}$$

Then
$$\frac{d^2w}{dx^2} = \sum_{n=1}^{\infty} -\frac{n^2\pi^2}{L^2} w_n \sin\frac{n\pi x}{L}$$

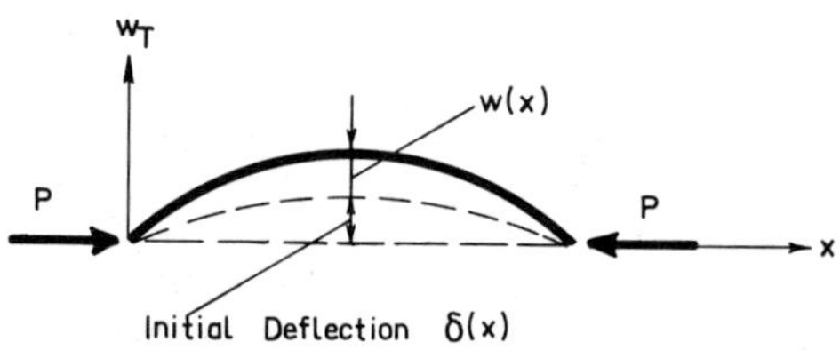

Figure 11.7 Eccentric column.

and from (11.1.9)

$$\sum_{n=1}^{\infty}\left[-\frac{n^2\pi^2}{L^2}w_n+\frac{P}{EI}(w_n+\delta_n)\right]\sin\frac{n\pi x}{L}=0$$

This must be true for any value of n as well as for the complete summation and therefore, since $\sin n\pi x/L$ cannot be zero, we must have

$$-\frac{n^2\pi^2}{L^2}(w_n)+\frac{P}{EI}(w_n+\delta_n)=0$$

and therefore

$$w_n=\frac{\delta_n}{\dfrac{n^2P_E}{P}-1},\quad \text{where } n=1,2,3\ldots \quad (11.1.10)$$

The condition when $(P_E/P)\rightarrow 1$ is of particular interest. An examination of (11.1.10) for various values of n gives the following:

n	w_n/δ_n
1	∞
2	$\frac{1}{3}$
3	$\frac{1}{8}$
⋮	⋮
∞	0

from which it is clear that the dominant term is $n = 1$.

Hence for any configuration of initial deflection δ, the extra deflection which P induces is given by

$$w=\frac{\delta}{\dfrac{P_E}{P}-1}$$

and the total or "magnified" deflection is

$$w_T=\delta+w=\frac{P_E}{P_E-P}\delta=\phi\delta \quad (11.1.11)$$

The factor

$$\phi=\frac{P_E}{P_E-P} \quad (11.1.12)$$

is called the *magnification factor*.

The foregoing load–deflection relationship for an elastic column is illustrated in Fig. 11.8. As the axial load approaches the Euler buckling load the deflection becomes very large—regardless of the value of the initial deflection. This implies that for an elastic column the maximum value of axial compressive load is close to the Euler buckling load, and hence the latter represents an upper bound for such a column. Strictly speaking, (11.1.12) is exact only for a sinusoidal initial deflection. However, the most serious shape for an initial deflection is a single half wave shape such as this, and the value of ϕ for other half wave shapes differs only negligibly from the value given by (11.1.12). Hence this equation can be used for nearly all types of initial eccentricity.

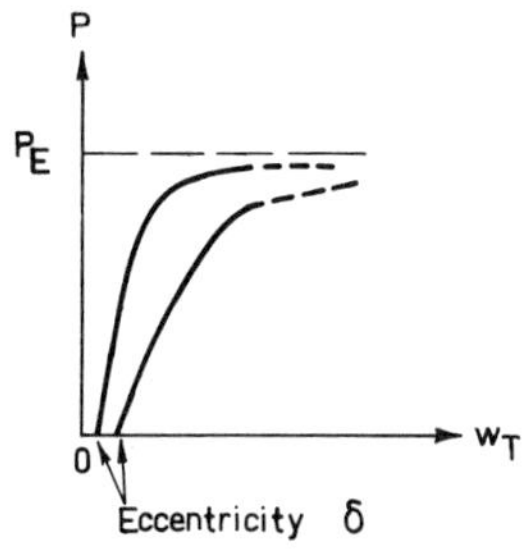

Figure 11.8 Load–deflection curves for eccentric columns.

Eccentricity effects may also arise due to eccentricity of load, as shown in Fig. 11.9. In this case it may be shown (see, for example, Ref. 6) that the magnification factor is given by

$$\phi=\sec\left(\frac{\pi}{2}\sqrt{\frac{P}{P_E}}\right) \quad (11.1.13)$$

The two types of eccentricity (load application and column geometry) can be combined linearly (superimposed) provided that the correct magnification factor is used for each. By plotting (11.1.12) and (11.1.13) it may be demonstrated that they give values of ϕ that never differ from each other by more than

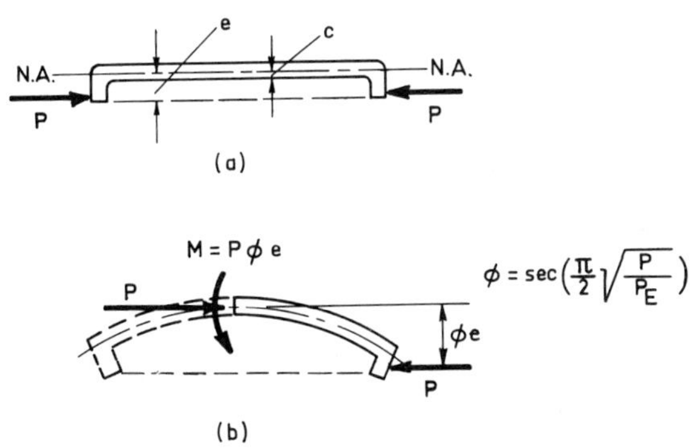

Figure 11.9 Column with eccentric load.

5%. Therefore, since it is simpler, (11.1.12) is usually used for eccentric loads as well as for initial deflections. This has the further advantage that the two effects may be combined directly, the total eccentricity $\Delta = \delta + e$ being used in place of δ in (11.1.11). The combination is illustrated in Fig. 11.10. The figure also shows that the maximum compressive stress in such a column is

$$\sigma_{\max} = \frac{P}{A} + \frac{P\phi\Delta}{Z} = \frac{P}{A}\left(1 + \phi\frac{\Delta}{r_c}\right) \tag{11.1.14}$$

in which Z is section modulus (on the compression side) and

$$r_c = \text{core radius} = \frac{Z}{A} = \frac{\rho^2}{c} \tag{11.1.15}$$

The ratio Δ/r_c is the *eccentricity ratio*. If the eccentricity of the load is large it will cause significant induced bending stresses, and these will tend to eclipse the residual stresses. The choice of a typical value for Δ for design purposes has long been a vexed question. Most authorities have related eccentricity to the slenderness ratio L/ρ, on the basis that a more slender column would be expected to have a larger eccentricity. If the relationship is assumed to be linear then

$$\frac{\Delta}{r_c} = \alpha\frac{L}{\rho} \tag{11.1.16}$$

On the basis of a series of tests of typical columns (with varying degrees of residual stress and eccentricity) Robertson [7] proposed a mean value of 0.003 for α, since this gave a lower bound to the test results.

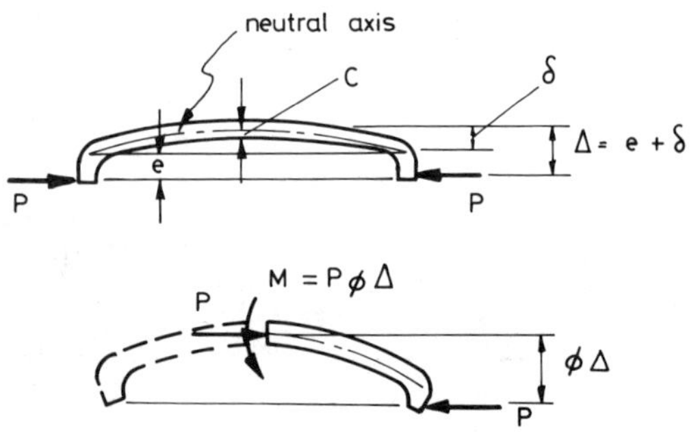

Figure 11.10 Combined eccentricity.

This approach of allowing for both eccentricity and residual stress by means of some effective eccentricity has formed the basis of many structural engineering design codes for some time.

11.2 COLUMN DESIGN FORMULAS

Perry-Robertson Formula

One of the simplest formulas for the ultimate load is the Perry-Robertson formula which adopts Robertson's value for α and assumes that the column will collapse when the maximum compressive stress reaches the yield stress. This gives, from (11.1.14) and (11.1.16)

$$\sigma_Y = \frac{P_{ult}}{A}\left(1 + \frac{\alpha\left(\frac{L}{\rho}\right)P_E}{P_E - P_{ult}}\right) \tag{11.2.1}$$

or, in terms of stress

$$\sigma_Y = \sigma_{ult}\left(1 + \frac{\frac{\alpha L}{\rho}}{1 - \frac{\sigma_{ult}}{\sigma_E}}\right) \tag{11.2.2}$$

with σ_E given by (11.1.2). This equation is a quadratic, and may be rewritten in a nondimensional, factored form

$$(1 - R)(1 - \lambda^2 R) = \eta R$$

in which R is the *strength ratio* of the column:

$$R = \frac{\sigma_{ult}}{\sigma_Y}$$

η is the *eccentricity ratio*, defined in (11.1.16):

$$\eta = \frac{\alpha L}{\rho}$$

and λ is the *column slenderness parameter*:

$$\lambda = \sqrt{\frac{\sigma_Y}{\sigma_E}} = \frac{L}{\pi\rho}\sqrt{\frac{\sigma_Y}{E}}$$

The solution is

$$R = \frac{1}{2}\left(1 + \frac{1+\eta}{\lambda^2}\right) - \sqrt{\frac{1}{4}\left(1 + \frac{1+\eta}{\lambda^2}\right)^2 - \frac{1}{\lambda^2}} \tag{11.2.3}$$

The Perry-Robertson formula, that is, the specific case of $\alpha = 0.003$, is plotted for mild steel ($E/\sigma_Y = 1000$) in Fig. 11.5, which shows that it lies significantly below the Johnson parabola, indicating the extent to which eccentricity degrades the column strength, quite apart from the effects of residual stress. For a perfectly straight column, having $\eta = 0$, (11.2.3) reduces to the Euler curve, $R = 1/\lambda^2$, providing that $\lambda > 1$.

Accurate Design Curves

Extensive experimental research [8] has shown that different types of sections have significantly different column collapse curves, partly due to geometry and partly to residual stresses. The results have shown that a single design curve such as the Perry-Robertson formula cannot accurately represent all types of sections. Instead, it is necessary to have different sets of curves for each type of section, each curve corresponding to a particular grade of steel. For instance, a Universal Column section has a quite different set of curves to a thin-wall tube, and should be treated accordingly. The newer codes have incorporated this approach. As shown by Dwight [9] the Perry-Robertson formula can still be used as a basis for these curves, with different values of α being used for different types of sections. Three such curves, corresponding to $\alpha = 0.0020$, 0.0035, and 0.0055, are given in Figs. 11.11, 11.12, and 11.13, and for rolled sections the curve to be chosen for each type of section is indicated in Table 11.1. In order to allow for strain hardening, the curves have a horizontal plateau at values of slenderness ratio less than some threshold value $(L/\rho)_0$. To achieve this the value of the eccentricity ratio, Δ/r_c, is defined as

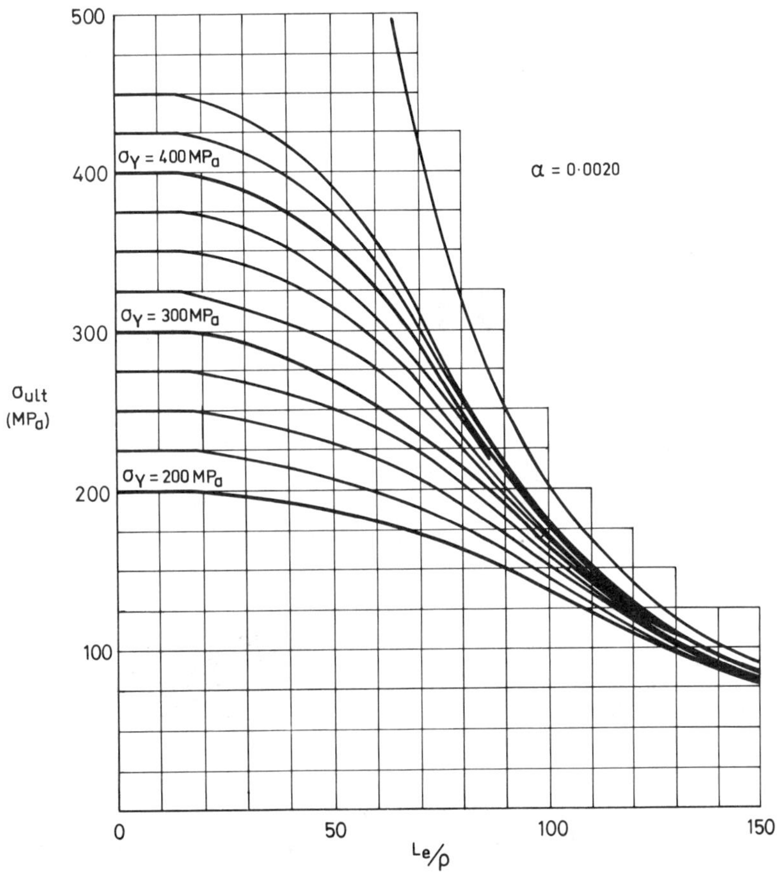

Figure 11.11 Column design curves for $\alpha = 0.002$.

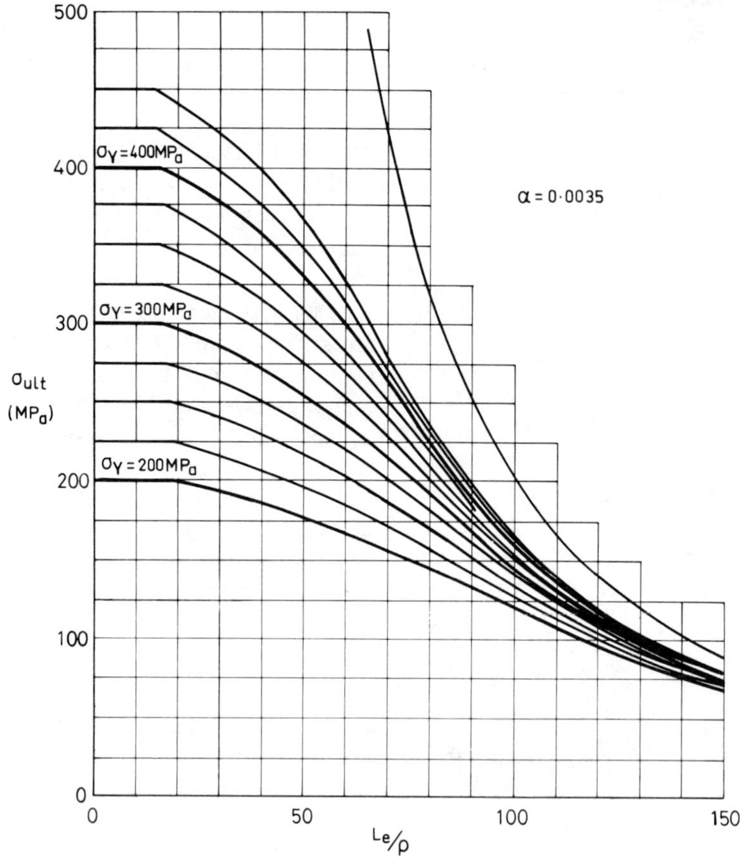

Figure 11.12 Column design curves for $\alpha = 0.0035$.

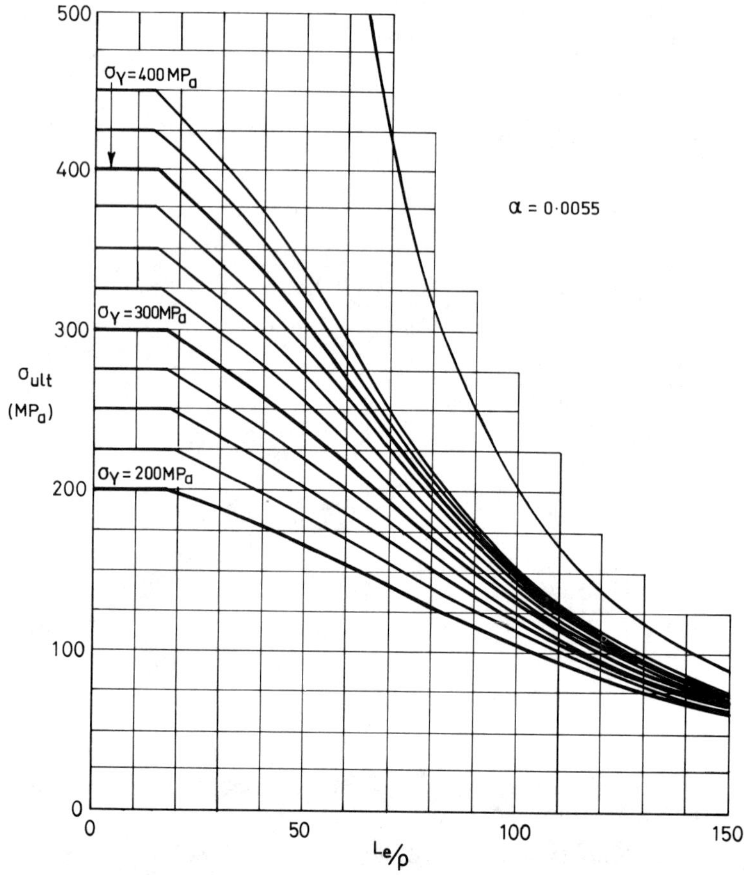

Figure 11.13 Column design curves for $\alpha = 0.0055$.

follows:

$$\left.\begin{array}{lll} \text{for} & \dfrac{L}{\rho} < \left(\dfrac{L}{\rho}\right)_{0'} & \dfrac{\Delta}{r_c} = 0 \\ \text{for} & \dfrac{L}{\rho} > \left(\dfrac{L}{\rho}\right)_{0'} & \dfrac{\Delta}{r_c} = \alpha\left[\dfrac{L}{\rho} - \left(\dfrac{L}{\rho}\right)_0\right] \end{array}\right\} \qquad (11.2.4)$$

where $(L/\rho)_0$ is given by:

$$\left(\frac{L}{\rho}\right)_0 = 0.2\pi\sqrt{\frac{E}{\sigma_Y}} \qquad (11.2.5)$$

For welded columns the residual stress σ_r in the flanges is generally larger than for rolled sections, and a further allowance must be made. This may be done to a satisfactory degree of accuracy by taking a reduced yield stress equal to about 95% of true yield stress.

TABLE 11.1 CURVE SELECTION TABLE FOR ROLLED AND WELDED SECTIONS (Note: For welded I- and box-sections the yield stress should be reduced by 5% to allow for welding residual stresses.)

Section	Axis of Buckling	α
Universal column	*xx*	0.0035
Universal column	*yy*	0.0055
Universal beam	*xx*	0.0020
Universal beam	*yy*	0.0035
UC or UB with cover-plates	*xx*	0.0035
UC or UB with cover-plates	*yy*	0.0020
Channel	*xx*	0.0055
Channel	*yy*	0.0055
Tee	*xx*	0.0055
Tee	*yy*	0.0055
Angle	any	0.0055
Round tube	any	0.0020
Rectangular hollow	*xx*	0.0020
Rectangular hollow	*yy*	0.0020
Welded I-sections	*xx*	0.0035
Welded I-sections	*yy*	0.0055
Welded box-sections		0.0035

11.3 EFFECT OF LATERAL LOAD: BEAM COLUMNS

Use of Magnification Factor

It may be shown that for a pinned column subjected to a uniform lateral load q and an axial load P, the exact solution for the maximum deflection is

$$w_{\max} = \frac{5qL^4}{384EI}\left[\frac{24}{5\xi^4}\left(\sec\xi - 1 - \frac{\xi^2}{2}\right)\right] \qquad (11.3.1)$$

in which $\xi = \left(\dfrac{L}{2}\right)\sqrt{\left(\dfrac{P}{EI}\right)}$

The first factor, $5qL^4/(384EI)$, is the central deflection of a laterally loaded member with pinned ends and with no axial loading. Hence the second factor gives the effect of the axial load in magnifying the central deflection; it is the magnification factor for the deflection for this particular loading and end condition. It may also be shown that the maximum bending moment is

$$M_{\max} = \frac{qL^2}{8}\left[\frac{2(1 - \sec\xi)}{\xi^2}\right] \qquad (11.3.2)$$

Once again the first factor, $qL^2/8$, is the central bending moment without the axial load, and the second factor is the magnification factor for the bending moment for this case. It is evident that the magnification factor for bending moment is different from that for deflection, and it may be shown that further differences occur for each combination of load and end condition.

Hence, in the interest of simplicity, the usual approach in dealing with laterally loaded columns (or "beam columns") is to use the simpler magnification factor which was derived for sinusoidal initial eccentricity:

$$\phi = \frac{P_E}{P_E - P} \qquad (11.3.3)$$

In a beam column the maximum bending moment is the sum of the bending moment due to the lateral load, M_0, plus that due to the eccentricity, which in this case includes the deflection δ_0 caused by the lateral load. That is

$$M_{\max} = M_0 + P\phi(\delta_0 + \Delta) \qquad (11.3.4)$$

The maximum compressive stress is then

$$\sigma_{\max} = \frac{P}{A} + \frac{M_{\max}}{Z}$$

To demonstrate the accuracy of this approach let us take the case of a beam column carrying a uniform lateral load q and an axial load $P = 0.5P_E$. We let

$\Delta = 0$ in order to be able to compare with the exact solution given by (11.3.2), which for this case is $M_{max} = 2.030M_0$. In (11.3.4) the value of δ_0 is $5qL^4/(384EI)$ and $\phi = 2$. Substituting these values gives

$$M_{max} = M_0 + 0.5\left(\frac{\pi^2 EI}{L^2}\right)2\frac{5qL^4}{384EI}$$

$$= M_0\left(1 + \frac{5\pi^2}{48}\right)$$

$$= 2.028M_0$$

which agrees with the exact result to three significant figures.

Because of the relatively large bending stress caused by the lateral load, the total bending stresses will be large and will tend to eclipse the residual stress. Hence we again make the simplifying (and only slightly conservative) assumption that the column will collapse when the maximum compressive stress reaches the yield stress. Therefore

$$\sigma_Y = \frac{P_{ult}}{A} + \frac{M_0}{Z} + \frac{P_{ult}(\delta_0 + \Delta)}{\left(1 - \dfrac{P_{ult}}{P_E}\right)Z} \qquad (11.3.5)$$

As before, this equation can be expressed in terms of nondimensional parameters:

$$R = \frac{\sigma_{ult}}{\sigma_Y}$$

$$\lambda = \sqrt{\frac{\sigma_Y}{\sigma_E}} = \frac{L}{\pi\rho}\sqrt{\frac{\sigma_Y}{E}}$$

$$\eta = \frac{(\delta_0 + \Delta)A}{Z}$$

$$\mu = \frac{\dfrac{M_0}{Z}}{\sigma_Y}$$

The resulting equation is the quadratic

$$(1 - R - \mu)(1 - \lambda^2 R) = \eta R \qquad (11.3.6)$$

for which the solution is

$$R = \frac{1}{2}\left(1 - \mu + \frac{1 + \eta}{\lambda^2}\right) - \sqrt{\frac{1}{4}\left(1 - \mu + \frac{1 + \eta}{\lambda^2}\right)^2 - \frac{1 - \mu}{\lambda^2}} \qquad (11.3.7)$$

This expression is plotted in Fig. 11.14 for selected values of η and μ. For a pinned beam column subjected to a specified lateral load, the axial load to cause collapse may be obtained by first calculating M_0 and δ_0, and then η and μ, and then interpolating for R in the figure.

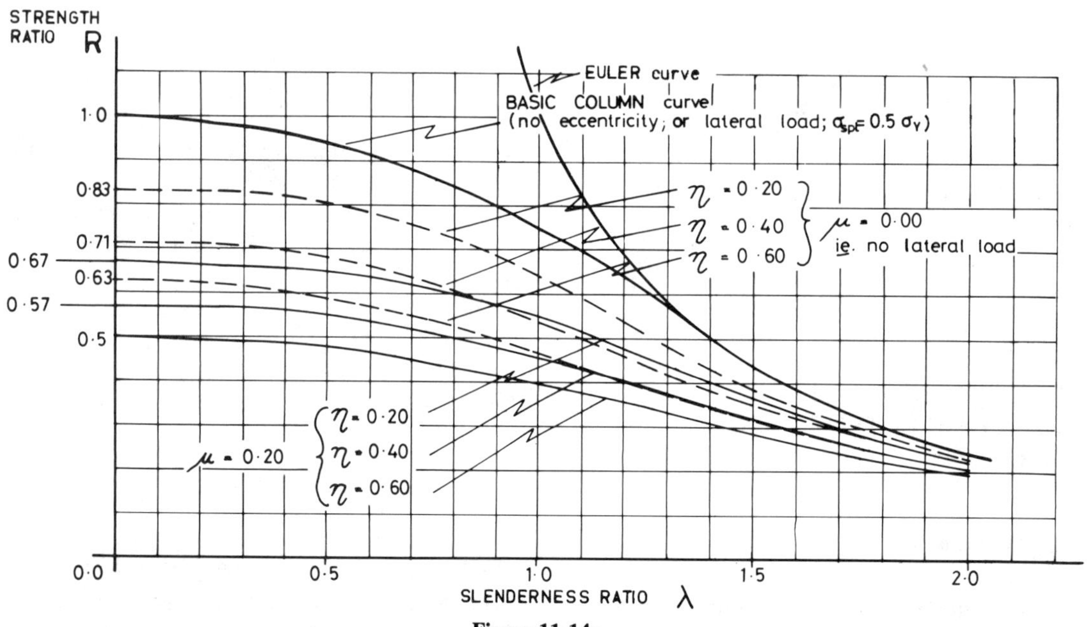

Figure 11.14a

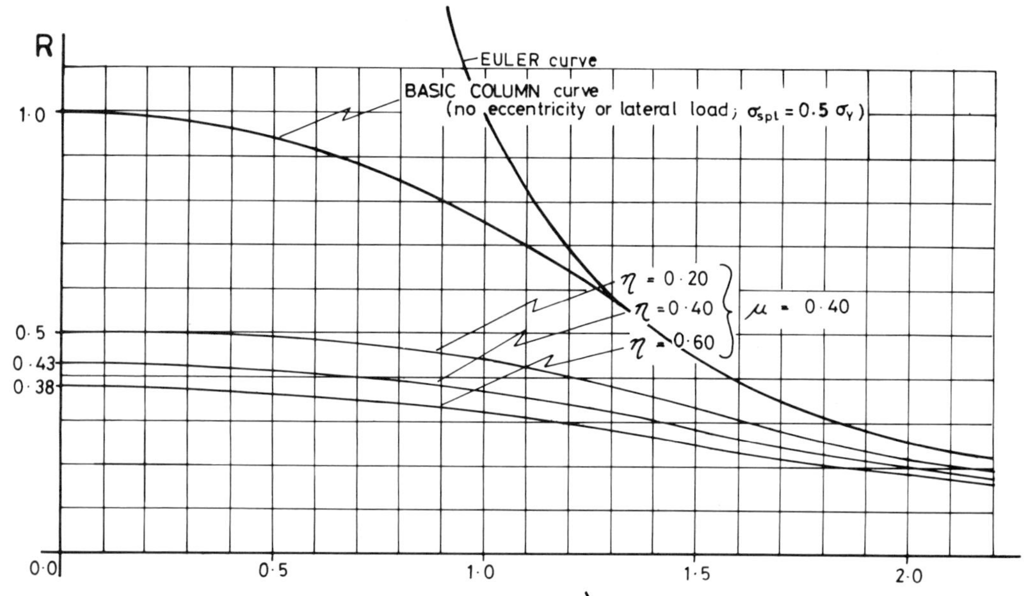

Figure 11.14b

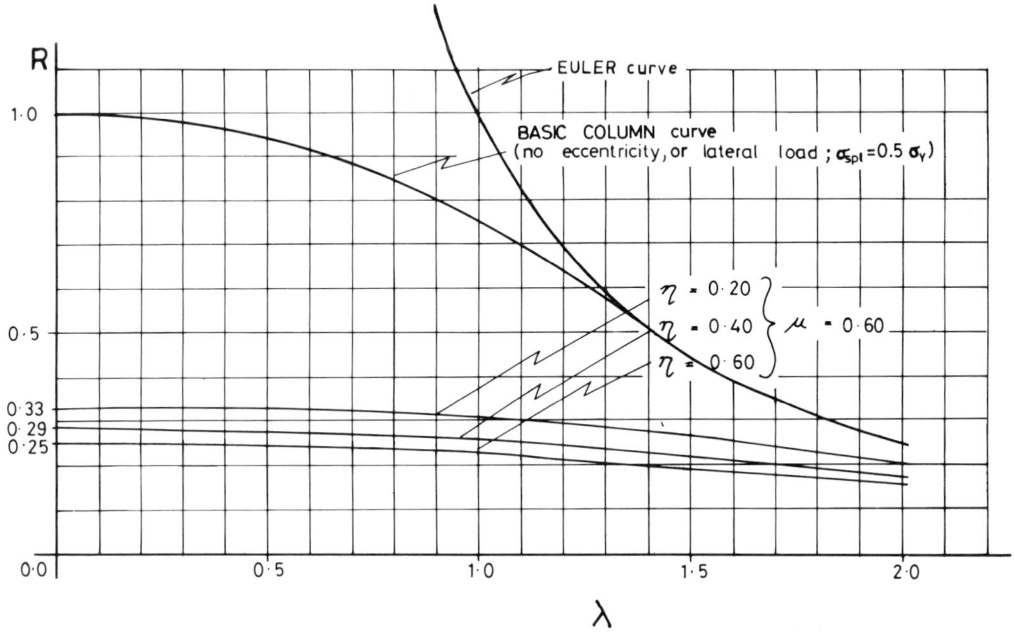

Figure 11.14c

Clamped Beam Column

If the ends of the beam column are rotationally restrained such that there is an end moment M_e, then the stress should be calculated at the end as well as at the center, because it is possible that yield in a compression flange may occur first at the ends. This is particularly likely with side frames, deck beams, and so on, in which one of the flanges is an effective width of plating. The other flange (i.e., the beam flange) is further from the neutral axis and therefore has larger stress, and this flange becomes the compression flange at the ends of the beam (assuming that the lateral load acts on the opposite side of the plating to the beam).

If the load is symmetric and if the ends are clamped then the beam column is equivalent to three pinned columns: the central portion, of length υL between the points of zero bending moment, and the two end portions, each of effective length $(1 - \upsilon)L$. The value of υ is 0.577 for a uniform load and 0.5 for a central point load. Figure 11.15 illustrates this for a uniformly loaded clamped beam column. The central portion is of length $0.577L$ and is subjected to a parabolic bending moment and the end portions are each equivalent to a pinned beam column of length $(1 - \upsilon)L = 0.423L$ and are subjected to an approximately triangular bending moment. Earlier in Section 11.3 it was mentioned that the magnification factor ϕ is relatively insensitive to the shape of the eccentric beam, and that the expression for ϕ for sinusoidal eccentric bending can also be used for a constant eccentricity. The same is true for the three portions of a clamped beam column. As shown in Fig. 11.15 the eccentricity of each end portion is equal to the deflection due to the lateral load (only) at the point of zero bending moment. This may be expressed as a fraction of δ_0, the maximum deflection of the beam due to the lateral load.

$$\delta_e = \gamma\delta_0 \tag{11.3.8}$$

The value of γ is 0.444 for a uniform load and 0.5 for a central point load. Although the eccentricity caused by the lateral load is not sinusoidal, the maximum eccentricity-induced bending moment in the end portion is approximately $P\phi_e\delta_e$, where the magnification factor ϕ_e is given by

$$\phi_e = \frac{P_{E,e}}{P_{E,e} - P} \tag{11.3.9}$$

in which

$$P_{E,e} = \frac{\pi^2 EI}{L_e^2}$$

and, as shown in Fig. 11.15

$$L_e = (1 - \upsilon)L$$

Similarly, the maximum eccentricity-induced bending moment in the center portion is approximately $P\phi_c\delta_c$, where δ_c and ϕ_c are the eccentricity and magnification factor of the center portion, given by

$$\delta_c = (1 - \gamma)\delta_0 \tag{11.3.10}$$

and

$$\phi_c = \frac{P_{E,c}}{P_{E,c} - P} \tag{11.3.11}$$

in which

$$P_{E,c} = \frac{\pi^2 EI}{L_c^2}$$

and

$$L_c = \upsilon L$$

Thus the maximum values of the extra, eccentricity-induced bending moments at the ends and at the center are $P\phi_e\gamma\delta_0$ and $P\phi_c(1 - \gamma)\delta_0$ respectively. Therefore, at the ends of a clamped beam column the total bending moment under the combined lateral and axial loads is $M_e + P[(1 - \upsilon)\Delta + \gamma\delta_0]\phi_e$, where M_e is the end moment due to the lateral load

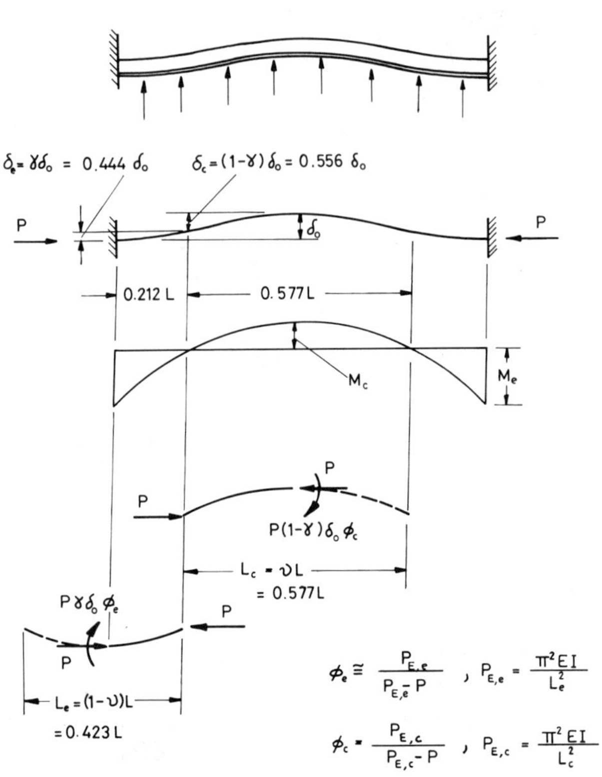

Figure 11.15 Clamped beam-column under uniform load.

(acting alone), while at the center of a clamped beam column the total bending moment is $M_c + P[v\Delta + (1 - \gamma)\delta_0]\phi_c$ where M_c is the central bending moment due to the lateral load, acting alone. The two portions of the beam are analyzed separately, and the effective length of each portion must be used in computing λ: vL for the middle portion and $(1 - v)L$ for the end portion.

This approach is much simpler than the exact beam column theory and yet retains sufficient accuracy for engineering use. The exact theory gives the following expression for ϕ_e for a uniform load:

$$\phi_e = \frac{3(\tan \xi - \xi)}{\xi^2 \tan \xi} \tag{11.3.12}$$

in which

$$\xi = \frac{\pi}{2}\sqrt{\frac{P}{P_E}} = \frac{L}{2}\sqrt{\frac{P}{EI}}$$

In order to assess the accuracy of the simplified approach let us take the case in which $P = 0.5P_E$. The end portion is of length $(1 - v)L = 0.423L$ and therefore

$$P_{E,e} = \frac{\pi^2 EI}{(0.423L)^2}$$

Hence

$$\frac{P}{P_{E,e}} = \frac{0.5P_E}{P_{E,e}} = 0.5(0.423)^2 = 0.0895$$

from which

$$\phi_e = 1.098$$

For $P = 0.5P_E$ the value of ξ in (11.3.12) is $\xi = 1.111$, from which $\phi_e = 1.090$. Thus the discrepancy is less than 1%.

The foregoing theory does not apply to intermediate degrees of end restraint. However, the idealized conditions of simply supported ends and clamped ends constitute the limits of possible variations and hence these two solutions provide useful information for the structural evaluation or design of the member.

The principal advantage of this approach is that it permits the solution for a clamped beam column to be obtained from the same direct and explicit formula as for simply supported beam columns, that is, (11.3.7) or, alternatively, Fig. 11.14. At the end of the beam column the eccentricity ratio and bending stress ratio to be used in (11.3.7) are, respectively,

$$\eta = \frac{[(1 - v)\Delta + \gamma\delta_0]A}{Z_e} \tag{11.3.13}$$

$$\mu = \frac{\dfrac{M_e}{Z_e}}{\sigma_Y} \tag{11.3.14}$$

where Z_e is the section modulus for whichever flange is in compression at the ends. Similarly, at the center of a clamped beam column these ratios are

$$\eta = \frac{[v\Delta + (1 - \gamma)\delta_0]A}{Z_c} \tag{11.3.15}$$

$$\mu = \frac{\dfrac{M_c}{Z_c}}{\sigma_Y} \tag{11.3.16}$$

where Z_c is the section modulus to the compression flange at the center.

REFERENCES

1. F. R. Shanley, "Inelastic Column Theory," *J. Aero. Sci.*, **14,** May 1947, pp. 261–267.
2. F. Bleich, "Buckling Strength of Metal Structures," McGraw-Hill, 1952.
3. Column Research Council, "Guide to Design Criteria for Metal Compression Members," Crosby Lockwood, London, 1960.
4. American Insititute of Steel Construction, "Specification for the Design, Fabrication, and Erection of Structural Steel for Building," New York, A.I.S.C., 1963.
5. L. Tall, "Recent Developments in the Study of Column Behaviour," *J. Inst. Eng., Aust.*, December 1964.
6. S. Timoshenko and J. Gere, *Theory of Elastic Stability,* McGraw-Hill.
7. A. Robertson, "The Strength of Struts," *Inst. Civ. Eng.*, Selected Paper No. 28, 1925.
8. L. S. Beedle, "Ductility as a Basis for Steel Design," *Conf. on Eng. Plasticity,* Cambridge, 1968, Cambridge University Press.
9. J. B. Dwight, "Use of Perry Formula to Represent the New European Strut Curves," Cambridge University, Dept. of Engineering Report TR.30, 1972.

Chapter Twelve

BUCKLING AND ULTIMATE STRENGTH OF PLATES

12.1 ELASTIC PLATES SUBJECTED TO UNIAXIAL COMPRESSION

Buckling of a Wide Column

As an introduction to the buckling of rectangular plating, we consider first the plate shown in Fig. 12.1. The plate is subjected to in-plane compression, with simple support along the loaded edges and with no support along the unloaded edges. Because of the lack of side support, the plate is acting more as a wide column than as a plate. The critical or buckling load for such a plate is the same as for a column except that the product EI is replaced by the plate flexural rigidity D. Thus, for a plate of length a the Euler formula for the critical load is

$$P_{cr} = \frac{\pi^2 Db}{a^2}$$

and the critical stress is

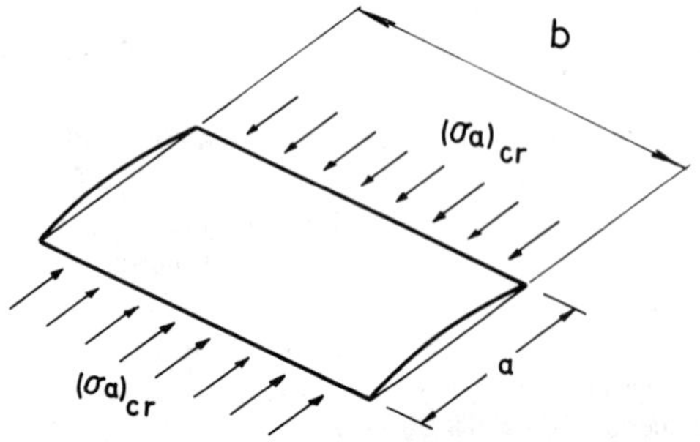

Figure 12.1 Buckling of wide column.

$$\sigma_{cr} = \frac{\pi^2 D}{a^2 t} = \frac{\pi^2 E}{12(1-\nu^2)}\left(\frac{t}{a}\right)^2 \qquad (12.1.1)$$

In (12.1.1) it is evident that the thickness/length ratio plays the same role as the slenderness ratio for columns. On the other hand, the width b of the plate plays no part whatever and it is clear that, from a buckling point of view, it is inefficient to use a plate in the role of a wide column, that is, without support along the unloaded edges.

Buckling of a Simply Supported Plate

Figure 12.2 shows a simply supported panel of plating subjected to a uniform in-plane compressive stress σ_a in the x-direction. Therefore, in the general plate equation (9.2.1) we substitute $N_x = -\sigma_a t$ and $p = N_y = N_{xy} = 0$ which yields

$$\nabla^4 w = -\frac{\sigma_a t}{D}\frac{\partial^2 w}{\partial x^2} \qquad (12.1.2)$$

Since the edges are simply supported, the deflected shape can be expressed in the form:

$$w = \sum_m \sum_n w_{mn} = \sum_m \sum_n C_{mn} \sin\frac{m\pi x}{a} \sin\frac{n\pi y}{b} \qquad (12.1.3)$$

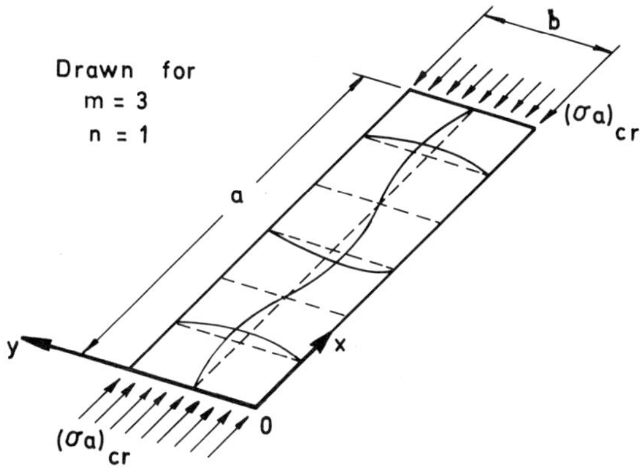

Figure 12.2 Buckled shape of a long plate.

which satisfies both the boundary conditions and the general biharmonic equation, of which (12.1.2) is a special case. We assume that the plate is free to move inward under the action of the in-plane compression (this should always be assumed for compressive loads) and hence there is no longitudinal strain in the mid-plane of the plate. Under these conditions the strain energy of deformation is due to bending only and is given by

$$U = \frac{D}{2}\int_0^a \int_0^b \left\{ \left(\frac{\partial^2 w}{\partial x^2} + \frac{\partial^2 w}{\partial y^2} \right)^2 - 2(1 - \nu) \left[\frac{\partial^2 w}{\partial x^2} \frac{\partial^2 w}{\partial y^2} - \left(\frac{\partial^2 w}{\partial x \, \partial y} \right)^2 \right] \right\} dx\, dy$$

Substitution of (12.1.3) yields

$$U = \frac{\pi^4 ab}{8} D \sum_m \sum_n C_{mn}^2 \left(\frac{m^2}{a^2} + \frac{n^2}{b^2} \right)^2$$

Likewise, the work done by the in-plane compressive stress is (with $N_y = N_{xy} = 0$)

$$W = \frac{\sigma_a t}{2} \int_0^a \int_0^b \left(\frac{\partial w}{\partial x} \right)^2 dx\, dy$$

and substitution of eq. (12.1.3) yields

$$W = \frac{\pi^2 b \sigma_a t}{8a} \sum_m \sum_n C_{mn}^2 m^2$$

From the principle of virtual work, $W = U$, and hence the equilibrium value of σ_a is

$$\sigma_a = \frac{\pi^2 a^2 D \sum_m \sum_n C_{mn}^2 \left(\frac{m^2}{a^2} + \frac{n^2}{b^2} \right)}{t \sum_m \sum_n m^2 C_{mn}^2}$$

The values of C_{mn} which give the minimum value of σ_a may be determined from the fact that an expression of the form $(c_1 + c_2 + c_3 \ldots)/(d_1 + d_2 + d_3 \ldots)$ has some intermediate value between the maximum and the minimum of the fractions $c_1/d_1, c_2/d_2, c_3/d_3, \ldots$ Thus there exists one fraction, c_j/d_j, which is less than any other fraction, c_i/d_i, and also less than any other sum or partial sum of fractions of the form shown. Therefore the minimum value of σ_a will be obtained by taking only one term of (12.1.3), say C_{mn}. Then the minimum value of σ_a is given by

$$(\sigma_a)_{cr} = \frac{\pi^2 a^2 D \left(\frac{m^2}{a^2} + \frac{n^2}{b^2} \right)^2}{t m^2}$$

The parameters m and n indicate the number of half-waves in each direction in the buckled shape. Both must be integers, and it can be seen that the value of n that gives the smallest value of σ_a is $n = 1$. Hence the plate will buckle into only one half-wave transversely, and the buckling or critical stress is

$$(\sigma_a)_{cr} = \frac{\pi^2 D}{a^2 t} \left[m + \frac{1}{m} \left(\frac{a}{b} \right)^2 \right]^2 \qquad (12.1.4)$$

This equation was derived by G. H. Bryan in 1891. It is usually written in a more general form in terms of a buckling coefficient k and the plate width b

$$(\sigma_a)_{cr} = k \frac{\pi^2 D}{b^2 t} \qquad (12.1.5)$$

The expression for the buckling coefficient k depends on the type of boundary support. From a comparison of (12.1.4) and (12.1.5) it is evident that for simply supported plates, k is given by

$$k = \left(\frac{mb}{a} + \frac{a}{mb} \right)^2 \qquad (12.1.6)$$

Expressions for k for other types of boundary conditions will be presented in Section 12.2. For design applications, in which the plate thickness is to be determined, (12.1.5) is usually written in the alternative form

$$\left.\begin{aligned}(\sigma_a)_{cr} &= KE\left(\frac{t}{b}\right)^2 \\ \text{in which} \quad K &= \frac{\pi^2 k}{12(1-\nu^2)}\end{aligned}\right\} \tag{12.1.7}$$

In Fig. 12.3 the coefficient k is plotted against aspect ratio a/b for various values of m. The figure shows that the lowest (and therefore truly critical) value of σ_a will occur for different values of m, depending on the aspect ratio. This can also be seen by setting $dk/dm = 0$, from which is obtained $m = a/b$; that is, the stress is lowest when the number of half-waves in the longitudinal direction is equal to the aspect ratio. Under these conditions, $k = 4$.

This indicates that a long panel of plating always tends to buckle into a number of square panels; if the length a is not an exact multiple of the width b, then the panel will buckle into the nearest whole number of half-waves which will make the critical stress a minimum. The value of k is therefore given by the solid (and discontinuous) curve in Fig. 12.3 from which it may be seen that, although the value of k is somewhat greater for noninteger aspect ratios, this effect is slight and diminishes as aspect ratio increases. Hence, for long simply supported plates it is usually assumed that $k = 4$. Equation (12.1.5) then becomes

$$(\sigma_a)_{cr} = 4\frac{\pi^2 D}{b^2 t} \tag{12.1.8}$$

and, assuming $\nu = 0.3$, (12.1.7) becomes

$$(\sigma_a)_{cr} = 3.62E\left(\frac{t}{b}\right)^2 \tag{12.1.9}$$

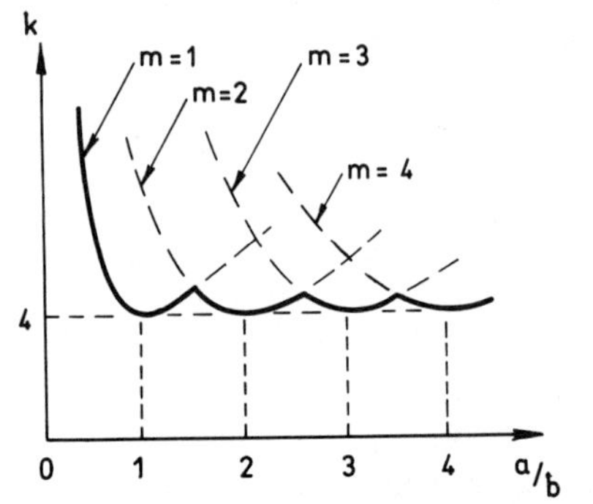

Figure 12.3

For a wide plate, in which the aspect ratio is less than 1.0, m will be equal to unity. Therefore, (12.1.4) becomes

$$(\sigma_a)_{cr} = \frac{\pi^2 D}{a^2 t}\left[1 + \left(\frac{a}{b}\right)^2\right]^2 \tag{12.1.10}$$

and this gives the critical stress for a simply supported wide plate. As before, a more general form of (12.1.10) can be written using a buckling coefficient. However, in the case of wide plates the buckling equations are usually written in terms of the length a, rather than the width b, because a wide plate more nearly resembles a column. The general "wide plate" equation is then

$$(\sigma_a)_{cr} = \bar{k}\frac{\pi^2 D}{a^2 t} \tag{12.1.11}$$

From (12.1.5) it may be seen that the wide plate buckling coefficient $\bar{k}$ is related to the usual coefficient k by

$$\bar{k} = \left(\frac{a}{b}\right)^2 k \tag{12.1.12}$$

The use of (12.1.11) in place of (12.1.5) overcomes the difficulty of dealing with very large values of k for low values of aspect ratio. It can be seen from (12.1.10) that for very wide plates, $\bar{k}$ approaches unity and the critical stress approaches the value for a wide column given in (12.1.1).

For design purposes (12.1.11) may be written as

$$(\sigma_a)_{cr} = \bar{K}E\left(\frac{t}{a}\right)^2 \tag{12.1.13}$$

in which

$$\bar{K} = \frac{\pi^2}{12(1-\nu^2)}\left[1 + \left(\frac{a}{b}\right)^2\right]^2$$

For $\nu = 0.30$,

$$\bar{K} = 0.905\left[1 + \left(\frac{a}{b}\right)^2\right]^2$$

The question of wide plates versus narrow plates leads to the consideration of the relative merits of stiffening a large sheet of plating in the longitudinal or in the transverse direction. Let the plating be of length

L and breadth B and subject to a uniform compressive stress σ_a. If longitudinal stiffeners are fitted at a spacing s as in Fig. 12.4a then the buckling stress is found from eq. (12.1.8) with $b = s$

$$(\sigma_a)_{cr} = \frac{4\pi^2 D}{s^2 t}$$

On the other hand, if the stiffeners are fitted transversely, at the same spacing, $(\sigma_a)_{cr}$ is obtained from (12.1.10) with $a = s$ and $b = B$

$$(\sigma_a)_{cr} = \frac{\pi^2 D}{s^2 t}\left[1 + \left(\frac{s}{B}\right)^2\right]^2$$

The term s/B is generally quite small for a ship—seldom greater than $\frac{1}{6}$—and so the term in brackets is approximately unity. Therefore, the buckling strength of longitudinally stiffened plating is nearly four times as great as that of transversely stiffened plating. This shows the great advantage of longitudinal over transverse stiffening in ship structures, and the former is used wherever possible (the deep web frames of a longitudinal framing system may make this system undesirable in some instances).

12.2 OTHER BOUNDARY CONDITIONS

The foregoing discussion has assumed that the sides of the plates are simply supported. If any other type of boundary condition exists along the plate edges then a new solution must be obtained for (12.1.2) which satisfies this condition. Many authors have presented solutions for a variety of cases, and the principal cases are summarized in Figs. 12.5a and b. The latter figure gives values of K in the "design" version of the Bryan formula, (12.1.7). Still other cases are covered in Refs. 1–4.

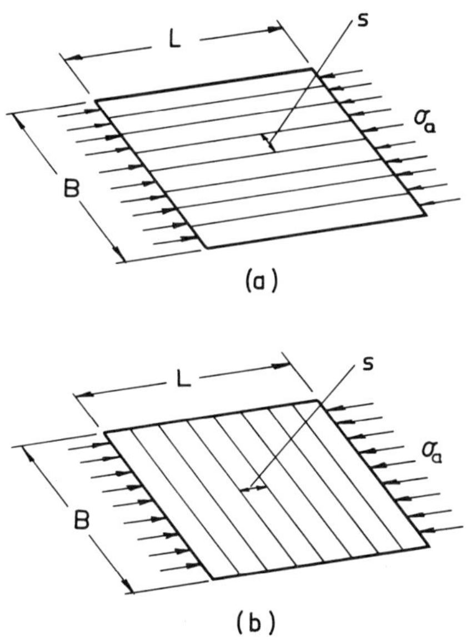

Figure 12.4

Clamped Edges

For in-plane loads, as in the case of lateral loads, it is not possible to obtain finite expressions for the solution of clamped plates. Numerical solutions have been obtained by Faxen [5] and Maulbetsch [6], the latter employing the energy (or Ritz) method. Levy [7] obtained a solution using infinite series which have good

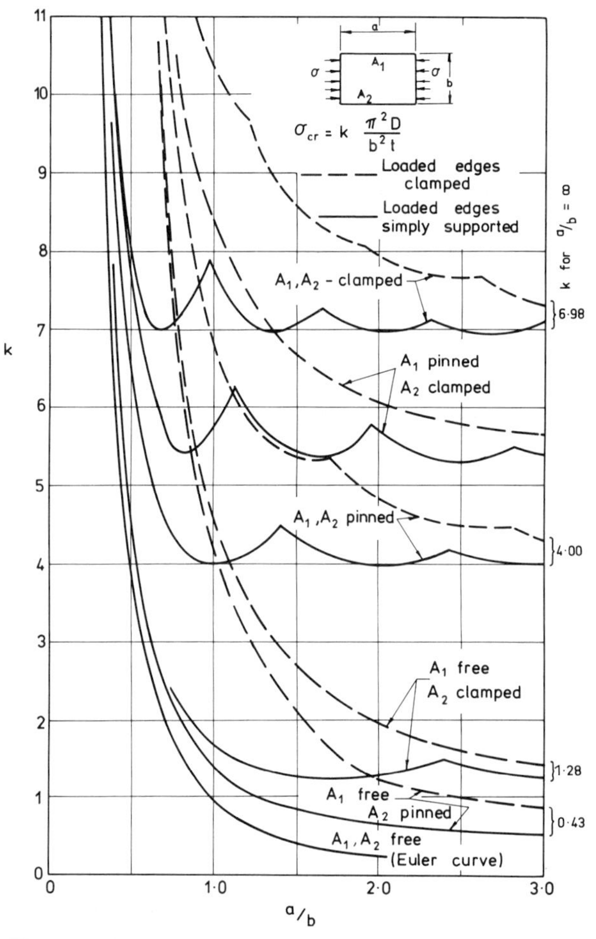

Figure 12.5a Buckling stress coefficient k for flat plates in uniaxial compression.

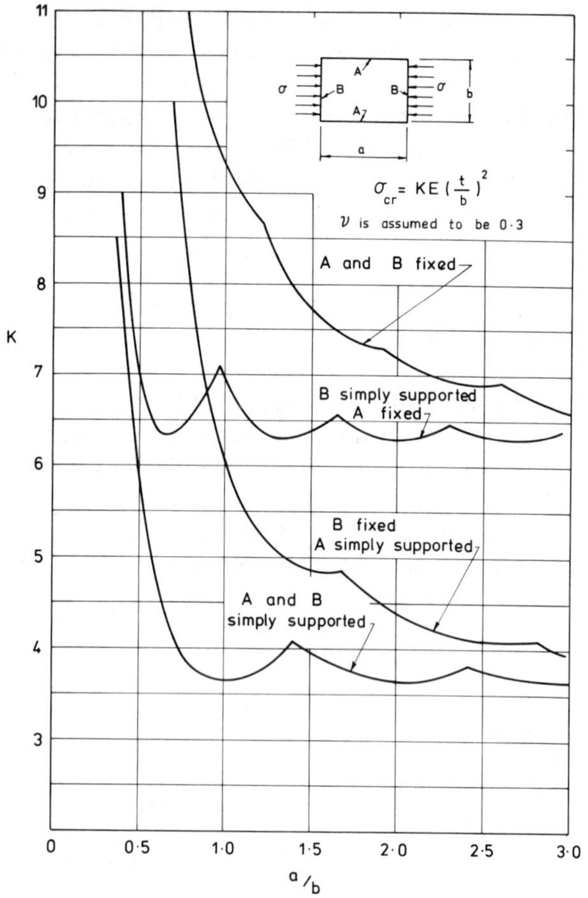

Figure 12.5b Buckling coefficient $k$in the design formula for flat plates in uniaxial compression.

TABLE 12.1 BUCKLING COEFFICIENT k FOR CLAMPED PLATES UNDER UNIAXIAL COMPRESSION

Aspect Ratio a/b	Number of Buckles	Faxen	Maulbetsch	Levy
0.50	1	18.30	—	—
0.75	1	11.39	12.77	11.66
1.00	1	10.07	10.48	10.07
1.25	2	9.2	9.38	9.25
1.50	2	8.30	8.45	8.33
1.75	2	8.18	8.17	8.11
2.00	3	7.87	8.06	7.88
2.25	3	—	7.96	7.63
2.50	3	—	7.99	7.57
2.75	4	—	7.76	7.44
3.00	4	—	7.59	7.37
3.25	4	—	7.86	7.35
3.50	5	—	7.37	7.27
3.75	5	—	7.40	7.24
4.00	5	—	7.45	7.23
∞	∞	—	—	—

convergence. Table 12.1 summarizes the results of the various authors.

Unloaded Edges Rotationally Restrained

Lundquist and Stowell [8] have investigated the case in which the support along the unloaded edges is intermediate between simply supported and clamped. The degree of rotational restraint is specified in terms of a coefficient of restraint ζ, defined as

$$\zeta = \frac{b}{D} C_y$$

in which C_y is the rotational stiffness of the supporting structure along the unloaded edge, that is, the ratio of edge moment to edge slope. Values of k for various values of ζ are given in Fig. 12.6. The value of ζ depends not only on the dimensions of the supporting structure, but also on the extent to which any com-

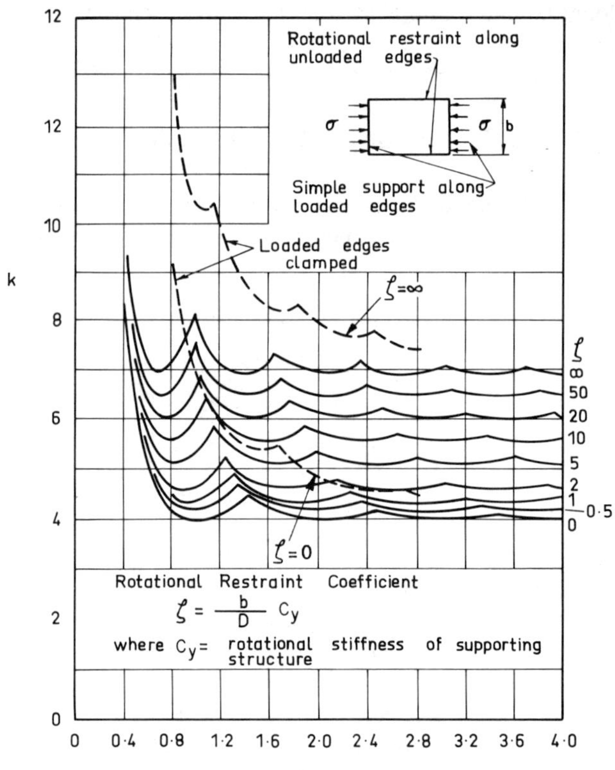

Figure 12.6 Buckling coefficient k for plates with loaded edges simply supported and longitudinal edges rotationally restrained [8].

pressive load acting on it approaches its own buckling value. Values of ζ for the webs and flanges of various structural sections are given in Section 92 of Ref. 1. Figure 12.6 can also be used when there are unequal rotational restraints along the unloaded edges. This is done by obtaining k from Fig. 12.6 for each value of ζ and taking the geometric mean of these two: $k = (k_1 k_2)^{1/2}$. Reference 8 shows that this approximation is sufficiently accurate for design.

Loaded Edges Rotationally Restrained

In general, the important boundary conditions are those along the longer edges of the plate. Thus, for short wide plates the edge restraint along the loaded edges becomes significant. The effect is similar to the effect of end conditions in a column and can be dealt with in the same way; that is, by using an effective length a_e in the "wide plate" formula, (12.1.10)

$$(\sigma_a)_{cr} = \frac{\pi^2 D}{a_e^2 t}\left[1 + \left(\frac{a_e}{b}\right)^2\right]^2 \qquad (12.2.1)$$

The effective length a_e for any combination of end constraints is the same as for a column. For example, for clamped ends $a_e = \frac{1}{2}a$, and for one end simply supported and the other clamped $a_e = 0.707$. Strictly speaking this approach applies only when the short or unloaded edges are simply supported. However, for wide plates the exact nature of the support along the short edges has little effect *as long as these edges are at least simply supported*. Hence the effective length approach can be used for all wide plates which fulfill this condition. An alternative approach is to use a coefficient of restraint ζ, as was done for long plates restrained along the unloaded edges. In this case we define the coefficient of restraint ζ as $\zeta = (a/D)C_x$, where C_x is the rotational stiffness of the supporting structure along the loaded edges. The solution to this case is obtained from

$$K_1 \tan\frac{\alpha k_1}{2} - K_2 \tan\frac{\alpha k_2}{2} + \frac{\alpha}{\zeta}(K_1^2 - K_2^2) = 0$$

in which the coefficients K_1 and K_2 are related to the buckling coefficient k by

$$K_{1,2} = \frac{\pi}{2}(\sqrt{k} \pm \sqrt{k-4})$$

TABLE 12.2 BUCKLING COEFFICIENT $\bar{k}$ FOR WIDE PLATES IN COMPRESSION ELASTICALLY RESTRAINED ON THE LOADED EDGES

Coefficient	Aspect Ratio a/b							
ζ	0.1	0.2	0.3	0.4	0.5	0.6	0.8	1.0
clamped	4.02	4.08	4.19	4.34	4.55	4.82	5.59	6.74
40	3.35	3.40	3.51	3.63	3.88	4.15	4.94	6.18
20	2.88	2.93	3.04	3.18	3.42	3.70	4.51	5.78
13.2	2.56	2.62	2.72	2.86	3.10	3.38	4.21	5.48
10	2.32	2.38	2.49	2.62	2.86	3.15	3.98	5.26
8	2.15	2.21	2.32	2.45	2.68	2.97	3.80	5.09
6.67	2.01	2.07	2.18	2.32	2.55	2.84	3.67	4.98
5.92	1.90	1.96	2.08	2.22	2.44	2.73	3.57	4.80
5	1.81	1.88	1.99	2.13	2.35	2.65	3.48	4.80
4.46	1.74	1.81	1.91	2.06	2.28	2.58	3.41	4.73
4	1.69	1.75	1.85	2.00	2.23	2.52	3.35	4.68
2	1.39	1.46	1.55	1.71	1.92	2.22	3.06	4.37
1	1.22	1.27	1.38	1.55	1.76	2.04	2.89	4.19
pinned	1.02	1.08	1.19	1.35	1.56	1.85	2.69	4.00

The results are given in Table 12.2 in terms of the wide plate buckling coefficient $\bar{k} = (a/b)^2 k$. The corresponding coefficient $\bar{K}$ in the "design" version of the wide plate formula

$$(\sigma_a)_{cr} = \bar{K}E\left(\frac{t}{a}\right)^2$$

is given in Fig. 12.7. As in the case of longitudinal edge restraint, the geometric mean values of $\bar{k}$ and $\bar{K}$ can be used for cases of unequal end restraint.

In ship structures the rotational restraint is usually provided by flange-and-web type of transverse stiffeners, such as "T" or "I" sections. Bleich [1] has shown that in this case ζ is given approximately by

$$\zeta = \frac{27a}{t^3 b^2}\left(\frac{\pi^2 I d^2}{b^2} + \frac{J}{2.6}\right) \qquad (12.2.2)$$

where d = depth of the web
I = second moment of area of the stiffener about the midthickness of the web
J = Saint-Venant's torsion constant for the stiffener

12.3 BIAXIAL COMPRESSION

All Edges Simply Supported

With two in-plane loads, we need to adopt a more comprehensive nomenclature. In order to be consistent

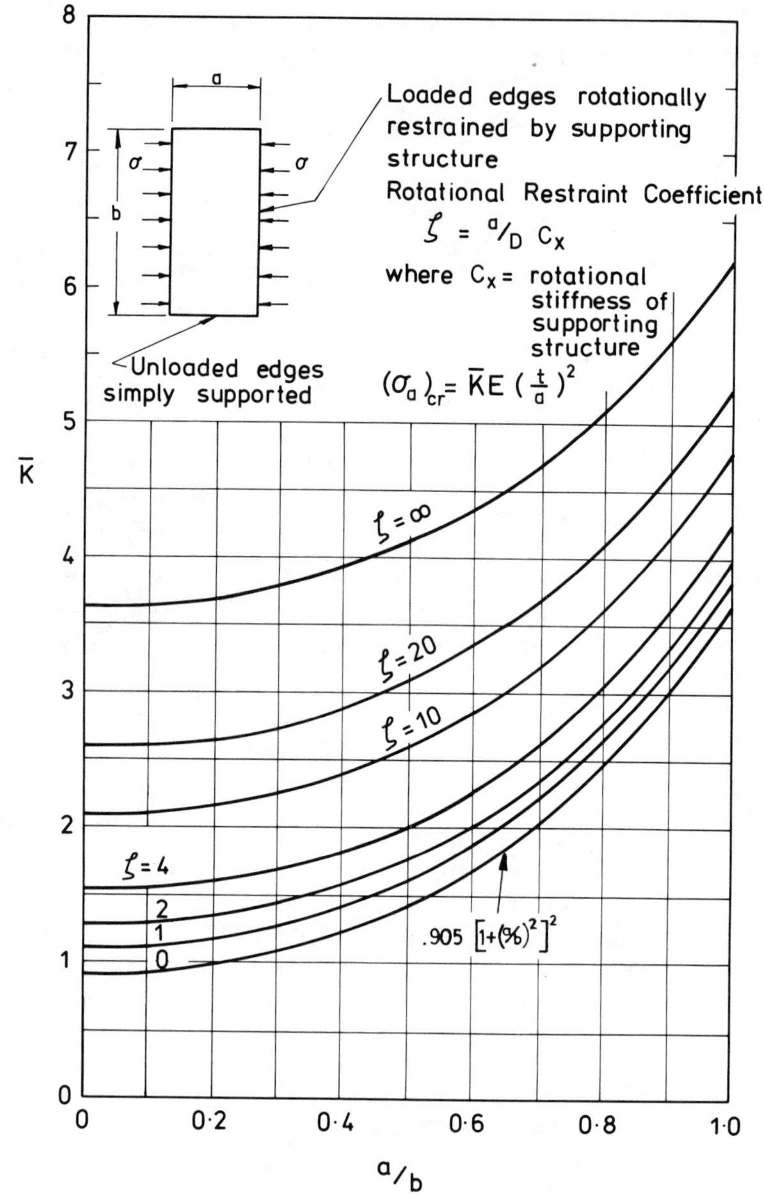

Figure 12.7 Buckling coefficient $\overline{K}$ in the design formula for flat plates (wide plate version).

with the convention for uniaxially loaded plates and yet allow for maximum flexibility, we shall say that the dimension a is the plate length in line with the principal load σ_{ax}, whichever that may be. That is, we shall take the x-axis as being in the direction of the principal load and designate this load as σ_{ax}, but the designer is free to choose whichever of the loads he wishes to regard as the principal load; it need not be the larger of the two. With this convention the dimension a is always parallel to σ_{ax} and the dimension b is always parallel to σ_{ay}, as shown subsequently in Fig. 12.9. The aspect ratio $\alpha(=a/b)$ may be any value.

In plating which is subjected to biaxial in-plane compression the critical values of σ_{ax} and σ_{ay} are interdependent. In the case of a simply supported plate the deflection can again be represented by the series of (12.1.3) and it can again be proved that only one term in the series need be taken. Applying the energy method as before yields the following expression for the critical combination of σ_{ax} and σ_{ay}:

$$\left[\left(\frac{m}{\alpha}\right)^2 \sigma_{ax} + n^2\sigma_{ay}\right]_{cr} = \frac{\pi^2 D}{b^2 t}\left[\left(\frac{m}{\alpha}\right)^2 + n^2\right]^2 \quad (12.3.1)$$

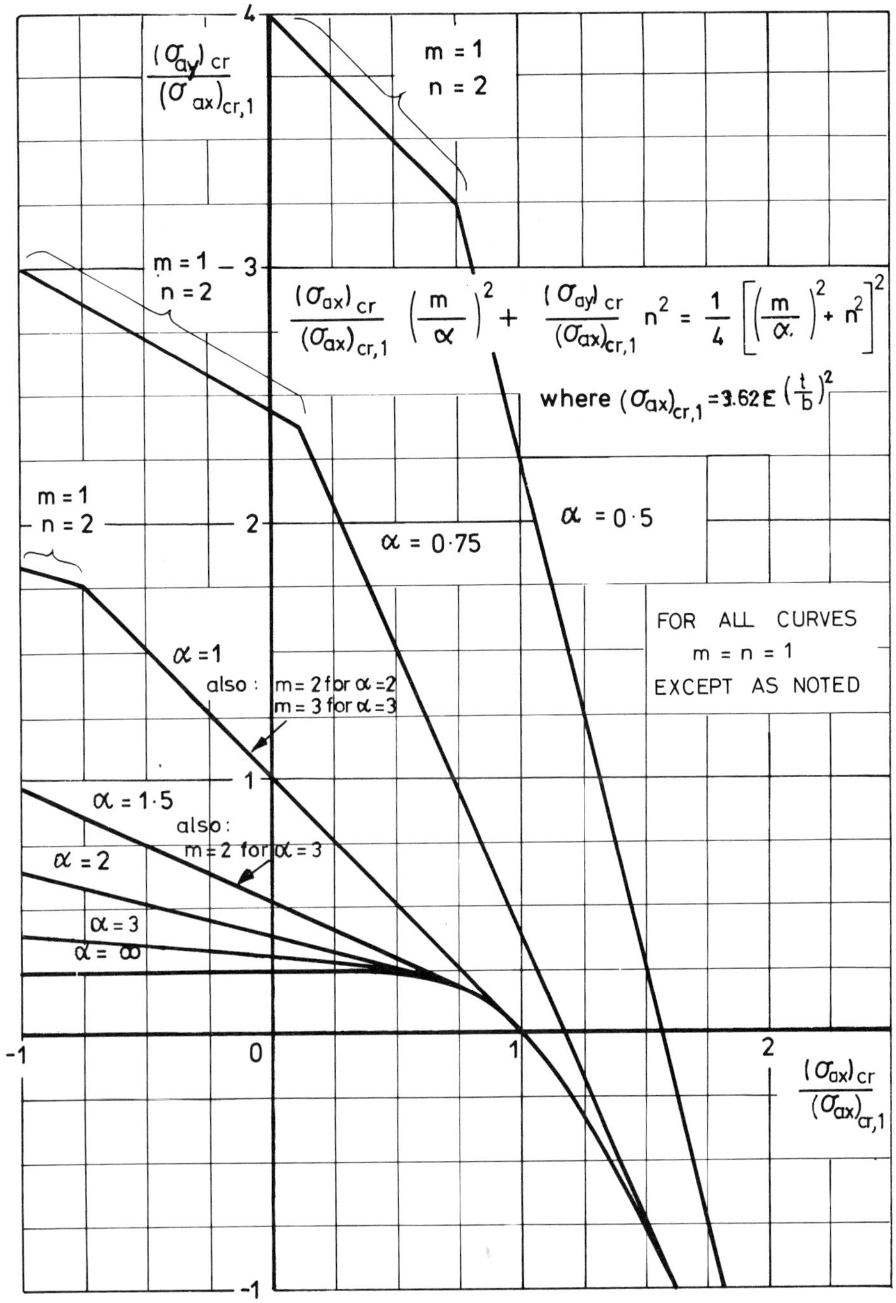

Figure 12.8 Buckling stresses of biaxially loaded simply supported plates.

If we denote the square plate critical stress as

$$(\sigma_{ax})_{cr,1} = 3.62E\left(\frac{t}{b}\right)^2$$

we can write this equation in nondimensional form:

$$\left[\left(\frac{m}{\alpha}\right)^2 \frac{\sigma_{ax}}{(\sigma_{ax})_{cr,1}} + n^2 \frac{\sigma_{ay}}{(\sigma_{ax})_{cr,1}}\right]_{cr} = \frac{1}{4}\left[\left(\frac{m}{\alpha}\right)^2 + n^2\right]^2 \tag{12.3.2}$$

In applying the formula, values of m and n must be chosen so as to give the smallest buckling stress. Figure 12.8 indicates the appropriate values of m and n, and the corresponding buckling stresses, for a range of aspect ratios. It is worth noting that whenever a decrease in aspect ratio would produce a smaller buckling stress, the plate will in fact buckle at this lower stress because the buckled shape can adopt this pattern over at least a portion of the plate. This effect is responsible for the curved segment of line in Fig. 12.8,

TABLE 12.3

σ_{ay}/σ_{ax}	1	$\frac{3}{4}$	$\frac{1}{2}$	$\frac{1}{3}$	$\frac{1}{4}$	0	$-\frac{1}{4}$	$-\frac{1}{2}$	-1	$-\frac{3}{2}$	-2
σ_{ax}/σ_e	5.61	6.41	7.48	8.40	8.80	10.65	13.12	15.20	17.4	18.75	20.0
σ_{ay}/σ_e	5.61	4.81	3.74	2.80	2.20	0	-3.28	-7.60	-17.4	-28.1	-40.0

$\sigma_e = (\pi^2 D)/(b^2 t) = 0.905E(t/b)^2$.

which is formed by a series of overlapping straight lines.

All Edges Clamped

For the case of all edges clamped, Timoshenko [2] has shown that for plates subjected to approximately equal compressive stresses the interaction formula is

$$(\sigma_{ax} + \alpha^2 \sigma_{ay})_{cr} = 1.20E\left(\frac{t}{b}\right)^2\left(\frac{3}{\alpha^2} + 3\alpha^2 + 2\right) \tag{12.3.3}$$

For square plates, critical combinations of σ_{ax} and σ_{ay} are given in Table 12.3 for particular values of σ_{ax}/σ_{ay}, including cases in which σ_{ay} is tensile.

σ_{ay} Edges Clamped: σ_{ax} Edges Simply Supported

There is no simple interaction formula available for this case, but Fig. 12.9 gives values of the buckling coefficient k to be used in calculating $(\sigma_{ax})_{cr}$. As before, the value of m—the number of half-waves in the x-direction—must be chosen so as to give the lowest value of k.

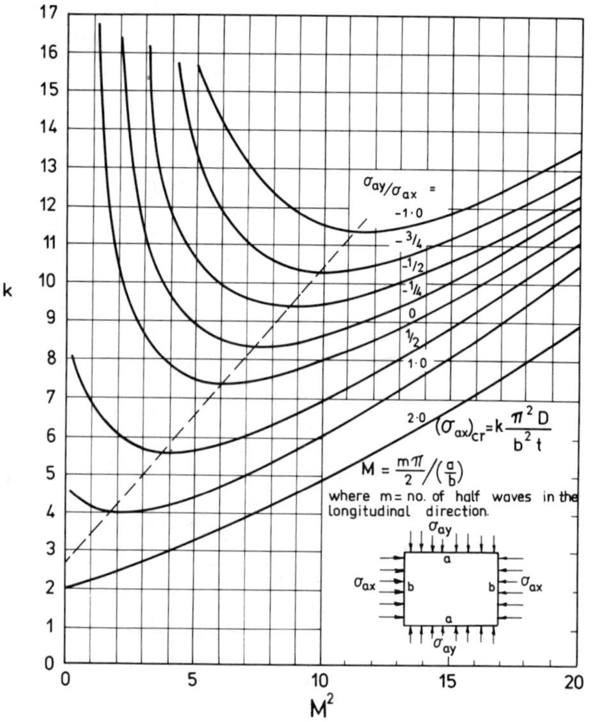

Figure 12.9 Buckling coefficient for biaxial edge compression with longitudinal edges clamped.

12.4 OTHER TYPES OF IN-PLANE LOADS

Pure Shear

In ship structures the plating is commonly subjected to large shear loads. For example, the side plating constitutes the web in the overall hull girder and therefore carries the greater part of the overall shear load. This shearing load can cause buckling since, as shown in Fig. 12.10*a* it gives rise to an in-plane compressive stress σ_a. For the case of pure shear, σ_a is equal to the shear stress τ and acts at 45° to the shear axis. The wave length and other details of the buckled shape will depend on aspect ratio and boundary conditions; the case of an infinitely long, simply supported plate is

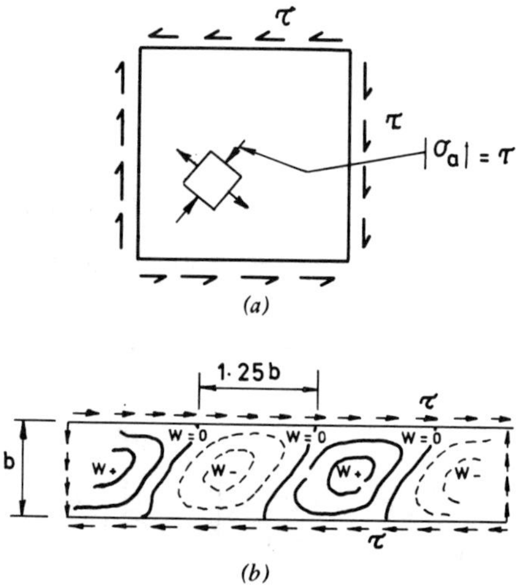

Figure 12.10

illustrated in Fig. 12.10*b*. The mathematical equation governing the above problem is the general plate equation (9.2.1) with $N_{xy} = \tau t$ and $N_x = N_y = p = 0$, which is

$$\nabla^4 w = \frac{2\tau t}{D} \frac{\partial^2 w}{\partial x\, \partial y} \tag{12.4.1}$$

Solutions to (12.4.1) for various boundary conditions are presented in Ref. 2; the methods are similar to those used for uniaxial compression and will not be reproduced here. As would be expected from the close relation between shear and in-plane compression, the resulting expressions for the critical shear stress τ_{cr} closely resemble equations (12.1.5) and (12.1.7) for in-plane compression. In fact, the only change is in the value of the buckling coefficients in these equations; in the case of shear buckling these coefficients are denoted as k_s and K_s, and the two equations are

$$\tau_{cr} = k_s \frac{\pi^2 D}{b^2 t} \tag{12.4.2a}$$

and

$$\tau_{cr} = K_s E \left(\frac{t}{b} \right)^2 \tag{12.4.2b}$$

Values of k_s and K_s are given in Figs. 12.11*a* and 12.11*b* for various types of boundary conditions. Because of the symmetry of the pure shear loading, the choice of a and b is independent of the load. The convention is to choose a as the longer side, and for this reason the figures are drawn only for $a/b > 1$ (note the use of the inverse aspect ratio). The following formulas can also be used to give approximate values for k_s in (12.4.2a):

$k_s = 5.35 + 4(b/a)^2$ for simply supported plates

$k_s = 8.98 + 5.6(b/a)^2$ for clamped plates

Values of k_s for several other cases are given in Ref. 4.

Biaxial Compression and Shear

For this combination Wagner [9] has shown that for long plates ($a/b > 2$) k_s is given approximately by:

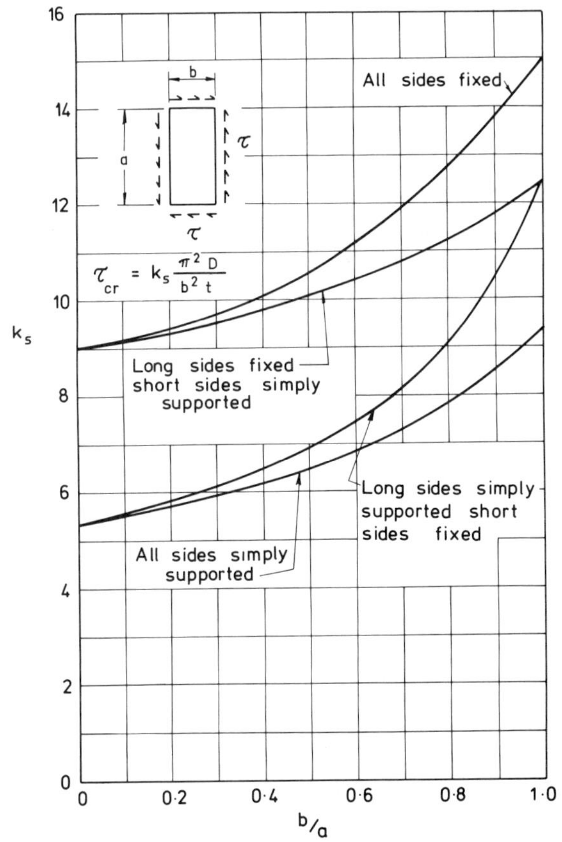

Figure 12.11a Buckling coefficient of flat plates in shear.

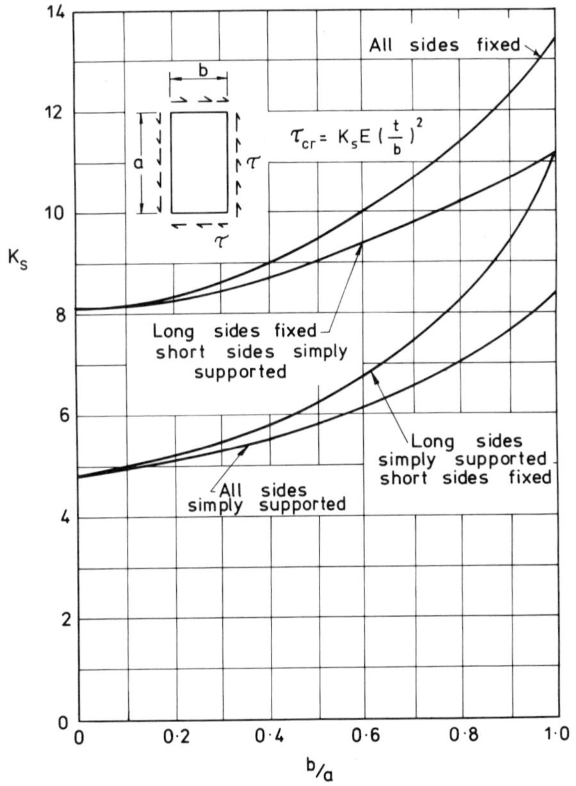

Figure 12.11b Buckling coefficient of flat plates in shear (design formula).

1. All edges simply supported:

$$k_s = \left[2\left(1 - \frac{\sigma_{ay}}{\sigma_e}\right)^{1/2} + 2 - \frac{\sigma_{ax}}{\sigma_e}\right]^{1/2} \times \left[2\left(1 - \frac{\sigma_{ay}}{\sigma_e}\right)^{1/2} + 6 - \frac{\sigma_{ax}}{\sigma_e}\right]^{1/2}$$

2. All edges clamped:

$$k_s = \left[\frac{4}{\sqrt{3}}\left(4 - \frac{\sigma_{ay}}{\sigma_e}\right)^{1/2} + \frac{8}{3} - \frac{\sigma_{ax}}{\sigma_e}\right]^{1/2} \times \left[\frac{4}{\sqrt{3}}\left(4 - \frac{\sigma_{ay}}{\sigma_e}\right)^{1/2} + 8 - \frac{\sigma_{ax}}{\sigma_e}\right]^{1/2}$$

where $\sigma_e = (\pi^2 D/b^2 t)$

In-plane Bending

Figure 12.12 shows the buckled pattern for plates acted upon by pure "in-plane bending," in which σ_b denotes the largest or edge value of the applied stress. The critical value of σ_b can be expressed in the same form as for uniaxial compression; viz:

$$(\sigma_b)_{cr} = k_b \frac{\pi^2 D}{b^2 t} \qquad (12.4.3)$$

Values of k_b are given in Table 12.4. Some approximate formulas are:

1. simply supported edges:

$$\text{for } a/b \le \tfrac{2}{3}: \quad k_b = 15.87 + 1.87(b/a)^2 + 8.6(a/b)^2$$

$$\text{for } a/b > \tfrac{2}{3}: \quad k_b = 23.9$$

2. clamped edges:

$$\text{for } a/b \ge 1: \quad k_b = 41.8$$

3. one unloaded edge clamped; other three edges simply supported:

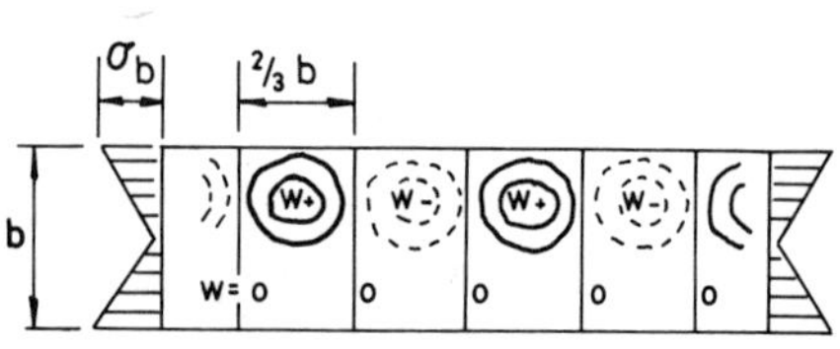

Figure 12.12

TABLE 12.4 BUCKLING COEFFICIENT k_b FOR PURE IN-PLANE BENDING

All edges simply supported					
a/b	0.3	0.4	0.5	0.6	0.67
k_b	37.4	29.1	25.5	24.1	23.9
a/b	0.8	0.9	1.0	1.2	1.5
k_b	24.4	25.6	25.5	24.1	24.1
One unloaded edge clamped; other three edges simply supported					
a/b	0.4	0.5	0.6	0.67	0.7
k_b	29.5	26.0	24.6	24.6	24.6
a/b	0.8	0.9	1.0	1.2	1.6	2.1
k_b	25.2	26.5	26.0	24.6	25.2	24.6
Unloaded edges clamped; loaded edges simply supported					
a/b	0.3	0.35	0.4	0.47	0.5	0.6
k_b	47.3	43.0	40.7	39.6	39.7	41.8
a/b	0.7	0.8	1.0	1.2	1.5	2.0
k_b	43.0	40.7	39.7	40.7	39.7	39.7

$$\text{for } a/b \ge \tfrac{1}{2}: \quad k_b = 25$$

4. unloaded edges clamped; loaded edges simply supported:

$$\text{for } a/b \ge 0.4: \quad k_b = 40$$

In this type of buckling the half-wave length is relatively short, being $2b/3$ for long simply supported plates, and for this reason the values of k are larger than those for uniaxial compression or shear.

Figure 12.13 illustrates the more general case in which the bending is unsymmetric. For simply supported edges the value of k_b is given approximately by:

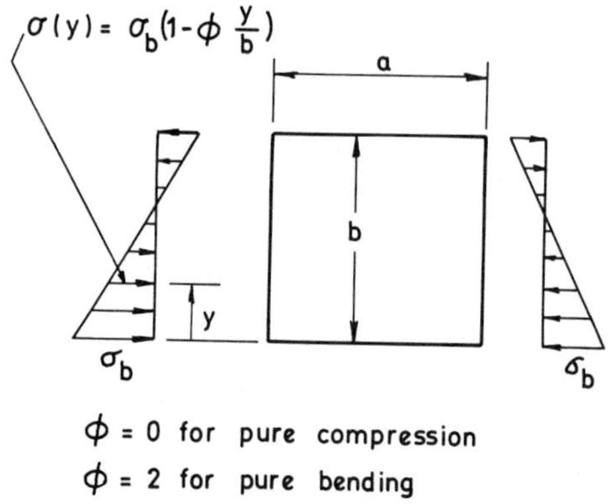

Figure 12.13

$$k_b = 5\phi^2 + 4 \qquad \left(\frac{a}{b} > \frac{2}{3}; \ \begin{matrix}\text{simply supported} \\ \text{edges only}\end{matrix}\right)$$

Combined In-plane Loads: Interaction Formulas

In many cases, ships' plating is subjected to combined in-plane loads. For example, in the hull side plating both shear and longitudinal stresses are present, and it is necessary to estimate the combination which would cause buckling. As can be seen from the various buckling formulas given previously, the critical value for buckling under one type of load depends on the magnitude of the other loads. Likewise, if these other loads are types which can cause buckling then each of their critical values may be similarly affected. Thus there is generally an extensive interdependence or interaction among the loads, and this interaction is usually nonlinear. One of the best ways of dealing with this complicated situation is to make use of interaction formulas, in which the variables are the ratios of each load to its critical value. If just one type of load acted, a value of unity for that load ratio would correspond to failure (i.e., buckling). In the case of several simultaneous loads the failure ratios would be less than unity, and the interaction formula combines the various load ratios in such a way that failure corresponds to the *sum* of the terms being equal to unity, thereby indicating quantitatively the way in which the various loads interact. If each load ratio is plotted along a coordinate axis then the interaction formula is the equation of the curve, or surface, or "hypersurface" (for 2, 3, or n separate loads) corresponding to failure.

One of the advantages of interaction formulas is that they may be derived empirically from experimental results or from numerical solutions. This is especially helpful in complicated cases for which the theory is inadequate or in which the analytical results are unwieldy. This is usually the case in plate buckling, and interaction formulas have been derived empirically for several principal load combinations.

UNIAXIAL COMPRESSION AND IN-PLANE BENDING

Buckling occurs when σ_a and σ_b satisfy the interaction formula:

$$\left(\frac{\sigma_a}{(\sigma_a)_{cr}}\right) + \left(\frac{\sigma_b}{(\sigma_b)_{cr}}\right)^{1.75} = 1 \qquad (12.4.4)$$

in which $(\sigma_a)_{cr}$ and $(\sigma_b)_{cr}$ are the critical values for these two types of loading taken separately, that is, the values given by either (12.1.9) for $\alpha \geq 1$ or (12.1.13) for $\alpha < 1$, and by (12.4.3).

UNIAXIAL LOAD (COMPRESSIVE OR TENSILE) AND SHEAR

This combination was investigated by Batdorf and Stein [10] who produced a set of interaction curves for various values of aspect ratio, α. For computer-aided applications a mathematical expression is more useful, and therefore an empirical formula has been obtained which gives a satisfactory representation of these curves. For convenience we adopt the symbol R to denote a critical load ratio (or "strength" ratio). In the present case the strength ratios are

$R_c = \dfrac{\sigma_a}{(\sigma_a)_{cr}}$, in which $(\sigma_a)_{cr}$ is the critical value of σ_a when it acts alone on the plate, given by (12.1.9) for $\alpha \geq 1$ and by (12.1.13) for $\alpha < 1$.

$R_s = \dfrac{\tau}{\tau_{cr}}$, in which τ_{cr} is the critical value of τ when it acts alone on the plate, given by (12.4.2) (substituting a in place of b if $\alpha < 1$).

The interaction formula is

$$\left.\begin{aligned} R_c + R_s^2 &= 1 \quad (\alpha \geq 1) \\ \left(\frac{1 + 0.6\alpha}{1.6}\right)R_c + R_s^2 &= 1 \quad (\alpha < 1) \end{aligned}\right\} \qquad (12.4.5)$$

This formula can be used for negative (i.e., tensile) values of σ_a as well as for the compressive case.

IN-PLANE BENDING AND SHEAR

For this case we define the strength ratio for in-plane bending as

$$R_b = \frac{\sigma_b}{(\sigma_b)_{cr}}$$

in which $(\sigma_b)_{cr}$ is given by (12.4.3).

A suitable interaction formula is

$$R_b^2 + R_s^2 = 1 \quad (\alpha > \tfrac{1}{2}) \qquad (12.4.6)$$

BIAXIAL COMPRESSION, IN-PLANE BENDING, AND SHEAR

Figure 12.14 gives a set of interaction curves for square plates, obtained from ref. [11]. The two compression strength ratios are

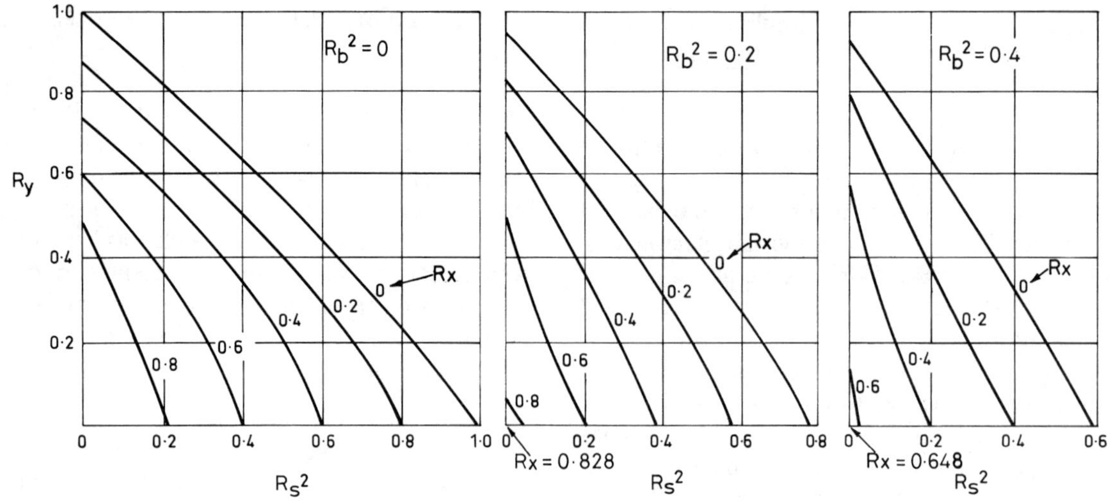

Figure 12.14a Interaction curves for biaxial compression, in-plane bending, and shear drawn for $\alpha = 2$.

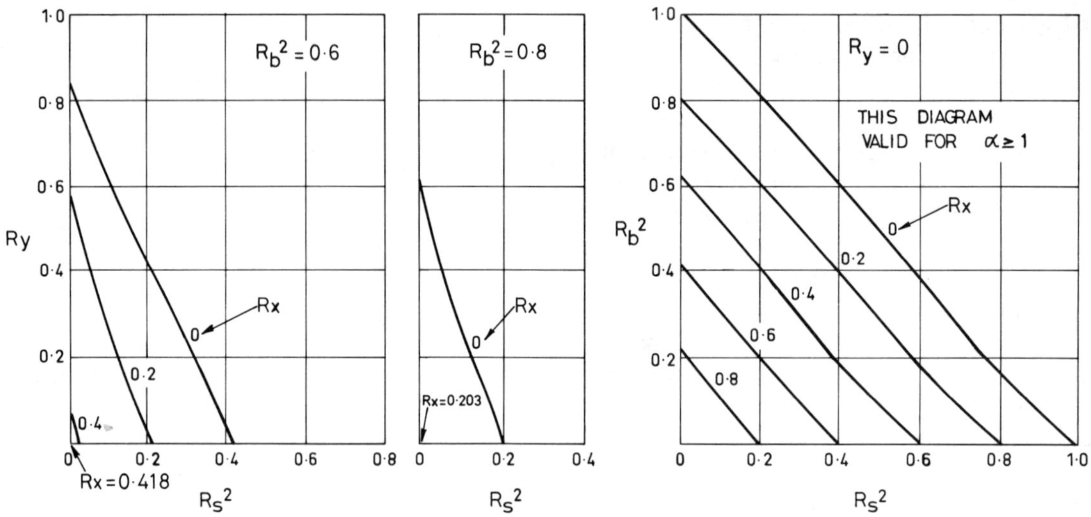

Figure 12.14b Interaction curves for biaxial compression, in-plane bending, and shear drawn for $\alpha = 2$.

$R_x = \dfrac{\sigma_{ax}}{(\sigma_{ax})_{cr}}$, in which $(\sigma_{ax})_{cr}$ is the critical value of σ_{ax} when it acts alone on the plate, given by (12.1.9).

$R_y = \dfrac{\sigma_{ay}}{(\sigma_{ay})_{cr}}$, in which $(\sigma_{ay})_{cr}$ is the critical value of σ_{ay} when it acts alone on the plate, given by (12.1.13), with a and b interchanged.

Here again, since a mathematical expression is more useful than a set of curves, an expression has been developed which fits these curves when $\alpha = 1$. Moreover, by performing a series of four-variable curve-fitting solutions it has been possible to develop the following formula, which is valid for all $\alpha \geq 1$.

$$\frac{0.625(1 + 0.6/\alpha)R_y}{(1 - 0.625R_x)\left[1 - \dfrac{R_b^4}{(1 - R_x)^2}\right]} + \frac{R_s^2}{1 - R_x} = 1 \qquad (\alpha \geq 1) \quad (12.4.7)$$

If $R_b = 0$ this formula can also be used for $\alpha < 1$ by interchanging R_x and R_y, and putting b/a in place of α. In this case $(\alpha_{ax})_{cr}$ is calculated from (12.1.13) and this equation is used as is; there is no interchange of a and b. The value of $(\sigma_{ay})_{cr}$ is calculated from (12.1.9), but with a in place of b.

12.5 COMBINED IN-PLANE COMPRESSION AND LATERAL LOAD

Chapter 9 dealt with laterally loaded plates and showed the restraining effect which in-plane tension has on lateral deflection, but did not consider the opposite effect which in-plane compression would have. From the analysis of columns it is apparent that the application of an in-plane compressive stress σ_a will magnify the initial lateral deflection of a plate. If the plate carries a lateral load and hence has initial bending stresses, these will also be magnified by the application of σ_a. In dealing with this phenomenon we must distinguish between two quite different cases:

1. moderate in-plane compression, such that σ_a is well below any value that would of itself cause any type of collapse; specifically, a value that is well below the elastic buckling stress $(\sigma_a)_{cr}$ and the yield stress. In this case the role of the in-plane load is secondary; it merely magnifies the deflections and stresses caused by the lateral load. An example would be the linear response analysis of deck or bottom plating, that is, the calculation of the deflection and stress under typical in-service values of lateral load and in-plane compression.
2. large in-plane compression, such that the possibility of collapse is present. In this case the in-plane load assumes primary importance and must be dealt with explicitly; it cannot be regarded as simply a magnifying effect. The calculation of the ultimate compressive strength of plating obviously falls into this category.

Case 1 is treated in this section and case 2 in the next section. Hence in the present section the word "buckling" implies *elastic* buckling, unless indicated otherwise. Elastic buckling only occurs for slender plates, for which the buckling stress is less than the yield stress. The buckling of such plates differs from column buckling in that it does not represent total failure or "collapse" of the member. When the applied in-plane compression σ_a acting on a slender plate reaches the elastic critical stress $(\sigma_a)_{cr}$ the deflection in the center of the plate becomes large, but when this happens some of the load is transferred toward the straighter regions at the sides of the plate, which are sturdier because the edges are restrained from deflecting. The load–deflection relationship is shown in Fig. 12.15. Of course, if the load is further increased it eventually reaches an ultimate or collapse value. The investigation of the post-buckling range and the ultimate strength of plates will be taken up in Section 12.6, but it is mentioned here in order that plate buckling may be seen in its proper perspective.

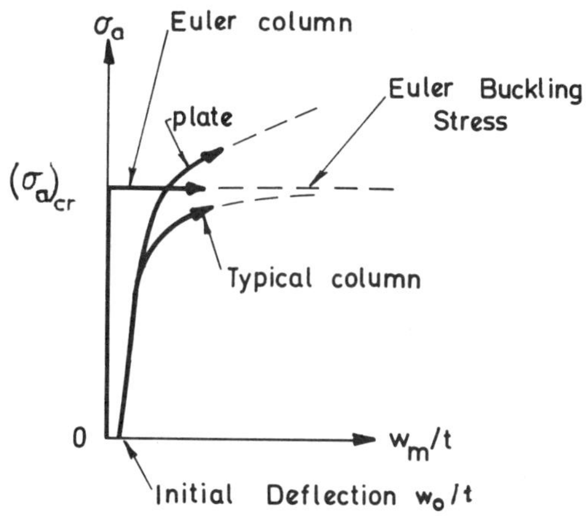

Figure 12.15

Interaction Between Lateral and In-Plane Loads

When in-plane compression and a lateral load act together, each of these loads can alter the effect which the other would have if it were acting alone. If the lateral load is sufficiently large it can influence both the deflected shape and the boundary conditions of a plate and thus influence the buckling stress. At the same time, the in-plane compression can alter the effects of the lateral load. The significance of this influence and the way it is dealt with depends entirely on whether the in-plane compression is moderate or large, as defined previously. If it is moderate then it merely magnifies the effects of the lateral load, and in this case the deflected shape and the boundary conditions which are used for the calculation of the elastic critical stress are those corresponding to the lateral load.

If the in-plane compression is large, then the analysis is concerned with the question of the actual buckling and/or collapse of the plate. In this case the important thing is to determine the "most critical" mode of buckling, that is, the shape of buckled deflection that corresponds to the lowest value of critical stress. We shall refer to this most critical mode as the "primary" buckling mode, and we note that the word "primary" is not meant to imply that either $m = 1$ or

$n = 1$. The primary mode of a plate is calculated taking into account whatever external circumstances, such as loads and boundary conditions, would result in the lowest buckling stress. For example, in a long plate the primary mode is alternating, square-panel half-waves. Therefore, in regard to plate distortion due to fabrication, which could be any one of a variety of shapes, the worst shape (and hence the shape to be assumed for ultimate strength purposes) is that in which the initial deflection has the same number of half-waves as the primary buckling mode. The corresponding load–deflection relationship is illustrated by curve (a) in Fig. 12.16.

If, on the other hand, a circumstance has the effect of raising the buckling stress, and if it is not a permanent and reliable feature of the structure, then it should be ignored when calculating either the buckling stress or the ultimate failure load of the plate. In many cases a lateral load is such a circumstance. Under a lateral load the initial deflection corresponds to the $m = n = 1$ mode. For wide plates this coincides with the primary buckling mode, but for long plates the primary buckling mode has $m > 1$. As σ_a is increased, a laterally loaded long plate will at first maintain the $m = 1$ shape [curve (b) in Fig. 12.16] and σ_a may approach or even exceed the primary buckling stress $(\sigma_a)_{cr}$ without buckling because the plate is being forced into a higher mode of buckling. However, as σ_a is further increased there are two possibilities:

1. The plate may buckle in this higher mode ($m = 1$).
2. The plate may "snap through" into its primary mode, in spite of the presence of the lateral load. If the applied stress has already exceeded the lowest buckling stress then the snap-through becomes a sudden, almost explosive, buckling.

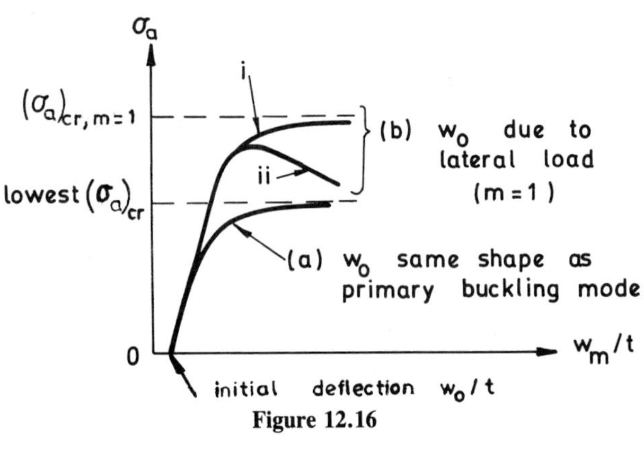

Figure 12.16

The occurrence of 1 or 2 depends mainly on the relative magnitudes of the lateral and the in-plane loads and on the plate properties.

In view of the unpredictability of many of the loads which act on ship plating, and the undesirability of snap-through buckling, it is best to ignore the increase in buckling stress corresponding to case 1. That is, when dealing with the question of actual buckling of plating (as opposed to the calculation of the elastic critical stress corresponding to some imposed deflection) *the lowest or primary buckling mode should be chosen, even in the presence of a lateral load,* unless there is some special reason to the contrary.

Magnification Effect of a Moderate In-Plane Load

For a plate that is subjected to lateral load and in-plane compression, the designer must be sure to investigate all possible loads and boundary conditions which may occur. For example, a load which acts on a single panel and not on the surrounding panels (such as a wheel load) may be a worse case than an even larger multipanel load (such as pressure) because it corresponds to simply supported panel boundaries rather than clamped.

In this section we present some expressions for the magnification factor for various boundary conditions for a plate subjected to an in-plane load $N_x = -\sigma_a t$ and a lateral pressure p. For these loads the general plate equation (9.2.1) becomes

$$\nabla^4 w = \frac{1}{D}\left(p - \sigma_a t \frac{\partial^2 w}{\partial x^2}\right) \qquad (12.5.1)$$

Unfortunately, there is no exact method of solution for this which is valid for all boundary conditions. The Navier method of solution, for instance, is suitable only for the simply supported case. Bleich [1] has presented solutions for a variety of boundary conditions, using an energy method and assumed deflected shapes. However, these solutions are only valid for wide plates. Also, they only give the magnification factor for the center of the plate, whereas for clamped wide plates the maximum plate bending stress occurs at the loaded edges, and the magnification factor there is different. Nevertheless Bleich's solution is useful for simply supported wide plates and so we present a brief summary of it.

Magnification Factor for the Center of Wide Plates

The strain energy resulting from the deformation of the plate is

$$U = \frac{1}{2}D\int_{-a/2}^{a/2}\int_{-b/2}^{b/2}\left\{\left(\frac{\partial^2 w}{\partial x^2}+\frac{\partial^2 w}{\partial y^2}\right)^2 - 2(1-\nu)\left[\frac{\partial^2 w}{\partial x^2}\frac{\partial^2 w}{\partial y^2}-\left(\frac{\partial^2 w}{\partial x\,\partial y}\right)^2\right]\right\}dx\,dy$$

in which the origin is taken at the center of the plate. The total in-plane shortening of the plate in the x-direction due to the buckled deflection is

$$\int_{-a/2}^{a/2}\frac{1}{2}\left(\frac{\partial w}{\partial x}\right)^2 dx$$

per unit width of plate. Therefore, the work done by the compressive stress is

$$W_\sigma = \int_{-a/2}^{a/2}\int_{-b/2}^{b/2}\frac{\sigma_a}{2}\left(\frac{\partial w}{\partial x}\right)^2 t\,dx\,dy$$

The work done by the lateral pressure p is

$$W_p = \int_{-a/2}^{a/2}\int_{-b/2}^{b/2}\frac{1}{2}pw\,dx\,dy$$

Equating the total work done by the loads to the strain energy of deformation gives

$$\int_{-a/2}^{a/2}\int_{-b/2}^{b/2}\left[\frac{1}{2}D\left\{\left(\frac{\partial^2 w}{\partial x^2}+\frac{\partial^2 w}{\partial y^2}\right)^2 - 2(1-\nu)\left[\frac{\partial^2 w}{\partial x^2}\frac{\partial^2 w}{\partial y^2}-\left(\frac{\partial w}{\partial x\,\partial y}\right)^2\right]\right\} - \frac{\sigma_a t}{2}\left(\frac{\partial w}{\partial x}\right)^2 - \frac{1}{2}pw\right]dx\,dy = 0$$

In order to apply this equation we must have some assumed expression $w(x,y)$ for the deflected shape of the plate, and this must satisfy the specified boundary conditions. The following expressions are suitable for plates in which a/b is such that the primary buckling mode is a single wave, that is, wide plates ($a/b < (2)^{1/2}$ for simply supported plates; $a/b < 1$ for clamped plates).

1. All edges clamped:

$$w(x,y) = w_0\left(1+\cos\frac{2\pi x}{a}\right)\left(1+\cos\frac{2\pi y}{b}\right)$$

2. All edges simply supported:

$$w(x,y) = w_0\cos\frac{\pi x}{a}\cos\frac{\pi y}{b}$$

3. Loaded edges clamped, others simply supported:

$$w(x,y) = w_0\left(1+\cos\frac{2\pi x}{a}\right)\cos\frac{\pi y}{b}$$

4. Loaded edges simply supported, others clamped:

$$w(x,y) = w_0\cos\frac{\pi x}{a}\left(1+\cos\frac{2\pi y}{b}\right)$$

For these expressions, it can be shown that

$$\int_{-a/2}^{a/2}\int_{-b/2}^{b/2}\left[\frac{\partial^2 w}{\partial x^2}\frac{\partial^2 w}{\partial y^2}-\left(\frac{\partial^2 w}{\partial x\,\partial y}\right)^2\right]dx\,dy = 0$$

in each case.

If $\phi w(x,y)$ represents the magnified deflection following the application of σ_a, it may be shown that for the foregoing four cases the magnification factor ϕ is

$$\phi = \frac{1}{1-\dfrac{\sigma_a t}{D}\dfrac{\iint\left(\dfrac{\partial w}{\partial x}\right)^2 dx\,dy}{\iint\left(\dfrac{\partial^2 w}{\partial x^2}+\dfrac{\partial^2 w}{\partial y^2}\right)^2 dx\,dy}} \tag{12.5.2}$$

For all four of the cases the result of the integration in (12.5.2) is of the form

$$\left.\begin{aligned}\phi &= \frac{(\sigma_a)_{cr}}{(\sigma_a)_{cr}-\sigma_a}\\ \text{where}\quad (\sigma_a)_{cr} &= k^*\frac{\pi^2 D}{b^2 t}\end{aligned}\right\} \tag{12.5.3}$$

and k^* is a constant which depends on the plate dimensions and the boundary conditions, as follows:

1. All edges clamped:

$$k^* = 4\frac{a^2}{b^2}+\frac{8}{3}+4\frac{b^2}{a^2} \tag{12.5.4a}$$

2. All edges simply supported:

$$k^* = \frac{a^2}{b^2} + 2 + \frac{b^2}{a^2} \qquad (12.5.4b)$$

3. Loaded edges clamped, others simply supported:

$$k^* = \frac{3a^2}{4b^2} + 2 + \frac{4b^2}{a^2} \qquad (12.5.4c)$$

4. Loaded edges simply supported, others clamped:

$$k^* = \frac{16a^2}{3b^2} + \frac{8}{3} + \frac{b^2}{a^2} \qquad (12.5.4d)$$

If the plate is sturdy the buckling stress from (12.5.3) may exceed the yield stress, in which case it is no longer a true buckling stress but simply an indicator that the plate proportions are such as to preclude buckling, and in such a case there is very little magnification. Hence it is permissible to use this artificially high value of $(\sigma_a)_{cr}$ to calculate ϕ because it gives a value close to unity, which is in fact the case. In other words, even though ϕ becomes less "exact" as $(\sigma_a)_{cr}$ increases, this causes no difficulty because ϕ is also approaching unity and hence becoming unnecessary.

It is emphasized that these formulas apply only when the plate aspect ratio is such that the primary buckling mode is a single wave. For clamped plates this means $a/b < 1$, and for simply supported plates $a/b < (2)^{1/2}$. The maximum values of a/b for other cases can be determined from Fig. 12.5*a*.

For small total deflection, the maximum bending stress in the center of the plate is linearly proportional to the deflection. Hence the magnification factor of (12.5.3) can also be applied to bending stress; that is, the total bending stress in the center of the plate under combined lateral and in-plane load is ϕ times the bending stress due to the lateral load alone.

The experimental work of Young [12] has verified this approach, showing that for deflections of less than $t/2$, the error in predicted stress is less than 6%.

Magnification Factor for Bending Moment at Clamped Loaded Edges

As noted earlier, if the plate's edges are clamped the maximum bending moment occurs at these edges rather than at the center. The case of greatest interest is a plate with its loaded edges clamped, because then the applied stress σ_a combines directly with the magnified plate bending stress along these edges (on the concave side of the plate). There seems to be no exact method which is sufficiently general for the case of a plate with clamped loaded edges and arbitrary boundary conditions on the sides. If the sides are simply supported a solution may be obtained by a generalized Navier method in which the deflection is represented as a series whose terms are sinusoidal in y and hyperbolic in x. This solution is valid for all aspect ratios but it is cumbersome, and we will show that it is wide plates which are of greatest interest. Therefore, after presenting the exact solution we present a simplified method for such plates, based on beam-column theory.

Solution for Clamped Loaded Edges and Simply Supported Sides*

We again take the origin at the center of the plate. The lateral pressure p_0 is represented as a Fourier series in y:

$$p(x,y) = \sum_{n=1,3,\ldots} p_n \cos\left(n\frac{2\pi y}{b}\right)$$

with

$$p_n = \frac{4(-1)^{(n-1)/2}}{n\pi} p_0$$

and the lateral deflection is represented as a series of terms which are trigonometric in y (thus satisfying the differential equation and the simply supported boundary condition in that direction) and complex exponential in x:

$$w(x,y) = \sum_{n=1,3,\ldots} w_n(x) \cos\frac{n\pi y}{b} \qquad (12.5.5)$$

with

$$w_n = e^{\zeta X}$$

where $X = (2x)/a$ and ζ is a complex parameter to be determined from the differential equation and the boundary conditions. The foregoing form of $w(x,y)$ makes the equation "separable," and the resulting differential equation for $w_n(x)$ is a "beam-column" type of equation:

*The author is indebted to Dr. Y. N. Chen of the American Bureau of Shipping for supplying this solution and for making several helpful suggestions regarding the approach and coverage of this section.

$$w_n'''' - 2(\lambda^2 - 2\mu^2)w_n'' + \lambda^4 w_n = \frac{p_n a^4}{16D} \qquad (12.5.6)$$

in which:

$$\lambda = \frac{n\pi a}{2b}$$

$$\mu = \frac{\pi}{2}\sqrt{\frac{\sigma_a}{\sigma_0}}$$

and

$$\sigma_0 = 4\frac{\pi^2 D}{a^2 t}$$ = the critical value of σ_a for a simply supported square plate of length a.

It may be seen that the parameter μ plays an equivalent role here to the parameter ξ in beam-column theory, where it is defined as $\pi/2(P/P_E)^{1/2}$, with P_E being the Euler buckling load. After solving for $w(x,y)$ the bending moment in the x-direction is obtained from (9.1.8a) which gives

$$M_n = \frac{4D}{a^2}(w_n'' - \nu\lambda^2 w_n)$$

The final result for the bending moment in the x-direction is

$$M_x(x,y) = \sum_{n=1,3,\ldots} M_n(x) \cos\frac{n\pi y}{b} \qquad (12.5.7)$$

The expression for $M_n(x)$ varies, depending on the relative magnitude of λ and μ. Three cases must be distinguished:

1. $\mu = 0$, *that is, lateral load only; no in-plane compression*

$$M_n(x) = p_0 a^2 \frac{(-1)^{(n-1)/2}}{n\pi\lambda^2}\left\{-\nu + \left[\frac{2A}{\lambda} - (1-\nu)B\right]\cosh\lambda X + (1-\nu)AX\sinh\lambda X\right\} \qquad (12.5.8)$$

where

$$A = \frac{\lambda\sinh\lambda}{\lambda + \sinh\lambda\cosh\lambda}$$

$$B = \frac{\sinh\lambda + \lambda\cosh\lambda}{\lambda + \sinh\lambda\cosh\lambda}$$

2. $\lambda > \mu > 0$

$$M_n(x) = p_0 a^2 \frac{(-1)^{(n-1)/2}}{n\pi\lambda^2}\left\{-\nu + \left[\frac{2\mu\gamma A}{\lambda^2} - \left(\frac{\gamma^2 - \mu^2}{\lambda^2} - \nu\right)\right]B\cos\mu X\cosh\gamma X \right.$$
$$\left. + \left[\frac{2\mu\gamma B}{\lambda^2} + \left(\frac{\gamma^2 - \mu^2}{\lambda^2} - \nu\right)A\right]\sin\mu X\sinh\gamma X\right\} \qquad (12.5.9)$$

where

$$\gamma = \sqrt{\lambda^2 - \mu^2}$$

$$A = \frac{\gamma\cos\mu\sinh\gamma - \mu\sin\mu\cosh\gamma}{\mu\sinh\gamma\cosh\gamma + \gamma\sin\mu\cos\mu}$$

$$B = \frac{\mu\cos\mu\sinh\gamma + \gamma\sin\mu\cosh\gamma}{\mu\sinh\gamma\cosh\gamma + \gamma\sin\mu\cos\mu}$$

3. $\mu > \lambda > 0$

$$M_n(x) = p_0 a^2 \frac{(-1)^{(n-1)/2}}{n\pi\lambda^2}\left[-\nu + A\left(\frac{\delta_1^2}{\lambda^2} + \nu\right)\cos\delta_1 X - B\left(\frac{\delta_2^2}{\lambda^2} + \nu\right)\cos\delta_2 X\right] \qquad (12.5.10)$$

where

$$\delta_{1,2} = \sqrt{2\mu^2 - \lambda^2 \pm \sqrt{4\mu^2(\mu^2 - \lambda^2)}}$$

$$(+ \text{ for } \delta_1, \; - \text{ for } \delta_2)$$

$$A,B = \frac{\delta_{2,1} \sin \delta_{2,1}}{\delta_2 \cos \delta_1 \sin \delta_2 - \delta_1 \sin \delta_1 \cos \delta_2}$$

The foregoing solution gives the total bending moment anywhere in the plate, and for any aspect ratio. For wide plates the bending moment is maximum at the midpoint of the loaded edges: $x \pm a/2$, $y = 0$. For purposes of comparison this can be put into the form of a magnification factor by dividing the expression for the total bending moment $M_x(a/2, 0)$ by the value corresponding to $\sigma_a = 0$, which is case 1 above. The result is shown in Fig. 12.17 for $a/b = 0, 0.2, 0.5$, and 1.0. In order to permit comparison with the simplified beam-column solution which is presented later, the results are plotted in terms of a modified beam-column parameter, ξ_p, defined in (12.5.13).

Long Plates

It may be demonstrated from the foregoing equations that for long plates ($a/b > 1.4$, approximately) the magnification factor for edge bending moment diminishes as aspect ratio increases, and even falls below 1.0 for some combinations of p_0 and σ_a. Also, for long plates the maximum bending moment for p_0 acting alone does not occur at the loaded edges but at the center of the plate, and in the y-direction instead of the

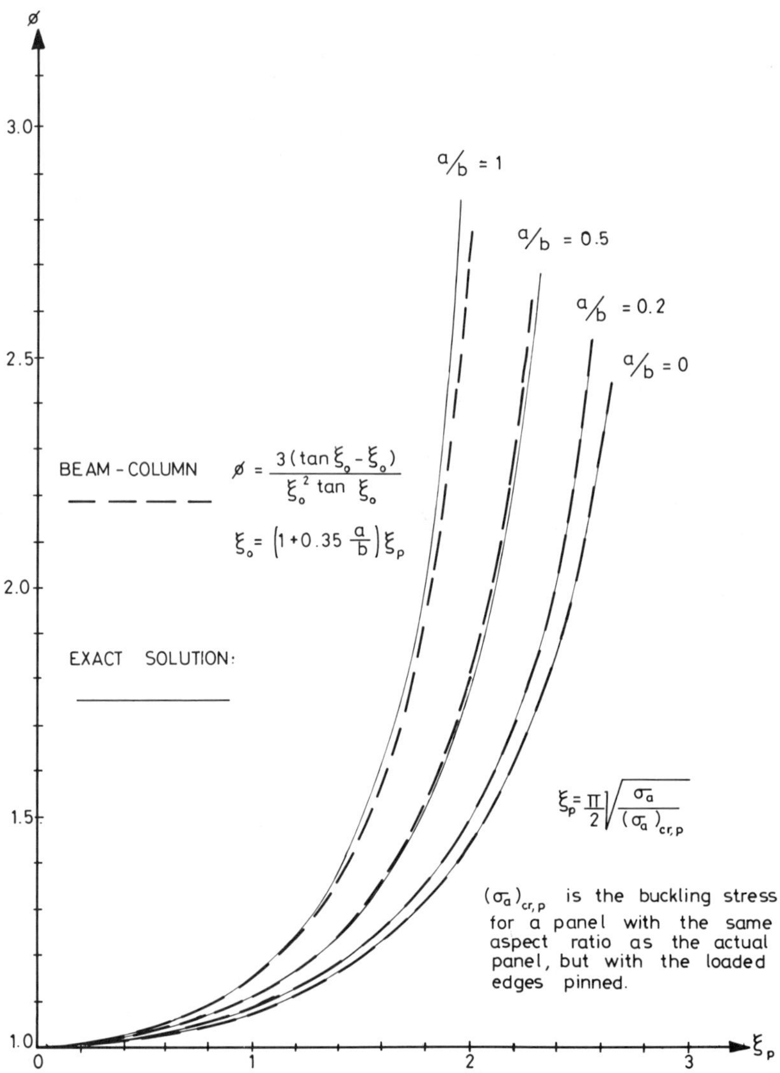

Figure 12.17 Magnification factors for clamped moment in wide plates (unloaded edges pinned).

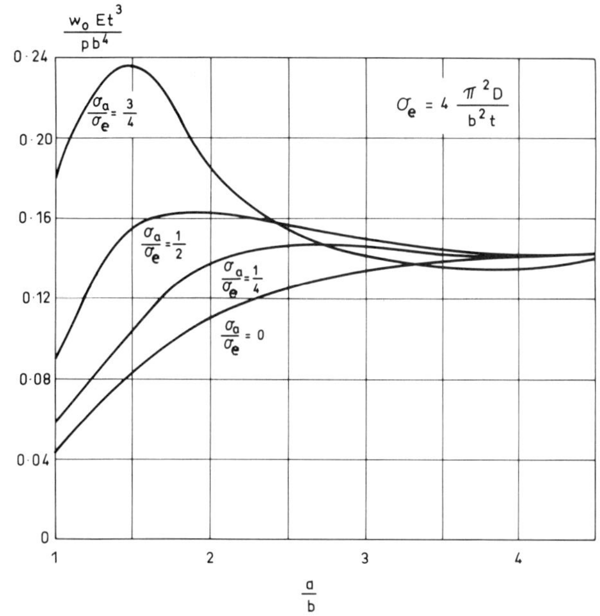

Figure 12.18 Maximum deflection due to uniaxial compression and lateral pressure; simply supported plate [13].

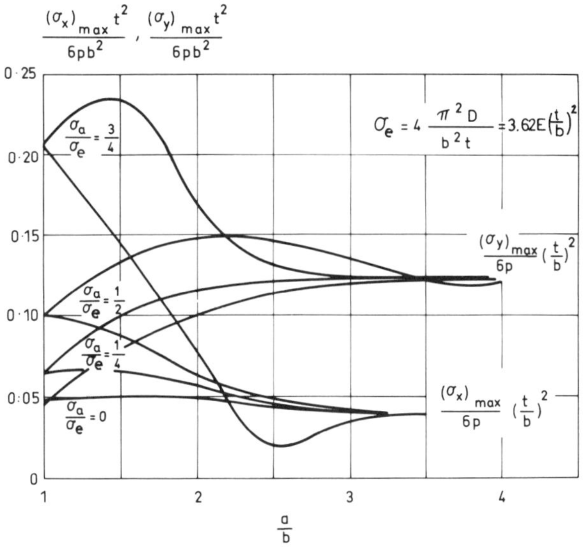

Figure 12.19 Maximum bending stress due to uniaxial compression and lateral pressure; simply supported plate [13].

x-direction. If the unloaded edges are clamped the maximum occurs there, again in the y-direction. In both cases the maximum value is not greatly affected by an in-plane compressive load in the x-direction. This is even true for plates which are simply supported on all sides. A general solution for this case has been obtained by Conway [13] using an energy method of the type originally developed by M. Levy [14]. Conway's results for the simply supported case are presented in Figs. 12.18 and 12.19.

If a long plate is subjected to transverse compression in addition to the primary longitudinal compression, then the transverse load will cause a magnification of the maximum bending moment M_y, the same as for a wide plate. Table 12.5 from Ref. 15 gives

TABLE 12.5 DEFLECTIONS ALONG THE X-AXIS AND MAXIMUM VALUES OF σ_X AND σ_y FOR CLAMPED RECTANGULAR PLATES UNDER LATERAL LOAD AND EQUAL BIAXIAL COMPRESSION OF $\frac{\sigma_{ax}}{\sigma_0} = \frac{\sigma_{ay}}{\sigma_0} = \frac{1}{2}$[a]

		$\frac{a}{b} = 1$	$\frac{a}{b} = 1.5$	$\frac{a}{b} = 2$	$\frac{a}{b} = 2.5$	$\frac{a}{b} = 3$	$\frac{a}{b} = 4$	$\frac{a}{b} = 5$
$\frac{w_0Et^3}{pb^4}$ $Y = 0, X =$	0	0.02866	0.05878	0.05956	0.05310	0.04961	0.05223	0.05427
	0.1	0.02643	0.05486	0.05818	0.05535	0.05298	0.05193	0.05336
	0.2	0.02027	0.04345	0.05179	0.05716	0.05932	0.05553	0.05267
	0.3	0.01178	0.02647	0.03642	0.04722	0.05538	0.06162	0.06045
	0.4	0.00371	0.00879	0.01387	0.02058	0.02734	0.03989	0.05036
	0.5	0	0	0	0	0	0	0
$\frac{(\sigma_x)_{max}}{6p}\left(\frac{t}{b}\right)^2$		0.05406	−0.09159	−0.09051	−0.09445	—	—	—
occurs at		$X = 0$, $Y = 0$	$X = 0.5$, $Y = 0$	$X = 0.5$, $Y = 0$	$X = 0.5$, $Y = 0$			
$\frac{(\sigma_y)_{max}}{6p}\left(\frac{t}{b}\right)^2$		0.05406	−0.15004	−0.14927	−0.14197	—	—	—
occurs at		$X = 0$, $Y = 0$	$X = 0$, $Y = 0.5$	$X = 0$, $Y = 0.5$	$X = 0.2$, $Y = 0.5$			

[a]Origin is at center of plate; coordinates X and Y are: $X = \frac{x}{a}$, $Y = \frac{y}{b}$; $\sigma_0 = \frac{\pi^2 D}{b^2 t} = 0.905E\left(\frac{t}{b}\right)^2$.

values of ϕ for long clamped plates subjected to biaxial compression, for the special case when σ_{ax} and σ_{ay} are each equal to half of the critical value.

Simplified Method for Wide Plates with Loaded Edges Clamped

For wide plates with clamped loaded edges an approximate expression for ϕ could be obtained by applying the solution for a clamped beam-column to a unit strip of plating in the x-direction. This would give

$$\phi = \frac{3(\tan \xi - \xi)}{\xi^2 \tan \xi}$$

where

$$\xi = \frac{\pi}{2}\sqrt{\frac{P}{P_E}} = \frac{\pi}{2}\sqrt{\frac{\sigma_a}{\sigma_E}}$$

Of course using the Euler stress gives a value for ξ which is too large since it does not allow for the restraining effect of aspect ratio. The logical replacement would be the value of $(\sigma_a)_{cr}$ for a plate having the same aspect ratio and side support as the actual plate, but with the loaded edges pinned. We shall denote this as $(\sigma_a)_{cr,p}$. Solutions using this quantity show that it gives values for ξ which are somewhat too small, and that good accuracy is achieved by applying a linear correction term, $(1 + 0.35a/b)$, to ξ. The solution is then

$$\phi = \frac{3(\tan \xi_0 - \xi_0)}{\xi_0^2 \tan \xi_0} \qquad (a/b \leq 1) \qquad (12.5.11)$$

where

$$\xi_0 = \left(1 + 0.35\frac{a}{b}\right)\xi_p \qquad (12.5.12)$$

and

$$\xi_p = \frac{\pi}{2}\sqrt{\frac{\sigma_a}{(\sigma_a)_{cr,p}}} \qquad (12.5.13)$$

If the sides are simply supported, $(\sigma_a)_{cr,p}$ may be obtained from the wide plate equation, that is, (12.1.11) or (12.1.13), or read from the bottom curve of Fig. 12.7. If the sides are clamped it is necessary to use the appropriate curve from Fig. 12.5*a* or *b*.

The accuracy of (12.5.11) is illustrated in Fig. 12.17. Although the correction term in (12.5.12) was obtained using the solution for a plate with the sides pinned, the foregoing three equations are also valid for a plate with clamped sides because the effect of the side support is reflected in $(\sigma_a)_{cr,p}$, which is evaluated for the actual boundary condition at the sides.

Example 1

A panel of deck plating is 1 m long, 0.5 m wide, and 10 mm thick, and has a yield stress of 250 MPa. In a certain cargo loading configuration it is subjected to a lateral pressure of 100 kPa, while adjacent panels have no lateral pressure. The maximum hull girder sagging stress is 150 MPa. Calculate the maximum deflection and also determine whether the plate will undergo any surface yielding.

Solution:
From the load description the plate should be treated as simply supported. Therefore we may use Figs. 12.18 and 12.19.

Calculate the reference stress:

$$\sigma_e = 3.62E\left(\frac{t}{b}\right)^2 = 304 \text{ MPa}$$

hence

$$\frac{\sigma_a}{\sigma_e} = \frac{150}{304} = 0.49$$

At 0.49 and $a/b = 2$ in Fig. 12.18, read

$$\frac{w_{max}}{t} \Big/ \frac{p}{E}\left(\frac{b}{t}\right)^4 = 0.162;$$

hence

$$w_{max} = 4.8 \text{ mm}$$

In Fig. 12.19 read

$$\frac{(\sigma_y)_{max}}{6p}\left(\frac{t}{b}\right)^2 = 0.145;$$

hence

$$(\sigma_y)_{max} = 218 \text{ Mpa}$$

and

$$\frac{(\sigma_x)_{max}}{6p}\left(\frac{t}{b}\right)^2 = 0.064;$$

hence

$$(\sigma_x)_{max} = 96 \text{ MPa}$$

Total σ_x stress:

$$(\sigma_x)_{total} = 96 + 150 = 246 \text{ MPa}$$

Hencky-von Mises equivalent stress:

$$\sigma_{eq} = 0.707\sqrt{(246 - 218)^2 + 246^2 + 218^2}$$
$$= 233 \text{ MPa}$$

$\sigma_{eq} < \sigma_Y$; therefore yielding does not occur.

Example 2

Suppose that the previous lateral load is doubled and that in order to cope with this increased load three transverse stiffeners are added to the panel, so that the length of each plate panel becomes $a = 0.25$ m. For one of the internal plate panels perform the same calculations as in the previous example. The lateral pressure continues to act on the entire overall panel.

Solution:
Since the pressure load now acts over several adjacent plate panels, the edges of the internal panels are approximately clamped. Since these are wide panels we may use a magnification factor approach. In clamped plates the maximum bending stress occurs at the edges and therefore we use the beam-column expression for ϕ given in (12.5.11), rather than (12.5.3) which applies only for the center of the plate. The buckling stress $(\sigma_a)_{cr,p}$ corresponds to a plate that is simply supported at the loaded edges. Since the sides are clamped the buckling coefficient K is given by the second highest curve of Fig. 12.5*b* ("B simply supported; A fixed"). For a plate aspect ratio of 0.5, K is approximately 7.0. Therefore

$$(\sigma_a)_{cr,p} = 7.0E\left(\frac{t}{b}\right)^2 = 588 \text{ MPa}$$

Hence

$$\xi_0 = \left(1 + 0.35 \times \frac{1}{2}\right)\frac{\pi}{2}\sqrt{\frac{\sigma_a}{(\sigma_a)_{cr,p}}} = 0.932$$

and

$$\phi = \frac{3(\tan \xi_0 - \xi_0)}{\xi_0^2 \tan \xi_0} = 1.063$$

The bending stress at the loaded edges when the lateral load acts alone is, from Fig. 9.6 (with a and b switched, in order to follow the convention for laterally loaded plates)

$$\sigma_{x0} = kp\left(\frac{b}{t}\right)^2 = (0.5)(200 \times 10^{-3})\left(\frac{0.25}{0.01}\right)^2 \text{ (MPa)}$$
$$= 62.5 \text{ MPa}$$

and $$\sigma_{y0} = \nu\sigma_{x0} = 18.76 \text{ MPa}$$

Hence the magnified stress is $\phi\sigma_{x0} = 66.44$ MPa and the total stress in the x-direction, on the inside surface of the plating

$$(\sigma_x)_{total} = 66.44 + 150 = 216.44$$

The total equivalent stress is

$$\sigma_{eq} = 0.707\sqrt{(216.44 - 18.76)^2 + (216.44)^2 + (18.76)^2} = 207.7 \text{ MPa}$$

Hence the relative margin below yielding is

$$\frac{250 - 207.7}{250} = 16.9\%$$

12.6 ULTIMATE STRENGTH OF PLATES

As shown in Chapter 11 the collapse of columns occurs either by elastic buckling if the column is slender, or by the commencement of yielding which so seriously decreases the column bending stiffness that collapse by inelastic buckling follows almost immediately. In the case of plating the mechanism of collapse is decidedly more complex. The principal parameters which can influence plate collapse are: type of loading, edge restraint, aspect ratio, initial distortion (out-of-flatness), and residual stress. We shall begin by describing the collapse behavior of a uniaxially loaded, simply supported square plate, with sides free to pull in, and having some typical initial distortion in the form of a half-wave in each direction. Later we shall examine the effect of altering these various parameters.

The two remaining parameters are the plate slenderness β, defined in (9.3.1), and the amount of residual stress in the plating due to welding.

Plates Without Residual Stress

The most useful way of describing plate collapse is to record the relationship between the applied load (which is in the form of an applied stress σ_a) and the axial shortening of the plate. The corresponding internal quantities are the average stress, which equals σ_a, and the average strain ε_a in the plate. Figure 12.20*a* shows some idealized load versus end shortening curves for stress-free plates (i.e., plates free of residual stress) having different values of slenderness ratio β. These curves differ in their shape and to explain them we define three ranges of β, as shown in Fig. 12.20*b*: slender plates ($\beta > 2.4$), intermediate plates ($1.0 < \beta < 2.4$), and sturdy plates ($\beta < 1.0$). We now consider each of these in turn.

Fig. 12.20*b* includes the elastic buckling stress curve, (12.1.9) and it may be seen that for $\beta > 2.4$ the buckling stress is well below the yield stress and also below the curve of collapse stress, which corresponds to the peaks of the σ_a–ε_a curves in Fig. 12.20*a*. Thus for a slender plate the first noticeable response to an applied load σ_a is that the plate buckles. As a result the center portion of the plating partially escapes from the axial shortening being imposed by the load, and a greater proportion of the load must be taken by the region of plating near the sides. As shown in Fig. 12.21, compressive stress within the plate is no longer uniform. Nevertheless, the plate is still supporting the load and so, although it has buckled, it has not yet collapsed.

However, because of the deflected shape of the buckled portion of the plate, the axial stiffness is substantially reduced in this region and hence the overall stiffness of the plate, $d\sigma_a/d\varepsilon_a$, is reduced. This explains the reduced slope of the two lowermost curves in Fig. 12.20*a*.

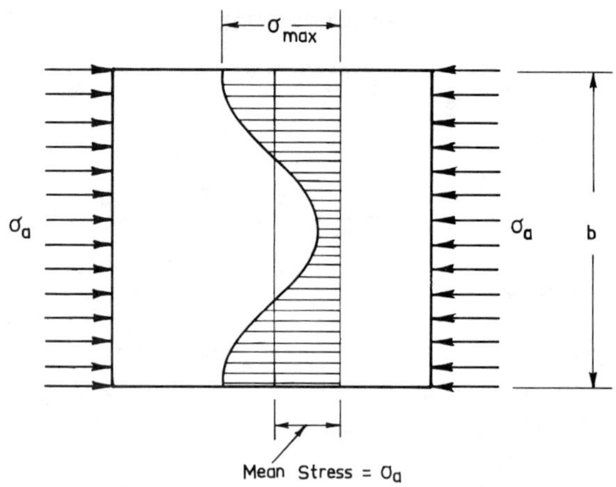

Figure 12.21 Post-buckling stress distribution.

As the load is further increased the shirking of the center region becomes more pronounced and the maximum stress at the sides increases ever more rapidly, until finally this stress reaches the yield stress and collapse occurs. The peak value of the applied load, that is, the value which causes collapse, is referred to as the *ultimate strength* of the plate and is denoted as $\sigma_{a,u}$. For a stress-free plate it is denoted $(\sigma_{a,u})_{sf}$. After

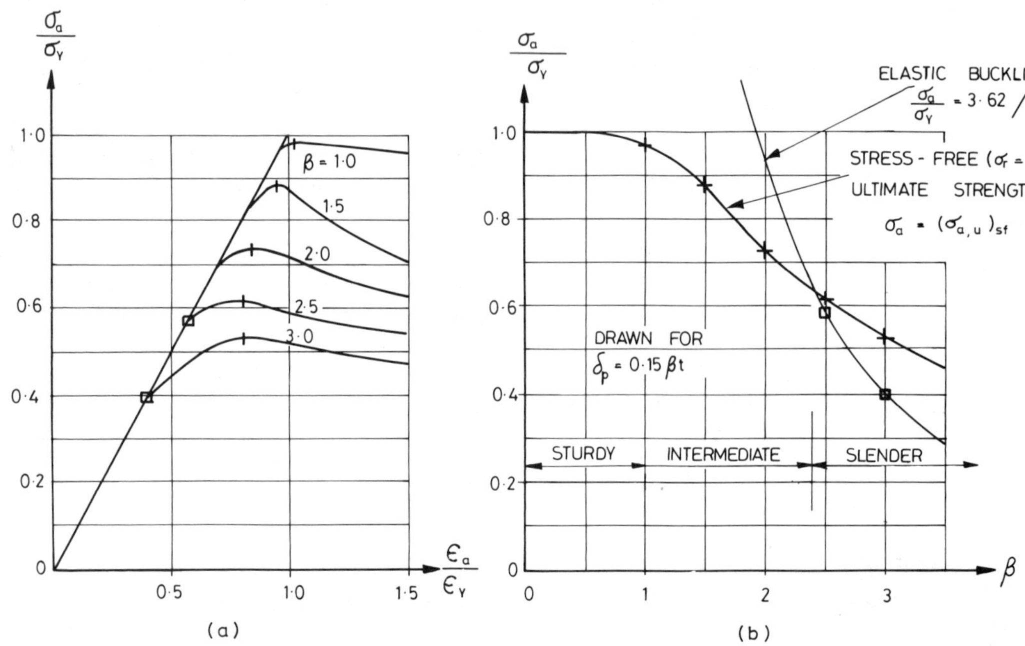

Figure 12.20 Plate strength without welding ($\sigma_r = 0$).

the collapse there is usually some reduction in load-carrying ability, which means that if the load is not taken over by some adjacent structure then collapse will be relatively sudden and will involve large axial deflection.

If the plate is very slender, as in aircraft and in some lightweight surface vessels, there is a significant margin between buckling and collapse. For example, for $\beta = 3$ the values of applied stress at buckling and collapse are $0.40\sigma_Y$ and $0.53\sigma_Y$ respectively. Under these conditions buckling can often be treated as merely an unserviceability type of failure, particularly since the buckling is elastic and disappears when the load diminishes. However, if the load is cyclic it would also be necessary to ensure that all fatigue requirements were satisfied.

For plates of intermediate slenderness, which are more typical of most large steel plated structures, the mechanism of collapse is somewhat different because the buckling stress is approximately equal to the yield stress. For a rigorous analysis of the collapse of such plates it is necessary to use elasto-plastic large deflection theory as in Refs. 16–19. These analyses show that, as the applied stress increases, the initial distortion (which is always present in real plates) is magnified and this causes some loss of stiffness and some local yielding, both of which give rise to some stress redistribution. The stress is no longer uniform in either the transverse or the longitudinal direction. Collapse occurs when the average equivalent stress along the sides of the plate reaches yield stress—the equivalent stress being that given by the Hencky-von Mises failure criterion. For these plates there is no preliminary buckling to give warning of the collapse. Moreover, the collapse is accompanied by a sharp reduction in load-carrying ability, which means that the collapse may be quite sudden. In most cases the collapse is accompanied by what is termed a "plastic mechanism" in the plating: the "pitched roof" configuration shown in Fig. 12.22. It may be shown that this configuration allows large axial shortening with minimum strain energy.

For sturdy plates ($\beta < 1$) the initial distortion is smaller in proportion to the thickness and the magnification is less because the elastic buckling stress is very large. Hence these plates can carry a load equal, or nearly equal, to the full "squash load," in which $\sigma_{a,u} = \sigma_Y$. Since the plate remains elastic up until collapse there is no readily apparent warning before the collapse. However, after the peak load is reached the load-carrying capacity remains approximately constant up to very large strains, of the order of $2\varepsilon_Y$, after which there is some load-shedding. Hence the collapse itself is a relatively gradual process.

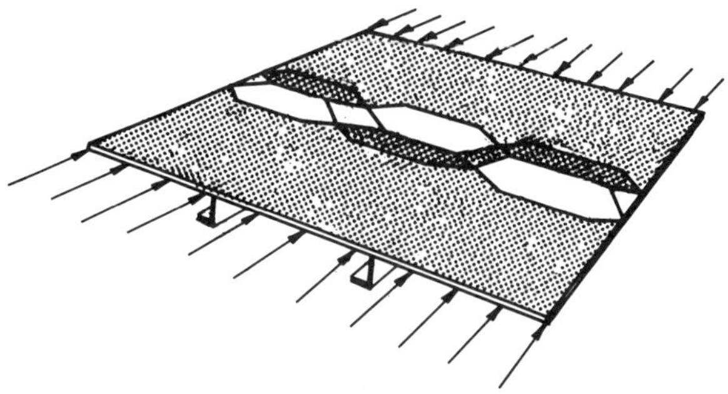

Figure 12.22 Typical post-buckling deformation.

Plates With Residual Stress

Figure 11.6 presented a typical distribution of residual stress in a plate due to welding, together with an idealization that has been found useful. The stress in the tension zone adjacent to the weld is equal to the yield stress and hence from equilibrium there is a compressive stress in the rest of the plating given by

$$\sigma_r = \frac{2\eta}{\frac{b}{t} - 2\eta}\,\sigma_Y \qquad (12.6.1)$$

For analytical purposes the total stress distribution is subdivided into a uniform compressive stress $\sigma_a + \sigma_r$ acting over the entire width and an augmented tensile stress of magnitude $\sigma_Y + \sigma_r$ acting in the two edge strips adjacent to the welds. Since the tensile zones are situated adjacent to the supported edges they can be assumed not to affect the stability of the plate. Therefore the middle part of the plate reaches its ultimate strength when the average equivalent stress on its edges, adjacent to the tensile zone, reaches compressive yield stress. The fact that the aspect ratio of this middle portion is slightly smaller has no appreciable effect and therefore the average stress-strain curve up to this point has the same shape as for a stress-free plate; the only difference is that everything happens sooner. That is, as shown in Fig. 12.23, the departure from linearity and all other points on the curve occur at a value of applied stress which is σ_r less than for a stress-free plate.

The collapse of the middle portion marks a new phase in the load-end shortening history, in which the edge strips continue to carry further loads until the

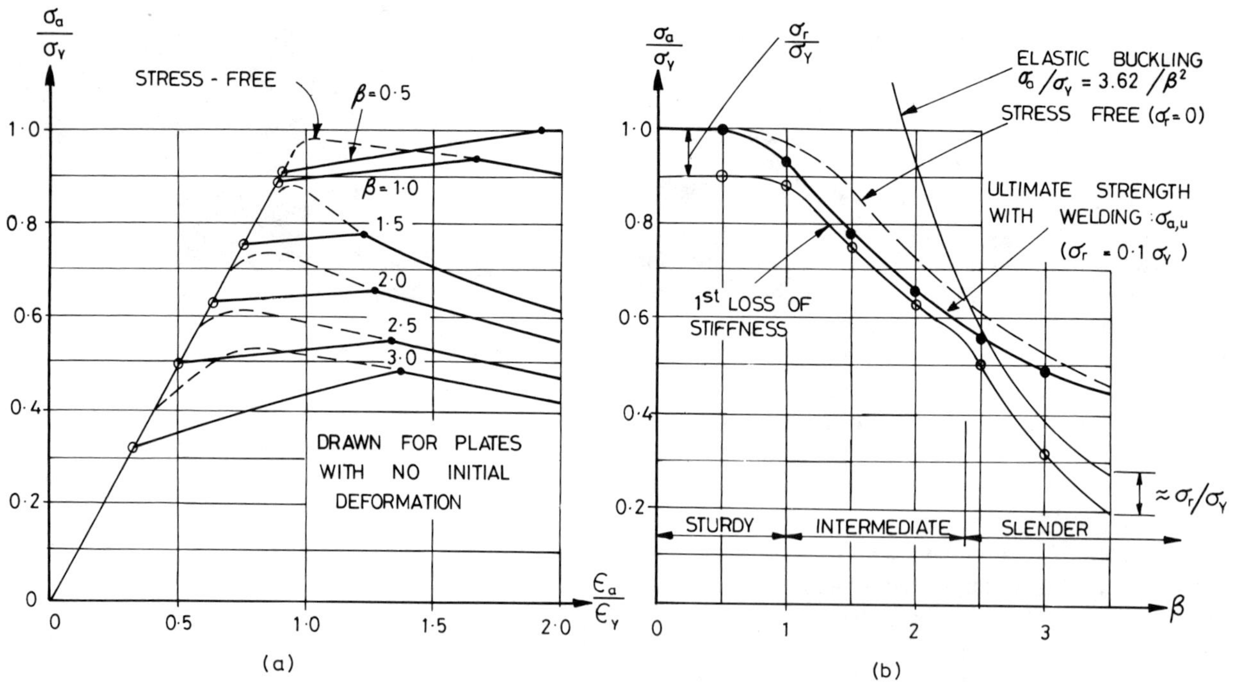

Figure 12.23 Plate strength with welding ($\sigma_r = 0.1\ \sigma_Y$).

initial tensile stress of $(\sigma_Y + \sigma_r)$ reaches compressive σ_Y. This additional compressive force can be shown to be $bt\sigma_r$. During this stage the middle part of the plate offers no resistance, though it continues to be strained by an amount $(\sigma_Y + \sigma_r)/E$. In fact, since the slope of the post-peak collapse curve of the stress-free plate is negative, some additional load is shed by the middle portion into the edge strips during this second phase of the collapse. Therefore during this second phase the stiffness of the plate is very low. Eventually, the tension zones reach compressive yield and the plate collapses. As shown in Fig. 12.23*a*, this corresponds to the point where the second phase of the stress-strain curve intersects the curve for the stress-free plate. At this point the self-equilibrating residual stress distribution has been "washed out" and the plate response again matches the stress-free case.

The figure shows that the overall effect of residual stress is to remove the peakiness of the stress-free plate collapse curves, thus making the collapse less sudden. Also, the collapse strain is two or three times larger than for the stress-free plate, which means that in most cases the structure supporting the plate (usually a stiffener) will begin taking over some of the load before the plate collapses.

As was the case for stress-free plates, the exact shape of the collapse curve depends on the plate slenderness. For sturdy plates ($\beta < 1$) there is no load-shedding in the stress-free condition and therefore the collapse load is the same with residual stress as it is for the stress-free plate. However, there is a very large reduction in plate stiffness when the applied stress reaches the value $\sigma_a = (\sigma_{a,u})_{sf} - \sigma_r$ (or, approximately, $\sigma_a = \sigma_Y - \sigma_r$) and it is safer and usually more realistic to regard this as collapse. Plates of intermediate slenderness exhibit the most severe load-shedding and hence they suffer the greatest loss of ultimate strength due to residual stress, the loss being almost equal to σ_r. For slender plates with residual stress the initial buckling occurs earlier, at $\sigma_a = (\sigma_a)_{cr} - \sigma_r$, making the transition to collapse become still more gradual.

For normal proportions of stiffeners and plating the magnitude of the residual welding stresses in stiffened plating can be estimated from

$$\sigma_r = \frac{c_1 a_w}{bt + A_s} c_2 \tag{12.6.2}$$

where a_w = cross-sectional area of weld
c_1 = coefficient for shrinkage force in the weld
c_2 = coefficient for stress reduction due to shakedown

The coefficient for shrinkage force in the weld, c_1, has been found experimentally to be approximately 1 ×

10^4 MPa for single pass fillet welds [20]. The second coefficient allows for the stress-relieving which occurs due to "shakedown," the process in which the alternating tension and compression of the plate under the action of the hull girder bending causes or permits small local plastic strains to occur. These strains smooth out the sharply peaked tension zones adjacent to the welds, thereby greatly reducing σ_r. On the basis of detailed calculations Sadden [21] has recommended a value of 0.5 for c_2. In normal stiffened panels the value of σ_r from (12.6.2) rarely exceeds 0.1 σ_Y, and this value has been used in Fig. 12.23. More detailed design curves, covering a range of σ_r values, are given subsequently, in Fig. 12.25.

Effects of Other Parameters

RESTRAINT AT SIDES

Although clamping the sides of a plate increases the elastic buckling stress by 75%, Moxham [22] has shown that the increase in the ultimate strength is never more than about 14%. Therefore, except in special applications where clamping can be guaranteed to be always present, it is best ignored. Similarly, in most plated structures there is little restraint against transverse contraction (pull-in) of the plate, and even when it exists Crisfield [23] and others have shown that the strengthening effect does not exceed 5%. Finally, there is the constraint that in stiffened plating the edges of each plate panel must at least remain straight, from conditions of symmetry. Reference 17 shows that this effect only becomes significant for slender plates and even then is generally less than 10%. Since plate panels near the outer edges of a stiffened panel would not have this restraint, and since their collapse might trigger the collapse of internal panels, it is probably best to ignore this restraint also.

ASPECT RATIO

As shown at the end of Section 12.1 longitudinal stiffening is more efficient and hence the panels of plating between stiffeners usually have an aspect ratio greater than unity. Ideally such plates would buckle into a series of alternating, dished square subpanels but in practice, because of initial out-of-plane deformations, there is usually some region along the plate which is more vulnerable and collapse involves only one buckled half-wave, as in Fig. 12.22. Experiments [22] and elasto-plastic analyses [17, 24] have verified that for aspect ratio $a/b \geq 1$ the maximum strength is virtually independent of aspect ratio. The only exception to this is plates with very large initial deformation ($\delta_0 > 1.5t$) in which case square plates have significantly less strength than long plates.

If the aspect ratio is less than 1.0 then the effect of the shorter length will be to strengthen the plate; that is, for a given β the collapse load will exceed the value of $\sigma_{a,u}$ given by Fig. 12.23. Note that this strengthening is for a *given* width, or a given value of β; if we compare plates of two *different* widths and of the same length, the wider plate always has *less* strength. The strength of wide plates is discussed later.

INITIAL DEFORMATION

Initial out-of-plane deformation of plating is of two types: transverse and longitudinal. Transverse distortion is caused mainly by the shrinkage effects of the fillets welds at the stiffeners, giving rise to the familiar "hungry horse" pattern of cylindrical deformation occurring between each stiffener. In Section 9.4 it was mentioned that for typical ship panels the central deflection is of the order of $0.065\beta^{1.65}t$, and hence the maximum amplitude can be as large as $0.5t$ for heavily welded slender plates. Nevertheless this type of initial deformation has little effect on plate strength.

The other type of initial out-of-plane deformation is rippling or waviness in the longitudinal direction, superimposed on the cylindrical welding deformation. We shall use the symbol δ_p to denote this type of deformation. Its magnitude depends mainly on the manner and circumstances of the fabrication of the structure. Typical upper and lower values are $0.04\beta^2t$ and $0.18\beta^2t$, the former corresponding to good workmanship and careful fabrication. The only significant effect of δ_p is that for intermediate and slender plates it softens or completely removes the sharp knuckle in the curve of σ_a versus ε_a, which was shown in Fig. 12.23 ("1st loss of stiffness"). In practice, nearly all plates have some initial deformation. As the applied stress increases, the deformation is magnified and the increasing lateral deflection causes a progressive reduction in the in-plane stiffness of the plate. That is, the slope of the σ_a versus ε_a curve begins to decrease sooner, with the result that the collapse is more gradual, as shown in Fig. 12.24. However, the maximum height of the curves—the ultimate strength—is only slightly decreased. Tests [25] on plating in practical welded structures have shown that, as long as the amplitude of δ_p is less than, say, $0.12\beta^2t$ (which is readily achieved by a reasonable and average quality

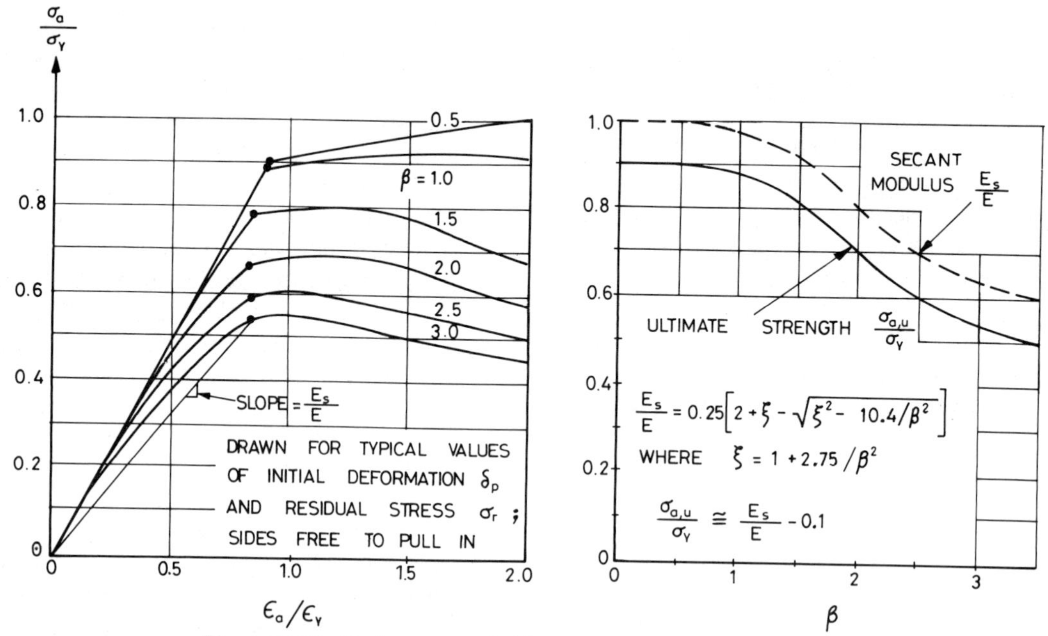

Figure 12.24 Design curves of ultimate strength and secant modulus.

of workmanship) then δ_p has no significant effect on collapse strength. An exception to this is panels in which stiffeners are only intermittently welded to the plating; Ref. 25 may be consulted for further details.

SHEAR STRESS

The presence of an in-plane shear stress τ in a plate tends to lower the resistance to longitudinal compression. Since the b/t values used in compression panels are far less than those at which pure shear buckling becomes a factor, the only effect of shear is to hasten yielding. This may be allowed for by using a reduced yield stress $r_\tau \sigma_Y$ in all plate calculations. From the Hencky-von Mises yield criterion, the plate strength reduction factor is

$$r_\tau = \sqrt{1 - 3\left(\frac{\tau}{\sigma_Y}\right)^2}$$

Dwight et al. [26] report that experiments seem to indicate that the strength reduction due to shear only becomes appreciable at high levels of shear, and that for lower levels (say $\tau < 0.175\sigma_Y$) the foregoing expression is overly conservative. They recommend that for $\tau < 0.175\sigma_Y$, the effect of shear stress should be ignored, and that for $\tau > 0.175\sigma_Y$ the reduction factor should be

$$r_\tau = 1.05\sqrt{1 - 3\left(\frac{\tau}{\sigma_Y}\right)^2} \qquad (12.6.3)$$

Ultimate Strength of Uniaxially Loaded Plates

Figure 12.24 presents some sample curves of σ_a versus ε_a, and the corresponding ultimate strength and secant modulus curves, for the plating of uniaxially loaded, longitudinally stiffened steel panels of the type used in ships and box girder bridges, with typical amounts of initial deformation ($\delta_p < 0.2\beta t$) and residual welding stress ($\sigma_r \approx 0.1\sigma_Y$) and with the sides constrained to remain straight but free to pull in. The curves are based on the experimental and numerical results presented in Ref. 16. For sturdy plates, which suffer a relatively sudden "first loss of stiffness," this point is taken as collapse; for plates of greater slenderness the loss of stiffness is more gradual and the point representing collapse is chosen closer to the peak of the curve. The curve of secant modulus is based on a curve presented by Chatterjee and Dowling [27], but since a mathematical expression is more useful for computer-aided design, the curve of Fig. 12.24 is the following expression, which gives a very close fit to the curve in Ref. 27

$$\left.\begin{aligned} T = \frac{E_s}{E} &= 0.25\left(2 + \xi - \sqrt{\xi^2 - \frac{10.4}{\beta^2}}\right) \\ \text{where} \\ \xi &= 1 + \frac{2.75}{\beta^2} \end{aligned}\right\} \qquad (12.6.4)$$

The symbol T is given to the secant modulus ratio because in the limit analysis of stiffened panels (Chapter 14) this ratio is used as a "transformation factor" for the plate sectional area, as demonstrated in Section 3.6, as a means of allowing for a portion of the cross section which has a different Young's modulus.

Chatterjee and Dowling observed that for typical panels of the type just described, the curve of plate ultimate strength is very similar in shape to the curve of T, and is approximately a distance σ_r/σ_Y (here being taken as 0.1) beneath it. The curves are not exactly alike, but the similarity is close enough to allow this simplifying approximation. Hence the mathematical expression for the ultimate strength of the plating in panels of the foregoing type is

$$\left.\begin{aligned} \frac{\sigma_{a,u}}{\sigma_Y} &= (T - 0.1) \\ &= 0.25\left(1.6 + \xi - \sqrt{\xi^2 - \frac{10.4}{\beta^2}}\right) \\ \text{where again}& \\ \xi &= 1 + \frac{2.75}{\beta^2} \end{aligned}\right\} \qquad (12.6.5)$$

If there is any significant amount of shear stress in the plating (say $\tau > 0.175\sigma_Y$) it may be accounted for by using a reduced yield stress $r_\tau \sigma_Y$ in (12.6.5) where r_τ is defined in (12.6.3). Note, however, that the full value of σ_Y should be used when calculating β.

Figure 12.25 gives values of $\sigma_{a,u}$ for a wider range of residual stress and side restraint ("restrained" means the sides remain straight and do not pull in; "unrestrained" means both types of transverse deformation can occur). The magnitude of σ_r is difficult to predict with any precision; it depends on details of fabrication procedures and welders' skills, and considerable differences have been recorded between different panels of the same grillage [28]. Equations 12.6.1 and 12.6.2 are only approximate formulas for the mean value of σ_r. Therefore, as suggested by Dwight [26], the approach which is generally taken is to define three broad classes of plates according to the approximate and average level of residual stress, as given by (12.6.1). The three classes are unwelded or stress-relieved ($\sigma_r = 0$), average welding ($\sigma_r \leq 0.1\sigma_Y$), and heavily welded ($\sigma_r \leq 0.33\sigma_Y$). Plate strength curves for these three classes are given in Fig. 12.25, taken from Ref. 17, but modified such that for welded plates with $\beta < 1$ $\sigma_{a,u} = \sigma_Y - \sigma_r$. If the average residual stress exceeds $0.33\sigma_Y$ by a moderate amount (say up to $\sigma_r = 0.4\sigma_Y$) recourse may be had to extrapolation. Beyond this it may be advisable to employ special measures (such as stress-relieving or special welding techniques) or possibly to undertake a more detailed analysis of the structure using a large deflection elasto-

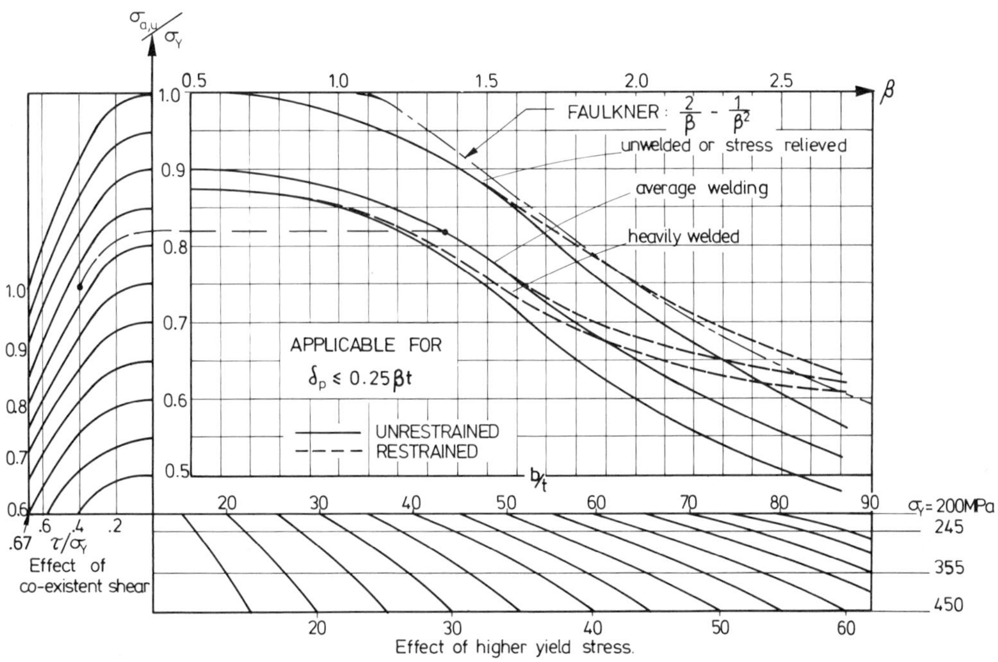

Figure 12.25 Curves for ultimate strength of plates.

plastic theory. Alternatively, in many cases it will be possible (and generally easier) to redesign the structure so as to decrease the level of residual stress. For example, it may be possible to increase the plate width b (i.e., decrease the number of stiffeners) by increasing the plate thickness. Another possibility is to use a higher strength steel so as to permit a larger b/t value. An important factor which tends to mitigate the effect of plate residual stresses is that collapse of a complete structural assembly must involve some averaging of properties over a number of plate panels. Therefore the very high variabilities of weld-induced residual stresses in plating do not cause quite such high variabilities in the aggregate strength of typical ship structures.

The curves in Fig. 12.25 are for an initial out-of-plane deformation $\delta_p = 0.145\beta t$ ($b/200$ for mild steel), which corresponds to an average level of workmanship and care in fabrication. They can be used for δ_p up to about $0.25\beta t$. Figure 12.25 also permits the effects of coexistent shear on collapse strength to be allowed for as in (12.6.3). For example, as indicated by the dashed line in the shear correction curves of the figure, an ultimate stress $\sigma_{a,u}$ of $0.82\sigma_Y$ is reduced to $0.75\sigma_Y$ in the presence of a coexisting shear stress $\tau = 0.4\sigma_Y$.

Figure 12.25 also shows Faulkner's simple and useful expression for the ultimate strength of unwelded plates which was proposed in 1964 [29] and subsequently shown [30, 31] to be in good agreement with later and more extensive experimental data. The expression is

$$\frac{\sigma_{a,u}}{\sigma_Y} = \frac{2}{\beta} - \frac{1}{\beta^2}$$

In Faulkner's approach the effect of residual stress is accounted for by a strength reduction factor R_r defined by

$$R_r = 1 - \frac{\sigma_r E_{ts}}{\sigma_Y E}$$

in which E_{ts} is the structural tangent modulus defined by (11.1.4). Faulkner has also presented [32] a simpler expression for E_{ts} for use in design

$$\frac{E_{ts}}{E} = \frac{2\beta - 1}{\beta} \qquad (1 < \beta < 2.5)$$

For $\beta > 2.5$, $E_{ts} \approx E$ and for $\beta < 1$, $E_{ts} \approx 0$.

Ultimate Strain of Uniaxially Loaded Plates

In a plate which forms part of the hull girder the longitudinal in-plane load is caused by overall bending of the hull girder, and thus the load on the plate can be taken as an applied uniform strain ε_a, instead of the average applied stress σ_a. For most applications it is more convenient to use stress. But in the hull girder ultimate strength analysis strain becomes the primary quantity because throughout the collapse the strain within each plate panel remains uniform and increases in direct proportion to the local bending curvature of the hull girder at the section in question. In contrast, as each plate panel collapses the stress becomes very nonuniform and is not so directly related to the overall response. Hence for the hull module limit analysis it is necessary to have an expression for the ultimate strain of plating, $\varepsilon_{a,u}$. This can be obtained directly from $\sigma_{a,u}$ by means of the secant modulus since $\sigma_{a,u} = E_s \varepsilon_{a,u}$. Thus

$$\varepsilon_{a,u} = \frac{\sigma_{a,u}}{TE} \qquad (12.6.6)$$

and if we again utilize the similarity of the curves of $\sigma_{a,u}$ and T this becomes

$$\varepsilon_{a,u} = \frac{(T - 0.1)\sigma_Y}{TE} = \left(1 - \frac{0.1}{T}\right)\varepsilon_Y \qquad (12.6.7)$$

with T given by (12.6.4). An increase in initial deflection δ_p reduces the plate's tangent stiffness and produces a larger value of ε_a for a given value of σ_a. Hence the effect of δ_p is to increase the value of $\varepsilon_{a,u}$. The foregoing expression applies for values of δ_p up to about $0.25\beta t$. Smith [33] has presented curves of σ_a versus ε_a for larger values of δ_p.

Ultimate Strength of Wide Plates

All of the foregoing is for plates that are long, or at least square. As the aspect ratio a/b decreases below 1.0 the ultimate strength decreases markedly because the middle portion of the plate receives less support from the sides and becomes more and more like a wide column, for which the buckling mode is a single half-wave pattern. The most common instance of wide plating is a transversely stiffened panel, and here again the plating is best regarded as simply supported and having the worst pattern of initial deflection, which is alternating half waves as shown in Fig. 12.26 (greatly exaggerated). Unlike a long plate, the ultimate strength of a wide plate depends on the magnitude of the initial deflection, δ_p. The center portion of the plate is similar to an eccentric wide column and the collapse is by nonbifurcation buckling, with the plate stiffness steadily decreasing as the applied stress increases. The stress distribution within the plate will be as shown in Fig. 12.27*a*. Figure 12.27*b* shows an idealized model of the stress distribution. The center portion of the plate is regarded as a wide column ($b = \infty$) and at collapse the stress in this portion will be equal to the ultimate strength of a wide column, $(\sigma_{a,u})_{wc}$. Smith [34] has presented numerically derived curves for $(\sigma_{a,u})_{wc}$; for computer applications the following expression gives a reasonable fit to the curves:

$$(\sigma_{a,u})_{wc} = \frac{0.63}{1 + 3.27\dfrac{\delta_p}{\beta_L^2 t}} \frac{\sigma_Y}{\beta_L^2} \quad (\beta_L > 1) \qquad (12.6.8)$$

in which, since $b = \infty$, we introduce a "longitudinal" plate slenderness parameter having a in place of b

$$\beta_L = \frac{a}{t}\sqrt{\frac{\sigma_Y}{E}} \qquad (12.6.9)$$

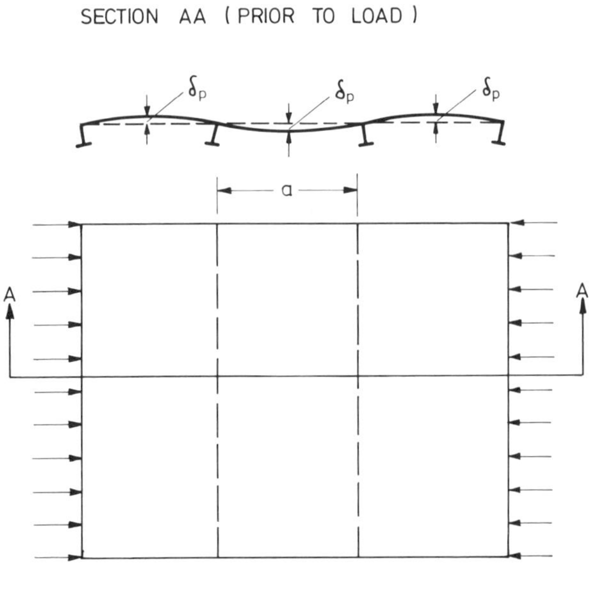

Figure 12.26 Initial deflection in a transversely stiffened panel.

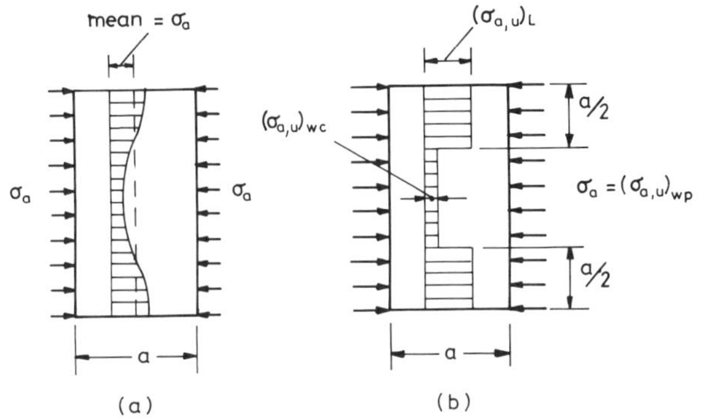

Figure 12.27 Wide plate ultimate strength.

When the applied load σ_a reaches the value given by (12.6.8) the center of the plate has collapsed and this region has an average value of stress equal to $(\sigma_{a,u})_{wc}$. As σ_a is further increased this center region becomes wider but the stress level in it remains constant because $(\sigma_{a,u})_{wc}$ is independent of the width of the wide column. With increasing load the stress in the side regions continues to increase and eventually these side regions undergo inelastic buckling and the plate collapses. It would be very difficult to predict the exact stress distribution in these side regions at collapse. However, this is not necessary if we employ an approximation suggested by Valsgård [35] that at collapse each of the side regions is similar to one-half of a square plate of length a. That is, we say that at collapse the average stress within these regions is approximately equal to the collapse strength of a square plate of width a, and this value may be obtained from the curve of Fig. 12.25 or (12.6.5). However, since the width is a it is necessary to use a value of β which corresponds to a, rather than to b; in other words, we must use the longitudinal slenderness β_L. We shall denote the corresponding value of ultimate strength as $(\sigma_{a,u})_L$.

We have now determined the average value of stress in the center region and in the side regions of a wide plate at collapse, as illustrated in Fig. 12.27*b*. From equilibrium between the applied load and the internal average stresses we have

$$b(\sigma_{a,u})_{wp} = a(\sigma_{a,u})_L + (b - a)(\sigma_{a,u})_{wc}$$

and so, finally, the ultimate strength of a wide plate is

$$(\sigma'_{a,u})_{wp} = \frac{a}{b}(\sigma'_{a,u})_L + \left(1 - \frac{a}{b}\right)(\sigma'_{a,u})_{wc} \qquad (12.6.10)$$

in which the two strength values on the right-hand side are obtained from either Fig. 12.25 or (12.6.5) evaluated at β_L, and from (12.6.8). A prime indicates a quantity which has been normalized by dividing by the yield stress; for example, $(\sigma'_{a,u})_{wc} = (\sigma_{a,u})_{wc}/\sigma_Y$.

Ultimate Strength under Biaxial and Lateral Loads

We have already seen, in Section 12.3, that if a plate is subjected to two orthogonal compressive stresses σ_{ax} and σ_{ay}, elastic buckling involves a strong interaction between these two. This is also true for inelastic buckling and for ultimate strength. As before, with two in-plane loads we adopt the convention that the more "important" load,or the reference load, as chosen by the designer, is designated σ_{ax}. Since this is only for reference purposes σ_{ax} need not be the larger of the two, although this would usually be the case.Then, to be consistent with the convention for uniaxially loaded plates, the dimension a is always taken as parallel to the principal load σ_{ax}. The aspect ratio a/b may have any value.

The interaction between the two loads depends mainly on aspect ratio and plate slenderness. With regard to aspect ratio, we have seen previously that for plates compressed in the long direction the ultimate strength $\sigma_{a,u}$ is independent of aspect ratio, whereas for plates compressed in the short direction the ultimate strength is $(\sigma_{a,u})_{wp}$ given by (12.6.10) which is a function of aspect ratio. Also, for a given plate the wide plate strength is always less than the strength in the long direction. When there are two loads, one of them will always be acting on a "wide plate." Usually this will be σ_{ay} because plating is usually stiffened parallel to the primary load, giving $a/b > 1$, as in Fig. 12.28*a*. In this case, the "length" of the wide plate is b, whereas in dealing with a uniaxially loaded wide plate we followed the convention that the length (i.e., the dimension in line with the load) is designated a, and so a/b was less than 1.0. Also, we introduced the longitudinal slenderness $\beta_L = (a/t)(\sigma_Y/E)^{1/2}$. Now, with biaxial loading, the possibility exists (and indeed is the more likely) that a/b will be greater than 1.0. Whenever this is so, a and b must be interchanged in the wide plate equations, (12.6.8) and (12.6.10), and the ordinary slenderness $\beta = (b/t)(\sigma_Y/E)^{1/2}$ must be used in place of β_L. For ease of reference and to avoid confusion, the complete procedure is set out here in the two forms which are appropriate, depending on a/b:

For $a/b > 1$, Fig. 12.28*a*:

$$\sigma'_{ax,u} = \sigma'_{a,u} \qquad \text{(see note, later)} \tag{12.6.11}$$

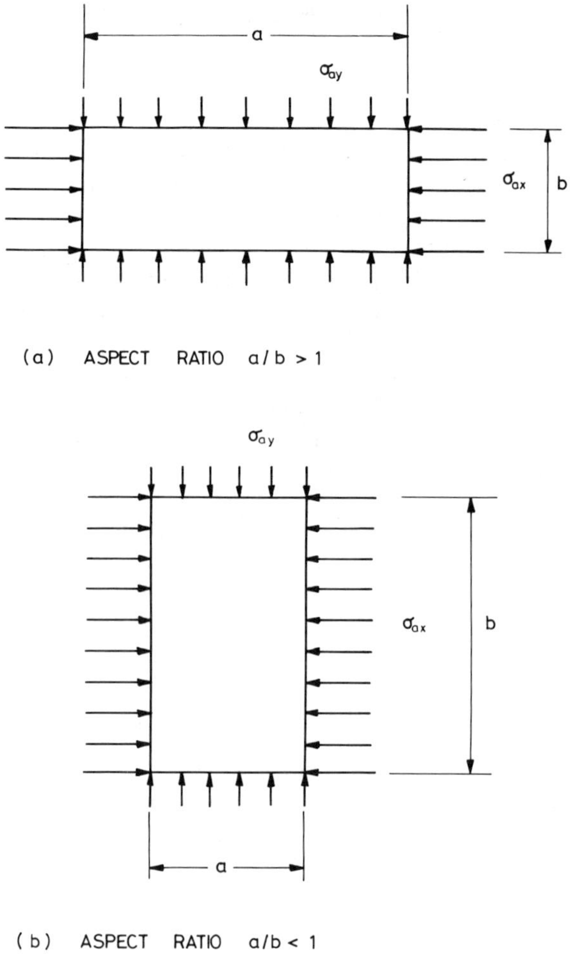

Figure 12.28 Aspect ratio for biaxially loaded plates.

$$\sigma'_{ay,u} = \frac{b}{a}\sigma'_{a,u} + \left(1 - \frac{b}{a}\right)(\sigma'_{a,u})_{wc} \tag{12.6.12}$$

in which $(\sigma'_{a,u})_{wc}$ is obtained from (12.6.8) with β in place of β_L.

For $a/b < 1$, Fig. 12.28*b*:

$$\sigma'_{ay,u} = (\sigma'_{a,u})_L \tag{12.6.13}$$

$$\sigma'_{ax,u} = \frac{a}{b}(\sigma'_{a,u})_L + \left(1 - \frac{a}{b}\right)(\sigma'_{a,u})_{wc} \tag{12.6.14}$$

in which $(\sigma'_{a,u})_{wc}$ is obtained from (12.6.8).

Note: 1. A prime (′) indicates normalization by dividing by the yield stress.

2. In all of the foregoing, both $\sigma'_{a,u}$ and $(\sigma'_{a,u})_L$ are obtained from either Fig. 12.25 or (12.6.5), using β for the first quantity and β_L for the second.

The interaction between σ_{ax} and σ_{ay} is best described in terms of the individual strength ratios for each load acting alone.

$$R_x = \frac{\sigma_{ax}}{\sigma_{ax,u}}; \qquad R_y = \frac{\sigma_{ay}}{\sigma_{ay,u}} \qquad (12.6.15)$$

The nature of the interaction depends mainly on the plate slenderness. For slender plates collapse begins with elastic buckling, leading to a loss of effectiveness of the middle portion of the plate. In each direction the effective width of the plate is made up of two strips of plating along the sides. Collapse ultimately occurs when one pair of plate edges reaches a yield condition. If the plate is slender the buckling is essentially elastic and so the interaction curve would be the appropriate elastic buckling interaction curve of Fig. 12.8, depending on the aspect ratio. For square plates the interaction curve is a straight line and so for square plates which are very slender ($\beta \to \infty$) the biaxial ultimate strength interaction equation is approaching the simple linear form

$$R_x + R_y = 1 \qquad (12.6.16)$$

If on the other hand the plate is sturdy, collapse will involve inelastic buckling, and if the plate is very thick collapse would be by crushing, that is, pure yield. Thus, for sturdy plates ($\beta \to 0$) collapse mainly involves yielding, and so the interaction curve for σ_{ax} and σ_{ay}, and hence also for R_x and R_y, would tend toward the Hencky-von Mises ellipse.

$$R_x^2 - R_x R_y + R_y^2 = 1 \qquad (12.6.17)$$

Square Plates

Since square plates have the simplest elastic buckling interaction formula, it is somewhat easier to develop an ultimate strength interaction formula for square plates. Even so, the problem is very complex and a rigorous solution requires a finely detailed elasto-plastic, large deflection numerical analysis. The results of such an analysis are presented in the form of a graph in Ref. 17. Since a mathematical expression is more useful for computer-aided design, the following expression has been developed based on the graph

$$R_x^2 - R_x R_y + R_y^2 + \left(\beta - \frac{\frac{4}{9}}{\beta}\right)(R_x + R_y - 1) = 1 \qquad (\beta \geq \tfrac{2}{3}) \qquad (12.6.18)$$

The results of the numerical solution showed that for $\beta < \frac{2}{3}$ failure is by crushing, just as for the uniaxial case (see Fig. 12.23*a*). Therefore the foregoing expression has been chosen such that it gives the Hencky-von Mises ellipse when $\beta = \frac{2}{3}$. For any value of β less than $\frac{2}{3}$ the expression should be evaluated at $\beta = \frac{2}{3}$. In order to illustrate the overall nature of the interaction and the effect of slenderness, (12.6.18) is plotted in Fig. 12.29.

Rectangular Plates

Rectangular plates are more difficult because of the infinite variety of possible aspect ratios. It would appear that no numerical solution has yet been published. However, from the information which we have about square plates we can formulate at least a qualitative explanation of the ultimate strength interaction of rectangular plates.

If the plate is very sturdy, such that failure is by crushing, then there will be no change from the square plate interaction. But for all other plates, collapse involves a combination of yielding and buckling, and because of the different lengths of side the two uniaxial collapse loads have different values. For the sake of discussion we shall take the case of $a/b > 1$; this is the most common case, and it will be shown that by a simple change of axis the interaction curves for this case can also be used for $a/b < 1$.

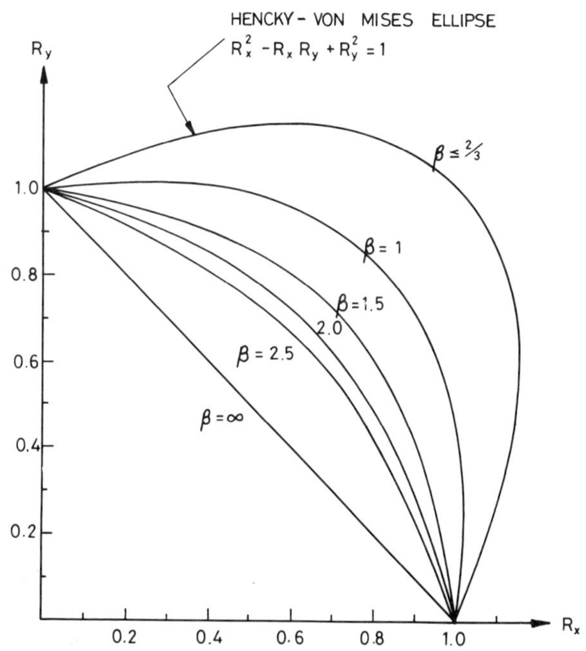

Figure 12.29 Interaction diagram for collapse of biaxially loaded square plates (Eq. 12.6.18 plotted for specific values of β).

To begin, we regard b as fixed and we consider various values of a, with $a > b$. As a increases, $\sigma_{ax,u}$ remains constant whereas $\sigma_{ay,u}$ decreases. Because of this change in $\sigma_{ay,u}$ it will facilitate the explanation if we take $\sigma_{ax,u}$ as the reference stress in both directions. That is, in the y-direction we use $\sigma_{ax,u}$ to nondimensionalize the applied stress, σ_{ay}, and so in place of R_y we have $\sigma_{ay}/\sigma_{ax,u}$. For square plates the two quantities are identical and so Fig. 12.29 remains unchanged. For $a/b > 1$ the uniaxial collapse load $\sigma_{ay,u}$ is smaller than for a square plate because in the y-direction the plate is a wide plate. Therefore, along the $\sigma_{ay}/\sigma_{ax,u}$ axis of the interaction diagram the collapse point is no longer 1.0 but some number less than 1. It will further facilitate the explanation if we take a specific value of a/b; since there are some experimental results available for $a/b = 3$ we choose that case. The interaction diagram for this case is shown—qualitatively only—in Fig. 12.30, and we now proceed to explain why the curves are of the form illustrated.

First, if σ_{ax} is large and σ_{ay} is small, then the interaction curve is no different from a square plate because:

1. The collapse is being caused mainly by σ_{ax}, for which the uniaxial collapse value $\sigma_{ax,u}$ is independent of a/b.
2. Even though the collapse load in the y-direction is less than for a square plate, if σ_{ay} is small then it is still well below this value and so it has no different effect here than it had for a square plate.

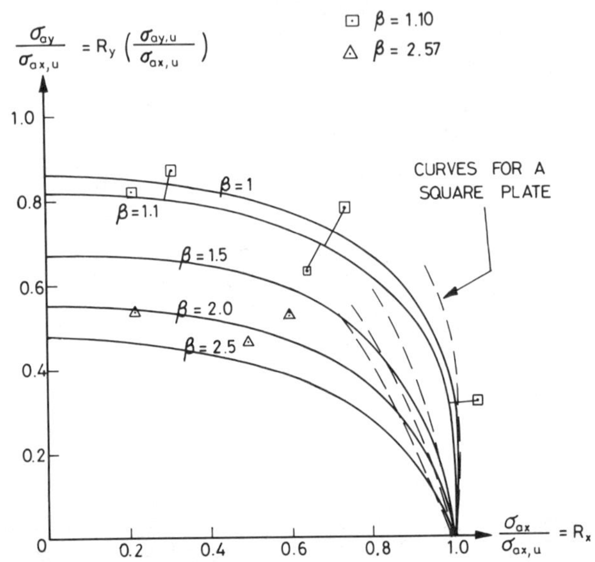

Figure 12.30 Interaction diagram for collapse of biaxially loaded plate with $a/b = 3$.

Therefore, as shown in Fig. 12.30, in this region of the interaction diagram the collapse curves for various values of β all resemble the curves for a square plate, which are shown dashed.

At the other extreme, when σ_{ax} is small and σ_{ay} is large, collapse is essentially that of a wide plate loaded uniaxially. Therefore the point at which each collapse curve meets the $\sigma_{ay}/\sigma_{ax,u}$ axis can be determined: σ_{ay} is equal to the uniaxial wide plate collapse load $\sigma_{ay,u}$, calculated from (12.6.12), and $\sigma_{ax,u}$ is calculated from (12.6.11).

If we start from each of these points and consider an increase in σ_{ax} we can safely assume that this increase will not be beneficial; that is, the curves will not rise, except in the extreme case of a very sturdy plate ($\beta < \frac{2}{3}$) in which the collapse is purely by yielding. Also, if σ_{ax} is only moderate it will have little detrimental effect and so the curves will be essentially horizontal in the left-hand portion of the diagram. Ultimately, as σ_{ax} becomes large, the curves tend toward the limiting shapes of the square plate case, discussed earlier.

Figure 12.30 includes some experimental results for plates with $a/b = 3$ and $\beta = 1.10$ and 2.57, from Ref. 36. Although there is some scatter, the values agree reasonably well with the qualitative curve for $\beta = 1.1$ and thus serve to substantiate the foregoing explanation.

For slender plates collapse is preceded by elastic buckling, and under certain conditions this can result in a strengthening of the plate. To illustrate this phenomenon the buckling interaction curve for $a/b = 3$ and $\beta = 3.3$ is drawn on Fig. 12.31. The curve consists of three straight line segments, corresponding to the formation of 1, 2, or 3 half-waves in the x-direction, as shown earlier in Fig. 12.8. Each segment corresponds to the mode which gives the lowest critical load combination. However, the actual occurrence of one mode or the other depends on the relative magnitude of the two loads prior to buckling; that is, it depends on the particular loading sequence. If σ_{ay} grows more rapidly than σ_{ax} (curve A in Fig. 12.31) then the plate first buckles into one half-wave and then collapses as expected, when the load reaches the collapse curve for that particular plate, that is, for that particular aspect ratio and slenderness. The collapse curve for the plate in question ($a/b = 3$, $\beta = 3.3$) is drawn in the figure, having been estimated from Fig. 12.30.

However, once the $m = 1$ buckling has occurred, the cylindrical shape of the buckled plate serves to strengthen the plate in the x-direction; that is, it in-

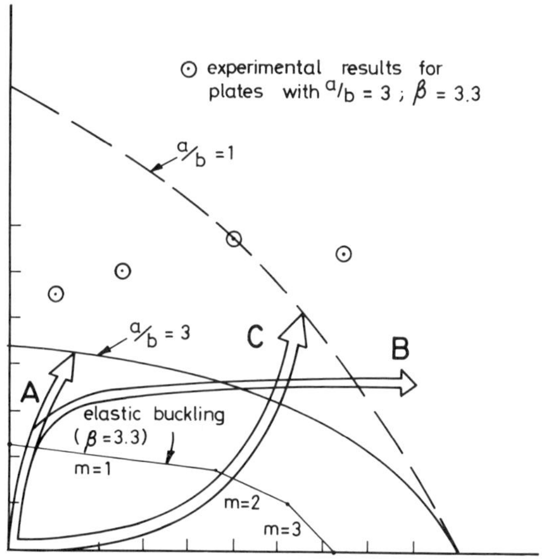

Figure 12.31 Illustration of the effect of load sequence on plate collapse.

creases $\sigma_{ax,u}$. In this case the collapse load can exceed even the square plate value, as in curve *B* of Fig. 12.31. This effect was deliberately ignored when dealing with uniaxial strength because the presence of such curvature cannot be relied on since it depends on loads. It is even less reliable here since it depends on the sequence of the loads—whether σ_{ay} is applied first. Therefore, although the occurrence of this strengthening has been verified experimentally, for design purposes it is ignored.

The third case to be considered is that in which σ_{ax} grows more rapidly, at least up to the point of buckling. In this case the plate buckles into two or three half-waves in the long direction, usually three. Because of this, the plate has essentially been converted into three square plates, at least in regard to the transverse load, σ_{ay}. Therefore, if σ_{ax} and σ_{ay} are further increased (curve C in the figure) the combination can now exceed the normal collapse value for $a/b = 3$. In principle, collapse will not occur until the load combination reaches the square plate collapse curve for the same value of β, because each subpanel is equivalent to a square plate of side b. The square plate collapse curve for the plate in question is shown in Fig. 12.31, having been obtained from Fig. 12.29.

The figure also shows the experimental results for a slender plate ($\beta = 3.3$) of aspect raito $a/b = 3$, from reference [36]. When the ratio σ_{ax}/σ_{ay} is small, the collapse load combination is close to, and somewhat above, the theoretical collapse curve for a plate of that aspect ratio, that is, the value predicted by Fig. 12.30. But when the load ratio is larger, the collapse occurs later, at loads which correspond approximately to the square-plate collapse curve. In the experiment the σ_{ax} load was applied first, which corresponds to path C in Fig. 12.31. Therefore it seems likely that the strengthening occurs because the plate first buckles into two or three half-waves, and this effectively converts it into a plate of aspect ratio 1.5 or 1.0. This explanation is also consistent with the fact that this phenomenon only occurred with the most slender group of plates ($\beta = 3.3$). For all other groups ($\beta = 1.10$, 1.84, 2.57) the collapse points are much closer to the theoretical collapse curves of Fig. 12.30. Some sample points are shown for $\beta = 1.1$ and $\beta = 2.57$.

Ultimate Strength under Shear

If the plating is subjected to very large shear loading this alone can cause collapse, quite apart from in-plane compression. The mechanism of collapse depends on the plate slenderness in much the same manner as for compressive collapse. In very slender plating an applied shear load will cause elastic buckling as described in Section 12.4. After buckling, the plating has some further load-carrying capacity because in one of the two diagonal directions there is a stable arrangement of tensile stresses (a "tension field") which can carry further shear load until yielding occurs, and this marks collapse. In very sturdy plates there is no buckling and collapse occurs purely by yielding. In plates of intermediate slenderness the initial buckling becomes inelastic and less distinct but the tension field mechanism still occurs.

An example of plating subjected to large shear stress is the web plating of a large long-span beam of uniform web depth (a "plate girder") near the support points, where there is large local shear. Because plate girders are so common there is a good deal of information available concerning the ultimate shear strength of plating and the design of plate girders; for example Refs. 37–40.

In ship hull and deck plating such large values of shear stress are generally not present because of the wide expanse of the plating and the distributed nature of the loads. And in ship frames and girders it is often possible to increase the web depth near intersections and points of support. Nevertheless there are some instances in which plate girders are used in ships, such as the transverse deck beams above the hanger deck in an aircraft carrier or above the vehicle decks in roll-on roll-off ships. For such applications adequate information may be found in the preceding references and other sources in the civil engineering literature.

REFERENCES

1. F. Bleich, *Buckling Strength of Metal Structures,* McGraw-Hill, 1952.
2. S. Timoshenko and J. M. Gere, *Theory of Elastic Stability,* McGraw-Hill, 1961.
3. L. G. Jaeger, *Elementary Theory of Elastic Plates,* Pergamon, London, 1963.
4. Column Research Committee of Japan, Ed., *Handbook of Structural Stability,* Corona Publishing Co. Ltd., Tokyo, 1971.
5. O. H. Faxen, "Die Knickfestigkeit rechtechkiger Platten," *Zeitschr. angewandte Math. Mech.* **15,** 1935, pp. 268–277.
6. J. L. Maulbetsch, "Buckling of Compressed Rectangular Plates with Built-In Edges," *Trans. ASME, J. Appl. Mech.,* **4,** 1936, pp. 52–62.
7. S. Levy, "Buckling of Rectangular Plates with Built-In Edges," *Trans. ASME,* **64,** 1942, p. A171.
8. E. Lundquist and E. Z. Stowell, "Critical Compressive Stress for Flat Rectangular Plates . . . Elastically Restrained . . . ," NACA Tech. Note 733, 1942.
9. H. Wagner, "Uber Konstruktion- und Berechnungs-fragen des Blechbaues," *Jb. wiss. Ges. Luftf.,* 1928, S. 113–124.
10. S. B. Batdorf and M. Stein, "Critical Combinations of Shear and Direct Stress for Simply Supported Rectangular Flat Plates," NACA Tech. Note 1223, 1947.
11. K. I. McKenzie, "The Buckling of a Rectangular Plate Under Combined Biaxial Compression, Bending and Shear," *Aero. Q.,* **15** (3), 1964, pp. 239–246.
12. A. G. Young, "Ship Plating Subjected to Loads Both Normal to and in the Plane of the Plate," *Trans. RINA,* **99,** 1957, pp. 578–600.
13. H. D. Conway, "Bending of Rectangular Plates Subjected to a Uniformly Distributed Lateral Load and to Tensile or Compressive Forces in the Plane of the Plate," *J. Appl. Mech.,* **16,** 1949, pp. 301–309.
14. M. Levy, "Sur l'equilibre elastique d'une plaque rectangulaire," *Compt. Rend.* **129,** 1899, pp. 535–539.
15. C. C. Chang and H. D. Conway, "The Marcus Method Applied to Solution of Uniformly Loaded and Clamped Rectangular Plate Subjected to Forces in its Plane," *J. Appl. Mech.,* **19,** 1952, pp. 179–184.
16. P. J. Dowling, J. E. Harding, and P. A. Frieze, Eds., *Steel Plated Structures,* Crosby Lockwood Staples, London, 1976.
17. P. A. Frieze, P. J. Dowling, and R. E. Hobbs, "Ultimate Load Behaviour of Plates in Compression," Chapter 2 of Ref. 16, pp. 24–50.
18. J. E. Harding, R. E. Hobbs, and B. G. Neal, "Ultimate Load Behaviour of Plates under Continued Direct and Shear In-Plane Loading," Chapter 15 of Ref. 16, pp. 369–403.
19. M. A. Crisfield, "Full-Range Analysis of Steel Plates and Stiffened Plating under Uniaxial Compression," *Proc. ICE,* **59,** Dec. 1975, pp. 595–624.
20. B. W. Young, P. Elliott, and G. Bowers, "Residual Stresses and Measurements of Tolerances," *Proc. Int'l. Conf. Steel Box Girder Bridges,* Inst. Civil Engineers, London, Feb. 1973.
21. J. A. Sadden, "Primary Strength Design for Destroyers and Frigates," Chapter 28 of Ref. 16, pp. 658–675.
22. K. E. Moxham, "Buckling Tests on Individual Welded Steel Plates in Compression," Report TR.3 (1971), Cambridge University Engineering Dept., Cambridge.
23. M. A. Crisfield, "Collapse Analysis of Box Girder Components using Finite Elements," Paper presented at Symposium on Non-Linear Techniques and Behaviour in Structural Analysis, Transport and Road Research Laboratory, Dec. 1974.
24. K. E. Moxham, "Theoretical Prediction of the Strength of Welded Steel Plates in Compression," Report TR.2 (1971), Cambridge University Engineering Dept., Cambridge.
25. M. R. Horne and R. Narayanan, "Ultimate Strength of Stiffened Panels under Uniaxial Compression," Chapter 1 of Ref. 16, pp. 1–23.
26. J. B. Dwight, G. H. Little, and N. A. Rogers, "An Approach to Stiffened Steel Compression Panels," Report TR.32 (1973), Cambridge University Engineering Dept., Cambridge.
27. S. Chatterjee and P. J. Dowling,"The Design of Box Girder Compression Flanges," Chapter 8 of Ref. 16, pp. 196–228.
28. C. S. Smith, "Compressive Strength of Welded Steel Ship Grillages," *Trans. RINA,* **117,** 1975, pp. 325–359.
29. J. B. Caldwell, "Ultimate Longitudinal Strength," *Trans. RINA,* **107,** 1965, p. 411; Discussion by D. Faulkner.
30. D. Faulkner, "A Review of Effective Plating for Use in the Analysis of Stiffened Plating in Bending and Compression," *J. Ship Res.* **19** (1), March 1975, pp. 1–17.
31. D. Faulkner, Chapters 21 and 22 of *Ship Structural Design Concepts,* J. H. Evans, Ed., Cornell Maritime Press, 1975.
32. D. Faulkner, "Design Against Collapse for Marine Structures," in *Advances in Marine Technology,* Vol. 2, Norwegian Institute of Technology, Trondheim, June 1979, pp. 275–308.
33. C. S. Smith, "Influence of Local Compressive Failure on Ultimate Longitudinal Strength of a Ship's Hull," Proceedings of International Symposium on Practical Design in Shipbuilding (PRADS), Tokyo, October 1977, pp. 73–80.
34. C. S. Smith, "Imperfection Effects and Design Tolerances in Ships and Offshore Structures," *Trans. Inst. Engrs. Shipbuilders Scotland,* February 1981.
35. S. Valsgård, "Ultimate Capacity of Plates in Transverse Compression," Det norske Veritas, Report no. 79-0104, 1979.
36. H. Becker and A. Colao, "Compressive Strength of Ship Hull Girders—Part III," Ship Structure Committee Report SSC-267, U.S. Coast Guard Headquarters, Washington, 1977.
37. K. Basler, "Strength of Plate Girders in Shear," *J. Struct. Div., ASCE,* **87** (ST7), Proc. Paper 2967, Oct. 1961.

38. A. Ostapenko, C. Chern, and S. Parsanejad, "Ultimate Strength Design of Plate Girders," in *Developments in Bridge Design and Construction,* Crosby, Lockwood and Son, Ltd., University College, Cardiff, 1971.

39. K. C. Rockey, H. R. Evans, and D. M. Porter, "A Design Method for Predicting the Collapse Behaviour of Plate Girders," *Proc. Inst Civil Eng.,* Part 2, **65,** March 1978, pp. 85–112.

40. K. C. Rockey, "The Design of Web Plates for Plate and Box Girders—A State of the Art Report," Chapter 19 of Ref. 16, pp. 459–485.

Chapter

Thirteen

ELASTIC BUCKLING OF STIFFENED PANELS

This chapter deals with the elastic buckling of flat rectangular panels that are stiffened in one or both directions, under various combinations of loading. It is not possible, in one chapter, to present solutions for all types of loading, stiffener arrangements and geometry, and boundary conditions. This type of information is available in structural stability handbooks. Reference 1, for example, contains an extensive collection of solutions to specific cases. The purpose of this chapter is to explain the principal features of stiffened panel buckling, to discuss its applications and limitations with respect to ship panels, to present solutions for the most common cases, and, in Section 13.6, to present a general method of solution which can be used with a wide variety of loads, stiffening arrangements, and boundary conditions.

Stiffened panels can buckle in essentially two different ways. In *overall* buckling, the stiffeners buckle along with the plating; in *local* buckling either the stiffeners buckle prematurely, because of inadequate rigidity or stability, or the plate panels buckle between the stiffeners, thus shedding extra load into the stiffeners so that eventually the stiffeners buckle in the manner of columns. For most ship panels the proportions are such that the buckling—of whichever type—is inelastic, and when dealing with that, the word "failure" is more properly used instead of "buckling": overall panel failure, local stiffener failure, local plate failure. Nevertheless, an elastic buckling analysis gives a good indication of the likely modes of failure, and also provides a foundation for the more complex question of the inelastic buckling and ultimate strength of stiffened panels, which is covered in the next chapter.

The available techniques for a buckling analysis of stiffened panel structures fall into two categories: analytical methods and numerical methods. Analytical methods of solution involve solving the governing differential equation, usually by means of a series representation for the solution in conjunction with an energy approach. There are two principal types of analytical methods: discrete beam and orthotropic plate, the same as for panel bending analysis. Most of the comments and recommendations which were presented in Section 10.1 regarding these alternative approaches apply equally well to buckling analysis. In general the discrete beam approach is more versatile and more accurate but involves more computation. In panels that are stiffened in only one direction, the stiffeners are more independent and, unless the stiffeners are very closely spaced, the orthotropic approach should not be used. Therefore in this chapter the orthotropic plate approach is demonstrated only for cross-stiffened panels, in Section 13.5. If desired the information and solutions presented there can easily be adapted to unidirectional stiffened panels.

Numerical methods currently in use are of two main types: finite difference and finite element. In both cases the solution is written in terms of a discrete number of unknowns located at points in the panel. This leads, in general, to a large-order matrix equation whose coefficients are numerically evaluated functions of the material properties, geometry, and applied in-plane loads. The buckling load is found by determining the load at which the value of the determinant of the coefficients is zero.

As noted earlier when dealing with unstiffened plates, the calculation of elastic buckling stress should usually assume simple support, irrespective of the presence of a lateral load, because in some cases such a load may be absent or may not be large enough to provide rotational restraint, and apart from this possible influence on boundary conditions, a lateral load has little effect on elastic buckling. Therefore, throughout this chapter, unless stated otherwise, it is assumed that the edges of the panel are simply supported. It is also assumed that individual elements of the stiffeners are not subject to instability.

The way in which a stiffened panel will buckle when subjected to in-plane longitudinal compressive forces depends mainly on the stiffeners. The two principal requirements are that they have sufficient torsional stability so that they do not buckle prematurely (i.e., before the plating) and that they have sufficient lateral rigidity so that the possibility of overall buckling is either eliminated or made sufficiently unlikely. For all practical purposes, stiffener buckling is synonymous with overall buckling, because if the stiffeners buckle the plating is left with almost no lateral rigidity whatever. Since overall buckling involves the entire panel, it is usually regarded as collapse rather than as unserviceability. Moreover, it is a quite sudden mode of collapse and therefore more undesirable than other modes. Thus, the first and most basic principle in regard to stiffeners is that they should be at least as strong as the plating; that is, they should be sufficiently rigid and stable so that neither overall buckling nor local stiffener buckling occurs before local plate buckling. Most ship panels must carry substantial lateral loads and this requirement usually produces stiffeners that are already larger and more rigid than the minimum sizes required by consideration of elastic overall buckling. Therefore in most cases an inelastic failure analysis is required. However, if the panel is *slender* (this term will be quantified later) as may occur with small lateral load and in light weight construction (small, closely spaced stiffeners) then elastic overall buckling becomes a possible collapse mode and an elastic buckling analysis becomes essential. Generally speaking, it is not possible, without a specific analysis, to know with certainty just what failure modes—elastic or inelastic—will be the governing requirements which determine the plate thickness and the stiffener sizes and spacing. The best approach is to first perform an elastic buckling analysis because: (1) it is relatively simple, consisting mostly of explicit formulas; (2) for slender panels, elastic buckling of one type or another may be possible and may be one of the governing failure modes; and (3) the elastic analysis indicates whether or not an inelastic analysis is required, and some of the elastic buckling parameters are needed for the inelastic analysis.

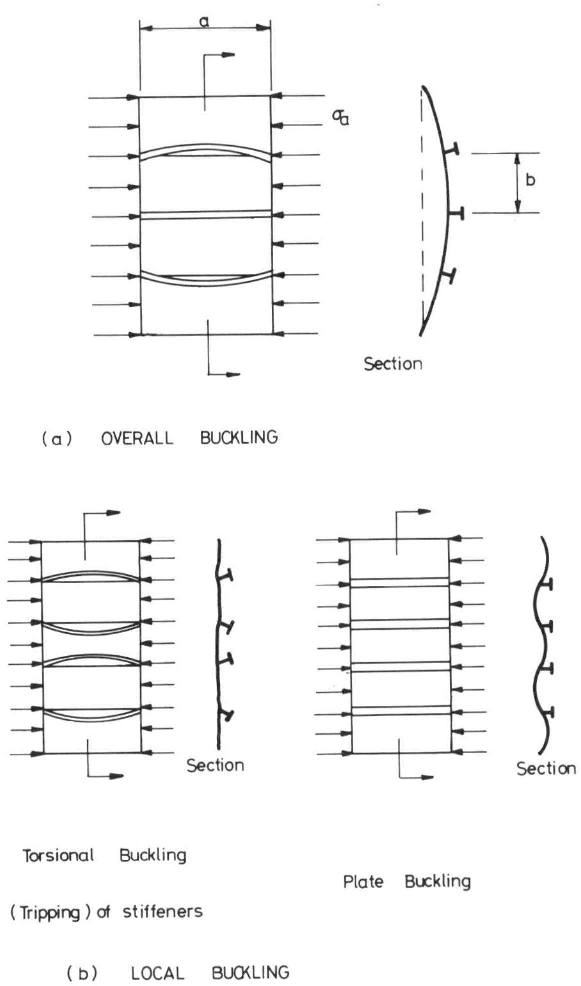

Figure 13.1 Buckling of longitudinally stiffened panels.

13.1 LONGITUDINALLY STIFFENED PANELS

Figure 13.1 illustrates overall buckling and the two types of local buckling for longitudinally stiffened panels.* We first investigate the minimum lateral rigidity to ensure that overall buckling does not precede plate buckling, and then the required torsional rigidity to prevent local stiffener buckling.

*The term "overall buckling" is used with reference to the type of panel which is being considered. Thus it is important to distinguish overall buckling of a longitudinally stiffened panel from overall buckling of a cross-stiffened panel, which is dealt with in Section 13.5. To assist in making this distinction, the term "gross panel buckling" will be used for the latter.

Overall Buckling vs. Plate Buckling

Minimum Flexural Rigidity to Avoid Overall Buckling

The minimum rigidity of longitudinal stiffeners necessary to ensure that overall buckling does not precede plate buckling has been investigated by various authors [2,3,4] for a panel containing one, two, or three equally spaced longitudinal stiffeners. For the first two cases Bleich [5] has presented approximate formulas which are reproduced here. As before, the symbol I_x denotes the moment of inertia of a section comprised of a stiffener together with a width b of plate. The minimum required rigidity is expressed in terms of a parameter γ_x which is the ratio of the flexural rigidity of the combined section to the flexural rigidity of the plating:

$$\gamma_x = \frac{EI_x}{Db} = \frac{12(1 - \nu^2)I_x}{bt^3} \qquad (13.1.1)$$

The other parameters are the panel aspect ratio Π

$$\Pi = \frac{L}{B} = \frac{a}{B} \qquad (13.1.2)$$

in which L is the panel length (which in this case equals the plate length a) and the area ratio δ_x

$$\delta_x = \frac{A_x}{bt} \qquad (13.1.3)$$

in which A_x is the cross-sectional area of the stiffener (only).

In terms of these parameters the minimum value of γ_x to ensure that stiffener buckling does not precede plate buckling is given by the following formulas:

(a) Panel with one central longitudinal stiffener:

$$\left.\begin{aligned} &\text{either} \\ &\gamma_x = 22.8\,\Pi + (2.5 + 16\,\delta_x)\Pi^2 - 10.8\sqrt{\Pi} \\ &\text{or} \\ &\gamma_x = 48.8 + 112\,\delta_x(1 + 0.5\,\delta_x) \end{aligned}\right\} \qquad (13.1.4)$$

whichever is less.

(b) Panel with two equally spaced longitudinal stiffeners:

$$\left.\begin{aligned} &\text{either} \\ &\gamma_x = 43.5\sqrt{\Pi^3} + 36\,\Pi^2\,\delta_x \\ &\text{or} \\ &\gamma_x = 288 + 610\,\delta_x + 325\,\delta_x^2 \end{aligned}\right\} \qquad (13.1.5)$$

whichever is less.

NOTE: The range of validity for the above formulas is $0 < \delta_x < 0.2$.

Cox and Riddel [4] discussed in detail plates with one, two, or three stiffeners, and their analysis is capable of extension to four, five, or more stiffeners. They investigated the smallest size of stiffeners necessary to prevent overall buckling of a flat panel before buckling of the plate between stiffeners. The effect of torsional stiffness of the stiffeners is included. The analysis was done using a strain energy method and the solution was given in closed form.

A more general solution, valid for any number of stiffeners, has been presented by Klitchieff [6]

$$\gamma_x = \delta_x(1 + N_B^2\Pi^2)^2 + \frac{4}{\pi}\Pi(1 + N_B^2\Pi^2)\sqrt{2 + N_B^2\Pi^2}$$

where N_B = number of panels = 1 + number of longitudinal stiffeners.

Calculation of Overall Buckling Stress

An alternative approach is to calculate the overall buckling stress $(\sigma_a)_{cr}$ and compare it to the plate buckling stress to ensure that it is larger. One method of calculating $(\sigma_a)_{cr}$ is to regard each stiffener and its associated width of plating as a column, having some equivalent slenderness ratio $(L/\rho)_{eq}$. The elastic buckling stress $(\sigma_a)_{cr}$ is then obtained from the Euler column formula and the requirement that overall buckling not precede plate buckling then appears in a more explicit form:

$$(\sigma_a)_{cr} \geqslant \sigma_o$$

where

$$\sigma_o = 3.62\,E\left(\frac{t}{b}\right)^2$$

In short panels, this approach is virtually exact; the stiffeners are in fact a row of parallel columns which are essentially identical and, as far as buckling is concerned, independent. The equivalent slenderness ratio of each column is the actual slenderness of the section

$$\left(\frac{L}{\rho}\right)_{eq} = \frac{a}{\rho} = \frac{a}{\sqrt{\dfrac{I_x}{(A_x + bt)}}} \tag{13.1.6a}$$

or, in terms of nondimensional parameters

$$\left(\frac{L}{\rho}\right)_{eq} = \frac{a}{t}\sqrt{\frac{12(1 - \nu^2)(1 + \delta_x)}{\gamma_x}} \tag{13.1.6b}$$

In long panels the stiffeners receive some lateral restraint from the sides of the panel and this may cause them to buckle in more than one half-wave. In this case the equivalent slenderness ratio is smaller than the value given by (13.1.6). The effect occurs for large values of the panel aspect ratio Π and for small values of stiffener rigidity relative to the plating (i.e., small values of γ_x). From the work of Sharp [9] it is possible to derive an aspect ratio coefficient C_Π which accounts for this effect as follows

$$\left(\frac{L}{\rho}\right)_{eq} = C_\Pi \frac{a}{\rho} = \frac{C_\Pi a}{\sqrt{\dfrac{I_x}{(A_x + bt)}}} \tag{13.1.7a}$$

in which C_Π is given by either

$$C_\Pi = \frac{1}{\Pi}\sqrt{\frac{\gamma_x}{2(1 + \sqrt{1 + \gamma_x})}}$$

or (13.1.7b)

$$C_\Pi = 1$$

whichever is less. The resulting value of $(L/\rho)_{eq}$ is then used in the standard column buckling formula

$$(\sigma_a)_{cr} = \frac{\pi^2 E}{\left(\dfrac{L}{\rho}\right)_{eq}^2} \tag{13.1.8}$$

As is stands, the purpose of this equation is simply to verify that the critical stress for the combined stiffener and plate is greater than the plate buckling stress σ_0, and for this purpose the plate flange should be taken at its full width b because a value less than b would be "begging the question," since it could only be less if the plate had already buckled. Of course, the above value of $(\sigma_a)_{cr}$ does not constitute a true overall buckling stress if it exceeds the yield stress. For convenience, we shall use the adjective "slender" to describe a panel in which the overall buckling stress calculated from elastic theory is less than the yield stress. That is, we say that a panel is slender if

$$\frac{\pi^2 E}{\left(\dfrac{L}{\rho}\right)_{eq}^2} < \sigma_Y$$

and so in terms of the column slenderness parameter $\lambda = (L/\pi\rho)(\sigma_Y/E)^{1/2}$ which was introduced in Chapter 11, the foregoing criterion becomes simply $\lambda_{eq} > 1$.

Since slender panels are normally designed such that plate buckling precedes overall buckling, when the latter occurs the plate flange of the stiffener will not be fully effective over the width b. Instead, it is necessary to take some reduced effective width, b_e. Note that this is not the same as the reduced effective *breadth* which was used in Chapter 3 in order to correct for shear lag effects. In that case the lack of effectiveness was due to in-plane deformation of the plating due to shear; in the present case it is due to out-of-plane deformation, caused by buckling.

The effective width due to buckling has long been a vexed question, mainly because in most cases it was being discussed and applied in the difficult context of the ultimate strength of panels which were not slender and which therefore did not buckle elastically. For elastic or near-elastic buckling a satisfactory formula was derived by von Karman [7] as early as 1924 (the reference quoted is a later paper, in English). Von Karman's approach was characteristically simple and practical. He idealized the state of stress within the buckled plate by assuming that, due to buckling, the center portion has no compressive stress, while the edge portions of the plate remain fully effective and carry a uniform stress σ_e, as shown in Fig. 13.2. In other words, the buckled center portion is discounted completely and the original plate of width b is replaced by a narrower unbuckled plate of effective width b_e. From statics it is clear that σ_e and σ_a are related by

$$\sigma_e = \frac{b}{b_e}\sigma_a \tag{13.1.9}$$

To simulate the progressive growth of the buckling, it is further assumed that the (as yet unbuckled) effec-

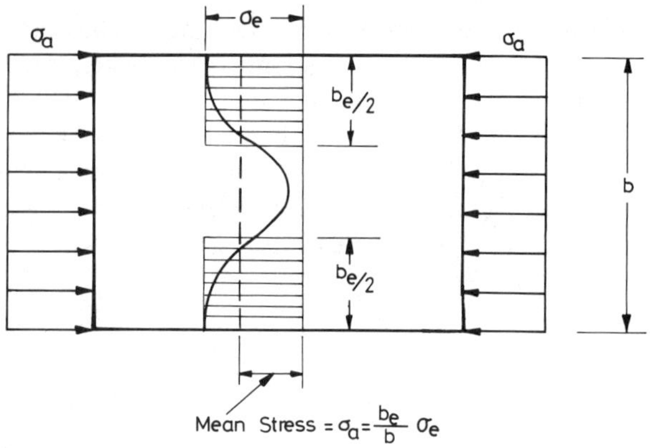

Figure 13.2 Post-buckling stress distribution; effective width.

tive plate is always on the verge of further buckling; that is, the effective width is taken to be the width at which the equivalent plate would buckle at an applied stress of σ_e. This implies that

$$\sigma_e = k\frac{\pi^2 D}{b_e^2 t}$$

For the original plate

$$(\sigma_a)_{cr} = k\frac{\pi^2 D}{b^2 t}$$

and if it is assumed that k is the same in both cases then

$$\frac{b_e}{b} = \sqrt{\frac{(\sigma_a)_{cr}}{\sigma_e}} \qquad (13.1.10)$$

The latter assumption is not strictly correct because although the boundary conditions can be regarded as similar in both cases, the aspect ratios are different. However, it was shown in Section 12.1 that for aspect ratios greater than 1.0, k can be taken as 4.0. Substitution of this value, together with $\nu = 0.3$, into the expression for $(\sigma_a)_{cr}$ converts (13.1.10) into

$$\frac{b_e}{b} = 1.9\frac{t}{b}\sqrt{\frac{E}{\sigma_e}} \qquad (13.1.11)$$

The effective width would reach its smallest possible value when σ_e reached yield. In this case the right-hand side becomes a simple inverse function of the plate slenderness parameter β:

$$\left(\frac{b_e}{b}\right)_{\min} = \frac{1.9}{\beta}$$

However, if the panel is truly slender, overall panel buckling would in most cases occur before σ_e reaches σ_Y and so we use the more general expression for b_e given by (13.1.11).

Now that we have an expression for b_e we can proceed to obtain an expression for the ultimate or collapse load of slender stiffened panels that is, panels for which collapse is by overall elastic buckling. The effective width is used in calculating the equivalent slenderness ratio from (13.1.7); that is, b_e is used as the plate flange width in calculating I_x, δ_x, and ρ. We shall denote these values as I_{xe}, δ_{xe}, and ρ_e, and the resulting value of equivalent slenderness ratio as $(L/\rho_e)_{eq}$. The axial stress in the stiffener "columns" is σ_e and the critical value of this stress is given by (13.1.8)

$$(\sigma_e)_{cr} = \frac{\pi^2 E}{\left(\dfrac{L}{\rho_e}\right)_{eq}^2} \qquad (13.1.13)$$

Note that this equation refers to σ_e rather than to σ_a; the axial stress in the stiffener is larger than the external applied stress σ_a because of the reduced width of the plate. The quantity of interest is the value of σ_a corresponding to $(\sigma_e)_{cr}$. From statics the two are related by $\sigma_a(bt + A_x) = \sigma_e(b_e t + A_x)$, and from this together with (13.1.13) we have

$$(\sigma_a)_{cr} = \left(\frac{b_e t + A_x}{bt + A_x}\right)\frac{\pi^2 E}{\left(\dfrac{L}{\rho_e}\right)_{eq}^2} \qquad (13.1.14)$$

Because of the presence of σ_e in (13.1.11) the foregoing sequence of calculations must be performed iteratively. A suitable procedure would be:

1. Assume some initial value of b_e (e.g., $0.8b$).
2. Calculate δ_{xe} and I_{xe}, and then evaluate $(L/\rho_e)_{eq}$ from (13.1.7), using b_e in place of b.
3. Calculate $(\sigma_e)_{cr}$ from (13.1.13).
4. Using this value of σ_e, recalculate b_e from (13.1.11).
5. Repeat from step 2 until b_e has converged.
6. Calculate $(\sigma_a)_{cr}$ from (13.1.14).

In computer-aided design, steps 1 to 5 would simply become an inner loop of the overall procedure. For short panels ($\Pi < 1$) in which $C_\Pi = 1$, steps 1 to 5 can be condensed into a single nonlinear equation for

b_e by expressing ρ_e in terms of the stiffener web and flange areas A_w and A_f, as in (8.3.6). The resulting equation is

$$\frac{\pi d}{1.9at} b_e \sqrt{\frac{A_w}{12}(A_w + 4A_f) + \left(\frac{A_w}{3} + A_f\right) b_e t} = A_w + A_f + b_e t$$

The preceding analysis only applies to slender panels, that is, panels for which the overall buckling stress is, at its lowest, less than the yield stress. The phrase "at its lowest" is required because the value of $(\sigma_e)_{cr}$ depends on ρ_e, which in turn depends on the value of b_e. As shown in Fig. 13.3a, if the plate flange area bt is substantially greater than the stiffener area A_x (say, $bt > 2A_x$) *as is usually the case,* then the progressive decrease in effective width causes an increase in ρ_e, making the panel *sturdier* (because L/ρ_e is smaller). If the plate area is less than $2A_x$, then the effect can be neutral or even opposite, but this is not common. Because of this somewhat paradoxical relationship between ρ_e and b_e, the lowest (hypothetical) value of $(\sigma_e)_{cr}$ corresponds to $b_e = b$; that is, plating fully effective. This is illustrated in Fig. 13.3b, which also shows that the same paradoxical relationship usually holds for $(\sigma_a)_{cr}$, but the effect is somewhat diminished because of the multiplication by the area ratio in (13.1.14). The full range of possible situations is shown in Fig. 13.4. As long as the value of $(\sigma_a)_{cr}$ corresponding to $b = b_e$ is greater than the plate buckling stress σ_0 then it remains merely hypothetical; it would not be the actual collapse stress because the plating would buckle first. But, if this value of $(\sigma_a)_{cr}$ were less than σ_0 then the panel would collapse by overall elastic buckling at an applied stress equal to $(\sigma_a)_{cr}$, with no prior plate buckling. As mentioned earlier, this is a highly undesirable mode of collapse and for this reason one of the basic steps in panel design is to ensure that $(\sigma_a)_{cr}$ for fully effective plating is greater than the plate buckling stress. We see now that if $bt > 2A_x$ (approximately) then in making a preliminary check as to whether or not a panel is slender (by calculating λ_{eq} and seeing if it is greater than 1.0) it is necessary to use the full plate width.

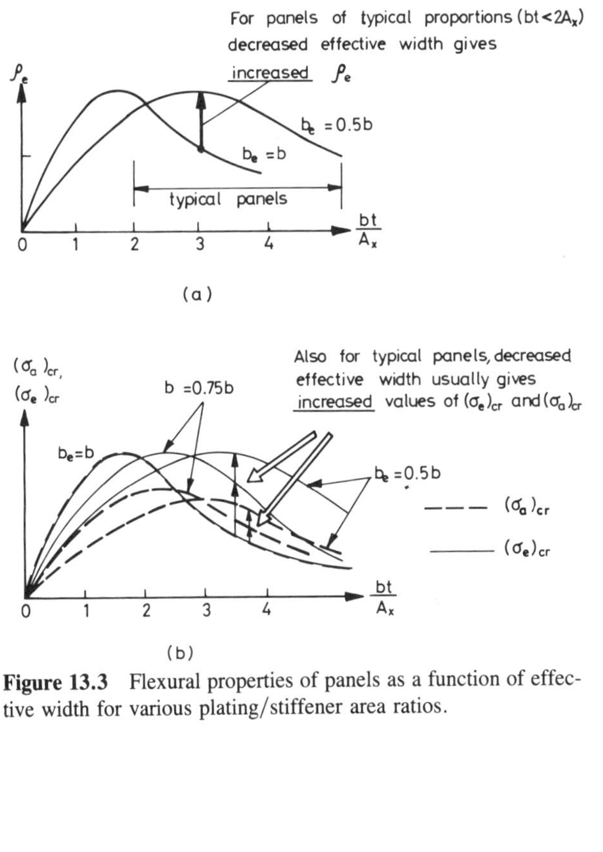

Figure 13.3 Flexural properties of panels as a function of effective width for various plating/stiffener area ratios.

Local Buckling of Stiffener (Tripping)

As shown in Fig. 13.5 a stiffener may buckle by twisting about its line of attachment to the plating. This is commonly referred to as tripping. The plate may rotate somewhat to accommodate the stiffener rotation, and the direction of rotation usually alternates as shown because this involves less elastic strain energy in the plating. However, this is not plate buckling. Tripping and plate buckling do interact, of course, but they can occur in either order, depending on the stiffener and plating proportions. As noted earlier, tripping failure is regarded as collapse because once tripping occurs the plating is left with no stiffening and so overall buckling follows immediately. Also, *elastic* tripping is a quite sudden phenomenon and hence it is a most undesirable mode of collapse, akin to elastic overall panel buckling. Since the open sections that are used as stiffeners in ship panels have relatively little torsional rigidity, such panels can be susceptible to tripping and so it is very important to consider this mode of buckling and to provide an adequate margin of safety. With stiffeners of closed cross section tripping cannot occur and the use of such stiffeners produces panels that have a larger strength-to-weight ratio. However, their use also increases the cost of panel fabrication and there may be difficulties regarding inspection and control of corrosion.

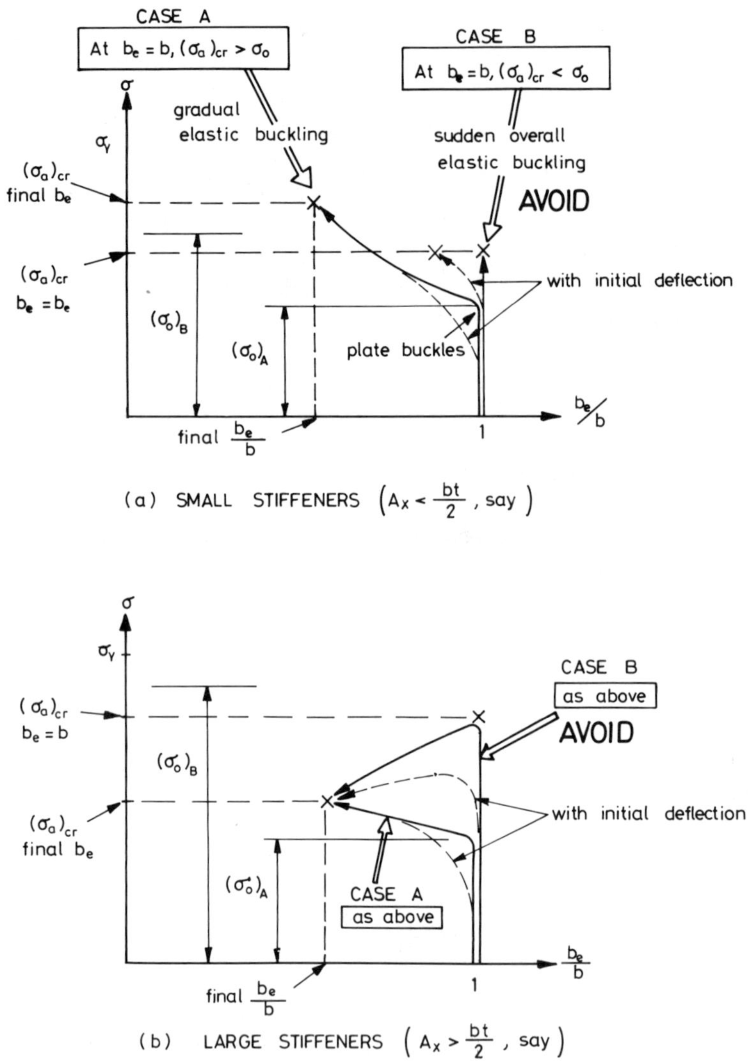

Figure 13.4 Possible modes of elastic buckling of stiffened panels.

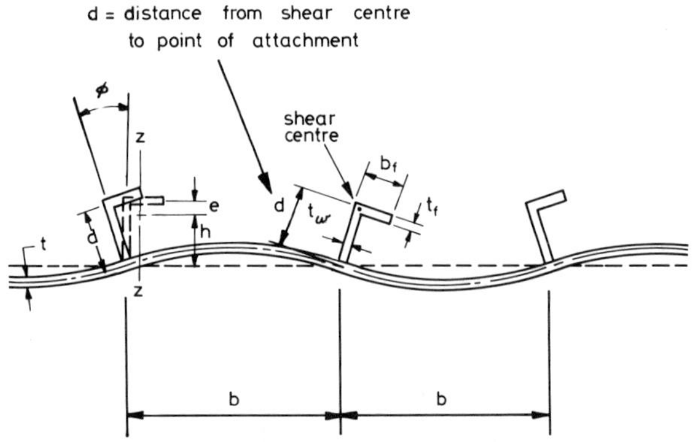

Figure 13.5 Stiffener tripping.

Types of Stiffener Buckling and Methods of Analysis

The theory presented in this section is for a stiffener subjected to axial compression. Torsional buckling of a stiffener may also be caused by a bending moment which puts the flange in compression; this is referred to as *flexural-torsional* buckling. It is decidedly more complicated than buckling under a purely axial load because of the strong influence of eccentricity and the inherently nonlinear coupling between the flexural and torsional response of the stiffener. There are two quite different methods for dealing with nonlinear elastic buckling: (1) "folded plate" analysis based on finite difference methods (this approach is summarized and applied to ship structural components in Ref. 8); and (2) nonlinear frame (or finite element) analysis, which accounts for nonlinearity by means of a "geometric stiffness matrix." The first method is simpler but is restricted to certain boundary conditions and to simple forms of structural geometry; also it is basically limited to elastic buckling of bifurcation type. The second method involves more computation and is necessarily computer-based, but is nevertheless quite economical. It is much more general and it can deal with other forms of nonlinearity such as yielding, large deflections, plastic hinge formation, and the development of a "mechanism" in the overall structure. Therefore only this second method is considered in this text. Because of its greater generality, its principal application in the design method presented herein (Fig. 1.1) is the ultimate strength analysis of the overall structure (hull module). Therefore, although the method also has many applications to smaller portions of structure, even down to the level of stiffener buckling, its presentation is deferred until later in the book (mainly Chapter 15) where its application to individual structural members comes as a natural and logical way of presenting its basic features. Thus for example flexural-torsional buckling is introduced in Section 15.4. The present chapter concentrates on the other forms of elastic panel buckling, for which analytical methods, and often explicit expressions, are available. Chapter 14 does the same for inelastic panel buckling.

Stiffener Buckling Due to Axial Compression

Under this type of load a stiffener acts essentially as a column, but its tripping or torsional buckling differs from that of a column in three ways: first, because the rotation occurs about an enforced axis—the line of attachment to the plating; second, because the plate offers some restraint against this rotation; and third, because it is not necessarily rigid body rotation (if the plating is sturdy there will be some distortion of the stiffener due to web bending, as shown in Fig. 13.6). The analysis that follows takes account of all these factors and it applies to both symmetric and unsymmetric stiffeners.*

*Note, however, that in regard to flexural-torsional buckling, unsymmetric stiffeners (angles, etc.) have a lower buckling load than symmetric stiffeners (tees) because they begin to rotate as soon as a lateral load is applied.

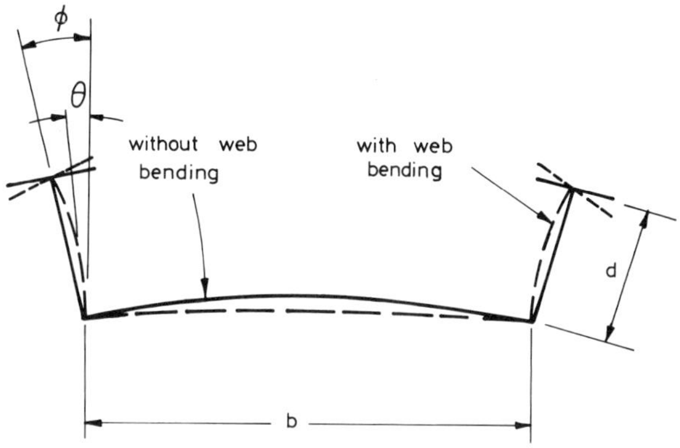

Figure 13.6 Effect of web bending.

The governing differential equation for the torsional buckling of a column about an enforced axis of rotation under the action of an applied axial stress is presented in Reference 2. In applying this to stiffeners we may take advantage of the fact, known from the basic theory of the torsion of thin-walled sections, that for all cross-sectional shapes that are composed of thin rectangles which meet at a common point, the shear center is at this point and the warping constant is zero. In this case it may be shown that the governing differential equation for the rotation ϕ is

$$EI_{sz}\, d^2 \frac{d^4\phi}{dx^4} - (GJ - \sigma_a I_{sp}) \frac{d^2\phi}{dx^2} + K_\phi\, \phi = 0$$

in which I_{sz} = moment of inertia of the stiffener (only) about an axis through the centroid of the stiffener and parallel to the web

d = stiffener web height

I_{sp} = polar moment of inertia of the stiffener about the center of rotation; it may be shown that for thin-wall open sections $I_{sp} \cong d^2(A_f + A_w/3)$, where A_f is the area of the flange and A_w is the area of the web

K_ϕ = distributed rotational restraint which the plating exerts on the stiffener.

If the ends of the stiffener are regarded as simply supported, the solution for $\phi(x)$ is a buckled shape in which the rotation ϕ varies sinusoidally in m half-waves over the length a. The elastic tripping stress, that is, the value of the applied in-plane stress σ_a that would cause tripping according to elastic theory, will be denoted as $\sigma_{a,T}$. From the foregoing equation it may be seen that $\sigma_{a,T}$ is the minimum value of σ_a that satisfies the following, in which m is a positive integer

$$EI_{sz}\, d^2 \frac{m^4\pi^4}{a^4} + (GJ - \sigma_a I_{sp})\frac{m^2\pi^2}{a^2} + K_\phi(\sigma_a, m) = 0 \qquad (13.1.15)$$

As with plate buckling, the critical mode of tripping corresponds to whichever integer value of m gives the minimum value of σ_a in the foregoing expression. In this expression K_ϕ is written as a function of σ_a and m because it is strongly dependent on both of these quantities. It is dependent on σ_a because the onset of plate buckling can diminish or eliminate K_ϕ, or even make it negative; and it depends on m because this diminishing of K_ϕ depends on whether m matches or approaches the number of half-waves in the plate's own critical buckling mode. Conversely, the value of m depends on the value of K_ϕ; if K_ϕ is small relative to EI_{sz} the critical mode will be $m = 1$, but for larger values of K_ϕ, m may be larger.

In the absence of other factors, the rotational restraint offered by the plating comes directly from the plate's flexural rigidity which causes, in response to the rotation ϕ of the stiffener, a total distributed restraining moment $M_R = 2M$ along the line of the stiffener attachment, as shown in Fig. 13.7. If the individual plate panels are long, that is, if $a >> b$, then we may ignore aspect ratio effects, and by considering a unit strip of plating across the span b it may be shown that $\phi = \frac{1}{2}Mb/D$. Therefore the rotational restraint coefficient, or rotational stiffness, is

$$K_\phi = \frac{M_R}{\phi} = \frac{4D}{b}$$

However, this assumes that the buckled displacement of the stiffener is entirely due to rigid body rotation. This is only accurate if the flexural rigidity of the stiffener web is much larger that of the plate. In practice, some of the sideways displacement of the stiffener flange occurs due to bending of the web, and this effect becomes important if the plating is sturdy or if the stiffener web is slender. Sharp [9] has presented an expression which accounts for this effect. Figure 13.6 shows the deflected shape when web bending does and does not occur, for the same amount of maximum sideways displacement of the stiffener flange. The rotational restraining moment M_R exerted by the plate (Fig. 13.7) is proportional to the angle of rotation of the plate along the line of attachment, and this angle is

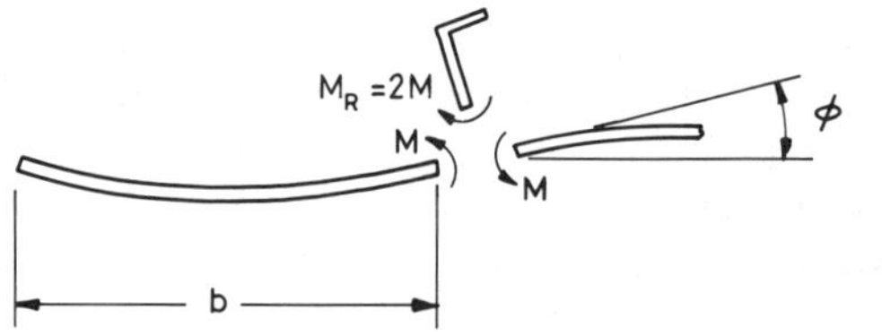

Figure 13.7 Restraining moment exerted by plating.

denoted as θ and ϕ in the two cases, respectively. It may readily be shown that the two angles are related by

$$\theta = C_r\,\phi$$

in which
$$C_r = \frac{1}{1 + \dfrac{2}{3}\left(\dfrac{t}{t_w}\right)^3 \dfrac{d}{b}}$$

That is, C_r is the factor by which the plate rotational restraint is reduced due to web bending. This correction does not account for all of the effects of web bending; for that it would be necessary to take web bending into account from the very beginning, in the derivation of the governing differential equation. Nevertheless, on comparing the results of the foregoing treatment with a more detailed computer-based solution, Sharp found good agreement, particularly if the $\frac{2}{3}$ factor was changed to 0.4. Also, the effect of plate aspect ratio may be accounted for by applying another correction factor $C_\alpha = (1 + m^2/\alpha^2)$. We then have

$$K_\phi = \frac{4D}{b} C_r\, C_\alpha \qquad (13.1.16)$$

with
$$C_r = \frac{1}{1 + 0.4\left(\dfrac{t}{t_w}\right)^3 \dfrac{d}{b}} \qquad (13.1.17)$$

and
$$C_\alpha = 1 + \frac{m^2}{\alpha^2} \quad (\text{for } \sigma_a = 0)$$

The qualification regarding σ_a is added because we have not yet allowed for the decrease in plate stiffness which occurs as the applied stress σ_a approaches the plate buckling stress σ_0 [which, for $\alpha > 1$, is $\alpha_0 = 3.62\, E(t/b)^2$]. Kroll [10] has calculated a systematic set of values of a plate stiffness factor which accounts for both aspect ratio and buckling effects, and these values can be used for C_α in place of $(1 + m^2/\alpha^2)$. However, Kroll's results are in tabular form, whereas for purposes of computer-aided design it is preferable to have an analytical expression for C_α as a function of σ_a, m, and α, even at the cost of some accuracy. The following expression is an adequate representation of the tabulated values, and the small discrepancies which are introduced occur on the conservative side.

$$C_\alpha = 1 - \left(\frac{2\sigma_a}{\sigma_0} - 1\right)\frac{m^2}{\alpha^2} \qquad (13.1.18)$$

From this expression it may be seen that for $\sigma_a = \sigma_0$, the factor C_α, and hence also the plate rotational restraint K_ϕ, is proportional to $1 - (m/\alpha)^2$; that is, the restraint disappears when $m = \alpha$. This reflects the fact that if the plate panel between the stiffeners is long, or at least square, it will buckle into a number (approximately α) of square subpanels. In most cases the number will exceed m, the number of half-waves of the stiffener tripping mode, and in such cases the plate buckling has little deleterious effect on K_ϕ. However, if the plate buckling pattern matches that of the stiffener ($\alpha = m$) then as σ_a approaches σ_0 the plate loses its ability to provide any rotational restraint. For stiffened panels of usual proportions, tripping occurs in a single half-wave, $m = 1$, and hence it is mainly square or short panels in which this loss of stiffness can occur. Applying the correction factors C_r and C_α gives

$$K_\phi = \frac{4D}{b}\left[\frac{1}{1 + 0.4\left(\dfrac{t}{t_w}\right)^3 \dfrac{d}{b}}\right]\left[1 - \left(\frac{2\sigma_a}{\sigma_0} - 1\right)\frac{m^2}{\alpha^2}\right] \qquad (13.1.19)$$

We have now seen that tripping involves three variables which are interrelated in a rather complex fashion: σ_a, m, and K_ϕ. The critical value of σ_a given by (13.1.15) depends on K_ϕ and m, the value of m depends on the magnitude of K_ϕ relative to EI_{sz}, and (13.1.19) shows that K_ϕ depends on both σ_a and m.

As mentioned previously, in stiffened panels of average proportions the critical tripping mode is usually the $m = 1$ mode, but of course this cannot be simply assumed; the correct value of m must be ascertained in each case. The only exception to this is the calculation of a lower bound solution in which the plating rotational restraint is deliberately ignored, in which case the critical mode is always $m = 1$. Also, in this case the solution for $\sigma_{a,T}$ is much simpler, as will be shown. Therefore, in the analysis of tripping we distinguish three different cases: $m \geqslant 2$, $m = 1$, and the lower bound case. Accordingly, it will be helpful and even necessary to have some method for predicting the likely value of m, even if only approximate, instead of having to use a purely trial-and-error approach. An estimate can be obtained by substituting (13.1.19) for K_ϕ in (13.1.15) and solving the latter for σ_a. The result is

$$\sigma_{a,T} = \underset{m = 1, 2, \ldots}{\text{Minimum}} \left\{\frac{1}{I_{sp} + \dfrac{2C_r b^3 t}{\pi^4}}\left[GJ + \frac{m^2\pi^2}{a^2}EI_{sz}d^2 + \frac{4DC_r}{\pi^2 b}\left(\frac{a^2}{m^2} + b^2\right)\right]\right\} \qquad (13.1.20)$$

in which C_r is defined in (13.1.17). We then temporarily regard m as a continuous variable, differentiate (13.1.20) with respect to m and set this equal to zero. The result is

$$m \cong \frac{a}{\pi}\sqrt[4]{\frac{4D\,C_r}{EI_{sz}\,d^2\,b}} \qquad (13.1.21)$$

After obtaining this estimate the strictly correct procedure would be to try the two integer values above and below it to see which value gives the lowest value of $\sigma_{a,T}$ in (13.1.20). However, if the estimated m from (13.1.21) is, say, 1.7 or more, indicating that m is definitely greater than 1, then it is sufficiently accurate (and slightly conservative) to use a new form of (13.1.20) in which the above expression for m has already been substituted. In fact, this type of approximation is particularly helpful, and even advisable, in computer-aided design applications because it helps to avoid the problems which can arise when variables have only discrete values.

On the other hand, if the estimated value of m from (13.1.21) indicates that m may be 1, then this case should be dealt with explicitly, substituting $m = 1$ in (13.1.20). And finally, if K_ϕ is being taken as zero, then there is no need to estimate m; it is necessarily 1. We thus have the three categories of solution mentioned earlier. If K_ϕ is not being taken as zero, the first step is to obtain an estimate for m from (13.1.21) in order to know which of the first two categories is relevant.

(a) Solution when $m \geqslant 2$

To provide for this case, (13.1.21) is substituted directly into (13.1.20) giving

$$\sigma_{a,T} = \frac{1}{I_{sp} + \dfrac{2C_r b^3 t}{\pi^4}}\left(GJ + 4\sqrt{\frac{DC_r EI_{sz} d^2}{b}} + \frac{4DC_r b}{\pi^2}\right) \qquad (13.1.22)$$

in which C_r is defined by (13.1.17). In some cases a simpler expression may be desired, even at the expense of some accuracy, such as for non-computer-aided design or for quick estimates. This is presented in part (c) below.

(b) Solution when $m = 1$

When the value of m from (13.1.21) lies approximately midway between 1 and 2 (say, when it is less than 1.7, to be safe) the analysis of part (a) should be supplemented by an explicit investigation of the $m = 1$ case. Also, when the estimated m is less than, say, 1.3 then part (a) is not required at all. For $m = 1$ (13.1.20) becomes

$$\sigma_{a,T} = \frac{1}{I_{sp} + \dfrac{2C_r b^3 t}{\pi^4}}\left[GJ + \frac{\pi^2 EI_{sz} d^2}{a^2} + \frac{4DC_r}{\pi^2 b}(a^2 + b^2)\right] \qquad (13.1.23)$$

in which, once again, C_r is defined in (13.1.17).

(c) Lower Bound Solution

If the plating is very light, or if a simple and direct formula for $\sigma_{a,T}$ is required, either for manual design or for obtaining a quick check regarding tripping, the solution for $\sigma_{a,T}$ can be simplified by ignoring the first and third terms, which are usually small relative to the term containing I_{sz}. The equation for $\sigma_{a,T}$ becomes

$$\sigma_{a,T} = \frac{1}{I_{sp}}\left[\pi^2 E\left(\frac{d}{a}\right)^2 I_{sz}\right] \qquad (13.1.24)$$

I_{sp} and I_{sz} may be expressed approximately in terms of the stiffener areas $A_f = b_f t_f$, $A_w = dt_w$, and $A_x = A_f + A_w$ as follows:

$$I_{sp} = d^2\left(A_f + \frac{A_w}{3}\right)$$

$$I_{sz} = b_f^2 \frac{A_f\left(\dfrac{A_x}{3} - \dfrac{A_f}{4}\right)}{A_x} \cong \frac{A_f A_w b_f^2}{3A_x}$$

Also, by defining a fractional flange area, $f = A_f/A_x$, (13.1.24) becomes

$$\sigma_{a,T} = \frac{\pi^2 Ef(1-f)}{1+2f}\left(\frac{b_f}{a}\right)^2 \qquad (13.1.25)$$

This equation reveals the strengthening effect of the flange width b_f, and the equally important adverse effect of the panel length a. Note that b_f cannot be increased indefinitely because of the possibility of flange buckling. This elemental buckling is avoided by maintaining an adequate thickness-to-width ratio. From Fig. 12.5 the buckling stress of a flange is approximately $0.43\pi^2 D/(b_f^2 t_f)$ or $0.39E(t_f/b_f)^2$. Hence, in order for this value not to fall below, say, $2\sigma_Y$, the flange proportion should not exceed $b_f/t_f = 14$ for mild steel, and smaller values for high yield steels. Most rolled sections are sufficiently sturdy to preclude elemental buckling, but careful attention is warranted for fabricated sections.

13.2 TRANSVERSELY STIFFENED PANELS

The panel geometry for this case is shown in Fig. 13.8. The minimum size of transverse stiffeners for plates loaded in uniaxial compression has been defined by Timoshenko and Gere [2] for one, two, or three equally spaced stiffeners, and by Klitchieff [11] for any number of stiffeners. The stiffeners as sized provide a nodal line for the buckled plate and thus prohibit overall buckling of the stiffened panel. The strength of the stiffened panel is then determined by the buckling strength of the plate between stiffeners. The required minimum value of γ_y given by Klitchieff is

$$\gamma_y = \frac{(4N_L^2 - 1)[(N_L^2 - 1)^2 - 2(N_L^2 + 1)\kappa + \kappa^2]}{2(5N_L^2 + 1 - \kappa)\Pi^4} \qquad (13.2.1)$$

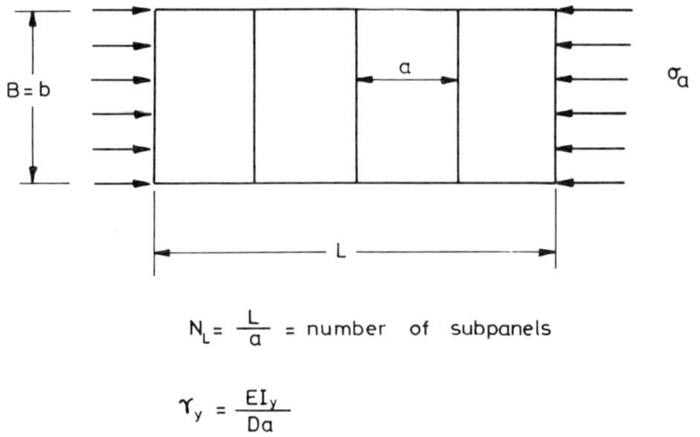

Figure 13.8 Transversely stiffened panel.

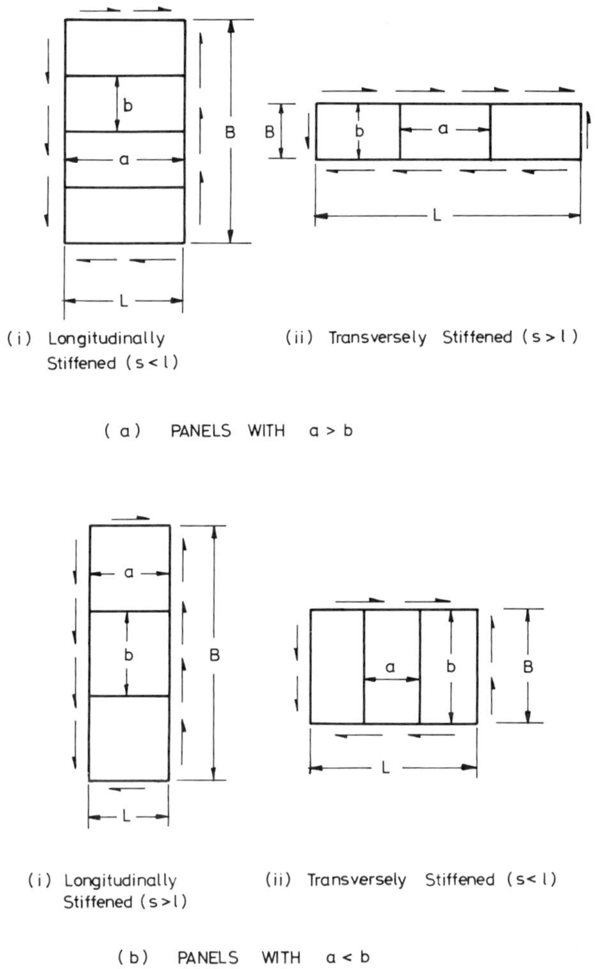

Figure 13.9 Variations in subpanel aspect ratio.

where
$$\gamma_y = \frac{EI_y}{Da}$$
$$\kappa = \alpha^2\Pi^2$$
$$\Pi = \frac{L}{B} = \frac{L}{b}$$
$$N_L = \text{number of plate panels} = \frac{L}{a}$$
$$N_L - 1 = \text{number of stiffeners}$$

and EI_y = flexural rigidity of one transverse stiffener, including a plate flange of full width a. The full width is used because the foregoing equation deals with the occurrence of overall buckling *before* local plate buckling; see also the discussion following (13.1.8).

13.3 BUCKLING DUE TO SHEAR

Figure 13.9 shows various types of stiffened panels subjected to a uniform shear loading. If s and l denote the stiffener spacing and stiffener length, then $s = b$ or a, and $l = a$ or b, depending on whether the panel is (1) longitudinally or (2) transversely stiffened, respectively. For the present we will assume that $a > b$, as in Fig. 13.9*a*. If the stiffeners have sufficient rigidity to remain straight until the plate elements buckle, then the critical shear stress is $\tau_{cr} = K_s E(t/b)^2$, where K_s is given in Fig. 12.11*b* as a function of the aspect ratio and the degree of restraint at the edges. If the stiffeners do not have sufficient rigidity then the panel will undergo overall buckling at a smaller value of applied shear stress. The expression for τ_{cr} can still have the foregoing form providing that we obtain a general expression for K_s that accounts for the rigidity of the stiffeners.

When the stiffeners are relatively weak flexurally and closely spaced, the buckle pattern becomes independent of stiffener spacing, and the analysis follows that for orthotropic plates in shear, in which the general form of the buckling stress equation is $\tau_{cr} = k_s(\pi^2 D/l^2 t)$. If we define the bending stiffness parameter γ_s as either EI_x/Db or EI_y/Da for (1) or (2) respectively, then the relationship between γ_s and the plate buckling coefficient k_s is given in the accompanying table from Ref. 12.

γ_s	0	2	5	20	50	100	200
k_s	5.34	10.34	16.07	37.14	68.99	112.2	184.6

The problem was investigated in more detail by Timoshenko and Gere [2] for the cases of one and two stiffeners, using an energy approach. For one stiffener (say in the transverse or y-direction, such that $a = L/2$) the critical value of shear stress is

$$\tau_{cr} = \frac{9\pi^2 D(1 + \Pi^2)^2}{32BLt}\sqrt{1 + \frac{2\gamma_y \Pi^4}{(1 + \Pi^2)^2}}$$

where $\gamma_s = \gamma_y = EI_y/Da$ and $\Pi = L/B$. For a single longitudinal stiffener, $\gamma_s = \gamma_x = EI_x/Db$ and $1/\Pi$ replaces Π.

Rockey and Cook [13] presented a solution for the opposite case, when the number of stiffeners is large. From these results and from some further numerical solutions it has been possible to obtain an empirical expression for K_s that covers the full range of panel geometry. The values from the expression do not depart from the numerical results by more than 8% and the discrepancy is on the conservative side. For the cases shown in Fig. 13.9*a*, in which the subpanel aspect ratio a/b is greater than 1.0, the complete expression for τ_{cr} is

$$\tau_{cr} = K_s E\left(\frac{t}{b}\right)^2 \qquad (a \geqslant b) \qquad (13.3.1)$$

in which

$$K_s = 4.5\left[\left(\frac{b}{a}\right)^2 + \frac{1}{N^2} + \left(\frac{N^2 - 1}{N^2}\right)\left(\frac{\omega}{1 + \omega}\right)^r\right]$$

in which N = number of subpanels

$$= \begin{cases} B/b \text{ for longitudinal stiffening} \\ L/a \text{ for transverse stiffening} \end{cases}$$

$$\omega = \frac{I_{se}}{lt^3}$$

I_{se} = moment of inertia of a section consisting of the stiffener and a plate flange of effective width $s/2$.

$$r = \begin{cases} 1 - 0.75(s/l) \text{ for longitudinal stiffening} \\ 0.25 \text{ for transverse stiffening} \end{cases}$$

The numerical solutions showed that when the stiffener spacing exceeds the stiffener length (i.e., when $s/l > 1$) the effect of stiffener size remains the same as for a square subpanel ($s = l$). Since we are here dealing with the case when $a \geqslant b$, transversely stiffened panels will have $s/l \geqslant 1$, and hence for such panels, r is set to the value corresponding to $s/l = 1$, namely 0.25.

For the panel geometry shown in Fig. 13.9*b*, in which the subpanel aspect ratio is less than 1.0, a and b must be interchanged in the foregoing equations because a shear load has no directionality and therefore in the analysis of shear buckling the subpanel aspect ratio is always taken as being greater than or equal to 1.0. Therefore, whenever $a < b$ the equation for the shear buckling stress is

$$\tau_{cr} = K_s E\left(\frac{t}{a}\right)^2 \qquad (a < b) \qquad (13.3.2)$$

in which

$$K_s = 4.5\left[\left(\frac{a}{b}\right)^2 + \frac{1}{N^2} + \left(\frac{N^2 - 1}{N^2}\right)\left(\frac{\omega}{1 + \omega}\right)^r\right]$$

All other quantities are as defined previously except that the two expressions for r must be reversed:

$$r = \begin{cases} 0.25 \text{ for longitudinal stiffening} \\ 1 - 0.75(s/l) \text{ for transverse stiffening} \end{cases}$$

13.4 BUCKLING OF LONGITUDINALLY STIFFENED PANELS UNDER COMBINED COMPRESSION AND SHEAR

Analytical and experimental results on the buckling behavior of stiffened plates under combined compression and shear are relatively scarce. Recourse is therefore usually had to data for unstiffened plates supplemented with whatever data are available for the type of longitudinally stiffened panels used in ships, such as that shown in Fig. 13.10. The case of unstiffened rectangular plates under combined compressive and shear stresses was dealt with in Section 12.4, which for $\alpha \geqslant 1$ gave the parabolic interaction formula of (12.4.5) which is

$$R_x + R_s^2 = 1 \quad (\alpha \geqslant 1) \qquad (13.4.1)$$

where $R_x = \dfrac{\sigma_a}{(\sigma_a)_{cr}}$

= ratio of compressive stress when buckling occurs in combined shear and direct stress to compressive stress when buckling occurs in pure compression

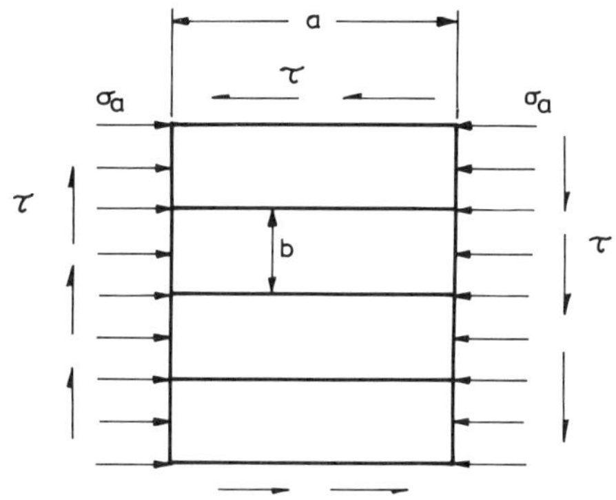

Figure 13.10 Longitudinally stiffened panel under combined compression and shear stresses.

and $R_s = \dfrac{\tau}{\tau_{cr}}$

= ratio of shear stress when buckling occurs in combined shear and direct stress to shear stress when buckling occurs in pure shear

Harris and Pifko [14] obtained a finite element solution for the buckling of infinitely wide longitudinally stiffened panels under compression and shearing stresses. The stiffeners were assumed to have both bending and torsional stiffness, and the grid refinement used was judged to be adequate to ensure accurate results. In Fig. 13.11 the results are compared to the parabolic expression of (13.4.1). Except for the case of assumed large torsional stiffness ratio ($GJ/bD = 10^6$) the analytical points follow the parabolic relationship very well.

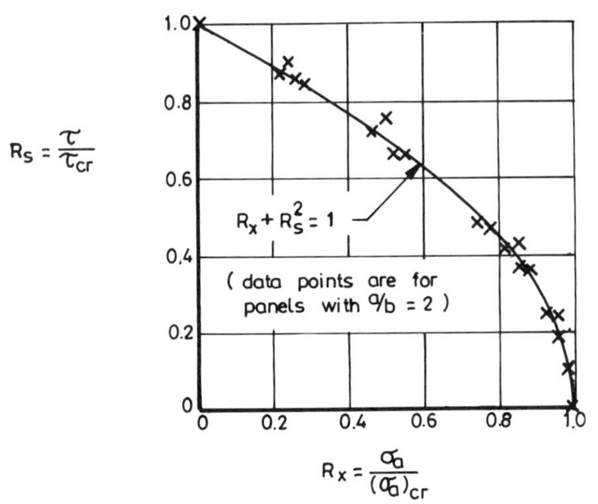

Figure 13.11 Interaction relation for longitudinally stiffened panel under combined compression and shearing stresses.

From the work of Section 13.1 we have (13.1.14) which, together with (13.1.7), gives $(\sigma_a)_{cr}$ for longitudinally stiffened panels of any panel aspect ratio Π, stiffener spacing b, and stiffener rigidity γ_x. Likewise, (13.3.1) and (13.3.2) give the value of τ_{cr} for uniaxially stiffened panels of any value of Π, b, and ω (which is an alternative measure of stiffener rigidity). Therefore, these general expressions can be used to calculate the denominators of the strength ratios in the interaction formula given in (13.4.1).

In fact, we can go a few steps further, and make some allowance for the presence of a transverse compressive stress σ_{ay} and in-plane bending stress σ_b. First, in regard to transverse compression, (13.2.1) gives the minimum stiffener rigidity at which plate buckling occurs before stiffener buckling under transverse compression. A numerical parametric study of this equation indicates that for most proportions of stiffened panels the stiffeners exceed this rigidity. This is nearly always true if the panel is intended to carry a primary longitudinal compressive stress, σ_{ax}, and it is virtually guaranteed if the panel must withstand a significant lateral pressure load. Therefore, the main effect of a transverse compressive stress is to promote plate buckling between the stiffeners. That is, the interaction of σ_{ay} and τ is mainly at the plate level and this form of interaction is described by (12.4.7), with the in-plane bending term omitted. The equation becomes

$$\frac{0.625\left(1+\dfrac{0.6}{\alpha}\right)R_y}{1-0.625R_x}+\frac{R_s^2}{1-R_x}=1 \quad (\alpha \geqslant 1) \tag{13.4.2}$$

in which $R_y = \dfrac{\sigma_{ay}}{(\sigma_{ay})_{cr}}$.

Since σ_{ay} interacts at the plate level the value of $(\sigma_{ay})_{cr}$ is obtained from (12.1.13), with a and b interchanged.

This interaction formula is a generalization of (13.4.1) which Harris and Pifko verified for stiffened panels, and it reduces to (13.4.1) when $R_y = 0$.

The full version of (12.4.7) includes the effect of in-plane bending stress, but this loading cannot be associated exclusively with plate buckling; in its compressive portion it is similar to σ_{ax}, the only difference being that it starts at a maximum value σ_b at one edge and attenuates linearly across the panel. In a uniaxially stiffened panel each stiffener is essentially an independent column. The column which has the largest compressive stress will buckle first; this will throw more load onto the others, causing the next most heavily loaded stiffener to buckle and continuing in a rapid chain reaction which can scarcely be distinguished from simultaneous buckling of all stiffeners. Therefore, the safest and easiest way of handling an in-plane bending stress of maximum value σ_b is to simply add σ_b to σ_{ax}. The interaction formula of (13.4.2) then becomes

$$\frac{0.625\left(1+\dfrac{0.6}{\alpha}\right)R_y}{1-0.625\,R_{x+b}}+\frac{R_s^2}{1-R_{x+b}}=1 \quad (\alpha \geqslant 1) \tag{13.4.3}$$

in which

$$R_{x+b}=\frac{\sigma_{ax}+\sigma_b}{(\sigma_{ax})_{cr}}$$

$$R_y=\frac{\sigma_{ay}}{(\sigma_{ay})_{cr}}$$

$$R_s=\frac{\tau}{\tau_{cr}}$$

The value of $(\sigma_{ax})_{cr}$ is obtained from (13.1.14) and (13.1.7), τ_{cr} is given by either (13.3.1) or (13.3.2), depending on a/b, and $(\sigma_{ay})_{cr}$ is given by

$$\left.\begin{aligned}(\sigma_{ay})_{cr}&=\overline{K}E\left(\frac{t}{b}\right)^2\\ \text{in which}\quad \overline{K}&=0.905\left[1+\frac{1}{\alpha^2}\right]^2\end{aligned}\right\} \tag{13.4.4}$$

The foregoing approach lacks experimental confirmation but seems reasonable and should be sufficiently accurate for panels of normal proportions, as long as R_{ay} and σ_b/σ_{ax} are not large (each less than, say, 0.3) as is usually the case.

13.5 BUCKLING OF CROSS-STIFFENED PANELS

When the panel is large it will generally be necessary to provide stiffening in both directions. In general the relative proportions of the stiffeners will depend on the overall panel aspect ratio, the direction and size of the primary in-plane compressive load σ_{ax}, and the size of the secondary compressive load σ_{ay}, if there is one. As mentioned earlier, if there is a lateral load it is likely that the scantlings which are required to provide adequate bending strength will preclude elastic buckling, and the ultimate compressive strength would be calculated by the methods to be presented in Chapter 14. But, as also mentioned earlier, the elastic analysis provides a relatively simple method for testing whether this is true.

In cross-stiffened panels which are designed to carry in-plane loads there is usually a hierarchy of structural elements. In the first place, because of its larger cross-sectional area, the plating takes most of the in-plane compressive load. Next are the stiffeners, which carry most of the lateral load and which also are required to stiffen and stabilize the plating so that it can carry the in-plane load. Therefore wherever practicable the stiffeners should be oriented parallel to the primary in-plane load, which is nearly always the longitudinal direction; such stiffeners are commonly referred to as "longitudinals." Finally there are the transverse members, whose main role is to provide nondeflecting intermediate supports for the longitudinals; the latter are acting as columns and the most efficient way of strengthening a column is to give it intermediate support, thus shortening its span. In general the transverse members are not expected to provide rotational restraint because this is much more difficult and the benefits are not commensurate. Because they must provide (nearly) undeflecting support to the longitudinals the transverse members must usually have a larger rigidity; that is, $I_y > I_x$. Therefore they usually have deeper webs than the longitudinals and for this reason they are often referred to as "web frames" or "deep webs." Another reason for making them deeper is to simplify the construction by permitting one set of members to pierce the webs of the other set, leaving the flanges of both undisturbed.

If the transverse members are not rigid enough the panel may undergo "gross panel" buckling, in which the transverses buckle with the longitudinals. This is also referred to as "grillage" buckling to indicate that both sets of stiffeners are involved. If, on the other hand, the transverses are sufficiently rigid then the stiffened panels between them are ordinary, simply supported, longitudinally stiffened panels, and can be analyzed by the methods of the previous sections. Thus the provision of cross-stiffening introduces two new quantities to be calculated:

1. The minimum rigidity of the transverses to provide nondeflecting support to the longitudinals.
2. Alternatively, if large transverse stiffeners are undesirable or impractical, the value of applied stress $(\sigma_a)_{cr,gp}$ at which gross panel buckling would occur.

Minimum Transverse Rigidity to Prevent Gross Panel Buckling (Uniaxial Load)

Johnston [15] derived an expression for estimating the required stiffness of transverse stiffeners from a consideration of the buckling of columns having elastic springs at evenly spaced intermediate points which restrain the lateral deflection at those points. Johnston's results are based on the work of Timoshenko and Gere [2] who showed that for a single column having multiple spring supports, the required spring constant K_y for the supports to remain undeflected and to constitute nodal points is given by:

$$K_y = \frac{NP}{CL} \tag{13.5.1}$$

where

$$P = \frac{N^2 \pi^2 EI_x}{L^2}$$

and EI_x = flexural stiffness of column
N = number of spans
L = total length of column

C is a parameter that depends on N, and that decreases from 0.5 for $N = 2$ to 0.25 for infinitely large N. It is given approximately by

$$C = 0.25 + \frac{2}{N^3} \tag{13.5.2}$$

In the case of a cross-stiffened panel (Fig. 13.12), the longitudinal stiffeners act as columns that are elastically restrained by the transverse stiffeners. Assuming that the loading from the longitudinal stiffener to the transverse stiffener is proportional to the deflection of the latter, the spring constant for each column support can be estimated. For example, for the limit case of an infinite number of small longitudinals, the deflected shape of the transverses is a half sine wave and the spring constant per unit width is

$$k_y = \frac{\pi^4 EI_y}{B^4} \tag{13.5.3}$$

For this case the spring constant K_y which is required at each stiffener must also be converted to a per-unit-width value by dividing by the stiffener spacing b. Thus from (13.5.1) the required stiffness per unit width is

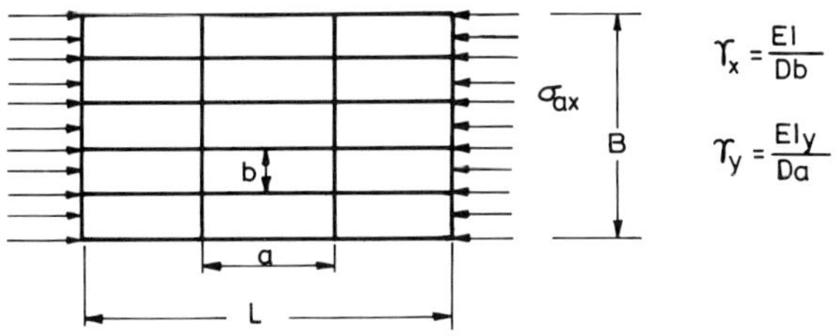

Figure 13.12 Cross-stiffened panel.

$$k_y = \frac{K_y}{b} = \frac{NP}{bCL} \tag{13.5.4}$$

Equating the two values of k_y and inserting the value given for P yields

$$\frac{b\,EI_y}{B\,EI_x} = \frac{N^3}{\pi^2 C \left(\frac{L}{B}\right)^3}$$

$$= \frac{B^3}{\pi^2 C a^3}$$

Therefore, in panels containing a large number of longitudinal stiffeners the minimum rigidity $\gamma_y (= EI_y/Da)$ of the transverse members to prevent gross panel buckling is

$$\frac{\gamma_y}{\gamma_x} = \frac{B^4}{\pi^2 C a^4} \tag{13.5.5}$$

in which C is given by (13.5.2). Timoshenko and Gere presented solutions for several particular cases involving a specific number of longitudinals and these are given in Table 13.1. Johnston showed that a general expression which covers all of these cases with satisfactory accuracy is

$$\frac{\gamma_y}{\gamma_x} = \frac{B^4}{\pi^2 C a^4}\left(1 + \frac{1}{p}\right) \tag{13.5.6}$$

where p is the number of longitudinal stiffeners, and the other parameters are as defined previously (see also Fig. 13.12). As long as the relative rigidity of the transverse stiffener exceeds the value given by

TABLE 13.1

Number of Longitudinal Stiffeners	Minimum Transverse Rigidity Ratio γ_y/γ_x to Prevent Gross Panel Buckling
1	$\frac{0.206}{C}\left(\frac{B}{a}\right)^4$
2	$\frac{0.152}{C}\left(\frac{B}{a}\right)^4$
4	$\frac{0.133}{C}\left(\frac{B}{a}\right)^4$

(13.5.6) the compressive strength of the cross-stiffened panel is determined by the compressive strength of the longitudinally stiffened panels between transverse stiffeners.

Gross Panel Buckling—Orthotropic Plate Approach

As shown in Chapter 10, panels that are stiffened in two orthogonal directions may be idealized as orthotropic plates by "smearing" the bending rigidity of the stiffeners over the region of the plating. Clearly this approach is satisfactory only when the stiffeners are uniform and closely spaced in each direction. As noted in the introduction, this approach can also be used for panels which are stiffened in only one direction, but in this case the stiffeners are much more independent and, unless they are very closely spaced the discrete beam approach is preferable. The equations and solutions for unidirectional stiffened panels can be obtained from those presented in this section simply by setting D_x or D_y equal to D.

In the first part of Chapter 10, in the discussion of the use of orthotropic plate theory for the analysis of stiffened panels, six main points were made and it is recommended that the reader review these before proceeding further. Also, of the three conclusions drawn from those six points, the first applies equally well here, for in-plane loads, as it did for lateral loads; that is, that for singly plated cross-stiffened panels the discrete beam approach is more versatile and more accurate than the orthotropic plate approach. The discrete beam method will be presented in the next section. But the orthotropic plate approach remains useful for hand calculations and for obtaining approximate solutions.

Much work has been done regarding the elastic buckling of stiffened panels, particularly in the aircraft field. Solutions have been achieved for a wide variety of cases and at all three levels of accuracy: linear small-deflection theory, corrected small-deflection theory, and large-deflection theory. However, as explained at the beginning of this chapter, because of the lateral loads and the general sturdiness of ship panels, the buckling of such panels is usually inelastic. Elastic gross panel buckling can occur only in lightweight panels (thin plating; slender, closely spaced stiffeners). For panels of this type, elastic gross panel buckling should be investigated as accurately as possible because it constitutes collapse (and usually a quite undesirable type of collapse). For best results a large deflection method which accounts for the interaction between lateral and in-plane loads should be used. For this purpose Mansour [16] has presented a comprehensive and convenient set of charts which give deflection, bending moment, buckling stress, and effective width. However, since this type of ultimate strength analysis is only relevant for lightweight panels it will not be discussed further, either here or in Chapter 14 which deals with the ultimate strength of stiffened panels.

Also, the present chapter deals only with singly plated panels because, as mentioned in point (2) of Section 10.1, for double wall panels the elastic buckling stress far exceeds the yield stress and therefore has little meaning, even as a reference parameter.

Because the elastic gross panel buckling stress for typical ship panels exceeds the yield stress and is therefore not the actual collapse stress, but merely a parameter which represents the panel characteristics, there is no need for great accuracy in its calculation. Hence it is usually sufficient to use small-deflection theory. As noted earlier, most ship panels are sufficiently sturdy that by the time a lateral load is large enough to produce any significant lateral deflection, it would already have produced extensive yielding in the panel, thus rendering an elastic analysis invalid. However, if the panel is slender enough to undergo a significant elastic lateral deflection, either due to panel eccentricity or to a large lateral load, then the corrected small-deflection approach is sufficient. Mansour [17], Falconer and Chapman [18], and Chapman and Slatford [19] have presented suitable methods.

In all of its basic aspects the buckling of orthotropic plates is similar to that of isotropic plates. For a uniform compressive distributed load N_x (positive if compressive) the governing differential equation of small deflection orthotropic plate theory is

$$D_x \frac{\partial^4 w}{\partial x^4} + 2H \frac{\partial^4 w}{\partial x^2 \partial y^2} + D_y \frac{\partial^4 w}{\partial y^4} = -N_x \frac{\partial^2 w}{\partial x^2} \tag{13.5.7}$$

It is often more appropriate or more convenient to express the load as an applied stress σ_{ax} (e.g., the hull girder stress). The relationship between the two is $N_x = \sigma_{ax}(t + A_x/b)$, where A_x is the area of each longitudinal stiffener.

In the foregoing equation D_x and D_y are the bending rigidities and H is the torsional rigidity of the orthotropic plate. As shown in Chapter 10, orthotropic plate solutions are most conveniently expressed in terms of two nondimensional parameters

$$\rho = \frac{L}{B}\left(\frac{D_y}{D_x}\right)^{1/4} = \text{virtual aspect ratio}$$

$$\eta = \frac{H}{\sqrt{D_x D_y}} = \text{torsional stiffness parameter} \tag{13.5.8}$$

For simply supported edges the solution to (13.5.7) is

$$(N_x)_{cr,gp} = k_0 \frac{\pi^2\sqrt{D_x D_y}}{B^2} \tag{13.5.9}$$

where
$$k_0 = \frac{m^2}{\rho^2} + 2\eta + \frac{\rho^2}{m^2} \tag{13.5.10}$$

and m is the number of half-waves in the x-direction. Note that all of the formulas presented in this section are based on the assumption that the panel buckles into only one half-wave in the y-direction.

In terms of an applied stress σ_{ax} the solution is

$$(\sigma_{ax})_{cr,gp} = k_0 \frac{\pi^2\sqrt{D_x D_y}}{B^2\left(t + \dfrac{A_x}{B}\right)} \tag{13.5.11}$$

The close parallel with the isotropic case can be seen by comparing these equations with (12.1.5) and (12.1.6). If η is set to unity (13.5.10) becomes

$$k_0 = \left(\frac{m}{\rho} + \frac{\rho}{m}\right)^2$$

which is the same as (12.1.6) with the virtual aspect ratio ρ in place of the isotropic aspect ratio a/b.

The primary buckling load corresponds to whichever value of m gives the smallest value of k_0 in (13.5.10). We can estimate this by temporarily regarding m as a continuous variable, differentiating (13.5.10) with respect to m and setting this equal to zero. The result is $m = \rho$; that is, the number of half-waves in the x-direction is the integer nearest to ρ, the virtual aspect ratio of the panel. This also is completely analogous to the situation with isotropic plates, for which the primary buckling mode is $m = \alpha$ (buckling into square subpanels). Thus, if $D_x \cong D_y$, a cross-stiffened panel will also tend to buckle into square subpanels, but if the two flexural rigidities are unequal the primary buckling mode will be some other pattern. Obviously, if $\rho < 1$, (a short panel and/or relatively rigid longitudinal stiffeners) the primary mode will be $m = 1$.

In orthotropic plate theory the bending rigidity of the stiffeners in both directions is "smeared" into the plating. This implicitly assumes that in the buckled deformation pattern all of the stiffeners are participating; that is, they are all undergoing flexure. If the primary mode is such that $m > 1$ and if N_L, the number of subpanels in the x-direction, is a multiple of m (i.e., if $N_L = m$, $2m$, $3m$, etc., where m is the nearest integer to ρ), then $m - 1$ of the transverse stiffeners are located at or near node lines and will not undergo any significant flexure. The extreme case is $N_L = m$ because then *all* of the transverse stiffeners occur at node lines. In any of these cases the elastic buckling stress predicted by orthotropic plate theory will be larger than the true value, and the error is important because it lies on the optimistic (nonconservative) side. Therefore when using the buckling formulas which are presented below it is very important to check whether N_L is a multiple of either of the integers m_1, m_2 above and below ρ. If so (if, for example, $N_L = jm_1$) then the subpanel of length L/m_1 should also be investigated, again using the formulas presented in this section. That is, ρ should be calculated for the reduced length and the elastic buckling stress evaluated for this value of ρ. The true buckling stress is then the lower of the two calculated values.

For singly plated panels the neutral axis in each direction is close to the plating and therefore, as shown by Schultz [20], the effect of Poisson's ratio may be ignored. Hence the flexural rigidities are

$$D_x = \frac{EI_x}{b} = Ei_x \qquad \text{and} \qquad D_y = \frac{EI_y}{a} = Ei_y \tag{13.5.12}$$

where I_x and I_y are the moments of inertia of the stiffeners, including a plate flange of width b and a respectively. The full width is used because we want to guard against overall buckling occurring before local plate buckling and therefore we want the value of $(\sigma_{ax})_{cr}$ with the plating fully effective.

The torsional rigidity is

$$H = D_{xy} = \frac{1}{6} Gt^3 + \frac{GJ_x}{b} + \frac{GJ_y}{a} \quad (13.5.13)$$

where J_x and J_y are the St. Venant torsion constants of the stiffeners which for open sections is given by

$$J = \frac{1}{3} \sum_i l_i t_i^3$$

in which t_i and l_i are the thickness and width of the stiffener flange and web.

The following cases are dealt with in this section:

1. Uniaxial load; ends simply supported, sides elastically restrained.
2. Uniaxial load; sides clamped, ends simply supported or clamped.
3. Biaxial load; sides simply supported, ends simply supported or clamped.

Case 1 is based on work by Schultz [20]; cases 2 and 3 are based on work by Wittrick [21].

Uniaxial Load; Ends Simply Supported, Sides Elastically Restrained

The buckling stress is given by:

$$(N_x)_{cr,gp} = (k + 2\eta) \frac{\pi^2 E \sqrt{i_x i_y}}{B^2} \quad (13.5.14)$$

The coefficient k is given by Bleich [5] in the form of curves. For computer-aided design a mathematical expression is more suitable, and for this purpose the following expression has been developed which gives a satisfactory representation of these curves.

$$k = \frac{\rho^2}{0.4 + \rho^2}\left(2 + \frac{0.5}{0.2 + C_{fy}}\right) \quad (13.5.15)$$

In this expression C_{fy} is a flexibility coefficient which accounts for the rotational restraint along the sides of the panel due to adjacent structure. If $\overline{K}_y$ denotes the rotational stiffness of the adjacent structure (such that a distributed edge moment m_y acting on the adjacent structure would cause it to rotate an angle of $m_y/\overline{K}_y$) then the flexibility coefficient is

$$C_{fy} = \frac{2D_y}{B\overline{K}_y}$$

Note that C_{fy} is defined inversely to the coefficient ζ for plates (page 408); for clamped edges $C_{fy} = 0$ and for simply supported edges $C_{fy} = \infty$. When the two sides are of differing flexibility, sufficient accuracy is achieved by averaging the two values of k that correspond to the values of C_{fy}.

In ship panels the adjacent structure along each side of the panel is usually another panel. For this case Schultz has derived the following expression for C_{fy}

$$C_{fy} = \frac{D_y}{\pi \overline{D}_y}\left(\frac{\overline{D}_x}{\overline{D}_y}\right)^{1/4} \xi r \quad (13.5.16)$$

where the barred symbol refers to the restraining plate. The effect of an in-plane load $\overline{N}_x$ acting on the restraining plate is reflected by r, which for a general, nonuniform compressive load is given by

$$r = \frac{1}{1 - \left(\frac{1+\psi}{2}\right)\left(\frac{1+\eta}{1+\overline{\eta}}\right)\frac{\overline{D}_x D_y}{\sqrt{D_x \overline{D}_y}}\left(\frac{\overline{B}}{B}\right)^2} \quad (13.5.17)$$

in which all barred symbols refer to the restraining panel. The parameter ψ describes the distribution of $\overline{N}_x$, as shown in Fig. 13.13. The factor ξ accounts for the torsional stiffness parameter $\overline{\eta}[=\overline{H}/(\overline{D}_x\overline{D}_y)^{1/2}]$ of the restraining panel. Its value may be obtained from Table 13.2, in which

$$\varepsilon = \frac{\pi \overline{B}}{B}\left(\frac{D_y}{\overline{D}_y}\right)^{1/4} \quad (13.5.18)$$

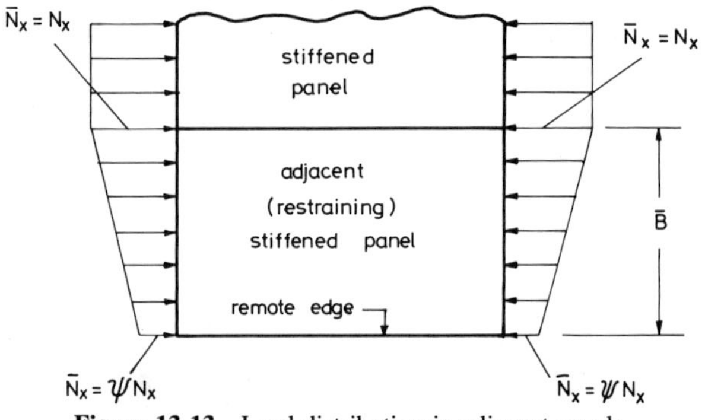

Figure 13.13 Load distribution in adjacent panels.

TABLE 13.2 EFFECT OF ORTHOTROPIC PARAMETERS AND REMOTE EDGE SUPPORT OF RESTRAINING PANEL

Remote edge simply supported

$\overline{\eta} > 1$

$$\xi = \frac{1}{\sqrt{\dfrac{\overline{H}^2}{\overline{D}_y^2} - \dfrac{\overline{D}_x}{\overline{D}_y}}} (\rho_1^* \coth \varepsilon\rho_1^* - \rho_2^* \coth \varepsilon\rho_2^*) \qquad \rho_1^* = \sqrt{\frac{\overline{H}}{\overline{D}_y} + \sqrt{\frac{\overline{H}^2}{\overline{D}_y^2} - \frac{\overline{D}_x}{\overline{D}_y}}} \qquad \rho_2^* = \sqrt{\frac{\overline{H}}{\overline{D}_y} - \sqrt{\frac{\overline{H}^2}{\overline{D}_y^2} - \frac{\overline{D}_x}{\overline{D}_y}}}$$

$\overline{\eta} = 1$

$$\xi = \frac{\cosh \varepsilon\rho^* \sinh \varepsilon\rho^* - \varepsilon\rho^*}{\rho^* \sinh^2 \varepsilon\rho^*} \qquad \rho^* = \sqrt[4]{\frac{\overline{D}_x}{\overline{D}_y}}$$

$\overline{\eta} < 1$

$$\xi = \frac{\rho_2^* \sinh \varepsilon\rho_1^* \cosh \varepsilon\rho_1^* - \rho_1^* \sin \varepsilon\rho_2^* \cos \varepsilon\rho_2^*}{\rho_1^* \rho_2^* (\sinh \varepsilon\rho_1^* + \sin^2 \varepsilon\rho_2^*)} \qquad \rho_1^* = \sqrt{\frac{1}{2}\left(\sqrt{\frac{\overline{D}_x}{\overline{D}_y}} + \frac{\overline{H}}{\overline{D}_y}\right)} \qquad \rho_2^* = \sqrt{\frac{1}{2}\left(\sqrt{\frac{\overline{D}_x}{\overline{D}_y}} - \frac{\overline{H}}{\overline{D}_y}\right)}$$

Remote edge clamped

$\overline{\eta} > 1$

$$\xi = \frac{2\rho_1^*\rho_2^* \cosh \varepsilon\rho_1^* \cosh \varepsilon\rho_2^* - \sinh \varepsilon\rho_1^* \sinh \varepsilon\rho_2^*(\rho_1^{*2} + \rho_2^{*2}) - 2\rho_1^*\rho_2^*}{\sqrt{\dfrac{\overline{H}^2}{\overline{D}_y^2} - \dfrac{\overline{D}_x}{\overline{D}_y}}(\rho_2^* \sinh \varepsilon\rho_1^* \cosh \varepsilon\rho_2^* - \rho_1^* \sinh \varepsilon\rho_2^* \cosh \varepsilon\rho_1^*)} \qquad \rho_1^* = \sqrt{\frac{\overline{H}}{\overline{D}_y} + \sqrt{\frac{\overline{H}^2}{\overline{D}_y^2} - \frac{\overline{D}_x}{\overline{D}_y}}} \qquad \rho_2^* = \sqrt{\frac{\overline{H}}{\overline{D}_y} - \sqrt{\frac{\overline{H}^2}{\overline{D}_y^2} - \frac{\overline{D}_x}{\overline{D}_y}}}$$

$\overline{\eta} = 1$

$$\xi = \frac{1}{\rho^*} \frac{\sinh^2 \varepsilon\rho^* - (\varepsilon\rho^*)^2}{\sinh \varepsilon\rho^* \cosh \varepsilon\rho^* - \varepsilon\rho^*} \qquad \rho^* = \sqrt[4]{\frac{\overline{D}_x}{\overline{D}_y}}$$

$\overline{\eta} < 1$

$$\xi = \frac{1}{\rho_1^*\rho_2^*} \frac{\rho_2^{*2} \sinh^2 \varepsilon\rho_1^* - \rho_1^{*2} \sin \varepsilon\rho_2^*}{\rho_2^* \sinh \varepsilon\rho_1^* \cosh \varepsilon\rho_1^* - \rho_1^* \sin \varepsilon\rho_2^* \cos \varepsilon\rho_2^*} \qquad \rho_1^* = \sqrt{\frac{1}{2}\left(\sqrt{\frac{\overline{D}_x}{\overline{D}_y}} + \frac{\overline{H}}{\overline{D}_y}\right)} \qquad \rho_2^* = \sqrt{\frac{1}{2}\left(\sqrt{\frac{\overline{D}_y}{\overline{D}_y}} - \frac{\overline{H}}{\overline{D}_y}\right)}$$

For a simply supported panel (13.5.14) reduces to

$$(N_x)_{cr,gp} = \left(\frac{2\rho^2}{0.4 + \rho^2} + 2\eta\right) \frac{\pi^2 E \sqrt{i_x i_y}}{B^2} \qquad (13.5.19)$$

UNIAXIAL LOAD; SIDES CLAMPED, ENDS SIMPLY SUPPORTED OR CLAMPED

Wittrick [21] has derived an approximate expression for $(N_x)_{cr,gp}$ utilizing the fact that the principal term, k, in the orthotropic buckling coefficient k_0 and the virtual aspect ratio ρ have exactly the same relationship as the buckling coefficient k and the ordinary aspect ratio $\alpha = a/b$ of isotropic plates. Therefore, all of the curves of Fig. 12.5*a* can be used to obtain k for an orthotropic panel, with the value of ρ being used in place of α. For clamped sides the expression for $(\sigma_{ax})_{cr}$ is

$$(N_x)_{cr,gp} = [k - c(1 - \eta)] \frac{\pi^2 E \sqrt{i_x i_y}}{B^2} \qquad (13.5.20)$$

in which k is obtained from one of the two curves in Fig. 12.5*a* labeled "A_1, A_2 clamped." The value of c is 2.40 for simply supported ends and 2.46 for clamped ends.

BIAXIAL LOAD; SIDES SIMPLY SUPPORTED, ENDS SIMPLY SUPPORTED OR CLAMPED

If the sides are simply supported the foregoing technique can be extended to the case when a tensile or compressive load N_y (positive for compression) acts on the sides of the plate. The effect of N_y is accounted for by modifying the virtual aspect ratio, which becomes

$$\rho = \frac{L}{B}\left[\frac{D_y}{D_x}\left(1 - \frac{B^2 N_y}{\pi^2 D_y}\right)\right]^{1/4} \qquad (13.5.21)$$

where N_y must be less than $\pi^2 D_y / B^2$. N_y is positive for compression. The buckling stress is then given by

$$(N_x)_{cr,gp} = \left\{\left[(k-2)\sqrt{1-\frac{B^2N_y}{\pi^2 D_y}}\right] + 2\eta\right\}\frac{\pi^2 E\sqrt{i_x i_y}}{B^2} \qquad (13.5.22)$$

with k being obtained from one of the two curves in Fig. 12.5*a* labeled "A_1, A_2 pinned."

13.6 DISCRETE BEAM METHOD FOR GROSS PANEL BUCKLING ANALYSIS

As mentioned in the introduction to this chapter, for singly plated cross-stiffened panels in which the stiffeners are of open section, an appropriate technique for buckling analysis is the discrete beam approach developed by Timoshenko and Gere [2], based on Rayleigh's energy principle that the strain energy V is always greater than or equal to the external work W done by the applied in-plane loads N_x and N_y. This means that an approximate "upper bound" buckling load can be obtained by equating the total strain energy and the external work. The buckled deformation is specified in the form of an assumed displacement function which satisfies the boundary conditions and which is able to represent the various modes of buckled deformation. Each mode is tried in turn, and the correct buckling mode is that which gives the lowest value of $(N_x)_{cr}$. Although not possessing the degree of generality of frame buckling analysis by matrix methods, the energy approach has the great advantage of producing an explicit analytical solution, and this is preferable to a numerical solution because it indicates the precise role and relative importance of each structural design variable. Moreover, from the computational standpoint the energy approach is much more economical, a factor which is of great importance in any structural optimization and/or computer-aided design scheme. The energy approach is also superior to orthotropic plate theory, primarily because the latter involves more difficulty in solving equations for the general case of partial rotational constraints along the boundaries of the stiffened panel. The orthotropic plate theory also suffers a significant loss of accuracy when the number of transverse stiffeners is about three or less, which is a serious limitation.

Rotational Restraint due to Adjacent Structure

Faulkner [22] extended the energy method to take into account the effect of torsional rigidity of stiffeners and elastic rotational restraint on pairs of opposite edges of the panel. In this section we present Faulkner's method for estimating the rotational restraint which is given to the panel by the adjacent structure. The method is quite general and includes the effects of compression on adjacent structure and even allows for the case where the adjacent structure becomes unstable before the panel under consideration. The rotational restraint at each end of each stiffener in the panel is considered to be a linear elastic rotational spring. If the spring constant of the supporting structure at the ends of the longitudinal stiffeners is denoted as C_x (in units of moment/radian per stiffener) and is the same for both ends, then a dimensionless restraint parameter ζ_x can be defined as

$$\zeta_x = \frac{C_x L}{EI_x} \qquad (13.6.1a)$$

In a similar manner, the dimensionless restraint parameter ζ_y is

$$\zeta_y = \frac{C_y B}{EI_y} \qquad (13.6.1b)$$

The spring constants C_x and C_y depend on a number of characteristics of the adjacent structure, but the flexural stiffnesses of adjacent beams which are connected to the stiffener ends are by far the most influential. Thus by concentrating on the flexural stiffnesses of adjacent beams, it is possible to quickly determine the restraint parameters ζ_x and ζ_y to a sufficient degree of accuracy.

Figure 13.14 illustrates the beams in the adjacent structure which lie in the plane of bending of a typical panel stiffener. If end i of the "second remove" beam ji (considered in isolation) is acted on by a moment M_i, it will rotate by an amount θ_i which is related to the beam characteristics I_j and l_j and the elastic rotational spring constant C_j (connected to end j) as given by

$$\frac{M_i}{\theta_i} = 4\bar{k}_j\frac{EI_j}{l_j} \qquad (13.6.2a)$$

where

$$\bar{k}_j = \frac{k_j+3}{k_j+4} \qquad (13.6.2b)$$

$$k_j = \frac{C_j l_j}{EI_j} \qquad (13.6.2c)$$

If the joint i is assumed rigid, which is reasonable for ship-type welded connections, then any rotation θ_i will be shared by all N_i second remove members meeting at joint i. It thus follows from statics that the total

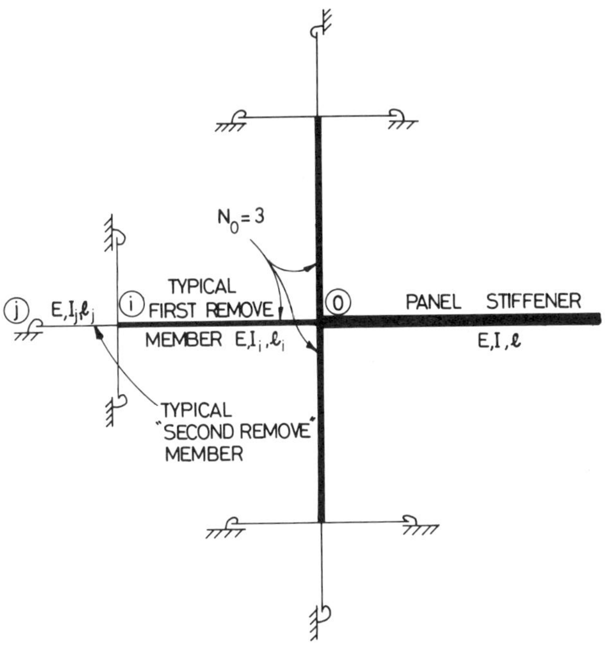

Figure 13.14

moment exerted by all N_i second remove members on the end i of the typical first remove beam iO is the sum

$$M_i = 4 \sum_{j=1}^{N_i} \bar{k}_j \frac{EI_j}{l_j} \theta_i$$

The rotational spring stiffness $C_i = M_i / \theta_i$ arising from the presence of structure connected to end i of beam iO is therefore given by

$$C_i = 4 \sum_{j=1}^{N_i} \bar{k}_j \frac{EI_j}{l_j} \tag{13.6.3}$$

This process can be extended to all of the beams connected to end O of the stiffener. Thus the rotational spring stiffness C_O provided by all of the structure adjoining end O of the stiffener is equal to

$$C_O = 4 \sum_{i=1}^{N_O} \bar{k}_i \frac{EI_i}{l_i} \tag{13.6.4a}$$

where

$$\bar{k}_i = \frac{k_i + 3}{k_i + 4} \tag{13.6.4b}$$

$$k_i = \frac{C_i l_i}{EI_i} \tag{13.6.4c}$$

In this expression, N_O is the total number of first remove members connected to end 0 and C_i is given by the summation, (13.6.3), for the second remove members.

This process is very general in nature and as such need not be limited to the two levels of "removed" members discussed here and illustrated on Fig. 13.14. More than two levels may well exist in actual structures. However, an examination of (13.6.2b) indicates that $\bar{k}_j$ has only a limited range, from 0.75 to 1.0, even though this represents a transition at end j from $C = 0$ (pinned) to $C_j = \infty$ (completely clamped). Since a very wide variation in constraint at one end of a single beam causes only a small difference in the rotation at the other end, it logically follows that the influence of the boundary conditions and physical characteristics of adjacent structure on the spring stiffness at the panel end becomes weaker and weaker as the structure becomes further and further removed from the panel itself. Thus only slight inaccuracies are likely to occur when it is assumed that the boundaries of an adjacent structure can be idealized as being one, or at most, two joints away from the panel. Note that however many levels are used, some specific numerical value must always be assumed for the rotational spring constant C_j of the outermost rotational spring support, in order to commence the calculation chain.

Faulkner [22] showed that for typical ship structures $\bar{k}_j$ is about 0.9 ($k_j = 6$). If this value (or a more precise value if one is known) is assumed at second remove joints, the rotational stiffness C_O can be determined with only negligible error by using (13.6.3) and (13.6.4). However, an even simpler assumption is sufficient for most purposes. Calculations have indicated that for moderately redundant frameworks, accuracy within 5% is attainable by taking the boundary of the adjacent structure at the first remove joints and using $\bar{k}_i = 0.9$. Inserting this value into (13.6.4a) gives

$$C_O = 3.6E \sum_{i=1}^{N_O} \frac{I_i}{l_i} \tag{13.6.5}$$

and for the dimensionless restraint parameter ζ,

$$\zeta = \frac{C_O l}{EI} = 3.6 \frac{l}{I} \sum_{i=1}^{N_O} \frac{I_i}{l_i} \tag{13.6.6}$$

Buckling Formula for Panels with Open-Section Stiffeners

Adamchak [23] further improved the application of the energy method to cross-stiffened panel buckling by providing a better assumed displacement function, and it is this result which is given here. The formula neglects shear deflection because for singly plated panels of typical ship proportions this effect is negligible. Also, the formula presented herein is intended for panels with stiffeners of open cross section, in which case the effects of torsion, sideways bending, and longitudinal warping of the stiffeners are negligible. The formula presented in Ref. 23 is a more general (and much more complex) solution which includes these three effects, but Adamchak shows that for typical ship panels with stiffeners of open cross section all three of these effects are insignificant. As an illustration, in his numerical example these three effects together contributed less than 0.05% to the elastic buckling load.

Because it explicitly accounts for the precise number of beams in each direction, the formula involves several alternative subexpressions depending on the number of stiffeners in each direction (p and q), the number of half-waves in each direction (m and n), and whether the latter numbers are odd or even. If either one (say m) is even, the formula also involves a parameter γ_m which is the $(m/2)$th positive root of the equation

$$\tan\left(\frac{\gamma_m}{2}\right) = \frac{\gamma_m}{2}$$

For example, for $m = 2$, $\gamma_m = 8.9868$. Values of γ_m and γ_n corresponding to the first eight even values of m and n are tabulated in Table 13.3.

TABLE 13.3

m (or n)	γ_m (or γ_n)
2	8.98682
4	15.45051
6	21.80825
8	28.13239
10	34.44151
12	40.74261
14	47.03891
16	53.33211

For selected values of m and n the formula gives the critical value of N_x, including the effect of a transverse in-plane load N_y. The formula is

$$(N_x)_{cr,gp} = \frac{\pi^2 E}{B^2}\sqrt{\frac{I_x I_y}{ab}}\left(\frac{H_{xm}}{H_{1xm}}\right)\left(\frac{m^2}{\rho^2}F_{xm} + \frac{n^4\rho^2}{m^2}F_{yn}\frac{H_{yn}S_{xm}}{S_{yn}H_{xm}}\right) - \left(\frac{nL}{mB}\right)^2 \frac{H_{1yn}S_{xm}}{S_{yn}H_{1xm}} N_y \tag{13.6.7}$$

in which the parameters are defined as follows:

$$G_{xm} = \begin{cases} \dfrac{m\zeta_x}{\pi(m+1)^2} & m \text{ odd} \\[2ex] \dfrac{m\pi\zeta_x}{\gamma_m^2} & m \text{ even} \end{cases}$$

$$H_{xm} = \begin{cases} 1 + \dfrac{8}{\pi^2(2m+1)}\zeta_x + \dfrac{1}{m^2\pi^2}\zeta_x^2 & m \text{ odd} \\[2ex] 1 + \dfrac{8}{\gamma_m^2 - m^2\pi^2}\zeta_x + \dfrac{1}{m^2\pi^2}\zeta_x^2 & m \text{ even} \end{cases}$$

$$H_{1xm} = \begin{cases} 1 + \dfrac{8}{\pi^2(2m+1)}\zeta_x + \dfrac{1}{(m+1)^2\pi^2}\zeta_x^2 & m \text{ odd} \\[2ex] 1 + \dfrac{8}{\gamma_m^2 - m^2\pi^2}\zeta_x + \dfrac{1}{\gamma_m^2}\zeta_x^2 & m \text{ even} \end{cases}$$

$$F_{xm} = 1 + \frac{4\zeta_x}{m^2\pi^2 H_{xm}} \qquad \text{All } m$$

There are several alternative expressions for the parameter S_{xm}; the choice depends on whether m or $m + 1$ is an integer multiple of N_L, the number of subpanels in the longitudinal direction. These expressions use the symbols θ_m and β which are

$$\theta_m = \frac{m\pi}{q+1} = \frac{m\pi}{N_L} \qquad \beta_m = \frac{\gamma_m}{q+1} = \frac{\gamma_m}{N_L}$$

Case A: $m/N_L \neq$ integer, $(m+1)/N_L \neq$ integer.

$$S_{xm} = \begin{cases} 1 + \dfrac{4(1-\cos\theta_{m+1})(\sin\theta_m)G_{xm}}{N_L(1-\cos\theta_m)(\cos\theta_m - \cos\theta_{m+1})} \\ \quad + 3G_{xm}^2 & m \text{ odd} \\ 1 + \dfrac{4(1-\cos\beta_m)(\sin\theta_m)G_{xm}}{N_L(1-\cos\theta_m)(\cos\theta_m - \cos\beta_m)} \\ \quad + \dfrac{G_{xm}^2}{N_L}\Bigg[\dfrac{2}{3}q\left(\dfrac{q+5}{q+1}\right) \\ \quad - \left(\dfrac{4}{N_L}\right)\left(\dfrac{1+\cos\beta_m}{1-\cos\beta_m}\right) \\ \quad + \left(1+\dfrac{4}{\gamma_m^2}\right)(q+\cos\gamma_m) \\ \quad + \dfrac{4}{\gamma_m}\left(\dfrac{2\sin\beta_m}{1-\cos\beta_m} - \dfrac{\cos\beta_m}{\sin\beta_m}\right)\Bigg] & m \text{ even} \end{cases}$$

Case B: $m/N_L =$ integer, $(m+1)/N_L \neq$ integer

$$S_{xm} = \begin{cases} 3G_{xm}^2 & m \text{ odd} \\ \dfrac{G_{xm}^2}{N_L}\Bigg[\dfrac{2}{3}\left(\dfrac{q+5}{q+1}\right)q - \dfrac{4}{N_L}\left(\dfrac{1+\cos\beta_m}{1-\cos\beta_m}\right) \\ \quad + \left(1+\dfrac{4}{\gamma_m^2}\right)(q+\cos\gamma_m) \\ \quad + \dfrac{4}{\gamma_m}\left(\dfrac{2\sin\beta_m}{1-\cos\beta_m} - \dfrac{\cos\beta_m}{\sin\beta_m}\right)\Bigg] & m \text{ even} \end{cases}$$

Case C: $m/N_L \neq$ integer, $(m+1)/N_L =$ *odd* integer

$$S_{xm} = \begin{cases} 1 + \dfrac{8G_{xm}}{N_L \sin\theta_m} + 4G_{xm}^2 & m \text{ odd} \\ \text{(Same as Case A)} & m \text{ even} \end{cases}$$

Case D: $m/N_L \neq$ integer, $(m+1)/N_L =$ *even* integer

$$S_{xm} = \begin{cases} 1 & m \text{ odd} \\ \text{(Same as Case A} & m \text{ even} \end{cases}$$

Case E: Approximate form for m/N_L and $(m+1)/N_L \ll 1$

$$S_{xm} = \begin{cases} 1 + \dfrac{8(m+1)^2 G_{xm}}{\pi m(2m+1)} + 3G_{xm}^2 & m \text{ odd} \\ 1 + \dfrac{8\gamma_m^2 G_{xm}}{\pi m(\gamma_m^2 - m^2\pi^2)} \\ \quad + \left[\dfrac{5}{3}q(q+2) - 1\right]\dfrac{G_{xm}^2}{N_L^2} & m \text{ even} \end{cases}$$

The corrcsponding expressions for G_{yn}, H_{yn}, H_{1yn}, F_{yn}, and S_{yn} can be obtained by substituting the variables y, n, p, and B for x, m, q, and L, respectively.

The above formula is a function of the mode numbers m and n. The mode numbers of interest are those corresponding to the minimum value of the critical buckling load. By analogy with isotropic plate buckling theory, it may be shown that n is usually equal to unity. It would be rather complex to prove this for the general case, but it is relatively easy to demonstrate (not rigorously) when one of the loads predominates. For ship panels the N_x loading is usually much greater than N_y. In this case, an examination of the first term of (13.6.7) and the behavior of the various coefficients contained within it will demonstrate that the predominant influence of n comes from its appearance in the numerators. Thus it follows that the value of mode number n will always be unity for small N_y.

Even with the value of mode number n determined, the choice of the critical value of m is somewhat complex. The coefficient terms (H_{xm}/H_{1xm}), (S_{xm}/H_{1xm}), and F_{xm} are all functions of m and tend to be most sensitive at the smaller values of m, those of most practical interest. However, because n has been determined and the critical value of m for most practical ship-type panels is likely to be small, the determination of this value can be achieved simply by starting with $m = 1$ and incrementing by one until the critical value is found. For automatic computation, for which this analysis was developed, this technique imposes very little time or cost penalties.

Unequal and/or Reduced Values of ζ_x and ζ_y

The foregoing formula for gross panel buckling assumes that the value of ζ_x (or ζ_y) is the same at both ends of each stiffener and is the same for all stiffeners in the x- (or y-) direction. That is, there is only one value of ζ_x and one value of ζ_y. If this is not the case then an approximate solution can be obtained by using the Lundquist-Stowell approximation, as outlined in Section 12.2. This involves evaluating the buckling load for each value of ζ and then taking the geometric mean.

If the adjacent structure is also subject to buckling due to in-plane loading, the constraint it can offer to the panel will diminish as the loading approaches the critical value for the adjacent structure. In this situation it is appropriate to reduce ζ_y to a value ζ_{yr} according to the linear interaction formula

$$\zeta_{yr} = \zeta_y\left(1 - \frac{\overline{N}_x}{(\overline{N}_x)_{cr}}\right) \qquad (13.6.8)$$

where $\overline{N}_x$ is the axial in-plane loading on the adjacent structure and $(\overline{N}_x)_{cr}$ is the corresponding critical value assuming edges pinned. Although the development of the buckling formula theory was made on the assumption of constant linear springs, this inconsistency is of no great consequence and the adoption of the conservative (13.6.8) will, if anything, result in an underestimation of the critical buckling load. This expression is also valid for negative ζ_{yr}, which corresponds to those cases where the tendency of the adjacent structures to buckle is so strong that they are restrained by their attachment to the panel.

When the rotational restraint ζ_x is being computed, the nature of the buckling phenomenon is such that it is generally not valid to assume that a similar structure lying in the plane of compression provides any restraint. In this case, contributions should be considered only from adjacent structures which lie essentially normal to this plane and which themselves have no tendency to buckle.

REFERENCES

1. Column Research Committee of Japan, *Handbook of Structural Stability*, Corona Publishing Co., Ltd., Tokyo, 1971.
2. S. P. Timoshenko and J. M. Gere, *Theory of Elastic Stability*, 2nd Ed., McGraw-Hill, New York, 1961.
3. P. Seide, The Effect of Longitudinal Stiffeners Located on One Side of a Plate on the Compressive Buckling Stress of the Plate-Stiffener Combination, NACA TN 2873, Jan. 1953.
4. H. L. Cox and J. R. Riddel, "Buckling of a Longitudinally Stiffened Flat Panel," *Aeronaut. Q.*, **I**, Nov. 1949, pp. 225–244.
5. F. Bleich, *Buckling Strength of Metal Structures*, McGraw-Hill, New York, 1952.
6. J. M. Klitchieff, "On the Stability of Plates Reinforced by Longitudinal Ribs," *J. Appl. Mech.*, **18** (4), Dec. 1951, pp. 364–366.
7. T. von Karman, E. Sechler, and L. H. Donnell, "The Strength of Thin Plates in Compression," *Trans. ASME*, **54**, 1932, p. 53.
8. C. S. Smith, "Bending, Buckling and Vibration of Orthotropic Plate-beam Structures," *J. Ship. Res.* **12** (4), 1968, pp. 249–268.
9. M. L. Sharp, "Longitudinal Stiffeners for Compression Members," *ASCE J. Struct. Div.*, **92,** ST5, Oct. 1966.
10. W. D. Kroll, "Tables of Stiffness and Carry-Over Factor for Flat Rectangular Plates Under Compression", NACA Advanced Report No. 3K27, Nov. 1943.
11. J. M. Klitchieff, "On the Stability of Plates Reinforced by Ribs," *J. Appl. Mech.* **16** (1), March 1949, pp. 74–76.
12. M. Stein and R. W. Fralich, "Critical Shear Stress of Infinitely Long, Simply Supported Plate with Transverse Stiffeners," NACA Tech. Note 1851, 1949.
13. K. C. Rockey and I. T. Cook, "Shear Buckling of Clamped and Simply-Supported Infinitely Long Plates Reinforced by Transverse Stiffeners," *Aeronaut. Q.*, **8,** Feb. 1962, p. 41.
14. H. G. Harris and A. B. Pifko, Elastic-Plastic Buckling of Stiffened Rectangular Plates, *Proc. Symp. Appl. Finite Element Meth. Civil Eng.*, Vanderbilt University, November 1969.
15. B. G. Johnston, Ed. *Guide to Stability Design Criteria for Metal Structures*, 3rd ed., Wiley, New York, 1976.
16. A. Mansour, "Charts for Buckling and Post-Buckling Analyses of Stiffened Plates under Combined Loading," *SNAME T & R* Bull., 2–22, 1976.
17. A. Mansour, "Ship Bottom Structure under Uniform Lateral and Inplane Loads," *Schiff und Hafen*, **5,** (1967) 19 Jahrgang, pp. 323–339.
18. B. H. Falconer and J. C. Chapman, Compressive Buckling of Stiffened Plates, *The Engineer*, **195,** 1953, pp. 789–791 and pp. 822–825.
19. J. C. Chapman and J. E. Slatford, "Design of Stiffened Plating in Compression," *The Engineer*, **207,** 1959, pp. 292–294.
20. H. G. Schultz, "Sum Stabilitatsproblem elastisch eingespannter orthotroper Platten," *Schiff und Hafen*, **14** (617), 1962, pp. 479–86 and pp. 569–76.
21. W. H. Wittrick, "Correlation between Some Stability Problems for Orthotropic and Isotropic Plates under Bi-axial and Uni-axial Direct Stress," *Aeronaut. Q.*, **4,** 1952, pp. 83–92.
22. D. Faulkner, "The Overall Compression Buckling of Partially Constrained Ship Grillages," M.I.T. Sea Grant Publication No. MITSG-73-10, 1973.
23. J. C. Adamchak, "A Rapid Technique for Predicting the General Instability of Ship Grillages with Elastically Constrained Edges," NSRDC Report 4604, 1975.

Chapter Fourteen

ULTIMATE STRENGTH OF STIFFENED PANELS

As explained at the beginning of Chapter 12, the stiffened panels in ship structures are usually sufficiently sturdy that $(\sigma_a)_{cr} > \sigma_Y$. Hence the mode of compressive collapse is not elastic panel buckling, but rather a more complicated inelastic process, for which we use the words "failure" or "collapse" instead of buckling. Nevertheless, there are several aspects of elastic buckling which have their counterpart in the inelastic compressive collapse of plating. For example, in parallel with elastic buckling there are several possible levels and modes of inelastic plate collapse. If the panel is transversely stiffened, then plate failure by itself constitutes collapse, and the theory and methods of Section 12.7 are sufficient. If the panel is cross-stiffened there are two possible levels of collapse:

1. Interframe collapse, that is, collapse of stiffened panels between transverse frames.
2. Gross panel collapse, involving both longitudinal and transverse stiffeners.

In general, cross-stiffened panels should be designed such that interframe collapse occurs before gross panel collapse because the latter involves a larger portion of structure and is likely to be more catastrophic. Also, the structural proportions of ship panels are usually such that gross panel collapse is inelastic, and therefore it is an extremely difficult and complex task to calculate the collapse load with sufficient accuracy for design purposes. Therefore the best procedure is to design the panel such that gross panel collapse cannot occur before interframe collapse. This can be achieved by ensuring that the elastic gross panel buckling stress calculated by the discrete beam method of Section 13.6 is substantially larger than the elastic interframe buckling stress given by (13.1.8), and that both of these buckling stresses are substantially larger than the yield stress.

In practice the scantlings that a panel must have in order to satisfy the requirements of other modes of failure are usually sufficient to meet both of these requirements, but they should always be checked nonetheless. With this design philosophy it is not necessary to consider inelastic gross panel collapse and therefore this chapter deals only with inelastic interframe collapse.

14.1 BOUNDARY CONDITIONS, LOAD TYPES, AND COLLAPSE MODES

Boundary Conditions for Ultimate Strength Analysis

In a deck or side it is clear that the rotational restraint provided to the panels by the transverse frames is relatively small and is best ignored. In the bottom and in other regions having a large lateral load each panel receives some rotational restraint from the adjacent panel, and in principle this would give a larger ultimate compressive strength than if the panel were simply supported. However, in most cases it is possible for the compressive load to occur with little or no lateral load. Also there are many situations in which a bottom panel is far from being clamped. Figure 14.1 shows a common example, in a tanker the end panel of an empty cargo tank adjacent to a full cargo tank has only moderate rotational restraint at one end and is subjected to a destabilizing moment at the other end.

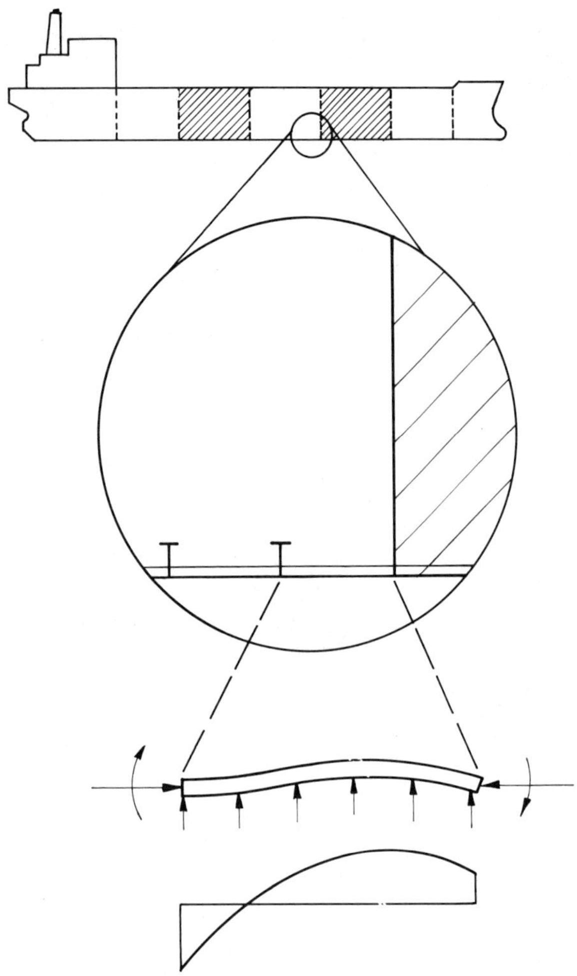

Figure 14.1 Occurrence of nonclamped conditions in bottom panels.

Although other panels may have more rotational restraint, it is obvious that this panel will govern the design of the bottom structure; it would be uneconomical and impractical for the panels to differ from one frame to another. In other ship types the same situation arises if a cargo hold is flooded due to underwater damage. It also arises in a double bottom in which one compartment is filled with fuel or ballast and the adjacent compartment is empty. Therefore unless a panel's rotational restraint is guaranteed to be permanently present under all conditions (which is quite rare) it should not be counted when calculating the ultimate strength of the panel. In general the safest and best procedure is to regard the panel as simply supported. However, the other effects of the lateral load—the maximum deflection and the maximum bending moment—should not be ignored since they decrease the panel's ultimate compressive strength. In this chapter the deflection and the bending moment due to the lateral load (acting alone) are denoted by δ_0 and M_0. Although the panel ends are being taken as simply supported the theory presented in this chapter does not require that δ_0 and M_0 be the simply supported values. They can be either these values or, if desired and if it is practicable to calculate them, they can be the largest values which will occur, allowing for the actual amount of rotational restraint that is present in that worst case.

Although simply supported boundary conditions generally correspond to small or moderate values of lateral load, this chapter covers the simply supported case over the full range of lateral loads, from zero up to the plastic hinge value, and acting on either side of the plating.

Basic Load Types and Associated Mechanisms of Collapse

For longitudinally stiffened panels with simply supported edges there are three basic types of loads:

1. Lateral load causing negative bending of the panel (i.e., of the plate-stiffener combination).
2. Lateral load causing positive bending of the panel.
3. In-plane compression.

Throughout this chapter, bending moment in the panel is positive when it causes compression in the plating, and in-plane loads are positive if compressive.

Each of these three basic loads has associated with it one or more possible mechanisms of collapse. Whenever in-plane compression occurs in combination with one of the first two load types there is interaction between the specific mechanisms of collapse. As we shall see, the various interactions give rise to three distinct modes of collapse. The specific collapse mechanisms associated with each of the basic load types follow (see Fig. 14.2).

Negative Bending

With this type of load the plating is in tension and hence cannot buckle. Also, since the neutral axis is close to the plating, it is the stiffener flange which first reaches yield (point A in Fig. 14.2). As the bending moment increases, yielding spreads through the stiffener. The plating then begins to yield in tension and, if there is no instability in the stiffener, collapse eventually occurs when the bending moment reaches M_P,

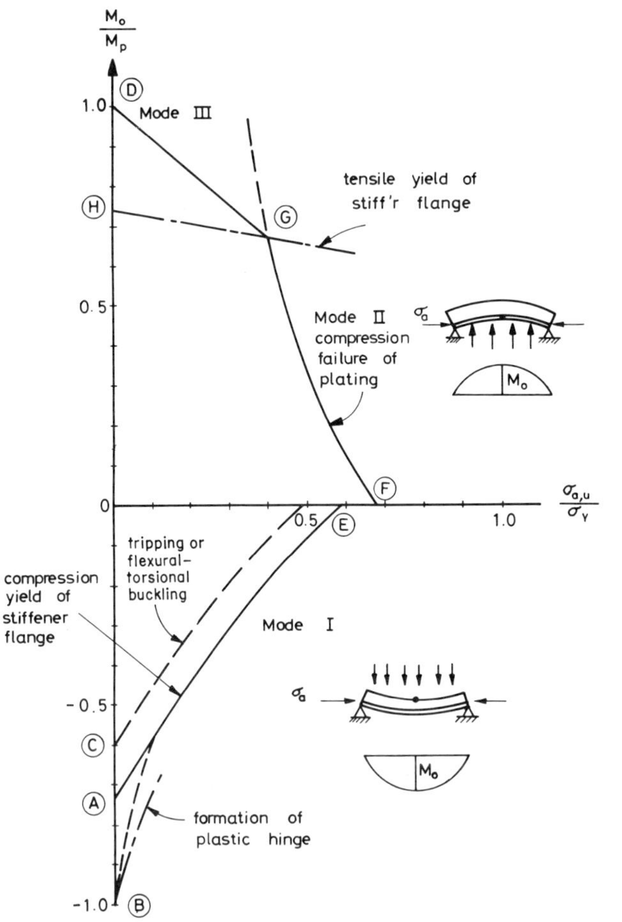

Figure 14.2 Collapse mechanisms in a stiffened panel under lateral and in-plane loads.

the value at which a plastic hinge is formed (see Section 16.1). This corresponds to point B in Fig. 14.2. However, since the stiffener is in compression it is possible that stiffener buckling might occur. Thus, for a negative bending load the plate-stiffener combination may not be capable of withstanding the full value of M_P for the section (point C in Fig. 14.2).

Positive Bending

With this type of load the plating is in compression, but since the neutral axis is very close to the plating the plate will not be heavily stressed until the load becomes very large, by which time the entire stiffener has yielded in tension and collapse is imminent. The plate then begins to yield in compression, and collapse finally occurs as the result of a plastic hinge, with the plate reaching compressive yield. In fact, for most ship panels the plastic neutral axis lies within the plating thickness, and so some portion of the plate's thickness is in tension. This stabilizes the plate and prevents it from buckling and hence for most ship panels the plate-stiffener combination is capable of absorbing almost the full value of the plastic bending moment M_P (point D in Fig. 14.2).

Uniform In-plane Compression

By "uniform compression" we mean an applied load which causes a uniform shortening of the panel, and it is assumed that the load is not (directly) influenced by the panel's response. The most common example is the hull girder stress. Strictly speaking, the loading on the panel is an *applied uniform strain* ε_a, but for the present we shall deal with the *average applied stress* σ_a (or average distributed load N).

Under this type of load the panel is essentially a group of identical (and almost independent) unsymmetric columns, each consisting of a stiffener and a plate flange of effective width b_e. If the panel is sufficiently slender these columns will buckle elastically and the ultimate strength of the panel can be calculated by the methods of Chapter 13. But for typical ship panels the plate's response becomes inelastic prior to collapse because the plate's load versus end shortening curve is the σ_a versus ε_a curves of Fig. 12.24. This makes the collapse mechanism decidedly more complicated, but a method is presented in this chapter which overcomes the complications and gives an accurate prediction of the collapse value of this type of load. Basically the method consists in regarding the plate portion of the column as being made of a different material, having a reduced elastic modulus, and then using the basic column approach of Section 11.2 to calculate the collapse load, making careful allowance for initial eccentricity. In this approach, "collapse" is considered to occur when the stress in the extreme fiber of the compression flange of the column reaches the failure value for the material of that flange. For the stiffener flange this is simply the yield stress, but for the plate flange it is some other value which allows for the nonlinearity of the σ_a versus ε_a curve. Thus there are two different values of in-plane collapse load, depending on which flange is the compression flange, that is, depending on the direction of the buckling. These two collapse loads correspond to points E and F on the interaction diagram of Fig. 14.2. In general these two loads are not the same and either of them can be the smaller. Since almost any panel can be in a zero lateral load condition, both values must be calculated and the lower value taken as the collapse load.

Combined Loads

In ship structures the stiffened panels are usually subjected to a combination of bending and in-plane loads. Also, in many ship panels the lateral bending can be either positive or negative. For example, a bottom panel in a tanker is subjected to an upward or a downward pressure, depending on the depth of cargo oil in that space relative to the draft. With combined loads there is an interaction among the three basic mechanisms outlined above. This interaction is rather complex and will be discussed fully in the next three sections. But at this point it will be helpful to present at least a qualitative overall view of the situation, and this will now be done with the aid of Fig. 14.2.

The first and most basic point is that with a lateral load the plate-stiffener "column" becomes a "beam column," and the basic column ultimate strength theory of Section 11.2 is replaced by the beam column theory of Section 11.3. In other words, collapse still occurs as the result of "failure of a flange," but the bending moment and deflection caused by the lateral load must be taken into account. Because of the bending, either flange could be the failure flange, depending on the sign of the bending moment, and the failure could be either tensile or compressive. Thus, in principle, there are four modes of collapse; in practice, one of the four—tensile failure of the plating—never occurs because the neutral axis is so close to the plating. Among the other three modes, the precise nature of the collapse mechanism differs, depending on which flange is the compression flange, that is, depending on the sign of the bending moment.

For greater generality we shall allow for the possibility of the stiffeners having a different value of yield stress from the plating. For this purppose we shall use the symbols σ_{Ys} and σ_{Yp}. We will also need to refer to the mean value of yield stress for the entire section, which is

$$\bar{\sigma}_Y = \frac{\sigma_{Ys}A_s + \sigma_{Yp}bt}{A_s + bt} \qquad (14.1.1)$$

Since the in-plane load usually arises from hull girder bending, it is usually more fundamental and more variable than the lateral load. Hence it is customary to define "ultimate strength" in terms of σ_a and to treat the value of the lateral load as an independent parameter. That is, the ultimate strength of a panel is the value of σ_a that causes collapse, in the presence of a specified value of lateral load. In an interactive collapse diagram such as Fig. 14.2 the ultimate strength is the horizontal distance to the curve at the height corresponding to the lateral load. Obviously this convention is not appropriate for panels in which the lateral load is always the dominant load. For such panels the ultimate strength is defined in terms of the lateral load and can be calculated by the plastic hinge theory of Section 16.1* But for the general case of a panel subjected to combined loads the ultimate strength theory must be sufficiently general to handle any combination of load. The theory presented in this chapter meets this requirement.

Compression Plus Negative Bending (Fig. 14.3)

Mode I: Compression Failure of the Stiffener. With this combination of loads the stiffener flange is the compression flange; therefore collapse occurs as the result of compressive failure of the stiffener flange, either by compressive yield or by buckling. Let us first consider compressive yield. When there is any appreciable amount of axial compressive stress σ_a, it directly increases the compressive bending stress in the stiffener and decreases the tensile bending stress in the plating. Therefore compressive yielding in the stiffener commences sooner and progresses more rapidly, whereas plate yielding is delayed. The result is that the section can no longer achieve a plastic hinge condition, as it could when the load was purely lateral (point B in Fig. 14.2). Instead, the stiffener reaches its limit of stress absorption soon after the yielding has penetrated through the stiffener flange, and when the stiffener is thus "neutralized" the effective moment of inertia of the section becomes very small, since only the plating is effective. The result is that once the stiffener flange is fully yielded collapse occurs soon afterward. Moreover, this mode of collapse is generally quite sudden. In most cases the precise cause of the failure is the formation of a local plastic mechanism in the flange such as that shown in Fig. 14.4. It is quite local because the flange becomes fully plastic at the point where the bending moment is a maximum. This mechanism involves large local sideways deflection and rotation, permitting the stiffener to shorten and thereby

*A cross-stiffened or gross panel subjected to lateral loads is termed a *grillage*. This type of structure can be designed in isolation, separate from the hull module, and is therefore not particular to ship structural design.

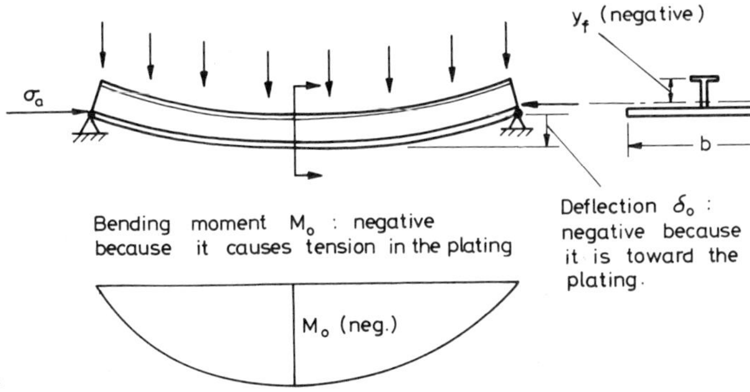

Figure 14.3 In-plane compression and negative bending.

escape from carrying the axial compressive load. The local sideways deflection of the flange is combined with rotation about the line of attachment and hence this type of failure is sometimes taken to be tripping, but this is not correct since the cause of the sideways deflection is a local plastic mechanism rather than overall flange buckling.

As noted earlier, when the loading is purely lateral the plate-stiffener combination can reach a plastic hinge condition before collapse occurs. Hence for load combinations which involve only a small amount of in-plane compression the collapse curve can depart from the "stiffener flange yield" curve and eventually, at zero axial load, it can reach the plastic hinge collapse point. However, the curve in this region is very steep, such that a small underestimate of the axial load would cause a serious overestimate of the collapse value of the lateral load. For this reason the plastic hinge mechanism of collapse should only be used when it is certain that in-plane compression cannot occur (e.g., platforms, ramps, machinery flats, and other laterally loaded stiffened panels). The calculation of ultimate strength under this type of loading is dealt with in Chapter 16. Therefore, for stiffened panels which are subject to in-plane compression it is best to use the "stiffener flange yield" mechanism for all of the Mode I region, even though for very small in-plane loads this approach is conservative. Hence the Mode I collapse curve begins at point A in Fig. 14.2.

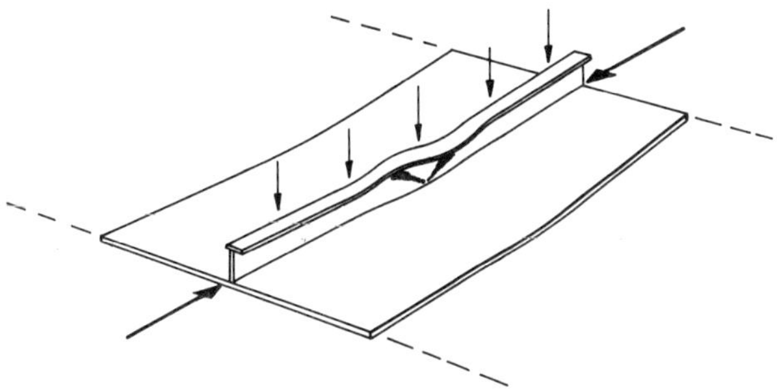

Figure 14.4 Local plastic collapse mechanism in a stiffener flange.

Alternatively, the stiffener may fail by tripping. If the amount of bending is small the elastic tripping stress $\sigma_{a,T}$ is given by the formulas of Section 13.1. If it is not small then the stiffener may undergo flexural-torsional buckling, and providing the stiffener remains elastic the value of $\sigma_{a,T}$ may be calculated by elastic nonlinear frame analysis, which is presented in Section 15.5. The analysis is simple and rapid because the buckling is of bifurcation type and requires only an eigenvalue analysis. However, as mentioned earlier, for most ship panels, buckling is not purely elastic and the simple bifurcation/eigenvalue approach cannot provide the true collapse value, $\sigma_{a,u}$, either with or without M_0. To obtain an accurate value of $\sigma_{a,u}$ would require the fully nonlinear finite element method (incremental and iterative), which is discussed in Section 15.6, but this involves too much computation for such a local and frequently occurring calculation (every panel) especially in a design procedure. Instead, there is a simple alternative suggested by Smith [1]. Since the maximum compressive stress in the flange must be kept below σ_{Ys} in order to avoid a local plastic mechanism, stiffener buckling can be avoided simply by requiring that the value of $\sigma_{a,T}$ obtained from the elastic theory (Section 13.1 or 15.5, as appropriate) must be well above σ_{Ys}.

Since collapse occurs when the stiffener flange stress reaches yield (or earlier if $\sigma_{a,T} < \sigma_{Ys}$) the beam-column remains fully elastic prior to collapse. Therefore the combination of compression and negative bending is relatively straightforward to analyze, requiring only the ordinary beam column theory of Section 11.3.

Compression Plus Positive Bending (Fig. 14.5)

With this combination of loads the plating is the compression flange and there are two distinct modes of collapse, depending on the relative magnitude of the bending and the in-plane loads:

Mode II: Compression Failure of the Plating. With small and moderate lateral loads, collapse occurs when the stress in the plating reaches the failure value for the "special material" which is used to represent the plating. As will be shown, the use of a secant modulus to represent the plate's stress-strain relationship allows the beam-column to be treated as if it remained entirely elastic prior to collapse.

Mode III: Combined Failure of Stiffener and Plating. With large positive lateral loads the bending causes large tensile stresses in the stiffener. These are reduced by the in-plane applied compressive stress, but for larger and larger lateral loads the stiffener undergoes more and more tensile yielding, and collapse, when it finally occurs, is due to a combination of compressive failure of the plating material and tensile yielding of the stiffener.

In the next section we investigate each of these three modes of collapse.

14.2 MODES OF COLLAPSE

Mode I Collapse: Compression Failure of the Stiffener

The combination of in-plane compression and negative bending is illustrated in Fig. 14.3. With this combination, collapse occurs due to compression failure of the stiffener flange. That is, collapse occurs when the total stress (σ_a plus the magnified bending stress) throughout the thickness of the flange has reached the "failure value" σ_F. The failure value is either the yield stress σ_{Ys} or the elastic tripping stress $\sigma_{a,T}$, whichever is less. Stated mathematically

$$\sigma_F = \text{MIN}\{\sigma_{Ys}, \sigma_{a,T}\} \tag{14.2.1}$$

where $\sigma_{a,T}$ is obtained from the method of Section 13.1 or 15.5.

Strictly speaking, as the yield zone penetrates from the outer surface to the midthickness of the flange there is some nonlinearity, but since the half-thickness is small relative to the web depth the effect is negligible. Therefore, the failure process can be regarded as entirely elastic and can be completely described by the beam-column theory of Section 11.3. Accordingly, the total stress in the midthickness of the flange is

$$\sigma_f = \sigma_a + \frac{M_0 y_f}{I} + \frac{\sigma_a A(\delta_0 + \Delta) y_f}{I}\phi \tag{14.2.2}$$

in which A and I = cross-sectional area and moment of inertia of the beam-column. (Note: Since the plating is in tension the plate flange is assumed to be of full width b. Therefore $A = A_s + bt$.)

M_0 and δ_0 = maximum bending moment and maximum deflection due to the lateral load acting alone. It is assumed that these maximum values occur at the midlength of the beam-column.

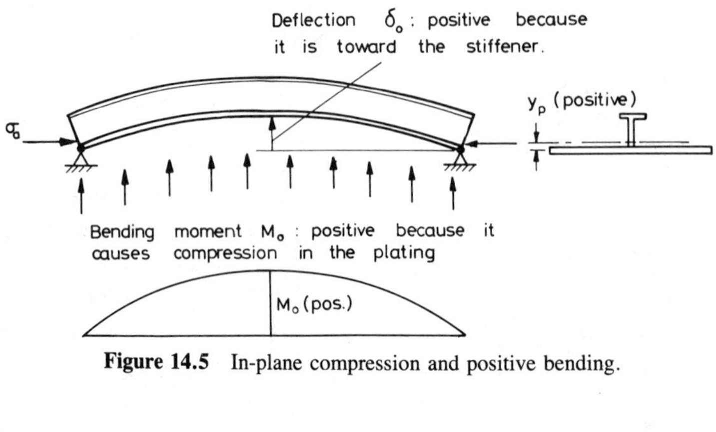

Figure 14.5 In-plane compression and positive bending.

Δ = initial eccentricity of the beam-column; a typical maximum permissible value for welded panels is $a/750$, where a is the length of the panel.

y_f = distance from the centroidal axis of the cross section to the mid-thickness of the stiffener flange.

ϕ = magnification factor due to the axial compressive stress σ_a:

$$\phi = \frac{1}{1 - \dfrac{\sigma_a}{\sigma_E}}$$

where σ_E is the Euler buckling stress for the beam-column:

$$\sigma_E = \frac{\pi^2 EI}{Aa^2}$$

The collapse or "ultimate" value of applied stress $\sigma_{a,u}$ is the value of σ_a at which $\sigma_f = \sigma_F$, and so we now proceed to make these substitutions and solve for $\sigma_{a,u}$. As in Section 11.3, we make use of the following nondimensional parameters:

$$\left.\begin{aligned} R &= \frac{\sigma_{a,u}}{\sigma_F} \\ \lambda &= \frac{a}{\pi\rho}\sqrt{\frac{\sigma_F}{E}} \qquad \text{where } \rho = \sqrt{\frac{I}{A}} \\ \eta &= \frac{(\delta_0 + \Delta)y_f}{\rho^2} \\ \mu &= \frac{M_0 y_f}{I\sigma_F} \end{aligned}\right\} \qquad (14.2.3)$$

In terms of these parameters the magnification factor is

$$\phi = \frac{1}{1 - \lambda^2 R}$$

Equation 14.2.2 becomes

$$(1 - R - \mu)(1 - \lambda^2 R) = \eta R \qquad (14.2.4)$$

and the solution for the strength ratio R is

$$R = \frac{\zeta}{2} - \sqrt{\frac{\zeta^2}{4} - \frac{1 - \mu}{\lambda^2}} \qquad (14.2.5)$$

where $\zeta = 1 - \mu + \dfrac{1 + \eta}{\lambda^2}$

Therefore, if $\sigma_F = \sigma_{Ys}$ the equation for the interactive collapse curve in the Mode I collapse region of Fig. 14.2 is

$$\left(\frac{\sigma_{a,u}}{\sigma_{Ys}}\right)_I = R \qquad (14.2.6a)$$

with R as given previously. The influence of M_0 is contained in the bending parameter μ. If $\sigma_F = \sigma_{a,T}$ then the equation becomes

$$\left(\frac{\sigma_{a,u}}{\sigma_{Ys}}\right)_I = R\frac{\sigma_{a,T}}{\sigma_{Ys}} \qquad (14.2.6b)$$

This equation is based on the following sign convention, which is used throughout this chapter.

Stress: Positive if compressive.

Bending moment: Positive when the plating is in compression.

Lateral deflection, Eccentricity: Positive toward the stiffeners (i.e., positive upward in Fig. 14.3). Note: For a simply supported panel, positive bending causes positive lateral deflection.

Distance from neutral axis: Positive toward the plating.

In simply supported panels that are subjected to negative bending, the collapse mode is always Mode I for any significant amount of negative bending, that is, throughout the lower half of the interaction diagram of Fig. 14.2. To make the diagram more useful the loads are nondimensionalized: values of maximum bending moment are normalized by dividing by M_P, and values of applied in-plane compressive stress are divided by $\bar{\sigma}_Y$. Thus the point corresponding to collapse under a pure bending load is the point -1 on the bending moment axis, providing that buckling does not precede flange yield. As the amount of in-plane compression increases, the amount of bending which is required to cause collapse is lessened. The exact shape of the interactive collapse curve depends on the scantlings and geometry of the panel. If the failure mechanism is buckling, the ultimate strength will be less than it would be for failure by yielding, and the entire curve will lie closer to the origin, as shown by the dashed line in Fig. 14.2.

Mode II Collapse: Compression Failure of the Plating

We next consider the combination of in-plane compression σ_a and positive bending moment M_0 as shown in Fig. 14.5. If the bending is small or moderate in magnitude, collapse occurs due to compression failure of the plate flange. If the plating remained perfectly elastic, with no buckling or nonlinearity, the total applied stress acting on the plating, σ_{pa}, would be given by the usual beam-column formula

$$\sigma_{pa} = \sigma_a + \frac{M_0 y_p}{I} + \frac{\sigma_a A(\delta + \delta_0) y_p}{I} \phi \tag{14.2.7}$$

As shown in Section 12.6, the compressive collapse of welded plating is a complex inelastic process. The relationship between load and end shortening, that is, between the applied stress σ_{pa} and the average strain ε_a, is described by the curves of Fig. 12.24. For our present application the applied stress in the plating is σ_{pa} and so this symbol replaces σ_a of Fig. 12.24. As an illustration, and for later reference, a typical curve of σ_{pa} versus ε_a from Fig. 12.24 is reproduced in Fig. 14.6. The earlier figure showed that for most plates ($\beta > 1$) the relationship between σ_{pa} and ε_a becomes nonlinear well before collapse. Both figures also illustrate what we mean by "plate failure," namely, the point in the σ_{pa} versus ε_a curve at which the plating has lost most of its in-plane stiffness. In geometric terms it is the point at which the tangent of the curve has become significantly less than the original value, E. Strictly speaking, the failure point is the peak of the curve. However, for plates of low and intermediate slenderness there is a discernible knee in the curve (unless the initial distortion δ_p is very large) and this is the simplest and safest choice of the failure point. For very slender plates the failure point is either the peak of the curve or the point at which the slope has decreased to some specified value (e.g., $\frac{1}{3}E$).

Since the plating is actually being subjected to an imposed uniform strain, collapse is, strictly speaking, measured by the value of average strain corresponding to this loss of stiffness: $\varepsilon_a = \varepsilon_{\text{ult}}$, as shown in Fig. 14.6. However, it is more convenient to deal in terms of stress. For later use we define σ_F as the equivalent elastic failure stress, that is, the value of σ_{pa} which

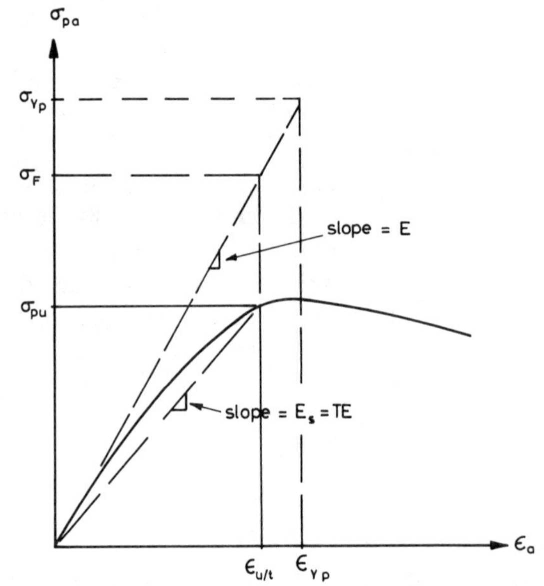

Figure 14.6 Typical curve of load versus end shortening for welded steel plating.

would correspond to ε_{ult} if the plate maintained a linear relationship between σ_{pa} and ε_a.

For most plates the slope of the curve begins to decrease well before the point of "plate failure." Therefore σ_{pu}, the true value of applied stress which corresponds to plate failure, is significantly less than σ_F. This also means that during the load sequence that leads to plate failure, the average plate stiffness is significantly less than the elastic or material stiffness E. The average stiffness over the entire range of loading is given by the secant modulus $E_s = TE$, which is the slope of the line joining the origin to the point of plate failure, as shown in Fig. 14.6. For our purpose this straight line can be used in place of the actual curve because we simply want to be able to predict plate failure, and we are not interested in the detailed plate response prior to that. With this approach we are, in effect, representing the plating as if it were made of a different material which has a smaller value of Young's modulus, and for which the *curve of σ_{pa} versus ε_a is perfectly linear, right up to failure*. As we shall see, the latter property is of very great advantage.

The mathematical expression for the secant modulus was given in (12.6.4) which for convenience is reproduced here

$$T = \frac{E_s}{E} = 0.25\left(2 + \xi - \sqrt{\xi^2 - \frac{10.4}{\beta^2}}\right) \tag{14.2.8}$$

where

$$\xi = 1 + \frac{2.75}{\beta^2}$$

Also, it was noted in Section 12.6 that the ultimate strength curve has the same shape as the curve for T and can be represented sufficiently accurately by simply subtracting 0.1 from T. That is

$$\frac{\sigma_{pu}}{\sigma_{Yp}} = T - 0.1$$

$$= 0.25\left(1.6 + \xi - \sqrt{\xi^2 - \frac{10.4}{\beta^2}}\right)$$

with ξ again given by (14.2.9)

$$\xi = 1 + \frac{2.75}{\beta^2}$$

We now consider the composite beam column comprised of a stiffener and a plate flange of width b, as shown in Fig. 14.7. The total cross-sectional area is $A = A_s + bt$. The loads are an applied compressive stress σ_a and a lateral load which causes a positive bending moment of maximum value M_0. The plate flange has a different elastic modulus than the stiffener and hence, before we can apply beam theory, we must transform this composite section into an equivalent section which has a uniform elastic modulus, as was done for the hull girder in Section 3.6. We choose the stiffener as the reference material, and hence the cross-sectional area of the plating is multiplied by the transformation factor T. In the present application, to avoid any confusion about plate thickness we apply T to the flange width b instead of to the plate thickness. Therefore, in the transformed section the plate flange has a transformed width $b_{tr} = Tb$. The total area is now $A_{tr} = A_s + b_{tr}t$, the neutral axis is at the centroid of the transformed section, and the effective moment of inertia of the beam-column is I_{tr}, which is the second moment of the transformed section. The strain distribution is linear and continuous but the stress distribution is discontinuous because in the plating the stress is $\sigma_{pa} = T\sigma^*_{pa}$, where σ^*_{pa} is the stress obtained by applying ordinary beam theory to the transformed section (see Fig. 14.7). It should be noted that σ_{pa}, the actual value of stress in the plating, acts across the entire width of the plate flange; the transformed width b_{tr} is merely a device for obtaining the neutral axis position and moment of inertia of the composite section.

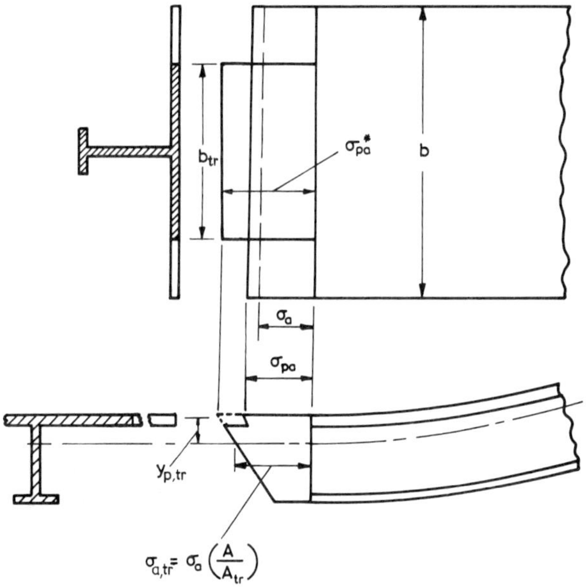

Figure 14.7 Use of transformed section to represent combined stiffener-and-plate as an equivalent elastic beam column.

The axial stress in the transformed section is

$$\sigma_{a,tr} = \sigma_a \frac{A}{A_{tr}} = \sigma_a \frac{A_s + bt}{A_s + b_{tr}t} \qquad (14.2.10)$$

and this is the value of stress at the neutral axis of the section. If we now apply the ordinary beam column theory as in (14.2.7) to the transformed section we obtain the uncorrected plate stress σ^*_{pa}

$$\sigma^*_{pa} = \sigma_{a,tr} + \frac{M_0 y_{p,tr}}{I_{tr}} + \frac{\sigma_{a,tr} A_{tr}(\Delta + \delta_0) y_{p,tr}}{I_{tr}} \phi \qquad (14.2.11)$$

and the true value of σ_{pa} is then

$$\sigma_{pa} = T\sigma^*_{pa} \qquad (14.2.12)$$

From beam-column theory we know that collapse of the beam-column occurs when the stress in the compression flange (here σ_{pa}) reaches the failure value for that material (σ_{pu}, as shown in Fig. 14.6). The ultimate strength of the beam-column is defined as the value of applied axial stress in the beam-column ($\sigma_{a,tr}$) which is sufficient to cause the above collapse condition. That is, $(\sigma_{a,tr})_{\text{ult}}$ is defined as the value of $\sigma_{a,tr}$

which satisfies

$$\sigma_{pu} = \sigma_{pa}$$

or

$$\sigma_{pu} = T\sigma_{pa}^* \tag{14.2.13}$$

where σ_{pa}^* is as given previously. Rather than multiplying every term of σ_{pa}^* by T, it will be more convenient to bring T to the left-hand side and define an equivalent plate failure stress σ_F:

$$\sigma_F = \frac{\sigma_{pu}}{T}$$

as was foreshadowed earlier (see Fig. 14.6).

Thus far we have considered only longitudinal stress in the plating. If there is a transverse stress σ_{ay} (positive compressive) and shear stress τ, the ultimate strength of the plating is less than the uniaxial value σ_{pu}. If these other two stresses are not large (less than, say, 0.3 σ_{ax}) the value of σ_{pu} can be corrected by applying a reduction factor to σ_{pu}. The reduction factor to account for τ is given by (12.6.3)

$$r_\tau = \sqrt{1 - 3\left(\frac{\tau}{\sigma_{Yp}}\right)^2}$$

The interaction between σ_{ax} and σ_{ay} is discussed in Section 12.6, and is shown to depend on the plate slenderness β and aspect ratio a/b. The interaction is quite complex, being actually beneficial in the case of sturdy plates ($\beta < 1$) and becoming linear, and detrimental, in the case of slender plates. For design purposes we want a simple relationship which covers the worst cases without being unduly pessimistic in other cases. The worst case is the linear interaction given by (12.6.16) corresponding to slender plates. If we adopt this then the reduction factor to be applied to σ_{pu} is

$$r_{ay} = 1 - \frac{\sigma_{ay}}{\sigma_{ay,u}}$$

in which $\sigma_{ay,u}$ is given by (12.6.12) assuming that $a/b \geqslant 1$. We now apply these two reduction factors to σ_{pu}, and for simplicity of notation we do so by redefining σ_F, the equivalent failure stress for the plating, such that it includes these factors. We also take advantage of the fact that σ_{pu} can be represented in terms of T, as in (14.2.9): $\sigma_{pu} = (T - 0.1)\sigma_{Yp}$. The definition of σ_F becomes

$$\sigma_F = \frac{T - 0.1}{T}\sigma_{Yp}\sqrt{1 - 3\left(\frac{\tau}{\sigma_{Yp}}\right)^2}\left(1 - \frac{\sigma_{ay}}{\sigma_{ay,u}}\right) \tag{14.2.14}$$

and σ_F is an artificial or equivalent failure stress for the plating, such that the plate fails when the uncorrected plate stress σ_{pa}^* reaches σ_F. Substituting for σ_{pa}^* from (14.2.11) we obtain the equation governing the collapse condition:

$$\sigma_F = (\sigma_{a,tr})_{\text{ult}} + \frac{M_0 y_{p,tr}}{I_{tr}} + \frac{(\sigma_{a,tr})_{\text{ult}} A_{tr}(\Delta + \delta_0) y_{p,tr}}{I_{tr}}\phi \tag{14.2.15}$$

In parallel with the procedure adopted for Mode I collapse, we define the following nondimensional parameters

$$R = \frac{(\sigma_{a,tr})_{\text{ult}}}{\sigma_F}$$

$$\lambda = \frac{a}{\pi\rho_{tr}}\sqrt{\frac{\sigma_F}{E}} \qquad \text{where } \rho_{tr} = \sqrt{\frac{I_{tr}}{A_{tr}}}$$

$$\eta = \frac{(\Delta + \delta_0) y_{p,tr}}{\rho_{tr}^2} \tag{14.2.16}$$

$$\mu = \frac{M_0 y_{p,tr}}{I_{tr}\sigma_F}$$

$$\phi = \frac{1}{1 - \lambda^2 R}$$

In terms of these parameters, (14.2.15) reduces to the same nondimensional beam-column equation as for Mode I, given in (14.2.4)

$$(1 - \mu - R)(1 - \lambda^2 R) = \eta R$$

and the solution for the strength ratio R or (R_{II}, to indicate the collapse mode) is

$$R_{\text{II}} = \frac{\zeta_{\text{II}}}{2} - \sqrt{\frac{\zeta_{\text{II}}^2}{4} - \frac{1 - \mu}{\lambda^2}} \tag{14.2.17}$$

where $$\zeta_{\text{II}} = 1 - \mu + \frac{1 + \eta}{\lambda^2}$$

This equation is based on the sign convention defined earlier and hence $y_{p,tr}$, δ_0, and Δ are all positive quantities.

Induced Load Eccentricity Due to Decreased Plate Stiffness

In investigating panel collapse of a stiffened panel structure we usually wish to find and analyze the first panel to collapse. Hence the panels which are immediately adjacent in the direction of the load should be treated as fully effective. But the main reason for treating them as such is that this is a more severe condition for the subject panel. The reduced stiffness of the plating has been allowed for by defining a transformed section in which the neutral axis is displaced slightly further from the plating. By definition the neutral axis is the line of action, within the cross section of the beam column, of the total axial force $P(P = A\sigma_a = A_{tr}\sigma_{a,tr})$ which is transmitted by the cross section. But the line of action of the externally applied force is the centroid of the total area A, which means that the applied load has a slight positive eccentricity, and this creates an extra positive moment in the section and therefore decreases the ultimate strength. The eccentricity is equal to the distance moved by the neutral axis, which is

$$\Delta_p = hA_s\left[\frac{1}{A_{tr}} - \frac{1}{A}\right] \qquad (14.2.18)$$

where h is the distance from the midplane of the plating to the centroid of the stiffener, and A_s is the sectional area of the stiffener.

The effect of Δ_p is to add another term to the right-hand side of (14.2.15). Unlike other forms of eccentricity, Δ_p does not undergo magnification since it is the final value of the neutral axis eccentricity, at the instant of failure. Therefore the additional term is

$$\frac{(\sigma_{a.tr})_{\text{ult}}A_{tr}\Delta_p y_{p,tr}}{I_{tr}}$$

Accordingly, we define an additional nondimensional parameter η_p, analogous to the geometric eccentricity parameter η

$$\eta_p = \frac{\Delta_p y_{p,tr}}{\rho_{tr}^2} \qquad (14.2.19)$$

The nondimensional equation governing collapse becomes

$$[1 - \mu - R(1 + \eta_p)](1 - \lambda^2 R) = \eta R$$

for which the solution is

$$R_{\text{II}} = \frac{\zeta_{\text{II}}}{2} - \sqrt{\frac{\zeta_{\text{II}}^2}{4} - \frac{1 - \mu}{(1 + \eta_p)\lambda^2}} \qquad (14.2.20)$$

where

$$\zeta_{\text{II}} = \frac{1 - \mu}{1 + \eta_p} + \frac{1 + \eta_p + \eta}{(1 + \eta_p)\lambda^2}$$

NONUNIFORM APPLIED COMPRESSIVE STRESS

If the panel forms part of a hull girder the overall depth of the hull girder is sufficiently large that the hull girder stress σ_a can be regarded as constant over the height of the panel. If, however, the overall structure is a shallower box girder, such as a double bottom or a barge or a simple box girder beam, then the applied stress σ_a has a trapezoidal distribution over the height of the panel and this also induces some additional eccentric bending. The eccentricity is again positive (providing that the stiffeners are on the inside of the box girder flanges) and is given by

$$\Delta_H = \frac{I}{AH} \qquad (14.2.21)$$

where A and I refer to the total panel cross section (not the transformed section) and H is the distance from the centroidal axis of the total panel cross section to the neutral axis of the overall box girder, that is, to the axis of zero stress in the box girder, taking into account any axial stress which may be present in it. Since Δ_H is an initial eccentricity it may simply be added to Δ, and there is no need for a new term as there was for Δ_p.

The steps in calculating the Mode II ultimate strength may be summarized as follows:

1. For the given lateral load, calculate M_0 and δ_0.
2. From (14.2.8) calculate T and then calculate σ_F from

$$\sigma_F = \frac{T - 0.1}{T}\sigma_{Yp}\sqrt{1 - 3\left(\frac{\tau}{\sigma_{Yp}}\right)^2}\left(1 - \frac{\sigma_{ay}}{\sigma_{ay,u}}\right)$$

with $\sigma_{ay,u}$ given by (12.6.12).

3. For a transformed section having a plate flange of width Tb, calculate A_{tr}, $y_{p,tr}$, I_{tr}, ρ_{tr}, and Δ_p. (Note: Eqs. 8.3.6 are helpful for this task.
4. Using (14.2.16) and (14.2.19) calculate the parameters λ, η, η_p, and μ.
5. Calculate the strength ratio R_{II} from (14.2.20).
6. Calculate the collapse load $\sigma_{a,u}$ from (14.2.10)

$$\sigma_{a,u} = (\sigma_{a,tr})_{\text{ult}}\left(\frac{A_{tr}}{A}\right)$$

which is, in terms of R_{II}

$$\sigma_{a,u} = \sigma_F R_{\mathrm{II}}\left(\frac{A_{tr}}{A}\right) \qquad (14.2.22a)$$

or in nondimensional terms

$$\left(\frac{\sigma_{a,u}}{\sigma_Y}\right)_{\mathrm{II}} = \frac{\sigma_F}{\sigma_Y} R_{\mathrm{II}}\left(\frac{A_{tr}}{A}\right) \qquad (14.2.22b)$$

R_{II} is a function of M_0 through the medium of μ

$$\mu = \frac{M_0 y_{p,tr}}{I_{tr}\sigma_F}$$

and therefore (14.2.22) is the equation for the Mode II interactive collapse curve of Fig. 14.2.

Mode III Collapse: Combined Failure of Stiffener and Plating

If the positive bending moment M_0 is large it creates a large tensile stress in the stiffener. This is partly canceled by the applied compressive stress, but if the bending moment is sufficiently large there will be tensile yielding in the stiffener flange as well as compressive failure of the plating. This further reduces the effectiveness of the section, and so the plate failure analysis of the previous section is no longer sufficient. The point on the Mode II curve where this simultaneous failure begins to occur is labeled G on the interaction diagram of Fig. 14.2. This simultaneous failure occurs in all combinations of collapse loads in which the bending moment exceeds $(M_0)_G$. Therefore in this region of the interaction diagram there is a new mode of collapse—Mode III—and a new segment of the collapse curve extending from point G up to point D, where the load is a pure bending load. For this loading, the stiffener can achieve full yield and as this occurs the effective neutral axis moves toward the plating. For most panel proportions it eventually enters the plating, putting it partly in tension. This stabilizes the plating such that it can absorb almost the full value of compressive yield stress. Hence for a purely lateral load collapse does not occur until the panel has reached a plastic hinge condition.

A precise calculation of the collapse loads for Mode III collapse would be extremely difficult because of the complicated interaction between the simultaneous stiffener tensile yielding and plate compressive failure. Experiments [2] show that the interaction curve for this collapse mode is convex upward and that it always intersects the Mode II curve at some distance above point G. That is, the Mode II collapse analysis becomes inaccurate only gradually, in proportion to the extent to which M_0 exceeds $(M_0)_G$, such that the collapse involves some yielding of the stiffener flange. Therefore we choose a straight line between points D and G as the interaction curve for Mode III collapse. This choice is safe since the actual collapse points always lie slightly above this line, and it greatly simplifies the analysis of Mode III collapse. The analysis only requires finding point G, the load combination which causes simultaneous compression failure of the plating and tensile yield of the stiffener flange. The latter occurs when the total flange stress reaches the tensile yield stress, which in the sign convention of this chapter is $-\sigma_{Ys}$. Therefore the condition for flange yield is

$$-\sigma_{Ys} = (\sigma_{a,tr})_{\mathrm{ult}} + \frac{M_0 y_{f,tr}}{I_{tr}} + \frac{(\sigma_{a,tr})_{\mathrm{ult}} A_{tr}(\delta_0 + \Delta) y_{f,tr}}{I_{tr}}\phi + \frac{(\sigma_{a,tr})_{\mathrm{ult}} A_{tr}\Delta_p y_{f,tr}}{I_{tr}} \qquad (14.2.23)$$

This equation corresponds to the curve GH in Fig. 14.2. To facilitate the solution for the value of $(\sigma_{a,tr})_{\mathrm{ult}}$ we again define a set of nondimensional parameters. These differ from those of Mode II in that the flange distance $y_{f,tr}$ replaces $y_{p,tr}$, which means that both η and η_p will have negative values. Also, since the failure stress is now negative ($\sigma_F = -\sigma_{Ys}$) the strength ratio R will likewise have a negative value, and this means that the expression for the magnification factor must be altered to

$$\phi = \frac{1}{1 + \lambda^2 R}$$

In order to distinguish the nondimensional parameters which correspond to the curve GH from those of Mode II, we shall use a subscript GH. Accordingly, the parameters are

$$R_{GH} = \frac{(\sigma_{a,tr})_{\mathrm{ult}}}{-\sigma_{Ys}}$$

$$\lambda_{GH} = \frac{a}{\pi\rho_{tr}}\sqrt{\frac{\sigma_{Ys}}{E}}$$

$$\eta_{GH} = \frac{(\Delta + \delta_0) y_{f,tr}}{\rho_{tr}^2}$$

$$\eta_{p,GH} = \frac{\Delta_p y_{f,tr}}{\rho_{tr}^2} \qquad (14.2.24)$$

$$\mu_{GH} = \frac{M_0 y_{f,tr}}{I_{tr}(-\sigma_{Ys})}$$

$$\phi_{GH} = \frac{1}{1 + \lambda^2 R_{GH}}$$

The condition for flange yield (i.e., the equation of curve GH) is then

$$[1 - \mu_{GH} - R_{GH}(1 + \eta_{p,GH})](1 + \lambda_{GH}^2 R_{GH}) = \eta_{GH} R_{GH}$$

and solving for R_{GH} gives*

$$R_{GH} = \frac{\zeta_{GH}}{2} \pm \sqrt{\frac{\zeta_{GH}^2}{4} + \frac{1 - \mu_{GH}}{(1 + \eta_{p,GH})\lambda_{GH}^2}}$$

where

$$\zeta_{GH} = \frac{1 - \mu_{GH}}{1 + \eta_{p,GH}} - \frac{1 + \eta_{p,GH} + \eta_{GH}}{(1 + \eta_{p,GH})\lambda_{GH}^2}$$

The other condition defining point G is the Mode II condition. Therefore, from (14.2.20) the second equation for determining point G is

$$R_{\text{II}} = \frac{\zeta_{\text{II}}}{2} - \sqrt{\frac{\zeta_{\text{II}}^2}{4} - \frac{1 - \mu_{\text{II}}}{(1 + \eta_{p,\text{II}})\lambda_{\text{II}}^2}}$$

where (14.2.26)

$$\zeta_{\text{II}} = \frac{1 - \mu_{\text{II}}}{1 + \eta_{p,\text{II}}} + \frac{1 + \eta_{p,\text{II}} + \eta_{\text{II}}}{(1 + \eta_{p,\text{II}})\lambda_{\text{II}}^2}$$

in which the nondimensional parameters are those of (14.2.16) and (14.2.19). From these expressions for R_{GH} and R_{II} we can obtain two equations for $(\sigma_{a,u})_G$ as a function of $(M_0)_G$. For the stiffener tensile yield condition the equation for $(\sigma_{a,u})_G$ in terms of R_{GH} is similar to (14.2.22), namely

$$\left(\frac{\sigma_{a,u}}{\overline{\sigma}_Y}\right)_G = \frac{-\sigma_{Ys}}{\overline{\sigma}_Y}\left(\frac{A_{tr}}{A}\right)R_{GH} \qquad (14.2.27)$$

*We here indicate both roots of the quadratic because the solution for point G can involve either root, depending on M_0 and the panel proportions.

The result will be positive since R_{GH} is negative. R_{GH} is a function of $(M_0)_G$ since

$$\mu_{GH} = \frac{(M_0)_G y_{f,tr}}{I_{tr}(-\sigma_{Ys})}$$

The second equation for $(\sigma_{a,u})_G$ and $(M_0)_G$ comes from the plate failure condition, for which (14.2.22) gives

$$\left(\frac{\sigma_{a,u}}{\overline{\sigma}_Y}\right)_G = \frac{\sigma_F}{\overline{\sigma}_Y}\left(\frac{A_{tr}}{A}\right)R_{\text{II}} \qquad (14.2.28)$$

in which R_{II} is a function of $(M_0)_G$, since

$$\mu_{\text{II}} = \frac{(M_0)_G y_{p,tr}}{I_{tr}\sigma_F}$$

The required collapse loads $(\sigma_{a,u})_G$ and $(M_0)_G$ are obtained by solving (14.2.27) and (14.2.28) simultaneously, using an appropriate numerical technique. By equating the right-hand sides of these equations we obtain a single implicit equation for $(M_0)_G$:

$$R_{\text{II}}\sigma_F = -R_{GH}\sigma_{Ys} \qquad (14.2.29)$$

The solution must be performed iteratively, and the best approach is to begin with a trial value of $(M_0)_G$, calculate the corresponding value of $(\delta_0)_G$ (which depends on the type of lateral load), then evaluate R_{II} and R_{GH}, and hence the left and right sides of (14.2.29). In general these will not agree and it will be necessary to choose another value of $(M_0)_G$ and repeat until the agreement is satisfactory. Some numerical technique (Newton's method, Secant method, etc.) may be used to obtain each new value of $(M_0)_G$.

Since the other end-point of the Mode III collapse line is $(\sigma_{a,u}/\overline{\sigma}_Y = 0; M_0/M_P = 1)$ the equation for the ultimate strength of a panel in the Mode III collapse regime is

$$\left(\frac{\sigma_{a,u}}{\overline{\sigma}_Y}\right)_{\text{III}} = \frac{M_P - M_0}{M_P - (M_0)_G}\left(\frac{(\sigma_{a,u})_G}{\overline{\sigma}_Y}\right) \qquad (14.2.30)$$

in which $(\sigma_{a,u})_G$ is obtained from either of two expressions:

$$(\sigma_{a,u})_G = \sigma_F\left(\frac{A_{tr}}{A}\right)R_{\text{II}} \qquad (14.2.31)$$

or

$$(\sigma_{a,u})_G = -\sigma_{Ys}\left(\frac{A_{tr}}{A}\right)R_{GH}$$

Effect of Initial Eccentricity

If there is no lateral load the initial eccentricity Δ becomes the crucial factor in determining which mode of collapse occurs. The load eccentricity Δ_p which is induced by the shift of the neutral axis at plate failure is always positive; that is, it is always such as to encourage a Mode II collapse. Therefore if the initial geometric eccentricity Δ is positive, then the theory necessarily predicts a Mode II collapse because both of the eccentricities, Δ and Δ_p, are positive. Of course there are other possible factors, such as stiffener imperfections and eccentricity of applied load, which can override this tendency and cause a Mode I collapse. But it is a fact that panels which have a positive initial eccentricity and are subjected to a purely in-plane load usually collapse in Mode II. The converse is not true; that is, a panel with a purely axial load and with a small or moderate negative initial eccentricity may collapse in either Mode I or Mode II. This is because the induced load eccentricity Δ_p is always positive, and hence it may eclipse a negative value of Δ.

Also, a negative initial eccentricity can give an apparent strengthening against a Mode II collapse. A negative deflection relieves the plate of some compression, and therefore a small or moderate negative initial eccentricity (not large enough to cause a Mode I collapse) may delay plate failure and hence give a small but measurable increase in the Mode II collapse strength.

This strengthening effect is not restricted to the case of pure in-plane loading. For any lateral load it will occur whenever Δ is of opposite sign to δ_0, whether positive or negative. However, the effect is significant only for small lateral loads, when δ_0 is of the same order of magnitude as Δ. In this case η is very small since it is proportional to the sum of these two eccentricities. The effect which this has on the collapse curves is shown in Fig. 14.8, exaggerated for clarity. Some examples of this effect in actual panels are presented in Section 14.4.

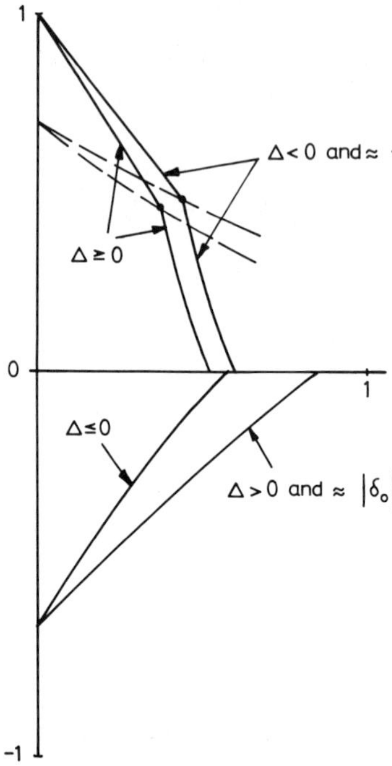

Figure 14.8 Effect of initial eccentricity.

14.3 ULTIMATE STRAIN OF STIFFENED PANELS

As mentioned earlier, in overall structural collapse due to hull girder bending the loading condition on each plate panel is actually a uniform applied strain, ε_a, rather than a uniform applied stress, and it is more correct to express plate collapse in terms of the value of strain in the plate when it collapses, $\varepsilon_{a,u}$. The same is true for the stiffener-plate combination and so it is necessary to have an expression for $\varepsilon_{a,u}$ for this type of member. The expression depends on the degree of yielding in the cross section, which in turn depends on the collapse mode. For Mode I the cross section is still elastic and so the ultimate strain is

$$\epsilon_{a,u} = \frac{\sigma_{a,u}}{E} \tag{14.3.1}$$

with $\sigma_{a,u}$ given by (14.2.6).

For Mode II the stiffener is again elastic since the upper end of the Mode II range is stiffener tensile yield. The plating is inelastic but it has been replaced by a pseudoelastic plate, having an effective elastic modulus E_s. Thus the stiffener and the plate are like two parallel elastic rods for which the axial stiffnesses are EA_s/L and E_sA_p/L. Since they are in parallel and must have the same axial shortening, the total stiffness is the sum of the individual stiffnesses. Expressing the total stiffness in terms of the total area $A = A_p + A_s$ and a mean value of secant modulus $\overline{E}_s$ gives

$$\frac{\overline{E}_s A}{L} = \frac{E A_s}{L} + \frac{E_S A_p}{L}$$

and by noting that $A_{tr} = A_s + T A_p$ we have

$$\overline{E}_s = \frac{A_{tr}}{A} E \qquad (14.3.2)$$

The ultimate strain is then

$$\varepsilon_{a,u} = \frac{A}{A_{tr}} \frac{\sigma_{a,u}}{E} \qquad (14.3.3)$$

in which $\sigma_{a,u}$ comes from the procedure given at the end of the discussion of Mode II collapse, during which A_{tr} is also evaluated.

For Mode III the degree of yielding in the stiffener is not known precisely and the safest and easiest course is to regard the stiffener as being fully yielded, with an effective tangent modulus of zero. The mean value of secant modulus for the panel is then

$$\overline{E}_s = \frac{A_p}{A} E_s \qquad (14.3.4)$$

and the ultimate strain is

$$\varepsilon_{a,u} = \frac{A}{A_{p,tr}} \frac{\sigma_{a,u}}{E} \qquad (14.3.5)$$

14.4 COMPARISON OF THEORY AND EXPERIMENT

Single Span Panels

In order to verify the beam column approach presented in the previous sections, we now compare the collapse loads predicted by this method with the measured collapse loads of a series of fifteen panels which were tested to destruction under various combinations of lateral and in-plane load by Michelutti and Murray at Monash University in Melbourne [2,3,4]. The experimental setup is shown in Fig. 14.9. The bending moment was provided by two point loads W applied to each stiffener at the $a/3$ positions, giving a bending moment $M_0 = Wa/3$ which is constant over the middle one-third of the panel. For this load the maximum deflection is

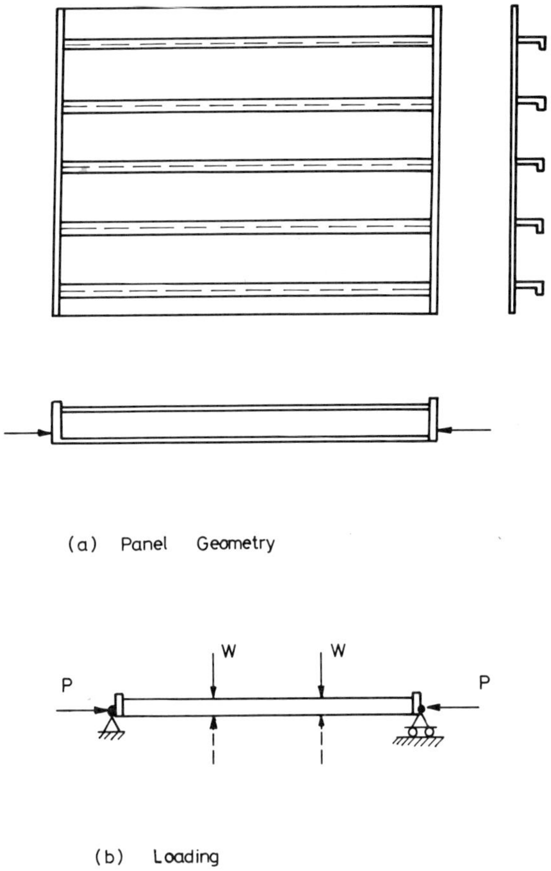

Figure 14.9 Experimental setup of Monash tests.

$$\delta_0 = \frac{23 W a^3}{648 EI} \qquad (14.4.1)$$

Also the plastic hinge load W_P is related to M_P by

$$M_P = \frac{W_P a}{3} \qquad (14.4.2)$$

Thirteen of the panels were of the same (nominal) overall dimensions, namely:

$a = 3.20$ m, $B = 2.44$ m, $b = 0.533$ m,

$h_w = 0.152$ m, $t_p = 9.7 \pm 0.3$ mm,

$t_w = 7.3 \pm 0.2$ mm

One panel (panel A4 in the terminology of Ref. 2) was slightly shorter (a = 3.05 m) but was geometrically similar to the others. Only one panel (panel A6) had any significant difference in geometry; panel A6 was 54% longer, having a = 4.93 m.

The principal difference among the panels was in regard to yield stress. Within each panel the stiffeners and plating had different yield stresses, and the ratio varied from panel to panel. In nine panels σ_{Ys} exceeded σ_{Yp}, the three largest ratios being 1.25 (panel Q) and 1.19 (panels A5 and N) and in six panels σ_{Ys} was less than σ_{Yp}, the two smallest ratios being 0.86 (panel L) and 0.91 (panel A3). Details are given in Table 14.1, including the value of initial eccentricity Δ/a and the values of various parameters. The table gives the experimental collapse loads and the corresponding normalized values $\sigma_{a,u}/\overline{\sigma}_Y$ and W_{ult}/W_P. This is followed by the theoretical normalized values obtained from the theory of the preceding sections.

Although fourteen of the panels were virtually identical in geometry, the variation in the ratio of yield stress means that each panel has a different collapse curve. For each panel the complete curve was calculated using the particular values of plate and stiffener thickness and of initial eccentricity of that panel, in order to make the comparison of theory and experiment more precise. Some examples are shown in Fig. 14.10. These have been chosen to cover the full range of collapse load combinations and to illustrate the variety among the curves. For clarity most of the curves are not plotted in full but only in the vicinity of the experimental collapse load for that panel.

Panel Q is one of the panels with a large value of σ_{Ys}/σ_{Yp} and hence its Mode III curve is higher. Panels A2, H, and Z are typical of most of the panels, with a yield stress ratio close to unity. Panels H and A6 were subjected to pure in-plane compression and they collapsed due to plate failure (Mode II). The behavior of stiffened panels that are subjected to a purely axial load is an important aspect of stiffened panel response and merits some discussion. Therefore we shall use these two panels as examples to illustrate and explain this behavior.

When the loading is pure in-plane compression the initial eccentricity has a strong influence on whether collapse occurs in Mode I or Mode II. Also, as discussed in the previous section, a negative initial eccentricity can give an apparent strengthening against a Mode II collapse. This phenomenon occurred in panel H. Unfortunately, neither the direction nor the magnitude of the initial eccentricity were recorded for panel H. However, it appears clear from the test results that the initial eccentricity was negative because the measured values of lateral deflection are negative at all stages of the loading, until just before collapse. At an applied load of σ_a = 232 MPa ($\sigma_{a,u}/\overline{\sigma}_Y$ =

TABLE 14.1 COMPARISON OF THEORY AND EXPERIMENT

	Panel Characteristics						Experiment				Theory[a]		Discrepancy		
Panel	σ_{yp}	σ_{Ys}	$\frac{\sigma_{ys}}{\sigma_{Yp}}$	$\overline{\sigma}_Y$	a/Δ	$\lambda_{\text{I,II,GH}}$	$\sigma_{a,u}$	W_{ult}	$\frac{\sigma_{a,u}}{\overline{\sigma}_Y}$	$\frac{W_{\text{ult}}}{W_P}$	$\frac{\sigma_{a,u}}{\overline{\sigma}_Y}$	$\frac{W_{\text{ult}}}{W_P}$	Value	%	Coll-apse Mode
L	315	270	0.86	305	−3150	0.79 (I)	0	−39.0	0	−0.89	0	−0.72	0.17	19.1	I
L1	396	379	0.96	391	1600	0.94 (I)	0	−39.6	0	−0.71	0	−0.71	0	0	I
Y	363	378	1.04	367	1580	0.94 (I)	62	−29.8	0.17	−0.65	0.15	−0.58	0.08	11.9	I
Z	367	370	1.01	368	6300	0.93 (I)	123	−14.6	0.33	−0.31	0.31	−0.30	0.02	4.4	I
N	332	396	1.19	346	930	0.97 (I)	184	−9.0	0.53	−0.19	0.51	−0.18	0.02	3.5	I
J	370	381	1.03	374	−1200[b]	0.95 (I)	235	0	0.63	0	0.64	0	0.01	1.6	I
H	378	389	1.03	381	not given	0.80 (II)	253	0	0.66	0	0.59	0	0.07	10.6	II
A6[c]	368	383	1.04	372	530	1.21[c] (II)	153	0	0.41	0	0.39	0	0.02	4.9	II
Q	318	397	1.25	335	890	0.75 (II)	172	11.1	0.51	0.24	0.51	0.24	0	0	II
A5	324	385	1.19	338	970	0.75 (II)	170	22.8	0.50	0.49	0.46	0.45	0.06	8.6	II
A4	372	361	0.97	369	1330	0.76 (II)	154	36.4	0.42	0.65	0.38	0.59	0.07	9.6	II
A3	370	338	0.91	363	1150	0.82 (GH)	139	37.0	0.38	0.69	0.33	0.59	0.12	15.2	III
A2	383	354	0.92	378	1050	0.84 (GH)	139	38.4	0.37	0.71	0.31	0.60	0.12	15.1	III
A1	378	371	0.98	376	1400	0.85 (GH)	93	42.0	0.25	0.79	0.24	0.74	0.05	6.1	III
P	324	369	1.14	334	800	0.88 (GH)	0	46.8	0	1.02	0	1.00	0.02	2.0	III

[a] All theoretical results are for $a/\Delta = \pm 750$ (− for Mode I; + for Modes II and III).
[b] In this panel the eccentricity was in the axial load, and was equivalent to $\Delta = -a/1200$.
[c] Long panel (a = 4.93 m); therefore λ is large.

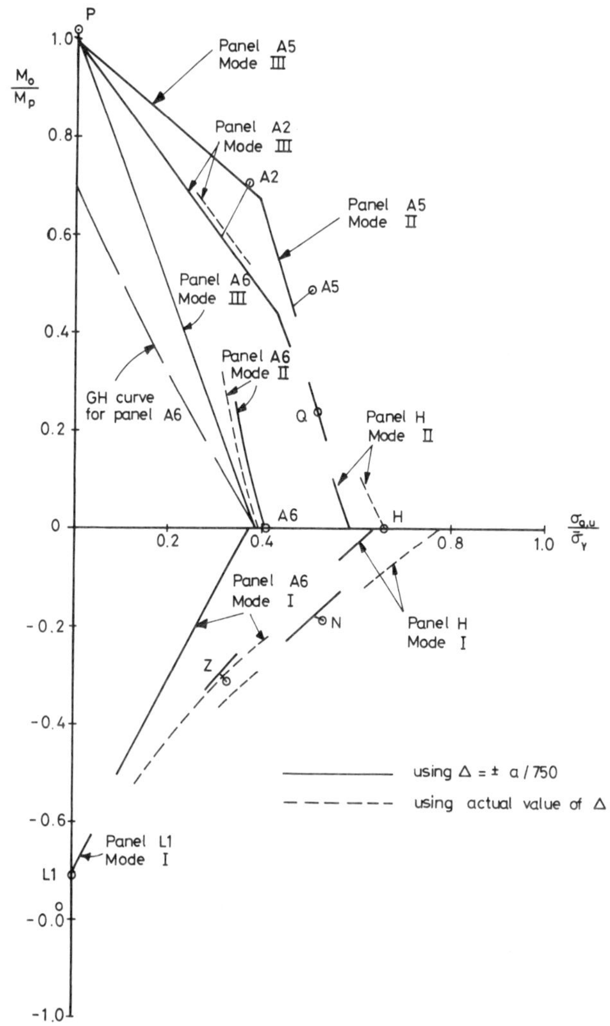

Figure 14.10 Comparison of theory and experiment for sample panels.

$232/381 = 0.61$) the plating began to fail and the panel immediately deflected in the positive direction (i.e., toward the stiffeners). On further increase of load, the positive deflection grew rapidly and a Mode II collapse occurred soon after, at $\sigma_{a,u}/\bar{\sigma}_Y = 0.66$. With a positive eccentricity (e.g., the standard design value of $+a/750$) the beam column theory for Mode II, as outlined previously, gives a collapse load ratio of 0.59. Therefore it appears that the panel had a negative initial eccentricity which artificially strengthened it. Obviously, one must ignore such strengthening when designing, but in comparing theory and experiment it is important to model the actual situation as closely as possible. Therefore a small negative initial eccentricity ($-a/2000$) was assumed and the theoretical collapse loads were calculated for Mode I and Mode II (with a small negative value of Δ, both equations have solutions). The resulting values of $\sigma_{a,u}/\bar{\sigma}_Y$ were 0.77 and 0.67 for Modes I and II, as shown in Fig. 14.10. Thus when the negative initial eccentricity which appears to have been present is represented by a typical value, the theory correctly predicts that Mode II is the collapse mode, and the predicted value agrees closely with the experimental value of 0.66. For comparison, Fig. 14.10 also shows the collapse curves corresponding to the standard design values of initial eccentricity ($-a/750$ for Mode I and $+a/750$ for Mode II). These are the curves which would in practice be taken as the predicted capability of the panel. Since they are design curves they must always give a predicted collapse load which is less than the actual collapse load for any panel whose initial eccentricity is within the permissible range: $-a/750 < \Delta < a/750$. At the same time they should not be excessively conservative. As shown in the figure, the curves for panel H fulfill both of these requirements.

Panel A6 is the long panel and consequently it has a much larger slenderness parameter: $\lambda_{\mathrm{I}} = 1.46$ and $\lambda_{\mathrm{II}} = 1.21$, compared to typical values of 0.95 and 0.75 for the other panels. This causes a substantial decrease in ultimate strength in both modes. In fact, the stiffener tensile yield curve for panel A6 (the curve GH in Fig. 14.10) is so close to the origin that it does not intersect the Mode II curve, and the Mode III collapse curve is a straight line from the point $M_0/M_P = 1$ to the point where the GH curve crosses the horizontal axis. As noted earlier, this is conservative because in such cases the collapse load combinations would follow the Mode II collapse curve upward for some distance (i.e., up to moderate values of M_0). But the conservativism is not excessive and it emphasizes the fact that such a panel does have generally low values of nondimensional collapse loads.

Panel A6 also provides another illustration of the crucial influence which the initial eccentricity has when there is no lateral load. For this panel, when the standard design value of $-a/750$ is assumed for Mode I and $+a/750$ for Mode II, the two values of $\sigma_{a,u}/\sigma_Y$ are 0.376 and 0.405 respectively, as shown in the figure. Thus if we were considering this panel in an objective design context, in which both positive and negative eccentricity must be allowed for, the panel could be said to be weaker in Mode I than in Mode II. But in this specific case the panel had a relatively large positive initial eccentricity ($+a/530$) and therefore Mode I collapse could not and did not occur. For a Mode I collapse to occur, Δ would have to be negative (and sufficiently negative to overcome the induced eccentricity effect). When the measured value of $+a/530$ is used for the Mode II calculation, the predicted collapse strength is 0.390, which agrees well with the experimental result: the panel in fact underwent a Mode II collapse at an applied load ratio of 0.41.

Since the tests covered a full range of load combinations, including purely lateral loads and purely inplane loads, the discrepancy between theory and experiment is best measured along a radial line from the origin as shown for panels A2, A5, N, and Z in Fig. 14.10. The relative error is then obtained by dividing this radial discrepancy by the length along the radial line from the origin to the experimental collapse point. This has been done for all fifteen cases and the results are given in Table 14.1. Also, to give a clearer overall comparison, Fig. 14.11 contains all of the radial discrepancies plotted in their correct positions relative to a single typical collapse curve. The calculations were performed using the standard design values of initial eccentricity, $\pm a/750$. The results indicate that the beam column equations presented in the previous sections are of satisfactory accuracy, and that the value of $\pm a/750$ for the initial eccentricity is suitable for design. In all panels except A6 (the long panel) the design value of eccentricity exceeded the actual value (see Table 14.1). The design value of $\pm a/750$ is a maximum permissible value; panels with larger eccentricity would either be forbidden by the fabrication specifications or would be designed as curved panels. For all panels in which the initial eccentricity was within these permissible limits the actual value of ultimate strength was never less than the design value, and the design value was never excessively conservative. Moreover, even in panel A6, in which the initial eccentricity was slightly over the permissible value, the collapse load was still slightly above the design value. These results demonstrate that the beam column approach is well suited for the design of stiffened panels, since it gives an accurate estimate of strength for all load combinations and for any typical amount of initial eccentricity.

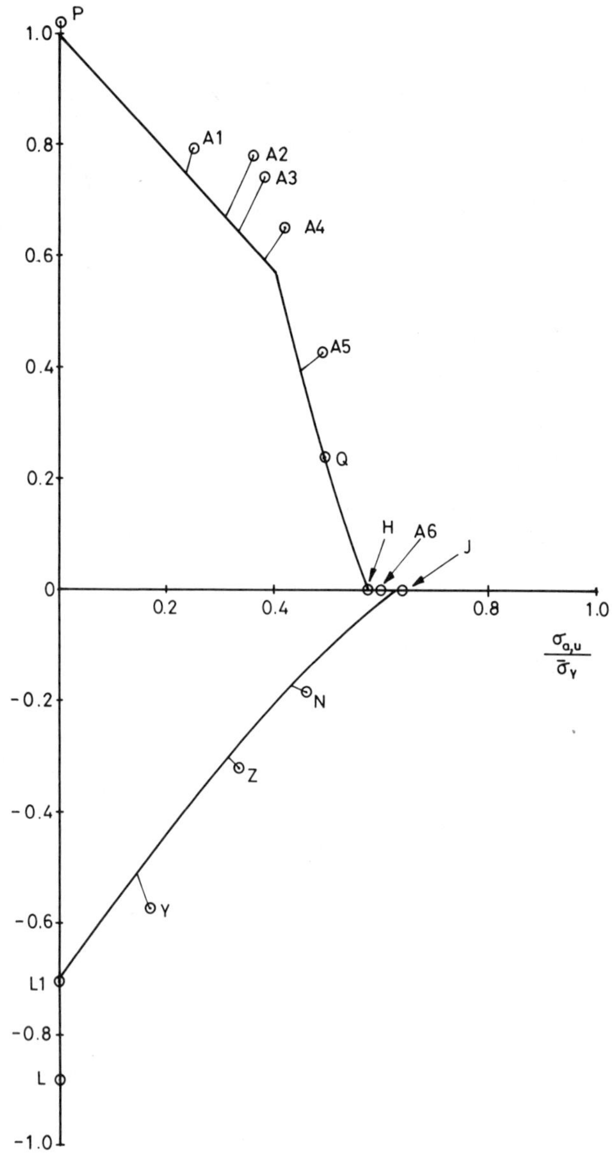

Figure 14.11 Composite plot of results.

Multispan Panels Without Lateral Load

To demonstrate the validity of the method for multispan panels we next compare its predicted values of $\sigma_{a,u}$ and $\varepsilon_{a,u}$ with the measured values obtained by Dowling et al. [5] in a test of a 3.9- m long box girder subjected to a sagging bending moment (model no. 2 in the terminology of Ref. 5). The cross section is shown in Fig. 14.12. The upper flange (the "strength

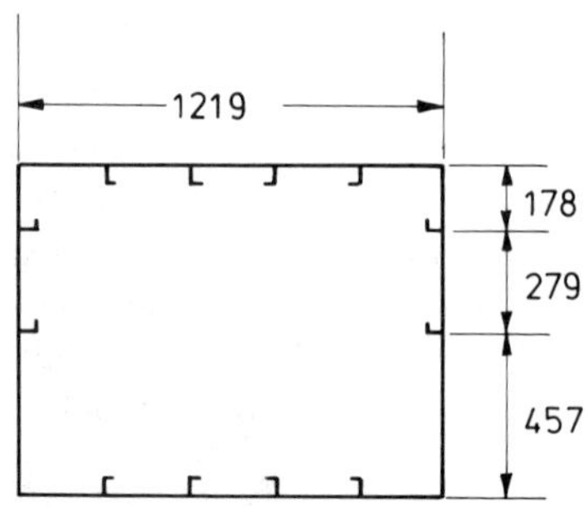

Figure 14.12 Box girder model section.

deck") consisted of five in-line panels 787 mm in length. The panels were typical in all respects to ship panels; details are given in the figure and in Table 14.2. From Ref. 5, in the bay where the collapse occurred the conditions before loading were:

maximum initial deflection of stiffeners:

upward (towards plating)

$$\Delta = -0.010 \text{ in.} = -0.25 \text{ mm} \quad (a/3100)$$

downward (towards stiffener)

$$\Delta = 0.055 \text{ in.} = 1.40 \text{ mm} \quad (a/563)$$

average longitudinal residual stress:

$$\sigma_r = 0.18\,\sigma_Y$$

Since there is no lateral load, both Mode I and Mode II collapse must be investigated. In order to provide a worked example of the method some details of the calculation will be given.

Mode I collapse

Beam-column parameters:

from (8.3.6): $C_1 = 0.083$, $I = 326700 \text{ mm}^4$,

$$\rho = 14.65 \text{ mm}$$

$$C_2 = 0.8775, \; y_f = -44.6 \text{ mm}$$

from (14.2.3): $\lambda = 0.623$, $\eta = 0.052$

from (14.2.5): $R = 0.925$

and hence the value of $\sigma_{a,u}$ for a Mode I collapse is $0.925\sigma_{Ys} = 255.8 \text{ N/mm}^2$. From (14.3.1) the value of $\varepsilon_{a,u}$ is 1.226×10^{-3}.

Mode II collapse

Plate slenderness: $\beta = 1.89$

from (14.2.8): $T = 0.825$,

TABLE 14.2 PROPERTIES OF BOX GIRDER MODEL

Component	Dimensions (mm)	σ_Y (N/mm²)	E (N/mm²)
Compression flange	4.88	298.0	208,500
Tension flange	4.88	298.0	208,500
Web	3.38	211.6	216,200
Stiffeners	50.8 × 15.9 × 4.8L	276.5	191,500

$$\sigma_F = \frac{T - 0.1}{T}\,\sigma_{Yp} = 262 \text{ N/mm}^2$$

from (8.3.6): $C_1 = 0.104$, $I_{tr} = 352000 \text{ mm}^4$,

$$\rho = 16.40 \text{ mm}$$

$$C_2 = 0.857, \; y_{p,tr} = 7.27 \text{ mm}$$

from (14.2.18): $\Delta_p = hA_s\left(\frac{1}{A_{tr}} - \frac{1}{A}\right)$

$$= (29.5)297\left(\frac{1}{1308} - \frac{1}{1523}\right)$$

$$= 0.946 \text{ mm}$$

from (14.2.21): $\Delta_H = \frac{I/A}{H} = \frac{\rho^2}{H} = \frac{(16.40)^2}{457 - 8}$

$$= 0.60 \text{ mm}$$

from (14.2.16): $\lambda = 0.542$, $\eta = 0.054$

from (14.2.19): $\eta_p = 0.031$

from (14.2.20): $R_{II} = 0.905$

from (14.2.22): $\sigma_{a,u} = 203.7 \text{ N/mm}^2$

and from (14.3.3): $\varepsilon_{a,u} = 1.138 \times 10^{-3}$

Since the Mode I value of $\sigma_{a,u}$ was larger than this Mode II value the latter is selected as the predicted value of ultimate strength.

As shown in Fig. 14.13 this predicted result is in quite good agreement with the experimental results:

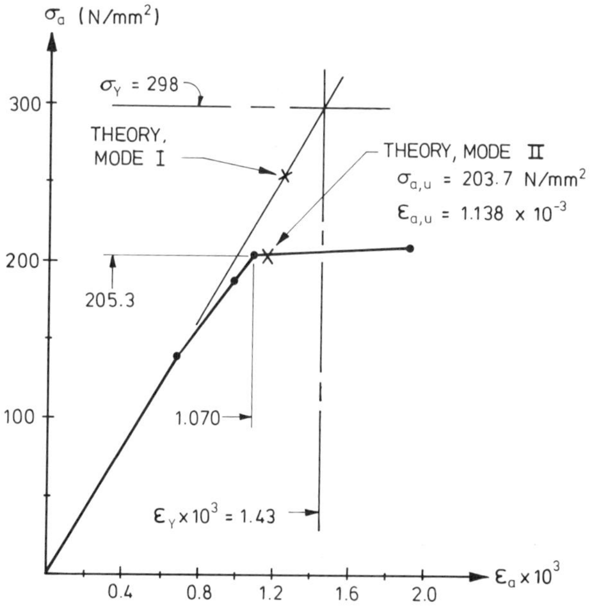

Figure 14.13 Collapse of upper flange of box girder model.

Mode II collapse at an average applied stress of 205.3 N/mm^2 and a strain of 1.070×10^{-3}.

It should be noted that the selection of the predicted mode of collapse must be based on $\sigma_{a,u}$ and not on $\varepsilon_{a,u}$. A smaller value of $\varepsilon_{a,u}$ for Mode I does not mean that Mode I collapse would occur first, because the Mode II collapse actually begins much earlier, when the plate begins to fail, and from this point onward the beam-column undergoes ever-increasing positive bending, opposite to that of Mode I. Thus a Mode I collapse can only occur if the value of $(\sigma_{a,u})_{\text{I}}$ is smaller than the load which initiates plate failure. Hence we are being conservative in saying that a Mode I collapse only requires that $(\sigma_{a,u})_{\text{I}}$ be less than $(\sigma_{a,u})_{\text{II}}$, and then taking this smaller value to be the ultimate strength of the panel. But as mentioned earlier, whenever the lateral load is small or is absent the occurrence of Mode I or Mode II depends not only on the properties and proportions of the panel but also on the particular circumstances and details such as initial deflections, load sequence, and residual stress, which cannot be known or predicted by the designer. Hence the designer must always be conservative and assume the worst, which means choosing whichever mode gives the lower value of strength.

Multispan Panels With Lateral Load

We next compare the method's predictions with the results of tests by Smith [6] on multispan panels subjected to a uniform lateral pressure in addition to axial compression. In comparing theory and experiment it is necessary to have full information about the initial conditions (residual stress, preload deflections, and the values of δ_0 and M_0 due to the lateral load) for the particular stiffener-plate combination which triggered the collapse. Of the 11 panels tested in Ref. 6 panels 3a and 3b are chose for the comparison because for these panels all of the required information was recorded and presented. Also these two panels provide valuable information about the effect of lateral pressure because they were a matching pair and one of them (3a) had a lateral pressure of 3 psi whereas the other had no lateral load.

PANEL 3a

This panel was subjected to a constant lateral pressure of 3 psi. The panel collapsed when the applied axial stress reached 0.69 $\overline{\sigma}_Y$, where $\overline{\sigma}_Y$ is the mean value of yield stress of the stiffeners and plating, averaged over the panel cross section.

This case is worth examining in some detail, especially in regard to the causal sequence of failures, because it provides an example of how a laterally loaded multispan panel, in spite of the rotational restraint at the ends of each span, can undergo the same type of collapse as a laterally loaded pinned panel.

As shown in Fig. 14.14 the collapse involved upward flexural buckling in one of the central bays and downward bucking of an adjacent bay, together with a stiffener flange mechanism of the type shown in Fig. 14.4, close to the central transverse frame. All three failures occurred more or less simultaneously, and to determine the precise interactions and the causal relationships we must examine the state of loading just prior to failure. Figure 14.15 illustrates the bending moment distribution in the stiffeners due to the lateral pressure. This was not given in Ref. 6 but was calculated by the author by a nonlinear finite element analysis of the gross panel. The total compressive stress in the stiffener flange is a maximum at the midlength of the gross panel, because of the peak value of bending moment there. The finite element analysis showed that when the applied axial stress reached the experimental collapse value the total stress in the stiffener flange at the midlength had just slightly exceeded the yield stress of the stiffener. At this time yield had not occurred anywhere else and all of the panels were still stable. This shows clearly that it was the stiffener

Figure 14.14 Panel 3a after collapse. (Photo courtesy of Admiralty Marine Technology Establishment, Dunfermline, Scotland).

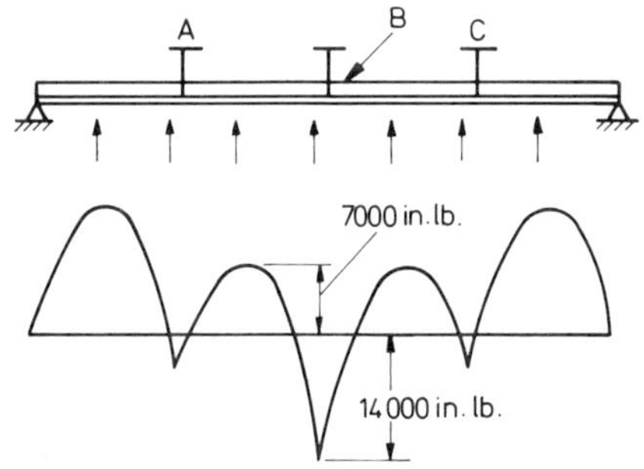

Figure 14.15 Bending moment in panel 3a.

flange mechanism which triggered the collapse, and also provides the explanation as to why the panels buckled at the same time as the formation of the stiffener flange mechanism. Such a mechanism is virtually the same as a hinge (an ordinary hinge, not a plastic hinge) because the flange has lost nearly all of its axial stiffness. Therefore the stiffeners rotated about this point and transferred the bending moment into the spans. The new bending moment in each span is approximately that for a beam which is pinned at one end and clamped at the other, with a midspan value of $pba^2/12$. The increased bending stress in the plating was sufficient to cause a Mode II collapse of span AB, and so this span collapsed immediately after the flange mechanism had occurred, with no further increase in axial load. Since the transverse frame provides very little rotational restraint and also permits at least a small lateral deflection, it is the hinge rather than the frame which constitutes the endpoint of the two beam-columns, and since AB is longer than BC the Mode II collapse occurred in the former.

This case demonstrates that in spite of the rotational restraint which exists at the ends of each span in a laterally loaded multispan panel, collapse can still occur in a manner very similar to that of a panel with pinned ends. Therefore the pinned beam-column approach, with allowance for the values of M_0 and δ_0 which exist within the span, is directly relevant to such panels.

In the rationally-based design process of Fig. 2.19, the hull module analysis, member limit analysis and constraint evaluation of the panels (steps 3, 4, and 5) will indicate if a stiffener flange mechanism can occur, and if so the use of any reasonable estimates of the post-mechanism values of M_0 and δ_0 will give a good estimate of $\sigma_{a,u}$. For example, using the approximate post-mechanism values of $pba^2/12$ for M_0 and $\frac{1}{2}(5pba^4)/384EI$ (half of the simply supported value) for δ_0, together with a typical design value of $a/750$ for Δ, gives a value of $0.654\overline{\sigma}_Y$ for $\sigma_{a,u}$, which differs from the actual value by only 5%, and on the conservative side.

The actual measured value of Δ for the span where the collapse occurred was $\Delta = 0.03$ in. ($a/2000$) and when this value is used the predicted value of $\sigma_{a,u}$ is $0.69\overline{\sigma}_Y$, the same as the experimental value.

Panel 3b

A second panel which was virtually identical to panel at an applied stress of $\sigma_{a,u} = 0.61\overline{\sigma}_Y$. Fig. 14.16 shows the panel after collapse. Since there is no lateral load, a Mode II collapse would also have been possible and one of the requirements of any method for estimating ultimate strength is that it should correctly identify the critical mode, that is, the mode with the lower value of $\sigma_{a,u}$. For this panel the initial deflection of the collapsed span was approximately 0.15 in. toward the plating ($\Delta = -a/400$). With the method presented herein the predicted values of $\sigma_{a,u}$ for Modes I and II are $0.57\overline{\sigma}_Y$ and $0.69\overline{\sigma}_Y$.* Thus the method correctly identifies the collapse mode and predicts a value for $\sigma_{a,u}$ which differs from the experimental value by only 7% on the conservative side. This 7% margin comes mainly from the conservative assumption that a stiffener flange mechanism forms as soon as the stress in the midthickness of the stiffener flange reaches yield, whereas in reality if forms only after the flange is fully yielded.

Figure 14.16 Panel 3b after collapse. (Photo courtesy of Admiralty Marine Technology Establishment, Dunfermline, Scotland).

* The latter is obtained using the largest positive value of Δ, which was 0.22 in. ($a/270$).

In the previous multispan examples the collapse has been Mode II, and hence this example demonstrates the accuracy of the method for a Mode I collapse of a multispan panel.

CONCLUSIONS

The analysis of panel 3b shows that when a multispan panel has no lateral load the ultimate strength is basically that of a simply supported subpanel. For reference purposes we shall denote this value as $(\sigma_{a,u})_0$. It is always the lesser of two values: either the Mode I value from (14.2.6) or the Mode II value from (14.2.22). If the lateral load is above a certain magnitude, each subpanel is effectively clamped and if this remains true then the ultimate strength is larger than for a simply supported panel. For panels 3a and 3b the increase was from $0.61\overline{\sigma}_Y$ to $0.69\overline{\sigma}_Y$, or 13%. But it is extremely difficult to predict the magnitude of lateral load that is required to constrain all subpanels to behave as clamped, first, because the behavior of each subpanel is strongly influenced by local factors (eccentricity, residual stress, nonuniformity of lateral load, etc.) and, second, because each subpanel interacts with its neighbors. Moreover, although the lateral load may be sufficient to keep each subpanel rotationally restrained for a time, such that the value of σ_a can exceed $(\sigma_{a,u})_0$, there is always the possibility of a stiffener flange mechanism, whereupon two or more of the subpanels suddenly change to the alternating wave shape and immediately undergo collapse, because σ_a already exceeds $(\sigma_{a,u})_0$. We thus have four factors to consider:

1. The possibility that the lateral load may become small.
2. The difficulty of calculating the magnitude of the lateral load which would guarantee end clamping.
3. The possibility of a stiffener flange mechanism, which becomes more likely with increasing lateral load.
4. The undesirability of sudden collapse.

For all of these reasons, any value of ultimate strength larger than $(\sigma_{a,u})_0$ should not be used for design purposes, even though larger values may in fact be achieved when there is a lateral load. That is, $(\sigma_{a,u})_0$ is an upper limit or "cutoff" value. In a rationally-based design process it is necessary to calculate $\sigma_{a,u}$ for each load case. For the various load cases involving a lateral load it would be possible to calculate accurate values of $\sigma_{a,u}$ by using the information from steps 3, 4, and 5 of the overall design procedure, as discussed previously. But for design purposes this is not necessary because these values will nearly always exceed the cutoff value, $(\sigma_{a,u})_0$. Hence for design purposes it is sufficient to use the simple and conservative pinned beam-column approach, with M_0 and δ_0 equal to their simply supported values, for estimating the values of $\sigma_{a,u}$ for the various load cases involving a lateral load. Although the true ultimate strength would be larger than this value it would usually be ruled out because it exceeded $(\sigma_{a,u})_0$. Therefore the value obtained by the pinned beam-column approach will not be unduly conservative. For example, for the load case involving lateral pressure (panel 3a) the use of the simply supported values of M_0 and δ_0 and a typical value of Δ $(a/750)$ gives a value of $0.56\overline{\sigma}_Y$. For this load case the true ultimate strength is $0.69\overline{\sigma}_Y$, but this cannot be used for design purposes because it exceeds the value for zero lateral load, $(\sigma_{a,u})_0 = 0.61\overline{\sigma}_Y$. Hence, taking $0.56\overline{\sigma}_Y$ as the ultimate strength for the laterally loaded case is not overly conservative since it is only 8% below the experimentally obtained value of $(\sigma_{a,u})_0$.

REFERENCES

1. C. S. Smith, "Influence of Local Compressive Failure on Ultimate Longitudinal Strength of a Ship's Hull," Proc. Int'l. Symp. on Practical Design in Shipbuilding (PRADS), Tokyo, October 1977.
2. N. W. Murray, "Analysis and Design of Stiffened Plates for Collapse Load," *Struct. Eng.*, **53** (3), March 1975, pp. 153–158.
3. W. M. Michelutti, *Stiffened Plates in Combined Loading*, Ph.D. thesis, Dept. of Civil Engineering, Monash University, Melbourne, 1976.
4. W. M. Michelutti and N. W. Murray, "The Collapse Behaviour of Stiffened Plates under Combined Axial and Bending Loads," 6th Australasian Conference on the Mechanics of Structures and Materials, Univ. of Canterbury, Christchurch, N. Z., 1977.
5. P. J. Dowling, S. Chatterjee, P. A. Frieze, and F. M. Moolani, "The Experimental and Predicted Collapse Behaviour of Rectangular Stiffened Steel Box Girders," Int'l. Conference on Steel Box Girder Bridges, Institute of Civil Engineers, London, 1973.
6. C. S. Smith, "Compressive Strength of Welded Steel Ship Grillages," *Trans. RINA*, **117**, 1975, pp. 325–359.

Chapter

Fifteen

STRUCTURAL INSTABILITY: NONLINEAR FRAME AND FINITE ELEMENT ANALYSIS

As mentioned in Section 2.6, the linear stiffness-based method of finite element analysis can be generalized and applied to nonlinear problems. In a structure, the two main types of nonlinearity are geometric nonlinearity and material nonlinearity. The former occurs when the loads on a structure cause a change in the structure geometry which is sufficient to alter its state of equilibrium. The most common example is buckling, which is covered in this chapter. Material nonlinearity is covered in Chapter 16.

For simplicity the only element considered in these chapters is the beam element and hence the precise topic of these two chapters is nonlinear *frame analysis*. But this demonstrates all of the basic principles of nonlinear finite element analysis, just as Chapter 5, dealing with linear frame analysis, demonstrated the basic principles of linear finite element analysis. Also, as shown in Chapter 17, it is possible to use a framework type of structural model for the hull module ultimate strength analysis.

Section 15.1 discusses the interaction that exists between axial load and lateral deflection—an interaction which is not considered in ordinary (linear) frame analysis. It is shown that this interaction gives rise to extra terms in the beam stiffness matrix which are dependent on the member forces, which are in turn dependent on the unknown nodal displacements; it is this which makes the analysis nonlinear. These extra terms are usually collected together to form a separate element stiffness matrix—the *geometric* stiffness matrix $\mathbf{k}_G$—which is additive to the ordinary stiffness matrix. Because of the nonlinearity the solution must be obtained iteratively, and Section 15.2 presents two computational algorithms for iterative finite element analysis. Section 15.3 presents $\mathbf{k}_G$ for a completely general beam element, and Section 15.4 deals with *elastic bifurcation buckling,* which is essentially an eigenvalue problem. Section 15.5 deals with the special case of a beam attached to plating. Section 15.6 considers nonbifurcation buckling, which requires an incremental load–deflection approach; in effect, a mathematical simulation of the buckling process. The methods that use this approach are referred to as *incremental frame analysis,* and *incremental finite element analysis.*

15.1 INTERACTION BETWEEN LOAD AND DEFLECTION: GEOMETRIC STIFFNESS MATRIX

As explained in Section 2.3, there are several instances in which the deflection caused by a load alters the equilibrium conditions of the member. We will now demonstrate this for a simple beam element, showing that there is an interaction between member force and lateral deflection, and introducing some corresponding new terms in the equilibrium equations that are ignored in the usual first order analysis. If higher-order terms are neglected, the new terms are linearly proportional to the deflection and hence they can be considered as an additional stiffness matrix.

In a beam of normal proportions the shear deformation is negligibly small and hence the strain is entirely longitudinal and is given by

$$\varepsilon_x = \frac{du}{dx} - y\left(\frac{d^2v}{dx^2}\right) + \frac{1}{2}\left(\frac{dv}{dx}\right)^2 \qquad (15.1.1)$$

The first and second terms are the familiar components of axial and flexural strain respectively, whereas the third term, which is nonlinear in the transverse displacement v, represents a strain component associated with the coupling of flexural and axial action. The origin of this term can be seen from Fig. 15.1, in which a differential length of fiber is shown in the displaced position. The ends of the displaced fiber are drawn directly beneath their undisplaced positions because the axial strain due to any horizontal stretching or shortening is accounted for by the first two components of strain, and we now wish to account for the effect of the lateral displacement only. In the displaced position the length ds is given by

$$ds = \sqrt{1 + \left(\frac{dv}{dx}\right)^2}\, dx$$

and since dv/dx is small this may be approximated by

$$ds \approx \left[1 + \frac{1}{2}\left(\frac{dv}{dx}\right)^2\right] dx$$

Hence the extra strain caused by the lateral deflection is

$$\frac{ds - dx}{dx} \approx \frac{1}{2}\left(\frac{dv}{dx}\right)^2$$

which is the third term of (15.1.1).

We now apply the virtual work theorem in order to derive the stiffness matrix for a beam element. Since the strain is nonlinear we use the variational form of the theorem, which states that $\delta W_{\text{ext}} = \delta U_{\text{int}}$. That is, if a variational change is made to the nodal displacements, $\boldsymbol{\delta\delta}$, then the variation of external work which is done by the actual nodal forces $\mathbf{f}$ (which are constant) is equal to the variation in internal strain energy, which involves the variation in strain $\delta\boldsymbol{\varepsilon}$ (expressed in terms of $\delta\boldsymbol{\delta}$) and the actual stress $\boldsymbol{\sigma}$ (which is constant and is expressed in terms of the actual displacements $\boldsymbol{\delta}$, corresponding to $\mathbf{f}$). The variation in external work is

$$\delta W_{\text{ext}} = \delta(\boldsymbol{\delta}^T\mathbf{f}) = \delta\boldsymbol{\delta}^T\mathbf{f}$$

and if we adopt a notation similar to (6.3.1) for nodal forces and displacements in the x- and y-direction this becomes

$$\delta W_{\text{ext}} = \delta\mathbf{u}^T\mathbf{f}_u + \delta\mathbf{v}^T\mathbf{f}_v \qquad (15.1.2)$$

where the nodal forces are

$$\mathbf{f}_u = \begin{Bmatrix} F_{x1} \\ F_{x2} \end{Bmatrix} \quad \text{and} \quad \mathbf{f}_v = \begin{Bmatrix} F_{y1} \\ M_{z1} \\ F_{y2} \\ M_{z2} \end{Bmatrix}$$

and the nodal displacements are

$$\mathbf{u} = \begin{Bmatrix} u_1 \\ u_2 \end{Bmatrix} \quad \text{and} \quad \mathbf{v} = \begin{Bmatrix} v_1 \\ \left(\frac{dv}{dx}\right)_1 \\ v_2 \\ \left(\frac{dv}{dx}\right)_2 \end{Bmatrix}$$

The variation in internal strain energy is

$$\delta U_{\text{int}} = \int_{\text{Vol}} \delta\varepsilon_x \sigma_x \, d\text{Vol}$$

Equation 15.1.1 expresses the strain ε_x in terms of the internal displacements u and v. If the corresponding expression for $\delta\varepsilon_x$ is substituted in the foregoing equation the result is

$$\delta U_{\text{int}} = \int_0^L \left\{ \int_A \delta\left(\frac{du}{dx}\right)\sigma_x \, dA - \int_A \delta\left(\frac{d^2v}{dx^2}\right)\sigma_x y \, dA + \frac{1}{2}\int_A \delta\left[\left(\frac{dv}{dx}\right)^2\right]\sigma_x \, dA \right\} dx$$

We next observe that the inner integrals on the right-hand side can be expressed in terms of internal member forces $\mathcal{F}_x$ and $\mathcal{M}_z$

$$\mathcal{F}_x = \int_A \sigma_x \, dA \quad \text{and} \quad \mathcal{M}_z = \int \sigma_x y \, dA$$

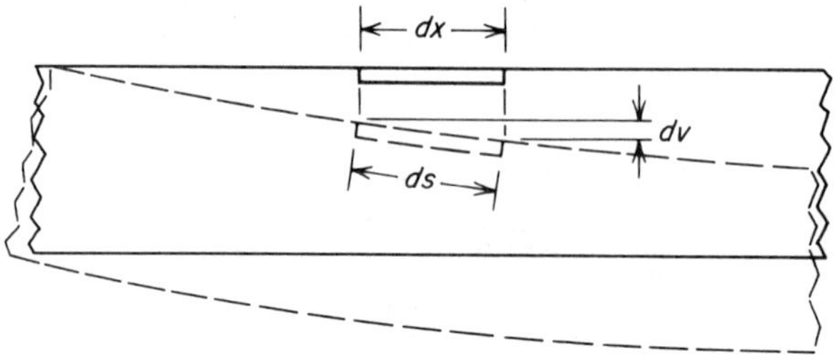

Figure 15.1 Axial strain due to lateral deflection.

Also the third term can be expanded. The result is

$$\delta U_{\text{int}} = \int_0^L \left[\delta\left(\frac{du}{dx}\right)\mathcal{F}_x - \delta\left(\frac{d^2v}{dx^2}\right)\mathcal{M}_z + \left(\frac{dv}{dx}\right)\delta\left(\frac{dv}{dx}\right)\mathcal{F}_x \right] dx$$

We next make use of the following relationships between the internal member forces and the internal member displacements

$$\mathcal{F}_x = EA\frac{du}{dx} \quad \text{and} \quad \mathcal{M}_z = EI_z\frac{d^2v}{dx^2}$$

Substitution of these in the first two terms of the right-hand side gives

$$\delta U_{\text{int}} = \int_0^L \left[\delta\left(\frac{du}{dx}\right)EA\left(\frac{du}{dx}\right) - \delta\left(\frac{d^2v}{dx^2}\right)EI_z\left(\frac{d^2v}{dx^2}\right) + \left(\frac{dv}{dx}\right)\delta\left(\frac{dv}{dx}\right)\mathcal{F}_x \right] dx \qquad (15.1.3)$$

We note that since $\mathcal{F}_x$ is proportional to du/dx and is not related to dv/dx there is no point in expressing $\mathcal{F}_x$ in terms of displacement in the third term of this equation. We next choose a *shape function* to express the internal member displacement in terms of nodal displacements. For convenience, we choose the same shape function as for the ordinary beam element.* For axial displacement the shape function is

$$u(x) = [(1-\xi) \quad \xi]\begin{Bmatrix} u_1 \\ u_2 \end{Bmatrix} = \mathbf{N}_1(\xi)\mathbf{u} \qquad (15.1.4)$$

and for the flexural displacement the shape function is, from (5.3.9b)

$$v(x) = [(1-3\xi^2+2\xi^3) \quad x(1-2\xi+\xi^2) \quad (3\xi^2-2\xi^3) \quad x(-\xi+\xi^2)]\begin{Bmatrix} v_1 \\ v_1' \\ v_2 \\ v_2' \end{Bmatrix} = \mathbf{N}_3(\xi)\mathbf{v} \qquad (15.1.5)$$

where $\xi = \dfrac{x}{L}$

In terms of these shape functions the derivatives of displacement are

$$\frac{du}{dx} = \frac{1}{L}\frac{du}{d\xi} = \frac{1}{L}\mathbf{N}_1'\mathbf{u}$$

$$\frac{dv}{dx} = \frac{1}{L}\frac{dv}{d\xi} = \frac{1}{L}\mathbf{N}_3'\mathbf{v}$$

$$\frac{d^2v}{dx^2} = \frac{1}{L^2}\frac{d^2v}{d\xi^2} = \frac{1}{L^2}\mathbf{N}_3''\mathbf{v}$$

Substitution of these in (15.1.3) gives

$$\delta U_{\text{int}} = \delta\mathbf{u}^T\left[\frac{EA}{L}\int_0^1 \mathbf{N}_1'^T\mathbf{N}_1'\,d\xi\right]\mathbf{u} - \delta\mathbf{v}^T\left[\frac{EI_z}{L^3}\int_0^1 \mathbf{N}_3''^T\mathbf{N}_3''\,d\xi\right]\mathbf{v} + \delta\mathbf{v}^T\left[\frac{\mathcal{F}_x}{L}\int_0^1 \mathbf{N}_3'^T\mathbf{N}_3'\,d\xi\right]\mathbf{v} \qquad (15.1.6)$$

Equating δW and δU, and noting that for corresponding components the variational terms cancel on the left and right, we obtain

$$\begin{Bmatrix} \mathbf{f}_u \\ \mathbf{f}_v \end{Bmatrix} = \begin{bmatrix} \mathbf{k}_u & \mathbf{0} \\ \mathbf{0} & \mathbf{k}_v + \mathbf{k}_{v,G} \end{bmatrix}\begin{Bmatrix} \mathbf{u} \\ \mathbf{v} \end{Bmatrix} \qquad (15.1.7)$$

where $\mathbf{k}_u$ is the usual stiffness matrix for axial displacement given in (6.3.2)

$$\mathbf{k}_u = \frac{AE}{L}\begin{bmatrix} 1 & -1 \\ -1 & 1 \end{bmatrix}$$

and $\mathbf{k}_v$ is the usual flexural stiffness matrix, given in (6.3.3). The additional terms in the element stiffness matrix, represented by the symbol $\mathbf{k}_{v,G}$, come from the third term of (15.1.6). That is

$$\mathbf{k}_{v,G} = \frac{\mathcal{F}_x}{L}\int_0^1 \mathbf{N}_3'^T\mathbf{N}_3'\,d\xi \qquad (15.1.8)$$

Substituting for $\mathbf{N}_3$ and integrating gives

*Note that the chosen shape function need not be the exact analytical solution to the differential equation governing the problem. In pure flexure, with no distributed loads, the exact solution is a cubic polynomial. For combined axial and flexural loads the exact solution involves trigonometric functions, but a cubic polynomial can still be used as the shape function.

$$\mathbf{k}_{v,G} = \frac{\mathcal{F}_x}{30L}\begin{bmatrix} 36 & & & \text{symm.} \\ 3L & 4L^2 & & \\ -36 & -3L & 36 & \\ -3L & -L^2 & -3L & 4L^2 \end{bmatrix} \tag{15.1.9}$$

The matrix $\mathbf{k}_{v,G}$ accounts for the interaction between the member axial force $\mathcal{F}_x$ and the flexural nodal displacements $\mathbf{v}$. It thus accounts for the elastic instability effects and may be viewed as a correction to the basic flexural stiffness matrix $\mathbf{k}_v$. It can be seen from (15.1.9) that the individual terms of this matrix are solely dependent on the geometric parameters—in this case, the length. This matrix is therefore usually referred to as the *geometric* stiffness matrix.

The global geometric stiffness matrix $\mathbf{K}_G$ is obtained in the same manner as the ordinary stiffness matrix, that is, by transforming each element geometric stiffness matrix to global coordinates and summing the terms which correspond to the same global degree of freedom. The resulting equilibrium equations are

$$\mathbf{F} = (\mathbf{K} + \mathbf{K}_G)\boldsymbol{\Delta} \tag{15.1.10}$$

15.2 ITERATIVE FINITE ELEMENT ANALYSIS

In the foregoing equation $\mathbf{K}_G$ depends on the internal member force and therefore on $\boldsymbol{\Delta}$, and the latter is the unknown which is being solved for. Therefore (15.1.10) is nonlinear and must be solved by iteration, using some numerical technique such as Newton-Raphson or modified Newton-Raphson to obtain progressively better values for the nodal displacements. These techniques are illustrated in Figs. 15.2 and 15.3 respectively and the procedure for both may be summarized as follows:

1. Obtain the first estimate of the nodal displacements $\boldsymbol{\Delta}_i$ by solving the equilibrium equations without $\mathbf{K}_G$

$$\boldsymbol{\Delta}_0 = \mathbf{K}_0^{-1}\mathbf{F}$$

(Note: $\mathbf{K}_0$ is not actually inverted but only triangularized).

The vector of nodal displacements $\boldsymbol{\Delta}$ will be calculated in a stepwise fashion and each step will be denoted as $\delta\boldsymbol{\Delta}_i$. To initialize set $\delta\boldsymbol{\Delta}_0 = \boldsymbol{\Delta}_0$. Then repeat the

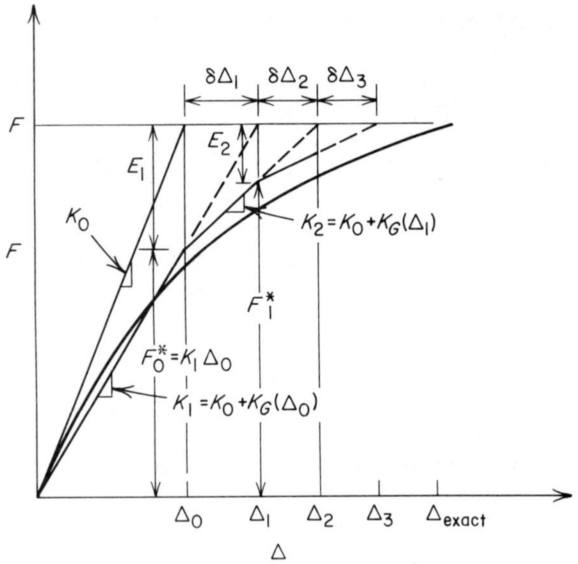

Figure 15.2 Iterative finite element analysis; repeated triangularizations.

following cycle of steps (i = cycle number = 1, 2, . . .).

2. Compute $\mathbf{K}_G(\boldsymbol{\Delta}_{i-1})$, which is the value of $\mathbf{K}_G$ based on $\boldsymbol{\Delta}_{i-1}$, and update the total stiffness matrix

$$\mathbf{K}_i = \mathbf{K}_0 + \mathbf{K}_G(\boldsymbol{\Delta}_{i-1})$$

3. Compute $\mathbf{F}^*_{i-1} = \Sigma_1^i \mathbf{K}_i\, \delta\boldsymbol{\Delta}_{i-1}$, which is the set of structure nodal forces corresponding to $\boldsymbol{\Delta}_{i-1}$.

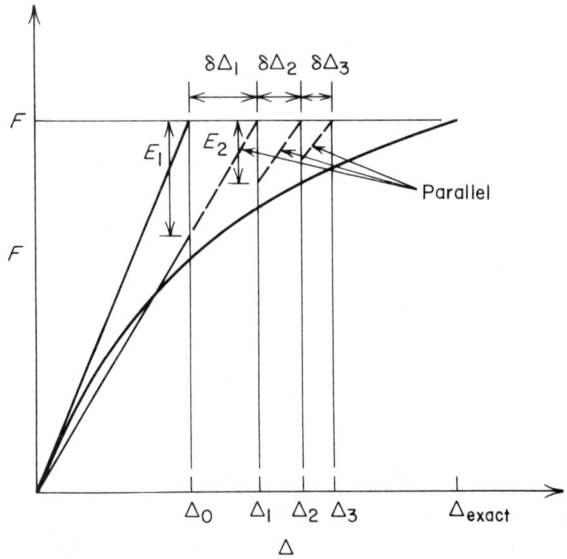

Figure 15.3 Iterative finite element analysis; two triangularizations only.

Note that these forces are only approximate values, based on a single updating of $\mathbf{K}_i$, and the updating was done using $\mathbf{\Delta}_{i-1}$.

4. Compute the error of the solution, which is the imbalance between the actual loads and the above set of nodal forces

$$\mathbf{E}_i = \mathbf{F} - \mathbf{F}^*_{i-1}$$

5. Solve for the correction to the displacements $\delta\mathbf{\Delta}_i = \mathbf{K}_i^{-1}\mathbf{E}_i$ using the ordinary Newton-Raphson method, or alternatively $\delta\mathbf{\Delta}_i = \mathbf{K}_1^{-1}\mathbf{E}_i$ if the modified Newton-Raphson is chosen.

6. Determine the new total displacement

$$\mathbf{\Delta}_i = \mathbf{\Delta}_{i-1} + \delta\mathbf{\Delta}_i$$

Steps 2 to 6 are repeated until the unbalanced force vector $\mathbf{E}_i$ is acceptably small; i.e. until the ratio of the norm of the unbalanced forces to that of the applied loads is less than the prescribed tolerance ε. That is, until

$$\frac{\mathbf{E}_i^T\mathbf{E}_i}{\mathbf{F}^T\mathbf{F}} < \varepsilon \tag{15.2.1}$$

As shown in Fig. 15.3 the modified Newton-Raphson method avoids the repeated triangularization of $\mathbf{K}_i$, but it requires more iterations to achieve the same accuracy.

15.3 GEOMETRIC STIFFNESS MATRIX FOR A GENERAL BEAM ELEMENT

Figure 15.4 shows a prismatic thin-walled beam element in which the x-axis passes through the centroid of the beam cross section and the y- and z-axes are the principal axes. The coordinates of the shear center are (y_0, z_0). The element nodal forces are shown for node 1, and a similar set exists at node 2. However, in the x-direction the force is the *internal member force*, $\mathcal{F}_x$ (written in script to distinguish it from a nodal force). This force is positive if tensile and acts through the point (e_y, e_z). Uniform distributed loads f_y and f_z act through the point $(y_0 + \bar{a}_y, z_0 + \bar{a}_z)$. The displacement of the shear center S is u, v, and w.

For the sake of brevity a detailed derivation is not given. Instead, expressions are presented for the total potential energy, $\Pi = U - W$, where U is the strain energy and W is the work done by the loads on the element, that is, by the element nodal forces. Then the variational condition for equilibrium, $\delta\Pi = 0$, is applied. References are cited in which further details may be obtained.

The strain energy U is given by:

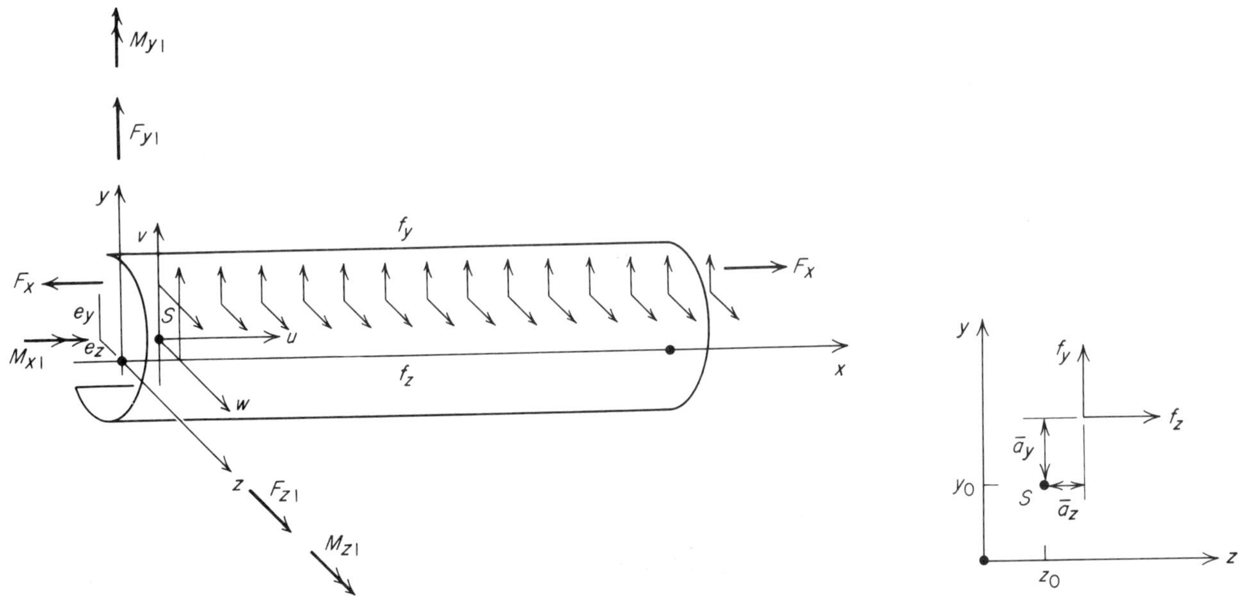

Figure 15.4

$$U = \frac{1}{2}\int_0^l [EI_z v''^2 + EI_y w''^2 + EI_\omega \theta_x''^2 + GJ\theta_x'^2 + EAu'^2]\,dx \tag{15.3.1}$$

where a prime represents d/dx, I_y and I_z are the moments of inertia of the cross section about the principal axes, I_ω is the warping torsional constant, J is the St. Venant torsional constant, and A is the cross-sectional area.

The work done by the nodal forces consists of two parts. The first part W_1 is the direct work done by the nodal forces acting through their corresponding displacements (prebuckling):

$$\begin{aligned} W_1 = {} & \mathcal{F}_x(u_2 - u_1) + F_{y1}v_1 + F_{y2}v_2 + F_{z1}w_1 + F_{z2}w_2 \\ & + M_{x1}\theta_{x1} + M_{x2}\theta_{x2} - M_{y1}w_1' - M_{y2}w_2' + M_{z1}v_1' + M_{z2}v_2' \\ & + \int_0^l f_y v\,dx + \int_0^l f_z w\,dx + \int_0^l f_z\bar{a}_y\theta_x\,dx - \int_0^l f_y\bar{a}_z\theta_x\,dx \end{aligned} \tag{15.3.2}$$

The second part W_2 is the work associated with the additional displacements due to buckling. These displacements may be grouped into three categories: axial displacement (shortening of the element), displacements due to lateral buckling, and displacements due to torsional buckling. We now consider each of these separately and derive expressions for the corresponding amounts of work: W_a, W_b, and W_t.

(a) Work done due to shortening of the element

$$W_a = \int_0^l (ds - dx)\left[\int_A \left(\frac{\mathcal{F}_x}{A} + \frac{M_z}{I_z}y - \frac{M_y}{I_y}z\right)dA\right] \tag{15.3.3}$$

where $(ds - dx)$ is the shortening of a differential element of original length ds and is given by (Ref. 1, p.158.)

$$ds - dx = \frac{1}{2}\left[v' + \theta_x'(y_0 - y)\right]^2 + \frac{1}{2}\left[w' - \theta_x'(z_0 - z)\right]^2 \tag{15.3.4}$$

Also

$$\left.\begin{aligned} M_y &= M_{y1} + F_{z1}x + \frac{1}{2}x^2 f_z + \mathcal{F}_x e_x \\ M_z &= M_{z1} - F_{y1}x - \frac{1}{2}x^2 f_y - \mathcal{F}_x e_y \end{aligned}\right\} \tag{15.3.5}$$

Substituting (15.3.4) and (15.3.5) into (15.3.3) and noting that

$$\int_A y\,dA = \int_A z\,dA = \int_A yz\,dA = 0$$

$$\int_A y^2\,dA = I_z, \qquad \int_A z^2\,dA = I_y,$$

we obtain the following expression

$$\begin{aligned} W_a = \frac{1}{2}\int_0^l \Big\{ & -\mathcal{F}_x(v'^2 + w'^2) - 2\theta_x' v'[\mathcal{F}_x(z_0 - e_z) + M_{y1}] + 2\theta_x' w'[\mathcal{F}_x(y_0 - e_y) - M_{z1}] \\ & - 2\theta_x' v' x F_{z1} - \theta_x' v' x^2 f_z + 2\theta_x' w' x F_{y1} + \theta_x' w' x^2 f_y \\ & + \theta_x'^2\left[-\mathcal{F}_x\left(\frac{I_0}{A} + \beta_1 e_z + \beta_2 e_y\right) + M_{y1}\beta_1 - M_{z1}\beta_2\right] \\ & + \theta_x' x(\beta_1 F_{z1} + \beta_2 F_{y1}) + \frac{1}{2}\theta_x'^2 x^2(\beta_1 f_z + \beta_2 f_y)\Big\}\,dx \end{aligned} \tag{15.3.6}$$

where I_0 is the polar moment of inertia about the shear center

$$I_0 = I_y + I_z + A(y_0^2 + z_0^2) \qquad (15.3.7)$$

and β_1 and β_2 are coefficients that account for the second-order effect in which an axial stress may produce a torsional moment if a warping deformation exists in the section [2, p.231]

$$\beta_1 = \frac{1}{I_y}\int_A (y^2 + z^2)z\,dA - 2z_0$$

$$\beta_2 = \frac{1}{I_z}\int_A (y^2 + z^2)y\,dA - 2y_0 \qquad (15.3.8)$$

(b) Work done due to lateral buckling

Consider the lateral deformation in the xy plane, as shown in Fig. 15.5. In a differential element of length dx the increase of displacement in the y-direction is $v'dx$. When a twist θ_x exists, there will be an additional displacement in the z-direction, $v'\,\theta_x\,dx$. It may be shown [3] that because of this displacement, the shear force $-F_z(x)$ will do work $-F_z(x)[v'\theta_x\,dx]$ where

$$F_z(x) = F_{z1} + xf_z$$

Likewise there will also be a component of work $F_y(x)[w'\,\theta_x\,dx]$ where

$$F_y(x) = F_{y1} + xf_y$$

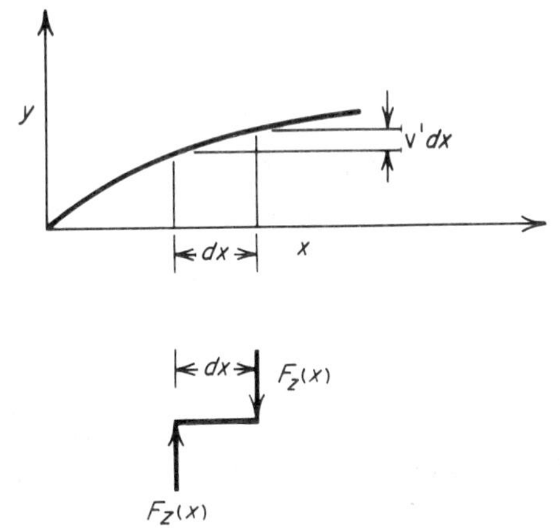

Figure 15.5

Therefore the total work W_{b1} done by the lateral shear forces and distributed forces is given by

$$W_{b1} = \int_0^l [(F_{y1} + xf_y)w' - (F_{z1} + xf_z)v']\theta_x\,dx \qquad (15.3.9)$$

Another component of work associated with lateral buckling is the work done by the twisting moment in the element, characterized by the value at node 1, M_{x1}. It may be shown [4] that this work is

$$W_{b2} = \frac{-1}{2}M_{x1}\int_0^l (v'w'' - w'v'')\,dx \qquad (15.3.10)$$

(c) Work done due to torsional buckling

Consider a distributed load f_y acting at a distance $\bar{a}_y$ above the shear center S, as shown in Fig. 15.6. When there is a torsional buckling f_y is moved a small distance and for a differential element of length dx a torque $f_y\bar{a}_y\theta_x\,dx$ about the shear center is produced. The work done over the full length of the element is

$$W_{c1} = -\frac{1}{2}f_y\bar{a}_y\int_0^l \theta_x^2\,dx$$

Similarly, for a uniform distributed force f_z

$$W_{c2} = -\frac{1}{2}f_z\bar{a}_z\int_0^l \theta_x^2\,dx$$

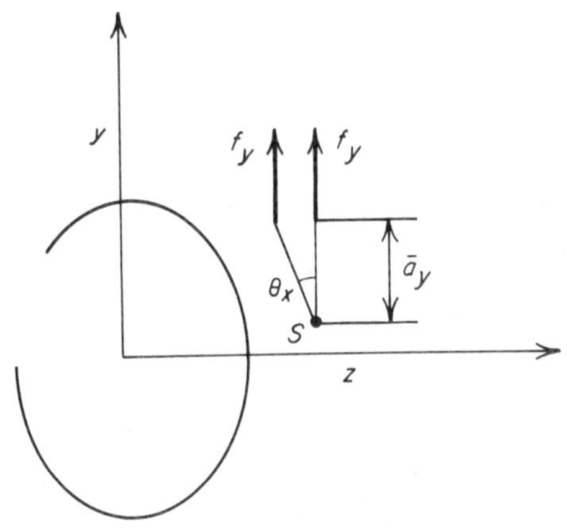

Figure 15.6

Therefore the total work done is

$$W_c = -\frac{1}{2}(f_y\bar{a}_y + f_z\bar{a}_z)\int_0^l \theta_x^2\,dx \quad (15.3.11)$$

Summing up the components of Π given by (15.3.1), (15.3.2), (15.3.6), and (15.3.9) through (15.3.11) we obtain

$$\begin{aligned}
\Pi = {} & \frac{1}{2}\int_0^l (EI_z v''^2 + EI_y w''^2 + EI_\omega \theta''^2_x + GJ\theta'^2_x + EAu'^2)\,dx \\
& - \mathcal{F}_x(u_2 - u_1) - F_{y1}v_1 - F_{y2}v_2 - F_{z1}w_1 - F_{z2}w_2 \\
& - M_{x1}\theta_{x1} - M_{x2}\theta_{x2} + M_{y1}w_1' + M_{y2}w_2' - M_{z1}v_1' - M_{z2}v_2' \\
& - \int_0^l f_y v\,dx - \int_0^l f_z w\,dx - \int_0^l f_z\bar{a}_y\theta_x\,dx + \int_0^l f_y\bar{a}_z\theta_x\,dx \\
& + \frac{1}{2}\int_0^l \Big\{ \mathcal{F}_x(v'^2 + w'^2) + 2\theta_x' v'[\mathcal{F}_x(z_0 - e_z) + M_{y1}] - 2\theta_x' w'[\mathcal{F}_x(y_0 - e_y) - M_{z1}] \\
& + 2\theta_x' v' x F_{z1} + \theta_x' v' x^2 f_z - 2\theta_x' w' x F_{y1} - \theta_x' w' x^2 f_y \\
& + \theta_x'^2\left[\mathcal{F}_x\left(\frac{I_0}{A} + \beta_1 e_z + \beta_2 e_y\right) - M_{y1}\beta_1 + M_{z1}\beta_2\right] \\
& - \theta_x' x(\beta_1 F_{z1} + \beta_2 F_{y1}) - \frac{1}{2}\theta_x' x^2(\beta_1 f_z + \beta_2 f_y)\Big\}\,dx \\
& - \int_0^l [(F_{y1} + xf_y)w' - (F_{z1} + xf_z)v']\theta_x\,dx + \frac{1}{2}M_{x1}\int_0^l (v'w'' - w'v'')\,dx \\
& + \frac{1}{2}(f_y\bar{a}_y + f_z\bar{a}_z)\int_0^l \theta_x^2\,dx
\end{aligned} \quad (15.3.12)$$

For this general case we can use the same shape functions as before—the linear function of (15.1.4) for the axial degree of freedom and the cubic function of (15.1.5) for the flexural and torsional degrees of freedom. This is the same combination that was used in Section 6.2. Substitution of these functions into (15.3.12) yields an expression for Π in terms of the element nodal displacements $\boldsymbol{\delta}$. For equilibrium the total energy must satisfy the variational expression $\delta\Pi = 0$, and in terms of nodal displacements this condition becomes

$$\delta\Pi = \frac{\partial\Pi}{\partial\delta}\,\delta\boldsymbol{\delta} = 0 \quad (15.3.13)$$

After performing the variation on the expression for Π the result may be put into the form

$$\mathbf{f} = (\mathbf{k}_S + \mathbf{k}_G)\,\boldsymbol{\delta}$$

where $\mathbf{f}$ is the vector of nodal forces as defined in (6.3.1), $\mathbf{k}_S$ is the ordinary stiffness matrix, and $\mathbf{k}_G$ is the geometric stiffness matrix. The former is the same matrix which was obtained in Section 6.3, but it is presented here in the more usual form, with the degrees of freedom grouped according to nodes. The geometric stiffness matrix $\mathbf{k}_G$ is comprised of all of the terms which contain an element nodal force or internal force as a multiplying factor. The derivation is not given here because the calculation is lengthy and also straightforward. The final expressions for $\mathbf{k}_S$ and $\mathbf{k}_G$ are given on the next page.

$$
\mathbf{K}_S = \begin{array}{c} \\ \end{array}
\begin{array}{cccccccccccccc}
u_1 & v_1 & w_1 & \theta_{x1} & \theta_{y1} & \theta_{z1} & \theta'_{x1} & u_2 & v_2 & w_2 & \theta_{x2} & \theta_{y2} & \theta_{z2} & \theta'_{x2}
\end{array}
$$

$$
\mathbf{K}_S = \begin{bmatrix}
\mathfrak{a} & & & & & & & -\mathfrak{a} & & & & & & \\
 & 12\mathfrak{b}_z & & & & 6l\mathfrak{b}_z & & & -12\mathfrak{b}_z & & & & 6l\mathfrak{b}_z & \\
 & & 12\mathfrak{b}_y & & -6l\mathfrak{b}_y & & & & & -12\mathfrak{b}_y & & -6l\mathfrak{b}_y & & \\
 & & & c_1 & & & c_3 & & & & -c_1 & & & c_3 \\
 & & & & 4l^2\mathfrak{b}_y & & & & & 6l\mathfrak{b}_y & & 2l^2\mathfrak{b}_y & & \\
 & & & & & 4l^2\mathfrak{b}_z & & & -6l\mathfrak{b}_z & & & & 2l^2\mathfrak{b}_z & \\
 & & & & & & c_2 & & & & -c_3 & & & c_4 \\
 & & & & & & & \mathfrak{a} & & & & & & \\
 & & & & & & & & 12\mathfrak{b}_z & & & & -6l\mathfrak{b}_z & \\
 & & & & & & & & & 12\mathfrak{b}_y & & 6l\mathfrak{b}_y & & \\
 & & & & & & & & & & c_1 & & & -c_3 \\
 & & & & & & & & & & & 4l^2\mathfrak{b}_y & & \\
 & & & & & & & & & & & & 4l^2\mathfrak{b}_z & \\
 & & & & & & & & & & & & & c_2
\end{bmatrix}
$$

Note: $\theta_y = -w'$

$$\mathfrak{a} = \frac{EA}{l} \qquad \mathfrak{b}_z = \frac{EI_z}{l^3} \qquad \mathfrak{b}_y = \frac{EI_y}{l^3} \qquad \mathfrak{b}_\omega = \frac{EI_\omega}{l^3}$$

$$c_1 = 1.2\,GJ/l + 12\mathfrak{b}_\omega \qquad c_2 = \frac{2}{15}GJl + 4l^2\mathfrak{b}_\omega \qquad c_3 = 0.1GJ + 6l\mathfrak{b}_\omega \qquad c_4 = -\frac{1}{30}GJl + 2l^2\mathfrak{b}_\omega \tag{15.3.14}$$

$$
\begin{array}{cccccccccccccc}
u_1 & v_1 & w_1 & \theta_{x1} & \theta_{y1} & \theta_{z1} & \theta'_{x1} & u_2 & v_2 & w_2 & \theta_{x2} & \theta_{y2} & \theta_{z2} & \theta'_{x2}
\end{array}
$$

$$
\mathbf{K}_G = \begin{bmatrix}
 & & & & & & & & & & & & & \\
 & a & & V_1 & -e & b & V_4 & & -a & & V_7 & e & b & V_{10} \\
 & & a & W_1 & -b & -e & W_4 & & & -a & W_7 & -b & e & W_{10} \\
 & & & \theta_1 & -W_2 & V_2 & \theta_2 & & -V_1 & -W_1 & \theta_3 & -W_3 & V_3 & \theta_4 \\
 & & & & c & & -W_5 & & e & b & -W_8 & -d & f & -W_{11} \\
 & & & & & c & V_5 & & -b & e & V_8 & f & -d & V_{11} \\
 & & & & & & \theta_5 & & -V_4 & -W_4 & \theta_6 & -W_6 & V_6 & \theta_7 \\
 & & & & & & & & & & & & & \\
 & & & & \text{Symmetric} & & & & a & & -V_7 & -e & -b & -V_{10} \\
 & & & & & & & & & a & -W_7 & b & -e & -W_{10} \\
 & & & & & & & & & & \theta_1 & -W_9 & V_9 & \theta_8 \\
 & & & & & & & & & & & c & & -W_{12} \\
 & & & & & & & & & & & & c & V_{12} \\
 & & & & & & & & & & & & & \theta_9
\end{bmatrix}
$$

$$a = 1.2\frac{\mathcal{F}_x}{l} \qquad b = 0.1\mathcal{F}_x \qquad c = \frac{2}{15}\mathcal{F}_x l \qquad d = \frac{1}{30}\mathcal{F}_x l \qquad e = \frac{M_{x1}}{l} \qquad f = \tfrac{1}{2}M_{x1} \tag{15.3.15}$$

$$V_1 = 1.2\mathcal{A} + 0.1F_{z1} - \frac{1}{70}f_z l \qquad \theta_1 = 1.2\frac{\mathcal{C}}{l} - 0.6\mathcal{D} - \frac{6}{35}l\mathcal{E} + \frac{13}{35}l\mathcal{F} \qquad \mathcal{A} = \mathcal{F}_x(z_0 - e_z) + M_{y1}$$

$$V_2 = 0.1\mathcal{A} + 0.2lF_{z1} + \frac{11}{420}f_z l^2 \qquad \theta_2 = 0.1\mathcal{C} - 0.1l\mathcal{D} - \frac{1}{28}l^2\mathcal{E} + \frac{11}{210}l^2\mathcal{F} \qquad \mathcal{B} = \mathcal{F}_x(e_y - y_0) + M_{z1}$$

$$V_3 = 0.1\mathcal{A} - 0.1lF_{z1} - \frac{17}{420}f_z l^2 \qquad \theta_3 = -1.2\frac{\mathcal{C}}{l} + 0.6\mathcal{D} + \frac{6}{35}l\mathcal{E} + \frac{9}{70}l\mathcal{F} \qquad \mathcal{C} = \mathcal{F}_x\left(\frac{I_0}{A} + e_z\beta_1 + e_y\beta_2\right) - \beta_1 M_{y1} + \beta_2 M_{z1}$$

$$V_4 = 0.1\mathcal{A} - \frac{1}{140}f_z l^2 \qquad \theta_4 = 0.1\mathcal{C} \qquad + \frac{1}{70}l^2\mathcal{E} - \frac{13}{420}l^2\mathcal{F} \qquad \mathcal{D} = \beta_1 F_{z1} + \beta_2 F_{y1}$$

$$V_5 = \frac{2}{15} l\mathcal{A} + \frac{1}{30} l^2 F_{z1} + \frac{1}{210} f_z l^3 \qquad \theta_5 = \frac{2}{15} l\mathcal{C} - \frac{1}{30} l^2 \mathcal{D} - \frac{1}{105} l^3 \mathcal{E} + \frac{1}{105} l^3 \mathcal{F} \qquad \mathcal{E} = \beta_1 f_z + \beta_2 f_y$$

$$V_6 = -\frac{1}{30} l\mathcal{A} - \frac{1}{30} l^2 F_{z1} - \frac{1}{84} f_z l^3 \qquad \theta_6 = -0.1\mathcal{C} + 0.1 l\mathcal{D} + \frac{1}{28} l^2 \mathcal{E} + \frac{13}{420} l^2 \mathcal{F} \qquad \mathcal{F} = \bar{a}_y f_y + \bar{a}_z f_z$$

$$V_7 = -1.2\frac{\mathcal{A}}{l} - \frac{11}{10} F_{z1} - \frac{17}{35} f_z l \qquad \theta_7 = \frac{-1}{30} l\mathcal{C} + \frac{1}{60} l^2 \mathcal{D} + \frac{1}{140} l^3 \mathcal{E} - \frac{1}{140} l^3 \mathcal{F} \qquad \beta_1 = \frac{1}{I_y} \int_A z(y^2 + z^2)\, dA - 2z_0$$

$$V_8 = -0.1\mathcal{A} - 0.2 l F_{z1} - \frac{23}{210} f_z l^2 \qquad \theta_8 = -0.1\mathcal{C} \qquad - \frac{1}{70} l^2 \mathcal{E} - \frac{11}{210} l^2 \mathcal{F} \qquad \beta_2 = \frac{1}{I_z} \int_A y(y^2 + z^2)\, dA - 2y_0$$

$$V_9 = -0.1\mathcal{A} + 0.1 l F_{z1} + \frac{13}{105} f_z l^2 \qquad \theta_9 = \frac{2}{15} l\mathcal{C} - 0.1 l^2 \mathcal{D} - \frac{3}{70} l^3 \mathcal{E} + \frac{1}{105} l^3 \mathcal{F}$$

$$V_{10} = 0.1\mathcal{A} + 0.1 l F_{z1} + \frac{3}{70} f_z l^2$$

$$V_{11} = -\frac{1}{30} l\mathcal{A} + \frac{1}{210} f_z l^3$$

$$V_{12} = \frac{2}{15} l\mathcal{A} + 0.1 l^2 F_{z1} + \frac{4}{105} f_z l^3$$

The W_i terms are obtained by replacing $\mathcal{A}$ by $\mathcal{B}$, F_{z1} by $-F_{y1}$, and f_z by $-f_y$ in the expressions for V_i.

15.4 BUCKLING OF BIFURCATION TYPE: EIGENVALUE METHOD

The equilibrium equations for a structure which are given in (15.1.10) can also be expressed in incremental form. That is, they are equally valid for an increment of load $\tilde{\mathbf{F}}$ and the corresponding increment in deflection $\tilde{\mathbf{\Delta}}$, providing that $\mathbf{K}_G$ is evaluated for the structure's current state; that is, the member forces must be calculated using the current total displacement $\mathbf{\Delta}_i$. This is illustrated in Fig. 15.7 for a single force and a single displacement. In incremental form the equilibrium equations are

$$\tilde{\mathbf{F}} = (\mathbf{K} + \mathbf{K}_G)\, \tilde{\mathbf{\Delta}} \tag{15.4.1}$$

In the eigenvalue approach to buckling the assumption is made that the structure remains undeformed until the load reaches a critical value, at which point the deflection becomes indefinitely large, as shown in Fig. 15.8. That is, the critical condition is defined as a condition in which there can be a nonzero deflection with no increase in load ($\tilde{\mathbf{F}} = 0$). In this situation the incremental equilibrium equation becomes

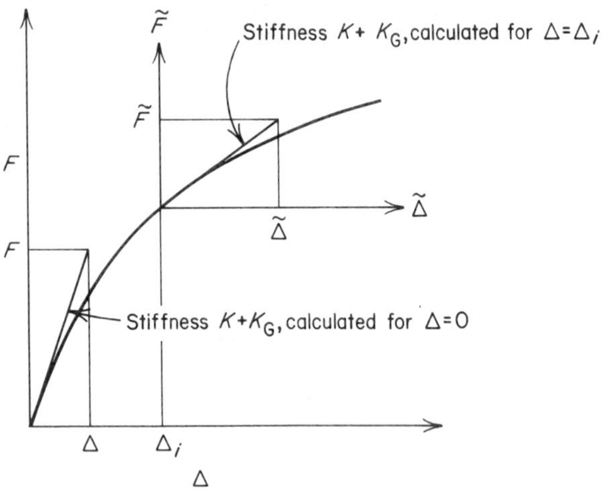

Figure 15.7 Incremental equilibrium (force–deflection) relationship.

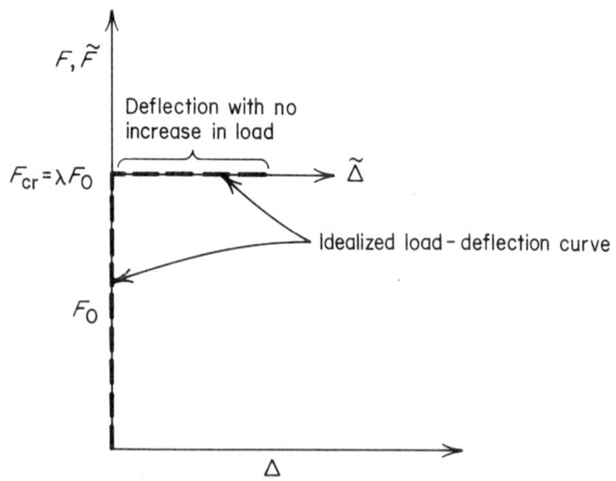

Figure 15.8 Idealized load–deflection curve for bifurcation buckling.

$$(\mathbf{K} + \mathbf{K}_G)\,\tilde{\mathbf{\Delta}} = 0 \qquad (15.4.2)$$

Since the displacement increment $\tilde{\mathbf{\Delta}}$ is nonzero the equilibrium can only be satisfied if the total stiffness matrix is singular. Hence the mathematical criterion for buckling is the vanishing of the determinant of the stiffness matrix; that is,

$$\det(\mathbf{K} + \mathbf{K}_G) = 0 \qquad (15.4.3)$$

The assumption is made that throughout the loading history the external loads maintain the same relative proportions. Therefore the load vector can be expressed as a scalar multiple of a representative or reference set of loads, $\mathbf{F}_0$: $\mathbf{F} = \lambda \mathbf{F}_0$. Then the general form of the geometric stiffness matrix is

$$\mathbf{K}_G = \lambda\, \mathbf{K}_{G,0} \qquad (15.4.4)$$

where $\mathbf{K}_{G,0}$ represents the geometric stiffness matrix computed for the reference level of load $\mathbf{F}_0$. Introducing (15.4.4) into (15.4.2) yields

$$\mathbf{K}\,\tilde{\mathbf{\Delta}} = -\lambda\, \mathbf{K}_{G,0}\, \tilde{\mathbf{\Delta}}$$

or

$$\frac{1}{\lambda}\tilde{\mathbf{\Delta}} = -\mathbf{K}^{-1}\mathbf{K}_G\, \tilde{\mathbf{\Delta}} \qquad (15.4.5)$$

The critical buckling load $\lambda \mathbf{F}_0$ is now obtained by solving (15.4.5) for the unknown eigenvalue λ. The eigenvalue solution can be obtained using any of the standard software library subroutines. The buckling mode can be identified as the eigenvector corresponding to the lowest eigenvalue.

In the following three sections we examine the three basic types of bifurcation buckling which can occur with beam members: flexural, torsional, and lateral-torsional. The aim of these sections is simply to illustrate the common and most basic aspects of the finite element approach to bifurcation buckling—the fundamental assumptions, the limitations which these entail, and the degree of accuracy to be expected. To keep the presentation as simple as possible we will take very simple examples which can be solved by hand.

Flexural Buckling

Figure 15.9 shows a simple Euler column represented by two beam elements. Because of symmetry we need only deal with one element. The boundary conditions are $v_1 = \theta_2 = 0$. In this simple example the member and the structure are one and the same, and therefore the global stiffness matrix $\mathbf{K}$ is the same as the member stiffness matrix $\mathbf{k}$ of (15.1.7). This means that in this particular "structure" the axial response is independent of the flexural response and can be obtained separately. We therefore take the axial portion of (15.1.7) and substitute (6.3.2) which gives

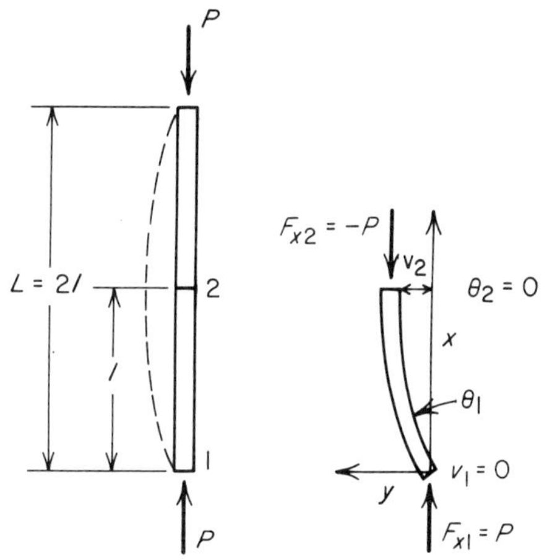

Figure 15.9 Buckling of a simple column.

$$\begin{Bmatrix} P \\ -P \end{Bmatrix} = \frac{AE}{L}\begin{bmatrix} 1 & -1 \\ -1 & 1 \end{bmatrix}\begin{Bmatrix} u_1 \\ u_2 \end{Bmatrix}$$

The solution of this, when substituted into the member internal force equation, gives the member internal force: $\mathcal{F}_x = -P$. This demonstrates that whenever the axial portion of the *global* stiffness matrix is independent of the flexural portion, it is possible to perform a separate solution for the internal axial forces in each of the members, expressed in terms of the external load. Knowing these internal axial forces greatly simplifies the solution of the overall problem, in which $\mathbf{K}_G$ is a function of these forces.

Thus bifurcation buckling theory is based on the assumption that in the assembled structure the distribution of axial force among the members is not influenced by the flexural response. This makes it possible to perform a separate and prior structural analysis using a representative set of loads $\mathbf{F}_0$, and thus to determine the corresponding member axial forces, say $\boldsymbol{F}_{x0}$.[†]

[†]For consistency this symbol should be a *script* boldface F because it is a vector of *internal* forces, but because of typesetting limitations an *italic* boldface F is used instead.

Then in the subsequent buckling analysis, with external loads $\mathbf{F} = \lambda \mathbf{F}_0$, the axial member forces are λF_{x0}, and the only unknown aspect of the member forces is the scalar multiplier λ. This converts the problem from a set of nonlinear equations to an eigenvalue problem thereby allowing a direct solution.

In the present example we are seeking the value of P which satisfies the zero determinant condition of (15.4.3). In this particular case the global stiffness matrix $\mathbf{K} + \mathbf{K}_G$ is so small that we can solve (15.4.3) directly, in its determinantal form. That is, although we are solving an eigenvalue type of problem there is here no need to use a formal eigenvalue approach. Because the element and global stiffness matrices are identical, (15.4.3) can be expressed in terms of the former. Also, because the axial response is independent we need write only the flexural portion. Equation 15.4.3 therefore becomes

$$\det\left(\mathbf{k}_v + \mathbf{k}_{v,G}\right) = 0$$

Substituting from (6.3.3) and (15.1.9), accounting for the boundary conditions by deleting the v_1 and θ_{x2} degrees of freedom, and setting $\mathcal{F}_x = -P$ gives

$$\det\left(\frac{-P}{30l}\begin{bmatrix} 4l^2 & -3l \\ -3l & 36 \end{bmatrix} + \frac{EI}{l^3}\begin{bmatrix} 4l & -6l \\ -6l & 12 \end{bmatrix}\right) = 0$$

Introducing the scalar parameter $\lambda = \dfrac{Pl^2}{30EI}$ gives

$$\det\left(\frac{EI}{l^3}\begin{bmatrix} 4l^2(1-\lambda) & 3l(-2+\lambda) \\ 3l(-2+\lambda) & 12(1-3\lambda) \end{bmatrix}\right) = 0$$

Expanding this gives the characteristic equation

$$48l^2(1-\lambda)(1-3\lambda) - 9l^2(-2+\lambda)^2 = 0$$

The smaller root of this quadratic equation is

$$\lambda = 0.08392$$

and hence

$$P = \lambda\frac{30EI}{l^2} = \lambda\frac{120EI}{L^2}$$

$$= 10.07\frac{EI}{L^2}$$

This result is only 2% above the exact value, π^2EI/L^2. Obviously with smaller and more numerous elements the accuracy would be even better. This shows clearly that the generalized finite element method, utilizing the geometric stiffness matrix, gives quite satisfactory solutions to instability problems. It also shows that for instability of bifurcation type the use of finite element theory results in an eigenvalue type of problem, the solution of which can be achieved directly, using standard numerical methods. The amount of computation is not much more than solving a linear system of the same size.

However, bifurcation buckling is an idealized phenomenon involving several important assumptions. Some of these relate to the individual members: that they are perfectly straight, that axial loads act through the centroid and transverse loads act through the shear center, and that no other type of failure occurs prior to buckling. It is these assumptions which make the axial portion of the member stiffness matrix of (15.1.7) independent of the flexural portion. In practice the members often possess some degree of eccentricity, either in their own geometry or in their attachment to other members or in the point of application of the loads. The result of these eccentricities is a coupling between the member's axial and flexural response, and the appearance of many new terms in the stiffness matrix which account for this coupling. The structural members undergo additional bending and twisting, but because the axial rigidity of most members is much larger than the flexural rigidity the member axial forces are not much affected. If the members are orthogonal to each other then the transformation from local to global coordinates will not alter this situation. Hence for an orthogonal structure (and only such a structure) a moderate amount of eccentricity does not substantially affect the accuracy of a bifurcation buckling solution and, if desired, the coupling terms can be omitted in the prebuckling analysis. They must, however, be included if it is desired to obtain realistic values for the bending and twisting moments in the prebuckling analysis.

It should be noted that in practice the axial degrees of freedom do not occur in just one block of the global stiffness matrix—they are scattered throughout the matrix. Hence the prebuckling analysis is performed using the complete stiffness matrix and not a purely axial submatrix as in this simple example.

The most serious limitation of bifurcation buckling theory is that it is based on the assumption that the independence of the axial response from the flexural response exists in the global stiffness matrix as well as in the individual member stiffness matrices. This assumption is only satisfied if the members are all orthogonal to each other, so that in the transformation to global coordinates the uncoupled status of the axial degrees of freedom is preserved. If the members are at various angles then the transformation will destroy this uncoupled status. Hence the member axial forces are affected by the lateral displacements and cannot be assumed to remain in the same proportions and to grow linearly with the external load. The instability analysis of this type of structure is inherently nonlinear and an iterative load–deflection analysis is required.

In Section 2.3 a structural response analysis was classified as *second order* (i.e., nonlinear) if there is interaction between member displacements and member forces. In this section we have seen that if the influence is only in one direction—axial force influencing the lateral displacement but not vice-versa—then the analysis becomes a "second-order linearized" analysis. If, however, the axial force is also influenced *by* the lateral displacement then the analysis is completely second order and iteration is required.

Torsional Buckling

Another type of bifurcation buckling is torsional buckling of a column subjected to axial compression applied at the centroid ($e_y = e_z = 0$), as illustrated in Fig. 15.10. Here again because of the simplicity of the structure the axial degree of freedom is independent from the others and the axial member force $\mathcal{F}_x$ can be obtained separately, and even by inspection: $\mathcal{F}_x = -P$. Therefore we need only examine the other degrees of freedom, which are Θ_x and Θ_x'. In order to have any rotation we must have at least one internal node, that is, two elements, and because of symmetry we need examine only one of these. The boundary conditions are $\Theta_{x1} = 0$ and $\Theta_{x2}' = 0$, which leaves only two degrees of freedom in the structure: Θ_{x1}' and Θ_{x2}. For simplicity let us take the St. Venant torsional constant J as zero. Since there are no nodal or internal forces other than $\mathcal{F}_x$ most of the terms in (15.3.15) are zero. From this equation and (15.3.14) the total structure stiffness matrix, flexural plus geometric, is

$$\mathbf{K}_T = \begin{bmatrix} 4\mathfrak{b}_\omega l^2 - 4PC_0 l^2 & 6\mathfrak{b}_\omega l - 3PC_0 l \\ 6\mathfrak{b}_\omega l - 3PC_0 l & 12\mathfrak{b}_\omega - 36PC_0 \end{bmatrix}$$

where $\mathfrak{b}_\omega = EI_\omega / l^3$ and $C_0 = I_0/30Al$.

The condition for bifurcation buckling is

$$\det(\mathbf{K}_T) = 0$$

which may be written as

$$\mathfrak{b}_\omega \det \begin{pmatrix} 4l^2(1-\lambda) & 3l(2-\lambda) \\ 3l(2-\lambda) & 12(1-3\lambda) \end{pmatrix} = 0$$

$$\text{where } \lambda = P(C_0/\mathfrak{b}_\omega) = PI_0 l^2/30EI_\omega A = PI_0 L^2/120 EI_\omega A \qquad (15.4.6)$$

This yields a quadratic equation for λ for which the solution is $\lambda = 0.08286$. Substituting from (15.4.6) and using the full length L of the column:

$$P = 9.944 \frac{EI_\omega A}{I_0 L^2}$$

The coefficient is only 0.75% above the exact value, π^2, which shows that a two-element modeling of the column is quite adequate for ultimate strength analysis and other structural design applications.

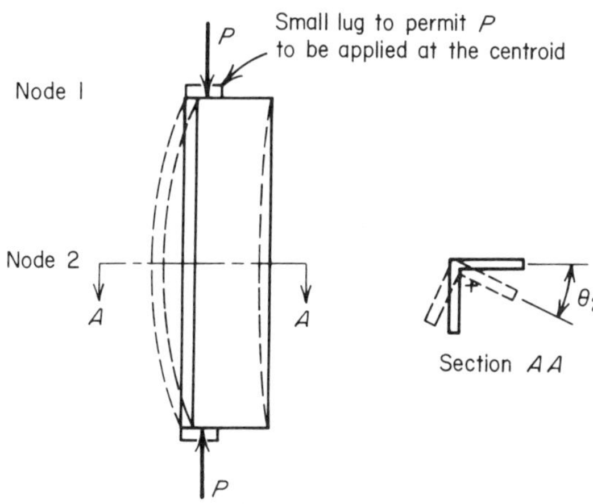

Figure 15.10 Torsional buckling of a column.

Flexural-Torsional Buckling: Pure Bending

A third type of bifurcation buckling involves the combination of a sideways deflection and a rotation of the cross section. This can occur in a slender thin-wall beam of open cross section. Figure 15.11 illustrates such a beam with the ends simply supported and prevented from rotating and subjected to applied moments M_Z and $-M_Z$ in the vertical plane. For simplicity the beam is assumed to be doubly symmetric, which means that prior to buckling it deflects only in the vertical plane; that is, the bending response involves only the V and Θ_Z degrees of freedom, and the bending moment within the elements can be obtained by a separate and prior calculation. Of course in this partic-

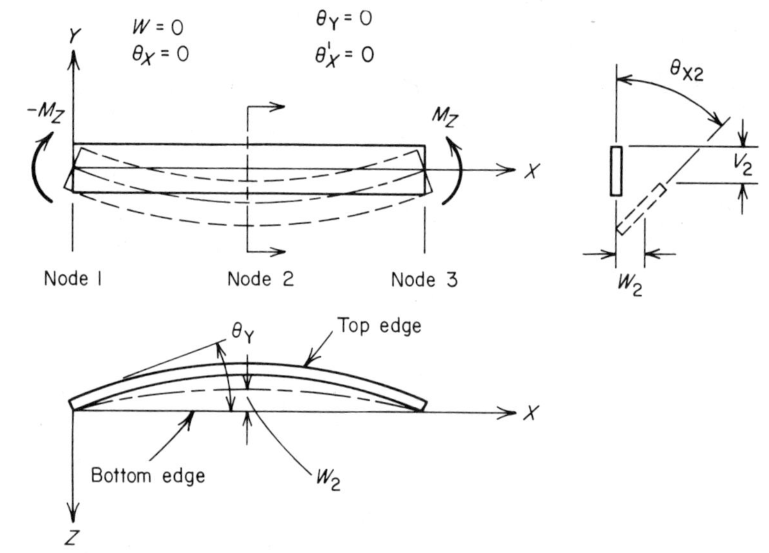

Figure 15.11 Flexural-torsional buckling; pure bending.

ular example the bending moment is obviously equal to M_Z. Since the V and Θ_Z degrees of freedom are independent of the others and have already been examined, we need only deal with the lateral and torsional degrees of freedom: W, Θ_Y, Θ_X, and Θ_X'.

We again model the structure using only two elements. For simplicity we will perform the analysis for two separate and idealized cases for which an exact solution is available: the case when the St. Venant torsional stiffness is negligible relative to the warping torsional stiffness ($J = 0$) and the opposite case when there is free warping of the cross section ($I_\omega = 0$).

SOLUTION FOR $J = 0$

From (15.3.14) and (15.3.15) the total structure stiffness matrix is

$$\mathbf{K} = \begin{array}{c} \begin{matrix} \quad\Theta_{Y1} & \qquad\qquad \Theta_{X1}' & \qquad\qquad W_2 & \qquad\qquad \Theta_{X2} \end{matrix} \\ \begin{bmatrix} 4\mathfrak{b}_y l^2 & & \text{symm.} & \\ -8M_Z l/60 & 4\mathfrak{b}_\omega l^2 & & \\ -6\mathfrak{b}_y l & -6M_Z/60 & 12\mathfrak{b}_y & \\ -6M_Z/60 & -6\mathfrak{b}_\omega l & -72M_Z/60l & 12\mathfrak{b}_\omega \end{bmatrix} \end{array}$$

where $\mathfrak{b}_y = EI_y/l^3$ and $\mathfrak{b}_\omega = EI_\omega/l^3$.

Setting det ($\mathbf{K}$) = 0 and performing the required multiplications gives

$$(M_Z)_{cr} = 9.9443\,\frac{E\sqrt{I_y I_\omega}}{L^2}$$

and the coefficient is only 0.76% above the exact value, π^2, which shows that the two-element representation is again quite sufficient.

SOLUTION FOR $I_\omega = 0$

Here the only change in the total stiffness matrix is in the following terms:

$$K_{22} = \frac{4GJl}{30}$$

$$K_{42} = \frac{GJ}{10}$$

$$K_{44} = \frac{12GJ}{10l}$$

The result of setting det ($\mathbf{K}$) = 0 is

$$(M_Z)_{cr} = 3.1534\,\frac{\sqrt{EI_y GJ}}{L}$$

in which the coefficient is only 0.375% above the exact value, π.

Flexural-Torsional Buckling: Uniform Lateral Load

We next examine the bifurcation buckling of a simply supported beam subjected to a uniform lateral load, which is a more typical loading for beams in a ship structure. Here again for simplicity the beam is assumed to be doubly symmetric and the lateral load q is directly in line with the centroidal axis, so that we need only consider the lateral and torsional degrees of freedom. The boundary conditions for a two-element model are illustrated in Fig. 15.12, which shows that the relevant degrees of freedom are θ_{y1}, θ'_{x1}, w_2, and θ_{x2}. From (15.3.14) the ordinary stiffness matrix is

$$\mathbf{K}_S = \begin{bmatrix} \theta_{y1} & \theta'_{x1} & w_2 & \theta_{x2} \\ 4l^2\mathfrak{b}_y & & & \text{symm.} \\ 0 & \dfrac{2GJl}{15} + 4l^2\mathfrak{b}_\omega & & \\ -6l\mathfrak{b}_y & 0 & 12\mathfrak{b}_y & \\ 0 & -\dfrac{GJ}{10} - 6l\mathfrak{b}_\omega & 0 & \dfrac{6GJ}{5l} + 12\mathfrak{b}_\omega \end{bmatrix}$$

and from (15.3.15) the geometric stiffness matrix is

$$\mathbf{K}_G = \begin{bmatrix} \theta_{y1} & \theta'_{x1} & w_2 & \theta_{x2} \\ 0 & & & \text{symm.} \\ -ql^3/35 & 0 & & \\ 0 & ql^2/140 & 0 & \\ 0.0905ql^2 & 0 & -0.6143ql & 0 \end{bmatrix}$$

Setting $\det(\mathbf{K}_S + \mathbf{K}_G) = 0$ and solving for q gives

$$q_{cr} = \frac{28.38\sqrt{EI_yGJ}}{L^3}\sqrt{1 + 10.01\frac{EI_\omega}{GJL^2}} \qquad (15.4.7)$$

The exact solution to this problem is given in Ref. 2 in the form

$$q_{cr} = \gamma\frac{\sqrt{EI_yGJ}}{L^3}$$

and the value of γ is given for a range of values of the torsional stiffness ratio $\rho = GJl^2/EI_\omega$. These are presented in Table 15.1 together with the values from (15.4.7). Once again it may be seen that the finite element solution agrees closely with the analytical solution and that a two-element model gives acceptable accuracy for engineering design.

TABLE 15.1

ρ	γ (Ref. 2)	γ (Eq. 15.4.7)
0.4	143	144.8
4	53.0	53.1
16	36.3	36.2
48	31.5	31.2
80	30.1	30.1
200	29.0	29.1
400	28.6	28.7
∞	28.3	28.38

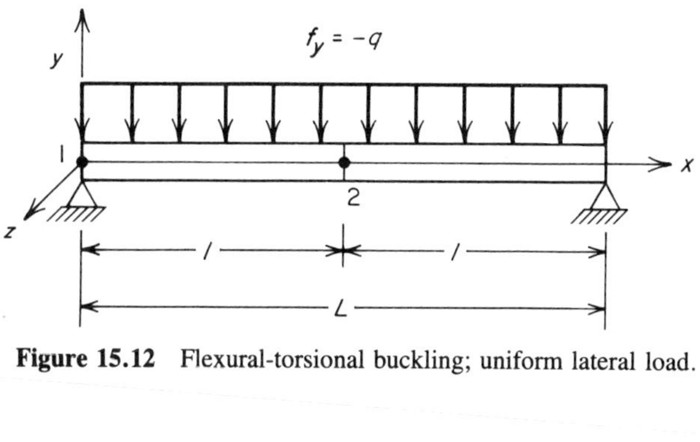

Figure 15.12 Flexural-torsional buckling; uniform lateral load.

15.5 BEAM ATTACHED TO PLATING

In a ship structure nearly all of the beams are attached to plating, and the plating's in-plane stiffness is so large compared to the beam's flexural and torsional stiffness that the beam may be regarded as prevented from undergoing any sideways translation along the line of attachment; that is, in the coordinates of Fig. 15.13 there is no w deflection in the plane of plating. This restraint is very important because it strongly influences the beam's response in regard to flexural-torsional buckling, and it also allows the development of a simplified geometric stiffness matrix for such beams. The greatest simplification is obtained for beams which are symmetric about the plane of the web (*monosymmetric* beams) and in a ship structure most of the larger or more important beams are given such symmetry whenever possible because this greatly increases the beam's resistance to flexural-torsional buckling. Therefore this section deals with monosymmetric beams, but it should be noted that the gen-

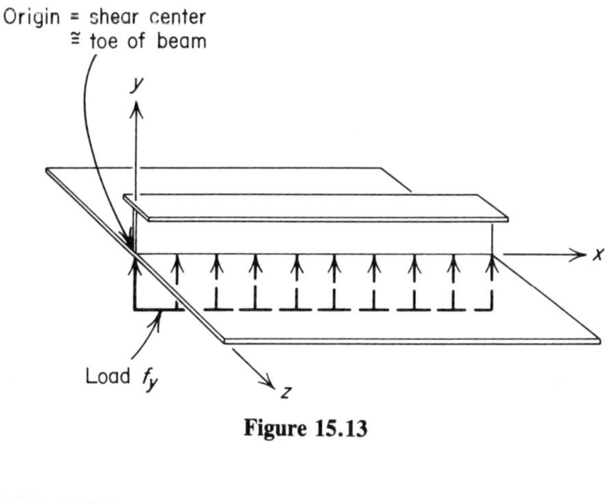

Figure 15.13

eral approach which is presented here applies to all beams which are attached to plating. The only difference is that for unsymmetric beams the vertical bending cannot be uncoupled from the lateral and torsional response, and so there are more terms to be evaluated in $\mathbf{K}_G$. In fact, for these beams the geometric stiffness matrix approach is particularly important, first because they are more susceptible to flexural-torsional buckling and second because there is no other method that can perform this type of analysis with the same scope and generality.

Figure 15.13 shows a typical beam in a ship structure. The beam is monosymmetric and is loaded only in the plane of the web; there are no sideways loads or bending moments. In the notation of Fig. 15.4, z_0, e_z, $\bar{a}_z$, β_1, F_z, f_z, and M_y are all zero. Under these conditions the vertical bending response (i.e., the v and θ_z degrees of freedom) of an *unrestrained* monosymmetric beam is independent of the lateral and torsional response (w, θ_y, θ_x, and θ_x'). The center of twist (or shear center) of such a beam is located at a distance y_0 from the large flange, given by Ref. 5 as

$$y_0 = \frac{I_{ys}}{I_y} h_w \tag{15.5.1}$$

where I_{ys} and I_y are the second moments of area of the smaller flange and of the complete beam, about the y-axis (see Fig. 15.14).

Due to the sideways restraint of the plating, the actual axis of rotation is at the line of attachment rather than at the shear center. But the ratio I_{ys}/I_y is approximately proportional to the cube of the flange width ratio and therefore the value of y_0 is negligibly small for a beam attached to plating. Therefore we may investigate the lateral and torsional response separately from the in-plane bending.

In Section 15.3 the nodal degrees of freedom v, w, θ_y, and θ_z were defined to be the displacements and rotations at the shear center. We are neglecting y_0 and placing the nodes in the plating, where $w = \theta_y = 0$. Thus the effect of the plating is to delete the w and θ_y degrees of freedom ($w = \theta_y = 0$) leaving only the torsional degrees of freedom: θ_x and θ_x'. Therefore in analyzing this type of beam for flexural-torsional buckling it is only necessary to consider these two degrees of freedom, and so both the ordinary and the geometric stiffness matrices of (15.3.14) and (15.3.15) are of dimension 4×4.

We next consider the effect which this special restraint has on the interpretation and evaluation of the torsional parameters J, I_ω, and β_2. This depends entirely on the extent to which the beam's torsional response is restricted by the plating. The plating does contribute some resistance because the twist of the beam involves some plate bending, as shown in Fig. 15.15. In practice, however, for typical thicknesses and overall proportions of beams and plates this restraint is relatively small, and ignoring it is both conservative and simplifying. We shall also assume that the beam cross section is not distorted by web bending, as shown in Fig. 15.16*a*. For slender beams, web bending, if allowed to occur, would cause a significant reduction in the flexural-torsional buckling load. Therefore it is prevented by fitting vertical web brackets at sufficiently close intervals along the beam, as shown in Fig. 15.16*b*. Since this arrangement is by far the most common we will obtain the most widely applicable result by assuming that web bending does not occur. With these assumptions it is only the beam which is involved in the torsional response, and so the values of J, I_ω, and β_2 are calculated using only the beam cross section. In terms of the beam dimensions of Fig. 15.17 the expressions for these parameters are

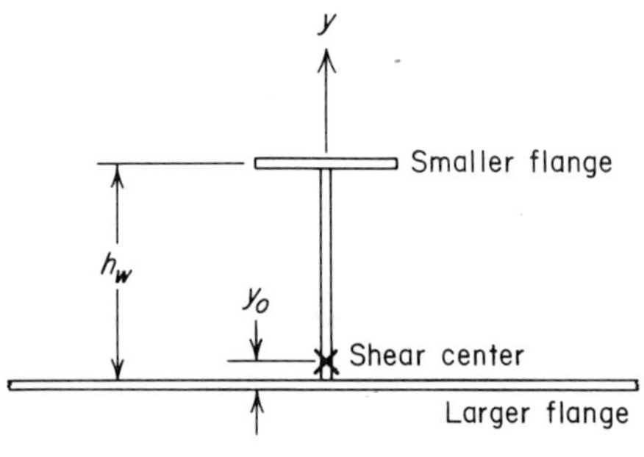

Figure 15.14 Shear center of a monosymmetric beam.

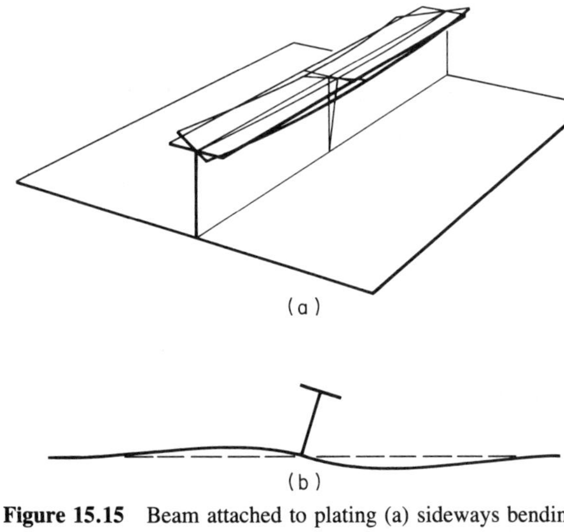

Figure 15.15 Beam attached to plating (a) sideways bending of flange, (b) section at midlength

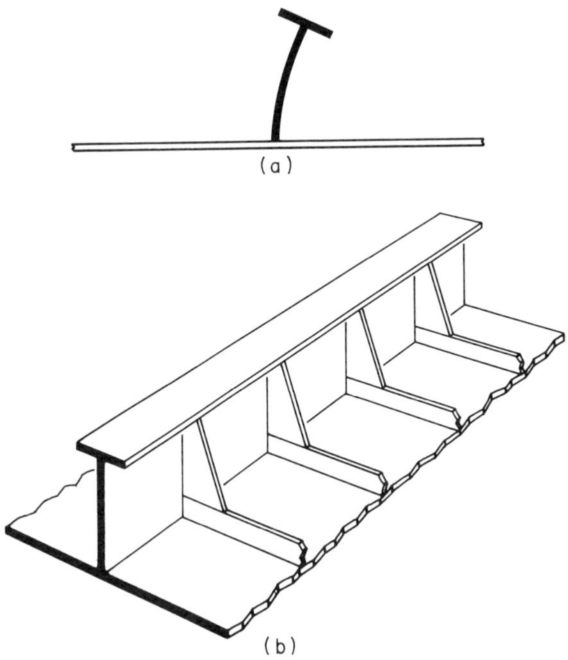

Figure 15.16 (*a*) web bending, (*b*) web brackets

$$J = \frac{1}{3}(h_w t_w^3 + b_f t_f^3) \tag{15.5.2}$$

$$I_\omega = \frac{1}{12} h_w^2 t_f b_f^3 \tag{15.5.3}$$

$$\beta_2 = \frac{1}{I_z}\int_{A_b} y(y^2 + z^2)\, dA$$

$$= \frac{1}{I_z}\left(\frac{1}{4}h_w^4 t_w + h_w^3 b_f t_f + \frac{1}{12} b_f^3 h_w t_f\right) \tag{15.5.4}$$

where A_b is the cross-sectional area of the beam and I_z is the moment of the inertia of the entire *composite* beam, including the plating.

The expressions for J and β_2 are seen to be the usual expressions, evaluated for the beam cross section

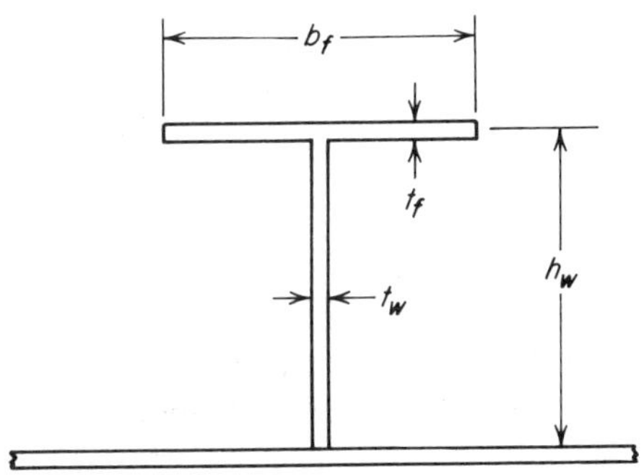

Figure 15.17 Beam dimensions.

only. Similarly the expression for I_ω may be obtained by evaluating (6.1.17) for the beam cross section. But it is instructive to show that this parameter relates entirely to the sideways bending of the beam flange. This may be seen by writing the expression for the strain energy of a beam undergoing flexural-torsional buckling

$$U = \frac{1}{2}E\int_L\int_A [\varepsilon(y,z)]^2\, dA\, dx$$

The strain in the flange is due to its curvature

$$\varepsilon = z\frac{\partial^2 w}{\partial x^2}$$

and the sideways deflection of the flange is

$$w = h_w \theta_x$$

The web has relatively little resistance to sideways bending and hence its contribution to the strain energy is negligible. Substituting into the expression for U we obtain

$$U = \frac{1}{2}E\int_L\int_A [h_w \theta_x'' z]^2\, dA\, dx$$

$$= \frac{1}{2}E\left[\frac{t_f b_f^3 h_w^2}{12}\right]\int_0^L (\theta_x'')^2\, dx$$

and comparison with (15.3.1) shows that the quantity in square brackets is I_ω.

Rotational Restraint due to Plate Stiffeners

Since the rotational stiffness of the plating is relatively small the web brackets should always be located in line with the plate stiffeners, as shown in Fig. 15.16b. With this arrangement the flexural stiffness of the plate stiffeners provides some restraint against rotation θ_x, and this should be allowed for. If, as is usual, there are several equally spaced stiffeners within the length L, it may be shown that the average rotational stiffness per unit length is

$$s_{\theta_x} = \frac{4EI_s}{ab}$$

where a, b, and I_s are the stiffener span, spacing, and moment of inertia (including the plate flange) respectively. The relationship between the internal moment $\mathcal{M}_x(x)$ and the rotation $\theta_x(x)$ is $\mathcal{M}_x(x) = s_{\theta_x}(x)\theta_x(x)$. If we again adopt the cubic shape function of (5.3.9b)–or (15.1.5)–to represent $\theta_x(x)$

$$\boldsymbol{\theta}_x(x) = \mathbf{N}_3(x)\boldsymbol{\theta} \qquad (15.5.5)$$

then it may be shown that the matrix of stiffness terms relating the nodal moments and bimoments to the four nodal degrees of freedom (θ_{x1}, θ'_{x1}, θ_{x2}, and θ'_{x2}) is given by

$$\mathbf{k}_{\theta x} = s_{\theta x} L \int_0^1 \mathbf{N}_3^T \mathbf{N}_3 \, d\xi$$

from which the final result is

$$\mathbf{k}_{\theta x} = \frac{EI_s L}{105\,ab} \begin{bmatrix} 156 & & & \text{symm.} \\ 22L & 120L^2 & & \\ 54 & 13L & 156 & \\ -13L & 120L^2 & -22L & 120L^2 \end{bmatrix} \qquad (15.5.6)$$

Example: Beam Subjected to Pure Bending

We next present a simple example of the use of the geometric stiffness matrix for analyzing the flexural-torsional buckling of a beam attached to plating. In order to assess the accuracy of a two-element model of the beam we choose the very simple case of pure bending, for which it is possible to obtain an exact solution. As shown in Fig. 15.18 the beam is simply supported and is subjected to end bending moments $-M$ and M which place the beam flange in compression. If the bending were applied in the opposite direction there would be no torsional buckling. Because of the lengthwise symmetry we need only consider the left-hand element, for which the boundary conditions are $\theta_{x1} = 0$ and $\theta'_{x2} = 0$. In the stiffness matrices only the θ'_{x1} and θ_{x2} terms are required. The value of M_{z1} is $-M$ and so the resulting matrices are

$$\mathbf{K}_S = \begin{bmatrix} \frac{2}{15} GJl + 4l_1^2 \mathfrak{b}_\omega & \text{symm.} \\ -\frac{1}{10} GJ - 6l\mathfrak{b}_\omega & \frac{6GJ}{5l} + 12\mathfrak{b}_\omega \end{bmatrix}$$

and

$$\mathbf{K}_G = \begin{bmatrix} -\frac{2}{15} l\beta_2 M & \text{symm.} \\ \frac{1}{10}\beta_2 M & -\frac{6}{5}\frac{\beta_2 M}{l} \end{bmatrix}$$

The buckling condition $\det(\mathbf{K}_S + \mathbf{K}_G) = 0$ leads to

$$M_{cr} = \frac{GJ + 9.944 \dfrac{EI_\omega}{L^2}}{\beta_2} \qquad (15.5.7)$$

The analytical solution to this problem can be obtained by setting $\theta_x = a \sin(\pi x/L)$, which satisfies the boundary conditions, and then imposing the buckling condition $\partial\Pi/\partial a = 0$, with Π being given by the θ_x terms of (15.3.12), in which $M_{z1} = -M$. The result is

$$M_{cr} = \frac{GJ + \pi^2 \dfrac{EI_\omega}{L^2}}{\beta_2}$$

which again shows that a two-element model is quite accurate.

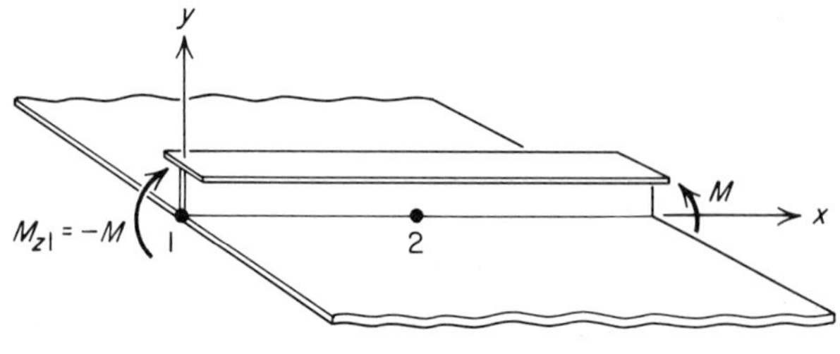

Figure 15.18 Beam attached to plating; pure bending.

15.6 NONBIFURCATION BUCKLING: NONLINEAR FINITE ELEMENT ANALYSIS

A structure is nonlinear whenever its properties or equilibrium relationships—as typified by its stiffness matrix **K**—change with increasing load. We have seen in Section 15.1 that if the change in **K** is due entirely to the small changes in structure geometry which are the displacements, then this gives rise to additional stiffness terms (the geometric stiffness matrix) which are linear functions of the internal member forces. Since the latter are dependent on the displacements, the analysis requires repeated solutions of the equilibrium equations until the internal member forces are consistent with the displacements. This iterative process was illustrated in Fig. 15.2.

However, there are many other ways in which **K** could change; for example, because of a change in a material property of an element or, in the case of a structure, a failure in one or more of the members. In such cases the changes in **K** may be a continuous nonlinear function of either the displacements (e.g., nonbifurcation buckling) or internal member forces (e.g., strain hardening). Alternatively, the changes in **K** may occur only at a certain level of internal member force (yielding, bifurcation buckling, and fracture). As a result there are two extra complications which are not present in simple geometric nonlinearity:

1. Since it is not known beforehand which terms in **K** will change, all elements or members must be examined explicitly.
2. Since the terms themselves are not linear and continuous functions of the member forces, they cannot be updated automatically simply by updating the member forces; instead they must be examined continuously and updated individually as required.

These complications make it necessary to use an incremental approach, which consists in applying a sequence of load increments, modifying **K** so that it always gives a correct portrayal of the structure stiffness, and updating and monitoring all of the internal member forces and deformations, comparing them with limit values which have been calculated for the various modes of failure, for each type of member.

There are two types of incremental finite element analysis, depending on whether or not there is a continuous change in **K**, as in Fig. 15.19, in addition to whatever discrete changes may occur. If there is continuous change then the analysis must be both *incremental and iterative*. If the change in **K** occurs only at discrete values of load, such that the load–deflection curve is piecewise linear as in Fig. 15.20, then the analysis need only be *incremental;* iteration is not required.

Continuous nonlinearity can occur either at the member level (e.g., the gradual spread of plasticity throughout a highly stressed element) or at the overall structure level (e.g., nonbifurcation buckling of a structure). Thus in a ship structure there are many possible applications of incremental and iterative finite element analysis. However, as mentioned in Section 2.6, this type of analysis requires a very large amount of computation, even for an individual structural member, while for a complete structure the computational cost is prohibitive, at least for present day computers. Hence for individual structure members the various approximate analysis methods presented in Chapters 9 to 14 are preferred because they involve much less computation and yet have sufficient accuracy. For a ship structure, overall structural collapse is due to the sequential failure (loss of stiffness) of individual members, generally due to some combination of local plastic deformation and bifurcation buckling accompanied by some yielding. Thus the load–deflection curve is piecewise linear, as in Fig. 15.20, and the much simpler noniterative type of incremental finite element analysis may be used. In this simpler method the next individual failure can be identified by checking all of the members and hence the magnitude of the next load increment can be determined. If two or more failures have approximately the same failure load they are taken together and any interactions are allowed for.

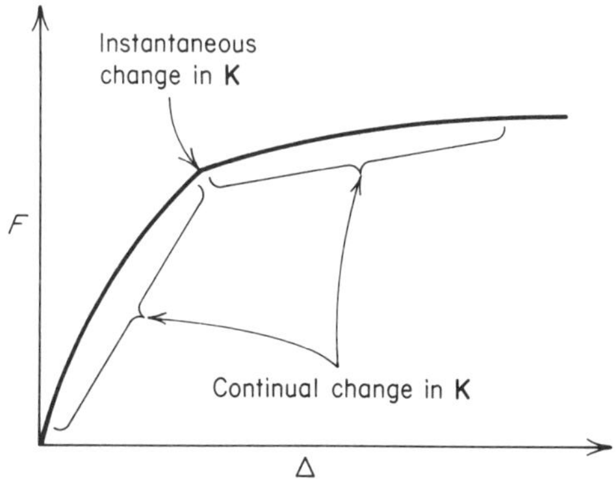

Figure 15.19 Structure with continously changing stiffness matrix.

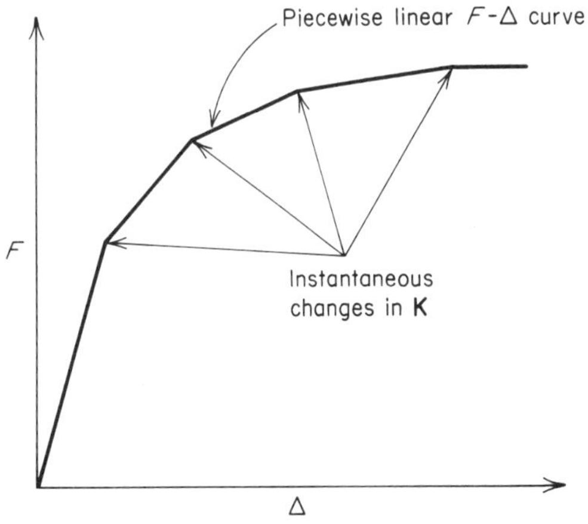

Figure 15.20 Structure with only discrete changes in stiffness matrix.

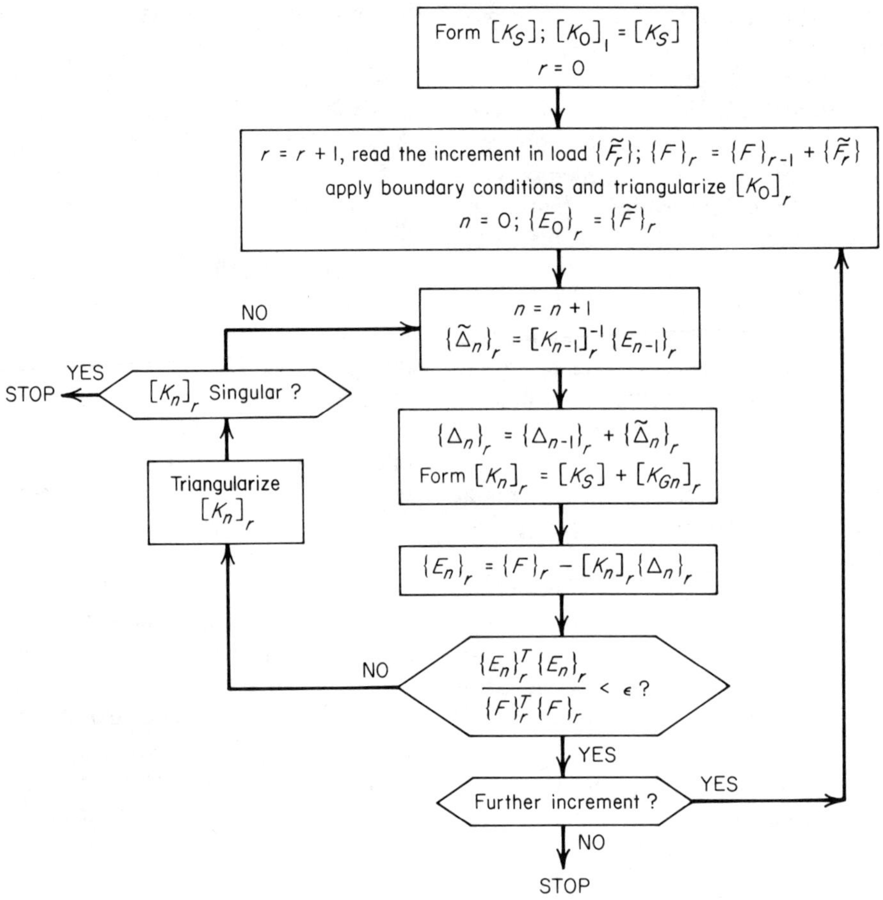

Figure 15.21 Flow chart for incremental and iterative finite element analysis.

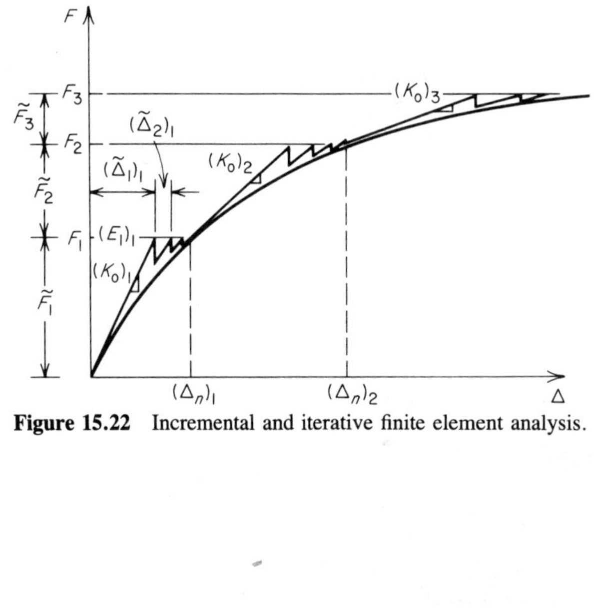

Figure 15.22 Incremental and iterative finite element analysis.

Each failure is accounted for by making the appropriate change to the structure stiffness matrix. In most cases this involves either altering or adding a few degrees of freedom and recalculating the member forces. Some of the basic features of this method are presented in the next chapter, which deals with the special case when all of the individual failures are of the same type: the formation of a plastic hinge. The complete version of the method is presented in Chapter 17.

Because of its large computational cost the iterative type of incremental finite element analysis is not presently used for ship design, but rather only for specially commissioned ultimate strength investigations and for research projects. Therefore in this text we give simply the basic algorithm, as summarized by the flow chart in Fig. 15.21 and the load-deflection curve in Fig. 15.22.

REFERENCES

1. T. V. Galambos, *Structural Members and Frames,* Prentice-Hall, 1968.
2. S. P. Timoshenko and J. M. Gere, *Theory of Elastic Stability,* McGraw-Hill, New York, 1961.
3. B. J. Davidson and I. C. Medland, "A Finite Element Approach to Stability Analysis in Frames (Including Warping Effects)," First Int'l. Conf. on Finite Element Methods in Eng'g., Sydney, Aug. 1974, pp. 621–637.
4. R. S. Barsoum and R. H.Gallagher, "Finite Element Analysis of Torsional and Torsional-Flexural Stability Problems," *Int'l. J. Num. Meth. Eng.,* **2,** 1970, pp. 335–352.
5. S. Kitipornchai and N. S. Trahair, "Elastic Behavior of Tapered Monosymmetric I-Beams," *J. Struct. Div., ASCE,* **101**(ST8), Aug. 1975, pp. 1661–1689.

Chapter

Sixteen

PLASTIC FRAME ANALYSIS

This chapter examines the collapse of a frame structure due to the occurrence of large plastic deformation. The first section reviews the basic aspects of such deformation, including the concept of a plastic hinge. The next section examines the collapse of an individual beam. The next two sections deal with *plastic frame analysis,* which assumes that collapse of a frame structure occurs when sufficient plastic hinges have been formed to convert some portion of the structure into a mechanism. It is shown that there are two alternative approaches: *direct* plastic frame analysis and *incremental* plastic frame analysis. The latter is a load–deflection approach which traces the formation of the hinges. Until now the direct approach has predominated because it involves much less computation, but Section 16.4 shows that because of its basic underlying assumptions, direct plastic frame analysis is unsuitable for hull module ultimate strength analysis.

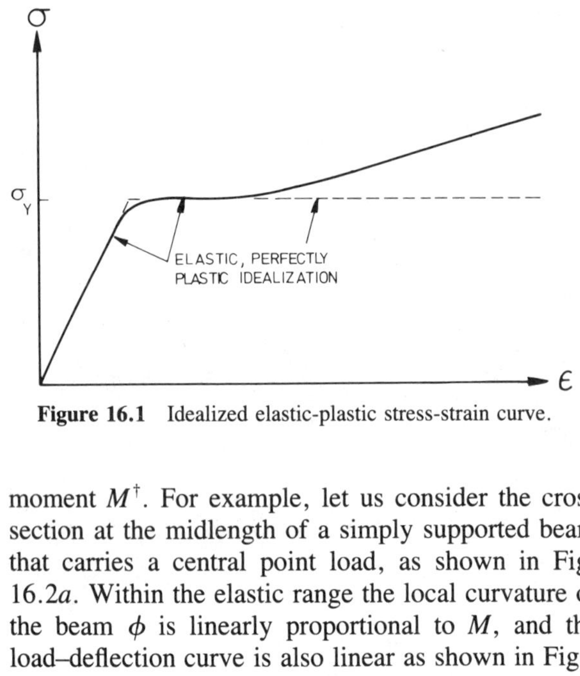

Figure 16.1 Idealized elastic-plastic stress-strain curve.

16.1 PLASTIC THEORY AND APPLICATION TO BEAMS

Plastic Hinge and Plastic Section Modulus

Plastic theory is based on an idealized "elastic–perfectly plastic" stress-strain curve. As shown in Fig. 16.1, this idealization is quite suitable for mild steel, with its definite yield point, and it is conservative since it ignores the subsequent strain-hardening of the material.

Let us consider a cross section of a beam that is subjected to a steadily increasing value of bending moment $M^{\dagger}$. For example, let us consider the cross section at the midlength of a simply supported beam that carries a central point load, as shown in Fig. 16.2*a*. Within the elastic range the local curvature of the beam ϕ is linearly proportional to M, and the load–deflection curve is also linear as shown in Figs. 16.2*b* and *c*. Figure 16.3 shows the resulting growth and distribution of the bending stresses for two different cross sections, one symmetric and the other nonsymmetric. For the present we shall assume that there is no axial force acting on the beam and we shall neglect the effect of shear. At some value of M, say M_Y, the maximum bending stress will reach the yield stress of the material. If the load, and therefore the bending moment, is further increased, plasticity will spread throughout the depth of the beam until the sec-

†In this chapter nodal moments are not involved, and the symbol M refers always to bending moment.

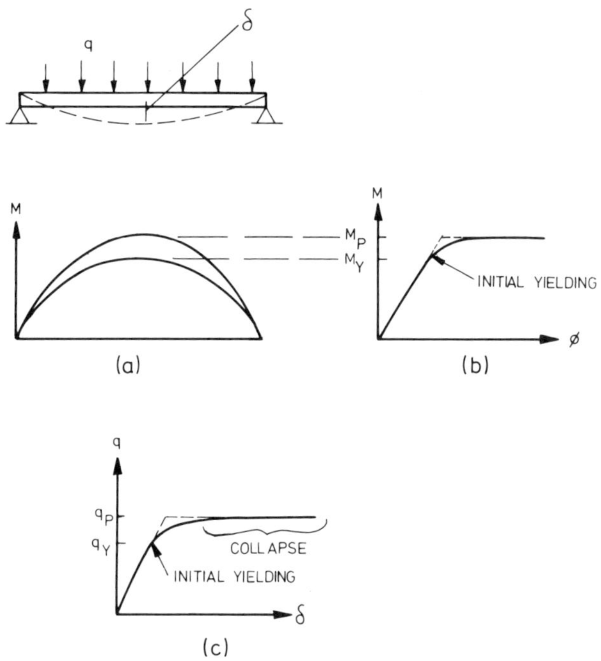

Figure 16.2 Plastic bending of a beam.

tion is fully plastic. The local bending moment corresponding to this condition is known as the *plastic moment* of the section, M_P. Because all of the fibers have now reached the limit of their load-carrying ability, the beam can absorb no further bending moment at this section. It is as if a hinge had been inserted in the beam at this point, and hence this condition in a beam is referred to as a "plastic hinge." If the beam is merely

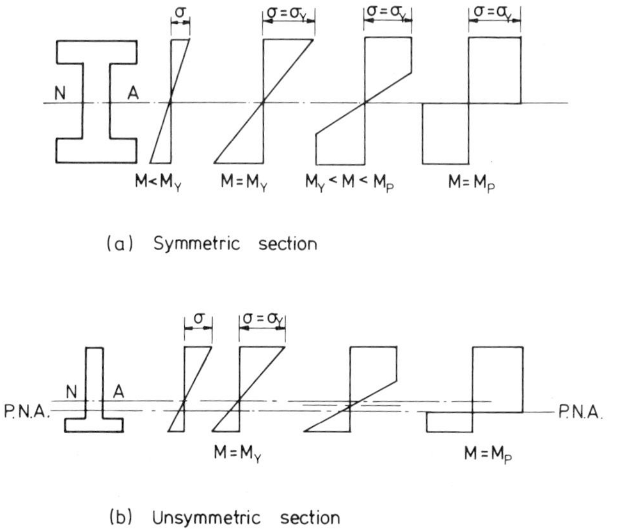

Figure 16.3 Plastic hinge formation.

simply supported as in Fig. 16.2, or if it is a cantilever, then the occurrence of such a hinge makes it completely incapable of carrying any further load and it would collapse. In practice the strain-hardening of the material would delay the collapse slightly, to some value of M slightly greater than M_P, but nevertheless at $M = M_P$ the deflection would already be so large as to constitute effective collapse.

Since the stresses in a plastic hinge are equal and opposite in the upper and lower portions of the cross section and since there is no axial force, equilibrium in the longitudinal direction requires that the areas of these two portions must be equal. Therefore, the "plastic neutral axis" (PNA)—that is, the axis or line in the fully plastic section where the stress reverses sign—is simply the horizontal line which divides the section into two equal areas. In a symmetric section this will coincide with the elastic neutral axis (axis of equal first moments) but in an unsymmetric section they will differ, as shown in Fig. 16.3*b*.

We now consider equilibrium of moments for the fully plastic section, which requires that the external bending moment M_P must be balanced by the moment of the stress forces in the section. Since the stresses are of constant magnitude the result is simply

$$M_P = \sigma_Y \frac{A}{2}(\bar{y}_1 + \bar{y}_2) \qquad (16.1.1)$$

where $\bar{y}_1$ and $\bar{y}_2$ are the vertical distances from the PNA to the centroids of the two half area portions, as shown in Fig. 16.4.

As a further analogy to elastic beam bending we define a "plastic section modulus" Z_P such that

$$M_P = \sigma_Y Z_P \qquad (16.1.2)$$

From (16.1.1) we obtain

$$Z_P = \frac{A}{2}(\bar{y}_1 + \bar{y}_2) \qquad (16.1.3)$$

and thus we see that Z_P is simply the sum of the first moments of area of the two half-areas of the fully

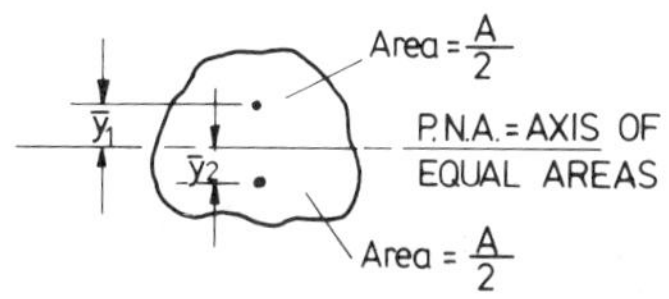

Figure 16.4

yielded section. In computing Z_P the two first moments must be combined fully, and hence the two moment arms $\bar{y}_1$ and $\bar{y}_2$ must each be treated as positive. Z_P is a purely geometric quantity and for standard sections such as I-beams, tees, angles, and so on, the value of Z_P may be obtained from structural handbooks. In ship structures the beams are usually nonstandard since they are attached to plating and have a "plate flange" (of effective breadth b_e) which must be included. Therefore Z_P must be calculated by first solving for the position of the PNA and then calculating and adding the two area moments. An alternative method, using formulas (and therefore suitable for computer applications) is given later in this section.

Both Z_P and Z are geometric quantities which depend on the shape of the section, and it may be shown that Z_P is always larger than Z. For instance, for a rectangular section of depth h and breadth b the two values are $Z_P = bh^2/4$ and $Z = bh^2/6$, giving a ratio of 1.5. This ratio is referred to as the *shape factor* ν_s

$$\nu_s = \frac{Z_P}{Z} \tag{16.1.4}$$

For thin-walled, flange-and-web beams the shape factor is generally in the range $1.1 < \nu_s < 1.2$. From the definition of M_Y (moment to cause initial yielding) we have $M_Y = \sigma_Y Z$ and hence from (16.1.2)

$$\nu_s = \frac{Z_P}{Z} = \frac{M_P}{M_Y} \tag{16.1.5}$$

Thus the shape factor has physical significance in that it is the ratio by which the plastic hinge moment exceeds the initial yield moment at any given section of the beam. If the beam is statically determinate, so that only one hinge is required for collapse, then ν_s also indicates the margin between yielding and collapse. However, beams in structures generally have more than this minimum degree of fixity, and consequently their true reserve of strength is somewhat larger than ν_s.

The fact that bulky sections have a larger value of ν_s than thin-walled sections does not mean that they are better; it arises simply because they have a smaller value of Z, and hence a smaller M_Y, due to a less efficient distribution of material. For loads within the elastic range most of the material is only lightly stressed, whereas thin-walled beams have a large portion of the material in the flanges, where it is highly stressed. It is only after yielding begins that the inner material of a bulky section becomes highly stressed, and so there is a relatively larger margin between yielding and collapse. The fact that the more efficient sections do not, of themselves, have as large a local margin emphasizes the need for the design to be accurate and rationally-based, to ensure that the structure as a whole, and each of its component beams, will have an adequate collapse margin of safety.

Because the local margin between M_Y and M_P is relatively small (about 15%) the rounded portion of the $M - \phi$ curve in Fig. 16.2*b* is often neglected so as to obtain the idealized bilinear moment–curvature relationship shown in the figure. That is, we often ignore the elasto-plastic transition range and the corresponding additional curvature; this is equivalent to saying that the cross section is unaffected until M reaches M_P, whereupon a plastic hinge suddenly forms.

Formulas for Plastic Section Modulus

A systematic method for calculating Z_P for the nonstandard stiffener-and-plate sections of ship structures is to compute the contributions of the web and the flanges separately and to add them. That is,

$$Z_P = Z_f + Z_w + Z_p \tag{16.1.6}$$

in which Z_f, Z_w, and Z_p are the contributions of flange, web, and plate flange respectively. These can be calculated using the formulas given in Table 16.1. As shown in Fig. 16.5, two possible cases arise, depending on whether the plate area A_p is greater or less than the stiffener area $(A_w + A_f)$.

If the section is subjected to an axial force F (positive tensile) the plastic neutral axis is not the axis of equal areas but rather the axis for which the areas of

TABLE 16.1

	Case 1 $A_p > A_f + A_w$	Case 2 $A_p < A_f + A_w$
C_1	$\dfrac{A_p - A_w - A_f}{2A_p}$	$\dfrac{A_w + A_f - A_p}{2A_w}$
C_2	$C_1^2 - C_1 + \frac{1}{2}$	Same as for Case 1
g	$C_1 t_p$	$C_1 h_w$
Z_f	$A_f(h_w + g + \frac{1}{2}t_f)$	$A_f(h_w - g + \frac{1}{2}t_f)$
Z_w	$A_w(\frac{1}{2}h_w + g)$	$A_w h_w C_2$
Z_p	$A_p t_p C_2$	$A_p(g + \frac{1}{2}t_p)$

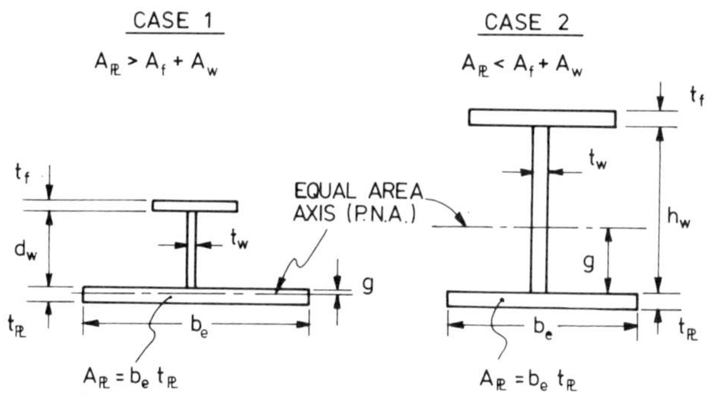

Figure 16.5

compression and tension, A_c and A_t, satisfy the relationship

$$A_t\sigma_Y - A_c\sigma_Y = F$$

from which A_t is

$$A_t = \frac{F + A\sigma_Y}{2\sigma_Y} \qquad (16.1.7)$$

where A is the total area. In this case the value of g is obtained by calculating A_t and then locating the horizontal axis that would isolate that much area on the tension side of the cross section; this is the plastic neutral axis of the section. Case 1 is when this axis lies within the plating; the relative magnitude of A_p and $A_w + A_f$ is no longer the criterion. As before, g is the distance from the plastic neutral axis to the inside surface of the plating.

The foregoing definition of a plastic hinge assumes that the tension and compression portions of the section can both reach a fully yielded condition. As shown in Section 14.1 this may not be true for a free beam flange in compression because it may develop a local S-shaped plastic mechanism. Hence in practice it is best to use two values of collapse bending moment for any section: the full plastic hinge value M_P when the flange is in tension, and the flange yield value M_{fY} when the flange is in compression.

Correction for Shear

The foregoing equations do not account for the effect of a shear force Q and the resulting shear stress within the beam cross section. The relative magnitude of Q and M, and hence τ and σ, depends mainly on the type of load and the slenderness (height/length) of the beam; the ratio τ/σ will be largest for a point load and for short deep beams. Taking, for example, a simply supported thin-walled beam of normal cross section proportions (either open or closed) of height h and of length L, it may be shown that the ratio of τ to σ is of the order of h/L. Hence for most beams the shear stress is much smaller than the bending stress, and it is often neglected in calculating ultimate strength. However, since it acts to diminish the ultimate strength, its effect should be allowed for. If its magnitude is small or moderate ($\tau/\sigma < 0.3$, say) the following approach may be used. If it is large then it is possible that web shear might in itself cause collapse of the beam. References 37 through 40 of Chapter 12 deal with the design of such beams.

The Hencky-Von Mises yield criterion states that yielding occurs in an element when σ and τ satisfy the relation

$$\sigma_Y = \sqrt{\sigma^2 + 3\tau^2} \qquad (16.1.8)$$

Hence, the effect of a shear stress τ is to reduce the failure value of σ, that is, the value of σ which will cause yield. The reduced value can be expressed in terms of a reduction factor r

$$\sigma_{\text{failure}} = r\sigma_Y = \sqrt{1 - 3\left(\frac{\tau}{\sigma_Y}\right)^2}\,\sigma_Y$$

and the effect of τ can be accounted for by using this reduced value in place of σ_Y and continuing to treat the stress distribution as uniaxial.

We may apply this to the present case by noting that for flanged, thin-walled beams the shear stress occurs mainly in the web and is given approximately by

$$\tau \cong \frac{Q}{A_w}$$

where A_w is the area of the web. Hence the reduction factor r may be approximated by:

$$r = \sqrt{1 - 3\left(\frac{Q}{A_w\sigma_Y}\right)^2} \qquad (16.1.9)$$

If we ignore the smaller values of shear stress in the flanges, the reduction factor need only be used in the web. Furthermore, at collapse the stress level is constant in magnitude throughout the web, and we may therefore apply the reduction factor to the web area instead of to the stress. That is, we calculate Z_P for a transformed section in which the web area A_w is reduced by the factor r. The use of this value of Z_P in all ultimate strength calculations automatically corrects for the effect of shear.

16.2 COLLAPSE OF A BEAM OR OF A LATERALLY LOADED STIFFENED PANEL

As explained in Section 8.1 the stiffened panels in a hull module usually have stiffeners in one direction only because the beams in the other direction are modeled explicitly. In such panels the stiffeners are virtually independent and can be analyzed as separate beams, with a plate flange of effective width b (or b_e if allowance for shear lag is desired; see Section 3.8 and Fig. 3.36). The flexural properties of such beams may be calculated by (8.3.6). Therefore all of the methods and results presented in this section may be used for laterally loaded stiffened panels as well as for beams. For simplicity the word "beam" is used throughout. In regard to the support conditions to be assumed at the ends of the panel, the comments in Section 8.1 also apply here. If desired, a more accurate analysis may be made simply by including some or all of the adjacent panels in the analysis and using the multibeam (frame) analysis methods presented in the subsequent sections of this chapter.

Basic Collapse Mechanism

An individual beam becomes a mechanism (a two-bar linkage) if it is given three pinned joints, providing that one end of the beam can move inward toward the other, as shown in Fig. 16.6. If the beam is simply supported at the ends, only one plastic hinge is required and it will form when the maximum bending moment reaches M_P. For a central point load $M_{max} = PL/4$, and hence the collapse or "ultimate" load for this case is

$$P_{ult} = \frac{4M_P}{L} \tag{16.2.1}$$

The ultimate load can be expressed in terms of beam properties by substituting $Z_P\sigma_Y$ in place of M_P. It can be seen from similar triangles that $u/2\delta = 2\delta/L$ (assuming that $\delta \ll L$). Therefore, for an unrestrained beam the axial displacement u is proportional to the square of the deflection. In dimensionless form the relationship is

$$\frac{u}{L} = 4\left(\frac{\delta}{L}\right)^2 \tag{16.2.2}$$

If, on the other hand, the ends are completely restrained from approaching, then a beam can continue to carry a load even with three pinned joints, through the action of axial tensile forces, as illustrated in Fig. 16.7. From geometry, assuming $\delta \ll L$, the tensile force is approximately

$$T \cong \frac{PL}{4\delta} \tag{16.2.3}$$

The elongation is related to T by

$$u = \frac{TL}{AE} \tag{16.2.4}$$

If the three pinned joints were present from the very beginning (i.e., a two-bar linkage with its ends fully restrained from moving in) then (16.2.3) would apply throughout all of the load history and from the foregoing three equations the load–deflection relationship would be

$$P = \frac{16AE}{L^3}\delta^3 \tag{16.2.5}$$

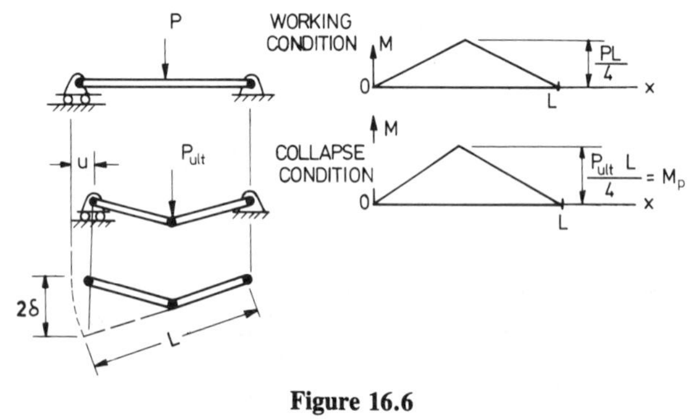

Figure 16.6

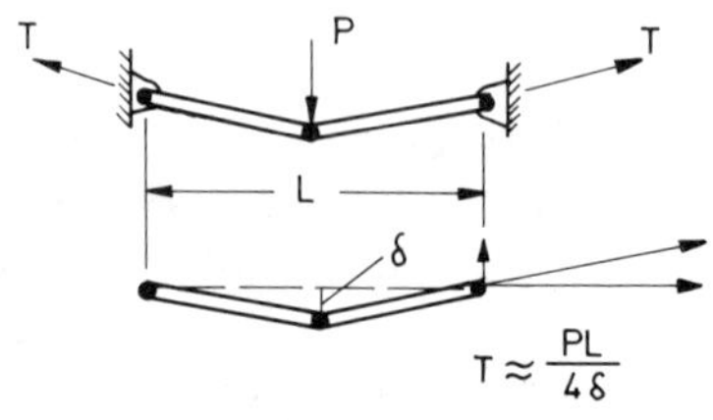

Figure 16.7

which is shown in Fig. 16.8. Collapse would finally occur when the axial stress reached the yield stress. By setting $T/A = \sigma_Y$ in (16.2.4) and substituting for u in (16.2.2) the corresponding deflection is found to be

$$\frac{\delta_{\rm ult}}{L} = \frac{1}{2}\sqrt{\frac{\sigma_Y}{E}}$$

and substitution of this into (16.2.5) gives the ultimate strength of a fully restrained two-bar linkage

$$P_{{\rm ult},r} = \frac{2A\sigma_Y^{3/2}}{\sqrt{E}} \qquad (16.2.6)$$

Figure 16.8 also shows the load–deflection relationship for a pinned beam if it has complete axial restraint. As the central plastic hinge forms, the deflection begins to increase sharply and the beam begins to carry the load by axial tension, which makes up for the loss of flexural rigidity. The beam thus resembles a two-bar linkage and its subsequent load–deflection curve is similar to that of (16.2.5) but displaced upward and to the right. Under the idealized conditions which we are considering, $P_{{\rm ult},r}$ for the beam would be the same as for the two-bar linkage.

Since the ratio δ/L is always very small (usually less than 1 or 2%) (16.2.3) indicates that the tensile restraining force T is many times larger than the load. Under the action of this very large force the supporting structure must undergo some deflection and/or deformation (and hence allow some pull-in) in proportion to its own structural stiffness and material properties. Also we have seen that in an unrestrained beam the pull-in is proportional to the square of the δ/L ratio, and hence it is very small. For example, for a beam having $\delta/L = \frac{1}{40}$, $u = L/400$. Therefore only a small amount of deflection and/or deformation of the supporting structure is enough to preclude axial restraint. For this reason axial restraint in free-standing structures is very difficult to provide, and is virtually impossible in frame structures, including ships, because in that case an axial displacement of one beam corresponds to transverse deflections of the supporting beams, and the transverse (or bending) stiffness of beams is much smaller than the axial stiffness. Therefore, for beams in rigid-jointed frame structures the basic mechanism for member collapse is *three hinges*.

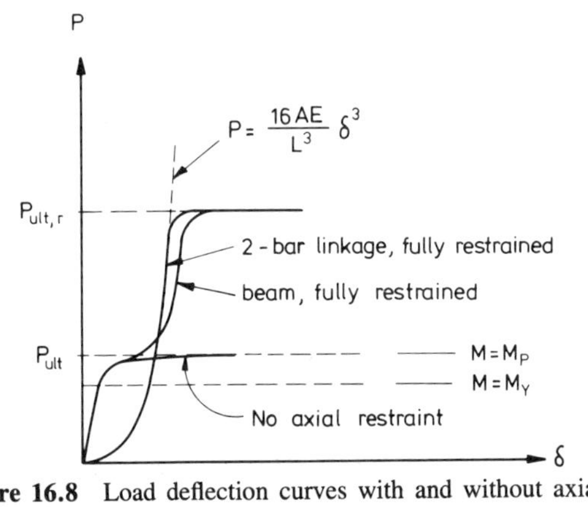

Figure 16.8 Load deflection curves with and without axial restraint.

Incremental or Load–Deflection Approach

In order to investigate the collapse of an individual beam in more detail let us take the case of a clamped beam which carries a uniform distributed load of intensity q. The bending moment diagram for the elastic range of loading is shown as curve (a) in Fig. 16.9. The value of bending moment at which yield first commences is $M_Y = \sigma_Y Z$. The corresponding load is

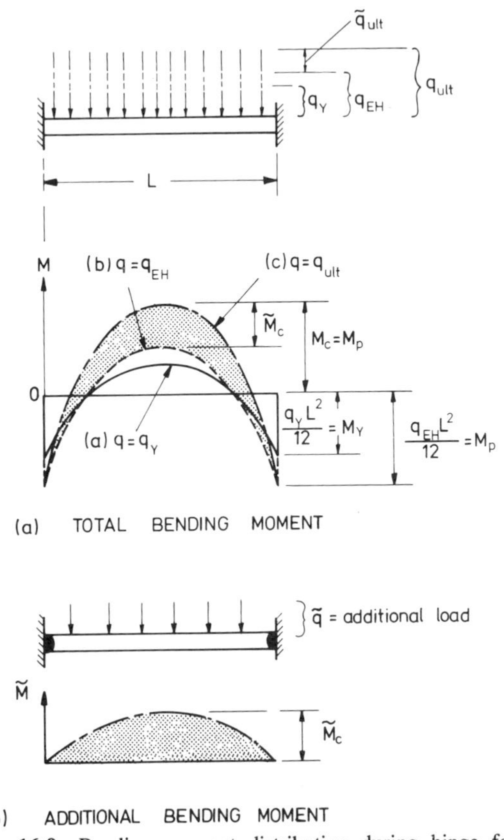

Figure 16.9 Bending moment distribution during hinge formation.

obtained by equating the maximum absolute value of the bending moment to M_Y, which gives

$$\left| -\frac{qL^2}{12} \right| = M_Y$$

or

$$q_Y = \frac{12M_Y}{L^2} \tag{16.2.7}$$

As the load increases beyond this value plastic hinges are formed at the ends, and they are fully formed when the end bending moment reaches M_P, as shown in curve (b). Therefore the load on the beam at the completion of the end hinges is

$$q_{EH} = \frac{12M_P}{L^2} \tag{16.2.8}$$

Figure 16.10 presents the load-deflection diagram, showing the elasto-plastic portion corresponding to the formation of the hinges and the idealized version in which the beam is assumed to behave elastically until the bending moment reaches M_P, at which point a plastic hinge suddenly appears.

In the elastic range of loading, the midspan deflection of the beam is given by

$$\delta = \frac{qL^4}{384EI} \tag{16.2.9}$$

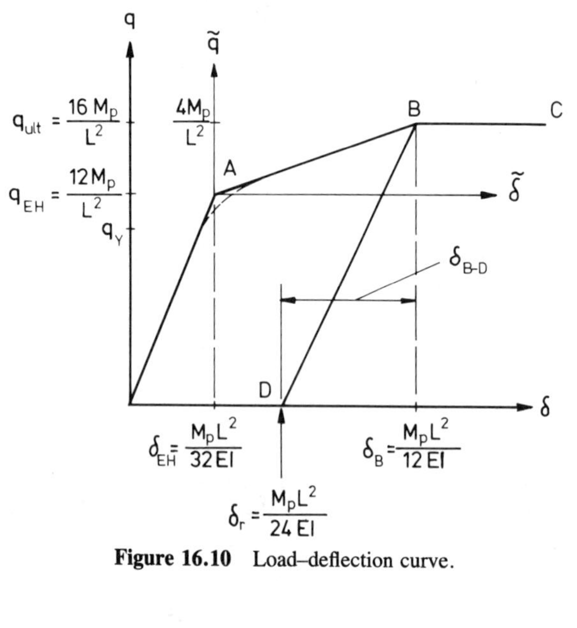

Figure 16.10 Load–deflection curve.

By adopting the simplified load–deflection behavior, in which the beam remains elastic until $q = q_{EH}$, we can estimate the corresponding midspan deflection by substituting q_{EH} for q in (16.2.9). The result is

$$\delta_{EH} = \frac{M_PL^2}{32EI}$$

From this point on, as the load is increased beyond the value q_{EH}, the bending moment at the ends remains constant at M_P. Therefore, with respect to the *additional* load $\tilde{q}$, over and above q_{EH}, the beam behaves as if it were simply supported at the ends. Hence to investigate this next stage in the beam's response we alter the representation of the beam, making it simply supported at the ends. This means that the additional bending moment at the center is related to the additional load $\tilde{q}$ by the usual expression for a simply supported beam

$$\tilde{M}_c = \frac{\tilde{q}L^2}{8}$$

and the total bending moment at the center is

$$M_c = \frac{q_{EH}L^2}{24} + \frac{\tilde{q}L^2}{8} \tag{16.2.10}$$

$$= \frac{M_P}{2} + \frac{\tilde{q}L^2}{8}$$

Similarly, the additional deflection is given by the expression for a simply supported beam

$$\tilde{\delta} = \frac{5\tilde{q}L^4}{384EI} \tag{16.2.11}$$

and the total deflection is

$$\delta = \delta_{EH} + \tilde{\delta} \tag{16.2.12}$$

Equations 16.2.9 and 16.2.11 indicate that the rate of increase of deflection in the modified beam is five times the rate for the original clamped beam, and therefore the slope of the load–deflection curve is only one-fifth of the value prior to the formation of the end hinges. That is, the occurrence of the end hinges reduces the beam's stiffness to only one-fifth of its original value.

Eventually, as $\tilde{q}$ is increased, the bending moment in the beam reaches the distribution shown in curve (c), in which M_c has reached the value M_P. Setting M_c

equal to M_P in (16.2.10) and solving for $\tilde{q}$ gives

$$\tilde{q}_{\text{ult}} = \frac{4M_P}{L^2} \qquad (16.2.13)$$

Therefore the total load at which collapse occurs is

$$\begin{aligned} q_{\text{ult}} &= q_{EH} + \tilde{q}_{\text{ult}} \\ &= \frac{12M_P}{L^2} + \frac{4M_P}{L^2} \\ &= \frac{16M_P}{L^2} \end{aligned} \qquad (16.2.14)$$

At this point the beam has become a mechanism and cannot sustain any further load. The load–displacement curve becomes horizontal and the beam undergoes very large (theoretically unrestricted) plastic deflection. The deflection just prior to collapse is obtained from (16.2.12) by substituting the expressions for δ_{EH} and $\tilde{\delta}$, using (16.2.11) and (16.2.13) for the latter. The result is

$$\delta_B = \frac{M_P L^2}{12EI}$$

This value is $\frac{8}{3}$ of the end hinge deflection, δ_{EH}, which in turn is only slightly larger than the maximum elastic deflection δ_Y. Therefore $\delta_B \approx 3\delta_Y$, which is not so large that it invalidates the basic assumptions of simple beam theory.

We see that the behavior of the clamped beam is quite different from that of the simply supported beam which was considered at the beginning of this chapter (Fig. 16.2). The latter reaches a collapse state as soon as the maximum bending moment reaches the value M_P. In the clamped beam, however, the formation of end hinges initiates a change in the distribution of bending moment, and there is a further reserve of strength from the formation of the end hinges to the final collapse of the beam. For the clamped uniformly loaded beam this reserve of strength is

$$\frac{q_{\text{ult}} - q_{EH}}{q_{EH}} \times 100\% = 33\tfrac{1}{3}\% \qquad (16.2.15)$$

Note that this figure does not include the margin of approximately 15% between q_Y and q_{EH}, corresponding to the gradual formation of the end hinges.

If the beam were to be completely unloaded after having just reached the collapse load (point B in Fig. 16.10) the load–deflection curve would decrease linearly, parallel to the original elastic portion. The deflection therefore decreases by

$$\delta_{B-D} = \frac{q_{\text{ult}} L^4}{384EI} = \frac{M_P L^2}{24EI}$$

and the residual deflection after complete unloading is

$$\delta_r = \delta_B - \delta_{B-D} = \frac{M_P L^2}{12EI} - \frac{M_P L^2}{24EI} = \frac{M_P L^2}{24EI} \qquad (16.2.16)$$

Similarly, after complete unloading of the beam, residual bending moments remain. The diagram of residual moments is the difference between curve (c) of the bending moment diagram of Fig. 16.9 and an elastic bending moment diagram similar to curve (a) but drawn for $q = q_{\text{ult}}$ (in which case the end value would be $-\frac{4}{3}M_P$ and the center value $\frac{2}{3}M_P$). The curve obtained by subtracting these two curves is a straight horizontal line at a value of $M_P/3$; that is, the unloaded beam has a constant residual bending moment. The constancy is due to the fact that it is produced by the two identical plastic rotations of the end sections. This can be verified by performing the calculations at selected points. For example, at the ends of the beam the residual bending moment is

$$M_r = -M_P - \left(-\frac{4}{3}M_P\right) = \frac{1}{3}M_P$$

and at midspan the value is

$$M_r = M_P - \frac{2}{3}M_P = \frac{1}{3}M_P$$

Figure 16.11 gives the collapse loads for a beam subjected to either a point load or a uniform distributed load, for a variety of end support conditions.

Direct Method of Collapse Analysis

Although in the foregoing analysis we calculated the deflection throughout the elastic and elasto-plastic portions of the response, we in fact did not use this information in calculating the collapse load. Instead the collapse load was obtained purely by considering what the bending moment distribution would have to be in order for specified hinges to occur. This was done in

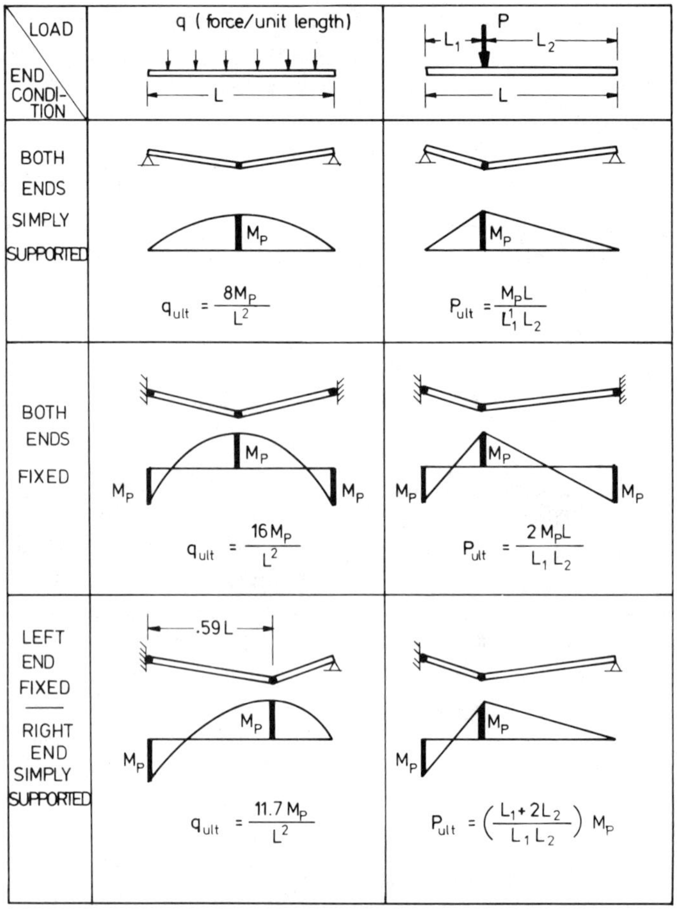

Figure 16.11 Collapse loads for single-span beams.

two stages, first for end hinges and then for a center hinge; but since we did not require any values of deflection or other information for this analysis we could just as well have done it with all three hinges present. That is, we could have examined the collapse state directly. This we will now do. The bending moment diagram for a uniformly loaded beam with equal but arbitrary end moments is shown in Fig. 16.12. It consists of the inverted parabola of the simply supported bending moment diagram, of height $qL^2/8$, displaced downward by an amount equal to the magnitude of the end moments. The height of the parabola and the magnitude of the end moments have been chosen so that the bending moment diagram satisfies the requirements for a three-hinge collapse mechanism. As can be seen this requires that the height of the parabola must be eqal to $2M_P$. Therefore

$$\frac{qL^2}{8} = 2M_P$$

or

$$q = q_{ult} = \frac{16M_P}{L^2}$$

which agrees exactly with the result obtained by gradually tracing the development of the mechanism. Clearly this second method, which looks immediately and directly at the collapse mechanism, involves much less computation and is far more efficient.

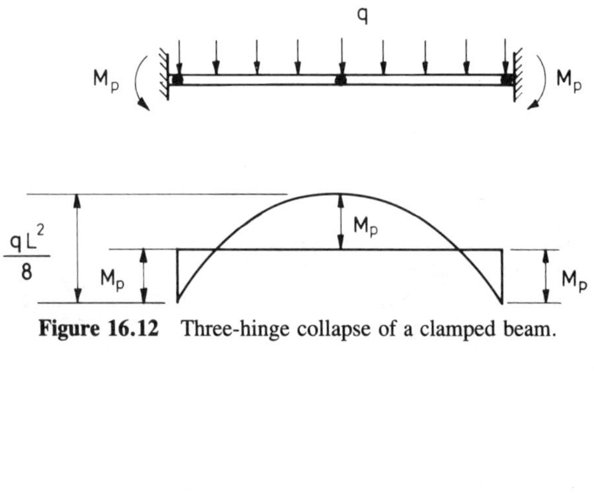

Figure 16.12 Three-hinge collapse of a clamped beam.

On the other hand there are some situations in which the collapse load is influenced by the deflection or by other aspects of the response prior to collapse. For example, if the beam is also subjected to axial compression the lateral deflection represents an eccentricity and this produces extra ending in the beam, thereby causing it to collapse at a lower value of load than that predicted by the direct method. In the gradual or incremental type of analysis the deflection is known at all stages of the loading and its effect can be allowed for.

Another limitation of the direct method is that it cannot account for the interaction between plastic hinge failure and instability failure. For example, the occurrence of the end hinges reduces the buckling strength of the beam and the beam might undergo a buckling collapse at this stage of the loading. But if this end-hinge stage is never examined, this fact will be missed.

Effect of Brackets

We have seen in Section 8.6 that in rigid-jointed frame structures the use of brackets is very beneficial in regard to the elastic response, because brackets allow the structure to withstand large local values of shear force and bending moment (which usually occur at joints) by providing locally increased values of section area and section modulus. They also shorten the effective span, thus reducing the bending moment.

Brackets also provide significant benefits in regard to the ultimate strength of frame structures, especially if the members are subjected to distributed loads. With such loads the peak values of bending moment nearly always occur at the joints. By providing a large value of Z_P in the joint region brackets delay the formation of the end hinges, and hence a collapse mechanism, thereby increasing the ultimate strength. For example, we have seen that for a clamped uniformly loaded beam the first hinges occur at the ends at a load of $12\,M_P/L^2$. The ultimate load, corresponding to the formation of a center hinge, occurs at $q_{ult} = 16\,M_P/L^2$. The load–deflection relationship was shown in Fig. 16.10. Figure 16.13*a* shows the same beam fitted with brackets, and Fig. 16.13*b* shows a composite plot of bending moment and of (negative) M_P along the length of the beam. A plastic hinge occurs wherever the bending moment curve touches the M_P curve. To provide for positive bending moment the central portion of the M_P is also drawn above the axis. The bending moment distribution is drawn for the load which would cause end hinges if there were no

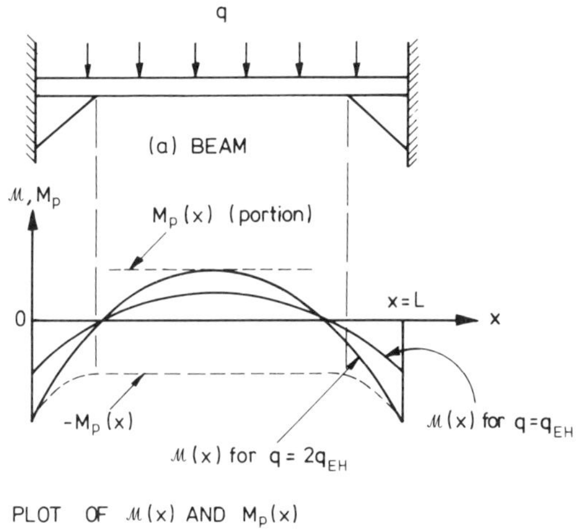

Figure 16.13 Plastic collapse of a bracketed beam.

brackets: $q = q_{EH} = 12M_P/L^2$. The brackets delay the formation of end hinges so that the bending moment distribution maintains the same parabolic shape and continues to grow proportionately throughout the beam. As shown in Fig. 16.14 the load–deflection curve continues along the elastic straight line. When the load becomes twice the value of q_{EH} the bending moment at the center reaches M_P and a hinge occurs there.

If the brackets are just large enough so that the three hinges form simultaneously then this load is the collapse load; that is,

$$q_{\text{ult}} = 24\frac{M_P}{L^2} \qquad (16.2.17)$$

compared to a value of only $16M_P/L^2$ without brackets. Thus the effect of brackets is to increase the ultimate strength by 50%—a dramatic improvement.

If the brackets were made still larger, the collapse load would be increased further but the gain would be

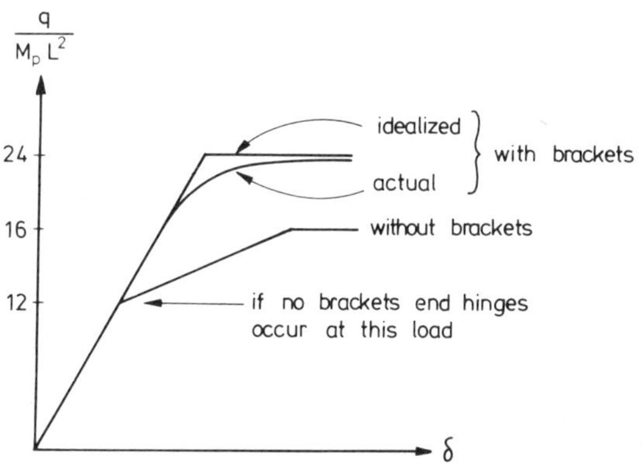

Figure 16.14 Load–deflection curve for a bracketed beam.

much less because once the central hinge has formed then with respect to any further load the beam is effectively two cantilevers, and the largest growth of bending moment occurs at the ends. The absolute optimum would be to design the brackets such that the curve of M_P exactly matched the curve of $\mathcal{M}$ within the length of the brackets, but the extra fabrication cost is rarely justified. Also, most structures are subjected to loads of various types and distributions and it is rare that the distribution of the worst load would be known with such precision.

The load–deflection diagram for this example (Fig. 16.14) reveals an important characteristic of the use of brackets to increase the ultimate strength of a beam, namely that it produces simultaneity of hinge formation and hence a more sudden type of collapse. This is another example of the general fact that whenever a structure is made more efficient by better utilization of material and the prevention of lesser precollapse failures, the collapse itself becomes more sudden (see Section 2.8).

In practice the collapse is not quite as sudden as the idealized plastic hinge theory predicts because the theory ignores the elasto-plastic transition phase of the moment–curvature relationship which was illustrated in Fig. 16.2*b*. As mentioned there the transition begins about 15% below the final value. A more realistic shape of the load–deflection curve for the present example is shown in Fig. 16.14.

16.3 INCREMENTAL PLASTIC FRAME ANALYSIS

Methods of Plastic Frame Analysis

Since a frame is an assemblage of beams the notions of a plastic hinge and of a failure mechanism due to a sufficient number of plastic hinges, have an immediate application in frame structures, and they constitute the basis for *plastic frame analysis,* which is a well-established method for calculating the ultimate strength of a frame structure.

Plastic frame analysis incorporates the basic assumptions of plastic hinge theory which were presented in Sections 16.1 and 16.2. These include the idealized moment–curvature relationship of Fig. 16.2*b* and the assumption that the material is sufficiently ductile to sustain the large local strains in the narrow zones of high curvature without fracture. In order to simplify the analysis it is also assumed that the beams are prismatic and that each joint of the structure is able to transfer the maximum plastic moment of the connected members. If necessary a more complicated form of plastic frame analysis could be developed which did not include these assumptions, but in practice nearly all frame structures satisfy both of them.

In a structure which has many members there is a wide variety of possible collapse mechanisms, depending on the extent of structure which is involved. This in turn depends on the number of plastic hinges and their locations. At one extreme there are some collapse mechanisms which cause large deflection (collapse) of the entire structure, as illustrated in Fig. 16.15*a*. It can be shown that the number of hinges

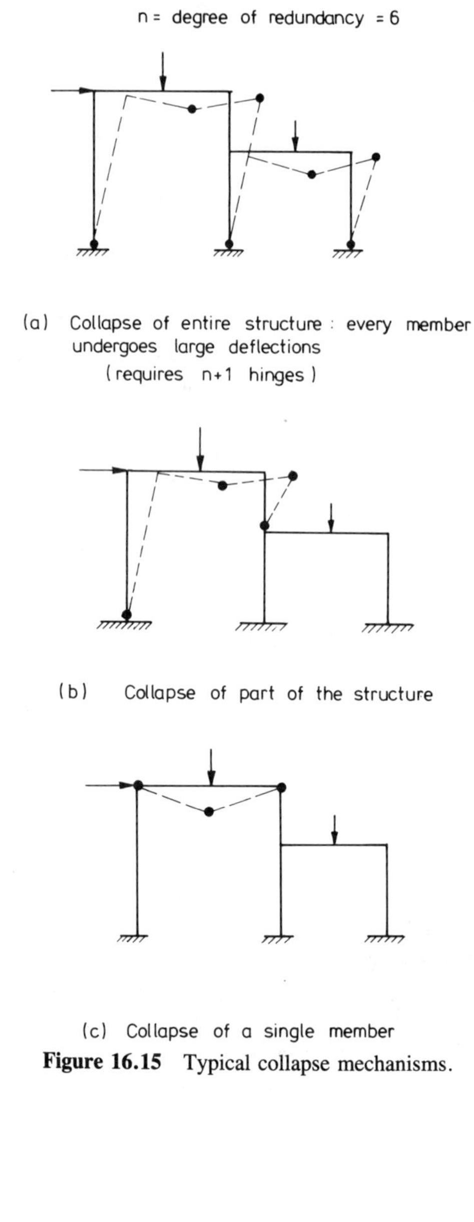

Figure 16.15 Typical collapse mechanisms.

required for this type of collapse is $n + 1$, where n is the degree of statical indeterminacy of the structure. But in addition to these special mechanisms which involve all of the members there are many more which involve fewer members, but which nonetheless would be considered as failure of the overall structure because of the seriousness of the consequences. For example the formation of four hinges in the locations shown in Fig. 16.15*b* causes a portion of the structure to become a mechanism. Finally there is also the possibility of a mechanism occurring within a single member, as shown in Fig. 16.15*c*. In this case the mechanism is a two-bar linkage and the number of hinges is always three (including any pinned joints which the member may already have). The collapse of an individual member may or may not be regarded as collapse of the structure, depending on the consequences. The best procedure is to analyze the behavior of the structure beyond the single-member collapse load in order to see what further mechanisms develop and to better decide what particular value of load should be taken as the collapse load of the structure.

Thus in a frame structure containing several members the collapse mechanism for the structure is not at all obvious beforehand. Moreover, the possibility of a given mechanism and the order of occurrence of the various mechanisms are both dependent on the particular combination of loads which are acting on the structure and on their sequence. There is a separate collapse mechanism for each separate "loadcase" (combination and sequence of loads) and each of these requires a separate ultimate strength analysis. Hence there is a need for an efficient and reliable method for calculating the collapse load of a structure for any nominated loadcase.

There are two types of plastic frame analysis (pfa): *direct* pfa and *incremental* pfa. These correspond to the two methods of analysis that were presented for an individual beam in the previous section. Direct pfa looks directly and exclusively at the collapse itself, with no regard for any aspect of the structural response prior to collapse; these are assumed to have no influence on either the collapse mechanism or the magnitude of the collapse load. This assumption is crucial to the method and it is the source of its conceptual simplicity and computational efficiency. But as we shall see, it also greatly limits the applicability of the method or, alternatively, it necessitates a good deal of additional calculation to ensure that the frame actually possesses the required and intended degree of safety.

In contrast, incremental pfa is a load–deflection approach. It is a step-by-step (or hinge-by-hinge) mathematical simulation of the collapse by means of a series of elastic frame analyses, each of which determines the load and the location of the next hinge, alters the global stiffness matrix to simulate a hinge at the location, and tests whether a mechanism has been created. If the mechanism involves only one member, the member is deleted and the series of hinge-by-hinge analyses is continued until the occurrence of a multimember collapse mechanism. This constitutes collapse of the structure and the current level of the loading is the collapse load for that loadcase.

A large thin-walled steel structure such as a hull module is subject to both instability failure and failure due to large plastic deformation (plastic hinge). Moreover, these two types of failure usually occur in combination, and this rules out the use of direct pfa. Incremental pfa, however, being a load–deflection approach, can be adapted so as to deal with combined modes of failure. In the previous chapter we considered instability failure and noted that when it occurs in combination with plastic deformation, a load–deflection approach is required. In this section we examine the same combined failure in terms of its other aspect—failure by large plastic deformation—and we will see that here again the problem requires a load–deflection approach. In fact there is a parallel situation regarding the analysis methods for these two basic types of failure. In both cases there is a direct method which is based on the assumption that there is only one type of failure, together with an idealized definition of the failure which makes it possible to calculate the collapse load directly, without tracing the development of the collapse. Because of the large computational requirements of the load–deflection methods, ultimate strength analysis has until now been based entirely on the two direct methods—eigenvalue analysis and direct pfa. These two methods have been used, separately, for virtually all calculations of ultimate strength, even though they both become invalid when the two types of failure occur together. But modern computers possess the computational power which is required by the load–deflection methods, and these methods can and should be used whenever the assumptions and idealizations of the direct methods are not fulfilled. The direct methods can often be used for the analysis of individual members, but they are not adequate for the ultimate strength analysis of a large complex structure such as a hull module. Therefore in this chapter the emphasis is on incremental pfa, with merely a brief summary of direct pfa.

Stiffness Matrix Representation of a Plastic Hinge

Figure 16.16 shows a simple portal frame subjected to a point load. Because of the load the horizontal member is represented by two beam elements, A and B. For simplicity we shall use the flexure-only beam element for which the stiffness matrix is given in (5.3.1). The maximum bending moment occurs at structure node j, directly under the load. Therefore as the load is increased the first plastic hinge forms at that point. It is important to note that although we commonly speak of a hinge occurring at a node, this is not fully precise because a node is a junction point whereas a hinge in fact occurs within a particular structural element. It occurs when the local value of bending moment in an element exceeds that element's value of M_P at that point.* At each node the maximum number of hinges which can occur is one less than the number of elements which meet at that node, because when that number have formed, the remaining intact element is simply supported at that node. Thus if two elements meet at a node there can be only one hinge, and it will form in whichever element has the smaller value of M_P. If the elements have the same value of M_P then it makes no difference which element is given the hinge.

In the present example a plastic hinge forms when the bending moment at j reaches the plastic moment at that point, M_P. Since the members have the same value of M_P we arbitrarily place the hinge at the right end (node 2) of member A. Let us denote the load at which the hinge occurs as P_1. If P is increased beyond P_1 the bending moment at j remains at the value M_P and the plastic hinge begins to rotate. Let us denote this additional load by the symbol $\tilde{P}$: $\tilde{P} = P - P_1$. Now by definition a hinge cannot sustain any value of bending moment, and the plastic hinge is only a hinge with respect to the *additional* load, $\tilde{P}$, over and above P_1. Thus to model the occurrence of a plastic hinge we adopt an incremental approach in which we deal separately with the pre-hinge and post-hinge stages of the loading, as shown in the figure. In the post-hinge stage the load is $\tilde{P}$ and the structure is different—it now contains a hinge. This is illustrated in Fig. 16.16*c*, which shows that in the new structure there is a discontinuity in the incremental nodal rotation at node j; the element rotations $(\tilde{\theta}_2)_A$ and $(\tilde{\theta}_1)_B$ are not equal. Since the plastic hinge occurs in element A the rotation $(\tilde{\theta}_1)_B$

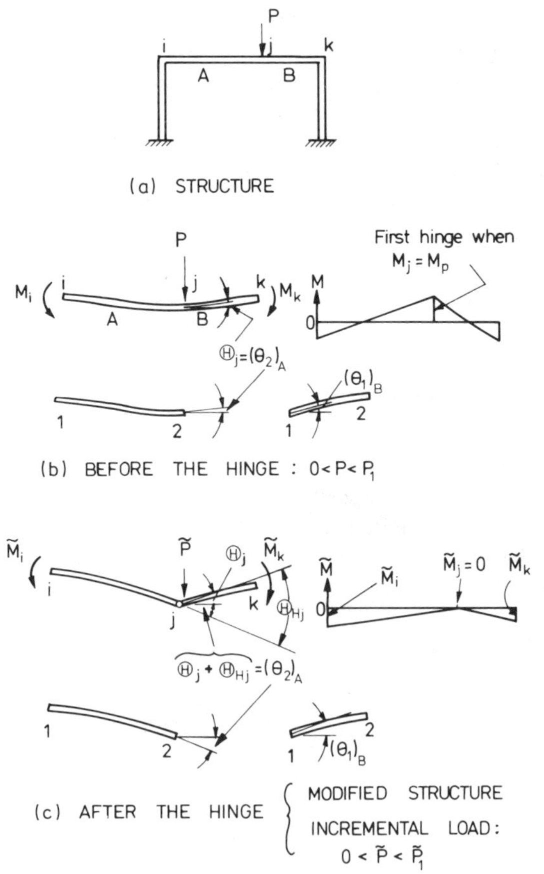

Figure 16.16 Idealization of a structure before and after a plastic hinge.

is still equal to the structure nodal rotation $\tilde{\Theta}_j$, but the rotation $(\tilde{\theta}_2)_A$ of element A is not. Instead it is given by

$$(\tilde{\theta}_2)_A = \tilde{\Theta}_j + \Theta_{Hj} \qquad (16.3.1)$$

where Θ_{Hj} is the hinge rotation.* The *incremental* structure displacement vector has five terms instead of four:

$$\tilde{\mathbf{\Delta}} = [\tilde{V}_i \quad \tilde{\Theta}_i \quad \tilde{V}_j \quad \tilde{\Theta}_j \quad \Theta_{Hj}]^T$$

Since the hinge rotation is an additional degree of freedom in the structure it necessitates an additional row and column in the *structure* stiffness matrix. However, there is no need to increase the size of the *element* stiffness matrix. Instead, the relationship of (16.3.1) between the element rotation and the structure

*As mentioned in Section 16.1, M_{fY} should be used in place of M_P when the beam flange is in compression. To simplify the explanation we here use only M_P.

*Since the rotation Θ_{Hj} is zero prior to the hinge formation there is no need to designate it as an increment by placing a ~ above it.

rotation can be incorporated into the transformation from element coordinates to structure coordinates. In other words we introduce the additional degree of freedom as part of the displacement transformation, $\tilde{\boldsymbol{\delta}} = \mathbf{T}\tilde{\boldsymbol{\Delta}}$. In this particular example the element coordinates are parallel to the structure coordinates and so before the hinge occurs the transformation matrix **T** is simply the unit matrix **I**. The occurrence of the hinge introduces an additional degree of freedom—Θ_{Hj}—into the displacement vector $\tilde{\boldsymbol{\Delta}}$. Hence in order to satisfy the relationship specified by (16.3.1) we must insert an additional column in **T**, after the Θ_j column, containing a 1 in the θ_2 position. The modified transformation is

$$\begin{Bmatrix} \tilde{v}_1 \\ \tilde{\theta}_1 \\ \tilde{v}_2 \\ \tilde{\theta}_2 \end{Bmatrix} = \left[\begin{array}{c|c} & 0 \\ \mathbf{I}_{4\times 4} & 0 \\ & 0 \\ & 1 \end{array} \right] \begin{Bmatrix} \tilde{V}_i \\ \tilde{\Theta}_i \\ \tilde{V}_j \\ \tilde{\Theta}_j \\ \tilde{\Theta}_{Hj} \end{Bmatrix} \qquad (16.3.2)$$

The element stiffness matrix in structure coordinates is obtained by the usual matrix product $\mathbf{K}^e = \mathbf{T}^T\mathbf{k}^e\mathbf{T}$. The result for the above case, in which the hinge occurs at element node 2, is

$$\mathbf{K}^e_{H2} = \frac{EI}{L^3} \begin{array}{c} \begin{matrix} V_i & \Theta_i & V_j & \Theta_j & \Theta_{Hj} \end{matrix} \\ \left[\begin{array}{ccccc} 12 & 6L & -12 & 6L & 6L \\ 6L & 4L^2 & -6L & 2L^2 & 2L^2 \\ -12 & -6L & 12 & -6L & -6L \\ 6L & 2L^2 & -6L & 4L^2 & 4L^2 \\ 6L & 2L^2 & -6L & 4L^2 & 4L^2 \end{array} \right] \end{array} \qquad (16.3.3)$$

Note that when the symbols v, θ, V, and Θ denote a *degree of freedom* there is no need for the tilde ($\tilde{\ }$) because the occurrence of a hinge does not change the degrees of freedom; it simply adds one new degree to the existing ones. It is only when these symbols represent actual values of displacements that there is any need to distinguish between the initial (pre-hinge) values and the incremental or additional values. Hence in labeling the columns of $\mathbf{K}^e_{H2}$ the ordinary symbols are used, without the tilde.

In this particular example the structure coordinates are the same as the element coordinates, and so if there were no hinge the stiffness matrix in structure coordinates, $\mathbf{K}^e$, would be the same as the stiffness matrix in element coordinates, as given in (5.3.1). A comparison of that matrix with $\mathbf{K}^e_{H2}$ shows that the entire upper left 4×4 portion of $\mathbf{K}^e_{H2}$—the portion corresponding to the original four degrees of freedom—is the same as $\mathbf{K}^e$. Hence the only change caused by the additional degree of freedom Θ_{Hj} is the addition of a new final row and column. Moreover, as shown by the dashed outlines, the new row and new column are the same as the row and column corresponding to the ordinary rotational degree of freedom at the node where the hinge is located, that is, the Θ_j row and column. The additional Θ_{Hj} diagonal term is identical to the Θ_j diagonal term.

In the analysis of the original structure, before the occurrence of the hinge, the ordinary element stiffness matrix $\mathbf{k}^e$ would have been calculated, transformed into $\mathbf{K}^e$ and inserted into the overall structure stiffness matrix **K**. We have just seen that after the hinge has occurred the portion of $\mathbf{K}^e_{H2}$ (and hence of **K**) corresponding to the original four degrees of freedom is the same as before. Thus there is no need to change any of the existing terms in **K**; it is only necessary to add the new Θ_{Hj} row and column. Moreover, even these terms are not new; they are copies of the existing terms in the ordinary element stiffness matrix $\mathbf{K}^e$. Hence *there is no need to create* $\mathbf{K}^e_{H2}$; its appearance here is simply to facilitate the explanation.

The structure load vector also has a new term corresponding to the new rotation Θ_{Hj}. Since a hinge cannot sustain any moment the value of the new term is set to zero.

If a hinge were to occur at structure node i, and hence at element node 1, the relationship between the incremental values of element nodal rotation and structure nodal rotation would be

$$\tilde{\theta}_1 = \tilde{\Theta}_i + \Theta_{Hi} \qquad (16.3.4)$$

and therefore the transformation between element and structure coordinates for a node 1 hinge is

$$\begin{Bmatrix} \tilde{v}_1 \\ \tilde{\theta}_1 \\ \tilde{v}_2 \\ \tilde{\theta}_2 \end{Bmatrix} = \begin{bmatrix} 1 & 0 & 0 & 0 & 0 \\ 0 & 1 & 1 & 0 & 0 \\ 0 & 0 & 0 & 1 & 0 \\ 0 & 0 & 0 & 0 & 1 \end{bmatrix} \begin{Bmatrix} \tilde{V}_i \\ \tilde{\Theta}_i \\ \tilde{\Theta}_{Hi} \\ \tilde{V}_j \\ \tilde{\Theta}_j \end{Bmatrix} \qquad (16.3.5)$$

The corresponding element stiffness matrix in structure coordinates is

$$\mathbf{K}^e_{H1} = \frac{EI}{L^3}\begin{array}{c} \begin{array}{ccccc} V_i & \Theta_i & V_j & \Theta_j & \Theta_{Hi} \end{array} \\ \left[\begin{array}{c|c|cc|c} 12 & 6L & -12 & 6L & 6L \\ \hline 6L & 4L^2 & -6L & 2L^2 & 4L^2 \\ \hline -12 & -6L & 12 & -6L & -6L \\ 6L & 2L^2 & -6L & 4L^2 & 2L^2 \\ \hline 6L & 4L^2 & -6L & 2L^2 & 4L^2 \end{array}\right] \end{array} \tag{16.3.6}$$

Here again the only effect of introducing the hinge at node i is to add some new terms to the structure stiffness matrix, which are copies of the terms in the Θ_i row and column of the transformed element stiffness matrix.

It can be shown that the allowance for a hinge continues to follow this simple rule even when the element is not parallel to the structure coordinates and the transformation matrix therefore has the general form given in (5.3.2). That is, if a hinge occurs in element e at structure node h, a new row and column is created in the structure stiffness matrix, corresponding to the hinge rotation Θ_{Hh}. The new row and column contain some terms which come from $\mathbf{K}^e$, the transformed element stiffness matrix for element e; namely the row and column of $\mathbf{K}^e$ corresponding to the Θ_h degree of freedom. This will be either the θ_1 or the θ_2 row and column, depending on whether node h corresponds to element node 1 or 2, respectively. Also the new diagonal term in $\mathbf{K}$ is set equal to the diagonal Θ_h term of the row and column in $\mathbf{K}^e$. Finally, the new term in the structure load vector corresponding to Θ_{Hh} is set to zero.

For a general (6 d.o.f.) beam element at an arbitrary orientation χ (see Fig. 5.9) if Θ_h corresponds to element node 1 the new row to be added to the structure stiffness matrix is the vector $\mathbf{V}_{H1}$ given by

$$\mathbf{V}_{H1} = \frac{EI}{L^3}\begin{array}{c}\begin{array}{ccccccc} U_h & V_h & \Theta_h & U_j & V_j & \Theta_j & \Theta_{Hh}\end{array}\\ [-6L\mu \quad 6L\lambda \quad 4L^2 \quad 6L\mu \quad -6L\lambda \quad 2L^2 \quad 2L^2]\end{array} \tag{16.3.7}$$

in which j is the structure node corresponding to element node 2. The new column to be added is $\mathbf{V}^T_{H1}$. (Note that the last term of each vector has been halved so that after the double insertion the new diagonal term has the correct value.)

If Θ_h corresponds to element node 2 the new row to be added is

$$\mathbf{V}_{H2} = \frac{EI}{L^3}\begin{array}{c}\begin{array}{ccccccc} U_i & V_i & \Theta_i & U_h & V_h & \Theta_h & \Theta_{Hh}\end{array}\\ [-6L\mu \quad 6L\lambda \quad 2L^2 \quad 6L\mu \quad -6L\lambda \quad 4L^2 \quad 2L^2]\end{array} \tag{16.3.8}$$

in which i is the structure node corresponding to element node 1. Again the new column to be added is $\mathbf{V}^T_{H2}$.

Practical Technique for Insertion of $\mathbf{V}_H$ into $\mathbf{K}$

In practice, since the structure stiffness matrix is stored internally as a vector, it is not convenient to add new rows and columns during the solution. Instead it is better to reserve several rows and columns for this purpose from the beginning. These should initially contain some nonzero number on the diagonal and zeros elsewhere, and the corresponding terms in the load vector should be zero. The extra storage required is not large because in plane frames the number of hinges required to cause collapse is usually less then eight and rarely more than 12 or so.

Since it is easier and conceptually clearer to speak of "adding new rows and columns" in order to insert plastic hinges, we shall continue to use this terminology.

Basic Features of Incremental Plastic Frame Analysis

We now demonstrate the basic features of incremental plastic frame analysis using, for convenience, an example from Majid [1], which is a basic and pioneering text in this field. The structure is illustrated in Fig. 16.17. The first step is to obtain the bending moment distribution in the structure in order to ascertain where the first hinge will occur. A plastic hinge can occur at any joint in the structure and also under any point load. Of their very nature both of these correspond to structure nodes and so if a structure has only point loads there is no need to add any extra nodes to allow for possible hinges. The procedure for distributed loads is presented subsequently.

To obtain the bending moment distribution we apply an arbitrary initial value of the load, referred to as

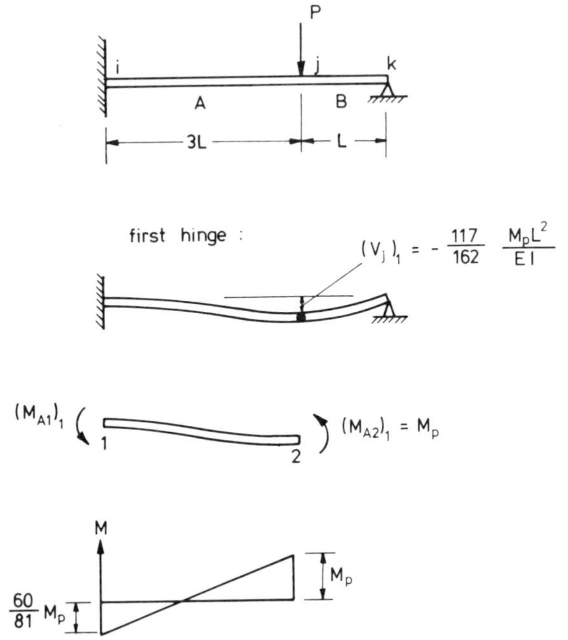

Figure 16.17. Two-member frame structure.

the *base load*, P_0. The base load can be of any magnitude because it is used only for a linear scaling of the bending moment distribution, to determine the first hinge load, and for this range of loading the structure is linear. (For the present we are considering only plastic hinge type of failure and not buckling.) Thus it would not matter even if the base load turned out to be larger than P_1, but in that case the bending moment values would not be useful for anything other than for scaling. For maximum usefulness the base load should be some magnitude of load below P_1 for which information about the structure response is desired. The most obvious example is the service load, that is, the mean value of the expected load, but any other choice would be equally valid.

At this stage of the loading there are no hinges and so the equilibrium equations for the structure are assembled in the usual manner. After allowing for the boundary conditions the resulting equations are

$$\begin{Bmatrix} -P_0 \\ 0 \\ 0 \end{Bmatrix} = \frac{EI}{L_B^3}\begin{bmatrix} \frac{12}{27} + 12 & \frac{-6L_A}{27} + 6L_B & 6L_B \\ \frac{-6L_A}{27} + 6L_B & \frac{4L_A^2}{27} + 4L_B^2 & 2L_B^2 \\ 6L_B & 2L_B^2 & 4L_B^2 \end{bmatrix}\begin{Bmatrix} (V_j)_0 \\ (\Theta_j)_0 \\ (\Theta_k)_0 \end{Bmatrix}$$

in which the factor of 27 arises because $(L_A/L_B)^3 = 27$. In symbolic form these equations are

$$\mathbf{F}_0 = \mathbf{K}_0\mathbf{\Delta}_0$$

The solution is

$$(V_j)_0 = \frac{-117P_0L^3}{256EI}$$
$$(\Theta_j)_0 = \frac{63P_0L^2}{256EI} \qquad (16.3.9)$$
$$(\Theta_k)_0 = \frac{144P_0L^2}{256EI}$$

where $L = L_B = L_A/3$.

The element bending moment values at the nodes are obtained from (5.3.12b)

$$(\mathcal{M}_{A1})_0 = \frac{EI}{L_A^3}[\ 6L_A(V_j)_0 - 2L_A^2(\Theta_j)_0] = -\frac{60}{128}P_0L$$
$$(\mathcal{M}_{A2})_0 = \frac{EI}{L_A^3}[-6L_A(V_j)_0 + 4L_A^2(\Theta_j)_0] = \frac{81}{128}P_0L$$

From these values it is evident that the first hinge will occur at node j of the structure, and since the elements have the same value of M_P we arbitrarily place the hinge at node 2 of element A.

For a more typical structure having many elements, each of which may have a different value of M_P, the location of the first hinge is found by evaluating the *plastic hinge ratio* at each structure node j. This ratio is defined as

$$\gamma_{P,j} = \overset{n_j}{\underset{\lambda=1}{\text{MIN}}}\left\{\left|\frac{(M_P)_\lambda}{\mathcal{M}_{j,\lambda}}\right|\right\} \qquad (16.3.10)$$

where n_j is the number of elements which are attached to node j, and $\mathcal{M}_{j,\lambda}$ is the nodal bending moment in each such element. The first hinge occurs at the location having the smallest value of plastic hinge ratio, and this smallest value is the *load scaling factor* μ:

$$\mu = \underset{j=1}{\overset{N}{\text{MIN}}} [\gamma_{P,j}]$$

where N is the number of nodes in the structure. The load required to cause the first hinge is then obtained by scaling the base load

$$P_1 = \mu P_0$$

In the present example the minimum plastic hinge ratio is

$$\mu = \frac{(\mathcal{M}_P)_A}{\mathcal{M}_{A2}} = \frac{128 M_P}{81 P_0 L}$$

and therefore the load to cause the first hinge is

$$P_1 = \mu P_0 = \frac{128}{81}\frac{M_P}{L}$$

and at this stage of the loading the deflection under the load is, from (16.3.9),

$$(V_j)_1 = \mu (V_j)_0$$

$$= -\frac{117}{162}\frac{M_P L^2}{EI} \qquad (16.3.11)$$

and the bending moment at the clamped support is

$$(\mathcal{M}_{A1})_1 = \mu(\mathcal{M}_{A1})_0 = \frac{128}{81}\frac{M_P}{P_0 L}\left[\frac{-60}{128} P_0 L\right] = -\frac{60}{81} M_P$$

These values of response are shown in Figs. 16.17 and 16.18.

Once a hinge is formed the structure is different from its previous condition and therefore the analysis must be performed in separate stages, each of which terminates with the formation of the next hinge. The loading is therefore applied in separate increments, from one hinge to the next, until collapse occurs. (In the present example only one more hinge is required for collapse.) Each new stage in the analysis begins with a modification of the structure stiffness matrix **K** to reflect the presence of the latest hinge at, say, structure node h. As we have seen, this is done by adding a new row and column to **K** consisting of the row and column of the transformed element stiffness matrix which correspond to the Θ_h degree of freedom. In the

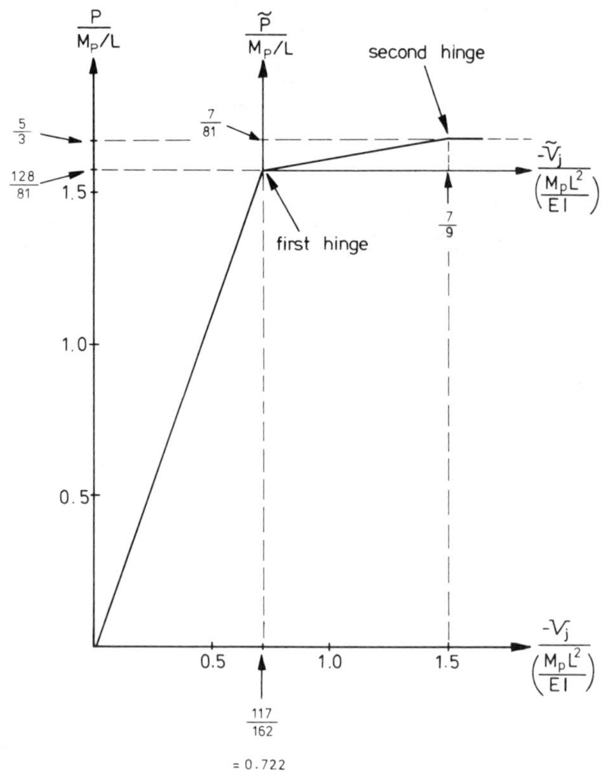

Figure 16.18 Load–deflection diagram for two-member structure.

present case the hinge forms in member A at structure node j, and this node corresponds to element node 2. Therefore the row and column come from the stiffness matrix of element A, and they are the fourth row and column, as illustrated in (16.3.3). As it happens, the first two rows and columns of $\mathbf{K}^A$ correspond to V_i and Θ_i, and these degrees of freedom have been deleted because of the clamped support at i. Therefore only the second two terms of the fourth row and column are actually inserted into **K**. The resulting augmented structure stiffness matrix is

$$\begin{matrix} & V_i & \Theta_j & \Theta_k & \Theta_{Hj} \end{matrix}$$

$$\mathbf{K}_1 = \frac{EI}{L_B^3}\begin{bmatrix} \frac{12}{27} + 12 & \frac{-6L_A}{27} + 6L_B & 6L_B & \frac{-6L_A}{27} \\ \frac{-6L_A}{27} + 6L_B & \frac{4L_A^2}{27} + 4L_B^2 & 2L_B^2 & \frac{4L_A^2}{27} \\ 6L_B & 2L_B^2 & 4L_B^2 & 0 \\ \frac{-6L_A}{27} & \frac{4L^2{}_A}{27} & 0 & \frac{4L_A^2}{27} \end{bmatrix}$$

(16.3.12)

The next step is to analyze the modified structure and determine the location of the next hinge and the value of the load increment $\tilde{P}$ which causes it. The analysis is performed using a *trial load increment* $\tilde{\mathbf{F}}^*$, and the resulting incremental bending moment $\tilde{\mathcal{M}}^*_{j,\lambda}$ is calculated in each of the n_j elements which are attached to each node. Then the plastic hinge ratio is calculated at each node. This ratio specifies, for each node, the relative amount by which the current load would have to be increased to form a plastic hinge at that node. In the loading process which forms the first hinge the bending moment in all members $\mathcal{M}_{j,\lambda}$ commences at a zero value, and the plastic hinge ratio is simply the lowest ratio of $(M_P)_\lambda$ to $\mathcal{M}_{j,\lambda}$, given by (16.3.10). But now the load involved is an *incremental* load and at the start of this first increment all of the members already have (known) values of nodal bending moment, given by $(\mathcal{M}_{j,\lambda})_1$. For example, the bending moment at node i (member A) is $-(60/81)M_P$. Therefore to determine the increase in the current load increment which would cause the next hinge, it is necessary to use the *remaining margin* of plastic hinge moment at each location, obtained by subtracting the local initial value of nodal bending moment, $(\mathcal{M}_{j,\lambda})_1$, from the corresponding local plastic hinge value $(M_P)_\lambda$. Also, in order to allow for both positive and negative values of $\tilde{\mathcal{M}}^*_{j,\lambda}$, including the case in which the increment is of opposite sign to the initial value, it is necessary to write the generalized version of (16.3.10) in two parts. And finally, to make the expression applicable for all load increments we replace the subscript 1 by i, so that $(\mathcal{M}_{j,\lambda})_i$ is the nodal bending moment in each element which is attached to node j when the ith hinge has just formed. The resulting generalized equation is

$$\text{for } \tilde{\mathcal{M}}^*_{j,\lambda} > 0: \qquad \gamma_{P,j} = \underset{\lambda=1}{\overset{n_j}{\text{MIN}}}\left\{\frac{(M_P)_\lambda - (\mathcal{M}_{j,\lambda})_i}{\tilde{\mathcal{M}}^*_{j,\lambda}}\right\}$$

$$\text{for } \tilde{\mathcal{M}}^*_{j,\lambda} < 0: \qquad \gamma_{P,j} = \underset{\lambda=1}{\overset{n_j}{\text{MIN}}}\left\{\frac{-(M_P)_\lambda - (\mathcal{M}_{j,\lambda})_i}{\tilde{\mathcal{M}}^*_{j,\lambda}}\right\} \tag{16.3.13}$$

Once the values of $\gamma_{P,j}$ have been calculated for all nodes, the load scaling factor can be determined.

$$\mu = \underset{j=1}{\overset{N}{\text{MIN}}}[\gamma_{P,j}] \tag{16.3.14}$$

Then the load increment to cause the second hinge is $\tilde{\mathbf{F}}_1 = \mu\tilde{\mathbf{F}}^*$.

In the present example there is only one load component and so the incremental load is a scalar

$$\tilde{\mathbf{F}}_1 = \begin{Bmatrix} -\tilde{P}_1 \\ 0 \\ 0 \\ 0 \end{Bmatrix} = -\tilde{P}_1$$

Hence there is no need to use a trial load increment—we can solve for $\tilde{P}_1$ explicitly. The equilibrium equations are

$$\tilde{\mathbf{F}}_1 = \mathbf{K}_1\tilde{\mathbf{\Delta}}_1$$

where $\mathbf{K}_1$ is the stiffness matrix for the modified structure containing a hinge at node j, given by (16.3.12). The solution is

$$\tilde{\mathbf{\Delta}}_1 = \begin{Bmatrix} \tilde{V}_j \\ \tilde{\Theta}_j \\ \tilde{\Theta}_k \\ \Theta_{Hj} \end{Bmatrix} = \frac{\tilde{P}_1 L_B^3}{EI}\begin{Bmatrix} -9 \\ \dfrac{9}{L_B} \\ \dfrac{9}{L_B} \\ -\dfrac{27}{2L_B} \end{Bmatrix} \tag{16.3.15}$$

The next step is to calculate the increment of bending moment in the members using (5.3.12b). At node 1 the expression for the incremental bending moment is

$$\tilde{\mathcal{M}}_{A1} = \frac{EI}{L_A^3}[6L_A\tilde{v}_2 - 2L_A^2\tilde{\theta}_2] \tag{16.3.16}$$

and to evaluate this we must transform the foregoing increments of nodal displacement back to element coordinates. Since the modified structure contains a hinge at node 2 of member A, the member rotation at that point, $\tilde{\theta}_2$, is not the same as the structure nodal rotation, $\tilde{\Theta}_j$. Instead they are related by

$$\tilde{\theta}_2 = \tilde{\Theta}_j + \Theta_{Hj}$$

Substituting into (16.3.16) and noting that $v_2 = V_j$ gives

$$\tilde{\mathcal{M}}_{A1} = \frac{EI}{L_A^3}[6L_A\tilde{V}_j - 2L_A^2(\tilde{\Theta}_j + \Theta_{Hj})]$$

and substitution from (16.3.15) gives

$$\tilde{\mathcal{M}}_{A1} = -3\tilde{P}_1 L$$

where again we utilize the reference length $L = L_B = L_A/3$. The total bending moment at the clamped support is the sum of the first hinge value and the incremental value:

$$\begin{aligned}\mathcal{M}_{A1} &= (\mathcal{M}_{A1})_1 + \tilde{\mathcal{M}}_{A1} \\ &= -\frac{60}{81}M_P - 3\tilde{P}_1 L\end{aligned}$$

The second hinge forms when this total bending moment reaches M_P. Therefore the value of incremental load $\tilde{P}_1$ which causes the second hinge is the value which satisfies the relationship

$$-M_P = -\frac{60}{81}M_P - 3\tilde{P}_1 L$$

which gives

$$\tilde{P}_1 = \frac{7}{81}\frac{M_P}{L}$$

Hence the total second hinge load is

$$\begin{aligned}P_2 &= P_1 + \tilde{P}_1 \\ &= \left(\frac{128}{81} + \frac{7}{81}\right)\frac{M_P}{L} \\ &= \frac{5}{3}\frac{M_P}{L}\end{aligned}$$

From (16.3.15) the incremental deflection is

$$\tilde{V}_j = -9\frac{\tilde{P}_1 L^3}{EI}$$

Substitution for $\tilde{P}_1$ gives

$$\tilde{V}_j = -\frac{63}{81}\frac{M_P L^2}{EI}$$

and adding this to the prehinge deflection of (16.3.11) gives the total deflection

$$\begin{aligned}V_j &= \left(-\frac{117}{162} - \frac{63}{81}\right)\frac{M_P L^2}{EI} \\ &= -1.5\frac{M_P L^2}{EI}\end{aligned}$$

These results are illustrated in Fig. 16.18.

This completes the first increment and for a large and complex structure a new cycle would now be commenced, to determine the incremental load $\tilde{P}_2$ that would cause the next hinge. Of course in this simple example we can see by inspection that a mechanism has now formed in the structure, and hence P_2 is the collapse load. But in general the achievement of a collapse mechanism is not so obvious and there is a need for a clear and unequivocal indication that the collapse state has been reached. The clearest and most reliable indication is that the structure stiffness matrix becomes singular. Thus in the analysis of a typical structure we would now proceed with the next cycle, the first step of which would be to alter the structure stiffness matrix to account for the new hinge at node i. Since the hinge corresponds to node 1 of element A the terms to be inserted into the new Θ_{Hi} row and column are the terms in the second row and column of the (transformed) stiffness matrix of element A, as shown in (16.3.6). In the present example node i is fully constrained and so only the last two terms of the row and column, corresponding to V_j and Θ_j are added to $\mathbf{K}_1$. Also, the new diagonal term is set equal to the diagonal term of the element stiffness matrix: $K_{\theta i,\theta i}[= (EI/L_A^3)(4L_A^2) = (EI/L_B^3)(4L_A^2/27)]$. The resulting structure stiffness matrix is

$$\mathbf{K}_2 = \frac{EI}{L_B^3}\left[\begin{array}{c|c} \frac{L_B^3}{EI}\mathbf{K}_1 & \begin{matrix}\frac{-6L_A}{27} \\ \frac{2L_A^2}{27} \\ 0 \\ 0\end{matrix} \\ \hline \begin{matrix}\frac{-6L_A}{27} & \frac{2L_A^2}{27} & 0 & 0\end{matrix} & \frac{4L_A^2}{27}\end{array}\right] \quad (\text{column } \Theta_{Hi})$$

where $\mathbf{K}_1$ is the previous (one hinge) structure stiffness matrix, given by (16.3.12). The new incremental load vector $\tilde{\mathbf{F}}_2$ contains an additional zero corresponding to the zero increment of moment at the new hinge. The equilibrium equations are

$$\tilde{\mathbf{F}}_2 = \mathbf{K}_2\tilde{\mathbf{\Delta}}_2$$

and the next step is the solution for $\tilde{\mathbf{\Delta}}_2$. Near the end of the equation solving process, in the triangularization of the second last (Θ_{Hj}) row and column of $\mathbf{K}_2$, a zero appears in the final (Θ_{Hi}) diagonal position, indicating that with two hinges the structure contains a mechanism. Hence the two-hinge load is the ultimate load.

$$P_{ult} = P_2 = \frac{5}{3}\frac{M_P}{L} \qquad (16.3.17)$$

Procedure for Distributed Loads

When the loading on a member is a distributed load, there are three possible locations for peak values of bending moment and hence for plastic hinges: at the ends and somewhere near the center. In general the internal peak value will not occur exactly at the midlength but in most cases the midlength value will not be much less than the peak value because of the smoothness of the bending moment distribution in this region. Thus for any member which has a distributed load it is necessary (and usually sufficient) to check for plastic hinges at these three locations. For example, Fig. 16.19 shows a portal frame for which Neal [3] gives 5.92 M_P/PL as the exact value of μ_{ult}, the load scaling factor at collapse. The internal hinge in the horizontal member occurred 0.415L from the left end. Placing the hinge at the midlength gives $6M_p/PL$, which differs by only 1.4%. Thus in the method of incremental pfa which is presented herein, and which forms part of the method for transverse ultimate strength analysis presented in Section 17.3, all members that carry distributed loads are represented by means of two half-length beam elements. Hence throughout the rest of this section and in Section 17.3 the word "member" refers to the complete (full span) member, and the word "element" refers to the basic beam finite element.

The representation of a distributed load involves an equivalent nodal moment. Once a plastic hinge has formed at one end of an element the equivalent nodal forces for subsequent load increments cannot contain a moment at that node. Instead the new set of equivalent nodal forces is equal and opposite to the reaction forces and moments for a beam carrying that same load, while being simply supported at the end corresponding to the plastic hinge and clamped at the other end. For a uniform distributed load q and a plastic hinge at element node 1 the reactions are: $R_1 = 3qL/8$, $M_1 = 0$, $R_2 = 5qL/8$, $M_2 = -qL^2/8$.

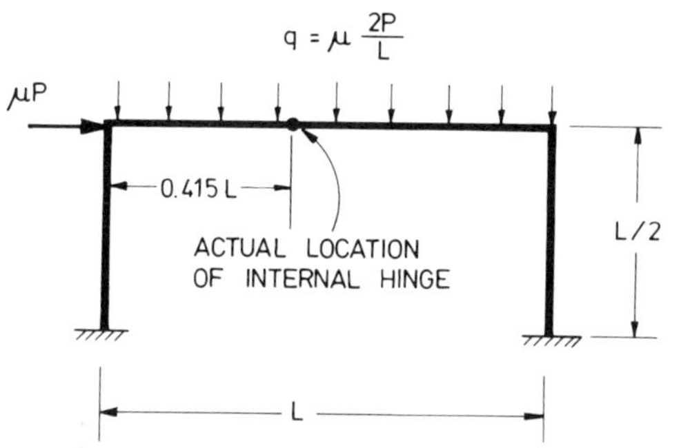

Figure 16.19 Internal plastic hinge.

Independent Loads: Loadcases

As mentioned in Section 2.3 and elsewhere, if some of the principal components of the loading are independent it is necessary to perform a separate ultimate strength analysis for each of several "loadcases;" that is, several different combinations, sequences, and relative magnitudes of the components. Incremental pfa has great flexibility in this regard because, if desired, new proportions of the various component loads of $\tilde{\mathbf{F}}$ can be specified at the beginning of each new increment. One of the components of the trial load increment $\mathbf{F}^*$ is selected as a reference value and given a unit magnitude, and the other components are then sized in proportion to it. If desired some components of the overall load can be held constant by setting the corresponding component of $\mathbf{F}^*$ to zero.

Local (Single Member) Collapse

If at any stage a member incurs a total of three hinges that member undergoes collapse, but this may or may not constitute collapse of the structure depending on the degree of indeterminacy (redundancy) of the structure and on the number and location of hinges which are already present in other members. But regardless of whether it does or does not cause collapse of the structure, this local mechanism causes singularity of the structure stiffness matrix. Hence to determine whether the structure has collapsed it is necessary to delete the collapsed member and then reexamine the structure. In order not to have to solve the structure equilibrium equations twice, it is best to make a continual check of the number of hinges in each member, so as to detect a member with three hinges and to delete it before commencing the solution. The deletion is achieved by subtracting the transformed element stiffness matrices of its two component elements from the structure stiffness matrix. Then if the solution reveals that the stiffness matrix is singular this indicates that the structure has collapsed.

Summary of Basic Steps

At this point it may be helpful to list the basic steps of incremental pfa, before considering more detailed aspects concerning deep webs and brackets.

1. Define the loadcase for the base load, $\mathbf{F}_0$. That is, choose a particular combination of individual loads and specify their relative magnitudes.

2. Assemble the structure stiffness matrix for the intact structure, $\mathbf{K}_0$, and add some extra rows and columns which have a 1 on the diagonal.

3. Solve $\mathbf{F}_0 = \mathbf{K}_0\boldsymbol{\Delta}_0$ and calculate the nodal bending moment $(\mathcal{M}_{j,\lambda})_0$ in each element, λ, at each structure node, j. Then calculate the plastic hinge ratio $\gamma_{P,j}$ at each of the N structure nodes using (16.3.10).

4. Determine the load scaling factor

$$\mu = \underset{j=1}{\overset{N}{\mathrm{MIN}}}[\gamma_{P,j}]$$

and hence the location of the first hinge; scale up the base load to obtain the first hinge load: $\mathbf{F}_1 = \mu\mathbf{F}_0$ and the corresponding response values: $(\mathcal{M}_j)_1 = \mu(\mathcal{M}_j)_0$, $\boldsymbol{\Delta}_1 = \mu\boldsymbol{\Delta}_0$, and so on.

Steps 5 to 11 are performed for each load increment, i, starting with $i = 1$.

5. Obtain the modified structure stiffness matrix $\mathbf{K}_i$ by adding the appropriate terms to the ith *extra* row and column of $\mathbf{K}_{i-1}$.

6. Select the proportions of the next trial load increment, $\tilde{\mathbf{F}}_i^*$.

7. Solve $\tilde{\mathbf{F}}_i^* = \mathbf{K}_i\tilde{\boldsymbol{\Delta}}_i^*$; if $\mathbf{K}_i$ is singular go to Step 12.

8. Calculate the nodal bending moment $(\mathcal{M}_{j,\lambda})_i^*$ at each structure node $j (j = 1, \ldots, N)$ for each element, λ, at that node $(\lambda = 1, \ldots, n_j)$.

9. Calculate the load scaling factor μ from (16.3.13) and (16.3.14) and use it to obtain the load increment for the next hinge

$$\tilde{\mathbf{F}}_i = \mu\tilde{\mathbf{F}}_i^*$$

10. Update all loads and responses

$$\mathbf{F}_{i+1} = \mathbf{F}_i + \tilde{\mathbf{F}}_i$$

$$(\mathcal{M}_{j,\lambda})_{i+1} = (\mathcal{M}_{j,\lambda})_i + \mu(\tilde{\mathcal{M}}_{j,\lambda})_i^* \qquad \text{for all } \lambda \text{ and } j$$

$$\boldsymbol{\Delta}_{i+1} = \boldsymbol{\Delta}_i + \mu\tilde{\boldsymbol{\Delta}}_i^*$$

11. Set $i = i + 1$ and return to Step 5.

12. If the latest hinge is the third hinge in any member, delete that member by subtracting the two corresponding transformed element stiffness matrices from $\mathbf{K}_i$, set $i = i + 1$ and return to Step 5. If it is not a third hinge then the structure has formed a collapse mechanism which involves more than one member. Hence $\mathbf{F}_{\text{ult}} = \mathbf{F}_i$.

Allowance for Web Height and Brackets

In a hull module the beams often have relatively deep webs and/or brackets, both of which increase the flexural rigidity at the beam ends, and neglecting this effect can cause considerable error. Section 8.6 showed how this effect can be accounted for by representing the end portions of the beam as flexurally rigid. It is just as important to account for these effects in plastic frame analysis as it is in elastic analysis, and so in this section we apply the rigid length transformation to $\mathbf{K}_{H1}^e$ and $\mathbf{K}_{H2}^e$—the stiffness matrices for elements with plastic hinges.

As mentioned earlier, the possibility of a plastic hinge occurring at the midlength of a member is allowed for by using two beam elements to represent each member. As shown in Fig. 16.20 this arrangement requires only one rigid length for each element, but for generality of application we will give the stiffness matrix for an element having a rigid length at each end, and either r_1 or r_2 can be set to zero, as required.

The stiffness matrix for a beam with rigid ends was presented in (8.6.5) in the form

$$\mathbf{k} = \check{\mathbf{k}} + \mathbf{k}_r$$

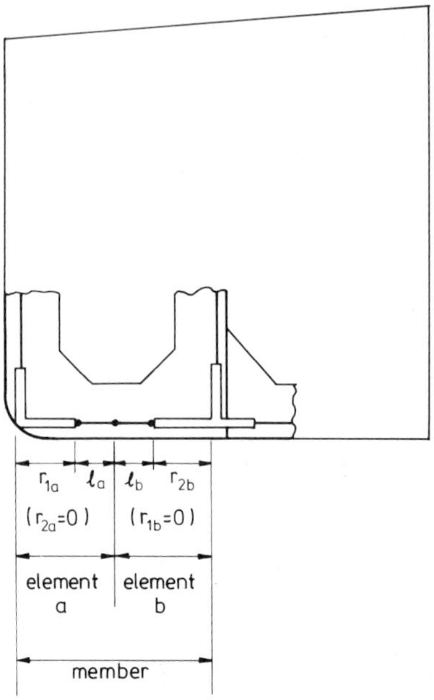

Figure 16.20 Two-element members for plastic frame analysis.

in which $\breve{\mathbf{k}}$ is the ordinary beam stiffness matrix for the flexural portion, of length $l = L - r_1 - r_2$, and $\mathbf{k}_r$ is the matrix which accounts for the rigid lengths, and is given by (8.6.6). For a beam at an arbitrary orientation χ (see Fig. 5.9) the corresponding expression in structure coordinates is

$$\mathbf{K} = \breve{\mathbf{K}} + \mathbf{K}_r \qquad (16.3.18)$$

in which $\breve{\mathbf{K}}$ is given by (5.3.16) (with l in place of L) and $\mathbf{K}_r$ is given by (8.6.7).

In the present application the flexural portion contains a hinge and the stiffness matrix of this portion is $\mathbf{K}^e_{H1}$ or $\mathbf{K}^e_{H2}$ from (16.3.3) or (16.3.6), depending on whether the hinge is at element node 1 or 2. Hence these two matrices here take the place of $\breve{\mathbf{K}}$, and so we write them as $\breve{\mathbf{K}}^e_{H1}$ and $\breve{\mathbf{K}}^e_{H2}$. The same symbols without the ˇ above them now denote the generalized stiffness matrices, after the rigid lengths have been added. These matrices are obtained by applying the transformation

$$\mathbf{K}^e_{H1} = \mathbf{T}_1^T \breve{\mathbf{K}}^e_{H1} \mathbf{T}_1$$

in which $\mathbf{T}_1$ is the transformation matrix from internal nodal displacements $\breve{\boldsymbol{\delta}}$ to external nodal displacements $\boldsymbol{\delta}$, corresponding to $\mathbf{T}$ of (8.6.2). For an element with a plastic hinge at node 1, $\breve{\boldsymbol{\delta}}$ includes an extra rotation $\breve{\theta}_{H1}$ and so the transformation is

$$\begin{Bmatrix} \breve{v}_1 \\ \breve{\theta}_1 \\ \breve{v}_2 \\ \breve{\theta}_2 \\ \breve{\theta}_{H1} \end{Bmatrix} = \begin{bmatrix} 1 & r_1 & & & \\ & 1 & & & \\ & & 1 & -r_2 & \\ & & & 1 & \\ & & & & 1 \end{bmatrix} \begin{Bmatrix} v_1 \\ \theta_1 \\ v_2 \\ \theta_2 \\ \theta_{H1} \end{Bmatrix}$$

or

$$\breve{\boldsymbol{\delta}} = \mathbf{T}_1 \boldsymbol{\delta} \qquad (16.3.19)$$

A similar transformation is required to obtain $\mathbf{K}^e_{H2}$.

The derivation is not shown here, first because it is straightforward and second because, as pointed out earlier there is no need to create $\mathbf{K}^e_{H1}$; all that is required is the vector $\mathbf{V}_{H1}$ containing the terms which are to be inserted in the new row and column of the structure stiffness matrix. It may be shown that for a beam having a flexural length l and rigid ends of lengths r_1 and r_2 and a plastic hinge at the lefthand internal node, $\mathbf{V}_{H1}$ is

$$\begin{matrix} & & U_h & V_h & \Theta_h & U_j & V_j & \Theta_j & \Theta_{Hh} \\ \mathbf{V}_{H1} = & \dfrac{EI}{l^3} [& -6l\mu & 6l\lambda & 4l^2 + 6lr_1 & 6l\mu & -6l\mu & 2l^2 + 6lr_2 & 2l^2] \end{matrix} \qquad (16.3.20)$$

The expression for $\mathbf{V}_{H2}$ is

$$\begin{matrix} & & U_i & V_i & \Theta_i & U_h & V_h & \Theta_h & \Theta_{Hh} \\ \mathbf{V}_{H2} = & \dfrac{EI}{l^3} [& -6l\mu & 6l\lambda & 2l^2 + 6lr_1 & 6l\mu & -6l\lambda & 4l^2 + 6lr_2 & 2l^2] \end{matrix} \qquad (16.3.21)$$

Allowance for Shear Deformation and Yield of Unbracketed Joints

Section 8.6 showed that in the elastic analysis of a frame containing deep members and/or brackets it is important to account for the shear deformation of the web regions of joints, and this also applies for incremental pfa. As shown therein this can be done by providing two separate rotational degrees of freedom at each joint and connecting them with an elastic rotational spring whose stiffness k_J is a function of the scantlings of the joint, given by (8.6.18) and (8.6.21).

In addition to elastic joint deformation, incremental pfa must also account for the shear yield of the inner web region of joints. In an unbracketed joint, or in a joint in which the bracket has already failed, this would constitute failure of the joint. For such joints the average shear strain in this region, γ_w, is given by (8.6.16). In each cycle of incremental pfa γ_w should be evaluated at each unbracketed joint and when it has reached the yield value the elastic rotational spring should be converted to a plastic hinge by setting k_J to zero for all subsequent increments.

Allowance for Yield of Bracket Flange

Because they occur at the most heavily loaded points, brackets have a crucial influence on the type and sequence of failures in a framed structure, far exceeding their proportional size, weight, and cost relative to the structure. In a bracketed joint the largest stress occurs in the bracket flange, shown as σ_f in Fig. 16.21. If the flange were to undergo tensile yield the bracket could not provide any significant further rotational stiffness to the joint. If the flange were to undergo compressive yield it would lose its axial and rotational stiffness (i.e., rotation of the flange within its own plane) and would collapse either axially, by means of a local plastic mechanism as in Fig. 14.4, or laterally as a two-bar mechanism. In either case the bracket web would immediately buckle and the bracket would collapse. In doing so it would lose most of its strength (load-carrying capability) and shed its load into the web region of the joint, which could cause γ_w to reach the yield value. Also the collapse of the bracket reduces the rigid lengths at the joint, since a_i and a_j are now zero.

Therefore in incremental pfa it is necessary to monitor σ_f and to account for the bracket failure in an appropriate manner if σ_f reaches positive or negative yield. The value of σ_f can be estimated from the following equation, adapted from Ref. 14 of Chapter 1

$$\sigma_f = \frac{M}{e\left(\frac{et}{3} + A_f\right)} \qquad (16.3.22)$$

in which e is the perpendicular distance from the mid-length of the flange to the neutral axis (see Fig. 16.21), M is the moment at this point, interpolated from the values at the two ends of the rigid length, t is the thickness of the bracket web, and A_f is the sectional area of the flange. Equation (16.3.22) is based on the assumption of a linear distribution of normal stress from the neutral axis to the line of the flange, as shown in the figure. For a T-shaped joint with two brackets the definition of M is given in Fig. 16.22.

Tensile failure of a bracket flange is accounted for simply by deleting the bracket (i.e., setting $a_1 = a_2 = 0$) for all subsequent load increments. Bracket failure due to compressive yield of the flange is more complicated, but it too can be accounted for. Because it involves buckling, this type of failure is relatively sudden and usually involves some load shedding (negative stiffness). That is, unlike tensile yield in which the bracket has zero stiffness but continues to carry its collapse load, the bracket stiffness has suddenly become negative and the bracket strength has diminished sharply. This disturbs the equilibrium of the structure, and it immediately undergoes further deflection such that the excess local load is shed onto the rest of the structure. A precise calculation of the new equilibrium state would require detailed incremental analysis because the bracket stiffness is nonlinear. This is simply not practical, at least not for design purposes.

However, for typical brackets the post-collapse strength is relatively small and hence one approach which is simple and not overly conservative is to regard the bracket as having lost *all* of its strength. This assumption makes it possible to account for compressive collapse of a bracket as part of incremental pfa, at a cost of only one additional analysis cycle (initial or incremental) for each such collapse. For example, let us assume that the initial analysis has revealed that the first failure in the structure is compressive collapse of a bracket. Figure 16.23*a* is the

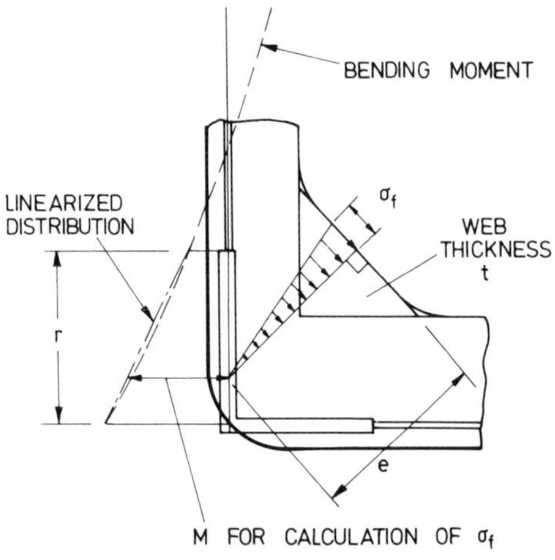

Figure 16.21 Stress in a bracket flange.

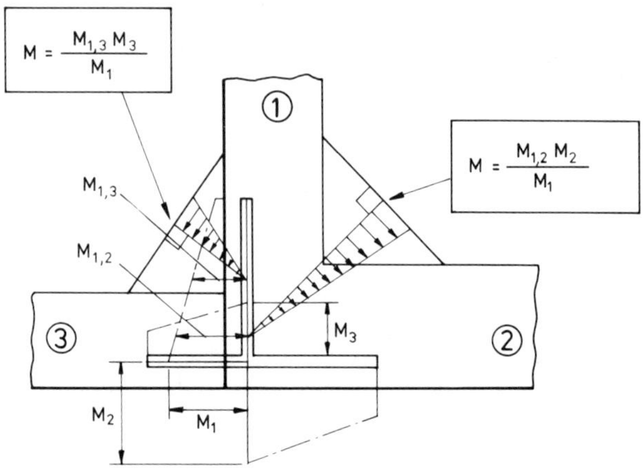

Figure 16.22 Definition of M for a bracketed T-shaped joint.

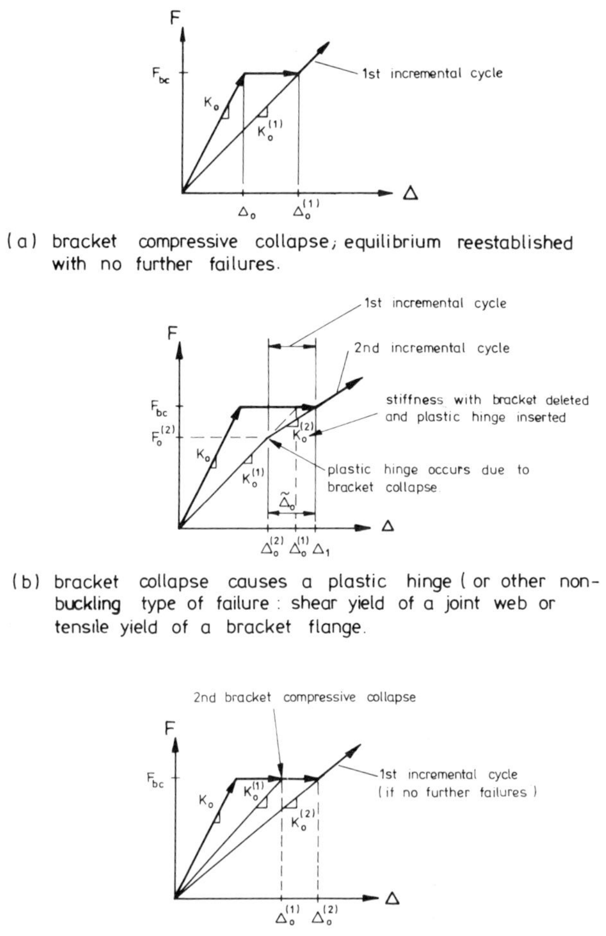

(a) bracket compressive collapse; equilibrium reestablished with no further failures.

(b) bracket collapse causes a plastic hinge (or other non-buckling type of failure: shear yield of a joint web or tensile yield of a bracket flange.

(c) bracket collapse causes a second bracket collapse

Figure 16.23 Procedure for compressive yield of bracket flange in incremental PFA.

load–deflection diagram for the structure. F_{bc} is the load which causes bracket collapse, and the slope K_0 is the stiffness of the initial undamaged structure. $K_0^{(1)}$ is the stiffness with the bracket deleted; the corresponding structure stiffness matrix $\mathbf{K}_0^{(1)}$ is obtained by reducing the rigid lengths (setting $a_1 = a_2 = 0$) and recalculating the relevant terms in $\mathbf{K}_0$. At collapse the bracket effectively disappears and the structure stiffness matrix suddenly becomes $\mathbf{K}_0^{(1)}$. The external load remains the same and to regain equilibrium the structure increases its deflection until the structure nodal forces $\mathbf{K}_0^{(1)}\mathbf{\Delta}_0^{(1)}$ are in equilibrium with the external loads. Thus the final equilibrium state of the structure is the same as if the bracket had been omitted from the beginning, and so $\mathbf{\Delta}_0^{(1)}$ can be determined by repeating the analysis with the bracket omitted.

After this reanalysis a check should be made as to whether any other form of failure has occurred in the altered structure; if so this means that the load shedding due to the bracket collapse has caused this other failure to occur immediately afterwards. For example, the end of the rigid length now occurs in line with the web height of the adjacent beam, and at this new location and with the new bending moment distribution it is possible that $M > M_P$; that is, the bracket collapse has led to a plastic hinge failure at this location. If this happens at *any* location in the structure then $\mathbf{K}_0^{(1)}$ must be further modified by inserting the vector $\mathbf{V}_H$ of (16.3.20) and (16.3.21) to represent the plastic hinge. This gives $\mathbf{K}_0^{(2)}$ and the load–deflection diagram now has a reduced slope $K_0^{(2)}$ starting from the point along $K_0^{(1)}$at which the plastic hinge occurred, as illustrated in Fig. 16.23*b*. Therefore the incremental analysis must begin at this point rather than at $\Delta_0^{(1)}$. The load vector $\mathbf{F}_0^{(2)}$ and the displacements $\mathbf{\Delta}_0^{(2)}$ corresponding to this point are obtained by linearly scaling $\mathbf{F}_{bc}$ and $\mathbf{\Delta}_0^{(1)}$ such that the plastic hinge has just formed. Then the equilibrium state is determined by solving $\mathbf{F}_{bc} - \mathbf{F}_0^{(2)} = \mathbf{K}_0^{(2)}\tilde{\mathbf{\Delta}}_0$ to obtain $\tilde{\mathbf{\Delta}}_0$, the amount of additional (post-hinge) deflection which is required to place the structure in equilibrium with the external load $\mathbf{F}_{bc}$. The equilibrium displacement is $\mathbf{\Delta}_1 = \mathbf{\Delta}_0^{(2)} + \tilde{\mathbf{\Delta}}_0$, and once this has been obtained the complete structure must again be checked for any other failures which may have occurred because of the larger total displacement which has resulted from the plastic hinge. If there are no further failures then the next incremental cycle can be commenced from the Δ_1 point.

Alternatively, the check following the first reanalysis (to obtain $\mathbf{\Delta}_0^{(1)}$) may reveal that because of the bracket collapse the web shear strain γ_w (which is now calculated with $a_1 = a_2 = 0$) now exceeds the yield value. If so, then the elastic rotational spring at that joint should be converted to a plastic hinge by setting k_J to zero. The rest of the analysis is similar to that described previously and illustrated in Fig. 16.23*b*: the loads and deflections are scaled, $\tilde{\mathbf{\Delta}}_0$ is calculated, and the structure is checked for further failures.

Another possibility is that the check following the first reanalysis may reveal that at some other bracket the new value of $|\sigma_f|$ exceeds σ_Y; that is, the first bracket collapse has led to the immediate failure of a second bracket. If this failure is tensile the bracket is deleted by reducing the rigid lengths and modifying the appropriate terms in $\mathbf{K}_0^{(1)}$ to produce $\mathbf{K}_0^{(2)}$. Then the first incremental analysis is performed commencing at the scaled displacements $\mathbf{\Delta}_0^{(2)}$, as in Fig. 16.23*b*.

If, however, the second bracket failure is compressive, then this bracket has also lost its strength, equilibrium is further disturbed and it is necessary to calculate the new equilibrium point. This is done by altering the stiffness matrix from $\mathbf{K}_0^{(1)}$ to $\mathbf{K}_0^{(2)}$ to reflect the reduced rigid lengths, and then solving for the new equilibrium deflections $\boldsymbol{\Delta}_0^{(2)}$ (see Fig. 16.23*c*). It is then necessary to check for any further failures. This process must be repeated until there are no further compressive bracket failures, or any other failures, for the current load. If there is any plastic (nonbuckling) type of failure, then the first incremental cycle is commenced from that point, which is obtained by linear scaling. Similarly, in all subsequent incremental cycles a bracket compressive collapse requires that the incremental analysis be repeated, without the bracket, using the current incremental load vector (that which caused the bracket collapse) and then a check must be made as to the occurrence of other collapse-induced failures.

Figure 17.8 gives a more detailed summary of the procedure for calculating the additional incremental displacement which occurs as the result of compressive yield of a bracket flange.

16.4 DIRECT PLASTIC FRAME ANALYSIS

In contrast to the step-by-step load–deflection procedure of incremental pfa, direct pfa deals directly and exclusively with the ultimate state of the structure; there is no need to examine earlier stages of the collapse or even to determine the order in which the hinges are formed. In essence, direct pfa is a systematic search for possible failure mechanisms and the use of some theorems, derived from equilibrium and virtual work, to determine which of the possible mechanisms corresponds to the smallest magnitude of the applied loading, the components of which are assumed to be of constant proportion. This smallest value of load is the ultimate or collapse load. This approach is simple in concept and it provides a relatively quick and easy method for calculating the collapse load of frame structures. These quite valuable attributes have stimulated the development and use of plastic frame analysis, and it is because of the existence of direct pfa that frame design was the first area of structural design in which the ultimate strength of the structure was examined and was used as a basis for the design.

The development of the method took place from about 1920 to 1950, with most of the development occurring in the third decade of this period. This development was the work of many people in several different countries, among whom may be mentioned G. V. Kazinczy, A. A. Gvozdev, S. M. Feinberg, J. F. Baker, M. R. Horne, H. J. Greenberg, W. Prager, J. A. van den Broek, R. A. Foulkes, and J. Heyman. The details of this development may be obtained from various textbooks on the method [2–6].

It will be shown subsequently that, although direct pfa is eminently suitable for relatively simple or idealized frame structures, it is not suitable for the calculation of the ultimate strength of a hull module, because it is based on several simplifying assumptions and idealizations which are not satisfied in this type of structure. Some of these discrepancies can be overcome at the cost of increased computation and complexity of the solution, but others are inherent and cannot be overcome. On some occasions the method has been used for such structures, but this is mainly because at that time the computation required by incremental pfa could not be provided conveniently and economically.

Nevertheless direct pfa continues to have a number of useful applications, and every structural designer should at least be aware of its basic features. In a ship structure, for example, there are several smaller portions of structure which can be idealized as a two-dimensional frame or grillage. The method is particularly suitable for simple grillages, that is, two-dimensional frames that carry only a lateral load. In such structures there are no large axial member forces, the loads are generally of fixed proportions, and there is no buckling, so that collapse is entirely due to the formation of plastic hinges.

Therefore this section presents the basic assumptions of direct pfa and a brief description of its basic features, using very simple frame and grillage examples. It does not illustrate all of the features and capabilities of the method, but a fuller coverage is not warranted here for the reasons just given. Also there are several basic texts available, such as Refs. 2–6.

Assumptions and Basic Features

In addition to the general assumptions of plastic frame analysis, direct pfa is based on the following special assumptions:

1. Prior to the appearance of the final plastic hinge the deflections are small and have negligible influence on the equilibrium conditions of the structure.
2. The value of M_P of each member is independent of the internal forces in the member.
3. The members do not undergo local buckling or any

other type of failure prior to the formation of a plastic hinge.

4. The members are not subject to column type of instability. Hence the bending moment in each member is essentially independent of the axial force in the member.

5. The individual components of the loading always maintain the same relative proportions.

6. The plastic hinge moment has the same magnitude regardless of the sign of M.

This is a very extensive set of assumptions and even without considering them in detail one is led to suspect that direct pfa is too idealized for the ultimate strength analysis of large, thin-walled steel structures. But there are some applications for which these assumptions are valid and so we will now show how they result in a very simple form of plastic frame analysis, which deals directly with the collapse state of the structure.

For most structures there are many possible collapse mechanisms and hence it is necessary to have some systematic and logical procedure for ascertaining what are the possible collapse mechanisms and which of them is the true collapse mechanism. The procedure that is used to achieve this is based on two theorems, a Static theorem which expresses the requirement of an equilibrium distribution of bending moment M which everywhere satisfies the condition $|M| \leq M_P$, and a Kinematic theorem which involves the geometry of the collapse mechanism and the Principle of Virtual Work. We shall present these theorems subsequently but first we shall demonstrate the basic features of direct pfa by considering a simple one-member structure, for which the collapse mechanism is immediately evident. In such a case, when the collapse mechanism is known, the theorems and the rigorous procedure are not required. To allow a comparison with incremental pfa we will use the same example as before—a propped cantilever carrying a single point load, as shown in Fig. 16.24. Because of the greater simplicity of direct pfa we can here obtain a solution for an arbitrary position of the load.

With point loads, the maximum values of bending moment can only occur at joints and at loads, and therefore the plastic hinges can only occur at these locations. For a single member a total of three pinned joints is required to form a mechanism. Since the right end of the cantilever is already pinned, only two plastic hinges are required to form the mechanism, and it is evident that they will occur at the clamped support and at the load. Therefore we can immediately draw

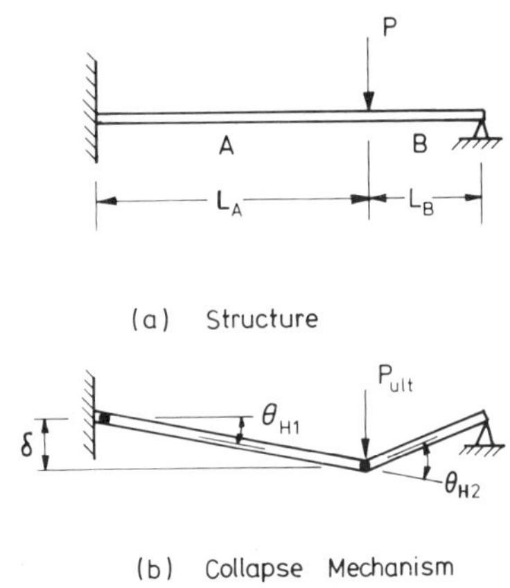

Figure 16.24 Two-member structure; direct plastic frame analysis.

the collapse mechanism as shown in Fig. 16.24 *b*. In this case, with the collapse mechanism already known, direct pfa is simply the application of the Principle of Virtual Work to the collapsing structure, taking the deflections of the structure as the virtual displacements. As the collapse occurs the external load moves through some distance and therefore some work W is done. By the Principle of Virtual Work this external work must equal the internal strain energy U which is generated during the collapse. By considering the geometry of the collapse mechanism we can obtain expressions for W and U in terms of the hinge rotations, the structure geometry and the (as yet unknown) collapse load. Then by equating these two expressions we can solve for the collapse load.

The first of the foregoing assumptions implies that until the collapse mechanism is fully formed, all deflections are negligibly small, even after some hinges have formed. That is, direct pfa assumes that the load–deflection relationship may be idealized as rigid-plastic, as shown in Fig. 16.25. This is equivalent to assuming that all of the hinges form suddenly and simultaneously. No account is taken of the se-

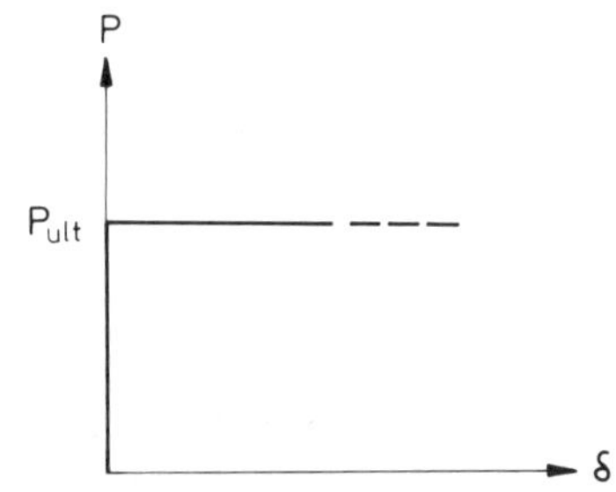

Figure 16.25 Rigid plastic load–deflection relationship.

quence of the hinges, or of anything else which happens prior to collapse. Since there is no internal deformation prior to collapse there is no internal strain energy generated during this period. Similarly, since there is no deflection of the structure prior to collapse there is no work done by the load. When the final plastic hinge occurs the structure collapses, and during the collapse the load remains constant. Therefore the work done by the load is simply the product of the load times the distance through which it moves. In the present example this is $W = P\delta$, where δ is the deflection under the load. From the geometry of the collapse mechanism this deflection can be expressed in terms of either of the hinge rotations. For example, in terms of Θ_{H1}, $\delta = L_A\Theta_{H1}$. Therefore the work done is

$$W = PL_A\Theta_{H1}$$

Notice that to obtain this result we ignored the elastic deflection of the beam and considered each portion to be perfectly straight.

The internal deformation is the rotation Θ_H of each plastic hinge. In each case the internal force which causes the rotation is a bending moment, and its value is equal to the local value of M_P. Since the bending moment is constant throughout the rotation, the internal strain energy at each hinge is simply $M_P\Theta_H$, and the total strain energy is the sum of the values for each hinge:

$$U = \sum_{h=1}^{n_h} (M_P)_h\Theta_{Hh} \tag{16.4.1}$$

where n_h is the total number of plastic hinges. From the geometry of the collapse mechanism the rotation Θ_{H2} of the internal hinge can be expressed in terms of Θ_{H1}

$$\Theta_{H2} = \Theta_{H1} + \frac{L_A}{L_B}\Theta_{H1}$$

Here again we have ignored the elastic deflection of the members. Substituting into (16.4.1) and equating the expressions for U and W we obtain

$$M_P\Theta_{H1} + M_P\left(1 + \frac{L_A}{L_B}\right)\Theta_{H1} = PL_A\Theta_{H1}$$

and we see that, true to its role here as a virtual displacement, the rotation Θ_{H1} drops out. Solving for P, which is now the ultimate load P_{ult}, we obtain

$$P_{\text{ult}} = \frac{L_A + 2L_B}{L_AL_B}M_P \tag{16.4.2}$$

In the specific example of the previous section the load was placed such that $L_B = L_A/3 = L$. In this case the above expression becomes

$$P_{\text{ult}} = \frac{5}{3}\frac{M_P}{L}$$

which agrees with the result obtained by incremental pfa, given in (16.3.17). The very small amount of computation required to achieve this result contrasts dramatically with that required by incremental pfa.

The exact agreement of the results is due to the fact that in the incremental analysis we implicitly followed or accepted all five of the assumptions on which direct pfa is based. For example, although we calculated the elastic deflection we did not use it in any way. We did not, for example, check whether it was affecting the equilibrium equations; instead we used ordinary small-deflection theory which ignores this effect. But the point is that we could have checked if we wanted to, and we could likewise have checked for member instability, or for the influence of axial or shear loads on M_P, or for the occurrence of local failures. In direct pfa none of this is possible.

Explicit Method

As mentioned earlier, most frame structures have a large number of possible collapse mechanisms and it is necessary to have a rigorous procedure for determining which of them is the true collapse mechanism. But for very simple frames, in which the collapse mechanisms are few and obvious, it is possible to obtain the solution using the foregoing method. This is referred to as the Explicit Method, because it involves the explicit calculation of the collapse load for various possible collapse mechanisms in turn, until the correct mechanism is found. It is suitable only for hand solutions, and this is another reason why it is only practical for very simple plane frames and grillages.

In the Explicit Method the procedure to be followed for each of the possible collapse mechanisms is as follows:

1. Draw the collapsed configuration of the structure, noting particularly the angle of rotation Θ_{Hh} at each of the n_h plastic hinges.
2. Choose one of these angles (say, Θ_{H1}) as a refer-

ence value and, from geometry, express the other angles in terms of Θ_{H1}.

3. Calculate the strain energy of deformation (rotation) at each hinge, $(M_P)_h\Theta_{Hh}$, and add to get the total strain energy of deformation in terms of Θ_{H1}:

$$U = \sum_{h=1}^{n_h} (M_P)_h\Theta_{Hh} \qquad (16.4.3)$$

If there is a discontinuity in Z_P at any of the hinge locations, the smallest value should be used in computing the value of M_P for that hinge.

4. If there are several loads (say, n_l in number) it is assumed that their relative proportions are fixed and are known beforehand. Each component load must be expressed in the form $P_l = \kappa_l P$, where κ_l is the proportion and P is a reference value. Also from the geometry of the collapse mechanism the deflection δ_l under each component load must be expressed in terms of Θ_{H1}. Then the work done is

$$W = P \sum_{l=1}^{n_l} \kappa_l\delta_l \qquad (16.4.4)$$

5. Equate the strain energy of deformation to the work done by the load, and solve for the collapse value of the reference load P for the assumed mechanism.

6. The foregoing steps can be repeated for the various possible collapse mechanisms, so as to determine which mechanism has the lowest value of collapse load. But it is possible that, through an oversight, the true mechanism might not be included among those examined. In that case the apparent collapse load—the lowest among those examined—would be too large, which is an error on the dangerous side. Therefore it is essential to have some means for checking whether a particular collapse mechanism is the true one. As will be shown later, this can be done by calculating the bending moment distribution corresponding to the trial collapse mechanism. If it does not exceed M_P at any point in the structure then the trial mechanism is the true collapse mechanism.

Load Sequence

It may be shown (see, for example, Ref. 3) that if assumptions 1 to 4 are satisfied then the magnitude of the collapse load is independent of the sequence in which its individual components are applied. In other words, if the first four assumptions are satisfied, then the collapse load obtained by direct pfa, which is based on proportional loading (assumption 5), is valid regardless of the sequence in which that final combination of loads is reached. Under these circumstances direct pfa can be used for the analysis of structures with random or independent load components. Unfortunately, however, these four assumptions are generally satisfied only in small simple frames or in special cases; it is rare for all four of them to be satisfied in a large and complex structure.

Application to Grillages

Direct pfa is commonly used for simple grillages because such structures generally satisfy the five assumptions of the method. We here illustrate this type of application, and for simplicity we again choose a structure which has only a few possible collapse mechanisms, so that we can use the Explicit Method. Figure 16.26 shows a simply supported doubly symmetric 3×3 grillage in which all of the members are identical. Because of the symmetry of structure and of load, the number of possible modes of collapse may be immediately reduced to three, depending on whether the collapse of each of the nine beams occurs wholly within the grillage, and therefore requires three plastic hinges in that beam, or extends to both ends of the grillage, in which case it requires only one plastic hinge in that beam (for each individual beam the mixed case is ruled out by symmetry). Although the latter requires fewer hinges per beam, the choice is not as simple as that. The expression for the strain energy in (16.4.3) shows that it is *hinge rotations* rather than hinges that are counted, and for hinges at centerlines, such as at point C, the hinge angle is twice as large. Also if some of the beams have different values of M_P the true collapse mechanism is by no means obvious.

In the present example we will calculate the collapse load for all three mechanisms and then apply Step 6 for the mechanism with the lowest value of load, to make sure that it is the true collapse load. For the first mechanism

Strain energy:

$$U = \overbrace{(\Theta + 2\Theta + \Theta)M_P}^{\text{beam BCD}} + \overbrace{(\Theta + 2\Theta + \Theta)M_P}^{\text{beam GCH}}$$

$$= 8\Theta M_P$$

Work done:

$$W = Pb\Theta$$

Therefore, from $U = W$:

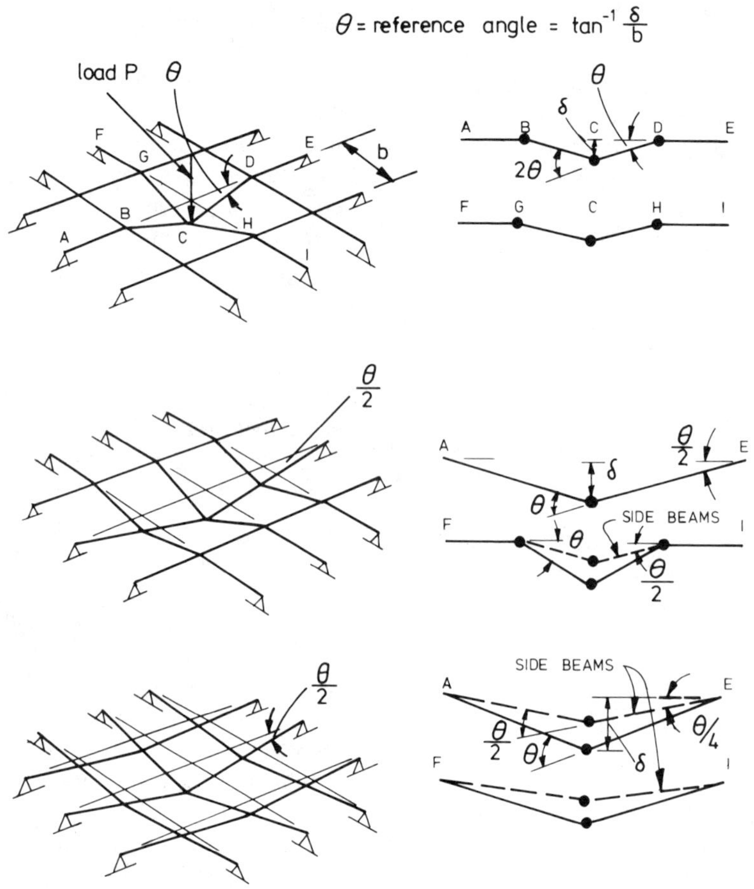

Figure 16.26 Use of Explicit Method of direct pfa for a grillage.

$$P = \frac{8M_p}{b}$$

For the second mechanism

$$U = \Big[\underset{\text{ACE}}{\Theta} + \underbrace{\Theta + 2\Theta + \Theta}_{\text{FCI}} + \underbrace{2\left(\frac{\Theta}{2} + \Theta + \frac{\Theta}{2}\right)}_{\text{side beams}}\Big]M_P$$

$$= 9\Theta M_P$$

$$W = Pb\Theta$$

Therefore, $P = \dfrac{9M_P}{b}$

For the third mechanism

$$U = \Big[\underset{\text{beam ACE}}{\Theta} + \underbrace{\frac{\Theta}{2} + \frac{\Theta}{2}}_{\text{side beams}} + \underset{\text{beam FCI}}{\Theta} + \underbrace{\frac{\Theta}{2} + \frac{\Theta}{2}}_{\text{side beams}}\Big]M_P$$

$$= 4\Theta M_P$$

$$W = P\left(2b\,\frac{\Theta}{2}\right)$$

$$P = \frac{4M_P}{b}$$

Since the third value is smallest we now check the bending moment distribution to verify that this is the true collapse mechanism. For this mechanism the bending moment diagram for each member is simply a triangle: zero at the ends and a peak value of M_P in the middle. Thus M_P is nowhere exceeded and we have therefore found the ultimate or collapse load:

$$P_{\text{ult}} = \frac{4M_P}{b}$$

Basic Theorems and Methods

The general methods of direct pfa are based on two fundamental theorems, called the Static theorem and the Kinematic theorem. These theorems were developed over many years, beginning as intuitive hypotheses and gradually being given a more rigorous foundation and formulation. Formal proofs of the theorems were supplied by Gvozdev [7] and later by Greenberg and Prager [8]. The first theorem furnishes a lower bound for the collapse load and the second an upper bound.

Since the individual loads are assumed to always maintain the same relative proportions (assumption 5) they can be characterized by a single representative load, P.

Static Theorem

The Static theorem is based on a consideration of the requirements of statical equilibrium for a frame. In general, there will exist many distributions of bending moment throughout a given redundant frame which satisfy all the conditions of statical equilibrium with a given external load. If in addition any such bending moment distribution does not exceed the local value of M_P anywhere in the frame it is said to be *statically admissible* for that magnitude of load. Using this terminology the Static theorem is:

> For a given frame and loading, if there exists any distribution of bending moment throughout the frame which is statically admissible for a load magnitude P, the value of P must be less than or equal to the collapse load.

An alternative form of the theorem is:

> The collapse load is the largest load for which a statically admissible bending moment distribution can be found.

Kinematic Theorem

The derivation of the second theorem involves the use of a statement known as Feinberg's lemma (see, for example, Ref. 5):

> If one strengthens a frame by modifying the dimensions of some member cross sections so as to increase the plastic moment of at least one section without decreasing it anywhere, this strengthening cannot decrease the collapse value of the load.

By the insertion of an adequate number of hinges, any frame structure can be transformed into a mechanism which corresponds to total or partial collapse. Such a mechanism is called *kinematically admissible* to indicate that it need not represent the actual mode of collapse of the structure.

When we regard the kinematically admissible mechanism as a (possible) collapse mechanism, we can compute the corresponding collapse value of the load from the Principle of Virtual Work, as shown in the previous section. Now it is unlikely that the trial mechanism is the actual collapse mechanism of the frame. However, it is easy to modify the frame so that the trial mechanism is the actual collapse mechanism. To do this we increase, by arbitrarily large amounts, the plastic moments of all cross sections except those in which plastic hinges have been assumed in the trial mechanism. This mechanism, which was merely kinematically admissible for the original beam, has now become the actual collapse mechanism of the strengthened frame. Because of Feinberg's lemma, the collapse load of the strengthened frame cannot be smaller than the collapse load of the original frame. We thus have what is known as the Kinematic theorem:

> The actual collapse load is the smallest load that can be computed by arbitrarily inserting an adequate number of plastic hinges in the structure and equating the energy dissipated in the hinges to the work done by the applied load.

Combined Theorem

The Static and Kinematic theorems can be combined into a single theorem. Indeed, since the first specifies a lower bound and the second specifies an upper bound, the magnitude of load corresponding simultaneously to a statically admissible bending moment diagram and to a kinematically admissible mechanism is the actual collapse load. This can be expressed in the following theorem:

> If for a given frame and loading a statically admissible bending moment diagram can be found such that the magnitude of the bending moment equals the plastic moment in sufficiently large number of sections to produce a kinematically admissible mechanism with plastic hinges in these sections, and if at each of these sections the sign of the bending moment agrees with the sense of the hinge rotation of this mechanism, the corresponding load is the collapse load of the structure.

The need for agreement between the sign of the bending moment and the sense of the rotation is important, since otherwise spurious solutions would be generated.

The Combined theorem is very useful because it provides positive confirmation as to whether or not a particular mechanism is the true collapse mechanism. This is the basis for step 6 of the Explicit Method.

METHODS OF DIRECT PFA

For a frame structure containing many members it is not practical to deal explicitly with each collapse mechanism. A method is needed which calculates the collapse load in a more systematic manner. Two such methods are listed here, one for hand solution and one for computer solution. Also, brief mention is made of the analogy between plastic design and minimum weight design. Information about these methods and about other aspects of direct pfa may be found in the references cited.

Method of Combining Mechanisms. This method was developed by Neal and Symonds [9]. It is essentially a hand method and requires considerable skill. Also it is intended mainly for point loads; distributed loads must be represented as equivalent point loads. The solution time depends on the degree of familiarity with the method; with practice a solution can be achieved relatively quickly. The method is described in various texts such as Refs. 3–6.

Method of Linear Programming. The most straightforward way of setting up the problem in a form which is suitable for automated calculation is to use the Kinematic theorem (in an inverse form) to set up a system of linear inequality constraints which must be satisfied and to then use linear programming to calculate the collapse load, which corresponds to the maximum value of load that will still satisfy these constraints. For every joint in the frame the Principle of Virtual Work produces one inequality, and the condition for a plastic hinge constitutes $2n_j$ further inequalities, where n_j is the number of members which meet at that joint. In Ref. 6 this approach is illustrated for a grillage subjected to point loads which occur only at joints. The method becomes more complicated for distributed loads, requiring internal joints and equivalent point loads.

Plastic Design and Least Weight Design. The task of minimizing the weight of a frame structure also constitutes a linear programming problem and hence it bears a close relationship with the calculation of the collapse load of a frame. As shown by several authors (see, for example, Refs. 3, 4, and 6) the analogy may be utilized to formulate an efficient method for achieving minimum weight frames governed by plastic hinge type of failure. However this is not pursued here because weight is seldom the exclusive measure of merit in ship structural design, and also there are many other types of failure besides plastic hinge formation.

Limitations of Plastic Frame Analysis

Because of the simplicity and computational efficiency of direct plastic frame analysis, and the much greater computation required by incremental pfa, there have been instances in which the former has been used, or its use has been proposed, even though some of its basic assumptions were not satisfied. The question therefore arises as to what are the limitations of the method and what are the circumstances, the means and the extent to which these limitations can be overcome. In particular, the question arises as to whether direct pfa can or should be used for the ultimate strength analysis of a hull module. The aim of this section is to provide some brief qualitative answers to these questions. In those cases where it *is* possible to overcome the limitations by performing extra computation, these questions are to some extent matters of opinion, because someone who is highly skilled in direct pfa may be willing to accept the extra effort required to overcome the limitations.

LIMITATION TO PLANAR STRUCTURES

The first point is that on practical grounds, direct pfa is effectively limited to plane frames and grillages or, equivalently, to three-dimensional frames that are constrained to deflect in only one plane. There are several reasons for this limitation. First, for three-dimensional frames the solution logic of direct pfa becomes extremely complicated and difficult to automate, while at the same time the amount of computation increases to the point where an automated solution method is essential. Second, in a free-standing structure (whether two-dimensional or three-dimensional) the loads will cause some "sway," that is, the translation of joints due to a deflection of the overall structure, as shown in Fig. 16.27. Sway begins to develop as soon as loads are applied. It represents a change of geometry of the structure and it sets up second-order bending moments

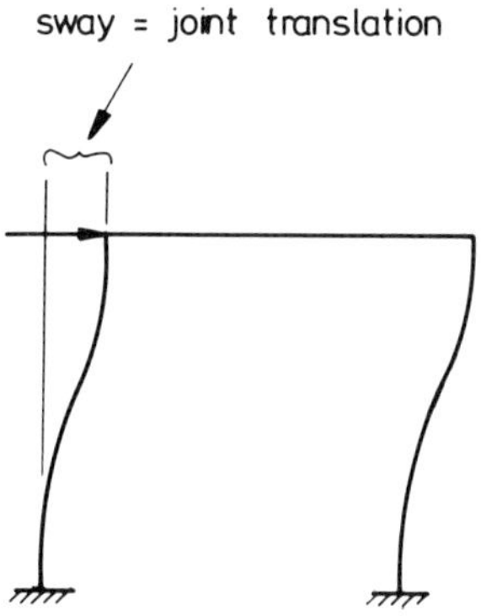

Figure 16.27 Sway in a frame structure.

in the various members. Although this additional bending moment can be allowed for in the direct plastic analysis of plane frames, it is much more difficult to do so in three-dimensional frames.

Moreover, for a three-dimensional *skeletal* structure which consists entirely of beam members, there are two further reasons why direct pfa is not suitable:

1. If the beams carry distributed loads, the loads in any two orthogonal members which meet at a joint will exert biaxial bending on a third orthogonal member at that joint as shown in Fig. 16.28. It may be shown [10] that in biaxial bending the relationship between load and deflection (or between bending moment and local curvature) is nonlinear even at small values of load, and not the bilinear elastic–perfectly plastic relationship of plastic hinge theory. Also the failure is due partly or even entirely to excessive deflection; it is not due purely to material yield. Hence the member cannot achieve the theoretical value of M_P, which is for bending in a single plane. More importantly, the failure level of bending moment cannot be calculated beforehand.

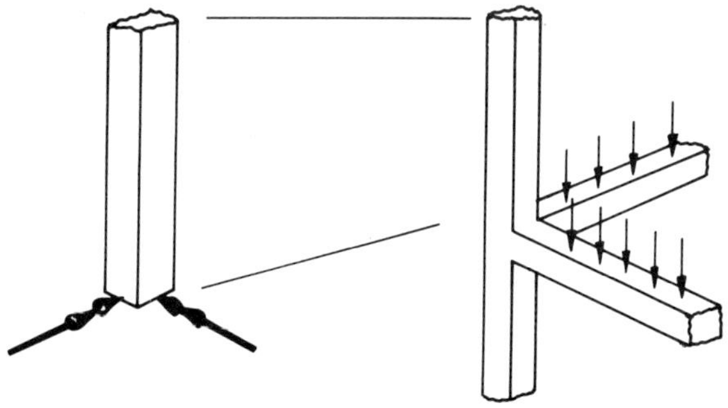

Figure 16.28 Biaxial bending in a three-dimensional frame structure.

2. Even if a beam is loaded in only one plane, if it is not constrained to remain in that plane it can undergo flexural-torsional buckling, as shown in Fig. 15.11. That is, in a beam which can deflect sideways as well as vertically, buckling can occur even though the bending occurs in only one plane, and *even in the absence of an axial load*. As we shall see, the possibility of any type of buckling makes direct pfa much less reliable and the predicted ultimate strength requires a good deal of checking.

For all of these reasons direct pfa is effectively limited to frame structures which are constrained to deflect in a single plane.

As discussed in Section 2.6 the hull module is a plated framework in which the transverse frames are constrained by the plating to deflect in the transverse plane, and this allows the transverse ultimate strength to be analyzed separately. For this reason direct pfa appears, at first glance, to be well suited for the transverse ultimate strength analysis of a hull module. However, there are still the problems and limitations imposed by the six basic assumptions of the method. The ultimate strength analysis of a hull module involves interactions between member loads, member deflections, and the two types of member failure: buckling and plastic hinge formation. As we will now show, this type of problem is beyond the capability of direct pfa. It can only be solved by means of a load–deflection approach, and in the next section we will present a generalized form of incremental pfa which is suitable for problems of this type.

Limitations of Direct PFA Arising from Its Basic Assumptions

Most steel frame structures satisfy the first of the basic assumptions listed earlier, but it is rare for all of the other five assumptions to be satisfied. Therefore if direct pfa is used for the design of a structure it is necessary to perform additional calculations to ensure that the structure has adequate strength, and to add local strength wherever required. For example, it is necessary to calculate the buckling loads for all possible types of buckling and all levels: local buckling, single member buckling, and multiple member buckling, and to alter the design in whatever manner is required to avoid premature failure due to buckling. These extra calculations involve about the same amount of computation as incremental pfa, thus canceling the computational advantage of direct pfa.

Moreover, this approach is "biased" in that it does not permit any buckling failure to be involved in the collapse mechanism, and this may very well preclude the optimum design from being achieved.

If the structure contains moderately large axial forces or large shear forces then their effect on M_P must be allowed for. This can be done approximately by means of formulas such as (16.1.9) but this then requires that the solution be performed iteratively, because to obtain an accurate value of the collapse load it is necessary to know the axial and shear forces in each member at this load.

If the structure contains large axial forces in some members then direct pfa simply cannot be used because such members are beam-columns; they involve the important and entirely distinct phenomenon of instability, and at any cross section there is complete interdependency between bending moment, axial force, and curvature. Thus, for example, the deflection due to the lateral load is magnified by the axial force, and the two together cause an increase in the bending moment.

But there is still another reason why large axial forces rule out the possibility of using direct pfa. Since the value of M_P in each member depends on the magnitude of the axial force in that member, the ultimate strength of the structure depends on the *order of application* of the axial loads and the lateral loads. If the various loads always maintained the same proportions (assumption 5) then the effect of large axial forces could be allowed for by performing an iterative solution, as mentioned previously. But in most structures, including ship structures, at least some of the loads are independent. In such cases direct pfa should not be used.

Finally, there is the assumption that M_P has the same value regardless of the sign of the bending moment. As shown in Section 14.1, the beams in ship structures can develop a local plastic mechanism in the flange when the stress there reaches yield in compression. Thus for bending in the direction of flange compression, the actual plastic hinge moment is markedly less than M_P.

REFERENCES

1. K. I. Majid, *Non-Linear Structures,* Butterworths, London, 1972.
2. J. A. van den Broek, *Theory of Limit Design,* John Wiley (N.Y.), Chapman & Hall (London), 1948.
3. B. G. Neal, *The Plastic Methods of Structural Analysis,* Chapman and Hall, London, 1956.
4. P. G. Hodge, Jr., *Plastic Analysis of Structures,* McGraw-Hill, New York, 1959.
5. C. E. Massonet and M. A. Save, *Plastic Analysis and Design,* Blaisdell, New York, 1965.
6. M. R. Horne, *Plastic Theory of Structures,* Thomas Nelson and Sons, London, 1971.
7. A. A. Gvozdev, "The Determination of the Value of the Collapse Load for Statically Indeterminate Systems Undergoing Plastic Deformation," *Proceedings of the Conference on Plastic Deformations,* December 1936, p. 19. Akademiia Nauk SSSR, Moscow-Leningrad, 1938. Translated by R. M. Haythornthwaite, *Int. J. Mech. Sci.,* **1**, 1960, p. 322.
8. H. J. Greenberg and W. Prager, "Limit Design of Beams and Frames," *Proc. Am. Soc. Civil Eng.,* **117**, 1952, p. 447.
9. B. G. Neal and P. S. Symonds, "The Rapid Calculation of the Plastic Collapse Load of a Framed Structure," *Proc. Inst. Civil Eng.,* London, **1**, 1952, pp. 58–71.
10. W. F. Chen and T. Atsuta, *Theory of Beam Columns,* McGraw-Hill, 1977.

Chapter

Seventeen

ULTIMATE STRENGTH OF A HULL MODULE

We come finally to that aspect of rationally-based design which is generally the most complicated and computationally demanding. The hull module is a three-dimensional structure containing many members, and its collapse involves various combinations of plastic deformation and inelastic bifurcation buckling of members. The most accurate and most general method for calculating the ultimate strength is to perform incremental finite element analysis of the entire hull module, but the computational requirements are too great, or too costly, with present day computing capability. It is therefore necessary to develop a simplified approach which retains sufficient accuracy but involves an acceptable amount of computation. As explained in Section 2.6, this can be achieved by making two simplifications:

1. Since the transverse structure is approximately orthogonal to the longitudinal structure, and since the shell and deck plating prevent sway in the longitudinal direction, there are only two independent modes of overall collapse: longitudinal collapse and transverse collapse.
2. By placing constraints on the relative sizes of the transverse frames and the longitudinal structure between these frames it is possible to ensure that longitudinal collapse will only occur between two adjacent transverse frames.

This means that in the hull module there will be some segment between adjacent transverse frames which will be a critical segment for hogging collapse and another for sagging collapse. Similarly for each substantially different set of transverse loads there will be a critical cross section of the hull module where transverse collapse first occurs. This chapter begins with a brief discussion of the identification of the critical segments and the factors which are involved in the question of striking a balance between accuracy and computational cost. The second part deals with longitudinal strength, giving a brief description of a rigorously accurate method and a more complete description of a simpler design-oriented method which involves much less computation. The method is based mainly on the panel strength analysis method of Chapter 14. The accuracy of the method is demonstrated by comparison with experimental results. The third part presents a method for transverse ultimate strength analysis which is also intended to provide a balance between accuracy and computational cost which is appropriate for preliminary design. The method is based on the theory and methods presented in Chapters 15 and 16.

17.1 BASIC ASPECTS OF HULL STRENGTH ANALYSIS

Identification of the Critical Segment

In some cases it will be obvious which segment is the critical segment. For example, if all segments are identical and the ship contains hatches, the critical segment will be at the middle of each hold in way of the hatch. But if the segments differ in geometry and/or loading, it will not be at all obvious which is the critical segment, and there is need for a procedure by which it can be identified, or at least by which the number of possible critical segments (all of which must be analyzed) is reduced to a manageable number.

The methods for overall ultimate strength analysis which are presented in this chapter are intended as part of the rationally-based overall design method outlined in Chapter 1 and summarized in Figs. 1.1 and 2.19. Therefore these methods are entirely computer based, as are all of the other aspects of the overall method, and they require and utilize information that is provided by three earlier steps:

1. The elastic hull module analysis provides the load effects $\mathbf{Q}$ in all of the members for hogging and sagging, using the characteristic extreme wave bending moment, $\hat{M}_{w,c}$, together with whatever stillwater loads are the worst for each condition.
2. The limit analysis of the principal members provides the limit values $\mathbf{Q}_L$ for each type of failure for each member, considered in isolation. The latter phrase is simply a reminder that multimember failures cannot be dealt with at the member level, but rather only as part of the overall ultimate strength analysis.
3. The formulation of the constraints for each submodule optimization (step 4 of Fig. 2.19) consists essentially in searching for and identifying the worst values of each constraint for each different member (i.e., the lowest value of $Q_L - Q$) and all associated information: which member, where, and which loadcase.

From the latter information it is a relatively simple matter to find the critical segments for hogging and sagging, both for longitudinal collapse and for transverse collapse (these may all be different segments). If in any one of these four cases there are two or more segments of different geometry with approximately the same degree of criticality, an ultimate strength analysis should be performed for each of them.

Choice of Method for Design Applications

As mentioned in Chapter 14, panel collapse involves not only the loss of in-plane stiffness—the ability to carry a further increase in load—but also a rapid and progressive loss of total load-carrying capacity. Consequently as a panel collapses it sheds its load onto other members. Under these conditions the overall ultimate strength, the speed of the collapse, and the amount of overall deformation all depend on the particular sequence of panel collapses and this can only be determined by an incremental type of analysis. In transverse structure load-shedding can also occur, in the form of a redistribution of bending moment, and so an incremental approach is also required for this part of the analysis.

Because of the complexity of hull module ultimate strength analysis there is a wide range of possible trade-offs between the degree of optimality of the final design and the computational cost. The computational cost depends mainly on the mathematical rigor of the analysis (in particular whether it involves nonlinear finite element analysis) and on whether it is included within the design process, and hence performed repeatedly, or used only for a post-design check on overall structural adequacy.

If an incremental method of sufficient accuracy is included in the design process the partial safety factor for overall strength can be reduced without decreasing the overall safety. That is, as shown in Fig. 17.1, if the ultimate strength is known with greater precision then the variance is smaller and hence the mean value of ultimate strength $\overline{Q}_L$ can be reduced while still keeping the characteristic value (and hence the probability of not achieving this value) the same. Reducing the mean value of the ultimate strength can mean significant savings in weight or initial cost or VCG of structure, or some combination of these. Thus a designer can choose one of two courses and still maintain a given degree of safety.

1. He can include an incremental method in the design process, requiring more time and computational cost, and be able to use a smaller value of $\overline{Q}_L$ and thereby achieve a savings in weight, initial cost, and so on.
2. He can use an approximate (even nonincremental) method, in conjunction with an accurate post-design check. This reduces the time and cost of the design process, but results in a larger value of $\overline{Q}_L$ and hence a greater weight, initial cost, and so on.

The savings obtained in the first approach can be quite important in naval vessels and some other high performance ships, and for these vessels there is good reason to include incremental overall ultimate strength analysis in the design process. However, since rationally-based design involves optimization and hence repeated ultimate strength analysis, it is necessary to have a design-oriented method which provides the appropriate level of accuracy for design at an economical computational cost. The truly rigorous incremental methods are too detailed and require too much computation for design. They are better suited

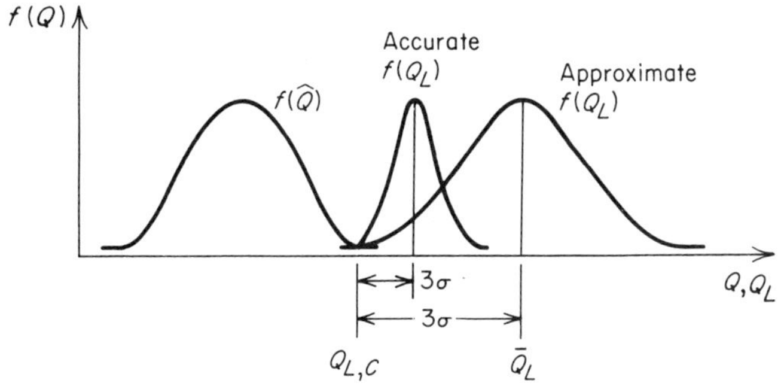

Figure 17.1

for making an accurate study of a particular mode of collapse, or for checking the results of a direct method, or for making an accurate ultimate strength evaluation of a finished design or of an existing ship.

On the other hand, in most commercial vessels, unless they are of the high performance type, the savings in initial cost which can be achieved by reducing the ultimate strength of the structure generally does not lead to increased profitability in the long term. On the contrary, in a commercial environment a more robust structure, even though not strictly necessary and somewhat more expensive initially, often means greater long term profitability because it reduces minor structural failures, with their consequent delays and repairs. Putting it another way, with commercial vessels it often happens that the scantlings which are required by the constraints against member failure, which are deliberately set at a conservative level, are themselves sufficient to give more than the required margin against overall collapse. (But even so, a post-design overall ultimate strength analysis should still be performed to ensure that this is so.)

17.2 LONGITUDINAL ULTIMATE STRENGTH ANALYSIS

Structural Members to be Included

The main type of structural member which is involved in the interframe collapse of the critical segment is a longitudinally stiffened panel. As mentioned in Section 12.6, transversely stiffened panels have relatively low ultimate strength, being at most equal to the ultimate strength of wide plates, providing that the stiffeners are sufficiently large to prevent overall panel buckling (Section 13.2). Therefore such panels should only be used in locations where the hull girder bending strain is relatively small, such as the portion of deck between hatches. In any case, regardless of their location, their contribution to the hull module ultimate strength is so small that it is usually best to ignore it.

A question also arises as to how much a longitudinal deep girder contributes to the ship's ultimate strength and how to calculate its contribution. This depends mainly on the proportions of the girder relative to the frames. As explained in Section 2.6, the design philosophy which is adopted in this text is to make the framing sufficiently strong to ensure that longitudinal collapse can only occur between two adjacent frames. For a deep girder the critical mode of buckling may be multiframe instead of interframe, and in estimating the ultimate strength of such a member the first step is to use the appropriate formulas of Sections 13.5 and 13.6 to establish the critical mode. Because of the complicated interaction with the frames the calculation of the girder's true ultimate strength would require nonlinear finite element analysis (both incremental and iterative). If this is not feasible, and if it is not possible to avoid such a large girder, then an alternative approach is to ensure that all of the relevant elastic buckling stresses obtained by the methods of Chapter 13 are well above the yield stress, and that at the load level corresponding to the first panel failure the actual maximum stress in the girder is still well below yield. Also, in making all of the required assumptions about boundary conditions, load combinations, and so on, the most conservative or "worst case" should always be selected.

Once this is done then in the analysis of the critical segment the web of the deep girder is regarded as a panel, either stiffened or unstiffened as the case may be. However, two further requirements must be met. Firstly the girder flange must be sized such that its lowest elastic buckling stress (flexural, torsional, or combined) is well above the yield stress so that the web panel has at least simple support along this edge, right up to collapse. Secondly, it is necessary to account for the bending stress in the girder due to the lateral load (maximum or design extreme value). This stress, which is obtained in the hull module analysis, must be converted to an applied strain and added to the strain arising from hull girder bending.

Thus even when there is a deep girder the longitudinal ultimate strength analysis eventually involves only longitudinally stiffened panels, and so the principal computational task is the calculation of the nonlinear response of such a panel under enforced in-plane shortening. As shown in Chapter 14, panel collapse can be initiated either by stiffener buckling (flexural or torsional and usually involving some local yielding and, for large lateral loads, the formation of a plastic hinge) or by plate buckling, which for ship panels is nearly always inelastic. Since each stiffener acts almost independently of the others the analysis may be confined to a single stiffener and its associated width of plating, as in Chapter 14.

Requirement for Panel Side Support

The methods for calculating panel ultimate strength are based on the assumption that the panel edges are approximately simply supported, and that the adjacent structure on each side of the panel is capable of preventing any significant amount of deflection along that side in the direction normal to the plate. Therefore this adjacent structure does not qualify as a panel side support unless it has sufficient stiffness against normal displacement and retains this stiffness until after panel collapse. Thus it must contain at least some plating which is approximately normal to the panel in question. For example, a panel of deck plating would need to be supported by either side plating or a longitudinal bulkhead or a longitudinal girder. The actual force which is required to provide restraint against normal deflection of the panel edges is not large and the restraint need not be total or perfect. Therefore even the vertical longitudinal panels of a passageway which extends for only a single deck height can be counted as a support. Also, providing that the vertical restraining panel is itself similarly restrained at its remote edge, it will continue to provide sufficient stiffness normal to the deck panel to satisfy the assumed boundary condition, even if it undergoes longitudinal collapse *before* the deck panel.

However, this is not true of a longitudinal girder, and therefore if a girder is to qualify as an adjacent support to a panel it must fulfill the conditions for a deep girder: all buckling stresses well above yield, and only a moderate maximum stress when other members are beginning to fail. The girder's elastic buckling stresses will have been calculated in the member limit analysis and therefore these conditions can be included among the constraints in the optimization of the submodule which contains the girder.

Rigorous Incremental Method

As explained in Section 2.6 and as illustrated in Fig. 17.2, the ultimate longitudinal strength is the value of bending moment M_{ult} at which the flexural stiffness of the hull girder (the slope of the M–ϕ curve) becomes zero. This occurs when sufficient panels within a particular segment of the hull girder (the critical segment) have themselves undergone collapse, either tensile or compressive. During the overall collapse the various panels are all at different stages of member collapse, and within each member the principal load effect is the elongation or shortening which the hull girder bending moment imposes as a result of curvature ϕ. Because the collapse is constrained to occur between frames, the cross section of the critical segment remains plane and the normal strain due to hull girder flexure varies linearly across the segment, in accordance with the Euler-Navier beam bending theory. That is

$$\varepsilon = y\phi$$

where $$y = Y - Y_{NA}$$

is the distance from the instantaneous elastic neutral axis of the cross section. Both vertical and horizontal bending can be accommodated, and the effects of transverse shear can be allowed for in the calculation of ε_{ult}. As shown by Smith [1] the linear distribution of strain greatly simplifies matters since it means that the principal load effect in each and every member is

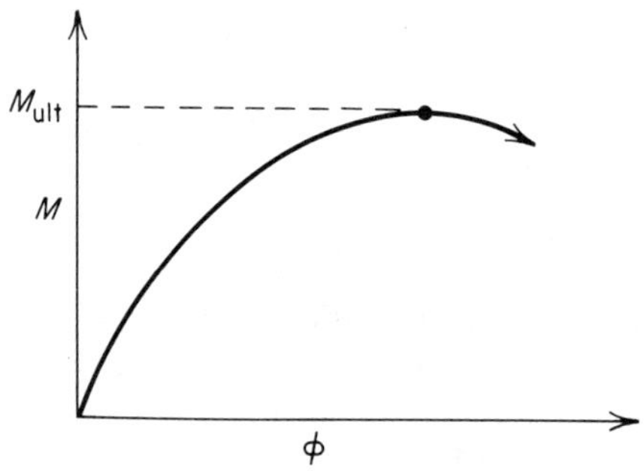

Figure 17.2

uniquely related to the overall response variable ϕ. In order to take advantage of this, ϕ is chosen as the independent variable and M_{ult} is calculated as a function of ϕ. That is, the value of ϕ is increased in increments $\Delta\phi$ and the corresponding strain increment in the ith member is

$$\Delta\varepsilon_i = y_i \Delta\phi \tag{17.2.1}$$

For each increment of strain it is necessary to calculate the increment in the load carried by each member, expressed in terms of an increment in the average stress $(\Delta\sigma_a)_i$. Since the member response is nonlinear a rigorous calculation of $(\Delta\sigma_a)_i$ requires a nonlinear (iterative and incremental) finite element analysis of each member following the procedure given in Figs. 15.21 and 15.22. In order to allow for the occurrence of yield it is necessary to check the stress level in each element at every iteration and to alter the value of E as required. An elastic-perfectly plastic stress-strain relationship (as in Fig. 16.1) is usually assumed. In order to account for the progressive spread of plasticity in the member it is necessary to use a relatively fine mesh and a large number of elements. Reference 1 presents an example of such an analysis for a representative longitudinal stiffener and its associated width of plating. The stiffener was divided lengthwise into 20 segments, each of which was further subdivided into 20 horizontal layers (the bottom layer being the plating) giving a total of 400 elements.

Once the stress increment has been calculated for each member the incremental tangent modulus E_t is evaluated (see Fig. 17.3)

$$E_{ti} = \frac{(\Delta\sigma_a)_i}{\Delta\varepsilon_i} \tag{17.2.2}$$

Then the condition of zero net axial force is used to calculate the new location of the instantaneous elastic

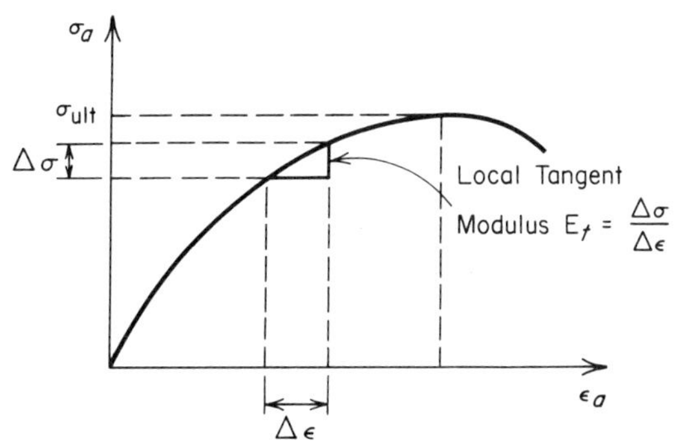

Figure 17.3

neutral axis Y_{NA}

$$\sum \Delta\varepsilon_i E_{ti} A_i = 0$$

or

$$\sum (Y_i - Y_{NA}) \Delta\phi E_{ti} A_i = 0$$

and the increment in hull girder bending moment is then obtained from

$$\Delta M = \sum (Y_i - Y_{NA})(\Delta\sigma_a)_i A_i \tag{17.2.3}$$

After completing each increment all of the cumulative or total values of nodal displacement, stress, strain, internal force, and so on, are updated. M_{ult} is the value of M when the slope of the M–ϕ curve has become effectively zero.

The complete calculation must be performed twice, once for hogging and once for sagging. In each case the other loads (stillwater bending moment and shear force, draft, etc.) are set at whichever are the worst values for that condition. The elastic portion of the M–ϕ curve is obtained by scaling the result of the hull module analysis up to the value of M at which the first member failure occurs.

Approximate Design-Oriented Method

Even for just one panel the method described in the previous section involves a great deal of computation. A typical critical segment contains some 15 or 20 panels, and there may be several potential critical segments which need to be analyzed. Also a rationally-based design process necessarily involves several overall design cycles, each of which requires a new hull module ultimate strength analysis. Therefore the computational cost of using nonlinear finite element analysis for each panel is too large for most design applications, at least for present day computers. As mentioned earlier its main use is for checking a final design, for investigating special cases, and for verifying alternative and less costly methods.

This section presents a simplified method which involves far less computation and yet possesses sufficient accuracy for most design applications. The method completely avoids the small interval incrementation of ϕ and the piecewise construction, by nonlinear finite element analysis, of the full curve of σ_a versus ε_a for each stiffener-plate combination, both before and after panel collapse. Instead the method utilizes the method of Chapter 14 to calculate $\sigma_{a,u}$ and $\varepsilon_{a,u}$ directly and immediately for each panel (or stiffener-plate combination), it proceeds directly from one collapse to the next, and it makes the (only slightly) conservative assumption that after each stiffener-plate combination has collapsed it has no strength whatever; that is, it sheds all of its load onto the other stiffeners.

Adamchak has developed a similar simplified method, and a computer program which implements it [2], using an alternative approach to panel ultimate strength, including flexural-torsional buckling [3].

In most panels the stiffeners are identical or differ only moderately in size or spacing. For all such panels, if the imposed strain is approximately uniform the collapse of the first stiffener marks collapse of the panel; there is no need to perform the calculations for each stiffener-plate combination. But if the imposed strain has a pronounced gradient, as in a side panel, or if the stiffeners are distinctly different, then in the method presented herein each stiffener-plate combination of that panel should be treated separately. To simplify the description of the method we shall refer only to panels.

The first step is to calculate the lowest value of ultimate strength $\sigma_{a,u}$ of each panel using the formulas derived in Section 14.2, and then for that lowest value calculate the corresponding ultimate strain $\varepsilon_{a,u}$ from Section 14.3. These values are then compared to the values of average elastic hull girder bending strain for each panel, $(\varepsilon_a)_i = y_i\phi$, to determine which panel will collapse first. For each panel in which $\sigma_{a,u}$ corresponds to a Mode II failure the neutral axis location should be calculated using $A_{p,tr}$ as the plate area of that panel, because during its failure its effective elastic modulus is the secant modulus, $E_s = TE$. The value of y_i is measured from this effective neutral axis to the furthest stiffener in the panel. Providing there is no tensile yield in the tension portion of the hull girder, this procedure gives the value of ϕ_0, the curvature of the hull girder at which the first panel collapses. Expressed mathematically

$$\phi_0 = \operatorname*{MIN}_{i=1}^{N_p}\left(\left|\frac{(\varepsilon_{a,u})_i}{y_i}\right|\right)$$

which signifies the smallest value (in magnitude) of the ratio $(\varepsilon_{a,u})_i/y_i$ for each panel, among the total number of panels N_p. However it is also necessary to check for panel failure by tensile yield. Thus the foregoing equation becomes

$$\phi_0 = \operatorname*{MIN}_{i=1}^{N_p}\left(\left|\frac{(\varepsilon_{a,u})_i}{y_i}\right|, \left|\frac{\varepsilon_Y}{y_i}\right|\right) \qquad (17.2.5)$$

in which a total of $2N_p$ quantities are being compared in order to find the one with the smallest magnitude.

The corresponding value of M_1 is

$$M_1 = EI_0\phi_0 \qquad (17.2.6)$$

in which I_0 is the hull girder moment of inertia with all of the panels intact, and with $A_{p,tr}$ being used as the plate area in the failed panel if the failure is Mode II. We have thus reached the point of first failure in the moment–curvature diagram of Fig. 17.4*a*.

At this point it is necessary to consider the difference between panel yield and panel compressive collapse. When a panel fails by tensile yield there is no immediate shedding of its load; rather, it continues to carry the load it had at failure. The only change is that it no longer has the ability to carry any further increase in load. Therefore if the first failure is tensile yield the only action required is to note that in all future increments that panel is to be ignored.

In contrast, when a panel fails by compressive collapse it loses some of its *current* load-carrying ability. This disturbs the equilibrium of the hull girder, and because of M_1 (which does not change) the hull girder increases its curvature until equilibrium is reestablished. During this process the panel is effectively absent from the hull girder, because of its negative stiffness. Therefore the flexural stiffness of the hull girder is now $EI_0^{(1)}$, where $I_0^{(1)}$ is the moment of inertia calculated without the panel. Also the axis of zero strain is now at $(Y_{NA})_0^{(1)}$, the height of the centroid of the new section, and the new distribution of imposed strain is in accordance with this new axis (see Fig. 17.4*b*).

As shown in Fig. 17.4*a* the new equilibrium point is the intersection of the new stiffness line $EI_0^{(1)}$ and M_1. Hence the new curvature is

$$\phi_0^{(1)} = \frac{M_1}{EI_0^{(1)}} \qquad (17.2.7)$$

and the new strain in the ith panel is $\phi_0^{(1)}(y_i)_0^{(1)}$, where $(y_i)_0^{(1)}$ is the distance to the new neutral axis.

In reality the panel does not shed all of its load immediately, and so our assumption is slightly conservative. Also, if the subsequent increase in imposed strain is large enough the panel can provide some further resistance, but it is very small in comparison with the unfailed panels.

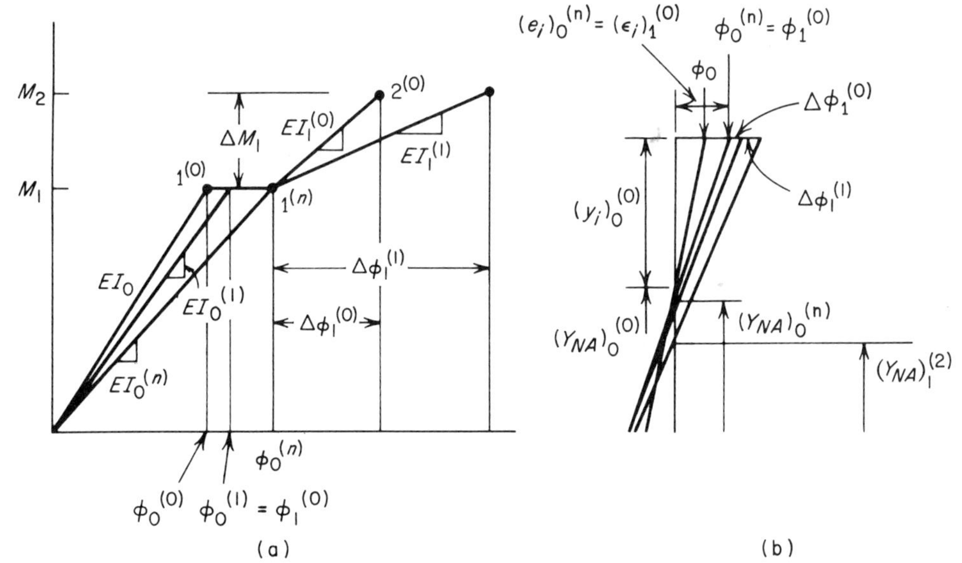

Figure 17.4

Besides the load shedding and the sudden increase in ϕ, the other important feature of panel collapse is that the increase in load on the other panels makes it quite possible that some others may also fail, even though the bending moment remains constant at M_1. It is therefore necessary to check whether $\phi_0^{(1)}(y_i)_0^{(1)}$, the strain caused by this new curvature, exceeds $\varepsilon_{a,u}$ of any other panel. If so then a new moment of inertia is calculated without that panel, say $I_0^{(2)}$, and the curvature is again increased to the point corresponding to the new stiffness $EI_0^{(2)}$

$$\phi_0^{(2)} = \frac{M_1}{EI_0^{(2)}}$$

This increased curvature raises the possibility that the corresponding strain in the panels, which is now $\phi_0^{(2)}(y_i)_0^{(2)}$, may now have reached or exceeded the ultimate strain of still more panels. This must be checked, and the entire process of deleting collapsed panels, updating the curvature and the new values of imposed strain, and rechecking for further collapse, must be continued until there are no panels in which the current updated strain exceeds $\varepsilon_{a,u}$. Only then is the structure in an equilibrium state, with $M = M_1$ and ϕ equal to its latest value, $\phi_0^{(n)}$, where n is the number of collapsed panels.

From the foregoing description it can be seen that the collapse of one panel increases the possibility of collapse of a second, and if that second collapse does occur the combined load shedding of two panels makes it even more likely that a third panel will collapse, and so forth in a chain reaction. Also, when a panel collapses in the compression flange of a box girder the axis of zero strain moves toward the tension flange of the girder, thus decreasing the possibility of tensile yield while greatly increasing the possibility of further panel collapse. Therefore in a box girder structure such as a ship, overall collapse can occur purely due to compressive collapse; tensile yield is not necessary for overall collapse. In fact, unless tensile yield occurs first (as it can in the deck in the hogging condition) it is unlikely to occur at all and if it does it will be of secondary importance. This is because if the tension flange has not yet yielded then in order for panel collapse to cause yield the new strain gradient would have to be very steep, and this would mean the collapse of many more panels, which is of far greater seriousness than tensile yield.

For this reason, the foregoing description of the procedure for establishing the first equilibrium failure point only mentioned the initial test for tensile yield, after which emphasis was placed on checking for the exceedence of $\varepsilon_{a,u}$. In practice the check should include a test for tensile yield, and in subsequent increments a yielded panel should not be counted when calculating incremental hull girder properties.

We have now reached the first equilibrium failure point. As shown in Fig. 17.4*b* the strain in the ith panel is

$$(\varepsilon_i)_0^{(n)} = \phi_0^{(n)}(y_i)_0^{(n)} \qquad (17.2.9)$$

We now begin to calculate the first increment in curvature, $\Delta\phi_1$, which will bring us to the second equilibrium failure point. Thus we again make a survey of all of the surviving panels to identify the one whose ultimate strain (either $\varepsilon_{a,u}$ or ε_Y) is closest to the strain currently imposed on that panel, $(\varepsilon_i)_1^{(1)}$, given by*

$$(\varepsilon_i)_1^{(1)} = [\phi_0^{(n)} + \Delta\phi_1^{(0)}](y_i)_0^{(n)}$$

Since we are commencing the first increment we must update the subscripts to 1 and re-zero the superscripts. Accordingly the above equation becomes

$$(\varepsilon_i)_1^{(1)} = [\phi_1^{(0)} + \Delta\phi_1^{(0)}](y_i)_1^{(0)} \quad (17.2.10)$$

In mathematical terms the increment of curvature which corresponds to the next failure is

$$\Delta\phi_1^{(0)} = \mathrm{MIN}\left(\left|\frac{(\varepsilon_{a,u})_i - (\varepsilon_i)_1^{(1)}}{(y_i)_1^{(0)}}\right|, \frac{|\varepsilon_{Yi}| - |(\varepsilon_i)_1^{(1)}|}{(y_i)_1^{(0)}}\right) \quad (17.2.11)$$

The corresponding increment in bending moment is

$$\Delta M_1 = EI_1^{(0)}\Delta\phi_1^{(0)} \quad (17.2.12)$$

in which $I_1^{(0)}$ is calculated using $A_{p,tr}$ as the area of the newly failed panel. If the failure is by panel collapse we must correct for load shedding and then determine whether there are any further failures. To do this we delete the panel, calculate the revised hull girder properties $(Y_{NA})_1^{(1)}$ and $I_1^{(1)}$, calculate $(y_i)_1^{(1)}$ for each panel, update the incremental curvature

$$\Delta\phi_1 = \frac{\Delta M_1}{EI_1^{(1)}}$$

and calculate the new strain in each panel due to load shedding:

$$(\varepsilon_i)_1^{(2)} = (\varepsilon_i)_1 + \Delta\phi_1^{(1)}(y_i)_1^{(1)}$$

If in any panel this new strain exceeds $\varepsilon_{a,u}$ that panel is deleted, $I_1^{(2)}$ and $(y_i)_1^{(2)}$ are calculated, the incremental curvature is updated

$$\Delta\phi_1^{(2)} = \frac{\Delta M_1}{EI_1^{(2)}}$$

and another check is made as to whether the new strain, $(\varepsilon_i)_1^{(3)} + \Delta\phi_1^{(2)}(y_i)_1^{(2)}$, exceeds $\varepsilon_{a,u}$ anywhere. As long as values of $\varepsilon_{a,u}$ continue to be exceeded this process is repeated. If the original survey of panels represented by (17.2.11) showed that the next failure was tensile yield, then the failure point lies along the EI_1 line, and the increment in curvature is

$$\Delta\phi_1 = \frac{|\varepsilon_{Yi}| - |(\varepsilon_i)_1^{(1)}|}{(y_i)_1} \quad (17.2.13)$$

in which i denotes the yielded panel. The increment in bending moment is

$$\Delta M_1 = EI_1\Delta\phi_1$$

In all subsequent increments the yielded panel is not counted when computing hull girder properties.

In nearly all cases overall collapse is caused by the occurrence of a runaway sequence of panel collapse within a particular increment, and in most cases the number of increments prior to collapse is quite small—often only one or two. But a hogging collapse often involves yield and may even be primarily caused by yield. In this case the collapse is much more gradual, and a suitable measure of collapse is the slope of the M–ϕ curve. Although strictly speaking collapse is the zero slope condition, in practice any small value (say 5% of the initial value, EI) represents collapse, and the cumulative value of M at this point constitutes the ultimate longitudinal strength M_{ult}.

The method just outlined is extremely rapid because it uses explicit equations only and it proceeds directly from one panel failure to the next. In most ships the majority of the panels in the strength deck will have roughly the same value of $(\varepsilon_{a,u})$ and the same is true of the panels in the bottom. Hence most of the $\Delta\phi$ increments are quite short and so the assumption of linearity within each segment does not introduce any substantial error.

*Throughout this section a numerical subscript indicates an equilibrium failure point, and it refers to conditions just after that failure has been reached. A superscript such as (m) refers to conditions within that same increment, after the collapse of m panels within that increment. Each increment terminates with either equilibrium or tensile yield. In calculating hull girder quantities such as $(y_i)_1^{(m)}$ the panels to be excluded are all those which have failed, by yield or collapse, up to and including failure point 1, together with the first m panels to collapse in the current increment.

Comparison with Rigorous Method and Experiment

Obtaining a full scale experimental value of M_{ult} requires testing a complete ship to destruction. The only values which appear to be available are three values which were obtained from 30 to 50 years ago, and as indicated in Ref. 4 they are not really suitable for verifying the theory because of their method of construction; two of them were transversely framed and all of them were riveted. Therefore it is necessary to use data from small scale tests on simple box girder models, such as those of Dowling et al. [5] and Reckling [6]. These experimental results have already been used by Dow et al. [4] to demonstrate the accuracy of the rigorous method of longitudinal ultimate strength analysis. Hence they are appropriate for assessing the accuracy of the simpler design-oriented method of the previous section. All of the models were basically similar and so we select Model 2 of Ref. 5, shown in Fig. 14.12, because the upper flange of this model was used to demonstrate the method for calculating panel ultimate strength. Although each side of the model contains just one panel the fact that each stiffener-plate combination acts separately makes it possible to allow for the different heights of the stiffeners.

The first step is to calculate $\sigma_{a,u}$ (two values) and then $\varepsilon_{a,u}$ for each stiffener-plate combination, in order to establish which is the first to fail. This has already been done for the upper flange in Section 14.4, and since the stiffeners in the side are all the same we only need to calculate $\sigma_{a,u}$ for one of them. The scantlings and material properties of the side (web) panel are given in Table 14.2. The calculations are similar to those for the upper (compression) flange in Section 14.4, and so only an abbreviated summary of the calculations is given here. Reference 5 does not give the initial deflection Δ of the stiffeners in the side panel, but since these are identical to those in the deck panel the same values may be used: $\Delta_{\text{I}} = -0.25$ mm and $\Delta_{\text{II}} = 1.40$ mm.

Mode I collapse

$A = 1070\ \text{mm}^2,\qquad I = 294100\ \text{mm}^4,$
$y_f = -41.4\ \text{mm}$

$\lambda = 0.575,\qquad \eta = 0.0377,\qquad R_{\text{I}} = 0.95$
Hence $(\sigma_{a,u})_{\text{I}} = 263$ MPa

Mode II collapse

$\beta = 2.116,\qquad T = 0.77,\qquad \sigma_F = 184$ MPa
$A_{tr} = 892\ \text{mm}^2,\qquad I_{tr} = 277800\ \text{mm}^4,$
$y_{p,tr} = 10.47$ mm
$\lambda = 0.415,\qquad \Delta_p = 1.63$ mm, $\qquad \Delta_H = 0.692$ mm
$\eta = 0.070,\qquad \eta_p = 0.055,\qquad R_{\text{II}} = 0.879$
Hence $(\sigma_{a,u})_{\text{II}} = 135$ MPa

Therefore Mode II governs and the ultimate strain is

$$\varepsilon_{a,u} = \frac{A}{A_{tr}}\frac{\sigma_{a,u}}{E} = 0.749 \times 10^{-3}$$

From Section 14.4, the ultimate strain of the stiffener-plate combination in the deck panel is $\varepsilon_{a,u} = 1.138 \times 10^{-3}$, and from Table 14.2 the yield strain in the bottom plating is 1.429×10^{-3}. The value of hull girder curvature ϕ_0 at the first failure is given by (17.2.5). In this example there are three possible causes of the first failure: compressive failure of the stiffener-plate combinations in the deck (all of which are identical), compressive failure of the upper side stiffener, and tensile yield of the bottom flange.

For failure of the deck panel the reduced area of plating is $A_{p,tr} = 0.825 \times 1219 \times 4.88 = 4908\ \text{mm}^2$ and the distance from the effective neutral axis to the upper flange is $y_i = 472$ mm. For failure of the uppermost stiffener-plate combination in the side the value of y_i is 294 mm. Substituting into (17.2.5)

$$\begin{aligned}\phi_0 &= \text{MIN}\left[\frac{1.138 \times 10^{-3}}{472}, \frac{0.749 \times 10^{-3}}{294}, \frac{1.429 \times 10^{-3}}{457}\right]\\ &= \text{MIN}[2.41, 2.55, 3.13] \times 10^{-6}\\ &= 2.41 \times 10^{-6}\ \text{mm}^{-1}\end{aligned}$$

and so the method correctly predicts that the upper flange fails first. The failure moment M_1 is given by (17.2.6) in which I_0, the moment of inertia with a plate area of 4908 mm^2 in the upper flange, is 3.17×10^9 mm^4. Since the value of E differs within the section (see Table 14.2) it is necessary to use the mean value, which is 208 kN/mm^2. The result is

$$M_1 = EI_0\phi_0 = 1.59 \times 10^9\ \text{Nmm}$$

Since the first failure involves *all* of the stiffener-plate combinations in the upper flange, the value of $I_0^{(1)}$ is quite small, the equilibrium curvature $\phi_0^{(1)}$ is very large, and the resulting value of strain which is imposed on the upper stiffener-plate combination in the side is far larger than its ultimate strain (0.749×10^{-3}). Hence equilibrium cannot be established and so the method says that M_1 is the collapse load: $M_{ult} = 1.59 \times 10^9$ Nmm.

This predicted value differs by only 4% from the experimental value of 1.53×10^9 Nmm from Ref. 5. In Ref. 4 the results of the rigorous method are given in terms of the plastic moment of the section, which is $M_P = 2.27 \times 10^9$ Nmm. The comparison is shown in Fig. 17.5. Because it is far more detailed the rigorous method gives slightly different results depending on what allowance is made for the "hard spot" in the upper corners of the structure. As shown in the figure, the conservative assumptions of the simplified method are in themselves sufficient to compensate for this slight strengthening effect.

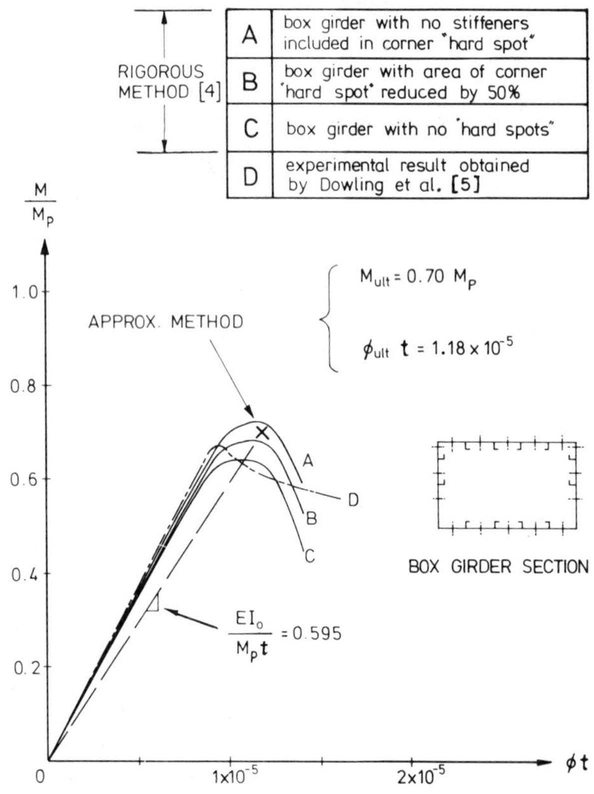

Figure 17.5 Ultimate strength of a box girder in bending; theory and experiment.

17.3 TRANSVERSE ULTIMATE STRENGTH ANALYSIS

Estimation of Transverse Loading

In nearly all ship types including naval vessels transverse loading derives mainly from local and easily quantified effects. In most ships it is simply the net effect of the transverse distribution of weight and fluid pressure, with a quasi-static allowance for dynamic effects. Thus it is essentially deterministic and the worst load combinations and the resulting maximum loads can be identified and quantified to a good degree of accuracy and reliability. This is in direct contrast with the longitudinal ultimate strength analysis, in which the largest uncertainty comes from the loading.

With naval vessels the maximum transverse loading is usually some form of external blast pressure, both above and below the waterline. But even here, although the load magnitude is not deterministic, it *is* known by the designer because the value of such pressure which the ship must be able to withstand is one of the design specifications which is selected beforehand, in the light of the ship's purpose, operating environment, and other factors.

There are two alternative ways in which the designer can derive benefit from the greater certainty of the transverse load, while still maintaining a satisfactory degree of safety: either the margin can be left the same and the calculation of the transverse ultimate strength can be simplified, or the transverse ultimate strength can be calculated to a similar degree of accuracy as the transverse load and the margin between the two can be reduced. As with longitudinal strength analysis there are some applications in which a more accurate calculation of the ultimate strength is to be preferred, and other applications in which an approximate calculation is preferred. The factors affecting the choice are much the same as those discussed at the beginning of this chapter.

Basic Features of the Analysis

Transverse ultimate strength analysis consists of incremental pfa (plastic frame analysis) of the beam members which lie in the transverse plane of the critical segment, together with an eigenvalue analysis for flexural-torsional buckling. Most of the beams are attached to plating and therefore have a plate flange whose width is equal, or nearly equal, to the ship's frame spacing. The geometric stiffness matrix of such beams is derived in Section 15.3. This section presents a brief summary of the basic aspects of transverse ultimate strength analysis. For the most part the details of each aspect have been presented in earlier chapters, and the relevant sections are indicated herein. Some further details are explained after presenting the basic features.

As with the longitudinal ultimate strength analysis, the critical segments are identified as such in the constraint formulation following the hull module analysis and the member limit analysis (step 4 of Fig. 2.19). If there are two or more segments of differing geometry and similar levels of criticality they should each be analyzed. Each critical segment is analyzed for the combination of loads which caused it to be selected as a critical segment. If there are two or more "worst combinations" of loads for a given critical segment a separate ultimate strength analysis should be performed for each such combination.

If any specific form of damage tolerance is to be provided, the damage should be simulated in the critical segment in an appropriate manner, for example by omitting the relevant member.

As with longitudinal collapse, transverse collapse involves a combination of beam-column buckling (flexural or flexural-torsional) and the formation of plastic hinges. However, there are several significant differences from longitudinal analysis.

1. Instead of being nearly independent there is a high degree of interaction among the members.
2. The members are generally all different; there are no groups of identical members as in stiffened panels.
3. The axial (i.e., girthwise) compression in the plating is relatively small and hence there is no prior plate failure by inelastic buckling. Therefore each member is essentially a beam and the formation of plastic hinges is governed by the basic theory of Section 16.1, with allowance for the effects of axial force and transverse shear.
4. Once a beam flange reaches compressive yield at some point along the beam length, then in all subsequent load increments the flange at that point has negligible stiffness against rotation in the plane of the flange. This greatly diminishes the member's resistance to flexural-torsional buckling, and it is essential to account for this.
5. The members are relatively short and deep, and are therefore highly resistant to flexural buckling, but are susceptible to flexural-torsional buckling. Hence transverse collapse is mainly instigated by the formation of plastic hinges and compressive yield of beam flanges, and/or flexural-torsional buckling of individual members. Since these failures mark the boundaries of the load increments, there is no basic change in the general state of the structure within an increment. Therefore within each increment it is not necessary to iteratively update the stiffness matrix; it is sufficient to update it once at the end of each increment.
6. Because of symmetry of both load and structure about the plane of the transverse frame, the lateral-torsional response can be uncoupled and analyzed separately from the in-plane bending, within each increment. This analysis is an eigenvalue problem.

Bracket Failure by Buckling and by Fatigue

As noted earlier, in a frame structure, brackets have a strong influence on the structure's ultimate strength, both in magnitude and in regard to the mode of collapse. Section 16.3 showed how bracket failure by yielding is accounted for in incremental pfa. Brackets can also fail by buckling and by fatigue, and the question arises as to whether and to what extent these other types of failure need to be explicitly included in the transverse ultimate strength analysis. This depends on whether the associated limit value can be calculated independently of the analysis and whether the failure can be totally prevented either by making local changes to the bracket or by a simple constraint on the limit value.

Failure by buckling is prevented by fitting stiffeners, and the number, size, and location of these is essentially a local matter since it depends on the particular bracket geometry and also on the nature and distribution of the loading on the bracket. Conversely as long as the stiffening is adequate its particular arrangement has no influence on the overall strength. Therefore the investigation and prevention of buckling of brackets can be done either as part of the overall ultimate strength analysis or as part of detail design.

Fatigue depends mainly on local factors, especially the amplitude of the cyclic stress and the magnitude of the static tensile stress, both of which can be largely controlled by altering the bracket scantlings and geometry and avoiding stress concentrations. In longitudinal brackets some of the stress may be hull girder bending stress; this is controlled by means of a constraint on section modulus, which then allows the other stresses to be suitably adjusted by making local changes. In bulk carriers, brackets contain a component of cyclic stress due .to the alternating loaded/ballasted conditions. This may be controlled by means of a constraint on the amplitude of the cyclic stress in the brackets, which is known from the hull module analysis.* Thus fatigue failure need not be explicitly included in the hull module ultimate strength analysis.

Members, Nodes, and Degrees of Freedom

The basic approach is that of incremental pfa as presented in Section 16.3, including the use of rigid ends to allow for the webs of adjacent beams and for brackets, the use of an elastic rotational spring to allow for the shear deformation of joints, and the checking for joint failure due to yield of the bracket flange (if relevant) followed by failure of the web region, either by shear or by the formation of a plastic hinge. As before, each member is represented by two beam elements, with a node between them at the midlength of the original member. Here again, to avoid confusion the word "member" in this section always means the original beam member, and the word "element" always means one of the two beam elements in the finite element model.

Figure 17.6 shows a typical portion of transverse structure and the structure coordinate system. The structure X-axis extends longitudinally and the transverse structural coordinates are Y and Z. The displacements in structure coordinates are V, W, and Θ_X; the U displacement is zero. It will be shown that flexural-torsional buckling can be analyzed separately and therefore the Θ_Y, Θ_Y', Θ_Z, and Θ_Z' degrees of freedom are not required in the structural model.

To allow for plastic hinges the structure has two independent values of Θ_X at each midlength node and at each internal node at the inside ends of the rigid lengths. The latter nodes do not occur in the final structural model; they are removed by transforming the beam element stiffness matrix as shown in Section 16.3. The U and V deflections at these locations do not qualify as degrees of freedom because they are totally

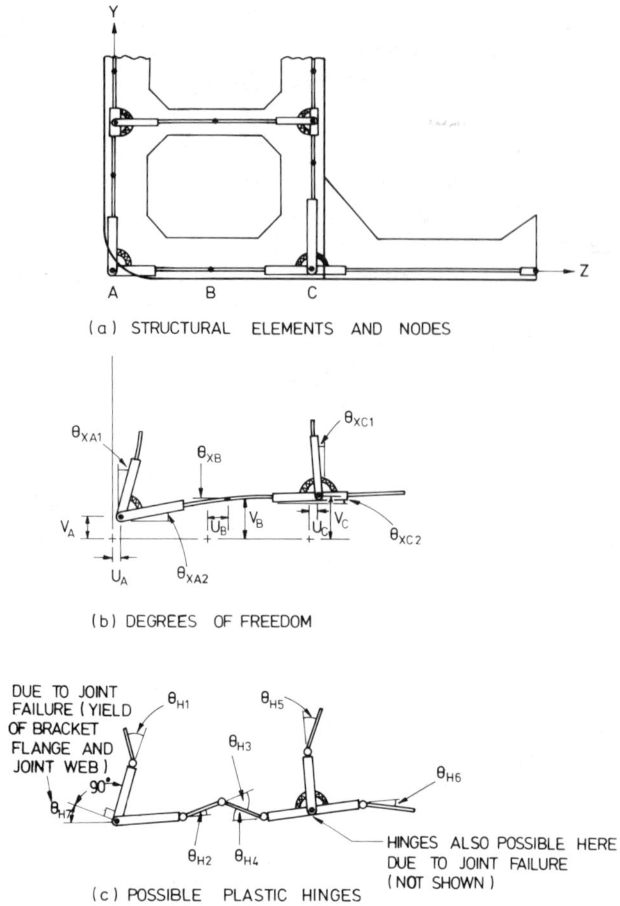

Figure 17.6 Structural model for analysis of transverse strength.

determined by the nodal values of U, V, the Θ_X value for that rigid length, and the length itself, r. The value of Θ_X is, of course, constant along the rigid length. Because of shear deformation the angle of rotation of one rigid length (such as Θ_{XA1}) is separate from that of the orthogonal rigid length (Θ_{XA2}). These two degrees of freedom are related only by the 2×2 stiffness matrix of the elastic rotational spring, given by (8.6.9) and (8.6.14). Since there is no bending along the entire rigid length, there are also only two values of Θ_X for T-shaped and X-shaped junctions (see also Fig. 8.48).

Constant Loads and Variable Loads

The loads that were used in the hull module analysis were the *design extreme value* loads (for those of probabilistic type) and *design maximum* loads; let us represent this set of design loads by the symbol $\mathbf{F}_d$. In the transverse ultimate strength analysis these loads are divided into two parts. The first part is the constant load $\mathbf{F}_c$ which consists of those loads which have a

* Hence for such vessels the hull module analysis should include at least an approximate calculation of the stress in the bracket flange. If the bracket is modeled as a superelement this is known directly; with rigid lengths (16.3.22) may be used.

fairly definite maximum value; this would include most of the stillwater loads. The other part is the variable load $\mathbf{F}_v$ which consists of those loads which could occur simultaneously and which could conceivably be large enough to cause transverse collapse. This would include loads such as acceleration forces and additional external pressure due to extreme wave height or slamming or underwater blast. It should also include overloads which might occur due to emergency operation, human error, or equipment failure. It is these variable loads which are increased incrementally, up to collapse. Each increment is defined in terms of a basic set of variable loads $\mathbf{F}_{v0}$ and a *load scaling factor* μ; thus for the ith increment

$$\tilde{\mathbf{F}}_{vi} = \mu_i \mathbf{F}_{v0}$$

The magnitude of each increment is controlled by μ, while the proportions of the individual component loads are determined by $\mathbf{F}_{v0}$. These proportions are constant within each increment but if desired they may be varied from one increment to the next. Thus at any stage in the analysis the total load is

$$\mathbf{F} = \mathbf{F}_c + \mathbf{F}_v$$

$$= \mathbf{F}_c + \sum_i \mu_i \mathbf{F}_{v0}$$

and the transverse ultimate strength of the ship, $\overline{F}_{\text{ult}}$, is the norm of the total load vector when overall collapse finally occurs

$$\overline{F}_{\text{ult}} = (\mathbf{F}^T\mathbf{F})^{1/2} \tag{17.3.1}$$

where $$\mathbf{F} = \mathbf{F}_c + \left(\sum \mu_i \mathbf{F}_{v0}\right)_{\text{collapse}}$$

As first discussed in Section 2.7, the value of $\overline{F}_{\text{ult}}$ depends on $Q_{L,hg}$, the limit values of those member load effects which are involved in the hull girder collapse. For transverse collapse the load effect is the bending moment $\mathcal{M}$ at all of the nodes in the structural model, and there are three types of limit values: M_P for plastic hinge formation, M_{fY} for compressive yield of a beam flange, and M_{cr} for flexural-torsional buckling. The constraint against transverse collapse is

$$\overline{F}_{\text{ult}}(\mathbf{Q}_{L,hg}) \geqslant \gamma_{s,0}\overline{F}_d$$

in which the parentheses indicate functional dependency. $\overline{F}_d$ is the norm of the design load $\mathbf{F}_d$, which consists of $\mathbf{F}_c$ and $\mathbf{F}_v$. This constraint is formulated in step 7 of the overall design procedure shown in Fig. 2.19.

Principal Steps of the Analysis

The overall procedure is illustrated in Fig. 17.7. The tilde (~) above any quantity indicates an incremental value, and an asterisk indicates a trial value, of arbitrary magnitude. The principal steps are:

1. Perform an elastic frame analysis with the constant load $\mathbf{F}_c$ to obtain the constant load effects $\mathbf{Q}_c$. In all subsequent increments these are added to the incremental load effects to get the total values.

Steps 2–11 are performed for each increment.

2. For a trial incremental load vector $\tilde{\mathbf{F}}^* = \mu^*\mathbf{F}_{v0}$ calculate the equivalent nodal forces, allowing for any plastic hinges which have already formed.
3. Assemble the structure stiffness matrix $\mathbf{K}$ and solve for the corresponding incremental displacements $\tilde{\mathbf{\Delta}}^*$. If $\mathbf{K}$ is singular, calculate $\overline{F}_{\text{ult}}$ as in (17.3.1) and stop.
4. At each element node calculate the nodal bending moments $\tilde{\boldsymbol{M}}^*$ and the axial force and shear force.*
5. For each element calculate the limit values $\mathbf{Q}_L$. In the main these consist of the nodal bending moments $\boldsymbol{M}$ which would produce a hinge due to: (a) $\mathcal{M}$ reaching M_P with the beam flange in tension, (b) $\mathcal{M}$ reaching M_{fY} at the midlength of a member with the beam flange in compression, (c) shear yield of the web region of a joint which has no bracket or in which the bracket has already failed. The effects of axial force and shear force should be allowed for in all cases.
6. Calculate the hinge ratios $\gamma_{L,j}$ at each structure node λ, using (16.3.13). Here γ_L is used in place of γ_P, and M_L in place of M_P because there are now three types of limit moment and approximately three times as many hinge ratios. Then determine the load scaling factor which would cause the next hinge.

$$\mu = \operatorname*{MIN}_{j=1}^{N}[\gamma_{L,j}]$$

7. For each bracket calculate the bracket yield ratio

$$\gamma_{bY,k} = \left[\frac{M_{bY,k} - |\mathcal{M}_k|}{\tilde{\mathcal{M}}_k^*}\right]$$

*Due to typesetting limitations, an italics boldface M is used in place of a script boldface M; see the footnote on page 497.

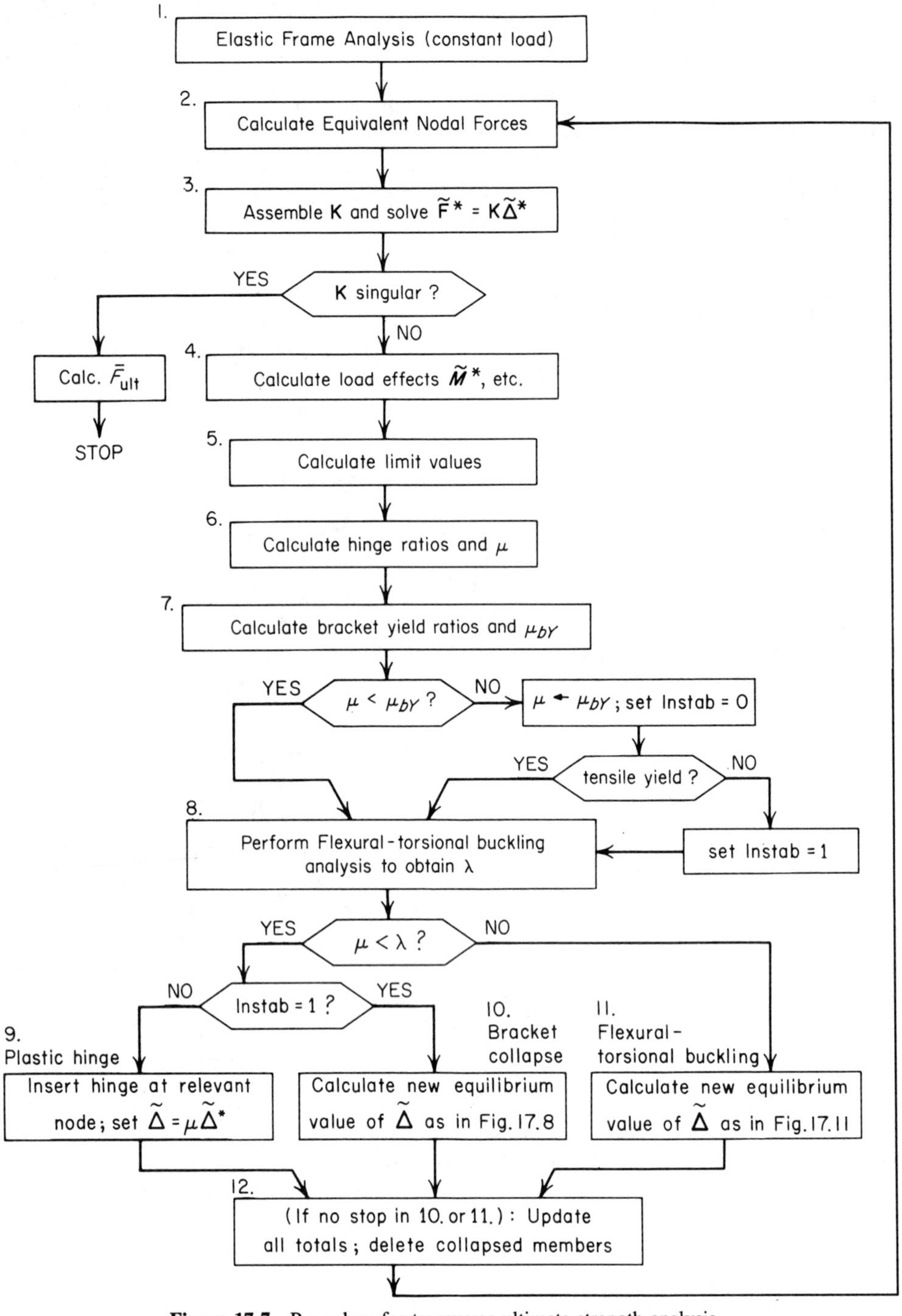

Figure 17.7 Procedure for transverse ultimate strength analysis.

where $\mathcal{M}_k$ is the joint moment in the kth bracket and $M_{bY,k}$ is the value of $\mathcal{M}_k$ that would cause σ_f to reach yield; this is obtained from (16.3.22) with $\sigma_f = \sigma_Y$. Then calculate the load scaling factor for bracket yield:

$$\mu_{bY} = \underset{k=1}{\overset{n_b}{\text{MIN}}}[\gamma_{bY,k}]$$

If $\mu < \mu_{bY}$ proceed to step 8; if $\mu \geqslant \mu_{bY}$ replace μ by μ_{bY}; if the yield is compressive set INSTAB to 1 to indicate this.

8. Perform a flexural-torsional analysis of the structure as explained in the next section and calculate the load scaling factor λ for buckling.

9. If $\mu < \lambda$ and INSTAB $\neq 1$ then for this incremental cycle the failure is due to hinge formation: modify the stiffness matrix and the load vector so as to insert a Θ_X hinge at the relevant node. Scale all trial values: $\tilde{\mathbf{\Delta}} = \mu\tilde{\mathbf{\Delta}}^*$, $\tilde{\mathbf{M}} = \mu\tilde{\mathbf{M}}^*$, $\tilde{\mathbf{F}} = \mu\tilde{\mathbf{F}}^*$, and so on, and go to Step 12.

10. If $\mu < \lambda$ and INSTAB $= 1$ then the failure is due to bracket collapse. Calculate the value of $\tilde{\mathbf{\Delta}}$ which is required for equilibrium to be reestablished, taking account of all further failures of any type which may occur during this process. The details of this step are given in a separate list and illustrated in Fig. 17.8. Then go to Step 12.

11. If $\mu > \lambda$ then the failure in this cycle is by flexural-torsional buckling. Calculate the value of $\tilde{\Delta}$ which is required for equilibrium to be reestablished, taking account of all further failures of any type which may occur during this process. The details of this step are given subsequently in a separate list.

12. Update all totals $\mathbf{F}_v = \mathbf{F}_v + \tilde{\mathbf{F}}$, and so on. If this is a third hinge in a member delete the member.

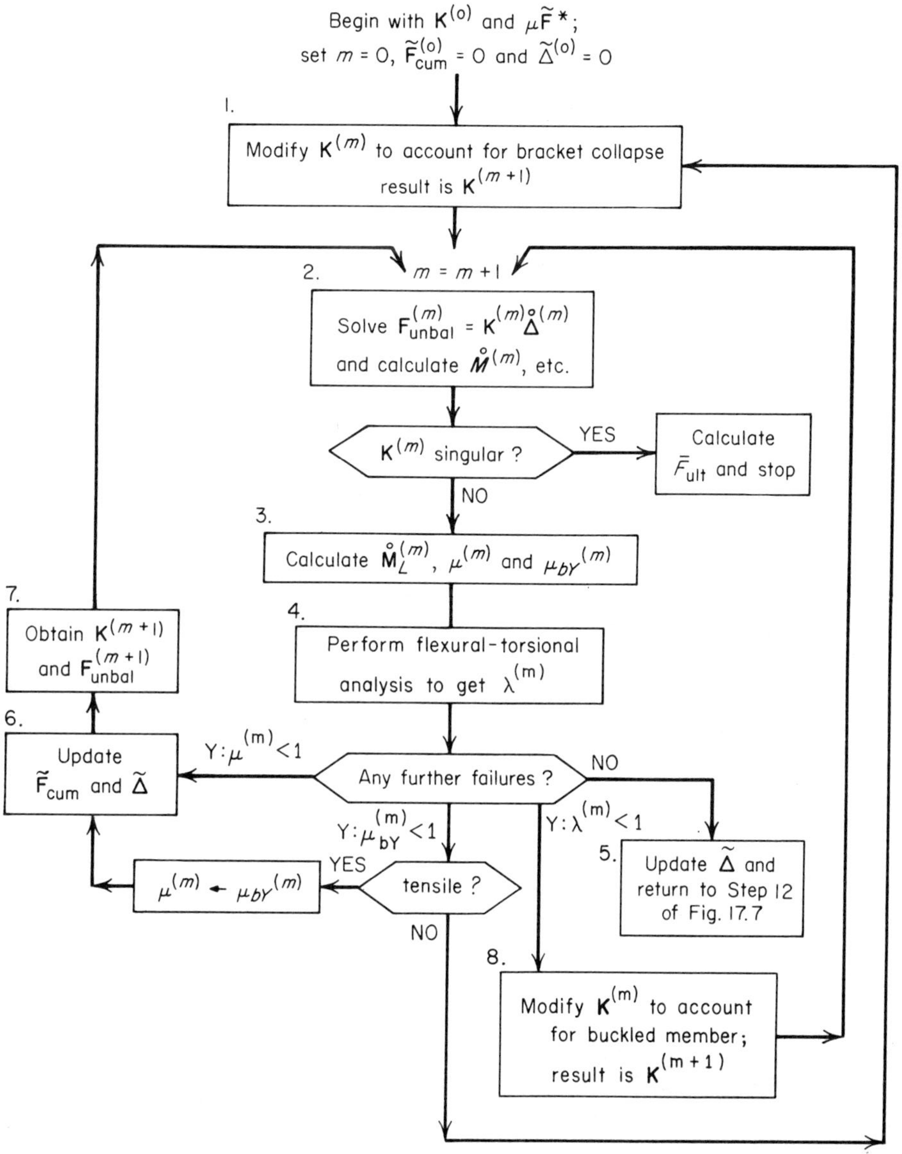

Figure 17.8 Calculation of $\tilde{\Delta}$ for bracket collapse (step 10 of Fig. 17.7).

Calculation of Incremental Deflection Due to Compressive Collapse of a Bracket Flange

The details of step 10 of the method for transverse ultimate strength analysis are given here. The procedure is given in the flow diagram of Fig. 17.8, and we here give the details of the numbered steps shown there. The procedure is cyclic and the cycle number is shown as a superscript in parentheses to distinguish between successive values; (0) indicates initial values, at the instant just prior to bracket collapse. The incremental load which causes the collapse is $\mu\tilde{\mathbf{F}}^*$, and at that time the incremental displacement is $\mu\tilde{\mathbf{\Delta}}^*$. The purpose of the analysis is to find the final value of $\tilde{\mathbf{\Delta}}$, at which the structure's internal forces are in equilibrium with the load, $\mu\tilde{\mathbf{F}}^*$ (which remains unchanged throughout the process; see Fig. 17.9). Since the bracket collapse may cause additional plastic hinges, the calculation of $\tilde{\mathbf{\Delta}}$ involves another level of incrementation, denoted by $\mathring{\mathbf{\Delta}}$. Figure 17.9*a* illustrates the case in which there are two further plastic hinges. The symbol $\tilde{\mathbf{F}}_{\text{cum}}^{(m)}$ denotes the cumulative value of the internal nodal forces as the incrementation proceeds. Hence at any stage during the recovery of equilibrium the remaining unbalance between internal force and external load is

$$\tilde{\mathbf{F}}_{\text{unbal}}^{(m)} = \mu\tilde{\mathbf{F}}^* - \tilde{\mathbf{F}}_{\text{cum}}^{(m)} \tag{17.3.2}$$

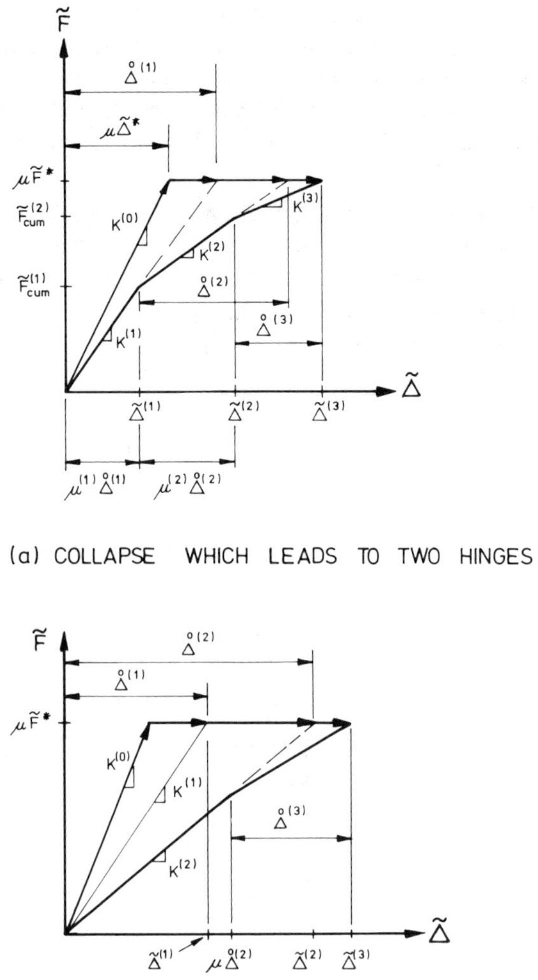

(a) COLLAPSE WHICH LEADS TO TWO HINGES

(b) COLLAPSE WHICH LEADS TO A SECOND COLLAPSE AND ONE HINGE

Figure 17.9 Additional deflection due to bracket collapse.

Procedure:

1. The collapse of a bracket (either the original collapse or another which has resulted from it) eliminates the bracket's contribution to the rigid lengths at that joint, and so it is necessary to modify the appropriate terms in the stiffness matrix (by subtracting the old values and adding the new values) thus obtaining $\mathbf{K}^{(1)}$ (or $\mathbf{K}^{(m+1)}$ in the general case). Then begin the next incremental cycle by updating the cycle number m (which was set to zero initially).

2. Calculate the remaining unbalanced forces from (17.3.2) and then solve

$$\tilde{\mathbf{F}}_{\text{unbal}}^{(m)} = \mathbf{K}^{(m)}\mathring{\mathbf{\Delta}}^{(m)}$$

which gives the increment in displacement to achieve equilibrium if there are no further failures. Then calculate the corresponding increment in bending moment, $\mathring{\mathbf{M}}^{(m)}$ and in the other load effects. If $\mathbf{K}^{(m)}$ is found to be singular it means that the structure has developed sufficient hinges (plastic hinges and/or shear yield of joint webs) to form a mechanism, and so the current total load is the collapse load. Calculate $\overline{F}_{\text{ult}}$ as in (17.3.1) and stop.

3. Calculate the limit values and the values of $\mu^{(m)}$ and $\mu_{bY}^{(m)}$, as in Steps 5, 6, and 7 of the overall procedure, with $\mathring{\mathbf{M}}^{(m)}$ in place of $\tilde{\mathbf{M}}^*$. Note that in the second and later cycles (i.e. when $m \geqslant 2$) the base value of bending moment that is subtracted from the limit value must include the accumulated increment, $\tilde{\mathbf{M}}^{(m-1)}$; that is, the value of bending moment corresponding to the accumulated incremental displacement, $\tilde{\mathbf{\Delta}}^{(m-1)}$.

4. Perform a flexural-torsional analysis on the modified structure (for which the stiffness matrix is $\mathbf{K}^{(m)}$ to obtain the new value of $\lambda^{(m)}$, the load scaling factor for the next flexural-torsional instability.

5. If there are no further failures ($\mu^{(m)} > 1$, $\mu_{bY}^{(m)} > 1$ and $\lambda^{(m)} > 1$) perform the final updating

$$\tilde{\Delta} = \tilde{\Delta}^{(m-1)} + \mathring{\Delta}^{(m)}$$

and return to Step 12 of the overall procedure (Fig. 17.7). If the failure is a plastic hinge proceed to Step 6. If it is tensile yield of a bracket flange replace $\mu^{(m)}$ by $\mu_{bY}^{(m)}$ and proceed to Step 6. If it is compressive yield return to Step 1. If it is flexural-torsional buckling go to Step 8.

6. Update the deflections

$$\tilde{\Delta}^{(m)} = \tilde{\Delta}^{(m-1)} + \mu^{(m)}\mathring{\Delta}^{(m)}$$

and the cumulative internal nodal forces

$$\tilde{\mathbf{F}}_{\text{cum}}^{(m+1)} = \tilde{\mathbf{F}}^{(m)} + \mu^{(m)}\tilde{\mathbf{F}}_{\text{unbal}}^{(m)}$$

7. Calculate the new unbalanced force vector $\mathbf{F}_{\text{unbal}}^{(m+1)}$ from (17.3.3). Modify $\mathbf{K}^{(m)}$ in the appropriate manner to account for the new plastic hinge or the bracket tensile yield; the result is $\mathbf{K}^{(m+1)}$. For a plastic hinge, also insert a zero for the corresponding moment term in the unbalanced force vector. Then update m and return to Step 2.

8. As discussed in the next section, flexural-torsional buckling of a beam member totally eliminates the beam's flexural rigidity in the transverse plane. It is the equivalent of installing an ordinary hinge (not a plastic hinge) within the length of the member, say at the midlength. Therefore it is accounted for by deleting from the global stiffness matrix the terms corresponding to θ_z at the member's midlength node, thus obtaining $\mathbf{K}^{(m+1)}$. Then update m and return to Step 2.

Figure 17.9*a* shows a case in which bracket collapse induces the formation of two hinges, either because the bending moment has exceeded M_P or M_{fY}, or because of shear yield of the web region of a joint at which the bracket, if present, has already undergone flange yield. Figure 17.9*b* shows a case in which bracket collapse induces the collapse of a second bracket, followed by a hinge, either at the same location or elsewhere.

Analysis for Flexural-Torsional Buckling

For transverse structure the most important mode of buckling is flexural-torsional buckling of the individual members, that is, the possibility that the beam flange may buckle sideways in order to escape from being shortened by the axial compressive force to which it is subjected due to the flexure of the beam. The frame is a plane frame, with $U = 0$ at every node, but the beam flange can deflect out of the *YZ* plane because of the beam's torsional degrees of freedom, which in member coordinates are: θ_{x1}, θ'_{x1}, θ_{x2} and θ'_{x2}. For this analysis the nodes are placed along the line of attachment to the plating because the plating constrains the beam to rotate about this axis. As shown in Section 15.5 the shear center y_0 is sufficiently close to the plating that the flexural-torsional response is effectively uncoupled from the deflection and bending in the *YZ* plane. This allows the check for flexural-torsional buckling to be performed as a separate step (step 11) within each incremental cycle of the overall analysis.

In essence the analysis is an eigenvalue analysis, similar to that demonstrated in Section 15.5 for a single two-element beam member carrying a lateral load. In the lateral-torsional response there are only two degrees of freedom per node, θ_x and θ'_x, because the w and θ_y deflections are prevented by the plating. Because direct axial compression in the members is relatively small it is not necessary to include the axial degree of freedom. Note however that if this were not so it would have to be included.

Since the load and the structure are both symmetric about the beam web, the quantities z_o, e_z, $\bar{a}_z$, β_1, F_z, f_z, and M_y (see Fig. 15.4) in (15.3.15) are all zero. Also, since the distributed load arises from pressure on the plating the point of application of the load is along the base of the web (regardless of which side the pressure acts on) which is also the shear center (since $y_o \approx 0$) and so $\bar{a}_y$ is also zero. All of this greatly simplifies the terms in $\mathbf{k}_G$.

Since the analysis involves rigid lengths it is necessary to include the rigid length terms of (8.6.7) in the ordinary stiffness matrix $\mathbf{k}_S$. Also for the general case it would be necessary to re-derive the geometric stiffness matrix of (15.3.15), taking account of the new distribution of $v(x)$. However, for the special case that is being considered here (symmetry about the beam web, both of load and of structure, and relatively small axial forces) all but two of the terms in (15.3.12) which involve v (or v' or v'') disappear. The only such terms which remain are the first, involving $(v'')^2$, and the sixteenth, involving $f_y v$. It may be shown that the change in the first term generates the identical rigid length stiffness terms as in (8.6.7). It may also be shown that the change in the sixteenth term does not affect $\mathbf{k}_S$ or $\mathbf{k}_G$, but instead leads to a modified set of equivalent nodal forces. For the element representing the left half of the member (see Fig. 17.10) the equivalent nodal forces are

$$\mathbf{f} = \begin{Bmatrix} f_y\left(r_1 + \dfrac{\ell_1}{2}\right) \\ f_y\left(\dfrac{r_1(r_1+\ell_1)}{2} + \dfrac{\ell_1^2}{12}\right) \\ f_y\dfrac{\ell_1}{2} \\ -f_y\dfrac{\ell_1^2}{12} \end{Bmatrix} \tag{17.3.3}$$

and for the element representing the right half the forces are

$$\mathbf{f} = \begin{Bmatrix} f_y\dfrac{\ell_2}{2} \\ f_y\dfrac{\ell_2^2}{12} \\ f_y\left(\dfrac{\ell_2}{2} + r_2\right) \\ -f_y\left(\dfrac{\ell_2^2}{12} + \dfrac{r_2(r_2+\ell_2)}{2}\right) \end{Bmatrix} \tag{17.3.4}$$

For greatest accuracy the analysis would be performed for the complete transverse frame contained within the critical segment. The principal steps are calculating the 4 × 4 element stiffness matrices $\mathbf{k}_S$ and $\mathbf{k}_G$ corresponding to θ_x and θ_x' for each element, assembling these, and solving the resulting eigenvalue problem. The main reason for dealing with the complete critical segment is to account for interactions between members. But in flexural-torsional buckling there is relatively little interaction between members that are at (approximately) right angles to each other, which is a common situation in a transverse frame. In such members the beam flange receives very little restraint from adjacent members in regard to rotation in its plane, that is, rotation about an axis parallel to the y-axis (see Fig. 15.13). Also, at the ends of a beam it is not common for the beam flange to be bracketed in its own plane. In many cases the flange itself does not reach the end of the beam but curves around to become the flange of another beam. In this case a stiffener may be fitted to act as a continuation of the flange, but there is still relatively little rotational restraint. Therefore if it is desired to reduce the amount of computation the beam flange may be regarded as simply supported at the ends of each orthogonal joint. This allows a separate analysis of each member, or group of in-line members, extending from one orthogonal intersection

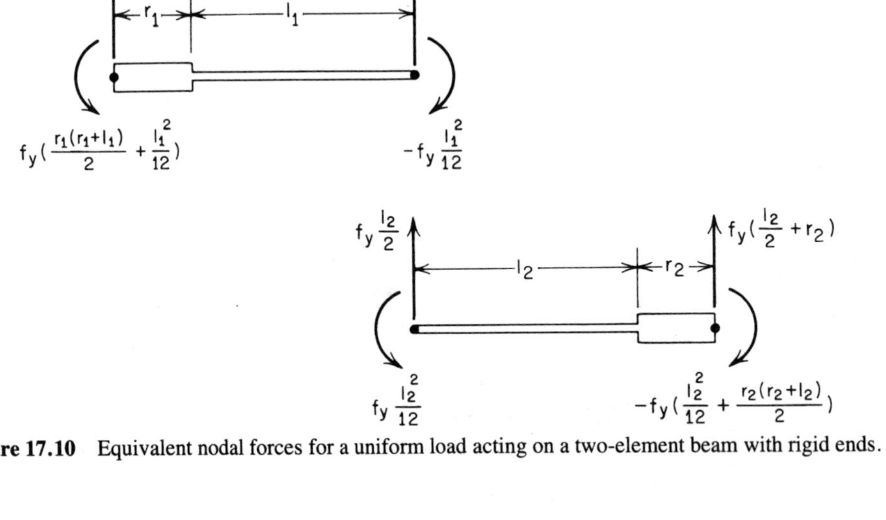

Figure 17.10 Equivalent nodal forces for a uniform load acting on a two-element beam with rigid ends.

to the next. This gives a substantial reduction in the amount of computation.

Representation of Prior Plastic Hinges

It is also necessary to account for plastic hinges that have already occurred in the structure in previous incremental cycles. Since a plastic hinge in the *YZ* plane involves yield of the beam flange it also constitutes a point of zero rotational stiffness in regard to bending in the plane of the flange, which corresponds to the θ_x' degree of freedom. Therefore in the flexural-torsional buckling analysis a plastic hinge is represented by inserting an additional value of θ_x' at that point, that is, by adding a new row and column to the structure stiffness matrix and a new term, equal to zero, in the structure load vector, as is done for bending in the *YZ* plane.

The principal reason why an accurate flexural-torsional analysis cannot be done as part of the isolated member limit analysis (step 3 of Fig. 2.19) is not because of member interaction but because the occurrence of plastic hinges causes the bending moment distribution to change, sometimes quite significantly, and there is no way of anticipating or allowing for this at the member level. In fact, since the member scantlings are changing as the design proceeds, not even the pattern or manner of change of the bending moment distribution can be predicted.

Calculation of Incremental Deflection due to Flexural-Torsional Buckling

The details of step 11 of the method for transverse ultimate strength analysis are shown in the flow diagram of Fig. 17.11, and we here give the details of the numbered steps shown there. The procedure is similar to that for bracket collapse and we use the same notation; a superscript in parentheses indicates the cycle number, and (0) indicates initial values, at the instant just prior to flexural-torsional buckling. Figure 17.9 again illustrates this process; the only difference is that the failure load is now $\lambda\tilde{\mathbf{F}}^*$ instead of $\mu\tilde{\mathbf{F}}^*$, and the corresponding initial incremental displacement is $\lambda\tilde{\mathbf{\Delta}}^*$ instead of $\mu\tilde{\mathbf{\Delta}}^*$. That is, the incremental load which causes the buckling is $\lambda\tilde{\mathbf{F}}^*$, and at that time the incremental displacement is $\lambda\tilde{\mathbf{\Delta}}^*$. The purpose of the analysis is to find the final value of $\tilde{\mathbf{\Delta}}$, at which the structure's internal forces are in equilibrium with the load, $\lambda\tilde{\mathbf{F}}^*$ (which remains unchanged throughout the process, as does $\mu\tilde{\mathbf{F}}^*$ in Fig. 17.9). The buckling may cause additional failures, either by plastic hinge or by bracket collapse or by further flexural-torsional buckling. Therefore, the calculation of $\tilde{\mathbf{\Delta}}$ again involves a second level of incrementation, denoted by $\mathring{\mathbf{\Delta}}$. Figure 17.9 illustrates typical combinations of additional failures. Again the symbol $\tilde{\mathbf{F}}_{\text{cum}}^{(m)}$ denotes the cumulative value of the internal nodal forces as the incrementation proceeds. Hence at any stage during the recovery of equilibrium the remaining unbalance between internal force and external load is

$$\tilde{\mathbf{F}}_{\text{unbal}}^{(m)} = \lambda\tilde{\mathbf{F}}^* - \tilde{\mathbf{F}}_{\text{cum}}^{(m)} \qquad (17.3.5)$$

Procedure

1. Flexural-torsional buckling of a beam member (either the original instability or another which has resulted from it) totally eliminates the beam's flexural rigidity in the transverse plane. It is the equivalent of installing an ordinary hinge (not a plastic hinge) within the length of the member, say at the midlength. Therefore the first step is to delete from the global stiffness matrix the terms corresponding to θ_z at the member's midlength node, thus obtaining $\mathbf{K}^{(1)}$ (or $\mathbf{K}^{(m+1)}$ in the general case). Then begin the next incremental cycle by updating the cycle number, *m*.
2. Calculate the remaining unbalanced forces from (17.3.5) and then solve

$$\tilde{\mathbf{F}}_{\text{unbal}}^{(m)} = \mathbf{K}^{(m)}\mathring{\mathbf{\Delta}}^{(m)}$$

which gives the increment in displacement to achieve equilibrium if there are no further failures. Then calculate the corresponding increment in bending moment, $\mathring{M}^{(m)}$ and in the other load effects. If $\mathbf{K}^{(m)}$ is found to be singular it means that the structure has developed sufficient hinges (plastic hinges and/or shear yield of joint webs) to form a mechanism, and so the current total load is the collapse load. Calculate $\bar{F}_{\text{ult}}$ as in (17.3.1) and stop.
3. Calculate the limit values and the values of $\mu^{(m)}$ and $\mu_{bY}^{(m)}$, as in steps 5, 6, and 7 of the overall procedure, with $\mathring{M}^{(m)}$ in place of $\tilde{M}^*$. Note that in the second and later cycles (i.e., when $m \geqslant 2$) the base value of bending moment that is subtracted from the limit value must include the accumulated increment, $\tilde{M}^{(m-1)}$; that is, the value of bending moment corresponding to the accumulated incremental displacement, $\tilde{\mathbf{\Delta}}^{(m-1)}$.
4. Perform a flexural-torsional analysis on the modified structure (for which the stiffness matrix is $\mathbf{K}^{(m)}$) to obtain the new value of $\lambda^{(m)}$, the load scaling factor for the next flexural-torsional instability.

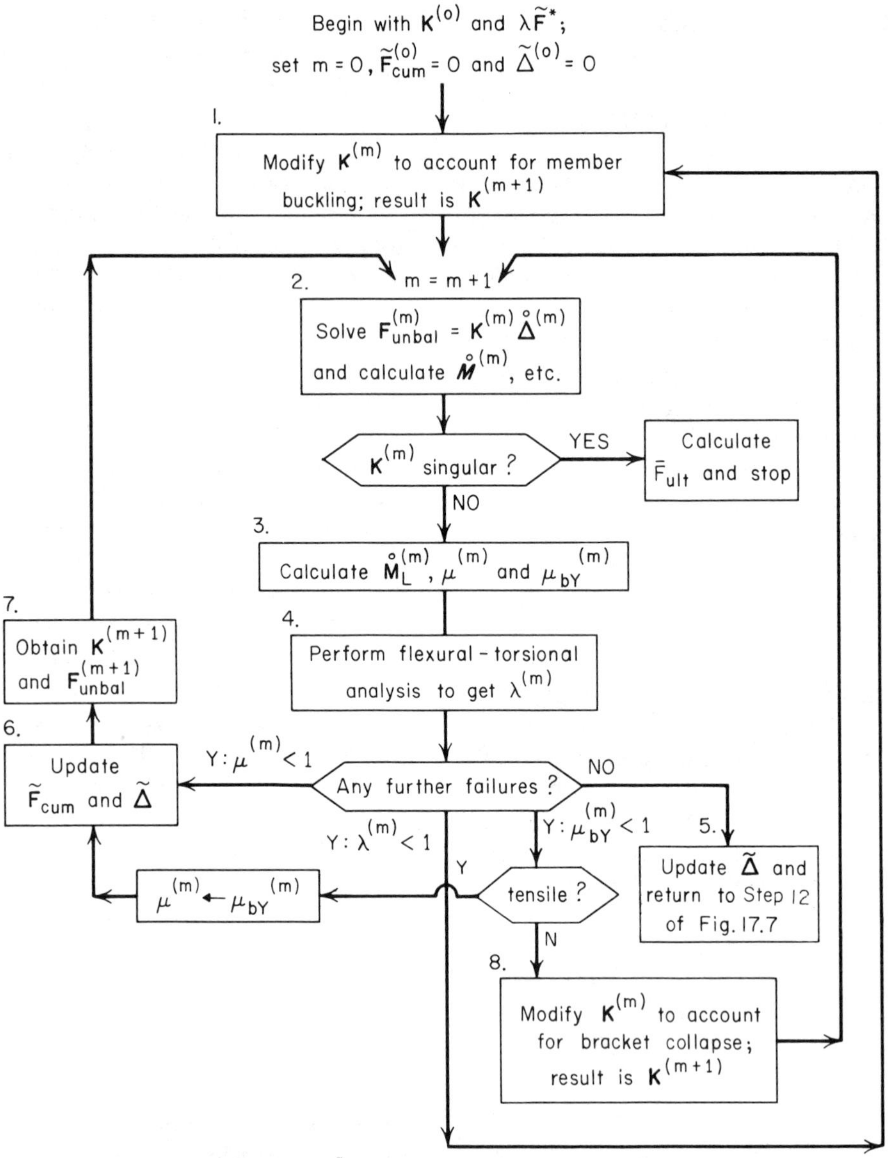

Figure 17.11 Calculation of $\tilde{\Delta}$ for flexural-torsional buckling (Step 11 of Fig. 17.7).

5. If there are no further failures ($\mu^{(m)} > 1$, $\mu^{(m)}_{bY} > 1$ and $\lambda^{(m)} > 1$) perform the final updating

$$\tilde{\mathbf{\Delta}} = \tilde{\mathbf{\Delta}}^{(m-1)} + \mathring{\mathbf{\Delta}}^{(m)}$$

and return to step 12 of the overall procedure (Fig. 17.7). If the failure is a plastic hinge proceed to step 6. If it is tensile yield of a bracket flange replace $\mu^{(m)}$ by $\mu^{(m)}_{bY}$ and proceed to step 6. If it is compressive yield go to step 8. If it is another flexural-torsional failure return to step 1.

6. Update the deflections

$$\tilde{\mathbf{\Delta}}^{(m)} = \tilde{\mathbf{\Delta}}^{(m-1)} + \mu^{(m)}\mathring{\mathbf{\Delta}}^{(m)}$$

and the cumulative internal nodal forces

$$\tilde{\mathbf{F}}^{(m+1)}_{cum} = \tilde{\mathbf{F}}^{(m)} + \mu^{(m)}\tilde{\mathbf{F}}^{(m)}_{unbal}$$

7. Calculate the new unbalanced force vector $\mathbf{F}^{(m+1)}_{unbal}$ from (17.3.5). Modify $\mathbf{K}^{(m)}$ in the appropriate manner to account for the new plastic hinge or the bracket tensile yield; the result is $\mathbf{K}^{(m+1)}$. For a plastic hinge also insert a zero for the corresponding moment term in the unbalanced force vector. Then update m and return to step 2.

8. Reduce the rigid lengths where the bracket collapsed and modify the stiffness matrix accordingly, obtaining $\mathbf{K}^{(m+1)}$. Then update m and return to step 2.

REFERENCES

1. C. S. Smith, "Influence of Local Compressive Failure on Ultimate Longitudinal Strength of a Ship's Hull," Proc. Int'l. Symp. on Practical Design in Shipbuilding (PRADS), Tokyo, October 1977, pp. 73–80.
2. J. C. Adamchak, "ULTSTR: A Program for Estimating the Collapse Moment of a Ship's Hull under Longitudinal Bending," DTNSRDC Report 82/076, October 1982.
3. J. C. Adamchak, "Design Equations for Tripping of Stiffeners under Inplane and Lateral Loads," DTNSRDC Report 79/064, October 1979.
4. R. S. Dow, R. C. Hugill, J. D. Clark, and C. S. Smith, "Evaluation of Ultimate Ship Hull Strength," Extreme Loads Response Symposium, SNAME, October, 1981, pp. 133–147.
5. P. J. Dowling, S. Chatterjee, P. A. Frieze, and F. M. Moolani, "Experimental and Predicted Collapse Behaviour of Rectangular Steel Box Girders," Int. Conf. on Steel Box Girder Bridges, London, 13–14 February 1973.
6. K. A. Reckling, Preliminary Report on Box Girder Experiments (included in Report of Committee II.2 on Nonlinear Structural Response, Proc. of 7th Internat. Ship Structures Congress, Paris, 1979).

INDEX